Index of Tables

Reviews, Techniques, and Mathematical Assistance

$$x = x_0 + v_0 t + \tfrac{1}{2} a t^2$$
$$v = v_0 + at$$
$$v^2 - v_0^2 = 2a(x - x_0)$$

$$y = y_0 + v_0 t - \tfrac{1}{2} g t^2$$
$$v = v_0 - gt$$
$$v^2 - v_0^2 = -2g(y - y_0)$$

Physics

p 51 ball up down → a and g

Richard L. Childers died on February 19, 2000, shortly after the update to the third edition of this book was completed. Richard's interest in physics and in teaching is evident on these pages. His interest and ability as a photographer can be seen in many of the photographs that appear in this volume.

Richard was also known for his work with high school teachers and students, especially for his role in establishing the Midway Physics Day at the S.C. State Fair.

He was a friend and colleague for 35 years and will be sorely missed. He is survived by his wife, Sigrid, two daughters, and five grandchildren. This update to the third edition of Contemporary College Physics is dedicated to his memory.

Edwin R. Jones
Columbia, SC
March 17, 2000

contemporary college

Physics

2001 UPDATE   THIRD EDITION

Dr. Edwin Jones
University of South Carolina

Dr. Richard Childers
University of South Carolina

Boston   Burr Ridge, IL   Dubuque, IA   Madison, WI   New York   San Francisco   St. Louis
Bangkok   Bogotá   Caracas   Lisbon   London   Madrid
Mexico City   Milan   New Delhi   Seoul   Singapore   Sydney   Taipei   Toronto

McGraw-Hill Higher Education

A Division of The **McGraw-Hill** *Companies*

CONTEMPORARY COLLEGE PHYSICS 2001 UPDATE
THIRD EDITION

Published by McGraw-Hill, an imprint of the McGraw-Hill companies, Inc., 1221 Avenue of the Americas, New York, NY 10020. Copyright © 2001 by Edwin R. Jones and Richard L. Childers. All rights reserved. Copyright © 1999 by The McGraw-Hill Companies, Inc. All rights reserved. Previous edition © 1993 by Addison–Wesley Publishing Company, Inc. All rights reserved. No part of this publication may be reproduced or distributed in any form or by any means, or stored in a database or retrieval system, without the prior written consent of The McGraw-Hill Companies, Inc., including, but not limited to, in any network or other electronic storage or transmission, or broadcast for distance learning.

Some ancillaries, including electronic and print components, may not be available to customers outside the United States.

This book is printed on acid-free paper.

1 2 3 4 5 6 7 8 9 0 VNH/VNH 0 9 8 7 6 5 4 3 2 1 0

ISBN 0–07–239911–2
ISBN 0–07–118236–5 (ISE)

Vice president and editor-in-chief: *Kevin T. Kane*
Publisher: *JP Lenney*
Sponsoring editor: *Daryl Bruflodt*
Editorial assistant: *Jenni Lang*
Marketing managers: *Mary K. Kittell/Debra A. Besler*
Project manager: *Jill R. Peter*
Media technology project manager: *Phillip Meek*
Production supervisor: *Laura Fuller*
Coordinator of freelance design: *David W. Hash*
Cover designer: *David W. Hash/Crispen Prebys*
Cover images: *©Corbis/©PhotoDisc*
Photo research coordinator: *John C. Leland*
Photo research: *Feldman & Associates*
Senior supplement coordinator: *Candy M. Kuster*
Compositor: *York Graphic Services, Inc.*
Typeface: *10/12 Times Roman*
Printer: *Von Hoffmann Press, Inc.*

The credits section for this book begins on page C–1 and is considered an extension of the copyright page.

The update includes 16 new photos and 13 new or revised pieces of line art along with approximately 30 new examples. There were also a number of minor corrections and updates made throughout the text.

Library of Congress Cataloging-in-Publication Data

Jones, Edwin R., 1938–
    Contemporary college physics / Edwin R. Jones, Richard L. Childers. — 3rd ed., 2001 update
        p.   cm.
    Includes index.
      ISBN 0–07–239911–2
      1. Physics.   I. Childers, Richard L.   II. Title.

QC21.2.J66   2001
530—dc21
                                      00–035478
                                         CIP

INTERNATIONAL EDITION   ISBN 0–07–118236–5
Copyright © 2001. Exclusive rights by The McGraw-Hill Companies, Inc., for manufacture and export. This book cannot be re-exported from the country to which it is sold by McGraw-Hill. The International Edition is not available in North America.

www.mhhe.com

Contents

* Different teachers will emphasize different parts of the text. To make the choices easier, we have designated some material as optional. Sections marked with an asterisk may be safely omitted without fear that their content will be needed in subsequent sections or chapters.

Preface

We are pleased to present the third edition of *Contemporary College Physics.* By building on the strengths of the first two editions of our non-calculus general physics text, we feel this edition introduces students to the beauty and usefulness of physics while teaching problem-solving skills that can help them throughout their studies and careers. With considerable help from students and instructors from across the country, we have updated and improved a wide variety of elements and introduced new features to create the third edition.

New to the Third Edition

- Consistent color coding of line art elements helps students identify and differentiate between force vectors, velocity vectors, acceleration vectors, magnetic fields, and positive and negative charges.

- *Master the Concept* boxes walk students through the principles that apply to a given situation to clarify the application of concepts.

- Strategy boxes have been added throughout to guide students through problem-solving issues.

- An enhanced art program features more photographic illustrations of concepts and principles.

- The number of worked examples in the chapters has been significantly increased to help students understand how to approach problem solving.

- Expanded end-of-chapter exercises include conceptual questions as well as computational problems.

- New interior design makes this text easier to use than ever before.

2001 Update

This 2001 Update of *Contemporary College Physics,* third edition, differs from the 1999 third edition in two ways. First, we have further highlighted biomedical applications of physics and added coverage of some recent technologies. For example, we introduce hydrostatic weighing (to determine percentage of body fat) as an application of Archimedes' principle, and our new discussion of recent advances in laser eye surgery is accompanied by a detailed illustration (see page 957). This focus on the life and health sciences, achieved through new topic coverage as well as photographs and illustrations added to the Update, should help to motivate the many students who take this course in preparation for a career in health-related fields. For a list of relevant applications, see page xix following this preface.

The second change to the third edition is the expansion and improvement of the Interactive Student Tutorial, the CD-ROM that accompanies the book. The CD supports problems-solving practice with a wealth of new examples and exercises. For details, see the description of the CD under "Supplements" on page xvi.

The organization and exercise sets of the 1999 third edition remain unchanged in the Update, so the solutions manuals and other supplements have not changed, nor will users of the third edition need to change their lesson plans in any way.

Goals

Our main goals are to increase student understanding of natural laws and to develop the analytical skills critical for success in both educational undertakings and lifetime decision making. We approach these goals by emphasizing basic principles and the unity of physics.

We have the additional goal of providing students a thorough coverage of modern physics so that students will better comprehend the important public policy issues facing them as citizens. We want students to see that physics is a dynamic, exciting field. We are now preparing students for the twenty-first century, when the need for scientific understanding will be greater than ever. Classical physics is presented from a contemporary perspective. Modern physics is treated thoroughly, as an integral part of the course. The entire book speaks to today's students, using the latest pedagogical aids.

We introduce the concept of a model in Chapter 1 and then point out throughout the text how physicists use models as part of the scientific process. We emphasize that the first part of developing any theory is to make a model of the physical situation and state its assumptions. Then we show how later observations serve to refine the model and improve our overall understanding. Examples include such fundamental models as the kinetic theory of gases, the free electron model of metals, the wave model of particles, and the quark model of matter.

Problem Solving

Solving physics problems has long been regarded by physics instructors as a key to learning. We are aware of the difficulty students have in developing good problem-solving skills and habits. For this reason, we have put special emphasis on helping students with problems.

■ **Examples** There are over 340 worked examples in the body of the text, and in most cases the solutions are divided into three sections: strategy, solution, and discussion. The strategy section shows the students a conceptual way of analyzing the problem in order to decide what to do. The solution section presents the analysis and computation, and the discussion section points out to the student the significance of the answer and analysis. This approach helps direct students to a more productive way of solving problems than merely grasping for equations.

Example 4.2

Safety regulations require that all cars traveling at a given speed be able to stop within a given distance. What must be the relationship between the minimum braking forces allowed by law for two cars whose masses have the ratio of 3 to 2?

Strategy We can use Newton's second law to relate the force to mass and acceleration. From your knowledge of kinematics, you know that the requirement for stopping in equal distances from the same initial speed for both cars means they must both have the same acceleration.

Solution We can write the two accelerations in terms of the braking forces F_1 and F_2 and the masses m_1 and m_2 as

$$a_1 = \frac{F_1}{m_1} \quad \text{and} \quad a_2 = \frac{F_2}{m_2}.$$

When we set the accelerations equal we find

$$\frac{F_1}{m_1} = \frac{F_2}{m_2},$$

so

$$\frac{F_1}{F_2} = \frac{m_1}{m_2} = \frac{3}{2}.$$

Discussion This result tells us that a larger braking force is needed to stop a larger (more massive) car in the same distance that a smaller (less massive) car is stopped.

Pedagogical Use of Color

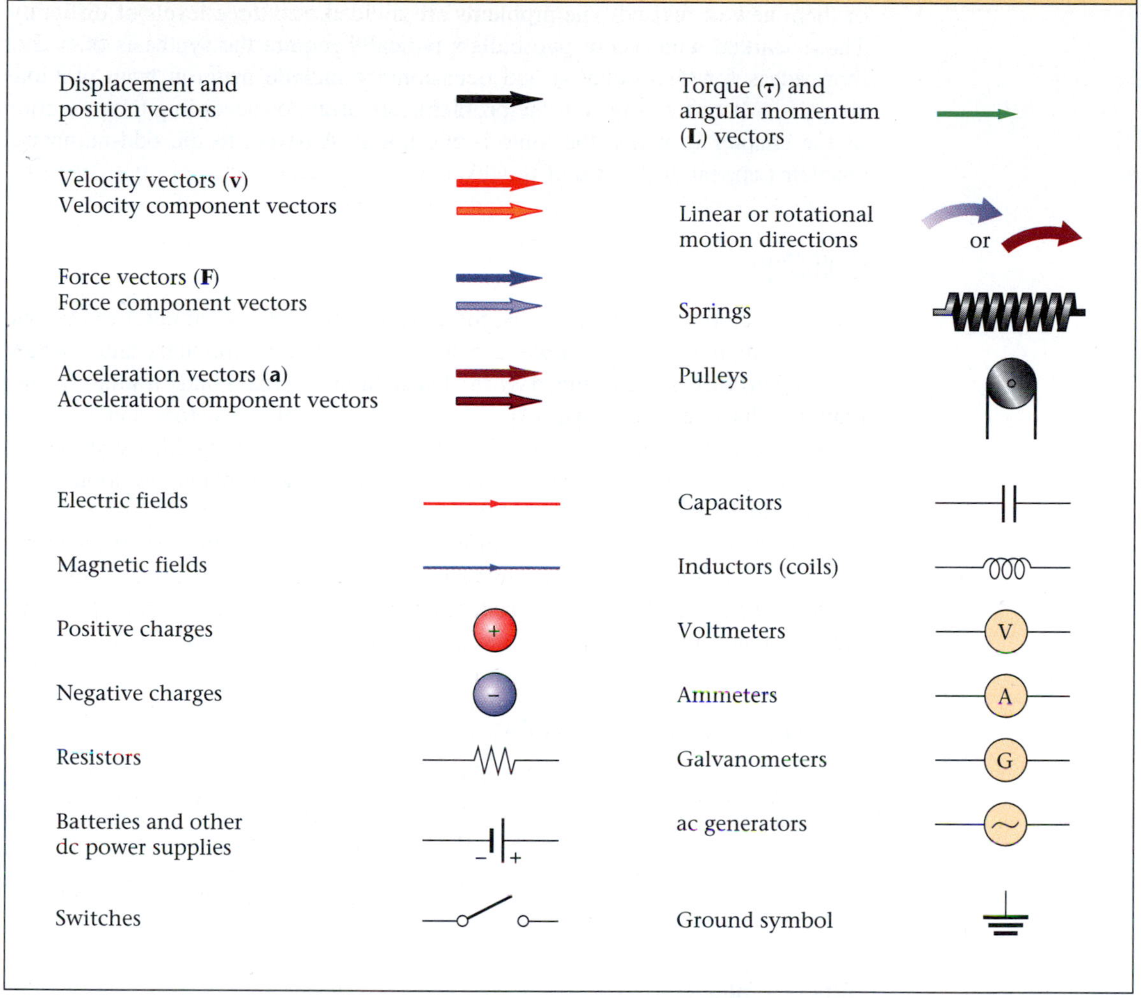

Displacement and position vectors	Torque (τ) and angular momentum (**L**) vectors
Velocity vectors (**v**) Velocity component vectors	Linear or rotational motion directions *or*
Force vectors (**F**) Force component vectors	Springs
Acceleration vectors (**a**) Acceleration component vectors	Pulleys
Electric fields	Capacitors
Magnetic fields	Inductors (coils)
Positive charges	Voltmeters
Negative charges	Ammeters
Resistors	Galvanometers
Batteries and other dc power supplies	ac generators
Switches	Ground symbol

■ **Problem-Solving Guidelines** A general step-by-step guide to problem solving is given in Section 1.7 (p. 19) and this approach is reinforced throughout the book: in the examples, in the **Hints for Solving Problems** distributed throughout the end-of-chapter problems, and in the **Problem-Solving Strategy** boxes included in the narrative.

■ **Master the Concepts** A step-by-step solution of conceptual questions is presented utilizing the basic principles. This nonnumerical analysis will help students visualize the concepts involved.

Color Key

We have implemented a color key to help students identify elements in the illustrations. (Top of page.)

Master the Concept

Force and Circular Motion

Question: The Wave Swinger ride at the fair has two circular rows of chairs suspended by chains of equal lengths. The outer row is along the outer edge of the top of the ride and the inner row has a smaller radius. Do riders in both rows swing out at the same angle when the ride is rotating?

Answer: When the ride is in motion, the passengers in the chairs swing outward. At one particular angle the chain provides the necessary horizontal force for circular motion and the vertical force needed to balance the gravitational force. Riders in both rows move with the same angular velocity, but those in the outer row move in a larger radius and thus have a larger angular acceleration. Because of this larger angular acceleration, a larger centripetal force is needed. Thus, for riders in the outer row, the horizontal component of the chain force is larger relative to their gravitational force than it is for those on the inner row. The result is that riders in the outer row swing out to the larger angle from the vertical direction than do the riders in the inner row as you can see in Fig. 5.22 on p. 170.

Problems

The end-of-chapter problem sets have been significantly expanded. There are now over six hundred conceptual questions and more than 2,250 problems, many of them new or revised. The problems are divided into three levels of difficulty. Those marked with one or two bullets typically require the synthesis of two or more ideas for their solution and occasionally include material from previous chapters. About two-thirds of the problems are arranged according to the section of the chapter in which the topic is discussed. Answers to the odd-numbered problems appear at the end of the text.

Coverage

As with the previous two editions, the coverage of topics is comprehensive, but not encyclopedic. We introduce a new section on measurements and models. Model building is introduced with discussions of blackbody radiation and Planck's discovery. We have expanded our treatment of vectors and included more material on vector addition. There is more coverage of Maxwell's equations and electromagnetic waves. Care has been taken to include all topics covered on the MCAT.

We understand that different teachers will emphasize different parts of the text. To make the choices easier, we have designated some material as optional. **Sections marked with an asterisk may be safely omitted** without fear that their content will be needed in subsequent sections or chapters.

Emphasis on Basic Principles

"The student can't see the forest for the trees," say our colleagues. Having heard this over and over again, we have made the emphasis of basic principles one of our highest priorities. A good example is our treatment of conservation laws. The ability to explain and predict observations using conserved quantities is emphasized conceptually as well as mathematically.

Unity of Physics

Our treatment of conservation laws also illustrates another of our goals: to show that physics is not just a collection of independent ideas but is an interconnected whole. We believe this approach reflects the spirit of physics today, and we also believe that it helps students retain more of what they've learned after they leave the course.

If you read a mystery novel all the way through in one sitting, you immediately have at your fingertips all the clues necessary to solve the puzzle. However, students generally read a physics textbook in small sections and cover groups of chapters over a period of time. As a result, they inevitably forget some of the clues and are less prepared to solve the puzzle—or in this case, to see the big picture and appreciate the beauty of physics. For this reason, we give frequent reminders in the text and examples of previously covered topics and of topics to be covered later.

Level

The text assumes that students have no previous background in physics. The basic mathematical working tools are algebra, and trigonometry, and a high school course in these subjects is certainly a prerequisite. One of the challenges in teaching this course is that the students' math preparation is often weaker than the teacher would like. Most students need a math refresher beyond the typical review stuck in the back of texts. To that end, we have included chapter appendices on key math topics in those chapters where they are first needed: quadratic equations (Chapter 2), basic trigonometry (Chapter 3), simultaneous equations (Chapter 4), and the exponential function (Chapter 12).

The exponential function is first used in Chapter 12 in describing the barometric formula and the distribution of molecular speeds. Subsequently, it appears in analyses of electric circuits, radioactive decay, and other topics. The addition of the exponential function to the usual mix of algebra and trigonometry affords students a better comprehension of the individual topics.

Motivation

Teachers frequently hear the complaint that the subject matter has no relevance to the students' subsequent courses and careers. To overcome this misconception, we have made a special effort to show applications of fundamental principles in everyday life as well as in biology, medicine, architecture, and technology.

■ **Physics in Practice** Applications can be found in the text and examples and in special essays called **Physics in Practice,** which deal with topics ranging from automobile tires to liquid-crystal displays. Great care has been taken to provide a diversity of applications that will appeal to the broadest range of students. A list of these applications follows on page xix.

■ **Back to the Future** Physics is a science based on the efforts of real men and women struggling to understand how the world works. In essays called **Back to the Future,** we present physics as a human activity in which new ideas are constantly being tried and in which scientific truth is never absolute. We generally introduce a new topic by describing the efforts of the scientists who made the breakthrough discoveries and advances. For example, in Chapter 1 we present the work of Arno Penzias and Robert W. Wilson, and discuss the problems they encountered on their way to the discovery of cosmic background radiation. This type of real-world illustration of the topics of measurements, models, and analysis is meant to bring to life the world of physics. Throughout, we have emphasized physics as a way of thinking, investigating, and understanding rather than as a body of facts and theories.

Accuracy

We have made a strenuous effort to ensure accuracy.

■ Realistic Examples: We have made a point to make the text correspond to reality. By this we mean that if we use an example of an airliner accelerating to takeoff, the numerical values given for mass, takeoff speed, and so on, are those of a real airliner. We have tried to introduce reality by referring to real objects such as baseballs, golfballs, automobiles, and animals with realistic masses moving with realistic speeds.

■ Adherence to Nature: We have taken care to correctly describe what is actually observed in nature as, for example, in the description of the temperature dependence of electrical resistivity of metals (Section 18.3) and to give correct information on friction (Section 4.8).

■ Fidelity to History: We have read original papers and the current history-of-science research in order to ensure the accuracy of the presentation. Discussions of experiments correspond to what was actually done and discussions of theories correspond to what the authors actually wrote.

■ Answers: We were also determined to have correct answers to the end-of-chapter problems. Both of the authors have independently worked each of the problems. University of South Carolina student Jeremy Thomason also worked the problems and provided insight into wording them more clearly. We hope that this process not only has confirmed the right answers, but also has eliminated problems that students might find confusing.

Supplements

For the Student:

■ **Study Guide/Solutions Manual** This study guide provides a variety of exercises designed to help the student master the important concepts of each chapter. Solutions for the odd-numbered end-of-chapter problems from the text are in a separate section. The manual was written by the authors and Professor John Safko of the University of South Carolina.

■ **Interactive CD-ROM** This is included free with each textbook. It is a powerful study tool for the student—a browser-based CD with interactive simulations, animated sequences, quizzes, glossary, links to related web-based tutorial material, and more. The 2001 Update adds Practice Problems and Practice Quizzes that greatly expand the opportunities for solving problems like those in the book. Many of these new problems are taken from *Schaum's Outline of College Physics,* Ninth Edition, by Bueche and Hecht (the complete *Schaum's Outline* is available from McGraw-Hill).

For the Instructor:

■ **Instructor's Solutions Manual** This manual provides the answers and solutions to all the problems in the text. Expanded explanations will help with classroom discussions, as will the graphs and diagrams.

■ **Test Bank** This compilation has been updated and expanded for the third edition by Professor Nancy Woods of Des Moines Area Community College. It contains two thousand testing questions. It is available as a bound book or in computerized format for either Windows or Macintosh systems.

■ **Overhead Transparencies** A collection of 275 four-color illustrations from the text is provided on acetate for use with overhead projectors.

■ **Visual Resource Library** A group of 150 four-color illustrations is provided on a CD-ROM. This CD-ROM allows you to quickly browse through the images and captions from the text, arrange images in your own custom-designed slide show, and enlarge images for greater projection. You can also print full- and half-size images and export images for use in word-processing programs. Additional features include:

(1) find and sort thumbnail image records by name, type, location, and user-defined keywords; (2) search using keywords or terms; (3) view hundreds of images at the same time with the Small Gallery View; (4) flip through hundreds of images quickly using a mouse; (5) display all important file information for easy identification; (6) drag and place images into virtually any graphics, desktop publishing, presentation, or multimedia application.

Special Text/MEPI Package

WCB/McGraw-Hill is offering a package of Jones/Childers, *Contemporary College Physics* 3/e plus *MEPI; Multimedia Enhanced Physics Instruction,* a two-CD boxed set produced by Maha Ashour-Abdalla at the UCLA Space Plasma Simulation Group.

Using a hyperlink interface, *MEPI* gives students hands-on experience in exploring the world of physics through interactive demonstrations and experiments. Each topic in *MEPI* is comprised of five tutorial components:

■ *Concepts.* In-depth topical explanations

■ *Videos.* 28 short (3–6 minute) videos enable students to view a wide variety of actual experiments being performed

■ *Simulations.* Students change variables to set up different scenarios or experiments in simulated physical phenomena (see below for list of *MEPI* simulations)

■ *Quizzes.* Approximately 8–10 conceptual multiple-choice questions per chapter

■ *Problems.* 3–5 computational problems are first animated, then carefully outlined for step-by-step solving; students who enter wrong answers receive hints for guidance, and the system "keeps score."

Website

A *Contemporary College Physics* website is available with additional information, course-presentation material, practice problems (including MCAT-type questions), chapter-by-chapter web links, and more. The website address is www.mhhe.com/jones.

Acknowledgments

We have been blessed with the continued strong support of our wives, Betty Jones and Sigrid Childers. We thank them for all their encouragement and patience. We have had many profitable discussions with our colleagues at the University of South Carolina. We again thank Stuart Johnson, David Chelton, and Jennifer Albanese for their help with earlier editions. We are grateful to Jill Birschbach, for her photo-research efforts, and to the professional staff at McGraw-Hill Higher Education for its efforts in bringing this project to completion. A special note of thanks goes to Jean Fornango, Jill Peter, J. P. Lenney, Lisa Gottschalk, David Dietz, Dave Edwards, Lloyd Black, and Jim Smith.

Reviewers

We would like to thank the many physics teachers whose reviews, focus-group discussions, and personal communications helped to shape this third edition:

David B. Aaron, *South Dakota State University*
B. N. Narahari Achar, *University of Memphis*
Zaven Altounian, *McGill University*
Michael S. Berger, *Indiana University*
John Berryman, *Palm Beach Community College*
Shane C. Brower, *Youngstown State University*
Michael E. Browne, *University of Idaho*
James J. Carroll, *Youngstown State University*
Neal M. Cason, *University of Notre Dame*
K. Kelvin Cheng, *Texas Tech University*
Marek Cieplak, *Rutgers University*
R. Kent Clark, *University of South Alabama*
Lawrence B. Coleman, *University of California, Davis*
Lawrence Corrado, *University of Wisconsin, Manitowoc*
Mark Davenport, *San Antonio College*
Don DeYoung, *Grace College*
Chaden Djalai, *University of South Carolina*
Paul Draper, *University of Texas, Arlington*
Miles J. Dresser, *Washington State University*
Andrew Duffy, *Boston University*
John J. Dykla, *Loyola University, Chicago*
Angelo M. Ferrari, *Santa Fe Community College, Florida*
Leonard N. Feuerhelm, *Oklahoma Christian University*
Lewis Ford, *Texas A&M University*
Charles Gale, *McGill University*
J. David Gavenda, *University of Texas, Austin*
Simon George, *California State University*
Peter K. Glanz, *Rhode Island College*
John L. Hubisz, *North Carolina State University*
Fred W. Inman, *Mankato State University*
Larry D. Johnson, *Northeast Louisiana University*
Alain Karma, *Northeastern University*
Sanford Kern, *Colorado State University*
Diandra Leslie-Pelecky, *University of Nebraska*
Bo Lou, *Ferris State University*

Alfredo Louro, *University of Calgary*
Robert March, *University of Wisconsin, Madison*
William E. McCorkle, *West Liberty State College*
Marles L. McCurdy, *Tarrant County Junior College, Northeast*
Daniel McLaughlin, *University of Hartford*
Charles R. Meitzler, *Sam Houston State University*
R. D. Murphy, *University of Missouri*
David Mylander, *California State Polytechnic University*
Melvyn J. Oremland, *Pace University*
B. E. Powell, *State University of West Georgia*
C. W. Price, *Millersville University*
Michele Rallis, *Ohio State University*
Richard W. Robinett, *Penn State University*
Ian Robinson, *University of Champaign, Urbana*
Charles W. Rogers, *Southwestern Oklahoma State University*
Lawrence G. Rowan, *University of North Carolina at Chapel Hill*
Michael Simon, *Houstonic Community Technical College*
Igor Strakovsky, *The George Washington University*
Michael G. Strauss, *University of Oklahoma*
George Strobel, *University of Georgia*
James F. Sullivan, *University of Cincinnati*
Jeffery Sundquist, *Palm Beach Community College*
Andrew Sustich, *Arkansas State University*
John A. Swez, *Indiana State University*
Colin Terry, *Ventura College*
Herman Trivilino, *College of the Mainland*
David Vesper, *Indiana State University*
Chris Viulle, *Embry-Riddle University*
Gail Welsh, *Salisbury State University*
Donald A. Whitney, *Hampton University*
Luc T. Wille, *Florida Atlantic University*
Anthony Zito, *Dutchess Community College*

List of Selected Applications

Geology

Architecture

Sports

Everyday Life

Measurement, Models, and Analysis

T he word *physics* comes from a Greek word meaning "knowledge of nature." Physics attempts to describe the fundamental nature of the universe and how it works, always striving for the simplest explanations common to the most diverse behavior. For example, physics explains why rainbows have colors, what keeps a satellite in orbit, and what atoms and nuclei are made of. The goal of physics is to explain as many things as possible using as few laws as possible, revealing nature's underlying simplicity and beauty.

In achieving their goal, physicists construct models to represent the world around us. To the physicist, a **model** is an idealized description of a physical system or natural phenomenon. Such a model forms a conceptual framework that permits us to reduce complex situations into simpler, more understandable forms. For example, although we cannot see atoms, we can construct useful models of them that enable us to understand their behavior. Usually, such models of physical systems take a mathematical form. It should be understood that these models are by nature incomplete and, therefore, imperfect. For instance, we can describe the main features of the motion of a baseball if we use a model that ignores air resistance. Such a

1

model has its limitations, however, because it does not accurately describe the path of a curve ball thrown by a major league pitcher. We can obtain better agreement with observations by using a model that includes air resistance.

Physics is an experimental science. By this we mean that the acceptance of any physical theory depends on its success in predicting and explaining reproducible observations. To understand physics we must be able to connect our theoretical description of nature with our experimental observations of nature. This connection is made through quantitative measurements. In part, a thorough understanding of physical theories rests on knowing how measurements are made and how reliable the measured information is.

Until this century, scientists assumed that a sufficiently clever observer, given enough time and money, could, in principle, measure any thing or set of things as accurately as necessary. Our understanding of the measurement process is now more refined. We know that we cannot make a measurement that does not in some way affect the system being measured, thereby limiting the precision of our measurement. This limitation is of little or no importance in the everyday measurements with which we are most familiar: the length of a board or the speed of an automobile. However, we will find that when we come to submolecular processes, the interaction of the observer with the measured quantity cannot be ignored. ■

1.1 Measurements and Models

The role of conceptual models in physics can be demonstrated by discussing one in detail. We illustrate here the interplay of measurement and modeling using an example chosen from physics near the end of the last century. This example exhibits the main features of a physical model discussed in the introduction. It is an important model that we will consider in more detail in later chapters for it was a key to the development of twentieth-century physics.

When a solid object is heated to several hundred degrees Celsius, it becomes incandescent; that is, it glows. The radiation emitted by an incandescent object, such as a very hot tungsten filament, forms a continuous range of wavelengths, part of which lies in the visible range. As the filament's temperature is increased, the relative intensities of the light across the visible spectrum change, shifting the observed color from dull red to almost white (Fig. 1.1). Redder (longer) wavelengths correspond to lower energies while whiter or bluer (shorter) wavelengths signal higher energies.

Experiments show that the light emerging from most incandescent objects depends only on the temperature of the object and not on its composition. An ideal object for producing incandescent light is called a blackbody. A good approximation of an ideal blackbody is an enclosed oven or kiln with only a tiny opening for radiation (i.e., light) to get out. When the oven is cold, light from outside that passes into the opening is not reflected back (Fig. 1.2a) and so the

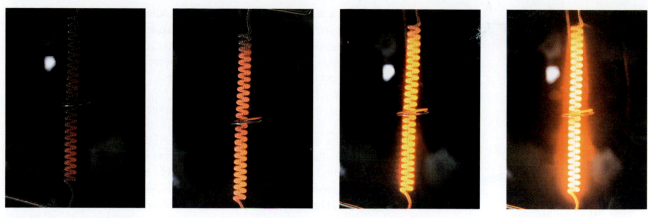

Figure 1.1

As the temperature of a hot filament increases, it emits larger amounts of radiation that peak at shorter and shorter wavelengths. The result is that the color shifts from dark red to bright white.

hole looks completely black. When the oven is very hot, light emitted inside passes out through the hole (Fig. 1.2b). A blackened lamp filament is also a good approximation of an ideal blackbody. Such a blackbody not only is a perfect radiator but also is a perfect absorber that absorbs all radiation falling on it.

During the latter part of the nineteenth century, scientists measured the intensity of the continuous spectrum of light emitted from hot objects. Figure 1.3 shows the radiant energy distribution for a blackbody at several temperatures. The intensity of the radiation is plotted along the vertical axis and wavelength (color) is plotted along the horizontal axis. As the temperature increases, the wavelength of the maximum intensity λ_m shifts to smaller values. The curve for 6000 K (about 5700°C or 10,300°F) corresponds to the approximate temperature

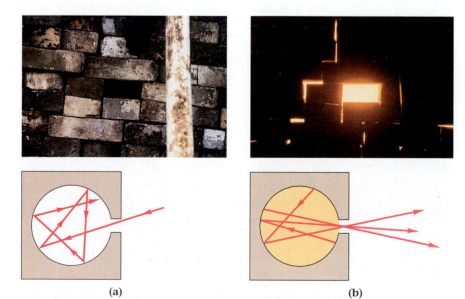

(a)   (b)

Figure 1.2

A small kiln of the type used to fire ceramics behaves nearly like an ideal blackbody cavity. (a) At a temperature of approximately 300 K, the kiln is very dark inside. Light entering from outside is absorbed. (b) When the kiln is heated to 1400 K, the interior glows brightly and the light emitted inside the cavity passes out through the opening.

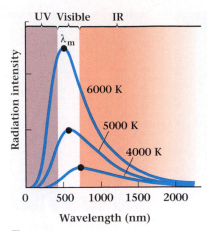

Figure 1.3

Blackbody spectra for several different temperatures. The peak of the curve (λ_m) shifts to shorter wavelengths at higher temperatures. The total amount of energy radiated is proportional to the area under the curve and increases with increasing temperature.

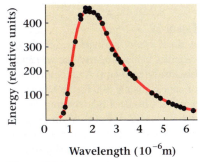

Figure 1.4

Spectral energy distribution of a blackbody at 1596 K. The line is computed from Planck's formula for blackbody radiation. The circles show observation by Coblenz published in *Bulletin of the National Bureau of Standards,* **13,** 436 (1916).

of the sun and has a maximum intensity in the middle (yellow-green) part of the visible spectrum. (The human eye is most sensitive to green light near the maximum of the sun's spectrum.)

In the 1880s and 1890s, numerous attempts were made to explain the shape of the blackbody spectrum. These explanations took the form of conceptual and mathematical models that not only would describe observed spectral curves such as we see in Fig. 1.3 but also would predict curves at wavelengths longer or shorter than visible light. A good model should explain what we can see and measure *and* predict things we have not yet observed. An important part of the scientific method is the construction of experiments to test the predictive value of models. A hundred years ago theoretical models based on the by then well-established—and presumed complete—laws of electromagnetism and thermodynamics failed to reproduce the experimental measurements adequately. The inability to explain blackbody radiation in terms of the framework of physics known in 1890 ultimately led to the development of a new branch of physics, commonly called *modern physics.*

The correct shape of the blackbody radiation curve was first found by Max Planck (1858–1947) as the result of a novel hypothesis. In 1900, Planck reported his discovery of an empirical formula that accurately described the shape of the blackbody spectrum for all observable wavelengths and temperatures. We know that Planck sought a model that would explain his formula (see Section 27.3). His initial model, based on classical theories, assumed that the energy produced through blackbody radiation behaved in a continuous manner. When this model failed to fit the observations, Planck postulated a model in which vibrational energy is *quantized,* that is, limited to certain discrete quantities.

Planck's quantum model was a modification of classical ideas that brought his theory into agreement with experimental observations. (The word *quantum* has the same origin as quantity, and means the smallest possible unit.) Physicists of the time had considerable doubt as to the validity of the quantization of electromagnetic radiation. Planck originally suspected that it was a mathematical trick that did not correspond to reality. However, his formula for blackbody radiation commanded attention because of its striking agreement with observations (Fig. 1.4). The replacement of the model in which energy flowed like a smooth unbroken stream of water, by one in which energy was thought of as coming in little packets, marked the beginning of quantum mechanics and the end of a time in which all physical explanations were in terms of continuous flows or motions.

A fascinating aspect of physics, and science in general, is that an important discovery—like the explanation of blackbody radiation—made years ago can have significance in the present. Though the idea that the universe started with a "big bang" was first proposed in the 1920s, it was not until the 1940s that serious attention was paid to establishing a functional Big Bang theory. Working through the equations that described a hot Big Bang, a few astronomers predicted that there should be surviving radiation from the early universe. Though this radiation initially was superhot and superdense, the expansion of the universe over billions of years should have resulted in the radiation becoming diffuse and cold. The initial prediction placed the temperature of this primordial radiation between 5 K and 50 K.* Later refinements in the early 1960s also indicated that the temperature would be near 5 K.

---

*The letter K is the symbol for the kelvin, the unit of temperature on the Kelvin scale. One kelvin is an interval the same size as the Celsius degree. The zero point of the Kelvin temperature scale is −273.15°C. The temperature of 0 K is called absolute zero.

In 1964, Arno Penzias and Robert W. Wilson discovered microwave radiation with a wavelength of 7.35 cm that was independent of where in the sky their antenna was aimed (see the box: Echoes of the Big Bang). These first measurements and others that soon followed at different wavelengths, were fitted to a Planck radiation curve corresponding to blackbody at a temperature of only 3 K above absolute zero. This radiation is called the cosmic background radiation and its existence is consistent with the prediction made by the Big Bang theory. For their discovery, Penzias and Wilson received the 1978 Nobel prize in physics.

Following Penzias and Wilson's original measurement, other researchers soon measured values at other radio wavelengths to verify the 3 K value. The peak of the blackbody curve for a temperature of 3 K occurs at a wavelength of 1.5 mm. Unfortunately, this wavelength is in the far infrared, which is blocked by the earth's atmosphere. In 1989, NASA launched the Cosmic Background Explorer (COBE) satellite, which proceeded to make a high-precision mapping of the sky (Fig. 1.5). Because the measurements have been so good, the average background temperature of the cosmic radiation is now known to be 2.735 ± 0.06 K.

We see in this example of blackbody radiation the interplay between measurement and analysis. Here, *analysis* means more than graphing the data to observe their essential features. In this context, it means fitting the data to some model (both conceptual and mathematical). Such analysis is almost always done against some background of preconceived ideas. The best models not only explain fully the observations that first led to the model but go on to predict new observations that were previously unexpected. In Planck's case, his original model had to be modified. Others later used his work to predict that the remnant of the radiation left over from the Big Bang would be found within a few

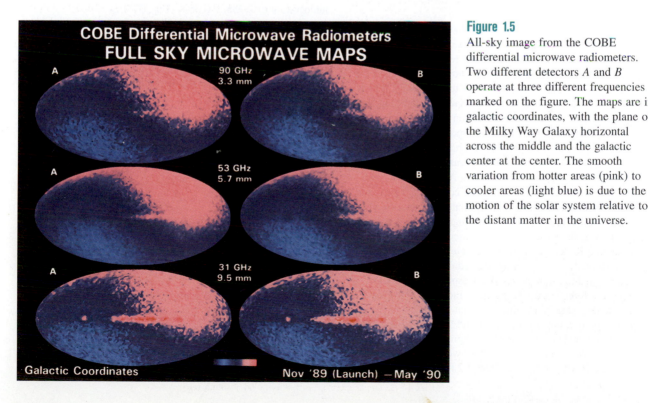

Figure 1.5

All-sky image from the COBE differential microwave radiometers. Two different detectors *A* and *B* operate at three different frequencies marked on the figure. The maps are in galactic coordinates, with the plane of the Milky Way Galaxy horizontal across the middle and the galactic center at the center. The smooth variation from hotter areas (pink) to cooler areas (light blue) is due to the motion of the solar system relative to the distant matter in the universe.

ECHOES OF THE BIG BANG

What is cosmic background radiation? How was it discovered? And what does it tell us about the universe?

In 1964, Arno Penzias and Robert W. Wilson of the Bell Telephone Laboratories (Fig. B1.1) began measuring the intensity of radio waves emitted from the gases that surround our galaxy. They used a special 20-foot antenna (called a radio telescope) built for communicating with the Echo and Telstar satellites. The antenna was much more directionally sensitive than were other radio telescopes of that era.

Penzias and Wilson set out to minimize the noise or background signals that had been seen in earlier measurements with the antenna. They considered obvious sources of noise including the Milky Way, the sun, poor antenna joints, even the pigeon droppings that had collected on the antenna. Despite their best efforts, a noticeable radio signal at a wavelength of 7.35 cm was still detected no matter where in the sky the antenna was pointed, the time of day, or the season of the year. The intensity of the signal at that wavelength corresponded to the intensity expected from a thermal radiator (blackbody) at a temperature of about 3 K, that is 3° above absolute zero. This radiation came to be known as the *cosmic background radiation.*

In the spring of 1965, Penzias and Wilson learned of the work of Robert Dicke and James Peebles of Princeton University, who had predicted a residual temperature of 10 K for the universe evolving from a particular point in time in the distant past. This model is referred to as the *Big Bang.* Meanwhile, David Wilkinson and Peter Roll had been building an antenna to look for the remanent radiation at a wavelength of 3.2 cm. The following year, they, too, reported finding radiation corresponding to a temperature of about 3 K.

The cosmic background radiation was first predicted in the late 1940s by George Gamow, Ralph Alpher, and Robert Herman who, in developing a Big-Bang cosmology to explain the formation of the elements, had estimated the necessary temperature and density. They showed that a consequence of their model was the existence of a cosmic background radiation at about 5 K. As the young universe expanded and cooled from its extremely hot beginning, the radiation emitted would survive, although by now it would correspond to a very low temperature. Moreover, the background radiation is isotropic, that is, it is the same in any direction that we observe.

In the years following Penzias's and Wilson's discovery, many measurements of the cosmic background radiation were made at other wavelengths. All of these were fit to the Planck law at the same temperature of about 2.7 K. In 1989, NASA

Figure B1.1 Arno Penzias (right) and Robert W. Wilson with their horn-shaped antenna in the background. Penzias and Wilson won the 1978 Nobel Prize in Physics for their discovery of the cosmic background radiation.

launched the Cosmic Background Explorer (the COBE satellite) designed to measure the cosmic background radiation over a wide range of wavelengths. Orbiting high above the earth's atmosphere, the COBE was able to make better observations than could be made with ground-based radio telescopes. The COBE operated for 10 months, taking many data points in all directions of space. When the data were finally analyzed, they fit precisely a Planck blackbody curve for a temperature of 2.735 ± 0.06 K (Fig. B1.2).

The noise that Penzias and Wilson could not eliminate turns out to be the residual thermal radiation left over from the beginning of the universe. This radiation, which shines on us from every direction, is a uniform glow coming from the most distant regions of the universe. It is the echo of the Big Bang.

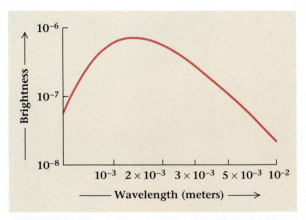

Figure B1.2 Intensity spectrum of the cosmic background radiation detected with the COBE satellite. The data points and their error bars are fit to a Planck curve for $T = 2.735 \pm 0.06$ K.

kelvins of absolute zero. Instruments were then built that could confirm the prediction.

The story of the Big Bang radiation shows how science works—an interplay between measurement and analysis. However, you should remember that even the best models are only as good as the assumptions that go into making them. For this reason we must be careful in applying models to complex situations and not forget the approximations that went into the models.

1.2 Units and Standards of Measurement

When the distance between two objects is measured, that measurement is reported as a number. However, the particular number given depends on the size of the basic unit of measurement. For example, suppose a surveyor needs to measure the distance between two stakes firmly stuck in the ground. If the distance is 25.4 meters, the surveyor must report both "25.4" and "meters." If he reports the distance in centimeters, he says "2540 cm." If he reports the distance in inches, he says "1000 inches."

We say that a measured quantity has **dimensions** when the size of the numerical result depends on the units chosen for measurement. The separation between the two stakes measured by the surveyor has the dimension of length. Time and mass are other familiar dimensions. In order for measurements made in different places and at different times to have meaning, we must first define **base,** or **fundamental, units.** Base units are sometimes referred to as standards.

There are only seven base units, corresponding to the quantities of length, time, mass, electric current, temperature, the amount of substance, and luminous intensity. All measured quantities are expressed in terms of these base units or combinations of them. For instance, a unit of area is meter^2, and a unit of speed is meters per second. Although various systems of units have been used over the years, scientists have generally agreed to use the **International System of Units.** The International System is also known by its abbreviation, SI, which comes from its French name, *Système International.* Table 1.1 lists the SI base units, which are largely based on what is often called the metric system. In the past, this system of units has also been called the mks system, in reference to the base units of meter, kilogram, and second.

Table 1.1 *SI Base Units*

Quantity	Name	Symbol
Length	meter	m
Mass	kilogram	kg
Time	second	s
Electric current	ampere	A
Thermodynamic temperature	kelvin	K
Amount of substance	mole	mol
Luminous intensity	candela	cd

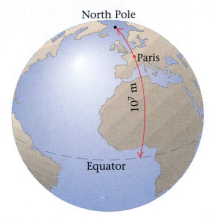

Figure 1.6

The meter was originally chosen to be one ten-millionth of the distance from the equator to the North Pole along a meridian through Paris.

Length

The basic unit of length, the meter, was approved by the first General Conference on Weights and Measures (CGPM) in 1889. It was defined as the distance between two fine lines engraved near the ends of a bar of platinum-iridium alloy when the bar was maintained at 0°C. The original international prototype meter, which is still kept at the International Bureau of Weights and Measures in Sèvres, France, was intended to be one ten-millionth of the distance from the equator to the North Pole along a line of longitude through Paris (Fig. 1.6). To give you some sense of the size of a meter, it is approximately the distance from the tip of your nose to the fingertips of your outstretched arm.

In 1983, the seventeenth General Conference on Weights and Measures redefined the meter in terms of the speed of light: The meter is the length of the path traveled by light in vacuum during a time interval of 1/299,792,458 of a second.

Mass

The base unit of mass, the kilogram, is defined as the mass of a prototype cylinder of platinum-iridium alloy that is kept at the International Bureau of Weights and Measures. Platinum-iridium was chosen because that alloy is particularly stable. The United States has a duplicate of the standard kilogram that is kept by the National Institute of Standards and Technology. In your everyday experience, the mass of a typical baseball bat is about one kilogram.

Time

The base unit for time is the second, which was originally defined as 1/86,400 of the mean solar day. In 1967, the CGPM adopted a standard based on the cesium-beam atomic clock, which can have a stability and accuracy of about one part in 10^{12} or even 10^{13}. The second is defined to be exactly 9,192,631,770 periods of a specified (microwave) radiation from cesium-133 atoms.

Understanding the nature of the atom allows us to use certain periodic properties of atoms as the characteristics that regulate the most modern timekeepers, such as the cesium-beam clocks. Over the years, improvements in timekeepers have often been related to ideas and developments that were in the forefront of science and technology. For example, advances in understanding the thermal properties of materials allowed for the development of temperature-compensated clocks. The use of electromagnets and electric switches brought about still further improvements in pendulum clocks. Studies of the electromechanical properties of quartz led eventually to crystal-controlled clocks that are many times more accurate than mechanical clocks. Figure 1.7 shows the improvement of timekeeping accuracy since the invention of the mechanical clock around the year 1300.

An estimate of the duration of one second is about the time required to say "one Mississippi." The time between human heartbeats is also approximately one second.

Symbols for Units

It is often more convenient to use the shorter symbol for a unit than to use its full name. Thus, we write 10 kg instead of 10 kilograms. Table 1.1 gives the

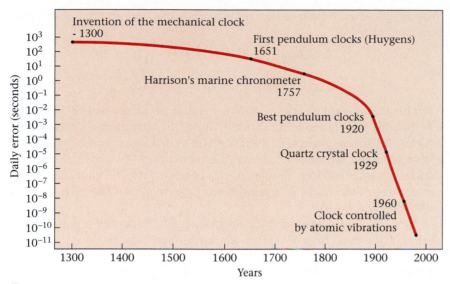

Figure 1.7

The diminishing daily error in clocks is shown as a function of the year in which the improvement was made.

symbols for the seven base units. Several rules govern the writing and use of these unit symbols: Roman (upright) type, generally lowercase, is used for symbols of units; however, if the symbols are derived from proper names, capital roman type is used for the first letter. Unit symbols are not followed by a period (except in. for inch) and do not change in the plural.

A product of two or more units may be indicated by

$$A \cdot m \quad \text{or} \quad A\,m.$$

A unit may also be raised to a power, as

$$m^3 \quad \text{or} \quad cm^2.$$

Units formed from two others by division may be indicated by a horizontal line, a solidus (an oblique line, /), or a negative power; for example,

$$\frac{m}{s}, \quad m/s, \quad \text{or} \quad m \cdot s^{-1}.$$

In complicated cases, negative powers or parentheses should be used; for example,

$$m \cdot kg/(s^3 \cdot A) \quad \text{or} \quad m \cdot kg \cdot s^{-3} \cdot A^{-1}.$$

Prefixes

In physics we frequently encounter numbers that are very large or very small. Such numbers are most easily expressed by the use of **scientific notation,** which uses powers of 10. This notation provides an easier way of writing large and small numbers and of doing arithmetic with them. To illustrate, note that we may write the number "one thousand" in the following way:

$$1000 = 10 \times 10 \times 10 = 10^3.$$

Table 1.2	SI Prefixes	
Factor	Prefix	Symbol
10^{18}	exa	E
10^{15}	peta	P
10^{12}	tera	T
10^{9}	giga	G
10^{6}	mega	M
10^{3}	kilo	k
10^{2}	hecto	h
10^{1}	deka	da
10^{-1}	deci	d
10^{-2}	centi	c
10^{-3}	milli	m
10^{-6}	micro	μ
10^{-9}	nano	n
10^{-12}	pico	p
10^{-15}	femto	f
10^{-18}	atto	a

With this notation, we can conveniently express very large and very small numbers. For example,

$$127,000,000 = 1.27 \times 100,000,000 = 1.27 \times 10^8$$

and

$$0.00037 = 3.7 \times 0.0001 = 3.7 \times 10^{-4}.$$

Using the scientific notation, 1670 meters can be written as 1.67×10^3 meters. But this same quantity can also be written 1.67 kilometers—abbreviated as 1.67 km. This is an example of using a prefix to indicate a decimal multiple of a base unit. Here the prefix kilo means one thousand or 10^3. Table 1.2 gives the prefixes for other multiples or submultiples of units. Table 1.3 gives the approximate values of various lengths, masses, and times in units with prefixes and expressed in powers of 10.

Numbers expressed in scientific notation may be multiplied and divided according to the usual rules of algebra. Remember that

$$10^n \times 10^m = 10^{n+m} \quad \text{and} \quad \frac{10^a}{10^b} = 10^{a-b}.$$

When using numbers with prefixes, they should be converted to scientific notation in the proper units before being used in calculations.

Example 1.1

According to the official rules for international soccer, the minimum size for the rectangular playing field is 100 m by 64 m. Compute the area of a minimum-sized soccer field in units of kilometers squared.

Solution The area of the field is the product of length and width. If we express the length and width in units of km, then the area will be given in km^2. Remember that 1 km equals 1000 m so that $1 \text{ m} = 10^{-3}$ km.

$$\text{Area} = (100 \text{ m})(64 \text{ m}) = (0.100 \text{ km})(0.064 \text{ km}) = 0.0064 \text{ km}^2$$

$$\text{Area} = 6.4 \times 10^{-3} \text{ km}^2.$$

Table 1.3	Approximate Values of Some Physical Quantities		
Quantity		Magnitude	
Radius of a proton	1 femtometer	1 fm	1×10^{-15} m
Distance from Atlanta to San Diego	3000 kilometers	3000 km	3×10^3 km $= 3 \times 10^6$ m
Mass of a small marble	5 grams	5 g	5×10^{-3} kg
Mass of a grain of salt	1 milligram	1 mg	1×10^{-3} g $= 1 \times 10^{-6}$ kg
Time for light to travel 0.3 m	1 nanosecond	1 ns	1×10^{-9} s
Time for sound to travel 1.0 m	3 milliseconds	3 ms	3×10^{-3} s

Example 1.2

What is the volume of a rectangular sheet of paper that is 8.60×10^{-3} cm thick, 21.6 cm wide, and 27.9 cm long?

Solution The volume V of a rectangular-shaped piece of paper is given by the product of the length, width, and thickness:

$$V = \text{length} \times \text{width} \times \text{thickness}$$
$$V = (27.9 \text{ cm})(21.6 \text{ cm})(8.60 \times 10^{-3} \text{ cm}) = 5.18 \text{ cm}^3.$$

1.3 Unit Conversions

Frequently you will need to change from one set of units to another. You may have to change a given number of seconds into minutes, or a given number of inches into centimeters or meters. A systematic method for doing this starts with writing a conversion factor, a fraction whose numerator in one set of units equals the denominator in the other set of units. For example, the inch is defined to be exactly 2.54 cm.

$$1 \text{ in.} = 2.54 \text{ cm.}$$

We can rewrite this equation as

$$1 = \frac{2.54 \text{ cm}}{1 \text{ in.}} \quad \text{or} \quad 1 = \frac{1 \text{ in.}}{2.54 \text{ cm}}.$$

Both of these fractions are pure numbers equal to 1. That is, the quantity on the right-hand side of the equation has the value of unity and has no dimensions, since it is equal to the left-hand side. Consequently, we can multiply by 1 in the form of either of these fractions in any equation where they are needed. To change units, you simply multiply the quantity to be converted by the conversion factor chosen so that the undesired units cancel out. Several brief examples of unit conversions will help illustrate the procedure.

Example 1.3

How many centimeters are there in one foot?

Strategy We know that the foot is defined to be exactly 12 in. and that the inch is defined to be exactly 2.54 cm. Thus we can express the foot as 12 in. and then use a conversion factor to take us from inches to centimeters.

Solution We start with the equation

$$1 \text{ ft} = 12 \text{ in.,}$$

and multiply the right-hand side by 1 in the form of 2.54 cm/1 in.:

$$1 \text{ ft} = (12 \text{ in.})(1) = (12 \text{ in.})\left(\frac{2.54 \text{ cm}}{1 \text{ in.}}\right) = 30.48 \text{ cm.}$$

Notice that when you use this procedure, the correctness of the units ensures the correctness of the answer.

Example 1.4

How many kilometers are in one mile?

Strategy We know that 1 km is 1000 m, 1 m is 100 cm, and 1 mile is 5280 ft. We can find the number of kilometers in one mile by using these relationships along with the result of Example 1.3 relating cm to ft.

Solution Using the outlined procedure, we obtain

$$1 \text{ mi} = (1 \text{ mi})\left(\frac{5280 \text{ ft}}{1 \text{ mi}}\right)\left(\frac{30.48 \text{ cm}}{1 \text{ ft}}\right)\left(\frac{1 \text{ m}}{100 \text{ cm}}\right)\left(\frac{1 \text{ km}}{1000 \text{ m}}\right)$$

$$1 \text{ mi} = 1.609 \text{ km}.$$

Example 1.5

An automobile has a speed of 30 mi/h. What is its speed in m/s?

Solution Following the method used in the preceding examples, we use a succession of conversion factors to change the units to the desired form:

$$30\frac{\text{mi}}{\text{hr}} = \left(\frac{30 \text{ mi}}{1 \text{ h}}\right)\left(\frac{1 \text{ h}}{60 \text{ min}}\right)\left(\frac{1 \text{ min}}{60 \text{ s}}\right)\left(\frac{1.609 \text{ km}}{1 \text{ mi}}\right)\left(\frac{1000 \text{ m}}{1 \text{ km}}\right)$$

$$\frac{30 \text{ mi}}{\text{hr}} = 13.4 \text{ m/s}.$$

1.4 Measurements, Calculations, and Uncertainties

There is an important difference between how we use numbers in arithmetic and algebra and how we use them in science. This difference arises from the nature of measurement. When we say that a length is 9.2 cm, we mean that this is a length that has been, or can be, measured. Implicit in our statement is information about the precision with which the measurement has been made. By 9.2 cm, we mean a length that is closer to 9.2 cm than to either 9.1 cm or 9.3 cm. Had we estimated between the 0.1-cm divisions, or used a ruler with finer divisions, we might have said the length was 9.23 cm (Fig. 1.8). This means a length closer to 9.23 cm than to 9.22 cm or 9.24 cm. The point is that the last figure in a quoted measurement is in some sense uncertain and lacks the same significance as the figures to its left.

The number of digits reported in a measurement, irrespective of the location of the decimal place, is called the number of **significant figures.** The number of significant figures is a reflection of how well you know a given quantity. For

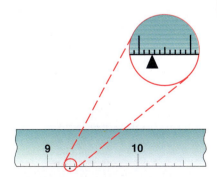

Figure 1.8
Position of a reference mark along a scale. The inset shows a magnified view, which permits greater precision in locating the position of the mark.

example, length is, in the classical sense, assumed to be infinitely divisible. This means that to specify a length "exactly" would require an infinite number of significant figures. Therefore, all numbers that represent lengths are inexact, but they do carry with them information about how well those lengths are known.

Measured values are often used to calculate other quantities. For example, we can determine the circumference C of a circle from a measurement of the radius r of the circle and the expression

$$C = 2\pi r. \tag{1.1}$$

The value of π is well known and has been determined to many significant figures. To seven significant figures, $\pi = 3.141593$. If we had measured the radius of a circle to be 1.60 cm, direct application of Eq. (1.1) would give a numerical answer of

$$C = 10.053098 \text{ cm.}$$

However, this is not a physically meaningful or sensible answer. The circumference can be known only to the same precision as the radius. In this case, the precision is three significant figures. Therefore we must *round off* the answer to the correct number of significant figures:

$$C = 10.1 \text{ cm.}$$

Note that the last figure has been rounded up. The rule we will follow in this book is that if the first digit beyond the last significant figure is 5 or greater, the last significant figure is to be increased by unity (1). All other figures beyond the last significant figure are dropped. If the digit beyond the last significant figure is less than 5, the last significant figure remains unchanged. Thus 10.05 rounds off to 10.1, but 10.04 rounds off to 10.0. (Most pocket calculators that round off automatically use this rule.)

Another numerical example will further illustrate the use of significant figures. Let us calculate the area of a paperback book cover. Suppose the cover is a rectangle whose sides we measure to be 10.6 cm and 17.9 cm. To find the area A of a rectangle, we multiply the length l times the width w:

$$A = l \times w.$$

The product of 10.6 and 17.9 is 189.74. However, it is *not* correct to give the area of the cover as 189.74 cm^2. To do so would imply that the area is known to be between 189.73 cm^2 and 189.75 cm^2, a precision that is unwarranted on the basis of the precision with which the lengths of the individual sides are known.

To see what we mean, notice that the length of the first edge of the cover is between 10.55 cm and 10.64 cm, and the length of the second edge is between 17.85 cm and 17.94 cm. The product of the least of these numbers is 188.3175, and the product of the greater is 190.8816. We see, then, that even the last place *before* the decimal is uncertain, so it would be a mistake to give an answer with more than three figures. Thus, we round off the computed area of 189.74 cm^2 to 190 cm^2. The general rule is that your answer must have no more significant figures than is warranted by the least precise of your values (that is, the value with the fewest significant figures).

When multiplying or dividing several quantities, the number of significant figures in the result is the same as the number of significant figures in the factor with the least number of significant figures.

The precision of the sum or difference of two quantities is no better than that of the least-precise quantity. For example,

$$14.1 \text{ cm} + 1.32 \text{ cm} = 15.4 \text{ cm}.$$

Even though both of these numbers have three significant figures, the first number is less precise as it is given only to tenths of cm while the second number is given to hundredths of cm. Consequently, the sum is given only to the nearest tenth of a centimeter. Thus, we have the rule:

When adding or subtracting two quantities, the least significant figure in the result occupies the same position relative to the decimal point as the position of the last significant figure in the number whose least significant figure is farthest to the left.

Sometimes it is not clear whether the final zero or zeros in a number are significant figures or are merely needed to locate the decimal point. In the preceding example, the result of 190 cm was correct to three figures so that the zero was a significant figure. To clearly indicate the number of significant figures, we often express a number in scientific notation. Thus, we write 190 cm as 1.90×10^2 cm. The presence of the final zero indicates that the number is known to three significant figures. We will assume integers of numeration to be precise. Thus, the answer to "What is the cost of 2 hamburgers at $1.89 each?" is $3.78 and not $4.

Example 1.6

|← 10.2 cm →|

18.4 cm

$V = hA = h\pi(d/2)^2$

Figure 1.9
Example 1.6: Calculating the volume of a cylindrical oatmeal box.

Calculate the volume of a cylindrical oatmeal box with a diameter of 10.2 cm and a height of 18.4 cm (Fig. 1.9).

Solution   The volume of a cylinder is the product of the height h and the area of the base,

$$V = hA = h\pi\left(\frac{d}{2}\right)^2,$$

where d is the diameter. Upon inserting the values for h, π, and d, we get

$$V = 18.4 \text{ cm} \times 3.14 \times \left(\frac{10.2}{2}\right)^2 \text{ cm}^2.$$

Direct computation with a hand calculator yields

$$V = 1502.75376 \text{ cm}^3 = 1.50275376 \times 10^3 \text{ cm}^3.$$

However, since the dimensions are known to only three significant figures, the answer must be rounded off to give

$$V = 1.50 \times 10^3 \text{ cm}^3.$$

| 1.5 | ## Estimates and Order-of-Magnitude Calculations |

As we have just seen, it is important to keep track of uncertainties in measurement when calculating answers to problems. Sometimes, in everyday life as well as in science, we need to solve a problem for which we don't have enough information for a precise answer. We can often obtain a useful answer by estimating the values of the appropriate quantities. These estimates, usually made to the nearest power of ten, are called **order-of-magnitude** estimates. The resulting order-of-magnitude calculation is not exact, but usually it is accurate to within a factor of ten. Just knowing the order of magnitude of physical quantities often provides enough information for us to gain a useful understanding of the physical situation and to be able to make judgments and rough calculations for constructing models. Figure 1.10 shows the range of the magnitudes of lengths encountered in physics.

Making order-of-magnitude estimates is often easy. For example, imagine you are going to college for the first time and that you want to estimate how much money you need for books. You know that the usual load for most students is five courses and that each course requires a textbook. You can estimate the cost of a single book using the following reasoning. You know from experience that $1 is too little and that $100 too much. Even $10 is low. A reasonable estimate might be $50. Thus, the estimated cost of books for one term is 5 × $50 = $250. Although the result is certainly not accurate, it is the right order of magnitude and provides a reasonable estimate to a real problem. The following example further illustrates the application of order-of-magnitude estimates.

When making computations of this sort we often make other approximations as well. Replacing π with 3 or replacing $\sqrt{2}$ with $\frac{3}{2}$ makes little difference in the order of magnitude, but doing so greatly simplifies the calculations. The following examples illustrate the technique.

Example 1.7

Estimate the volume of rubber worn from automobile tires each year in the United States. The average radial tire has a useful tread depth of $\frac{5}{16}$ in. and can be driven 35,000 mi before it is worn out.

Strategy We begin by estimating the number of automobiles in the United States and the average number of miles driven per car each year so that we can find the total number of miles driven per year. From that result we can compute the equivalent number of tires worn out each year. Finally, we compute the volume of rubber lost from each worn out tire and multiply by the number of worn out tires to get the total volume of rubber.

Solution The number of cars can be estimated from the size of the population, which is about 260 million people. We know that not every person has a car and that many families have more than one. If we estimate that there is about one car for every four people, then we get approximately 65 million automobiles. The average number of miles driven/year is about 15,000 mi/year.

10⁷ meters
The earth. The average radius of the earth is roughly 6×10^6 m. This is almost twice the radius of Mars, but less than one-tenth the radius of Jupiter.

10²¹ meters
The Andromeda galaxy. This spiral galaxy is our close neighbor within the so-called "Local Group" of galaxies, lying about 2×10^{22} m away from the earth. It is similar in form to our Milky Way galaxy, which contains about 10^{11} individual stars.

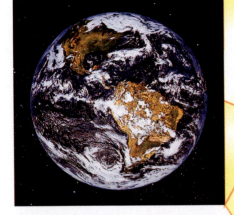

10⁰ meters
People. Most humans are 1.5–2.0 m tall. The present population of the United States is about 2.6×10^8 people, so if we were all laid out end to end in a straight line, we would cover a distance about 30 times greater than the diameter of the earth.

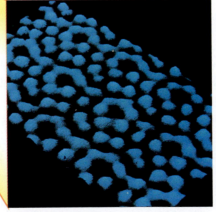

10⁻¹⁰ meters
An atom. This image from an electron tunneling microscope shows individual atoms of silicon, located at the corners of the hexagonal pattern. The nucleus of the atom is only 10^{-14} m in size, so the atom as a whole consists mostly of empty space.

10⁻⁵ meters
A red blood cell, or erythrocyte. A human contains about 2.5×10^{13} erythrocytes, which function primarily to transport oxygen within the body. New cells form at a rate of about 10^6 per second to replace dying cells and keep a constant number in the blood.

Figure 1.10
The range of sizes in physics is from the very large to the very small. Here we see a range of sizes covering 31 orders of magnitude. Physicists have directly observed distances spanning nearly 41 orders of magnitude, from the diameter of a proton (10^{-15} m) to the most distant galaxies seen by the Hubble Space Telescope (10^{26} m or 12 billion light-years). Extending this range further, particle physicists theorize that quarks, which have not been "seen," have dimensions on the order of 10^{-18} m.

(Surely 1500 mi/year is too low and 150,000 mi/year is too high.) From these numbers we get the number of tire miles per year.

Total tire miles/year ≈ number of cars × 4 tires/car × 15,000 mi/year.

(The symbol ≈ means that the quantity on the left is *approximately* equal to the quantity on the right.) The number N of worn out tires each year is given by

$$N = \frac{\text{total tire miles/year}}{\text{miles/worn out tire}}$$

$$N \approx \frac{(6.5 \times 10^7 \text{ cars})(4 \text{ tires/car})(1.5 \times 10^4 \text{ mi/year})}{3.5 \times 10^4 \text{ mi}} \approx 1.1 \times 10^8 \text{ tires/year.}$$

The volume of rubber worn off each tire is given by

$$V_1 = 2\pi r \times \text{width} \times \text{thickness of tread worn off,}$$

where r = radius of the tire ≈ 15 in. ≈ 38 cm, width ≈ 6 in. ≈ 15 cm, and thickness ≈ $\frac{5}{16}$ in. ≈ 0.8 cm. The total volume of rubber worn from all tires in one year is

$$V_T = NV_1 \times 1 \text{ year}$$
$$V_T \approx (1.1 \times 10^8 \text{ tires/year})(2\pi)(38 \text{ cm})(15 \text{ cm})(0.8 \text{ cm})(1 \text{ year})$$
$$V_T \approx 3.2 \times 10^{11} \text{ cm}^3 \approx 3 \times 10^5 \text{ m}^3.$$

Thus, we estimate that the volume of rubber worn each year from automobiles in the United States is about $3 \times 10^5 \text{ m}^3$.

Discussion Just how big is a pile of rubber $3 \times 10^5 \text{ m}^3$? If it was piled 1 m deep across two lanes of an interstate highway (width ≈ 7 m), it would reach more than 40 km.

Example 1.8

A record store offers a prize to the customer with the guess closest to the correct number of jelly beans that fill a liter jar on a display counter in the store. (One liter is equal to 1000 cm³.) Estimate what that number should be.

Strategy A careful look at the jar (Fig. 1.11) reveals several things. The jelly beans can be roughly approximated by little cylinders about 2 cm long by about 1.5 cm in diameter. Furthermore, the jelly beans are not tightly packed; perhaps only as much as 80% of the volume of the jar is filled. We can use these observations to estimate the number of beans in the jar.

Solution The number of jelly beans is the occupied volume of the jar divided by the volume of a single jelly bean,

$$\text{number of beans} = \frac{\text{occupied volume of jar}}{\text{volume of 1 bean}}.$$

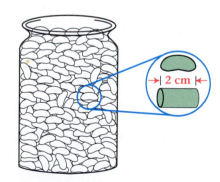

Figure 1.11
Example 1.8: How many jelly beans fill a one-liter jar?

The volume of one jelly bean is approximated by the volume of a small cylinder,

$$\text{volume of 1 bean} = h\pi\left(\frac{d}{2}\right)^2 \approx 2\ \text{cm} \times 3\left(\frac{\frac{3}{2}\ \text{cm}}{2}\right)^2 = \frac{27}{8}\ \text{cm}^3.$$

Thus the approximate number of beans in the jar is

$$\text{number of beans} \approx \frac{0.80 \times 1000\ \text{cm}^3}{\frac{27}{8}\ \text{cm}^3} \approx 240.$$

An actual count, by one of the authors, of the jelly beans filling a quart jar (0.95 liter) gave 255 beans.

1.6 How to Study Physics

In the first five sections of this chapter, you have seen only a little of what physics is about. Even so, you have probably noticed one thing: words are important. Words are frequently given a more precise meaning in the physical sciences than in ordinary use. To make headway in learning physics, you need to know the meaning of these key words, or terms, as they are used in physics. The key terms in this book have been written in **boldface type** when they are first introduced and defined. As a help in studying, these same terms are gathered together at the end of each chapter in the summary. You should write out the definition of each term, just as you write out vocabulary items when learning a new language. Doing so will help ensure that you learn their precise meanings. Learning the precise meaning of the terms here is just as important as learning the meaning of new words in a language.

In fact, you should always study with a pen or pencil in your hand. (What we are talking about is entirely different from underlining or highlighting part of the text.) You should outline the ideas as they are presented and work out the algebraic steps for yourself. Then you should go on to an example and work through it. By writing out the material you will notice details that reading alone may not convey. Just as words and ideas are important, symbols are important. A lowercase letter, say t, will not stand for the same thing as a capital letter, in this case T. Thus, mA does not mean the same thing as MA. Likewise—as we will see in Chapter 3—the boldface letter **A** will not mean the same thing as A. Be careful when writing the Greek letters so that they are not confused with other letters or symbols.

When working for yourself—as well as when working problems to turn in or on a test—you should always help yourself by arranging your work in an orderly way. It is no accident that the equations in a textbook are placed one under another, as in Example 1.7 and elsewhere. This layout makes it easier for you to understand the text. Likewise such a procedure will help you as you try to understand the material. Don't scatter equations around the page, but start at the top and work down going from left to right as you proceed. Furthermore, using a systematic and orderly procedure will help you as you try to tell others—perhaps the graders of your work—about your solutions. Do not make learning more difficult by ignoring purely mechanical procedures.

The mathematical formulas that you will encounter represent more than algebraic equations. The formulas are shorthand for physical situations and therefore represent ideas. You must understand where and when the ideas and the accompanying formulas apply. Always include the units when you substitute a value for a symbol in an equation. With almost no exceptions, all the symbols in equations stand for physical quantities with units. When you are carrying out mathematical manipulations for your private study, for homework, or on a test, it is a good idea to make some comments in words to indicate what you are doing and why. Doing so will make it easier for you when studying and reviewing and will help your grader understand what you did.

When you reach the end of a section in the text, you should ask yourself what key ideas and terms have been introduced. Write them down and compare what you have with the items listed in the summary. After mastering the ideas and concepts of the text, you should work some of the problems at the end of the chapter. Doing so will reinforce the ideas that you have been learning.

1.7 Problem Solving

Learning physics requires more than just learning new terms and definitions and stating the concepts and laws. To find out what physics is really about, you must learn how to apply these concepts and laws to real or hypothetical situations. Experience has shown that this kind of learning cannot occur without practice. For this reason, in a physics course you may spend more time working problems than you do reading the text.

In order to be productive in your study and to make the best use of your time, you must recognize that problem solving is much more than merely substituting numbers in a formula or fitting together the pieces of a jigsaw puzzle. Thumbing through the book to find a formula that seems to fit or a worked-out example that resembles your particular problem is a waste of time and effort. You should begin by studying the ideas, the concepts, and their relationships first. Then you attempt the problems as a way to find out for yourself whether or not you understand the material.

Even though the worked-out examples and problems in this text may sometimes represent an oversimplification of nature—for example, we might ignore the effects of air resistance on a moving body—they illustrate basic ideas, concepts, and techniques that you must master before going on to more complicated situations. The most helpful advice we can offer you is to use the working of problems as a method of studying. When you begin studying a chapter, go over the text examples with pencil and paper, work them out carefully in more detail than the text does, and then proceed to the problems. Although practice is the most important thing, you may find the following general rules to be helpful.

Problem-Solving Guidelines

1. **Read the entire problem carefully.** Then read it again to try and find out what you are being told. Don't worry about the question at first; focus more on what information you are being given.

2. Whenever possible, **draw a diagram of the physical situation.** Label the diagram with the information given in the problem. Be sure to include units, such as meters or kilograms, with the quantities. If some standard symbols have been introduced, label the parts of the diagram with them, too. For example, you might label r for radius, h for height, or l for distance. You might also make a list of the known and unknown quantities involved.

3. Only after you are sure you understand what is given and after you have labeled the diagram should you **tackle the question.** It is often a good idea to briefly write down the question, using symbols.

4. The next step is to **find a mathematical relationship between the known and unknown quantities.** For example, if asked how many dimes make $1.30, you call on your knowledge of the value of a dime. Or, if asked how far an auto goes in 2 hours at 30 kilometers per hour, you call on another type of knowledge. In most cases, you will need to write the relationship between the known and unknown quantities in the form of an equation, or perhaps several equations.

5. Next you should **solve the equation, or equations, for the unknown quantity or quantities.** This means rearranging the formulas in accord with the rules of algebra so that you have an equation with the unknown on the left-hand side of the equals sign and all of the known quantities and constants on the right-hand side.

6. Now, and only now, should you **substitute numerical values into the equation.** Do not substitute just "bare numbers," but substitute both the numerical values *and* the units. As indicated in Example 1.7, units are multiplied and divided as if they were algebraic quantities. Your answer should then come out in the appropriate units. If the units do not come out correctly, you have probably made some basic error. On the other hand, if the units are correct, there is a higher probability that your work is correct. Remember that in giving your final answer you must give the proper number of significant figures.

7. As a final check, you should **consider whether your answer is reasonable.** Does your result have the proper order of magnitude? You may even carry out a quick order-of-magnitude estimate as a way of confirming your work.

As an example of these ideas, let us work a sample problem.

Example 1.9

Suppose you get into your car and travel 25 km due north. Then you turn and travel an unknown distance east. Finally, you travel 32 km directly back to your starting point. How long was the eastern leg of your trip?

Strategy and Solution   By reading the problem carefully and drawing a diagram similar to Fig. 1.12, you will have covered steps 1, 2, and 3.

For step 4, recall the relationship between the sides of a right triangle, which is what the diagram shows. This relationship is the Pythagorean theorem: The square of the hypotenuse is equal to the sum of the squares of the other two sides. In terms of the symbols on the diagram, this theorem can be written

$$c^2 = a^2 + b^2.$$

In this equation, the unknown quantity is b and the known quantities are a and c. We may isolate the unknown quantity by subtracting a^2 from both sides to get

$$b^2 = c^2 - a^2.$$

To complete the solution, we take the square root of both sides,

$$b = \sqrt{c^2 - a^2}.$$

The negative solution can be discarded because a negative length is not physically meaningful.

Now, in accord with step 6, we substitute the values and the units for the algebraic symbols:

$$b = \sqrt{(32 \text{ km})^2 - (25 \text{ km})^2}$$
$$b = \sqrt{1024 \text{ km}^2 - 625 \text{ km}^2} = \sqrt{399 \text{ km}^2}$$
$$b = 20 \text{ km}.$$

Discussion Notice how the units were handled and how many significant figures were kept in the answer. The rounding off was done last after all of the calculations were made.

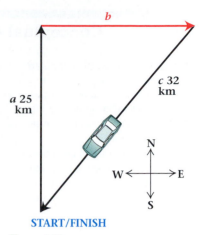

START/FINISH

Figure 1.12
Example 1.9: How long is the distance b?

Summary

Useful Concepts

■ A model is an idealized description of a physical system that enables us to form a conceptual framework for understanding that system.

■ There are seven SI base units or standards. The names of the base units for mass, length, and time are kilogram, meter, and second, respectively.

■ Numbers expressed in scientific notation can be multiplied and divided in the following way:

$$(A \times 10^n) \times (B \times 10^m) = (A \times B) \times 10^{n+m}$$
$$\frac{C \times 10^n}{D \times 10^m} = \left[\frac{C}{D}\right] \times 10^{n-m}.$$

■ To convert from one system of units to another, first write the conversion factor as a fraction whose numerator and denominator are physically equal and then multiply this fraction by the quantity to be converted.

■ Numbers arising from measurement convey the precision of the measurement by the number of significant figures used.

■ The results of a calculation may not have more significant figures than is warranted by the least precise of the values used. When multiplying or dividing, the number of significant figures in the result is limited by the number of significant figures in the least-precise value. When adding or subtracting numbers, the number of decimal places in the result is the same as that of the value with the fewest decimal places.

Important Terms

You should be able to write the definition or meaning of each of the following:

model	scientific notation
dimensions	significant figures
base, or fundamental, unit	order of magnitude
International System of Units	

Conceptual Questions

1.1 Suppose you are planning a trip by car to another city and you estimate the time required to get there. Show how this estimate depends on a model, as described in the text. What does the model depend on and how reliable is it?

1.2 Give a personal example of the use of a model for analysis of measured data.

1.3 Make a list of those quantities for which you think it might be helpful to establish standard units. Why?

1.4 Explain the difference between dimensions and units.

1.5 What problems are there in maintaining and using a metal bar as the primary length standard as was done before the meter was redefined in 1983?

1.6 What problems are there in maintaining and using a primary time standard in which the second is defined as 1/86,400 of a mean solar day?

1.7 Explain the basic idea behind unit conversion.

1.8 Discuss the difference in meaning of the three quantities 10 m, 10.0 m, and 10.00 m.

1.9 Estimate the precision of your watch or a clock. You may do this by using radio or television time signals spaced 24 or 48 hours apart.

1.10 Which of the following numbers is given to three significant figures: 0.003 m, 0.32 cm, 0.320 cm, 3.21 mm, or 3.213 mm?

1.11 A student measures a rectangle with a meterstick that measures no better than ±1 mm. She finds the height to be 37 mm and the width to be 46 mm. Why does she then report the area of the rectangle to be 1700 mm^2 instead of 1702 mm^2?

1.12 Estimate the total weight of the population of the United States.

Problems

Section 1.1 Measurements and Models

> **Hints for Solving Problems**
>
> The problems in this section ask you to devise a physical model that explains a given set of observations. In conceiving your model, you may assume a "perfect" world in which friction and other perturbing effects do not influence the working of your model. The model may be conceptual or mathematical.

1.1 What model most simply describes the following observations? (a) A ball placed anywhere on the floor remains at rest. (b) A ball placed anywhere on the floor begins to roll. (c) Give other, less-simple models for these observations.

1.2 A handful of small pebbles is dropped down a large, vertical tube. Only one-half of the pebbles fall out of the lower end. (a) Devise a simple model for a device within the tube that accounts for the observations. (b) Devise another, less-simple model.

1.3 A die is thrown many times with the following results for the number showing on the top face: 1, 63 times; 2, 58 times; 3, 62 times; 4, 63 times; 5, 75 times, and 6, 61 times. What model can you make for the die?

1.4 An insect travels between two fixed points 0.50 meters apart in 4.0 seconds. Another insect travels between fixed points 1.0 meters apart in 2.0 seconds. (a) What is the simplest model for the relationship between the way the insects move? (b) Propose another possible model.

1.5 A metal cube floats in a liquid. What is the simplest model of the cube and liquid? Are there other models?

Section 1.2 Units and Standards of Measurement

1.6 Express 2500 m in kilometers and in centimeters.

1.7 One liter (L) is a volume of 10^3 cm^3. How many cubic centimeters are in 2.5 milliliters?

1.8 How many picoseconds are there in 9.2 microseconds? 9.2×10^6

1.9 How far will light go in a vacuum in 1.0 nanosecond? (Speed of light = 3.0 × 10^8 m/s)

1.10 What fraction of a kilogram is (a) a milligram and (b) a microgram? a) $\overline{1\,000\,000}$ b) $1/1\,000\,000\,000$

1.11 The black grains in some types of photographic films are about 0.8 μm across. Assume that the grains have a square cross section and that they all lie in a single plane in the film. How many grains are required to completely obscure 1 square centimeter of film?

1.12 The hard-drive memory of a desktop computer used for word processing has a capacity of 2 GByte (GB). (A byte is the amount of memory needed to store one alphabetic character.) A single page of double-spaced text requires about 5 kB of memory and a simple line drawing takes about 15 kB. How many pages of text and drawings can be stored on the hard drive if there is one drawing for every two pages of text material?

1.13 A formula reads $y = \frac{1}{2}at^2$ where y is in meters and t is in seconds. What are the dimensions of a?

1.14• During the total eclipse of the sun, the moon just obscures the sun. The distance from the earth to the moon is 3.84 × 10^8 m and the distance from the earth to the sun is

1.50×10^{11} m. Estimate the relative diameter of the sun and moon. $\frac{1.5 \times 10^{11} m}{3.84 \times 10^{9} m}$

Section 1.3 Unit Conversions

1.15 What is the height in centimeters of a person who is 5′11″ tall?

1.16 How many minutes make a microcentury?

1.17 What is 40.2 mi expressed in kilometers?

1.18 A cord is a volume of cut wood equal to a stack 4 ft high, 4 ft wide, and 8 ft long. How many cubic meters equals one cord?

1.19 Express 130 km/h in terms of miles per hour.

1.20 (a) Express 75 miles per hour in units of km/h. (b) In units of m/s.

1.21 A store advertises carpet that costs $18.95 per square yard. How much does that carpet cost per square meter?

1.22 As an estimate, you say that your mass in kilograms is numerically equal to one-half your weight in pounds. What is the percent error in this estimate relative to the value computed with the proper conversion factor? (The conversion factor is found in a table in the end papers. The percent error is the ratio of the error to the correct value expressed as a percent.)

1.23 When gasoline sells for $1.069 per gallon, what is the price in dollars per liter? (1 gal = 3.7853 L)

1.24 A student traveling in Hamburg, Germany, finds a radio for sale. The price of the radio is given as 547 marks (DM547). If the exchange rate on that day is $1.00 = DM1.4495, what is the cost of the radio in dollars and cents? $\frac{547}{1.4495}$

1.25 What is the area in square centimeters of an $8\frac{1}{2}″ \times 14″$ piece of paper?

1.26 The blade on a wood saw has 12 teeth per inch. What is the separation between adjacent teeth in units of mm?

1.27 Wooden slats in a picket fence are spaced 6.0 inches apart, center to center. How many slats are contained in one meter of fence?

1.28 Compute the number of liters in one gallon, given that one gallon is exactly 231 in³.

1.29 The moon turns on its axis once every $27\frac{1}{3}$ days so that the same face is always toward the earth. Through how many degrees does the moon rotate about its own axis in one hour?

1.30• A clock loses 3.0 s per day. By how many minutes will it be off at the end of one year (365 days)?

1.31• How many revolutions does the second hand of a clock make in three years? Assume no leap years in the interval.

1.32• The AU, the distance from the earth to the sun, is 1.50×10^{11} m, and the speed of light is 3.00×10^8 m/s. (a) How long does it take for light to come from the sun? $\frac{1.5 \times 10^{11}}{3.0 \times 10^8}$ (b) A light-year is the distance light travels in one year. How far is a light-year in m? (c) How far is a light-year in AU?

1.33• The earth has a mass of 5.98×10^{24} kg and a radius of 6.38×10^6 m. (a) What is the mass per unit volume of the earth in kg/m³? (b) What is the mass per unit volume of a gold nucleus that has a mass of 3.27×10^{-25} kg and a radius of 6.98×10^{-15} m? (c) What would be the radius of the earth

if its mass were unchanged but it had the same mass per unit volume as the gold nucleus?

Section 1.4 Measurements, Calculations, and Uncertainties

1.34 Calculate the volume of a rectangular cereal box of height 27.5 cm, width 19.2 cm, and depth 5.9 cm. Remember the rule regarding significant figures.

1.35 Calculate the volume of the rectangular board (Fig. 1.13) with height 17.5 mm, width 29.4 cm, and length 115.4 cm. Remember the rule regarding significant figures.

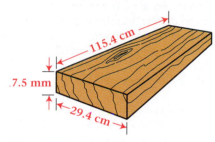

Figure 1.13
Problem 1.35.

1.36 A sphere has a surface area of 0.683 cm². What is the volume of the sphere? $A = 4\pi r^2$

1.37 If you measure the sides of a square to be ten centimeters with an accuracy of ± one percent, what is the area of the square and what is the uncertainty?

1.38 Find the number of seconds in a year and express your answer to two significant places using scientific notation.

1.39 Add the following numbers: 3.57×10^2, 2.43×10^3, and 4.865×10^2.

1.40 A rectangular file cabinet has a height of 133 cm, a width of 37.5 cm, and a length of 72.0 cm. Express its volume in scientific notation. Give your answer in units of cm³ and m³.

1.41 The area of the United States is about 9.4×10^6 square kilometers. In early 1996, there were 2.6×10^8 people living in the United States. What was the population density in people per km² at that time?

1.42 What error do you make by approximating the length of a football field to be 100 m instead of 100 yards? Give your answer as the percent error, which is defined as % error = $\frac{\text{error}}{\text{correct value}} \times 100\%$.

1.43• A ream of copy paper is 2.00 in. thick. What is the thickness of a single sheet of the paper? Express your answer in m and in mm.

1.44• (a) If the height of the cereal box in Example 1.6 were increased 5%, by what percentage would the volume be increased? (b) If the radius were increased 5%, by what percentage would the volume be increased?

1.45• The speed of an automobile is said to be 103.2784 mi/h on a one-quarter-mile course. How well must the distance be measured for this statement to be accurate?

1.46• The inner edge of a race track is in the form of a rectangle with two semicircular ends on the shorter sides. The rectangle is 87.0 m wide and 137 m long. The race track is 5.25 m wide. What is the difference between the longest path around the track taken by following along the outer edge and the shortest path taken by following along the inner edge?

1.47•• The rectangular floor of a gymnasium has sides of length $x \pm \Delta x$ by $y \pm \Delta y$, where Δx and Δy are the estimated measurement uncertainties and are small compared to x and y. Show by direct computation that the area of the floor and the uncertainty in that area are given by $xy \pm xy\left(\frac{\Delta x}{x} + \frac{\Delta y}{y}\right)$ when very small terms, of order $(\Delta x)^2$, are ignored. (In most cases, this result overestimates the uncertainty in the area, because it does not take into account that the uncertainties in the lengths, Δx and Δy, come from a series of measurements that have a natural spread in their values.)

Section 1.5 Estimates and Order-of-Magnitude Calculations

1.48 How high would the stack reach if you piled one trillion dollar bills in a single stack?

1.49 Estimate the thickness of the pages in this book. Give your result in millimeters.

1.50 How high can you count out loud in half an hour?

1.51 About how many bricks does it take to build a shoulder-high brick wall 100 ft long? Standard bricks are 8 in. long by 2 1/4 in. high and are separated by 3/8 in. of mortar.

1.52 A large power plant burns a 100-car trainload of coal every day. If the coal is 10% ash, estimate the volume of ash generated each year by the power plant.

1.53• Approximately how many gallons of gasoline are used by passenger cars each year in the United States?

1.54•• Approximately what fraction of the area of the continental United States is covered by automobiles?

Additional Problems

1.55 What is the volume in cubic millimeters of a cube 1.00 in. on a side?

1.56 How many square kilometers are there in 10 acres? (1 acre = 43,560 ft² or 1/640 mi²)

1.57 In some countries, the gasoline consumption of an automobile is expressed in liters consumed per 100 km of travel. If an automobile gets 27 miles/gallon, what is its fuel consumption in liters per 100 km? (1 gal = 3.7853 L)

1.58 The equatorial radius of the earth is 6.38×10^6 m. If the time zones are equally spaced, how wide is a time zone at the equator?

1.59 The speed of sound at room temperature is 340 m/s. Express the speed of sound in units of miles per hour.

1.60 The radius of a circle is given as 1.300 cm. When asked to find the area of the circle, a student uses her calculator to arrive at an answer of 5.30998 cm². (a) What did she use for the value of π? (b) To how many significant figures was her value of π correct? (c) What is the area, including the correct number of significant figures, when using the best value of π?

1.61• (a) How many milliseconds are there in one minute? (b) How many gigaseconds are there in a century?

1.62• A spherical ball will just pass through a circular hole with an area of 38.2 cm². What is the volume of the ball?

1.63• (a) Calculate the height of a cylinder of radius R that has the same volume as a sphere of radius R. (b) Show that the cylinder has a larger surface area than the sphere.

1.64• (a) Calculate the edge length of a cube that has the same volume as a sphere of radius R. (b) Show that the surface area of the cube is greater than the surface area of the sphere.

1.65• Consider a sphere that fits exactly inside a cube. What is the ratio of the volume of the sphere to the volume of the cube?

1.66• (a) By what percentage is the area of a 12.5-cm-radius circle increased if the radius is increased 1.0 cm? (b) By what percentage does the surface area of the side (not the ends) of a 12.5-cm-radius cylinder increase if the radius is increased 1.0 cm?

1.67•• A cylindrical milk shake cup has a measured inside radius of $r \pm \Delta r$ and a height of $h \pm \Delta h$. Show that the volume of the cup is

$$V = \pi r^2 h \pm 2\pi rh\Delta r \pm \pi r^2 \Delta h$$

if very small terms of order $(\Delta r)^2$ are ignored.

Motion in One Dimension

This chapter begins our study of motion. You already know a lot about motion from your everyday experience. For example, when you walk briskly along a straight line at a rate of 2 meters per second, you know that in 5 seconds you will travel a distance of 10 meters. We will build on your knowledge to develop methods for accurately describing the positions and motions of objects. To do this we will give precise meaning to terms in everyday use, such as velocity and displacement, and find relationships between them. This study of motion is called kinematics.

The word *kinematics* comes from the Greek word *kinema,* meaning "motion"—the same root from which we get the word *cinema.* **Kinematics** describes the positions and motions of objects in space as a function of time but does not consider the causes of motion. The study of the causes of motion is called **dynamics.** Separating the study of motion into kinematics and dynamics is a great aid in understanding **mechanics,** which is the branch of physics that deals with large-scale (macroscopic) objects. In this chapter, and the next several chapters, we present the basic principles of mechanics.

Kinematics provides the means for describing the motions of such varied things as planets, golf balls, and subatomic particles. Because of its precision and generality, mathematics is the natural language for kinematics. The ideas and techniques of kinematics that you learn here are used throughout the text. The causes and effects of motion are discussed for many seemingly different situations, but the techniques used are the same. The range of these applications runs from gravity to electricity and magnetism to nuclear physics.

We begin our study of kinematics by considering motion in only one dimension. This restriction has the advantage of introducing almost all of the necessary concepts in their simplest form. A mastery of these ideas is necessary for understanding the following chapters, which discuss the motion of objects in two and three dimensions. We start by building on something you already know about motion. ■

2.1 Reference Frames, Coordinate Systems, and Displacement

Your experience teaches you that motion is a relative term. For example, when you ride in a smoothly moving car and toss a ball into the air, the ball goes up and comes back down to your hand. You see only an up-and-down motion (Fig. 2.1a). However, a person standing beside the moving car sees the ball move forward with the car, as well as up-and-down (Fig. 2.1b). Only one thing actually happens, yet different observers see different motions. We can avoid confusion if we carefully specify the location and motion of each observer. We say that the different observers are in different reference frames. A **reference frame** is a physical entity, such as the ground, a room, or a moving car, to which we refer the position and motion of objects.

To adequately describe motion, we must be able to say where something is located within a given reference frame. For example, we can locate the ball in the car by saying it is 0.75 m from the left door, 0.53 m above the floor, and 0.23 m in front of the seat. This choice is not a particularly convenient reference system. However, it does remind us of something: When we say space is three-dimensional, we mean that we need three numbers to completely locate the

Figure 2.1

(a) Motion of a ball thrown vertically by a passenger in a car moving along a straight line at a constant rate, as seen by an observer in the car. The ball's motion appears the same as if the car were not moving. (b) Motion of a ball thrown vertically by a passenger in the moving car, as seen by an outside observer at rest relative to the road. Now the ball has a forward motion equal to that of the car.

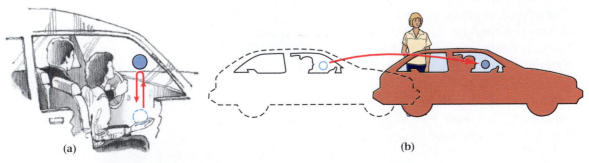

(a)   (b)

position of an object or point. A system for assigning these three numbers, or co-ordinates, to the location of a point in a reference frame is called a **coordinate system.** Most frequently, we will use a *Cartesian* (x, y, z) coordinate system (Fig. 2.2). The Cartesian coordinates in space are often called rectangular coordinates because the axes that define them meet at right angles. Figure 2.3 depicts several reference frames with the coordinate axes shown explicitly. In this chapter, we will only consider motion in one dimension, and so we will align the axis system so that the motion is along the direction of one of the axes, ordinarily the *x* axis.

Because a coordinate system is a mathematical construction, you are free to choose the system that you want, orient it however you wish, and place its origin wherever you prefer. The physical locations of the two people shown in Fig. 2.4, for instance, do not depend on where you place the origin of the coordinate system. However, their positions as *measured* within a coordinate system do depend on where the origin is placed. Two points are worth noting. First, when faced with a physical problem, try to place the origin of the coordinates in a position that will make the problem as easy to solve as possible. Second, having chosen a coordinate system for working the problem, do not change it during the solution.

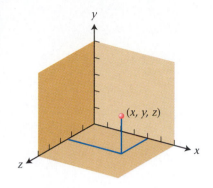

Figure 2.2
A Cartesian coordinate system. The system shown is right-handed.

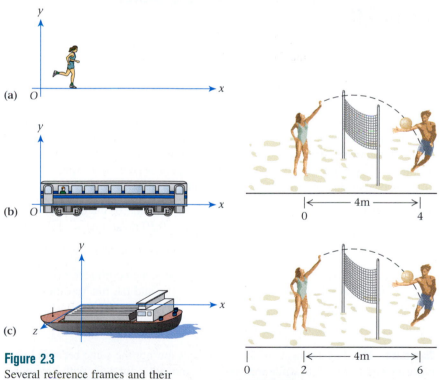

Figure 2.3
Several reference frames and their associated coordinate systems. (a) A runner moves in the reference frame of the earth. (b) A passenger walks in the reference frame of a moving railroad car. (c) A reference frame associated with a floating barge.

Figure 2.4
The location of the two people and, hence, the distance between them does not depend on the location of the origin of the coordinate system used to make the measurements.

To avoid mistakes later on, you need to be sure of the difference between reference frames and coordinate systems. Although the terms have been defined individually, it is helpful to compare the two ideas. The reference frame is the local surroundings from which the changes in positions of other objects can be observed. Within such a frame, the coordinate system permits the measurement of the positions.

In one dimension, the position of an object is given by saying how far it is from the origin of the coordinate system. The change in position, including the sign of the change, is called the **displacement.** When something moves from one location to another, we say it undergoes a displacement. Thus, the position of an object is also referred to as its displacement from the origin. Suppose an object moves along a line from an initial position x_i to a final position x_f, then the net change in position (its displacement) is given by

$$\Delta x \equiv x_f - x_i. \tag{2.1}$$

The symbol Δ (the Greek capital letter delta) is used to indicate a change in something—in this case, the position x. The symbol Δx (delta ex) represents one quantity, the change in x. Three lines instead of two in an equals sign mean that the quantity on the left is defined by that equation.

Example 2.1

(a) What is the displacement of the black chess piece if it is moved to the right all the way across the board as shown in Fig. 2.5? (b) What would be the displacement of the white chess piece if it were moved to the left all the way across the board?

Strategy   The chessboard can be used as our frame of reference. A coordinate system may be chosen with its x axis aligned parallel to the edge of the board. The solution of the problem can then be found directly from the definition of displacement given in Eq. (2.1). However, we must be careful to specify the initial and final positions correctly in this and all other cases.

Solution   (a) We see from Fig. 2.5 that the initial position of the black chess piece is $x_i = 0.00$ m and its final position is $x_f = 0.30$ m, so that we have

$$\Delta x = x_f - x_i = 0.30 \text{ m} - 0.00 \text{ m} = 0.30 \text{ m}.$$

(b) In the case of the white chess piece, using the same coordinate system, we find the initial position of that piece is $x_i = 0.30$ m and the final position is $x_f = 0.00$ m, so that

$$\Delta x = x_f - x_i = 0.00 \text{ m} - 0.30 \text{ m} = -0.30 \text{ m}.$$

Discussion   The displacements of the two pieces are not the same because one piece moved to the right while the other one moved to the left. With the coordinate system that we chose, motion to the right is positive while motion to the left is negative. Even though both pieces are moved through the same distance, their displacements are different. Be careful to distinguish between distance and displacement. Also notice that the position of the origin of coordinates does not affect the value or sign of the displacement. You should show this for

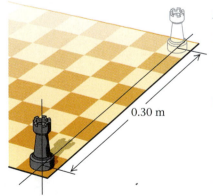

0.30 m

Figure 2.5
Chess pieces are moved on a chessboard. The chessboard provides the frame of reference. A coordinate system is fixed relative to the chessboard with its x axis parallel to the edge of the board as shown. The origin of the coordinate system is chosen to lie under the center of the black chess piece.

yourself by repeating this example with the origin of the coordinates located at some other position along the edge of the chessboard.

Example 2.2

You walk along a long straight sidewalk for 45 m, then you turn around and walk 25 m in the opposite direction. Finally, you turn again and walk 37 m in the original direction and stop. What is your displacement from your starting point?

Strategy Your trip consists of several segments, each corresponding to a displacement defined by the starting and ending positions of that segment. The total displacement can be found from the sum of the individual displacements.

Solution If we call the initial direction positive, the first segment is a displacement of 45 m. Similarly, the displacement of the second segment is -25 m and the displacement of the final segment is 37 m. The total displacement is

$$\Delta x = x_f - x_i = 45 \text{ m} - 25 \text{ m} + 37 \text{ m} = 57 \text{ m}.$$

Discussion For motion in one dimension, the total displacement is just the algebraic sum of the individual displacements. Thus, the final position is 57 m from the initial or starting point.

2.2   Average Speed and Average Velocity

You are already familiar with the concept of speed and know that the speed of an object is measured in units such as miles per hour, kilometers per hour, or meters per second. The speed is the ratio of the distance traveled to the time required for the travel. We define the **average speed** as the total distance, s, traveled during a particular time divided by that time interval, t:

$$\text{average speed} \equiv \frac{\text{total distance traveled}}{\text{time interval for travel}} = \frac{s}{t}. \tag{2.2}$$

It is important to recognize that this definition refers to neither the size nor shape nor mass nor any other property of the moving body, nor to how the body is influenced by its surroundings. The definition deals only with the motion itself. In the same way, other definitions in kinematics are restricted to properties of the motion only. If the average speed is the same for all parts of a trip, then the speed is constant.

Example 2.3

On a clear October day, two students take a three-hour automobile trip to enjoy the fall foliage. In the first two hours, they travel 100 km at a constant speed. In the third hour they travel another 80 km, at a different constant speed. What is the average speed for each segment and for the entire trip?

Strategy For each segment we can find the average speed as the distance traveled in that segment divided by the elapsed time. To find the average speed for the whole trip, we must find the total distance and divide it by the total elapsed time.

Solution For the first portion of the trip, the average speed is:

$$\text{average speed} = \frac{\text{distance traveled}}{\text{time elapsed}}$$

$$\text{average speed (1)} = \frac{100\ \text{km}}{2\ \text{h}} = 50\ \text{km/h}.$$

For the second portion of the trip, the average speed is given by the same formula but now with a distance of 80 km and a time of 1 h:

$$\text{average speed (2)} = \frac{80\ \text{km}}{1\ \text{h}} = 80\ \text{km/h}.$$

For the entire trip, the average speed is the *total* distance divided by the *total* time interval:

$$\text{average speed (total)} = \frac{180\ \text{km}}{3\ \text{h}} = 60\ \text{km/h}.$$

Discussion Note that the average speed for the entire trip, in this case 60 km/h, is *not,* in general, the same as the direct average of the individual speeds, which in this case is 65 km/h.

■

As noted before, in this chapter we limit our study to motion along a line or in one dimension. This limitation makes our definition of speed and related concepts easier to understand. In reality, however, motion is usually not restricted to one dimension, and we must take account of the direction as well as the speed of an object's motion. The name for the quantity that describes both the direction and the speed of motion is **velocity.** Even though we are considering only one-dimensional motion in this chapter, we must still take account of direction (for example, positive versus negative, or east versus west), so we will use the term *velocity.* We will consider two- and three-dimensional motion in the later chapters.

Suppose a car is located at point x_1 at a time t_1, and at another point x_2 at a later time t_2 (Fig. 2.6). Then the car's **average velocity** $\bar{v}$ over the time interval is

$$\bar{v} = \frac{\text{final position} - \text{initial position}}{\text{final time} - \text{initial time}} = \frac{x_2 - x_1}{t_2 - t_1}. \tag{2.3}$$

Equation (2.3) can be put in words: The average velocity is the displacement divided by the time elapsed during that displacement. In general, a bar over a symbol (as in $\bar{v}$) indicates the average value of that quantity, in this case the average velocity. Note that the average velocity can be either positive or negative. The

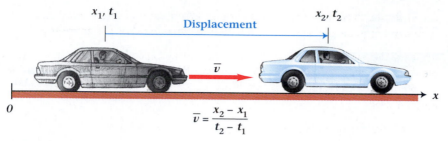

Figure 2.6

When the car moves from point x_1 to point x_2, its displacement is given by $x_2 - x_1$. The time interval from t_1 to t_2 is $t_2 - t_1$.

difference between speed and velocity is more than just an algebraic sign; it involves the difference between the total distance traveled (for speed) and the net change in position (for velocity). These two quantities are not necessarily the same. For example, a trip of 30 km away and 30 km back gives a total distance traveled of 60 km, but a net change in position (displacement) of zero.

In the preceding paragraphs, we used the terms *initial position* and *final position,* measuring them from the zero point of our reference system of coordinates. So far we have used a coordinate system that is just the x or y axis with the origin (zero) placed at some convenient point. Most often we have chosen the zero to be the starting point for the motion described. Notice that we can have both positive and negative positions measured from the origin. For example, in one dimension, if we define positions measured to the right of the origin to be positive, then positions measured to the left of the origin are negative. In this case, a positive velocity indicates motion to the right; a negative velocity indicates motion to the left.

The same argument may be made for time, the other basic quantity needed for kinematics. We often say that we "start the clock running" at the moment a process starts or when some particular event happens. In that case we choose the initial time to be zero. Times before that would be negative and all times after the beginning would be positive.

Example 2.4

(a) What is the average velocity of a helicopter: (i) If it takes off from its pad at the hospital and travels 150 km due east in one hour (Fig. 2.7a)? (ii) If, instead, it travels 150 km due west from the hospital in one hour? (iii) If it starts from a location 20 km east of the hospital, travels to a point 50 km due east, then turns around and travels to a spot 80 km west of the hospital in one hour? (b) What would the average speed be in each of these cases?

Strategy (a) From the statement of the problem, we see that all of the motion is confined to a straight line along the east-west direction. We begin by making a simple diagram to help visualize the situation (Fig. 2.7b). After making the drawing, we must fix the coordinate origin and choose which direction is positive. Here we choose the origin to be at the hospital and choose east to be the

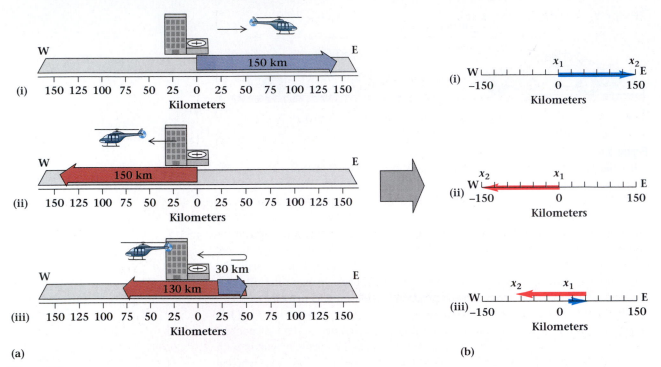

(a)   (b)

Figure 2.7

Example 2.4: (a) Starting at hospital helicopter pad, (i) the helicopter flies 150 km due east and (ii) the helicopter flies 150 km due west. (iii) Starting 20 km east of the hospital, the helicopter flies first to a position 50 km east of the hospital and then to a position 80 km west of the hospital. (b) Simplified drawing for analysis. We have chosen east as positive and marked the positions of x_1 and x_2 for all three situations.

positive direction and west to be negative. To find the average velocity in each case, we use

$$\overline{v} = \frac{\text{final position} - \text{initial position}}{\text{final time} - \text{initial time}} = \frac{x_2 - x_1}{t_2 - t_1}.$$

Notice that the duration given by the difference between initial and final times is the elapsed time:

$$\overline{v} = \frac{\text{final position} - \text{initial position}}{\text{time elapsed}}.$$

Solution (a)   So, for the three cases,

$$\text{(i)} \quad \overline{v} = \frac{150 \text{ km} - 0 \text{ km}}{1 \text{ h}} = +150 \text{ km/h}.$$

$$\text{(ii)} \quad \overline{v} = \frac{-150 \text{ km} - 0 \text{ km}}{1 \text{ h}} = -150 \text{ km/h}.$$

$$\text{(iii)} \quad \overline{v} = \frac{-80 \text{ km} - 20 \text{ km}}{1 \text{ h}} = -100 \text{ km/h}.$$

Strategy (b)   Recalling that speed is not the same as velocity and depends on the total distance traveled, we write

$$\text{average speed} = \frac{\text{distance traveled}}{\text{time elapsed}}.$$

Solution (b) For the first two cases,

$$\text{(i) average speed} = \frac{150\ \text{km}}{1\ \text{h}} = 150\ \text{km/h}.$$

$$\text{(ii) average speed} = \frac{150\ \text{km}}{1\ \text{h}} = 150\ \text{km/h}.$$

For the third case, the distance traveled is 30 km east plus 130 km west, for a total distance of 160 km. Thus,

$$\text{(iii) average speed} = \frac{160\ \text{km}}{1\ \text{h}} = 160\ \text{km/h}.$$

Discussion Notice that the average speed is always a positive number. In situation (iii) the magnitude of the displacement and the total distance traveled are not the same. Consequently the numerical value of the average speed is not the same as that of the average velocity.

2.3 Graphical Interpretation of Velocity

We now supplement our algebraic definition of velocity in Eq. (2.3) with a graphical interpretation of velocity. Consider two people, one running and one walking with constant velocity (Fig. 2.8). We plot the elapsed time along the horizontal axis, or **abscissa,** and the velocity along the vertical axis, or **ordinate.** (It is conventional to plot the independent variable along the abscissa and the dependent variable along the ordinate, as we have done here.) The point on the graph that represents the velocity at any time traces out a smooth line as time goes by. In this case, the horizontal line v_A represents the particular constant velocity of the runner. A slower-moving person, but one with constant velocity also, gives rise to the horizontal line v_B.

In Fig. 2.9, we have plotted the positions of the people (their displacement from the origin) against the elapsed time. Our definitions of both velocity and speed show that the displacement at constant velocity or speed is directly proportional to the time. When one variable is directly proportional to the other, as *displacement* is to *time* in this example, a straight line results. For this reason, such relationships are called *linear.* Plotting displacement against time for constant velocity, therefore, will result in a straight line such as A in Fig. 2.9, where we have started at zero displacement at $t = 0$. A graph due to a smaller constant velocity gives rise to another straight line, such as the line B.

In Fig. 2.10 we have plotted displacement against time for a case in which the velocity is not constant. Over the portion of the trip between times t_1 and t_2, the line is straight and the velocity is constant. Between times t_2 and t_3, the velocity remains the same and is still constant. Between times t_4 and t_5, the velocity is constant, but is not the same as between times t_1 and t_2 or between t_2 and t_3. The velocity is not constant for the time interval between t_3 and t_4, nor is it constant overall between times t_1 and t_5. However, we can use Eq. (2.3) to define the average velocity for the time interval t_3 to t_4, t_1 to t_5, or any other interval.

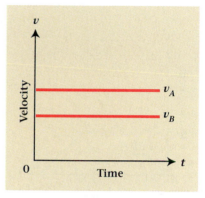

Figure 2.8
Velocity graphed against time for two people moving with different constant velocities. Person A is moving faster than B.

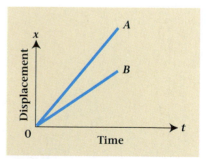

Figure 2.9
Displacement graphed against time for two people moving at different constant velocities. Both people were at zero displacement when the time was zero.

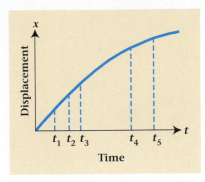

Figure 2.10
Displacement graphed against time for varying velocity.

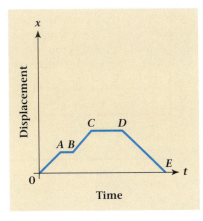

Figure 2.11
Example 2.5: Displacement against time for a drive from home to the store and back.

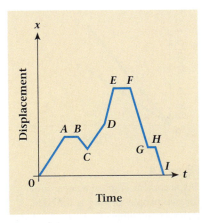

Figure 2.12
Example 2.6: Graph of the displacement against time for a child walking along a path.

Example 2.5

A woman drives her car to the store that is located straight down the street from her home. Halfway there she pauses for a traffic light, then continues to the store. After making her purchase she returns directly home, this time catching the green light so that she doesn't have to stop for the traffic light. Make a graph of the displacement of her car from home against time.

Strategy We can make a simple graph by considering her motion to be one-dimensional and by treating the velocity in each segment to be constant. Then the graph will consist of only straight lines.

Solution Let the origin of the graph be at her home. The first segment (Fig. 2.11) is a straight line showing increasing displacement with increasing time. At point A she stops for the traffic light, so the displacement is constant with time until point B where she resumes moving toward the store. At C she stops upon reaching the store. After completing her shopping (D), she drives toward her home (E); thus the displacement decreases as the time increases until she is again at home.

Example 2.6

The movement of a child walking along a straight path is represented by the graph of displacement against time shown in Fig. 2.12. Describe the motion of the child in words.

Strategy We can interpret the graph if we break it into sections with simple features and analyze each section separately.

Solution The child starts from the origin and walks with constant velocity to point A where she stops. After a short time (B), she begins walking back toward the origin. At C she reverses and starts walking away again. From D to E she walks faster or perhaps runs, stopping at E for a while. Then at the time corresponding to point F, she begins walking back toward home (the origin) until at the point labeled G she stops. At the time corresponding to H, she resumes her return reaching home (I).

■

To use our graphical method further, we need a more formal definition for the slope of a line. Suppose an object moves with constant velocity so that its position-time graph looks like Fig. 2.13. If the object was at point x_1 at time t_1, and at another point x_2 at a later time t_2, then the net change in position is given by $\Delta x = x_2 - x_1$. Similarly, the change in time is given as Δt (delta tee) $= t_2 - t_1$. The average velocity of the object during the time interval from t_1 to t_2 is then written

$$\overline{v} = \frac{x_2 - x_1}{t_2 - t_1} = \frac{\Delta x}{\Delta t}.$$

As we said earlier, the change in position $\Delta x = x_2 - x_1$ is the displacement of the moving object, including the magnitude of the change and its direction.

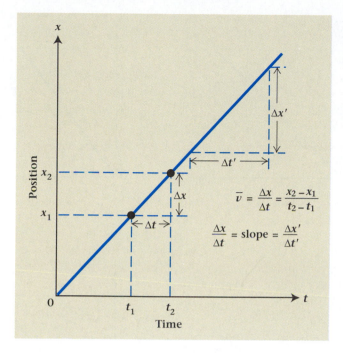

Figure 2.13
The velocity is the slope of the position-time curve. For a straight line, the slope $\Delta x/\Delta t$ is constant and is independent of the particular choice of time interval. This corresponds to constant velocity.

Notice that the position of an object is the same as its displacement from the origin, or zero, of the coordinate system.

An alternative way of defining velocity is to say that velocity is the slope of the position-time curve. The **slope of a line** is defined to be the ratio of the change in the line's ordinate to the corresponding change in the abscissa. For the case of constant velocity (Fig. 2.13), the position-time curve is a straight line and the slope is $\Delta x/\Delta t$. Thus, we can determine the velocity from the slope of the line. Since the distance traveled is proportional to the time, we get the same numerical value for this ratio, $\Delta x/\Delta t$, no matter what interval of time we choose to consider. Thus, in Fig. 2.13 the ratio $\Delta x/\Delta t$ is equal to the ratio $\Delta x'/\Delta t'$. This behavior is characteristic of constant velocity. Notice that the slope, and therefore the velocity, has the dimensions of length/time.

Example 2.7

Find the velocities corresponding to the displacement-time graphs in Fig. 2.14. Then list them in decreasing order starting with the largest velocity.

Strategy We can find the magnitude and sign of the velocity for each distance-time graph by finding the slope of the line using the techniques illustrated in Fig. 2.13.

Solution For the line shown in (a), the velocity is the rise over the run, which is $v = \dfrac{\Delta x}{\Delta t} = \dfrac{8\ \text{m} - 0\ \text{m}}{4\ \text{s} - 0\ \text{s}} = 2\ \text{m/s}$.

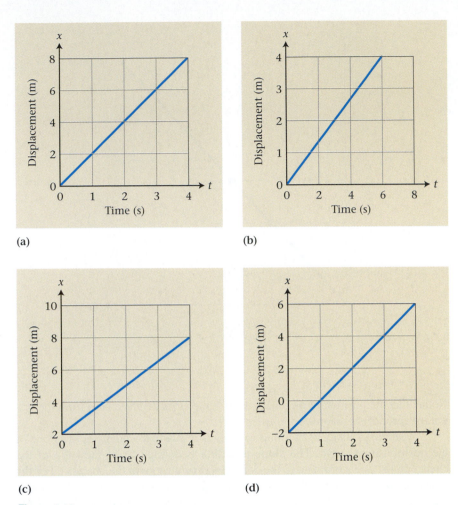

(a)

(b)

(c)

(d)

Figure 2.14

Example 2.7: Four graphs of displacement versus time.

For (b) the velocity is $v = \dfrac{\Delta x}{\Delta t} = \dfrac{4\text{ m} - 0\text{ m}}{6\text{ s} - 0\text{ s}} = \dfrac{2}{3}$ m/s.

For (c) the velocity is $v = \dfrac{\Delta x}{\Delta t} = \dfrac{8\text{ m} - 2\text{ m}}{4\text{ s} - 0\text{ s}} = 1.5$ m/s.

For (d) the velocity is $v = \dfrac{\Delta x}{\Delta t} = \dfrac{6\text{ m} - (-2\text{ m})}{4\text{ s} - 0\text{ s}} = 2$ m/s.

To list them in decreasing order, we have (a) and (d) are equal and are the largest, (c) is next, and then (b) is the smallest.

2.4 Instantaneous Velocity

Suppose that a runner's velocity is not constant, but changes as illustrated in the displacement-time graph in Fig. 2.15. Now that the line is curved rather than straight, how do we find the velocity at a particular instant of time? To determine

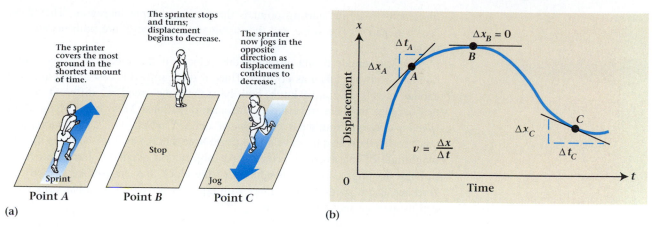

The sprinter covers the most ground in the shortest amount of time.

The sprinter stops and turns; displacement begins to decrease.

The sprinter now jogs in the opposite direction as displacement continues to decrease.

Sprint — Point A

Stop — Point B

Jog — Point C

(a)

(b)

the velocity at point A, for instance, we draw a tangent to the curve at that point. Then the slope of that tangent, which is a straight line, is determined as previously described; it is given by $v_A = \Delta x_A / \Delta t_A$. The subscripts refer to the line tangent to the curve at point A. The velocity measured at any given moment (for example, at point A) is called the **instantaneous velocity.**

At the point of contact, the tangent line is parallel to the displacement-time curve. The average velocity over the time interval Δt that contains A approaches the value of the instantaneous velocity as the interval Δt becomes smaller (Fig. 2.16). We can therefore define the instantaneous velocity as the limiting value of $\Delta x / \Delta t$ as Δt becomes vanishingly small. In symbols, the idea is

$$v = \lim_{\Delta t \to 0} \frac{\Delta x}{\Delta t}. \tag{2.4}$$

The symbol $\lim_{\Delta t \to 0}$ tells us to evaluate the ratio $\Delta x / \Delta t$ in the limiting case of Δt approaching zero. Note that we don't simply set $\Delta t = 0$; rather, we examine the ratio $\Delta x / \Delta t$ as both Δx and Δt get smaller. This ratio goes to a definite value as Δt goes to zero.

The slopes at other points, such as B and C in Fig. 2.15, are determined in the same way. Notice that in the immediate neighborhood of point C the dis-

Figure 2.15
The slope of a curved line is determined at any point by the slope of the line tangent to the curve at that point. The steepness of the slope corresponds to the magnitude of the velocity. Notice that since Δx_C is negative, the slope at point C is negative, corresponding to velocity in a negative direction (back toward the starting point).

Figure 2.16
The instantaneous value of the velocity is obtained from the average value as the interval Δt becomes vanishingly small.

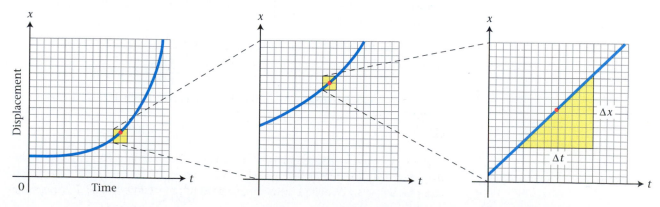

placement from the starting point is decreasing as time increases. Therefore, Δx_C is negative. The slope and the instantaneous velocity are both negative at point C.

The magnitude of the instantaneous velocity is the **speed;** it has no direction and is always taken as positive. Thus, if one car travels east at 60 km/h and another travels west at 60 km/h, both have the same speed but different velocities because they are traveling in different directions. This usage of the term *speed* is different from the *average speed,* defined in Section 2.2 as the total distance traveled divided by the elapsed time. Note that the average speed is not the magnitude of the average velocity. An object that oscillates back and forth through the origin, like the pendulum of the clock, has an average velocity of zero because it spends equal times with equal positive and negative velocities. But its average speed is not zero.

Example 2.8

Estimate the relative size and sign of the velocity corresponding to the displacement-time graph shown in Fig. 2.17(a). Then make a sketch of the velocity-time graph.

Strategy We can estimate the relative size of the velocity at the points marked from the slope of the curve at each point.

Solution The curve seems to be starting at point A with zero slope. At point B the slope is the most negative and at C it is zero again. Point D has a positive slope and point E has an even greater slope. The corresponding graph of velocity versus time should look like Fig. 2.17(b)

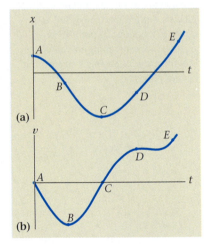

Figure 2.17
Example 2.8: (a) Graph of displacement versus time. (b) Graph of velocity versus time corresponding to the slope of (a).

Example 2.9

At the start of a 100-m race, a sprinter is poised for action. When the starter fires the gun, the sprinter pushes off the starting block and quickly reaches maximum velocity. The graph in Fig. 2.18 shows a sprinter's displacement as a function of the time elapsed after the starting signal. Determine the instantaneous velocity of the sprinter at 0.50 s, 1.50 s, and 2.50 s.

Solution The graph in Fig. 2.18 shows tangent lines drawn through the points corresponding to $t = 0.50$ s, 1.50 s, and 2.50 s. Therefore, Δt and Δx can be read directly from the graph.

At $t = 0.50$ s we have

$$v = \frac{\Delta x}{\Delta t} = \frac{2.0 \text{ m}}{0.50 \text{ s}} = 4.0 \text{ m/s}.$$

At $t = 1.50$ s,

$$v = \frac{4.0 \text{ m}}{0.50 \text{ s}} = 8.0 \text{ m/s}.$$

At $t = 2.50$ s,

$$v = \frac{4.8 \text{ m}}{0.50 \text{ s}} = 9.6 \text{ m/s}.$$

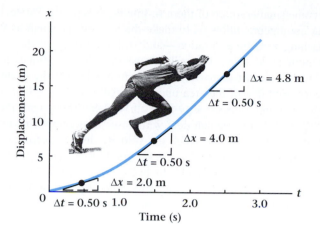

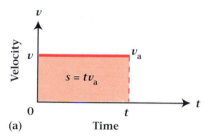

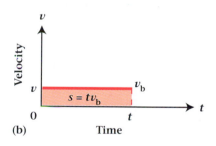

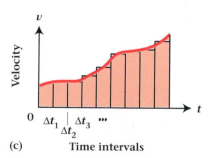

Figure 2.18

Example 2.9: Graph of displacement versus time for a sprinter. The slope of the curve at any time is measured from the line tangent to the curve at that time. The tangents shown correspond to times of 0.50 s, 1.50 s, and 2.50 s.

Discussion In this example, we see that the velocity changes with time. The velocity increases rapidly as the runner springs from the starting block but begins to level off as he approaches maximum velocity.

In addition to being used for specific examples, graphical techniques can also help us obtain general relationships between variables. Let us consider one example of such a procedure. The example is specific to kinematics, but we will use the idea and technique many times again.

Figure 2.19(a and b) shows velocity-time graphs for an object moving at constant velocity. The length of a vertical line from the horizontal axis to the line v_a (or v_b) is equal to the velocity. The line v_a is higher than the line v_b, indicating a greater velocity for the object represented in Fig. 2.19(a). Note that the area of the rectangle under the line v is given by tv, where t is the elapsed time and v is the velocity. However, this quantity is also the net distance traveled, s, which is the displacement. We can extend this observation to state a general principle: *The area under any velocity-time curve between two times is equivalent to the displacement during that time interval.*

To see that this last statement is true, consider the velocity-time graph of Fig. 2.19(c), for which the velocity is not constant. We can divide the total time into small time intervals of duration Δt. Over each of these intervals, we may approximate the velocity by a constant value, given by the average of the initial and final velocities for the interval. Then for each interval, we have, as before, a rectangle whose area is equivalent to the displacement during that interval. The total displacement is then represented by the sum of all the areas of these rectangular strips—that is, the total area under the velocity-time curve. Thus

$$\text{total displacement} = \bar{v}_1 \Delta t_1 + \bar{v}_2 \Delta t_2 + \bar{v}_3 \Delta t_3 + \cdots,$$

where $\bar{v}_1$ is the average velocity during the first interval, $\bar{v}_2$ is the average velocity during the second interval, and so on.

If the intervals Δt are too long, or if the velocity changes too much during each interval, this approximation is unsatisfactory because we assume that the velocity is constant for each time interval Δt. However, if we make the intervals smaller and smaller, this approximation becomes more accurate. Also, as the

Figure 2.19

Graphs of velocity versus time for three moving objects: (a), and (b) Constant velocity; (c) Time-varying velocity. The area under the velocity curve is equivalent to the displacement of the object during the time interval.

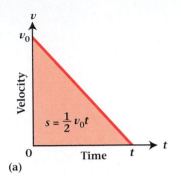

(a)

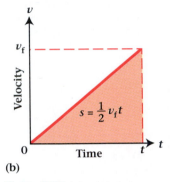

(b)

Figure 2.20
Calculating displacement from velocity-time graphs. (a) Velocity decreasing uniformly with time. (b) Velocity increasing uniformly with time.

intervals become smaller, more of them fit into a given range. The techniques of calculus, but not algebra, allow us to make the intervals as small as we want— even approaching zero time duration—and still keep track of the areas of the tremendous number of strips. We will not use calculus techniques in this text, but we will make use of two basic concepts from calculus: the meaning of the slope of a curve and that of the area under a curve.

Since the area under the velocity-time curve always represents the displacement of the moving body, we can deal with cases of nonuniform velocity. Figure 2.20(a) represents a case in which the velocity decreases from some initial value, v_0, to zero in a linear way. This graph might represent the slowing down and stopping of a car when you apply the brakes. The area under the curve, and therefore the distance traveled, is easily determined by noting that the required area is that of a right triangle. The area of the triangle is one-half the height times the base, so distance traveled $= \frac{1}{2}v_0t$.

Instead of starting with a car in motion and smoothly bringing it to rest, we could consider a car initially at rest and smoothly increase its velocity to some final value v_f (Fig. 2.20b). The distance traveled during the time the velocity is increasing is given by $s = \frac{1}{2}v_ft$.

Example 2.10

Traveling on the German Autobahn in your BMW at a speed of 166 km/h, you smoothly brake to a halt in 12 s. How far do you travel after first applying the brakes?

Solution Using the result from Fig. 2.20(a), we have

$$s = \tfrac{1}{2}v_0t = \tfrac{1}{2}(166 \text{ km/h})(12 \text{ s})$$

$$s = \tfrac{1}{2}\left(\frac{166 \text{ km} \times 12 \text{ s}}{\text{h}}\right)\left(\frac{1 \text{ h}}{60 \text{ min}}\right)\left(\frac{1 \text{ min}}{60 \text{ s}}\right)$$

$$s = 0.277 \text{ km}.$$

Because the time is known to only two significant figures, we round the answer to

$$s = 0.28 \text{ km}.$$

2.5 Acceleration

In Section 2.2 we defined the average velocity of an object as its change in position divided by the time elapsed, $\bar{v} = \Delta x/\Delta t$. This tells us how the object's position changes with time. From our discussion in the last section, it is reasonable to define a quantity that indicates how the object's velocity changes with time. We define the **average acceleration, $\bar{a}$,** as the change in velocity divided by the time required for the change. The average acceleration can be written as

$$\bar{a} \equiv \frac{v_2 - v_1}{t_2 - t_1} = \frac{\Delta v}{\Delta t}. \tag{2.5}$$

The dimensions of acceleration are velocity divided by time, which is the same as length divided by time squared. In the SI system, the unit of acceleration is m/s². Acceleration, like velocity, has a direction as well as a magnitude.

The following two examples illustrate the meaning of average acceleration and the use of Eq. (2.5).

Example 2.11

A bicyclist starts from rest and increases his velocity at a constant rate until she reaches a speed of 4.0 m/s in 5.0 s (Fig. 2.21). What is his average acceleration?

Strategy   We can use the definition, Eq. (2.5), to find the acceleration because the statement of the problem gives us the initial and final velocities and the time interval. Since the bicycle starts from rest, the initial velocity is 0. The final velocity is 4.0 m/s and the time interval is 5.0 s.

Solution   The correct relationship is

$$a = \frac{v_2 - v_1}{t}.$$

Inserting the numerical values gives

$$a = \frac{4.0 \text{ m/s} - 0 \text{ m/s}}{5.0 \text{ s}} = 0.80 \frac{\text{m/s}}{\text{s}},$$

or

$$a = 0.80 \text{ m/s}^2.$$

Figure 2.21
Example 2.11: A bicyclist accelerating from rest at a constant rate reaches a speed of 4.0 m/s in 5.0 s.

Discussion Notice that the units of acceleration are m/s². We have a new unit that is derived from the previously defined base units for length and time.

Example 2.12

A motorcyclist starts from rest and accelerates in one direction with a constant acceleration of 4.0 m/s² for 12 s. What is the rider's velocity at the end of the 12 s?

Strategy Here we are given a constant acceleration and a time interval and need to find the change in velocity. We can rewrite Eq. (2.5) in the form

$$v_2 = \bar{a}(t_2 - t_1) + v_1,$$

which will give us the final velocity (v_2) in terms of the known quantities.

Solution Inserting the values of $\bar{a} = 4.0$ m/s², $t_2 - t_1 = 12$ s, and $v_1 = 0$ into this equation gives

$$v_2 = 4.0 \text{ m/s}^2 \times 12 \text{ s} + 0 = 48 \text{ m/s}.$$

Discussion Again notice the units. In working problems you should always insert the units and perform the algebraic operations on them as well as on the numbers. After all, a number without any units is just a number. But if it represents a physical quantity, it should be accompanied by the appropriate units. Furthermore, if the units have the correct form, you probably have the correct answer. For example, if you get units of m/s for velocity, you have probably worked the problem correctly, but if you get units of m/s², then you know that something has gone amiss.

A graphical approach offers further insight into the relationships between acceleration, velocity, time, and distance. From the definition of average acceleration in Eq. (2.5), we see that acceleration may also be described as the slope of the curve of velocity versus time. In Fig. 2.22(a), the horizontal line (zero slope) corresponds to an object moving with constant velocity and therefore zero acceleration. In Fig. 2.22(b), the slope of the line is everywhere the same, indicating that the velocity is increasing at a constant rate. Therefore, the acceleration is positive and constant. But suppose the velocity changes as shown in Fig. 2.22(c). How can we determine the acceleration at any particular instant of time? We can define the **instantaneous acceleration** in a manner similar to the way we defined the instantaneous velocity. The instantaneous acceleration is the limiting value of $\Delta v/\Delta t$ as the time interval Δt becomes vanishingly small. In symbols, the instantaneous acceleration is

$$a = \lim_{\Delta t \to 0} \frac{\Delta v}{\Delta t}. \tag{2.6}$$

In the limit that the time interval Δt approaches zero, the ratio $\Delta v/\Delta t$ becomes the value of the instantaneous acceleration. Recall that the slope of a curved line

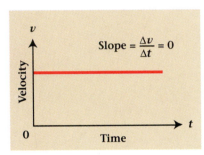

(a)

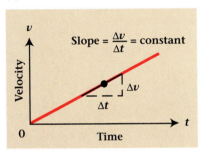

(b)

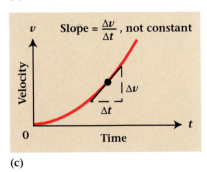

(c)

Figure 2.22

Acceleration is the slope of the curve of velocity versus time. (a) Zero acceleration. (b) Constant nonzero acceleration. (c) Variable acceleration.

at any point is determined by the line tangent to the curve at that point. Thus, the slope of a velocity-time curve at any point of time equals the instantaneous acceleration at that point.

Example 2.13

A sprinter, at rest ($v = 0$) at the start of a race, quickly accelerates to maximum velocity (Fig. 2.23). Determine the instantaneous acceleration of the sprinter at 0.50 s, 1.50 s, and 2.50 s. Notice that the data here are related to those in Example 2.9.

Solution The results can be read directly from the graph of Fig. 2.23, where we have drawn the tangent lines through the points corresponding to $t = 0.50$ s, 1.50 s, and 2.50 s.

At $t = 0.50$ s we have

$$a = \frac{\Delta v}{\Delta t} = \frac{3.0 \text{ m/s}}{0.50 \text{ s}} = 6.0 \text{ m/s}^2.$$

At $t = 1.50$ s we have

$$a = \frac{1.25 \text{ m/s}}{0.50 \text{ s}} = 2.5 \text{ m/s}^2.$$

At $t = 2.50$ s we have

$$a = \frac{0.40 \text{ m/s}}{0.50 \text{ s}} = 0.80 \text{ m/s}^2.$$

Discussion As expected, we find that the acceleration decreases with time as the sprinter approaches maximum velocity (see the inset). Compare this

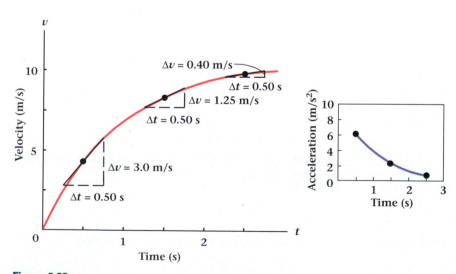

Figure 2.23
Example 2.13: Graph of the velocity of a sprinter versus time.

example with Example 2.9 and Fig. 2.18. Be sure you clearly understand how the shapes of these two curves are related.

2.6 Motion with Constant Acceleration

Many situations occur in which an object's acceleration is constant. The most common case is the motion of a freely falling object when air resistance is negligible. Because of the numerous motions that can be described by constant acceleration, we will assume constant acceleration for the remainder of this chapter. This restriction to constant acceleration allows us to develop some simple relationships among four kinematic quantities—displacement, velocity, acceleration, and time.

First, let us find an expression for the average velocity of an object moving with constant acceleration. Figure 2.24 shows a velocity-time graph of this situation. The velocity is increasing from v_1 to v_2, as discussed in Section 2.4. The displacement is the total area under the velocity-time curve and is equal to the sum of the areas of rectangle $ABCD$ and triangle BCE. The average velocity is then

$$\overline{v} = \frac{\text{displacement}}{\text{time elapsed}} = \frac{\text{area of } ABCD + \text{area of } BCE}{\Delta t}$$

$$\overline{v} = \frac{v_1 \Delta t + \frac{1}{2}(v_2 - v_1)\Delta t}{\Delta t} = \frac{v_1 + v_2}{2}.$$

We see that for the special case of constant acceleration, and *only* for this case, the average velocity is one-half the sum of the initial and final velocities.

We can use this result, along with the definition of average velocity (Eq. 2.3) and the definition of acceleration (Eq. 2.5), to get another useful relationship. [We can use Eq. (2.5) because the instantaneous and average accelerations are the same when the acceleration is constant.] The resulting equations are simpler if we measure time in terms of elapsed time. In this case we set the initial time $t_1 = 0$, and t_2 becomes any later time t. Thus,

$$t_2 - t_1 = t - 0 = t.$$

We also let x_0 represent the initial position (at time $t = 0$) and let x be the position at time t. Using these quantities, we can represent the distance x in terms of the average velocity (from Eq. 2.3) as

$$x - x_0 = \overline{v}t.$$

If we also let v_0 represent the initial velocity when $t = 0$ and let v be the velocity at time t, then the average velocity is

$$\overline{v} = \frac{v_0 + v}{2}.$$

Inserting this value of $\overline{v}$ into the equation for displacement, we find

$$x - x_0 = \frac{1}{2}(v_0 + v)t. \tag{2.7}$$

This equation expresses the change in displacement (distance traveled) of an object in terms of its initial and final velocities and the elapsed time of motion.

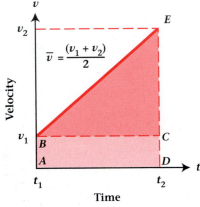

Figure 2.24
Velocity-time graph for constant acceleration, starting with an initial velocity $v_1 \neq 0$.

To express the distance traveled in terms of the acceleration rather than the final velocity v, we rewrite Eq. (2.5) in terms of v, v_0, and t, solving for v:

$$v = v_0 + at.$$

Then we substitute this result into Eq. (2.7), getting

$$x = x_0 + v_0 t + \tfrac{1}{2}at^2. \qquad (2.8)$$

You should carry out these steps to assure yourself that Eq. (2.8) is correct.

Equation (2.8) can also be obtained by examining Fig. 2.24. The term $v_0 t$ represents the area of rectangle $ABCD$, and the term $\tfrac{1}{2}at^2$ represents the area of the triangle BEC. Equation (2.8) is one of the more important and useful kinematic expressions, as we shall see in Chapter 3. The following examples illustrate the use of this relationship.

Example 2.14

A Boeing 777 airliner, initially at rest, undergoes a constant acceleration of 2.3 m/s^2 down the runway for 34 s before it lifts off (Fig. 2.25). How far does it travel down the runway before taking off?

Strategy   The distance the airplane travels is given by Eq. (2.8):

$$x = x_0 + v_0 t + \tfrac{1}{2}at^2.$$

If we choose the coordinates so that the airliner is at rest at the origin at $t = 0$ and the direction of the acceleration is along the positive x axis, then the numerical values are $x_0 = 0$, $v_0 = 0$, $a = 2.3$ m/s^2, and $t = 34$ s.

Solution   Using these values the distance is then

$$x = \tfrac{1}{2} \times 2.3 \text{ m/s}^2 \times (34 \text{ s})^2 = 1329.4 \text{ m}.$$

Because the acceleration and time are known to only two significant figures, we round off the answer to

$$x = 1.3 \times 10^3 \text{ m} = 1.3 \text{ km}.$$

Figure 2.25
Example 2.14: An airliner accelerates down the runway to gain enough speed to become airborne.

Example 2.15

A drag racer travels one-quarter mile in 10 s from a standing start. Calculate the acceleration in ft/s^2, under the assumption that it is constant. Choose the origin to be at the starting point and the direction of travel along the positive x axis.

Solution   Setting $v_0 = 0$ in Eq. (2.8) and solving for the acceleration gives

$$a = \frac{2(x - x_0)}{t^2}.$$

Next, we insert the numerical values for the distance $x - x_0$ and t.

$$a = \frac{2(0.25 \text{ mi} - 0 \text{ mi})}{(10 \text{ s})^2} = 0.0050 \text{ mi/s}^2.$$

Since we are asked to calculate the acceleration in units of ft/s^2, we need to convert the units from miles to feet. We do so by multiplying a by the conversion factor of 5280 ft/mi:

$$a = 0.0050 \text{ mi/s}^2 \times 5280 \text{ ft/mi} = 26 \text{ ft/s}^2.$$

Example 2.16

Suppose a child driving a go-cart is traveling 4.0 m/s when she crosses a line 4.0 m from her starting point. She continues, with a steady acceleration of 0.40 m/s^2, until she crosses a mark 40 m from the starting point. How long does it take for her to go from the 4.0-m mark to the 40-m mark?

Strategy The problem gives us the initial velocity v_0, the acceleration a, and the displacement $x - x_0$, and it asks us to compute the time. These quantities are all related in Eq. (2.8),

$$x = x_0 + v_0 t + \tfrac{1}{2}at^2.$$

We can rearrange this equation into the general form of a quadratic equation in the variable t:

$$\tfrac{1}{2}at^2 + v_0 t + (x_0 - x) = 0.$$

The values of the coefficients can now be inserted into the quadratic formula to find t. (See the appendix to this chapter for a review of using the quadratic formula.)

Solution On comparing the last equation with the quadratic formula, we get

$$t = \frac{-v_0 \pm \sqrt{v_0^2 - 4(\tfrac{1}{2}a)(x_0 - x)}}{2(\tfrac{1}{2}a)}.$$

Notice that the a in Eq. (2.7) is not the same as the a of the quadratic formula. Inserting the numerical values leads to

$$t = \frac{-4.0 \text{ m/s} \pm \sqrt{(4.0 \text{ m/s})^2 - 4(0.20 \text{ m/s}^2)(4.0 \text{ m} - 40 \text{ m})}}{0.40 \text{ m/s}^2}$$

$$t = \frac{-4.0 \text{ m/s} \pm \sqrt{44.8 \text{ m}^2/\text{s}^2}}{0.40 \text{ m/s}^2} = \frac{-4.0 \pm 6.69}{0.40} \text{ s}.$$

Discussion The quadratic equation gives two answers: $t = 6.7$ s and $t = -27$ s, where we have rounded the answer to two significant figures. We discard the negative time because, although it does satisfy the algebraic equation, it has no physical meaning in this case. The meaningful answer is that the child took 6.7 s to reach the 40-m mark after she passed the 4.0-m mark. Notice that units are handled just like numbers or symbols.

We can derive another useful expression relating velocities and distances for motion with constant acceleration by combining Eqs. (2.5) and (2.7). Again using t to stand for elapsed time, we may rewrite Eq. (2.5) as

$$t = (v - v_0)/a.$$

Inserting this into Eq. (2.7) and rearranging, we get

$$v^2 = v_0^2 + 2a(x - x_0). \tag{2.9}$$

This equation gives us a way of calculating distances, velocities, or acceleration without needing to know the elapsed time involved. (It is another basic equation of constant-acceleration kinematics that shows how our definitions of the quantities fit together consistently.)

For convenience, we summarize the important kinematic equations for straight-line motion with constant acceleration in Table 2.1.

Table 2.1	Summary of Kinematic Equations for Constant Acceleration In One Dimension
$x = x_0 + \bar{v}t$ $x = x_0 + \frac{1}{2}(v_0 + v)t$ $x = x_0 + v_0 t + \frac{1}{2}at^2$ $v = v_0 + at$ $v^2 = v_0^2 + 2a(x - x_0)$	

In these equations we have taken the initial values of time, position, and velocity to be 0, x_0, and v_0, respectively.

Example 2.17

You are driving your new sports car at a velocity of 90 km/h, when you suddenly see a dog step into the road 50 m ahead (Fig. 2.26). You hit the brakes hard to get maximum deceleration of 7.5 m/s^2: that is, $a = -7.5$ m/s^2. How far will you go before stopping? Can you avoid hitting the dog?

Strategy In this problem, we choose the positive x axis as the direction of travel. We are looking for distance in terms of velocity and acceleration. To find the distance we can rearrange Eq. (2.9) to give the stopping distance:

$$x - x_0 = \frac{v^2 - v_0^2}{2a}.$$

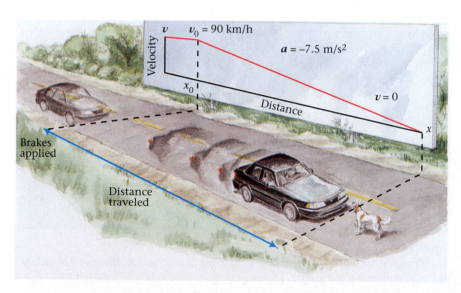

Figure 2.26
Example 2.17: Can you stop quickly enough to avoid hitting the dog?

Solution The values of the known quantities are $v_0 = 90$ km/h, $v = 0$, and $a = -7.5$ m/s^2. Before inserting the known quantities into the equation, we should express v_0 in units of m/s:

$$v_0 = 90 \text{ km/h} \times 1000 \text{ m/km} \times 1 \text{ h/3600 s} = 25 \text{ m/s}.$$

The stopping distance can now be evaluated:

$$x - x_0 = \frac{0^2 - (25 \text{ m/s})^2}{2 \times (-7.5 \text{ m/s}^2)} = 42 \text{ m}.$$

Fortunately, you are able to stop without hitting the dog.

Discussion In solving this problem we have ignored the very real effects of perception time (the time required for the driver to perceive a hazard) and reaction time (the time required to react by moving the foot to the brake pedal). Including these times would change the results of the calculation.

2.7 Galileo and Free Fall

We now discuss in some detail a familiar and important example of constant acceleration: we examine the work by Galileo Galilei on **freely falling bodies,** objects that are moving freely under the influence of gravity. One of Galileo's earliest scientific studies of motion is illustrated in Fig. 2.27. The data for this experiment are recorded in Galileo's notes. Galileo held a ball at the top of an inclined, grooved board and marked its position. Releasing the ball, he marked its position at the end of equal intervals of time. This is much like dropping a ball from a height, except that the effect of gravity has been "reduced" by allowing the ball to roll slowly down the inclined board rather than fall straight down. The positions as measured by Galileo are given in Table 2.2. The observations show what was already known qualitatively to Galileo and others of his time—that a rolling (or falling) object picks up speed as it continues to roll (or

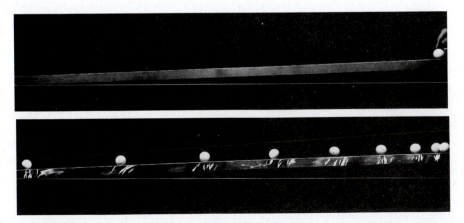

Figure 2.27

Reconstruction of Galileo's experiment on accelerated motion. The distance traveled by the ball rolling down the plane is proportional to the square of the elapsed time.

Table 2.2	Galileo's Results for a Ball Rolled Down an Inclined Plane ▼		
Time t (Equal Intervals)	t^2	Distance x (Points)	x/t^2
1	1	33	33.0
2	4	130	32.5
3	9	298	33.1
4	16	526	32.9
5	25	824	33.0
6	36	1192	33.1
7	49	1620	33.1
8	64	2104	32.9

Distances were measured in points, a unit that equals 29/30 mm.

Source: This experiment is described by Stillman Drake, "The Role of Music in Galileo's Experiments," *Scientific American,* June 1975, p. 98.

(a) In air   **(b) In airless space**

Figure 2.28
An apple and a feather are dropped simultaneously from the same height. (a) The feather falls more slowly than the apple due to the effects of the air. (b) The feather and the apple fall at the same rate when dropped in the airless space on the moon.

fall). However, the debt we owe to Galileo is for his careful measurements and his quantitative (mathematical) interpretation of the data.

Galileo's object was to find a general rule describing how distances increase with increasing time of fall. After some trial and error, and with considerable insight, Galileo realized that the distance traveled was proportional to the square of the elapsed time. That is, with $x_0 = 0$,

$$x \propto t^2,$$

where the symbol $\propto$ means "is proportional to." The fourth column in Table 2.2 lists the ratio of distance to the square of the time, which is seen to be essentially constant. It was experiments like this, rather than the legendary Tower of Pisa experiment, that led to Galileo's conclusion that the distance a falling body moves is proportional to the square of the elapsed time.

Galileo further deduced from his observations that heavy objects fall in the same way that light objects do. Some thirty years later, Robert Boyle, in a series of experiments made possible by his new vacuum pump, showed that this observation is strictly true for bodies falling without the retarding effect of air resistance. This experiment was also demonstrated in 1971 by an astronaut on the moon. There a hammer and a feather both fell with the same acceleration when they were dropped from rest in the airless space around the moon. Figure 2.28 illustrates this effect.

Therefore, we can write the relation between distance and time squared as an equality, where the proportionality constant k does not depend on the nature of the falling object:

$$x = kt^2.$$

For the case of an object released from rest at $x_0 = 0$, Eq. (2.8) reduces to $x = \frac{1}{2}at^2$. (For an object initially at rest $v_0 = 0$.) Thus, the constant $k = \frac{1}{2}a$, and we say that all bodies fall with the same acceleration. This acceleration is called the **acceleration of gravity** and is usually denoted by g. Its standard value is 9.80665 m/s^2, or approximately 9.81 m/s^2. In British units, g is about 32.2 ft/s^2.

Up to now, the relationships between kinematic quantities such as velocity and acceleration were not dependent upon any property of nature, but rather on how they were defined. Here, for the first time, we have introduced a quantity, the acceleration of gravity, which reflects a property of nature. We cannot calculate the acceleration of gravity from just our knowledge of the kinematical relationships, but rather it must be measured. The value we measure depends on the coordinate system and, hence, the units of measurement. But the fact that all things fall with the same acceleration (in the absence of air friction) is a consequence of natural law.

As we will see later in Chapter 5, the acceleration of gravity near the earth's surface is slightly different at different locations on earth. The acceleration depends on latitude because of the earth's rotation. It also depends on altitude. But for any given location, the acceleration there is the same for all objects.

Galileo knew that there was an effect on motion due to air resistance. However, his statement, even neglecting air resistance, was in much better agreement with his observations and measurements than was the generally believed (but untested) conclusion from Aristotle, namely, that a body ten times the weight of another would fall to the ground in one-tenth the time. We see here the importance of extracting from a series of observations the essence of what is important and setting aside or holding for later the less important aspects. It was the genius of Galileo that enabled him to see what was important and what was of secondary concern.

The measurements quoted here are not the ones Galileo referred to in *Discourses on Two New Sciences,* published in 1638, but are taken from his unpublished notes. For the published presentation of his ideas, Galileo discussed experiments in which he marked off distances in equal intervals and measured the time for those distances. He measured the short time intervals with a water clock. The result was that x/t^2 was constant, no matter which method was used. Galileo publicly described an experiment that differed from his real discovery experiment for just the same reason that scientists often do so today: to make the results more easily understood and to give the appearance that a logical mode of inquiry has been followed throughout. That a logical sequence of inquiry is often not the case is one of the important lessons in the history of science.

An extension of Galileo's experiments with inclined planes led to another extremely important qualitative result. Galileo positioned two inclined planes facing each other so that after a ball rolled down one, it would roll up the other. From his experiments with this setup, Galileo concluded that under ideal conditions (lack of friction and air resistance), the ball would roll up the second plane to a height above the base equal to the height from which it had been released on the first plane. If the upper end of the second track was lowered, the ball would roll farther along it in order to reach the same height. As the track was lowered more, the ball would travel farther, with less deceleration each time. Thus, on a level plane the ball would have no deceleration and its speed would be constant (in the absence of friction). By this reasoning, Galileo discovered the essence of what is called the law of inertia, later stated in full by Newton (see Chapter 4). Galileo realized that his experiment showed, at least for bodies on the earth, that it was not necessary to constantly apply a force in order to move an object. He concluded that rest and motion are both "natural" states of a body.

This conclusion was in direct and complete opposition to most beliefs of the time. The need for a force to keep a body in motion was accepted by some as a proof of the existence of God. The conclusions from Galileo's work in mechanics were not published for many years, in part because of the contradictions they presented with Aristotle's views.

Master the Concept

Velocity and Acceleration in Free Fall

Question: A ball thrown vertically upward rises to a maximum height and then falls to the ground. What are the ball's velocity and acceleration at the instant it reaches its maximum height?

Answer: When the ball is released, it has an initial upward velocity. It also has a constant downward acceleration due to gravity. As it moves up its (upward) velocity steadily decreases because of the downward acceleration until at one instant the velocity is zero. Afterward, the velocity becomes increasingly negative (that is, directed downward) and the ball moves down. At the instant the velocity becomes zero the ball has reached its maximum height; thereafter it falls back toward the ground. Thus, at the maximum height the ball has zero velocity and a downward acceleration of gravity g.

Example 2.18

At Six Flags Over Georgia near Atlanta, Free Fall riders seated four abreast in a padded gondola are taken to the top of a 10-story tower (Fig. 2.29a). Then the gondola is dropped 30 m down a vertical track that curves near the bottom, where the gondola slows to a stop. (a) How long does it take to fall from top to bottom? (b) What maximum speed is reached?

Strategy (a)   First we make a simplified sketch of the situation (Fig. 2.29b). We approximate the situation to be entirely vertical motion along an x coordinate that is vertically oriented with positive direction down. In this case we have chosen the origin at the top of the ride so that $x_0 = 0$. Then with the direction of travel chosen as positive, the gondola drops from rest and falls with a positive acceleration almost equal to that of all freely falling objects, 9.81 m/s^2. The relation between time, acceleration, and distance is Eq. (2.8) with $v_0 = 0$,

$$x - x_0 = \tfrac{1}{2}at^2.$$

This equation can be rearranged to give

$$t^2 = \frac{2(x - x_0)}{a}.$$

Solution (a) If we insert the values for x, x_0, and a (30 m, 0 m, and 9.81 m/s^2, respectively), we get the time to fall:

$$t = \sqrt{\frac{2(x - x_0)}{a}} = \sqrt{\frac{2(30 \text{ m})}{9.81 \text{ m/s}^2}} = \sqrt{6.116 \text{ s}^2} = 2.5 \text{ s}.$$

Strategy (b) The maximum speed is reached near the bottom of the ride. We can determine the speed from Eq. (2.9), which relates the final velocity to the initial velocity, the acceleration, and the displacement. In this case, the initial velocity is zero and the equation simplifies to

$$v^2 = 2a(x - x_0),$$

or

$$v = \sqrt{2a(x - x_0)}.$$

Solution (b) Inserting the numerical values, we find

$$v = \sqrt{2(9.81 \text{ m/s}^2)(30 \text{ m})} = 24 \text{ m/s}.$$

This speed is equivalent to 54 mi/h. The riders attain this speed because the gondolas are in free fall until slowed at the bottom of the ride.

Discussion In computing both of these solutions we had to find the square root. Mathematically, we should allow for both positive and negative values. However, since we chose the initial time to be zero and we can't make time run backward, only the positive value was reported for the time. Similarly, for positive displacement down the track with constant positive acceleration, the velocity must also be positive. Note also that the results for t and v are limited to two significant figures because the displacement was given only to two significant figures. ■

Figure 2.29
Example 2.18: (a) Free Fall, an experience that's like falling off a 10-story building. (b) Simplified diagram for analysis of the problem.

(a)   (b)

GALILEO AND EXPERIMENTAL SCIENCE

What makes science different from other human activities? Perhaps the principal distinction is the role of experiments that can be reproduced by other scientists around the world. Although people had previously tested their ideas of nature with observations, Galileo Galilei (Fig. B2.1) was one of the first to make the experimental process central to science. His influence, which persists to the present day, is in large measure due to his literary skill in describing his theories and experiments so clearly and beautifully that quantitative methods became attractive and fashionable. Galileo, as he is universally known, played a uniquely pivotal role in history. His breakthrough way of linking theory with experiment influenced all subsequent scientific thought, and, to some extent, nonscientific thought as well.

Galileo's work on falling bodies is an important early study of how nature works. This work also illustrates the modern view of how one should arrive at scientific conclusions. It is an example of what is sometimes called the scientific method.

An experiment believed to have been first conducted by Galileo in 1604 was reconstructed for *Scientific American* in 1975 by photographer Ben Rose according to specifications supplied by Stillman Drake. The objective of this modernized test was to measure the distances traveled from rest by a ball rolling down an inclined plane at the ends of eight equal times (in this case, at 0.55-s intervals). The grooved inclined plane was fitted with a stop at the higher end, against which a 2-in. steel ball could be held.

Galileo lived before the development of accurate clocks, and could not

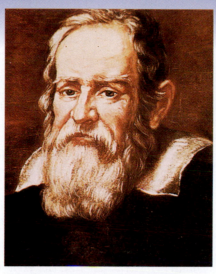

Figure B2.1   Galileo Galilei (1564–1642).

Figure B2.2   High-speed multiflash photograph of a freely falling billiard ball.

have measured time accurately if it had been necessary to measure the elapsed time. However, he needed to determine only equal intervals of time, not total elapsed time. Galileo was especially suited to do this because of his known ability as a musician. Even people who have no training in music are able to detect small differences in timing.

In the modern reconstruction of Galileo's experiment, the time intervals were established by singing the song "Onward Christian Soldiers" at a tempo of about two beats per second. At one beat the ball was released, and the positions of the ball at subsequent beats were marked with chalk; for comparison, the exact 0.55-s positions were also captured by multiple-flash photography (Fig. 2.27). A rubber band was then put around the plane at each chalk mark and the positions adjusted so that the audible bump made by the ball in passing each band would always come exactly at a beat of the march. The ratios of the successive measured distances were found to agree closely with a set of figures recorded by Galileo.

You may easily repeat his experiment by using a long board with rubber bands stretched around it, and tilting it at about two degrees to the horizontal. Hold a heavy ball lightly in place with your finger until you are ready to release it. You will hear a slight thump as the ball rolls over each of the rubber bands. When the thumps are regular, the time intervals are of equal length. Sing a song with a good strong beat and place the rubber bands so as to keep time with the music. The results depend only on whether the beat is regular.

Today's experimental techniques are far more sophisticated than anything Galileo could have imagined. For example, the development of high-speed photography, pioneered by Dr. Harold Edgerton, has enabled us to measure time intervals as small as 10^{-6} s. These techniques have led to calculations of speed and distance not otherwise obtainable (Fig. B2.2).

Summary

Useful Concepts

■ A reference frame is a physical entity to which we refer the position and motion of objects.

■ A coordinate system is used to specify the location of a point in a reference frame.

■ Average speed is the total distance traveled divided by the elapsed time interval.

■ Average velocity is given by

$$\overline{v} = \frac{\Delta x}{\Delta t},$$

where $\Delta x = x_f - x_i$ is the object's displacement. Velocity has direction as well as magnitude.

■ Instantaneous velocity is the limiting value of $\frac{\Delta x}{\Delta t}$ as Δt becomes vanishingly small; that is,

$$v = \lim_{\Delta t \to 0} \frac{\Delta x}{\Delta t}.$$

Alternatively, it is the slope of the tangent to the displacement-time curve at a point.

■ The area under the velocity-time curve between two time intervals is equivalent to the displacement during that time interval.

■ Average acceleration is the change in velocity divided by the time required for the change,

$$\overline{a} = \frac{\Delta v}{\Delta t}.$$

■ Instantaneous acceleration is the limiting value of $\frac{\Delta v}{\Delta t}$ as Δt becomes small; that is,

$$a = \lim_{\Delta t \to 0} \frac{\Delta v}{\Delta t}.$$

Alternatively, it is the slope of the tangent to the velocity-time curve at a point.

■ All bodies near the earth's surface fall with the same acceleration due to gravity, $g = 9.81$ m/s². If released from rest and if air resistance is not important, they traverse the same distance in the same amount of time.

■ The most frequently used kinematic equations for motion with constant acceleration are

$$x = x_0 + \overline{v}t \qquad v = v_0 + at$$
$$x = x_0 + v_0 t + \tfrac{1}{2}at^2 \qquad v^2 = v_0^2 + 2a(x - x_0)$$

Important Terms

You should be able to write the definition or meaning of each of the following terms:

kinematics	abscissa
dynamics	ordinate
mechanics	slope of a line
reference frame	instantaneous velocity
coordinate system	speed
displacement	average acceleration
average speed	instantaneous acceleration
velocity	freely falling bodies
average velocity	acceleration of gravity

Conceptual Questions

2.1 Carefully distinguish between speed and velocity.
2.2 Can an object have zero velocity but nonzero acceleration? Can it have zero acceleration and nonzero velocity? Give examples of each if the answer is yes.
2.3 Sketch a curve of velocity versus time for the displacement-time curve of Fig. 2.15. Sketch the acceleration-time curve also.
2.4 Sketch graphs to represent the following situations: (a) A car driven for 1 hour at a constant speed of 37 km/h (b) A person runs as fast as possible to the corner mailbox and immediately runs back as fast as possible (c) The motion of your hand as you wave it up and down
2.5 Discuss the qualitative relationship between the total na-

tional debt, the increase per year, and the rate of change of the increase. Use a graphical method. Which would you judge to be affected more by national and world events?
2.6 The distance-time curve for a hypothetical journey has the shape of an equilateral triangle with the base along the time axis. Discuss the velocity *and* acceleration necessary to bring about such a journey. Comment on whether or not this is a realistic journey.
2.7 What is meant by the word *slope* in the term *ski slope?*
2.8 Use a graphical method to answer the following question. The initial population of snowy egrets on a barrier island was 100 on January 1. Each January 1 thereafter, the population was 5% more than for the previous year. What was the popu-

lation at the end of 15 years? Is the rate at which the population grew a constant? (Recall the graphical meaning of constant velocity.)

2.9 In a common parlor trick, one person holds a dollar bill so that it hangs vertically. A second person places his hand so that the dollar is between but does not touch his thumb and forefinger. When the dollar is dropped, the second person tries to catch it but fails. Estimate the minimum reaction time of the second person.

2.10 Summarize your concept of the scientific method as you understand it from considering the work of Galileo.

2.11 Describe another observation or experiment that shows Galileo was correct in concluding that both straight-line motion and rest are "natural states" for a body (that is, they do not require the application of an external force).

2.12 A car's speedometer is correctly calibrated for tires of a specific size. If larger-diameter tires are substituted, what will be the effect on the speedometer reading?

$$s = vt \qquad v_2 = v_1 + at \qquad s = v_1t + \tfrac{1}{2}at \qquad v^2 = v_0^2 + 2as$$

Problems

Hints for Solving Problems

You may find it helpful to review the steps for solving problems that were given in Chapter 1 on pp. 19 and 20. In addition, be sure to choose the positive direction for the coordinate system used in each problem and apply the signs consistently to displacement, velocity, and acceleration. Be careful in converting units. (A conversion table is included inside the front cover.) Check your answers to see whether they are reasonable.

Section 2.1 Reference Frames, Coordinate Systems, and Displacement

2.1 A helicopter leaves its base and travels 20.0 km north. After a brief stop, it flies 35.7 km south, pauses briefly and then flies 17.0 km north. Finally it flies 6.0 km south and lands. At the end of the trip, what is the displacement of the helicopter from its base?

2.2 A scale is marked off in cm on a long table. A coin initially at the position labeled 100 cm is moved to the position marked −30 cm. What is the displacement of the coin?

2.3 The cursor on a computer screen is moved with a "mouse" resting on a 6.0-cm-wide pad. Two "swipes" completely across the pad move the cursor from the extreme left to the extreme right of the 25-cm-wide screen. Suppose that initially the cursor is at the left of the screen and the mouse is in the center of the mouse pad. Then the mouse is moved to the right edge of the pad, picked up and placed on the left edge of the pad and then moved all the way across the pad to the right edge. Where is the cursor?

2.4 Figure 2.30 shows the position of an object as a function of time. (a) At which point is the displacement from zero greatest? (b) At what point is the distance from zero greatest? Explain your answer.

2.5 Figure 2.31 shows position-time graphs for three different objects A, B, and C. (a) Which object is farthest from the origin at time $t = 0$? (b) Which object is at rest at time $t = 2.5$ s? (c) Which object moves the greatest distance away from the

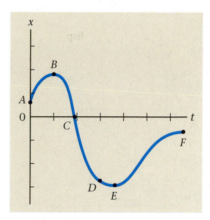

Figure 2.30
Problem 2.4.

origin? (d) How far from the origin is object C at $t = 2.0$ s? (e) Which object is farthest from the origin at $t = 3.0$ s? (f) Where is object B at 1.5 s?

2.6• You move 3.27 m ahead, 2.00 m ahead, 7.95 backward, 2.34 m ahead, 4.56 m backward, and 4.90 m ahead. (a) What is your final displacement? (b) What is your maximum displacement? (c) What is your minimum displacement?

2.7• A bike has a displacement of 1.27 km after a trip consisting of three segments x_1, x_2, and x_3 along a straight line. The sum of the first two segments is 3.79 km, the sum of the last two segments is −7.82 km. How long is each segment?

Section 2.2 Average Speed and Average Velocity

2.8 (a) What is the speed in kilometers per hour of a car traveling at a constant speed of 60 mi/h? (b) What is the speed of the car in meters per second? (1 mi = 1.609 km.)

2.9 What is the average speed for a trip of 157 km that requires 2.45 h?

2.10 How far will an automobile go in 3.5 h at a constant speed of 95 km/h?

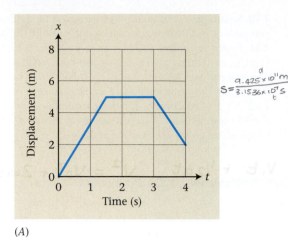

(A)

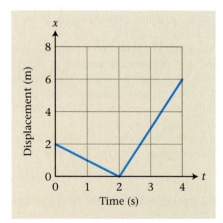

(B)

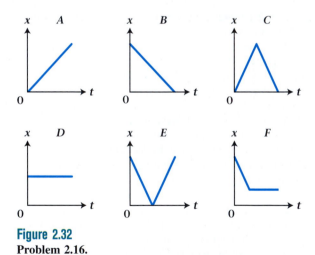

(C)

Figure 2.31
Problem 2.5.

$$S = \frac{9.425 \times 10^{11} \, m}{3.1536 \times 10^{7} \, s}$$

2.11 How long can you afford to stop for lunch if you can drive a steady 104 km/h on the highway and must make a 260-km trip in $3\frac{1}{2}$ h?

2.12 What is the linear speed of the earth in its orbit around the sun? Give your answer in meters per second. (Consider the earth's orbit to be a circle of radius 1.5×10^{11} m.)

2.13 It takes 2.51 s for a laser signal to go from the earth's surface to the moon and back. How far is the lunar surface from the earth's surface? The light travels at a speed of 3.00×10^8 m/s.

2.14 The speed of light is 3.0×10^8 m/s and the speed of sound is 340 m/s. Find the value of the integer n in the following statement: "If you start counting seconds when you see something happen and stop when you hear it happen, for every n seconds counted the event was about 1 km away."

2.15 When a batter hits a baseball its velocity is changed from 128 km/h due west to 136 km/h due east. (a) What is the change in speed? (b) What is the change in velocity?

2.16 Which of the displacement-time graphs in Fig. 2.32 correspond to each of the physical situations listed? (a) A ball rolled along the floor toward the origin, (b) a book at rest on a table, (c) a ball rolling along the floor away from the origin hits a wall and bounces straight back, (d) a ball rolling away from the origin at constant speed, (e) a ball rolled along the floor toward a wall at the origin, the ball then rebounds, and (f) an object rolling toward the origin that suddenly stops.

Figure 2.32
Problem 2.16.

2.17 A medivac helicopter travels 78 km due south of its base to pick up a patient. The helicopter then travels 93 km due north to a hospital. The entire trip takes 1.22 h. (a) What is the average velocity? (b) What is the average speed?

2.18 A long-distance runner starts at a given place and runs around a circular track of 50 m radius at a speed of 6.0 m/s in the clockwise direction for 60 s. Then the runner reverses direction and runs in the counterclockwise direction at 4 m/s for 120 s. At the end, how far around the track is the runner from the starting point?

2.19 Two children cross a starting line at the same time, one running with a velocity of $+3.5$ m/s and the other with a velocity of -4.0 m/s. How far apart are they after 12 s?

2.20● A commuting student leaves home and drives to school at an average speed of 40 km/h. After 24 min he realizes that he has forgotten his homework and returns home to get it at the same average speed. It takes 10 min to find the report, after which the trip to school 40-km away to the east is resumed at the same speed as before. (a) What is the average speed for the entire trip? (b) What is the average velocity for the entire trip?

2.21● If a greyhound runs in a straight line for 3.0 min at a velocity of 40 m/s, what must its velocity be in order to return to its starting point in 2.5 min?

2.22● Runner A, who runs with an average speed of 3.0 m/s, starts out at 3:00 P.M. Runner B, who runs with an average speed of 4.0 m/s, starts after A from the same place exactly 5 min later. (a) At what time will runner B catch up with runner A? (b) If the runners stop when B catches A, how far do they run?

2.23● A 9.0-h trip is made at an average speed of 50 km/h. If the first half of the distance is covered at an average speed of 45 km/h, what is the average speed for the second half of the trip?

2.24● Two automobiles travel in opposite directions from the same starting point. If the speed of one is twice the speed of the other and they are 200 km apart at the end of 1 hour, what is the speed of each car?

Section 2.3 Graphical Interpretation of Velocity

Hints for Solving Problems

Remember that the slope of a line is the ratio of the rise to the run, or $\Delta y/\Delta x$.

2.25 Plot the points given in the following table and sketch a curve through them. Determine the slope of the curve at $x = 5$ and at $x = 10$.

x	0	1	2	3	4	5	6	7
y	6.7	6.3	6.0	5.9	6.0	6.5	7.3	8.5

x	8	9	10	11	12	13
y	10.2	12.0	14.2	15.0	14.4	12.4

2.26 You are driving a car with initial speed v_0. At time $t = 0$, you begin to increase your speed at a constant rate. Twenty seconds later you are traveling at a speed of 80 km/h. At $t = 60$ s you are traveling at a speed of 140 km/h. (a) What was your initial speed v_0? (b) What was your speed at $t = 40$ s? Use a graphical method to arrive at the answers.

2.27 Figure 2.33 is an observed position-time graph. Sketch the corresponding velocity-time curve.

2.28 Match the velocity time graphs in the second row of Fig. 2.34 to the distance-time graphs in the first row.

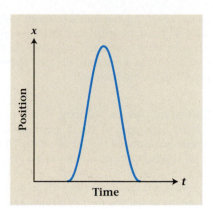

Figure 2.33
Problem 2.27.

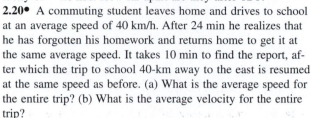

(A) (B) (C) (D)

(E) (F) (G) (H)

Figure 2.34
Problem 2.28.

2.29 Figure 2.35 contains pairs of distance-time and velocity-time graphs. In each of the four cases, identify which is the distance-time and which is the velocity-time graph.

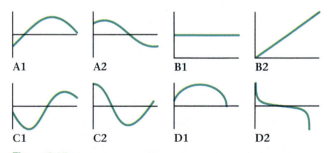

A1 A2 B1 B2

C1 C2 D1 D2

Figure 2.35
Problem 2.29.

2.30● Use the following data to plot a distance-time curve. Determine the velocity (slope) at each second and plot a velocity-time graph.

Time (s)	1	2	3	4	5	6	7	8	9
Distance (m)	2	4	6	8	9.2	9.8	10	10	10

2.31● A test car driver, starting with zero speed at time zero, drove in such a way that the speed-time graph is approximately an isosceles triangle with the base along the time axis. The maximum speed was 30 m/s, and the total elapsed time was 50 s. What distance did she travel?

Section 2.4 Instantaneous Velocity

2.32 The velocity-time graph of part of a cyclist's trip consists of a straight line between $v = 20$ m/s at $t = 10$ s and $v = 0$ at $t = 20$ s. What distance was covered during this portion of the trip?

2.33 Figure 2.36 shows the displacement of an object as a function of time. At which time is the velocity greatest? Explain your answer.

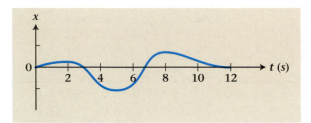

Figure 2.36
Problem 2.33.

2.34● Sketch an approximate distance-time curve for an object that has a velocity given by

$$v = (4.0 + 4.0t^{1/2})\text{m/s}$$

for the first five seconds of motion, where t is given in seconds.
2.35● The velocity-time graph of a jogger's trip is approximated by a triangle that starts at $v = 0$ at $t = 0$, rises to a maximum at $t = 6$ s, and then returns to $v = 0$ at $t = 10$ s. If the maximum speed was 6 m/s, how far did the jogger go?
2.36● (a) Plot the curve corresponding to $y = 3x^2 + 7$ and find the slope at $x = 5$ by graphical means. (b) Find the area under the curve between $x = 0$ and $x = 10$.

Section 2.5 Acceleration

2.37 A motorist traveling at 90 km/h applied the brakes for 5.0 s. If the braking acceleration was -2.0 m/s^2, what was her final speed?
2.38 The 542 hp Jaguar XJ220 can accelerate from 0 to 60 mi/h (26.8 m/s) in 4.0 s. What is the average acceleration during this time interval?
2.39 A Dodge Stealth turbo can accelerate from 0 to 60 mi/h (26.8 m/s) in 5.0 s. What is the average acceleration during this time interval?
2.40 A motorcycle rider moving with an initial velocity of 8.0 m/s uniformly accelerates to a speed of 17 m/s in a distance of 30 m. (a) What is the acceleration? (b) How long does this take?

2.41 Figure 2.37 shows the speed-time graph for a compact car under acceleration from a standing start. Use graphical techniques to determine the acceleration at 10 s and 30 s.

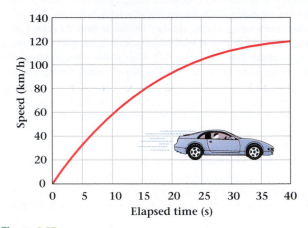

Figure 2.37
Problem 2.41: Graph of speed versus elapsed time for a compact car under acceleration from rest.

2.42 Describe in words the three types of motion represented in Fig. 2.38(a).

Displacement-time

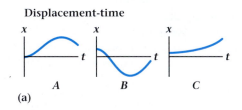

(a)

Velocity-time

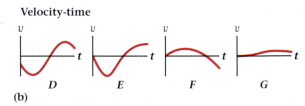

(b)

Acceleration-time

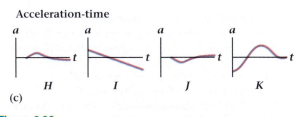

(c)

Figure 2.38
Problems 2.42 and 2.43: (a) Displacement-time curves. (b) Velocity-time curves. (c) Acceleration-time curves.

2.43● Figure 2.38 shows displacement-time curves for three different straight-line motions (A, B, C). Figure 2.38 also shows velocity-time curves for four different motions (D, E, F, G), three of which correspond to the motion shown in the displacement curves. (a) Match the three displacement curves to the appropriate velocity curves. (b) Match the three displacement curves to the appropriate acceleration curves shown as parts H, I, J, and K. (*Hint:* Use the velocity curves when looking for the accelerations.)

2.44● Starting from rest, a Mazda MX-6 reaches 48 km/h in 3.2 s and 96 km/h in 8.4 s. (a) Calculate the average acceleration needed to reach 48 km/h. (b) Calculate the average acceleration during the time it takes to go from 48 to 96 km/h. (c) Calculate the average acceleration during the time it takes to go from 0 to 96 km/h.

Section 2.6 Motion with Constant Acceleration

> ### Hints for Solving Problems
>
> You may find it helpful to refer to Table 2.1 to select the proper relationship between the quantities given in a problem and what you need to find. Remember that the equations in the table apply only to motion with constant acceleration. Include units in your work.

2.45 A motorcyclist moving with an initial velocity of 8.0 m/s undergoes a constant acceleration for 3.0 s, at which time his velocity is 17.0 m/s. (a) What is the acceleration? (b) How far does he travel during that 3.0-s interval?

2.46 The nominal stopping distances of a Nissan Sentra SE Sport Coupe are (a) 147 ft from 60.0 mi/h and (b) 264 ft from 80.0 mi/h. Determine the value of the acceleration for each case, assuming that it is constant during each event. Give your answer in units of m/s^2.

2.47 An object at rest is subject to a constant acceleration of 2.00 m/s^2 for 10.0 s. For the next 10.0 s there is no acceleration. Finally the object undergoes an acceleration of −2.00 for 10.0 s. (a) What is the final speed? (b) How far did the object go during the 30.0 s?

2.48● A soccer ball is released from the top of a smooth incline. After 4.22 s the ball travels 10.0 m. One second later it has reached the bottom of the incline. (a) Assume the ball's acceleration is constant and determine its value. (b) How long is the incline?

2.49● A speeding motorist passes a stopped police car. At the moment the car passes, the police car starts from rest with a constant acceleration of 4.28 m/s^2. The speeding motorist continues with constant velocity until caught by the police car 14.8 s later. How fast is the speeding car going?

2.50● Most state driver's handbooks contain a table of stopping distances for cars with good brakes and tires. The tables frequently include the braking distance for several speeds. The braking distance is the distance to stop after the brakes have been applied. Typical values of braking distances for a number of initial speeds are given in the following table. (a) Do these stopping distances represent approximately the same acceleration for each initial speed? (b) If the accelerations are constant, determine the stopping distance from 60 mi/h.

Initial speed (mi/h)	20	30	40	50
Stopping distance (ft)	20	45	80	125

2.51● You are driving down the street when you suddenly see a child dart out in front of you. Assuming that it takes 0.75 s for you to apply the brakes, compute the total stopping distance (braking distance plus distance traveled while moving your foot to the brake pedal) for initial speeds of 20 mi/h, 40 mi/h, and 50 mi/h. Use the information about braking distances given in Problem 2.50.

2.52● A Nissan Sentra can accelerate from 0 to 48 km/h in 3.6 s and from 0 to 96 km/h in 10.2 s. In addition, under constant acceleration from rest it crosses the 0.40-km marker at a speed of 130 km/h. (a) Calculate the average acceleration needed to reach 48 km/h. (b) Calculate the average acceleration during the time it takes to go from 48 to 96 km/h. (c) What constant acceleration would be required to reach a speed of 130 km/h over the 0.40-km course when starting from rest?

2.53● A motorcycle traveling with a constant acceleration of 2.00 m/s^2 crosses a 100-m-long bridge in 4.23 s. (a) What was the velocity at the beginning of the bridge? (b) What was the velocity at the end of the bridge?

2.54●● (a) A slowly moving train with 12-m-long flatcars is passing a station at 10 km/h. A person on the station platform tosses rocks onto the moving flatcars at the rate of once every second. (a) If the first rock just hits the front edge of one car, how many rocks will fall onto that car? (b) How many rocks will fall onto that car if the train begins to accelerate at 0.50 m/s^2 just as the first rock hits the car?

2.55●● Two motocross bikes start from one corner of a square field and go to the corner diagonally opposite in the same time t. They both start from the same place and take different routes. One travels along the diagonal with constant acceleration a, and the other accelerates momentarily and then travels along the edge of the field with constant speed v. What is the relationship between a and v? Assume a negligible acceleration time for the biker traveling along the edge.

Section 2.7 Galileo and Free Fall

2.56 A cannonball is dropped from the top of a building. If the point of release is 32.0 m above the ground, what is the speed of the cannonball just before it strikes the ground?

2.57 How long would it take an object to fall to the ground from the top of the Leaning Tower of Pisa (height = 54.6 m)?

2.58 Make a table of the velocity and total distance fallen at the end of each half-second during the first 3 s for an apple dropped from rest from the top of a very tall building. Make a graph of distance versus time and velocity versus time.

2.59 What is the acceleration of gravity on a planet where an object released from rest falls 54.2 cm in 1.08 s.

2.60• A ball is allowed to roll from rest down an inclined plane, and the distances are marked every 2.0 s. If the second mark is made 1.60 m from the starting point, where are the first and fourth marks?

2.61• A ball is thrown straight up so that it reaches a height of 25 m. How fast was it going when it was 5 m high? (*Hint:* Use the symmetry of the upward and downward paths.)

2.62• A small parachute dropped from a 30-m-high cliff falls with a constant velocity of 1.2 m/s. Twenty seconds after the parachute is dropped, a stone is dropped from the cliff. Will the stone catch up with the parachute before it reaches the ground?

2.63• The hollow cylinder shown in Fig. 2.39 is free to rotate about a horizontal axis. One hole is cut in the side of the cylinder. The object of a game is to spin the cylinder so fast that an object dropped through the hole when it is in the uppermost position will fall through the same hole when it has rotated to the bottom position. If the diameter of the cylinder is 0.50 m, how many revolutions per second must it make? (*Hint:* First calculate the time for the object to fall the appropriate distance, then use that result to determine the number of revolutions per second.)

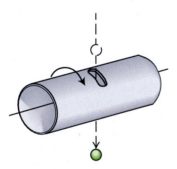

Figure 2.39
Problem 2.63.

2.64•• A professor drops one lead sinker each second from a very high window. (a) How far has the first sinker gone when the second one is dropped? (b) Does the distance between the first and second sinker remain constant? Explain your answer.

2.65•• A loose bolt falls from a high-flying helicopter that is rising at a constant 8.76 m/s. How far is the bolt below the helicopter 3.05 s later?

2.66•• A ball is dropped from a height h directly above a base toward which a fielder is running with speed v. When the ball is dropped, the glove on the fielder's outstretched hand is a distance d from the base. Find an expression for v so that the fielder just catches the ball if her glove is a height y above the base.

Additional Problems

2.67 A 3.0-h trip was made at an average speed of 75 km/h. For the first hour, the average speed was 90 km/h. What was the average speed for the remainder of the trip?

2.68 Figure 2.40 shows the speed curve for a sports coupe under maximum acceleration from a standing start. From the graph, determine the acceleration at a number of points, and make your own graph of acceleration versus time over the range from 0 to 40 s.

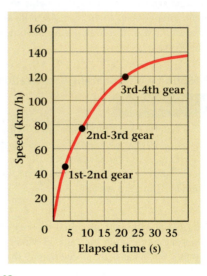

Figure 2.40
Problem 2.68: Graph of speed versus elapsed time for a sports coupe under maximum acceleration from rest.

2.69 (a) The flying time of the Concorde airplane from New York to London (5564 km) is approximately 3 h. What is the Concorde's average speed in km/h? (b) The Wright brothers' first sustained powered flight at Kitty Hawk, North Carolina, lasted 12 s and covered approximately 120 ft. What was their average speed in kilometers per hour? (c) How many times faster is the Concorde than the Wrights' first powered flight?

2.70• Starting from rest, you move with a constant acceleration of 2.0 ft/s² for 12 s and then move with an acceleration of −2.0 ft/s² for another 12 s. (a) What is your maximum speed attained? (b) How far do you go during the whole trip? (c) What is your average speed?

2.71• With what initial velocity must you throw a ball from a second-story window ($h = 4.0$ m) in order for it to reach the ground in half the time it would have taken if it had been dropped and not thrown?

2.72• A rock is dropped from rest from a height above a strange planet, and a strobe-light photograph is taken. The image is damaged in transmission to earth so that an unknown part of the top of the picture is lost. However, five successive images of the falling rock can be seen. The spacing between the remaining images corresponds to 0.70, 0.90, 1.10, and 1.30 m, and the flash rate is 4.0 flashes per second. Calculate the acceleration of gravity on that planet.

2.73•• Two cyclists start in opposite directions at the northernmost point of a circular track of 25 m radius. (a) If the speed of each bike is constant and is 10 km/h for cyclist A and 15 km/h for cyclist B, where will they meet? (b) Where will the cyclists meet the second time? Assume that they go around the track on essentially the same path, with cyclist A initially headed west and cyclist B initially headed east.

2.74•• A Boeing 767 is initially moving down the runway at 4.5 m/s preparing for takeoff. The pilot pulls on the throttle so that the engines give the plane a constant acceleration of 1.8 m/s². The plane then travels a distance of 1700 m down the runway before lifting off. How long does it take from the application of the acceleration until the plane lifts off, becoming airborne?

2.75•• A person can throw a ball with an upward velocity v_0 so that it will just reach the top of a 20-m-tall building. If the same person stands on top of the building and throws the ball downward with velocity v_0, how much sooner does it reach the ground than if it were merely dropped from the same height?

2.76•• A model rocket is fired upward. The rocket's average initial acceleration is 37.5 m/s² until the fuel burns out in 0.845 s. How high does the rocket go? Ignore the effects of air friction.

2.77•• A stone is dropped from rest from a height of 20 m. At the same time, a stone is thrown upward from the ground with a speed of 17 m/s. At what height do their paths intersect?

2.78•• A ball bearing is dropped from rest at a point A. At the instant it passes a mark 10 m below A, another ball bearing is released from rest from a position 11 m below A. (a) At what time after its release will the second ball bearing be overtaken by the first? (b) How far does the second bearing fall in that time?

2.79•• An electric-powered model car starts from rest and moves in a straight line for 4 s. The relationship between time and position is approximated by

$$x^2 + 2t^2 = 32,$$

where x is in centimeters and t is in seconds. Using graphical methods, find the velocity at $t = 1$ s and at $t = 3$ s.

2.80•• A ball dropped from the top of a tower reaches a velocity v_f just before it reaches the ground. An automobile traveling toward the tower with constant speed v_f just reaches the tower when the ball strikes the ground. Show that at the instant the ball was released, the distance of the car from the tower was twice the height of the tower.

2.81•• A ball is dropped from the top of a tower at the same time that one is thrown upward from the ground below. They collide at the top of the second ball's trajectory 2.0 s after the ball is thrown upward. How tall is the tower? Neglect the height of the thrower. (*Hint:* A ball thrown upward with speed v acquires the same speed v when it returns again to the ground.)

Solving Quadratic Equations

Any quadratic equation can be put in the form

$$ax^2 + bx + c = 0,$$

where a, b, and c are constants. An equation in this form has solutions for x given by the quadratic formula:

$$x = \frac{-b \pm \sqrt{b^2 - 4ac}}{2a}.$$

Thus, a particular quadratic equation such as

$$7x^2 + 4x - 3 = 0,$$

has values of $a = 7$, $b = 4$, and $c = -3$, which can be substituted into the quadratic formula to give

$$x = \frac{-4 \pm \sqrt{(4)^2 - (4)(7)(-3)}}{2(7)}$$

$$x = \frac{-4 \pm 10}{14}.$$

There are two values of x: the first one corresponds to the + sign and the second one corresponds to the − sign. Our final answers are

$$x = 0.429 \quad \text{and} \quad x = -1.00.$$

Both values satisfy the original equation.

Example 2.19

A student riding a skateboard down an incline is traveling 3.0 m/s when she crosses a mark on the sidewalk. She continues down the incline, with a steady acceleration of 0.50 m/s^2, until she crosses another mark 20 m from the first one. How long does it take for her to go from the first mark to the second one?

Strategy The position of the student is given by Eq. (2.8):

$$x = x_0 + v_0 t + \tfrac{1}{2}at^2.$$

Subtracting x from both sides of the equation and rearranging the order gives a quadratic equation in the variable t:

$$\tfrac{1}{2}at^2 + v_0 t + (x_0 - x) = 0.$$

On comparing this with the quadratic formula we get

$$x = \frac{-v_0 \pm \sqrt{v_0^2 - 4(\tfrac{1}{2}a)(x_0 - x)}}{2(\tfrac{1}{2}a)}.$$

Notice that the a in Eq. (2.8) is not the same as the a of the quadratic formula.

Solution If we let the first mark represent the origin, then the numerical values become $x_0 = 0$, $x = 20$ m, $v_0 = 3.0$ m/s, and $a = 0.50$ m/s^2. Inserting these numerical values gives

$$t = \frac{-3.0 \text{ m/s} \pm \sqrt{(3.0 \text{ m/s})^2 - 4(0.25 \text{ m/s}^2)(0 \text{ m} - 20 \text{ m})}}{0.50 \text{ m/s}^2}$$

$$t = \frac{-3.0 \text{ m/s} \pm \sqrt{29 \text{ m}^2/\text{s}^2}}{0.50 \text{ m/s}^2} = \frac{-3.0 \pm 5.39}{0.50} \text{s}.$$

This gives two answers: $t = 4.8$ s and $t = -17$ s, where we have rounded the answer to two significant figures. We discard the negative time because, although it does satisfy the algebraic equation, it has no physical meaning in this case. (We assumed $t = 0$ when $x = x_0$, and that time only increases in one direction. That is, t must be positive.) Thus, the meaningful answer is that the student took 4.8 s to go from the first mark to the second mark. Notice that units are handled just like numbers or symbols.

Motion in Two Dimensions

Galileo's studies of motion went beyond the case of free fall in one dimension to explore the two-dimensional motion of projectiles, such as shells fired from a cannon. However, before we can analyze projectiles, we need to understand more about the description of motion in two and three dimensions.

The motion of a runner in an open field is two dimensional. In order to describe the runner's velocity, we need to know how fast and in what direction the runner is going. Quantities, such as velocity, that have both direction and magnitude are called vectors. The vector techniques presented here are useful in physics and will be used often throughout the remainder of the text.

We will use the example of a projectile, such as a thrown baseball, as an illustration to show that it is possible to treat independently the horizontal and vertical components of the motion of a body. These ideas will appear again in the chapters that follow, especially those on mechanics, electricity, and magnetism. ∎

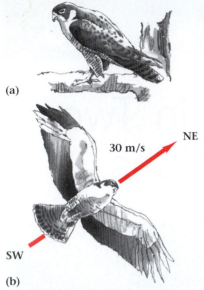

(a)

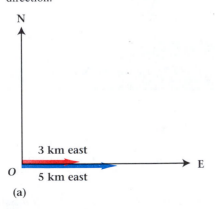

30 m/s

NE

SW

(b)

Figure 3.1

Direction and magnitude. (a) A quantity with magnitude but no direction, such as the falcon's mass, is a *scalar*. (b) The speed and direction of the falcon's flight result in a quantity with both magnitude and direction, called a *vector*.

Figure 3.2

(a) The magnitude of a vector is indicated by the length of the arrow representing the vector. (b) Arrows representing displacements of 5 km east and 5 km north. The vectors have the same magnitude but different directions. (c) The three vectors are the same because they all have the same magnitude and the same direction.

3.1 Vectors

Many physical quantities can be completely specified by their magnitude alone. Such quantities are called **scalars.** Examples include such diverse things as distance, time, speed, mass, and temperature. Another physically important class of quantities is that of **vectors,** which *have direction as well as magnitude.* For example, if we say a grocery store is ten miles from home, we have not completely specified its location unless we state its direction from us—north, south, east, or west. The distance and the direction together constitute a vector called the displacement. Without knowing the direction, we do not know the displacement even when we know the distance.

Many other physical quantities are vectors besides displacement, including velocity and acceleration, which were introduced in Chapter 2. Force and momentum, which will be defined in later chapters, are also vector quantities. Often an object (such as the falcon shown in Fig. 3.1) can have both scalar properties (mass) and vector properties (velocity) at the same time. Because we live in a three-dimensional world, we need vectors to describe the motions of objects. In this chapter, as in Chapter 2, we will consider only the way in which objects move. In the next and subsequent chapters, we will inquire into the causes of motion and the relationships between the various motions and their causes.

In printed materials such as this text, we represent a vector quantity in boldface roman type, **A**. If we are referring only to the magnitude of that quantity, we use the same letter in lightface italic type, *A*. For your own purposes in working problems and taking notes, you may find that a letter with an arrow drawn over it ($\vec{A}$) is a useful symbol for a vector.

In diagrams we frequently use an arrow to represent a vector. The arrow is drawn so that it points in the direction of the vector and its length is proportional to the magnitude of the vector. For instance, we might represent a displacement of 5 km east by an arrow 25 mm long (Fig. 3.2a). To represent a displacement of 3 km east, we would draw a second arrow parallel to the first but shorter, in the ratio of 3 : 5 or 15 mm. A vector representing 5 km north is directed 90° counterclockwise from a vector representing 5 km east (Fig. 3.2b). Because a vector is denoted by its length and direction, it remains the same vector when translated to a new starting point (Fig. 3.2c).

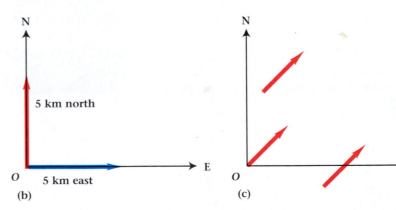

N

3 km east

O   5 km east

E

(a)

N

5 km north

O   5 km east

E

(b)

N

O

E

(c)

The **magnitude of a vector** quantity is represented by the same letter used for the vector, but in lightface italic type instead of boldface type. An alternative notation for the magnitude is the vector symbol with vertical bars on both sides. Thus

$$\text{magnitude of } \mathbf{A} = A = |\mathbf{A}|.$$

By definition, the *magnitude of a vector quantity is a scalar and is always positive.*

3.2   Addition of Vectors

Addition of scalars uses just simple arithmetic: 3 kg + 5 kg = 8 kg, for example. Addition of vectors, however, must be different to take account of the directions of the quantities. To illustrate vector addition, let us first take a simple example, after which we will state the general rule. Consider the following problem: If a woman walks 4 km north and 3 km east, how far and in what direction is she from the starting point?

Figure 3.3 is a scale map of this walk. The initial northward walk is a displacement and hence is represented by a vector, which we have labeled **A** in the figure. The eastward journey results in a displacement represented by vector **B**. After traveling north and east, the walker arrives at point *P*. However, a more direct way to reach the same point would be to walk along the straight line *OP*. This line, which represents the resultant displacement, is represented by vector **C** and measures 5 km at an angle of 36.9° east of north. (The distance and angle may be determined either by measuring on a scale drawing or by calculating with geometry and trigonometry.)

Whether the walker takes the indirect or the direct route, she ends up in the same place. Thus, the displacement represented by the sum of the two vectors **A** + **B** equals the displacement represented by the vector **C**:

$$\mathbf{A} + \mathbf{B} = \mathbf{C}.$$

We say that the vector **C** is the **vector sum,** or **resultant,** of the two vectors **A** and **B**.

Two important points are worth mentioning before proceeding. First, note that although the sum of the magnitudes of the individual parts of the trip is 7, this is not the magnitude of the resultant vector. The sum of two vectors depends on their directions as well as their magnitudes (Fig. 3.4). The second point is that

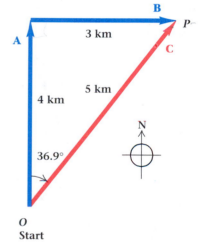

Figure 3.3
Addition of vectors. The displacement that results from walking along vectors **A** and **B** is equal to that of vector **C**. However, the sum of the magnitudes of **A** and **B** is not the same as the magnitude of **C**.

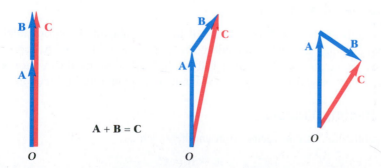

$$\mathbf{A} + \mathbf{B} = \mathbf{C}$$

Figure 3.4
Variations in magnitude and direction. If vectors **A** and **B** are of constant magnitude but their direction changes, then the magnitude and direction of **C** will also change.

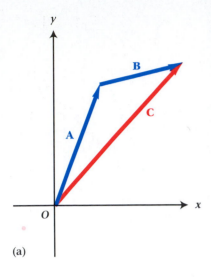

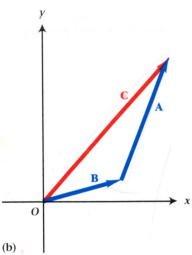

Figure 3.5
(a) Vector **C** is the vector sum **A** + **B**.
(b) Vector **C** is also the sum **B** + **A**.

the technique for adding vectors is the same for all vectors, whether they represent displacement or any other vector quantity—just as the rules for adding all scalars are the same, no matter whether they represent time, money, mass, or pure numbers.

Graphical Methods

To add two vectors graphically, make a scale drawing and place the vectors "head to tail"—that is, with the tail (origin) of the second vector starting from the head (end point) of the first. Then draw the resultant vector from the origin of the first vector to the end point of the second vector. Finally, measure the length (magnitude) and direction of the resultant vector directly from the scale drawing. This method is called the *triangle method of addition.*

Now consider an object that undergoes a displacement **A**, followed by a second displacement **B** (Fig. 3.5a). If we make the displacements in reverse order, as in Fig. 3.5b, with displacement **B** followed by **A**, the resultant is the same. We conclude that the order in which we add the vectors may be reversed without changing the result:

$$\mathbf{A} + \mathbf{B} = \mathbf{B} + \mathbf{A}.$$

We express this behavior by saying that vector addition is commutative. Moreover, commutation is a general rule for the addition of any two vectors.

In the previous example of a woman walking 4 km north and then 3 km east, she would reach the same end point if she reversed the order in which she made the two parts of the trip. The final distance and direction between her ending point and her starting point are the same in both cases.

Vectors may be moved (translated) without changing their value, so long as their directions and magnitudes are not changed. The vector **A** in Fig. 3.5a is identical to the vector **A** in Fig. 3.5b. Similarly the two vectors **B** are identical. Their resultants **C** are also identical in both direction and magnitude.

An alternative graphical method for adding two vectors is shown in Fig. 3.6. If the two vectors **A** and **B** are drawn with their tails joined together, their resultant **C** is the diagonal (bisector) of the parallelogram constructed with **A** and **B** as its sides. This method is known as the *parallelogram method of addition.*

To add more than two vectors together, we repeat the rule by adding successive vectors head to tail (Fig. 3.7). If we wish to add three vectors **A**, **B**, and **C**, we first add **A** and **B** to get a resultant **D**. Then we add **C** to **D** to give **E**. Additional vectors could also be added one at a time. It is not necessary to draw the intermediate sum **D**; we may simply add **C** by placing it next to **B**, as shown in the figure.

Negative of a Vector

The negative of vector **A** is a vector of the same magnitude and parallel to **A** but pointing in the opposite direction (Fig. 3.8). The negative of **A**, represented as −**A**, can be obtained by adding 180° to the angle that specifies vector **A**. The vector and its negative are equal in magnitude and opposite in direction.

Subtracting Vectors

The subtraction of one vector from another, such as

$$\mathbf{A} - \mathbf{B} = \mathbf{C},$$

can be considered as the addition of the first vector to the negative of the second, as

$$A + (-B) = C.$$

Figure 3.9 illustrates the procedure.

Multiplication of a Vector by a Scalar

The result of multiplying a vector by a scalar is another vector. For example, the vector 2**A** has a magnitude twice as great as that of **A** and points in the same direction as **A**. The vector −2**A** is also twice as great as **A**, but it points in the opposite direction.

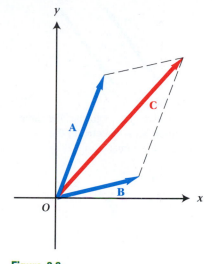

Figure 3.6
The sum of two vectors may be found from the diagonal of the parallelogram formed with the vectors for its sides.

Example 3.1

A vector **A** has a magnitude of 7 units and a direction of 30° measured counterclockwise from the positive x axis. The vector **B** has a magnitude of 11 units and a direction of 140° measured counterclockwise from the positive x axis. What is the vector sum of **A** and **B**?

Strategy First we draw a careful scale diagram. Then using the triangle method of addition, the vectors **A** and **B** are drawn head to tail (Fig. 3.10a). Then the resultant vector **C** is drawn from the tail of **A** to the head of **B**. This diagram corresponds to

$$A + B = C.$$

Solution After making the diagram, we determine the magnitude of **C** by measuring its length on the drawing and converting it to the appropriate number of units by using the scale indicated. Always indicate on the diagram the scale you use. In this case, the magnitude of **C** is 10.8 units. The direction of **C** can be measured with a protractor and is 103° from the positive x axis.

Alternate strategy We can also find the sum by using the parallelogram method of addition (Fig. 3.10b). Again we make a careful scale drawing, but this time both vectors are drawn with their tails at the origin of the coordinates.

Solution We complete the parallelogram and draw the diagonal from the origin to the opposite corner. This diagonal is the resultant **C**. As before, we measure the length of the **C** with a ruler and its angle from the x axis with a protractor to find a magnitude of 10.8 units and an angle of 103°.

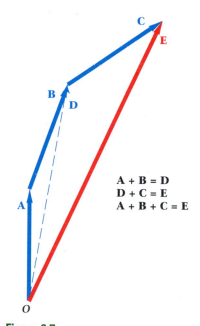

$$A + B = D$$
$$D + C = E$$
$$A + B + C = E$$

Figure 3.7
Addition of three vectors.

Example 3.2

Your friend standing at point P wishes to join you at point Q located a distance of 200 m due north across a lake (Fig. 3.11). Your friend decides to go around the lake by traveling along two straight-line paths, **A** and **B**. If the first path **A** is 150 m at an angle of 25° east of north, what are the distance and the direction of the second path?

Strategy The distance across the lake is the **C**, the sum of two contributing vectors **A** and **B**. We know **C** and one vector **A**, so to find the other vector **B**, we must subtract **A** from **C**.

Figure 3.8
The negative of a vector has the same magnitude but is directed in the opposite direction.

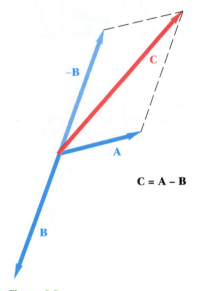

Figure 3.9
Subtraction of vectors. We can obtain the difference between two vectors such as **A** and **B** by adding the first vector to the negative of the second.

Figure 3.10
Example 3.1: Addition of vectors by (a) the triangle method and (b) the parallelogram method.

Solution   We may formulate this vector relation as

$$B = C - A,$$

where **C** is 200 m to the north and **A** is 150 m at 25° to north. Figure 3.11 is a scale diagram from which we may determine the answer by direct measurement. The second path **B** is found to be 90 m at an angle of 45° west of north.

3.3   Resolution of Vectors

It is often useful to think of a vector as the sum of two or three other vectors. We call these other vectors *components*. Usually we choose components at right angles to each other. Resolving vectors into their components makes it easier to carry out mathematical manipulations such as addition and subtraction.

In two dimensions, we frequently choose the component vectors to lie along the *x* and *y* axes of a rectangular (Cartesian) coordinate system. For example, consider the vector **A** lying in the *xy* plane (Fig. 3.12). We can construct two component vectors by drawing lines from the end of **A** perpendicular to the *x* and *y* axes. The two vectors that lie along the *x* and *y* directions add to form **A**. When we find the magnitude of these two vectors, we say that we have *resolved* the vector **A** into its *x* and *y* components.

Since the known vector and its component vectors form a right triangle, we may use trigonometry to resolve the vector into its components. (A review of basic trigonometry is found in the appendix to this chapter.) This is why components at right angles are particularly convenient. Thus, for the vector of magnitude *A* that makes an angle θ with the *x* axis, the component A_x along the *x* direction and the component A_y along the *y* direction are given by

$$A_x = A \cos \theta, \qquad A_y = A \sin \theta. \tag{3.1}$$

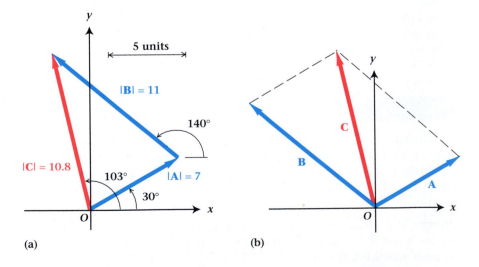

(a)

(b)

If we know the components A_x and A_y of a vector, then we can obtain the magnitude and direction of the vector by applying trigonometry. The magnitude A of the vector is computed from the Pythagorean theorem as

$$A = \sqrt{A_x^2 + A_y^2}. \tag{3.2}$$

The angle θ between the vector **A** and the x axis is determined from

$$\tan \theta = \frac{A_y}{A_x}. \tag{3.3}$$

Note that for a given ratio A_y/A_x there are two possible values for the angle. Your calculator will give an angle θ between $-90°$ and $+90°$. The other value is $\theta + 180°$. You must inspect each situation to know which value is correct.

Example 3.3

Swimming at an angle of $27°$ from the horizontal, an angelfish has a velocity vector **v** with a magnitude of 25 cm/s (Fig. 3.13). Find the x and y components of **v**.

Strategy   Choose a coordinate system with the x axis along the horizontal. Then we can resolve the velocity into its x and y components by applying Eq. (3.1).

Solution   To resolve vector **v** into its x and y components, we use Eq. (3.1):

$$v_x = v \cos \theta = (25 \text{ cm/s}) \cos 27° = 25 \text{ cm/s} \times 0.891 = 22 \text{ cm/s},$$
$$v_y = v \sin \theta = (25 \text{ cm/s}) \sin 27° = 25 \text{ cm/s} \times 0.454 = 11 \text{ cm/s}.$$

■

Adding Vector by Components

Adding or subtracting vectors by resolving them into perpendicular components simplifies the mathematics by enabling us to add vectors in the same directions. This way we avoid having to make careful scale drawings or using trigonometry for every manipulation of vectors. The way to add or subtract two or more vectors is first to resolve each vector into components. Then add the components

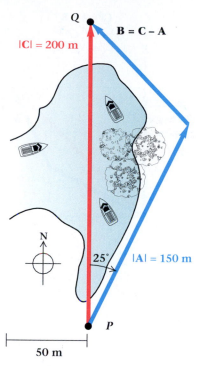

Figure 3.11
Example 3.2: Subtraction of vectors.

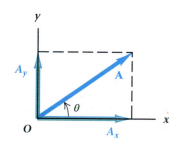

Figure 3.12
A vector **A** lying in the xy plane has components A_x along the x direction and A_y along the y direction. From trigonometry we see that $A_x = A \cos \theta$ and $A_y = A \sin \theta$.

Figure 3.13
Example 3.3: Resolution of a vector into components along x and y.

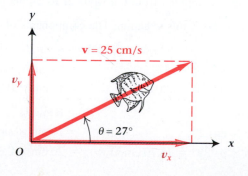

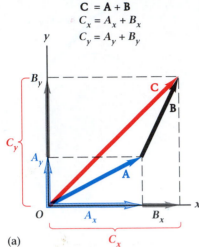

$$C = A + B$$
$$C_x = A_x + B_x$$
$$C_y = A_y + B_y$$

(a)

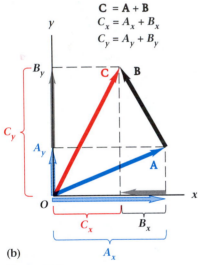

$$C = A + B$$
$$C_x = A_x + B_x$$
$$C_y = A_y + B_y$$

(b)

Figure 3.14

Adding two vectors **A** and **B** to form a new vector **C**. (a) We add the x components to form C_x and the y components to form C_y. (b) If the x component of B is directed along the negative direction, the x component of C is less than A_x.

along the x direction to form the x component of the resultant vector. Similarly, add the y components to get the y component of the resultant vector. For example, the vectors **A** and **B** add together to form a new vector **C** (Fig. 3.14a) whose components are

$$C_x = A_x + B_x, \qquad C_y = A_y + B_y.$$

We can obtain the magnitude and direction of the new vector **C** from Eqs. (3.2) and (3.3) using the new components C_x and C_y. In the case illustrated in Fig. 3.14(b), the x component of vector **B** points in the negative direction, so the sum of the x components is less than A_x.

If we want to subtract two vectors **A** and **B**, we can use the same procedure of resolving each vector into components and subtracting them in proper order. Then we use the resulting components to find the magnitude and direction of the new vector **C**.

Problem-Solving Strategy

Adding Vectors

1. Start by choosing a coordinate system and sketching the vectors. Use graphical techniques to get a qualitative estimate of the resultant.
2. Resolve the vectors into x and y components and be careful to keep their algebraic signs.
3. Add the x components algebraically to find the resultant x value and add the y components algebraically to find the resultant y value.
4. Then find the magnitude of the resultant vector using Eq. (3.2) and find the angle θ from the x axis using Eq. (3.3).

Example 3.4

Vector **A** has a length of 14 cm at 60° with respect to the x axis, and vector **B** has a length of 20 cm at 20° with respect to the x axis. Add the vectors by first resolving them into components and then adding the components.

Strategy We find the x and y components of each vector. Then we add the x components to get the x component of the resultant and add the y components to get the y component of the resultant. Then the magnitude and direction of the resultant is computed from the value of its components.

Solution Figure 3.15 shows this situation. The components of the vectors are

$$A_x = A \cos \theta_A$$
$$= 14 \text{ cm} \times \cos 60°$$
$$= 14 \text{ cm} \times 0.500 = 7.00 \text{ cm};$$

$$A_y = A \sin \theta_A$$
$$= 14 \text{ cm} \times \sin 60°$$
$$= 14 \text{ cm} \times 0.8660 = 12.12 \text{ cm};$$

$$B_x = B \cos \theta_B$$
$$= 20 \text{ cm} \times \cos 20°$$
$$= 20 \text{ cm} \times 0.9397 = 18.79 \text{ cm};$$

$$B_y = B \sin \theta_B$$
$$= 20 \text{ cm} \times \sin 20°$$
$$= 20 \text{ cm} \times 0.3420 = 6.840 \text{ cm}.$$

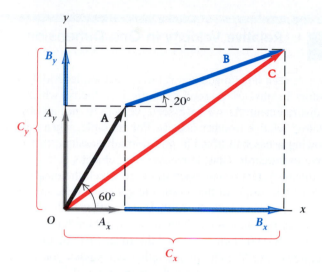

Figure 3.15
Example 3.4: Adding **A** and **B** by components.

The sum is given by

$$\mathbf{C} = \mathbf{A} + \mathbf{B},$$

where

$$C_x = A_x + B_x = 25.79 \text{ cm}, \qquad C_y = A_y + B_y = 18.96 \text{ cm}.$$

The magnitude of **C** is then found from the Pythagorean theorem as

$$C = \sqrt{C_x^2 + C_y^2},$$
$$C = \sqrt{(25.79 \text{ cm})^2 + (18.96 \text{ cm})^2} = 32 \text{ cm}.$$

The angle of **C** with respect to the x axis is given by

$$\tan \theta = \frac{C_y}{C_x} = \frac{18.96}{25.79} = 0.735$$
$$\theta = 36°.$$

Discussion Notice that we carried out all the computations before rounding off the answers to two significant figures. This is consistent with the rules given in Chapter 1.

●● ■

To this point we have considered the addition or subtraction of two vectors that lie in the same plane. Other cases occur in nature in which the vectors occupy three dimensions, rather than two. Although we will not deal with such cases very often in this text, we will state the basic principle for adding and subtracting vectors in three dimensions. The vectors to be added (or subtracted) are resolved into components along the x, y, and z axes, and the components are added (or subtracted). This procedure is the same as that used for vectors lying in a plane except that there are three, rather than two, dimensions. Figure 3.16 shows the resolution of a vector into components in three dimensions.

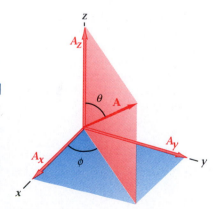

Figure 3.16
Components of a vector in three dimensions.

| 3.4 | **Relative Velocity in One Dimension** |

Now that we've seen how to add and subtract vectors, let's look at some examples of vectors in physics. A good example is velocity, which is the rate of change of displacement. As we have seen, velocity is not an absolute quantity but is measured relative to other objects. For example, when you hear about a pitcher throwing a baseball 85 mi/h, you normally assume that it means a velocity relative to the earth. Thus, to measure any object's velocity we must specify the coordinate system or reference frame in which the measurement is to be made. Ordinarily the origin of the coordinate system is fixed in some other body. In the example of the baseball, the earth was the other body.

When we say that an automobile is traveling 90 km/h (55 mi/h), we usually mean that it is going 90 km/h relative to the road. Imagine that you are driving down the highway at 90 km/h when another car passes you at 100 km/h. Although both cars are moving rapidly down the road, the faster car appears to overtake you very slowly. Relative to a coordinate system fixed on your car, the passing car is going only 100 km/h − 90 km/h = 10 km/h.

When two objects are traveling along the same line, the **relative velocity** of one to the other is obtained simply by ordinary subtraction. (Recall that the negative of a vector is a vector of the same magnitude pointing in the opposite direction.) To clarify the frame of reference of a particular object, we will use the notation v_{AB} to indicate the velocity of object A with respect to object B. Then if we have a velocity v_{AB} of A with respect to B and a velocity v_{BC} of B with respect to C, the velocity of A relative to C is found by vector addition,

$$v_{AC} = v_{AB} + v_{BC}.$$

| **Example 3.5** |

A freight train pulling several flatcars is slowly passing a highway intersection at 10 km/h. A hobo on one of the flatcars is walking toward the engine at 5 km/h (Fig. 3.17). What is the velocity of the hobo relative to an observer waiting in a truck stopped at the crossing?

Strategy In this example both velocities are in the same direction. We choose a coordinate system oriented with the positive direction in the direction of the train's motion. The hobo walks toward the engine with a velocity v_{he} = 5 km/h. At the same time, the train moves past the observer with velocity v_{eo} = 10 km/h. We want to find v_{ho}, the velocity of the hobo relative to the observer.

Solution We can write the expression for the velocity of the hobo relative to the observer as

$$v_{ho} = v_{he} + v_{eo}.$$

Since v_{he} and v_{eo} are in the same direction, we can simply add their magnitudes to get the magnitude of v_{ho}.

$$v_{ho} = v_{he} + v_{eo}.$$

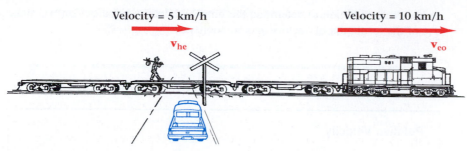

Velocity = 5 km/h   Velocity = 10 km/h

$\mathbf{v}_{he}$ $\mathbf{v}_{eo}$

Figure 3.17
Example 3.5: A hobo walking with velocity $\mathbf{v}_{he}$ on a train moving with velocity $\mathbf{v}_{eo}$ passes a stationary observer with a velocity equal to the sum of the two velocities.

When we insert the numerical values we get

$$v_{ho} = 5 \text{ km/h} + 10 \text{ km/h} = 15 \text{ km/h}.$$

The direction of $\mathbf{v}_{ho}$ is positive.

Example 3.6

Suppose the hobo in Fig. 3.17 turns around and walks away from the engine at 5 km/h. What is his velocity with respect to the observer then?

Strategy   If we choose the direction of the engine's velocity as positive, then for this case the hobo's velocity must be negative.

Solution   The velocity of the hobo relative to the observer is computed just as before.

$$\mathbf{v}_{ho} = \mathbf{v}_{he} + \mathbf{v}_{eo}.$$

Because the direction of $\mathbf{v}_{he}$ is opposite to that of $\mathbf{v}_{eo}$, the indicated sum of the magnitudes is

$$v_{ho} = -v_{he} + v_{eo}.$$

Upon inserting the numerical values we get

$$v_{ho} = -5 \text{ km/h} + 10 \text{ km/h} = 5 \text{ km/h}.$$

Discussion   This example illustrates the need for choosing a coordinate system and sticking with it throughout the problem. The difference between this example and the previous one is that the hobo is walking away from the engine. Since we choose the engine's velocity to be positive, the hobo's velocity relative to the engine was necessarily negative.

3.5   Relative Velocity in Two Dimensions

We have seen that when two objects are traveling along the same straight line, the relative velocity of one to the other is obtained by ordinary addition or subtraction. However, if the two velocities are not along the same line, then we need to use vector addition to determine the relative velocity. In particular, we make use of the resolution of vectors into components.

▼
Problem-Solving Strategy

Relative Velocity

1. Identify all velocities and make a vector diagram.
2. Use double subscripts to specify the velocities. For example, use $\mathbf{v}_{AB}$ to indicate the velocity of A relative to B.
3. Then use the techniques of vector addition to find the needed velocity.

▼
Example 3.7

A person who can row a boat at 5.0 km/h in still water tries to cross a river whose current moves at a rate of 3.0 km/h (Fig. 3.18a). The boat is pointed straight across the river, but its progress includes a downstream motion due to the river current. (a) What is the velocity of the boat with respect to the bank? (b) If the river is 200 m wide, how far downstream does the rower land?

Strategy (a) Velocity is the rate of change of displacement, a vector quantity. Hence, velocity is also a vector and has both direction and magnitude. Thus, the rule for adding relative velocities is a rule of vector addition. First make a scale

Figure 3.18

Example 3.7: (a) A boat crossing a stream with a current. (b) Vector diagram of the velocities. The boat travels downstream as it crosses the river.

(a)

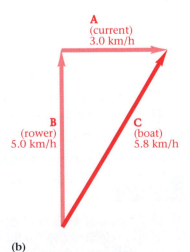

(b)

drawing of the vector diagram that corresponds to the physical situation. It should look like Fig. 3.18(b).

Solution (a) Once you have the scale diagram, it is relatively simple to use a ruler and protractor to get the answer of 5.8 km/h at an angle of 59° with respect to the river's motion. It is equally correct to give the direction as 31° with respect to a line straight across the river. You should do this example for yourself on paper, using a scale different from the one used in the text.

Strategy/Solution (b) The displacement of the boat is proportional to its velocity. Consequently, the ratio of the distance that the boat drifts downstream to the width across the river is the same as the ratio of the magnitude of the river's velocity to the magnitude of the velocity of the boat in still water:

$$\frac{\text{distance along bank}}{\text{width of river}} = \frac{\text{downstream current}}{\text{velocity across river}} = \frac{3.0 \text{ km/h}}{5.0 \text{ km/h}} = \frac{3.0}{5.0}.$$

Thus, the boat lands downstream a distance given by

$$\text{distance} = \frac{3.0}{5.0} \times \text{width} = \frac{3.0}{5.0} (200 \text{ m})$$

$$\text{distance} = 120 \text{ m}.$$

Example 3.8

A small airplane flies with an air speed of 200 km/h. A novice pilot wishing to fly from Columbia to Charlotte heads along a path that is due north. The wind is blowing from northwest to southeast at 28 km/h. What is the resultant ground speed of the plane and what is the direction in which the plane actually travels?

Strategy We begin by sketching a vector diagram indicating the velocity of the plane when no wind is present and indicating the velocity of the wind. For simplicity we choose a coordinate system with the y axis oriented north. The plane's velocity $\mathbf{v}_p$ in still air is along the y axis (Fig. 3.19a). The wind velocity $\mathbf{v}_w$ is

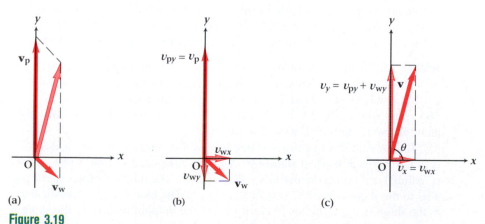

(a)   (b)   (c)

Figure 3.19

Example 3.8: (a) Vector diagram showing the velocity of the plane in still air and the velocity of the wind. (b) Resolving the wind velocity into components. (c) Finding the resulting velocity.

directed to the southeast, that is, at a 45° angle below the $+x$ axis. We can then use the parallelogram method to get a sense of the magnitude and direction of the resultant vector. The numerical values can be computed by resolving the wind velocity into x and y components (Fig. 3.19b) and then adding to get the components of the vector that represents the net motion of the plane.

Solution The components of the wind velocity are

$$v_{wx} = v_w \cos \theta = (28 \text{ km/h})\cos(-45°) = 19.8 \text{ km/h},$$
$$v_{wy} = v_w \sin \theta = (28 \text{ km/h})\sin(-45°) = -19.8 \text{ km/h}.$$

The plane's velocity $\mathbf{v}_p$ lies along the y axis so that

$$v_{py} = v_p = 200 \text{ km/h},$$
$$v_{px} = 0.$$

The y component of the resultant velocity is (Fig. 3.19c)

$$v_y = v_{wy} + v_{py} = -19.8 \text{ km/h} + 200 \text{ km/h} = 180 \text{ km/h}.$$

The x component is

$$v_x = v_{wx} = 19.8 \text{ km/h}.$$

The resultant velocity has a magnitude

$$v = \sqrt{v_x^2 + v_y^2} = \sqrt{(19.8 \text{ km/h})^2 + (180 \text{ km/h})^2} = 181 \text{ km/h},$$

The angle is found from

$$\tan \theta = \frac{v_y}{v_x} = \frac{180 \text{ km/h}}{19.8 \text{ km/h}} = 9.09$$

$$\theta = 83.7° \approx 84°.$$

Example 3.9

A dolphin traveling at 10 km/h in still water enters a tidal current at an angle of 30° and swims in that direction. The current is moving parallel to the shore at a speed of 3.0 km/h. What is the dolphin's velocity relative to the shore?

Strategy Let us begin by looking at a diagram of the situation (Fig. 3.20a). The velocity of the water relative to the shore is shown as $\mathbf{v}_{ws}$. The velocity of the dolphin relative to the water is $\mathbf{v}_{dw}$, which makes an angle of 30° with $\mathbf{v}_{ws}$. The vector sum of these two velocities ($\mathbf{v}_{dw} + \mathbf{v}_{ws}$) is the velocity of the dolphin relative to the shore, which we call $\mathbf{v}_{ds}$.

To calculate this quantity, we first choose a coordinate system so that $\mathbf{v}_{ws}$ is in the x direction. Then we resolve $\mathbf{v}_{dw}$ into components along x and y, that is, parallel and perpendicular to $\mathbf{v}_{ws}$ (Fig. 3.20b). The dolphin's total speed parallel to the shore is just the sum of $\mathbf{v}_{ws} + v_{dwx}$. The total speed perpendicular to the shore is $\mathbf{v}_{dwy}$. The total velocity $\mathbf{v}_{ds}$ of the dolphin relative to the shore is the vector sum of these components.

Solution The parallel component of the dolphin's velocity relative to the water is

$$v_{dw_x} = v_{dw} \cos(-30°) = (10 \text{ km/h}) \cos(-30°) = 8.66 \text{ km/h}.$$

The perpendicular component is

$$v_{dw_y} = v_{dw} \sin(-30°) = (10 \text{ km/h}) \sin(-30°) = -5.00 \text{ km/h}.$$

The dolphin's total speed parallel to the shore is $v_{ws} + v_{dw_x} = 11.66$ km/h and its speed perpendicular to the shore is -5.00 km/h. The magnitude of $\mathbf{v}_{ds}$ is given by the Pythagorean theorem as

$$v_{ds} = \sqrt{(11.66 \text{ km/h})^2 + (-5.00 \text{ km/h})^2}$$
$$v_{ds} = 13 \text{ km/h}.$$

The magnitude of $\mathbf{v}_{ds}$ is the speed of the dolphin relative to the shore. The resulting direction is

$$\tan \theta = \frac{v_{ds_y}}{v_{ds_x}} = \frac{-5.00}{11.66},$$

$$\theta = -23°.$$

Discussion The direction of the dolphin's velocity is $-23°$ away from the direction of the current (Fig. 3.20c) because one of its components is in the negative y direction. You should verify these results both by repeating the calculations yourself and by making a scale drawing and measuring with a ruler and protractor.

Figure 3.20

Example 3.9: (a) A dolphin swims in a tidal current at an angle of 30° from the direction in which the water is flowing. (b) Components of the dolphin's velocity relative to the water velocity $\mathbf{v}_{ws}$. (c) The velocity of the dolphin relative to the shore is $\mathbf{v}_{ds}$, the vector sum of $\mathbf{v}_{ws}$ and $\mathbf{v}_{dw}$.

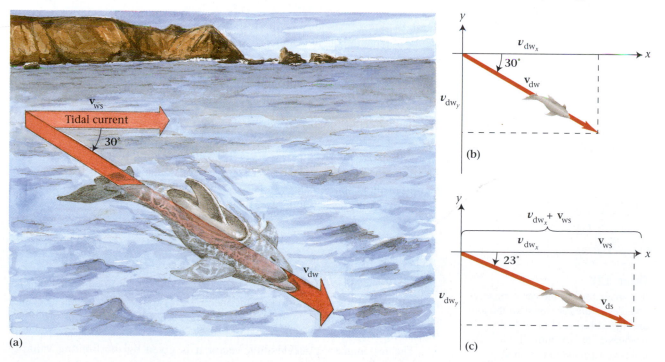

3.6 Kinematics in Two Dimensions

We can now rephrase our earlier kinematic definitions in vector form. The one-dimensional kinematic equations developed in Chapter 2 are for motion along a straight line. The vector equation defining average velocity is

$$\overline{\mathbf{v}} \equiv \frac{\Delta \mathbf{r}}{\Delta t} = \frac{\mathbf{r} - \mathbf{r}_0}{t - 0},$$ (3.4)

where $\mathbf{r}_0$ and $\mathbf{r}$ are the initial and final position vectors, t is the final time, and the initial time is set equal to zero. Here $\mathbf{r}$, $\mathbf{r}_0$, and $\mathbf{v}$ represent vectors that can have components along the directions of x, y, and z.

Now we can see that the average velocity is distinct from the average speed, which depends on the total distance traveled. For example, imagine a round trip in which a bee moves a distance of 5 cm and then returns to its starting point in 1 s. During that time interval, the bee travels a total distance of 10 cm and has an average speed of 10 cm/s. However, its net change in vector displacement is zero, since $\mathbf{r}_0$ (the position at the start) and $\mathbf{r}$ (the position at the end) are the same. Therefore, the bee has an average velocity of zero.

As in the case for one dimension, the instantaneous velocity $\mathbf{v}$ is given by the limiting value of $\Delta \mathbf{r}/\Delta t$ as Δt becomes vanishingly small:

$$\mathbf{v} \equiv \lim_{\Delta t \to 0} \frac{\Delta \mathbf{r}}{\Delta t}.$$ (3.5)

Similarly, the components of the instantaneous velocity vector are the limiting values of the component displacements divided by Δt as Δt becomes vanishingly small:

$$v_x = \lim_{\Delta t \to 0} \frac{\Delta x}{\Delta t}, \qquad v_y = \lim_{\Delta t \to 0} \frac{\Delta y}{\Delta t}.$$

The magnitude of the instantaneous velocity is given by

$$v = \sqrt{v_x^2 + v_y^2}.$$

The direction of motion is the direction of the instantaneous velocity, *not* the direction of the displacement.

The average acceleration is defined by

$$\overline{\mathbf{a}} \equiv \frac{\Delta \mathbf{v}}{\Delta t} = \frac{\mathbf{v} - \mathbf{v}_0}{t - t_0}.$$ (3.6)

Notice that an acceleration can arise from a change in the velocity's direction as well as from a change in its magnitude or speed (Fig. 3.21). In particular, a body moving in a circle may have constant speed, yet it will be accelerating because its velocity is continuously changing direction. We will study uniform circular motion in Chapter 5.

The instantaneous acceleration vector $\mathbf{a}$ is given by the limiting value of $\Delta \mathbf{v}/\Delta t$ as Δt becomes vanishingly small:

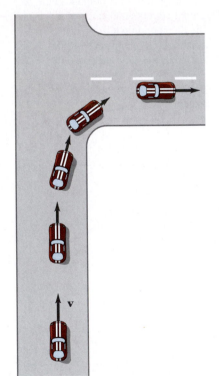

Figure 3.21
The magnitude of the car's velocity changes as the car slows for the turn and the direction of the velocity changes as the car turns. Both of these effects contribute to the car's acceleration.

$$\mathbf{a} \equiv \lim_{\Delta t \to 0} \frac{\Delta \mathbf{v}}{\Delta t}. \qquad (3.7)$$

As in the case of velocity, the instantaneous acceleration can be resolved into components of acceleration a_x and a_y,

$$a_x = \lim_{\Delta t \to 0} \frac{\Delta v_x}{\Delta t}, \qquad a_y = \lim_{\Delta t \to 0} \frac{\Delta v_y}{\Delta t}.$$

The magnitude of the instantaneous acceleration is given by

$$a = \sqrt{a_x^2 + a_y^2}.$$

The direction of the acceleration depends on the direction of change of velocity, *not* on the direction of the velocity.

The kinematic equations for motion with constant acceleration can be extended to vector form. Galileo found that the horizontal and vertical components of motion can be treated separately. Thus, we can express the vector equations of motion in terms of their separate components. For example, Eq. (2.8) written for two dimensions becomes

$$x = x_0 + v_{0x}t + \tfrac{1}{2}a_x t^2, \qquad (3.8a)$$

$$y = y_0 + v_{0y}t + \tfrac{1}{2}a_y t^2, \qquad (3.8b)$$

where x_0 and y_0 are the components of initial position, v_{0x} and v_{0y} are the components of initial velocity, and a_x and a_y are the components of acceleration. We can apply the same reasoning to extend these equations to the case of three dimensions.

Example 3.10

A particle is confined to move in a horizontal plane. Its location at time $t = 0$ is chosen as the origin of a coordinate system. The particle has an initial velocity along the x direction of 10 cm/s and is subject to a constant acceleration along the y direction of 2 cm/s^2. (a) Determine the path of the particle by computing its position at $t = 1, 2, 3, 4,$ and 5 s and graphing the result. (b) What is the displacement of the particle from the origin at time $t = 5$ s, and (c) what is its velocity?

Strategy (a) Since the acceleration is constant, we can use Eq. (3.8a,b) to find the x and y coordinates of the particle at the times specified:

$$x = x_0 + v_{0x}t + \tfrac{1}{2}a_x t^2 \quad \text{and} \quad y = y_0 + v_{0y}t + \tfrac{1}{2}a_y t^2.$$

A graph of these values gives the path of the particle.

Solution (a) Substituting the values $x_0 = 0$, $v_{0x} = 10$ cm/s, $a_x = 0$, $y_0 = 0$, $v_{0y} = 0$ and $a_y = 2$ cm/s^2 into the equations for x and y, we obtain the equations

$$x = 0 + (10 \text{ cm/s})t + 0 \quad \text{and} \quad y = 0 + 0 + \tfrac{1}{2}(2 \text{ cm/s}^2)t^2.$$

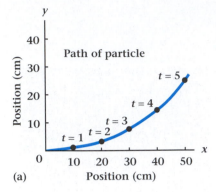

(a)

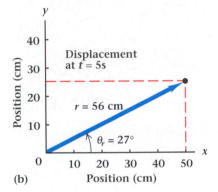

(b)

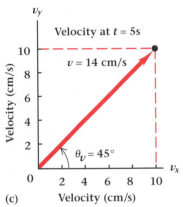

(c)

Figure 3.22

Example 3.10: (a) Path of the particle described. (b) Displacement of the particle at time $t = 5$ s. (c) Velocity of the particle at $t = 5$ s.

We can compute the coordinates x and y of the particle at each second along its path by inserting the values for the time in these equations:

t	x	y
0	0	0
1	10 cm	1 cm
2	20 cm	4 cm
3	30 cm	9 cm
4	40 cm	16 cm
5	50 cm	25 cm

These points, plotted on the graph in Fig. 3.22(a), show the path taken by the particle.

Solution (b) The displacement of the particle from the origin at $t = 5$ s is the vector $\Delta \mathbf{r} = \mathbf{r} - 0$ whose coordinates are $x = 50$ cm and $y = 25$ cm. Its magnitude is

$$r = \sqrt{x^2 + y^2} = \sqrt{(50)^2 + (25)^2} \text{ cm} = 56 \text{ cm.}$$

The direction of the displacement with respect to the x axis is given by the angle θ_r, where

$$\tan \theta_r = \frac{y}{x} = \frac{25}{50}.$$

This corresponds to an angle $\theta_r = 27°$ (Fig. 3.22b).

Strategy (c) The equation for velocity as a function of acceleration and time can be expressed in component form as

$$v_x = v_{xo} + a_x t \quad \text{and} \quad v_y = v_{yo} + a_y t.$$

For the case at hand we are seeking v_x and v_y at $t = 5$ s.

Solution (c) Upon inserting the numerical values we get

$$v_x = 10 \text{ cm/s} + 0 \quad \text{and} \quad v_y = 0 + (2 \text{ cm/s}^2)(5 \text{ s})$$
$$v_x = 10 \text{ cm/s} \qquad\qquad v_y = 10 \text{ cm/s.}$$

The velocity vector at $t = 5$ s, shown in Fig. 3.22c, has a magnitude

$$v = \sqrt{v_x^2 + v_y^2} = \sqrt{(10)^2 + (10)^2} \text{ cm/s} = 14 \text{ cm/s.}$$

The velocity vector is directed at an angle θ_v given by

$$\tan \theta_v = \frac{v_y}{v_x} = \frac{10 \text{ cm/s}}{10 \text{ cm/s}} = 1.0$$
$$\theta_v = 45°.$$

Discussion The direction in which the particle is traveling is the direction of its instantaneous velocity. Notice that this is not the same as the direction of the displacement from the origin. To help understand two-dimensional motion, you should extend this computation out to $t = 10$ s.

3.7 Projectile Motion

In the last part of his work *Discourses on Two New Sciences,* Galileo published a particularly useful idea that arose in connection with the motion of projectiles. Galileo observed that we could think of a projectile's motion as consisting of a horizontal part with constant speed and a vertical part with constant downward acceleration. Each of these motions is independent of the other, but their combination describes the overall motion of the projectile. For example, a ball projected horizontally falls with the same downward acceleration as a ball that is simply dropped (Fig. 3.23). The downward motion is unaffected by the horizontal motion. Thus, if air resistance is negligible, a ball projected horizontally reaches the floor at the same time as a ball dropped vertically, a result predicted by Galileo.

Galileo also used the example of a stone dropped from the mast of a ship moving with constant speed. To a shipboard observer the stone appears to travel straight down alongside the mast, falling with constant acceleration. However, to an observer on the shore, the path of the falling stone on the moving ship is equivalent to projecting it horizontally from a stationary point with an initial velocity equal to the velocity of the ship.

It is important to understand that the motion seen from the shore consists of two independent parts: the downward motion with constant acceleration and the horizontal motion with constant velocity. (In this discussion we neglect the effect of air resistance.) Experiments have confirmed the independence of the horizontal and vertical components. The reason for this will become clear in Chapter 4. To clarify the idea of independent components of motion, study the following example.

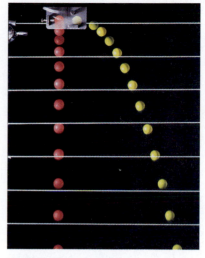

Figure 3.23
Photograph of two balls illuminated with a flashing strobe light. The balls are released at the same time. The ball projected horizontally has the same downward acceleration as the ball that is simply dropped.

Problem-Solving Strategy

Projectile Motion

1. Choose a coordinate system and stick with it while solving the problem.
2. Identify the initial position, initial velocity, and acceleration and resolve these quantities into x and y components.
3. Identify the quantities that you know and those that you need to find.
4. Write the appropriate kinematic equations in component form and solve them separately making use of the fact that the time of flight is the same for both components of motion.

Example 3.11

A ball is thrown horizontally from the Leaning Tower of Pisa (Fig. 3.24) with a velocity of 22 m/s. If the ball is thrown from a height of 49 m above ground,

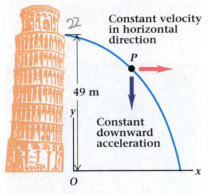

Figure 3.24

Example 3.11: A ball thrown from the Leaning Tower of Pisa.

how far from the point on the ground directly below the launch point does the ball strike the ground?

Strategy We can treat the x and y motions independently, and we know that the time it takes the ball to reach the ground will be the same for both motions. So, knowing the vertical height through which the ball falls, we can solve the equation of vertical motion for the time of fall and then use this value in the equation for horizontal motion to determine the horizontal distance.

Solution Choose a coordinate system with x along the horizontal and the positive y axis up. The time for the ball to reach the ground depends on the height h and the constant vertical acceleration a_y. (In Section 2.7 we discussed the motion of a body falling freely under the acceleration of gravity.) For motion in the vertical direction, we may use Eq. (3.8b),

$$y = y_0 + v_{0y}t + \tfrac{1}{2}a_yt^2,$$

where the initial y coordinate position is $y_0 = h$ and the acceleration in the y direction is $a_y = -g$. The negative sign is used since the positive y direction is upward and the acceleration of gravity is downward. Because the initial velocity is horizontal, the initial velocity in the y direction is $v_{0y} = 0$. The time t for the ball to reach the ground is given by this equation with $y = 0$ (the height of the ground) and the values for y_0, v_{0y}, and a_y inserted. The equation then becomes

$$0 = h + 0 - \tfrac{1}{2}gt^2.$$

Solving for t and inserting the numerical values gives

$$t = \sqrt{\frac{2h}{g}} = \sqrt{\frac{2 \times 49 \text{ m}}{9.81 \text{ m/s}^2}}$$

$$t = 3.162 \text{ s} \approx 3.2 \text{ s}.$$

This is the elapsed time before the ball strikes the ground after it has been thrown horizontally. We have chosen the positive square root because time increases as the ball falls.

The gravitational acceleration acts downward and does not affect the horizontal component motion. Thus, the horizontal component of acceleration is zero ($a_x = 0$) and the ball continues to travel with a constant x component of velocity until it strikes the ground. The horizontal distance traveled in time t at constant velocity v_{0x} is

$$x = v_{0x}t = 22 \text{ m/s} \times 3.162 \text{ s}$$

$$x = 70 \text{ m}.$$

Thus, the ball strikes the ground 70 m from the point directly below the launch point.

■

The combined effect of horizontal motion with constant velocity and vertical motion with constant acceleration is that the body moves in a parabolic path. In the following discussion, we prove that the path of a projectile thrown horizontally (as in Example 3.11) is a parabola. We then extend our discussion to cases in which the body is projected at any angle, not just horizontally. If you need to review some of the properties of a parabola, refer to page A–3 of Appendix A.

Consider an object, like a golf ball, projected horizontally (Fig. 3.25). For the y (vertical) component of motion, the initial velocity is zero and the acceleration is that of gravity, giving

$$y = -\frac{1}{2}gt^2.$$

We use the minus sign because the acceleration of gravity is downward and we have chosen the upward direction to be positive in the figure. The projected object starts at $y = 0$ and falls to negative values of y.

The horizontal component of motion has an initial velocity but is not accelerated, so

$$x = v_0 t.$$

Solving this equation for t and inserting the value of t into the equation for y, we get

$$y = -\frac{g}{2v_0^2}x^2. \tag{3.9}$$

Equation (3.9) has the same form as the equation for a parabola. In both cases the factor that is multiplied by x^2 on the right-hand side is a constant for a particular problem. Thus, we conclude that projectile motion is parabolic.

A more detailed analysis would take account of air resistance, which causes a departure from a true parabolic path in many real situations. In this book, we will consider air resistance to be negligible unless stated otherwise. Although the assumption does not correspond to physical reality, it often gives a good approximation. The mathematical model needed to include the effects of air resistance is more complicated than we wish to analyze here.

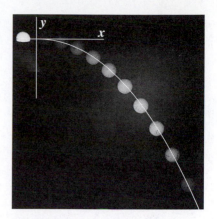

Figure 3.25
A photograph of a golf ball illuminated with a flashing strobe light and given an initial horizontal velocity. Superimposed on the photograph is a section of a parabola drawn from Eq. (3.9). Whether an object is thrown horizontally, upward, or downward, the principle remains the same: the object follows a parabolic path.

Master the Concept

Projectile Motion

Question: Will a ball dropped from rest reach the ground quicker than one launched from the same height but with an initial horizontal velocity?

Answer: In both cases, the balls have zero initial velocity in the vertical direction and are accelerated downward with gravitational acceleration g. The vertical motion is unaffected by the horizontal motion. The time to fall depends only on the initial height and g. Thus the time to reach the ground is the same for both balls.

Example 3.12

A boy aims his slingshot directly at an apple hanging in a tree. At the moment he shoots a pebble, the apple drops. Show that the pebble hits the target.

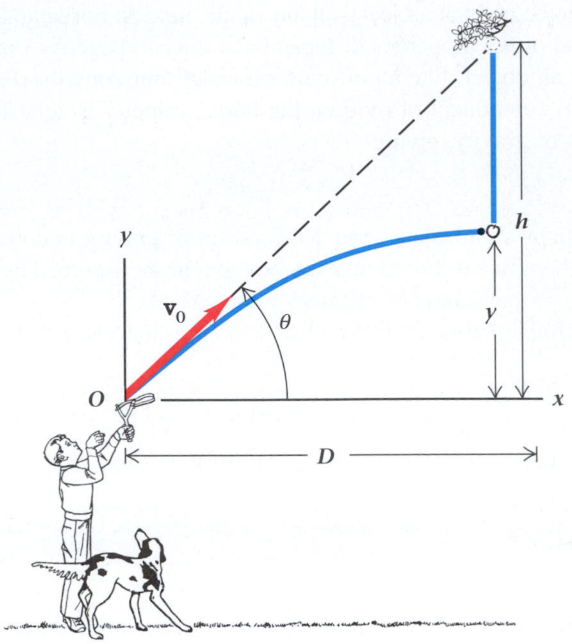

Figure 3.26

Example 3.12: Shooting a pebble at an apple that drops at the instant the pebble is fired.

Strategy For the pebble to hit the apple, they both need to be at the same place at the same time (Fig. 3.26). We know the horizontal distance the pebble must travel, namely D, so we need to calculate the time t required for that to happen. Then we can use that value of time in the equation for the pebble's vertical motion to determine its height when it reaches the horizontal distance D. Finally, we substitute this same time into the apple's equation of motion to see if it's at the same height as the pebble.

Solution For convenience, we place an x-y coordinate system with its origin at the slingshot (Fig. 3.26). We first consider the motion of the projectile. Let θ be the firing angle measured from the horizontal. Then the horizontal component of the pebble's velocity is $v_0 (\cos \theta)$. In a time t the pebble travels the horizontal distance D given by

$$D = v_0 (\cos \theta)\, t,$$

or

$$t = \frac{D}{v_0 \cos \theta}.$$

The vertical component of the pebble's initial velocity is $v_0(\sin \theta)$, so the vertical distance traveled by the pebble during the same period of time is given by Eq. (3.8b):

$$y = v_0(\sin \theta)t - \tfrac{1}{2}gt^2.$$

By substituting the value of t from the equation above, we get

$$y = D \tan \theta - \tfrac{1}{2}gt^2.$$

Examination of the diagram shows that $D \tan \theta$ is the initial height of the apple h, so that the height reached by the pebble in time t is

$$y = h - \tfrac{1}{2}gt^2.$$

Next we find the position of the apple at time t, which is given by

$$y = y_0 + v_{0y} + \tfrac{1}{2}a_y t^2.$$

Since $y_0 = h$, $v_{0y} = 0$, and $a_y = -g$, the position of the apple at time t is

$$y = h - \tfrac{1}{2}gt^2.$$

This height is just exactly the point that the pebble reached in time t. So, the pebble hits the apple.

Discussion This result is independent of the speed v_0, so collision occurs for any projectile (pebble, bullet, arrow, or whatever) as long as the projectile travels a distance D before hitting the ground. For the case shown in Fig. 3.26, the projectile was still rising when it hit the target. However, the collision would still occur even if the projectile were past its maximum height. This example is often the basis for a classroom demonstration experiment.

Example 3.13

A baseball thrown upward leaves the player's hand at a height of 1.60 m above a level playing field (Fig. 3.27). The ball has an initial speed of 28.0 m/s at an angle of 45.0° above the horizontal. How far from a point on the ground directly below the point of release will the baseball strike the field?

Strategy Again we can treat the x and y motions independently. We know that the time it takes the ball to reach the ground is the same for both motions. We can solve the equation of vertical motion for the time of fall and then use this value of time in the equation for horizontal motion to determine the horizontal distance. This is the same strategy that we used in Example 3.11, only in this case the ball's initial velocity has an upward component.

Solution The ball has an initial component of velocity in the vertical (y) direction given by $v_{0y} = v_0 \sin 45.0° = 28.0 \text{ m/s} \times 0.707 = 19.8$ m/s. We can use Eq. (3.8b) to find the time of flight of the ball since we know the initial

Figure 3.27
Example 3.13: A baseball thrown with an initial velocity making an angle of 45° with the horizontal is released at a height of 1.60 m above the ground.

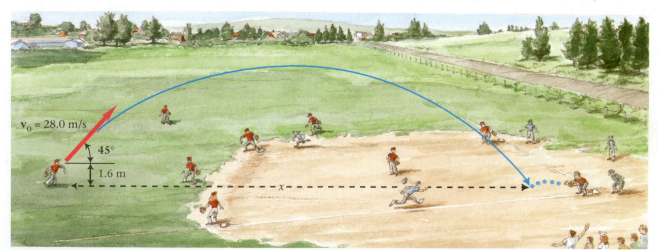

$v_0 = 28.0$ m/s

45°

1.6 m

x

velocity, the initial position $y_0 = 1.60$ m, the final position $y = 0$, and the vertical acceleration $a_y = -g = -9.81$ m/s². The equation may be put in the form

$$\tfrac{1}{2}a_y t^2 + v_{0y}t + (y_0 - y) = 0,$$

which is quadratic in t. When the proper values are inserted for the coefficients, we get

$$(-4.91 \text{ m/s}^2)t^2 + (19.8 \text{ m/s})t + 1.60 \text{ m} = 0.$$

Upon applying the quadratic formula, we obtain two values for the time t: $+4.11$ s and -0.0792 s. Since the elapsed time is positive, we choose the positive value, $t = 4.11$ s.

Now that we know the time of flight of the ball, we can compute the horizontal distance traveled from the product of the horizontal velocity component with the time:

$$x = v_{0x}t = v_0 \cos 45.0° \, t$$
$$x = (28.0 \text{ m/s})(0.707)(4.11 \text{ s}) = 81.4 \text{ m}.$$

The ball travels a horizontal distance of 81.4 m before it hits the ground.

Example 3.14

Find an expression for the *horizontal range* of a projectile, the horizontal distance it travels during the time it rises and returns to its initial height. Ignoring the effects of air resistance, show that the range depends on the initial velocity (both the magnitude and the launch angle) in a relatively simple way.

Strategy Figure 3.28 depicts the parabolic path of a football projected upward. The football's initial velocity $\mathbf{v}_0$ makes an angle θ with the horizontal. The horizontal component of velocity is $v_0 \cos \theta$ and is constant because there is no horizontal acceleration. (As usual, we have neglected air resistance.) The initial vertical or y component of velocity is $v_0 \sin \theta$. The vertical component of velocity changes with time as a result of gravitational acceleration. As the

Figure 3.28

Range of a projectile (football) with initial velocity $\mathbf{v}_0$ directed at an angle θ with respect to the horizontal.

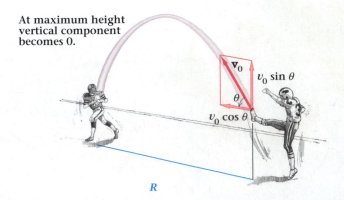

At maximum height vertical component becomes 0.

$\mathbf{v}_0$

$v_0 \sin \theta$

θ

$v_0 \cos \theta$

R

object rises, it slows down and its vertical component of velocity v_y at any time t is

$$v_y = v_0 \sin \theta - gt,$$

where the initial vertical velocity is $v_0 \sin \theta$ and the acceleration is $-g$.

The time for an object thrown upward from the ground to reach its maximum height and the time for it to fall to the ground again are the same, provided the initial and final positions are on the same level. This symmetry in the upward and downward parts of the path results from the fact that the acceleration of gravity, g, is constant and is the same for upward and downward motion at any height. The time required for the object to reach its maximum height is then half the total time T that it is in the air. Also, at the maximum height, the vertical component of velocity becomes zero; that is, the projectile stops rising before it begins to fall down. (It continues moving horizontally at this point.) Thus, at the maximum height, $v_y = 0$, and $t = T/2$.

Solution At the maximum height the velocity equation in the vertical direction becomes

$$0 = v_0 \sin \theta - g\frac{T}{2}.$$

Upon rearranging to find the time T,

$$T = \frac{2v_0 \sin \theta}{g}.$$

The horizontal range R, indicated in Fig. 3.28, is the horizontal distance traveled in the time T. It is given by $v_x t$, with $v_x = v_0 \cos \theta$ and $t = T$:

$$R = v_0(\cos \theta)T.$$

If the previous equation is used for the time T, the expression for the range becomes

$$R = \frac{2v_0^2}{g} \sin \theta \cos \theta.$$

Making use of the trigonometric identity $2 \sin \theta \cos \theta = \sin 2\theta$, we can express the range as

$$R = \frac{v_0^2}{g} \sin 2\theta.$$

Discussion The horizontal range derived here does not include the effects of air resistance. In cases where air resistance plays a role, the range is usually less than the amount given by this expression. An obvious exception is the Frisbee, a light disk that acquires a lifting force from the air passing over its surface.

Figure 3.29 shows the paths of an object projected upward at various angles with the same initial speed. Figure 3.30 is a graphical representation of the expression for the range. From either figure we can see that the maximum range occurs when $\theta = 45°$. (At this value of θ, $\sin 2\theta$ has its maximum value of 1.) The expression for the maximum range $R_{\max}$ is then

$$R_{\max} = \frac{v_0^2}{g}.$$

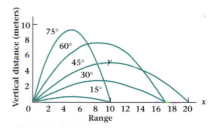

Figure 3.29
Trajectories of an object thrown upward with the same initial speed at various angles of inclination.

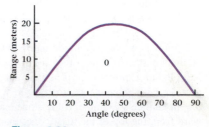

Figure 3.30
Range versus angle with respect to the horizontal for the same initial speed.

Summary

Useful Concepts

■ We can determine the sum of two vectors by using geometry, algebra, or trigonometry. However, it is a good idea, especially at the outset, to always make a scale drawing.

■ Vectors are added graphically by making a scale drawing. In the triangle method, the vectors are placed head to tail and the resultant is drawn from the origin (tail) of the first vector to the head of the last vector. In the parallelogram method, two vectors are placed with their tails joined together. A parallelogram is constructed and the diagonal is the resultant.

■ In two dimensions, a vector $\mathbf{A}$ may be resolved into its components:

$$A_x = A \cos \theta, \qquad A_y = A \sin \theta.$$

■ The angle θ is measured counterclockwise from the positive x axis.

■ Vectors may be added together by adding their components. The magnitude and direction of a vector $\mathbf{A}$ in terms of its components are

$$A = \sqrt{A_x^2 + A_y^2} \quad \text{and} \quad \tan \theta = \frac{A_y}{A_x}.$$

■ The instantaneous velocity vector is

$$\mathbf{v} \equiv \lim_{\Delta t \to 0} \frac{\Delta \mathbf{r}}{\Delta t}.$$

■ The instantaneous acceleration vector is

$$\mathbf{a} \equiv \lim_{\Delta t \to 0} \frac{\Delta \mathbf{v}}{\Delta t}.$$

■ A projectile launched with an initial velocity $\mathbf{v}_0$ at an angle θ with the horizontal has initial x and y components of velocity

$$v_x = v_0 \cos \theta \quad \text{and} \quad v_y = v_0 \sin \theta.$$

■ The path of a projectile is a parabola (neglecting air resistance).

Important Terms

You should be able to write out the definitions or meanings of the following terms:

scalar	vector sum
vector	resultant
magnitude of a vector	relative velocity

Conceptual Questions

3.1 List as many vector quantities as you can think of.

3.2 How is the meaning of the word *vectorcardiography* connected with our definition of a vector? Can you give any other compound words containing *vector* or any uses of the word outside of mathematics and physics?

3.3 What do we mean when we say that a weather map indicating temperatures represents a scalar field but a map indicating winds represents a vector field?

3.4 Show how you would add the following three vectors: 10 units north, 10 units south, and 10 units straight up.

3.5 Find an expression or procedure for finding the length of a vector with components along the x, y, and z axes.

3.6 In Galileo's example, if the stone had been thrown downward from the top of the mast rather than dropped from rest, what would the path have looked like to observers on the ship and on the shore?

3.7 If the object had been thrown horizontally from the mast of the ship in Galileo's example, what type of path would a person on the ship have observed? What type of path would a person on shore have observed? Be as explicit as you can for both cases.

3.8 A waiter in the dining car of a smoothly riding train pours a glass of water for a passenger. How would his task be affected if the velocity of the train is zero, if the velocity is nonzero but constant, and if the train is accelerating?

3.9 Can a group play volleyball on the deck of a ship moving at 15 km/h without taking into account the motion of the ship?

3.10 A child on a moving train rolls a ball down the aisle toward the back of the train. The ball travels 9.0 m along the floor in 3.0 s while the train moves forward at a constant speed of 18 m/s. Discuss the displacement of the ball during the 3-second interval from the point of view of the child and from the point of view of someone standing on the ground beside the track.

Problems

Section 3.2 Addition of Vectors

Hints for Solving Problems

Draw scale diagrams carefully. Choose a scale that gives you plenty of room to work. Indicate your scale on the diagram.

3.1 (a) If you walk three city blocks east and then four blocks north, how many blocks are you from your starting place? (b) What direction are you from the starting point? Give your answer as an angle measured from due east.

3.2 A jogger runs directly east for 3.0 km, then turns and goes northwest for 5.0 km. He then travels directly south for 2.0 km. How far and in what direction is he from the starting point? (Northwest is the direction lying exactly half way between north and west.)

3.3 A motorist travels directly north for 36 km, then turns to the west and goes 18 km. How far is she from the starting point and in what direction? Give the angle with respect to east.

3.4 Vector **A** has a magnitude of 13 units at a direction of 250°, and vector **B** has a magnitude of 27 units at a direction of 330°, both measured with respect to the positive x axis. (a) What is the vector sum of **A** and **B**? (b) What is the vector **C** = **A** − **B**? Find your answers by graphical analysis.

3.5 Add the following vectors graphically in the order given, then add them in reverse order on a separate diagram, thereby testing that vector addition is commutative: **A** = 5 units at 60° and **B** = 7 units at 180°.

3.6 A vector **A** has a magnitude of 25 units at 200°, and a vector **B** has a magnitude of 30 units at 45°. What is the vector difference **A** − **B**?

3.7● The vectors diagrammed in Fig. 3.31 have the same magnitude, A. Find: (a) **B** + **C**, (b) **A** + **B** + **C**, (c) **B** − **C**, (d) **A** − **B** − **C**.

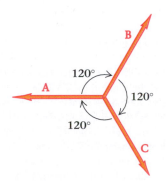

Figure 3.31
Problem 3.7.

3.8● Four ropes are tied to a stake, and each is pulled toward a compass direction, N, S, E, or W. A force of 10 lb is applied to the rope pulled toward the east. Forces of 20, 30, and 40 lb are applied toward the south, west, and north, respectively. What is the net force exerted on the stake by the ropes and in what direction is it? (Forces are vector quantities. The net force is the vector sum of the individual forces.)

3.9● A motorcyclist rides a given distance due north, then rides twice that far to the west. At the end of the trip, the direct distance from the starting point is 112 m. What is the length of each leg of the trip and in what direction is the end point with respect to due north?

3.10● A salesperson made a trip starting from the home office. The trip was made in two straight-line parts, one of which was 20 km long to the east. The end point of the trip was 80 km from the starting point at a direction of 45° southeast. Describe the other leg of the trip. Work this as a problem in vector subtraction.

Section 3.3 Resolution of Vectors

Hints for Solving Problems

Solve the following vector problems using trigonometry. A review of trigonometry is found in the appendix at the end of this chapter.

3.11 A vector with a magnitude $V = 16$ makes an angle of 50° with the x axis. Resolve the vector into its x and y components.

3.12 A vector with magnitude $V = 24$ makes an angle of −36° with the x axis. Find the x and y components of the vector.

3.13 What are the magnitude and direction of the vector with $A_x = 4.8$ and $A_y = -6.2$?

3.14 A vector has components of 9.0 units along the x axis and 12 units along the y axis. (a) What is the magnitude of the vector? (b) What is the angle between the vector and the x axis?

3.15 A helicopter takes off from pad #1 at ground level and rises straight up to an altitude of 327 m. The helicopter then flies in a straight level flight to a point directly above pad #2 on top of a building that is 873 m horizontally away from pad #1. The helicopter then descends in a straight line to pad #2. What is the helicopter's final displacement from pad #1 if pad #2 is 180 m above the ground level?

3.16 A person rides a motorbike 150 m along a road that slants upward at 4.50° from the horizontal. (a) How much higher will the rider be at the end of the trip? (b) How far has the rider gone in the horizontal direction?

3.17 The direction of a vector is 118° from the positive x axis, and its y component is 18.0. Find the x component and the magnitude of the vector.

3.18 Vector **A** has a magnitude of 8.0 at an angle of 60° from the *x* axis. Vector **B** has a magnitude of 6.0 at an angle of −30° from the *x* axis. Determine the vector sum **C = A + B** by resolution into components.

3.19 Find the difference **D = A − B** for the vectors given in Problem 3.18.

3.20 Find the vector **V = 2A − B** for **A** and **B** given in Problem 3.18.

3.21 (a) Find the sum **C** of the two vectors **A** and **B** given by $A_x = 3.0$, $A_y = 7.0$ and $B_x = 10.0$, $B_y = −9.0$. (b) Find the difference **D = A − B**.

3.22• If a vector has a magnitude of 18 and an *x* component of −7.0, what are the two possibilities for its *y* component and direction?

3.23•• A woman traveling due north makes a detour around a large lake in the following manner (Fig. 3.32). At point *A* she turns 46° toward the west and travels 4.91 km to *B*. At *B* she turns to the north and travels a distance of 8.27 km to *C*. At *C* she turns 60° to the east and travels a distance of 4.08 km to *D*. At *D* she turns to the north and continues her trip. (a) Are the initial and final northward paths along the same line? (b) What is the distance between *A* and *D*.

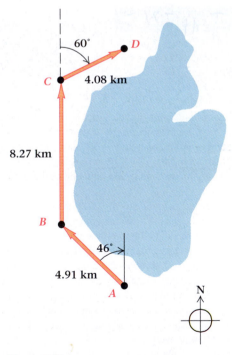

Figure 3.32
Problem 3.23.

Section 3.4 Relative Velocity in One Dimension

Hints for Solving Problems

You will find it helpful to use subscripts on the velocities similar to those used in Example 3.5.

3.24 A vendor on a train moving in the forward direction at 2.00 m/s pushes her cart toward the rear of the train at 0.47 m/s while an ant on a sandwich crawls toward the front of the train at 0.01 m/s. What is the velocity of the train station with respect to the ant.

3.25 Two airplanes are flying side by side at the same altitude. Plane A is slowly overtaking plane B at 4 km/h. A flight attendant in plane A is walking toward the rear of the plane at a speed of 2 km/h and a passenger is walking toward the front at 2 km/h. What are their speeds relative to a passenger watching them from plane B?

3.26 A boat is traveling in a river with a current of 3.0 km/h. The boat is capable of traveling at 10.0 km/h in still water. (a) How long will it take the boat to travel 7.0 km upstream? (b) How long will it take to travel 7.0 km downstream?

3.27 A boat capable of making 9.0 km/h in still water is traveling upstream in a river flowing at 4.0 km/h (Fig. 3.33). An object lost overboard is not missed for 30 minutes. (a) If the boat is then turned around, how long will it take to overtake the floating lost object? (b) What total distance will the boat travel relative to the shore from the point of turnaround to the point of overtaking the object?

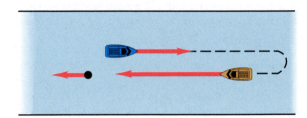

Figure 3.33
Problem 3.27.

3.28• A passenger rushing to catch a plane at the airport walks on a moving sidewalk at a speed of 3.0 km/h relative to the sidewalk in the direction that the sidewalk is moving. The sidewalk is 100 m long and moves with a steady velocity of 1.0 km/h. (a) How long does it take for the passenger to get from one end of the sidewalk to the other, that is, to cover the 100 m? (b) How much time does the passenger save by taking the moving sidewalk instead of just walking beside it? (c) Through what distance does the passenger walk relative to the moving sidewalk? (d) If the passenger's stride is 80 cm, how many steps are taken in going from one end of the moving sidewalk to the other?

3.29• Two cars approach each other on the highway. Car A moves north at 90 km/h, car B moves south at 70 km/h. (a) What is the velocity of car A as seen from car B? (b) What is the velocity of car B as seen from car A? (c) What are their velocities relative to car C, which is traveling north at 100 km/h?

Section 3.5 Relative Velocity in Two Dimensions

3.30 A small airplane flies with a speed relative to the ground (ground speed) of 208 km/h in a direction 18.0° to the east of north. If the plane is headed due north and the deviation from that direction is due to a crosswind blowing from west to east (Fig. 3.34), what is the speed of the wind?

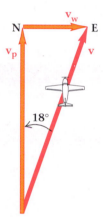

Figure 3.34
Problem 3.30: An airplane headed due north is blown off course by a crosswind blowing due east.

3.31 A small airplane has a cruising speed of 260 km/h in still air. The pilot heads the plane in an easterly direction on a day when the wind is blowing at 25 km/h in a direction 60° north of east. In what direction will the plane move and what will be its ground speed?

3.32• An airplane with a maximum air speed of 320 km/h takes off and heads to its destination, which is due east of its starting point, when a 70 km/h crosswind starts blowing from the north. The schedule calls for the plane to travel a distance of 1590 km between airports in a time of 5.0 h. Will the plane arrive on schedule?

3.33• An airplane heading south with an air speed of 200 km/h is in a cross wind of 10 km/h blowing toward the west. How far does the airplane go in two hours and in what direction?

3.34• A boat capable of making 9.0 km/h in still water is used to cross a river flowing at a speed of 4.0 km/h. (a) At what angle must the boat be directed so that its motion will be straight across the river? (b) What is its resultant speed relative to the shore?

Sections 3.6 and 3.7 Kinematics in Two Dimensions and Projectile Motion

> **Hints for Solving Problems**
>
> Choose a coordinate system and stick with it throughout the problem. Remember that horizontal and vertical components of motion are independent of each other, but they take place during the same time interval. Pay careful attention to signs and apply them consistently throughout the problem. Assume that air resistance can be ignored for these problems.

3.35 A student touring Japan accidentally drops a 500-yen coin from a height of 1.20 m above the floor of the Bullet Train, which is traveling at 250 km/h (Fig. 3.35). (a) How long does it take for the coin to hit the floor? (b) Where does the coin land with respect to the floor of the train? (c) How far along the track does the train go during the time it takes for the coin to fall?

Figure 3.35
Problem 3.35: The Bullet Train.

3.36 A rock thrown horizontally from the top of a radio tower lands 17.0 m from the base of the tower. If the speed at which the object was projected was 9.50 m/s, how high is the tower?

3.37 A monkey on a cliff throws a coconut horizontally from a height of 17 m with a speed of 2.1 m/s. If the ground below the cliff is level, how far from the base of the cliff does the coconut strike the ground?

> **Hints for Solving Problems**
>
> Problems 3.38 to 3.42 can be solved using the techniques and results of Example 3.14.

3.38 A rock thrown with an initial velocity of 32.5 m/s at an angle of 50° with respect to the horizontal has a range of 985 m on a certain planet. What is the acceleration of gravity on this planet?

3.39 An astronaut playing golf on earth consistently hits a golf ball 170 m. Using the same swing, how far could the astronaut hit golf balls on the moon? (*Hint:* The gravitational acceleration on the moon is approximately 1/6 that on the earth.)

3.40 (a) How far will a stone travel over level ground if it is thrown upward at an angle of 30.0° with respect to the horizontal and with a speed of 12.0 m/s? (b) What is the maximum range that could be achieved with the same initial speed?

3.41 Locusts have been observed to jump horizontal distances up to 80 cm on a level floor. Photographs of their jump show that they usually take off at an angle of about 55° from the horizontal. Calculate the initial velocity of a locust making a jump of 80 cm with a takeoff angle of 55°.

3.42• If you can throw a ball vertically upward to a height $h = 20$ m, what is the maximum horizontal range over which you can throw the same ball, assuming you throw it at the same initial speed?

3.43• Plot the path of an object in the xy plane that moves according to

$$x = (2 + 3t + 2t^2) \text{ cm},$$
$$y = (2t + 5t^2) \text{ cm}$$

from $t = 0$ to $t = 4$ s.

3.44• A cougar leaps horizontally from the top of a cliff with an initial velocity of 8.25 m/s. The cliff is 6.43 m tall. (a) Sketch the path of the cougar. (b) What are the magnitude and direction of the velocity when the cougar is halfway to the ground?

3.45• A ball thrown horizontally from a 13-m-high building strikes the ground 5.0 m from the building. With what velocity was the ball thrown?

3.46• A ball is thrown upward from a platform 5.2 m high with a speed of 15 m/s at an angle of 40° from the horizontal. What is the magnitude of its velocity when it hits the ground?

3.47• A monkey throws a coconut horizontally out of a tree at 5.36 m/s. The coconut leaves his hand 13.4 m above the level ground. The coconut lands on the ground by falling directly into a cylindrical basket tilted so that the coconut goes to the bottom without touching the sides. (a) How far is the basket horizontally from the point directly below the release point for the coconut? (b) At what angle is the basket tilted?

3.48• A third baseman makes a throw to first base 39.0 m away. The ball leaves his hand with a speed of 38.0 m/s at a height of 1.50 m from the ground and making an angle of 20.0° with the horizontal. How high will the ball be when it gets to first base?

3.49•• A third baseman makes a throw to first base 39.0 m away. (a) If the ball leaves his hand traveling horizontally with a speed of 38.0 m/s at a height of 1.20 m from the ground, how far will it go before striking the ground? (b) At what angle must he throw the ball so that it reaches the first baseman's glove at a height of 1.20 m above the ground?

3.50•• A cannon sitting on the battlement of a castle is aimed away from the castle at an angle θ above the horizontal. Find an equation for the time for the cannonball to hit the level ground below in terms of the height of the battlement, the magnitude and direction of the initial velocity, and the acceleration of gravity.

3.51•• An object on a horizontal plane starts at the origin and moves with velocity $v_x = 10.0$ cm/s, $v_y = -5.0$ cm/s, and acceleration $a_x = 0$, $a_y = 4.0$ cm/s². (a) How far is it from the origin after 5.0 s? (b) In what direction is it from the origin? (c) In what direction is it moving?

3.52•• A Hollywood daredevil plans to jump the canyon shown in Fig. 3.36 on a motorcycle. If he desires a 3.0-second flight time, what is (a) the correct angle for his launch ramp, (b) his correct launch speed, (c) the correct angle for his landing ramp, and (d) his predicted landing speed? (Neglect air resistance.)

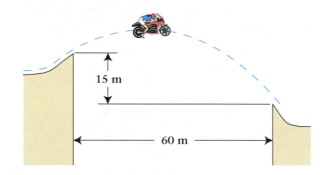

Figure 3.36
Problem 3.52: Motorcycle daredevil jumping canyon.

3.53•• An intramural quarterback throws a football with an initial speed of 9.7 m/s at an upward angle of 38° above a horizontal playing field. (a) How long will the ball remain in the air before falling to the ground if it leaves his hand at a point 1.5 m above the ground? (b) What horizontal distance will the ball travel before striking the ground?

Additional Problems

3.54 Five vectors of equal magnitude radiate outward from the center of a regular pentagon, one vector pointing toward each vertex (Fig. 3.37). Show by a graphical treatment that the sum of all five vectors is zero.

3.55 A ladder leans against a house at an angle of 76° from the horizontal. The base of the ladder is 1.48 m from the

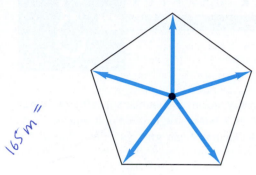

165 m =

Figure 3.37
Problem 3.54.

house. (a) How long is the ladder? (b) How far up the wall does the ladder touch the house?

3.56 A guy wire bracing a power pole makes an angle of 53° with the ground. How long is the guy wire if it attaches to the pole at a height of 5.50 m above the ground?

3.57 A vector **A** has a magnitude of two units at 30° with respect to the horizontal. Vector **B** has a magnitude of five units at 90°. Calculate **C = 3A + 2B**.

3.58 A physics student standing on the top floor of a building sights downward to a newspaper box on the ground below. The line of sight to the box makes an angle of 75° with respect to the horizontal. After coming down, the student measures the distance from the building to the box to be 40.5 m. How high above the ground were the student's eyes when he was looking at the box from the top floor?

3.59 Observations are made of a distant television tower from each end of a 100-m base line. From each position the angle between the line of sight and the base line is 85.5°. How far is the tower from the center of the line?

3.60 Three vectors have the same magnitude of 14 units. They make angles of θ, 2θ and 3θ with respect to the x axis, where $\theta = 20°$. What is their vector sum?

3.61 (a) How far will a stone travel over level ground if it is thrown upward at an angle of 30.0° with respect to the horizontal and with a speed of 12.0 m/s? Assume the stone is launched at a height of 2.0 m above the ground. (b) What is the maximum horizontal range that could be achieved with the same initial speed? Refer to Example 3.14.

3.62● A ladder that is 4.00 m long is leaning against a wall at an angle of 64° with respect to the ground. If the base of the

ladder is moved 0.30 m away from the wall, how far will the top of the ladder go down?

3.63● An arrow was fired horizontally from a platform above the ground. Exactly 3 s after it was released, it struck the ground at an angle of 45° from the horizontal. (a) With what speed and (b) from what height was it launched?

3.64● This problem helps to illustrate how important the effects of aerodynamic forces can be. A golf ball hit at a speed of 67 m/s will carry (achieve its maximum flight distance of) 183 meters when driven at a launch angle of 11°. What would be its horizontal range in a vacuum if launched at the same speed and (a) at the same angle and (b) at a 45° angle?

3.65●● A baseball player standing on a platform throws a baseball out over a level playing field. The ball is released from a point 3.50 m above the field with an initial speed of 14.3 m/s at an upward angle of 27° from the horizontal. How far from a point on the ground directly below the point of release will the baseball strike the field?

3.66●● A spring-loaded cannon aimed at 47° above the horizontal is on the last car of a long train of flat cars. The train has an initial velocity of 54.3 km/h. At the moment the train begins to accelerate forward at 0.325 m/s², the cannon fires a projectile at 180 m/s. The cannon points in the direction that the train is moving. (a) What is the horizontal range observed by a person standing on the ground? (b) How far on the train from the cannon does the projectile land? Neglect air resistance.

3.67●● Show that the maximum height h to which a football rises when kicked from the ground at an angle θ is given by $h = \frac{1}{4}R \tan \theta$, where R is the range.

3.68●● A cannon is adjusted for maximum range R_{max} on level ground. How high is the cannonball when its horizontal distance from the cannon is $\frac{3}{4}R_{max}$?

3.69●● How high does a golf ball rise when projected for maximum range? Express your answer in terms of the range.

3.70●● Galileo's great-great-great-...grandchild drops a Chianti bottle from the top of a vertical tower 54.6 m tall.
(a) How long does it take for the bottle to fall to the ground?
(b) What is the velocity of the bottle as it hits the ground?
(c) If the bottle is thrown straight out horizontally from the tower with a speed of 12.3 m/s, how far does it land from a point on the ground directly beneath the point from which it was launched? (d) What is the bottle's horizontal component of velocity, and (e) what is the magnitude of its velocity just before it strikes the ground for the situation in (c)?

Review of Trigonometry

This brief outline of basic trigonometry, together with a few other relationships introduced later, includes most of the mathematics you need to know for this text.

It may be intuitively obvious to you (or you may remember a formal proof from geometry) that in the right triangle ABC of Fig. A3.1, the *ratio* of side AC to the hypotenuse AB does not depend on the physical size of the triangle, but only on the angles. For instance, in the two similar triangles ABC and ADE in Fig. A3.1, we see that

$$\frac{BC}{AB} = \frac{DE}{AD}.$$

This ratio uniquely characterizes the angle marked θ in the diagram. This ratio is called the sine of the angle θ and is written

$$\sin \theta = \frac{BC}{AB}.$$

More generally, for any right triangle,

$$\sin \theta = \frac{\text{length of the side opposite the angle } \theta}{\text{length of the hypotenuse}}.$$

Two other trigonometric functions, the cosine of the angle and the tangent of the angle, are defined as

$$\cos \theta = \frac{\text{length of the side adjacent to the angle}}{\text{length of the hypotenuse}}.$$

and

$$\tan \theta = \frac{\text{length of opposite side}}{\text{length of adjacent side}}.$$

The tangent may seem somewhat superfluous, since $\tan \theta = \sin \theta / \cos \theta$, but its use is often convenient. A scientific pocket calculator can give you the value of the sine, cosine, or tangent of any given angle. There are other trigonometric functions, but they will not be needed in this text.

We always measure angles in the counterclockwise direction from the positive x axis (Fig. A3.2). Angles measured in the clockwise direction are negative. Our definitions of sine, cosine, and tangent do not change for angles greater than 90°; however, the sign of the trigonometric function changes from quadrant to quadrant. For 150° the hypotenuse is positive (as always), the side opposite the angle is positive, but the adjacent side (in the negative x-direction) is negative. Thus, the sine of an angle in the second quadrant (that is, an angle between 90° and 180°) is positive, but the cosine is negative. The signs of the sine and cosine functions are marked in each quadrant of Fig. A3.2.

If you know the value of the sine of an angle, you can also use your calculator to find the value of the an-

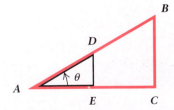

Figure A3.1
Two similar right triangles, ABC and ADE, with a common angle θ.

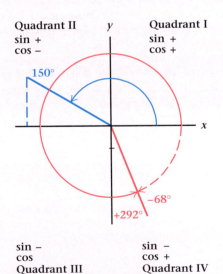

Figure A3.2
Angles are measured relative to the positive x axis. The signs of the sine and cosine functions are indicated for each quadrant.

gle. The angle θ whose sine is equal to A is written as $\theta = \sin^{-1}A$, where $\sin^{-1}$ is read as the inverse sine. The values of $\cos^{-1}$ and $\tan^{-1}$ can be obtained in a similar manner.

You can use your calculator to verify numerically that, for any angle,

$$\sin^2\theta + \cos^2\theta = 1.$$

We can show from Fig. A3.1 that this equation is true in general. The Pythagorean theorem gives

$$(BC)^2 + (AC)^2 = (AB)^2.$$

Dividing both sides by $(AB)^2$, we get

$$\frac{(BC)^2}{(AB)^2} + \frac{(AC)^2}{(AB)^2} = 1$$
$$\left(\frac{BC}{AB}\right)^2 + \left(\frac{AC}{AB}\right)^2 = 1,$$

or

$$\sin^2\theta + \cos^2\theta = 1.$$

Example 3.15

Standing on top of a tall building, you see a friend walking at a distance. The angle between your line of sight and the horizontal is 15°. How far from you is your friend? Assume the road is level and your eyes are 40.8 m above the ground.

Solution Let BC be the distance from the ground to your eyes and AB the distance from you to your friend. The angle θ in Fig. A3.3 is given as 15°. The sine of θ is

$$\sin\theta = \frac{BC}{AB},$$

or

$$AB = \frac{BC}{\sin\theta}.$$

Inserting the numerical values gives

$$AB = \frac{40.8 \text{ m}}{0.259} = 158 \text{ m}.$$

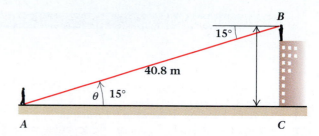

Figure A3.3
Example 3.15: How far away is your friend?

Example 3.16

Three narrow strips of wood, having lengths of 30, 40, and 50 cm, are arranged on a table top to form a right triangle (Fig. A3.4). What is the angle between the 40-cm side and the 50-cm side?

Solution Using the definition of the cosine, we write

$$\cos\theta = \frac{40 \text{ cm}}{50 \text{ cm}} = 0.80.$$

The value of the angle θ is

$$\theta = \cos^{-1}(0.80) = 37°.$$

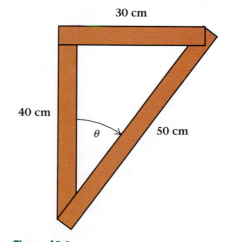

Figure A3.4
Example 3.16: How large is the angle θ?

Force and Motion

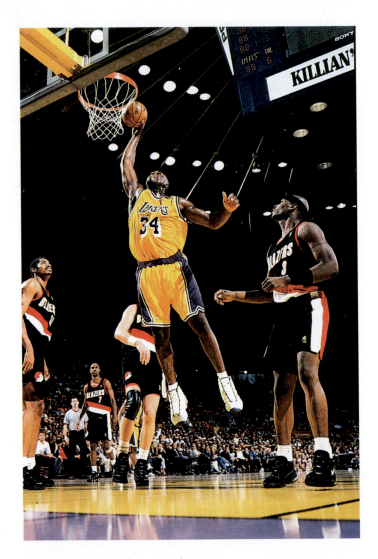

In the last three chapters you learned to describe motion. In this chapter we discuss the underlying causes of motion, a study called *dynamics*. **Newtonian mechanics,** or classical mechanics as it is also called, is the branch of physics that deals with the kinematics and dynamics of large-scale (macroscopic) objects. The essence of Newtonian mechanics, which is widely used to describe the motion of objects as different as golf balls and spacecraft, is found in Newton's three laws of motion. These laws, first clearly stated by Isaac Newton more than 300 years ago in his book the *Principia,* tell us how forces affect the motion of objects. The same laws apply whether the object is a mosquito zigzagging through the air or a planet moving about the sun. Although Newton's laws do not apply in the domain of the very small (that is, to atoms and molecules) and the very fast (to objects moving at speeds near the speed of light), they apply to almost everything else. Thus, the concepts introduced in this chapter form the basis for our understanding of much of physics and are essential to almost all of wh · at follows in this course. ■

Events Leading to Newton's *Principia*

In the sixteenth century, the idea that the earth was the fixed center of the universe was firmly ingrained in astronomical thought and had become an article of religious faith. Nicolas Copernicus (1473–1543) published in his last year a workable scheme for the solar system with the sun at the center and the planets in orbits about it. His idea was initially rejected, but the work of Galileo and Johannes Kepler (1571–1630) helped to bring about its adoption. Such a model is called a heliocentric theory, from the Greek word *helios,* for sun.

By Kepler's time, it was known that the orbits of the planets were not exact circles and that the sun was not at the exact center of the orbits. Kepler analyzed Tycho Brahe's (1546–1601) astronomical observations and found that the planets closer to the sun moved faster than those farther away, and that a single planet moved faster when it was closer to the sun than when it was farther away. He proposed that the sun was the cause of the planet's motion and that the sun's influence might decrease with increasing distance from it. With these ideas as guidelines and Tycho's detailed observations as his raw material, Kepler constructed a model of the solar system that was compatible with both ideas.

Kepler's laws of planetary motion were among the first "laws of nature" in the modern sense. The first of these laws, published in 1609, states that the planets move in elliptical orbits with the sun at one focus (Fig. 4.1).* (This law and Kepler's other two laws are discussed in detail in Chapter 5.) Though Kepler's laws could be considered statements about the geometry and kinematics of the solar system, Kepler himself was not content with this and thought that one should look for the cause of planetary motion. Many, including Kepler, held that the planets' orbits were due to some influence of the sun.

By the 1660s and 1670s, there was some reason to believe that the sun attracted each planet with a force that depended in a simple way on the distance between them: The force was thought to be proportional to the reciprocal of the square of this distance. That is,

$$F \propto \frac{1}{r^2},$$

where r is the distance between the sun and planet and F is the attractive force. Such a force is called an **inverse-square force.** This terminology means that if the force were, say, 36 units for a distance of 6000 m, then doubling the distance to 12,000 m would make the force become one quarter as much, or 9 units. Similarly, if the distance were halved to 3000 m, the force would increase by a factor of 4, to 144 units. The reasons for accepting the idea of an inverse-square force for the sun-planet attraction are developed later in Chapter 5.

One of the principal events that led to the writing of the *Principia* was Newton's deduction that an inverse-square force was responsible for the observed

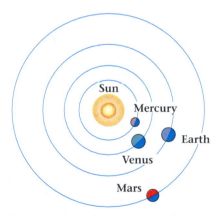

Figure 4.1
The orbits of the four innermost planets of the solar system. Relative distances from the sun are correct. On this scale the difference between the true ellipse and a circle cannot be detected. The size of the planets is exaggerated.

---

*A review of the properties of the ellipse is found on page A–2 of Appendix A.

orbits of the planets. Further, he identified the action of the inverse-square force as the law of gravitation and showed it to be universal by applying it to the moon, the planets, and objects on earth.

Newton's ideas about the way the universe worked, as expressed in the *Principia,* transcended science and influenced Western thought in general. The idea that a mechanical universe obeyed a single set of laws suggested that all observed behavior could be explained in mechanical terms. Within the last hundred years, physicists have found that it is not possible to take this deterministic viewpoint and that nature is much more subtle than it was imagined to be in Newton's day. We will discuss the contemporary viewpoint later in this book, when we get to relativity and quantum theory.

4.2   What Is a Force?

Understanding the concept of force forms the basis for understanding Newton's laws. In everyday language, a **force** is a "push" or a "pull." When we push a lawnmower or a grocery cart, we exert a force on it. When we pull open a drawer, we exert a force on the drawer. When we drop something, it falls, and we say that the force, or pull, of gravity made it fall. A roller coaster plunges downhill because of the force of gravity and moves through turns and loops because of the forces exerted on it by the track (Fig. 4.2). However, forces do not always cause motion. A book resting on a table experiences the downward force of gravity even though the book does not move.

Notice that force is not a property of an object, like mass, but rather it is an interaction of the object with an external agent. Consequently, for a force to act

Figure 4.2
A roller coaster moves in response to forces acting on it.

THE WRITING OF THE *PRINCIPIA*

The writing of the *Principia* came about from an argument over physics. The discussion was on the nature of gravitational force, and took place in January 1684 among three men: Christopher Wren (1632–1723), remembered today as an architect of churches and public buildings, especially St. Paul's Cathedral in London; Robert Hooke (1635–1703), who was curator of the Royal Society* at that time; and Edmund Halley (1656–1742), a young astronomer and mathematician. It was Halley (rhymes with Sally) who in 1705 predicted the return of the comet of 1682, which, upon its next appearance in 1758, was given his name.

Wren offered a prize to whichever of the other two could produce a proof that the force between the sun and the planets obeyed an inverse-square law. Hooke maintained that he had done so, but failed to produce evidence within a reasonable span of time. So Halley sought the help of Isaac Newton (Fig. B4.1), at that time a professor at Cambridge University. During the visit, Halley asked Newton what would be the orbit of planets attracted to the sun in an inverse-square manner. Newton replied that they would be ellipses and further said that he had produced a mathematical proof of this. Newton promised to repeat the calculation and send it to Halley, and the manuscript arrived in November. This so excited Halley that he suggested that Newton make his results public by sending them to the Royal Society. Newton did so in the form of a tract, *De Motu,* based in part on earlier calculations. In this work he proved that if "a body revolves in an ellipse . . . the law of the centripetal force . . . is reciprocally as the square of the distance." Here was the solution to the physical problem of planetary orbits, the first real answer to the question, "What *must* be the force that gives rise to the observed planetary orbits if the sun is the attracting body?"

Figure B4.1 Isaac Newton (1642–1727). As an old man, Newton recollected that his peak years were 1665 and 1666. While the black plague swept through England, Newton retreated to his family's farm. There he invented calculus, completed much of the preparatory work for his theory of gravity, and discovered that white light is composed of many colored rays—all in the span of 18 months!

Figure B4.2 Saturn's rings as seen by *Voyager* 2 in August 1981. The colors are computer enhanced.

After presenting *De Motu* to the Royal Society, Halley, with the support of the society, invited Newton to write a more comprehensive version. Newton finished the complete work within 18 months, an incredibly short time for such a monumental work. The first edition of the *Principia* was available by July 5, 1687.

Newton wrote the *Principia* for an elite audience who understood science and mathematics. In many ways he tried to make it as difficult and inaccessible as possible. He is said to have boasted to a friend that he made the *Principia* "abstruse" so that he would not have to argue with those of lesser learning. In contrast to Galileo's use of everyday language, Newton wrote the *Principia* in Latin, the international language of learning. Newton never attempted to produce an English version, though he lived forty more years and brought out revised editions in 1713 and 1723. The first, and only, complete English translation was published by Andrew Motte in 1729.

The *Principia* was and is a difficult book, so difficult that one historian has said "it is doubtful whether any work of comparable influence can ever have been read by so few persons." Like major scientific works in other times, it had strong proponents as well as strong detractors. However, the Newtonian concepts were admitted to the general body of scientific knowledge because of their utility and precision.

Newtonian mechanics is still important today. For example, when coupled with modern computer technology, Newtonian physics allows us to make the calculations necessary to guide probes to the outer planets of the solar system. In the 1970s and 1980s, Voyager spacecraft sent back dramatic pictures of Saturn, Jupiter, Uranus, and Neptune (Fig. B4.2).

---

*The Royal Society of London is one of the oldest scientific societies in the world. Founded in 1660, the society is still active publishing scientific journals.

on an object there must be something external other than the object itself that causes the force. For example, the force that moves a grocery cart is the result of an interaction between a person and the cart, and the force of gravity that keeps the moon in its orbit is an interaction between the earth and the moon.

Because force has direction as well as magnitude, it is a vector quantity. The **net force** is the *vector sum* of all forces acting simultaneously on an object. We often refer to the net force as the *resultant force* or the *unbalanced force*. *It is the net force that determines the motion of the object.*

Example 4.1

A patient with a fractured femur is placed in Russell's traction to help in the proper healing of the bone (Fig. 4.3a). The ropes and pulleys are positioned so that the resultant force is parallel to the upper leg. The downward force F exerted by gravity on the hanging mass is applied to the knee by a flexible rope that passes through the pulleys. The force on the knee has magnitude F and is directed upward along the rope. The force on the heel has magnitude $2F$ because there are two ropes leading to the heel pulley, each pulling with a force F. This force is directed horizontally away from the heel. The angle between the horizontal force and the upward force is 122°. What is the magnitude of the net, or resultant, force that pulls parallel to the upper leg?

Strategy We first draw a vector diagram (Fig. 4.3b) with the force on the knee represented by the vector **A** and the force on the heel by the vector **B**. If we measure the angles with respect to the positive x axis as shown, then vector **A** makes an angle of 58°. From the statement of the problem, we know that

Figure 4.3

(a) Russell's traction. (b) Vector diagram for Russell's traction.

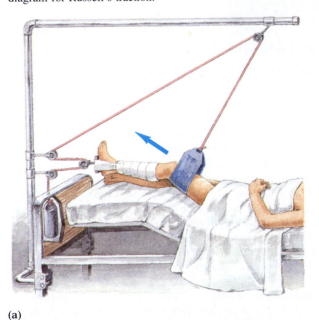

(a)

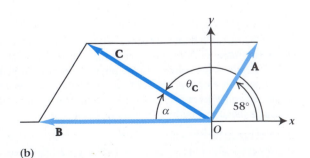

(b)

$A = F$ and $B = 2F$. Because forces are vector quantities, we use the vector addition techniques of Chapter 3 to find the magnitude of the resultant force. We resolve each force **A** and **B** into horizontal (x) and vertical (y) components. Then we add the x components together to get the x component C_x of the resultant vector and add the y components to get C_y. Finally, we use the Pythagorean theorem to obtain the magnitude C of the resultant. As a check of our work, we will also calculate the direction and verify that our answer agrees with the diagram.

Solution The components of the vectors are

$$C_x = A_x + B_x \qquad\qquad C_y = A_y + B_y$$
$$C_x = A \cos \theta_A + B \cos \theta_B \qquad C_y = A \sin \theta_A + B \sin \theta_B$$
$$C_x = F \cos 58 + 2F \cos 180 \qquad C_y = F \sin 58 + 2F \sin 180$$
$$C_x = F(0.530) + 2F(-1.00) = -1.47F \qquad C_y = F(0.848) + 0 = 0.848F.$$

The magnitude of the force vector **C** is then

$$C = \sqrt{C_x^2 + C_y^2}$$
$$C = \sqrt{(-1.47F)^2 + (0.848F)^2} = 1.70F.$$

Thus, the force applied parallel to the upper leg is $1.70F$, where F is the magnitude of the force due to the action of gravity on M. Measurement of the length of **C** on the diagram confirms that this is the magnitude of the applied force.

We can check our work by calculating the direction of the resultant force **C**. Because C_y is a positive quantity and C_x is a negative quantity, the head of the vector **C** will lie in the second quadrant. That is, the angle θ_C lies between 90° and 180°. The angle of **C** with respect to the horizontal direction is given by

$$\tan \theta_C = \frac{C_y}{C_x} = \frac{0.848F}{-1.47F} = -0.577.$$

Direct computation with a pocket calculator gives $\theta = -30°$. However we know that the vector lies in the second quadrant, so we must add 180°, making the correct answer

$$\theta_C = 150°.$$

Discussion When you use a calculator to determine the angle whose tangent is -0.577, you will get an answer of $-30.0°$. This happens for the following reason. Because the tangent is the ratio y/x, the value of the tangent can be negative in two situations, either because x is positive and y is negative or because y is positive and x is negative. For positive x and negative y, the angle is in the fourth quadrant, but in the case of negative x and positive y, the angle is in the second quadrant, as we know our answer to be in this instance. However, a calculator cannot distinguish between the two cases and is designed to give you the smaller angle. This is another reason why you must always draw a diagram and use it as part of your calculational procedure.

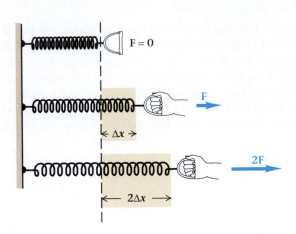

Figure 4.4
A spring attached to a fixed wall stretches through a distance proportional to the force applied to the ends. If a force **F** stretches the spring a distance Δx, then a force of 2**F** extends the spring a distance $2\Delta x$.

How can you measure a force? Consider that when you pull on the ends of a coiled spring (Fig. 4.4), the spring stretches. The harder you pull, the more the spring is extended. When you pull with a force **F**, the spring extends a length Δx; when you pull with twice the force (2**F**), the spring extends a length $2\Delta x$. If you calibrate the extension of the spring, you can use the distance the spring stretches to measure the magnitude of the applied force. Spring scales that use this principle are commonly available. Once the scale is calibrated, you can use it to measure forces of different origins, such as forces resulting from muscular effort, gravitation, or magnetism.

When you pull or push a spring, you exert a *contact force,* a force in which two bodies interact directly by contact between their surfaces. The force you feel on your hand when you push a door and the force you feel on your feet when you walk are familiar contact forces. Although contact forces are common, they are not considered fundamental forces in the physical world.

A dropped stone falls because of the force of gravity. In this case the stone falls freely without being in contact with anything. The force involved is an example of a field force; the force is exerted without actual contact between the bodies but acts through space. The gravitational field is discussed in the next chapter.

The gravitational force is one of the four *fundamental forces* in nature (Table 4.1). The other fundamental forces are the electromagnetic force, the strong force, and the weak force. The gravitational force is responsible for the weight of bodies on the earth as well as for the motion of the planets. It acts between all masses. We will study gravitation in detail in Chapter 5.

The electromagnetic force includes both electric and magnetic forces and is relatively strong. It is the electric force that you observe if you run a comb through your hair and then use the comb to pick up bits of paper. You see the magnetic force act when you pick up a pin with a magnet. The electromagnetic force is responsible for holding atoms and molecules together and for the structure of matter. Contact forces are actually large-scale manifestations of fundamental electromagnetic forces. We discuss electromagnetic forces in Chapters 16–21.

The strong force holds together the constituents of the atomic nucleus. It is sometimes called the nuclear force. Chapters 29 and 32 include material on the strong force.

Table 4.1	Fundamental Forces
Gravitational	
Electromagnetic	
Strong	
Weak	

The weak force acts between all matter, but is so weak that it plays no direct part in ordinary observable behavior. It is important, however, in the interactions between subnuclear particles.

Current theories have been partially successful in unifying the basic forces of nature. We now understand the electric and weak force to be separate manifestations of one force. However, efforts to fully unify the forces of nature under a single theory have not yet been successful.

4.3 Newton's First Law—Inertia

Because of our past experience, we can readily believe that a body at rest remains at rest unless some action is taken. However, it is not so easily seen from everyday experience that once a body is moving with nonzero constant velocity (constant speed in a straight line), it continues to do so without any additional outside effort. Your first reaction may be that this idea goes against common sense. Let's consider a glider moving along an air track with very little friction shown in the time exposure photograph in Fig. 4.5(a). While holding the glider at the left end of the track, the photographer sets off a regularly flashing light, opens the camera shutter, and gives the glider a push to the right. The constant spacing of the glider in the picture shows that the velocity of the glider remained essentially constant.

(a)

Newton understood this idea of constant velocity and incorporated it in his laws of mechanics. A paraphrase of an example given by one of Newton's acquaintances is understandable. Suppose you exert a certain force to roll a bowling ball on an unmowed lawn, and it rolls 20 yards. (In Newton's time, bowling took place on lawns, called bowling greens.) With the same force, you might roll it 30 yards if the grass were cut. Further smoothing of the grass surface would increase the ball's range. Since the range increases as obstacles are removed, we conclude that the removal of *all* retarding influences (forces) would allow the ball to continue rolling forever, with neither its speed nor direction changed. Once the ball is given a speed and direction by the bowler, the ball continues in its original state until acted on and retarded. In many cases, the major retarding force is friction.

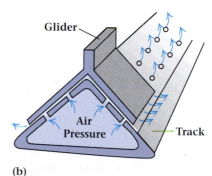

(b)

Newton's first law of motion is his embodiment of the idea we have just described that motion does not require a force. A paraphrase of Newton's first law is: *Every object continues in its state of rest, or of uniform motion in a straight line, unless compelled to change that state by forces acting on it.* Notice that this law is a cause-and-effect statement relating force and motion. A force causes a change in the state of motion. A statement of **Newton's first law** in more concise language, is: **An object has a constant velocity unless there is a net force acting on it.** A velocity of zero corresponds to the case of an object at rest.

Newton's first law is also known as the *law of inertia.* The word *inertia* is from the Latin word for *sluggish* or *inactive.* In modern terms, **inertia** is the property of matter that causes objects to resist changes in motion. You experience the effect of inertia when you set a bowling ball in motion. Likewise, you experience inertia when you try to stop the ball from moving. By the time Newton wrote the *Principia,* it was fairly well established that the kind of motion that

Figure 4.5

(a) A glider moving with a constant velocity on an air track. The regular spacing of the glider's successive positions indicates a constant velocity, which means that the horizontal force on it is negligible. (b) A typical linear air track is made from a hollow triangular rail that has tiny holes drilled in its upper surfaces. When it is pressurized, a cushion of escaping air supports a carefully fitted glider on the track. The glider floats nearly friction free along a straight line on this air cushion.

would continue indefinitely without any additional "push" was linear motion. The first law is a clear statement of this principle.

Inertial Frames of Reference

In Chapter 2 we discussed the necessity for a frame of reference to which we refer the position and motion of objects. The concept of a reference frame is central to Newton's laws of motion. Imagine yourself in an airliner in level flight at a constant velocity (Fig. 4.6a). You are at rest relative to the seat and to the other parts of the aircraft. A ball placed in the aisle will remain motionless relative to the airplane. A coin dropped from your hand will fall straight down. If we take the airplane as a frame of reference, everything within the cabin will behave as expected by Newton's first law. Such a reference frame in which Newton's first law is valid is called an **inertial reference frame.**

Now consider the very different circumstances of an airliner accelerating down a level runway for takeoff (Fig. 4.6b). You are at rest relative to the seat and the rest of the aircraft. However, you feel the force of the seat back that is accelerating you forward along with the plane. If the same ball is placed at rest in the aisle, it would begin rolling toward the back of the plane. Likewise, a coin dropped from your hand would not appear to fall straight down. Such motion is not consistent with the law of inertia (Newton's first law), so in this case the airplane is *not* an inertial frame. Reference frames such as the accelerating airplane, in which the law of inertia (Newton's first law) does not hold, are called *noninertial* reference frames.

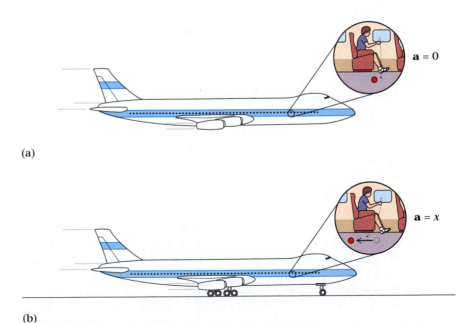

(a)

(b)

Figure 4.6

(a) An airliner in level flight at constant velocity is an inertial reference frame.
(b) An airliner accelerating down a level runway is not an inertial reference frame.

We usually consider Newton's laws to be valid on the face of the earth, even though, because it rotates, the earth is not truly an inertial reference frame. The effects of the rotation are not large and for most purposes Newton's laws apply. The effects of rotation can be calculated by using Newton's laws and an external inertial reference frame. Such calculations help explain the motion of the atmosphere and the formation of hurricanes and tornadoes.

4.4 Newton's Second Law

Newton's first law describes what happens when the net force acting on an object is zero. In that case, the object either remains at rest or continues in motion with constant speed in a straight line. Newton's second law describes the change of motion that occurs when a nonzero net force acts on the object.

To get a sense of what Newton's second law is about, imagine that you pull a child in a wagon along a smooth, level sidewalk. If you pull with a constant force, the wagon accelerates, moving faster as you continue to pull. If a friend helps pull so that the force is greater, the rate at which the speed increases is also greater; the larger force gives rise to a larger acceleration. This effect is predicted by Newton's second law.

The original translation of **Newton's second law** was, *The alteration of motion is ever proportional to the motive force impressed; and is made in the direction of the right line in which that force is impressed.* Elsewhere in the *Principia* Newton was clear that by "motion" he meant the product of the velocity and the mass. For the moment, it is sufficient to use Newton's identification of mass as the "quantity of matter." The modern name of the product of mass m and velocity $\mathbf{v}$ is **momentum.** (We consider momentum more fully in Chapter 7.) Using this term, we restate the second law:

The rate of change of momentum with time is proportional to the net applied force and is in the same direction:

$$\frac{\Delta(m\mathbf{v})}{\Delta t} \propto \Sigma \, \mathbf{F}, \tag{4.1}$$

where $\Sigma \, \mathbf{F}$ is the net force—that is, the vector sum of all forces acting on a body*—and the change in the momentum $\Delta(m\mathbf{v})$ is in the direction of $\Sigma \, \mathbf{F}$.

If we include a proportionality constant, the proportion becomes an equation. The value of the proportionality constant depends on the choice of units for force, mass, velocity, and time. If we choose the units appropriately, the proportionality constant can have the value 1. In that case the equation simply becomes

$$\frac{\Delta(m\mathbf{v})}{\Delta t} = \Sigma \, \mathbf{F}. \tag{4.2}$$

---

*The symbol Σ (the Greek letter sigma) indicates a summation. In Eq. (4.1), $\Sigma \, \mathbf{F}$ indicates the sum of all forces $\mathbf{F}$ acting on the body of mass m.

In the majority of real situations, the mass of an object does not change appreciably, so the change in momentum is just the mass times the change in velocity. Then

$$\frac{\Delta(m\mathbf{v})}{\Delta t} = m\frac{\Delta \mathbf{v}}{\Delta t} = m\mathbf{a}.$$

Thus Newton's second law may be expressed as

$$\mathbf{a} = \frac{\Sigma \mathbf{F}}{m}. \qquad (4.3)$$

Rearranging, we have the most common statement of the second law in the form

$$\boxed{\Sigma \mathbf{F} = m\mathbf{a}.} \qquad (4.4)$$

Although this last equation is mathematically correct, you should use care in interpreting it. *It is the force that causes the acceleration, not vice versa.*

Equation (4.4) is a vector equation. It is equivalent to a set of three equations, one for each component:

$$\Sigma F_x = ma_x, \qquad \Sigma F_y = ma_y, \qquad \Sigma F_z = ma_z. \qquad (4.5)$$

These equations relate three components of acceleration of a mass m to the three components of the net force causing the mass to accelerate.

Once we know the velocity and acceleration of an object, we can predict its motion. The ability to predict the motion of a body from a knowledge of the forces acting on it is one of the most useful accomplishments of physics. Not only is it useful for predicting the motion of bodies under mechanical or gravitational forces, but also it allows us to calculate the motion when the force is due to other causes, such as magnetism or electricity. Conversely, when the force is not known it may sometimes be deduced from a knowledge of the motion.

Before we go on to examples of the second law, it is worth noting that Newton's meaning of the second law is properly represented by Eq. (4.1). Equation (4.4), or even $\mathbf{F} \propto \mathbf{a}$, as it is sometimes stated, is a restricted special case for objects with constant mass. In this case, *the acceleration is proportional to the force, and the direction of the acceleration is the same as the direction of the force.*

Compare the glider moving with no horizontal force on it (Fig. 4.5) to the photo in Fig. 4.7(a). In this case, a constant horizontal force has been applied to the same glider in the manner shown in Fig. 4.7(b). The increased spacings between positions recorded by flashes at equal time intervals indicate that the glider accelerates as it moves to the right.

Measurements taken from a photograph similar to Fig. 4.7 are listed in Table 4.2. They show that over each equal interval of time the velocity increases by the same amount; that is, the acceleration is constant. In this demonstration, a constant force has given rise to a constant acceleration. In another experiment similar to this one, the horizontal force on the glider was increased by hanging another mass alongside the first. Measurements on the photograph again showed motion with constant acceleration.

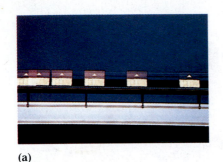

(a)

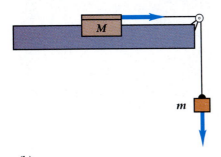

(b)

Figure 4.7

(a) An air-track glider moving under the influence of a small hanging mass. The photograph was made by the same technique used in Fig. 4.5. The increased spacing between the positions of the glider for successive strobe flashes indicates acceleration. (b) A simplified drawing of the air track showing the hanging mass. As the mass falls, it pulls the glider along with a constant force.

Table 4.2		*The Motion of an Air-track Glider Subjected to a Constant Force*	
Time	x (cm)	$\Delta x/\Delta t$ (cm/time interval)	$\Delta(\Delta x)/(\Delta t)^2$ [cm/(time interval)^2]
0	131.4		
		8.4	
1	139.8		3.0
		11.4	
2	151.2		3.2
		14.6	
3	165.8		3.1
		17.7	
4	183.5		3.0
		20.7	
5	204.2		3.1
		23.8	
6	228.0		

Time intervals were uniformly spaced. The position given is x, the location of the front of the glider. There is some uncertainty in the fourth digit. The change in x is Δx, which is proportional to the velocity. The acceleration is proportional to the change in $\Delta x/\Delta t$ and is given by $\Delta(\Delta x)/(\Delta t)^2$. From the table we see that the acceleration is constant.

Indeed, using a range of accelerating forces and a number of different gliders, we always observe that for a given total mass (glider plus hanging mass), the acceleration is proportional to the net force. Rearrangement of the mathematical statement of Newton's second law gives

$$m = \frac{F}{a}.$$

Then we can call on common experience and this equation to tell us that what we have called **mass** is a quantitative measure of the inertia of a body. Finally, we say that the more massive a body, the larger the force necessary to give it a particular acceleration.

In the SI system of units, the unit of mass is the kilogram. The present standard of mass is a platinum-iridium cylinder whose mass is defined to be, by international agreement, one kilogram. This standard is kept at the International Bureau of Weights and Measures near Paris. Other bodies are assigned their mass value by a comparison with the standard. If, when Eq. (4.4) is used, the unit of mass is the kilogram and the unit of acceleration is the meter/second^2, then the unit of force becomes the kilogram · meter/second^2. This combination of units is given the name **newton** (N). We say: **A force of one newton acting on a one-kilogram mass gives it an acceleration of one meter/second^2.** Some familiar forces and their approximate magnitudes in newtons are shown in Fig. 4.8.

Although we shall use SI units almost exclusively in this text, other systems of units are still in use. Principal among these are the British system and the centimeter-gram-second (CGS) system of units. The units of force, mass, and acceleration in these three systems are summarized in Table 4.3.

Force exerted by a weightlifter: 2000 N

Gravitational force (weight) on a person: 600 N

Force to flip a light switch: 3 N

Figure 4.8
Some familiar forces.

Table 4.3	Units of Force, Mass, and Acceleration		
System of Units	Force	Mass	Acceleration
SI	newton (N)	kilogram (kg)	m/s^2
CGS	dyne (dyn)	gram (g)	cm/s^2
British*	pound (lb)	slug	ft/s^2

*1 lb = 1 slug · ft/s^2 = 4.448 N

Example 4.2

Safety regulations require that all cars traveling at a given speed be able to stop within a given distance. What must be the relationship between the minimum braking forces allowed by law for two cars whose masses have the ratio of 3 to 2?

Strategy We can use Newton's second law to relate the force to mass and acceleration. From your knowledge of kinematics, you know that the requirement for stopping in equal distances from the same initial speed for both cars means they must both have the same acceleration.

Solution We can write the two accelerations in terms of the braking forces F_1 and F_2 and the masses m_1 and m_2 as

$$a_1 = \frac{F_1}{m_1} \quad \text{and} \quad a_2 = \frac{F_2}{m_2}.$$

When we set the accelerations equal we find

$$\frac{F_1}{m_1} = \frac{F_2}{m_2},$$

so

$$\frac{F_1}{F_2} = \frac{m_1}{m_2} = \frac{3}{2}.$$

Discussion This result tells us that a larger braking force is needed to stop a larger (more massive) car in the same distance that a smaller (less massive) car is stopped.

Example 4.3

(a) What force is necessary to give a 0.80-kg air-track glider a horizontal acceleration of 1.5 m/s^2 if the force is directed horizontally along the length of the level air track? (b) What is the acceleration of the glider when the applied force is 1/3 the value found in part (a)?

Solution (a) We can find the magnitude of the force from Newton's second law:

$$F = ma = 0.80 \text{ kg} \times 1.5 \text{ m/s}^2 = 1.2 \text{ N}.$$

(b) We can also use the second law to find the acceleration when the force applied is 1/3 the value just found in part (a). Then the magnitude of the acceleration is

$$a = \frac{F}{m} = \frac{\frac{1}{3}(1.2 \text{ N})}{0.80 \text{ kg}} = 0.50 \text{ m/s}^2.$$

Example 4.4

A fully loaded Lockheed L-1011 with a mass of 2.17×10^5 kg accelerates at full throttle down a level runway (Fig. 4.9). The engines push with a combined constant horizontal net force of 753 kN. If the plane starts from rest, how far will it go during the 33.5 s that it takes to reach liftoff velocity?

Strategy The problem asks us to find the distance traveled by an airplane accelerating from rest. We can use our knowledge of kinematics to find that distance if we know the acceleration. Because we are given the mass of the plane and the force of the engines, we can use Newton's second law to find the acceleration.

Remember to put all numerical values in SI units. For example, the force of the engines is

$$F = 753 \text{ kN}\left(\frac{1000 \text{ N}}{1 \text{ kN}}\right) = 753000 \text{ N} = 7.53 \times 10^5 \text{ N}.$$

Solution The acceleration is found from Newton's second law in the form of Eq. (4.3):

$$a = \frac{F}{m} = \frac{7.53 \times 10^5 \text{ N}}{2.17 \times 10^5 \text{ kg}} = \frac{7.53 \times 10^5 \text{ kg} \cdot \text{m/s}^2}{2.17 \times 10^5 \text{ kg}} = 3.47 \text{ m/s}^2.$$

This value of acceleration is then used with the kinematic equation for displacement,

$$x = v_0 t + \tfrac{1}{2}at^2.$$

Since the plane starts from rest, $v_0 = 0$. Inserting the values for a and t we get

$$x = \frac{1}{2}\left(\frac{7.53 \text{ m}}{2.17 \text{ s}^2}\right)(33.5 \text{ s})^2 = \frac{1}{2}(3.47)(1122 \text{ m})$$

$$x = 1947 \text{ m} \approx 1.95 \text{ km}$$

Figure 4.9
Example 4.4: An airliner accelerates under constant force.

4.5 Weight

One important example of Newton's second law is the expression for an object's weight. The **weight** of an object on earth is the gravitational force exerted on it by the earth. In Galileo's time, most scholars held Aristotle's belief that heavy (weightier) objects fall faster than light ones, just because they are heavier. As we saw in Chapter 2, Galileo showed that reasoning to be incorrect. Some objects do fall less swiftly than others because of air resistance, which slows the fall of objects such as feathers and leaves. More compact objects such as stones fall faster because, for a given mass, they offer a smaller area to the air, thus reducing the effect of air resistance to a negligible amount. In a vacuum, all objects fall with the same acceleration, regardless of their weight.

As a result, we know that near the earth's surface, when we neglect air resistance, the acceleration is the same for all falling bodies. This constant acceleration is known as the acceleration of gravity, **g**, and has the standard value 9.807 m/s^2.* When an object is dropped near the earth's surface, it is accelerated by the gravitational force (equal to its weight) with an acceleration **g**. Thus, by Newton's second law, the weight **w** becomes

$$\mathbf{w} = m\mathbf{g}. \tag{4.6}$$

We see in this equation the relation between mass and weight: *Weight is a force proportional to the mass of a body and g is the constant of proportionality.*

Example 4.5

What is the weight of a textbook whose mass is 1.85 kg?

Solution The weight may be computed from Eq. (4.6),

$$w = mg,$$

where the numerical value of g is 9.81 m/s^2. The weight becomes

$$w = 1.85 \text{ kg} \times 9.81 \text{ m/s}^2 = 18.1 \text{ kg} \cdot \text{m/s}^2 = 18.1 \text{ N}.$$

Discussion The weight of the book is 18.1 N. Notice that the unit of weight is the unit of force. Using the information in Table 4.3, you should show for yourself that this SI force of 18.1 N corresponds to 4.07 lb.

When you stand on a scale, the scale reading gives the magnitude of your weight. Because most of the force that comprises your weight is due to the gravitational attraction between you and the earth, we have described weight as the force of gravity. However, because the earth rotates, your weight is slightly less than it would be if the earth were not rotating. Consequently, we need a more precise definition of weight. It is: *The weight of a body in a specified reference frame is the force which, when applied to the body, would give it an acceleration equal to the local acceleration of free fall in that reference frame.* On the earth, the local acceleration of free fall is **g**. We still use Eq. (4.6) to express the relationship between mass and weight.

If we take an object to the moon, the force of gravity exerted on it by the moon is less than the force of gravity on it when it was on the earth. Thus, its

*Although the value of g is the same for all objects in a given locality, it is measurably different at other localities. There is about 0.5% variation with latitude, with g being smaller at the equator than at the poles. An additional variation depends on altitude and on the underlying geological formation. For example, the value of g in Denver, Colorado, is 9.796 m/s^2, while the value in Greenwich, England, is 9.812 m/s^2.

weight on the moon is different from its weight on the earth, even though its mass remains constant.

The Normal Force

When an object such as a brick rests on the ground, the gravitational force continues to act on the brick, even though it is not accelerating. According to Newton's second law, the net force on the brick at rest must be zero. There must be another force acting on the brick that opposes the gravitational force. This force is provided by the ground (Fig. 4.10a). The force provided by the ground is perpendicular to the surface of contact and is known as the **normal force**. (In geometry, the word *normal* means "perpendicular.") The normal force is simply the resistance of the ground to the motion of the brick acted upon by gravity. It is this normal force that keeps the brick from sinking into the ground.

If the brick rests on an inclined surface, the gravitational force $m\mathbf{g}$ acting on the brick is still directed downward. The normal force $\mathbf{N}$ acts perpendicular to the surface, and since the surface is inclined, the normal force must also be inclined (Fig. 4.10b). If we introduce a coordinate system with components parallel and perpendicular to the surface, then we can resolve the gravitational force into components along these directions. The component of gravitational force perpendicular to the surface must be opposed by an equal but opposite normal force if the brick is not to sink into the surface. If there are no other forces acting on the brick, then the component of gravitational force parallel to the surface will be unopposed and will cause the brick to slide down the slope. Notice that in this case the component of gravitational force parallel to the surface is the net force. That is, it is the vector sum of $m\mathbf{g}$ and $\mathbf{N}$. The brick will then accelerate down the incline at a rate determined by this net force and the brick's mass.

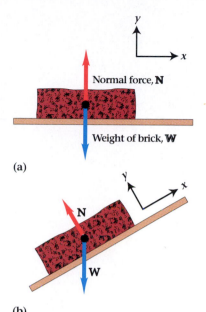

Figure 4.10
The normal force is always perpendicular to the contact surface. (a) The normal force is equal to the weight of the brick. (b) The normal force is less than the weight of the brick.

Example 4.6

A furniture van has a smooth ramp for making deliveries. The ramp makes an angle θ with the horizontal. A large crate of mass m is placed at the top of the ramp. Assuming the ramp is a frictionless plane, what is the acceleration of the crate as it moves down the ramp?

Strategy To find the acceleration of the crate we must first find the net force acting on it. The gravitational force $m\mathbf{g}$ acts downward on the crate (Fig. 4.11). The plane supports the crate with a normal force $\mathbf{N}$ perpendicular to the plane. We assume that the ramp is rigid and that the crate can slide freely over the surface, so there are no other forces acting on the crate. The net force is the vector sum of the two forces $m\mathbf{g}$ and $\mathbf{N}$. We can find the net force and divide by the mass of the crate to obtain the acceleration of the crate.

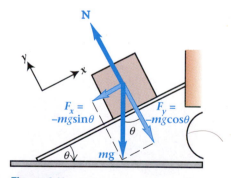

Figure 4.11
Example 4.6: A large crate slides down a slippery incline. The weight of the crate is $m\mathbf{g}$, the normal force is $\mathbf{N}$, which is perpendicular to the surface of the incline.

Solution In this case it is convenient to choose a coordinate system with one axis parallel to the surface of the inclined plane. Then we can resolve the gravitational force vector into components parallel and perpendicular to the surface. We choose the positive x direction to be upward along the plane, and the positive y direction perpendicular to it as indicated in the figure. The resulting components of the gravitational force along the x and y directions are

$$F_x = -mg \sin \theta$$

21 × 9.81 sin 34

8 × 9.81 sin 69 = 73.26739 N

$a = 3.3552176$

$a = 0.3$ $diff = 3.0552176$

parallel to the surface of the plane and

$$F_y = -mg \cos \theta$$

perpendicular to the plane.

There is no acceleration perpendicular to the plane, so the net force in the y direction must be 0:

$$F_{y\,net} = N - mg \cos \theta = 0.$$

The only force in the x direction is a component of the gravitational force,

$$F_{x\,net} = -mg \sin \theta.$$

Because there are no other forces along the x direction, this force is the net force along x that causes the crate to move down the plane. The acceleration is given by Newton's second law:

$$a_x = \frac{F_{x\,net}}{m} = \frac{-mg \sin \theta}{m} = -g \sin \theta.$$

Discussion The negative sign indicates that the crate accelerates to the left in the figure. Notice that the acceleration of the crate depends only on the acceleration of gravity (g) and on the angle of the inclined plane, and not on the mass of the crate. The crate moves with the same acceleration if it contains heavy furniture or if it is empty.

$a = -g \sin 20° =$

4.6 Newton's Third Law

Newton's **third law** of motion may be stated as: **For every action (force) there is a reaction force and the action and reaction forces are equal in magnitude, opposite in direction, and act upon different bodies.** Newton's own words help us understand the third law a little better. "Whatever draws or presses another is as much drawn or pressed by that other. If you press a stone with your finger, the finger is also pressed by the stone. If a horse draws a stone tied to a rope, the horse will be equally drawn back towards the stone; for the distended rope, by the same endeavor to relax or unbend itself, will draw the horse as much toward the stone as it does the stone toward the horse . . ."

Restating the third law in more quantitative terms, we have (Fig. 4.12)

**If body A exerts a force $\mathbf{F}_{AB}$ on body B,
then B exerts a force $\mathbf{F}_{BA}$ on A, so $\mathbf{F}_{AB} = -\mathbf{F}_{BA}$.**

The action and reaction forces of Newton's third law are equal in magnitude, opposite in direction, and *act upon different bodies.* Newton's third law tells us that forces always occur in pairs.

We should point out that when applying Newton's *second* law we consider only one of these forces at a time. Body A is subjected to the applied force $\mathbf{F}_{BA}$. The acceleration of A is thus proportional to $\mathbf{F}_{BA}$ and inversely related to its own mass m_A. Similarly, body B is subjected to the force $\mathbf{F}_{AB}$ and is accelerated by it.

It is important to remember that the third-law action-reaction pair of forces have equal magnitudes. Thus, when a large person pulls on a much smaller per-

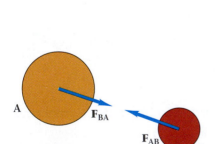

Figure 4.12

Body A exerts a force $\mathbf{F}_{AB}$ on body B, then B exerts a force $\mathbf{F}_{BA}$ on A. The two forces are equal in magnitude and opposite in direction.

(a)

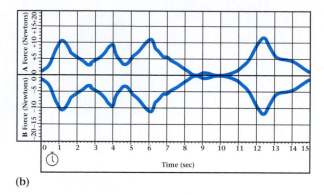

(b)

Figure 4.13
(a) Two people holding force sensors pull against each other. (b) Graph showing that the measured forces on the two people are equal in magnitude and opposite in direction.

son, the action-reaction forces are exactly the same size even though they act in the opposite directions (Fig. 4.13).

Example 4.7

An apple rests on an upturned basket (Fig. 4.14). What forces act on the apple and on the basket? Identify all of the action-reaction pairs.

Strategy There are two types of forces involved here, gravitational and contact forces. We can isolate each object and examine the forces on it. (We will consider the earth to be an inertial frame so that the gravitational force and the weight are the same.)

Solution There are two forces that act on the apple (Fig. 4.14b), a downward gravitational force $\mathbf{F}_{EA}$ of the earth (E) acting on the apple (A) and a contact force $\mathbf{F}_{BA}$, the normal force due to the basket (B) that pushes the apple up. For the apple to be at rest, those two forces must be equal and opposite so that their sum is zero.

$$\mathbf{F}_{EA} + \mathbf{F}_{BA} = 0.$$

The basket is subject to three forces (Fig. 4.14c), the contact force $\mathbf{F}_{AB}$ due to the apple pushing down, the gravitational force $\mathbf{F}_{EB}$ of the earth pulling it down, and a contact force $\mathbf{F}_{GB}$, the normal force of the ground (G) pushing it up. For the basket to be at rest, the sum of these three forces must also be zero.

$$\mathbf{F}_{AB} + \mathbf{F}_{EB} + \mathbf{F}_{GB} = 0.$$

Now, which forces are action-reaction pairs? Start with the apple. The two forces $\mathbf{F}_{EA}$ and $\mathbf{F}_{BA}$ cannot be an action-reaction pair because they act on the same object. Since the force $\mathbf{F}_{EA}$ is due to the earth's attraction for the apple, there must be an equal and opposite reaction force $\mathbf{F}_{AE}$ due to the apple acting on the earth (Fig. 4.14b). So, $\mathbf{F}_{AE} = -\mathbf{F}_{EA}$. The force $\mathbf{F}_{BA}$ of the basket on the apple and the force $\mathbf{F}_{AB}$ of the apple on the basket are an action-reaction pair, so

$$\mathbf{F}_{BA} = -\mathbf{F}_{AB}.$$

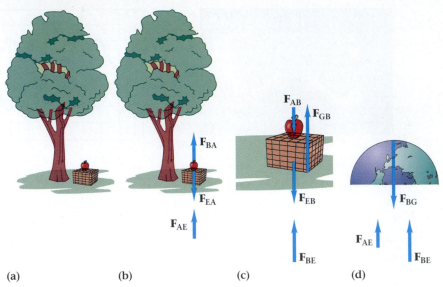

(a) (b) (c) (d)

Figure 4.14

Example 4.7: (a) An apple resting on a basket resting on the ground. (b) Two forces act on the apple and (c) three forces act on the basket. (d) Three forces act on the earth, two gravitational forces upward and a contact force downward.

Similarly, the gravitational force $\mathbf{F}_{EB}$ of the earth acting on the basket and the force $\mathbf{F}_{BE}$ of the basket attracting the earth (Fig. 4.14c) are an action-reaction pair:

$$\mathbf{F}_{EB} = -\mathbf{F}_{BE}.$$

Finally, the reaction force to the contact force $\mathbf{F}_{GB}$ of the earth's surface pushing up on the basket is a force $\mathbf{F}_{BG}$ of the basket pushing down on the earth. Thus

$$\mathbf{F}_{GB} = -\mathbf{F}_{BG}.$$

Discussion Notice that action-reaction forces always act on different bodies. The reaction to the contact force of the ground on the basket is the force of the basket on the ground. Even though the two forces $\mathbf{F}_{EA}$ and $\mathbf{F}_{BA}$ acting on the apple are equal and opposite, they are not an action-reaction pair because they act on the same object, the apple. You can show from the equations here that the vector sum of the three forces acting on the earth is zero as it should be (Fig. 4.14d).

Example 4.8

A 68-kg passenger rides in an elevator that is accelerating upward at 1.0 m/s^2 because of external forces. What is the force exerted by the passenger on the floor of the elevator?

Strategy Only two forces act on the passenger (Fig. 4.15a): the gravitational force $m\mathbf{g}$, pulling down, and the force $\mathbf{F}_{FP}$ of the floor on the passenger, pushing up. The force $\mathbf{F}_{FP}$ of the floor on the passenger and the force $\mathbf{F}_{PF}$ exerted by the passenger on the floor are an action-reaction pair of forces (Fig. 4.15b). From the third law we know that the action-reaction forces are equal and opposite. We can find the force $\mathbf{F}_{FP}$ from Newton's second law by observing that the vector sum of the forces $\mathbf{F}_{FP}$ and $m\mathbf{g}$ is the net force that provides the upward acceleration of the passenger, which is the same as the acceleration of the elevator. From our knowledge of the mass and acceleration of the passenger, we can find the force $\mathbf{F}_{FP}$. Then the force exerted by the passenger on the elevator floor is obtained from the third law.

Solution By Newton's second law, the net force on the passenger is

$$\mathbf{F}_{net} = \mathbf{F}_{FP} + m\mathbf{g} = m\mathbf{a}.$$

Taking into account the directions of the forces and choosing up as the positive direction, the magnitude of the net force is

$$F_{net} = F_{FP} - mg = ma.$$

Upon rearranging, we find the upward force of the floor on the passenger,

$$F_{FP} = ma + mg = m(a + g)$$

$$F = 68 \text{ kg} \times (1.0 + 9.81) \text{ m/s}^2 = 740 \text{ N}.$$

From Newton's third law, the passenger exerts a downward force of 740 N on the floor of the elevator.

Discussion If the passenger were standing on a scale placed on the elevator floor, the 740-N force would be the force read on the scale. This force is the passenger's weight in the reference frame of the elevator. If the elevator were at rest or moving with a constant speed, the scale would read

$$F = mg = 68 \text{ kg} \times 9.81 \text{ m/s}^2 = 670 \text{ N}.$$

If the elevator were accelerating downward, the weight of the passenger would be less than normal:

$$F = m(g - a).$$

You can feel this change in weight when riding in an elevator. If you walk around in the elevator, when it accelerates as it starts to ascend (or comes to a stop during a descent), you will notice how much more effort is needed to walk naturally. On the other hand, as the elevator begins to descend (or comes to a stop when moving upward), you will feel lighter and sense an additional spring in your step as you walk.

According to Example 4.8 and our definition of weight, when an elevator moves with a downward acceleration a, the weight of a passenger decreases as measured in the reference frame of the elevator. If you were riding in the elevator, you would say that your weight was less than normal. If the downward acceleration increased, your weight would also decrease. If the downward acceleration of the elevator reached g, the condition of free fall, your weight measured by a scale would become zero:

$$F = m(g - a) = m(g - g) = 0.$$

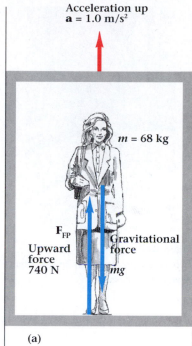

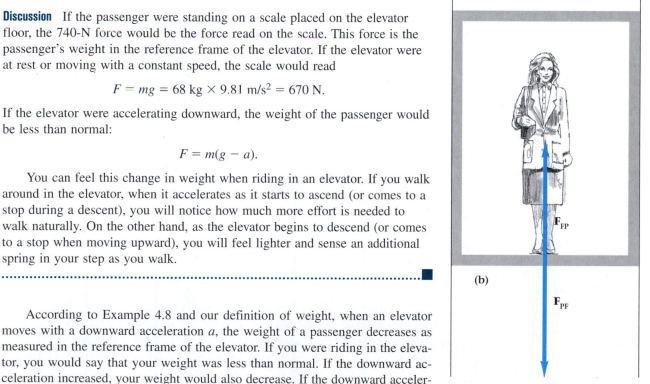

Figure 4.15
Example 4.8: A passenger in an accelerating elevator has an apparent weight different from her real weight.

Figure 4.16
Two astronauts demonstrate the effects of weightlessness.

This condition is called **weightlessness.** Notice that your mass is unchanged. The gravitational force on you is also unchanged, even though your weight becomes zero. Because you and the elevator are both falling with the same acceleration, you are free to float within the elevator. A slight push against the floor sends you to the ceiling. Because there is no net force between you and the floor, the concepts of "up" and "down" with respect to the elevator no longer apply. You are just as comfortable with your head toward the floor or lying on your side as with your head toward the ceiling.

Weightlessness is experienced by the occupants of orbiting artificial satellites like the space shuttle. In that case, the satellite and its occupants are both in free fall, just as in the case of the accelerating elevator. During weightlessness the occupants of artificial satellites move around freely (Fig. 4.16). Remember, the explanation is that the satellite and its occupants have the same acceleration because of gravity. There is no force acting between them, but the force of gravity still acts on all of them.

4.7 Some Applications of Newton's Laws

A useful device for illustrating Newton's second law was devised by George Atwood (1746–1807), nearly a century after the *Principia* appeared. His original device had an elaborate wheel work to reduce frictional forces and a pendulum to measure time. In the simplified diagram (Fig. 4.17a), Atwood's machine is a flexible cord connecting two masses, m_1 and m_2, that runs over a pulley that turns freely without friction. Since the masses of the pulley and the cord are very small in comparison with the masses m_1 and m_2, we can disregard them in our analysis.

We include the force provided by the cord, which we have labeled with the symbol **T**. The magnitude of this force is called the **tension** in the cord. (Recall

Figure 4.17
(a) A simplified version of an Atwood's machine in which two masses are suspended by a cord passing over a frictionless pulley. (b) Free-body diagrams, showing the forces acting on the two single masses, m_1 and m_2. The downward forces are due to the weight of the masses and the upward forces are due to the tension (**T**) in the cord.

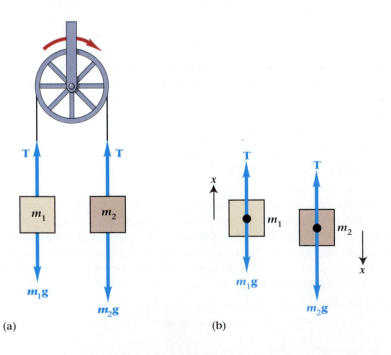

(a)　　　(b)

Newton's example of a horse pulling a stone with a rope, mentioned in the previous section.) We begin by making a separate vector diagram for each body, showing all the forces that act on that body. Such a diagram is called a **free-body diagram.** For example, in Fig. 4.17(b) we see that two forces act upon mass m_2: a downward force m_2g and an upward tension force T provided by the cord. We choose the downward motion of mass m_2 as the positive direction (that is, we assume $m_2 > m_1$). Since m_1 must move up when m_2 moves down, the corresponding positive direction for m_1 is upward. The net downward force on m_2 is therefore

$$F_2 = m_2g - T = m_2a.$$

The acceleration is positive (downward) when m_2g exceeds T. The net upward force on m_1 (as seen in Fig. 4.17b) is

$$F_1 = T - m_1g = m_1a.$$

The acceleration a must be the same for m_1 and m_2, since they are joined by the cord and we assume that the cord does not stretch. The tension T in the cord is the same at each end if we neglect all effects of the pulley.

We now have two equations involving two unknown quantities T and a. Both of these equations must be satisfied at the same time. For this reason they are called *simultaneous equations*. The appendix to this chapter reviews how to solve simultaneous equations. Note that you cannot solve either equation alone for either of the two variables.

Upon rearranging the two equations, we get

$$T = m_2g - m_2a \quad \text{and} \quad T = m_1g + m_1a.$$

▼

Problem-Solving Strategy

Free-body Diagrams

Free-body diagrams show all of the forces acting on an object. When there is more than one object involved, use of free-body diagrams helps us to isolate the forces acting on each object separately. Thus, the free-body diagram becomes a very useful tool for analyzing the motion of a physical situation. The following three steps illustrate the method.

1. Choose the object you wish to isolate and draw it along with whatever geometry and dimensions are important to solving the problem. Show objects as simple particles or blocks and keep your diagrams simple and uncluttered.
2. Draw all forces acting on the object as vector arrows, in approximately correct size and direction. Label all forces clearly.
3. Indicate a coordinate system and show the positive direction of displacement, velocity, or acceleration, depending on the problem. If you resolve vectors into components, mark out the original vector so that you don't count it twice.

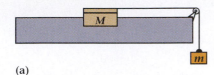

(a)

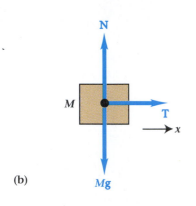

(b)

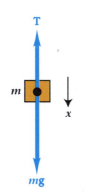

(c)

Figure 4.18

(a) An air-track glider of mass M connected to a small mass m by a light cord passed over a frictionless pulley. (b) A free-body diagram for the glider. The forces are the downward force of gravity $M\mathbf{g}$, the upward support $\mathbf{N}$ due to the air track (equal and opposite to $M\mathbf{g}$), and the horizontal force $\mathbf{T}$ due to the tension in the cord. (c) A free-body diagram for the small mass m.

The right-hand sides of these equations may be set equal and solved to find the acceleration,

$$a = \frac{m_2 - m_1}{m_2 + m_1} g.$$

We can now compute the tension T from the acceleration by inserting the expression for a into either of the equations for T,

$$T = \frac{2m_1 m_2}{m_1 + m_2} g.$$

Notice that when $m_1 = m_2$, the tension T equals the weight mg and there is no acceleration.

As another example of analysis using free-body diagrams, let's consider an air-track glider of large mass M moving on a frictionless air track (Fig. 4.18a). A small mass m is attached to M by a light cord that passes over a frictionless pulley of negligible mass. The forces on M are shown in the free-body diagram of Fig. 4.18(b). The gravitational force Mg is equal and opposite to the supporting force N provided by the air track because there is no vertical acceleration. An unbalanced horizontal force T, exerted by the string, accelerates the mass to the right:

$$T = Ma.$$

The forces acting on the smaller mass m (Fig. 4.18c) are a downward gravitational force mg and the upward force T, the tension in the string:

$$mg - T = ma.$$

Combining these two simultaneous equations, we find that

$$a = \frac{m}{m + M} g.$$

We can find the tension T by substituting for a in either of the force equations above,

$$T = \frac{mM}{m + M} g.$$

The tension T is always less than the downward force of gravity on the hanging mass m, and the acceleration is always less than g.

Example 4.9

Suppose an air-track glider of 1.000-kg mass is connected to a mass of 0.015 kg as in Fig. 4.18(a). What is the acceleration of the glider?

Strategy Because the two masses are joined by the light cord, they must move with identical speed and acceleration. We assume that the cord is so light that its mass may be neglected. Then the only force causing the masses to move is the gravitational force on the mass m. But, because of the cord, that force causes both masses to move.

Solution The acceleration of the glider is obtained from Newton's second law,

$$a = \frac{F_{\text{net}}}{m_{\text{total}}},$$

where F_{net} is given by mg and the total mass is the sum of the masses, $m + M$. Thus

$$a = \frac{mg}{m + M}$$

$$a = \frac{0.015 \text{ kg} \times 9.81 \text{ m/s}^2}{1.015 \text{ kg}} = 0.14 \text{ m/s}^2.$$

■

▼

Problem-Solving Strategy

Newton's Laws

You will need to work problems in order to deepen your understanding of the laws of motion. Developing your skills for problem solving requires practice. To help you solve problems, we summarize here the problem-solving guidelines from Chapter 1 as they apply to problems involving Newton's laws.

1. Read the entire problem carefully. Then read it again, focusing on what you are being told.
2. Draw and label a diagram of the physical situation. Draw a free-body diagram where appropriate. Choose a coordinate system and indicate it on your drawing. Include units, such as meters or kilograms, with the quantities. The diagram is more than a simplified picture, it is part of the solution. In complicated situations, drawing several free-body diagrams separates the problem into manageable pieces so that you can find the appropriate equations.
3. After you understand what is given and after you have labeled the diagram, then tackle the question. Briefly restate the question, perhaps in symbols, on your paper. It may help to make a list of the known quantities given in the problem as well as the unknown quantities being sought.
4. State the basic principles or concepts that apply. Find a mathematical relationship between the known and unknown quantities and write it in the form of an equation, or perhaps several equations.
5. Solve the equation for the unknown quantity (or quantities) so that you have an equation with only the unknown on the left-hand side of the equals sign and all of the known quantities and constants on the right-hand side.
6. Now substitute the numerical values into the equation if the problem has a numerical solution. Include both the numerical value *and* the units for each quantity. Then compute the numerical answer.
7. As a final check you should ask whether your answer is reasonable.

Example 4.10

Suppose that a block of mass M on an inclined plane is joined to a mass m by a cord over a pulley (Fig. 4.19a). The block slides on a frictionless surface and the effects of the pulley are negligible. What are the magnitude and direction of the acceleration of the block if the surface is inclined at 20° and $m = \frac{1}{2}M$?

Strategy We choose a coordinate system for block M with the positive x direction up the ramp, and we resolve the gravitational force on the block into components parallel and perpendicular to the surface of the inclined plane, as we did in Example 4.6. We choose a coordinate system for the mass m with the positive x coordinate directed downward so that when m moves in the positive x direction, M will also move in the positive x direction.

For block M, the perpendicular component of the weight is balanced by the normal force N of the plane supporting the block (Fig. 4.19b):

$$N - Mg \cos \theta = 0.$$

Two forces act on the block in the direction parallel to the plane: the gravitational component $-Mg \sin \theta$ and the tension force T due to the cord. Applying Newton's second law to the mass of the block, M, we find the net force along x to be

$$F_{\text{net(block)}} = T - Mg \sin \theta = Ma.$$

From the free-body diagram of the hanging mass (Fig. 4.19c) the net force is

$$F_{\text{net(mass } m)} = mg - T = ma.$$

Figure 4.19
Example 4.10: (a) External forces acting on mass M cause it to move on a frictionless surface inclined at angle θ. (b), (c) Free-body diagrams for masses M and m, respectively.

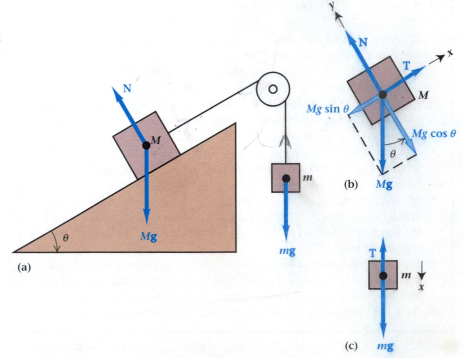

(a)

(b)

(c)

Handwritten annotations:

$Mg - T = Ma$

$3.6 \times 9.81 - T = 3.6 \times 3.27$

$T = 35.316 - 11.772$

$=$

where a positive net force corresponds to the mass m accelerating downward and M accelerating up the incline. We now have two equations that must be solved simultaneously for the two unknowns a and T.

Solution   We can add these two equations to eliminate T, resulting in an equation for the acceleration in terms of the masses and the gravitational acceleration g:

$$mg - Mg \sin \theta = (m + M)a.$$

The acceleration is then

$$a = \frac{(m - M \sin \theta)g}{m + M}.$$

If we insert the values $m = \frac{1}{2}M$ and $\theta = 20°$, we get

$$a = \frac{Mg(\frac{1}{2} - \sin 20°)}{\frac{3}{2}M},$$

[handwritten: $a = \dfrac{(3.6 - 8.0 \sin 69)\,9.81}{11.6}$

$= -3.27167$]

or

$$a = \tfrac{2}{3}g(0.5 - 0.34) = 0.11\,g = 1.0\ \text{m/s}^2.$$

The direction of the acceleration of block M is up the incline.

Example 4.11

Two blocks of masses m_1 and m_2 are connected by a light string passing over a pulley (Fig. 4.20). The blocks are at rest on inclined frictionless surfaces, and the effects of the pulley are negligible. Which way and how far do the blocks move in 0.750 s after being released?

Strategy   We can draw free-body diagrams for the two blocks and use them to find an equation for their acceleration. Once we have the acceleration, we can use the kinematic relationship between acceleration, time, and distance to determine how far they slide in 0.750 s.

Starting with block m_1, we resolve the gravitational force into components parallel and perpendicular to the inclined plane on which m_1 moves, just as we did in the previous example. We choose a reference coordinate system as indicated in Fig. 4.20(b), with the positive x axis directed up the left-hand plane. The component of the weight perpendicular to the plane is balanced by the normal force N_1. The component of the weight parallel to the plane is directed down the plane, and the tension T in the string is directed up the plane. Applying Newton's second law to m_1, we find the net force to be along the x axis,

$$F_{1\text{net}} = T - m_1g \sin \theta_1 = m_1a.$$

For the second block, we see again that the normal component is equal and opposite to the component of the weight perpendicular to the surface (Fig. 4.20c). Using the indicated coordinate system so that motion of the block m_2 down the plane is positive, we see that the component of the weight parallel to the plane is in the positive x direction and the tension T in the string is in the negative x direction. Applying Newton's second law to

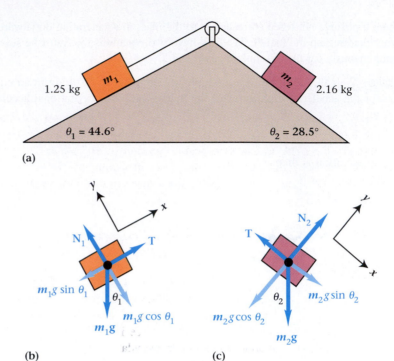

Figure 4.20
Example 4.11: (a) Two blocks rest on frictionless surfaces. (b), (c) Free-body diagrams for the two blocks.

m_2, we find the net force to be

$$F_{2\text{net}} = m_2 g \sin \theta_2 - T = m_2 a.$$

We now have two equations with two unknowns. We can solve them simultaneously to find the acceleration a.

Solution We eliminate T by adding the two equations from the free-body analysis to get

$$(m_2 \sin \theta_2 - m_1 \sin \theta_1)g = (m_1 + m_2)a.$$

The acceleration then becomes

$$a = \frac{(m_2 \sin \theta_2 - m_1 \sin \theta_1)g}{(m_1 + m_2)}.$$

We now insert the values $m_1 = 1.25$ kg, $m_2 = 2.16$ kg, $\theta_1 = 44.6°$, and $\theta_2 = 28.5°$ to get

$$a = \frac{[(2.16 \text{ kg})(\sin 28.5°) - (1.25 \text{ kg})(\sin 44.6°)]9.81 \text{ m/s}^2}{2.16 \text{ kg} + 1.25 \text{ kg}}$$

$$a = +0.440 \text{ m/s}^2.$$

The positive value of the acceleration tells us that the motion of the two blocks is in the positive x direction (to the right in the diagram). That is, block m_1 slides up its incline while block m_2 moves down.

Now that we know the acceleration, we can compute the distance the blocks move in the time $t = 0.750$ s by applying the kinematics expression

$$x = x_0 + v_0 t + \tfrac{1}{2}at^2$$

to either block. Taking mass m_1, we can choose the initial position to be $x_0 = 0$, and since they start from rest, $v_0 = 0$. The distance traveled becomes

$$x = \tfrac{1}{2}at^2 = \tfrac{1}{2}(+0.440 \text{ m/s}^2)(0.750 \text{ s})^2 = +0.124 \text{ m}.$$

Discussion The positive value for x indicates that the motion of the mass m_1 is up the plane, as we expected from the sign of the acceleration. In making our free-body diagrams, we arbitrarily chose the positive direction for m_1 to be up the plane on the left. We then chose the positive x direction for m_2 to be down the plane on the right, so that their positive motions were both to the right. Finally, with careful attention to these directions in writing down the equations and with careful algebra, we computed the acceleration and the displacement, complete with sign.

| 4.8 | **Friction** |

So far we have been careful to ignore the effects of friction. However, when one object slides over the surface of another, its motion is always opposed by a retarding force that resists this motion. This force is called **friction.** Frictional forces are especially important to us in our everyday lives, for without them we could not walk or hold things with our hands; cars would be unable to start or stop; nails and screws would be useless. We first examine frictional forces and show how to work problems that include friction. Then we present a description of the causes of friction. Frictional forces are not fundamental forces like gravity or electromagnetism, but arise as reaction to other applied forces.

Consider the motion of a solid object in contact with a horizontal surface. The object might be a brick on the floor or a telephone on a tabletop (Fig. 4.21). Initially the telephone of weight **w** rests on the horizontal surface. If we pull on the telephone's cord with a small horizontal force **T** parallel to the surface, the telephone does not slide; instead it remains at rest. According to Newton's second law, there must be another force acting on the telephone that is equal and opposite to **T** so that the net force is zero and the telephone remains stationary. This force is the frictional force **F**$_{fr}$ exerted on the telephone by the surface. If **T** is made smaller, the frictional force must also decrease so that the two remain equal but opposite. When the applied force **T** becomes large enough, the telephone begins to slide in the direction of the applied force. The frictional force is still present and directed opposite to the applied force, but is no longer equal in magnitude to **T**. The net difference in these forces causes the telephone to accelerate.

The general principles of frictional behavior have been known for nearly 500 years. (1) For objects in relative motion—that is, sliding or rolling—the force of friction always acts in a direction opposite to the direction of motion. (2) The frictional force is proportional to the perpendicular (or normal) force between the two surfaces in contact. (3) For solid objects, the frictional force is approximately independent of the area of contact between the surfaces. (4) The frictional force depends on the particular materials that make up the surfaces. These empirical rules usually hold to a good approximation. However, they are not "laws" in the same sense as Newton's laws; they simply sum up the approximate behavior under some simple conditions. Yet, in some situations the usual rules of friction do not apply. We consider one such situation on p. 127.

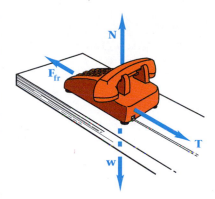

Figure 4.21
A telephone of weight **w** rests on a horizontal surface. When a horizontal force **T** is applied through its cord, the telephone's motion is opposed by a frictional force **F**$_{fr}$.

Let's put these statements about friction in quantitative form. For the static case, in which there is no relative motion between the surfaces, the magnitude of the frictional force is

$$F_{\text{fr}} \leq \mu N, \qquad [\text{static friction}] \qquad (4.7a)$$

where N is the magnitude of the normal force and the proportionality constant μ is the (dimensionless) **coefficient of friction.** The value of μ depends on the objects involved and on the condition of their surfaces. Table 4.4 lists typical values of μ for a few cases.

Equation (4.7a) actually sets an upper limit for the frictional force $\mathbf{F}_{\text{fr}}$ since, as we have seen, when an object is at rest the frictional force must be equal and opposite to the applied tangential force, that is, to the force parallel to the surface. If no tangential force is applied, there is no opposing frictional force. As the tangential force increases, the frictional force increases to oppose it. Ultimately, the frictional force reaches the maximum value expressed by Eq. (4.7a). If a still greater external force is applied, the object no longer remains at rest but begins to slide. For this case, in which the surfaces slide against each other, the magnitude of the frictional force is

$$F_{\text{fr}} = \mu N \qquad [\text{kinetic friction}] \qquad (4.7b)$$

Notice that the expression of Eq. (4.7b) is an equality and not an inequality as given in Eq. (4.7a). Although the coefficient of friction is not truly constant, Eqs. (4.7a) and (4.7b) are good empirical rules for approximating the force needed in many practical situations.

The magnitude of the frictional force depends on whether the two surfaces are in relative motion. For identical surface conditions and constant pressures, the coefficient of friction generally decreases slowly with increasing relative speed. If the normal force or the speed becomes too large, Eq. (4.7b) no longer applies. It is important to realize that an empirical law such as this has its limitations, beyond which it does not work.

Sometimes two frictional coefficients are given: μ_s for static friction and μ_k for kinetic, or sliding, friction. However, because the coefficient of friction depends on speed, and because it varies greatly as a result of conditions such as

Table 4.4	Coefficients of Friction	
Materials	**Conditions**	μ
Glass on glass	Clean	0.9–1.0
Wood on wood	Clean and dry	0.25–0.5
Wood on wood	Wet	0.2
Steel on steel	Clean	0.58
Steel on steel	Motor oil lubricant	0.2
Rubber on solids	Dry	1–4
Teflon on steel	Clean	0.04
Waxed hickory on dry snow		0.03–0.06
Brass on ice		0.02–0.08

Values are approximate. Frictional coefficients vary with surface conditions and cleanliness.

surface moisture, cleanliness, and wear, such coefficients are poorly known and not very reproducible. For these reasons we have not drawn a distinction between static and kinetic coefficients in the examples and in the problems. You should bear in mind that the values of the coefficients given in the table are only approximate. Ordinarily, for identical surface conditions, μ is slightly greater for static friction than it is for sliding (kinetic) friction. The situation is very complicated. As Richard Feynman observed,

> Many people believe that the friction to be overcome to get something started (static friction) exceeds the force required to keep it sliding (sliding friction), but with dry metals it is very hard to show any difference. The opinion probably arises from experiences where small bits of oil or lubricant are present, or where blocks, for example, are supported by springs or other flexible supports so that they appear to bind.*

Example 4.12

A horizontal force **T** of 100 N is applied to a box of books of mass 20 kg resting on a wooden table (Fig. 4.22). Does the box slide if the coefficient of friction of the box on the table is 0.40? If the box moves, find its acceleration.

Strategy   We can compute the maximum frictional force and compare it to the horizontal force **T**. If the maximum friction force exceeds T, the box stays at rest. If the friction force is less than T, the box will accelerate.

Solution   The normal force between the box and the table is just equal to the weight of the box, mg = 196 N. The maximum frictional force is

$$F_{fr} = \mu N = 0.40 \times 196 \text{ N} \approx 78 \text{ N}.$$

This force is less than the applied force, so the box slides in the direction of **T**.

The box is accelerated by the net force $F_{net} = T - F_{fr}$. Thus from Newton's second law we get

$$a = \frac{F_{net}}{m} = \frac{(T - F_{fr})}{m}$$

$$a = \frac{(100 - 78)\text{N}}{20 \text{ kg}} = 1.1 \text{ m/s}^2.$$

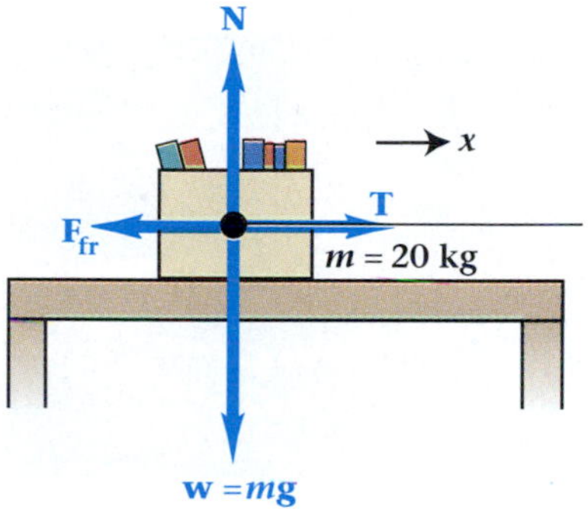

Figure 4.22
Example 4.12: A box of books moves only if the applied force **T** is greater than the force of friction between the books and the table.

What are the causes of frictional forces and how do we know about them? It is often supposed that frictional effects originate in the roughness of the surfaces in contact with one another. In fact, experiments have shown that friction does not generally increase with roughness. For example, two pieces of smooth, flat glass show much more frictional drag than two pieces of rough, ground glass. The frictional forces arise primarily from molecular forces in the regions of real contact. Thus, friction is determined not so much by the effect of the roughness

*R. P. Feynman, R. P. Leighton, and M. Sands, *The Feynman Lectures on Physics* (Reading, Mass.: Addison-Wesley, 1964), Vol. I, p. 12-5.

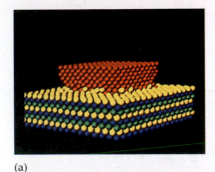

(a)

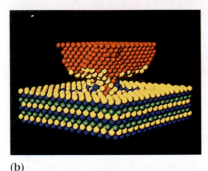

(b)

(c)

Figure 4.23

Friction at the atomic level as modeled by computer simulation for a nickel tip (red) and a surface of gold. (a) The tip is pressed into the surface. (b) The tip is slowly raised, forming an intermetallic bond. (c) Gold atoms continue to cling to the tip after it is well above the surface.

Master the Concept

Friction and Newton's Laws

Question: According to Newton's laws, an external force is needed to stop a car when the brakes are applied. Where is this force and what is its origin? Be careful to distinguish between internal forces and external forces.

Answer: You know that a car is slowed by applying pressure to the brake pedal causing a frictional force between the brake pads and the wheel. But these forces are internal forces and cannot stop the car. If the car is rolling smoothly in a straight line with no brake applied, the forces between the tires and the road are essentially only normal forces. The operation of the brakes retards the rotation of the wheels resulting in frictional forces between the tires and the road. These friction forces are parallel to the road surface. The tires push forward on the road and the road pushes backward on the tires. This behavior not only is true for skidding when the wheels are locked, but also is true even when the wheels are rolling. This backward or retarding force of the road on the tires is an external force acting on the car. Except for air resistance, there are no other external forces along the line of motion, so the net force opposes the forward motion of the car causing it to slow to a stop.

or smoothness of the surfaces as by the molecular forces in the area of actual contact.

Studies of blocks sliding on blocks show that friction is generally independent of the contact area of the blocks. This area, which is determined by multiplying length times width, is more properly thought of as the apparent area of coverage. Because objects that appear smooth may be microscopically rough, the apparent area of coverage is usually much larger than the actual area of contact. As the normal force increases, the actual contact area also increases because of deformations of the two surfaces at their interface. It is this increase in contact area with increasing load that gives the apparent connection between friction and normal force.

In dealing with surfaces that are easily deformed, an increase in apparent area may closely approximate an increase in the actual area of contact. For this reason the width (and hence area) of tires can make a significant difference in how an automobile drives. This effect (of dependence on area) stands in sharp contrast to that seen in experiments with metal blocks sliding over metal surfaces, where changes in the apparent area of the block make little difference in the observed friction.

Whenever you measure the friction of one metal block sliding against another block of the same metal, it is not really a case of pure metal sliding on pure metal. The surfaces of each block contain oxides and other impurities. If the surfaces are carefully cleaned in a high vacuum and are touched together, they will stick, forming a cold weld (Fig. 4.23). This surprising result happens when microscopically clean surfaces touch, since the bonding at the interface is then the same as it is anywhere else within the metal. Thus, the friction that we normally

THE FRICTION OF AUTOMOBILE TIRES

How do tires affect your safety when you drive your car along the highway? What factors help to prevent skidding and allow you to control your car when turning and stopping? What does friction have to do with this?

The tread pattern of rubber tires plays a major role in determining their friction, or skid resistance. Under dry conditions on paved roads, a smooth tire gives better traction than a grooved or patterned tread because a larger area of contact is available to develop the frictional forces. For this reason, the tires used for auto racing on the tracks at Darlington, Indianapolis, Talladega, and elsewhere have a smooth surface with no tread design (Fig. B4.3). Unfortunately, a smooth tire develops very little traction under wet conditions because the frictional mechanism is reduced by a lubricating film of water between the tire and the road. A patterned tire provides grooves or channels into which the water can squeeze as the tire rolls along the road, thus again providing a region of direct contact between tire and road (Fig. B4.4). A patterned tire gives typical dry and wet frictional coefficients of about 0.7 and 0.4, respectively. These values represent a compromise between the extreme values of about 0.9 (dry) and 0.1 (wet) obtained with a smooth tire.

Classical friction theory must be modified for tires because of their structural flexibility and the stretch of the tread rubber. Instead of depending solely on the coefficient of friction at the tire-road interface (which is determined by the nature of the road surface and the tread rubber compound), maximum stopping ability also depends on the resistance of the tread to tearing under the forces that occur during braking.

When a car is braked to a hard stop on a dry road, the maximum frictional force developed can be greater than the strength of the tread. The result is that instead of the tire

Figure B4.4   Tires used on racing cars driven only on dry tracks have a solid contact area, like that of the race tires shown. Grooved tires designed for general use provide traction under wet conditions by channeling water away from the tire. Because it has no similar tread pattern, the racing tire cannot be driven on a wet track.

merely sliding along the road, rubber is torn off the tread at the tire-road interface. Undoubtedly the tread resistance to this tearing is a combination of the rubber strength and the grooves and slots that make up the tread design.

The weight of the car is unevenly distributed over the tire-road contact area, creating areas of high and low pressure. (This is much like what you feel when you step on a pebble while walking in thin-soled shoes.) The resistance of the tread to tearing increases in the areas of higher pressure, where the tread is more compressed, causing an effective increase in traction.

Further, the size of the contact area is very important in car tires because the traction is dynamic rather than static; that is, it changes as the tire rolls along. The maximum coefficient of friction can occur anywhere in the contact area, so that the greater the area, the greater the likelihood of maximum traction. Thus, under identical load and on the same dry surface, the wider tire has a greater contact area and develops higher traction, resulting in greater stopping ability.

Next time you need to buy tires, think about what kind of climate you live in, what kind of roads you drive on, and what speeds you drive. If you live in a region with good paved roads, you may not need tires with extra tread. If you drive in areas with mud or snow, you need a tread designed for those conditions.

Figure B4.3   Race cars driven on the superspeedways are equipped with wide, smooth tires known as "racing slicks."

Figure 4.24
(a) Hovercraft vehicles use powerful blowers to maintain a cushion of air beneath them to float over land and sea with little frictional drag. (b) Fan *A* provides the air cushion and fan *B* provides the horizontal thrust.

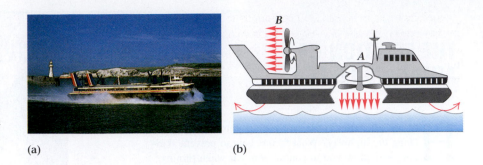

(a)

(b)

observe is due principally to the surface layer of contaminants that is always present and that serves to reduce the intermolecular forces at the interface.

A thin layer of oil is often placed between surfaces to make motion easier. This idea is not new; lubrication with fats and oils has been common for thousands of years. The principal effect of a lubricating film is to diminish the attractive forces of the sliding surfaces and thereby reduce their cohesion. Even an invisible layer of oil can reduce the friction of dry surfaces by several orders of magnitude.

A different effect occurs when the lubricating film separating two sliding surfaces is thick compared to the dimensions of molecules. Then the friction depends on the properties of the film rather than on the surfaces. A striking example of this is the use of a thin layer of air to support heavy objects. Because the friction of the air is so small, the objects appear to glide along almost without friction. This effect has been put to use in the air track of Fig. 4.5. Its application to air bearings has made possible the modern high-speed drills used by dentists. Hovercraft vehicles that float on a cushion of air are in regular service across the English Channel (Fig. 4.24).

4.9   Static Equilibrium

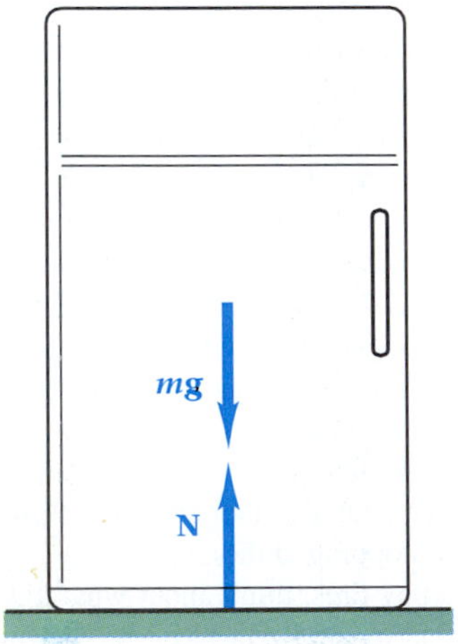

We have seen that a body subjected to a net force has an acceleration proportional to that force. But what if the vector sum of the forces acting on the body is zero? This is the condition of translational **equilibrium,** a state of motion in which the velocity of the body is constant. If the body is in motion with constant velocity, then we say it is in **dynamic equilibrium.** If the velocity of the body is zero, then the body is at rest and is said to be in **static equilibrium.**

In addition to translational motion, a body can also have rotational motion. However, these two types of motion can be separated and treated independently. We shall defer discussion of rotational motion and rotational equilibrium until Chapter 9. Our immediate concern is with static equilibrium of translational motion.

We begin by examining the forces on a refrigerator resting on a horizontal floor (Fig. 4.25). We know that for a refrigerator of mass m, a gravitational force $m\mathbf{g}$ pulls down on it, as illustrated in the figure. But from observation and Newton's laws, we have come to expect the refrigerator to remain at rest. Therefore, another force must be present, equal and opposite to the gravitational force. If this were not so, a net unbalanced force would be acting on the refrigerator and

*m***g**

N

Figure 4.25
A refrigerator rests in static equilibrium on a horizontal floor. Therefore, its weight must be balanced by an upward normal force.

it would accelerate. This equal and opposite force is provided by the floor, which pushes up on the refrigerator with a normal force **N**.

The condition of translational equilibrium for any object is given mathematically by the statement that the vector sum of all forces acting on that object must be zero. That is,

$$\mathbf{F}_{net} = \mathbf{F}_1 + \mathbf{F}_2 + \mathbf{F}_3 + \cdots = 0.$$

We can represent this summation more compactly as

$$\mathbf{F}_{net} = \sum_i \mathbf{F}_i = 0, \tag{4.8}$$

where i stands for the indices 1, 2, 3, 4, . . . and the symbol Σ represents the sum over all values of i. Here the forces $\mathbf{F}_i$ all act on the same object. This study of objects and forces in equilibrium is a special case of dynamics called *statics*. The vector Eq. (4.8) can be easily handled by resolving the forces into components along the vertical (y) direction and the horizontal (x) direction. If the forces are in equilibrium, then their x components and their y components must also be in equilibrium separately. The resulting equilibrium conditions are

$$\sum_i F_{ix} = 0 \quad \text{and} \quad \sum_i F_{iy} = 0.$$

Example 4.13

A child sits on a sled that rests on a snow-covered hill making an angle θ with the horizontal. If the coefficient of friction is 0.10, what is the maximum angle at which the sled remains at rest? (Assume the hill can be approximated by an inclined plane.)

Strategy We begin by making a sketch of the situation (Fig. 4.26). Then we treat the child and sled as a single free body and make a diagram showing the three forces that act on this "free body." These forces are the weight acting vertically downward, the normal force due to the surface and the frictional force that acts parallel to the surface and opposite to the direction of motion. At the maximum angle at which the sled remains stationary, the frictional force is maximum and equals the coefficient of friction times the normal force. We can use the equilibrium equations in directions parallel and perpendicular to the incline to express the friction force and the normal force in terms of the sled's weight and the angle of the hill. Then by combining these equations we can find a simple expression for the angle of the hill in terms of the coefficient of friction.

Solution We can resolve the weight of the sled plus child, $m\mathbf{g}$, into components parallel and perpendicular to the incline (Fig. 4.26). The perpendicular component of the weight is equal and opposite to the normal force **N**. The parallel component F_x down the incline is opposed by the friction force $\mathbf{F}_{fr}$. From Fig. 4.26 we see that the relations between **N**, F_x, and $m\mathbf{g}$ are

$$N = mg \cos \theta \text{ and } F_x = mg \sin \theta.$$

For the sled to be at rest, the condition of equilibrium requires that the vector sum of the forces must be zero in each direction. Thus, the magnitude

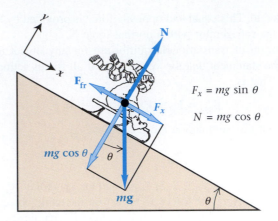

$$F_x = mg \sin \theta$$

$$N = mg \cos \theta$$

Figure 4.26
Example 4.13: How steep can the hill be before the sled slides down?

of $\mathbf{F}_x$ must be equal to the magnitude of the force of friction $\mathbf{F}_{\mathrm{fr}}$. The maximum value for F_{fr} is μN. Thus

$$F_x = \mu N$$

at the maximum angle. Inserting the relations for F_x and N into the equation $F_x = \mu N$, we obtain

$$F_x = mg \sin \theta = \mu N = \mu mg \cos \theta,$$

or

$$\mu = \tan \theta.$$

When $\mu = 0.10$, $\theta = 5.7°$.

For angles less than $5.7°$, the sled remains stationary on the incline, regardless of its weight. For angles greater than $5.7°$, the sled slides down.

Example 4.14

A lantern of mass m is suspended by a string that is joined to two other strings as shown in Fig. 4.27(a). What is the tension in each string if they make equal angles of $35°$ from the support beam as shown? Ignore the mass of the string.

Strategy Since we assume that the lantern and strings are at rest, we have static equilibrium. We can apply the equilibrium condition, Eq. (4.8), to obtain three equations containing the three unknown tensions.

Solution First make a free-body diagram of the mass m, which is acted on by a downward gravitational force $m\mathbf{g}$ and by an upward force of tension in the string $\mathbf{T}_1$ (Fig. 4.27b). For the lantern to be in static equilibrium,

$$T_1 = mg.$$

This equation gives the tension in the lower string, but how do we find the relationships for the other strings? The answer is to make a free-body diagram for the knot connecting the three strings (Fig. 4.27c). At the knot, the tension $\mathbf{T}_1$ in the lower string is downward. The tensions in the other two strings are $\mathbf{T}_2$ and $\mathbf{T}_3$ in the directions indicated. According to Eq. (4.8), the condition for the equilibrium can be expressed as

$$\mathbf{T}_1 + \mathbf{T}_2 + \mathbf{T}_3 = 0.$$

Treating the horizontal components first, we get

$$\Sigma\, T_x = T_{2x} + T_{3x} = 0.$$

Taking the positive direction to the right, we see that

$$T_{3x} = T_3 \cos 35° = 0.819 T_3$$

and

$$T_{2x} = -T_2 \cos 35° = -0.819 T_2.$$

Upon inserting these values in the equilibrium equation, we find that $T_2 = T_3$, something we might have suspected from the symmetry of the situation.

We now have enough information to evaluate T_2 (and T_3) in terms of T_1 by summing the vertical components in equilibrium:

$$\Sigma\, T_y = T_{1y} + T_{2y} + T_{3y} = 0.$$

If we take the positive direction to be upward, then

$$T_{3y} = T_3 \sin 35° = 0.574 T_3,$$
$$T_{2y} = T_2 \sin 35° = 0.574 T_2 = 0.574 T_3,$$

and

$$T_1 = -mg,$$

so

$$\Sigma\, T_y = -mg + 0.574 T_3 + 0.574 T_3 = 0.$$

This gives

$$T_3 = T_2 = \frac{mg}{1.15} = 0.87\, mg.$$

In summary, the three tensions are $T_1 = mg$, $T_2 = 0.87\, mg$, and $T_3 = 0.87\, mg$.

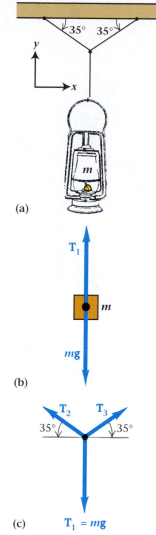

(a)

(b)

(c) $\mathbf{T}_1 = mg$

Figure 4.27
Example 4.14: (a) A lantern of mass m suspended by one string joined to two other strings. (b) A free-body diagram for the mass m. (c) A free-body diagram for the knot joining the strings.

Example 4.15

A child of mass M sits in a light swing suspended by a rope of negligible mass. His sister pushes him forward by a horizontal force until the rope makes an angle θ with the vertical (Fig. 4.28). What is the tension in the rope and how much horizontal force is required to hold the child in that position?

Solution We solve the problem by essentially the same method used in Example 4.14. First we sum the vertical components of force,

$$\Sigma\, F_y = T \cos \theta - Mg = 0.$$

There are only two vertical components, since the force $\mathbf{F}$ is purely horizontal. We obtain the tension immediately as

$$T = \frac{Mg}{\cos \theta}.$$

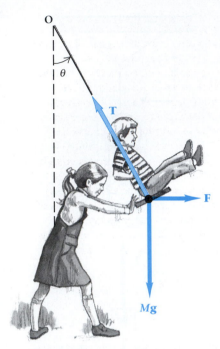

Figure 4.28
Example 4.15: A horizontal force **F** pushes a child in a swing until the rope makes an angle θ with the vertical.

The condition of equilibrium for the horizontal forces gives

$$\Sigma F_x = F - T \sin \theta = 0,$$

or

$$F = T \sin \theta.$$

The horizontal force F can be expressed in terms of Mg by inserting the value for T found above:

$$F = \frac{Mg \sin \theta}{\cos \theta} = Mg \tan \theta.$$

Discussion Notice that for $\theta = 0$, $F = 0$. For $\theta = 45°$, the force becomes $F = Mg$. As θ approaches $90°$, F must increase toward infinity; that is, a purely horizontal force can never make the rope become perfectly horizontal as long as it has to support the downward weight.

4.10 The Laws of Motion as a Whole

We have introduced Newton's laws and some of their limitations. But the real content of these laws is that forces have some independent properties in addition to that expressed in the law $F = ma$. Not only can we use the laws to calculate acceleration when a given force acts on a mass, but we can use Newton's laws to investigate the forces observed in nature. By studying accelerations, we can find how forces depend on other quantities, such as distance. In this way Newton's laws become a tool for understanding nature.

In Newton's *Principia* the laws of motion are in a section called "Axioms, or Laws of Motion." As with axioms in geometry, we may postulate a set of axioms for dynamics that give rise to an abstractly satisfactory scheme of dynamics, just so long as the axioms are not contradictory. However, Newton wanted to explain nature as it was observed. This desire determined his choice of axioms. Strictly speaking, the success of Newtonian dynamics rests not on verification of individual axioms, but on the success of the entire scheme in predicting what we observe.

In this chapter, "what we observe" has meant "ordinary-sized things moving with ordinary velocities." If, instead, we want to explain observations that include a much wider range of sizes and velocities, alternative or modified axioms are needed. In particular, if the range of observed velocities is to include those comparable to the velocity of light, then we need the postulates of Einstein's relativity (Chapter 25). If we go to small dimensions comparable to atomic sizes, then the postulates of quantum mechanics are required (Chapter 28). Figure 4.29 illustrates the domain of Newtonian physics. The overall success of Newtonian dynamics in its proper domain is evident from the great variety of situations covered and the enormous range of sizes, from the motion of the solar system to the motion of helium atoms in a gas.

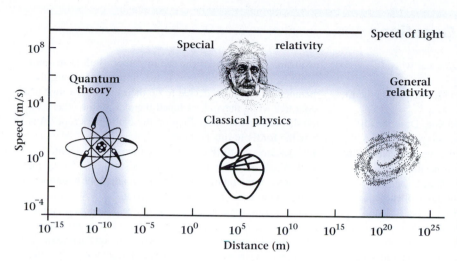

Figure 4.29
(A speed-distance diagram illustrating the range of applicability of Newtonian mechanics. (Note that the scales chosen are not linear, but logarithmic.) The laws of classical physics (center area) are consistent with observations of everyday life. However, when dealing with very small or very large distances or with very high speeds, these laws do not adequately describe what we observe. In those regions, we need the laws of quantum mechanics and relativity instead of Newton's laws to describe and predict the physical observations.

Summary

Useful Concepts

■ Newton's laws tell us that a body in motion stays in motion at constant velocity along a straight line unless acted upon by an external force; that a net force applied to a body causes it to change its motion by accelerating, and similarly, an accelerating body does so because a net force is applied; and that every action has an equal and opposite reaction, applied to different bodies. The mathematical statements of Newton's laws of motion are usually given as follows:

1. A body has a constant velocity unless there is a net force acting on it.
2. The rate of change of momentum with time is proportional to the net applied force and is in the same direction,

$$\frac{\Delta(m\mathbf{v})}{\Delta t} = \Sigma \mathbf{F}.$$

For the case of constant mass, Newton's second law is

$$\Sigma \mathbf{F} = m\mathbf{a}.$$

3. If a body A exerts a force $\mathbf{F}_{AB}$ on a body B, then B exerts a force $\mathbf{F}_{BA}$ on A, so that $\mathbf{F}_{AB} = -\mathbf{F}_{BA}$.

■ One newton is the force acting on one kilogram to give it an acceleration of one meter/second².

■ The weight of an object is proportional to its mass and to the free-fall acceleration,

$$\mathbf{w} = m\mathbf{g}.$$

■ The frictional force of one object sliding on another is

$$F_{\text{fr(static friction)}} \leq \mu N; \qquad F_{\text{fr(kinetic friction)}} = \mu N$$

■ For a body to be in translational equilibrium, the sum of all the forces acting on it must be zero:

$$\Sigma \mathbf{F}_i = 0.$$

Important Terms

You should be able to write the definition or meaning of each of the following terms:

Newtonian mechanics	normal force
inverse-square force	weightlessness
force	tension
net force	free-body diagram
inertia	friction
inertial reference frame	coefficient of friction
momentum	equilibrium
mass	static equilibrium
newton	dynamic equilibrium
weight	

Conceptual Questions

4.1 Why do packages slide off the seat of a car when the brakes are applied quickly and forcefully?

4.2 A book rests motionless on a table. Does that mean there are no forces acting upon it?

4.3 A child sits in a swing that is not moving. Describe all forces present and identify all action-reaction pairs.

4.4 A tennis ball thrown against the wall bounces back toward the thrower. Where does the force come from that sends it back?

4.5 Imagine a large adult and a small child on roller skates. Describe their motions if they push off each other.

4.6 Given the existence of a standard mass, devise a way of dynamically comparing other masses with the standard.

4.7 Suspend a heavy weight from a light string and attach a similar string below it (Fig. 4.30). If you pull on the lower string with steadily increasing force, the upper string will break; if you pull the lower string with a jerk, the lower string will break. Explain both cases.

Figure 4.30
Question 4.7.

4.8 Two teams of students are having a tug-of-war. The rope passes through a small hole in a high fence that separates the two teams. Neither team can see the other. Both teams pull mightily, but neither budges. As lunchtime approaches, the members of one team decide to tie their end of the rope to a stout tree while they take a lunch break. Can the other team tell that the first is not pulling on the motionless rope? Analyze the forces in this problem.

4.9 Describe the difference between mass and weight.

4.10 Would a 5.0-kg mass on the earth still be a 5.0-kg mass if it were on the moon? If the mass weighed 49 N on the earth, would it weigh 49 N on the moon?

4.11 A spring scale is used to weigh beans on an elevator. How will the readings for a given amount of beans change when the elevator is (a) going down with constant velocity, (b) moving with a constant downward acceleration less than g, (c) moving upward with a constant velocity, (d) accelerating upward with an acceleration a?

4.12 A man goes over Niagara Falls in a barrel with windows in the side. During the descent, the man takes out an apple, holds it up in front of his face and releases it. Describe what is seen by (a) an observer on the bank looking through the window and (b) the man in the barrel.

4.13 A person on an upward-moving elevator is throwing darts at a target on the elevator wall. How should she aim the dart if the elevator has (a) constant velocity, (b) constant upward acceleration, (c) constant downward acceleration?

4.14 Describe some of your everyday activities that would be seriously hindered, if not impossible, if there were no friction.

4.15 According to Newton's laws, an external force is needed to stop a car when the brakes are applied. Where is this force and what is its origin? Be careful to distinguish between internal forces and external forces.

4.16 When a moving car is slowed to a stop with its brakes, what is the direction of its acceleration vector? Describe the path of a ball dropped by a passenger during the time the car is slowing down.

Problems

Section 4.2 What Is a Force?

4.1 A man pushing a lawn mower exerts a force of 436 N on the handle (Fig. 4.31). The handle makes an angle of 40° with the horizontal. What is the horizontal force applied to the mower?

4.2 Two horses pull horizontally on ropes attached to a tree stump (Fig. 4.32). Each horse pulls with a force of magnitude F. If the resultant force is 1.79 F, what is the angle between the two ropes?

4.3 Two horses pull horizontally on ropes attached to a tree stump (Fig. 4.32). Each horse pulls with a force of magnitude F. If the angle between the two ropes is 126°, what is the resultant force?

4.4• Three coplanar forces act on a 7.0-kg mass: 14 N directed at 0°, 14 N at 138°, and 18 N at 275°. What are the magnitude and direction of the force?

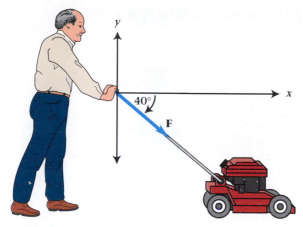

Figure 4.31
Problem 4.1.

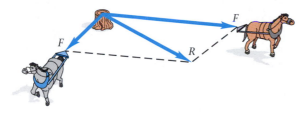

Figure 4.32
Problems 4.2 and 4.3.

Section 4.4 Newton's Second Law

Hints for Solving Problems

An object's acceleration is proportional to the net applied force. Because forces are vectors, they should be added vectorially to find the net force. Unless directed otherwise take all acceleration and all forces in this section to be constant and all motions to be along a straight line.

4.5 A force of 45.6 N is applied to a 2.00-kg discus. What is the acceleration of the discus?

4.6 (a) Find the net force that produces an acceleration of 6.4 m/s² for an 0.50-kg cantaloupe. (b) If the same force is applied to a 20-kg watermelon, what will its acceleration be?

4.7 A German Inter City Express train initially at rest accelerates to a speed of 200 km/h in 3 min 20 s (Fig. 4.33). What steady net force must be exerted by the engine if the total mass of the train is 8.63×10^5 kg?

4.8 An Inter City Express train traveling at 250 km/h is braked to a stop in a distance of 4820 m. If the mass of the train is 8.63×10^5 kg, what is the average braking force?

4.9● The total horizontal force exerted between the tires of a 1500-kg automobile and the ground is 980 N. If the car starts from rest, how far will it go in 5.0 s?

Figure 4.33
Problem 4.7.

4.10● A 400-kg ice boat moves on runners on essentially frictionless ice. A steady wind blows, applying a constant force to the sail. At the end of an 8.0-s run, the acceleration is 0.50 m/s². (a) What was the acceleration at the beginning of the run? (b) What was the force due to the wind? (c) What retarding force must be applied at the end of 4.0 s to bring the ice boat to rest by the end of the next 4 s? (The wind is still blowing. Assume the boat was at rest at time $t = 0$.)

4.11● A toy truck of 1.0 kg is at rest on a horizontal frictionless surface. At time $t = 0$ there are no horizontal forces acting on the truck. At $t = 1.0$ s the truck is suddenly acted upon by a force $F = 1.0$ N. This force is maintained until $t = 2.0$ s, when the force becomes 2.0 N. At $t = 3.0$ s the force becomes 3.0 N, and so on. Construct a graph of the velocity as a function of time. From this make a graph of displacement versus time.

4.12● A Lufthansa A320 accelerates from rest to liftoff speed of 73.7 m/s in 27.1 s. Each of the plane's two jet engines provides a forward force (thrust) of 111 kN. (a) What is the mass of the plane and (b) how far does it travel down the runway before liftoff?

4.13● A net force of 5.34 N acting for 4.23 s on a mass initially at rest causes it to travel 4.75 m in a straight line. What is the mass?

4.14●● A 7.31-g bullet is moving at 579 m/s as it leaves the 0.610-m-long barrel of a rifle. What is the average force on the bullet as it moves down the barrel? Assume that the acceleration is constant.

Section 4.5 Weight

Hints for Solving Problems

Assume the value of $g = 9.81$ m/s² unless stated otherwise in the problem.

4.15 What is the mass in kilograms of a bag of sugar that weighs 5.00 lb? (1 N = 0.2248 lb.)

4.16 A force of 1 newton is equal to 0.2248 lb. (a) Compute the weight in newtons of a 150-lb man. (b) Compute the mass of the man.

4.17 What is the weight of a 48.9-kg girl?

4.18 A brass block of mass m resting on a horizontal frictionless surface is given a horizontal acceleration of 4.5 m/s² by a force of 8.7 N. (a) What is the mass of the block? (b) What is the weight of the block?

4.19 A standard kilogram mass was prepared in Paris, where $g = 9.81$ m/s², and sent to Washington, where $g = 9.80$ m/s². What was the percentage change in the weight of the standard?

4.20 A 24-kg block is pushed up a frictionless inclined plane that makes an angle of 23° with the horizontal direction. What force is needed to move the block with constant speed? Assume that the force is parallel to the surface of the plane.

4.21• The free-fall acceleration on earth is about 9.81 m/s². On the moon the same quantity is 1.62 m/s². An astronaut in a space suit has a mass of 145 kg. (a) What is the astronaut's weight on earth? (b) On the moon? (c) What is the astronaut's mass on the moon?

4.22• A Saturn-Apollo launch rocket has a mass of 5.40×10^5 kg. What is the initial acceleration of the rocket if the thrust at liftoff is 7.40×10^6 N?

4.23• A force of 8.7 N is applied to a steel block initially at rest on a horizontal frictionless surface. The force, which is directed at an angle of 30° below the horizontal, gives the block a horizontal acceleration of 5.3 m/s². (a) What is the mass of the block? (b) What is the normal force of the surface acting on the block?

Section 4.6 Newton's Third Law

Hints for Solving Problems

Remember that action-reaction pairs act on different objects.

4.24 A large 500-kg magnet exerts a constant force of 3.00 N on a 0.250-kg bar magnet. What magnitude force does the bar magnet exert on the big magnet?

4.25 What force must be exerted at A to give an upward force of 32 N on the cervical traction device in Fig. 4.34?

4.26 The starship Enterprise is on a mission into deep space, where no human has gone before. It encounters an alien spacecraft, and Captain Kirk orders activation of the tractor beam to pull the alien craft to the starship. The alien ship is made of a super-dense alloy so its mass is 8 times the mass of the starship. Assuming the two craft have zero relative ve-

Figure 4.34
Problem 4.25.

locity at the time the tractor beam is activated, describe their motions by comparing the acceleration of the Enterprise to that of the alien craft as seen by a distant stationary observer.

4.27 An 59-kg woman stands in an elevator. What force does she exert on the floor of the elevator under the following conditions? (a) The elevator rises with a constant velocity of 2.0 m/s. (b) The elevator accelerates upward at 1.8 m/s². (c) The elevator goes down with a constant velocity of 4.0 m/s. (d) The elevator descends with a downward acceleration 2.8 m/s². (e) While going down, the elevator decelerates at 1.5 m/s².

4.28 An 75-kg man stands in an elevator. What force does he exert on the floor of the elevator under the following conditions? (a) The elevator is stationary. (b) The elevator accelerates upward at 2.0 m/s². (c) The elevator rises with constant velocity of 4.0 m/s. (d) While going up, the elevator accelerates downward at 1.5 m/s². (e) The elevator goes down with constant velocity of 7.0 m/s.

4.29 An unabridged dictionary of mass m rests on a weak table. (a) What is the reaction force to the force of the book on the table? (b) What is the reaction force to the force of gravity on the book? (c) The tabletop collapses under the book's weight. Answer (a) and (b) for the conditions while the table is collapsing.

4.30• A bird is placed on a perch in a large closed box that is sitting on a spring scale. What will happen to the scale reading when (a) the bird jumps off the perch, (b) the bird is flying, (c) the bird lights on the perch again?

4.31• A person weighing 650 N stands on spring scales in an elevator that is moving downward with constant speed of 3.6 m/s. The brakes suddenly grab, bringing the elevator to a stop in 1.8 s. Describe the scale readings from just before the brakes grab until after the elevator is at rest.

Section 4.7 Some Applications of Newton's Laws

Hints for Solving Problems

Be sure to draw a properly labeled free-body diagram for each problem.

4.32 Figure 4.35 shows a graph of distance along a straight course as a function of time for a 1000-kg dragster. Construct a graph of the horizontal force between the tires and ground (no slipping) as a function of time. (*Hint:* First construct a plot of velocity, taking points, say, every 0.5 s.)

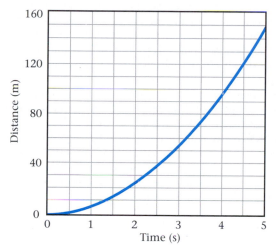

Figure 4.35
Problem 4.32.

4.33 (a) A 17.0-kg bucket is lowered by a rope with constant velocity of 0.500 m/s. What is the tension in the rope? (b) A 17.0-kg bucket is lowered with a constant downward acceleration of 1.00 m/s². What is the tension in the rope? (c) A 10.6-kg bucket is raised with a constant upward acceleration of 1.00 m/s². What is the tension in the rope?

4.34● A 0.0050-kg bullet traveling with a speed of 200 m/s penetrates a large wooden fence post to a depth of 0.030 m. What was the average resisting force exerted on the bullet?

4.35● A 0.0048-kg bullet traveling with a speed of 400 m/s penetrates into a large wooden fence post. If the average resisting force exerted on the bullet was 4.5 × 10³ N, how far did the bullet penetrate?

4.36● Two air-track gliders m_1 and m_2 are joined together with a light string (Fig. 4.36). A constant horizontal force of 4.0 N to the right is applied to mass m_2. (a) If $m_1 = 1.5$ kg and $m_2 = 0.50$ kg, what is the acceleration of the gliders? (b) What is the tension in the cord joining them?

4.37● Find the accelerations and the tension for the situation of Problem 4.36 given that $m_1 = 0.50$ kg and $m_2 = 1.5$ kg.

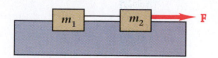

Figure 4.36
Problems 4.36 and 4.37.

4.38●● An 8.0-kg mass rests on an inclined frictionless surface as shown in Fig. 4.37. A light string runs parallel to the surface from the mass over a light, frictionless pulley to a 3.6-kg mass. Find (a) the acceleration of the masses and (b) the tension in the string.

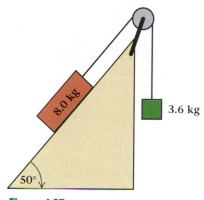

Figure 4.37
Problem 4.38.

4.39●● A simple Atwood machine composed of a single pulley and two masses m_1 and m_2 is on an elevator. When $m_1 = 44.7$ kg and $m_2 = 45.3$ kg, it takes 5.00 s for mass m_2 to descend exactly one meter from rest relative to the elevator. What is the elevator's motion? (That is, is it moving with constant velocity or accelerating up or down?)

4.40●● A classroom demonstration is done with an Atwood machine. The masses are $m_1 = 1.00$ kg and $m_2 = 1.10$ kg (Fig. 4.38). If the larger mass descends a distance of 3.00 m from rest in 3.6 s, what is the acceleration of gravity at that place? (Ignore the effects of pulley mass and friction.)

Section 4.8 Friction

4.41 A 9.75-kg lead brick rests on a level wooden table. If a force of 46.4 N is required to slide the brick across the table at a constant speed, what is the coefficient of friction?

4.42 A horizontal force pulls a 50.0-kg bag of fertilizer across the floor. What is the minimum force required if the coefficient of friction is 0.37?

4.43 A 5.00-g nickel coin sliding along a level table top with an initial velocity of 200 cm/s comes to a stop after traveling

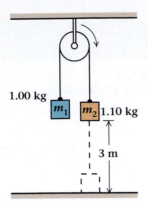

Figure 4.38
Problem 4.40.

50.0 cm. (a) What is the acceleration of the coin? (b) What is the coefficient of friction between the coin and the table?

4.44 A horizontal force of 3.00 N is applied to a 1.00-kg radio, initially at rest on a table with a level surface. (a) Will the radio move if the coefficient of friction is 0.45? (b) What is the coefficient of friction if it just begins to move? (c) What is its acceleration if the coefficient of friction is only 0.20?

4.45 A 5.00-kg concrete block rests on a level table. The coefficient of friction between the block and the table is 0.55. A 4.00-kg weight is attached to the block by a string of negligible mass passed over a light frictionless pulley (Fig. 4.39). What is the acceleration of the block when the 4.00-kg weight is released?

over a light, frictionless pulley (Fig. 4.39). If the acceleration of the block is measured to be 1.00 m/s², what is the coefficient of friction between the block and the table?

4.47 A crate starts from rest and slides 8.35 m down a ramp. When it reaches the bottom it is traveling at a speed of 5.25 m/s. If the ramp makes an angle of 20.0° with the horizontal, what is the coefficient of friction between the crate and the ramp?

4.48 A block of mass 4.7 kg slides 20 m from rest down an inclined plane making an angle of 30° with the horizontal. If the block takes 10 s to slide down the plane, what is the retarding force due to friction?

4.49 A 4.0-kg wooden block rests on a level table. The coefficient of friction between the block and the table is 0.40. A 5.0-kg mass is attached to the block by a horizontal string passed over a frictionless pulley of negligible mass. (a) What is the acceleration of the block when the 5.0-kg mass is released? (b) What is the tension in the string during the acceleration?

4.50 Determine the acceleration and the tension for the situation in Problem 4.49 when the coefficient of friction between the block and the table is 0.25.

4.51 Two clay pots joined together by a light string rest on a table (Fig. 4.40). The frictional coefficient between the pots and the table is 0.35. The pots are also joined to a 4.0-kg mass by a string of negligible mass passed over an ideal pulley as shown in the figure. (a) Calculate the acceleration of the system when the 4.0-kg mass is released. (b) Also find the tensions T_1 and T_2 in the strings during acceleration.

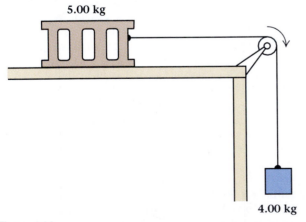

Figure 4.39
Problems 4.45 and 4.46.

4.46 A 5.00-kg concrete block rests on a level table. A 4.00-kg mass is attached to the block by a string passing

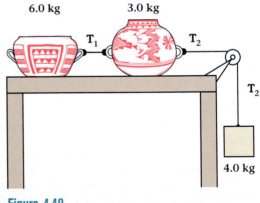

Figure 4.40
Problems 4.51 and 4.52.

4.52 Consider the situation shown in Fig. 4.40 but with the 4.0-kg mass replaced by a 6.5-kg mass. Find the acceleration of the system and the tensions T_1 and T_2 assuming that the frictional coefficient is 0.35.

Section 4.9 Static Equilibrium

Hints for Solving Problems

In static equilibrium problems, the equilibrium condition applies independently in each direction. In inclined-plane problems, it is often helpful to resolve forces into components parallel and perpendicular to the plane.

4.53 A steel paperweight rests on a clean dry steel incline making an angle θ with the horizontal. Find the maximum angle θ for the paperweight to remain at rest.

4.54 A glass cube rests on a glass incline making an angle θ with the horizontal. The coefficient of friction between the cube and the incline is 0.92. Find the maximum angle θ for the cube to remain at rest.

4.55 When the archer in Fig. 4.41 pulls on the bow string with a force of 267 N, the bow string makes angles of 62° with the arrow. What is the tension in the string?

Figure 4.41
Problem 4.55.

4.56 A 52.6-kg high-school student hangs from an overhead bar with both hands. (a) What is the tension in each arm if the bar is gripped with both arms raised vertically overhead? (b) What is the tension in each arm when the arms make an angle of 33° with respect to the vertical?

4.57 A 60-lb child is seated in a swing of negligible mass. How much horizontal force is required to pull the child and swing aside so that the support rope makes an angle of 30° with the vertical? See Fig. 4.28. (p. 132)

4.58 Three equal masses are suspended from frictionless pulleys as shown in Fig. 4.42. What are the angles of the strings with respect to the horizontal when the system comes to equilibrium?

4.59● Suppose that the masses in Problem 4.58 are not identical. What are the angles θ_1 and θ_2 if $m_1 = m_2 = 2.0$ kg and $m_3 = 3.0$ kg?

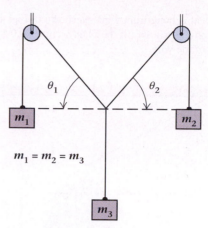

Figure 4.42
Problems 4.58 and 4.59.

4.60● A physicist finds that her car is stuck in the sand and cannot be driven out. Unable to push it out, she ties a strong rope from the front of the car to a large tree 30 m away and directly in front of the car. She then pushes on the middle of the rope with a force of 400 N in a direction perpendicular to the length of the rope. If the midpoint of the rope is displaced by 3.0 m, what is the force applied to the car?

4.61● A fish of mass m is suspended by a string as shown in Fig. 4.43. The string is fastened securely at point C but will pull loose from the wall at A when the string tension exceeds 22 N. What is the maximum mass of the fish that can be supported by the string?

4.62● What is the force exerted by the string on the wall at point A in Fig. 4.43 if the suspended fish has mass $m = 0.35$ kg?

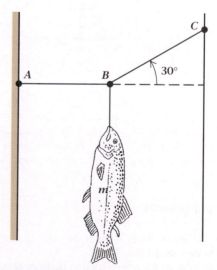

Figure 4.43
Problems 4.61 and 4.62.

4.63• A plant is hung from wires as shown in Fig. 4.44. What is the tension in each wire if the plant weighs 20.0 N? Ignore the weight of the wire.

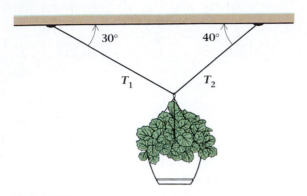

Figure 4.44
Problem 4.63.

4.64• Suppose that the weight w_2 in Fig. 4.45 is 400 N. What must be the values of the weights w_1 and w_3?

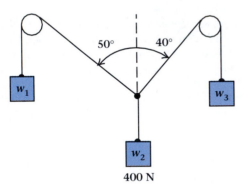

Figure 4.45
Problem 4.64.

4.65•• A 10.0-kg block is placed on a frictionless inclined plane and connected to a 5.0-kg block as shown in Fig. 4.46. (a) What would the angle θ have to be for the blocks to remain motionless? (b) What would be the acceleration of the blocks if $\theta = 37°$?

Additional Problems

4.66 A 1000-kg car is moving at 30 m/s. A braking force of 6000 N is applied for 4.0 s. What is the velocity of the car when the brakes are released?

4.67 What is the coefficient of friction between a sled and a plane inclined at 30° from the horizontal if the sled just slides without accelerating when given an initial push?

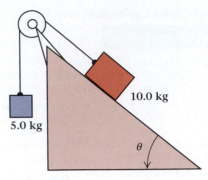

Figure 4.46
Problem 4.65.

4.68 A single cable supports an 837-kg elevator car. What is the tension in the cable when the car is moving with (a) constant speed, (b) an upward acceleration of 3.21 m/s², and (c) a downward acceleration of 3.21 m/s²?

4.69 A truck loaded with heavy cartons is forced to stop suddenly with a deceleration of 5.0 m/s². Calculate the minimum coefficient of friction between the cartons and the truck bed given that the cartons do not slide.

4.70 Assume that the Atwood's machine of Fig. 4.17 (a) (p. 116) is in static equilibrium. Make a diagram showing all forces. What must be the relationship of mass m_1 to mass m_2?

4.71 A 10-kg block is placed on a frictionless table and connected to a 5.0-kg block by a string that extends horizontally across the table over a pulley and down to the 5.0-kg block. What is the acceleration of the blocks? Ignore friction in the pulley.

4.72• For the accompanying diagram (Fig. 4.47), describe what will happen and what the spring scale S will read if (a) $m_1 = m_2 = 1$ kg; (b) $m_1 = 1.2$, $m_2 = 1$ kg. (Ignore the mass of the scale.)

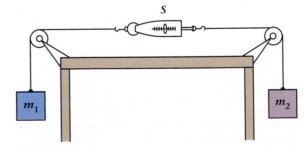

Figure 4.47
Problem 4.72.

4.73• A 500-kg trailer being pulled behind a car is subject to a 100-N retarding force due to friction. What force must the car exert on the trailer if (a) the trailer is to move forward at a constant speed of 25 km/h; (b) the trailer is to move forward with an acceleration of 2.0 m/s²; (c) starting from rest, the trailer and car are to travel 150 m in 10 s?

4.74● An Inter City Express locomotive pulls fourteen cars, each with a mass of 53,000 kg. What is the tension in the coupling between the third and fourth cars when the acceleration of the train is 0.26 m/s²?

4.75● (a) What is the minimum time in which one can hoist a 1.00-kg rock a height of 10.0 m if the string used to pull the rock up has a breaking strength of 10.8 N? Assume the rock to be initially at rest. (b) If the string is replaced by one that is 50% stronger, by what percentage will the minimum time for the hoist be reduced?

4.76● What minimum force is required to drag a carton of books across the floor at constant speed if the force is applied at an angle of 45° to the horizontal (Fig. 4.48)? Take the mass of the carton as 40 kg and the coefficient of friction as 0.60.

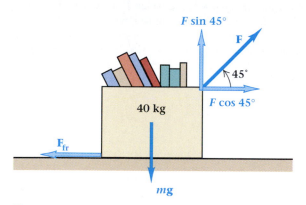

Figure 4.48
Problem 4.76.

4.77● A 65-kg skier goes down a 30° slope. (a) What would be the acceleration of the skier if friction could be neglected? (b) A skier continues to accelerate until gravitational force is balanced by the normal force and by the frictional forces due to the skis on the snow and the air resistance. What must be the frictional forces if the skier is no longer accelerating?

4.78● An elevator with a mass of 2500 kg carries four passengers with a combined mass of 260 kg. (a) What is the tension in the supporting cable when the elevator starts from rest and moves upward with constant acceleration until reaching a speed of 5.0 m/s in 2.0 s? (b) What is the tension in the cable when the elevator starts down with the same load and acceleration of the same magnitude?

4.79● Two blocks are connected by a light string passing over a pulley (Fig. 4.49). The inclined surfaces are frictionless and the effects of the pulley can be ignored. If the values are mass $m_2 = m_1 = 1.00$ kg, $\theta_1 = 46°$, and $\theta_2 = 34°$, what is the acceleration of the blocks?

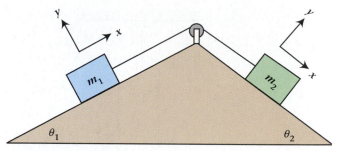

Figure 4.49
Problems 4.79, 4.85, and 4.86.

4.80●● A 400-N kangaroo exerts a constant force on the ground during the first 0.60 m of a vertical jump. After the kangaroo's feet leave the ground it rises an additional 1.8 m. When the kangaroo carries a baby kangaroo in its pouch and jumps with the same force, it can rise only 1.65 m higher after its feet leave the ground. What is the weight of the baby kangaroo?

4.81●● A 0.840-kg glider on a level air track is joined by strings to two hanging masses (Fig. 4.50). The strings have negligible mass and pass over light, frictionless pulleys. (a) Find the acceleration of the masses and (b) the tension in the strings.

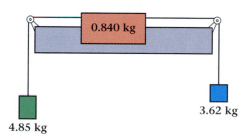

Figure 4.50
Problems 4.81 and 4.82.

4.82●● A 0.840-kg glider on a level air track is joined by strings to two hanging masses (Fig. 4.50). The strings have negligible mass and pass over light, frictionless pulleys. (a) Find the acceleration of the masses and (b) the tension in the strings when the air flow is turned off and the coefficient of friction between the glider and the track is 0.47.

4.83●● A monkey clinging to one end of a rope that passes over a frictionless pulley is balanced by a mirror of exactly equal weight on the other end of the rope (Fig. 4.51). When the monkey begins to climb the rope, it sees itself in the mirror. (a) Can the monkey climb up or down to escape the image? (b) After the monkey has traversed 4.0 m of rope, he stops. How high has he climbed? (c) What will happen if the monkey lets go of the rope? If you need additional information, make a specific assumption and answer the question in that light.

$M_m = M_b$

M_b

M_m

Figure 4.51
Problem 4.83.

4.84•• A 0.50 kg air-track glider has an initial speed of 0.25 m/s as it passes through a photoelectric gate that starts a timer. As it passes through, a constant force of 0.40 N is applied in the direction of motion. (a) What is the acceleration of the glider? (b) The glider then passes through a second gate that stops the timer at 1.3 s. What is the distance between the two gates? (c) The 0.40-N force is applied by means of a string attached to the glider. The other end of the string passes over a frictionless pulley and is attached to a hanging mass m. How big is the mass m? (d) Derive an expression for the tension T in the string as a function of the mass M of the glider, the mass m of the hanging mass, and the acceleration of gravity g.

4.85•• Two blocks are connected by a light string passing over a pulley (Fig. 4.49). The inclined surfaces are frictionless and the effects of the pulley can be ignored. The value of $m_1 = 1.05$ kg, $\theta_1 = 30°$, and $\theta_2 = 40°$. If the blocks accelerate to the left with acceleration $a = 0.010$ m/s^2, what is the value of m_2?

4.86•• Two blocks are connected by a light string passing over a pulley (Fig. 4.49). The inclined surfaces are frictionless, and the effects of the pulley can be ignored. The value of $m_1 = m_2 = 1.00$ kg and $\theta_2 = 40°$. If the blocks accelerate to the right with acceleration $a = 0.206$ m/s^2, what is the value of θ_1?

Solving Simultaneous Equations

Physics problems often lead to two or more equations that need to be solved at the same time to get the required answers. We call these equations simultaneous equations. The following problem illustrates how we solve simultaneous equations: A bicyclist heads north at a steady speed of 15 km/h from a point 3 km from a park. Three hours before the cyclist departed, a walker started walking north from the park along the same path at 5 km/h. When and where does the bicyclist overtake the walker?

Let y be the distance from the park where the bicyclist overtakes the walker. Then for the bicyclist

$$y = (15 \text{ km/h})t + 3 \text{ km},$$

where t is the cyclist's elapsed time in hours. The walker, having traveled for $t + 3$ h, goes the same distance,

$$y = (t + 3 \text{ h})5 \text{ km/h} = (5 \text{ km/h})t + 15 \text{ km}.$$

Each of these equations contains two unknown quantities, y and t. Neither equation alone can give us the answers, but we can solve them simultaneously to obtain both unknowns.

There are several techniques for solving simultaneous equations, all using only simple algebra and all giving the same result for a given problem. When you use these techniques to solve physics problems, be sure you take proper care of the units. However, because our intent here is to review the mathematical techniques, we will simplify them by omitting the units.

To solve the two equations just developed, let's write them again without units. We can eliminate the variable y by subtracting the second equation from the first. They become

$$
\begin{array}{r}
y = 15t + 3 \\
- \quad y = 5t + 15 \\
\hline
0 = 10t - 12.
\end{array}
$$

This new equation involves only t and may be rearranged to give

$$10t = 12.$$

Thus

$$t = \frac{12}{10} = 1.2.$$

Now that we have a value for t, it may be substituted into either of the original equations to give

$$y = 15t + 3 = 15(1.2) + 3 = 21$$
$$y = 5t + 15 = 5(1.2) + 15 = 21.$$

Notice that we get the same answer in both cases. Only one is necessary, but you may want to compute both as a check on your work. The physical answer to the original problem is that the cyclist overtakes the walker 1.2 h after the cyclist started. They meet at a position 21 km from the park.

We can solve the same pair of equations by eliminating t first. To do so we multiply the first equation by 5 and the second equation by 15. This procedure gives two new equations with the same coefficient of t. If we subtract the lower equation from the upper one, we get

$$
\begin{array}{r}
5y = 75t + 15 \\
- \quad 15y = 75t + 225 \\
\hline
-10y = -210.
\end{array}
$$

Thus

$$y = 21.$$

This is the same result as before. If you substitute this value of y into either original equation and solve for t, you will again get $t = 1.2$.

These examples are called the *addition and subtraction* method. Strictly speaking we used subtraction in both examples. Suppose the physical process had given equations of the type

$$z = 12x + 20$$

and

$$z = -2x + 7.$$

Then we could solve them by multiplying the second equation by 6 and adding the two equations:

$$
\begin{aligned}
z &= 12x + 20 \\
+\ 6z &= -12x + 42 \\
\hline
7z &= 62.
\end{aligned}
$$

The unknown quantity z is then

$$z = \frac{62}{7} = 8.86.$$

The value for x can be found by substituting z into one of the original equations.

Another technique for solving simultaneous equations makes use of substitution. For example, in the problem of the cyclist and walker, we could rewrite the first equation to express t in terms of y:

$$t = \frac{1}{15}y - \frac{1}{5}.$$

We then substitute this value of t into the second equation to get

$$y = 5\left(\frac{1}{15}y - \frac{1}{5}\right) + 15$$

$$y = \frac{1}{3}y - 1 + 15$$

$$y\left(1 - \frac{1}{3}\right) = \frac{2}{3}y = 14$$

$$y = \frac{14}{2/3} = 21.$$

As before, we can substitute this value of y into either original equation to get the value for t.

If your simultaneous equations include the square of an unknown quantity, you should use these same techniques to eliminate all but one variable. If the resulting equation still contains a squared term, it may be necessary to use the quadratic formula to find the solution. In some cases—for instance, the two-dimensional elastic collision of two objects (Chapter 8)—you will have three simultaneous equations to be solved for three unknown quantities. By applying any of the procedures we have just described, you can sequentially eliminate variables and thus reduce the number of equations to a single equation in one variable. The resulting solution can then be used to determine the other unknowns. In general, the number of independent equations you will need must be at least as great as the number of unknowns.

Uniform Circular Motion and Gravitation

We saw in Chapter 4 how we can use Newton's laws to calculate the motion of an object once we know the forces acting on it. The ability to do so does not depend on the nature or origin of the forces involved. We will see that in addition to the contact forces we have already discussed, there are other forces arising from gravity, electricity, magnetism, etc. Once we know the details of the particular force, we may then use Newton's laws to predict the motion of the object.

In this chapter we introduce Newton's law of universal gravitation, which was the first example of a fundamental force of nature expressed in mathematical form. The second force to be described mathematically was the electrostatic force, almost 100 years later (see Chapter 16). This process of mathematical development continues today for nuclear and subnuclear forces (see Chapters 29 and 32).

We observe that planets move in elliptical paths around the sun. In many cases these ellipses are almost circles. Therefore, we can improve our understanding of the motion of planets and satellites in their orbits if we first examine the dynamics of circular motion. Circular motion is quite common in nature and

can be found in varied settings from orbiting satellites to charged particles moving in a magnetic field to fair rides to cars rounding a curve. Furthermore, circular motion is closely related to vibrations and waves. Thus, the concepts learned here will be put to use in other areas. ■

5.1 Uniform Circular Motion

You know from experience that when you tie an object to a string and swing it in a circle around your head, you exert a force on the string. Because the string is flexible, the force that acts on the object is directed along the string toward your hand. According to Newton's second law, there is an acceleration in the same direction as this force. In this section we will show that any object moving in a circle at constant speed has an acceleration toward the center of the circle.

Before we begin the derivation, let us introduce a unit of angular measure, the radian. If we measure the length of an arc along the circumference of a circle (Fig. 5.1), we find that the arc length s is proportional to the angle θ subtended by the arc.* For a given angle, the arc length increases in direct proportion to increases in the radius of the circle. We define the angle in **radians,** θ, to be the ratio of the arc length s to the radius r:

$$\theta \equiv \frac{s}{r}. \tag{5.1}$$

Thus, one radian is the angle subtended by an arc length equal to the radius of the circle.

If we let the arc length become the entire circumference of the circle, then $s = 2\pi r$. (Recall that π is just a number, equal to 3.14159 . . .) Thus the angle in radians generated in going around a complete circle is 2π. We can say that the angular change in a full rotation is either 360° or 2π radians, depending on which unit of angular measure is used.

The relationship between the radian and the degree is

$$2\pi\,\text{rad} = 360°,$$

where we have used the symbol rad for the unit of radian. If we divide both sides by 2π, we find that

$$1\,\text{rad} = \frac{360°}{2\pi} = 57.3°.$$

We could have divided both sides by 360 and found that

$$1° = \frac{2\pi}{360} = 0.0175\,\text{rad}.$$

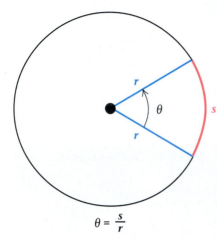

$\theta = \dfrac{s}{r}$

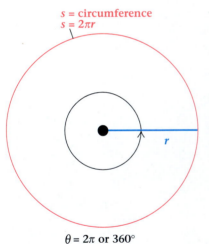

$s = \text{circumference}$
$s = 2\pi r$

$\theta = 2\pi$ or 360°

Figure 5.1
Angular measure in radians. The radian measure of an angle is the ratio of the subtended arc length to the radius. If the arc length is the entire circumference of a circle, then the angle in radians is 2π.

---

*In a triangle or sector of a circle, a line *subtends* the angle opposite to it; that is, the line extends across the entire angle.

Example 5.1

Eratosthenes (ca. 276–196 BC) knew that when the sun was directly overhead at Syene (modern Aswan) in southern Egypt, the sun was about 7° away from directly overhead in Alexandria (Fig. 5.2). Using the known distance between the two observing points, he was able to determine the circumference of the earth. Use a figure of 770 km for the distance to perform the same calculation.

Strategy From the figure we see that the distance from Syene to Alexandria is an arc length that subtends an angle of 7°. If we express the angle in radians, we can use the definition of Eq. (5.1) to determine the radius of the earth and, hence, the circumference.

Solution The angle measured in radians is

$$\theta = (7°)\left(\frac{2\pi\,\text{rad}}{360°}\right) = 0.122\ \text{rad}.$$

The radius of the earth, given by $r = s/\theta$, where s is the arc length of 770 m, is

$$r = \frac{s}{\theta} = \frac{770\ \text{km}}{0.122\ \text{rad}} = 6300\ \text{km}.$$

The circumference of the earth is found from

$$C = 2\pi r = 2\pi(6300\ \text{km}) = 40{,}000\ \text{km}.$$

Discussion The modern value for the average radius of the earth is 6380 km, barely 1% greater than Eratosthenes's measurement.

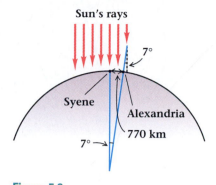

Figure 5.2
When the sun is directly overhead at Syene, it is about 7° away from directly overhead in Alexandria.

Now, consider a point particle moving along a circular path of radius r with constant speed v. A particle moving in this manner is said to undergo **uniform circular motion.** By a "particle" we mean an object of negligible size and constant mass. We use the idea of a particle as a way of creating a simplified model of a real physical situation. However, the resulting equations apply to real situations, such as a dot of paint on the rim of a wheel rotating at a steady rate. To a good approximation, we can extend our description of a point in uniform circular motion to a merry-go-round ride or to a satellite revolving around the earth.

As we learned in our study of kinematics (Chapter 2), a particle's speed is determined by measuring the distance traveled along its path and dividing by the elapsed time. Although the dot moves along its circular path with a constant speed v, its instantaneous velocity vector is constantly changing because the direction of its motion is constantly changing. (Remember that the speed is the magnitude of the instantaneous velocity.) At any moment of time, the instantaneous velocity is tangential to the circle. In a time interval Δt, the object moves along the circular path from one point, say P in Fig. 5.3(a), to another point, Q. The instantaneous velocity vectors at points P and Q, given by $\mathbf{v}_P$ and $\mathbf{v}_Q$, respectively, have the same magnitude but differ in direction. Each instantaneous velocity is tangent to the circle and perpendicular to the radius r at the point in question.

Because the velocity is constantly changing, there must be an acceleration. (Remember, acceleration occurs whenever the velocity changes in magnitude or

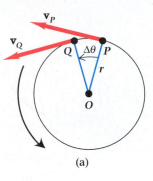

(a)

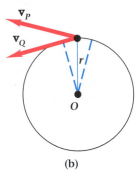

(b)

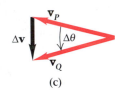

(c)

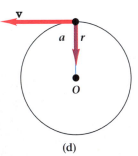

(d)

Figure 5.3

(a) The instantaneous velocity vector at points P and Q for a particle moving in a circular path. The velocity vector at a later time is shown at point Q.

(b) Vectors $\mathbf{v}_P$ and $\mathbf{v}_Q$ are translated to a common origin for comparison.

(c) The vector $\Delta\mathbf{v} = \mathbf{v}_Q - \mathbf{v}_P$. As $\Delta\theta$ becomes very small, $\Delta\mathbf{v}$ becomes perpendicular to both $\mathbf{v}_Q$ and $\mathbf{v}_P$.

(d) The moving particle has an instantaneous velocity tangential to the circle and an acceleration directed toward the center of the circle.

direction.) During the time interval Δt, the velocity changes by an amount $\Delta\mathbf{v} = \mathbf{v}_Q - \mathbf{v}_P$. The definition of average acceleration is

$$\bar{\mathbf{a}} \equiv \frac{\Delta\mathbf{v}}{\Delta t}. \tag{5.2}$$

To evaluate the acceleration we translate the vectors $\mathbf{v}_P$ and $\mathbf{v}_Q$ to a common origin (Fig. 5.3b). The change in velocity $\Delta\mathbf{v}$ is shown in Fig. 5.3(c). As the time interval Δt is made smaller, points P and Q are found closer together, reducing $\Delta\theta$, the angle between them. You can see that this angle is also the angle between the two velocity vectors. Eventually the angle $\Delta\theta$ becomes so small that $\mathbf{v}_P$ and $\mathbf{v}_Q$ are almost parallel and their difference $\Delta\mathbf{v}$ is almost perpendicular to both of them. In the limit that Δt goes to zero, $\Delta\mathbf{v}$ is exactly perpendicular to $\mathbf{v}$. Hence, the instantaneous acceleration, which is in the same direction as $\Delta\mathbf{v}$, is directed radially toward the center of the circular path. Therefore *a particle moving with constant speed around a circle is always accelerated toward the center* (Fig. 5.3d). In this special case, the particle has uniform circular motion, and its acceleration is always perpendicular to the velocity. This acceleration is called the **centripetal** (center-seeking) **acceleration.**

The centripetal acceleration can be readily evaluated. The angle $\Delta\theta$ between $\mathbf{v}_P$ and $\mathbf{v}_Q$, when measured in radians, is the ratio of the arc length to the radius. The arc length is the product of the speed and the time interval, so the angle becomes

$$\Delta\theta = \frac{\text{arc length}}{\text{radius}} = \frac{v\Delta t}{r}. \tag{5.3}$$

From the geometry of Fig. 5.3(c), we see that as $\Delta\theta$ gets smaller, the magnitude of $\Delta\mathbf{v}$ (indicated by $|\Delta\mathbf{v}|$) is approximately the same as the arc length made by turning a vector of magnitude $|\mathbf{v}_P|$ through the angle $\Delta\theta$. Thus $\Delta\theta$ may also be expressed as the ratio

$$\Delta\theta \approx \frac{|\Delta\mathbf{v}|}{|\mathbf{v}_P|}.$$

Since the magnitude of the velocity is constant, $|\mathbf{v}_P| = |\mathbf{v}_Q| = v$ and we can write

$$\Delta\theta \approx \frac{|\Delta v|}{v}.$$

The magnitude of the average acceleration becomes

$$|\bar{a}| = \frac{|\Delta\mathbf{v}|}{\Delta t} \approx \frac{v\Delta\theta}{\Delta t}.$$

Upon substituting Eq. (5.3) for $\Delta\theta$, we find that, in the limit of very small angles, we have

$$\boxed{a_{\text{c}} = \frac{v^2}{r},} \tag{5.4}$$

where the subscript c denotes centripetal acceleration. The velocity of a point in uniform circular motion is always tangential to the circle, and the acceleration always points to the center of the circle.

Example 5.2

A bicycle racer rides with constant speed around a circular track 25 m in diameter (Fig. 5.4). What is the acceleration of the bicycle toward the center of the track if its speed is 6.0 m/s?

Solution Since the speed around the circle is constant, we can compute the acceleration directly from Eq. (5.4):

$$a_c = \frac{v^2}{r} = \frac{(6.0 \text{ m/s})^2}{12.5 \text{ m}} = 2.9 \text{ m/s}^2.$$

Remember that this acceleration is directed toward the center of the circle as the bicycle moves at a constant speed around the circular track.

Sometimes it is more convenient to describe circular motion in terms of other quantities. For example, suppose that we know an object's **period** T, which is the time it takes the object to complete one revolution around its circular path. During this time, the object travels with constant speed v along a distance equal to the circumference of the circle, $vT = 2\pi r$. We can rearrange this last equation to find v, and then insert the result into Eq. (5.4) to get the centripetal acceleration in terms of r and T,

$$a_c = \frac{4\pi^2 r}{T^2}$$

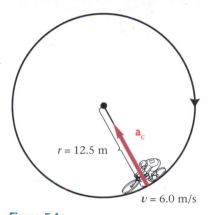

Figure 5.4
Example 5.2: Bicycle racer on a circular track.

The **frequency** f is the number of complete revolutions, or cycles, an object makes per unit of time. The frequency f is the reciprocal of the period T. If an object takes a time T to complete one revolution around the circle, then the number of revolutions per unit time, the frequency, is

$$f = \frac{1}{T}. \tag{5.5}$$

Frequency has the dimension of inverse time, that is, 1/time. The SI unit of frequency is the **hertz,** abbreviated Hz:

$$1 \text{ Hz} = 1/\text{s} = 1 \text{ s}^{-1}.$$

Example 5.3

An industrial grinding wheel with a 25.4-cm diameter spins at a rate of 1910 revolutions per minute (Fig. 5.5). What is the linear speed of a point on the rim?

Solution The speed of a point on the rim is the distance traveled, $2\pi r$, divided by T, the time for one revolution. However, we are given the frequency, which is the reciprocal of the period. Thus, the speed of a point on the rim, a distance

Figure 5.5
Example 5.3: A grinding wheel.

r from the axis of rotation, is

$$v = \frac{2\pi r}{T} = 2\pi rf.$$

Inserting the values of *r* and *f*, we get

$$v = (2\pi)\left(\frac{25.4 \text{ cm}}{2}\right)\left(\frac{1910}{\text{min}}\right)\left(\frac{1 \text{ min}}{60 \text{ s}}\right)$$

$$v = 2540 \text{ cm/s} = 25.4 \text{ m/s}.$$

▼

Problem-Solving Strategy

Comparing Quantities

When asked to compare two quantities, you should do so by computing the quantities and finding their ratio. In the following example we compare the acceleration of the moon to that of a body falling near the earth by forming the ratio a_{moon}/g.

▼

Example 5.4

Determine the centripetal acceleration of the moon as it circles the earth, and compare that acceleration with the acceleration of bodies falling on the earth. The period of the moon's orbit is 27.3 days.

Strategy According to Newton's first law, the moon would move with constant velocity in a straight line unless it were acted on by a force. We can infer the presence of a force from the fact that the moon moves with approximately uniform circular motion around the earth (Fig. 5.6). The acceleration toward the earth can be calculated from the period and the orbital radius. The mean center-to-center, earth-moon distance, given in a table in the end pages, is 3.84×10^8 m. The period of 27.3 days is not in SI units and must be converted.

Solution The acceleration of the moon toward the earth is

$$a_c = \frac{v^2}{r} = \frac{4\pi^2 r}{T^2} = \frac{4\pi^2(3.84 \times 10^8 \text{ m})}{(27.3 \text{ days} \times 24 \text{ h/day} \times 3600 \text{ s/h})^2}$$

$$a_c = 2.72 \times 10^{-3} \text{ m/s}^2.$$

The ratio of the moon's acceleration to that of an object falling near the earth is

$$\frac{a_c}{g} = \frac{2.72 \times 10^{-3} \text{ m/s}^2}{9.81 \text{ m/s}^2} = \frac{1}{3600}.$$

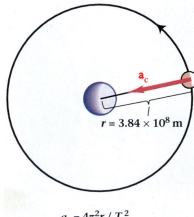

$a_c = 4\pi^2 r / T^2$

$T = 27.3$ days

Figure 5.6
Example 5.4: The moon orbiting the earth.

Discussion The ratio of the earth-moon distance to the earth's radius is about 60. According to Newton's law of gravitation, the ratio of the attracting forces, and thus the accelerations, is the inverse ratio of the square of the distances and is approximately $1/(60)^2 = 1/3600$. This predicted ratio and the ratio just calculated are the same. Considerations of this kind helped convince Newton and others of his time of the inverse-square nature of the gravitational force.

■

The position of a point in circular motion with tangential velocity v at a constant radial distance r from the center of the circle is given by the angle θ shown in Fig. 5.7. This angle in radians is given in terms of the arc length s and the radius r ($\theta = s/r$), as defined earlier. The rate of change of this angle is the **angular velocity** ω. The average angular velocity $\overline{\omega}$ is

$$\overline{\omega} \equiv \frac{\Delta\theta}{\Delta t} = \frac{\Delta s}{\Delta t} \cdot \frac{1}{r}.$$

In the limit that the time interval goes to zero, $\Delta s/\Delta t$ becomes the instantaneous speed v. Then the magnitude of the instantaneous angular velocity ω becomes

$$\boxed{\omega = \frac{v}{r}.} \qquad (5.6)$$

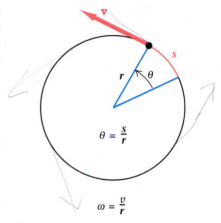

Figure 5.7
A point moving with a constant speed v along a circle of radius r is described in terms of its angular position θ. Its angular velocity (also called the angular frequency) is $\omega = v/r$.

We see from the definition of angular velocity that the dimension of ω is the reciprocal of time (that is, time^{-1}) and the units are radians per second (rad/s). Remember that, since the angle in radians was defined as a ratio of lengths, it has no dimension. The unit of radian is carried as a reminder that the angles are measured in radians and not in degrees.

When the change in time Δt is one period, the change in angle corresponds to one complete revolution, or 2π rad. Thus, we can also express the angular velocity as

$$\omega = \frac{\Delta\theta}{\Delta t} = \frac{2\pi}{T},$$

so that

$$\boxed{\omega = 2\pi f.} \qquad (5.7)$$

Since ω is directly proportional to f and has the dimension of inverse time, it is often called the **angular frequency.** The two names for ω may be used interchangeably. We will use these terms again in describing rotations (Chapter 9) and oscillations (Chapter 14).

We can also express the centripetal acceleration in terms of the angular velocity by combining Eqs. (5.4) and (5.6) to get

$$a_c = \omega^2 r.$$

Uniform circular motion represents the special case of two-dimensional motion in which the acceleration is always perpendicular to the velocity. In that case, the acceleration changes only the direction of the velocity, and not its magnitude (speed). We found earlier that for projectile motion the acceleration is generally neither parallel nor perpendicular to the velocity, so that both magnitude and direction of the velocity change. Figure 5.8 depicts these two cases.

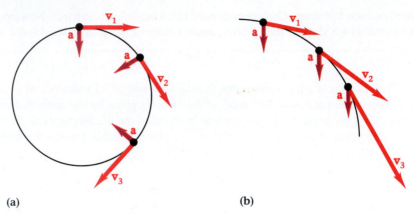

Figure 5.8
(a) In uniform circular motion, the acceleration is always perpendicular to the velocity, so that only the direction of the velocity changes. (b) In other two-dimensional motions, the acceleration is neither perpendicular nor parallel to the velocity. Then both direction and magnitude of the velocity change.

Figure 5.9
Example 5.5: The Wheelie.

Example 5.5

At Six Flags Over Georgia near Atlanta, the Wheelie carries passengers in a circular path with a radius of 7.7 m (Fig. 5.9). The ride makes a complete rotation every 4.0 s. (a) What is a passenger's angular velocity due to the circular motion? (b) What acceleration does a passenger experience?

Solution (a) The ride has a period $T = 4.0$ s. We can use it to compute the angular velocity as

$$\omega = \frac{2\pi}{T} = \frac{2\pi \text{ rad}}{4.0 \text{ s}} = \frac{\pi}{2.0} \text{ rad/s} \approx 1.6 \text{ rad/s}.$$

(b) Because the riders travel in a circle, they undergo a centripetal acceleration given by

$$a_c = \omega^2 r = \left(\frac{\pi}{2} \text{ rad/s}\right)^2 (7.7 \text{ m}) = 19 \text{ m/s}^2.$$

Notice that this is almost twice the acceleration of a body in free fall.

5.2 Force Needed for Circular Motion

We have just seen that an object of mass m moving in a circular path with a uniform speed v is accelerated because the direction of its instantaneous velocity is continuously changing. For example, a toy airplane whirled in a circle by a string is accelerated. By Newton's second law, the net force acting on the object is in the same direction as the observed acceleration. This net force is called the **centripetal force** because it is directed toward the center of the circle. It is this net force that causes the motion of the object to be circular; without the centripetal force, the object would travel in a straight line and not in a circle. The magnitude of the centripetal force is obtained from Newton's second law as

$$F_c = ma_c$$

$$F_c = \frac{mv^2}{r}.$$

(5.8)

Substituting for v in terms of T or ω also gives a valid expression. For example, centripetal force is also given by $F_c = m\omega^2 r$.

As we have seen, circular motion requires acceleration and the acceleration is the result of a net force. Thus, any object undergoing circular motion necessarily experiences a force that causes the object to move in a circular path. It is this force that we call the centripetal force. Note that the centripetal force is not a fundamental force in the same sense that gravity is a fundamental force. It is just the name we give to the net force—whatever its origin—that causes an object to move in a circle. Computations of centripetal force are illustrated by Examples 5.6 and 5.7.

Master the Concept

Force and Circular Motion

Question: The Wave Swinger ride at the fair has two circular rows of chairs suspended by chains of equal lengths. The outer row is along the outer edge of the top of the ride and the inner row has a smaller radius. Do riders in both rows swing out at the same angle when the ride is rotating?

Answer: When the ride is in motion, the passengers in the chairs swing outward. At one particular angle the chain provides the necessary horizontal force for circular motion and the vertical force needed to balance the gravitational force. Riders in both rows move with the same angular velocity, but those in the outer row move in a larger radius and thus have a larger angular acceleration. Because of this larger angular acceleration, a larger centripetal force is needed. Thus, for riders in the outer row, the horizontal component of the chain force is larger relative to their gravitational force than it is for those on the inner row. The result is that riders in the outer row swing out to the larger angle from the vertical direction than do the riders in the inner row as you can see in Fig. 5.22 on p. 170.

Example 5.6

Approximately how much force does the earth exert on the moon?

Strategy Assume the moon's orbit to be circular about a stationary earth. The force can be found from $F = ma_c$.

$$F = ma_c = m\frac{v^2}{r} = m\frac{4\pi^2 r}{T^2}.$$

Use the data given in Fig. 5.6 for r and T. The mass of the moon is given in the end sheets as 7.35×10^{22} kg.

Solution The numerical values can be inserted into the equation to give

$$F = \frac{(7.35 \times 10^{22} \text{ kg}) 4\pi^2 (3.84 \times 10^8 \text{ m})}{(27.3 \text{ day} \times 8.64 \times 10^4 \text{ s/day})^2} = 2.00 \times 10^{20} \text{ N}.$$

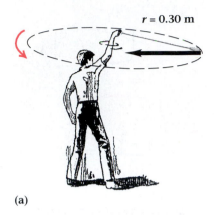

$r = 0.30$ m

(a)

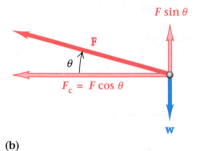

$F \sin \theta$

F

θ

$F_c = F \cos \theta$

w

(b)

Figure 5.10
Example 5.7: (a) A fishing weight is tied to a string and whirled in a horizontal circle. (b) The free-body diagram for (a). The force **F** along the string must provide both the horizontal centripetal force and the upward vertical force that opposes the gravitational force **w**.

Example 5.7

A student ties a 0.060-kg lead fishing weight to the end of a piece of string and whirls it around in a horizontal circle. If the radius of the circle is 0.30 m and the object moves with a speed of 2.0 m/s, what is the horizontal component of force that directs the lead weight toward the center of the circle (Fig. 5.10a)? What is the tension in the string?

Strategy To begin, we draw the free-body diagram of Fig. 5.10(b). The tension **F** along the string provides both a horizontal and a vertical force. The centripetal acceleration is provided by the horizontal component F_c. The weight will move up or down, changing the angle θ, until the vertical component of **F** is equal and opposite to the gravitational force **w**.

Solution The horizontal force component is

$$F_c = ma_c = \frac{mv^2}{r}$$

$$F_c = \frac{(0.060 \text{ kg})(2.0 \text{ m/s})^2}{0.30 \text{ m}} = 0.80 \text{ N}.$$

The tension in the string is the vector sum of F_c and the vertical force opposing **w**. Its magnitude is

$$F = \sqrt{F_c^2 + w^2}$$
$$F = \sqrt{(0.80 \text{ N})^2 + (0.060 \text{ kg} \times 9.81 \text{ m/s}^2)^2} = 0.99 \text{ N}.$$

A car moving in a circle with constant speed must be acted on by a force in order to execute circular rather than straight-line motion. This centripetal force is provided by the friction between the tires and the road. Passengers inside the car must also be subject to a centripetal force or they will not travel in the same path as the car. They experience forces exerted by the seat or the door of the car that cause them to move along the same path. This description is in accord with what is seen by an observer at rest outside the car.

Inside the noninertial frame of the car rounding the curve, the driver may actually believe she experiences an outward force that pushes her against the car door. In reality, however, it is the car seat and door that press inward on the driver, causing her to move in a circular path (Fig. 5.11). The occupant of the car exerts a force on the car that is equal and opposite to the centripetal force. This apparent outward force, which is experienced only by the person in the car moving in a curve (a rotating reference frame), is called a centrifugal force. However, an observer at rest outside the car would say that there is no centrifugal force. The words centripetal and centrifugal were used in their present-day context in the *Principia*.

Example 5.8

Imagine a giant donut-shaped space station located so far from all heavenly bodies that the force of gravity may be neglected. To enable the occupants to live a "normal" life, the donut rotates and the inhabitants live on the part of the donut farthest from the center (Fig. 5.12). If the outside diameter of the space station is 1.5 km, what must be its period of rotation so that the passengers at the periphery will perceive an artificial gravity equal to the normal gravity at the earth's surface?

Solution   The weight of a person of mass m on the earth is a force

$$F = mg.$$

The centripetal force required to carry the person around a circle of radius r is

$$F = ma_c = \frac{m4\pi^2 r}{T^2}.$$

We may equate these two force expressions and solve for the period T:

$$mg = \frac{m4\pi^2 r}{T^2},$$

$$T = 2\pi\sqrt{\frac{r}{g}} = 2\pi\sqrt{\frac{750 \text{ m}}{9.81 \text{ m/s}^2}} = 55 \text{ s} = 0.92 \text{ min.}$$

A common laboratory tool that operates by the principle of centripetal force is the centrifuge. This device is primarily used to increase the sedimentation rate of small particles suspended in a liquid or to separate slightly dissimilar liquids. One type of centrifuge is shown in Fig. 5.13(a). When the rotor spins, the tubes swing outward until at very high speeds they are virtually horizontal. In this position the liquid is unable to exert the centripetal force required to keep the small suspended particles moving in a circle. Consequently, the particles move outward toward the end (bottom) of the tubes. At high speeds, the resulting forces on the particles may be many times greater than the gravitational force, so that the small particles collect at the bottom of the tubes more quickly than if left to settle by gravity alone. Further details will be found in the discussion of Stokes's law in Chapter 10.

Example 5.9

A centrifuge used to separate blood cells from blood plasma rotates at 55 rotations per second. What is the acceleration at the center of a centrifuge tube 8.0 cm from the axis of rotation?

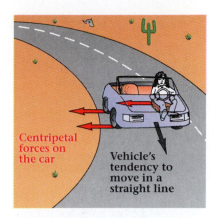

Figure 5.11
Centripetal forces cause both driver and car to move together in a circular path. Friction between the tires and the road exerts an inward force on the car; the car seat and door exert an inward force on the driver. By Newton's third law, the driver at rest inside the car also exerts an outward force on the car seat, equal in magnitude and opposite in direction to the centripetal force.

Figure 5.12
Example 5.8: A rotating space station. Occupants are kept on the outer wall by an "artificial gravity" produced by the rotational motion.

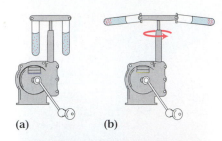

(a) (b)

(c)

Figure 5.13

(a) A simple centrifuge. At rest the tubes hang down as shown. (b) As the shaft rotates at high speed, the tubes swing out. (c) Commercial laboratory centrifuges with a fixed tube angle commonly rotate at speeds of 3400 rpm.

Figure 5.14

(a) A car rounding a flat curve. The car's weight is balanced by the sum of the normal forces $\mathbf{F}_N$. Frictional forces $\mathbf{F}_{fr}$ between the tires and the road provide the centripetal acceleration. (b) On a banked road, the normal force may be resolved into a vertical component and a horizontal component directed toward the center of the curve. For each bank angle there is one particular speed for which the normal force provides the necessary centripetal force.

Solution Since we are given a frequency of rotation, we choose the acceleration formula

$$a_c = \omega^2 r,$$

where $\omega = 2\pi f$. We then obtain

$$a_c = (2\pi f)^2 r = (2\pi\, 55/\text{s})^2 (0.080\text{ m})$$
$$a_c = 9.6 \times 10^3 \text{ m/s}^2.$$

We may compare this acceleration with the acceleration of gravity, $g = 9.81$ m/s^2, to get

$$a_c = 970\, g.$$

Discussion It is common to compare accelerations with the acceleration of gravity as we have done here. It becomes a way of gauging how much force acts on an object. An acceleration of one g corresponds to a force equal to the object's weight, $2g$ corresponds to a force twice the object's weight, and so on. Sometimes people refer to forces in terms of g. In that case they are not giving the force itself, rather they are giving the ratio of force on an object to its weight.

■

As we saw earlier, on a flat curve a car is turned by the frictional force exerted on the tires by the road. If the frictional force is not large enough, the car does not travel around the proper curve, but instead skids toward the outside. For high-speed turns on highways and race tracks, it is common practice to bank the curves to compensate for the tendency of vehicles to skid outward. Banking the curves also reduces the sideways force on the passengers and thus makes them more comfortable.

Figure 5.14(a) shows the force vectors acting on a car rounding an unbanked curve. The weight of the car is shown as the downward vector $m\mathbf{g}$. The normal forces $\mathbf{F}_N$ of the road supporting the car act on each of the tires. The sum of these four normal forces just equals the weight of the car. The frictional forces $\mathbf{F}_{fr}$ acting on the tires provide the unbalanced force that gives the car its centripetal acceleration and causes it to turn.

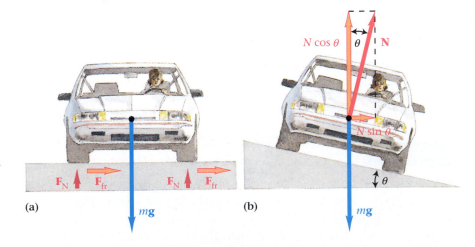

If the curved road is banked at an angle θ, the normal force has a horizontal component toward the center of the circle (Fig. 5.14b). When the car is moving, this inward component of the normal force still points to the center of the circle. At one particular speed, this force component by itself provides just the necessary force for turning the car without skidding, even if there are no sideways frictional forces between the tires and the road. At this speed the car will not skid, even on an ice-covered road.

At this nonskidding speed, all of the forces between the car and the road are perpendicular or normal to the road surface. The road exerts a normal force $\mathbf{N} = \Sigma \mathbf{F}_N$ on the car. The vertical component $N \cos \theta$ must be equal and opposite to the gravitational force $m\mathbf{g}$. The horizontal component of $\mathbf{N}$ provides the centripetal force that turns the car. Thus, for the vertical component we get

$$N \cos \theta = mg,$$

and for the horizontal component we get

$$N \sin \theta = \frac{mv^2}{r}.$$

We can divide the second equation by the first to get

$$\tan \theta = \frac{v^2}{gr}. \tag{5.9}$$

This equation gives the banking angle θ for a curve of a given radius r to be negotiated at a speed v without tending to slide out or in, away from the circular path.

Example 5.10

A race track designed for average speeds of 240 km/h (66.7 m/s) is to have a turn with a radius of 975 m. To what angle must the track be banked so that cars traveling 240 km/h have no tendency to slip sideways?

Strategy If there is no tendency to slip sideways, the vertical component of the normal force must equal the weight and the horizontal component must provide the centripetal force. These conditions were used to derive Eq. (5.9), which can be applied here.

Solution We can determine θ from Eq. (5.9):

$$\tan \theta = \frac{v^2}{gr} = \frac{(66.7 \text{ m/s})^2}{(9.81 \text{ m/s}^2)(975 \text{ m})} = 0.465,$$

$$\theta = 24.9°.$$

| 5.3 | **Kepler's Laws of Planetary Motion** |

As we mentioned in Chapter 4, Johannes Kepler and others of his time knew that the orbits of the earth and other planets about the sun are not exactly circular. Kepler made use of Tycho Brahe's observational data to deduce that planetary

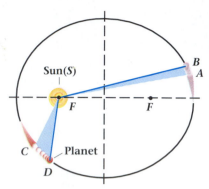

Figure 5.15
Diagram of Kepler's three laws of planetary motion. The ellipse has been exaggerated so that it has a much greater eccentricity than any actual planetary orbit. Note that in accordance with Kepler's first law, the sun is located at one of the foci, *F*. According to Kepler's second law, the time it takes a planet to move from *A* to *B* and from *C* to *D* is the same if the areas of the wedges *ABS* and *CDS* are the same. Kepler's third law says that the ratio of the orbital period *T* squared to the cube of the average radius *R* is the same for all the planets.

orbits are actually ellipses. Eventually his analysis of Tycho's data was expressed as three rules or laws describing the motion of the planets. Later, Newton relied on Kepler's results in formulating his theories about the motion of the planets. Here we state the three Kepler laws. Later, in this and other chapters, we show how Kepler's laws can be derived from Newtonian mechanics and the law of universal gravitation.

Kepler's laws of planetary motion give unambiguous predictions that are subject to verification. They are usually stated in the following way (refer to Fig. 5.15):

1. The orbit of each planet is an ellipse* and the sun is at one focus.

2. An imaginary line from the sun to a moving planet sweeps out equal areas in equal intervals of time.

3. The ratio of the square of a planet's period of revolution to the cube of its average distance from the sun is a constant.** This constant is the same for all planets. (The period of revolution is the time for one complete orbit.)

That Kepler worked out these laws without knowing their underlying cause is a remarkable example of perseverance. The cause of planetary motion did not become clear until nearly 70 years later, when Isaac Newton published the laws of gravitation and of motion.

Like all "laws" of nature, Kepler's laws represent the distillation or summation of many observations. They are not "fundamental" in the same way as some other laws that we will study later; that is, they describe observations without giving their causes. However, Kepler's laws are similar in an important way to almost all of the other laws that we will encounter: They are expressed in mathematical terms. Although these laws can be put in words alone, their most succinct and useful form is as a mathematical statement. For example, contrast the statement that "planets move faster when closer to the sun" with Kepler's second law. The second law includes not only the information conveyed by those words but also the ability to make exact predictions. Theories in the physical sciences, and physics in particular, are partly judged on their ability to predict what will occur, as well as to explain what has happened. This ability usually requires that they have a mathematical form. Likewise, compare the statement that "planetary orbits are elongated ovals" with the more precise and meaningful statement that "planetary orbits are ellipses."

Kepler's *first law* has wider applicability than given in the statement above. We now know that the orbit of the moon around the earth, the orbits of other moons around other planets, the orbits of comets, and the orbits of artificial satellites are all ellipses. Contrary to the exaggerated way in which Fig. 5.15 depicts the orbits, planetary orbits are almost circular. Table 5.1 lists the characteristics of the planetary orbits. Although we can usually approximate planetary orbits with circles, this approximation is not always valid for artificial satellites.

---

*A review of the properties of the ellipse is found on page A–2 of Appendix A.
**For Keplerian orbits, the average distance of the planet from the sun is the length of the semi-major axis of its orbit.

Table 5.1	The Orbits and Periods of the Planets	
Planet	**Semimajor Axis of Orbit in AU***	**Orbital Period in Years**
Mercury	0.387	0.241
Venus	0.723	0.615
Earth	1.000	1.000
Mars	1.524	1.881
Jupiter	5.203	11.862
Saturn	9.516	29.458
Uranus	19.166	84.013
Neptune	30.012	164.793
Pluto	39.557	248.530

Source: The Astronomical Almanac, for the Year 1991. Washington, D.C.: U.S. Government Printing Office, 1989.

*The distances in this table are given in astronomical units, AU. One AU is a distance equal to the semimajor axis of the earth's orbit around the sun. That is, 1 AU = 1.50×10^8 km = 9.3×10^7 mi.

Kepler's *second law* can be used to predict the speed of a planet in one part of its orbit if we know its speed in another part. Assume, for example, that a planet takes one month to go from A to B (Fig. 5.15). If we draw pie-shaped sectors ASB and CSD so that each sector has the same area, then according to the second law, the planet will also take one month to go from C to D. Since the arc CD is longer than the arc AB and the times of travel are the same, the speed along CD is greater than the speed along AB.

Kepler's *third law* deals with a relationship between planets, rather than predicting the behavior of a single planet alone. Stated in symbols, it becomes

$$\frac{T^2}{R^3} = k, \tag{5.10}$$

where the period T is the time for one complete orbit, R is the average distance of the planet from the sun, and k is a constant that is the same for all planets circling the sun.

Example 5.11

Compare the prediction of Kepler's third law for the distance of Mars from the sun with the measured value of the semimajor axis of its orbit given in Table 5.1.

Solution Since the ratio T^2/R^3 is the same for all planets, we may equate the value of the ratio for Mars to that for earth, using subscripts M to denote Mars and E to denote earth:

$$\frac{T_M^2}{R_M^3} = \frac{T_E^2}{R_E^3}$$

or

$$R_M^3 = \frac{T_M^2 R_E^3}{T_E^2},$$

$$R_M = \left(\frac{T_M^2 R_E^3}{T_E^2}\right)^{\frac{1}{3}}.$$

We may use the average distance of the earth from the sun as our unit of distance. This distance is called the astronomical unit, abbreviated AU. One astronomical unit is equal to 1.50×10^{11} m. We may likewise take the year as our unit of time. From Table 5.1, the period of Mars is 1.881 years, and the period of the earth is 1.000 year. These values give

$$R_M = \left(\frac{(1.881 \text{ year})^2 (1\text{AU})^3}{(1 \text{ year})^2}\right)^{\frac{1}{3}} = 1.524 \text{ AU}.$$

This result is in agreement with the observed value listed in Table 5.1.

5.4 The Law of Universal Gravitation

Now that we have examined uniform circular motion and Kepler's laws, let us return to the subject that prompted Newton's writing of the *Principia:* gravitation. In the *Principia,* Newton showed that the gravitational force acting on a body moving in an elliptical orbit—as a planet does—is inversely proportional to the square of the distance from the body to the center of force; that is, $F \propto 1/r^2$. We have seen in Example 5.4 why this was a reasonable conjecture, at least for a circular orbit.

The gravitational force between objects depends not only on the distance between them, but also on their masses. We have seen that the gravitational force on an object near the earth's surface is directly proportional to that object's mass. Furthermore, from Newton's third law we know that the same object exerts an equal and opposite force on the earth. From such reasoning, Newton proposed that the magnitude of the gravitational force between two objects is proportional to *both* their masses. Thus, the force between any two bodies with masses m_1 and m_2 has the form

$$F \propto \frac{m_1 m_2}{r^2},$$

where r is the distance between them. Strictly speaking, the law applies to what we call point masses, which are objects that have no size. However, for the sun, planets, and other bodies with spherical symmetry, the distance r is measured as the distance between their respective centers.

When we insert a constant of proportionality, this statement becomes

$$F = G\frac{m_1 m_2}{r^2}. \tag{5.11}$$

JOHANNES KEPLER

Today, children routinely learn the names of all the planets and their order from the sun, and travel throughout the solar system does not seem impossible. It's hard to imagine a time when people didn't know about the planets and how they move through space. An early step toward understanding the solar system occurred almost 400 years ago, with Johannes Kepler's analysis of planetary orbits.

Kepler (Fig. B5.1) proposed a scheme for explaining the radii of the orbits of the five known planets that was based on the geometry of the only five regular geometrical solids. This mystical scheme, which accidentally gave fairly good agreement with what was then known (but ruled out the existence of any more planets), gained him considerable publicity and led to his association with Tycho Brahe in 1600. Upon Tycho's death 18 months later, Kepler took possession of his data.

Kepler spent almost ten years trying to fit Tycho's observations of the position of Mars to a circular orbit, or to some combination of circles. He even came to a point where the disagreement between his calculations and Tycho's observations was only 8 minutes of arc. This is the angle covered by a dime held on edge and viewed from a distance of about 22 in. (56 cm). But Tycho's measurements were at least twice this good, equivalent to moving the dime to 44 in. Kepler had so much faith in the precision of Tycho's observations that he knew his own analysis must be in error. He discarded his work and started over many times, arriving finally at what are now called Kepler's laws of planetary motion. Kepler discovered the first two of his three laws in the attempt to understand the orbit of the planet Mars. They appeared in Kepler's *Astronomia Nova* in 1609. The third law appeared in 1619 in *Harmonices Mundi* (The Harmony of the Worlds). Kepler achieved a prominence in history for believing that the world worked according to logical principles that could be discovered and understood.

Figure B5.1   Johannes Kepler (1571–1630). In his first academic position at Graz, Austria, Kepler was both astronomer and astrologer.

Incidentally, Kepler was not the only person to benefit from the great precision of Tycho's measurements. Not only was there a general improvement in the quality of observations but in 1582 the new calendar of Pope Gregory XIII, called the Gregorian calendar, was instituted, in part because of the more accurate observations. The basic principle of this calendar system has not been changed since its introduction.

Because Kepler's laws rest on more general laws of nature, including Newton's universal law of gravitation, they apply to more systems than just the planets moving about the sun. The moon in its orbit about the earth obeys Kepler's laws with an adjusted constant k. Other objects in orbit about the earth also obey the same basic laws. The orbits and periods of artificial satellites used for observations and communications can likewise be understood in terms of Kepler's laws. Figure B5.2 shows a communication satellite of the type used to distribute television programs. When boosted to the proper altitude in an equatorial orbit, these satellites remain over the same place on earth as they move in synchronism with the earth's rotation on its axis. (They are called geosynchronous satellites.) Television companies and home users can receive the signals by aiming their dish antennas at the satellite.

Figure B5.2   A communication satellite of the type used to transmit television signals around the earth.

Equation (5.11) expresses Newton's **law of universal gravitation.** It states that every particle in the universe attracts every other particle with a force that is directly proportional to the product of their masses and inversely proportional to the square of the distance between them. Note that there is an attractive force between *any* pair of objects, whether they be the sun and the earth, the earth and you, or a leaf and a grain of sand.

Newton had some difficulty in showing to his own satisfaction that the law of universal gravitation holds when applied to spherical bodies of uniform density* if *r* is measured from the center of one sphere to the center of the other (Fig. 5.16). Newton's mathematical work on this problem eventually led to his invention of calculus, with which he proved that the mass of a symmetrical object of uniform density behaves under the law of universal gravitation exactly as if it were concentrated at the point of the object's center of symmetry. Such a point is called the *center of gravity.*

During the next two centuries, astronomers showed that the law of universal gravitation accounts for the motion of the entire solar system with great precision. The one exception was the motion of Mercury, a problem solved only in the twentieth century by the introduction of Einstein's theory of general relativity. Today's astronomers view gravitation as the force shaping the structure of the universe. No other fundamental force acts over such enormous distances and is always an attractive force, never balanced by repulsive forces. We believe gravitational forces are responsible for the formation of stars from clouds of gas and the formation of galaxies from millions of stars. When huge masses like these are involved, gravitational forces can be awesome in magnitude.

The **universal gravitational constant** *G* must be determined by experiment. Its numerical value depends on the system of units in which we make the measurement. Before describing the determination of *G*, let us consider two examples that allow us to use the law of gravitation without knowing the value of *G*. The first gives a result for things here on the surface of the earth; the second makes a prediction about the solar system.

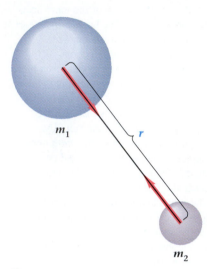

Figure 5.16

For spherical bodies of uniform density, the gravitational force between them is found from Newton's law with the distance *r* measured from the center of one to the center of the other.

Example 5.12

Consider a mass *m* falling near the earth's surface. Find its acceleration *g* in terms of the universal gravitational constant *G*, and draw some conclusions from the form of the answer.

Solution The gravitational force on the body is

$$F = \frac{GmM_E}{r^2},$$

where M_E is the mass of the earth and *r*, the distance of the mass from the center of the earth, is essentially the earth's radius (Fig. 5.17).

We have already noted that the gravitational force on a body at the earth's surface is

$$F = mg.$$

Figure 5.17

Example 5.12: A small object near the earth's surface is attracted by a gravitational force $F = GmM_E/r^2$.

---

*The **density**, ρ, of an object is defined as its mass per unit volume: $\rho = m/V$, where *m* is the mass of the object whose volume is *V*.

Setting the two expressions for the gravitational force on m equal to each other, we get

$$mg = \frac{GmM_{\mathrm{E}}}{r^2},$$

or

$$g = \frac{GM_{\mathrm{E}}}{r^2}.$$

Both G and M_{E} are constant, and r does not change significantly for small variations in height near the surface of the earth. Thus, the right-hand side of this equation does not change appreciably with position on the earth's surface. For this reason we may replace r with the average radius of the earth R_{E} to get

$$g = \frac{GM_{\mathrm{E}}}{R_{\mathrm{E}}^2}.$$

Discussion The law of gravitation predicts that the acceleration due to gravity of an object at the earth's surface is approximately constant and does not depend on the mass of the object. Experimentally we know that g does not vary appreciably from one place to another. This constancy of g is just what Galileo found (Section 2.7). Thus, the law of universal gravitation not only describes the forces that hold the planets in their orbits, but also describes the forces on objects close to the earth.

Example 5.13

Show that Kepler's third law follows from the law of universal gravitation.

Strategy We make the approximation that the orbits of the planets are circles. This approximation is essentially correct for those planets visible to the unaided eye. Because the mass of the sun is so much larger than the mass of the planet, we can assume, as Kepler did, that the sun lies at the center of the planetary orbit. It is also approximately true that the orbital speed is constant.

Solution Using the symbols in Fig. 5.18, we can say that the sun's gravitational force on any planet of mass m is

$$F = G\frac{mM}{r^2},$$

where M is the mass of the sun and r is the center-to-center distance between the planet and the sun. The circular orbit implies a centripetal force that may be expressed in terms of the orbital period as

$$F_{\mathrm{c}} = \frac{4\pi^2 mr}{T^2}.$$

This net force for circular motion is provided by the gravitational force. Equating these two forces, we get

$$\frac{GmM}{r^2} = \frac{4\pi^2 mr}{T^2}.$$

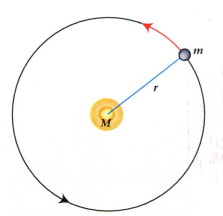

Figure 5.18
Example 5.13: A planet of mass m makes a circular orbit of radius r about the sun (mass M) with a period T. (Figure is not drawn to scale.)

Rearranging gives

$$\frac{T^2}{r^3} = \frac{4\pi^2}{GM}.$$

Discussion The right-hand side of this equation depends on the mass of the sun and on the universal gravitational constant, but not on any property of the planet. Therefore the ratio on the left-hand side is identical for all planets, just as Kepler observed. However, now we see that this result is not an independent physical law, but a consequence of a more fundamental law—the law of universal gravitation. We have derived Kepler's third law only for uniform circular motion of the planet, but the result is true for elliptical orbits if we use the average distance from the sun for r.

.5.5   The Universal Gravitational Constant *G*

After publication of Newton's *Principia,* experimenters tried to make some independent test of the law of universal gravitation, despite its obvious success in explaining the main features of planetary and lunar motion. Perhaps the most satisfying type of experiment would consist of placing known masses a known distance apart and measuring the attractive force between them. Under such conditions, the validity of the law could be checked and G determined in a straightforward manner from Eq. (5.11). However, common experience tells us that the gravitational force between ordinary bodies is extremely small. To measure this slight force, a sensitive device called a torsion balance was invented independently by Charles Coulomb in France and John Michell in England in the 1770s. Coulomb used his balance to measure the force between electrical charges (see Chapter 16). Michell, a professor of geology at Cambridge, designed his balance to measure gravitational forces and so to "weigh the earth." He did not complete the work before his death, but the apparatus eventually passed into the hands of Henry Cavendish, who refined and made use of it.

In 1798, 71 years after Newton's death, Cavendish first measured the force between small masses on earth. His interest, like that of many of his contemporaries, was in finding the earth's density. Only much later were his experimental results interpreted by others to give a value for G. Thus, although Cavendish is often remembered for determining the value of G, in fact he never did so.

Since the time of Cavendish, the value of G has been determined in a number of ways. The presently accepted value is

$$G = 6.673 \times 10^{-11} \, \text{N} \cdot \text{m}^2/\text{kg}^2.$$

Using this value of G, we can apply the law of universal gravitation to many different situations. Examples 5.14 and 5.15 illustrate a few of these.

Example 5.14

Use the law of universal gravitation and the measured value of the acceleration of gravity g to determine the average density of the earth.

Strategy We start with the result of Example 5.12 that

$$g = \frac{GM_\text{E}}{R_\text{E}^2},$$

where M_E is the mass of the earth, R_E is its radius, and g is the acceleration of gravity at the earth's surface. Then we substitute for M_E an expression involving ρ, the average density of the earth, defined as the ratio of the earth's mass to its volume, $\rho = M_\text{E}/V$. The result gives us ρ as a combination of numerical values.

Solution If we take the earth to be a sphere of radius R_E, then

$$\rho = \frac{M_\text{E}}{\frac{4}{3}\pi R_\text{E}^3}.$$

The equation for g can then be rewritten in terms of the density as

$$g = \frac{G\left(\frac{4}{3}\pi R_\text{E}^3 \rho\right)}{R_\text{E}^2} = \frac{4}{3}G\pi R_\text{E}\rho.$$

Upon rearranging, we find the density to be

$$\rho = \frac{3g}{4\pi R_\text{E}G}.$$

Inserting the numerical values, we get

$$\rho = \frac{3(9.81 \text{ m/s}^2)}{4\pi(6.38 \times 10^6 \text{ m})(6.67 \times 10^{-11} \text{ N} \cdot \text{m}^2\text{kg}^{-2})}$$
$$\rho = 5.50 \times 10^3 \text{ kg/m}^3.$$

This is the average density of the entire earth and is 5.5 times the density of water. Because the average density is greater than that of soil and rock, we know that the interior of the earth must be very dense compared to the material in the earth's crust. (See Henry Cavendish and the Density of the Earth, p. 167.)

Example 5.15

Estimate the period of an artificial earth-orbiting satellite that passes just above the earth's surface (Fig. 5.19).

Strategy In reality, a satellite's orbit must be high enough so that the satellite is above most of the earth's atmosphere. However, the height at which low-orbiting satellites operate is small compared to the radius of the earth (about 6.4×10^6 m). Therefore, for estimating purposes, we can approximate the radius of the orbit by saying that it's the same as the radius of the earth.

Solution We set the force required to give a circular orbit—the centripetal force—equal to the gravitational force. We let the mass of the satellite be m, the mass of the earth M_E, the radius of the orbit R_E, and the satellite's period T. Then

$$\frac{m4\pi^2 R_\text{E}}{T^2} = \frac{GmM_\text{E}}{R_\text{E}^2}, \quad \text{or} \quad T = \sqrt{\frac{4\pi^2 R_\text{E}^3}{GM_\text{E}}}.$$

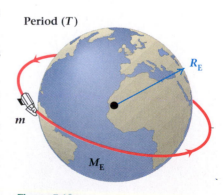

Period (T)

Figure 5.19
Example 5.15: Finding the period of an earth-orbiting satellite.

Rather than substituting the numerical values now, we use the result of Example 5.12 that $gR_E^2 = GM_E$, so that the above expression for the period becomes

$$T = \sqrt{\frac{4\pi^2 R_E^3}{gR_E^2}} = \sqrt{\frac{4\pi^2 R_E}{g}}.$$

Notice that the period depends only on the radius of the earth and the acceleration of gravity. To complete our estimate, we insert the approximate values of $\pi^2 \approx 10$, $g \approx 10$ m/s², and $R_E \approx 6.4 \times 10^6$ m,

$$T \approx \sqrt{\frac{4(10)(6.4 \times 10^6 \text{ m})}{10 \text{ m/s}^2}} \approx 5100 \text{ s} \approx 85 \text{ min.}$$

This value compares favorably with the observed periods of low-orbiting satellites. For example, the space shuttle typically has an orbital period of 90 min.

*5.6 Gravitational Field Strength

We can gain additional insight into gravitation by considering gravitational forces from a different but related point of view. Recall that gravitational forces are not contact forces but instead act over distances in space. The gravitational force on an object at some point in space can be described in terms of a property of that space. We consider that a mass, such as the earth, influences the surrounding space in such a way that another mass, such as the moon, in that space will experience a force in the direction of the first mass. This property of the space is called the gravitational field. The field due to the first mass exists even when the second mass is not present.

We define the **gravitational field strength** *at any point in space to be the gravitational force per unit mass on a test mass* m_0. Thus, at a point in space where a test mass m_0 experiences a gravitational force **F**, the gravitational field strength is

$$\Gamma \equiv \frac{\mathbf{F}}{m_0}. \tag{5.12}$$

Note that the gravitational field strength is just the acceleration that a unit mass would experience at that point in space. The test mass must be small so that its gravitational field does not modify the field that is being measured.

Later, we will use the concept of a field when studying electricity and magnetism. Even though the electric field does not have the same units as the gravitational field, the concepts of field apply in both cases, allowing us to draw useful analogies between gravitation and electricity when they are considered from the standpoint of fields.

Since the gravitational force is a vector, the field must also be a vector, having both magnitude and direction. If the gravitational force arises from the attraction of the test mass by a mass M located a distance r from the test mass, then the magnitude of the field strength is

$$\Gamma = \frac{F}{m_0} = \frac{GMm_0}{r^2 m_0} = \frac{GM}{r^2}.$$

HENRY CAVENDISH AND THE DENSITY OF THE EARTH

Like most eighteenth-century English scientists, Henry Cavendish (1731–1810) was influenced by the questions in Newton's *Principia* and *Optics*. This influence led him to investigate gravitational forces. Because gravitational forces between ordinary objects are so very small, Cavendish had to use a special balance to measure them (Fig. B5.3). His balance was based on a design by the Cambridge geologist John Michell (1724–1793)

Cavendish's 1798 paper, *Experiment to Determine the Density of the Earth,* contained a drawing of the torsion balance used in his experiment. In Cavendish's own words,

Figure B5.4 Details from a gravity map of South Carolina. The contour lines are spaced 5 mgal apart. The *gal,* a unit named after Galileo, is 1 cm/s². Thus, 1 mgal is approximately 10^{-6} *g.* From South Carolina Geological Survey MS 21, "Simple Bouguer Anomaly Map of South Carolina."

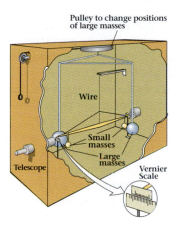

Figure B5.3 The Cavendish Balance: A torsional balance constructed of two small masses balanced on a light rod suspended by a thin fiber. Two large masses (lead spheres) are arranged symmetrically on either side of the rod so as to give it a rotational force. The restoring force is supplied by the fiber.

Labels in figure: Pulley to change positions of large masses; Wire; Small masses; Large masses; Telescope; Vernier Scale

"The apparatus is very simple; it consists of a wooden arm, 6 feet long, made so as to unite great strength with little weight. This arm is suspended in an horizontal position, by a slender wire 40 inches long, and to each extremity is hung a leaden ball about 2 inches in diameter; and the whole is enclosed in a narrow wooden case, to defend it from the wind.

As no more force is required to make this turn round on its center, than what is necessary to twist the suspending wire, it is plain, that if the wire is sufficiently slender, the most minute force, such as the attraction of a leaden weight a few inches in diameter, will be sufficient to draw the arm sensibly aside. The weights which Mr. Michell intended to use were 8 inches in diameter. One of these was to be placed on one side of the case, opposite to one of the balls, and as near it as could conveniently be done, and the other on the other

side, opposite to the other ball, so that the attraction of both the weights would conspire in drawing the arm aside; and when its position, as affected by these weights, was ascertained, the weights were to be removed to the other side of the case, so as to draw the arm the contrary way, and the position of the arm was to be again determined; and, consequently, half the difference of these positions would show how much the arm was drawn aside by the attraction of the weights."

By improving on Michell's apparatus and by using the utmost care, Cavendish performed this delicate experiment in a series of 17 trials whose results clustered around an average value of 5.48 for the density of the earth as compared with water. (See Example 5.14.) Recall that the density of any material is the ratio of its mass to its volume. Cavendish reckoned that his measurement was correct to within about 7%. His result corresponds to a value for G of $(6.70 \pm 0.48) \times 10^{-11}$ N · m²/kg², which differs little from the modern value of

$$G = 6.673 \times 10^{-11} \, \text{N} \cdot \text{m}^2/\text{kg}^2.$$

Cavendish's interest in the density of the earth has its counterpart today. Local variations in the density of the earth's crust can yield information about mineral and oil deposits. Consequently, geologists have developed instruments to measure the acceleration of gravity with great precision. One type of gravity meter uses a very sensitive spring scale in which a mass is pulled down more where the force of gravity (and therefore g) is larger. Another type uses the connection between the period of a pendulum and the acceleration of gravity, a topic discussed in Chapter 14.

These small variations from place to place can be used to map the underlying geological structure. Modern gravity meters can easily measure variations in g as small as 10^{-6} m/s². Figure B5.4 is a map showing the variations in the acceleration of gravity in South Carolina. The contours for constant values of g are spaced at intervals of 5×10^{-5} m/s². A trained geologist can use such a map to help infer information about the mineral content and seismic structure of the underlying terrain.

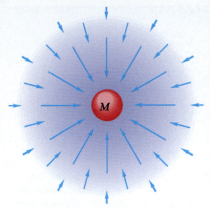

Figure 5.20

A representation of the gravitational field around a point mass M. Each arrow indicates the magnitude and direction of the field at the base of the arrow.

The field vector lies along the line from m to M and is directed toward the mass M. For example, the gravitational field at the earth's surface is a vector directed toward the center of the earth with magnitude 9.81 m/s^2.

Once the gravitational field at any point in space has been determined (either by measurement or by calculation), then we can compute the force on any other mass m placed at that point by using $\mathbf{F} = \mathbf{\Gamma}m$. In fact, we may say that the gravitational field is a property of space. Instead of focusing on the mass of an object and how the gravitational force depends on distance, we focus on the space itself and how a property of the space (the field) is affected by the presence of objects near and far. Masses can then be treated as sources of the gravitational field, and the force on some particular mass is determined by the field present at the location of that mass. We can represent this field visually with the aid of arrows representing the direction and magnitude of the field at different points in space. This representation is shown in Fig. 5.20 to illustrate the field around a point mass M. Each arrow represents the field at the base of the arrow. The lengths of the arrows are proportional to the magnitude of the field at each point. (We have shown a two-dimensional drawing, but the gravitational field itself extends outward from the source in all directions in three-dimensional space.)

Another way to help visualize the gravitational field is to diagram **lines of force,** also called **field lines.** These continuous lines are drawn in the direction of the force on a test mass (Fig. 5.21). The relative number of lines is proportional to the strength of the force and hence proportional to the field. Such a representation helps show that the field strength diminishes as the distance from the mass increases. The farther away from the source of the gravitational field, the farther apart the lines are, and the weaker the field becomes.

We will draw lines of force later, in the chapters on electricity and magnetism, as an aid in understanding the forces encountered there. The field concept was introduced by Michael Faraday (Chapter 20) in connection with his experiments on electricity and has become the contemporary way of describing the effects and interactions of the fundamental forces of nature.

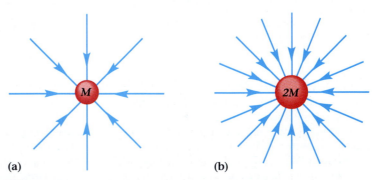

(a)   (b)

Figure 5.21

Lines of force (a) around a mass m and (b) around a mass $2M$.

Summary

Useful Concepts

■ Angles can be measured in degrees or radians, where the angle in radians is given by

$$\theta = \frac{s}{r} \quad \text{and} \quad 1 \text{ radian} = 57.3°.$$

■ Uniform circular motion can be described by an angular velocity and a centripetal (center-seeking) acceleration. The centripetal acceleration always points to the center of the circular path of motion and is perpendicular to the linear velocity:

$$a_c = \frac{v^2}{r}.$$

■ The frequency of revolution is the reciprocal of the period, which is the time required for one revolution:

$$f = \frac{1}{T}.$$

■ The angular velocity (also called angular frequency) is

$$\omega = \frac{v}{r}$$

and is proportional to the rotational frequency f,

$$\omega = 2\pi f.$$

■ The centripetal acceleration used in Newton's second law gives the centripetal force,

$$F_c = \frac{mv^2}{r}.$$

The centripetal force is not a fundamental force, but describes the effect of any net force applied to a body that keeps it moving in uniform circular motion.

■ Kepler's laws of planetary motion are

1. The orbit of each planet is an ellipse with the sun at one focus.

2. An imaginary line from the sun to a moving planet sweeps out equal areas in equal intervals of time.

3. The ratio of the square of a planet's period of revolution to the cube of its average distance from the sun is a constant. This constant is the same for all planets:

$$\frac{T^2}{R^3} = k,$$

■ Newton's law of universal gravitation gives the force between *any* two point masses, no matter what their mass and no matter where they are located,

$$F = \frac{Gm_1m_2}{r^2}.$$

■ We can visualize the gravitational field by drawing continuous lines of force (field lines) showing the direction of the gravitational force at any point.

Important Terms

You should be able to write the definition or meaning of each of the following terms:

radian	Kepler's laws of planetary
uniform circular motion	motion
centripetal acceleration	law of universal gravitation
period	density
frequency	universal gravitational constant
hertz	gravitational field strength
angular velocity	lines of force
(angular frequency)	field lines
centripetal force	

Conceptual Questions

5.1 Can you think of any reasons for dividing a full circle into 360°?

5.2 Draw a diagram showing how you would measure the angle that a distant person subtends at your eye.

5.3 Describe the forces acting on a passenger riding on a Ferris wheel. Explain any difference between the situation near the top of the ride and that near the bottom.

5.4 When a fenderless bicycle is ridden along a wet street, the rider gets a wet stripe down his back from water droplets thrown off the rear wheel. Explain this observation.

5.5 Explain the action of the spin cycle in removing water from clothes in an automatic washing machine.

5.6 What complications would you encounter in playing catch with a baseball if you and your partner were standing on a rotating carousel?

5.7 How can an object in uniform circular motion be moving at a constant speed and at the same time be constantly accelerated?

5.8 Make an order-of-magnitude estimate of the linear speed of a point on the earth's equator due to the earth's rotational motion. Compare this speed with the linear speed of the earth's orbital motion.

5.9 Explain what causes the riders in the swing of Fig. 5.22 to move outward when the ride begins to rotate.

5.10 What are the restrictions on the orbit of a communications satellite if it is to remain stationary above a given location on the earth?

5.11 If the average distance between the earth and the moon were suddenly cut in half, how would the lunar period be affected?

5.12 What is universal about the law of universal gravitation?

5.13 Assume that you are inside a train moving along a level track at constant speed. All of the window shades are drawn so that you cannot see out. You have with you a sensitive spring balance with a mass hanging from it and a finely ruled protractor. How can you use these instruments to tell whether the train is traveling along a straight track or going around a curve?

Figure 5.22
Question 5.9.

Problems

Section 5.1 Uniform Circular Motion

Hints for Solving Problems

In uniform circular motion, the speed is constant and the centripetal acceleration, which is perpendicular to the velocity, points to the center of the circle. Be sure you know how to express centripetal acceleration in terms of different rotational quantities.

5.1 A protractor is made so that the edge of its scale is 7.5 cm from the center point. If the scale is marked in degrees, how far apart are the marks along the edge?

5.2 The moon subtends an angle of 9.06×10^{-3} radians and is 3.84×10^8 meters from the earth. What is the approximate diameter of the moon?

5.3 A 12-in.-diameter phonograph record rotates about its center by one-quarter turn. (a) Through how many radians has it turned? (b) How far has a point on the rim moved?

5.4 What is the centripetal acceleration of an automobile driving at 40 km/h on a circular track of radius 20 m?

5.5 The earth is 1.5×10^{11} m from the sun and has a period of about 365 days. Assume the earth's orbit to be circular and determine the magnitude and direction of its radial acceleration in m/s².

5.6 The centripetal acceleration at a point in a sample tube 5.60 cm away from the axis of rotation of a centrifuge is 4300 m/s². What is the instantaneous speed at that point?

5.7 Show that v^2/r has the dimensions of acceleration.

5.8 The centripetal acceleration at the equator is about 3.4 cm/s². Use that information and the length of a day to estimate the radius of the earth.

5.9 A bicycle tire is 66 cm in diameter. (a) At what frequency f does the tire rotate when the bicycle is traveling at a speed of 30 km/h? (b) What is the angular frequency ω?

5.10• What is the magnitude and direction of the centripetal acceleration due to the earth's rotation at a location near Kansas City at 38° latitude?

5.11• The period of a stone swung in a horizontal circle on a 2.00-m radius is 1.00 s. (a) What is its angular velocity in rad/s? (b) What is its linear speed in m/s? (c) What is its radial acceleration in m/s²?

5.12• During 0.19 s, a wheel rotates through an angle of 2.36 rad as a point on the periphery of the wheel moves with a constant speed of 2.87 m/s. What is the radius of the wheel?

5.13• A 32.5-cm-radius tire on a moving car turns through 3π rad in 0.27 s. What is the speed of the car in kilometers per hour?

5.14• Jupiter's moon Europa has an average orbital radius of 6.67×10^8 m and a period of 85.2 h. Calculate the magnitude of (a) its average orbital speed, (b) the angular velocity, and (c) the centripetal acceleration of Europa.

Section 5.2 Force Needed for Circular Motion

Hints for Solving Problems

The formulas for centripetal force describe the force necessary to make an object move in a circle. In this chapter the applied force is often gravity. Remember that objects in circular motion obey Newton's three laws. Also, remember to draw and label a free-body diagram before attempting a problem.

5.15 Calculate the centripetal force on a 2000-kg automobile rounding a curve of 175 m radius at a speed of 50 km/h.

5.16 A stunt driver drives a car so fast that it leaves the ground as it tops a hill. If the hill can be approximated by a 165-m-radius vertical circle, what speed must the car exceed if it is to leave the ground?

5.17 A race track curve has a radius of 100 m and is banked at an angle of 68°. For what speed was the curve designed?

5.18 A velodrome track is banked so that a bicycle traveling at 62.3 km/h will have no tendency to slip to either side when traveling on the path that has a radius of curvature of 77.0 m. What is the banking angle?

5.19● What angle does a plumb bob line make with the vertical in a train rounding a 300-m-radius curve at 27 m/s?

5.20● A spring scale on a rotating platform indicates that the horizontal force on a 0.452 kg mass is 1.34 N when the mass is 2.37 m from the axis of rotation. How long does it take for the platform to make one revolution?

5.21● A 0.208-kg toy whistle can be whirled in a horizontal circle of 1.00 m radius at a maximum of 3.00 rev/s before the string breaks (see Fig. 5.10). What is the force needed to break the string?

5.22● A 0.237-kg block slides down the inside of a 0.213-m-radius circular track, reaching the lowest point with a speed of 1.37 m/s. If the coefficient of friction between the block and track is 0.28, what is the frictional retarding force on the block at the lowest point?

5.23● A 0.436-kg ball is suspended on a 0.452-m cord from a fixed point. The ball swings in a horizontal circular path at 0.811 revolutions per second. (a) What is the tension in the cord? (b) What is the angle θ between the cord and the vertical?

5.24● A coin placed on a turntable rotating at 33.3 rev/min will stay there if its center is placed no further than 8.5 cm from the axis of rotation. (a) Find an expression for the maximum distance the center of the coin can be placed from the axis if the turntable rotates at 45 rev/min? (b) What is the coefficient of friction between the coin and turntable?

5.25● A 0.255-kg ball tethered to a tall pole on a 1.37-m rope is thrown so that it travels in a horizontal circle with the rope making an angle $\theta = 40°$ with the vertical pole (Fig. 5.23). (a) What is the speed of the ball? (b) What is the tension in the rope?

5.26● An electron with mass 9.11×10^{-31} kg moves with a speed of 2.00×10^{6} m/s in a circle of 2.85 cm radius under the influence of a magnetic field. A proton of mass 1.67×10^{-27} kg, moving in the same plane with the same speed, experiences the same centripetal force. What is the radius of the proton's orbit?

5.27● A stunt pilot in an airplane diving vertically downward at a speed of 220 km/h turns vertically upward by following an approximately semicircular path with a radius of 180 m (Fig. 5.24). (a) How many g's does the pilot experience due to his motion alone? (b) By what factor does the pilot's weight appear to increase at the bottom of the dive?

5.28●● A highway curve with a radius of 750 m is banked properly for a car traveling 120 km/h. If a 1590-kg Porsche

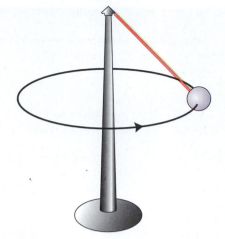

Figure 5.23
Problem 5.25.

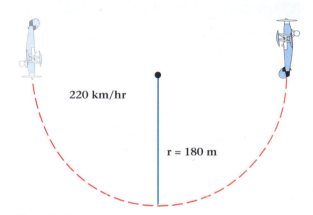

220 km/hr

r = 180 m

Figure 5.24
Problem 5.27.

928S rounds the curve at 230 km/h, how much sideways force must the tires exert against the road if the car does not skid?

5.29●● A small dog slides with constant speed down a metal sliding board at 30° to the horizontal. The dog then sits 1.3 m from the center of a rotating turntable made of the same metal as the sliding board. What is the maximum angular frequency at which the turntable may turn without the dog sliding off?

5.30●● Passengers riding in the Great Six Flags Air Racer are spun around a tall steel tower (Fig. 5.25). At top speed the planes fly at a 56° bank approximately 14 m from the tower. In this position the support chains make an angle of 56° with the vertical. Calculate the speed of the planes.

Section 5.3 Kepler's Laws of Planetary Motion

5.31 Calculate the distance from the sun to Saturn, given the information that Saturn's period of revolution about the sun is 29.46 years.

5.32 An amateur astronomer claims to see a new planet beyond the orbit of Pluto, with a calculated period of 230 years. Can this claim be true? Assume the orbit to be nearly circular.

Figure 5.25
Problem 5.30: The Air Racer.

5.33• Use the known period of $27\frac{1}{3}$ days for the motion of the moon about the earth and the distance from the earth to the moon of 3.84×10^8 m to calculate the radius of the orbit of an earth satellite that stays above the same point on the equator. (*Hint:* Use Kepler's third law.)

5.34• In *Gulliver's Travels,* the Lilliputians claim "They have likewise discovered two lesser stars, or satellites, which revolve about Mars, whereof the innermost is a distance from the centre of the primary planet exactly three of his diameters, and the outermost five; the former revolves in the space of ten hours, and the latter in twenty-one and a half; . . ." Could the claim of the Lilliputians be true, if time were measured to within the nearest one-quarter hour?

5.35• Calculate the ratio of T^2/R^3 for each planet in the solar system. Use the data from Table 5.1. What is the percentage difference between the highest and lowest values? The percentage difference is the ratio of the difference in the values to the average of the two values expressed as a percentage.

5.36•• Assuming that the orbits of the planets are circular, show that the product of the orbital radius with the square of the speed of a planet is the same for all planets. The speed is the distance traveled divided by the time required to go that distance.

Section 5.4 The Law of Universal Gravitation

5.37 What is the ratio of the acceleration of gravity on the surface of the moon to the acceleration of gravity on the surface of the earth? (The radius of the moon is 0.273 times the radius of the earth, and the mass of the moon is 0.0123 times the mass of the earth.)

5.38• Saturn's mass is 95 times that of earth, and its radius is 9.0 times that of earth. (a) Calculate the acceleration of gravity at the surface of Saturn. Express your answer in g's. (b) Calculate the average density of Saturn.

5.39• At what fraction of the center-to-center distance from the earth to the moon will their opposing gravitational forces on a spaceship traveling between them be equal in magnitude? (The mass of the moon is 0.0123 times the mass of the earth.)

5.40•• You drive an automobile of mass 2.10×10^3 kg from sea level to the top of a mountain 2.05 km high. What is the automobile's change in weight? (*Hint:* You will need to use the binomial expansion given in Appendix A.)

5.41•• Assuming the earth to be an oblate spheroid, so that the distance from the center to the equator is 27 mi (43.5 km) greater than the distance from the center to the poles, calculate the approximate percentage change in a person's weight at the poles and at the equator. Take into account the fact that the earth is turning. (*Hint:* You will need to use the binomial expansion given in Appendix A.)

Section 5.5 The Universal Gravitational Constant G

5.42 An apple ($m = 0.20$ kg) falls to the earth. Determine (a) the apple's acceleration toward the earth; (b) the earth's acceleration toward the apple. (c) Discuss the appropriate reference frames in which to determine the accelerations in (a) and (b).

5.43 Calculate the mass of a lead brick of dimensions 5.0 cm × 10 cm × 30 cm given that the density of lead is 1.13×10^4 kg/m^3.

5.44 Gold has a density of 1.93×10^4 kg/m^3. (a) Calculate the volume of 0.500 kg of gold. (b) If this amount were shaped into a cube, what would be the length of one edge of that cube?

5.45 Suppose the earth and moon were held at rest at their present separation and then released to move under their mutual gravitational attraction. What would be the initial acceleration of each body? Be specific about your choice of coordinate frames.

5.46 Compute the gravitational force between the sun ($M = 1.99 \times 10^{30}$ kg) and the planet Uranus ($m = 14.5M_E$, orbital radius $r = 19.2$ AU).

5.47 Calculate the mass of Jupiter from the knowledge that its satellite Io (Fig. 5.26) orbits at an average distance of 4.22×10^5 km from its center with an orbital period of 42.5 h.

5.48 Calculate the orbital period of Venus from knowledge of G, the mass of the sun ($M = 1.99 \times 10^{30}$ kg), and the Venusian orbit radius of 1.08×10^{11} m.

5.49 In Henry Cavendish's famous experiment, we can determine that the force between the ball and weight in each pair was about 1.53×10^{-7} N when the torsion pendulum was in its equilibrium position. From the data given in Fig. 5.27, calculate the universal constant of gravitation.

5.50• If the earth revolved rapidly enough, the weight of objects at the equator would be zero. What would be the length of the day in that case?

Figure 5.26
Problem 5.47: Jupiter's moon Io.

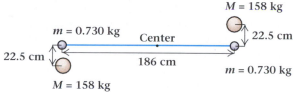

Figure 5.27
Problem 5.49: Diagram of a torsion pendulum in its equilibrium position as viewed from above.

5.51● (a) Determine the force the sun exerts on a kilogram of water on the earth's surface at a point nearest the sun and at a point farthest from the sun. (b) Do the same for the force exerted by the moon. (c) Explain why the tides are associated with the motion of the moon.

5.52● (a) Calculate the mass of the sun from the radius of the earth's orbit (1.5×10^{11} m), the earth's period in its orbit, and the gravitational constant G. (b) What is the density of the sun and how does it compare with the density of the earth? (The sun's radius is 6.96×10^8 m.)

5.53● (a) What is the radius of the orbit of a communications relay satellite that always remains above one point on the earth's surface? Such an orbit is called a geosynchronous orbit. (b) Can such a satellite be placed in geosynchronous orbit over *any* point on the earth's surface? Why?

5.54● Assume that the orbit of the moon about the earth is a perfect circle with a center-to-center distance from earth to moon of 3.84×10^8 m. (a) What is the gravitational force between the earth and the moon? (b) What is the speed of the moon in its orbit? Give your answer in meters per second.

5.55● An astronaut weighing 700 N on earth travels to the planet Mars. What does the astronaut weigh on Mars? (The mass of Mars is 0.107 of the earth's mass. The radius of Mars is 0.530 of the earth's radius.)

5.56● If an astronaut dropped a small rock near the surface of Mars, how far would the rock fall in 1.00 s? (The mass of Mars is 0.107 the mass of earth, and the radius of Mars is 0.530 the radius of the earth.)

5.57●● (a) Show that the speed of a satellite in orbit just above the surface of a planet of density ρ is proportional to the radius of the planet. (b) Imagine a small planet with the same density as the earth. What would be its radius if a sprinter capable of running 100 m in 10 s could launch herself into orbit just by running?

5.58●● Before Cavendish, Robert Hooke attempted to measure G by weighing the same object at heights differing by 300 ft (91.4 m). (a) What fractional change in weight would be predicted by the law of universal gravitation? (b) What difference in height would he have had to use to detect a change in the weight of a 1.0000-kg mass equivalent to the weight of 1.00-g mass?

*Section 5.6 Gravitational Field Strength

5.59 What is the gravitational field strength on the surface of the earth due to the earth alone? In your answer, give magnitude, direction, and proper units to agree with Eq. (5.12).

5.60 Compare the magnitude of the gravitational field at the surface of the earth due to the moon with that due to the sun. The mass of the sun is 1.99×10^{30} kg, the mass of the moon is 7.35×10^{22} kg, and the distance from the surface of the earth to the moon is 3.78×10^8 m.

5.61● Two 1.0-kg masses are located so that one is at each of two corners of an equilateral triangle with sides 1.0 m long. Determine the magnitude and direction of the gravitational field at the third corner due to these masses. Neglect the effects of all other masses, including the earth.

Additional Problems

5.62 The disk in a CD player does not rotate at a constant rate, but spins at a rate determined by a control circuit so that the linear speed of the track being read is constant. The laser beam used to read the data on the disk starts at an inner radius of 2.5 cm and continues to read until reaching an outer radius of 5.8 cm. If the disk turns at the rate of 490 rev/min at the start, what will be its rotation rate at the end?

5.63 Comets, like planets, are part of the solar system. Ericke's comet is a comet with the shortest known period, 3.3 years. What is its average distance from the sun?

5.64● A popular amusement park ride consists of a broad short cylinder arranged so that it rotates around its vertical axis (Fig. 5.28). People stand inside the cylinder with their backs to the outer wall and "feel pushed back" when the cylinder rotates. When the cylinder is rotating fast enough, it is tipped so that its axis of rotation is almost horizontal. If the radius of the cylinder is 4.5 m, how fast must it rotate so that the riders do not fall away from the walls at the topmost position? Give your answer in hertz.

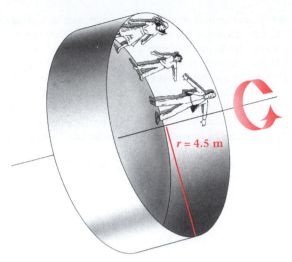

Figure 5.28
Problem 5.64.

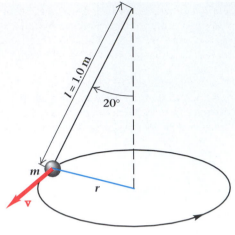

Figure 5.29
Problem 5.70.

5.65• The gravitational attraction due to a spherical mass is the same as that of a point mass of the same magnitude located at the center of the sphere. Use this fact to calculate the radius of a lead sphere that would attract you with a force 10^{-3} times your own weight if you stood right next to it. (The density of lead is 1.14×10^4 kg/m^3.)

5.66• The average density of the planet Mercury is 5.61×10^3 kg/m^3. The acceleration of gravity on its surface is 3.92 m/s^2. Calculate the average radius of Mercury.

5.67• You swing a bucket of water at constant speed in a vertical circle at arm's length (0.70 m). What is the minimum number of revolutions per second you must maintain to keep the water from spilling out of the bucket?

5.68• A strange new planet that has no atmosphere has a satellite that orbits very close to the planet's surface with a period of 1.63 h. What is the approximate density of the planet? (*Hint:* Approximate the orbital radius with the planet's radius.)

5.69• (a) Compute the mass of the earth from knowledge of the earth-moon distance (3.84×10^8 m) and of the lunar period (27.3 days). (b) Then calculate the average density of the earth. The average radius of the earth is 6.38×10^6 m.

5.70• A bob of mass m is whirled in a circular path on the end of a string 1.0 m long. If the string makes an angle of 20° with the vertical (Fig. 5.29), what is the tangential speed of the bob?

5.71• The Wheelie, described in Example 5.5 and shown in Fig. 5.9, can be tilted until its plane of rotation makes an angle of 89° with the horizontal. Show that when it is in this position, the force exerted by the wheel on the riders at the top is equivalent to only 0.94 g, while at the bottom it is 2.9 g. You may approximate by assuming the tilt angle to be 90°.

5.72•• Halley's comet has a period of about 75 years and comes relatively close to the sun. Estimate how far it is from the sun when it is at the farthest point of its elliptical orbit.

5.73•• A long string tied to an overhead support has a weight hanger attached to its lower end. The string can support a maximum weight Mg without breaking. A piece of the same string is doubled and a mass M is attached to one end of the doubled string. The mass is then swung in a vertical circle of 0.75 m radius. What is the maximum frequency of rotation that can be maintained without breaking the string?

5.74•• A coin placed on a flat stationary phonograph record just begins to slide when the record is tilted to an angle of 45° from the horizontal. The 30-cm-diameter record is placed on a horizontal turntable, and the coin is placed at the edge of the record. Will the coin slide off when the record is rotated at the rate of $33\frac{1}{3}$ rev/min?

5.75•• An automatic tumble dryer has a 0.65-m-diameter basket that rotates about a horizontal axis. As the basket turns, the clothes fall away from the basket's edge and tumble over. If the clothes fall away from the basket at a point 60° from the vertical (Fig. 5.30), what is the rate of rotation in units of revolutions per minute?

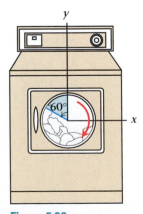

Figure 5.30
Problem 5.75.

5.76•• A newly found planet with a density of 3.90×10^3 kg/m^3 has no atmosphere and is orbited by a low altitude satellite with an orbital speed of 3.55 km/s. What is the mass of the planet? (*Hint:* Approximate the orbital radius with the planet's radius.)

5.77•• What is the gravitational acceleration on a rocket ship of mass m moving along a line between the earth and the moon at distances from the earth of $\frac{1}{4}$, $\frac{1}{2}$, and $\frac{3}{4}$ the earth-moon distance. Include the gravitational effects of both the moon and the earth. The mass of the moon is 0.0123 times the mass of the earth. Express your answer in terms of the acceleration of gravity g at the earth's surface.

5.78•• A light spring with spring constant k is attached to a peg in a level, frictionless surface. A mass m is attached to the other end of the spring. The spring is free to rotate about the peg. The mass is given a push so that it moves in a circular orbit about the peg. Find an expression for the extension in the spring when the mass orbits with a period T.

Work and Energy

Perhaps the most generally useful idea in all of science is the concept of energy and its conservation. Energy is a vital part of our daily lives. The food we eat gives our bodies energy for movement; electrical energy lights our homes and streets; oil and gas propel our cars and keep us warm. These are all examples of using energy. In this chapter we define work and mechanical energy and arrive at quantitative relationships between them. We will extend and apply these principles in the following chapters.

The terms *force* and *energy* were not always clearly defined. Before the midnineteenth century, they were often used interchangeably. However, progress in mechanics and in thermal physics helped clarify these ideas and the distinction between them. In 1807 the English scientist Thomas Young (1773–1829) introduced the word *energy* to denote the quantity of work that a system can do. Later, the Scottish engineer and thermal physicist W. J. M. Rankine (1820–1872) popularized this definition and coined the terms *potential energy* and *conservation of energy*. As is often the case in science, this clarification of terms and definitions led to greater insight and understanding of natural laws and their consequences.

Today the principle of conservation of energy is part of the framework of physical theory. Our faith in this principle is based on years of experience. We will encounter this idea in one form or another throughout the rest of the book and use it to derive other results. Indeed, we will see that many laws in various areas of physics are simply alternative versions of the law of conservation of energy, stated in different terms. ■

6.1   Work

The word *work* means many different things to us in our daily lives. We say that we work when we rake the yard, or buy groceries, or drive a truck. We also do work when we push a box across the floor. How much work we do depends on both how hard we push and how far we move the box. In the physical sciences, the meaning of work is more precise and restricted than in everyday usage. If we exert a constant force **F** on an object (Fig. 6.1), causing it to move a distance x parallel to **F**, then the **work** W done by the force is defined to be the product of the magnitude of the force times the distance through which it acts as the object is moved.

There are two important conditions in our definition of work. First, the force must be exerted on the object through a distance. In other words, *the force must move the object*. Consider the ancient Greek myth about Atlas, who held up the sky. Atlas is often depicted as a stooped but powerful figure, bearing the earth on his shoulders. Atlas soon tired of his terrible burden. But according to our definition, as long as he held the earth stationary, he actually did no work on it. In a similar fashion, you could push with all your might against a stationary wall until your muscles ached with the effort. Nevertheless if the wall did not move, you would not have done any work on the wall. Note, however, that work is being done within your body as muscle fibers repeatedly contract.

Second, for work to be done, the force must have a component parallel to the direction of motion. If an applied force is not along the direction of motion, we can resolve it into components parallel to and perpendicular to the displacement (Fig. 6.2). *Only the component of force that is parallel to the displacement contributes to the work.* Thus, if the force **F** makes an angle θ with the line of motion, chosen as the x direction in the figure, then the component of force that contributes to the work is $F_x = F \cos \theta$. Mathematically, the work done is defined to be

$$W \equiv F_x x = Fx \cos \theta. \qquad (6.1)$$

When **F** is along the direction of the displacement **x** (as shown in Fig. 6.1), then $\theta = 0$ and $\cos \theta = 1$. For that special case the work becomes

$$W = Fx.$$

Notice that, although the work done depends on the force, work itself is a scalar, not a vector, quantity. It has magnitude but no direction. We can add

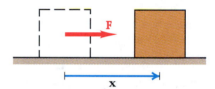

Figure 6.1
A force **F** acting in a direction parallel to the displacement **x** of an object does an amount of work $W = Fx$.

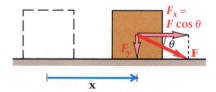

Figure 6.2
A force **F** making an angle θ with the horizontal pushes a box through the displacement **x** and does an amount of work $W = Fx \cos \theta$.

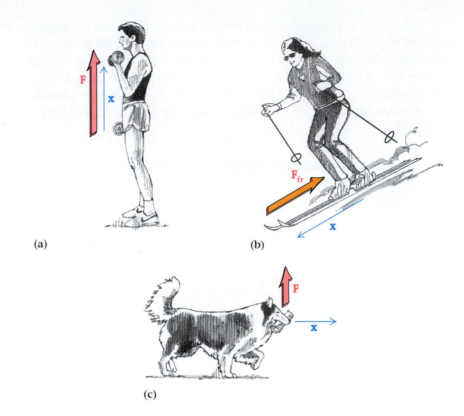

Figure 6.3

(a) A person does positive work by lifting the weight with an applied force **F** in the same direction as the displacement **x**. (b) The frictional force $\mathbf{F}_{fr}$ does negative work by acting in a direction opposite the displacement. (c) When the applied force is perpendicular to the displacement, zero work is done.

amounts of work directly, just like any other scalar quantities. When the component of the force is in the same direction as the displacement, the work done on the object is positive. When the component of the force is opposite to the displacement, the work done on the object is negative. If the force is perpendicular to the displacement, the work is zero (Fig. 6.3).

The SI unit for work is the newton-meter or $\mathrm{kg \cdot m^2/s^2}$. This combination of units has also been given the name **joule** (J), in honor of James Prescott Joule (1818–1889), one of the great contributors to our understanding of energy (see Chapter 11):

$$1 \text{ joule (J)} = 1 \text{ N} \cdot \text{m} = 1 \text{ kg} \cdot \text{m}^2/\text{s}^2.$$

In the British system, the unit of work is the foot-pound:

$$1 \text{ ft-lb} = 1.356 \text{ J}.$$

You can get some feeling for the size of the joule by looking ahead at Table 6.2 on page 182.

Example 6.1

A child pulls a toy 2.0 m across the floor by a string, applying a force of constant magnitude 0.80 N (Fig. 6.4). During the first meter the string is parallel to the floor. During the second meter the string makes an angle of 30° with the horizontal direction. What is the total work done by the child on the toy?

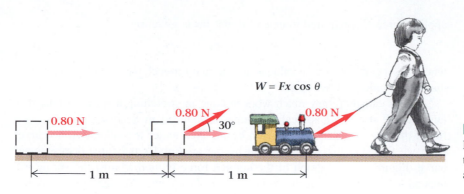

$W = Fx \cos \theta$

0.80 N

0.80 N

30°

0.80 N

1 m

1 m

Figure 6.4

Example 6.1: A child pulls a toy, with the string first parallel to the motion and then at an angle of 30°.

Strategy We must calculate the work separately for the first and second parts of the motion and then add them together. Since work is a scalar, we can add directly and do not need to use vector addition.

Solution For the first part, the work W_1 is

$$W_1 = F_1 x_1 \cos \theta_1, \qquad \text{where } \theta_1 = 0°$$
$$W_1 = (0.80 \text{ N})(1.0 \text{ m})(1.0) = 0.80 \text{ J}.$$

For the second part of the motion, the work W_2 is

$$W_2 = F_2 x_2 \cos \theta_2, \qquad \text{where } \theta_2 = 30°$$
$$W_2 = (0.80 \text{ N})(1.0 \text{ m})(0.87) = 0.69 \text{ J}.$$

The total work W is then

$$W = W_1 + W_2 = 0.80 \text{ J} + 0.69 \text{ J} = 1.5 \text{ J}.$$

6.2 Work Done by a Varying Force

If the force exerted on a moving object is constant, then we can calculate the work by the simple application of Eq. (6.1). In Example 6.1, the force changed after the first meter of displacement, but then remained constant for the remainder of the motion. In this case, we applied Eq. (6.1) to each part of the motion separately. However, often the force exerted on an object changes continuously. In these situations we must calculate the small amount of work ΔW for each small displacement Δx, over which the force is constant or approximately constant:

$$\Delta W = F \Delta x \cos \theta.$$

Then we can add together these amounts of work ΔW to give the total work for the entire process.

An important example of this idea is a spring that obeys what is called Hooke's law. We know that the more force we apply to a spring, the more it stretches. For a spring that obeys Hooke's law, the extension of the spring (its displacement from equilibrium) is proportional to the applied force. In Fig. 6.5, the displacement of the spring from its equilibrium position is denoted by x. The

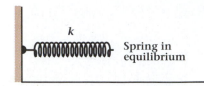

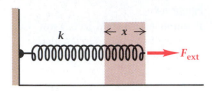

Figure 6.5
A spring that obeys Hooke's law. F_{ext} is the force applied to the spring to give an extension x.

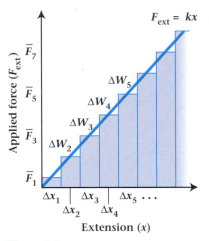

Figure 6.6
The force F_{ext} plotted against the extension of the spring of Figure 6.5. The work done in extending the spring equals the area under the force-displacement curve.

applied force F_{ext} required to extend it that far is given by

$$F_{ext} = kx, \tag{6.2}$$

where k is a constant whose value is determined for each particular spring.* Note that the SI units of k are N/m.

Let us calculate how much work is done in stretching a spring a total distance x. We use the graphical technique introduced in Section 2.4, where we found that the displacement is given by the area under the velocity-time curve. Here we find that the work is given by the area under the force-displacement curve. Figure 6.6 is a graph of the applied force plotted against the spring extension. (The extension of the spring is a displacement.) We mark off small displacements Δx along the abscissa of the graph and draw vertical lines from them up to the force-displacement curve. Then, much as we did in Chapter 2, we form a series of rectangles whose width is Δx and whose height is halfway between the applied force at the beginning and at the end of Δx. This height corresponds to the average applied force $\overline{F}$ over each displacement Δx.

We wish to calculate the small amounts of work ΔW corresponding to the small displacements Δx and then add them together:

$$W = \Delta W_1 + \Delta W_2 + \Delta W_3 + \cdots.$$

In each small displacement Δx, the work equals the product of Δx with the average force over that displacement. This is just the area of each rectangular strip in Fig. 6.6. The total work then corresponds to the total area of all the strips. As we decrease the size of each displacement Δx, thus increasing the number of strips, their total area approaches the area of the triangle with base x and altitude F_{ext}. Therefore

$$W = \tfrac{1}{2}xF_{ext} = \tfrac{1}{2}x(kx),$$

or

$$\boxed{W = \tfrac{1}{2}kx^2.} \tag{6.3}$$

The work done in stretching a spring by an amount x is given by $\tfrac{1}{2}kx^2$.

▼

Example 6.2

(a) How much work is required to extend an exercise spring (Fig. 6.7) by 45 cm if the spring constant k has the value 310 N/m? (b) What force is required to extend the spring by 45 cm?

Strategy We are given the spring constant k and the extension x of the spring from its equilibrium position. With these two quantities we can compute the work from Eq. (6.3) and the force from Eq. (6.2).

---

*Further discussion of Hooke's law is found in Chapter 14. Usually, Hooke's law is written as $F = -kx$, where F is the force exerted *by* the spring and is in the opposite direction from the spring extension x.

Solution (a) We can calculate the work by substituting directly into the expression for the work done in stretching a spring. We find

$$W = \tfrac{1}{2}kx^2 = \tfrac{1}{2}(310 \text{ N/m})(0.45 \text{ m})^2 = 31 \text{ J}.$$

(b) The force is found from Hooke's law,

$$F_{ext} = kx = (310 \text{ N/m})(0.45 \text{ m}) = 140 \text{ N}.$$

Discussion Notice that before inserting the numerical value for the extension x we converted the units from cm to m.

Figure 6.7
Example 6.2: Force is exerted to stretch the exercise spring. Work is done as the spring stretches.

6.3 Energy

Now that we understand the concept of work, we can use it to define energy. Because energy appears in so many different forms, such as mechanical, chemical, or electrical, it is difficult to give a single, brief definition. As a start, we can define **energy** by saying: *Energy is the ability to do work.* A compressed spring has energy because it may do work in returning to its uncompressed state. A falling body has energy because it may drive a stake into the ground upon striking it. Gunpowder, which may do work on exploding, has energy. An electrical battery has energy because it can turn an electric motor that does work.

An ocean wave or a falling rock has mechanical energy (Fig. 6.8), whereas gunpowder and gasoline have chemical energy, and steam has thermal energy.

Figure 6.8
An ocean wave possesses energy associated with its motion. This is a form of mechanical energy.

Table 6.1	The Values of Some Common Energy Units in Joules
1 ft-lb	$= 1.356$ J
1 Btu	$= 1.055 \times 10^3$ J
1 kWh	$= 3.600 \times 10^6$ J
1 calorie*	$= 4.187$ J

*The unit ordinarily used in nutrition is the Calorie or kilocalorie, which is equal to 10^3 calories. This unit is also called a "large calorie." Note that the larger unit is distinguished from the smaller unit by being written with a capital letter.

Energy may be transformed from one form to another. Chemical energy changes to mechanical energy when an automobile engine burns gasoline; mechanical energy changes to electrical energy when water from a dam turns the turbine of an electric generator; and mechanical energy turns into thermal energy when you rub your hands together to warm them.

Energy, in all its forms, is measured in the same units as work. Other energy units in common use, in addition to the joule and the foot-pound, arise out of convenience in different situations. The British thermal unit (Btu) and calorie (cal) are often useful in discussing thermal energy, and the kilowatt-hour (kWh) is most frequently used in the case of electrical energy. These units will be discussed again and are mentioned here only to emphasize that they all measure the same thing. Their values in joules are listed in Table 6.1. Table 6.2 gives the energy values in joules of a range of phenomena.

6.4 Kinetic Energy

A body in motion possesses energy associated with its motion because it can do work upon impact with another object. This energy of motion is called **kinetic energy.**

To derive an expression for kinetic energy, let's consider a particle* subjected to a constant force **F** directed along the x direction. This force may or may not be gravitational in origin; we only require that it be constant. Under the influence of this constant force, the particle moves a distance x. The work done on

---

* By a particle we mean an idealized object so small that we can imagine it as a single point in space with no size and no internal structure. For simplicity, we will use this model to approximate the behavior of real objects.

Table 6.2	Approximate Energy Values	
Source		**Approximate Energy (in J)**
Total U.S. use in one year (from all sources, 1995 est.)		9.6×10^{19}
Generated by Grand Coulee Dam in one year		6.5×10^{16}
Generated by Hoover Dam in one year		2×10^{16}
Burning 1 ton of coal		30×10^9
Burning 1000 ft^3 of natural gas		1×10^9
Burning 1 gallon of gasoline		2×10^8
Kinetic energy of a car at 60 mi/h		1×10^6
Person running 10 km/h		3×10^2
Bowling ball dropped from waist		70
1 calorie		4
Penny dropped 10 cm		2.5×10^{-3}
Fission of one atom of uranium		1.8×10^{-11}
Kinetic energy of a molecule of air		6×10^{-21}

the particle is $W = Fx$. If F is the net force on the particle, then we can then use Newton's second law to replace the force by the product of mass times acceleration, obtaining

$$W = (ma)(x).$$

If the particle was initially moving in the direction of **F** with a speed v_1, then after moving through the distance x it will have a speed v_2 given by the kinematic expression from Chapter 2:

$$v_2^2 = v_1^2 + 2ax.$$

If we rearrange this last expression and multiply by $m/2$, we get

$$(ma)(x) = \tfrac{1}{2}mv_2^2 - \tfrac{1}{2}mv_1^2,$$

or

$$W = \tfrac{1}{2}mv_2^2 - \tfrac{1}{2}mv_1^2. \tag{6.4}$$

In applying a force to the particle, we performed an amount of work $W = max$. The effect of the work done on the particle has been to change its motion. The quantity $\tfrac{1}{2}mv^2$ is given the name kinetic energy (KE). More specifically, this quantity is called the **translational kinetic energy.** A particle of mass m moving with a speed v possesses a kinetic energy due to its translational motion that is given by

$$\boxed{\text{KE} \equiv \tfrac{1}{2}mv^2.} \tag{6.5}$$

The SI units for kinetic energy are kg · m²/s² or joules, the same as the units for work.

The concept of kinetic energy is not limited to particles only. Any object with mass m moving with velocity v has a translational kinetic energy given by Eq. (6.5), whether it is a particle or an extended body (Fig. 6.9).

Using the definition of kinetic energy, we see that the quantity on the right-hand side of Eq. (6.4) is the difference between the final and initial kinetic energies:

$$\boxed{W = \Delta\text{KE} = \tfrac{1}{2}mv_2^2 - \tfrac{1}{2}mv_1^2.} \tag{6.6}$$

Figure 6.9
Even when moving at a low speed, a train has a large kinetic energy because of its large mass.

Equation (6.6) is known as the **work-energy theorem:** *the work done on a particle by the net force acting on it is equal to the change in kinetic energy of the particle.* The left-hand term represents the net work done on the particle. The right-hand side of the equation is the difference between the final and initial kinetic energies. If the work done on the particle is positive, then its kinetic energy increases. The work-energy theorem emphasizes that work, or equivalently energy, is needed to set a particle in motion. The theorem is valid even for a force that is not constant.

Master the Concept

Energy and Automobiles

Question: Why do cars get better mileage in freeway driving than in city traffic?

Answer: The fuel burned in a car engine provides the energy to propel the car. Much of this energy is expended in accelerating the car from rest or from slower speeds to higher speeds. When driving on the freeway, the speed can be held essentially constant so that the energy required is primarily used to overcome friction and air drag. However, in the stop-and-go driving of city traffic much more energy is needed to repeatedly accelerate the car up to speed. Consequently, city driving requires more fuel per mile than does freeway driving.

Example 6.3

A baseball player throws a 0.17-kg baseball at a speed of 36 m/s. Find the ball's translational kinetic energy.

Solution From the definition of kinetic energy we have

$$KE = \tfrac{1}{2}mv^2,$$

$$KE = \tfrac{1}{2}(0.17 \text{ kg})(36 \text{ m/s})^2 = 110 \text{ J}.$$

Discussion Remember that if you put the mass and the velocity in SI units of kg and m/s, the kinetic energy will be given in the SI unit of joules.

Example 6.4

Sometimes we can measure the kinetic energy of a particle emitted during radioactive decay by determining how far it travels in matter before stopping. Using this technique, a physicist determines that an alpha particle had an initial kinetic energy of 8.0×10^{-14} J. The mass of an alpha particle is known to be 6.65×10^{-27} kg. What was the initial speed of the alpha particle in m/s? What was the speed when expressed as a fraction of the speed of light ($c = 3.00 \times 10^8$ m/s)?

Strategy Because we are given the kinetic energy and the mass, we can calculate the particle's speed from our definition of kinetic energy even if we don't know what an alpha particle is. (We will describe it later, in Chapter 26.)

Solution From our definition of kinetic energy we have

$$KE = \tfrac{1}{2}mv^2,$$

which we can solve for v to give

$$v = \sqrt{\frac{2KE}{m}}.$$

Inserting the numerical values gives

$$v = \sqrt{\frac{2(8.0 \times 10^{-14} \text{ J})}{6.65 \times 10^{-27} \text{ kg}}} = 4.9 \times 10^6 \text{ m/s}.$$

Comparing the speed of the alpha particle to the speed of light c gives

$$\frac{v}{c} = \frac{4.9 \times 10^6 \text{ m/s}}{3.00 \times 10^8 \text{ m/s}} = 0.016,$$

or

$$v = 0.016c.$$

Example 6.5

(a) How much work is done to move a 1840-kg Jaguar XJ6 automobile from rest to 27.0 m/s (60 mi/h) on a level road? (b) If this takes place over a distance of 117 m, what is the average net force?

Solution (a) We can use the work-energy theorem to find the work,

$$W = \tfrac{1}{2}mv_2^2 - \tfrac{1}{2}mv_1^2.$$

We set v_1 equal to zero, $v_2 = 27.0$ m/s, and $m = 1840$ kg. Then the work is

$$W = \tfrac{1}{2}(1840 \text{ kg})(27 \text{ m/s})^2 = 6.71 \times 10^5 \text{ J}.$$

(b) The average net force can be found from

$$W = Fx,$$

$$F = \frac{W}{x} = \frac{6.71 \times 10^5 \text{ J}}{117 \text{ m}} = 5.74 \times 10^3 \text{ N}.$$

Discussion Note that we could have computed the car's average acceleration and then used Newton's second law to find the force. However, the present method of solution uses only scalar quantities, rather than vectors, and for more complicated situations is the easier method to use.

6.5 Potential Energy

We have seen that an object in motion has kinetic energy. Objects may also have energy in other forms. When a spring is stretched, it acquires energy called *potential energy*. For example, when you do work to wind the spring of a toy car, you give the spring potential energy; it has the ability to move the car when it unwinds and therefore to do work. Other examples include a jack-in-the-box and the compressed gas in an aerosol spray can. A mass lifted from the ground also gains potential energy. If it is dropped, it loses potential energy and gains kinetic energy. If it strikes a stake on the ground, the mass can do work by driving the stake farther into the ground.

In all of these examples, work must be done to increase an object's potential energy, and in all cases the object acquires the capability of doing work when released. We can think of **potential energy** as stored energy. These examples illustrate mechanical potential energy; other forms of potential energy include chemical and electrical potential energy, which we will discuss in later chapters.

The term *potential energy* does not mean that the energy is not real energy. Rather, it means that the energy is stored and is available to be converted into work or some other form of energy.

Gravitational Potential Energy

Gravitational potential energy is one of the most familiar forms of potential energy. Figure 6.10 shows a log of mass m initially at rest on the ground. It is then lifted slowly at a uniform velocity with a constant upward force just strong enough to equal the downward force of gravity mg. If the log is raised from the ground to a height h, the work done on the log is the net lifting force times the distance traveled, or

$$W = Fh = mgh.$$

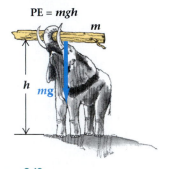

PE = *mgh*

m

h

mg

Figure 6.10
The gravitational potential energy of a log of mass *m* raised to a height *h* above the earth's surface is given by *mgh*.

If the log is released, it will fall. As it falls it accelerates, gaining velocity and kinetic energy, and thereby the ability to do work. Because the log at height h is capable of doing work if it is released, we say it has potential energy due to its position. More specifically, we say in this case that it has **gravitational potential energy.** The change in the log's gravitational potential energy is equal to the work required to raise the log to that height.

Thus, near the earth's surface an object's gravitational potential energy with respect to some reference level is

$$PE = mgh, \tag{6.7}$$

where h is the height above the reference level. Again, the units are joules. Note specifically that the gravitational potential energy depends on the position of the body and is not a property of the body alone.

Gravitational potential energy must always be referred to a specific reference level. For instance, in Fig. 6.11, the potential energy of the vase with respect to the floor is greater than its potential energy with respect to the table. Yet there is only one physical situation, not two. This example does not imply any ambiguity in the concept of potential energy; it only points out that the reference level is arbitrary. However, once you have chosen the reference level from which to measure potential energy in a given physical situation, you must keep that same reference throughout your analysis and calculation of that situation. Remember, *the physically important quantity is the difference in potential energy between two levels.*

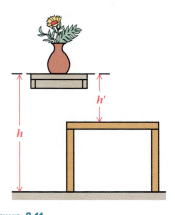

h

h'

Figure 6.11
The potential energy of a vase with respect to the table (*mgh'*) is less than its potential energy with respect to the floor (*mgh*).

Figure 6.12 shows several different paths by which a stack of boards can be lifted from point A to point B. Though some pathways are longer than others, *the same amount of work is done in each case.* Remember that only the component of the displacement in the direction of the force contributes to the work done on an object. In this case the direction of the gravitational force is vertical and the net vertical displacement is the same for each pathway. Thus, we conclude that the gravitational potential energy depends only on the difference between the heights of A and B. We will discuss this further in Section 6.6.

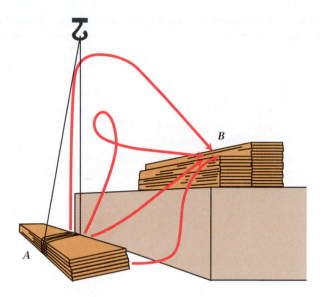

Figure 6.12
The work done in lifting a load between two heights in a uniform gravitational field is independent of the path traveled by the load.

▼

Problem-Solving Strategy

Gravitational Potential Energy

In working problems with gravitational potential energy, you must choose a reference coordinate system and stick with it throughout the problem. The particular choice is arbitrary, but a clever choice will simplify your efforts. The important quantity is the change in vertical position.

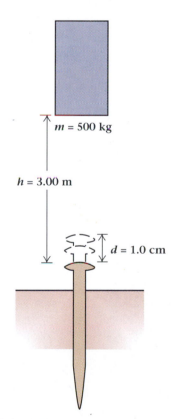

▼

Example 6.6

How much potential energy does a 7.5-kg ceiling fan have with respect to the floor when it is 3.0 m above it?

Strategy   We are looking for the potential energy with respect to the floor, so it is natural to choose the floor as the zero level of PE.

Solution   If we let h be the height above the floor, we find

$$PE = mgh$$
$$PE = (7.5 \text{ kg})(9.81 \text{ m/s}^2)(3.0 \text{ m}) = 220 \text{ J}.$$

▼

Example 6.7

In Fig. 6.13, the 500-kg mass of a pile driver is dropped from a height of 3.00 m onto a piling in the ground. The impact drives the piling 1.0 cm deeper into the ground. If all the original potential energy of the mass is converted

Figure 6.13
Example 6.7: A pile driver drops a 500-kg mass onto a piling, driving it deeper into the ground.

into work in driving the piling into the ground, what is the frictional force acting on the piling? Assume the frictional force to be constant over the 1.0-cm travel.

Strategy   We can solve this problem by using the concept of energy. This method has the advantage that we do not need to know the exact nature of the interaction, only that all of the energy is converted into work. We know that the driver falls through a distance h, striking the piling and driving it into the ground. We assume all of the potential energy goes into causing this motion. In driving the piling, the driver must overcome the frictional force between the ground and the piling. The frictional force will be opposite in direction to the force exerted by the driver.

Solution   The work needed to drive the piling through a distance d is

$$W = Fd,$$

where F is the force needed to overcome friction. The initial potential energy relative to the final position is

$$PE = mgh.$$

The final potential energy is 0. We equate the work done and the change in potential energy to give

$$Fd = mgh - 0,$$

$$F = \frac{mgh}{d}$$

$$F = \frac{(500 \text{ kg})(9.81 \text{ m/s})(3.00 \text{ m})}{0.010 \text{ m}} = 1.5 \times 10^6 \text{ N}.$$

Discussion   In reality, some of the original potential energy will go into radiated sound and into heating of the point of impact. Most of the original potential energy goes into moving the stake against the frictional force, energy that is eventually dissipated as heat. (We will discuss this in more detail in Chapter 11.) Also note that, because the height is large compared with the displacement of the piling, we approximate the distance through which the heavy mass moves as simply h.

Elastic Potential Energy

Just as we can store energy by raising a mass in a gravitational field, we can store energy in a spring by stretching or compressing it. We have seen in Section 6.2 that the work required to stretch a spring a distance x from its equilibrium position is $\frac{1}{2}kx^2$. When the external force is removed, the spring returns to its original length converting its stored energy into kinetic energy. Thus, in the same way that we defined gravitational potential energy, we can define a potential energy of the spring that we call **elastic potential energy**, PE$_s$, given by

$$PE_s = \tfrac{1}{2}kx^2. \tag{6.8}$$

Notice that when the spring is compressed corresponding to a negative value of x, the elastic potential energy is positive just as it is when the spring is stretched.

In either case, extension or compression, the spring exerts a force in a direction that would move the spring back to its equilibrium length.

Example 6.8

A 1550-kg Pontiac Gran Prix is supported by four coil springs, each with a spring constant of 7.00×10^4 N/m. (a) By how much are the springs compressed beyond their normal length? (b) How much energy is stored in the springs?

Strategy For simplicity we assume that the four springs support an equal portion of the car's weight. The amount of compression of each spring can then be found from the force on the spring and the spring constant. Finally, from the compression of the spring we can compute the stored energy from the definition of elastic potential energy given in Eq. (6.8).

Solution (a) Assuming the weight of the car is equally distributed, the force on each spring is one-fourth the weight of the car. The compression of each spring may be computed from the Hooke's law relationship $F_{\text{ext}} = kx$.

$$x = \frac{F_{\text{ext}}}{k} = \frac{mg/4}{k} = \frac{(1550 \text{ kg})(9.81 \text{ m/s}^2)/4}{7.00 \times 10^4 \text{ N/m}} = 5.43 \times 10^{-2} \text{ m}.$$

(b) The energy stored in one spring is given by $PE_s = \frac{1}{2}kx^2$. The total energy stored is four times the energy stored in one spring.

$$PE_{s \text{ tot}} = 4\left(\tfrac{1}{2}kx^2\right) = (2)(7.00 \times 10^4 \text{N/m})(5.43 \times 10^{-2} \text{ m})^2 = 413 \text{ J}.$$

6.6 Conservation of Mechanical Energy

When you apply a force to a spring and stretch it, the spring returns to its original length when released. On the other hand, if you apply a force to a book and push it across a table against the frictional force, the book does not return to its original position when you release it. In both cases you do work against a resisting force, but the nature of the forces is different. The spring force represents what is called a conservative force; friction is a dissipative, or nonconservative, force.

If the work done by a force on an object depends only on the initial and final positions of the object, then that force is a **conservative force.** The work is independent of the path taken. If you move an object against a conservative force and return it to the starting place, the total work done is zero. For example, if you do an amount of work to lift a barbell from the floor, the same amount of work is done on you if you lower the barbell to its initial location (Fig. 6.14). The total work you've done on the barbell is zero. (This does not mean that a weightlifter does no work in lifting and lowering a barbell. Energy losses occur within the body.)

As we saw earlier, the gravitational force is a conservative force. The expression for the work given in Eq. (6.7) depends on the difference in initial and final positions (see Fig. 6.12), not on the path taken between them. A spring with a restoring force proportional to its extension is another example of a

Figure 6.14
The work done in raising a barbell is equal to the work done on the weightlifter by the barbell if it is lowered to its initial position.

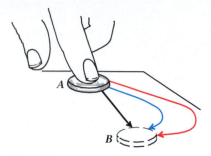

Figure 6.15

The work done to push a coin from A to B depends on the path taken.

conservative force. We conclude, from the requirement that the work done be independent of the path taken, that conservative forces must be forces that depend on position, rather than forces that may vary with time, speed of the object, path taken, or some other parameter.

The work done against a frictional force depends on the path. If you push a coin from A to B (Fig. 6.15) with constant speed, the work done is different for different paths. Thus, friction is a nonconservative force. The resistance of air or water to the motion of a body through it is another example of a nonconservative force.

An important distinction between conservative and nonconservative forces is that we can write an expression for the potential energy for conservative forces. You have seen this for gravitation. You have seen also that the energy needed to compress a spring that obeys Hooke's law is the stored, or potential, energy. No such expression is possible for frictional forces. The term *conservative force* is appropriate because a conservative force corresponds to the conservation, or constancy, of the sum of the kinetic energy and the potential energy.

Let's return to the case of a body that is raised to a small height h above the ground (Fig. 6.16). Over this distance the earth's gravitational field is essentially constant. The gravitational potential energy of the body is PE = mgh. If the rope breaks, the body is released from rest and falls. As it falls it gains speed as a result of the acceleration of gravity, and at any instant it has a kinetic energy corresponding to its instantaneous speed, given by

$$\text{KE} = \tfrac{1}{2}mv^2.$$

If the speed at height h_1 is v_1, then we can determine the speed v_2 at height h_2 by using the kinematic expression from Chapter 2,

$$v_2^2 = v_1^2 + 2a(x_2 - x_1).$$

Figure 6.16

A body is released from height h above the ground. It falls to height h_1, where its speed is v_1, and to height h_2, where it has a new speed v_2. Its total kinetic and potential energy stays the same.

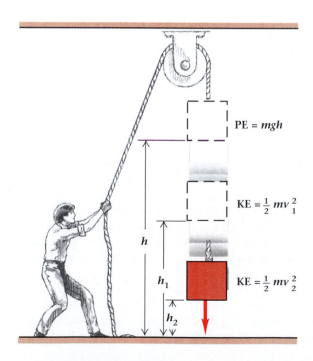

If we choose a coordinate system in which up is positive and down is negative, then the acceleration a is $-g$ and $x_2 - x_1$ becomes $h_2 - h_1$. The kinetic energy at h_2 is then

$$\tfrac{1}{2}mv_2^2 = \tfrac{1}{2}mv_1^2 - mg(h_2 - h_1).$$

This equation may be rearranged to give

$$\tfrac{1}{2}mv_2^2 + mgh_2 = \tfrac{1}{2}mv_1^2 + mgh_1. \tag{6.9}$$

The sum of the body's kinetic energy and potential energy is its total **mechanical energy.** According to Eq. (6.9), the mechanical energy at height h_2 is equal to the mechanical energy at height h_1. Thus, an object's total mechanical energy is constant for motion in a constant gravitational field provided that no other forces are introduced. For this special case we say that the mechanical energy is *conserved*. The total energy is the same at the top (h) as it is at any other height. The total mechanical energy E can be written in the form

$$E = \text{KE} + \text{PE}. \tag{6.10}$$

The value of the potential energy (PE) and kinetic energy (KE) may change, but their sum, the total energy (E), is a constant and does not change.

Equations (6.9) and (6.10), along with additional observations and experiments, leads to the generalization that *if the forces are all conservative, the sum of the kinetic and potential energies is a constant.* This statement is the **law of conservation of mechanical energy,** which can be written as

$$\boxed{\text{KE}_2 + \text{PE}_2 = \text{KE}_1 + \text{PE}_1.} \tag{6.11}$$

Alternatively, we can rearrange Eq. (6.11) to write it in the form

$$\Delta\text{KE} + \Delta\text{PE} = 0. \tag{6.12}$$

Sometimes this last equation is a useful way to state the law of conservation of mechanical energy.

We can combine Eq. (6.12) with the work-energy theorem Eq. (6.6) to get

$$W_c = \Delta\text{KE} = -\Delta\text{PE} \tag{6.13}$$

or

$$W_c = \text{PE}_i - \text{PE}_f,$$

where we have used W_c to represent the work done by conservative forces. Thus, the work done on an object by a conservative force is equal to the object's initial potential energy minus its final potential energy.

The principle of conservation of energy is extremely useful in a wide variety of cases. Often the effects of friction are small enough that we can ignore them and use the law of conservation of mechanical energy. In later chapters we will study other forms of energy, such as thermal energy and electrical energy. We will find that energy can be transformed from one form to another and that we can extend the principle of conservation of energy to include those other forms. Example 6.9, which follows, demonstrates an application of conservation of mechanical energy.

Significance of Conservation Laws

Those laws of nature that state that some quantity is the same before and after an event or interaction are called **conservation laws.** They reflect one of our most basic ways of describing nature. The law of conservation of mechanical energy is, in some sense, a more fundamental principle than Newton's mechanics, which we used to derive it here. Because conservation laws allow us to consider quantities, such as energy, that do not change during an event, we do not need to know the details of the interaction. We simply deal with the value of the conserved quantity before and after the interaction. Moreover, the fact that some quantities are conserved, and some are not, tells us something about nature itself.

In all of physics there are only a relatively small number of conservation laws. We will introduce the conservation of momentum in Chapter 7 and only a few more conservation laws in the remainder of the text. It can be argued that conservation laws represent the most powerful and, at the same time, simplest view of nature. They allow us not only to understand how much electrical energy it takes to keep our house warm or cool, but also to discover new particles and new behavior on the atomic and subatomic scale.

Example 6.9

A student accidentally knocks a plant off a window sill, where it falls from rest to the ground 5.27 m below (Fig. 6.17). Use the principle of conservation of energy to determine its speed just before it strikes the ground.

Strategy The total mechanical energy E of the plant is a constant. It consists of two parts: kinetic energy KE and potential energy PE, which are not individually constant, but whose sum is constant:

$$E = KE + PE.$$

By expressing these energies in terms of the height h and the final velocity v, we can determine v in terms of h.

Solution Use subscripts T for top and 0 for ground, we equate the total energy at the top of the fall to the total energy at the bottom,

$$E_T = E_0$$
$$KE_T + PE_T = KE_0 + PE_0.$$

But $KE_T = 0$ because the initial speed is zero, and $PE_0 = 0$ because $h = 0$. Thus we have

$$PE_T = KE_0,$$

or

$$mgh = \tfrac{1}{2}mv^2.$$

Upon solving for v, we get

$$v = \sqrt{2gh}$$
$$v = \sqrt{2(9.81 \text{ m/s}^2)(5.27 \text{ m})} = 10.2 \text{ m/s}.$$

Discussion Note that we could have obtained the same result using the methods described in Chapter 2. However, the principle of energy conservation is

Figure 6.17
Example 6.9: A plant falls from a height of 5.27 m above ground.

PE = mgh, KE = 0

5.27 m

PE = 0, KE = ½ mv^2

important because it can be applied in much more complicated situations where the methods of Chapter 2 are impractical.

Potential Energy Diagrams

Let us consider what happens to a roller coaster as it moves along a track (Fig. 6.18a). The height of the track above some reference level determines the potential energy at that point. In part (a) a car of mass m leaves point A with no initial velocity ($v_A = 0$). We can determine its speed at any other location B using the principle of conservation of energy:

$$KE_A + PE_A = KE_B + PE_B.$$

At A the kinetic energy of the coaster is zero because its speed is zero. Thus the equation for conservation of mechanical energy becomes

$$0 + mgh_A = \tfrac{1}{2}mv_B^2 + mgh_B$$
$$\tfrac{1}{2}mv_B^2 = mgh_A - mgh_B.$$

We rearrange to find the speed

$$v_B = \sqrt{2g(h_A - h_B)}.$$

This result suggests a diagrammatic way of viewing a situation in which energy is conserved. In Fig. 6.18(b) we have plotted the potential energy of the car as a function of its horizontal displacement x. Since the potential energy is

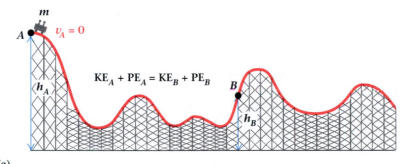

(a)

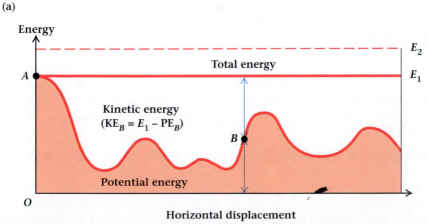

(b)

Figure 6.18

(a) A roller coaster on a track. The total mechanical energy $E = KE + PE$ is conserved. (b) The potential-energy diagram for the roller coaster on the track. We choose the zero energy level to be ground level. The kinetic energy is the difference between total energy E_1 and the potential energy.

directly proportional to the height h, the potential energy curve has the same shape as the track itself. Such a graph of potential energy against displacement is called a **potential-energy diagram.** The total energy is represented here by E_1 and is constant. In this case, since $\mathrm{KE}_A = 0$, E_1 is equal to the initial potential energy PE_A. On the diagram you can see that the kinetic energy is given by the distance between the total-energy curve and the potential-energy curve. Thus, the kinetic energy at point B is

$$\mathrm{KE}_B = E_1 - \mathrm{PE}_B.$$

When these energies are expressed in terms of position and speed, we get

$$\tfrac{1}{2}mv_B^2 = E_1 - mgh_B,$$

or

$$v_B = \sqrt{2\left(\frac{E_1}{m} - gh_B\right)}.$$

However, since for this case the total energy E_1 is equal to the initial potential energy mgh_A, this expression becomes

$$v_B = \sqrt{2g(h_A - h_B)}.$$

This result is the same as our earlier finding.

If the roller coaster in part (a) had been moving at point A, it would have had an initial kinetic energy, as well as an initial potential energy. Then the total energy would have been greater than E_1, as shown by the dotted line E_2 in part (b). In this case the speed at any point B would be

$$v_B = \sqrt{2\left(\frac{E_2}{m} - gh_B\right)}.$$

This is just the same as the previous expression, with E_1 replaced by E_2. In either case the value of the total energy would be known from the initial conditions.

We emphasize that the roller coaster's kinetic energy, and therefore its speed, is determined by the difference between the total-energy line and the potential-energy curve. As indicated by the potential-energy diagram, at a point where the potential energy is smaller, the kinetic energy, and therefore the speed, is larger. From such a diagram you can determine the speed if you know the position.

Example 6.10

A block of mass m is released from rest and slides down a frictionless track of height h (Fig. 6.19) At the bottom of the track the block slides freely along a horizontal surface until it hits a spring of spring constant k attached to a heavy, immovable wall. How far is the spring compressed at the maximum point of compression?

Strategy At the beginning of the problem, the mass is at rest at a height h above the horizontal surface. If we choose the horizontal surface as our origin for measuring gravitational potential energy, then the mass m has an initial potential energy of mgh and zero kinetic energy. As the mass slides down the track it loses potential energy and gains kinetic energy until at the bottom all

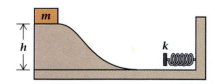

Figure 6.19
Example 6.10: A mass moves down a frictionless ramp until it strikes a spring.

of its energy is kinetic. Finally, the mass hits the spring and begins to compress it, exchanging kinetic energy for spring potential energy until it stops and all of the energy is potential energy.

Solution   We can equate the initial mechanical energy to the final mechanical energy to get

$$KE_i + PE_{gi} + PE_{si} = KE_f + PE_{gf} + PE_{sf},$$

where we have used the g and s to denote gravitational and spring energies and i and f to denote initial and final energies. Inserting the values gives

$$0 + mgh + 0 = 0 + 0 + \tfrac{1}{2}kx^2.$$

Solving for x, the maximum amount that the spring is compressed, we find

$$x = \sqrt{\frac{2mgh}{k}}.$$

Discussion   At the instant that the spring is compressed a maximum amount, the speed of the mass is zero. However, this is not a stable condition, for the spring is pushing outward with a force $F = kx$, which begins to accelerate the mass until all of the spring potential energy is converted into kinetic energy of the mass. Then, if there are no losses of energy, the mass will make its way back up the track to its starting point.

Example 6.11

Students from a physics class visit an amusement park to test their understanding of Newton's laws. They ride a roller coaster (Fig. 6.20) that is pulled up to the top of a 48-m-tall hill. The coaster then moves over the crest at an average

Figure 6.20
Example 6.11: A roller coaster.

speed of 0.50 m/s before it plunges to a low point 3 m above the ground. From there it climbs over a smaller hill only 16 m high. One of the students on the roller coaster records the trip with a 1.5-kg video camera. (a) What is the coaster's speed as it goes over the top of the 16-m hill? (b) What force (magnitude and direction) must the student exert on the camera to hold it steady as the car passes through the crest of the 16-m hill? The path of the track over the hill is approximately circular with a radius of 20 m. Assume that the student holds the camera 1.0 m above the track and that friction is small enough to be ignored.

Strategy The speed of the coaster at any point along the track depends on its height. We may determine the speed by using the law of conservation of mechanical energy. Once we know the speed we can find the centripetal force required to move the camera along with the coaster. The centripetal force is the net force exerted on the camera. It is the vector sum of the force exerted by the student and the gravitational force.

Solution (a) We find the speed of the coaster from the conservation of mechanical energy.

$$\text{KE}_{\text{top}} + \text{PE}_{\text{top}} = \text{KE}_B + \text{PE}_B,$$

where the subscript B represents any point along the track. In this case we let point B be the crest of the 16-m hill. Then the equation becomes

$$\tfrac{1}{2}mv_{\text{top}}^2 + mgh_{\text{top}} = \tfrac{1}{2}mv_B^2 + mgh_B.$$

After dividing out the common factor m, we get

$$\tfrac{1}{2}v_{\text{top}}^2 + gh_{\text{top}} = \tfrac{1}{2}v_B^2 + gh_B.$$

Rearrange and solve for v_B.

$$v_B^2 = v_{\text{top}}^2 + 2g(h_{\text{top}} - h_B),$$
$$v_B = \sqrt{v_{\text{top}}^2 + 2g(h_{\text{top}} - h_B)}.$$

We then insert the numerical values to find v_B.

$$v_B = \sqrt{(0.50 \text{ m/s})^2 + 2(9.81 \text{ m/s}^2)(48 \text{ m} - 16 \text{ m})} = 25.06 \text{ m/s} \approx 25 \text{ m/s}.$$

(b) Next we find $\mathbf{F}_s$ the force that the student exerts on the camera. The force of gravity $\mathbf{F}_g$ also acts on the camera. The vector sum of these forces provides the centripetal force that carries the camera around the curved path of the track. Thus the centripetal force is

$$\mathbf{F}_c = \mathbf{F}_g + \mathbf{F}_s.$$

We know that the magnitude F_g is mg and that it is directed vertically downward.

$$F_g = mg = (1.5 \text{ kg})(9.81 \text{ m/s}^2) = 14.72 \text{ N}.$$

We can compute the magnitude of the centripetal force from the speed and the radius of the path of the camera. Because the coaster is going over a hill, the centripetal force at the top must also be downward.

$$F_c = \frac{mv^2}{r} = \frac{(1.5 \text{ kg})(25.06 \text{ m/s})^2}{21 \text{ m}} = 44.86 \text{ N}.$$

The force provided by the student is

$$\mathbf{F}_s = \mathbf{F}_c - \mathbf{F}_g.$$

If we let up be positive and down negative, then the force becomes

$$F_s = -F_c - (-F_g)$$
$$F_s = -44.86 \text{ N} + 14.72 \text{ N} = -30.14 \text{ N} \approx -30 \text{ N}.$$

The student must pull down on the camera with a force of 30 N to keep the camera in the coaster.

Discussion Notice that the force of 30 N exerted on the camera by the student is about twice the normal weight of the camera. To the student riding in the coaster, the camera seems to be flying upward with a force twice its weight.

What would be the direction of the required force when the coaster passes through the lowest point of the track? The track curves upward at the lowest point so the center of the effective curve is above the track. The resulting centripetal force must be upward. Thus the rider would have to exert an upward force to overcome the gravitational force and provide the neccessary centripetal force. In that case the camera would seem heavier than normal.

| *6.7 | **Energy Conservation with Nonconservative Forces** |

We know that friction plays a role in most real situations. For example, a box sliding down a ramp is slowed by the retarding force of friction. Consequently, we expect that its mechanical energy when it reaches the bottom of the ramp will be less than it was at the top. Thus, in the presence of a frictional (or other nonconservative) force, mechanical energy is not conserved. Even so, we can still account for the effects of nonconservative forces.

Let us express the work in the work-energy theorem of Eq. (6.6) in two parts, the work W_c due to conservative forces and the work W_{nc} due to the nonconservative forces:

$$W_c + W_{nc} = \Delta \text{KE}. \tag{6.14}$$

If we substitute Eq. (6.13) for the work done by conservative forces, we find upon rearranging that

$$W_{nc} = \Delta \text{KE} + \Delta \text{PE} = \Delta E,$$

or

$$W_{nc} = E_{\text{final}} - E_{\text{initial}}. \tag{6.15}$$

As before, E is the total mechanical energy, so Eq. (6.15) means that the work done by the nonconservative forces equals the change in total mechanical energy. That is, it is equal to the change in kinetic energy plus the change in potential energy.

Friction is a nonconservative force. The direction of the frictional force acts to retard the motion. Therefore the work done by friction is negative. In that case, the final mechanical energy is less than the initial mechanical energy. Then we can write Eq. (6.15) as

$$E_{\text{final}} - E_{\text{initial}} = W_{\text{friction}}, \qquad (6.16)$$

where W_{friction} is the product of the frictional force F_{fr} with the distance d through which it acts. In order to make it clear that the final energy is less than the initial energy, we rewrite Eq. (6.16) as

$$E_{\text{final}} - E_{\text{initial}} = -|W_{\text{friction}}|,$$

or

$$\boxed{E_{\text{initial}} = E_{\text{final}} + |W_{\text{friction}}|.} \qquad (6.17)$$

The effect of a frictional force is to decrease the final mechanical energy of the system. For this reason frictional forces are known as **dissipative forces.** When dissipative forces are present, they act to remove mechanical energy from the system. This energy is not lost, however, but is transformed into a different form. As we will see in Chapter 11, during the nineteenth century careful observers found that the mechanical energy lost to friction is transformed into thermal energy.

▼

Example 6.12

A 55-kg carton of bananas with an initial speed of 0.45 m/s slides down a ramp inclined at an angle of 23° with the horizontal. If the coefficient of friction between the carton and the ramp is 0.24, how fast will the carton be moving after it has traveled a distance of 2.1 m down the ramp?

Strategy The carton is acted upon by the conservative force of gravity drawing it down the ramp and by the nonconservative force of friction retarding its motion, as shown in Fig. 6.21. We can find the speed of the carton by applying the law of conservation of energy in the form of Eq. (6.17):

$$E_{\text{initial}} = E_{\text{final}} + |W_{\text{friction}}|$$

$$\tfrac{1}{2}mv_i^2 + mgh_i = \tfrac{1}{2}mv_f^2 + mgh_f + F_{\text{fr}}d,$$

where d is the distance traveled down the ramp.

Solution The distance traveled by the carton is $d = 2.1$ m. The difference in height from the initial position to the final position is $h_i - h_f = d \sin 23°$, and the frictional force is $\mu mg \cos 23°$. Inserting these values into the last equation we get

$$\tfrac{1}{2}mv_f^2 = \tfrac{1}{2}mv_i^2 + mg(h_i - h_f) - F_{\text{fr}}d$$

$$\tfrac{1}{2}mv_f^2 = \tfrac{1}{2}mv_i^2 + mgd \sin 23° - \mu mgd \cos 23°.$$

We can multiply each term by $2/m$ to get

$$v_f^2 = v_i^2 + 2gd(\sin 23° - \mu \cos 23°),$$

$$v_f = \sqrt{v_i^2 + 2gd(\sin 23°) - \mu \cos 23°)}.$$

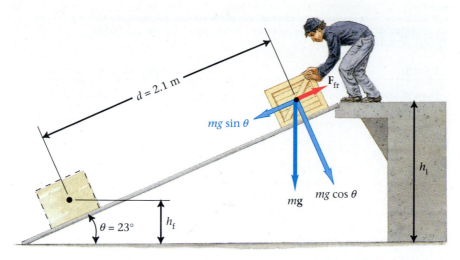

Figure 6.21
Example 6.12: A carton slides down a ramp.

Now we can insert the numerical values for v_i, g, d, and μ to get

$$v_f = \sqrt{(0.45 \text{ m/s})^2 + 2(9.81 \text{ m/s}^2)(2.1 \text{ m})(\sin 23° - 0.24 \cos 23°)}$$
$$v_f = 2.7 \text{ m/s}.$$

6.8 Power

In many cases it is useful to know not just the total amount of work being done, but how rapidly work is being done. For instance, if you have a motor that can provide only a certain amount of work in one day and you wish to accomplish twice that much work, then you must either take two days for the job or get an additional motor. We define **power** as the time rate of doing work; that is,

$$P \equiv \frac{\Delta W}{\Delta t}, \tag{6.18}$$

where ΔW is the amount of work done in the time interval Δt. In SI units, work is measured in joules and time in seconds. The unit of power is the joule/second, a combination that has been given the name **watt** (abbreviated W):

$$1 \text{ joule/second} = 1 \text{ watt}.$$

The measurement of power grew out of the need of early steam-engine builders to specify the properties of their engines. James Watt (1736–1819), the most inventive of these engine builders, developed the steam engine into an efficient, versatile engine that could be used to drive machinery. As part of his efforts to measure the power of his steam engine, Watt made systematic measurements on the work a horse could perform in a given time. From these measurements he defined a horsepower as 550 foot-pounds per second. The relationship between the watt and the horsepower (hp) is

$$1 \text{ hp} = 746 \text{ W} = 0.746 \text{ kW}.$$

Table 6.3	Approximate Power Production and Consumption ▼	
Device		**Approximate Power (in W)**
Hoover Dam		1.92×10^9
Jumbo jet aircraft		1.3×10^8
Automobile use of chemical energy at 60 mph		1.1×10^5
Electric stove		1.2×10^4
Clothes dryer		5.6×10^3
Average per capita use of electricity (U.S.)		1.5×10^3
Available solar power per square meter (average U.S. over 24-h day)		180
Solid-state color TV		120
Two-cell flashlight, halogen lamp		1.5
Pocket calculator (LCD display)		7.5×10^{-4}

Table 6.3 gives the approximate power produced or consumed in a number of situations.

Example 6.13: A jogger expends energy when running on a treadmill.

Example 6.13

Using a calibrated treadmill, a jogger measures her energy output as 4.8×10^5 J for an hour's run. What average output power did she develop?

Solution From the definition of power as the rate of doing work, we have

$$P = \frac{\Delta W}{\Delta t} = \frac{4.8 \times 10^5 \text{ J}}{3600 \text{ s}}$$

$$P = 1.33 \times 10^2 \text{ W} \approx 130 \text{ W}.$$

Discussion Compare this with the power consumption of a household electric lamp. Note that this is the output power—that is, the power expended to move the jogger along. The total power expended, some of which goes into work against the internal friction in the jogger's body, is greater than the output power.

Example 6.14

A 70-kg person runs up a staircase 3.0 m high in 3.5 s. How much power does he develop in climbing the steps?

Solution If we assume that the work is done at a constant rate, then the definition of Eq. (6.18) becomes

$$P = \frac{W}{t}.$$

In this case, the work done is the change in gravitational potential energy, mgh, so the power is

$$P = \frac{mgh}{t} = \frac{(70 \text{ kg})(9.81 \text{ m/s})(3.0 \text{ m})}{3.5 \text{ s}} = 590 \text{ W}.$$

Discussion Note that we have actually computed the average power, which is the total work done divided by the total time. The power may also be given in units of horsepower as

$$P = 590 \text{ W}\left(\frac{1 \text{ hp}}{746 \text{ W}}\right) = 0.79 \text{ hp}.$$

This result is consistent with the general observation that humans are capable of power outputs in the range of 0.5 to 1 hp for 30 s. For longer periods of time, human power output is much decreased. For steady work over 8 h, human power is of the order of 0.1 to 0.2 hp.

Our definition of power in Eq. (6.18) applies to all types of work, whether mechanical, electrical, or thermal. However, we can rewrite the definition in a special way for mechanical work by simply rearranging terms. When a force acts on an object so that it moves with a speed v, we can calculate the power from the force and the speed. If we consider the force to be constant, the change in work is $\Delta W = F \Delta x$. Then power becomes

$$P = F \frac{\Delta x}{\Delta t},$$

or

$$P = Fv. \tag{6.19}$$

It can be shown that Eq. (6.19) is true for instantaneous power even when the force is not constant.

Example 6.15

The heart may be regarded as an intermittent pump that forces about 70 cm^3 of blood into the 1.0-cm-radius aorta about 75 times a minute (Fig. 6.22). Measurements show that the average force with which the blood is pushed into the aorta is about 5.0 N. What is the approximate power used in moving the blood to the aorta?

Strategy Because we know the force, we can use Eq. (6.19) to determine the power if we can find the speed of the blood. The average speed can be computed from the volume of blood that is pumped and the pumping rate. Assume the aorta to be a cylinder and that a length s of it is filled with blood each time the heart beats. We start with the definition of speed and modify it by multiplying the numerator and denominator of the fraction by the cross-sectional area of the aorta, to give

$$v = \frac{s}{t} = \frac{s\pi r^2}{t\pi r^2} = \frac{\text{volume}}{t\pi r^2}.$$

Here r is the radius of the aorta and t is the time for one heartbeat.

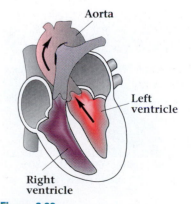

Figure 6.22
Example 6.15: The human heart.

HUMAN ENERGY

Does energy conservation apply to your body? The answer is yes. The food you eat is the fuel that provides your body with energy to maintain its functions and to do external work. Catalytic agents, called enzymes, allow this fuel to burn at body temperatures, converting chemical energy into thermal and other energy. If you take in too much fuel, some of the energy is stored in the form of body mass and your weight increases. If you take in too little energy, you lose weight. Thus, if you want to maintain your weight, the energy you take in must be equal to the energy your body uses. This equilibrium energy is sometimes called the sustaining energy.

The exact value of the sustaining energy intake depends on body mass and activity level. Figure B6.1 shows the approximate range of sustaining energy intake per day as a function of body mass. The lower boundary of the shaded area corresponds to inactive people; the upper boundary corresponds to very active people. In the United States, most people take in between 2000 and 3000 Calories (8400 – 12600 kJ) per day.

Most food energy goes into just running your body and keeping it warm. When you first wake up in the morning and are lying quietly in bed, your body is using energy at the lowest rate for the day. The average energy use in that condition, called the basal metabolic rate, is about 1400 Calories per day for women and about 1600 Calories per day for men. This rate of use corresponds to an average power output

of about 75 watts. Most of this energy goes into repairing cells. The waste energy shows up as heat to maintain your body temperature.

As you increase your activities during the day, your energy needs increase also. For example, Fig. B6.2 shows how the rate of energy use increases with increasing walking speed. The increased rate for faster walking implies that when you move your legs or arms, some energy is needed to overcome the internal friction in your body. Furthermore, your muscles are inefficient at converting chemical energy into mechanical motion. Only about one fifth of the chemical energy used by your muscles is converted into mechanical work; the remainder is dissipated as heat, explaining in part why you get hotter when you run than when you walk. The average daily expenditure of energy for inactive men is about 2800 Calories; for inactive women it is about 2000 Calories. People who engage in very strenuous work have much higher energy needs. For example, athletes in training for the triathlon need as much as 8000 Calories per day.

How much energy do you use in sports and other activities? Table B6.1 gives you some typical values for a 150-lb person. To get a value for your own weight, multiply the value in the table by your weight in pounds and divide the result by 150. The values in the table correspond to sustained activity. When you jump, throw, or bat, your peak effort is expended over a shorter time and the rate of energy use can be larger than for sustained activity. In short bursts of a second or so duration, 0.60 Calories *per second* have been measured in golf, weight lifting, and high jumping. An expenditure of 0.60 Calories per second is approximately two-and-one half times as great as the rate of energy use when running an 8-min mile.

Table B6.1	*Calories used by a 150-lb Person in 10 Minutes*	
Activity	**Energy Used (Calories)**	**(Joules)**
Volleyball	34	142,000
Walking (3 mph)	40	167,000
Walking (4 mph)	58	242,000
Jogging (11 min-mile)	91	380,000
Running (8 min-mile)	141	590,000
Bicycling (5.5 mph)	47	197,000
Bicycling (10 mph)	81	339,000
Swimming (breaststroke)	72	301,000
Swimming (crawl)	87	364,000
Calisthenics	49	205,000
Tennis	68	285,000
Handball or racquetball	95	398,000
Skiing (downhill)	95	398,000
Skiing (cross country)	108	452,000
Skating (moderate)	54	226,000
Canoeing (4 mph)	70	293,000
Mountain climbing	100	420,000
Golf	54	226,000

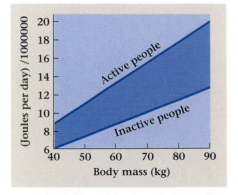

Figure B6.1 Sustaining energy versus body mass.

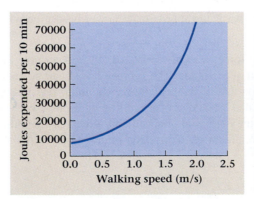

Figure B6.2 Rate of energy use as a function of walking speed.

Figure B6.3 Aerobic exercise burns quite a few Calories. This energy comes from the food you eat and store as fat in your body.

Solution   We know from the statement of the problem that the volume per unit time is 70 cm^3 per 0.80 s (75 beats per minute is 0.80 second per beat). Using the expression just found for the speed, we may rewrite Eq. (6.19) as

$$P = Fv = F \frac{\text{volume}}{t\pi r^2}.$$

Substituting the numerical values gives

$$P = 5.0 \text{ N} \frac{(70 \text{ cm}^3)(10^{-6} \text{ m}^3/\text{cm}^3)}{(0.80 \text{ s})\pi(0.010 \text{ m})^2} = 1.4 \text{ N} \cdot \text{m/s}$$

$$P = 1.4 \text{ W}.$$

Summary

Useful Concepts

■ The work done on an object by a force F acting over a distance x at an angle θ to the displacement is

$$W = Fx \cos \theta.$$

■ The work required to stretch a spring a distance x, and the energy stored in it, is

$$W = \tfrac{1}{2}kx^2.$$

■ Energy is the ability to do work.

■ An object's kinetic energy of translation is

$$\text{KE} = \tfrac{1}{2}mv^2.$$

■ According to the work-energy theorem, the work done on a particle by the net force acting on it is equal to the change in kinetic energy of the particle.

$$W = \Delta \text{KE} = \tfrac{1}{2}mv_2^2 - \tfrac{1}{2}mv_1^2.$$

■ The level from which you choose to measure an object's gravitational potential energy is arbitrary, but you must not change it while working a problem. The gravitational potential energy of a mass m at a height h near the earth's surface is given by

$$\text{PE} = mgh.$$

■ The potential energy stored in a spring (the elastic potential energy) is given by

$$\text{PE}_s = \tfrac{1}{2}kx^2,$$

where x is the displacement of the spring from its equilibrium position and k is the spring constant.

■ The work done by a conservative force acting on an object depends only on the initial and final positions of the object, not on the path taken. The sum of kinetic energy and potential energy is a constant if the forces are conservative. This is the law of conservation of mechanical energy.

$$\text{KE}_2 + \text{PE}_2 = \text{KE}_1 + \text{PE}_1$$

■ In the presence of frictional forces, the final mechanical energy is less than the initial mechanical energy. They differ by the work lost to friction:

$$E_{\text{initial}} = E_{\text{final}} + |W_{\text{friction}}|.$$

■ Power is defined as the rate of doing work:

$$P \equiv \frac{\Delta W}{\Delta t}.$$

Important Terms

You should be able to write the definition or meaning of each of the following terms:

work	conservative force
joule	mechanical energy
energy	conservation of
kinetic energy	mechanical energy
translational kinetic energy	conservation laws
work-energy theorem	potential-energy diagram
potential energy	dissipative forces
gravitational potential energy	power
elastic potential energy	watt

▼
Conceptual Questions

6.1 Why is no work done on an object when a force acting on the object does not move it?

6.2 Does the sun do work on the earth as it moves in its orbit?

6.3 A student rows upstream just fast enough to stay at rest with respect to the bank. Does the rower do work?

6.4 A car moves along the highway at a constant velocity, leading us to conclude that the net force on the car is zero. Discuss where, if anywhere, work is done.

6.5 What happens to the work done in stretching a spring?

6.6 A baseball and a Ping-Pong ball are thrown with the same velocity. Which one has the greater kinetic energy? Why?

6.7 Explain how a pile driver works.

6.8 Do you do the same work to lift a 1-kg weight through a vertical height of 1 m everywhere on the face of the earth?

6.9 Explain the basic ideas that govern the design and operation of a roller coaster. Under what conditions can successive hills be as high as or higher than the initial one?

6.10 Time yourself to see how long it takes you to do 10 deep knee bends. Then estimate how much work you do in lifting yourself back up each time. From these numbers, estimate the rate at which you were doing work.

6.11 Why do you shift to a lower gear to pedal a multispeed bicycle uphill? Do you save any energy in doing so?

6.12 Explain the observation that smaller cars generally get better fuel mileage than larger cars.

▼
Problems

Sections 6.1 and 6.2 Work and Work Done by a Varying Force

> **Hints for Solving Problems**
>
> Be careful about signs: Work done on an object is positive, work done by the object is negative. A force in the direction of displacement does positive work; a force opposite to the direction of displacement does negative work; a force perpendicular to the displacement does zero work.

6.1 How much work does a 52-kg woman do against gravity when climbing from the bottom to the top of a 2.8-m-high staircase?

6.2 A gardener pushes a box of tools across a driveway by applying a 32.5-N force that pushes downward at an angle of 22.4° with respect to the horizontal. How much work is done if the box moves 2.25 m in the horizontal direction?

6.3 A 50-kg sled is pulled 20 m over the ice at a constant speed. The coefficient of friction between sled and ice is 0.13. (a) What is the frictional force? (b) How much work is done in pulling the sled the 20 m?

6.4 Two forces parallel to the x axis do 14.7 J of work on a small tray while moving it 20.7 m in the x direction across a gym floor. One of the forces has a value of $+3.89$ N in the x direction. What is the other force?

6.5 A worker does 300 J of work against a frictional retarding force of 15 N in pushing a power sweeper across a floor in 3.0 s. If the sweeper moves with constant speed, how fast is it going?

6.6 A person finds that he can stretch a spring exercise device 1.25 times as far as a friend can. (a) What is the ratio of the forces they can apply? (b) What is the ratio of the work they each do in stretching the spring?

6.7● A boy uses 75 J in pushing a sled a distance of 3.0 m across the snow, applying the force in a horizontal direction. (a) How much force is needed to push the same sled through the same distance if the force is applied in a downward direction of 45° with respect to the horizontal? (b) How much force is required if it is applied in an upward direction of 45°? Assume that the work required in parts (a) and (b) is the same as that required when the force is applied horizontally.

6.8● A spring requires 46.1 J of work to extend it 12.0 cm and 278 J of work to extend it 27.0 cm. Does the spring obey Hooke's law?

6.9● A constant horizontal force of 2.29 N is applied to a 1.82-kg block initially at rest on a friction-free surface. How much work is done on the block in the first 1.35 s after the force is applied?

6.10● (a) How much work is needed to push a 132-kg packing crate a distance of 2.65 m up a frictionless inclined plane that makes an angle of 20.0° with the horizontal? (b) How much work would be required to move the crate the same distance if the coefficient of friction were 0.20?

6.11● How much work is done to move the 8.0-kg block 10 cm to the right if the spring is initially relaxed (Fig. 6.23)? The spring constant is 20 N/m, and the coefficient of friction between the block and the floor is 0.50.

6.12●● (a) A 50-kg gymnast stretches a vertical spring by 0.50 m when she hangs from it. How much energy is stored in the spring? (b) The spring is cut into two equal lengths, and the gymnast hangs from one section. In this case the spring stretches by 0.25 m. How much energy is stored in the spring this time?

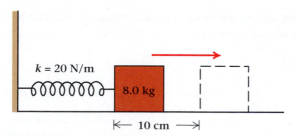

Figure 6.23
Problem 6.11.

6.13•• A net force acting on a large dictionary does 46.3 J of work in moving it a distance of 13.3 m along a frictionless surface that makes an angle of 36° with the horizontal *x* axis. The *x* component of the force is twice as large as the *y* (vertical) component. What are the magnitude and direction of the force vector?

Section 6.3 Energy

6.14 How many electricity-generating plants the size of Hoover Dam would be required to supply the total energy consumption of the United States? (Use the data in Table 6.2, p. 182.)

6.15 How many gallons of gasoline would it take to produce the electrical energy equal to that generated by Hoover Dam in one year? Assume that one third of the energy of the gasoline is converted to electrical energy and two thirds is lost.

6.16 (a) Approximately how many kilowatt-hours of electrical energy does one ton of coal produce if one third of the energy available is converted to electrical energy? (b) How many gallons of gasoline would be required to produce the same electrical energy? Again assume only one third of the available energy is converted to electrical energy. (Refer to the data in Table 6.2.)

6.17• The area of the United States (excluding Alaska and Hawaii) is approximately 3.5×10^6 mi². (a) Use the data in Table 6.3 to estimate the total solar power falling on this part of the United States. (b) Assuming a population of approximately 2.7×10^8 people, what area would have to be devoted to solar collectors with a 10% conversion efficiency in order to get all of the needed power from solar power? (c) What percentage of the area of the United States would be covered with solar collectors? (d) How does the answer to (b) compare with the area of the state in which you live?

Section 6.4 Kinetic Energy

6.18 (a) What is the kinetic energy of an 1800-kg car moving at 25 m/s? (b) At 120 km/h?

6.19 A 2.5-g Ping-Pong ball at rest is set in motion by the use of 1.8 J of energy. If all of the energy goes into the motion of the ball, what is the ball's maximum speed?

6.20 A 0.324-kg air-track glider moves linearly with an initial speed of 1.37 m/s. In order to increase the speed of the mass, 7.31 J of work are done. What is the final speed of the glider?

6.21 Sam can throw a baseball twice as fast as can his little brother, Bill. How many times as much kinetic energy can Sam give the baseball as Bill can?

6.22• How far does a 1.58-kg stone with a kinetic energy of 3.11 J go in 1.86 s if it is moving in a straight line?

6.23• (a) How much work is required to increase the speed of a 1200-kg automobile from 10 km/h to 30 km/h? (b) How much work is required to further increase the speed by the same amount, this time from 30 km/h to 50 km/h? Neglect the effects of friction.

6.24• A 1.03-kg hammer moving at 1.25 m/s drives a nail 0.752 cm into a board. What is the average resisting force?

6.25• Show for a satellite moving in a circular orbit about a center of gravitational attraction, such as the sun or earth, that the product of the orbital radius and the satellite's translational kinetic energy is a constant.

Section 6.5 Potential Energy

Hints for Solving Problems

Once you choose a reference level for gravitational potential energy, do not change it while solving the problem.

6.26 Two blocks of 6.00 kg and 2.00 kg are hung over a pulley on a rope, with the 6.00-kg block resting on the floor. What is the change in the potential energy of the system if the 6.00-kg block is raised 0.800 m?

6.27 How much potential energy with respect to ground level does a 10.0-kg lead weight have when it is 2.00 m above the surface of the ground?

6.28 A 0.302-kg coffee mug rests on a table top 0.740 m above the floor. (a) What is the potential energy of the mug with respect to the floor? (b) What is its potential energy with respect to a counter top 1.100 m above the floor?

6.29 A spring compressed by 0.080 m stores 150 J as elastic potential energy. What is the value of the spring constant *k*?

6.30 A spring hangs vertically from the ceiling. The spring extends an amount Δx when a mass *m* is hung from the spring's free end. Find an expression for the energy stored in the spring in terms of *m*, the acceleration of gravity *g*, and Δx.

6.31 A 750-g air-track glider slides along at a speed of 0.85 m/s until it hits a spring attached to the end of the very heavy track. The spring is compressed by 1.35 cm at the maximum compression. What is the value of the spring constant *k*?

6.32• By calculating the work done along each part of the path and adding the amounts (taking account of the signs), show that the work done in moving a 1.00-kg mass to a point 1.00 m over from and 1.00 m above its initial point is the same for all four paths shown in Fig. 6.24.

6.33• A constant force of 15.3 N acts upward on a iron block that weighs 11.7 N. (a) If the block is initially at rest, what is its kinetic energy 6.53 s after the force is applied? (b) What is the increase in the potential energy of the body 6.53 s after the force is applied?

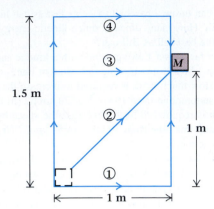

Figure 6.24
Problem 6.32.

6.34• A spring is extended by 2.35 cm when a mass of 0.250 kg is suspended from it. A second spring is extended twice as much (4.70 cm) when a mass M is suspended from it. If the energy stored in the two springs is the same (a) what is the value of the mass M and (b) how much energy is stored in each spring?

Section 6.6 Conservation of Mechanical Energy

Hints for Solving Problems

You should be careful to distinguish the initial and final states of a situation involving energy. Determine the potential energy and kinetic energy for each state. Then equate the initial mechanical energy to the final mechanical energy. Use energy conservation rather than kinematic methods to solve these problems.

6.35 A carpenter drops a hammer off the roof of a house. If the hammer falls a distance of 6.46 m, what is its speed just before striking the ground? Neglect air friction.
6.36 An ice cube slides from rest without friction down a long inclined ramp that makes an angle of 37.5° with the horizontal. What is the speed of the cube after it slides 1.27 m down the plane?
6.37 A block slides on a semicircular frictionless track (Fig. 6.25). If it starts from rest at position A, what is its speed at the point marked B?

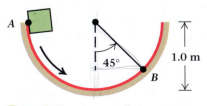

Figure 6.25
Problem 6.37.

6.38 An egg falls from a nest at a height of 3.08 m. What speed will it have when it is 0.50 m from the ground? Neglect air friction.
6.39 A baseball is thrown almost straight up at a speed of 12.3 m/s and falls back on the roof of a building 5.42 m above the height from which the ball was thrown. What is the speed of the ball just before it reaches the roof?
6.40 A block of mass $m = 750$ g is released from rest and slides down a frictionless track of height $h = 55.2$ cm. At the bottom of the track the block slides freely along a horizontal table until it hits a spring attached to a heavy, immovable wall (Fig. 6.26). The spring compressed by 2.64 cm at the maximum compression. What is the value of the spring constant k?

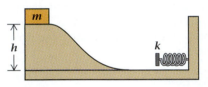

Figure 6.26
Problem 6.40.

6.41• Suppose the roller coaster car in Fig. 6.27 starts from rest at point A and moves without friction. (a) How fast is it going at points B, C, and D? (b) What constant deceleration must be applied at D to have it stop at E?

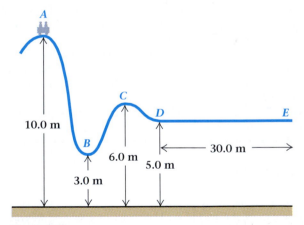

Figure 6.27
Problem 6.41.

6.42• A box of mass M is pushed at a constant speed for a distance s up an incline by a force parallel to the incline. The incline makes an angle θ above the horizontal and the coefficient of friction between the box and the surface is μ. (a) Draw a diagram of the situation. (b) Draw a free-body diagram. (c) Derive the equation for the work done.
6.43• A classroom demonstration is performed with an Atwood machine. The masses are $m_1 = 1.00$ kg and $m_2 = 1.10$ kg (Fig. 6.28). If the larger mass descends a distance of

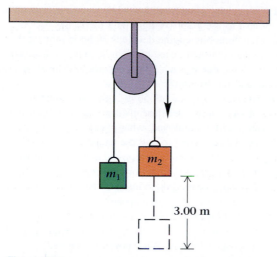

Figure 6.28
Problem 6.43.

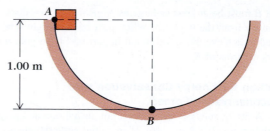

Figure 6.30
Problem 6.47.

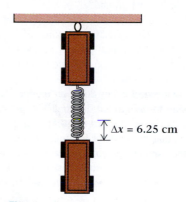

Figure 6.31
Problem 6.48.

3.00 m to the floor, what is the speed of the mass just before it hits?

6.44• From what height must a car be dropped to give it the same kinetic energy just before impact that it has when traveling at 60 km/h?

6.45• A 500-kg roller coaster starts from rest at point *A* (Fig. 6.29) and rolls freely (no friction) to point *B* where the brakes are applied and it slides along horizontally with a frictional force of 440 N. How far does the coaster slide past point *B* before coming to rest?

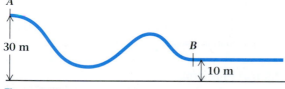

Figure 6.29
Problem 6.45.

6.46• A hammer dropped from rest by an astronaut from a height of 1.47 m above the surface of a planet has a speed of 4.1 m/s when it reaches a height of 0.32 m. Is the planet earth?

6.47• A block slides on a semicircular frictionless track (Fig. 6.30). (a) If it starts from rest at point *A*, what is its speed when it reaches the bottom of the track at point *B*? (b) Draw a free-body diagram of the block at the instant it is at point *B*. (c) What force does the track exert on the block when it passes through point *B* if the block's mass is 1.00 kg?

6.48•• Two low-friction carts of equal mass are joined by a spring. When suspended vertically the spring is extended 6.25 cm beyond its normal length (Fig. 6.31). The carts are then placed on a track and pulled together by a string until the spring is compressed by 6.25 cm from its normal length.

When the string is cut, the carts move off in opposite directions. What are the speeds of the carts when the distance between them is equal to the unextended length of the spring? Ignore the mass of the spring and the friction in the carts.

6.49•• The block on the loop-the-loop in Fig. 6.32 slides without friction. From what height must it start at *A* so that it presses against the track at *B* with a net upward force equal to its own weight? Give your answer in terms of *R*.

6.50•• A small mass slides without friction down the loop-the-loop track shown in Fig. 6.32. (a) Show that the speed at

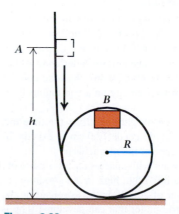

Figure 6.32
Problems 6.49 and 6.50.

point B must be at least as large as $\sqrt{gR}$ if the mass does not fall away from the track. (b) What must be the height h required to achieve the speed found in part (a)? Give your answers in terms of R.

*Section 6.7 Energy Conservation with Nonconservative Forces

6.51 A 30-kg suitcase falls from a hot-air balloon at a height of 1000 m. (a) If it loses 90% of its initial potential energy through friction with the air, what kinetic energy does it have just before it strikes the ground? (b) What speed does it have just before it strikes the ground?

6.52• A 4.20-kg box of books is pulled up a 4.06-m-long inclined plank to a height of 1.50 m. A force of 21.6 N parallel to the incline is needed to pull the box up the plank. (a) How much work is done by the 21.6-N force and (b) what percentage of that is lost to friction?

6.53• A 5.00-kg object slides down an inclined board from a height of 2.05 m to the ground, losing 10.0% of its energy to friction. What is its speed when it reaches the ground?

6.54• Assume that the force of friction between a shuffleboard puck and the playing surface is constant. Use the principle of energy conservation to show that a puck with an initial speed v will go a distance d before coming to a stop, where

$$d = \frac{v^2}{2\mu g}$$

and μ is the coefficient of friction.

6.55•• A wooden block placed at the end of a flat wooden board of length L begins to slide down the board when it is tilted to an angle of 34° from the horizontal. The block slides at a constant speed all the way to the other end of the board. (a) What is the change in potential energy of the block when it goes from the high end to the low end? (b) What is the work done against the force of friction? (c) Explain why your answers to (a) and (b) are the same or different. What assumptions have you made in working the problem?

Section 6.8 Power

6.56 (a) What is the engine power in kilowatts of a 108-hp Honda Civic CRX? (b) Of a 210-hp Ford Thunderbird SC? (c) Of Kyle Petty's 640-hp Winston Cup Pontiac race car?

6.57 About how many storage batteries would an electric car need in order to develop 80 hp? (Assume conventional electric batteries can deliver energy on a continuous basis at a rate of about 300 W.)

6.58 The 4.21-kg weight that drives the time mechanism of a grandfather clock descends 17.0 cm in exactly 24 hours. What power is delivered to the mechanism?

6.59 An astronaut with space suit has a mass of 110 kg. Climbing up a hill 7.3 m high in 7.2 s requires the astronaut to expend a power of 200 W. Is the astronaut on the earth?

6.60 An electric motor that can develop 1.0 hp is used to lift a mass of 25 kg through a distance of 10 m. What is the minimum time in which it can do this?

6.61 If electricity costs $0.083 per kilowatt-hour, how much does it cost to use a 250-W light bulb for 12 h?

6.62 An automobile engine develops 30 hp in moving the automobile at a constant speed of 50 mi/h. What is the average retarding force due to such things as wind resistance, internal friction, and tire friction?

6.63• Niagara Falls is about 53 m high. An estimated 6.0×10^6 kg of water pass over the falls every second. If all this energy were usefully employed, what power could be produced?

6.64• The force needed to pull the tape through an audio cassette player is 0.98 N. In operation the tape travels at a constant speed of $1\frac{7}{8}$ in./s. The motor consumes a power of 1.8 W. What percentage of the power input to the motor is required to pull the tape at its operating speed?

6.65• A 2150-kg loaded elevator moving with a constant speed rises 28 m in 15 s. The frictional force with the guide rails is a constant 1534 N. What power is required?

6.66•• A force of 125 N is needed to keep a small boat moving at 2.14 m/s. (a) What is the power required to keep the boat moving at the steady speed? (b) If the resistive force of the water increases with the square of the speed, what power is required if the speed is increased by 50%?

6.67•• A 73-kg person expends 400 W when walking on a level treadmill at a speed of 7.2 km/h. When the treadmill is inclined without changing the speed, the person's expended power increases to 600 W. Estimate the angle of incline of the treadmill by assuming that all of the increased output power goes into overcoming the force of gravity.

6.68•• Show that the height h to which an animal of mass m can jump is given approximately by

$$h = \frac{1}{2g}\left(\frac{4sP}{m}\right)^{\frac{2}{3}},$$

where P is the power expended and s is the distance over which the animal accelerates.

Additional Problems

6.69 A force of 24.3 N is needed to hold a spring extended 5.66 cm from its equilibrium position. How much work is done in extending the spring?

6.70 How many horsepower are developed when a 2.25-kg book is lifted 0.520 m in 2.00 s?

6.71• A 30-cm-long crank is attached to a simple machine that lifts a 210-kg load. The efficiency of a machine is the ratio of the output work to the input work, and its mechanical advantage is defined to be the ratio of the output force to the input force. When the crank turns through 400 complete revolutions by the application of a 12-N force perpendicular to the crank arm, the load is raised by 4.0 m. (a) What is the efficiency of the machine? (b) What is the mechanical advantage of the machine?

6.72• Simple machines such as levers, wedges, and pulleys allow a small force moved through a large distance to be transformed into a large force that acts through a small distance. The efficiency of a machine is the ratio of the output work to the input work, and its mechanical advantage is defined to be the ratio of the output force to the input force. Calculate the mechanical advantage of a crowbar used as a 100% efficient lever if the lifting end is 3.0 cm from the pivot and the input force is applied 50 cm from the pivot.

6.73• A stone thrown downward with a speed of 15.7 m/s from a height of 12.7 m above the ground has a kinetic energy of 293 J when it is 1.29 m above the ground. What is the mass of the stone?

6.74• A 40-kg child sits in a swing suspended with 2.5-m-long ropes. The swing is held aside so that the ropes make an angle of 15° with the vertical. Use conservation of energy to determine the speed the child will have at the bottom of the arc when she is let go.

6.75•• A force is given by $F = kx^2$, where x is in meters and $k = 10$ N/m². What is the work done by this force when it acts from $x = 0$ to $x = 0.1$ m? (Try using a graphical technique.)

6.76•• A 0.039-kg ball swings on the end of a 1.27-m-long string. On one swing the tension in the string is 0.435 N at the lowest point. By the second swing the ball has lost 3.1% of its energy. What is the ball's speed at the lowest point on the second swing?

6.77•• A 0.437-kg croquet ball depresses the pan of a spring balance 10.0 mm when resting upon it. If the croquet ball is then dropped from a height of 20.0 cm above the empty pan, how far will the pan be depressed if all the energy of the croquet ball goes into compressing the spring? (The energy in a compressed or stretched spring is the same for the same displacement.)

6.78•• The block in Fig. 6.33 is initially at rest on an inclined plane at the equilibrium position that it would have if there were no friction between the block and the plane. How much work is required to move the block 10 cm down the plane (a) if the frictional coefficient is $\mu = 0$ and (b) if the frictional coefficient is $\mu = 0.17$?

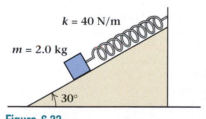

$k = 40$ N/m

$m = 2.0$ kg

30°

Figure 6.33
Problem 6.78.

6.79•• The 5.0-kg mass in Fig. 6.34 is released from rest 1.0 m above the floor. If the coefficient of friction between the

2.0-kg mass and the table is 0.28, what is the speed of the 5.0-kg mass just before it strikes the floor?

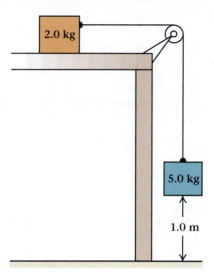

2.0 kg

5.0 kg

1.0 m

Figure 6.34
Problem 6.79.

6.80•• A 1300-kg car uses 15% more horsepower to travel 20 km/hr up a 5% grade than to travel at the same speed on a level surface. What is the average resisting force due to friction?

6.81•• Two masses are joined by a light cord that passes over frictionless pulleys as shown in Fig. 6.35. The 0.50-kg mass is pulled aside so that the cord makes an angle θ with the vertical and then it is released. What must be the angle θ for the 1.00 kg mass to just be lifted from its resting place by the motion of the smaller mass?

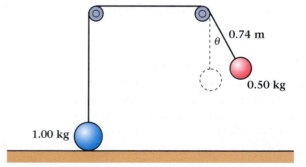

θ 0.74 m

0.50 kg

1.00 kg

Figure 6.35
Problem 6.81.

Linear Momentum

Nearly two decades before Newton's *Principia* was published, the Royal Society of London called for experimental studies of the behavior of colliding objects. Responses were received from several of Newton's contemporaries, including Sir Christopher Wren and Christiaan Huygens (1629–1695). Their observations led to the discovery of laws governing the exchange of momentum and energy between two colliding objects. These ideas were known to Newton and influenced his work. Their most important result was the law of conservation of linear momentum. According to this law, the total momentum after a collision is the same as the total momentum before the collision. This law made a key contribution to the growing understanding of mechanics.

Momentum, as stated in Newton's second law, is often called linear momentum to distinguish it from the angular momentum associated with rotational motion, which we will discuss in Chapter 9. The independent laws concerning the conservation of energy and of linear momentum are among the most basic laws in contemporary physics. Although we will derive these laws from Newton's laws of motion, in some respects they are even more fundamental and far-reaching than Newton's laws. For example, even in situations where Newton's laws do not apply, such as speeds approaching the speed of light or dimensions on atomic scales, these conservation laws are still valid. The use of conservation laws is one of the most fundamental ways of describing nature.

For simplicity, we focus on one conservation law at a time. Here we want to emphasize the conservation of momentum. However, in some cases we will first apply conservation of momentum and then use conservation of mechanical energy. In the next chapter we will examine collisions in which the laws of conservation of momentum and of conservation of kinetic energy are applied simultaneously. ■

7.1 Linear Momentum

The concept of momentum, which we first introduced in Chapter 4, Section 4.4, is extremely important in physics. Whenever we examine a moving object, we must consider both its mass and its velocity. The **linear momentum** of a body with mass m, traveling with velocity $\mathbf{v}$, is defined to be the product of the mass and the velocity. Since mass is a scalar quantity and velocity is a vector quantity, their product, momentum (which we designate with the letter $\mathbf{p}$) is a vector quantity:

$$\mathbf{p} \equiv m\mathbf{v}. \tag{7.1}$$

The word *momentum* (pl. *momenta*) is Latin and means "movement" or "moving power."

Most of us are intuitively aware of the importance of momentum in understanding the behavior of moving objects. For example, consider the difference between being hit by a bicycle traveling at 10 m/s (22 mi/h) and being hit by a locomotive traveling at the same velocity. We see that the mass of an object is an important consideration! Velocity is also important; just think of the difference between being hit by a baseball thrown by a child and being hit by a baseball thrown by a major league pitcher.

7.2 Impulse

When a baseball hits a bat (Fig. 7.1) or when two billiard balls collide, they exert forces on each other over a very short time interval. Forces of this type, which exist only over a very short time, are often called impulsive forces. Let's examine such forces to see how they are related to momentum.

In Chapter 4, we introduced Newton's second law in the form

$$\mathbf{F} = \frac{\Delta(m\mathbf{v})}{\Delta t} = \frac{\Delta \mathbf{p}}{\Delta t},$$

where $\mathbf{F}$ is the net force applied to an object and $\Delta \mathbf{p}$ is its change in momentum during a time Δt. If we multiply both sides of this equation by the time interval Δt, we get

$$\mathbf{F}\Delta t = \Delta \mathbf{p}. \tag{7.2}$$

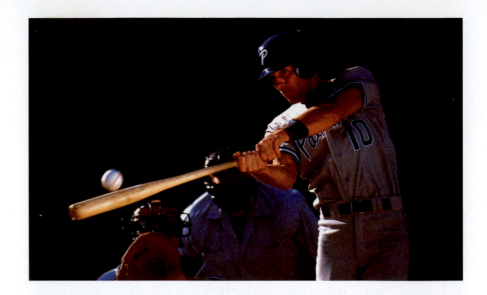

Figure 7.1
A bat striking a baseball. The interaction takes place over a very small time interval.

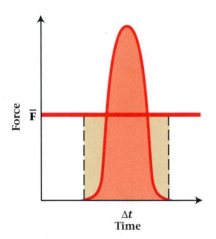

Figure 7.2
Average and instantaneous force during a typical brief collision between two moving bodies. The area under the curve of force versus time is equal to the impulse. Since the area of the rectangle whose height is the average force equals the area under the curve, we can replace the instantaneous force by the average force to obtain the impulse.

The quantity on the left, $\mathbf{F}\Delta t$, is called the **impulse.** It is the product of the force $\mathbf{F}$ and the time interval Δt over which the force acts. Even though the force occurs very briefly, the force is not usually constant over the time interval (Fig. 7.2). Nevertheless, the impulse is equal to the area under the force-time curve. Even if we don't know the exact shape of the force curve, we can replace the force in Eq. (7.2) with the average force $\overline{\mathbf{F}}$ over the time of interaction Δt:

$$\overline{\mathbf{F}}\Delta t = \Delta \mathbf{p}. \tag{7.3}$$

The usefulness of the impulse concept is illustrated in the two following examples.

Example 7.1

Using the following data, determine the average force on a baseball hit by a bat. The baseball has a mass of 0.14 kg and an initial speed of 30 m/s. It rebounds from the bat with a speed of 40 m/s in the opposite direction and is in contact with the bat for 0.0020 s. (High-speed photographs can be used to determine the contact time.)

Solution   Since the mass of the ball is constant, we can rewrite Eq. (7.3) as

$$\overline{F} = \frac{\Delta p}{\Delta t} = \frac{\Delta(mv)}{\Delta t} = \frac{m\,\Delta v}{\Delta t}$$

$$\overline{F} = \frac{m(v_{\text{final}} - v_{\text{initial}})}{\Delta t}.$$

We choose the direction of v_{final} as positive; then v_{initial} must be negative. The change in momentum becomes

$$m\,\Delta v = m(v_{\text{final}} - v_{\text{initial}}) = 0.14 \text{ kg}[40 \text{ m/s} - (-30 \text{ m/s})]$$
$$= (0.14 \text{ kg})(70 \text{ m/s}).$$

The average force is then

$$\overline{F} = \frac{(0.14 \text{ kg})(70 \text{ m/s})}{0.0020 \text{ s}} = 4900 \text{ kg} \cdot \text{m/s}^2 = 4900 \text{ N}.$$

Typically, the maximum force is much greater than the average force, as shown in Fig. 7.2.

Example 7.2

A 51-kg teenager jumps to the ground from a chair 0.34 m high (Fig. 7.3). If she bends her knees slightly on landing, lowering herself by only 8.0 cm, what is the average force with which her feet hit the ground?

Strategy   First, if we know the girl's momentum just before striking the ground, we can compute her change in momentum and thus the impulse involved. To determine her momentum, we need to know her speed just before she hits the ground. We obtain it from the kinematic relation in Eq. (2.9),

$$v^2 = v_0^2 + 2gh,$$

where g is the gravitational acceleration, h is the height of the chair, v is the speed just before striking the ground, and v_0 is her initial vertical speed. If we take $v_0 = 0$, the speed v becomes $v = \sqrt{2gh}$.

Second, if we know the time interval during which the impulse acts, we can determine the average force. The force of landing acts over the time during which the girl bends her knees to absorb the shock. If we assume the deceleration to be constant, then the slowing-down distance d is given by the product of the average speed and the time:

$$d = \frac{(v_i + v_f)}{2} \Delta t.$$

Here the initial speed when her feet contact the ground is $v_i = v$ and the final speed v_f is zero, so $d = (v/2)\Delta t$. The time over which the slowing-down, or knee-bending, process takes place is $\Delta t = 2d/v$. Because we can determine both the time interval and the change in momentum, we can compute the average force.

Solution   The girl's momentum changes from its initial value at the moment her feet strike the ground to a value of zero at the end of the knee flexing. So we may write the change of momentum as

$$\Delta p = p_{\text{final}} - p_{\text{initial}} = 0 - mv = -mv.$$

The average impulsive force is

$$\overline{F} = \frac{\Delta p}{\Delta t}.$$

Upon substituting $-mv$ for Δp and $2d/v$ for Δt, we get

$$\overline{F} = \frac{-mv}{2d/v} = \frac{-mv^2}{2d}.$$

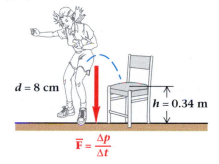

$$d = 8 \text{ cm} \qquad h = 0.34 \text{ m}$$

$$\overline{\mathbf{F}} = \frac{\Delta p}{\Delta t}$$

Figure 7.3
Example 7.2: A person jumping from a chair to the ground bends her knees to absorb the shock.

Using $v = \sqrt{2gh}$, we find

$$\overline{F} = \frac{-m(\sqrt{2gh})^2}{2d} = -\frac{mgh}{d}.$$

Now insert the numerical values, including $g = 9.81$ m/s^2, to get

$$\overline{F} = \frac{-(51 \text{ kg})(9.81 \text{ m/s}^2)(0.34 \text{ m})}{0.080 \text{ m}} = -2100 \text{ N (about 480 lb)}.$$

Discussion We see from the expression $\overline{F} = mgh/d$ that the impulsive force is inversely proportional to the knee-bending distance d. A greater amount of bending reduces the average force, whereas a smaller amount increases the force. If the girl had landed with her knees locked, the force would have been so large that she might have suffered one or more broken bones. The force we have just calculated is the impulsive force that changes her momentum. In addition, there is a contribution due to her weight (even when she is standing still), but we have neglected it in this calculation.

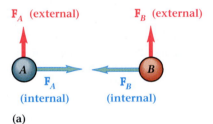

(a)

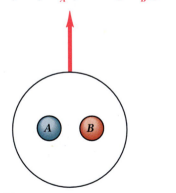

(b)

Figure 7.4

A system of two interacting bodies. (a) Considered individually, each body is subject to a net force composed of two forces: one external and one internal due to the other body. (b) Considered as a system, the net force on the two bodies is just the sum of the external forces.

7.3 Newton's Laws and the Conservation of Momentum

You already know that the velocity of a single particle or object does not change unless there is a net force acting on it. If there is no force acting on the object, its momentum is constant with time. We express this fact by saying that the object's momentum is conserved. In this section we first consider the motion of a single particle or object. Then we examine the motion of a system of interacting objects.

If a single object of mass m is subject to zero net force, Newton's second law states that the rate of change of momentum with time is zero. This is equivalent to saying that *the object's momentum remains constant when the net force acting on it is zero.* That is, if

$$\frac{\Delta \mathbf{p}}{\Delta t} = \mathbf{F} = 0,$$

then

$$\mathbf{p} = m\mathbf{v} = \text{constant.}$$

Most interesting physical systems consist of two or more objects interacting with one another. For such systems we can apply Newton's laws to each individual object. In doing so we find that while the momenta of the individual objects may change, *the total momentum of the system is constant whenever the net external force on the system is zero.* The effect of the internal forces of interaction among the objects is to exchange momentum among them in such a way that the total momentum is conserved.

To see how this result arises, consider the behavior of the two interacting bodies in Fig. 7.4. We are free to consider each of them independently (part a) or to regard them together as a system (part b). We can write the total force on one body as the sum of the forces exerted from outside the system (the external force) plus the force exerted on it by the other body inside the system (the internal force). Thus the total force on one body, say body A, is

$$\mathbf{F}_A \text{ (total)} = \mathbf{F}_A \text{ (internal)} + \mathbf{F}_A \text{ (external)}.$$

The force on the other body is

$$\mathbf{F}_B \text{ (total)} = \mathbf{F}_B \text{ (internal)} + \mathbf{F}_B \text{ (external)}.$$

The internal forces make up an action-reaction pair and, by Newton's third law, are equal in magnitude but opposite in direction. As a result, the sum of the internal forces is zero. Consequently, if we add up these two equations, we find that the net force on the system is just the sum of the external forces on the two bodies:

$$\mathbf{F} \text{ (total)} = \mathbf{F}_A \text{ (external)} + \mathbf{F}_B \text{ (external)}$$

The total rate of change of the system's momentum is equal to the sum of the external forces and is independent of the internal forces. We can express this result as

$$\mathbf{F}_{\text{net}} = \frac{\Delta \mathbf{p}}{\Delta t}, \tag{7.4}$$

where $\mathbf{F}_{\text{net}}$ is the net external force and $\mathbf{p}$ is the total momentum of the system. Equation (7.4) is the extension of Newton's second law to a system of two

Master the Concept

Momentum Conservation

Question: When a basketball player leaps high for the ball, he suddenly gets an upward momentum. How do you reconcile this observation with the law of conservation of momentum?

Answer: In order to jump, the player must push against the floor. In doing so the earth recoils as the player moves up. The momentum imparted to the earth has the same magnitude but opposite direction as that of the player. However, because the mass of the earth is so very much larger than the mass of the player, the earth's recoil velocity is too small to be noticed. Nevertheless, the total momentum is conserved.

Question: At the top of the leap the player's momentum changes direction. How do you reconcile this observation with the law of conservation of momentum?

Answer: The force of gravity acts on the player throughout the jump. It is this force that alters the player's momentum during the jump. If we only consider the motion of the player, then the gravitational force is an external force and the law of momentum conservation is not appropriate. However, let us consider the system that includes both the player and the earth. Gravity acts on each as equal-but-opposite internal forces in accord with Newton's third law. The combined momentum of the player and the earth remains constant as expected from the law of conservation of momentum: the total momentum of a system is constant when the net external force on the system is zero.

bodies. By similar reasoning we can extend it to include systems of three, four, or any number of bodies. In all cases, **when the net external force on a system is zero, the total momentum of that system is constant.** This is a statement of the law of **conservation of linear momentum.**

As we said earlier, conservation of linear momentum is one of the most useful laws in physics, enabling us to determine the momentum of a system after an interaction without needing to know all the details of the interaction. We will illustrate the application of momentum conservation with examples in the next few sections.

7.4 Conservation of Momentum in One-Dimensional Collisions

Christiaan Huygens and others knew of momentum conservation even before Newton published the *Principia*. The results of Huygens's investigations of collisions independently led to the following statement of the law of conservation of linear momentum: *Provided there are no external forces acting on a system, the total momentum before collision equals the total momentum after collision.* For a collision involving two bodies, the conservation law can be expressed symbolically as

$$m_1\mathbf{v}_1 + m_2\mathbf{v}_2 = m_1\mathbf{v}_1' + m_2\mathbf{v}_2'. \tag{7.5}$$

The subscripts indicate which of the two bodies is referred to; unprimed quantities stand for values before the collision, and primed quantities stand for values after the collision (Fig. 7.5). Although we have derived this general statement from theoretical considerations, it has been shown to be true experimentally as well.

Equation (7.5) is a statement of conservation of linear momentum that follows directly from Eq. (7.4). We will find it an especially useful version in problems involving collisions. Note that we do not need to know anything about the details of the collision mechanism itself. The rule holds for collisions between hard elastic bodies, such as billiard balls, as well as for collisions between soft bodies that do not "bounce" upon colliding, such as blobs of putty. In fact, in the absence of external forces, the law of conservation of momentum is observed to hold for any kind of collision on any scale, from subatomic to galactic.

Even though momentum is always conserved in collisions, kinetic energy may or may not be conserved. For this reason, we usually classify collisions according to whether kinetic energy is conserved or not. If kinetic energy is conserved during a collision, we call it an **elastic collision,** a situation approximated by the collisions between pool balls or glass marbles (Fig. 7.6). At the other extreme are collisions in which two objects stick together after impact. These collisions are known as **perfectly inelastic collisions.** A collision between two railroad cars that couple together upon impact is an example of a perfectly inelastic collision. In between are the **inelastic collisions** that are neither elastic nor perfectly inelastic. Because we can analyze perfectly inelastic collisions using only the law of conservation of momentum, we will consider them now. In the next chapter we will examine the somewhat more complicated case of elastic collisions in which both momentum and kinetic energy are conserved simultaneously.

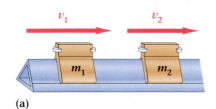

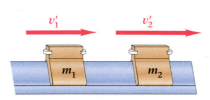

(a)

(b)

Figure 7.5
A one-dimensional collision of two gliders on an air track. (a) Before collision. (b) After collision.

Figure 7.6
Collisions between pool balls are approximately elastic.

First, let's simplify our discussion to just head-on, or one-dimensional, collisions. Suppose, for example, that we consider the *perfectly inelastic collision* of two gliders on a linear air track (Fig. 7.7). The air track allows us to control the gliders' motions so that they move only in one direction. This means that the vector equation for momentum conservation (Eq. 7.5) reduces to a single one-dimensional algebraic equation:

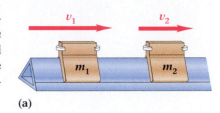

(a)

$$m_1v_1 + m_2v_2 = m_1v_1' + m_2v_2'.$$

If the collision is perfectly inelastic, the two bodies stick together and move off with the same velocity, $v_1' = v_2'$. We can immediately determine the final velocity in terms of the masses and the initial velocities.

Suppose that the glider of mass m_1 has an initial velocity v_1 and that glider m_2 has an initial velocity v_2 in the same direction. If $v_1 > v_2$, a collision occurs when glider m_1 overtakes glider m_2. If the gliders stick together after the collision, the final velocities m, are identical, so $v_1' = v_2' = v'$. Then the momentum equation becomes

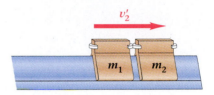

(b)

Figure 7.7
A one-dimensional collision of two gliders on an air track. (a) Before collision. (b) If putty or sticky tape is attached to their bumpers, the two gliders stick together. The resulting collision is perfectly inelastic.

$$m_1v_1 + m_2v_2 = (m_1 + m_2)v'.$$

Solving for the final velocity v', we get

$$v' = \frac{m_1v_1 + m_2v_2}{m_1 + m_2}$$

$$v' = \frac{m_1}{m_1 + m_2}v_1 + \frac{m_2}{m_1 + m_2}v_2.$$

This last equation reveals the significance of the restriction to perfectly inelastic collisions. If we know the masses and initial velocities, we can compute the final velocity readily from momentum considerations alone. If the collision is not perfectly inelastic, so that the two bodies do not stick together and therefore have different velocities after collision, then the more general equation (Eq. 7.5) applies. For collisions that are not perfectly inelastic, knowing the masses and initial velocities is not enough to determine the final velocities—we need more information. We consider this point again in Chapter 8.

▼

Problem-Solving Strategy

Conservation of Linear Momentum in One Dimension

Linear momentum is always conserved in any collision or event that is free from external forces. When applying the law of conservation of momentum the following steps are helpful.

1. Define a coordinate system and identify the velocities both in magnitude and direction.
2. Make a sketch of the situation including both "before" and "after" the collision.
3. Write the equation for the total momentum before and after the collision.
4. If the collision is perfectly inelastic, remember there is only one final velocity.

Example 7.3

In a safety test of automobile equipment, two cars of unequal mass undergo a head-on collision in which they stick together after the collision (Fig. 7.8). A Buick Park Avenue with a mass of 1660 kg and an initial velocity of 8.0 km/h strikes a 830-kg Geo Metro with a velocity of 10.0 km/h toward the first car. (a) What is the velocity of the combination immediately after collision? (b) How do the accelerations of the two cars during collision compare?

Strategy (a) We first choose a coordinate system. Let's take the direction of the Buick as the positive direction, and call its velocity v_1. Then $v_1 = 8.0$ km/h. Because the cars are traveling in opposite directions, the direction of the Geo is negative, $v_2 = -10.0$ km/h. The mass of the Buick is twice the mass of the Geo, so we can use m for the mass of the small car and $2m$ for the mass of the larger car. After impact, the cars stick together and thus move off with the same velocity. So, we can use the law of momentum conservation to find that single final velocity.

Solution (a) Because the collision is perfectly inelastic, the equation of momentum conservation becomes

$$m_1 v_1 + m_2 v_2 = (m_1 + m_2)v'.$$

If we set $m_1 = 2m$ and $m_2 = m$, the equation becomes

$$2mv_1 + mv_2 = (2m + m)v' = 3mv'.$$

The common factor of m can be eliminated to give

$$2v_1 + v_2 = 3v',$$

or

$$v' = \frac{2v_1 + v_2}{3}$$

$$v' = \frac{2(+8.0 \text{ km/h}) - 10.0 \text{ km/h}}{3} = +2.0 \text{ km/h}.$$

Figure 7.8

Example 7.3: A head-on collision between two cars of unequal mass.

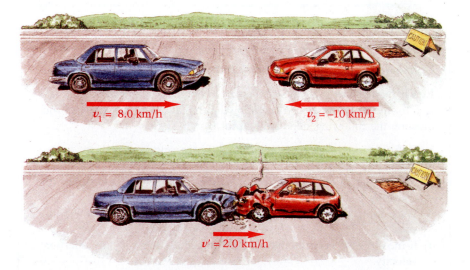

$v_1 = 8.0$ km/h $v_2 = -10$ km/h

$v' = 2.0$ km/h

The positive sign indicates that the final motion is in the same direction as the initial motion of the larger car.

(b) The change in velocity of the Buick is

$$\Delta v_{\text{large car}} = v_{\text{final}} - v_{\text{initial}} = +2.0 \text{ km/h} - 8.0 \text{ km/h} = -6.0 \text{ km/h}.$$

The change in velocity of the small car is

$$\Delta v_{\text{small car}} = v_{\text{final}} - v_{\text{initial}} = +2.0 \text{ km/h} - (-10.0 \text{ km/h}) = +12.0 \text{ km/h}.$$

The accelerations are given by $\Delta v/\Delta t$, where the collision time Δt is the same for both. Thus, the ratio of the accelerations is the ratio of the change in velocities. From our results we see that the magnitude of the small car's average acceleration is twice as great as the magnitude of the large car's acceleration.

Discussion   Because total momentum is conserved, the change in momentum of the two cars is equal in magnitude but opposite in direction. The forces on the two cars must also be equal and opposite. What about the forces on the occupants of the cars? Are they the same? No, they are not. The force on a passenger is $\mathbf{F} = m\mathbf{a}$, where m is the mass of the passenger. The acceleration is proportional to the change in velocities. Thus, the occupants of the smaller car, who undergo an acceleration that is twice as great as that of the passengers of the larger car, experience greater forces. Consequently, the occupants of the smaller car are much more likely to experience serious injuries than are the occupants of the larger car.

Example 7.4

A 60-kg ice skater is standing at rest on a frozen lake. The friction between his skates and the surface of the ice is negligible. If he throws a 2.0-kg block of ice horizontally with a velocity of 12 m/s, what is his recoil velocity?

Strategy   This situation is essentially that of a perfectly inelastic collision run in reverse; that is, a system separates into two bodies with no net force acting on the system as a whole. (We can neglect the force of gravity because it acts along a direction perpendicular to the direction of separation.) Therefore, we can apply the law of conservation of momentum to find the skater's recoil velocity.

Solution   The total momentum before the interaction is zero. Therefore, the total momentum afterward must also be zero. Writing the equation for momentum conservation, we get

$$0 = m_1 v_1' + m_2 v_2'.$$

If $m_1 v_1'$ is the momentum of the block of ice, then the skater recoils with a velocity

$$v_2' = -\frac{m_1 v_1'}{m_2}$$

$$v_2' = -\frac{(2.0 \text{ kg} \times 12 \text{ m/s})}{60 \text{ kg}} = -0.40 \text{ m/s}.$$

The negative sign indicates that the skater's motion is in the direction opposite to that in which the block of ice was thrown.

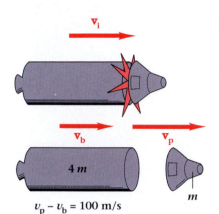

Figure 7.9
Example 7.5: A booster rocket separates from its payload.

Example 7.5

A booster rocket and its payload are traveling at a speed of 900 m/s. An explosion separates the booster from the payload with a relative speed of 100 m/s, and the payload is thrown forward along the initial direction of motion. Find the velocity of the payload and of the booster immediately after they separate. Assume that the mass of the booster is four times that of the payload and that the effects of gravity are negligible.

Strategy We first draw a sketch of the rocket before and after the explosion (Fig. 7.9). If the mass of the payload is m, the mass of the booster is $4m$. Before the explosion, both pieces are attached and travel with the same velocity $v_i = 900$ m/s. After the explosion, the payload is thrown forward, so momentum transfer to the booster must be in the backward direction. Since the explosion provides internal forces, but not external force on the system of payload and booster together, we can solve this problem by applying the principle of momentum conservation.

Solution The initial momentum is

$$\text{momentum before} = (m + 4m)v_i = (5m)v_i.$$

After the separation, the momentum is

$$\text{momentum after} = mv_p + (4m)v_b,$$

where v_p = payload velocity and v_b = booster velocity. Conservation of momentum tells us that

$$\text{momentum before} = \text{momentum after},$$

or

$$(5m)v_i = mv_p + 4mv_b.$$

We have two unknown quantities, v_b and v_p, so we need another equation relating them. This is the statement of relative speed after the explosion:

$$\text{relative speed} = 100 \text{ m/s} = v_p - v_b.$$

Solving for v_p and inserting the result into the equation of momentum conservation, we get

$$5mv_i = m(v_b + 100 \text{ m/s}) + 4mv_b$$
$$= 5mv_b + m(100 \text{ m/s}),$$

which may be solved for the booster velocity:

$$v_b = v_i - 20 \text{ m/s} = 880 \text{ m/s}.$$

The payload velocity is 100 m/s faster, or

$$v_p = 980 \text{ m/s}.$$

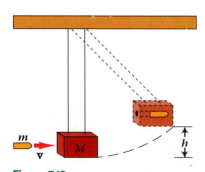

Figure 7.10
Example 7.6: A ballistic pendulum. The height to which the block rises may be used to find the speed of a bullet fired into the block.

Example 7.6

Figure 7.10 shows a *ballistic pendulum,* a device invented by Benjamin Robins in 1742 to measure the speed of a bullet. The bullet is fired into a block of wood (or other material) suspended from two light strings or wires. The bullet is stopped by the block, making a perfectly inelastic collision. The block containing the bullet then swings until it reaches a height h. Show that the initial velocity of the bullet can be determined from a knowledge of the mass of the bullet, the mass of the block, and the height of the swing.

Strategy We can analyze this situation in two parts; the first is the inelastic collision of the bullet with the block, and the second is the swinging of the block on its strings. First we use the law of conservation of momentum to analyze the collision. Then we use the law of conservation of mechanical energy to examine the swing. Combining these analyses will allow us to determine the initial speed of the bullet.

Solution Initially the bullet of mass m has a velocity v while the block of mass M is at rest. If the collision time is short, the bullet comes to rest with respect to the block before the block moves appreciably. Thus, there are no external forces, so momentum is conserved. If we let V represent the velocity of the block and bullet immediately after the collision, we get

$$mv = (m + M)V.$$

After the collision, the block and bullet have a kinetic energy given by

$$KE = \tfrac{1}{2}(m + M)V^2.$$

The pendulum swings out, rising as it goes. As it does so, it exchanges kinetic energy for gravitational potential energy. When it reaches the maximum height h, it will have a potential energy $(m + M)gh$ that just equals the kinetic energy it had immediately after the collision. By equating this potential energy with the kinetic energy, we can evaluate V and in turn find the bullet velocity v:

$$\tfrac{1}{2}(m + M)V^2 = (m + M)gh,$$

so

$$V = \sqrt{2gh}.$$

When we put this expression for V in the equation for momentum conservation and solve for the bullet velocity v, we get

$$v = \frac{m + M}{m}\sqrt{2gh}.$$

Discussion Looking back over the solution, you can see that we applied separate conservation laws to separate parts of the problem. In the first part (the collision), momentum was conserved, but mechanical energy was not. In the second part (the swinging of the pendulum), mechanical energy was conserved. Nevertheless, when we combine the two parts we have enough information to uniquely determine the initial velocity of the bullet.

7.5 · Conservation of Momentum in Two- and Three-Dimensional Collisions

In the previous examples, we limited ourselves to one-dimensional situations. However, in many common situations, such as collisions between billiard balls or air molecules, the objects move in different directions after the collision and it is necessary to consider two or three dimensions. Then the vector aspect of momentum becomes important. We first write out the general equations, simplify to two dimensions, and then consider a specific example. As before, provided that the net external force on the system is zero, we write the conservation rule as: *The momentum before collision equals the momentum after collision.*

In the general case, we could resolve each velocity vector in Eq. (7.5) into three mutually perpendicular components (v_x, v_y, v_z). For the vector equation to be true, the component equations must also be satisfied separately. As a consequence, the vector equation contains three independent statements, corresponding to its three components:

$$x \text{ component:} \quad m_1 v_{1x} + m_2 v_{2x} = m_1 v'_{1x} + m_2 v'_{2x}, \tag{7.6a}$$

$$y \text{ component:} \quad m_1 v_{1y} + m_2 v_{2y} = m_1 v'_{1y} + m_2 v'_{2y}, \tag{7.6b}$$

$$z \text{ component:} \quad m_1 v_{1z} + m_2 v_{2z} = m_1 v'_{1z} + m_2 v'_{2z}. \tag{7.6c}$$

We have used two subscripts on each velocity component. The first subscript indicates whether the velocity is for mass 1 or mass 2, and the second subscript indicates which component (x, y, or z) of velocity is being considered. An example follows of a two-dimensional collision in which the x and y components are analyzed separately.

Example 7.7

Two cars approaching each other along streets that meet at a right angle collide at the intersection. After the crash, they stick together. If one car has a mass of 1450 kg and an initial speed of 11.5 m/s and the other has a mass of 1750 kg and an initial speed of 15.5 m/s, what will be their speed and direction immediately after impact?

Strategy The collision is perfectly inelastic. We can use the law of conservation of momentum to find the speed just after impact. We must choose a coordinate system. Call the lighter car number 1 with velocity $\mathbf{v}_1$ in the x direction, and let car number 2 have velocity $\mathbf{v}_2$ in the y direction (Fig. 7.11). Then we can find the final velocity $\mathbf{v}$ in terms of the initial velocities and masses.

Solution From conservation of momentum we get

$$m_1 \mathbf{v}_1 + m_2 \mathbf{v}_2 = (m_1 + m_2)\mathbf{v}.$$

If we separate this vector equation into components, we get

$$x \text{ component:} \quad m_1 v_1 = (m_1 + m_2)v_x,$$

$$v_x = \frac{m_1 v_1}{m_1 + m_2} = \frac{1450 \text{ kg} \times 11.5 \text{ m/s}}{(1450 + 1750) \text{ kg}} = 5.21 \text{ m/s},$$

and

$$y \text{ component:} \quad m_2 v_2 = (m_1 + m_2)v_y,$$

$$v_y = \frac{m_2 v_2}{m_1 + m_2} = \frac{1750 \text{ kg} \times 15.5 \text{ m/s}}{(1450 + 1750) \text{ kg}} = 8.48 \text{ m/s}.$$

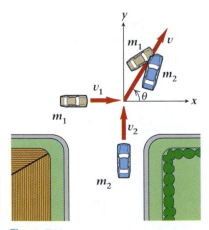

Figure 7.11
Example 7.7: A perfectly inelastic collision in two dimensions. The two cars stick together after impact.

The magnitude of the final velocity is

$$v = \sqrt{v_x^2 + v_y^2} = 9.95 \text{ m/s.}$$

The velocity $\mathbf{v}$ makes an angle with the x direction given by

$$\theta = \tan^{-1} \frac{v_y}{v_x} = 58.4°.$$

Thus, the two cars move off with a speed of 9.95 m/s at an angle of 58.4° from the initial direction of travel of the 1450-kg car.

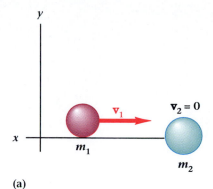

(a)

Consider the situation shown in Fig. 7.12, in which two objects collide but do not stick together after impact. The object m_1 with an initial velocity $\mathbf{v}_1$ strikes object m_2, initially at rest. These objects could be billiard balls, subatomic particles, whatever you like, and the two need not have the same mass. After the collision, we observe that object 1, which was initially in motion along the x direction, is now traveling with a velocity $\mathbf{v}_1'$ at an angle θ_1 with respect to the x direction. Can we determine the speed and direction of the second object?

The directions of the velocity vectors $\mathbf{v}_1$ and $\mathbf{v}_1'$ define a plane, which we may choose as the xy plane. There is no initial momentum in the z direction and $\mathbf{v}_1'$ has no z component, so there can be no z component of velocity associated with $\mathbf{v}_2'$. Thus $\mathbf{v}_2'$ must also lie in the xy plane. Because all the motion is in the xy plane—that is, over a flat surface—the problem is two-dimensional only. If we choose the x direction as the initial direction of $\mathbf{v}_1$, then the equations of momentum conservation reduce to

$$(p_x) \qquad m_1 v_{1x} = m_1 v_{1x}' + m_2 v_{2x}' \qquad (7.7a)$$

and

$$(p_y) \qquad 0 = m_1 v_{1y}' + m_2 v_{2y}'. \qquad (7.7b)$$

From these two equations we see that if we know the masses of the two objects, the velocity of the incident object, and the velocity components of one of the outgoing objects, then we can find the velocity components of the other object. Thus, we can indeed determine the speed and direction of the second object. An application of these equations is illustrated in the following example.

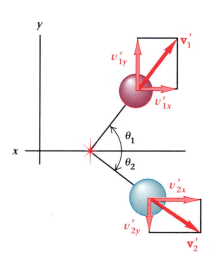

(b)

Figure 7.12

A collision in two dimensions. (a) Object m_1 moving in the x direction, object m_2 at rest. (b) Velocities after collision. Momentum is conserved in both the x and y directions.

▼

Problem-Solving Strategy

Conservation of Linear Momentum in Two Dimensions

The strategy that applies to momentum conservation in one dimension also applies in two dimensions. However, in two dimensions the velocities of each object must be resolved into x and y components. Then the equations for momentum conservation are applied separately for each direction.

Example 7.8

A billiard ball moving at 10.0 m/s along the positive x axis collides with a second billiard ball at rest. The balls have identical masses. After the collision, the incoming (or incident) ball moves on with a speed of 7.7 m/s at an angle of $40°$ from the x axis. What are the speed and direction of motion of the struck ball?

Solution We first determine the individual components of the final velocity of the second body, then calculate its direction of motion. In this example, $m_1 = m_2$, so the mass may be eliminated in Eq. (7.7). Referring to the diagram in Fig. 7.12(b), we see that the x component of velocity of the incident ball after collision is

$$v'_{1x} = v'_1 \cos \theta_1.$$

Substituting this equation into Eq. (7.7a), we obtain the x component of velocity of the struck ball,

$$v'_{2x} = (v_1 - v'_1 \cos \theta_1).$$

The y component of motion of the second ball is determined in a like manner. First, from the diagram we have

$$v'_{1y} = v'_1 \sin \theta_1,$$

which gives, upon substitution into Eq. (7.7b),

$$v'_{2y} = -v'_1 \sin \theta_1.$$

The minus sign indicates that the motion is in the negative y direction. The final speed and direction of the second body are now

$$v'_2 = \sqrt{(v'_{2x})^2 + (v'_{2y})^2} = \sqrt{(v_1 - v'_1 \cos \theta_1)^2 + (-v'_1 \sin \theta_1)^2},$$

$$\tan \theta_2 = \frac{v'_{2y}}{v'_{2x}} = \frac{-v'_1 \sin \theta_1}{v_1 - v'_1 \cos \theta_1}.$$

We can evaluate the speed and direction of mass 2 when the values for v_1, v'_1, and θ_1 are inserted into these equations. They give $v'_2 = 6.4$ m/s at $\theta_2 = 50°$ below the x axis. Note that we cannot determine the final velocities and angles from a knowledge of the initial conditions only, although it can be done in some special cases, as we will show in Chapter 8.

(a)

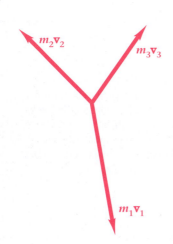

(b)

Figure 7.13

(a) Two occupants of a stationary ice sled simultaneously throw objects from the sled in different directions. The view is from directly overhead.
(b) Momentum diagram for the situation of (a) after the objects have been thrown.

For a situation involving more than two bodies, we still have the same general principles. Figure 7.13 shows one such case, which is two-dimensional and involves the movement of three bodies. Two occupants of an ice sled simultaneously throw objects horizontally from the sled in different directions. The accompanying vector diagram shows the final momenta of the objects and the sled. To analyze the situation, we would write x and y components for the momenta, as before. Only now we would have three bodies in the equations, which would be extensions of Eq. (7.6).

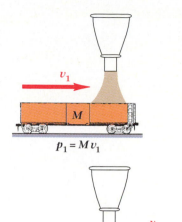

*7.6   Changing Mass

In Chapter 4 we considered the dynamics of bodies with constant mass. Because the mass was constant, it was relatively easy, when given the force, to find the acceleration and then the subsequent motion of the body. However, in many practical cases the mass is not constant.

    An example of a system with changing mass is the railroad car of Fig. 7.14. The car is in motion while taking on sand at a constant rate. For simplicity, we assume that the car has some initial velocity, v_1, and rolls freely without any friction. It is not connected to a locomotive.

    The initial momentum p_1 is the product of the mass of the car M and its initial velocity v_1,

$$p_1 = Mv_1.$$

Although the falling sand has zero initial speed in the horizontal direction, it is accelerated horizontally when it hits the moving car. The resulting action-reaction pair speeds up the sand while slowing down the car. After a time Δt, an amount of sand Δm has been added and the system (car plus sand) has a new velocity v_2. The corresponding momentum of the system is then

$$p_2 = (M + \Delta m)\, v_2.$$

The change in momentum is

$$p_2 - p_1 = Mv_2 + \Delta m v_2 - Mv_1 = M(v_2 - v_1) + \Delta m v_2,$$

which can be written as

$$\Delta p = M\Delta v + v_2 \Delta m.$$

If we divide through by Δt, we find the system's change in momentum per unit time is

$$\frac{\Delta p}{\Delta t} = M\frac{\Delta v}{\Delta t} + v_2 \frac{\Delta m}{\Delta t}.$$

But the change in momentum per unit time is equal to the net external force applied. Since there is no external force along the direction of motion, $\Delta p/\Delta t = 0$ and

$$M\frac{\Delta v}{\Delta t} = -v_2 \frac{\Delta m}{\Delta t}.$$

We see that $\Delta v/\Delta t$ is negative, as expected; the car is slowing down. If we wanted to keep the car moving with constant speed so that $v_1 = v_2$, then we would have to supply an external force $F = v_1 \Delta m/\Delta t$.

    A second example of motion with changing mass is that of a rocket, whose mass diminishes as it consumes its fuel. Figure 7.15 represents the basic principle of rocket propulsion. Gases are expelled at high velocity from the rear of the rocket. Conservation of momentum for the rocket-gas system then requires a forward velocity of the rocket.

Figure 7.14

A railroad car in motion takes on sand from a stationary hopper. Momentum in the horizontal direction is conserved despite the system's changing mass.

Figure 7.15

The basic principle of rocket propulsion: Gas expelled from the rear of the rocket causes it to move forward, because of the conservation of momentum.

Because the force of gravity acts on an earthbound rocket, we cannot properly apply the principle of conservation of momentum without considering the earth-rocket system. For simplicity, let us imagine a rocket so far from the earth that we can neglect external forces. The rocket engine ejects gases at the rate of $\Delta m/\Delta t$ (where m is the mass of the fuel) at a velocity v_r with respect to the rocket. The velocity v_r is taken to be constant here. Thus, the magnitude of the reaction force on the rocket is

$$F_R = v_r \frac{\Delta m}{\Delta t}.$$ (7.8)

This force is called the *thrust* of the rocket motor. The direction of $\mathbf{F}_R$ is opposite to the direction in which the gases are expelled.

Figure 7.16
Example 7.9: The acceleration of a rocket depends on the net force acting on it.

Example 7.9

A fully fueled rocket with a total mass of 5000 kg is set to be fired vertically. If the rocket engine ejects its exhaust gases at a speed of 3.0×10^3 m/s and burns fuel at the rate of 50 kg/s, what is the rocket's initial upward acceleration?

Strategy There are two forces that act on the rocket, the thrust of the engines that pushes upward and the gravitational attraction that pulls the rocket down (Fig. 7.16). The acceleration is a result of the net force that acts on the rocket.

Solution The thrust F_R generated by the rocket engine is the burn rate times the exhaust velocity:

$$F_R = v_r \frac{\Delta m}{\Delta t} = (3.0 \times 10^3 \text{ m/s})(50 \text{ kg/s}) = 1.5 \times 10^5 \text{ N}.$$

If we call the upward direction positive, then the force, $\mathbf{F}_R$, of the rocket engine is positive and the force, mg, due to gravity is negative. The initial acceleration of the rocket is given by the net force divided by the initial mass:

$$a = \frac{F_{net}}{m} = \frac{F_R - mg}{m} = \frac{F_R}{m} - g = \frac{1.5 \times 10^5 \text{ N}}{5 \times 10^3 \text{ kg}} - 9.81 \text{ m/s}^2$$

$$a = 20 \text{ m/s}^2 \text{ upward.}$$

Summary

Useful Concepts

This chapter introduces a fundamental conservation law of physics.

■ *Conservation of linear momentum:* The total linear momentum of a system is constant whenever the net external force on the system is zero.

■ Linear momentum is given by

$$\mathbf{p} = m\mathbf{v}.$$

■ An impulsive force is related to the change in momentum by

$$\overline{\mathbf{F}}\Delta t = \Delta \mathbf{p}.$$

■ For two bodies in collision, the statement of conservation of momentum is

$$m_1\mathbf{v}_1 + m_2\mathbf{v}_2 = m_1\mathbf{v}_1' + m_2\mathbf{v}_2'.$$

■ In the case of a rocket engine with a changing mass m of burning fuel, the thrust is

$$F_R = v_r \frac{\Delta m}{\Delta t}.$$

This is a form of Newton's second law with variable mass, $F = \Delta(mv)/\Delta t$.

Important Terms

You should be able to write the definition or meaning of each of the following terms:

linear momentum	elastic collision
impulse	perfectly inelastic collision
conservation of linear momentum	inelastic collision

Conceptual Questions

7.1 Discuss the usage of the word *momentum* in everyday conversation and compare it with the physical definition of momentum.

7.2 Explain the conservation of momentum that occurs when a fast-moving baseball is struck by a bat.

7.3 A golf club strikes a stationary golf ball and sends it flying. Which has the greater momentum change? Why?

7.4 You are in a cart loaded with bricks rolling without friction along a smooth straight track. What happens when you throw bricks (a) to the rear, (b) to the side, (c) or straight ahead?

7.5 A golf ball is thrown hard at a brick wall. A lump of soft clay with the same mass as the golf ball is thrown against the wall with the same initial velocity. Which of these events delivers the greater impulse to the wall?

7.6 What is the function of seat belts and air bags in automobiles? How do they reduce injuries?

7.7 When a balloon is blown up and released, it flies about as the air escapes. What makes it go?

7.8 The velocity of a bullet fired from a rifle held against the shooter's shoulder is measured very carefully. The rifle is then clamped to a massive bench so that it has no measurable recoil. How does that affect the velocity of the bullet?

7.9 Suppose you are holding the far end of a long garden hose of uniform diameter that extends straight from a faucet. Do you feel any force when the water is turned on? Would it make any difference if you bent the end of the hose through a 90° angle?

7.10 Why are the passengers in a bus less likely to be injured than passengers in a car in a collision of a bus with a car?

7.11 Sometimes when extinguishing a fire on a burning ship, a fireboat will have some of its nozzles pointing away from the fire. Why?

7.12 If the momentum of a system can only be changed by forces external to the system, how is it possible that a "Mexican jumping bean" can sometimes jump up from a table top?

7.13 Is it possible for two objects to simultaneously have identical kinetic energies and identical momenta? Explain your answer.

Problems

Section 7.1 Linear Momentum

Hints for Solving Problems

Momentum is the product of mass and velocity. Although SI units are usually preferred, many of the problems in this section can be solved using mixed units such as kg · km/h.

7.1 What is the momentum of a 1500-kg Mercedes-Benz 300E traveling at 115.0 km/h?

7.2 (a) What is the momentum of a 109-kg football player running at a top speed of 9.86 m/s? (b) What is the momentum of a 9.72-g rifle bullet traveling at 728 m/s?

7.3 Which is greater, the momentum of a 1645-kg Cadillac DeVille traveling at 32 km/h or a 1061-kg Mazda Miata traveling at 47 km/h?

7.4 An 1.73-kg physics book flies through the air with a momentum of 18.8 kg · m/s. What is its speed?

7.5 (a) What is the ratio of the momentum of a 2.3×10^5-kg jet passenger airplane flying at 960 km/hr to that of a 1.1-kg pitching horseshoe moving at 11.3 m/s? (b) What is the ratio of their kinetic energies?

7.6 The momentum of an object traveling at 5.3 m/s is determined to be 350 kg · m/s. Could the moving object be an automobile?

7.7 A detector of subatomic particles measures the momentum of a particle directly, without the mass being known. In one experiment, the particle's momentum was determined to be 1.82×10^{-26} kg · m/s. (a) What was the particle's speed if it was an electron? (b) What was its speed if it was a proton? The masses can be found in the end sheets.

7.8 Show that the expression for kinetic energy can be written in terms of the momentum as

$$KE = \frac{p^2}{2m}.$$

7.9● What is the momentum of a 46.0-g golf ball just before it hits the ground when it is dropped from a height of 1.36 m?

7.10● A stone of mass m is dropped from rest at a height of 1.84 m. From what height would a stone of mass $m/2$ have to be dropped to have the same momentum upon striking the ground?

7.11● A 0.145-kg baseball is thrown horizontally at 3.38 m/s from a height of 3.67 m above level ground. (a) What is the ball's momentum immediately after it is thrown? (b) What is the ball's momentum just before it strikes the ground?

7.12●● A 0.500-kg stone is thrown upward in a parabolic path. The magnitude of the stone's momentum at the top of the path is one-half its initial value. (a) At what angle with respect to the horizontal direction is the stone thrown? (b) If the stone reaches a maximum height of 3.60 m above the point from which it started, what is the initial momentum?

7.13●● A 0.564-kg block and a 1.54-kg block are held 2.38 m above the ground. The 1.54-kg block is allowed to fall from rest. (a) With what downward speed must the 0.564-kg block be thrown so that just before it strikes the ground it has the same momentum as the heavier block just as it strikes the ground? (b) With what upward speed must the 0.564-kg block be thrown so that it will have the same momentum as does the heavier block just before it strikes the ground in part (a)?

Section 7.2 Impulse

Hints for Solving Problems

When calculating a change in momentum, be sure to be consistent in choice of signs and directions before and after the interaction.

7.14 A croquet mallet delivers an impulse of 8.83 N · s to a 0.44-kg croquet ball initially at rest. What is the speed of the ball immediately after being struck?

7.15 The engine of a model rocket is rated with a total impulse of 5.00 N · s and a thrust duration of 1.20 s. What is the average force exerted by the engine?

7.16 A 0.145-kg baseball traveling 35.2 m/s is stopped in 0.163 s by a catcher's mitt. (a) What is the average accelera-

tion of the ball? (b) What is the average force on the catcher's mitt?

7.17 A 68-kg soccer player kicks a stationary 0.425-kg ball giving it a speed of 13.7 m/s. The player's foot is in contact with the ball for 0.097 s. (a) What is the average force on the ball? (b) What is the average force on the player's foot?

7.18 A 0.14-kg baseball with an initial speed of 28 m/s rebounds with a speed of 34 m/s after being struck with a bat. If the duration of contact between ball and bat was 2.1 ms, what was the average force between ball and bat?

7.19 By expressing each quantity in terms of SI base units, show that the product of force and time has the same dimensions as momentum. (*Hint:* You may want to refer to Section 1.3, Unit Conversions.)

7.20● A person about to jump from a 1.60-m-high platform wants to limit the average stopping force on landing to 12 times her weight. By how much will it be necessary to lower herself by flexing her knees as she lands? (By stopping force we mean the force in excess of her weight.)

7.21● How much work did the girl in Example 7.2 (p. 213) do in stopping by bending her knees? Give an algebraic rather than a numeric answer. Compare that work with the original potential energy she had before she jumped.

7.22● A 1.2-kg hammer hits a nail at a speed of 20 m/s and rebounds at 80% of that speed. The resisting force of the nail is 8000 N. Approximately how long is the hammer in contact with the head of the nail?

7.23● A 0.437-kg croquet ball rolls without friction on a smooth surface with a speed of 1.39 m/s toward a mallet-wielding player. (a) What impulse is required to just stop the ball? (b) What impulse is required to send it in the opposite direction with the same speed? (c) What is the average force in each of the previous parts if the mallet and ball are in contact for 2.00 ms?

7.24● When a 0.64-kg ball is dropped on your hand from 0.73 m above it, your hand recoils 2.4 cm before stopping. What is the average total force on your hand while stopping the ball?

7.25●● A machine gun fires 12 bullets/s into a target. The speed of the 0.014-kg bullets is 731 m/s. What is the average force necessary to hold the gun still?

7.26●● A 0.140-kg block on a wooden table is given a sharp blow with a hammer. The block then slides across the table to a stop. The coefficient of friction between the block and table is 0.32. The hammer's force on the block as a function of time is well approximated by an isosceles triangle with its base on the horizontal time axis. The length of the base is 0.0072 s and the maximum force is 35.7 N. How far does the block go? Use a graphical technique.

7.27●● A plastic ball dropped onto a hard surface from a height of 1.4 m rebounds to 60% of its original height. Measurements of the average force during the time the ball is in contact with the surface show the total average force to be 10 times the weight of the ball. How long was the ball in contact with the surface?

Section 7.4 Conservation of Momentum in One-Dimensional Collisions

Hints for Solving Problems

Collisions can be treated in terms of the final and initial states. When there is no net force, the final momentum equals the initial momentum.

7.28 A 78-kg ice hockey player standing on a frictionless sheet of ice throws a 6.0-kg bowling ball horizontally with a speed of 3.0 m/s. With what speed does the hockey player recoil?

7.29 A 3.51-kg rifle fires a 9.72-g bullet with a velocity of 891 m/s. What is the recoil velocity of the rifle?

7.30 A child playing marbles shoots a marble directly at another marble at rest. The first marble stops, and the second marble continues in a straight line with the same speed that the first marble had initially. What is the ratio of the masses of the two marbles?

7.31 A 0.20-kg model railroad car moving with a speed of 0.24 m/s is struck from behind by an 0.42-kg model locomotive moving along the same line with a speed of 0.52 m/s. If they stick together after the collision, what is their velocity?

7.32• A light spring-gun projectile launcher is mounted on an 0.443 kg air-track glider. The gun points upward at an angle of 28° with the horizontal track. With the glider at rest, a 74.3-g projectile is fired from the gun with a speed of 2.96 m/s. What is the speed of the air-track glider after the gun is fired?

7.33• A projectile launcher mounted horizontally on a stationary air-track glider with total mass of 235 g fires a 54.8-g projectile that hits and sticks to an adjacent 347-g glider initially at rest? The speed of the projectile is 96.4 cm/s. What are the final speeds of both gliders?

7.34• A perfectly inelastic collision takes place between two objects, one of which is initially at rest. Does the ratio of the final total kinetic energy to the initial total kinetic energy depend on the speed of the initially moving object?

7.35• The 3.56-g bullet from a 22-250 rifle is fired into the 1.174-kg block of a ballistic pendulum (Fig. 7.10). The bullet sticks within the block, which swings back, rising 0.595 m. What was the speed of the bullet just before impact?

7.36• A ballistic pendulum is used to measure the speed of a 9.72-g rifle bullet. The 2.27-kg block of the pendulum is measured to rise 14 cm. However, because of the difficulty of determining the position of the block, there is an uncertainty of ±0.5 cm in the measurement of the height. What range of bullet velocity is consistent with these measurements?

7.37• A 3000-kg rocket and its 500-kg payload are traveling at a speed of 2000 m/s. The payload is thrown forward by an explosion that separates it from the rocket with a relative velocity of 140 m/s. (a) What is the velocity of the payload after separation? (b) What is the velocity of the rocket?

7.38• Two toy locomotives approach each other along the same line, and upon collision both stop dead still. If one locomotive has three times the speed of the other and the sum of their masses is 2.88 kg, what is the mass of each locomotive?

7.39•• A 25-g dart is thrown with a speed of 32 m/s at a 1.60-kg target mounted on a spring as shown in Fig. 7.17. The target recoils 2.4 cm. Assume the spring obeys Hooke's law and determine the spring constant.

Figure 7.17
Problem 7.39.

7.40•• A 7.45-g bullet from a 9-mm pistol has a velocity of 353 m/s. It strikes the 0.725-kg block of a ballistic pendulum and passes completely through the block. If the block rises through a distance $h = 12.1$ cm, what was the velocity of the bullet as it emerged from the block?

7.41•• Two equal masses m are hung over a frictionless pulley as shown in Fig. 7.18. Another mass m is dropped onto one of the suspended masses from a height h above it. With what initial speed do the masses move immediately after the collision?

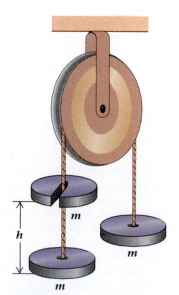

Figure 7.18
Problem 7.41.

7.42•• A ball of mass m is fired with velocity v_0 into the barrel of a spring gun of mass M initially at rest on a frictionless surface (Fig. 7.19). The ball sticks in the barrel at the point of maximum compression of the spring. No energy is lost to friction. What fraction of the ball's initial kinetic energy is stored in the spring?

Figure 7.19
Problem 7.42.

Section 7.5 Conservation of Momentum in Two- and Three-Dimensional Collisions

Hints for Solving Problems

Treat conservation of momentum in the x direction independently from that in the y direction.

7.43 Two cars approach each other along streets that meet at a right angle. They collide at the intersection. After the collision they stick together. If one car has a mass of 1300 kg and an initial speed of 2.25 m/s and the other has a mass of 1800 kg and an initial speed of 4.50 m/s, what will be their speed and direction immediately after impact?

7.44 An Escort and a Camaro traveling at right angles collide and stick together. The Escort has a mass of 1200 kg and a speed of 30 km/h in the positive x direction before the collision. The Camaro has a mass of 1500 kg and was traveling in the positive y direction. After the collision, the two move off at an angle of 64° to the x axis. What was the speed of the Camaro?

7.45• A railroad track lies alongside a frozen lake. A railroad train moves along the track with a constant speed of 7.0 m/s. A boy on a frictionless ice sled is initially moving parallel to the train track with the same speed as the train and 10 m away from it. The boy and sled together weigh 100 kg, and the boy carries a 5.0-kg bag of sugar with him. At some time the boy tosses the sugar to a girl on the train with a velocity of 2.0 m/s perpendicular to the direction of his motion. The girl on the train catches the bag of sugar and immediately throws it back to the boy, who catches it. The girl on the train throws the bag of sugar perpendicular to her motion with the same speed at which she caught it. What are the final direction and speed of the boy and sled?

7.46• Two pendulums are hung adjacent to each other as shown in Fig. 7.20. Bob 1 has a mass m, and bob 2 has a mass $2m$. Bob 1 is pulled aside until the support string makes an angle of 45° with the vertical direction and then released. When the bobs collide, they stick together. What is the maxi-

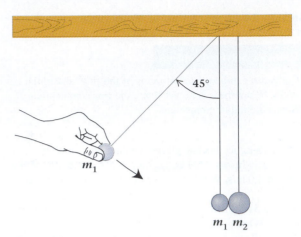

Figure 7.20
Problem 7.46.

mum angle made by the support strings with respect to the vertical after the collision?

7.47• A station wagon and a small car approaching each other at an angle, collide, and stick together after the collision. The station wagon has a mass 1820 kg and a speed of 30 km/h in the positive x direction. The car has a mass of 1060 kg and a speed of 25 km/h in a direction 135° from the x axis. With what velocity do they move off after the collision?

7.48• The ice sled in Fig. 7.13 is initially at rest. The two children on the sled each throw stones horizontally at the same time. The left-hand child throws a 0.87-kg stone with a speed of 2.1 m/s and the right-hand child throws her 0.23-kg stone with a speed of 5.4 m/s. The angle between the stones' paths is 75°. In what direction with respect to the 0.23-kg stone's path does the sled move? Assume that the sled moves without friction on the ice. (*Hint:* Choose a coordinate system with x along the direction of motion of the 0.23-kg stone.)

7.49•• A 7.2-kg bowling ball moving at 2.74 m/s strikes an identical ball that is originally at rest. After the collision the path of the initially moving ball makes an angle of 27° with respect to its original path. The path of the second ball makes an angle of −53° with respect to the same direction. What is the speed of each ball immediately after the collision?

7.50•• A 1000-kg car collides with a 1200-kg car that was initially at rest at the origin of an x-y coordinate system (Fig. 7.21). After the collision, the lighter car moves at 20 km/h in a direction of 30° with respect to the positive x axis. The heavier car moves at 12 km/h at −44° with respect to the positive x axis. What were the initial speed and direction of the lighter car?

*Section 7.6 Changing Mass

7.51 The initial mass of a rocket is 2.6×10^6 kg. A fuel-burning rate of 1.0×10^4 kg/s gives an initial acceleration of 1.5 m/s². What is the velocity of the exhaust gases? Assume the rocket is in free space.

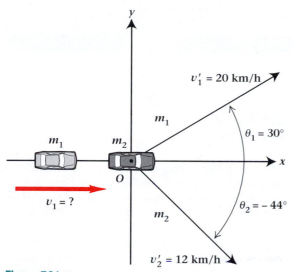

Figure 7.21
Problem 7.50.

7.52 What is the thrust of a rocket that burns fuel at a rate of 1.3×10^4 kg/s if the exhaust gases have a velocity of 2.5×10^3 m/s with respect to the rocket?

7.53 A 24-ton freight car rolls along a straight track with a speed of 10 mi/h. A 2.0-ton automobile falls vertically and lands on it from above and then falls off after a few moments. What is the final speed of the railroad car?

7.54 A 28,000-kg railroad car with an open top rolls along a straight track with a speed of 4.61 km/h. It passes under a waterfall that pours water straight into the car at a rate of 500 kg/s. How much force is necessary to keep the car traveling at constant speed? Assume that friction can be neglected.

7.55● A rocket with initial mass of 8.0×10^3 kg is fired vertically. Its exhaust gases have a relative velocity of 2.5×10^3 m/s and are ejected at a rate of 40 kg/s. (a) What is the initial acceleration of the rocket? (b) What is its acceleration after 20 s have elapsed?

7.56● A rocket with initial mass of 8.0×10^3 kg is fired in the vertical direction. Its exhaust gases are ejected at the rate of 45 kg/s with a relative velocity of 2.4×10^3 m/s. (a) What is the initial acceleration of the rocket? (b) What is the acceleration after 25 s have elapsed?

Additional Problems

7.57 A 0.14-kg baseball thrown with a speed of 25 m/s was hit with an average force of 4500 N. Afterward it had a velocity of 32 m/s in the opposite direction. How long was it in contact with the bat?

7.58 A proton with a speed of 2.36×10^6 m/s makes a signal in a modern particle detector that depends only on the proton's momentum. A subatomic particle called a D^+ meson is observed to leave the same momentum-dependent signal.

What is the speed of the D^+ meson? (You need only know the ratio of the masses. See Table 32.1, p. 1010.)

7.59 A 2.0-kg mass with a speed of 0.50 m/s collides head-on with a 1.5-kg mass moving with a speed of 0.30 m/s toward the first mass. After the collision, the 2.0-kg mass stops. What is the speed of the second mass after the collision?

7.60 In a crash test of automobiles, two cars of equal mass undergo a head-on collision in which they stick together after the collision. If the initial velocity of one car was 5.0 km/h and that of the second car was 8.0 km/h toward the first, what is the velocity of the combination immediately after collision?

7.61 A 0.46 kg golf ball dropped from rest at a height of 1.27 m rebounds to a height of 0.87 m. What is the change of the golf ball's momentum upon rebounding?

7.62● A baseball, thrown against a target that is free to move, rebounds from it with a speed 0.80 of its initial speed. The mass of the target is 20 times that of the baseball. If the baseball had an initial speed of 28 m/s, how far will the target move in 0.1 s?

7.63● A 9.72-g bullet is fired from a 30-30 rifle at a speed of 728 m/s into the 1.250-kg block of a ballistic pendulum suspended by strings 3.9 m long. (a) Through what vertical distance does the block rise? (b) How far does it swing horizontally?

7.64● A 0.046-kg golf ball was hit with an impulsive force that averaged 8000 N. If the ball was in contact with the club head for 5.0×10^{-4} s, what was the speed of the ball immediately after impact?

7.65● A bowling ball with initial velocity of v_0 strikes a stationary bowling ball a glancing blow. The first ball goes off in a direction $30°$ from the initial direction with a speed of 4.0 m/s. The second ball recoils in a direction $-45°$ from the initial direction with a speed of 3.0 m/s (Fig. 7.22). What is the ratio of the mass of the struck ball to that of the incident ball?

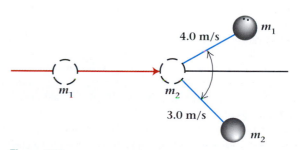

Figure 7.22
Problem 7.65.

7.66● A 900-kg car collides with a 1400-kg car that was initially at rest at the origin of an x-y coordinate system. After the collision, the lighter car moves at 20 km/h in a direction $40°$ with respect to the positive x axis. The heavier car moves at 10 km/h along the positive x axis. What were the initial speed and direction of the lighter car?

7.67• A bowling ball collides head-on with another bowling ball initially at rest. The first ball rebounds with a speed one-twentieth its original speed. The second ball moves off in the direction of the first ball's initial motion with a speed that is 95% of the original speed of the first ball. What is the ratio of the masses of the balls?

7.68•• A 1.0-kg toy car moves freely along a model railroad track with a speed of 0.50 m/s. An additional 0.20-kg mass is carried on the car in a device that projects the mass horizontally away from the car with a speed of 0.20 m/s relative to the initial motion of the car. Calculate the final velocity of the car along the track when the 0.20-kg mass is projected (a) in the same direction as the car's motion, (b) in the opposite direction from the car's motion, and (c) at right angles to the car's motion.

7.69•• In an introductory laboratory experiment, a 56.7-g steel ball is shot from a spring gun into the 203-g pendulum of a ballistic pendulum. The pendulum swings aside and is kept at its maximum height by mechanical means (Fig. 7.23). The change in height is measured to be 13.1 cm. (a) What is the velocity of the steel ball when fired from the spring gun? (b) The pendulum is then swung aside, and the gun is placed on a table so that it can be aimed horizontally to fire the ball into the room. If the ball leaves the gun at a height of 1.12 m from the floor, how far does it travel horizontally before hitting the floor?

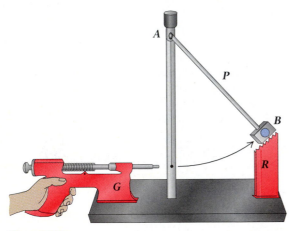

Figure 7.23
Problem 7.69: A ballistic pendulum consisting of a spring gun (G), a projectile (B), a pendulum (P) that pivots about axis (A), and a rack (R).

7.70•• A 15.2-g bullet hits a 0.463-kg block from below (Fig. 7.24). The initial speed of the bullet is 624 m/s and it emerges from the block at 131 m/s. (a) How high does the block rise? (b) If the block is 2.34-cm thick, estimate the average force on the block. Assume that the bullet passes completely through before the block moves appreciably.

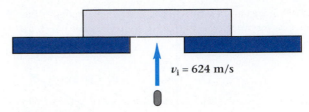

Figure 7.24
Problem 7.70.

7.71•• Hans Brinker, with a mass of 59 kg, is standing in the middle of a frozen lake of frictionless ice. He wants to reach the shore in the shortest possible time without skating. Hans has two 3.0-kg snowballs. He can throw snowballs with a relative speed of 10 m/s, regardless of their mass. Determine whether Hans should throw both masses at one time or throw one mass and then throw the other mass one second later. (If this is not a realistic question, state what is wrong with the assumptions made.)

7.72•• A 0.70-kg piece of modeling clay is held lightly in the extended fingers of the right hand 1.00 m above the floor. An identical piece of clay is dropped from the left hand held 0.80 m directly above the right hand. The falling piece hits and sticks to the other, and they continue together to the floor. (a) How long does it take for them to reach the floor? Call time zero the moment when the left-hand piece is released. (b) Is this more or less time than it would take the left-hand piece to fall straight to the floor if it had not struck the right-hand piece?

7.73•• A 0.350-kg air-track glider is set in motion by an electromechanical "pusher" doing 1.47 J of work on it. The moving glider then strikes another glider of the same mass that was at rest and sticks to it. (a) With what velocity does the pair of gliders move off? (b) What is the kinetic energy of the system after the collision? (c) Discuss your answer to part (b) and compare it with the original 1.47 J.

7.74•• Two cars approaching each other collide and stick together after the collision. One car has a mass 1820 kg and a speed of 30 km/h in the positive x direction. The second car has a mass of 1060 kg. After the collision they both move off with a speed of 18 km/h in a direction +37° from the x axis. What was the initial velocity of the second car?

7.75•• A 0.43-kg soccer ball resting on the ground is kicked so that it travels the maximum distance for the kicker's effort. It lands 42 m away from its initial position. If the kicker's shoe is in contact with the ball for 0.038 s, estimate the average force of the kick.

7.76•• A rocket with initial mass of 5.0×10^3 kg is fired vertically. Its exhaust gases are ejected at the rate of 30 kg/s with a relative velocity of 3.0×10^3 m/s. (a) Find the initial acceleration of the rocket. (b) How high does the rocket climb in 10 s? (*Hint:* Ignore the change in mass to perform this calculation.) (c) How would the answer to (b) change if you included the change in mass of the rocket?

7.77•• A model rocket handbook gives an approximate formula for the velocity v of a rocket at the time t at which the fuel burns out:

$$v = \left(\frac{F_R}{w_{ave}} - 1\right)gt,$$

where F_R is the thrust of the rocket motor and w_{ave} is the average weight of the rocket. Verify the formula and state the assumptions made in the derivation.

7.78•• A 0.74-kg apple is tossed straight up from 1.3 m above the ground with an initial speed of 7.3 m/s. When it has traveled 1.5 m upward it is struck by a 0.15-kg arrow, which stays in the apple. Just before impact the arrow was traveling at a speed of 30 m/s at an upward angle of 57° from the horizontal. (a) What speed did the apple have just before the arrow struck it? (b) What were the speed and direction of the apple-arrow combination immediately after impact? (c) How far from a spot directly below the apple's initial location does the apple-arrow combination hit the ground? (You may wish to refer to Chapter 3, Section 3.7.)

Applying the Conservation Laws

We have emphasized several times in the past two chapters that the application of conservation laws can solve an immense range of physical problems. Indeed, their importance becomes evident as you see how many different situations from all areas of natural science are governed by the same conservation laws. The collision of two subatomic particles and the orbiting of the sun by a comet obey the same conservation rules for energy and momentum, even though the important force is different in the two cases. The transformation of energy between living cells and the transformation of the chemical energy of fuel to warm your home obey similar principles, though we will have to extend our ideas of energy to understand thermal processes (Chapters 11 and 13).

In this chapter we analyze in some detail several different physical situations. The common threads that bind these analyses together are the principles of conservation of energy and conservation of momentum. Because we frequently apply these principles simultaneously, we often have to solve simultaneous equations and manipulate these equations algebraically. However, we never go beyond routine algebra and elementary trigonometry. The mathematics may sometimes become tedious, but there are no principles that you have not already used. Even so, simple collisions between billiard balls can give rise to a great deal of algebraic manipulation. For this reason, you should be particularly careful to study and fill in all the mathematical steps in the examples. In that way you will be able to handle the assigned problems.

The type of analysis we show here is similar in some respects to how physicists approach a new problem. The starting point is a simplified model of the physical situation (with assumptions and restrictions stated). Conserved quantities, such as energy or momentum, are carefully noted. The conservation laws are then used to get a solution. In fact, this type of analysis from conservation laws has led to the discovery of new subatomic particles, as we will see in Chapters 29 and 32. ■

8.1 Definition of Elastic Collisions

We have already encountered examples of momentum conservation in Chapter 7. There, for simplicity, we mainly considered perfectly inelastic collisions, in which two colliding objects stick together after impact. We now wish to broaden our understanding to include collisions in which the colliding objects rebound from each other.

Consider a ball of mass m dropped straight down onto a hard surface so that it rebounds straight up. Figure 8.1 shows the path of the ball slightly offset from vertical for clarity. Let us label the initial height from which the ball was dropped as h, and the height to which it rebounds as h'. We will show that the ratio of the magnitude of the ball's velocity just after impact to its magnitude immediately before impact can be expressed solely in terms of h' and h.

If the ball is dropped from rest, its total energy immediately before impact with the surface is equal to the total energy it had at the top, provided that we can neglect air resistance:

$$PE_{bottom} + KE_{bottom} = PE_{top} + KE_{top}.$$

The kinetic energy at the top is zero because the ball starts from rest. The kinetic energy at the bottom is then equal to the difference in potential energy between the top and the bottom:

$$KE_{bottom} = PE_{top} - PE_{bottom}.$$

If we choose the hard surface to be the zero level of potential energy, then the potential energy at the bottom is zero and the potential energy at the point of release—that is, at the height h—is mgh. The kinetic energy at the bottom thus becomes

$$\tfrac{1}{2}mv^2 = mgh.$$

Rearranging this equation, we obtain

$$v = \sqrt{2gh}.$$

The speed v is the speed immediately before impact.

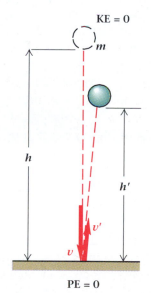

Figure 8.1

A ball of mass m dropped onto a hard surface from a height h rebounds to a height h'. If $h' = h$, the collision is elastic and kinetic energy is conserved.

Similar analysis for the upward path gives the speed v' immediately after impact in terms of the maximum rebound height h' as

$$v' = \sqrt{2gh'}.$$

The ratio of the speed immediately after impact to the speed just before impact is

$$\frac{v'}{v} = \sqrt{\frac{h'}{h}}.$$

If the ball were to rebound to its initial height (that is, to $h' = h$), then we would say that the collision was elastic. In this case v' would be the same as v and the kinetic energy immediately after the collision would be the same as the kinetic energy immediately before the collision. We call a collision **elastic** when the kinetic energy is conserved. If the ball does not rebound to its initial height, then kinetic energy is not conserved and the collision is called **inelastic.***** Note that inelastic collisions do not have to be perfectly inelastic. The colliding objects may have some recoil and loss of kinetic energy at the same time, as is often the case when two cars collide. In inelastic collisions, the total energy is still conserved, but some of the initial kinetic energy is converted into other forms, such as thermal energy.

When we studied momentum conservation in Chapter 7, we primarily considered perfectly inelastic collisions in which the colliding objects did not rebound at all, but stuck together after the collisions. Momentum was conserved in these inelastic collisions, but kinetic energy was not. At the other extreme are the elastic collisions, in which *both* kinetic energy and momentum are conserved simultaneously. Most collisions are neither elastic nor perfectly inelastic. Bouncing balls are prime examples of such collisions (Fig. 8.2). However, many situations can be approximated as either elastic or perfectly inelastic. *The collisions of hard spheres, such as billiard balls, are nearly elastic. A blob of modeling clay thrown against the wall is an example of a perfectly inelastic collision.*

The laws of conservation of energy and momentum have been validated countless times over the years. Today they have become such a fundamental part of our belief about nature that they are used as analytical tools in experiments, rather than being the subjects of experiments to test their validity. However, conservation of kinetic energy alone is not always sufficient. Kinetic energy may be exchanged for potential energy, for thermal energy, or, as we shall see in Chapter 25, for mass. The rule for the conservation of total energy includes energy in all its forms. Because we believe that energy and momentum are conserved, we can often obtain information about a collision or other interaction without observing it directly. Instead, we measure the energy and momentum of objects before and after the interaction and determine facts about the nature of the interaction itself.

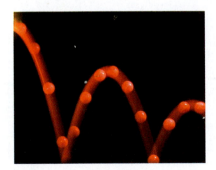

Figure 8.2
Successive bounces of a ball on a hard surface, showing the loss of energy with each bounce.

| 8.2 | **Elastic Collisions in One Dimension** |

We begin our analysis of elastic collisions by considering a collision in one dimension and applying the laws of conservation of momentum and conservation of kinetic energy to this situation. We will expand our analysis to two dimensions in the next section.

---

*Remember that in the absence of external forces, momentum is conserved whether or not kinetic energy is conserved.

Figure 8.3 shows two objects constrained to move along the x axis only. The first object, of mass m_1, is traveling in such a way that it collides head-on with the second object, of mass m_2, which may or may not be at rest. We choose the positive x direction to be the direction of travel of the first object. Since there are no external forces acting in this direction on the system of the two objects, we can use the law of conservation of momentum as given in Eq. (7.5), p. 216,

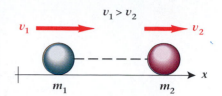

$$m_1 v_1 + m_2 v_2 = m_1 v_1' + m_2 v_2'. \tag{8.1}$$

Figure 8.3

An object of mass m_1 travels with velocity v_1 prior to a head-on collision with an object of mass m_2 and velocity $v_2 < v_1$.

Here v_1 and v_2 are the initial velocities of the first and second objects, respectively, and v_1' and v_2' are their velocities immediately after the collision. If the collision is elastic, we can also use conservation of kinetic energy to get

$$KE_1 + KE_2 = KE_1' + KE_2'. \tag{8.2}$$

Expressing the kinetic energy in terms of mass and velocity, we get

$$\tfrac{1}{2}m_1 v_1^2 + \tfrac{1}{2}m_2 v_2^2 = \tfrac{1}{2}m_1 v_1'^2 + \tfrac{1}{2}m_2 v_2'^2 \tag{8.3}$$

We can then use Eqs. (8.1) and (8.3) to predict the results of any straight-line, elastic collision between two bodies.

Identical Masses, One Object at Rest

A case of special interest, and one well known to billiard players, is the one-dimensional elastic collision between objects of equal mass in which one object (m_2) is at rest before the collision (Fig. 8.4). Because collisions between billiard balls are very nearly elastic, we describe them by assuming elastic conditions. For simplicity, we also neglect the effects of rotations. For the situation $m_1 = m_2$ and $v_2 = 0$, the momentum conservation equation simplifies to

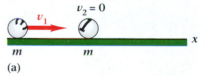

(a)

$$v_1 = v_1' + v_2' \tag{8.4}$$

and the kinetic energy equation becomes

$$v_1^2 = v_1'^2 + v_2'^2. \tag{8.5}$$

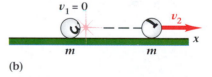

(b)

We can solve these two equations simultaneously by substitution, giving the final velocities v_1' and v_2' in terms of the initial velocity. You should show for yourself that one result is

Figure 8.4

A head-on collision of two billiard balls. (a) Before impact. (b) After impact.

$$v_1' = 0,$$
$$v_2' = v_1.$$

This result says that the first ball stops and the second ball moves with the same velocity that the first ball had initially. Equations (8.4) and (8.5) are also satisfied by the values $v_1' = v_1$ and $v_2' = 0$. These values are not useful, however, because they are the same as the initial conditions and do not correspond to a collision.

Unequal Masses, Both Moving

Now that we have seen the application of the conservation laws to the special case of a collision between a moving object with a stationary object of equal mass, let's consider the general situation in which both objects may be moving and their masses are unequal. If we know the masses and initial velocities of the two objects, then we have two unknown quantities, the final velocities after the collision. To find them we must solve the two Eqs. (8.1) and (8.3) simultaneously.

We begin by rewriting the equation for momentum conservation as

$$m_1(v_1 - v_1') = -m_2(v_2 - v_2').$$ (8.6)

Next we remove the common factor of $\frac{1}{2}$ from the kinetic energy equation, which we then rearrange and factor into

$$m_1(v_1 - v_1')(v_1 + v_1') = -m_2(v_2 - v_2')(v_2 + v_2').$$ (8.7)

Upon dividing Eq. (8.7) by Eq. (8.6), we get

$$v_1 - v_2 = -(v_1' - v_2').$$ (8.8)

This last equation shows that the magnitude of the relative velocity before the collision is equal to the magnitude of the relative velocity after the collision. In other words, the relative speed of approach of object 1 to object 2 before the collision is the same as their relative speed of separation after the collision. Notice that this relation is independent of the masses. It does not matter if the objects have the same mass or not and if they are different it does not matter which is more massive.

We can combine Eq. (8.8) with Eq. (8.6) to eliminate v_2' and get the final velocity of object 1 in terms of the two initial velocities:

$$v_1' = \frac{m_1 - m_2}{m_1 + m_2}v_1 + \frac{2m_2}{m_1 + m_2}v_2.$$ (8.9)

Similarly, we can solve for v_2', finding

$$v_2' = \frac{2m_1}{m_1 + m_2}v_1 - \frac{m_1 - m_2}{m_1 + m_2}v_2.$$ (8.10)

Unequal Masses, One Object at Rest

Now that we have the two general solutions for the final velocities, let's examine them in the special case in which object 2 is initially at rest. Then, since $v_2 = 0$, the second term on the right of the equal sign vanishes in the equations for v_1' and v_2'. The outcomes are

$$v_1' = \frac{m_1 - m_2}{m_1 + m_2}v_1$$ (8.11)

and

$$v_2' = \frac{2m_1}{m_1 + m_2}v_1.$$ (8.12)

These results can be evaluated in terms of the relative sizes of the masses. There are three possibilities: $m_1 < m_2$, $m_1 = m_2$, and $m_1 > m_2$.

1. Suppose m_1 is less than m_2, then v_1' is negative indicating that object m_1 recoils in the direction from which it came (Fig. 8.5). Object m_2 moves off in the original direction of m_1, but with a velocity smaller than v_1.

2. When the two masses are equal, we see that $v_1' = 0$ and $v_2' = v_1$. This is the same result that we obtained previously by direct computation.

3. Finally, suppose that m_1 is greater than m_2. Then v_1' and v_2' are both in the original direction of m_1, but v_1' is smaller than v_1 and v_2' is greater than v_1 (Fig. 8.6).

It is important to understand that these results depend on the simultaneous application of the law of conservation of momentum (Eq. 8.1) and the law of conservation of kinetic energy in elastic collisions (Eq. 8.3). From knowledge of the masses and the initial velocities, we have calculated both final velocities. Neither law alone is sufficient to predict the results for both velocities. (Because there are two unknowns, we must have two equations to solve for them.) The following examples illustrate the application of these conservation laws.

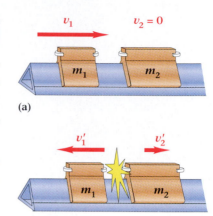

(a)

(b)

Figure 8.5

Head-on collision of two air-track gliders. The heavier glider is initially at rest. (a) Before impact. (b) After impact.

Problem-Solving Strategy

One-Dimensional Elastic Collisions

Remember that in elastic collisions both momentum and kinetic energy are conserved. For a one-dimensional collision between two objects, there are two separate equations that must be satisfied simultaneously. These equations contain six separate quantities: two masses, two initial velocities, and two final velocities. If any four of the quantities are known, the other two are uniquely determined by the equations. The following steps are helpful in applying the conservation laws to elastic collisions.

1. Define a coordinate system and identify the masses and velocities. Determine which quantities are given and which need to be determined.
2. Make sketches of the situations for both "before" and "after" the collision.
3. Write the equation for the total momentum before and after the collision.
4. Write the equation for the kinetic energy before and after the collision.
5. Solve the equations simultaneously.

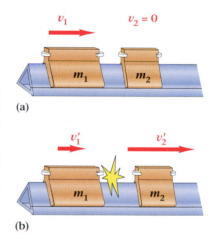

(a)

(b)

Figure 8.6

Head-on collision of two air-track gliders. The lighter glider is initially at rest. (a) Before impact. (b) After impact.

Example 8.1

Two railroad boxcars are initially in motion to the right along the same straight track, which we denote as the x axis with the positive direction to the right (Fig. 8.7). The ratio of the masses of the two cars is $m_1/m_2 = 1/2$. The rearmost boxcar, m_1, has a velocity twice as great as the forward boxcar, m_2. The faster-moving car, m_1, overtakes the other car and collides with it. After the collision, what is the velocity of each boxcar if the collision is elastic?

Strategy Although we could compute the answers from the general solutions given by Eqs. (8.9) and (8.10), we will start at the beginning by applying the laws of conservation of energy and momentum. This example illustrates a common situation. We will get two sets of answers that satisfy our mathematical formulation of the conservation laws. We must then take a careful look at the physical meaning of the solutions in order to select the answer that corresponds to reality. You should fill in the missing algebraic steps for yourself.

Solution If the initial velocity of the second boxcar, m_2, is v, then the initial velocity of the first boxcar, m_1, is $2v$. If we write $m_1 = m$ and $m_2 = 2m$, the law of conservation of momentum becomes

$$(m)(2v) + (2m)(v) = mv_1' + (2m)v_2'.$$

This equation may be simplified as

$$4v = v_1' + 2v_2'.$$

Conservation of kinetic energy gives

$$\tfrac{1}{2}(m)(2v)^2 + \tfrac{1}{2}(2m)(v)^2 = \tfrac{1}{2}mv_1'^2 + \tfrac{1}{2}(2m)v_2'^2,$$

or

$$6v^2 = v_1'^2 + 2v_2'^2.$$

We can rearrange the momentum equation to give v_1' in terms of the other velocities. Then we substitute v_1' into the energy equation to give

$$3v_2'^2 - 8vv_2' + 5v^2 = 0.$$

Figure 8.7

Example 8.1: Elastic collision between two boxcars of unequal mass.

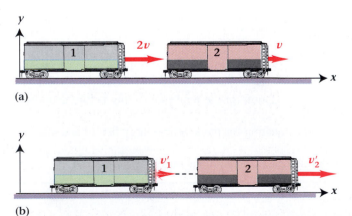

(a)

(b)

We now have a quadratic equation in v_2'. (The initial velocity v is a constant.) Upon substitution into the quadratic formula,* we find

$$v_2' = \frac{+8v \pm \sqrt{(-8v)^2 - 4(3)(5v^2)}}{2(3)} = \frac{8v \pm 2v}{6}.$$

Both solutions must be considered. If we choose the plus sign, we find

$$v_2' = \tfrac{5}{3}v.$$

If we choose the minus sign, we find

$$v_2' = v.$$

Both of these values cannot represent the physical situation, since only one event actually happens. However, we must keep both until we can eliminate one of them by physical considerations. Upon substituting $v_2' = v$ into the momentum equation, we get

$$v_1' = 2v.$$

Upon substitution of $v_2' = \tfrac{5}{3}v$ into the momentum equation, we get

$$v_1' = \tfrac{2}{3}v.$$

Thus, from the purely algebraic standpoint we have two pairs of possible answers:

$$\{v_1' = \tfrac{2}{3}v \quad \text{and} \quad v_2' = \tfrac{5}{3}v\} \quad \text{or} \quad \{v_1' = 2v \quad \text{and} \quad v_2' = v\},$$

both of which satisfy the conservation equations.

The resolution of the problem comes from a careful look at what the answers mean. Remember that this is a straight-line collision and that boxcar 1 is the rearmost car. It must always remain the rearmost car because it cannot pass through the other car. This means that after the collision, its velocity in its initial direction must be less than the velocity of car 2. Therefore the correct answer is $v_1' = \tfrac{2}{3}v$ and $v_2' = \tfrac{5}{3}v$. Note that the other answer—corresponding to the case in which both cars maintain their initial velocities, but with boxcar 1 passing through boxcar 2 without collision—is not physically possible. Because the equation for conservation of kinetic energy is quadratic, we always obtain two sets of answers; however, one set corresponds to the initial conditions and can be discarded.

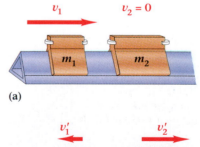

(a)

Example 8.2

In an experiment on a linear air track, a glider of mass $m_1 = m$ moving with initial speed v_1 to the right collides in an elastic collision with a glider of mass $m_2 = 2m$ initially at rest (Fig. 8.8). What are the final velocities of the two gliders?

Strategy We can obtain the final velocities by the simultaneous application of conservation of momentum and energy. Again we obtain two sets of solutions.

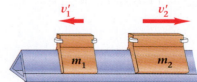

(b)

Figure 8.8
Example 8.2: A glider of mass m collides with a glider of mass $2m$ initially at rest.

---

*You may want to refer to the appendix for Chapter 2, Solving Quadratic Equations.

You should sketch a diagram for yourself before proceeding to help sort out which of the two solutions is physically meaningful.

Solution We apply conservation of momentum to get

$$mv_1 = mv_1' + 2mv_2',$$

where v_1' is the final velocity of the light glider and v_2' is the final velocity of the heavy glider. Conservation of energy gives

$$\tfrac{1}{2}(m)v_1^2 = \tfrac{1}{2}(m)v_1'^2 + \tfrac{1}{2}(2m)v_2'^2.$$

We may eliminate the common factors of $\tfrac{1}{2}m$ to get two equations that must be solved simultaneously for the two unknown velocities:

$$v_1 - v_1' = 2v_2', \qquad \text{(momentum)}$$
$$v_1^2 - v_1'^2 = 2v_2'^2. \qquad \text{(energy)}$$

The momentum equation may be squared to get

$$(v_1^2 - 2v_1v_1' + v_1'^2) = 4v_2'^2.$$

Now we multiply the energy equation by 2 and subtract the above equation from it, obtaining

$$v_1^2 + 2v_1v_1' - 3v_1'^2 = 0.$$

This equation may be factored, giving us

$$(v_1 - v_1')(v_1 + 3v_1') = 0.$$

As before, we have two solutions,

$$v_1' = v_1 \quad \text{or} \quad v_1' = -\tfrac{1}{3}v_1.$$

The solution $v_1' = v_1$ indicates that the body of mass m has the same velocity after the collision that it had before the collision; that is, it implies that there was no collision. We choose instead the physically meaningful solution of $v_1' = -\tfrac{1}{3}v_1$. The negative sign indicates that the light glider recoils in the direction opposite to its initial motion.

If we insert $v_1' = -\tfrac{1}{3}v_1$ into the momentum equation, we obtain

$$v_2' = \tfrac{2}{3}v_1.$$

You may wish to insert these values into the energy equation to verify their correctness.

Discussion We could have obtained the final velocities by direct substitution into Eqs. (8.11) and (8.12), which were derived for this situation, in which one mass is initially at rest. You should compute the values from those equations for yourself to check.

---

*8.3 Elastic Collisions in Two Dimensions

When two billiard balls collide on the surface of a billiard table, the general result is that they move off in different directions. Since all velocities lie in the plane of the surface, the collision is two-dimensional. We can analyze elastic collisions in two dimensions by using the fact that momentum is a vector quantity.

There are two equations for conservation of momentum—one for the x components and one for the y components of momentum—and one equation for conservation of energy. (Remember that energy is a scalar quantity.) However, *four* quantities must be determined to specify completely the final outcome of a two-dimensional collision: the two velocity components of each body, or equivalently, the magnitudes and directions of the two vector velocities. These four quantities cannot be determined from three simultaneous equations, even if we know all of the initial velocity components. However, if one of these quantities can be specified, either from observation or from some physical consideration, then the other three quantities can be uniquely determined.

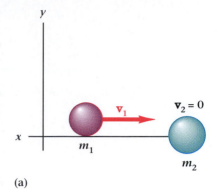

(a)

Figure 8.9 depicts an elastic collision between two spheres of mass m_1 and m_2. One of the spheres, m_2, is initially at rest. To simplify our work, we choose a coordinate system in which the initial velocity is along the x axis. This choice guarantees that there is no initial momentum in the y direction. Using the law of conservation of momentum, we obtain an equation for the x component of momentum:

$$m_1 v_{1x} = m_1 v'_{1x} + m_2 v'_{2x}. \qquad \text{(momentum along } x) \qquad (8.13)$$

Conservation of the y component of momentum gives

$$0 = m_1 v'_{1y} + m_2 v'_{2y}. \qquad \text{(momentum along } y) \qquad (8.14)$$

Conservation of energy gives

$$\tfrac{1}{2} m_1 v_1^2 = \tfrac{1}{2} m_1 v_1'^2 + \tfrac{1}{2} m_2 v_2'^2. \qquad \text{(kinetic energy)} \qquad (8.15)$$

Simultaneous solution of these three equations is necessary to determine the subsequent motion of the spheres. However, we still require prior knowledge of one postcollision velocity component or angle before we can find the other quantities. The following example illustrates the use of conservation laws in two dimensions.

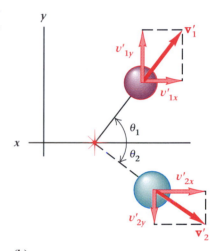

(b)

Figure 8.9
A glancing collision in two dimensions, viewed from above. (a) Before the collision. (b) After the collision, the two spheres move off in different directions.

Example 8.3

Consider two billiard balls of equal mass that collide elastically. One of them has an initial speed v_1, and the other is at rest. After the collision, the ball that was initially in motion is deflected at an observed angle θ_1 with respect to its original direction. Find the final speed of each ball and the direction in which the other ball is moving.

Solution Referring to Fig. 8.9, we can rewrite the equation for the x component of momentum (Eq. 8.13) in terms of the angles θ_1 and θ_2,

$$v_1 = v'_1 \cos \theta_1 + v'_2 \cos \theta_2, \qquad (x \text{ momentum})$$

where θ_1 and θ_2 are both defined as positive in the diagram. Notice that since both balls have the same mass, m is a common factor to all terms and has been eliminated from the equation. Similarly, the equation for conservation of the y component of momentum (Eq. 8.14) may be expressed as

$$v'_1 \sin \theta_1 = v'_2 \sin \theta_2, \qquad (y \text{ momentum})$$

and the kinetic energy conservation equation (Eq. 8.15) becomes

$$v_1^2 = v_1'^2 + v_2'^2. \qquad \text{(kinetic energy)}$$

The x-component momentum equation may be rearranged and squared to give

$$v_1^2 - 2v_1v_1' \cos\theta_1 + v_1'^2 \cos^2\theta_1 = v_2'^2 \cos^2\theta_2.$$

The y momentum equation may also be squared to give

$$v_1'^2 \sin^2\theta_1 = v_2'^2 \sin^2\theta_2.$$

When these two equations are added, we have

$$v_1^2 - 2v_1v_1' \cos\theta_1 + v_1'^2 = v_2'^2,$$

where we have made use of the trigonometric identity $\cos^2\theta + \sin^2\theta = 1$.

At this point we can combine the kinetic energy equation and the last equation to eliminate v_2'. The result expresses v_1' in terms of the initial speed v_1 and the angle θ_1:

$$v_1^2 - 2v_1v_1' \cos\theta_1 + v_1'^2 = v_1^2 - v_1'^2,$$
$$v_1'^2 = 2v_1v_1' \cos\theta_1 - v_1'^2$$
$$v_1'^2 = v_1v_1' \cos\theta_1.$$

The solution is

$$v_1' = v_1 \cos\theta_1,$$

corresponding to ball 1 emerging at angle θ_1. This value of v_1' may, in turn, be inserted into the kinetic energy equation to yield v_2':

$$v_2' = v_1 \sin\theta_1.$$

To get this result, you need to use the trigonometric identity

$$\cos^2\theta + \sin^2\theta = 1.$$

Finally, the recoil angle of the second ball can be obtained from the y momentum equation as

$$\sin\theta_2 = \frac{v_1'}{v_2'} \sin\theta_1 = \frac{v_1'}{v_1}.$$

Discussion We see that knowledge of the initial speed v_1 and the recoil angle θ_1 is sufficient to determine the magnitude of the other speeds v_1' and v_2' as well as the angle θ_2. This example is similar to Example 7.8. In both examples we were asked to find the speed and direction of the struck ball. In the earlier example we were given the speed and direction of the outgoing incident ball. Here we are given only the direction. Since we have three equations (x momentum, y momentum, and kinetic energy), we can solve for three unknowns. Thus, when kinetic energy is conserved we can make predictions about the results of collisions that cannot be made from momentum conservation alone.

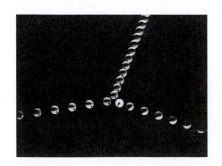

Figure 8.10
Example 8.4: A multiflash photograph of an elastic collision between two balls of equal mass. After the collision, the two balls move off at a right angle to each other.

Example 8.4

A ball of mass m moving with a speed of 2.00 m/s collides with a stationary ball of the same mass in an elastic collision (Fig. 8.10). The incident ball is scattered at an angle of 58° from its original direction. Find the final speed of each ball and the direction of recoil of the struck ball.

Strategy This problem is a numerical example of the collision described in Example 8.3. We are given that $v_1 = 2.00$ m/s and $\theta_1 = 58°$. We may use the equation developed in Example 8.3 to find the speed v_1' of the scattered ball, and conservation of kinetic energy to find the speed v_2' of the struck ball.

Solution The speed v_1' is

$$v_1' = v_1 \cos \theta_1 = (2.00 \text{ m/s}) \cos 58° = 1.06 \text{ m/s}.$$

We obtain the speed of the struck ball from the kinetic energy equation as

$$v_2' = \sqrt{v_1^2 - v_1'^2} = \sqrt{2.00^2 - 1.06^2} = 1.70 \text{ m/s}.$$

Finally, the recoil angle of the second ball is obtained from

$$\sin \theta_2 = \frac{v_1'}{v_1} = 0.530,$$

$$\theta_2 = 32°.$$

Discussion Notice that θ_1 and θ_2 add to 90°. The two objects move off at right angles to each other's motion. This behavior is characteristic of elastic collisions between two objects of equal mass; the struck object always moves off at a right angle to the recoil direction of the incident object.

*8.4 General Form of Gravitational Potential Energy

In Chapter 6 we considered gravitational potential energy near the surface of the earth. If we wish to determine the potential energy of an object that is far removed from the earth's surface, then the equation we found,

$$\text{PE} = mgh, \tag{8.16}$$

is inadequate. We derived this result under the assumption that the gravitational force on an object is constant, an assumption that is approximately correct for bodies near the earth's surface, but fails as the distances involved get larger. We can't expect Eq. (8.16) to be appropriate for describing the potential energy of the earth-moon system or even of an artificial satellite orbiting the earth.

Suppose we wish to calculate the potential energy difference between points A and B, where A is at a distance r_A from the earth's center and B is at a distance r_B from the earth's center (Fig. 8.11). We can determine this energy difference by computing the work required to move an object of mass m from A to B. First we move the object along the path of constant radius from point A to a point C. Remember, work is done only when we exert a force through a distance to overcome the gravitational force. Since the direction of the gravitational force is radial, no work is done in going from point A to point C because the motion is perpendicular to the force. To move from C to B, we must apply a force to move the mass outward against the gravitational force. The total work expended in going from A to B is then just the work done against the gravitational force in going from C to B.

This work W_{AB} equals the change in potential energy ΔPE_{AB} when the mass m is moved from A to B. We would like to express this as the difference between the potential energy at B and the potential energy at A,

$$\Delta \text{PE}_{AB} = \text{PE}_B - \text{PE}_A.$$

Physics in Practice

SYMMETRY AND CONSERVATION LAWS

What is symmetry? We see symmetry all around us: A butterfly is symmetric because the right side is a reflection of the left side (Fig. B8.1). We say that the butterfly has a mirror symmetry through it midline. Snowflakes have even more symmetry. In addition to mirror symmetries, they have rotational symmetry. A snowflake rotated through any multiple of 60° looks the same as before it was rotated (Fig. B8.2). Our technical definition of symmetry is much the same as our ordinary understanding: something is symmetric if the way it looks is unchanged by a reflection, rotation, or translation. Furthermore, any operation or procedure

Figure B8.1 A butterfly has a mirror symmetry through its midline.

has symmetry if performing the operation leaves an object unchanged. For example, if we examine one region of a honeycomb and then move over a few cells to a new region, that new region of the honeycomb looks the same as the first region (Fig. B8.3). We say the honeycomb has symmetry of translation. If we rotate the honeycomb by 60°, it still looks the same. The honeycomb has symmetry of rotation by 60°, 120°, or 180°, but not symmetry of rotation through 30°, 45°, or 90°. In this case, as in the case of the snowflake, the symmetry is discrete or discontinuous. Just as the butterfly and snowflake have mirror symmetries, so too does the honeycomb. These symmetries are all discontinuous; there is only a finite number of mirror axes.

Early in the history of science, crystals were classified according to the symmetry of their external shape. Eventually, it was discovered that these external shapes were due to the actual arrangement of the atoms and molecules that make up the crystals. These atomic alignments themselves are symmetric. After the discovery of x rays, crystallographers switched their attention from the shapes of crystals to their atomic structures. More recently, the invention of the scanning tunneling microscope has allowed us to visualize the atoms in the surfaces of materials and to see their symmetry (Fig. B8.4).

Other symmetries are continuous. For example, a uniform sphere such as a white billiard ball (cue ball) looks the same after rotation through any angle about any axis passing through its center. We say that the cue ball is invariant under rotation about any central axis. In contrast, a baseball or soccer ball has rotational symmetry about only a few axes, and the symmetries are discrete about those axes.

Symmetry is often seen in the work of architects, designers, and artists in addition to appearing in nature. Much of what we perceive as beauty in bridges and buildings is related to the symmetries they possess. The symmetrical shape of the Gateway Arch in St. Louis is important not only for its beauty but also for its strength (Fig. B8.5).

Is symmetry limited to geometry? No! There is much more. The mathematician Amalie "Emmy" Noether (1882–1935) recognized the physical significance of the

Figure B8.2 (a), (b) Snowflakes have mirror symmetries as well as (c)

Figure B8.3 A honeycomb has translational symmetry of as well as rotational symmetry

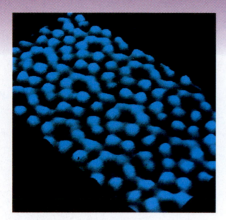

Figure B8.4 Atomic surface of a crystal.

Symmetries associated with the more abstract parameters found in condensed matter physics and particle physics are important to their understanding. A good example of how symmetry can lead to new ideas is found in the discovery of the Ω^- particle. The basic constituents of matter, such as protons, neutrons, electrons, and other less well known particles

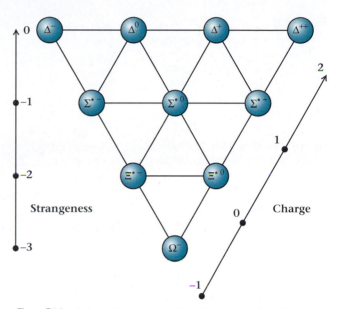

Figure B8.6 Geometric representation of the properties of elementary particles.

invariances associated with symmetries. Noether's Principle states that *for every continuous symmetry there is a corresponding conservation law and vice versa.* Both the law of conservation of momentum and the law of conservation of energy are consequences of symmetries. That the results of a mechanics experiment are independent of the location of the origin of the coordinate system indicates that space is symmetric (or uniform) under translation. The law of conservation of momentum corresponds to the translational symmetry of space. Conservation of angular momentum, charge, and other more abstract quantities also result from observed symmetries in nature.

Figure B8.5 The Gateway Arch in St. Louis has a mirror symmetry along a vertical line through its center.

were—at one time—called elementary particles. In 1962, the American physicist Murray Gell-Mann (1929–) and the Israeli physicist Yuval Ne'eman (1925–) independently used a property of particles previously developed in 1953 by Gell-Mann and Nisijima (1926–) to sort elementary particles into groups that display symmetry. Even though the properties are not spatial quantities and have no geometric meaning, these groups can be represented geometrically (Fig. B8.6). The test of the idea came when it was noted that one of the groupings contained a place for a particle yet unknown. A search was begun for a particle with the properties predicted by the theory. After careful experiments, the particle known as Ω^- was found in less than two years. The measured mass of the newly discovered particle was within 0.2% of the value predicted by the theory. Never before had the properties of a yet-unseen particle been so closely predicted. The prediction of this new particle was based on considerations of symmetry and its discovery confirms the fundamental correctness of the use of symmetry.

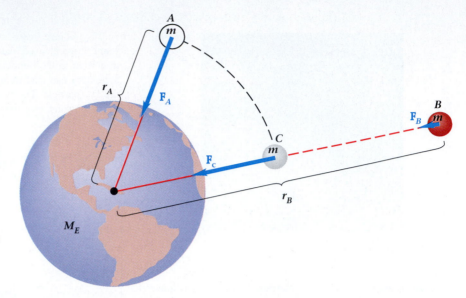

Figure 8.11

The gravitational potential energy of a body at point A is less than it is at point B, since work must be done to move the body from A to B. No work is done on moving from A to C because the motion is perpendicular to the direction of the force. $\mathbf{F}_A$, $\mathbf{F}_B$, and $\mathbf{F}_C$ are the gravitational forces acting on the body at points A, B, and C.

Since the gravitational force is not constant and varies as $1/r^2$, the work needs to be evaluated using the methods of calculus. The result for the work to move a mass m from A to B—that is, the difference in potential energy—is

$$\Delta PE_{AB} = \frac{-GM_Em}{r_B} - \frac{-GM_Em}{r_A}, \tag{8.17}$$

where M_E is the mass of the earth, and the potential energies at r_A and r_B are given by

$$PE_A = \frac{-GM_Em}{r_A} \quad \text{and} \quad PE_B = \frac{-GM_Em}{r_B}.$$

When we considered potential energy in Section 6.5, it was necessary to choose a reference, or zero. The choice of reference was based on convenience, the important physical quantity being the difference in potential energy between two positions. When the difference in gravitational potential energy is given by Eq. (8.17), we can again choose a reference on the basis of convenience. We will choose $r = \infty$ (infinity) to be the zero reference. This choice has the advantage of making the potential energy at the reference distance equal to zero. Then the potential energy at any distance is

$$PE = -G\frac{M_Em}{r}. \tag{8.18}$$

Equation (8.18) gives the potential energy at a distance r from the earth's center and is in agreement with our previous observation that objects have more potential energy as they are moved farther away from the earth. According to Eq. (8.18), a body's potential energy is negative near the earth's surface and becomes less negative (i.e., greater) as it moves away from the earth. The maximum potential energy is zero, the value obtained at an infinite distance from the

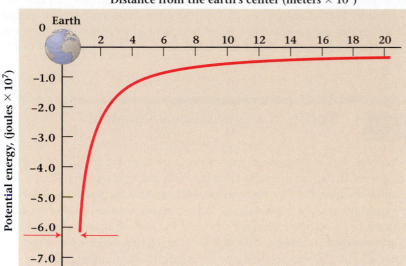

Figure 8.12
The potential energy of a 1-kg mass as a function of distance from the earth's center. The graph starts at the earth's surface.

earth's center. Figure 8.12 shows a graph of the potential energy of a 1-kg mass as a function of distance from the center of the earth.

It might seem as if we have two different expressions for the change in potential energy of a body lifted from the earth's surface. Certainly, Eqs. (8.16) and (8.18) do not appear to be the same. We can show, however, that they give essentially the same answer for small changes in distance from the earth's surface. Let us imagine that a mass m is initially at rest on the surface of the earth. We then lift it to a height h. What is its change in potential energy? If R_E is the radius of the earth, then Eq. (8.17) gives

$$\Delta PE = \frac{-GM_E m}{R_E + h} - \frac{-GM_E m}{R_E}$$

$$\Delta PE = GM_E m \left(\frac{1}{R_E} - \frac{1}{R_E + h} \right)$$

$$\Delta PE = GM_E m \left(\frac{h}{R_E(R_E + h)} \right).$$

If R_E is much greater than h, then the quantity in the brackets becomes approximately h/R_E^2 and the change in potential energy is given by*

$$\Delta PE = \frac{GM_E m h}{R_E^2}.$$

In Chapter 5 we found that the acceleration of gravity g is given by $g = GM_E/R_E^2$, so that the above equation can be written as

$$\Delta PE = mgh.$$

---

*If h is 1 km above the earth's surface, the approximation of $h/R_E(R_E + h)$ by h/R_E^2 gives an error of only 0.015%.

This equation is the same as Eq. (8.16). Thus, for changes in position near the earth's surface, Eqs. (8.16) and (8.18) both give the same answer. For motion over distances that are not small compared with the earth's radius, you must use Eq. (8.18). In all cases remember that it is the *change* in potential energy that is important.

*8.5   Motion in a Gravitational Potential

In this and the next section, we discuss some aspects of the motion of objects, such as spacecraft, that are projected from the earth and orbit around it (Fig. 8.13). These topics are included here because they demonstrate the power of conservation laws and because the mathematical techniques are similar to those we have already used in this chapter. Here we use only the laws of universal gravitation and of conservation of energy. Although a complete treatment of satellite and spacecraft motion also requires the law of conservation of angular momentum, which we will discuss in Chapter 9, we have chosen to restrict ourselves here to situations where angular momentum is zero and need not be included.

An object thrown straight up reaches a maximum height h that depends on its initial upward velocity v_0. We can determine the relation between h and v_0 from the principle of energy conservation by equating the potential energy at height h to the kinetic energy at height 0 (see Example 6.9). The result is $v_0 = \sqrt{2gh}$, where g is the gravitational acceleration. In arriving at this relationship, we assumed that the distance h is small enough to allow us to consider the gravitational force to be constant. If we consider objects such as rockets and satellites, which are projected to great distances above the earth's surface, the assumption of a uniform gravitational force is no longer acceptable. It is possible, though tedious, to derive the correct relationship between speed and distance from a consideration of the actual forces involved. However, it is much easier to analyze such problems from the standpoint of energy. This approach has the additional advantage of introducing some important techniques that are of general usefulness.

The gravitational potential energy of an object of mass m at a distance r from the center of the earth was given in the last section. For the case in which the potential energy is set equal to zero at $r = \infty$, it is

$$PE = -G\frac{M_E m}{r},$$

where G is the universal gravitational constant and M_E is the earth's mass. The potential-energy diagram for this case is the curve in Fig. 8.14, which plots potential energy against radial distance from the center of the earth. Because of our choice of reference level, the potential energy of the object is negative for any finite distance r.

Let the object be projected straight upward with a total energy E_0. As it rises it loses kinetic energy and gains potential energy. Its total energy remains constant as it moves in the earth's gravitational field if we neglect air resistance. The kinetic and potential energies at any point P are related by

$$KE_P + PE_P = E_0.$$

Figure 8.13
Booster rockets are used to launch the space shuttle into orbit around the earth.

If the total energy E_0 is known, then the kinetic energy at any point is the difference between the total energy and the potential energy. The total energy E_0 is indicated in Fig. 8.14 for a particular case in which E_0 is negative. For any radial distance $r < A$, where A depends on the total energy E_0, the object has a positive kinetic energy, as indicated at P. As the object moves away from the earth toward A, its kinetic energy decreases and its potential energy increases. When it reaches A, its kinetic energy becomes zero. At this point, the object stops and falls back toward the earth. When it passes the point P on its way back to the earth's surface, it has the same kinetic energy that it had at that same point on the way up. In other words, the object is moving with the same speed but now headed down instead of up.

A simple way to visualize the object's motion is to imagine it represented by a bead sliding without friction along a wire bent into the shape of the potential-energy curve. An upward push sends the bead along the wire from r_e to a, where it stops and slides back down to r_e. The bead loses speed as it goes from r_e to a, and regains speed as it returns. This analogy can be quite useful, but you must remember that the actual motion of the object is along a straight line directed radially away from the earth.

Let us return to the question of finding the relationship between the object's initial upward speed v_0 and the maximum height h to which it can ascend. We take the initial location to correspond to the earth's surface at $r = R_E$. Applying the rule of conservation of energy gives

$$KE(R_E) + PE(R_E) = KE(r) + PE(r).$$

At maximum height h ($r = R_E + h$), the kinetic energy of the object is zero and the energy equation becomes

$$\tfrac{1}{2}mv_0^2 - \frac{GM_Em}{R_E} = 0 - \frac{GM_Em}{r}.$$

We can rearrange this to give

$$v_0 = \sqrt{2GM_E\left(\frac{1}{R_E} - \frac{1}{r}\right)}.$$

This equation gives the initial upward speed required for an object at the earth's surface if it is to rise to a distance r from the center of the earth.

Compare this equation for v_0 with the simpler form given earlier for a uniform gravitational field, $v_0 = \sqrt{2gh}$. To make the comparison easier, we express the gravitational constant G in terms of the acceleration g (see Example 5.12):

$$GM_E = gR_E^2.$$

The result of inserting this expression into the above equation for speed is

$$v_0 = \sqrt{2gR_E^2\left(\frac{1}{R_E} - \frac{1}{r}\right)}.$$

We may also replace r by $R_E + h$ to get

$$v_0 = \sqrt{\frac{2ghR_E}{R_E + h}}. \tag{8.19}$$

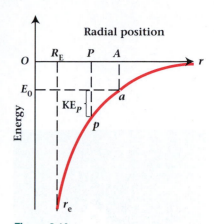

Figure 8.14
Potential-energy diagram of an object projected straight up. Energy is shown along the vertical axis and radial position from the center of the earth along the horizontal axis.

If h is small, so that $R_E + h \approx R_E$, then Eq. (8.19) reverts to the simpler form of the case for constant gravitational force. However, Eq. (8.19) is correct for all heights. The value of g remains the value measured at the earth's surface.

Example 8.5

A rocket is projected upward from the earth's surface (Fig. 8.15) to a height of $0.0100\, R_E$, where R_E is the mean radius of the earth. (a) What initial speed is necessary? For simplicity, assume that the rocket reaches its initial speed over a very short interval and that effects of air resistance can be ignored. (b) What fractional error would occur if the calculation were made under the assumption that the earth's gravitational field were uniform at all heights?

Solution (a) For the correct answer we use

$$
\begin{aligned}
v_0 &= \sqrt{\frac{2ghR_E}{R_E + h}} \\
&= \sqrt{\frac{2g(0.0100\,R_E)(R_E)}{R_E + 0.0100\,R_E}} = \sqrt{2gR_E(0.00990)} \\
&= \sqrt{2(9.81\ \text{m/s}^2)(6.38 \times 10^6\text{m})(0.00990)} \\
&= 1113\ \text{m/s} \approx 2490\ \text{mi/h}.
\end{aligned}
$$

(b) Let us take the ratio of the correct to the approximate expressions for the initial speed:

$$
\frac{v_0\ (\text{correct})}{v_0\ (\text{approximate})} = \frac{\sqrt{\dfrac{2ghR_E}{R_E + h}}}{\sqrt{2gh}} = \sqrt{\frac{R_E}{R_E + h}} = \sqrt{\frac{1}{1 + 0.01}}
$$

$$
\frac{v_0\ (\text{correct})}{v_0\ (\text{approximate})} = 0.995.
$$

Discussion As we expect, since the strength of the gravitational field decreases rather than remains constant with increasing height, the correct initial speed is less than the one calculated from the approximation. Nevertheless, for a distance of nearly 600 km, the difference in the calculations is only 0.5%.

*8.6 **Escape Speed**

Look at the potential-energy diagram of an object projected straight up (Fig. 8.14) and remember the analogy of the bead on the wire. We see that for a projectile to escape the earth—that is, for the bead not to slide back down again—its initial energy must be enough to move the point a to infinity. Thus, the total initial energy must be at least zero ($E_0 \geq 0$). For zero total energy, the projectile will "reach infinity" with zero speed. Rockets (or other objects) with total energy less than zero cannot escape the earth. Bodies with total energy greater than zero, as E_1 in Fig. 8.16, are unbound. For a projectile that is just able to escape from the earth's gravitational field, the total energy is zero.

Figure 8.15
A rocket is projected straight up from the earth's surface.

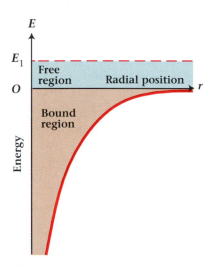

Figure 8.16
Potential-energy diagram for a mass attracted to the earth. Objects with total energy E less than 0 are bound to the earth. Those with E greater than zero are free to escape from it. Particles with $E = 0$ are just barely free and approach $r = \infty$ with zero velocity.

A rocket that is fired for only a short time behaves essentially like a projectile that is given an initial upward speed. The minimum initial upward speed necessary for the rocket not to fall back to earth again is called the **escape speed,** v_{esc}, and is found by setting the projectile's total energy equal to zero:

$$\frac{1}{2}mv_{esc}^2 - \frac{GM_E m}{R_E} = 0.$$

We can rearrange to find that

$$v_{esc} = \sqrt{\frac{2GM_E}{R_E}}. \tag{8.20}$$

Note that the escape speed does not depend on the mass of the rocket. The value of the escape speed for any object on the earth can be calculated to be 11.2 km/s. It is interesting to note that the speed needed to completely escape from the earth's gravitational field is only $\sqrt{2}$ times as much as the speed of an object in orbit just above the earth's surface. (See Problem 8.51.)

Example 8.6

What is the escape speed for a rocket on the surface of Mars?

Solution We can obtain the escape speed from the surface of Mars from the law of conservation of energy in the same way that we obtained Eq. (8.20):

$$\frac{1}{2}mv_{esc}^2 - \frac{GM_M m}{R_M} = 0.$$

The escape speed is then

$$v_{esc} = \sqrt{\frac{2GM_M}{R_M}}.$$

When the mass of Mars, 6.58×10^{23} kg, and its radius, 3.38×10^6 m, are inserted into the equation for v_{esc}, we get

$$v_{esc} = \sqrt{\frac{2(6.67 \times 10^{-11} \text{ N} \cdot \text{m}^2/\text{kg})(6.58 \times 10^{23} \text{ kg})}{3.38 \times 10^6 \text{ m}}},$$

$$v_{esc} = 5.10 \times 10^3 \text{ m/s}.$$

The escape speed from the surface of Mars is less than one-half the escape speed from the earth's surface.

Example 8.7

Calculate the minimum initial speed needed to project a rocket from the earth to the moon. For simplicity, use a model in which the positions of the earth and the moon are fixed and that ignores the effects of air friction in the earth's atmosphere.

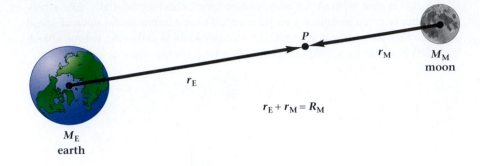

Figure 8.17
Example 8.7: Point P along a line from the earth to the moon is a distance r_E from the earth and a distance r_M from the moon.

Strategy The gravitational potential energy of the rocket due to both the earth and the moon has a maximum somewhere between the earth and the moon. We use a stationary model with a radial rocket velocity. The minimum initial speed needed corresponds to the minimum kinetic energy to get the rocket over this maximum potential energy.

Solution Consider points along a line between the earth and the moon (Fig. 8.17). Then the distance to the earth r_E plus the distance to the moon r_M equals R_M the center-to-center distance from earth to moon: $r_E + r_M = R_M$. The potential energy of the rocket is

$$PE = -G\frac{M_E m}{r_E} - G\frac{M_M m}{R_M - r_E},$$

where M_E is the mass of the earth, M_M is the mass of the moon, and m is the mass of the rocket.

This potential energy is shown in Fig. 8.18. Point A located a distance r_A from the earth is the position of the maximum potential energy along the path. It is the point at which the gravitational force on the rocket due to the earth is equal and opposite to the force on the rocket due to the moon:

$$\frac{GM_E m}{r_A^2} = \frac{GM_M m}{(R_M - r_A)^2}.$$

The rocket must have a total energy of at least E_A if it is to get "over the hump" at A and reach the moon along the direct path. A rocket leaving the earth with initial energy slightly greater than E_A will slow as it approaches A, where its velocity will be nearly zero. Past the point A, the rocket speeds up as it approaches the moon.

Upon solving the force equation for r_A we find

$$r_A = \frac{R_M}{1 + \sqrt{\dfrac{M_M}{M_E}}} = 0.90\,R_M.$$

The potential energy PE_A of the rocket at point A can be found in terms of the earth-moon distance by inserting $r_A = 0.90\,R_M$ for r_E in the potential energy equation. The result is:

$$PE_A = -\frac{1.234\,GM_E m}{R_M}.$$

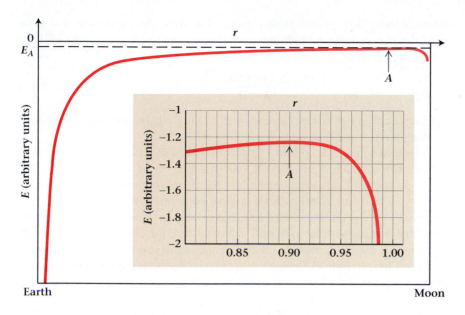

Figure 8.18
Example 8.7: Gravitational potential energy for a mass along a straight line between the earth and the moon.

To reach the moon along the direct path, a rocket must have an initial total energy KE + PE of at least PE$_A$. At the earth's surface, the rocket's total energy is

$$\text{KE} + \text{PE} = \tfrac{1}{2}mv_0^2 - \frac{GM_{\text{E}}m}{R_{\text{E}}} - \frac{GM_{\text{M}}m}{R_{\text{M}} - R_{\text{E}}} = \text{PE}_A.$$

Because the mass of the moon is 0.0123 M_{E} and $R_{\text{M}} \approx 60R_{\text{E}}$, the potential energy due to the moon at the earth's surface is only 1/6000 as big as that due to the earth. Thus, to a good approximation we may drop the term due to the moon, insert the value for PE$_A$, and solve to find the initial speed v_0.

$$v_0 = \sqrt{2GM_{\text{E}}\left(\frac{1}{R_{\text{E}}} - \frac{1.234}{R_{\text{M}}}\right)} = 11.1 \text{ km/s}.$$

The minimum initial speed required at the earth's surface is $v_0 = 11.1$ km/s or nearly 40,000 km/h.

Discussion It is interesting to note that the minimum initial upward speed to just project a rocket to the moon is only about 1% smaller than the speed needed to completely escape from the earth's gravitational field. Since travelers to the moon need to cross the potential hump in a reasonable time, they must necessarily have speeds greater than that found here. Consequently, great care is needed in steering the spacecraft to ensure that it reaches the moon and is not lost in space. Actual lunar voyages are not made along a straight line joining the earth and moon and the spacecraft are not simply fired from a giant gun. However, it is still necessary to give the spacecraft sufficient total energy to overcome the potential barrier between earth and moon.

The expression for the escape speed in the previous example leads to an interesting consequence that is outside the realm of Newtonian mechanics. Astronomical objects with greater mass to radius ratio than the earth (that is to say, denser objects) have greater escape speeds. For stars, the escape speed is quite high. If a star is dense enough, the escape speed approaches the speed of light c, the limiting speed for all motion. When

$$\sqrt{\frac{2GM}{R}} = c,$$

nothing, not even light itself, can escape. Because no light can escape from such a dense object, it is called a *black hole*.

The idea of a black hole was first proposed by P. S. Laplace (1749–1827) in 1796, before much of our present understanding of the nature of light or the stars was known. Current theories allow for the existence of black holes, although very different from Laplace's original conception. We know that stars are not permanent objects, but evolve and change over long periods of time. At the end of the lifetime of some extremely massive stars, the core can undergo a collapse; in this case it is possible for the central density to become large enough for a black hole to form. Although a proper understanding of black holes requires astronomy, general relativity, and quantum mechanics, the numerical result is the same as Laplace obtained.

We can obtain the radius of a black hole of mass M from the last equation. This radius, called the *Schwarzschild radius,* is given by

$$R_s = \frac{2GM}{c^2}. \tag{8.21}$$

The Schwarzschild radius indicates the size that an object of mass M must have if it is to have the enormous density of a black hole. For a black hole with the mass of the sun, the Schwarzschild radius is only about 3 km.

Even though black holes give off no light from within, they can be detected by their gravitational effects. For example, a black hole can combine with a cloud of gas, a star, or a galaxy of stars to form an orbiting system, whose

Figure 8.19

An image from the Wide Field and Planetary Camera II on NASA's Hubble Space Telescope shows a spiral-shaped disk of hot gas in the core of the galaxy M87. Measurements by the telescope reveal that disk rotates so rapidly that it contains a black hole at its hub.

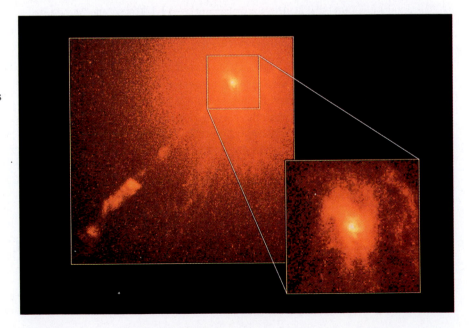

combined motions conform approximately to Kepler's laws. If we can infer the motion of the stars from astronomical observations, then we can detect the presence and mass of the black hole. One of the many observed objects thought to be a black hole is found in the core of the galaxy M87 located about 50 million light-years away in the constellation Virgo. Measurements from the Hubble Space Telescope reveal a spiral-shaped disk of hot gas rotating about a central hub (Fig. 8.19). Calculations show that the object at the hub has a mass three billion times the mass of the sun, yet it occupies a space no larger than our solar system. It must be a black hole.

Summary

Useful Concepts

In this chapter we applied the laws of conservation of energy and momentum to analyze elastic collisions, and motion in a gravitational field. These examples indicate the wide variety of problems solvable by conservation laws.

■ In one dimension the law of conservation of momentum for collision of two objects is

$$m_1 v_1 + m_2 v_2 = m_1 v_1' + m_2 v_2'.$$

■ In an elastic collision, kinetic energy is conserved, leading to the equation

$$\tfrac{1}{2}m_1 v_1^2 + \tfrac{1}{2}m_2 v_2^2 = \tfrac{1}{2}m_1 v_1'^2 + \tfrac{1}{2}m_2 v_2'^2$$

■ For collisions in two dimensions, we take account of the vector nature of momentum, determining separate conservation equations in the x and y directions.

■ The gravitational potential energy of an object of mass m at a distance r from the center of the earth and above the earth's surface is

$$\text{PE} = -G\frac{M_E m}{r}.$$

■ We can use the general form of the gravitational potential energy and the conservation of energy to determine the escape speed from the earth,

$$v_{\text{esc}} = \sqrt{\frac{2GM_E}{R_E}}.$$

Important Terms

You should be able to write the definition or meaning of each of the following terms:

elastic collision   escape speed
inelastic collision

Conceptual Questions

8.1 A stationary firecracker is hung by a thread from a tree limb and then explodes into three pieces. Do the paths of the three pieces lie in a plane? Explain your answer with care.
8.2 Give some examples of elastic, inelastic, and perfectly inelastic collisions.
8.3 Two bodies collide elastically in midair. The only restriction on their initial motions is that they must collide. Explain how this situation can be viewed as only a two-dimensional problem.
8.4 Can an object have, at the same time, more kinetic energy but less momentum than another object?
8.5 Are there any combinations of masses and initial velocities in a one-dimensional elastic collision so that the velocity of one of the bodies is unchanged by the collision? Are there

any conditions under which the speed is the same after the collision as it was before? In either case, explain your answer.
8.6 What would be the effect on the game of billiards if the cue ball were twice as massive as the other balls?
8.7 The Newtonian Demonstrator (Fig. 8.20) is a popular item in novelty shops as well as physics departments. When one ball is pulled aside, released, and allowed to strike the others, one ball pops out from the opposite side. When two balls are pulled aside and released, two balls pop out on the other side as a result of the ensuing collision. Explain why this is so.
8.8 A cosmonaut wishes to dock his spacecraft with another craft several hundred meters ahead of him in the same orbit. The two spacecraft are moving with the same speed at the same radius in the same circular orbit. The cosmonaut can use

Figure 8.20
Question 8.7: A Newtonian demonstrator.

thruster rockets directed fore, aft, up, or down. Which should he use and in what order? Describe the subsequent motion of his craft.

8.9 A projectile fired from the earth's surface needs to be given an upward speed equal to the escape speed v_{esc} in order to escape from the earth. Would a rocket fired from the earth need that same speed in order to escape? Why?

8.10 Why is gravitational potential energy (Eq. 8.18) negative?

8.11 What happens to a spacecraft that leaves the earth with a speed greater than the escape speed?

Problems

Section 8.1 Definition of Elastic Collisions

Hints for Solving Problems

Remember that for elastic collisions, both kinetic energy and momentum are conserved. You will find it helpful to draw two diagrams of a collision, one before and one after impact.

8.1 A steel ball is dropped from a height of 2.37 m onto a flat stone slab. On rebounding, the speed of the ball is 3.94 m/s when it is 1.52 m above the stone slab. Is the collision elastic?

8.2 A glass marble is dropped onto a steel slab from a height of 2.0 m. If the marble rebounds to a height of 1.6 m, what fraction of its initial energy was lost? Where has the energy gone?

8.3 A ball bearing is dropped from a height of 2.0 m onto a steel slab. If the ball bounces to a height of 1.4 m, what is its upward speed as it passes the 1.0-m mark?

8.4 A ball tossed upward with a speed of 4.27 m/s from 1.52 m above the floor falls to the floor and bounces to a height of 2.38 m. Is the collision with the floor elastic?

8.5 A ball thrown downward at 2.32 m/s from a height of 1.07 m above the floor bounces up to a height of 1.34 m. Is the collision with the floor elastic?

8.6• A tennis ball dropped from a height of 3.00 m loses 50% of its mechanical energy at each bounce. To what height does it rise after the third bounce?

8.7• A marble dropped from 1.37 m onto a flat surface bounces to a height of 1.27 m on its first bounce. If the marble loses the same fraction of mechanical energy on each bounce, how high does it rise after the fourth bounce?

Section 8.2 Elastic Collisions in One Dimension

8.8 A 0.400-kg toy truck moving at an initial speed of 0.100 m/s collides head-on with a 0.300-kg toy car at rest. The collision is elastic. Find their final speeds and directions.

8.9 Two billiard balls travel toward each other along a straight line with the same speed. What are their speeds and directions after an elastic collision?

8.10 Two air-track gliders of equal mass are initially moving in the same direction along a straight line. The rearmost glider has an initial speed of 3.0 m/s and the forward glider has a speed of 2.0 m/s. If the collision is elastic, what are the speeds and directions of the gliders after the collision?

8.11 An air-track glider with an initial speed of 4.0 m/s has a head-on collision with another glider at rest that is three times as massive. What are the final speeds and directions of the gliders if the collision is elastic?

8.12• Two glass marbles moving along a straight line toward each other undergo an elastic collision. The speed of one marble is $2v$ and its mass is m; the speed of the other marble is v and its mass is $2m$. What are the speeds and directions of the marbles after the collision?

8.13• A 1200-kg car traveling 27 m/s crashes into the rear of a 9000-kg truck moving in the same direction at 22 m/s. Immediately after the collision the car is moving at 20 m/s. (a) How fast was the truck moving immediately after the collision? (b) Is kinetic energy conserved in the collision?

8.14• A 326-g stationary air-track glider is attached to the end of an air track by a compressible spring with spring constant $k = 5.32$ N/m (Fig. 8.21). A 163-g glider moving at 1.27 m/s collides elastically with the stationary glider. How

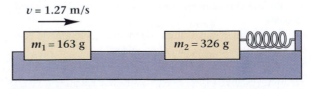

Figure 8.21
Problem 8.14.

far does the spring compress? Assume that the air track is very much heavier than the gliders.

8.15• Is it possible for a moving air-track glider to collide elastically with a stationary glider in such a way that both of them move in the same direction with the same speed after the collision? Use the conservation laws to explain your answer.

8.16• After an elastic collision between two pool balls of equal mass, one is observed to have a speed of 3.0 m/s along the positive *x* axis and the other has a speed of 2.0 m/s along the negative *x* axis. What were the original speeds and directions of the two balls?

8.17• A 1000-kg automobile going 10 m/s collides head-on with a 1200-kg automobile traveling in the opposite direction at 4.0 m/s. (a) What is the maximum kinetic energy that can be dissipated in damaging the two cars? (b) What limits this amount of energy?

8.18•• A 283-g air-track glider moving at 0.69 m/s on a 2.4-m long air track collides elastically with a 467-g glider at rest in the middle of the horizontal track. The end of the track over which the struck glider moves is not frictionless, and the glider moves with a coefficient of friction $\mu = 0.02$ with respect to the track. Will the glider reach the end of the track? Neglect the length of the gliders.

8.19•• A ball of mass 2 *m* is projected upward with speed v_0 from the floor (Fig. 8.22). Another ball of mass *m* is hung from the ceiling by a light string at a height *h* directly above the first ball, so that the projected ball collides with it. Derive an expression for the height above the floor to which the second ball will rise as a function of v_0, *h*, and *g*, assuming that the collision is elastic.

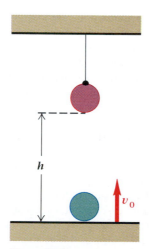

Figure 8.22
Problem 8.19.

8.20•• A 237-g air-track glider moving at 0.80 m/s on a 2.4-m long air track collides elastically with a 513-g glider at rest in the middle of the track. The end of the track over which the struck glider moves is not level, but slants upward at an an-

gle of 0.70° with respect to the horizontal. Will the glider reach the end of the track? Neglect the length of the gliders.

8.21•• Show that for a one-dimensional elastic collision between mass m_1 moving with initial velocity v_1 and mass m_2 with initial velocity v_2, the final velocities are

$$v_1' = \frac{m_1 - m_2}{m_1 + m_2}v_1 + \frac{2\,m_2}{m_1 + m_2}v_2 \quad \text{and}$$

$$v_2' = \frac{2\,m_1}{m_1 + m_2}v_1 - \frac{m_1 - m_2}{m_1 + m_2}v_2$$

8.22•• Blocks of mass *m* and 2 *m* are positioned on a semi-circular frictionless track at a height of *R*/4 above the lowest point (Fig. 8.23). The blocks are released simultaneously and collide elastically. How high does each block rise after the collision?

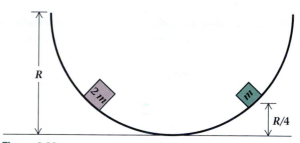

Figure 8.23
Problem 8.22.

*Section 8.3 Elastic Collisions in Two Dimensions

Hints for Solving Problems

In two dimensions, conservation of momentum provides independent equations in the *x* and *y* directions. A third equation comes from conservation of energy. For two objects in collision, if we know their masses and initial velocities, there will be four quantities in a typical two-dimensional collision that are required to define the final state. They might be the final velocity components v_{1x}, v_{1y}, v_{2x}, and v_{2y}. Since you can only determine three unknown quantities from three equations, you need one other quantity from the given information in order to uniquely determine the other three quantities. In elastic collisions between a moving object and a stationary object of the same mass, the angle between the two outgoing objects is 90°.

8.23 Use the results of Example 8.3 and the trigonometric relationship $\cos \theta = \sin(90° - \theta)$ to show that for a two-dimensional elastic collision between two balls of identical mass the relationship between the two outgoing angles is $\theta_1 + \theta_2 = 90°$.

8.24 A proton traveling with speed v_1 collides elastically with another proton initially at rest. After the collision the protons move off, making angles of $+45°$ and $-45°$ with respect to the direction of motion of the incident proton. What are the final speeds of the protons after the collision in terms of the initial speed v_1?

8.25• A ball of mass m moving with a speed v_1 collides elastically with a stationary ball of the same mass. The incident ball is scattered at an angle of $60°$ from its incident direction. What fraction of the initial kinetic energy is imparted to the struck ball?

8.26• An elastic collision occurs between two air hockey pucks in which one puck is at rest and the other is moving with a speed of 0.50 m/s. After the collision, the puck initially in motion makes an angle of $20°$ with its original direction, and the struck puck moves at an angle of $70°$ on the other side of the original direction. What is the final speed of each puck?

8.27• A deuteron collides elastically with another deuteron initially at rest. After the collision, one deuteron is observed to have half of the original kinetic energy. In what direction must it be moving? What is the direction of motion of the struck deuteron?

8.28•• A tennis ball collides with an identical ball at rest. The collision occurs in such a way that one-fourth the initial kinetic energy is lost in deformation of the balls. The outgoing balls leave the impact point, making equal angles with respect to the direction of the original ball. Determine the angle θ between the direction of one of the balls and the initial direction of the first ball.

8.29•• A billiard ball traveling at speed v_0 scatters elastically from an identical ball making an angle of $27°$ with respect to its original direction. The incident ball subsequently scatters from another identical ball so that the path of the incoming ball becomes parallel to its original direction. What is the speed of the incident ball after being scattered twice?

*Section 8.4 General Form of Gravitational Potential Energy

8.30 What is the value of the gravitational potential energy of a 1.00-kg mass on the surface of the earth if the zero of potential energy is taken at $r = \infty$?

8.31 What is the gravitational potential energy of the moon with respect to the earth if the zero of potential energy is taken at $r = \infty$?

8.32 What is the change in gravitational potential energy of a 1.00-kg mass that is carried from the surface of the earth to a distance of one earth radius above the surface?

8.33 What is the change in the gravitational potential energy of a 5.00-kg mass that is carried from the earth's surface to a height of $\frac{1}{4}$ the earth's radius above the earth?

8.34• A metal slug is dropped from a height of $1/20\ r_M$ above the moon's surface. Find the speed with which the slug strikes the moon's surface.

8.35•• How high from the surface of the earth must an object be raised so that the increase in potential energy as given by

PE $= mgh$ and by PE $= -GM_Em/r$ will differ by 2%? Express this distance as a multiple of the earth's radius.

8.36•• (a) What is the increase in potential energy of a mass m moved from the surface of the earth (R_E) to a distance of $2R_E$ from the center of the earth? (b) Is this the same value obtained by using ΔPE $= mg'\Delta R$ where g' is the simple average acceleration of gravity, $g' = (g_{RE} + g_{2RE})/2$? (c) If the average in (b) is not correct, determine the correct average g' by setting the actual change in potential energy equal to $mg'\Delta R$ and solving for g'.

*Section 8.5 Motion in a Gravitational Potential

Hints for Solving Problems

In a gravitational field, the potential energy is always negative if the reference level is zero at infinity.

8.37 With what minimum upward speed must a rocket be projected vertically to reach a height above the earth equal to the earth's radius? Ignore air friction.

8.38 How fast must you project an object for it to reach a height equal to the moon's distance from the earth? Ignore the gravitational attraction of the moon.

8.39 A rocket is projected upward from the earth's surface $(r = R_E)$ with an initial speed v_0 that carries it to a distance $r = 4R_E$ from the center of the earth. What is the launch speed v_0? Assume that air friction can be ignored and give your answer in terms of G, R_E, and the earth's mass M_E.

8.40 A rocket is projected upward from the earth's surface $(r = R_E)$ with an initial speed v_0 that carries it to a distance $r = 1.5R_E$ from the center of the earth. What is the launch speed v_0? Assume that air friction can be ignored and give your answer in terms of g and R_E.

8.41 What is the potential energy of a 1.0-kg mass at a distance three-quarters of the way from the earth to the moon? Ignore the effects of the sun, but not of the moon.

8.42• A spaceship lies along a line from the sun to Jupiter. At what distance from Jupiter $(1.90 \times 10^{27}$ kg$)$ is the gravitational attraction of the spaceship to the planet equal to the attraction of the spaceship to the sun $(1.99 \times 10^{30}$ kg$)$? Give your answer as a fraction of the distance between the sun and Jupiter.

8.43• A meteor headed straight toward the earth has an approach speed of 8.0×10^3 m/s at a distance of $4R_E$ from the center of the earth, where R_E is the earth's radius. How far from the center of the earth will it be when its speed increases to 1.2×10^4 m/s? Express your answer in terms of R_E.

8.44• Imagine that a meteor headed straight toward the earth has an approach speed of 8.0×10^3 m/s at a distance of $3R_E$ from the center of the earth, where R_E is the earth's radius. What will its speed be when it reaches the atmosphere at $R \approx R_E$?

*Section 8.6 Escape Speed

8.45 Compute the escape speed from the surface of Mars, which has a radius of 3.37×10^6 m and a mass of 6.42×10^{23} kg.

8.46 A distant planet has a mass of $0.82\ M_E$ and a radius of $0.95\ R_E$. What is the ratio of the escape speed from this planet to the escape speed from the earth?

8.47• The radius of Uranus is approximately 3.69 times the radius of the earth. If the escape speed from Uranus is 22 km/s, what is the ratio of the acceleration of gravity on Uranus to its value on earth?

8.48• The escape speed from a distant planet is 35 km/s. If the acceleration of gravity on the planet surface is 11.1 m/s², what is the radius of the planet?

8.49• (a) What is the Schwarzschild radius of a body with the mass of the earth? (b) What would be the average density of that body if its radius were the same as its Schwarzschild radius?

8.50•• Calculate the minimum initial speed needed to project a rocket from the moon to the earth. For simplicity, assume a straight line path as in Example 8.7.

8.51•• Show that the escape speed from the earth is $\sqrt{2}$ times as large as the orbital speed of an object in orbit just above the earth's surface.

Additional Problems

8.52 A 0.400-kg toy truck moving at an initial speed of 0.100 m/s collides head-on with a 0.300-kg toy car at rest. The collision is perfectly inelastic, so the two toys stick together. (a) Find their final speed and (b) calculate how much kinetic energy was lost in the collision.

8.53 Show that the escape speed from earth can be expressed as $v_{esc} = \sqrt{2gR_E}$, where g is the gravitational acceleration at the earth's surface and R_E is the earth's radius.

8.54 A 30-caliber rifle with a mass of 3.14 kg fires a 7.13-g bullet with a speed of 606 m/s with respect to the ground. What kinetic energy is released by the explosion of the gunpowder that fires the bullet?

8.55• A 58-kg girl skis down a slope from a height of 9.0 m above the bottom of a hill. At the bottom she plows into a snowdrift that stops her in 2.0 s. What average force does she exert on the snow? How far does she go into the snowdrift?

8.56• A 1200-kg spacecraft is separated from its 4800-kg booster stage by an explosion (Fig. 8.24). The two parts move away from each other with a relative speed of 100 m/s. Calculate the energy imparted to the two masses by the explosion.

8.57• A rocket moving at 1000 m/s consists of a 1200-kg space capsule and a 6000-kg booster stage. An explosion that separates them throws the capsule forward so that it has a speed of 1100 m/s. How much energy was released in the explosion?

8.58• A pendulum with a bob of mass $3\ m$ is lifted to a height h above its lowest point and allowed to fall so that it

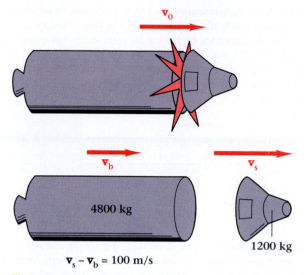

Figure 8.24
Problem 8.56.

collides elastically with a block of mass $2\ m$ that rests on a frictionless horizontal surface (Fig. 8.25). The struck mass then travels a distance s and collides elastically with another block of mass m. Find an expression for the speed of the block with mass m as a function of the height h to which the pendulum was raised.

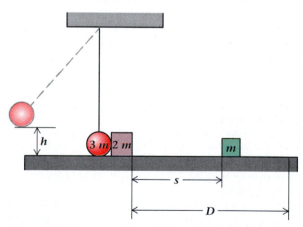

Figure 8.25
Problem 8.58.

8.59• Two air-track gliders of mass $m_1 = 234$ and $m_2 = 543$ g are tied together with thread, with a coil spring compressed between them. The gliders are at rest on an air track. The spring, with spring constant $k = 200$ N/m, is compressed by 4.70 cm from its equilibrium length. The thread is burned in two, and the gliders move apart. What is the speed of each glider?

8.60• A 2.5-kg block collides with a horizontal spring of negligible mass and spring constant $k = 320$ N/m. The block compresses the spring by 8.5 cm from its rest position (Fig. 8.26). How fast was the block going when it hit the spring if the frictional coefficient between the block and the horizontal surface is 0.40?

Figure 8.26
Problem 8.60.

8.61•• (a) Graph the potential energy for a 1-kg mass at positions along a straight line from the earth to the moon. Include the gravitational effects of the earth and the moon, but not of the sun. (b) Find the point along a straight line from the earth to the moon where the potential energy is a maximum by equating the force on the mass due to the earth with the force on it due to the moon.

8.62•• A puck collides elastically with identical stationary puck. The coefficient of friction between the pucks and the surface they slide on is μ. (a) Is the angle between the two outgoing pucks 90°? (b) If the angle is not 90°, is it always either greater than 90° or less than 90°?

8.63•• A ball is thrown upward with a speed v from a height h at $t = 0$. If it rebounds from the floor elastically, find an expression for the time t between bounces.

8.64•• A 36-g bullet with a speed of 350 m/s strikes a 8-cm-thick fence post. The bullet is retarded by an average force of 3.6×10^3 N while traveling all the way through the board. (a) What speed does the bullet have when it emerges? (b) How many such boards could the bullet penetrate?

8.65•• An air-track glider with an initial speed of 4.0 m/s collides head-on with another glider of equal mass initially at rest. The struck glider has a ball of wax on its bumper that deforms during the collision. What fraction of the initial energy is used in deforming the wax if the final speed of the struck glider is 3.0 m/s?

8.66•• A glass marble is dropped down an elevator shaft and hits a thick glass plate on top of an elevator that is descending at a speed of 2.0 m/s. The marble hits the glass plate 3.0 m below the point from which it was dropped. If the collision is elastic, how high will the marble rise, relative to the point from which it was dropped?

8.67•• After colliding elastically with a stationary puck of the same mass, an air hockey puck is observed to have a momentum whose magnitude is only half the magnitude of its original momentum. What angle does it make with its original direction?

8.68•• Two cars hit head-on in an inelastic collision in which they lose 40% of their kinetic energy. One car has a mass of 1230 kg and an initial speed of 37.3 km/h in the positive x direction. The second car has a mass of 1734 kg and an initial speed of 26.5 km/h in the negative x direction. (a) What happens after the collision? Do the cars separate or stick together? (b) What are the velocities of the two cars immediately after the collision? (*Hint:* Compare the net momentum with the momenta of the individual cars.)

8.69•• Two identical bumper cars collide elastically at right angles, with one going twice as fast as the other. If the faster car is deflected 45° away from its original direction of travel, by how much is the slower car deflected away from its original direction of travel?

8.70•• Two cars collide at right angles and stick together as they skid to a stop. Car 1 has a mass of 1280 kg and a speed of 25 km/h to the east; car 2 has a mass of 950 kg and a speed of 37 km/h to the north. The average coefficient of friction between the tires and the road is 0.65. How far from the collision point, and in what direction, do the cars stop?

8.71•• A 5.00-g glass marble moving at 1.2 m/s collides head-on with an identical stationary marble. After the collision, both marbles then move in the direction of the oncoming marble. The struck marble moves with a speed 100/99 of the initial speed of the oncoming marble. After the collision, a small piece of the struck marble is found directly below the point of impact. What is the mass of the small piece? (This problem is a one-dimensional classical mechanics example of a technique used in nuclear and elementary particle physics— determining the mass of an unobserved particle by measuring the initial and final kinematic quantities and then using the conservation laws to find the missing mass. Assume in this case that momentum and kinetic energy are conserved.)

8.72•• An object is shot directly upward from the earth's surface with an initial speed v_0 sufficient for it to reach a height of 50 km. (a) Do not assume the earth's gravitational field to be constant and show that the initial speed is given approximately by $v_0 \approx \sqrt{2gh}$. (b) Calculate the initial speed v_0 and express your answer as a fraction of the escape speed from earth.

Rigid Bodies and Rotational Motion

Our study of mechanics has included the revolution of the planets around the sun and the circular motion of an object as you swing it about your head on a string. A slightly different kind of angular motion is the daily turning of the earth on its axis or the spinning of a baseball when thrown. This latter type of motion is called rotational motion, and is defined more precisely in this chapter. Although we will primarily consider the motion of solid objects turning about a fixed axis, we will also examine moving objects that combine translational and rotational motion, as does a rolling wheel. In contrast with Chapter 5, we will not restrict ourselves to uniform circular motion, but will also include angular accelerations.

The material in this chapter is based on ideas with which you are already familiar. Topics will be developed in essentially the same order in which they were introduced for translational motion. For instance, we will begin by introducing kinematic equations to describe rotational motion, then go on to get an expression analogous to Newton's second law that will tell us what causes changes in rotational motion.

Almost all of the ideas of rotational motion can be developed in analogy with translational motion. Along the way, we will find a conservation law that is the rotational analog of the law of conservation of linear momentum, and we will find an additional expression for the kinetic energy of a rotating object. ■

9.1 Angular Velocity and Angular Acceleration

So far we have studied the translational motion of objects in one and two dimensions, including circular motion. Now we want to broaden our understanding of mechanics to include rotational motion. Any real object that has a definite shape can be made to rotate. If the object rotates with no deformation, so that all parts of the object remain at constant distances from every other part, we call the object a **rigid body.** The complex motion of a rigid body can be separated into a purely translational motion and a purely rotational motion.

In addition to the concepts of displacement, velocity, and acceleration that we use to describe linear motion, we need the corresponding quantities for angular motion. These quantities are angular displacement, angular velocity, and angular acceleration. With them we can describe the motion of a thrown, spinning ball or the start up of a merry-go-round.

Let's begin by thinking about a rigid body that rotates about a fixed axis, like the wheel of a bicycle turned upside down (Fig. 9.1). The angle θ describes the rotational position of a point on the wheel. In Section 5.1 we introduced

Figure 9.1
The wheel of an upturned bicycle spins freely about a fixed axis. When the angular velocity of the wheel changes, we say the wheel has angular acceleration.

angular measure in radians by defining an angle θ to be the ratio of the arc length s to the radius r:

$$\theta = \frac{s}{r}. \tag{9.1}$$

When the angle changes with time, the wheel has an *angular velocity*. We introduced angular velocity earlier when discussing uniform circular motion in Chapter 5. There we defined the average angular velocity $\overline{\omega}$ as the ratio of the change in angle to the change in time,

$$\overline{\omega} = \frac{\Delta\theta}{\Delta t}.$$

The instantaneous angular velocity ω is the limit of that ratio as the time interval goes to zero:

$$\omega = \lim_{\Delta t \to 0} \frac{\Delta\theta}{\Delta t}. \tag{9.2}$$

For a rigid body in rotation, all points on the body rotate with the same angular velocity ω (Fig. 9.2). Recall that the units of angular velocity are $\text{rad} \cdot \text{s}^{-1}$. (We could just use s^{-1} since the radian is dimensionless, but it is helpful to carry along the unit of radian as a reminder that the angles are measured in radians and not degrees.)

When the angular velocity of the rigid body changes, it has an angular acceleration. We define the **average angular acceleration** $\overline{\alpha}$ as the ratio of the change in angular velocity to the change in time,

$$\overline{\alpha} \equiv \frac{\omega_2 - \omega_1}{t_2 - t_1} = \frac{\Delta\omega}{\Delta t}. \tag{9.3}$$

The **instantaneous angular acceleration** α is the limit of the ratio $\Delta\omega/\Delta t$ as Δt approaches zero:

$$\alpha \equiv \lim_{\Delta t \to 0} \frac{\Delta\omega}{\Delta t}. \tag{9.4}$$

Since ω is the same for all points on a rotating rigid body, the angular acceleration will also be the same for all points on the body. The units of angular acceleration are $\text{rad} \cdot \text{s}^{-2}$.

As we saw in Chapter 5, the instantaneous tangential speed of a point on the rotating body depends on its radial distance r from the axis of rotation and on the angular velocity,

$$v = r\omega. \tag{9.5}$$

Similarly, there is a connection between the instantaneous tangential acceleration (linear motion) and the angular acceleration (rotational motion). The tangential acceleration associated with the motion of a point moving in a circular path of radius r is related to the instantaneous angular acceleration through

$$a_{\text{t}} = \alpha r. \tag{9.6}$$

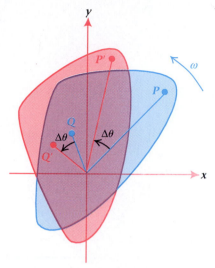

Figure 9.2
Points P and Q on the rigid body rotate with the same angular velocity, because they move through the same angle $\Delta\theta$ in the same time Δt.

In addition, for a point moving in a circular path with instantaneous angular velocity ω, we have seen that there is a centripetal acceleration a_c directed toward the rotation axis. The value of a_c is

$$a_c = \omega^2 r. \qquad (9.7)$$

Because the centripetal acceleration is along the radial line from the point to the rotation axis, it is a radial acceleration. The tangential acceleration is perpendicular to the line from the point to the axis (Fig. 9.3). Thus these two accelerations are at right angles to each other.

It is important to remember that the equations we have developed here are appropriate only for circular motion of a particle moving at a constant radius about an axis or for a rigid body in rotation about an axis. If the radial distance is allowed to change, there will be additional terms required for the description of both the tangential acceleration and the radial acceleration.

Be sure that you clearly understand the difference between angular acceleration, which we have just defined in Eq. (9.6), and centripetal acceleration given in Eq. (9.7). Centripetal acceleration occurs whenever an object moves in a curved path, even if it moves with constant speed. The centripetal acceleration acts radially toward the center of rotation. Angular acceleration occurs when there is a change in the angular speed of the object. When you swing a stone in a circle around your head at a constant rate, there is no angular acceleration, but there is a centripetal acceleration. On the other hand, when you increase or decrease the angular velocity, there is an angular acceleration as well as a centripetal acceleration. For rotational motion, there is *always* a centripetal acceleration, but there is angular acceleration *only* when the angular velocity is changing. When both accelerations are present, the instantaneous acceleration of a point on the rotating object is given by the vector sum of the two accelerations (Fig. 9.3).

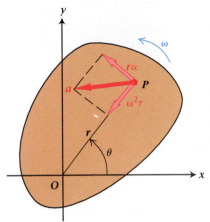

Figure 9.3

The acceleration **a** of a point P located a distance r from the axis of an object rotating with angular acceleration is the vector sum of a tangential component $a_t = r\alpha$ and a radial component $a_c = \omega^2 r$.

Figure 9.4

Example 9.1: An athlete swings the hammer in a circle to give it large velocity before letting it go.

Example 9.1

In the hammer throw, a 7.25-kg shot is swung in a circle five times and then released (Fig. 9.4). The shot moves with an average radius of 2.1 m and an average angular acceleration of 2.3 rad/s^2 reaching a maximum tangential speed of 25 m/s. (a) What is the average tangential force and (b) what is the maximum centripetal force exerted on the hammer?

Strategy (a) The average tangential force is given by the product of the average tangential acceleration with the mass of the hammer. The tangential acceleration is found from the relationship between angular acceleration and tangential acceleration given in Eq. (9.6). (b) The centripetal force is found from the product of centripetal acceleration with the mass.

Solution (a) The tangential acceleration is given by $a_t = \alpha r$. The tangential force is

$$F_t = ma_t = m\alpha r = (7.25 \text{ kg})(2.3 \text{ rad/s}^2)(2.1 \text{ m}) = 35 \text{ N}.$$

(b) The centripetal force is

$$F_c = m\omega^2 r = \frac{mv^2}{r} = \frac{(7.25 \text{ kg})(25 \text{ m/s})^2}{2.1 \text{ m}} = 2.2 \times 10^3 \text{ N}.$$

Discussion The tangential force is modest; approximately equivalent to the weight of a 3.6-kg object (the weight of two large textbooks). However, the centripetal force is quite large; nearly double the weight of a large athlete.

9.2 Rotational Kinematics

We now have three angular kinematic quantities θ, ω, and α, which are completely analogous to the linear kinematic quantities x, v, and a. If we restrict ourselves to cases in which the angular acceleration is constant, then we can find relationships between the angular kinematic variables just as we did for the translational kinematic variables. The definitions of angular displacement, angular velocity, and angular acceleration all differ from the related linear quantities by a factor of r. Because these corresponding quantities are analogous, we can simply write down the rotational kinematic equations from our knowledge of the translational ones. For example, starting with Eq. (2.8) for position,

$$x = x_0 + v_0 t + \tfrac{1}{2}at^2,$$

we can replace all the linear variables with the analogous angular variables to get

$$\theta = \theta_0 + \omega_0 t + \tfrac{1}{2}\alpha t^2. \tag{9.8}$$

In the same manner we could take Eq. (2.9),

$$v^2 = v_0^2 + 2a(x - x_0),$$

and by replacing the linear variables with the corresponding angular variables get

$$\omega^2 = \omega_0^2 + 2\alpha(\theta - \theta_0). \tag{9.9}$$

Table 9.1 lists the principal kinematic equations for constant acceleration for translational and rotational motion.

Table 9.1   *Summary of the Linear and Rotational Kinematic Equations*

Linear		Rotational	
$\bar{v} = \Delta x/\Delta t$	(Ch. 2)	$\omega = \Delta\theta/\Delta t$	(Ch. 5)
$\bar{a} = \Delta v/\Delta t$	(Ch. 2)	$\bar{\alpha} = \Delta\omega/\Delta t$	(Ch. 9)
$v^2 = v_0^2 + 2a(x - x_0)$	(Ch. 2)	$\omega^2 = \omega_0^2 + 2\alpha(\theta - \theta_0)$	(Ch. 9)
$v = v_0 + at$	(Ch. 2)	$\omega = \omega_0 + \alpha t$	
$x = x_0 + \tfrac{1}{2}(v_0 + v)t$	(Ch. 2)	$\theta = \theta_0 + \tfrac{1}{2}(\omega + \omega_0)t$	
$x = x_0 + v_0 t + \tfrac{1}{2}at^2$	(Ch. 2)	$\theta = \theta_0 + \omega_0 t + \tfrac{1}{2}\alpha t^2$	(Ch. 9)

This table contains analogous linear (translational) and rotational equations. They were first introduced in the chapters indicated. Some expressions that were not derived in the text have been included because you may find them useful in other applications.

Example 9.2

The wheel on a moving car slows uniformly from 70 rad/s to 42 rad/s in 4.2 s (Fig. 9.5). (a) What is the angular acceleration of the wheel? (b) What angle does the wheel turn through in the 4.2 s? (c) How far does the car go if the radius of the wheel is 0.32 m?

Solution (a) To find the acceleration, we substitute directly into the definition Eq. (9.3):

$$\alpha = \frac{\Delta\omega}{\Delta t} = \frac{42 \text{ rad/s} - 70 \text{ rad/s}}{4.2 \text{ s}}$$

$$\alpha = -6.67 \text{ rad/s}^2 \approx -6.7 \text{ rad/s}^2.$$

The negative sign corresponds to the decreasing angular speed.
(b) The wheel will turn through an angle given by Eq. (9.8). Because we are only interested in the rotation starting at the beginning of the 4.2 s, we set $\theta_0 = 0$.

$$\theta = \omega_0 t + \frac{1}{2}\alpha t^2$$

$$\theta = (70 \text{ rad/s})(4.2 \text{ s}) + \frac{1}{2}(-6.67 \text{ rad/s}^2)(4.2 \text{ s})^2$$

$$\theta = 294.0 \text{ rad} - 58.8 \text{ rad}$$

$$\theta = 235.2 \text{ rad} \approx 240 \text{ rad}.$$

Note that, because there are 2π rad in a full revolution, 240 rad is equivalent to $240/2\pi \approx 38$ revolutions of the wheel.
(c) The distance traveled by the car (Fig. 9.5) is the same as the path length of a point on the edge of the wheel as it turns through the angle θ calculated above. We can find this distance from Eq. (9.1), the definition of an angle in radians:

$$\theta = \frac{s}{r}.$$

Upon rearranging, we find the distance s to be

$$s = r\theta$$

$$s = (0.32 \text{ m})(235 \text{ rad})$$

$$s = 75 \text{ m}.$$

Figure 9.5

Example 9.2: When the car's wheels turn through an angle θ, the car travels a distance $s = r\theta$, where r is the radius of the wheels.

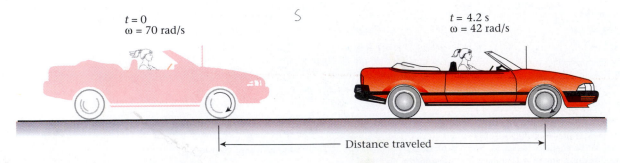

$t = 0$
$\omega = 70$ rad/s

$t = 4.2$ s
$\omega = 42$ rad/s

Distance traveled

Example 9.3

A bicycle tire turning at 0.21 rad/s is brought to rest by the brakes in exactly two revolutions. What is the angular acceleration of the wheel?

Solution   Equation (9.9) may be solved to give the angular acceleration:

$$\alpha = \frac{\omega^2 - \omega_0^2}{2(\theta - \theta_0)}.$$

The initial conditions are $\theta_0 = 0$ and $\omega_0 = 0.21$ rad/s. Two revolutions is 2 times 2π radians, so the final conditions are $\theta = 4\pi$ and $\omega = 0$. Substituting these into the equation for α gives

$$\alpha = \frac{0^2 - (0.21 \text{ rad/s})^2}{2(4\pi - 0)}$$

$$\alpha = -1.8 \times 10^{-3} \text{ rad/s}^2.$$

9.3   Torque

So far, we have seen that an object at rest, such as a book on a table, remain motionless when two equal but oppositely directed forces are applied to it. That is always true if the object is a point mass. It is even true for an extended object like the book when the two forces are applied in opposite directions along the same line. However, when the forces are not along the same line, the book may turn or rotate. This motion occurs even though the net force is zero. The manner in which the book turns depends on its size and shape, as well as on its mass. What causes the book to rotate? The answer to this question is found in this and later sections of this chapter.

We begin by considering a familiar rotating object, a door hinged at one edge. We can pull on it in several ways (Fig. 9.6). However, the most effective way to open the door is to grab the edge of the door farthest from the hinges and pull at right angles to the door. Intuitively, we realize that the distance from the hinge to the point of application of the force as well as the magnitude and direction of the force are important factors affecting the tendency of the door to rotate.

The quantity measuring how effectively a force causes rotation is called **torque.** The greater the distance from the axis of rotation (door hinges) to the point where we apply the force (door handle), the greater the torque. Also, maximum torque occurs when the direction of the applied force is perpendicular to a line drawn between the axis and the point where the force is applied. By contrast, when the line and the force are in the same direction, so that the force acts directly toward or away from the axis of rotation, there is no torque.

Figure 9.7 illustrates the application of a force causing a door to rotate about an axis at point O, looking down from above the door. The torque τ about point O is defined as

$$\tau \equiv rF \sin \theta, \tag{9.10}$$

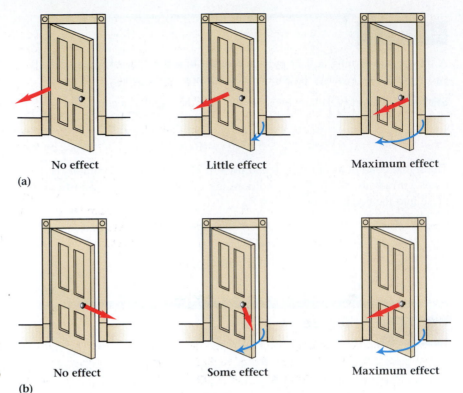

Figure 9.6
A door free to rotate about its hinges.
(a) The force applied farthest from the hinges produces the greatest torque.
(b) The force applied at right angles to the door produces the greatest effect.

(a) No effect   Little effect   Maximum effect

(b) No effect   Some effect   Maximum effect

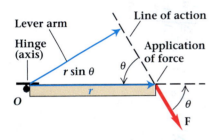

Figure 9.7
A force **F** applied at a distance r causes a torque $\tau = rF \sin \theta$ about the point O. The effective lever arm is $r \sin \theta$.

where r is the magnitude of the displacement from the axis to the point of application of the force **F** and θ is the angle between the direction of **r** and the direction of the force. The maximum torque occurs when θ is 90°—that is, when **r** and **F** are perpendicular. When **r** and **F** are in the same direction, θ becomes zero and there is no torque.

We see from the figure that applying a torque is equivalent to applying a force perpendicular to a lever. The lever arm, or moment arm, is defined as the perpendicular distance from the axis of rotation to the line of action of the force. From the geometry, we see that the lever arm is $r \sin \theta$. Therefore, saying that the torque is the product of force times the lever arm ($r \sin \theta$) is consistent with Eq. (9.10). Thinking of torque as the product of a force and a lever arm often makes a problem easier to analyze.

Figure 9.8 shows two forces generating torques. In both cases the forces are at right angles to **r**. Although the resulting torques have the same magnitude, they

Figure 9.8
(a) Force $\mathbf{F}_1$ produces a torque causing a counterclockwise rotation about O.
(b) Force $\mathbf{F}_2$ produces a clockwise rotation about O.

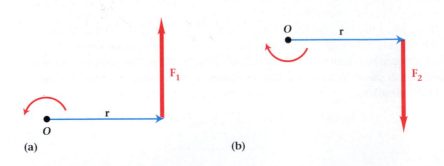

(a)   (b)

tend to cause rotations in opposite directions. In this book, we adopt the sign convention that a torque tending to produce a counterclockwise motion is positive; a torque tending to produce a clockwise motion is negative.

Torque is really a vector quantity, with both magnitude and direction. The magnitude is given by Eq. (9.10), and the direction is along the axis of rotation. This direction is perpendicular to the plane containing both the line of action of the force and the line from the axis to the point of application (Fig. 9.9a). The direction of the torque points along the direction a right-handed screw will move if **r** is rotated by **F**. Figure 9.9(b) shows an alternative way of establishing this direction.

The units of torque are the units of force multiplied by the units of length. Thus, the SI unit for torque is the newton-meter (N · m). In the British system, the unit for torque is the foot-pound (ft-lb). The units of torque are the same as those for work and energy; however, torque and work represent very different physical quantities. Moreover, remember that torque is a vector, whereas work is a scalar.

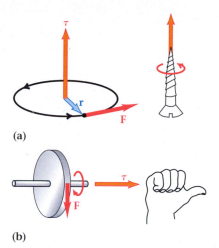

(a)

(b)

Figure 9.9
Direction of the torque vector. (a) In the direction of advance of a right-hand screw rotated by the force. (b) The right-hand rule: The curled fingers point in the direction of the force. The thumb then points in the direction of the torque.

Example 9.4

The instructions for replacing the head gasket on an automobile engine say that the bolts should be "torqued down" to 90 N · m. If you use a wrench that is 45 cm long, how much force must you apply in a direction perpendicular to the wrench handle to accomplish this? (Too much torque will pinch the gasket so that it will not seal properly.)

Solution   The physical situation is shown in Fig. 9.10. We are told that the angle between the direction of the force and the line from the axis to the point of application is 90°. Thus, Eq. (9.10) becomes

$$\tau = rF,$$

and

$$F = \frac{\tau}{r}$$

$$F = \frac{90 \text{ N} \cdot \text{m}}{45 \text{ cm}} \left(\frac{100 \text{ cm}}{1 \text{ m}} \right) = 200 \text{ N}.$$

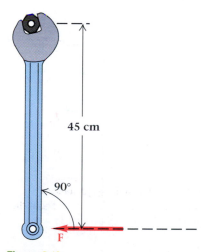

Figure 9.10
Example 9.4: Using a wrench to tighten the nut on a bolt. A force **F** is applied near the end of the handle.

Example 9.5

The crank arm of a bicycle pedal is 16.5 cm long. If a 52.0-kg woman puts all her weight on one pedal, how much torque is developed (a) when the crank is horizontal and (b) when the pedal is 15° from the top?

Solution   (a) For the case where the crank arm is horizontal (Fig. 9.11a), the angle between the downward force of the woman's weight and the crank arm is 90°. The lever arm is the full extent of the crank arm, and the torque is

$$\tau = rF = r(mg) = (16.5 \text{ cm})(52.0 \text{ kg})(9.81 \text{ m/s}^2)(10^{-2} \text{ m/cm})$$
$$\tau = 84.2 \text{ N} \cdot \text{m}.$$

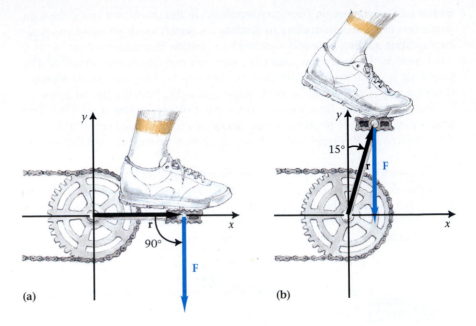

Figure 9.11
Example 9.5: Torque applied to a bicycle pedal. (a) The crank is horizontal. (b) The crank makes an angle of 15° with the vertical.

(a)

(b)

(b) For the case where the pedal is 15° from the top (Fig. 9.11b), the angle between the direction of the force and the direction of the crank arm is 165°, so the lever arm is (16.5 cm)(sin 165°) = 4.27 cm = 0.0427 m. The resulting torque is

$$\tau = \text{(lever arm)(force)}$$
$$\tau = (0.0427 \text{ m})(52.0 \text{ kg})(9.81 \text{ m/s}^2) = 21.8 \text{ N} \cdot \text{m}.$$

Have you ever noticed this difference when pedaling a bicycle up a hill?

9.4   **Static Equilibrium**

We saw in Section 4.9 that an object acted on by two forces, equal in magnitude but opposite in direction, has no linear acceleration. We say that the object is in translational equilibrium. The first condition for equilibrium (given in Section 4.9) was that the vector sum of the forces on a body be equal to zero:

$$\sum_{i=1}^{N} \mathbf{F}_i = 0, \qquad (9.11)$$

where there are N individual forces $\mathbf{F}_i$. This condition ensures that there is no translational acceleration.

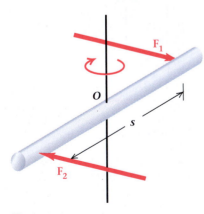

Figure 9.12
A couple. Two forces of equal magnitude act opposite and parallel to each other, but not along the same line. The rod is in translational equilibrium, but can still rotate.

However, an object in translational equilibrium may still rotate. For example, the wheel of a stationary exercise bicycle does not move relative to the floor, but it turns when you apply a torque to it. A pair of forces, such as $\mathbf{F}_1$ and $\mathbf{F}_2$ of Fig. 9.12, that are equal in magnitude but opposite in direction *and not lying along the same line* is called a **couple**. The couple applies a torque about O equal to the sum of the torques due to the individual forces. Note that although the forces are in opposite directions, each tends to rotate the body in the same direction about O. If the distance between the lines of action of the forces is s, then each force has a lever arm $s/2$, and the torque produced by the two equal forces is

$$\tau = \frac{s}{2}F_1 + \frac{s}{2}F_2.$$

However, since $F_1 = F_2 = F$, we may write

$$\tau = sF.$$

If we want to keep the object from rotating, we must subject it to another torque of the same magnitude but in the opposite direction. We conclude that for a body to be in **rotational equilibrium,** the sum of the torques must be zero, or

$$\sum_{i=1}^{N} \tau_i = 0, \qquad (9.12)$$

where there are N individual torques τ_i.

The idea expressed in Eq. (9.12) is sometimes called the *second condition for equilibrium.* The second condition ensures that there is no rotational, or angular, acceleration. An object satisfying both conditions of equilibrium is said to be in *equilibrium.* If such an object is stationary, it is in *static equilibrium;* if it is moving with neither translational nor rotational acceleration, it is in *dynamic equilibrium.*

Figure 9.13(a) shows a uniform rod, such as a meterstick, placed on a fulcrum. It balances at its midpoint. A nonuniform rod, like a baseball bat (Fig. 9.13b), will also balance at some point, though not at its midpoint. In each case the upward force due to the fulcrum is equal to the entire weight of the object, as required by the first condition for equilibrium. In addition, the sum of the torques about the fulcrum due to gravity equals zero if the rod is in rotational equilibrium. In other words, the object behaves as if its mass were concentrated at a point lying directly above the fulcrum. This point is called the **center of mass** of the body.*

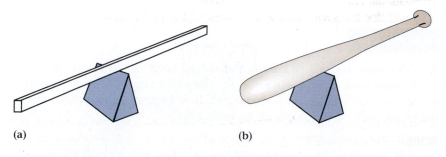

Figure 9.13
(a) A uniform rod balanced at its center of mass. (b) A nonuniform rod balanced at its center of mass.

(a)

(b)

---

*Strictly speaking, the *center of gravity* is the point at which the force of gravity can be considered to act. For spherical bodies, like the planets, that point is their geometric center. Frequently the center of gravity and the center of mass are the same. They are different only when the external gravitational field is not uniform over the body.

(a)

(b)

Figure 9.14
(a) The center of mass (c.m.) of a person standing erect lies above her feet. (b) When the person bends over, her hips move backward to keep her center of mass over her feet so that she does not fall over.

An object is in stable equilibrium if a small displacement causes a restoring torque to return it to its original position. If the resulting torque moves the object away from its original position, the equilibrium is unstable. We often refer to these conditions as being balanced or unbalanced. A person standing on a level floor will be balanced (stable equilibrium) if her center of mass is located over the area of support defined by the position of her feet (Fig. 9.14a). If she leans over, her legs and hips must move back as her torso moves forward so that her center of mass remains over her feet (Fig. 9.14b). Otherwise her center of mass would move out beyond her feet and she would topple over.

If a body is in static equilibrium, not only does it not rotate about some particular axis, but it does not rotate about any axis at all. We are free, therefore, to choose any possible rotation axis for the purpose of computing torques. Proper choice of the rotation axis can usually simplify the computations, since any force acting through the axis of rotation produces zero torque. Examples 9.6 and 9.7 illustrate this point.

Problem-Solving Strategy

Statics Problems

1. Start by drawing a diagram of the system.
2. Isolate the object to be analyzed and make a free-body diagram showing the forces acting *on* the object. Be careful to show the point where each force acts.
3. Choose coordinate axes and specify the positive sense of rotation. Then resolve the forces into components along the axes.
4. Write the equilibrium equations, $\Sigma F_x = 0$, $\Sigma F_y = 0$, and $\Sigma \tau = 0$.
5. Choose a convenient origin for the point to compute torques. Remember that if a force acts along a line passing through the point, the torque due to that force is zero. A clever choice of origin can often simplify a problem.
6. Solve the equilibrium equations simultaneously. Remember that you will need as many equations as you have unknowns to solve for.

Example 9.6

A 5.0-kg mass (m_2) and an unknown mass (m_3) hang from a 1.0-m rod of 2.0-kg mass (m_1), as shown in Fig. 9.15. The rod is supported on a knife-edge fulcrum at a distance 35 cm from one end. How large is the mass m_3 if the rod and masses are to balance on the knife edge?

Strategy We may consider all the weight of the rod to act downward at its center of mass, which in this case is at 50 cm. We then proceed as if a single mass m_1 equal to 2.0 kg (the mass of the rod) were hung from the rod at that point. Forces act at four points along the bar: the downward forces $m_1 g$, $m_2 g$, and $m_3 g$ at the points indicated and an upward force F at the fulcrum. The

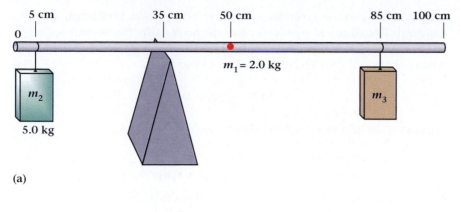

$m_1 = 2.0$ kg

m_2

5.0 kg

(a)

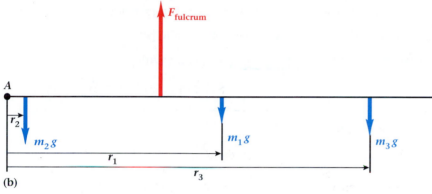

F_{fulcrum}

A

r_2

$m_2 g$

$m_1 g$

$m_3 g$

r_1

r_3

(b)

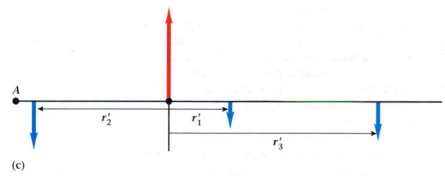

A

r'_2

r'_1

r'_3

(c)

Figure 9.15

Examples 9.6 and 9.7: A rod and three masses in static equilibrium. (a) Physical situation. (b) Free-body diagram for the rod indicating radii from point of rotation A. (c) Free-body diagram for indicating radii from the fulcrum.

upward force must equal the total downward force in order to have translational equilibrium:

$$F_{\text{fulcrum}} = m_1 g + m_2 g + m_3 g.$$

We have two unknowns in this equation, so we consider the second condition of equilibrium, setting the sum of the torques equal to zero to obtain a second equation. As we said above, if the rod is in static equilibrium, not only does it not rotate about some particular axis, but it does not rotate about any axis at all. We are free to pick any point for the axis to go through. For this example, we pick the left-hand end of the rod (point A on Fig. 9.15b) and calculate torques about this point.

Solution We can use Eq. (9.12) by calling torques that tend to rotate things in a counterclockwise direction positive and those that tend to rotate things in a clockwise direction negative. We have $m_1 = 2.0$ kg, $m_2 = 5.0$ kg, and m_3 is unknown. The distances from point A to the point where the weights act are $r_1 = 0.50$ m, $r_2 = 0.05$ m, and $r_3 = 0.85$ m. The distance from A to the fulcrum is $r_F = 0.35$ cm. According to Eq. (9.12),

$$\Sigma\tau = \sum_i r_i F_i = 0.$$

Therefore, the sum of the torques about point A is as follows:

$$-r_2 m_2 g + r_F F_{fulcrum} - r_1 m_1 g - r_3 m_3 g = 0,$$
$$-r_2 m_2 g + r_F(m_1 g + m_2 g + m_3 g) - r_1 m_1 g - r_3 m_3 g = 0.$$

Since g is a common factor, it can be divided out to give

$$-r_2 m_2 + r_F(m_1 + m_2 + m_3) - r_1 m_1 - r_3 m_3 = 0.$$

Next we insert the appropriate numerical values to get

$$-(0.05 \text{ m})(5.0 \text{ kg}) + (0.35 \text{ m})(5.0 \text{ kg} + 2.0 \text{ kg} + m_3)$$
$$-(0.50 \text{ m})(2.0 \text{ kg}) - (0.85 \text{ kg})(m_3) = 0,$$
$$-0.25 \text{ m} \cdot \text{kg} + 2.45 \text{ m} \cdot \text{kg} + (0.35 \text{ m})(m_3) - 1.0 \text{ m} \cdot \text{kg} - (0.85 \text{ m})(m_3) = 0.$$

Gathering terms, we get

$$(0.50 \text{ m})(m_3) - 1.20 \text{ m} \cdot \text{kg} = 0,$$
$$m_3 = 2.4 \text{ kg}.$$

Example 9.7

Calculate the mass m_3 for Example 9.6 by computing torques about the fulcrum.

Strategy If we choose the axis of rotation to be about the fulcrum, then we have only three torques to consider: the counterclockwise torque due to m_2 and the clockwise torques due to m_1 and m_3. The torque due to the fulcrum force becomes zero because the moment arm is now zero.

Solution The sum of the torques becomes

$$+r'_2 m_2 g - r'_1 m_1 g - r'_3 m_3 g = 0,$$

where we have used primes to indicate that the distance values are not the same as in Example 9.6. Upon rearranging, we find

$$m_3 = \frac{r'_2 m_2 - r'_1 m_1}{r'_3}.$$

From Fig. 9.15(c) we see that $r'_2 = 30$ cm, $r'_1 = 15$ cm, and $r'_3 = 50$ cm. When the numerical values are inserted into the equation for m_3, we get

$$m_3 = \frac{(30 \text{ cm} \times 5.0 \text{ kg}) - (15 \text{ cm} \times 2.0 \text{ kg})}{50 \text{ cm}}$$

$$m_3 = 2.4 \text{ kg}.$$

Discussion As expected, we get the same answer here as in Example 9.6. This time the equation was simpler and the computation reduced because we only had to consider three torques. The force acting through the point of rotation produced no torque and did not have to be considered.

Example 9.8

A sign weighing 400 N is suspended at the end of a 350-N uniform rod (Fig. 9.16a). (a) What is the tension in the support cable if it makes an angle $\theta = 35°$ with the rod? (b) What would be the tension if the upper end of the cable were moved so that $\theta = 55°$?

Strategy Again we have a static equilibrium problem. We can represent the forces on the rod by a diagram like Fig. 9.16(b). The tension in the cable is resolved into components along and perpendicular to the rod. The force of the wall on the rod is eliminated from the analysis by taking the point where the rod touches the wall to be the axis of rotation for computing the torques. Then we can apply the second condition for equilibrium (Eq. 9.12) to determine the component of the tension perpendicular to the rod.

Solution (a) Choose counterclockwise torques positive and clockwise torques negative. Let L be the length of the rod, w its weight, W the weight of the sign, and T the tension in the cable. Since the rod is uniform, the weight w acts at its center of mass, a distance $L/2$ from the wall. The sum of the torques becomes

$$(T \sin \theta)L - w(\tfrac{1}{2} L) - WL = 0.$$

We can remove the common factor of L and rearrange this equation to get the tension T as

$$T = \frac{\tfrac{1}{2}w + W}{\sin \theta}.$$

When we substitute the numerical values, we get

$$T = \frac{\tfrac{1}{2}(350 \text{ N}) + 400 \text{ N}}{\sin 35°} = 1000 \text{ N}.$$

(b) If the cable is moved so that the angle it makes with the rod is 55°, the analysis in part (a) doesn't change. However, when we substitute the new numerical value for the angle, the tension becomes

$$T = \frac{\tfrac{1}{2}(350 \text{ N}) + 400 \text{ N}}{\sin 55°} = 700 \text{ N}.$$

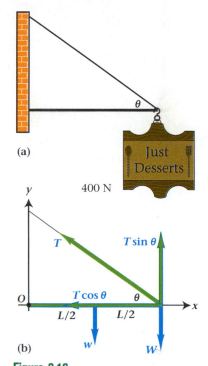

(a)

(b)

Figure 9.16
Example 9.8: (a) A 400-N sign is hung from the end of a long rod. (b) Force diagram for the sign and rod.

Example 9.9

The average mass of a woman's arm is approximately 5% of her whole body mass. When the arm is lifted as shown in Fig. 9.17, the distance between the center of mass of the arm and the center of rotation of the shoulder is

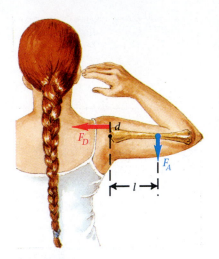

Figure 9.17
Example 9.9: The deltoid muscle supplies the force needed to hold the arm in equilibrium.

approximately 0.15 m. In a simplified model, we consider the deltoid muscle to pull horizontally with a lever arm of about 3 cm. With what force, in terms of the woman's body weight, must the deltoid muscle pull in order to hold the arm in equilibrium in the position shown in Fig. 9.17?

Strategy When the woman holds the arm steady, it is in a condition of static equilibrium for both translation and rotation. We can use the second condition for equilibrium to calculate torques about the center of rotation of the shoulder. Although we do not know a numerical value for the downward force on the arm, we can express it as a percentage of the woman's total weight $w = mg$.

Solution Remembering our sign convention for torques, we can apply Eq. (9.12) directly. Using the symbols shown in Fig. 9.17, we get

$$\sum_{i=1}^{N} \tau_i = F_D d - l F_A = 0,$$

where $F_A = 0.05w$. Upon rearranging, we find

$$F_D = \frac{l F_A}{d} = \frac{(0.15 \text{ m})(0.05w)}{0.03 \text{ m}} = 0.25w.$$

Discussion According to our model, the deltoid muscle must pull with a force of about one quarter of the woman's weight in order to hold the arm in the position shown. Although the model has oversimplified the structure of the shoulder, it does allow us to gain insight into the large forces that our muscles exert under the most ordinary conditions. If the woman extended her entire arm horizontally and held it outstretched, the center of mass would then be about 30 cm from the center of rotation. Then the force provided by the deltoid would be approximately half her body weight.

*9.5 Elasticity: Stress and Strain

In the previous section, we analyzed the forces on objects in static equilibrium. In doing so we assumed that the objects were completely rigid and did not deform under the applied forces. Yet a seemingly rigid solid object such as a steel bar will deform when large forces are applied to it. When more modest forces are applied, the bar will still deform, but the amount of the deformation may be quite small.

Suppose that we pull on the ends of a bar with a force F, as shown in Fig. 9.18. We say that the bar is in tension. The internal forces in the bar resist the tension forces and hold the bar together. Even so, the bar deforms and the equilibrium length of the bar will be greater when the external forces are applied than without them. If the bar is in equilibrium with the applied forces, then every cross section of the bar must be subject to the same internal forces that resist stretching. We define the **tensile stress** as the ratio of the magnitude of the applied force F to the cross-sectional area A:

$$\text{stress} \equiv \frac{\text{force}}{\text{area}} = \frac{F}{A}.$$

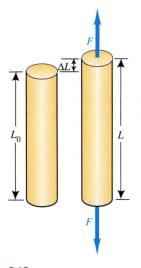

Figure 9.18
A bar with initial length L_0 is stretched by an amount ΔL when a force F is applied to its ends.

The **tensile strain** is defined as the ratio of the change in length ΔL to the initial length L_0 before the force was applied:

$$\text{strain} \equiv \frac{\text{change in length}}{\text{initial length}} = \frac{\Delta L}{L_0}.$$

The strain is the fractional change in length. As such, it is dimensionless. The bar would deform also if it were under compressive force, but this time it would be compressed rather than stretched. The **compressive strain** is defined just like the tensile strain; it is the ratio of the decrease in length to the initial length.

The amount of strain an object undergoes depends on the amount of stress applied to it. If the stress is not too great, the strain is observed to be proportional to the stress. The ratio of a stress to the corresponding strain is called an *elastic modulus*. For the tensile (or compressive) stress and strain that we have been describing, this ratio is called **Young's modulus,** denoted by Y:

$$Y = \frac{\text{stress}}{\text{strain}} = \frac{F/A}{\Delta L/L_0}. \tag{9.13}$$

Young's modulus has the same units as does stress, N/m^2. Some typical values are found in Table 9.2. Notice also that Eq. (9.13) is equivalent to Hooke's law, which we saw in Chapter 6. For a given initial length, cross-sectional area, and Young's modulus, we get $F = kx$, where $k = YA/L_0$.

What happens when the stress on an object gets very large? You know the answer already: the object breaks. If we plot a graph of tensile stress versus strain, we find a curve with the general shape seen in Fig. 9.19. When the stress is small, the relationship is linear and the material is characterized by a unique Young's modulus. This region, shown as Oa on the graph, is known as the *elastic* region. Between the points labeled a and b, stress and strain are no longer proportional, but if the stress is removed, the material returns to its original length. For strains less than that at b, the deformation is reversible. The stress at point b is known as the *elastic limit* (or yield point) because

Table 9.2	*Elastic Moduli*
Material	**Young's Modulus (GPa)**
Aluminum	70
Brass	91
Bone (compression)	7
Bone (tension)	18
Cast iron	120
Crown glass	60
Gold	80
Granite	46
Lead	16
Nylon	5
Steel	200
Tungsten	360

These are approximate values. The elastic behavior of a particular specimen depends on its past history and treatment.

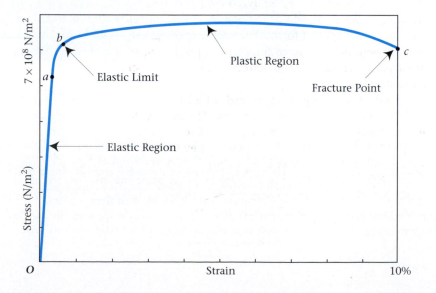

Figure 9.19
Stress-strain curve for an elastic solid, cold finished steel.

BRIDGES

How do you cross a river? Probably the first answer that comes to mind is to build a bridge. People have been building bridges for centuries, and continue today to design and construct longer and more elegant bridges.

The earliest bridges were tree trunks or stone slabs supported at both ends. The distance spanned by such beams was relatively short and depended on the strength and weight of the material used. The development of the truss, a combination of beams joined so that each piece shares part of the bridge's weight, increased the ratio of strength to weight. The members of the truss are straight pieces joined together to form a series of triangles. The resulting structure is lighter and more rigid than the equivalent simple beam and can support an external load over a much greater distance. In modern terms, the truss design required a knowledge of the strength of materials.

The earliest truss bridges were made of wood. Later trusses were reinforced with iron or even made entirely from iron. By the late 1800s, the common material for building truss bridges had become steel. Most of the railroad and highway bridges built in North America from 1890 to the middle of the twentieth century were steel truss bridges, especially for spans of 200 to 400 m.

Longer spans can be reached with arch bridges whose basic design was perfected centuries ago by the Romans. The secret of the arch is that the forces of its own weight and any added load are compressional forces, which allow the use of stone as a building material. Some of the stone bridges built by the Romans are still standing. The design of the arch results in a force that is downward and outward at the base of the arch. When the base is properly anchored, the arch bridge can span hundreds of meters. For example, the steel arch bridge over Sydney harbor spans 503 m and the New River Gorge Bridge in West Virginia spans 518 m. Both of these steel bridges utilize truss reinforcement of the basic arch. A maximum distance for steel arch bridges has been estimated at about 900 m.

The longest spans are achieved with suspension bridges that hang on steel cables stretched between tall towers. The ends of the cables are held in place on opposite shores by massive concrete anchorages. Because of the large strength-to-weight ratio of steel-wire cables, suspension bridges can be much longer than other types of bridges. The Akashi-Kaikyo Bridge in Japan is the longest span in the world, measuring 1990 m between the towers.

The modern suspension bridges of today owe much to the designers of the Brooklyn Bridge, John Roebling and his son Washington Roebling. In 1866 the elder Roebling, who

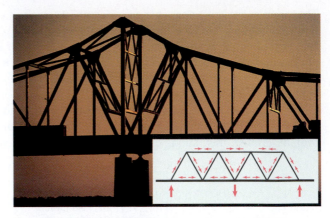

Figure B9.1 A truss bridge transmits its load to its end supports by a combination of compression (force vectors point toward each other) and tension (force vectors point away from each other).

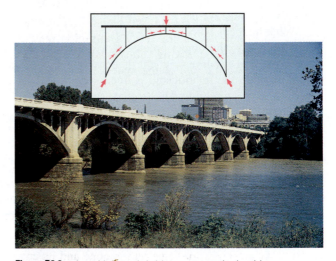

Figure B9.2 A multiple arch bridge supports its load by transmitting compressive forces along the arch to the end supports.

had already pioneered a method of spinning wires from one anchorage to the other through the top of the supporting towers, took on the task of designing and building a bridge to connect Brooklyn to Manhattan. Three years later, the design complete, John Roebling died from a tetanus infection as a result of an accident at the site. His son then took over as chief engineer for the bridge and supervised the construction. He adapted the use of water-tight working chambers, called caissons, to the high pressures at the depths of the bridge's foundations. While working in a caisson, Washington Roebling was struck with the bends caused by a sudden decompression. He became paralyzed and supervised the remainder of the construction from his bedroom window, with his wife Emily carrying out his instructions and dealing with the workers. When the bridge opened in May of 1883, it was the longest suspension span in the world with a main span of 486 m.

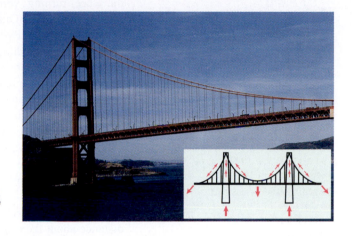

Figure B9.3 A suspension bridge supports its load by transmitting tension forces along the suspension cable to the main towers. The towers transmit compressive forces to the ground.

---

beyond *b* the distortion is not reversible and the object does not return to its original shape. If the stress is increased still more, the maximum elongation is reached and the material fractures. The region between the elastic limit and the fracture point (or breaking point) is known as the *plastic region*. The stress required to cause fracture of the material is known as the *breaking stress,* or ultimate strength.

Example 9.10

A weightlifter raises a weight of 600 N over her head. Assuming that each of her legs supports the same weight and that the legs are parallel and vertical, determine the amount by which each femur (thighbone) compresses. Each femur has an affective cross-sectional area of 7.5×10^{-4} m^2 and a length of 0.52 m.

Strategy We are given the force on the bone and its cross-sectional area and we want to find the resulting compression ΔL. We can use Young's modulus for bone (Table 9.2) to relate the strain ($\Delta L/L$) to the stress (F/A). Because the legs are vertical and share the load, the additional force on each femur is half the 600 N.

Solution We rearrange the defining equation for Young's modulus (Eq. 9.13) to get the compression as

$$\Delta L = \frac{FL_0}{YA} = \frac{(300 \text{ N})(0.52 \text{ m})}{(7 \times 10^9 \text{ N/m}^2)(7.5 \times 10^{-4} \text{ m}^2)} = 3 \times 10^{-5} \text{ m}.$$

Each femur compresses a small amount, 0.03 mm.

Example 9.10: A weightlifter's load puts additional stress on her legs.

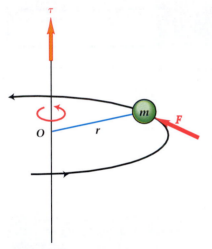

Figure 9.20
A mass constrained to move about a fixed point O at a distance r is subjected to a force $\mathbf{F}$. The resulting torque τ changes the angular velocity of the mass.

9.6 Torque and Moment of Inertia

Suppose you have a wheel mounted on an axle on which it is free to turn. To set the wheel turning, you need to apply a torque to it about its axle to overcome the wheel's inertia. Once the wheel is set in motion, it continues to rotate at constant angular velocity until another torque is applied. (We neglect the retarding effects of a torque due to friction.) In both cases, a torque causes the angular velocity to change. This situation is analogous to the application of a force to an object to change its linear velocity. We will find that the object undergoes an angular acceleration and that the connection between torque and angular acceleration is much like the connection between force and translational acceleration in Newton's second law. For simplicity we will introduce the ideas using only a single point mass, but the concepts apply to extended solid objects also. We start by applying a force to an object at a distance from its axis of revolution—that is, applying a torque—and then calculate the body's motion using Newton's laws.

Consider a single particle of mass m that is free to move in a plane at a fixed distance about the origin O (Fig. 9.20). This model might represent the physical situation of a ball bearing glued to one end of a soda straw of negligible mass. A force F applied to the ball bearing in a direction perpendicular to the straw causes the bearing to accelerate. The tangential acceleration a can be determined from Newton's second law written as

$$F = ma.$$

Multiplying both sides of this equation by r gives

$$Fr = mar.$$

The left-hand side of this equation is just the torque τ. The right-hand side may be rewritten in terms of the angular acceleration about O, $\alpha = a/r$. The result is

$$\tau = (mr^2)\alpha.$$

The quantity in parentheses is called the **moment of inertia, I**, and is a property of the mass and the radius r. In this instance, the value of I was fixed when we picked the mass and the length of the straw. However, the acceleration is the result of the torque acting on the system. Therefore, we write

$$\boxed{\tau = I\alpha.} \tag{9.14}$$

We have just computed the relationship between torque and angular acceleration for a single point of mass m moving in a circle of radius r. However, it can be shown that Eq. (9.14) is more general and may be used to find the angular acceleration produced by a net torque on any extended rigid body, provided we use the proper moment of inertia I. The moment of inertia plays the same role in the formulas for rotational motion that mass does in the equations for linear motion. However, the moment of inertia depends both on the mass and on the geometry, or shape, of the body, as well as on our choice of axis of rotation.

We can calculate the moment of inertia of an extended body by summing the moments of inertia of each small element of the body, obtaining the moment of inertia of the whole. The equation that expresses this idea is the definition of the moment of inertia:

$$I = \Sigma m_i r_i^2. \tag{9.15}$$

Except for a few simple cases, the techniques of calculus are needed to perform the indicated summation. Table 9.3 gives the moments of inertia, for several bodies about specific axes. Inserting these expressions into Eq. (9.14), we can readily compute the angular acceleration when we know the torque.

In order to better understand the meaning of the moment of inertia, you can do the following experiment. Find two identical long rods (perhaps two metersticks) and hold one in each hand. Grasp the one in your left hand at the center and the other in your right hand at its end, a situation corresponding to two of the entries in Table 9.3. Now try to twist the rods so that they both have the same rotational motion—that is, the same angular acceleration. You will quickly find that the torques you must apply are not the same for both rods. Instead, a noticeably larger torque is required to move the rod held by its end. This is not surprising because the distribution of mass relative to your hand is different, even though the rods have identical shape and mass. The mass is farther away from the axis of rotation for the rod held at the end, thus it has a larger moment of inertia. As you can see from Table 9.3, its moment of inertia is four times as great as for the rod held at its center.

Equation (9.14) is the rotational analog of Newton's second law. Notice the similarity between it and the relation $F = ma$. Equation (9.14) may be expressed in words as follows: *An object's angular acceleration is proportional to the applied torque, and the proportionality constant is the moment of inertia of the object.* The torque corresponds to the force in Newton's law, the angular acceleration corresponds to the linear acceleration, and the moment of inertia plays the same role as the mass.

Example 9.11

A cylindrical winch of radius R and moment of inertia I is free to rotate without friction about an axis (Fig 9.21a). A cord of negligible mass is wrapped about the shaft and attached to a bucket of mass m. When the bucket is released, it accelerates downward as a result of gravitational attraction. Find the acceleration of the bucket.

Strategy   We first examine the forces on the bucket, shown as a free-body diagram in Fig. 9.21(b). The downward gravitational force is mg, and the upward force T is the tension in the cord. If we choose the downward direction for the positive acceleration, then by Newton's second law

$$ma = mg - T,$$

and a is positive when mg exceeds T.

| Table 9.3 | *Moments of Inertia for Some Regular Objects of Mass m* |

Dumbbell about axis through center perpendicular to length; connecting rod of negligible mass

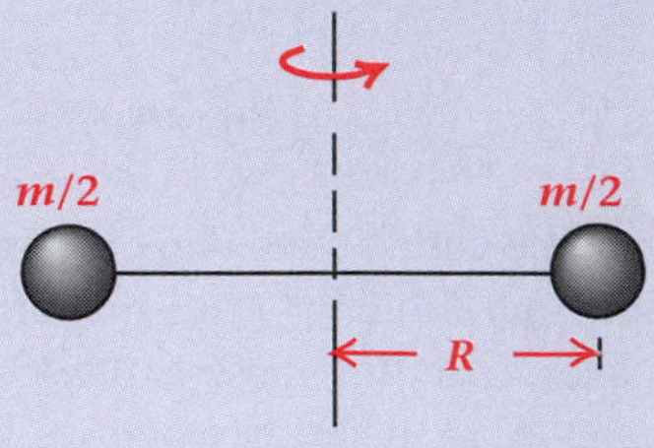

$I = mR^2$

Thin ring about axis through center

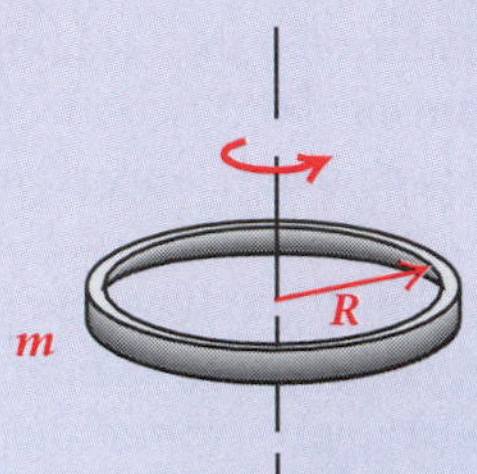

$I = mR^2$

Disk about axis through center

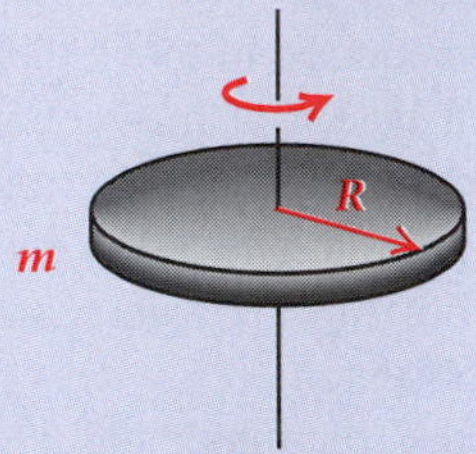

$I = \frac{1}{2}mR^2$

Solid sphere about any diameter

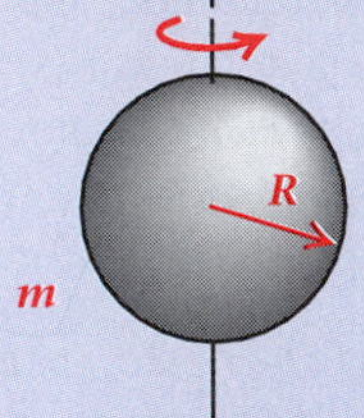

$I = \frac{2}{5}mR^2$

Thin rod about axis through one end perpendicular to length

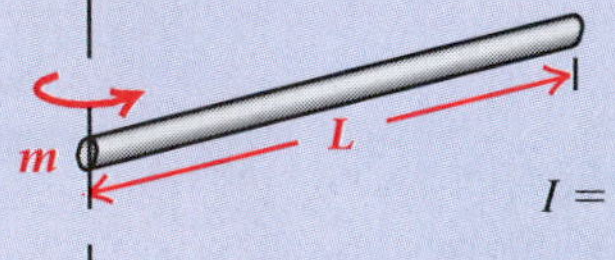

$I = \frac{1}{3}mL^2$

Thin rod about axis through center perpendicular to length

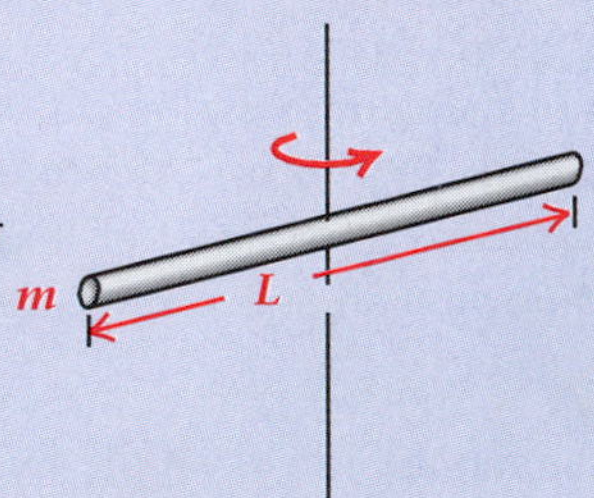

$I = \frac{1}{12}mL^2$

Disk about axis through one edge parallel to the symmetry axis

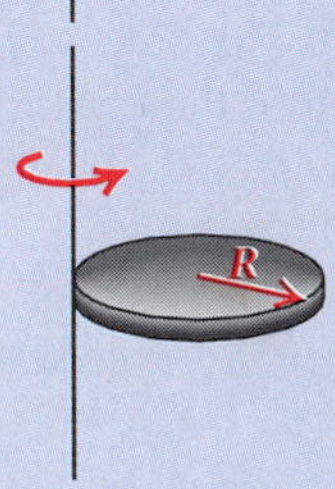

$I = \frac{3}{2}mR^2$

Next we observe that the cylinder is subject to a torque about its axis, given by RT (Fig. 9.21c). Its angular acceleration is given by

$$\tau = RT = I\alpha.$$

When the bucket moves down a distance s, a point on the circumference of the cylinder must also move through a distance s. Similarly, if the bucket accelerates an amount a, so does a point on the edge of the cylinder. The tangential acceleration a of a point on the circumference of the cylinder and the angular acceleration of the cylinder are related in the same way as the tangential velocity and the angular velocity are related. That is,

$$a = R\alpha.$$

From these equations we can determine the acceleration a.

Solution We may use the relationship for a to eliminate α in the torque equation:

$$RT = \frac{Ia}{R},$$

or

$$T = \frac{Ia}{R^2}.$$

This last expression may be inserted into the equation for the acceleration of the bucket:

$$ma = mg - \frac{Ia}{R^2}.$$

Figure 9.21

Example 9.11: (a) A cylinder rotates about its symmetry axis. A cord is wound around the cylinder, and a bucket is suspended from the end of the cord. (b) Free-body diagram for the bucket of mass m. (c) Free-body diagram for the torque about the axis of the cylinder.

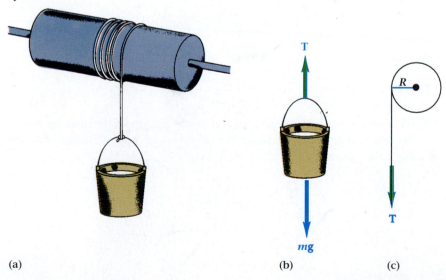

(a)   (b)   (c)

Upon rearranging we get

$$\left(m + \frac{I}{R^2}\right)a = mg,$$

or

$$a = \frac{mg}{m + \dfrac{I}{R^2}} = \left(\frac{1}{1 + \dfrac{I}{mR^2}}\right)g.$$

This last equation gives the acceleration of the bucket.

Discussion When the moment of inertia I becomes very large, a becomes very small. When I becomes small, a approaches g. The acceleration reverts to that for free fall in the limit that the inertia of the cylinder approaches zero. From this we see that the effect of the cylinder's inertia is to reduce the acceleration of the bucket.

9.7 Angular Momentum

Suppose you have a wheel mounted on an axle on which it is free to turn. To set the wheel turning, you need to apply a torque to it about its axle to overcome the wheel's inertia. Once the wheel is set in motion, it continues to rotate at constant angular velocity until another torque is applied. (We neglect the retarding effects of a torque due to friction.) In both cases, a torque causes the angular velocity to change. This situation is analogous to the application of a force to a body to change its linear velocity. In this section we develop this analogy further.

Consider a single particle of mass m that is free to move in a plane at a fixed distance r about the origin O, as shown in Fig. 9.22. A force F applied to the mass in a direction tangent to its path produces a torque that causes the mass to accelerate. The torque may be expressed as

$$\tau = rF = r\left(\frac{\Delta p}{\Delta t}\right) = rm\frac{\Delta v}{\Delta t}.$$

However, because r and m are constant, we can write this equation as

$$\tau = \frac{\Delta(rmv)}{\Delta t}.$$

Thus, the application of a torque causes a rate of change of the quantity rmv.

We can rewrite the equation for the torque as the rate of change of a new quantity called the angular momentum L:

$$\boxed{\tau = \frac{\Delta L}{\Delta t}.} \tag{9.16}$$

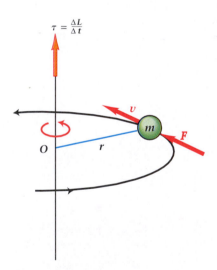

Figure 9.22

A mass constrained to move about a fixed point O at a distance r is subjected to a force **F.** The torque τ due to the force **F** causes a change in the angular momentum of the mass moving at a distance r about a fixed point O.

The **angular momentum** is the product of the linear momentum mv and the radius r:

$$L \equiv rmv. \qquad (9.17)$$

The quantity $L = rmv$ is the magnitude of the angular momentum vector **L**. Its direction is defined in a manner similar to the way in which we defined the direction of the torque in Fig. 9.9. That is, the angular momentum vector is at right angles to the plane containing **r** and **v**. The direction of the angular momentum is given by the right-hand rule and is directed to the right for the wheel in Fig. 9.23.

We may use the definition of angular velocity, $\omega = v/r$, to write the magnitude of the angular momentum in terms of the angular velocity ω:

$$L = mr^2\omega. \qquad (9.18)$$

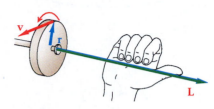

Figure 9.23
A wheel rotating on an axle. The direction of the angular momentum **L** is found from the right-hand rule. The fingers are curled so that they point along the direction of rotation; then the thumb points along the direction of the angular momentum.

Then we can separate the angular momentum into two parts: one term that depends on the properties of the body—that is, its moment of inertia—and another term that is the angular velocity. For a single point mass m at a fixed distance r from the axis of rotation, we can write

$$L = I\omega, \qquad (9.19)$$

where $I = mr^2$. For more general objects, Eq. (9.19) still holds with the appropriate moment of inertia $I = \Sigma m_i r_i^2$ (Table 9.3). Thus, we see that angular momentum is the product of moment of inertia and angular velocity in the same way that linear momentum is the product of mass and linear velocity.

9.8 Conservation of Angular Momentum

If the torque applied to a body is zero, the change in its angular momentum with respect to time is also zero. Thus, the angular momentum is constant. This situation is analogous to the conservation of linear momentum. That is, if

$$\tau = \frac{\Delta L}{\Delta t} = 0,$$

then

$$\Delta L = 0$$

and

$$L = I_i\omega_i = I_f\omega_f = \text{constant},$$

where the subscripts i and f stand for initial and final values, respectively.

When the net applied torque on an object is zero, its angular momentum is conserved. This statement of the law of **conservation of angular**

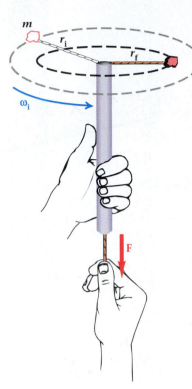

Figure 9.24
Example 9.12: A person whirls a stone around in a horizontal plane.

momentum, like that of the conservation of linear momentum, expresses an extremely important physical principle. It applies not only to large-scale phenomena, but also to atomic and nuclear phenomena. It applies to objects moving in a curved path or simply spinning about an axis.

If an object rotates with a large angular momentum, its axis of rotation remains relatively stationary in space unless a large torque is applied transverse to the initial rotation axis. The upright stability of a spinning top is a consequence of the fact that the torque due to gravity about the support point produces a changing angular momentum at right angles to the initial angular momentum. The vector sum of these two momenta causes the vertical axis of the top to move in a nearly circular path. As the top slows down, its angular momentum becomes smaller and the torque causes it to fall over.

Example 9.12

A stone attached to a string is whirled around in a horizontal circle (Fig. 9.24). If the stone is originally moving at a rate of 0.5 rad/s, what will its rate of revolution (i.e., its angular velocity) become if the radius of the circle is halved?

Strategy When the string is pulled in, the force acts through the axis of rotation. Thus, there is no torque applied to the stone and the angular momentum is unchanged.

Solution The equation for conservation of angular momentum, with subscripts i and f standing for initial and final values, respectively, becomes

$$mr_i^2 \omega_i = mr_f^2 \omega_f,$$

or

$$\omega_f = \frac{r_i^2}{r_f^2} \omega_i.$$

Substituting the initial angular velocity of 0.5 rad/s and the final radius $\frac{1}{2}r_i$ into the equation gives

$$\omega_f = \left(\frac{r_i}{0.5r_i}\right)^2 (0.5 \text{ rad/s})$$

$$\omega_f = 2 \text{ rad/s}.$$

Note that when angular momentum is constant, decreasing the radius increases the angular velocity.

In many situations, an object's moment of inertia I may change. This is especially true in some sports activities. For example, in the case of a high diver, the only external force is due to gravity, which acts on the diver's center of mass. Thus, there is no external torque acting on the diver and angular momentum is conserved. Since the product $I\omega$ remains constant, if I is made smaller, ω must increase. From the definition of I as $I = \Sigma m_i r_i^2$ we can see that the moment of inertia decreases when an object's mass is brought closer to the axis of rotation (smaller r). The diver does this by "tucking in" (Fig. 9.25). The diver's change in moment of inertia is accompanied by a corresponding increase in angular speed, allowing him to flip rapidly. The diver comes out of the tucked position to slow down his angular speed and enter the water vertically.

The motion of the diver combines rotational motion with translational motion. Careful observation shows that the diver's center of mass travels in a parabolic arc. This is the same path it would follow if it were a point mass. This is so because the net force (due to gravity in this case) acting on the diver can be considered to act on his center of mass. The motion of the diver about the center of mass is further complicated because location of the body's center of mass is affected by the positions of the limbs. This effect can be seen in Fig. 9.25(b), where the center of mass is represented by the dot.

Master the Concept

Conservation of Angular Momentum

Question: A diver goes off the high board and tucks into a flip. She then straightens out and enters the water vertically. What happened to her angular momentum just before entering the water?

Answer: There are no external torques acting on the diver after she leaves the board, so her angular momentum must be constant. When she straightens out, her moment of inertia is so much greater than when she was tucked that for a brief moment just before striking the water she appears to rotate no longer. However, careful attention to her motion reveals that she does not stop rotating, only that her angular velocity is very small. Her angular momentum is still the same as it was during the tuck.

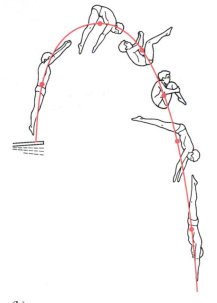

(a)

(b)

Figure 9.25

Motion of a diver. (a) In the tucked position, a diver has a smaller moment of inertia and thus a greater angular velocity than at any other point in the dive. (b) The center of mass of the diver follows a parabolic path, even though the location of the center of mass relative to the body changes as the positions of the arms and legs change.

Example 9.13

An ice skater starts spinning at a rate of 1.5 rev/s with arms extended. He then pulls his arms in close to his body, resulting in a decrease of his moment of inertia to three-quarters of the initial value (Fig. 9.26). What is the skater's final angular velocity?

Solution The skater's inward arm motion produces no external torque. Therefore, if we use the subscript i to denote initial values and f to denote final values, we have, from conservation of angular momentum,

$$I_i \omega_i = I_f \omega_f,$$

or

$$\omega_f = \frac{I_i}{I_f} \omega_i.$$

(a)

(b)

Figure 9.26

Example 9.13: An ice skater changes his angular velocity by changing his moment of inertia.

We are told that $I_f = \frac{3}{4}I_i$ and that the skater's initial frequency is $f = 1.5$ rev/s. Since the angular velocity is $\omega = 2\pi f$, we have

$$\omega_f = \frac{I_i}{\frac{3}{4}I_i\omega_i} = \frac{4}{3}2\pi f_i = \frac{4}{3}(2\pi \text{ rad/rev})(1.5 \text{ rev/s})$$
$$\omega_f = 4\pi \text{ rad/s},$$

which corresponds to a rotational rate of two rotations per second.

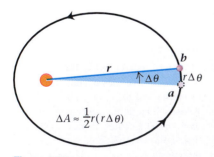

$$\Delta A \approx \frac{1}{2}r(r\Delta\theta)$$

Figure 9.27

Example 9.14: A portion of the orbit of a planet moving about the sun. During a time Δt the planet moves, making an angle $\Delta\theta$ with the sun.

Example 9.14

Show that Kepler's law of equal areas is equivalent to the law of conservation of angular momentum. (This result was, in fact, obtained by Newton in the early part of the *Principia*.)

Solution Figure 9.27 shows a planet of mass m in an orbit about the sun. We do not need to know anything about the force on this planet other than that it is directed along the line joining the planet and the sun, as required by the law of universal gravitation. As the planet moves from point a to point b in a time Δt, it sweeps out an area ΔA, which is approximately triangular. The area A of a triangle is given by

$$A = \tfrac{1}{2}(\text{base})(\text{altitude}).$$

For the triangle of Fig. 9.27 we have

$$\Delta A \approx \tfrac{1}{2}(r)(r\Delta\theta),$$

or

$$\Delta A \approx \tfrac{1}{2}r^2\Delta\theta.$$

The rate at which the area is swept out, $\Delta A/\Delta t$, is then given by

$$\frac{\Delta A}{\Delta t} \approx \tfrac{1}{2}r^2\frac{\Delta\theta}{\Delta t},$$

which becomes an equality in the limit of small Δt. In this limit, the quantity $\Delta\theta/\Delta t$ becomes the angular velocity ω. Thus

$$\frac{\Delta A}{\Delta t} = \tfrac{1}{2}r^2\omega,$$

which may be rewritten in the form

$$\frac{\Delta A}{\Delta t} = \frac{mr^2\omega}{2m}.$$

Since $mr^2\omega = L$, the angular momentum, we have

$$\frac{\Delta A}{\Delta t} = \frac{L}{2m}.$$

The rate at which area is swept out is constant because the angular momentum L is constant.

Discussion The dominant force on each planet is the gravitational attraction of the sun, which is a radially directed force that produces no torque. Forces between the planets are very much smaller. Thus, the torque on the planets in their orbits is essentially zero, and the angular momentum L of any planet is constant. Our result predicts that for a given planet, $\Delta A/\Delta t$ is a constant: that is, the rate at which the planet sweeps out an area is constant. We have just derived Kepler's second law (equal areas in equal time) from the law of conservation of angular momentum.

An additional observation may be made. The conservation of angular momentum means not only that the magnitude of the angular momentum is constant, but also that its direction in space is constant. The constancy of direction leads to the conclusion that the plane of a planet's orbit is constant.

9.9 Rotational Kinetic Energy

An important part of a moving object's total kinetic energy is its energy due to rotation. Sometimes an object's rotational kinetic energy even exceeds its translational kinetic energy. In order to present a more complete picture of conservation of energy, we discuss rotational kinetic energy here.

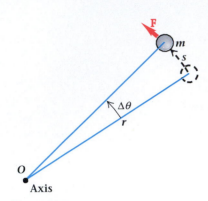

Figure 9.28

If a mass moving at a set distance r about a fixed axis is subjected to a force parallel to the direction of its motion, an increase in its kinetic energy results. This rotational kinetic energy is $\frac{1}{2}mr^2\omega^2$.

X - linear position   θ - angular position
v - linear velocity   ω - angular velocity
a - linear acceleration   α - ang. acceleration

F - force   τ - torque
m - mass   I - mom. of inertia
p - linear mom.   L - angular momentum
KE - lin.   KE - rot.

$\dfrac{\Delta\theta}{\Delta t} = \omega$

$\dfrac{\Delta\omega}{\Delta t} = \alpha$

$\theta = \theta_0 + \omega_0 t + \frac{1}{2}\alpha t^2$

$\omega = \omega_0 + \alpha t$

$\omega^2 - \omega_0^2 = 2\alpha(\theta - \theta_0)$

When an applied torque sets an object in rotation, it does work on the object. For example, you do work when you start a top or gyroscope spinning. The rotating body has kinetic energy due to its rotary motion, and this kinetic energy equals the work done in causing the rotation. How do we write an expression for the kinetic energy of rotation? You would be correct if you used the analogies we have developed in this chapter and guessed that the **rotational kinetic energy** is

$$\boxed{KE_{rot} = \tfrac{1}{2}I\omega^2,} \tag{9.20}$$

where I is the moment of inertia and ω is the angular velocity. Even though it is easy to guess correctly in this case, we will briefly present a discussion to support this definition.

We start with Eq. (9.9)—the kinematic expression for the relationship between angular speed, acceleration, and displacement. Multiply the equation by $\frac{1}{2}I$, where I is the moment of inertia. The result is

$$\tfrac{1}{2}I\omega_2^2 - \tfrac{1}{2}I\omega_1^2 = I\alpha\theta,$$

where we set the initial angular displacement θ_0 to zero. Recall that $I\alpha$ is the torque τ and that the angle $\theta = s/r$ (Fig. 9.28). Rearranging to put the torque on the left-hand side, we have

$$\tau\frac{s}{r} = \tfrac{1}{2}I\omega_2^2 - \tfrac{1}{2}I\omega_1^2.$$

But the torque divided by r is the force, and force times the distance through which it acts is work. So we have

$$W = \tfrac{1}{2}I\omega_2^2 - \tfrac{1}{2}I\omega_1^2. \tag{9.21}$$

This expression has the form of the work-energy theorem introduced in Chapter 6, except that now we have only rotational motion. Just as we did for translational motion, we identify the terms on the right-hand side as the rotational kinetic energies of a body with moment of inertia I.

Example 9.15

How much energy is required to set a 12-in. phonograph record into rotation at $33\frac{1}{3}$ revolutions per minute? The record has a mass of 0.115 kg.

Solution From Table 9.3, the moment of inertia of a disk of mass m and radius R rotating about an axis through its center is $I = \frac{1}{2}mR^2$. Substituting this value for the moment of inertia I into the equation for rotational kinetic energy, we get

$$KE_{rot} = \tfrac{1}{2}I\omega^2 = \tfrac{1}{2}(\tfrac{1}{2}mR^2)\omega^2 = \tfrac{1}{4}mR^2\omega^2.$$

We now insert the numerical values, noting that

$$R = 6 \text{ in.} = (6 \text{ in.})(0.0254 \text{ m/in.}) = 0.152 \text{ m}$$

and

$$\omega = 2\pi f = 2\pi(33\tfrac{1}{3}/\text{min})(1 \text{ min}/60 \text{ s}) = 3.49 \text{ rad/s}.$$

Thus

$$KE_{rot} = \tfrac{1}{4}(0.115 \text{ kg})(0.152 \text{ m})^2(3.49 \text{ rad/s})^2$$
$$KE_{rot} = 8.09 \times 10^{-3} \text{ J}.$$

9.10 Conservation of Energy: Translations and Rotations

In the previous section, we introduced rotational kinetic energy, but for the sake of keeping the mathematics simple, we did not discuss any examples that included both translational and rotational kinetic energy. Now we extend the law of conservation of mechanical energy to include, at the same time, translational and rotational kinetic energy. Both considerations need to be taken into account when dealing with such things as rolling tires and hoops. They are even important when describing the motion of molecules in a gas.

Figure 9.29 shows a disk of mass m and radius r at the top of an inclined plane. The axis of the disk is parallel to the top edge of the plane so that the disk, when released, rolls straight down the plane. If the frictional force is great enough, there is no sliding and the disk rolls without slipping. The thickness of the disk is not important here, except that for a given material and radius, the total mass of the disk depends upon its thickness.

At the top of the plane the disk has a potential energy mgh relative to its position at the bottom. Here h is the vertical distance through which the center of mass moves from the top to the bottom of the plane. If the disk rolls down to the bottom of the plane without slipping, then all of the initial potential energy is completely transformed into kinetic energies of rotation and translation at the bottom. Because the disk rolls without slipping, we can neglect energy loss due to friction. Then we may extend the idea of conservation of mechanical energy to include rotational as well as translational kinetic energy. Consequently,

$$\Delta PE + \Delta KE_{trans} + \Delta KE_{rot} = 0,$$

or

$$-\Delta PE = \Delta KE_{trans} + \Delta KE_{rot}.$$

Here KE_{trans} is the kinetic energy due to translation of the center of mass of the disk and KE_{rot} is the kinetic energy of rotation about its center of mass. The sum of these two terms is the total kinetic energy of the rolling disk:

$$KE_{tot} = KE_{trans} + KE_{rot} = \tfrac{1}{2}mv^2 + \tfrac{1}{2}I\omega^2. \tag{9.22}$$

We see that the magnitude of the decrease in potential energy is equal to the gain in total kinetic energy, or

$$mgh = \tfrac{1}{2}mv^2 + \tfrac{1}{2}I\omega^2, \tag{9.23}$$

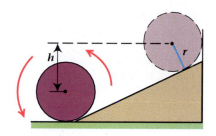

Figure 9.29
A disk rolling down an inclined plane. Its potential energy at the top is transformed into translational and rotational kinetic energy at the bottom.

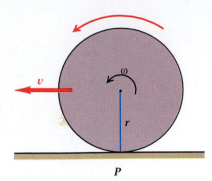

Figure 9.30
A disk rolling with speed v rotates with an angular velocity $\omega = v/r$.

where v and ω are the linear and angular speeds of the disk when it reaches the bottom of the plane and I is the moment of inertia of the disk about its center of mass. Table 9.3 lists moments of inertia for several objects of different shapes. Equation (9.23) is valid for any round object, whether it be a disk, a hoop, or a wheel. However, you must use the correct moment of inertia for the case at hand.

Because the disk rolls without slipping, there is a direct relationship between its linear and angular motions. If the disk is rolling with a speed v, then the instantaneous motion of the center of mass about the point of contact P (Fig. 9.30) is a rotation with angular velocity $\omega = v/r$. Since the center of mass is moving steadily in a straight line, a point on the rim must be in rotation about it at the same angular velocity. Thus, the angular velocity and the linear speed are related through $v = r\omega$. Then we can express Eq. (9.23) solely in terms of either v or ω. Let's choose to eliminate ω, obtaining

$$mgh = \tfrac{1}{2}mv^2 + \tfrac{1}{2}I\frac{v^2}{r^2}.$$

We can determine the speed of the disk at the bottom of the incline by solving this equation for v. The result is that after the body has rolled through a vertical height h, its speed is

$$v = \sqrt{\frac{2gh}{1 + \dfrac{I}{mr^2}}}. \tag{9.24}$$

We see that the body's speed at the bottom of the plane depends on the height of the plane and on the moment of inertia of the disk or other round object. Since the moment of inertia always depends linearly on the mass, the term I/mr^2 is independent of m and depends only on the geometry of the body. The following example illustrates a consequence of that point.

Example 9.16

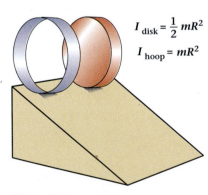

Figure 9.31
Example 9.16: A solid disk and a hoop of the same mass and radius on an inclined plane. Which one rolls to the bottom faster?

$$I_{\text{disk}} = \tfrac{1}{2}mR^2$$
$$I_{\text{hoop}} = mR^2$$

A uniform solid disk of radius R and mass m and a hoop of the same radius and mass are released from rest from the top of an incline (Fig. 9.31). Which object is moving more rapidly at the bottom?

Solution From Table 9.3 we find the moment of inertia of the disk to be

$$I_{\text{disk}} = \tfrac{1}{2}mR^2.$$

The moment of inertia of the hoop is

$$I_{\text{hoop}} = mR^2.$$

Inserting these values for the moment of inertia into Eq. (9.24), we get

$$v_{\text{disk}} = \sqrt{\frac{2gh}{1 + \tfrac{1}{2}}} = \sqrt{\frac{4gh}{3}}$$

and

$$v_{\text{hoop}} = \sqrt{\frac{2gh}{1 + 1}} = \sqrt{gh}.$$

Discussion The disk, because of its smaller moment of inertia, has a greater speed than the hoop at any point along the inclined plane, including the bottom. Consequently, if they are released simultaneously, the disk reaches the bottom first. Notice that this result is independent of the mass of each object.

Example 9.17

The Atwood's machine of Fig. 9.32 consists of a 0.400-kg pulley, having a diameter of 6.0 cm, and masses $m_1 = 1.20$ kg and $m_2 = 1.00$ kg. When released from rest, what is the speed of the mass m_1 after it has fallen a distance of 1.25 m? Assume that friction may be neglected.

Strategy We can solve this problem by use of the law of conservation of mechanical energy. This problem differs from those done earlier, because we now include the effects of the pulley's inertia.

Solution If we call the initial potential energy of masses m_1 and m_2 zero, then for the system at rest with zero kinetic energy, the total mechanical energy is 0. After the mass m_1 falls a distance of $h = 1.25$ m, the potential energy of the two masses is

$$\text{PE} = -m_1gh + m_2gh.$$

The kinetic energy of the masses and the pulley is

$$\text{KE} = \tfrac{1}{2}m_1v^2 + \tfrac{1}{2}m_2v^2 + \tfrac{1}{2}I\omega^2,$$

where I is the moment of inertia of the pulley, $I = \tfrac{1}{2}mr^2$, r is the pulley's radius, and m is its mass. Notice that both hanging masses move with the same speed v and that the angular velocity of the pulley is related to v through $\omega = v/r$. Thus, the total mechanical energy is

$$E = \text{PE} + \text{KE} = -m_1gh + m_2gh + \tfrac{1}{2}m_1v^2 + \tfrac{1}{2}m_2v^2 + \tfrac{1}{2}I\omega^2 = 0.$$

If we substitute for I and ω, we get

$$\tfrac{1}{2}m_1v^2 + \tfrac{1}{2}m_2v^2 + \tfrac{1}{2}(\tfrac{1}{2}mr^2)\left(\frac{v}{r}\right)^2 = m_1gh - m_2gh,$$

$$\tfrac{1}{2}(m_1 + m_2 + \tfrac{1}{2}m)v^2 = (m_1 - m_2)gh.$$

Solving for v, we get

$$v = \sqrt{\frac{2(m_1 - m_2)gh}{(m_1 + m_2 + \tfrac{1}{2}m)}}.$$

Substituting the numerical values and units, we find

$$v = \sqrt{\frac{[2(1.20 - 1.00)\text{kg}](9.81 \text{ m/s}^2)(1.25 \text{ m})}{[1.20 + 1.00 + \tfrac{1}{2}(0.400)]\text{kg}}} = 1.43 \text{ m/s.}$$

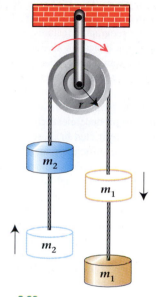

Figure 9.32
Example 9.17: Two masses on an Atwood's machine with a massive pulley.

THE EARTH, THE MOON, AND THE TIDES

The periodic rise and fall of the ocean on the beach—the tides—are familiar to everyone who has spent time at the seashore (Fig. B9.4). In the open ocean the tides are approximately a half meter high. As the tides approach the shore, the geographic features of the shoreline often channel the water so that typical shore tides are about two meters. These tides vary from place to place. In some areas they are smaller, while in a few locations they are much greater. In Canada's Bay of Fundy, the tidal level varies by as much as 15 m.

People have often dreamed of harnessing the motion of the tides to produce electricity. However, the possibility of doing so is restricted to those few places where the tidal variations are sufficiently large and where a dam can be constructed across the channel. At the present time, the expense of building such facilities has rendered them impractical in comparison with other means of generating electric power.

The tides are primarily caused by the gravitational pull of the moon.* In addition to the ocean tides, the moon also causes tides in the solid body of the earth, but these earth tides are harder to observe. As the moon moves in its orbit, the earth also moves, because each moves about the center of mass of the earth-moon system. Due to the inverse-square nature of the gravitational force, the water on the side of the earth near the moon is pulled toward the moon with a greater-than-average force, while the water on the far side is pulled with a less-than-average force. Moreover, the motion of the earth about the center of mass also helps raise a tidal bulge on the side away from the moon. As a result, two bulges appear in the water, on opposite sides of the earth.

Because the rotation of the earth about its axis is faster than the motion of the moon about the earth, and because of the frictional forces between the ocean currents and the sea floor, the earth drags the tidal bulges ahead of the position they would otherwise have (Fig. B9.5). This asymmetrical position of the bulges relative to the line joining the centers of the earth and the moon produces a net torque on the moon. This torque acts to increase the moon's angular momentum. By Newton's third law, a torque of equal magnitude acts to slow the rotation of the earth.

Although the total angular momentum of the earth-moon system is conserved, angular momentum is transferred from the earth to the moon. The total mechanical energy decreases as a result of the frictional losses of the tides. Consequently, the length of the day steadily increases as the earth's rate of rotation slows, and the length of the month decreases as the

---

*The sun also produces a tidal effect, but it is less than half that of the moon.

Figure B9.4   Ocean tides. (a) Low tide and (b) high tide.

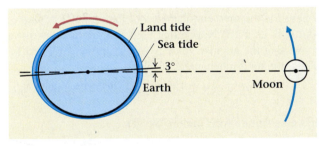

Figure B9.5   Tidal bulges occur 3° ahead of the line between the centers of the earth and the moon because of the earth's rotation. The view is from above the polar axis. (The sizes of the tides are greatly exaggerated for the sake of clarity.)

moon speeds up. Because of this increase in speed, and therefore energy, the distance of the moon from the earth also increases. These effects have been measured; the length of the day is gradually increasing at a rate of about 20 μs per year. (Thus, 200 million years ago in the Jurassic period, the length of a day was approximately 23 hours.) In addition, the moon is slowly moving away at approximately 3 cm per year. Calculations show that the moon will continue to move away from the earth until it reaches a distance of about 75 earth radii. Then the length of the day will equal the length of the month and the motion of the earth and moon will be synchronized. The earth will then keep the same face toward the moon, just as the moon now keeps the same face toward the earth.

Summary

Useful Concepts

■ Angles, measured in radians, are defined to be the ratio of the arc length s to the radius r:

$$\theta = \frac{s}{r}.$$

■ The instantaneous angular velocity is given by

$$\omega = \lim_{\Delta t \to 0} \frac{\Delta \theta}{\Delta t}.$$

■ The instantaneous angular acceleration is given by

$$\alpha \equiv \lim_{\Delta t \to 0} \frac{\Delta \omega}{\Delta t}.$$

■ The tangential acceleration of a point moving in a circle is related to the angular acceleration through

$$a_t = \alpha r.$$

The moving point also has a centripetal acceleration

$$a_c = \omega^2 r.$$

■ For motion with constant angular acceleration, two frequently used kinematic equations are

$$\theta = \theta_0 + \omega_0 t + \tfrac{1}{2}\alpha t^2 \quad \text{and} \quad \omega^2 = \omega_0^2 + 2\alpha(\theta - \theta_0).$$

$$\omega = \omega_0 + \alpha t$$

■ The torque about a point is

$$\tau = rF \sin \theta. \qquad F = \frac{\tau}{r}$$

Counterclockwise torques are positive; clockwise torques are negative. $\tau = fR \quad F = NF \quad \tau = \mu N R$

■ The conditions for equilibrium are $\tau = I\alpha$

1. The vector sum of all the external forces must be zero:

$$\sum_{i=1}^{N} \mathbf{F}_i = 0.$$

2. The vector sum of all the external torques must be zero:

$$\sum_{i=1}^{N} \tau_i = 0.$$

■ An object at rest, neither translating nor rotating, is in static equilibrium. An object in motion, but not accelerating linearly or rotationally, is in dynamic equilibrium.

■ A stress on an object causes a strain. The ratio of tensile stress to tensile strain is called Young's modulus:

$$Y = \frac{\text{stress}}{\text{strain}} = \frac{F/A}{\Delta L/L_0}.$$

■ For rotation about a fixed axis of a body with a constant moment of inertia, the torque is related to the angular acceleration by

$$\tau = I\alpha,$$

where I is the moment of inertia. For an extended body the moment of inertia is $I = \Sigma m_i r_i^2$.

■ A torque produces a change in an object's angular momentum L,

$$a = v^2/2ad$$
$$\tau = \frac{I\alpha}{r}$$
$$\tau = \frac{\Delta L}{\Delta t}. \qquad \tau = rT = I\alpha$$
$$mg - ma = T \qquad \alpha = \frac{a}{r}$$

■ The angular momentum of a point mass m moving in a circle of radius r with angular velocity ω is

$$L = mr^2\omega.$$

■ For an extended body, the angular momentum is

$$L = I\omega,$$

where I is the moment of inertia.

■ This chapter introduces a fundamental conservation law of physics, *conservation of angular momentum:* The total angular momentum of a system is constant whenever the net external torque on the system is zero.

■ The rotational kinetic energy of an object with moment of inertia I rotating with angular velocity ω is

$$KE_{rot} \equiv \tfrac{1}{2}I\omega^2.$$

■ The total kinetic energy of a rolling body separates into the translational kinetic energy of the center of mass and rotational kinetic energy about the center of mass,

$$\Delta KE_{tot} = KE_{trans} + KE_{rot} = \tfrac{1}{2}mv^2 + \tfrac{1}{2}I\omega^2.$$

Important Terms

You should be able to write the definition or meaning of each of the following:

rigid body	tensile strain
average angular acceleration	compressive strain
instantaneous angular	Young's modulus
acceleration	moment of inertia
torque	angular momentum
couple	conservation of angular
rotational equilibrium	momentum
center of mass	rotational kinetic energy
tensile stress	

Conceptual Questions

9.1 Carefully distinguish between force and torque. Give examples of forces without torques and forces that produce torques.

9.2 Explain how a yo-yo works.

9.3 Why do car owners go to the trouble to balance automobile tires? What happens when car wheels are unbalanced? Why is it better to balance the wheel on a rotating machine rather than by a static method?

9.4 Can a diver pull into a tuck and rotate while diving if he leaves the diving board with no angular velocity? Why?

9.5 What would happen to the planets if the gravitational force had a tangential component as well as a radial component?

9.6 A cat held upside down and dropped can right itself before it hits the floor (Fig. 9.33). Explain how the cat does so, since there are no external torques present. (*Hint:* Study the figure.)

9.7 Explain how ice skaters can quickly go from a slow to a fast spin and vice versa. What happens to their angular velocity and their moment of inertia? Is an external torque required?

9.8 How does a balancing pole help a tight-rope walker?

9.9 A passenger in the gondola of a hot-air balloon carries a motor-driven flywheel with its axis of rotation perpendicular to the earth's surface. Before lifting off, the wheel is at rest with respect to the gondola. After lifting off, the wheel is set into motion by an electric motor. A few minutes later the passenger inverts the wheel, then the wheel is allowed to come to rest. What would a person on the ground observe while all this was happening?

9.10 A ball rolls across the floor. Is it possible for its translational and rotational kinetic energies to be the same?

9.11 The large wheels in Fig. 9.34 have the same radius and mass, and they turn without friction. Initially θ_1 and θ_2 are equal, and the weight W, which is suspended by cords wrapped around the wheels, is allowed to fall from rest. Which angle will decrease more rapidly, θ_1 or θ_2?

9.12 An engineer desires to store energy in a rotating flywheel of a given mass and radius. Should he select a flywheel in the shape of a uniform solid disk or one with most of the mass on the rim? What is the ratio of the energies that can be stored in these two cases if both wheels rotate with the same angular frequency ω?

9.13 Two spheres of equal mass are released from rest at the top of an inclined plane. One sphere is solid and of uniform density. The other sphere is a shell of uniform density.
(a) Which sphere reaches the bottom of the plane first?
(b) Which sphere will have the greatest translational kinetic energy at the bottom?

9.14 How do the performers turn the Wheel of Death (Fig. 9.35)? When they are on the wheel, they are part of the rotating system. There is no motor to apply a torque to the axle.

9.15 An advertisement in a golfing magazine claims a "putter with the highest moment of inertia." How do you interpret such a claim and why is it an advantage?

Figure 9.33
Question 9.6.

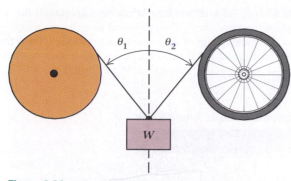

Figure 9.34
Question 9.11.

Figure 9.35
Question 9.14: The Wheel of Death.

9.16 The rolling motion of ocean liners is sometimes reduced by the use of a large flywheel within the ship. Explain how a flywheel spinning on a horizontal axis perpendicular to the length of the ship helps reduce the motion of waves striking the ship broadside.

9.17 A handbook says that to make a delicate balance you should (1) make the arms of the balance beam long, (2) make the beam as light as possible, (3) bring the center of mass of the beam very close under the point of support. What does each of these procedures do to make a balance delicate?

9.18 When the supporting stick *S* is jerked out from the apparatus shown in Fig. 9.36, the board falls down about the hinged end *H*. The ball *B* is caught by the cup *C*. Explain how the cup *C* can reach the ground before the ball, even though the ball is in free fall.

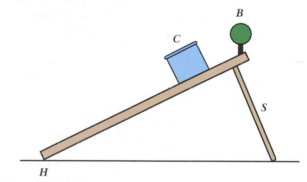

Figure 9.36
Question 9.18.

Problems

Sections 9.1 and 9.2 Angular Velocity and Angular Acceleration and Rotational Kinematics

Hints for Solving Problems

Remember the correspondence between linear quantities and rotational quantities. One revolution corresponds to 2π radians.

9.1 What is the average angular velocity of each of the three hands of a clock?

9.2 What are the initial and final speeds of the car in Example 9.2?

9.3 (a) A dentist's drill turns a 0.75-mm-diameter bit at 5.0×10^5 rev/min. What is the linear speed of the cutting edge? (b) When the drill is first turned on it takes about 1/4 s to come up to full speed. What is the average angular acceleration of the bit?

9.4 A wheel starts from rest and rotates about its axis with constant angular acceleration. After 6.8 s have elapsed, it has rotated through an angle of 25 radians. (a) What is the angular acceleration of the wheel? (b) What is the angular velocity when the time $t = 6.8$ s? (c) What is the centripetal acceleration of a point on the wheel a distance $r = 0.45$ m from the axis at $t = 6.8$ s?

9.5 A pneumatic high-speed cutter with a 7.50-cm-diameter cutting disc is advertised to have a rotation rate of between 5000 and 18,000 rev/min. (a) What is the range of angular speeds in radians per second? (b) What is the range of linear speeds of the edge of the disk? (c) What is the average angular acceleration if, starting from rest, the cutter comes up to its fastest speed in 2.8 s?

9.6• A wheel initially rotating at an angular speed of 1.6 rad/s turns through 36 revolutions during the time that it is subject to an angular acceleration of 0.32 rad/s². How long did the acceleration last?

9.7• The brakes are applied steadily on a car initially traveling at 86.5 km/hr. The braking gives the 15.3-cm-radius wheels an angular acceleration of −2.13 rad/s². (a) What is the average angular velocity of the wheels during the 16.9 s that the brakes are applied? (b) How far does the car go in that 16.9 s?

9.8• A screw with 20 threads per centimeter is driven 1.37 cm into its fixture in 12.8 s with a cordless electric screwdriver. What is the average angular speed of the screwdriver? Give your answer in radians per second (rad/s) and in rotations per minute (rpm).

9.9• A ball is tossed straight up with a spinning motion. The ball spins at 31 rad/s and makes 7.2 revolutions before it returns to its starting level. How high did the ball go?

9.10• A 3.60-cm-radius ball rolls down an inclined plane from rest at the top. The angular acceleration of the rolling ball about its center is 155 rad/s², and its angular speed at the bottom is 46.4 rad/s. How long is the plane?

9.11• A car engine idles at 800 rev/min. With the car in neutral, you depress the accelerator. After 1.2 s the tachometer indicates an engine speed of 3400 rev/min. (a) What are the initial and final angular velocities in rad/s? (b) What is the average angular acceleration during the 1.20-s interval? (c) How many revolutions does the engine make during that 1.20 s?

9.12•• A 0.50-m-radius wheel runs on a circular 1.75-m-radius horizontal track with a constant linear speed of 0.37 m/s. At $t = 0$, a dot painted on the rim of the wheel is in contact with a dot painted on the track. (a) What is the angular speed of the wheel about its axis? (b) What is the angular speed of the center of the wheel as it goes around the track? (c) Where is the dot on the wheel after the wheel makes one complete trip around the track? (d) Will the dots ever be in contact again? (e) If the dots will coincide again, how long will it take?

Section 9.3 Torque

9.13 If a force of 4.0 N is needed to open a 81-cm-wide door when applied at the edge opposite the hinges, what force must be applied to open the door if you push against the door 10 cm from the hinged side?

9.14 If a person can apply a maximum force of 50 lb, what is the minimum length of a wrench needed to apply a 35-ft-lb torque to the bolts on a motorcycle engine?

9.15 A force of 303 N is exerted on the end of a wrench in order to apply a torque of 43.2 N · m to a bolt head. The point of application of the force is 21.5 cm from the center of the bolt. What angle does the force make with respect to the wrench handle?

9.16 A clock has a second hand whose tip rubs against the inside of the glass cover. If the frictional force between the glass cover and the tip of the second hand is 0.0020 N and the length of the hand is 8.0 cm, what is the minimum torque that must be applied to the second hand if the clock is not to be stopped?

Section 9.4 Static Equilibrium

Hints for Solving Problems

Be sure to start with a diagram of the situation. Include the coordinate axes and indicate the direction of positive torques. Choose counterclockwise torques to be positive and clockwise torques to be negative.

9.17 A claw hammer is used to pry up a nail. Approximately how much force is applied to the nail when a force of 100 N is applied to the handle, as shown in Fig. 9.37?

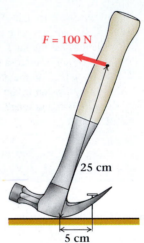

$F = 100$ N

25 cm

5 cm

Figure 9.37
Problem 9.17.

9.18 Two children are playing on a balanced seesaw. One child with a mass of 42 kg sits 1.4 m from the center. Where on the other side must the second child sit if her mass is 34 kg?

9.19 A person with upper arm vertical and forearm horizontal holds a 4.5-kg iron cannon ball (Fig. 9.38). Assume the mass of the forearm and hand is 1.5 kg, with a center of mass 15 cm from the elbow. The center of the cannonball is 32 cm from the elbow, and the force of the biceps is applied 5.0 cm from the elbow. (a) What force is exerted on the forearm by the biceps muscle? (b) What force is exerted on the upper arm at the elbow contact? Assume that all horizontal components are negligible.

9.20• A trucker needs to weigh a truck that is too long to fit on a platform scale. When the front wheels of the truck are run onto the scale, the scale reads W_1. When the rear wheels

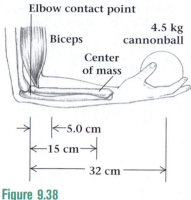

Figure 9.38
Problem 9.19.

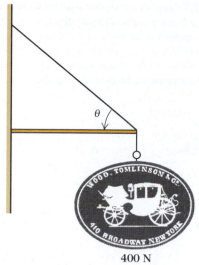

400 N

Figure 9.40
Problems 9.23 and 9.24.

are run onto the scale so that the front wheels are off, it reads W_2. (a) Prove that the total weight of the truck is $W_1 + W_2$. (b) Prove that if the truck is loaded so that its center of gravity is halfway between the front and rear wheels, the total weight is $2W_1$.

9.21● A bar balances 30.0 cm from one end. When a 0.75-kg mass is hung from that end, the balance point moves 8.0 cm toward that end. What is the mass of the bar?

9.22● A 4.0-m-long iron bar of uniform cross section is held perpendicular to the wall by a wire from the end of the bar to the wall (Fig. 9.39). What is the tension in the wire if the iron bar weighs 400 N and the wire is 5.0 m long?

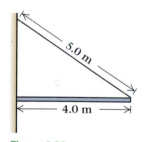

Figure 9.39
Problem 9.22.

9.23● A sign weighing 400 N is suspended at the end of a uniform rod 4.00 m long weighing 500 N (Fig. 9.40). (a) What is the tension in the support cable if it makes an angle $\theta = 40°$ with the rod? (b) What would be the tension if the cable were attached higher on the wall so that $\theta = 55°$?

9.24● Suppose the sign in Fig. 9.40 is suspended from the center of the boom rather than at the end. The cable is still connected to the end of the boom. What would be the tension in the cable if it made an angle $\theta = 40°$ with the bar?

9.25● A 75-kg Marine does pushups as shown in Fig. 9.41. What are the forces on his hands and feet?

9.26●● A bicycle carrier attached to a car by two horizontal straps carries a 11.3-kg bicycle (Fig. 9.42). What is the tension in each of the two straps marked S? Ignore the mass of the carrier.

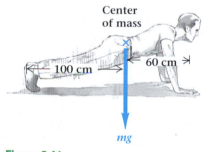

Figure 9.41
Problem 9.25.

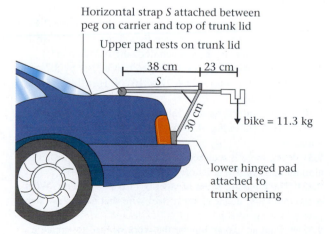

The distance between the center of the upper pad and the center of the lower pad is 31 cm.

Figure 9.42
Problem 9.26: A bicycle carrier mounted on a car.

9.27•• Figure 9.43(a) shows a person's outstretched arm holding a 5.0-kg dumbbell. The deltoid muscle is attached so that the force is applied at an angle of 17° from the horizontal at a point halfway between the shoulder joint and the center of mass of the arm (Fig. 9.43b). If the mass of the arm is 3.5 kg, what must be the tension in the deltoid muscle?

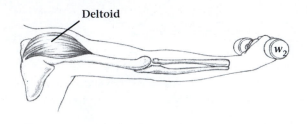

Deltoid

(a)

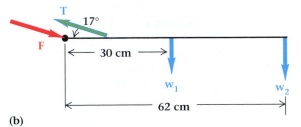

(b)

Figure 9.43
Problem 9.27.

*Section 9.5 Elasticity: Stress and Strain

Hints for Solving Problems

Remember that stress has units of N/m²; strain is dimensionless.

9.28 A 1.7-m long brass bar has a square cross section 2.4 cm on an edge. The bar is compressed by a force of 5.0×10^3 N applied to its ends. By how much does the bar shorten?

9.29 A steel cable, 1.27 cm in diameter, stretches by 8.7 mm when subjected to a force of 6.27×10^4 N. How long is the unstretched cable?

9.30 Use the information in Table 9.2 to plot a figure similar to the region Oa in Fig. 9.19 for aluminum, steel, nylon, and tungsten.

9.31 What force is required to compress a 2.50-cm cube of aluminum to 99.9% of its original height (Fig. 9.44)?

9.32• The breaking stress of steel is 11.0×10^8 N/m². What is the minimum diameter for a steel wire that can safely support a 70.0-kg person?

9.33•• Find an expression for the work needed to stretch a circular wire by an amount ΔL as a function of the applied force F when the wire has a Young's modulus Y.

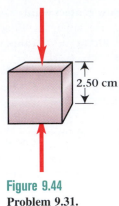

2.50 cm

Figure 9.44
Problem 9.31.

Section 9.6 Torque and Moment of Inertia

9.34 A torque of 12 N · m is applied to a heavy wheel whose moment of inertia is $I = 36$ kg · m². (a) What is the angular acceleration of the wheel? (b) If the wheel was initially at rest and the torque is applied for 10 s, what will be the rotational frequency ω of the wheel at the end of the 10 s?

9.35 A wheel whose moment of inertia is 32 kg · m² is subjected to a torque of 12 N · m. If the wheel is initially moving with an angular velocity $\omega = 6.0$ rad/s when the torque is applied, what will be its angular velocity if the torque is applied for 9.0 s?

9.36 An iron disk has a radius of 0.515 m and a mass of 307 kg. The disk is mounted on its axis so that it is free to spin. (a) What torque is required to give it an angular acceleration $\alpha = 1.00$ rad/s²? (b) If the torque is applied at the edge of the disk, how much force is required? (c) How much force is required if the force is applied at a distance 10.5 cm from the axis of rotation?

9.37 A wheel of radius R and moment of inertia I is subjected to a torque τ for a time t. (a) If the wheel was initially at rest at $t = 0$, what is its angular velocity at time t? (b) Calculate the tangential speed of the outer edge of the wheel.

9.38• A shaft extending from the side of an enclosed box is said to be connected to a flywheel inside. It is known that the mass of the flywheel is 1.37 kg and its radius is 7.50 cm. A 0.130-N · m-constant-torque motor connected directly to the shaft goes from rest to 500 rev/min in 3.0 s. Is the flywheel a uniform disk or is it some other shape? If it is some other simple shape, what might it be?

9.39• A student holding a rod by the center subjects it to a torque of 1.4 N · m about an axis perpendicular to the rod, turning it through 1.3 radians in 0.75 s. When the student holds the rod by the end and applies the same torque to the rod, through how many radians will the rod turn in 1.00 s?

9.40• A uniform 2.45-kg spinning disk with a radius of 0.243 m is brought to rest by a brake-pad pressed against the side 0.183 m from the rotation axis. The wheel is brought to rest from an angular velocity of 15.4 rad/s in a time of 24.3 s. What frictional force was applied by the brake pad?

9.41• In a student experiment, a torque of 2.0×10^{-2} N · m is applied to a rigid aluminum pipe, causing it to move about an axis through its center and perpendicular to its length with an acceleration $\alpha = 0.43$ rad/s². (a) What is the moment of inertia of the pipe? (b) The rigid pipe is made from light tubing with a mass of 0.20 kg inserted in each end. How far apart are the two masses? (Refer to Table 9.3.)

9.42•• A piece of metal is pressed against the rim of a 1.6-kg, 19-cm-diameter grinding wheel that is turning at 2400 rev/min. The metal has a coefficient of friction of 0.85 with respect to the wheel. When the motor is cut off, with how much force must you press to stop the wheel in 20 s?

Sections 9.7 and 9.8 Angular Momentum and Conservation of Angular Momentum

Hints for Solving Problems

A net torque causes a change in angular momentum with time, which is equivalent to angular acceleration times the moment of inertia. When there is no net torque, the final angular momentum equals the initial angular momentum.

9.43 A toy airplane moves at the end of a string about a fixed point in a horizontal circle of 1.00-m radius. If the linear speed is 4.86 m/s, what will the speed become if the string is pulled in to give a radius of 0.750 m? Assume there is no torque.

9.44 A 0.5-kg flashlight is swung at the end of a string in a horizontal circle of 0.80-m radius with a constant angular speed. If no torque is applied, what must the radius become if the angular speed of the flashlight is to be halved?

9.45 What is the angular momentum of the earth as it rotates about its axis? Approximate the earth by a sphere of uniform density.

9.46• A sphere with a mass of 4.37 kg and a radius of 6.29 cm is spun at 37.3 rad/s around an axis through its center. Independent measurements show that the angular momentum of the sphere is 0.186 kg · m²/s. Is the sphere of uniform density? Explain your answer.

9.47• A 12-in. record is dropped on a large 12-in. phonograph turntable that is freely turning at $33\frac{1}{3}$ rev/min. The mass of the record is 0.150 kg, and the mass of the turntable is 1.00 kg. What is the final speed of the turntable in revolutions per minute? (In this case you may add the angular momenta in a manner analogous to the way you add linear momenta.)

9.48•• A playground merry-go-round has a disk-shaped platform that rotates with negligible friction about a vertical axis. The disk has a mass of 200 kg and a radius of 1.8 m. A 36-kg child rides at the center of the merry-go-round while a playmate sets it turning at 0.25 rev/s. If the child then walks along a radius to the outer edge of the disk, how fast will the disk be turning?

Section 9.9 Rotational Kinetic Energy

Hints for Solving Problems

Refer to Table 9.3 for the moments of inertia of solid objects of assorted shapes.

9.49 What is the kinetic energy of a 0.145-kg, 12-in. phonograph record when rotated at 45 rev/min?

9.50 A solid iron cylinder has a radius of 0.250 m and a mass of 145 kg. Calculate the kinetic energy of this cylinder when it is rotating about its axis at a rate of 13.5 rev/s.

9.51• What is the approximate rotational kinetic energy of a 66-cm diameter bicycle wheel of mass 4.0 kg when the bicycle is traveling at 15 km/h?

9.52• (a) A solid 4.0-kg wheel of radius 0.23 m is initially at rest. How much work is required to make it rotate at 3.0 rev/s about its axis? (b) If the energy of the rotating wheel is doubled, how many revolutions per second will it make?

9.53•• A 400-g, 16-cm-diameter disk is suspended in a horizontal position from a helical spring as shown in Fig. 9.45. The disk is rotated through a certain number of turns, thereby storing energy in the spring. When released, the disk is observed to have an angular speed of 18 rad/s when the spring returns to its equilibrium position. The disk is removed and replaced with a 400-g, 16-cm rod attached at the center. The rod is rotated through the same number of revolutions as was the disk to wind up the spring, and then it is released. What is the angular speed of the rod as it passes through the equilibrium position?

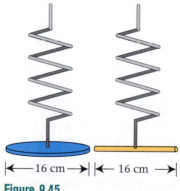

|←— 16 cm —→|←— 16 cm —→|

Figure 9.45
Problem 9.53.

Section 9.10 Conservation of Energy: Translations and Rotations

9.54 A hoop and a disk, both of 0.50-m radius and 2.0-kg mass, are released from the top of an inclined plane 3.0 m high and 8.0 m long. What is the speed of each when it reaches the bottom? Assume that they both roll without slipping.

9.55 You simultaneously release a 1.0-kg hoop of 0.50-m radius and a 1.0-kg disk of 0.25-m radius from the upper end of an inclined plane of 2.5-m height. Calculate the speed of each at the bottom of the plane, assuming that they roll without slipping.

9.56 A disk released from the top of an inclined plane has a speed of 4.52 m/s at the bottom. How high is the upper end of the inclined plane? Assume that the disk rolled without slipping.

9.57● In a demonstration an iron hoop rolls without slipping down an inclined plane from a height h to the bottom. In another demonstration the plane is lubricated and the hoop slides down the plane from the same height without rolling. What is the ratio of the speeds at the bottom?

9.58● A sphere is released from rest at the top of an inclined plane. Derive an expression for the speed of the sphere at a point a distance h below its starting point. Assume that the sphere rolls without slipping.

9.59● A hoop of mass m and radius r is released from rest and rolls without slipping down a hill to a point that is a distance h lower than the starting point. Show that at this time the hoop will be rotating with an angular velocity

$$\omega = \sqrt{\frac{gh}{r^2}}.$$

9.60● A solid bowling ball with a radius of 10.9 cm and a mass of 7.0 kg rolls along a bowling alley at a linear speed of 2.0 m/s. (a) What is its translational kinetic energy and (b) what is its rotational kinetic energy?

9.61● A solid 0.558-kg disk rolls without slipping down an inclined plane that makes an angle of 30° with the horizontal direction. The disk is released from rest a distance 0.834 m from the lower end of the plane. (a) How fast is the disk moving as it reaches the end of the plane? (b) What fraction of the total kinetic energy of the disk is rotational kinetic energy?

9.62● A physics teacher stands on a freely rotating platform (Fig. 9.46). He holds a dumbbell in each hand of his outstretched arms while a student gives him a push until his angular velocity reaches 1.5 rad/s. When the freely spinning professor pulls his hands in close to his body, his angular velocity increases to 5.0 rad/s. What is the ratio of his final kinetic energy to his initial kinetic energy? How do you account for the change in energy?

9.63●● An Atwood's machine of the type shown in Fig. 9.47 has hanging masses of mass m_1 and m_2 and a disk-shaped pulley of mass m_p. Show that if $m_1 = m_p = M$ and $m_2 = 2M$, the acceleration of the hanging masses is $\frac{2}{7} g$. Use the principle of the conservation of energy to solve this problem.

9.64●● The Atwood's machine of Fig. 9.47 consists of masses $m_1 = 1.35$ kg and $m_2 = 1.15$ kg and a solid pulley with a diameter of 6.70 cm and a mass $m_p = 0.546$ kg. When released from rest, what is the speed of the heavier mass after it has fallen a distance of 1.45 m? Treat the pulley as a uniform disk.

9.65●● Use the principle of conservation of energy to find the acceleration sought in Example 9.11. First find the speed of the bucket after traveling a distance s. Then find the acceleration necessary to give this velocity.

Figure 9.46
Problem 9.62.

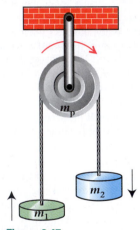

Figure 9.47
Problems 9.63 and 9.64.

9.66●● A 1.00-kg mass is attached to a string wrapped around a shaft of negligible mass and having a 6.0-cm radius. A dumbbell-shaped "flywheel" made from two 0.500-kg masses is attached to one end of the shaft and perpendicular to its axis (Fig. 9.48). The mass is released from rest and allowed

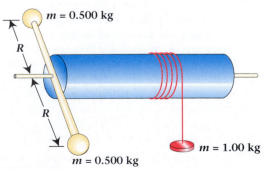

Figure 9.48
Problem 9.66.

to fall 1.00 m to the floor. It reaches a speed of 2.68 m/s just before striking the floor. How far apart are the masses of the dumbbell?

Additional Problems

9.67 An electric motor is used to lift a 20-kg bucket of water at constant speed from a well. If the diameter of the pulley is 10.0 cm, what must be the torque of the motor?

9.68● An unpowered flywheel is slowed by a constant frictional torque. At time $t = 0$ it has an angular velocity of 200 rad/s. Ten seconds later its velocity has decreased by 15%. What is its angular velocity at (a) time $t = 50$ s and (b) $t = 100$ s?

9.69● In Example 9.6, determine the mass m_3 by calculating torques about a point 0.50 m to the right of the left-hand end of the bar.

9.70● In Example 9.6, determine the mass m_3 by calculating torques about a point 0.10 m to the left of the right-hand end of the bar.

9.71● Figure 9.49 is a diagram of the jaw. Chewing is accomplished by the force of the masseter muscle closing the jaw about the fulcrum A. The distance $x_2 = 3x_1$. If the muscle exerts a force of 400 N, what force is applied by the front teeth?

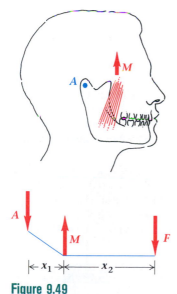

Figure 9.49
Problem 9.71.

9.72● A uniform solid disk of radius R_D and mass m and a hoop of radius R_H and mass m are simultaneously released from rest from the top of an incline. Can the ratio of the radii of the two objects be fixed so that they both reach the bottom of the plane at the same time?

9.73● Find the vertical and horizontal components of force exerted by the wall on the bar suspended as shown in Fig. 9.50. The weight of the bar is 300 N.

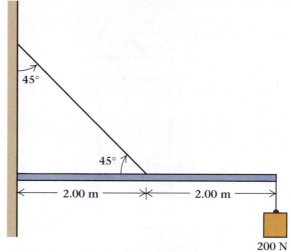

Figure 9.50
Problem 9.73.

9.74● A $\frac{1}{4}$-hp motor rotates at 1200 rpm. How much torque does it develop? (*Hint:* You may want to make an analogy with Eq. 6.19.)

9.75● A yo-yo resting on a horizontal table is free to roll (Fig. 9.51). When the string is pulled horizontally to the left, the yo-yo rolls to the left. When the string is pulled vertically, the yo-yo rolls to the right. Prove that the yo-yo will slide without rolling when the string is pulled at an angle θ given by $\sin \theta = r/R$.

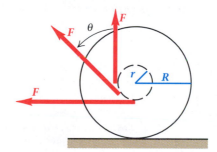

Figure 9.51
Problem 9.75.

9.76●● A cylinder of mass m and radius R is rigidly mounted to the same shaft as a lightweight cylinder of radius $r = R/2$. The shaft is free to turn with negligible friction. Cords are wound in opposite directions about the cylinders, and identical masses m are hung from the cords as shown in Fig. 9.52. (a) Which way do the cylinders turn? (b) What is the acceleration of the right-hand mass?

9.77●● A 1-m-radius flywheel is to be made from steel in the form of a solid disk. If the flywheel, when turning at 60 rev/min, is to store as much energy as a 100-watt lamp uses in 1.0 min, how thick must the flywheel be? The density of steel is 7.8 g · cm^{-3}.

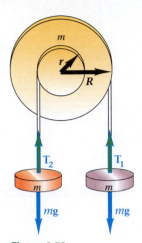

Figure 9.52
Problem 9.76.

9.78•• Two masses, one of value m and the other of $2m$, hang from light strings wrapped around a uniform solid cylinder of mass M and radius R that is free to rotate about a horizontal axis (Fig. 9.53). Find the acceleration of the masses when the cylinder is released from rest. Neglect effects of friction and the mass of the string.

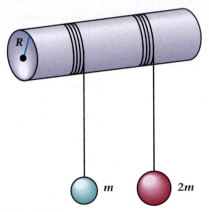

Figure 9.53
Problem 9.78.

9.79•• (a) Show that the kinetic energy of a satellite in circular orbit of radius r about the earth is one half the magnitude of its gravitational potential energy. (b) What is its total mechanical energy?

9.80•• A solid cylinder rolls without slipping on a horizontal surface. It collides with a resting hollow cylinder of the same mass and radius in a collision in which kinetic energy and momentum are conserved. If the initial speed of the rolling cylinder is 0.25 m/s, what is the final speed of the struck cylinder if it also rolls without slipping?

9.81•• When fitted over a narrow inclined plane and released, a yo-yo rolls slowly down the plane and then speeds up when

it hits the floor (Fig. 9.54). The diameter of the center rod of the yo-yo is one-fifth of the diameter of the outer disks. (a) Show that the speed at the bottom of the plane is

$$v = \sqrt{\frac{4gh}{27}},$$ where h is the vertical distance through which

the center of mass descends. (b) What is the speed on the floor? Neglect the mass of the rod that joins the disks.

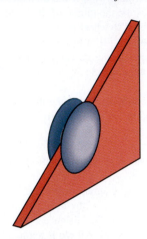

Figure 9.54
Problem 9.81.

9.82•• Find an expression for the speed with which a bowling ball must roll on a horizontal surface in order to roll up a ramp through a vertical distance equal to the ball's radius. (*Hint:* Include translational and rotational energy when in motion.)

9.83•• If the 0.70-kg block in Fig. 9.55 is released from rest, what speed will it have just before it hits the floor if there is no friction at the wheel's axis? Use conservation of energy and consider both translational and rotational kinetic energy.

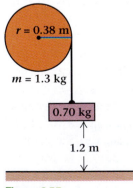

Figure 9.55
Problem 9.83.

9.84•• Assume that the axis of the wheel in Problem 9.83 is not frictionless. What is the frictional torque if the speed just before the 0.70-kg block hits the floor is half of the speed it would have without friction?

Fluids

T he study of fluids dates back to some of the earliest discoveries in physics. Many of the principles we examine in this chapter are associated with great scientists of the past, such as Archimedes (third century B.C.), Pascal (seventeenth century), Bernoulli (eighteenth century), and Stokes (nineteenth century). However, the study of fluid flow remains an active area of research. For example, the principles of fluid flow are used to minimize the aerodynamic resistance of a moving car or plane. Weather forecasters use computer simulation methods to model the fluid flow of our atmosphere, trying to understand the origin of hurricanes as well as of stable air currents like the jet stream. The occurrence of stable patterns of fluid flow from seemingly random initial conditions has become an exciting area of contemporary research in physical model building and computer science.

In this chapter, we discuss the fundamental properties of fluids at rest (the study of *hydrostatics*) and in motion (the study of *hydrodynamics*). The treatment here is somewhat

307

different from that of other chapters in that we present most results without derivation. Instead, we rely on physical intuition to show that these results are reasonable. We do this because in many cases the derivations are prohibitively long. However, these ideas are all developed from classical mechanics, primarily Newton's laws and conservation of energy.

After studying this chapter, you should be able to recognize the proper conditions for applying the various laws of fluid behavior, particularly those of fluid flow. These laws are strictly valid only for certain types of fluids and particular types of flow. They cannot give reliable results beyond their proper domain of applicability. ■

10.1 Hydrostatic Pressure

A **fluid** is any substance that cannot maintain its own shape; in other words, it is a substance that has no rigidity. It can flow and alter its shape to conform to the outlines of its container. This definition includes liquids, such as water; gases, such as air; very slowly flowing substances, such as tar and some plastics; and even some mixtures of solids and liquids that can flow, such as mud. Gases are easily compressible and have no natural volume; that is, they expand to uniformly fill the container in which they are held. Liquids, on the other hand, are practically incompressible, and a given mass of liquid has a characteristic volume. If the liquid volume is less than that of its container, the liquid will have a well-defined surface bounding its volume.

Unless otherwise stated, we deal in this chapter with fluids that cannot be compressed, that is, with liquids. However, we consider fluids to have viscosity, which is the friction that resists the motion of objects through the fluid.

Consider an upright cylinder containing a liquid (Fig. 10.1). The weight of the liquid exerts a force on the bottom of the cylinder. This force produces a pressure on the bottom of the cylinder, where the **pressure** P is defined as the mag-

Figure 10.1

A fluid exerts a pressure on the bottom of its cylindrical container equal to the total weight of the fluid divided by the area of the bottom of the container.

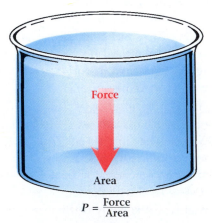

Force

Area

$$P = \frac{\text{Force}}{\text{Area}}$$

nitude of the perpendicular force acting on an area divided by that area. Thus, pressure is the *force per unit area:*

$$P \equiv \frac{\text{force}}{\text{area}}.$$ (10.1)

If we insert the weight of the liquid for the force in Eq. (10.1), we find that

$$P = \frac{\text{weight}}{\text{area}} = \frac{mg}{A},$$

where m is the mass of the liquid, A is the area of the bottom of the cylinder, and g is the acceleration of gravity. Note that pressure is a scalar quantity.

The different conditions under which pressure measurements are made have led to the development of a variety of commonly used units. The SI unit of pressure, N/m^2, is given the name **pascal** (Pa). Table 10.1 lists several other common units of pressure, along with their relationship to the pascal.

A tire gauge is a familiar pressure-measuring device (Fig. 10.2). One form consists of a hollow cylinder fitted with a spring-loaded piston. When you press the gauge against a tire valve stem, the pressurized air from the tire enters the cylinder. The air exerts a force on the piston that is equal to the pressure in the cylinder times the area of the face of the piston ($F = P \cdot A$). If the spring obeys Hooke's law, the piston is pushed a distance that is proportional to the force and therefore to the pressure. The rod on the end of the piston can be calibrated to read the pressure directly. We will discuss some other pressure-measuring devices later.

It is common to calibrate pressure gauges, such as the tire gauge, so that they indicate only the pressure in excess of atmospheric pressure. This pressure is called *gauge pressure.* As we will discuss in more detail in Chapter 12, the atmosphere exerts a pressure of about $1.013 \times 10^5 \ N/m^2$ (101.3 kPa or 14.7 lb/in.^{2}) at sea level, and this value must be added to the gauge pressure to give the total pressure.

Master the Concept

Air Pressure

Question: A driver with a flat tire measures its pressure with a tire-pressure gauge. The gauge gives a reading of zero. Does this reading indicate that there is no air in the tire?

Answer: The gauge for measuring tire pressure registers the difference between the pressure inside the tire and the atmospheric pressure outside the tire. Thus, a reading of zero only indicates that the pressure inside is equal to the pressure outside. There is still air within the tire, but the air present is at the same pressure as the outside air. There is no pressure difference to hold up the walls of the tire.

Table 10.1	Some Common Units of Pressure	
Name	**Value (N/m^2 = Pa)**	
1 pascal (Pa)	1	
1 bar	1.00×10^5	
1 atmosphere (atm)	1.01×10^5	
1 mm Hg	1.33×10^2	
1 torr	1.33×10^2	
1 lb/in.^{2} (psi)	6.89×10^3	

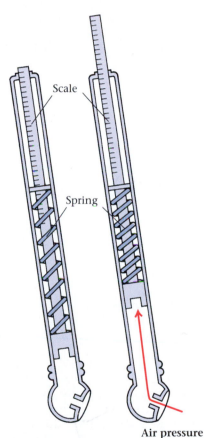

Scale

Spring

Air pressure

Figure 10.2
A tire-pressure gauge. The extension of the scale is proportional to the force on the spring (Hooke's law), which is proportional to the air pressure in the tire. When the gauge is removed from the tire, the scale, which is not attached to the spring, remains extended, held in position by friction with the sleeve that holds it.

It is convenient to use the concept of density when discussing pressure. The **density** of a substance, ρ, was defined in Chapter 5 as its mass per unit volume,

$$\rho = \frac{m}{V}, \tag{10.2}$$

The SI unit for density is kg/m^3. Table 10.2 gives the densities of several gases, liquids, and solids. A related quantity, the **specific gravity** of a substance, is defined as the ratio of its density to the density of water at 4°C, which is 1.00×10^3 kg/m^3. Thus lead, which has a density of 11.4×10^3 kg/m^3, has a specific

▼

Table 10.2	*Densities of Some Common Materials*	
Material	**Temperature (°C)**	**Density (kg/m^3)**
Gases*		
Hydrogen	0	0.09
Helium	0	0.18
Nitrogen	0	1.25
Air	0	1.29
Oxygen	0	1.43
Carbon dioxide	0	1.98
Liquids		
Gasoline	20	0.68×10^3
Methyl alcohol	20	0.791×10^3
Water	20	0.998×10^3
Sea water	20	1.03×10^3
Glycerin	20	1.26×10^3
Mercury	20	13.6×10^3
Solids		
Wood, balsa	—	$(0.12–0.20) \times 10^3$
Wood, pine	—	$(0.37–0.64) \times 10^3$
Wood, oak	—	$(0.67–0.79) \times 10^3$
Butter	—	$(0.86–0.87) \times 10^3$
Ice	—	0.92×10^3
Brick	—	$(1.4–2.2) \times 10^3$
Bone	—	$(1.7–2.0) \times 10^3$
Glass, common	—	$(2.4–2.8) \times 10^3$
Granite	—	$(2.64–2.76) \times 10^3$
Aluminum	20	2.70×10^3
Iron	20	7.87×10^3
Brass	—	$(8.41–8.86) \times 10^3$
Lead	20	11.4×10^3
Uranium	20	18.95×10^3
Gold	20	19.3×10^3

*The density given here is for a pressure P_0 of one standard atmosphere = 1.01325×10^5 Pa. The density ρ of a gas at other temperatures and pressures is given by

$$\rho = \rho_0 \left(\frac{P \times 273.15}{P_0 \times (T + 273.15)} \right),$$

where P is the pressure and T is the temperature in degrees Celsius.

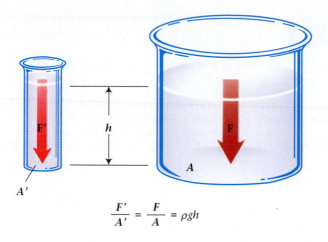

$$\frac{F'}{A'} = \frac{F}{A} = \rho g h$$

Figure 10.3
Containers of different size, filled to the same depth with identical fluids of uniform density, have equal pressure at the bottom.

gravity of 11.4. Because the specific gravity is a ratio of two quantities with the same units, it is a pure number without units or dimensions.

Equation (10.1) for pressure can be written in terms of density if we write the mass as the product of density and volume:

$$P = \frac{mg}{A} = \frac{\rho V g}{A} = \frac{\rho h A g}{A},$$

or

$$P = \rho g h, \qquad (10.3)$$

where ρ is the density and h is the depth of the liquid (Fig. 10.3).

We can now see that pressure is directly proportional to both the density and the depth of the liquid. Thus, separate containers of different size, holding identical liquids of uniform density, have equal pressure at equal depth. If the two containers are filled to the same height, they have equal pressure at the bottom, even though the total force on the bottom surface due to the liquid is greater at the bottom of the larger container (see Fig. 10.3). The pressure depends on the depth and not on the cross section. A thin upright tube filled to a height of 3 m with water has the same pressure at its bottom as does a large lake that is 3 m deep.

A force also is exerted on the sides of the container, and a corresponding pressure is present there. Moreover, a pressure exists at any point within the body of the liquid. If the liquid is at rest, this pressure is independent of direction. Thus, the pressure on the small rubber membrane in Fig. 10.4 is the same regardless of the orientation of its surface, provided that the center of the membrane is kept at a constant depth. If the pressures in opposite directions were not the same, a pressure difference would arise, resulting in an unbalanced force acting on the liquid. This force would cause the liquid to flow, which contradicts our original assumption that the liquid is at rest. So the pressure must depend only on the height of the liquid above the point in question, according to Eq. (10.3). Thus, the difference in pressure ΔP between two points that differ in depth by Δh is

$$\Delta P = \rho g \, \Delta h. \qquad (10.4)$$

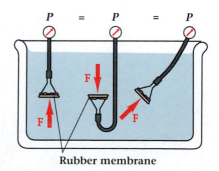

Rubber membrane

Figure 10.4
At constant depth, the pressure exerted by the fluid is the same in all directions.

Example 10.1

A tank is filled with water to a depth of 1.5 m. What is the pressure at the bottom of the tank due to the water alone?

Solution We can compute the pressure directly from Eq. (10.3). The density of water is approximately 10^3 kg/m^3, and the height h is 1.5 m. So

$$P = \rho g\,\Delta h = (10^3 \text{ kg/m}^3)(9.81 \text{ m/s}^2)(1.5 \text{ m}) = 1.5 \times 10^4 \text{ N/m}^2.$$

The combination of units N/m^2 is the pascal (Pa). Thus, the pressure at the bottom of the tank due to the water is 1.5×10^4 Pa, or 15 kPa.

Discussion Note that the pressure due to the water is completely independent of the size and shape of the tank. Note also that, if we wish to know the total pressure on the bottom of the tank, we must add atmospheric pressure to our answer here.

Example 10.2

A nurse administers medication in a saline solution to a patient by infusion into a vein in the patient's arm (Fig. 10.5). The density of the solution is 1.0×10^3 kg/m^3, and the gauge pressure inside the vein is 2.4×10^3 Pa. How high above the insertion point must the container be hung so that there is sufficient pressure to force the fluid into the patient?

Solution The container must be hung high enough that the gauge pressure due to the liquid in the tube and container is at least as great as the gauge pressure inside the vein:

$$P_{\text{liquid}} = \rho g h = 2.4 \times 10^3 \text{ Pa}.$$

Solving for the height h yields

$$h = \frac{2.4 \times 10^3 \text{ Pa}}{\rho g} = \frac{2.4 \times 10^3 \text{ Pa}}{9.81 \text{ m/s}^2 \times 1.0 \times 10^3 \text{ kg/m}^3}$$

$$h = 0.24 \text{ m} = 24 \text{ cm}.$$

To actually establish a flow through the needle, the container would need to be higher than this result.

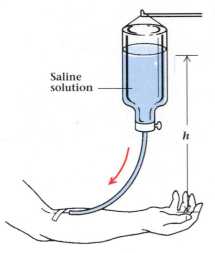

Saline solution

h

Figure 10.5
Example 10.2: A fluid is fed into a patient's arm from a suspended container. The pressure of the fluid must exceed the pressure in the arm.

10.2 Pascal's Principle

It is often desirable to know the pressure at one point in a fluid when we know it at another. The pressure at any point in the liquid in Fig. 10.6 depends on its depth below the surface h and on any additional pressure (such as atmospheric pressure, P_{atm}) exerted on the liquid above that point. If we know the pressure on the surface of the liquid (P_{atm}) and wish to determine the pressure at point B, our equation reads

$$P_B = P_{\text{atm}} + \rho g h. \tag{10.5}$$

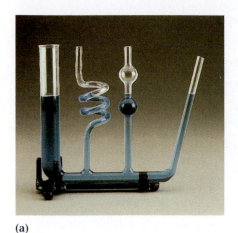

(a)

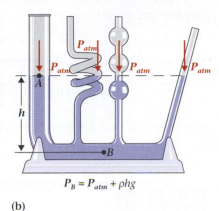

(b)

$$P_B = P_{atm} + \rho h g$$

Figure 10.6
(a) The height of the liquid surface is independent of the tube shapes showing that the pressure at any depth is the same for all tubes. (b) The pressure at point B is the pressure due to the liquid plus the atmospheric pressure P_{atm} on the surface.

If the pressure at A (that is, P_{atm}) is increased, then the pressure at B is correspondingly increased by the same amount. This fact was recognized by Blaise Pascal (1623–1662) and is embodied in the statement known as **Pascal's principle:** *The pressure applied at one point in an enclosed fluid is transmitted undiminished to every part of the fluid and to the walls of the container.* Pascal's principle holds for gases as well as for liquids, with some minor modifications due to the change in volume of a gas when the pressure is changed.

Hyperbaric medicine treats many physical problems through the application of high-pressure air or air-oxygen mixtures. Patients are enclosed in chambers pressurized at up to 6 atmospheres. By Pascal's principle, the pressure is distributed throughout the hyperbaric chamber. Patients breathing air at 6 atmospheres take in six times the amount of oxygen with each breath. The increased oxygen intake is useful in treating a variety of problems such as carbon-monoxide poisioning, slow-healing wounds, and burns.

A patient being placed in a hyperbaric chamber. The chamber provides a demonstration of Pascal's principle.

Example 10.3

A scuba diver searches for treasure at a depth of 20.0 m below the surface of the sea. At what pressure must the scuba (self-contained underwater breathing apparatus) device deliver air to the diver?

Strategy The pressure at the diver's depth is greater than atmospheric pressure because of the weight of the water above the diver. If the air breathed in is not at the same pressure as the external pressure on the diver's chest, the excess pressure will collapse the chest. Thus, the breathing apparatus must deliver air to the diver at the pressure of the surrounding water. We can calculate this pressure from Eq. (10.5).

Solution The pressure on the diver is

$$P = P_{atm} + \rho g h,$$

where P_{atm} is the pressure due to the atmosphere pressing down on the sea, ρ is the density of sea water, and h is the depth of the diver below the surface. The numerical values for P_{atm} (101.3 kPa) and ρ (approximately 1030 kg/m^3)

are found in Table 10.2. When the values for P_{atm}, ρ, and h are inserted into the equation, we get

$$P = 101.3 \times 10^3 \text{ Pa} + (1030 \text{ kg/m}^3)(9.81 \text{ m/s}^2)(20.0) = 303 \text{ kPa.}$$

When expressed in terms of atmospheres, the pressure becomes

$$P = 303 \text{ kPa} \left(\frac{1 \text{ atm}}{101.3 \text{ kPa}} \right) = 2.99 \text{ atm} \approx 3 \text{ atm.}$$

In other words, at a depth of 20.0 m, the pressure is three times the pressure at the surface.

Example 10.4

You can make a simple hydraulic lift by fitting a piston attached to a handle into a 3-cm-diameter cylinder, which is connected to a larger cylinder of 24-cm diameter (Fig. 10.7). If a 50-kg (110-lb) woman puts all her weight on the handle of the smaller piston, how much weight can be lifted by the larger one?

Strategy By Pascal's principle, the same applied pressure is transmitted everywhere within the enclosed liquid system. In particular, if the heights of the pistons a and b are the same, the pressure on the pistons must be the same, including the pressures due to the applied force. We can set the pressure on the pistons equal, express the pressures in terms of the forces and areas, and then solve for the force on the larger piston.

Solution Using subscripts a and b to denote the quantities at each place, we can write $P_a = P_b$. But the pressure is the force per unit area: $F_a / A_a = F_b / A_b$. The area of the circular pistons is πr^2, so

$$\frac{F_a}{\pi r_a^2} = \frac{F_b}{\pi r_b^2}.$$

Solving for F_b gives

Figure 10.7

Example 10.3: A woman pushes down on the piston a. The pressure is transmitted undiminished to piston b.

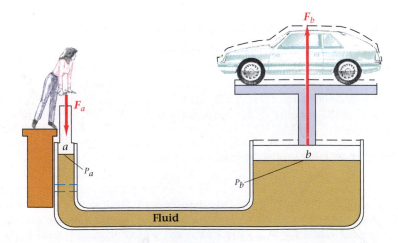

$$F_b = F_a \left(\frac{r_b}{r_a} \right)^2.$$

We see that the force at b is the applied force multiplied by the square of the ratio of the radii of the cylinders. In this case, the applied force is $F_a = mg$, so the answer becomes

$$F_b = mg \left(\frac{r_b}{r_a} \right)^2 = 50 \text{ kg} \times 9.81 \text{ m/s}^2 \left(\frac{12 \text{ cm}}{1.5 \text{ cm}} \right)^2 = 3.14 \times 10^4 \text{ N}.$$

This is enough force to lift two Jeeps, which together weigh 3.10×10^4 N. ■

In addition to the tire gauge mentioned earlier, there are several other pressure-measuring devices. One of these is the *manometer,* which measures a pressure by balancing the force due to the pressure with the weight of a column of liquid (Fig. 10.8). The pressure to be measured by this open-tube manometer is the pressure in the chamber C. If the density of air is neglected or the height h' is not too great, then Pascal's principle says that the pressure at A is the same as at C. Pascal's principle further says that the pressure at B is the same as at A. Therefore we can determine the gauge pressure at C by measuring the difference in the levels (heights) of the liquid h. This, as we saw in Section 10.1, is ρgh, where ρ is the density of the liquid. In some cases, it is common to state the pressure in terms of the height h. Thus, blood pressures are given as so many millimeters of mercury (mm Hg).

A *barometer* is a type of manometer used for measuring atmospheric pressure. A long tube is filled with mercury and inverted in a bowl of mercury (Fig. 10.9). The mercury drops in the tube until it is about 76 cm above the level of the mercury in the bowl. The atmosphere exerts a pressure on the mercury in the bowl that just balances the column of mercury. Small changes in atmospheric pressure, which are associated with changes in the weather, can be read from the barometer. Aneroid (without liquid) barometers are also commonly used to measure atmospheric pressure. The basic component of an aneroid barometer is a partially evacuated, sealed chamber made of thin metal with corrugated sides. The chamber expands and contracts in response to changes in atmospheric pressure. This movement of the chamber walls is transmitted to a pointer by a mechanical linkage.

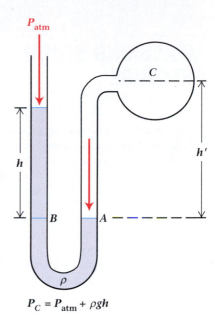

$$P_C = P_{atm} + \rho gh$$

Figure 10.8
An open U-tube manometer. The difference in the heights of the liquid indicates the pressure in the chamber C.

Archimedes' Principle

10.3

The story has been told that Archimedes (287–212 B.C.) conceived of the principle that bears his name after King Hiero of Syracuse asked Archimedes to determine the actual composition of the king's crown, which was alleged to be pure gold. Archimedes was ordered to do so without damaging the crown. According to legend, the Greek scientist's inspiration came to him as he lay partially submerged in his bath. On getting into the tub, he observed that the more his body sank into the tub, the more water ran out over the top. He immediately jumped out of the tub and rushed through the streets naked, shouting excitely in a loud voice "Eureka" ("I have found it"). **Archimedes' principle** says: *A body, whether completely or partially submerged in a fluid, is buoyed upward by a force that*

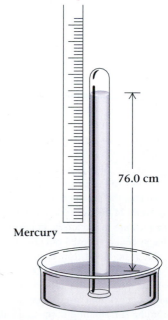

Figure 10.9
Mercury barometer.

MEASURING BLOOD PRESSURE

Almost any time you get a medical checkup, someone measures your blood pressure. The procedure is one of the most common in medicine: Someone wraps a cuff around your arm, inflates the cuff until it's tight, then listens through a stethoscope held to your arm while letting the cuff slowly deflate. What is happening during this procedure? The person is measuring the pressure in a fluid, your blood.

The heart is a large muscle, responsible for pumping oxygen-supplying blood to all parts of the body. The blood returns from the body through the veins to the right side of the heart, which pumps the blood to the lungs. The lungs remove carbon dioxide from the blood and add oxygen. The left side of the heart receives the oxygenated blood from the lungs and pumps it throughout the body by way of the arteries. The blood flows from the arteries to the veins through capillary beds.

Two pressures in the heart's action are of particular medical interest: the *systolic* pressure, when the heart is contracted, and the *diastolic* pressure, when the heart is relaxed between beats. Normal heart action causes arterial blood pressure to oscillate between these two values. Abnormally high or low arterial blood pressure can sometimes indicate physical and mental conditions of varying degrees of seriousness.

The most direct way of measuring blood pressure is to insert a fluid-filled tube into the artery and connect it to a pressure gauge. Though this is sometimes done, it is neither comfortable nor convenient. The commonly used indirect method involves a device called a sphygmomanometer. A nonelastic cuff that has an inflatable bag within it is placed around the upper arm, roughly at the same vertical level as the heart. The cuff is connected directly to some pressure gauge, such as a manometer (Fig. B10.1). When the cuff is inflated, the tissue in the arm is compressed; if sufficient pressure is applied, the flow of arterial blood in the arm stops. If the cuff is long enough and if it is applied snugly, the pressure in the tissues in the arm is the same as the pressure in the inflated part of the cuff, and is also the same as the pressure in the artery. In effect, Pascal's principle holds for the system composed of cuff, arm, and artery.

After the blood flow has been cut off, the pressure in the cuff is reduced by releasing some of the air. The falling pressure corresponds to the dashed line in Fig. B10.1. At some point, the maximum arterial pressure slightly exceeds the pressure in the surrounding tissue and cuff, allowing the blood to resume flowing. The acceleration of the blood through the arteries gives rise to a characteristic sound, which can be

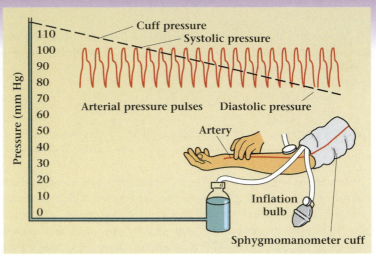

Figure B10.1 Measuring blood pressure with a sphygmomanometer. Identifiable sounds occur in the arm when the cuff pressure falls below the systolic and diastolic pressures.

identified by means of a stethoscope. When this sound occurs, the manometer indicates the maximum, or systolic, pressure. As the pressure in the cuff falls further, a second change in the sound is heard, characteristic of the drop below diastolic pressure. The readings shown in Fig. B10.1 correspond to the two pressures and are reported as "100 over 75," which is a typical value of the pressures for a healthy person.

The measurements made by this technique may vary because of the need to fit the cuff properly and the need to reliably estimate the point at which the sound changes. The condition of the manometer, the size of the arm, and the rate at which the cuff is inflated and deflated can have an effect. Figure B10.2 shows a comparison between pressures measured directly in the artery and pressures measured indirectly by a type of sphygmomanometer.

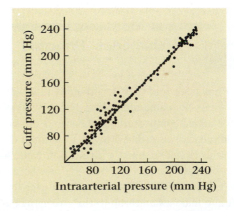

Figure B10.2 Comparison of blood pressure measurements by sphygmomanometry with direct measurements of arterial pressure.

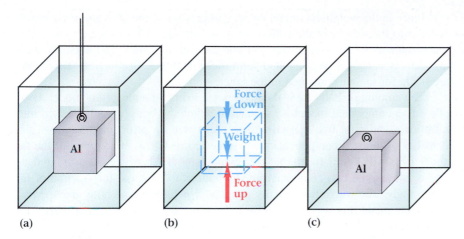

(a)   (b)   (c)

Figure 10.10
(a) An aluminum cube suspended in water. (b) An imaginary volume in the water, equal to the volume of the cube. The difference between the upward force and the downward force on the volume is equal to the weight of the water in the volume. (c) If the string is cut, the cube sinks because it is denser than water.

is equal to the weight of the displaced fluid. How this principle allowed Archimedes to solve the problem of the king's crown is seen in Problem 10.30.

Archimedes' principle can be obtained by a simple nonmathematical argument. Consider an object of arbitrary but known shape, such as an aluminum cube, that hangs completely submerged in a fluid, such as water (Fig. 10.10a). Now imagine that you remove the cube and replace it with a volume of water having the same size and shape as the "missing cube." (This is the amount of fluid displaced by the object.) Since this volume of fluid does not sink to the bottom of the container, the surrounding fluid must provide a net upward force, which counteracts the downward force due to the weight of fluid in the volume (Fig. 10.10b).

This upward force due to the surrounding water is the **buoyant force** and is equal to the weight of fluid in the volume displaced by the object. If you replace the aluminum cube with a lead cube of the same size, the buoyant force remains the same: It depends only on the volume of the submerged object, not on its mass or density. Of course, since aluminum is denser than water, the buoyant force alone cannot support the cube, which sinks to the bottom of the container if the supporting string is cut (Fig. 10.10c). If the fluid were mercury, which has a density greater than that of aluminum, the aluminum cube would float. We can generalize to a case in which the body is partially submerged, such as that of the iceberg in Example 10.6.

Blimps, hot-air balloons, and other lighter-than-air craft furnish another example of Archimedes' principle. They float in the air just as a submerged fish floats in the water. The blimp in Fig. 10.11 obviously has individual parts that will not float in air. However, if the *average* density of the whole craft, including passengers, is less than the density of air, the craft will take off without power. To meet this requirement in a practical way, the blimp contains a large volume of helium. The same type of observation can be made about large ships, which float even though they are made of steel and carry dense objects. The explanation is that the average density of the entire ship, including all of the air spaces, is less than the density of water.

Health professionals use Archimedes' principle to determine the percentage of fat in a person's body composition. The technique used, commonly known as the *hydrostatic weighing* method, requires two measurements of weight. A person is first weighed in air and is then weighed when submerged in water. The

Figure 10.11
The Goodyear airship *Europa* above the British research vessel *Eye of the Wind*. Although both the airship and the boat contain materials of high density, their average density is less than that of the fluid in which they float.

difference in the weight in air and the weight in water is the buoyant force, which is equal to the weight of the volume of water displaced. From these two weight measurements the person's mass and volume can be found. The person's average density can then be computed from these numbers. In practice, corrections are made for the air contained in the lungs.

The amount of body fat is then computed using an appropriate model. Usually a two-component model is assumed in which the body mass is divided into a fat mass and a fat-free mass. Furthermore, based on prior measurements, the densities of the two components are assumed to be 0.9×10^3 kg/m^3 for the fat and 1.10×10^3 kg/m^3 for the fat-free component. Then, since the average density is known, the two-component model is used to compute the fraction of fat in the body. The fraction is usually given as a percentage of body mass. Measurements made this way are precise to within a few percent of the body's fat composition.

Master the Concept

Buoyancy

Question: A bucket of water rests on a scale. Does the scale reading change when a lead block is suspended from a thread and lowered into the water where it is held submerged without touching the bottom or sides of the bucket?

Answer: The lead block displaces a volume of water equal to its own volume. Consequently, the water pushes the block upward with a buoyant force equal to the weight of the displaced water. By Newton's third law, there must be an equal but opposite force pushing down. The scale reading will increase by an amount equal to the buoyant force.

Question: If the lead block just described was suspended from a spring scale, what happens to the reading of that scale when the block is submerged in the water?

Answer: If the lead block is not accelerating, the upward forces acting on it must equal the downward force of gravity on it. When the block is in the air, the spring scale provides all of the upward force. When the block is submerged in the water, the water provides part of the upward force, so the spring scale reading is decreased by an amount equal to the buoyant force.

Example 10.5

An object of density ρ and mass m is submerged in a liquid with a smaller density ρ_0. Show that the effective weight of the submerged object is

$$w_{\text{eff}} = mg\left(1 - \frac{\rho_0}{\rho}\right).$$

Strategy Because the density of the object is greater than that of the surrounding liquid, the object will sink when dropped in the liquid. However, if we suspended it with a thread, we find that the force needed to hold it up (its effective weight) is less than its weight in air by the amount of buoyant force exerted by the liquid. It is this difference in forces that we need to find.

Solution The weight of the object is mg. The buoyant force is equal to the weight of the displaced liquid.

$$\text{buoyant force} = \rho_0 V g,$$

where V, the volume of the liquid displaced, is equal to the volume of the submerged object. The volume of the object can be expressed in terms of its density as

$$V = \frac{m}{\rho}.$$

The effective weight w_{eff} is the gravitational force less the buoyant force:

$$w_{\text{eff}} = mg - \rho_0 V g = mg - \rho_0 \left(\frac{m}{\rho}\right) g$$

$$w_{\text{eff}} = mg \left(1 - \frac{\rho_0}{\rho}\right).$$

Example 10.6

Icebergs are made of fresh-water ice, which has a density ρ_i of 0.92×10^3 kg/m^3 at 0°C. Ocean water, largely because of the dissolved salt, has a density ρ_w of about 1.03×10^3 kg/m^3. What fraction of an iceberg lies below the surface?

Strategy For simplicity, assume the iceberg to be a cube of volume L^3. The downward force acting on it is due to gravity:

$$F_{\text{grav}} = mg = \text{volume} \times \text{density} \times g = L^3 \times \rho_i \times g.$$

The buoyant force, according to Archimedes' principle, depends on the volume of ice submerged in the water. If the bottom of the floating cube is at a depth d below the surface of the water, then the upward buoyant force is

$$F_{\text{buoyant}} = d \times L^2 \times \rho_w \times g.$$

When the iceberg is floating in equilibrium, these forces are equal in magnitude. We can solve for the ratio of d to L to obtain the desired fraction.

Solution In equilibrium, the upward buoyant force must be equal in magnitude to the downward gravitational force.

$$F_{\text{buoyant}} = F_{\text{grav}},$$

or

$$d\,L^2\,g\rho_w = L^3 g \rho_i.$$

Rearrange the equation and insert the numerical values to get

$$\frac{d}{L} = \frac{\rho_i}{\rho_w} = \frac{0.92}{1.03} = 0.89.$$

Thus, 89% of the iceberg lies below the surface.

Discussion We have just computed how much of an iceberg lies below the surface for the particular case of a cubic iceberg. What happens if the iceberg is not cubic? The volume of the iceberg below the surface will be 89% of the total volume regardless of the shape of the iceberg. However, the ratio of the linear extent of berg below the surface to its total height will not, in general, be 0.89, but will depend on its exact shape.

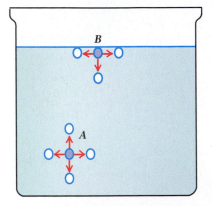

Figure 10.12
A molecule inside a liquid (point *A*) is attracted to other molecules on all sides. A molecule on the surface (point *B*) has attractive forces only to other molecules in the surface and below the surface. As a result, the surface acts like a membrane under tension.

*10.4 Surface Tension

Most people have observed the phenomenon called surface tension. You have seen spherical water drops hanging from a spider's web; or objects denser than water, like a needle or a razor blade, floating on the surface; or a water bug skittering over a pond. In each case, the liquid surface exhibits some of the characteristics of a stretched membrane under tension.

The molecules that compose a liquid attract one another; otherwise the liquid would not have a definite volume. Most liquids are nearly incompressible because the intermolecular forces are repulsive at distances less than their normal separations. At any point *A* inside a liquid at rest, the net molecular force on a molecule of the liquid is zero because the molecules surrounding it provide opposing forces in balance (Fig. 10.12). However, as shown, at a point *B* on the surface of the liquid, the situation changes. A molecule experiences attractive forces only in the direction of the other molecules of the liquid, and since at the surface these forces are not balanced by attractive forces in the outward direction, there is a net inward force. This inward force makes the surface act like a stretched drumhead. The surface of a liquid resists any effort to increase its area.

We can determine values of surface tension by measuring the force necessary to lift a ring out of the liquid (Fig. 10.13). The required force is proportional to *C*, the circumference of the ring. As the ring rises, a thin film of liquid clings to it until the weight of the film exceeds the attractive forces holding the film together. The film has two surfaces, one inside and one outside of the ring. Thus, the total length along which the lifting force acts is 2*C*. The **surface tension** γ is the ratio of the surface force to the length along which it acts:

$$\gamma = \frac{F}{2C}. \tag{10.6}$$

Notice that the surface tension is not simply a force, but a force divided by a length. The units of surface tension are N/m. Table 10.3 lists the surface tensions of some common liquids.

An equivalent way of thinking about surface tension is to think of it as the energy per unit area of the surface. That this is reasonable can be seen by writing the units N/m as $N \cdot m/m^2$, which is J/m^2. The equilibrium configuration of a surface corresponds to its lowest possible energy. One consequence is that liquids try to minimize their surface area. As a result, liquids tend to assume the shape of a sphere, which is the shape with the smallest surface area (and therefore energy) for a given volume. Figure 10.14 is a view of a splash of milk, showing the detached spheres in the process of separating from the body of liquid. For the same reason, soap bubbles tend to take the shape of spheres.

The attractive force between like molecules that acts to hold a liquid together is called cohesion. The attractive force between unlike molecules—say, between

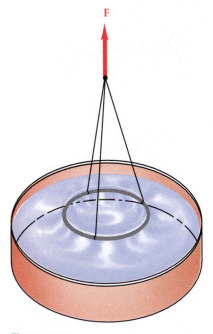

Figure 10.13
A force is required to lift a ring out of the liquid.

Table 10.3	Measured Values of Surface Tension	
Liquid in Contact with Air	Temperature (°C)	Surface Tension (10^{-3} N/m)
Water	0	75.6
Water	25	72.0
Water	80	62.6
Ethyl alcohol	20	22.8
Acetone	20	23.7
Glycerin	20	63.4
Mercury	20	435

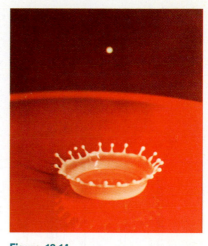

Figure 10.14
Close-up of the splash of a milk drop falling on a hard surface.

those of water and glass—is called adhesion. The adhesive forces may be as great as, or even greater than, cohesive forces. For example, the adhesion of water to clean glass is greater than the cohesion of water to itself, so the water wets the glass. By contrast, the adhesion of water to a newly waxed surface is less than the cohesion, so water does not wet the surface, but beads up into drops.

The surface tension of a liquid, such as water or oil, may be greatly reduced by the addition of certain chemicals called surfactants, or surface-active agents. Detergents are one type of surface-active agent. The lowering of the surface tension of water by detergents increases the ability of water to wet a surface. This improved wetting allows more water molecules to penetrate cloth fibers and wash away particles of soil.

If a tube of relatively small internal diameter (a capillary) is placed in a liquid that wets it, the liquid rises up in the tube (Fig. 10.15). This effect, known as *capillary action,* is due to the cohesive forces of surface tension and the adhesive forces between the liquid and the glass tube. The liquid rises until the upward force due to surface tension is equaled by the weight of the liquid in the tube.

In the weightless environment of an orbiting spacecraft, surface tension can pull liquids from their containers. Astronauts attempting to use a straw to drink a liquid find that the liquid climbs up the straw and collects as a spherical drop at the open end because there is no force (weight) opposing the capillary forces. The straws used aboard the space shuttle come with clamps to squeeze the straws shut lest the drinks come oozing out the top.

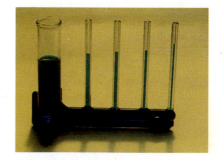

Figure 10.15
The internal diameters of the four tubes decrease from left to right.

Example 10.7

Show that the height to which a liquid rises in a narrow tube is given by $h = 2\gamma \cos \theta / r\rho g$, where r is the radius of the column of liquid, ρ is its density, g is the acceleration of gravity, γ is the surface tension, and θ is the angle the liquid surface makes with the wall at the line of contact (Fig. 10.16).

Solution First note that, for equilibrium, the weight of the liquid in the tube must be equal to the upward component of the adhesive force due to surface tension. The weight, or downward force F_d, is given by the density of the liquid times the volume of the column ($h\pi r^2$) times the acceleration of gravity:

$$F_d = \rho g V = \rho g h \pi r^2.$$

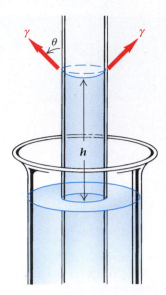

Figure 10.16

Example 10.7: Analysis of the height to which a liquid rises in a capillary tube.

Figure 10.17

The flow of the river is laminar in some regions and turbulent in others.

The upward force F_u, due to the surface tension acting all around the inside of the cylindrical surface, is the product of the upward force per unit length times the inner circumference of the cylinder. The upward force per unit length is the component of the surface tension along the vertical direction, $\gamma \cos \theta$. Thus, the upward force is

$$F_u = 2\pi r \gamma \cos \theta.$$

Equating the upward and downward forces gives

$$\rho g h \pi r^2 = 2\pi r \gamma \cos \theta,$$

or

$$h = \frac{2\gamma \cos \theta}{r\rho g}.$$

10.5 Fluid Flow: Streamlines and the Equation of Continuity

Let us now turn our attention to fluids in motion. The two main types of fluid flow are easily recognized in the river shown in Fig. 10.17. One type is the orderly flow of neighboring layers of fluid moving past each other smoothly. Each small element of fluid follows a path called a **streamline,** which does not cross over or become tangled with other streamlines. This smooth streamline flow is known as **laminar** flow. The other type of flow occurs when the fluid exceeds a certain critical velocity. Then the flow no longer is laminar but becomes **turbulent** and is characterized by an irregular, complex motion. (Turbulent flow will be considered in Section 10.9.)

Let us look at the steady laminar flow of an incompressible fluid moving through a tube or pipe. How does the flow change when the diameter of the pipe changes? Consider the case of a fluid moving from a region of cross-sectional area A_1 to a region of area A_2 (Fig. 10.18). Because the fluid is incompressible, the same amount of it leaves each region toward the right as enters from the left during the same time interval. The volume of fluid that flows into the tube across A_1 in a time interval Δt is $\Delta V_1 = A_1 v_1 \Delta t$, where v_1 is the velocity of the fluid at A_1. If the density of the fluid is ρ, then the mass of fluid that flows into the tube in time Δt is $\rho A_1 v_1 \Delta t$. Similarly, the mass of fluid that flows out of the tube through A_2 in the same time Δt is $\rho A_2 v_2 \Delta t$. Since the mass of fluid entering is the same as the mass leaving, we get

$$\rho A_1 v_1 \Delta t = \rho A_2 v_2 \Delta t.$$

We can divide out the density because it is constant for an incompressible fluid. We then get

$$v_1 A_1 = v_2 A_2. \tag{10.7}$$

This equation is called the **equation of continuity** and will be useful throughout our discussion of fluids in motion. It says that the flow of material (mass) through a tube of changing cross section is constant when the density of the fluid does not change. That is, the equation of continuity is a statement of conservation

SURFACE TENSION AND THE LUNGS

Take a deep breath. You've probably never thought about it, but some interesting physics goes on in your lungs every time you breathe. It all takes place without your having to think about it, but the interplay is fascinating.

The air passages into the two lungs branch and branch again until they end in tiny air sacs called *alveoli* (Fig. B10.3). It is here that the exchange of gases with the blood takes place. An adult's lungs contain on the order of 300 million alveoli, each with an average radius of 120 μm. Specialized cells in the walls of the alveoli produce a detergentlike material called surfactant that coats the inside of the alveoli, thus reducing the surface tension.

First, consider two soap bubbles connected to a pipe that contains a valve (Fig. B10.4). This arrangement is sometimes shown to a class and the question asked: "What will happen if the valve is opened?" When the valve is opened, the larger bubble grows and the smaller bubble becomes smaller until it completely goes away. We can understand this observation from the standpoint of energy. Just as a free drop of liquid takes on a spherical shape to minimize its total surface energy, the bubbles change relative sizes in such a way as to minimize their combined surface area, and therefore the surface energy. The surface area of a single bubble is about 30% less than the surface area of two smaller bubbles of equal size that together have the same volume as the larger bubble. So, when the valve is opened, the two bubbles become one system and form a single bubble.

If the effect of large bubbles absorbing smaller ones were to take place in the lungs, the smaller alveoli would all collapse and the larger ones would grow. This does not happen because of the pulmonary surfactant that coats the inside of

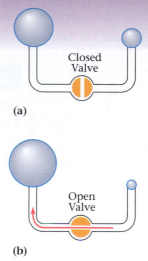

(a)

(b)

Figure B10.4 (a) Two soap bubbles of different radius are connected through a pipe. (b) When the valve is opened, the larger bubble grows as the smaller one shrinks.

the lungs. Experiments have shown that the surface tension of the pulmonary surfactant *increases* with its area, in contrast with the behavior of water and most other liquids. This means that the surface energy of larger alveoli, in spite of the smaller surface-to-volume ratio, can be the same as that of the smaller alveoli. Therefore, the larger and smaller air sacs can exist in equilibrium.

The experimental results that show the variation of surface tension with area also explain another observation about the lungs. If you take a big breath and then relax your chest muscles, the air is expelled from your lungs. Part of the reason that the air flows out is that the lungs have been blown up, much like a balloon, and the elasticity of the tissues causes them to contract and force the air out. However, tests have shown that the elasticity of the lung tissue itself is not sufficient to completely explain what happens. A large part of the effect is due to the surface tension in the alveoli, which causes them to contract and expel the air. Furthermore, the effect depends on a material for which the surface tension increases with area, or else the tendency of the lungs to expel air when inflated would be much less than it is.

Normal breathing is possible only when the pulmonary surfactant is present in the proper amount and has the proper surface tension. If the surface tension is greater than normal, the lungs tend to collapse and expansion of the alveoli is difficult. Some newborn babies, especially premature ones, do not have enough surfactant, making lung expansion difficult. Unless they receive immediate treatment, these infants die soon after birth because of inadequate ventilation. This condition is known as respiratory distress syndrome.

Figure B10.3 A plastic cast of the air passages in the lungs. The passages end in tiny sacs called alveoli.

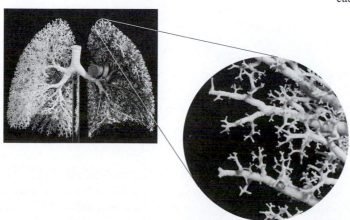

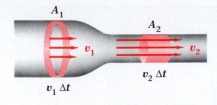

$$v_1 A_1 = v_2 A_2$$

Figure 10.18
Fluid passing cross section A_1 with speed v_1 passes through cross section A_2 with a new speed v_2, as required by the equation of continuity.

of mass. Notice that the product Av is the volume rate of flow, that is, the volume of fluid passing a given cross section per unit time. In SI units, the volume rate of flow is measured in m^3/s.

Example 10.8

A horizontal pipe of 25-cm^2 cross section carries water at a velocity of 3.0 m/s. The pipe feeds into a smaller pipe with a cross section of only 15 cm^2. What is the velocity of water in the smaller pipe?

Solution   We can determine the velocity of water in the smaller pipe from the equation of continuity. The velocity and area in the large pipe are $v_1 = 3.0$ m/s and $A_1 = 25$ cm^2. The area of the smaller pipe is $A_2 = 15$ cm^2. Thus, we have

$$v_2 = \frac{v_1 A_1}{A_2} = \frac{3.0 \text{ m/s} \times 25 \text{ cm}^2}{15 \text{ cm}^2} = 5.0 \text{ m/s.}$$

10.6   Bernoulli's Equation

We wish to find a relationship among the variables describing the steady laminar flow of a fluid, assuming the fluid not only is incompressible but also has no internal friction, or viscosity. (We will introduce the effects of viscosity in Section 10.7.) The result, called Bernoulli's equation,* describes the relationship of a fluid's pressure, velocity, and height as it moves along a pipe or other tube of flow.

The fluid flowing smoothly from region A to region B in Fig. 10.19 need not be constrained to a real pipe. Think, for example, of a portion of the water flowing in a river. However, if we draw all of the streamlines from the boundary of region A surrounding a portion of the fluid to their later positions when the fluid reaches B, we outline a *tube of flow*. The equation of continuity (Eq. 10.7) applies to such a tube of flow.

Figure 10.19
Fluid passing through area A later passes through area B. The streamlines mark the paths of small elements of the fluid. The tube that connects A to B along the streamlines is the tube of flow.

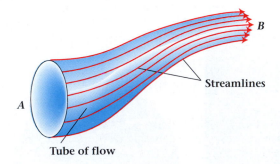

*This problem was first solved by the Swiss mathematician and physicist Daniel Bernoulli (1700–1782), who published the result in a book on fluid flow in 1738.

We can find the relationship that we seek by making use of the principle of conservation of energy. In particular, we calculate the work done on a small element of fluid moving along a tube of flow and then use the work-energy theorem (Section 6.4) to equate the change in kinetic energy to this work. To move a small element of fluid through a distance of Δx_1 at region 1 (Fig. 10.20) requires an amount of work $P_1 A_1 \Delta x_1$. At the same time, the same amount of fluid (given by $A_2 \Delta x_2$) moves a distance Δx_2 at region 2. The work in this case is $-P_2 A_2 \Delta x_2$. The negative sign indicates that the element of fluid at region 2 moves against the force due to the pressure of the fluid to its right.

Because the fluid is incompressible, the volume of the fluid displaced at region 1 is equal to the volume of the fluid displaced at region 2; $A_1 \Delta x_1 = A_2 \Delta x_2$. The work done by gravity in the net motion of fluid from region 1 to region 2 is $-mg(h_2 - h_1)$, where the mass $m = \rho A_1 \Delta x_1 = \rho A_2 \Delta x_2 = \rho \Delta V$. Thus, the work done by gravity is $-g\rho \Delta V(h_2 - h_1)$. The net work done is

$$W = P_1 \Delta V - P_2 \Delta V - g\rho \Delta V(h_2 - h_1).$$

According to the work-energy theorem, the change in kinetic energy of the mass $\rho \Delta V$ is equal to this work, so

$$\Delta KE = W,$$

or

$$\tfrac{1}{2}\rho \Delta V v_2^2 - \tfrac{1}{2}\rho \Delta V v_1^2 = P_1 \Delta V - P_2 \Delta V - \rho g \Delta V h_2 + \rho g \Delta V h_1.$$

Dividing through by ΔV and rearranging terms gives

$$P_1 + \rho g h_1 + \tfrac{1}{2}\rho v_1^2 = P_2 + \rho g h_2 + \tfrac{1}{2}\rho v_2^2 = \text{constant.} \qquad (10.8)$$

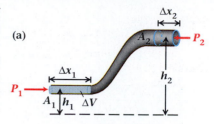

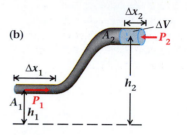

Figure 10.20
A volume ΔV of incompressible fluid is moved along a tube of flow from (a) region 1 to (b) region 2.

Equation (10.8) is called **Bernoulli's equation,** for the steady, nonviscous flow of an incompressible fluid. Under these conditions, Bernoulli's equation expresses conservation of energy in a moving fluid. To some extent, we can also apply Bernoulli's equation to compressible fluids of negligible viscosity in laminar flow.

If we consider a horizontal pipe ($h_1 = h_2$), the equation of continuity tells us that the fluid flows more rapidly in a constricted region of the pipe. If we combine the equation of continuity [$v_1 = (A_2/A_1)v_2$] with Bernoulli's equation we get

$$P_2 = P_1 + \frac{\rho v_2^2 (A_2^2 - A_1^2)}{2A_1^2}. \qquad (10.9)$$

The second term on the right-hand side is negative if A_1 is greater than A_2. In that case, the pressure P_2 is less than the pressure P_1. This result, which arises from the principles of conservation of energy and of mass, tells us that the pressure is less in the constricted region of the pipe. Similarly, if A_2 is greater than A_1, then P_2 is larger than P_1. In other words, when a moving fluid enters a narrower section of pipe (or artery), its speed increases but the pressure on the fluid decreases.

Equation (10.9) holds strictly only for incompressible nonviscous fluids. But the general qualitative conclusion above applies to both gases and liquids. If we consider an incompressible but viscous fluid, such as water or blood, we get

$$P_2 < P_1 + \frac{\rho v_2^2 (A_2^2 - A_1^2)}{2A_1^2}.$$

The inequality arises from noticing that in a viscous fluid, some of the work done is dissipated by the internal frictional forces in the liquid.

Example 10.9

Determine the pressure change that occurs on going from the larger-diameter pipe to the smaller pipe for the conditions of Example 10.8; that is, take $A_1 = 25$ cm², $A_2 = 15$ cm², and $v_2 = 5.0$ m/s.

Solution Since the pipe is horizontal, we can find the pressure change $P_2 - P_1$ from Eq. (10.9). It is

$$\Delta P = P_2 - P_1 = \rho v_2^2 \frac{(A_2^2 - A_1^2)}{2A_1^2}.$$

Here ρ is the density of water $= 10^3$ kg/m³.

$$\Delta P = (10^3 \text{ kg/m}^3)(5.0 \text{ m/s})^2 \left(\frac{(15 \text{ cm}^2)^2 - (25 \text{ cm}^2)^2}{2(25 \text{ cm}^2)^2} \right)$$

$$\Delta P = -8 \times 10^3 \text{ Pa}.$$

The negative result tells us that the pressure is smaller in the smaller pipe.

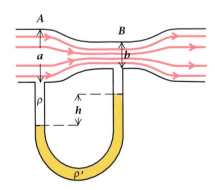

Figure 10.21

A Venturi meter enables us to measure fluid speed by observing height differences in a U-shaped tube.

The speed of a fluid flowing in a pipe can be measured with a device called a *Venturi meter* (Fig. 10.21). If the fluid in the pipe is flowing, the pressure at B is lower than the pressure at A. The difference in pressures is a function of the rate at which the fluid flows. The Venturi meter measures this difference in pressure with the U-shaped tube. Using Bernoulli's equation (Eq. 10.8) and the equation of continuity (Eq. 10.7), we can derive the following relationship for the velocity of the fluid at position A:

$$v = b \left(\frac{2(\rho' - \rho)gh}{\rho(a^2 - b^2)} \right)^{1/2},$$

where a and b are the cross-sectional areas at A and B, respectively, ρ is the density of the fluid flowing in the pipe, and ρ' is the density of the liquid in the U-tube. The difference in height of the liquid in the two arms of the U-tube is h, and the gravitational acceleration is denoted by g. Thus, the Venturi meter enables us to determine the flow velocity from a measurement of the height difference h.

Viscosity and Poiseuille's Law

Bernoulli's equation predicts that for a horizontal pipe of uniform cross section, the pressure in a moving fluid is constant. If there were no fluid friction, or viscosity, this would indeed be the case. However, for a viscous fluid flowing through a horizontal pipe of uniform cross section, the fluid pressure decreases with distance along the direction of flow.

Viscosity is that property of a fluid that indicates its internal friction. The more viscous a fluid, the greater the force required to cause one layer of fluid to slide past another. Viscosity is what prevents objects from moving freely through a fluid or a fluid from flowing freely in a pipe. The viscosity of gases is less than that of liquids, and the viscosity of water and light oils is less than that of molasses and heavy oils. Your experience with liquids such as motor oils and syrups tells you that viscosity increases with decreasing temperature. Thus, it is hard to start a car engine in subzero weather when the oil is thick and flows slowly, but it is easy to start the same car on a hot summer day when the oil is warm and flows readily.

Let us return to the situation of a fluid moving through a horizontal pipe. The walls of the pipe exert a resistive force, or drag, on the adjacent layers of fluid. These layers, in turn, slow down the next adjacent layers, and so on. As a result, the rate of flow is slowest near the pipe walls and fastest in the center of the pipe. Therefore, for a given rate of flow the pressure difference between two points along the length of the pipe depends on the radius of the pipe (Fig. 10.22). The pressure difference between the two points is also related to a quantity known as the coefficient of viscosity or simply the viscosity of the fluid. The exact relation is given by the following equation, called **Poiseuille's law:**[*]

$$P_1 - P_2 = 8\frac{Q\eta L}{\pi R^4}, \qquad (10.10)$$

where Q is the flow rate in m³/s, η is the coefficient of viscosity, R is the radius of the pipe, and L is the separation between the test points. If R and L are given in meters and the pressure is given in pascals, the unit of the coefficient of viscosity is the pascal-second (Pa · s). This equation is often used experimentally to determine the coefficient of viscosity of a liquid. Table 10.4 lists the coefficients of viscosity for a number of fluids.

An important observation about the behavior of real fluids is easily made from Eq. (10.10). A viscous fluid will not flow through a pipe unless there is a pressure difference between the ends.

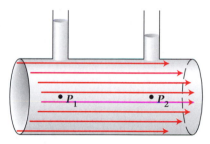

Figure 10.22
The pressure in a viscous fluid along a horizontal pipe of uniform cross section diminishes with distance along the direction of flow. Thus, pressure P_2 is less than pressure P_1. Note, too, that fluid flows fastest in the center of the pipe.

---

[*]Poiseuille is pronounced "pwa-zoy'."

Table 10.4	Viscosities of Some Common Fluids	
Fluid	Temperature (°C) (10^{-3} Pa · s)	Viscosity
Air	20	0.018
Water	40	0.656
Water	20	1.005
Motor oil (SAE 10)	30	200
Glycerin	20	1490
Castor oil	20	986
Mercury	20	1.550

Example 10.10

Blood in the extremities of the body is carried by arterioles, small vessels with an average diameter of about 0.1 mm. The muscles in the walls can contract and change the diameter of the "pipe," thereby decreasing the flow of the blood, a viscous fluid. Sometimes great strain or shock causes a severe reduction in blood flow. By approximately what amount would the arterioles have to contract to reduce the blood flow to 30% of its former value if we assume that the pressure drop remains constant?

Solution We use Poiseuille's law, writing subscripts 1 for the normal condition and 2 for the reduced-flow case. Then, because the pressure drop is constant, we have

$$8\frac{Q_1 \eta_1 L}{\pi R_1^4} = 8\frac{Q_2 \eta_2 L}{\pi R_2^4}.$$

Though the viscosity of the blood is not independent of its velocity or of the diameter of the tube through which it flows, we assume it to be constant over the range considered here. Then we have

$$\frac{R_2^4}{R_1^4} = \frac{Q_2}{Q_1},$$

or

$$\frac{R_2}{R_1} = \left(\frac{Q_2}{Q_1}\right)^{1/4}.$$

If the flow is reduced to 30% of its initial value, then $Q_2/Q_1 = 0.30$. Inserting this value in the above equation, we find that

$$\frac{R_2}{R_1} = (0.30)^{1/4} = 0.74.$$

The arteriole has constricted to 74% of its original diameter. For example, if the diameter of the arteriole were 0.10 mm originally, then its diameter after constriction would be 0.074 mm. The relative change in flow is greater than the relative change in the diameter of the tube.

Discussion In general, the diameter of blood vessels, unlike that of glass or metal tubes, increases as the internal pressure increases, because the blood vessels are distensible. Moreover, in blood vessels as small as capillaries, the viscosity of whole blood is as little as half of what it is in large vessels. This effect is due to the alignment of the red blood cells as they pass through the narrow vessels.

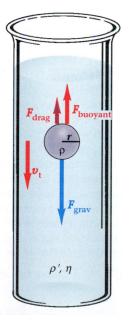

*10.8   Stokes's Law and Terminal Speed

Viscous Drag

An object moving through a fluid experiences a resistive force, or drag, that is proportional to the viscosity of the fluid. If the object is moving slowly enough, the drag force is proportional to its speed v. If the object is a sphere of radius r, the force is

$$F = 6\pi\eta r v,$$ (10.11)

where η is again the coefficient of viscosity. This equation is known as **Stokes's law**, after Sir George Stokes (1819–1903), who first conceived it in 1845. Stokes's law can be used to relate the speed of a sphere falling in a liquid to the viscosity of that liquid.

Consider a solid sphere of radius r dropped into the top of a column of liquid (Fig. 10.23). At the top of the column, the sphere accelerates downward under the influence of gravity. However, there are two additional forces, both acting upward: the constant buoyant force and a speed-dependent retarding force given by Stokes's law. When the sum of the upward forces is equal to the gravitational force, the sphere travels with a constant speed v_t, called the **terminal speed**. To determine this speed, we write the equation for the equilibrium of forces:

$$F_{\text{grav}} = F_{\text{buoyant}} + F_{\text{drag}}.$$

We can express the gravitational force in terms of the density ρ of the sphere, its volume $\frac{4}{3}\pi r^3$, and g:

$$F_{\text{grav}} = \frac{4}{3}\pi r^3 \rho g.$$

The buoyant force is equal to the weight of the displaced liquid, which has a density ρ':

$$F_{\text{buoyant}} = \frac{4}{3}\pi r^3 \rho' g.$$

The retarding force is expressed by Stokes's law with the speed v_t:

$$F_{\text{drag}} = 6\pi\eta r v_t.$$

Figure 10.23

A sphere falling in a viscous liquid reaches a terminal speed v_t that depends upon the radius and density of the sphere and the density and viscosity of the liquid.

HOW AIRPLANES FLY

Air travel is one of the great triumphs of the twentieth century. Every day hundreds of thousands of people are carried through the air to destinations all around the world. In every case the flight of heavier-than-air craft results from the flow of air around their wings.

Before their first powered flight in December 1903, the Wright brothers tested many different wing shapes in a wind tunnel to find the shape that produced the most lifting force. This shape, often called an airfoil, is shown in Fig. B10.5 along with the streamlines of the air moving past. The fluid moving over the top travels a greater distance than that moving just under the bottom of the wing. Consequently, the fluid moving over the top must travel faster in order to conform with the shape of the wing and still maintain the natural streamline. The shape of the wing also crowds the streamlines together above the wing, just as in the case of a constricting pipe. The result is that the region immediately above the wing experiences reduced pressure relative to the region immediately below the wing. Because the downward force on the top of the wing is less than the upward force on the bottom, a net upward force, or lift, arises from the air flow. (Beyond the airfoil the flowing air has a downward component of velocity. By Newton's third law, the reaction force to the net downward force exerted on the air is the lift.) Note that for lift to occur, a flow of air is required relative to the wing. The lift occurs equally well for a wing moving through stationary air or for air moving past a stationary wing.

You can demonstrate this effect with a small piece of paper, about 10 × 15 cm. Hold the short edge close below your lower lip and blow vigorously across the top of the paper (Fig. B10.6). The motion of the air above the paper will cause it to rise. This same effect helps to lift a plane into the air.

In addition, the angle of attack, or tilt of the wing relative to the air flow, can be changed to get additional lift from the deflection of the air stream (see Fig. B10.5). If the leading edge of the wing is higher than the trailing edge, the force of air against the underside of the wing is greater than its force

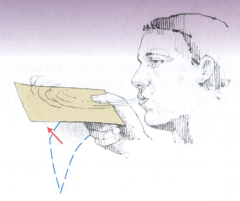

Figure B10.6 A demonstration of lift. Blowing across the paper causes it to rise.

against the upper side. In this case lift occurs even for a flat wing. However, if the angle of attack becomes too great, the streamline flow gives way to turbulence and the pressure difference is reduced. If the turbulence is great enough, the lift diminishes and the plane stalls.

In general, as the flow of air past the wing increases, both the lift force and the drag force (the resistance to forward motion) increase. Aircraft wings are designed so that pilots can change the wing shape during flight, producing greater lift for the slower speeds of takeoff and landing and producing less drag at cruising speeds. During takeoff and landing, flaps are extended backward and downward from the trailing edge of the wing (Fig. B10.7), increasing lift by imparting a greater downward velocity to the air. On some planes, extending the flaps increases the wing area as much as 25%, resulting in a much increased drag. At the same time, the leading edge of the wing may be moved forward, creating a slot that directs a high-speed layer of air over the top surface of the wing to reduce turbulence and increase lift. At higher speeds, the pilot closes the slot and retracts the flaps to reduce the drag forces. Passengers in commercial aircraft can easily see these changes in the wing during flight.

We should point out that lift is not in strict accord with Bernoulli's equation. The reason is that the Bernoulli equation holds exactly only for incompressible nonviscous fluids, yet air is both compressible and viscous. However, the pressure difference, and hence the lift, does occur in air, even if the amount is not in exact agreement with Eq. (10.8).

Figure B10.7 Flaps are extended during takeoff and landing to increase the lift.

Figure B10.5 An airfoil in a moving fluid.

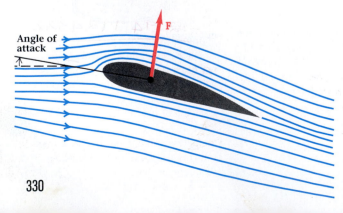

Angle of attack

F

Combining these equations, we get an expression for the terminal speed:

$$v_t = \frac{2r^2 g}{9\eta}(\rho - \rho').$$ (10.12)

The terminal speed is also called the sedimentation speed by biologists and geologists.

Example 10.11

An aluminum sphere of radius 1.0 mm is dropped into a bottle of glycerin at 20°C. What is the terminal speed of the sphere?

Solution Using Eq. (10.12), we calculate the terminal speed directly as

$$v_t = \frac{2r^2 g}{9\eta}(\rho - \rho').$$

The radius in meters is 1.0×10^{-3} m. The densities, from Table 10.2, are $\rho = 2.7 \times 10^3$ kg/m³ and $\rho' = 1.26 \times 10^3$ kg/m³. The viscosity, from Table 10.4, is 1.49 Pa · s. Thus,

$$v_t = \left(\frac{2 \times 1.0 \times 10^{-6} \text{ m}^2 \times 9.81 \text{ m} \cdot \text{s}^{-2}}{9 \times 1.49 \text{ N} \cdot \text{m}^{-2} \cdot \text{s}}\right)(2.7 - 1.26) \times 10^3 \text{ kg} \cdot \text{m}^{-3}$$

$$v_t = 2.1 \times 10^{-3} \text{ m}^2 \cdot \text{s}^{-2} \cdot \text{kg} / \text{N} \cdot \text{s}$$

But the units of N are kg · m · s⁻², so the speed is in m/s:

$$v_t = 2.1 \times 10^{-3} \text{ m/s, or 2.1 mm/s.}$$

Stokes's law suggests a method of measuring the size of small particles. If the rate at which material settles from a suspension is measured and the other parameters are known, then Eq. (10.12) gives the particle size. In the case of particles whose radius is less than about 5×10^{-6} m, the settling rate is prohibitively slow, even when the densities of the particles and fluid are quite different. However, we can increase the rate by using a centrifuge (see Section 5.2). The centripetal force per unit mass in an ordinary centrifuge may be 100 g, while that in a modern ultracentrifuge may go as high as 5×10^5 g.

Stokes's law is also useful in the consideration of geological processes in which the rate of sedimentation is important. Modifications are made to take account of nonspherical particles.

Form Drag

Stokes's law applies for situations in which the fluid flow is laminar, but not when the flow becomes turbulent. An important class that shows the effects of turbulence is illustrated by the retarding force of air on a moving car (Fig. 10.24), a falling raindrop, or a skydiver. In these cases it is observed that whenever an object moves rapidly enough, the retarding force F depends not on the speed (Stokes's law), but on the square of the speed:

$$F = bv^2,$$

where b is a constant determined for each different case.

Figure 10.24
Computer modeling of air flow around a moving car showing both laminar and turbulent flow.

An object falling from rest through the air falls with increasing speed until, at the terminal speed v_t, the retarding force of the air is equal in magnitude to the gravitational force:

$$mg = bv_t^2.$$

Thus, the terminal speed can be written as

$$v_t = \sqrt{\frac{mg}{b}}.$$

Elementary analysis (see Problem 10.59) shows that the constant b depends on the density ρ of the air and the area A of the body presented to the air flow. Then the equation for the terminal speed is

$$v_t = \sqrt{\frac{mg}{C_D \frac{\rho}{2} A}},$$

where C_D is called the **drag coefficient.** This equation also holds for objects moving horizontally through the air at any speed if mg is replaced by the re-tarding, or drag, force on the object. Thus, the aerodynamic drag on a moving object, such as a car, becomes approximately

$$F_{drag} = 0.65 C_D A v^2. \qquad (10.13)$$

One of the objects of modern automobile design is to reduce the drag in or-der to improve fuel economy. This is especially true for electric vehicles such as the General Motors EV_1. The small drag coefficient of the EV_1 is due in part to its smooth belly pan, its rear wheel skirts, and its teardrop shape resulting from a rear wheel track some nine inches smaller than that of the front wheels. (Some drag coefficients are given in Table 10.5.) On the other hand, the object of para-chute design is to have a large value of both C_D and A, so that descent is slow (Fig. 10.25).

Table 10.5	Aerodynamic Drag Coefficients	
Shape		C_D
1997 General Motors EV_1		0.19
1995 Lexus LS 400		0.28
1997 BMW 850ci		0.29
1997 BMW 750iL		0.32
1997 Dodge Intrepid		0.33
1997 Dodge Caravan		0.35
Typical 1970 U.S. auto		0.50
Typical 1970 U.S. station wagon		0.60
Small truck		0.70

For a skydiver falling through air, the terminal speed is approximately 60 m/s (about 120 mi/h); for a feather, it may be as small as 0.1 m/s. For a 2-mm-diameter raindrop, the terminal speed is about 7 m/s. Without air resistance, such a raindrop starting from rest would reach a speed of 7 m/s in less than three-quarters of a second, while it fell a distance of only 2.5 m. In this case the effects of air resistance are very important.

In our discussion of projectile range in Chapter 3, we used a simplified model that neglected the air resistance. Our model predicted that a thrown base-ball would follow a parabolic path. In a more realistic model including air resistance, the path of the ball is nonparabolic and the ball lands short of the range predicted with the simple model. Furthermore, the launch angle for maximum range is less than 45° and depends on the initial speed of the ball. If the ball spins, additional forces arise from the resulting turbulence that can also affect its path. These additional forces are described in the next section.

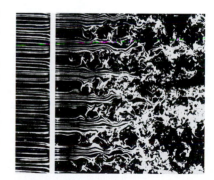

Figure 10.25
A parachutist falling with terminal speed. Parachutes are designed with large drag coefficients and large areas.

*10.9 Turbulent Flow

When you move your finger slowly through a liquid, such as water, you feel only a moderate force of resistance. This resistance arises from two sources: an inertial resistance to the acceleration of the water being displaced and a viscous drag force. As you move your finger faster, the resistance becomes larger because you are moving more water in the same time. The ratio of the inertial force to the viscous force, called the **Reynolds number,*** is a useful parameter for describing fluid flow and for determining the onset of turbulence. It is given by

$$\text{Re} = \frac{\rho v L}{\eta}, \qquad (10.14)$$

where ρ and η are the density and viscosity of the fluid, v is the speed of the object, and L is a length characteristic of the object. In this case, L is the length of your finger.

A large value of Re indicates large motion of the fluid. When motion through the fluid exceeds a certain critical speed, the laminar flow of the fluid around the object becomes turbulent and is characterized by an irregular, complex motion. It is the motion of the object relative to the fluid that is important; that is, a stationary object and a moving fluid give the same results. Similar behavior is observed in the flow of a fluid through a pipe. Laminar and turbulent flow are illustrated in Fig. 10.26.

For fluid flow through a tube, the Reynolds number becomes

$$\text{Re} = \frac{\rho v D}{\eta}, \qquad (10.15)$$

Figure 10.26
An example of turbulent flow. A laminar flow of smoke trails moving from left to right passes through a grid, which induces turbulent flow.

---

*The Irish-born engineer Osborne Reynolds (1842–1912) discovered in 1883 that laminar flow can become turbulent if the speed is sufficiently large.

where D is the diameter of the tube. The Reynolds number is dimensionless and has the same value in any consistent system of units. Observations show that for flow through a pipe, Reynolds numbers of less than about 2000 correspond to laminar flow and Reynolds numbers greater than 3000 correspond to turbulent flow. At values between 2000 and 3000, the flow is unstable and may change back and forth from one type of flow to another.

The frictional forces are much greater in turbulent flow than in nonturbulent flow. It is therefore often desirable to maintain laminar flow, whether in the case of water in pipes or blood in arteries and veins. In rigid pipes, this is accomplished primarily through using large pipes and low velocities. Blood vessels, however, are flexible and may expand with an increase in pressure. Turbulence in pipes can also be reduced by the appropriate placement of deflecting vanes and wires.

As a fluid moves past an object, it interacts with the object's surface forming a thin *boundary layer*. The interaction of the object with the boundary layer is the source of viscous drag. For very low Reynolds number (Re < 1), the flow is laminar and the drag is entirely viscous. At larger flow rates, the flow is not completely laminar. The boundary layer begins to separate from the surface leaving a turbulent wake in the trailing fluid. In this region (Re < 200,000), the resistive forces are partly due to viscous drag and partly due to form drag. As the flow increases further, the boundary layer separation begins to move closer to the front of the object—that is, the boundary separation moves in the upstream direction. The result is a larger wake and greater drag. At still larger flow velocities (Re > 200,000), the boundary layer becomes completely turbulent and the drag is entirely form drag.

At the transition to fully turbulent flow, there is a drop in the form drag, which then grows larger as the flow velocity is increased still further. The velocity at which the transition to fully turbulent flow occurs can be reduced by roughening the surface. For an object (a ball) moving through a fluid (air) at a speed near the transition to fully turbulent flow, the result of increasing the surface roughness is to lower the drag. Surface roughness affects the behavior of balls used in many sports. The dimples on a golf ball play a significant role in reducing the drag and thus increase the ball's time of flight. The fuzzy surface of tennis balls and the seams on baseballs play a role similar to the golf ball's dimples and contribute to the fully turbulent flow about them.

A related effect of turbulent air flow, this time around a spinning ball, causes the curve of a baseball or golf ball. To see this effect, imagine a stationary ball subjected to a flow of air (Fig. 10.27a). The motion of the air past the ball departs from streamline flow at the velocities normally attained by a thrown baseball, and some turbulence occurs, as shown. However, because of the symmetry of the ball, there is no net force on the ball perpendicular to the direction of flow.

When the ball is spinning, the friction between the ball's surface and the air drags a layer of air around with the ball (Fig. 10.27b). The pattern of turbulence becomes asymmetric, and the streamlines become more crowded at the top of the figure than at the bottom. The net result is a lowering of the pressure at the top and a force transverse to both the direction of flow and the axis of spin. This transverse force is the force that causes the ball to curve. By Newton's third law, there must also be a net force acting on the air whose effect is to deflect the air downward (Fig. 10.27c).

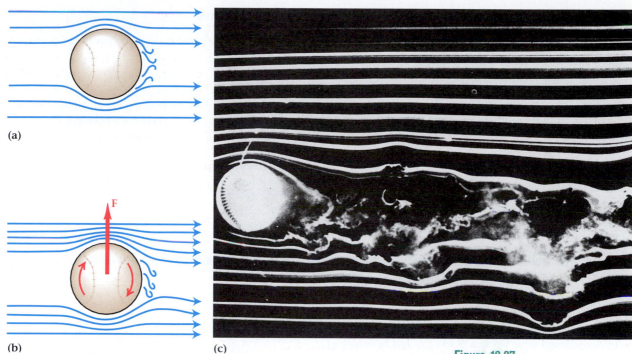

(a)

(b)

(c)

Figure 10.27
(a) Air flow about a baseball that is not spinning. (b) Air flow about a spinning baseball. (c) Photograph of smoke trails around a spinning baseball.

In golf, the impact of the slanted club head gives the ball spin around a horizontal axis (similar to Fig. 10.27c). The force resulting from the spin gives an additional lift to the ball, allowing it to travel farther. When the club does not strike the ball squarely, it gives the ball additional spin about a vertical axis, causing the ball to "hook" or "slice" to one side.

Summary

Useful Concepts

■ Pressure is defined as the force per unit area,

$$P = \frac{F}{A}.$$

The gauge pressure due to the weight of a fluid of density ρ at a depth h is

$$P = \rho g h.$$

■ Pascal's principle: The pressure applied at one point in an enclosed fluid is transmitted undiminished to every part of the fluid and to the walls of the container.

■ Archimedes' principle: A body, whether completely or partially submerged in a fluid, is buoyed upward by a force that is equal to the weight of the displaced liquid.

■ The equation of continuity,

$$v_1 A_1 = v_2 A_2,$$

is a statement of the principle of conservation of mass for fluids of constant density.

■ Bernoulli's equation is a statement of conservation of energy for fluids of constant density. For an incompressible nonviscous fluid in laminar flow, Bernoulli's equation is

$$P + \rho g h + \tfrac{1}{2}\rho v^2 = \text{constant}.$$

■ For a viscous fluid flowing in a horizontal pipe, Poiseuille's law gives

$$P_1 - P_2 = \frac{8Q\eta L}{\pi R^4},$$

where Q is the flow rate, L is the distance along the direction of flow, η is the coefficient of viscosity, and R is the radius of the pipe.

■ The force on a sphere moving in a viscous liquid is given by Stokes's law:

$$F = 6\pi\eta r v.$$

■ Aerodynamic drag is given by

$$F_{\text{drag}} = 0.65 \, C_{\text{D}} A v^2,$$

where C_{D} is the drag coefficient.

■ The Reynolds number is used to characterize turbulent flow, and for a tube is given by

$$\text{Re} = \rho v D / \eta$$

Important Terms

You should be able to write the definition or meaning of each of the following:

fluid	laminar flow
pressure	turbulent flow
pascal	equation of continuity
density	Bernoulli's equation
specific gravity	viscosity
Pascal's principle	Poiseuille's law
Archimedes' principle	Stokes's law
buoyant force	terminal speed
surface tension	drag coefficient
streamline	Reynolds number

Conceptual Questions

10.1 You are floating on a rubber raft in a small swimming pool in which the water level has been carefully measured. You throw overboard some wooden blocks that had been on the raft and watch the blocks float on the water. What happens to the water level as measured on the edge of the pool? Would the water level behave differently if the blocks were concrete and sank to the bottom of the pool?

10.2 Explain how a siphon works.

10.3 Three containers have the same base area but not the same shape. According to Pascal's principle, when they are filled to the same height with water, the same force acts on the base of each one. Yet when they are weighed on a scale, they do not weigh the same. Explain this apparent contradiction, which is sometimes called the *hydrostatic paradox.*

10.4 A block of wood floats half submerged in a container of water. If the same container were in an earth-orbiting satellite, how would the block float? Explain your reasoning.

10.5 The port of Hamburg is about 60 mi inland from the North Sea on the river Elbe. When a container ship left the port, a sailor noticed that there was a paint spot just at the water line. Where was the paint spot when the ship reached the open sea? Explain.

10.6 A sealed hollow glass tube is weighted at one end so that it floats upright when placed in a liquid. The level at which the tube floats depends on its weight and the density of the liquid. When calibrated, such a device can be used to measure fluid densities and is called a hydrometer. Discuss the effect of adhesive and cohesive forces on the level at which the tube floats.

10.7 A thin piece of wood is cut into the shape shown in Fig. 10.28, and a small piece of soap is placed at the point marked *A*. When the "boat" is placed in a pan of water, it moves in the direction indicated by the arrow. Explain.

10.8 A Ping-Pong ball can be suspended in a vertical stream of air, such as that from the exhaust pipe of a vacuum cleaner.

Figure 10.28
Question 10.7.

If the ball is given a small impulse to the side, it will return to the center of the stream rather than being ejected from the stream. Explain.

10.9 The card in the diagram in Fig. 10.29 will not fall away from the spool as long as one blows through the hole. Explain this effect. (The pin in the center is only to eliminate sideways motion and does not exert any vertical force.)

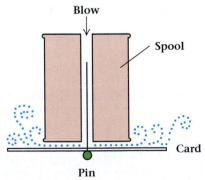

Figure 10.29
Question 10.9.

10.10 The viscosity of liquids decreases with increasing temperature. Give as many examples as you can of situations in which this effect can be encountered in everyday experience.

10.11 An upright cylindrical tank filled with water to a height *h* has three holes in its side through which the water escapes. The holes are at distances $h/4$, $h/2$, and $3h/4$ above the bottom of the tank. Which of the drawings in Fig. 10.30 corresponds to the streams of water that emerge? Explain your choice.

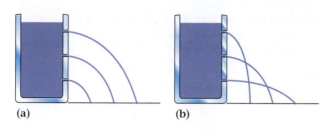

(a)   (b)

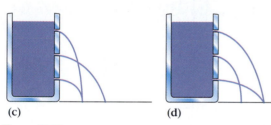

(c)   (d)

Figure 10.30
Question 10.11.

10.12 List some sports in which it is desirable to reduce turbulent fluid flow, and give some of the ways in which this is done.

10.13 Are there sports in which it is desirable to maximize turbulent fluid flow? If there are, give examples and indicate how turbulence is maximized.

10.14 How does the external design of a car affect its high-speed behavior? Why do race cars have spoilers (Fig. 10.31) on the back?

Figure 10.31
Question 10.14: The spoiler can be seen on the rear of the race car.

10.15 The explanation often given of how airplanes fly depends on the fact that the cross section of the wings usually is not symmetric but is curved on the upper surface and flat on the lower surface. How, then, do stunt flyers manage to fly upside down for considerable distances?

10.16 When syrup or oil is slowly poured from a container, the diameter of the stream decreases for a distance below the point at which it leaves the container. Explain this observation.

10.17 (a) If your mass is 62 kg and you float in water with only a negligible amount of your body above the surface, what is your approximate volume? (b) Determine your own volume by using either this technique or an improvement on it that takes account of how deep you really float.

Problems

Section 10.1 Hydrostatic Pressure

Hints for Solving Problems

Pressure is a scalar, not a vector. Remember that gauge pressure measures pressure above atmospheric pressure.

10.1 A block of wood 10 cm $\times$ 30 cm $\times$ 5.5 cm thick has a mass of 1240 g. (a) What is the density of this wood? (b) Is the wood balsa, oak, or pine?

10.2 A liter of corn oil has a mass of 0.925 kg. What is the (a) density and (b) specific gravity of the oil?

10.3 At the base of the Hoover Dam on the Colorado River, the depth of the water is 726 ft. What is the pressure at the base of the dam? Neglect the pressure due to the atmosphere.

10.4 A swimming pool is 50 m long by 23 m wide and is less deep at one end than at the other. The depth at the shallow end is 1.22 m, and the depth at the deep end is 4.35 m. The slope is continuous (smooth) from one end to the other. What is the difference in pressure on the bottom at opposite ends of the pool?

10.5 Organisms have been found living in the oceans where the pressure is as high as 1000 atm. To what depth does this pressure correspond? (Take the density of seawater to be 1.026×10^3 kg/m^3.)

10.6 An automobile tire is properly inflated at a pressure of 32.0 psi. What is its pressure expressed in kPa?

10.7 A 1350-kg automobile is supported by four tires inflated to a gauge pressure of 220 kPa. Ignoring the effects of tread thickness, calculate the area of contact between each tire and the road.

10.8 The gauge pressure at the bottom of a reservoir is four times what it is at a depth of 1.2 m. How deep is the reservoir?

10.9• A woman wearing high-heeled shoes places about 50% of her full weight on a single heel when walking. (a) Assuming the woman weighs 530 N, what is the pressure on the ground under one heel if the area of contact is 6.5 cm²? 1.0 cm²? (b) How does this compare with the pressure underneath an elephant's foot? For computation, assume that a full-grown elephant weighs 37,000 N and is standing evenly on four feet. Approximate the feet as circles 38 cm in diameter.

10.10• A rectangular fish tank measures 30 cm by 65 cm by 40 cm high. (a) If the tank is filled with water to a depth of 37 cm, what is the pressure at the bottom due to the water? (b) What is the total force of the water on the bottom?

10.11• A 1.000-m-tall pipe is filled to the halfway level with glycerin and then to the top with water. What is the gauge pressure at the bottom?

Section 10.2 Pascal's Principle

Hints for Solving Problems

In using Pascal's principle at some point in a fluid, make sure you know the pressure due to depth at that point before you consider the effect of an applied pressure.

10.12 A hydraulic jack is made with a small piston 1.2 cm in diameter that is used to move a large piston 5.4 cm in diameter. If a man can exert a force of 280 N on the small piston, how heavy a load can he lift with the jack?

10.13 A hydraulic press has a large piston with a cross-sectional area of 420 cm² and a small piston with a cross-sectional area of 5.00 cm². What is the force on the large piston when a force of 1.50 kN is applied to the small piston?

10.14 An air compressor maintains a pressure of 700 kPa over the hydraulic fluid in a tank (Fig. 10.32). The large piston that lifts the car has a cross-sectional area of 0.280 m². What is the maximum weight that it can lift?

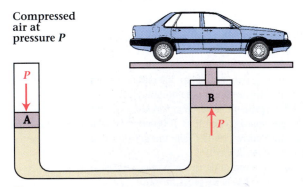

Figure 10.32
Problems 10.14 and 10.15: An air compressor provides a pressure at piston *A*. The pressure is transmitted through the hydraulic fluid to piston *B*.

10.15 A hydraulic lift of the type shown in Fig. 10.32 is used to raise a car weighing 15,000 N. The piston that supports the car has a diameter of 36 cm. What pressure of air within the system is required to just hold the car in place?

10.16 The column of mercury in a barometer stands 76.0 cm high. How tall would the barometer have to be if the mercury were replaced by water?

10.17• A U-shaped tube is partially filled with equal volumes of water and mercury (Fig. 10.33). If each liquid fills a 20-cm-long section of the tube, what will be the difference in the levels of the upper surfaces?

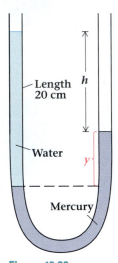

Figure 10.33
Problem 10.17.

10.18•• Corrosive liquids can be moved from containers by means of siphons rather than by pumps (Fig. 10.34). Over how high a wall can sulfuric acid (specific gravity 1.84) be siphoned?

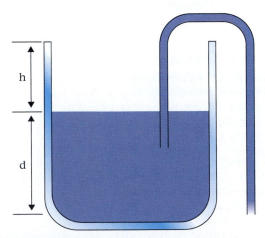

Figure 10.34
Problem 10.18.

Section 10.3 Archimedes' Principle

Hints for Solving Problems

Remember that the buoyant force equals the weight of the *displaced fluid* and is independent of the *weight* of the object. When working problems that use Archimedes' principle with gases, remember that the density of a gas depends on the pressure and the temperature through the relationship given in the footnote to Table 10.2.

10.19• A solid cube of unknown composition is seen floating upright in water with 30% of it above the surface. Calculate the density of the material.

10.20• A block of iron is suspended from one end of an equal-arm balance by a thin wire. To balance the scales, 2.35 kg are needed on the scale pan at the other end. (a) What is the volume of the block? (b) Next, a beaker of water is placed so that the iron block, suspended as in part (a), is submerged in the beaker but not touching the bottom (Fig. 10.35). What mass is now necessary to balance the scales?

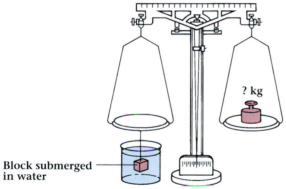

? kg

Block submerged in water

Figure 10.35
Problems 10.20 and 10.30.

10.21• A Goodyear blimp has a volume of 5750 m³ and a mass of 4300 kg when empty. What additional load is it able to lift when the entire volume is filled with helium at a temperature of 20°C? (Your answer will differ from the actual load because the entire volume is not filled with helium, the helium is not at atmospheric pressure, and the temperature of the air and helium may not be the same.)

10.22• The helium gas capacity of the dirigible Macon was 184,000 m³. The weight of the dirigible (less gas) was about 1,092,000 N. (a) Approximately how much additional load could the dirigible lift from the ground? Assume a temperature of 21°C. (b) What percent more load could it lift if hydrogen were used for the gas?

10.23• A plastic bag is filled with helium at atmospheric pressure and 21°C. How large a volume of helium is required to lift a 50-kg girl off her feet? Assume that the mass of the bag is negligible. (See Table 10.2.)

10.24• If, in Problem 10.23, hot air is used instead of helium, what is the required volume for the balloon if the air inside can be maintained at a temperature of 44°C? Assume the outside air is at 21°C.

10.25• A wooden dowel is placed in a test tube containing water. The dowel floats with 60% of its length below the water surface. How much of the dowel is submerged when it floats in methyl alcohol?

10.26• A 1.0-kg container of water sits on a scale A. A piece of aluminum 10 cm × 10 cm × 10 cm is suspended from a spring scale B so that half of the block is submerged in the water. (a) What is the reading on spring scale B? (b) What is the reading on scale A?

10.27• A 0.0132-kg seashell of density $\rho = 3.54 \times 10^3$ kg/m³ is suspended by a thread from a spring scale. The seashell is then lowered into seawater until it is completely submerged. If the scale is calibrated in units of newtons, what is the reading of the scale?

10.28•• A block of pine wood ($\rho = 0.40$ g/cm³) is floating on a pond. The block is 10 cm × 40 cm × 5 cm thick. (a) How much of the block protrudes above the water? (b) If the block is made to carry a load by placing additional mass on top of it, how much mass must be added to just submerge the block?

10.29•• A wooden block 20 cm × 20 cm × 10 cm has a density of 0.60 g/cm³. (a) How much iron ($\rho = 7.86$ g/cm³) can be placed on top of the block if the top of the block is to be level with the water around it? (b) If iron were attached to the bottom of the block instead, what mass of iron would it take to bring the top of the wooden block down to the level of the water? Why?

10.30•• A king's crown is said to be solid gold but may be made of lead and covered with gold. When it is weighed in air, the scale reads 0.475 kg. When it is submerged in water, the scale reads 0.437 kg. (a) Is it solid gold? (b) If not, what percentage by mass is gold? (Refer to Fig. 10.35.)

10.31•• An inverted Atwood machine is constructed by fixing a pulley P near the bottom of a container filled with a liquid of density ρ (Fig. 10.36). Two floats with the same mass

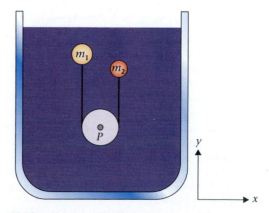

Figure 10.36
Problem 10.31.

($m_1 = m_2$) but different density are joined together by a string passing around the pulley. Because the masses are not the same size, the upward forces on them are unequal. The upward force F_1 on mass m_1 is greater than the upward force F_2 on mass m_2. The floats are released from rest. (a) Draw a free-body diagram for each float including all of the forces that act on the float. (b) What is the net force on each float? (c) Find an expression for the initial acceleration of each float. Ignore the pulley's mass, any friction in the pulley, and any fluid friction.

*Section 10.4 Surface Tension

10.32 What force is required to overcome surface tension when raising a horizontal 8.0-cm-diameter ring out of water at 25°C?

10.33 A 5.0-cm-diameter ring is used to determine the surface tension of a liquid. What is the liquid's surface tension if, in addition to the ring's weight, a force of 2.3×10^{-2} N is required to lift the ring from the liquid?

10.34 A wire frame with a sliding crosspiece is dipped into a soap solution and held vertically (Fig. 10.37). The crosspiece is 5.0 cm long and has a mass of 0.265 g. What is the value of the surface tension of the soap solution if the weight of the crosspiece is just balanced by the surface tension force?

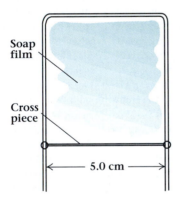

Soap film

Cross piece

|← 5.0 cm →|

Figure 10.37
Problem 10.34.

10.35 How high will water rise in a capillary tube with inside diameter of 0.10 mm? Assume the contact angle to be $\theta = 0°$. Take the temperature to be 25°C.

10.36• Water is poured into an upright U-shaped tube in which the legs have different internal diameters. If the diameter of one is 0.6 mm and the diameter of the other is 1.2 mm, what will be the difference in the height of the water in the two legs? Take the contact angle to be 0°.

10.37• By what factor is the surface energy of a 2.0-cm-diameter soap bubble increased when it is blown up to a diameter of 6.0 cm?

Sections 10.5 and 10.6 Fluid Flow: Streamlines and the Equation of Continuity; Bernoulli's Equation

Hints for Solving Problems

Remember that Bernoulli's principle applies to incompressible, nonviscous fluids in laminar flow. Be especially careful with units, since many of the quantities used in fluid flow involve derived units rather than fundamental units.

10.38• What is the pressure change in water going from a 4.0-cm-diameter pipe to a 2.0-cm-diameter pipe if the velocity in the smaller pipe is 3.0 m/s?

10.39• What is the pressure change in methyl alcohol flowing from a 4.0 cm diameter pipe to a 1.5 cm diameter pipe if the velocity in the larger pipe is 0.40 m/s?

10.40• Water is flowing in a horizontal pipe of variable cross section. Where the cross-sectional area is 1.0×10^{-2} m², the pressure is 5.0×10^5 Pa and the velocity 0.50 m/s. In a constricted region where the area is 4.0×10^{-4} m², what are the pressure and velocity?

10.41• A large storage tank is filled with water. Neglecting viscosity, show that the speed of water emerging through a hole in the side a distance h below the surface is $v = \sqrt{2gh}$. This result is known as *Torricelli's theorem*. Try using Bernoulli's equation.

10.42• Suppose you blow air at a speed of 10 m/s across one end of a U-shaped tube containing water. What will be the difference in the heights of the water surfaces on the two sides? Which one will be higher?

10.43•• Use the results of Problem 10.41 to show that you obtain the maximum range for water leaving a hole in the side of a tank resting on the ground when the hole is halfway between the top and bottom surfaces of the liquid.

10.44•• Assume that you wish to find the speed of a moving fluid that obeys Bernoulli's equation. Use the equation of continuity and Bernoulli's equation to derive the equation for the speed measured by a Venturi meter (refer to Fig. 10.21).

*Section 10.7 Viscosity and Poiseuille's Law

10.45 A horizontal garden hose 15 m long with an interior diameter of 1.25 cm is used to deliver water at the rate of 150 cm³/s. What is the pressure drop from one end of the hose to the other? Assume a temperature of 20°C.

10.46 Mercury flows through a horizontal pipe 4.0 cm in diameter and 0.50 m long. If the pressure drop from one end of the pipe to the other is 1.0×10^4 Pa (about 1/10 atm), what is the rate of flow through the tube?

10.47 By what fraction would the blood flow be reduced if an arteriole were reduced to 0.95 of its former diameter? Assume the pressure and viscosity to be constant.

10.48• How high above the point of injection must a container of blood plasma be if the plasma is to enter the patient's arm at a rate of 3.0 cm³/min through a needle that is 50 mm long and has an inside diameter of 0.55 mm? Assume the pressure in the vein to be 15 mm Hg. (Assume also that the density of blood plasma is 1.05 g/cm³ and its viscosity is about 1.5×10^{-3} Pa · s.)

*Section 10.8 Stokes's Law and Terminal Speed

10.49 Compare the sedimentation rates for a mixture of spherical particles that are all of the same material but have diameters that differ in the ratio of 1:2:3.

10.50 A steel ball bearing 8.00 mm in diameter is dropped into a cylinder of glycerin. The densities of steel and glycerin are 7.80×10^3 and 1.26×10^3 kg/m³, respectively. What is the terminal speed of the ball bearing?

10.51 (a) A bottle of corn syrup is taken from a refrigerator ($T = 5°C$), and a glass marble of density 2.5×10^3 kg/m³ is dropped into it. The marble takes 45 s to sink to the bottom. The diameter of the marble is 1.57 cm, the depth of the liquid is 12.1 cm, and its density is 1.2×10^3 kg/m³. What is the viscosity of the syrup at that temperature? (b) If the bottle is kept out of the refrigerator for several hours and the experiment is done again, the marble takes 5.0 s to fall through the liquid. What is the viscosity of the syrup at room temperature?

10.52 Measurements of falling coffee filters show that they experience a drag force proportional to the square of their speed. If three filters are nested together so that their effective cross-sectional area is the same as that of a single filter, what would be the ratio of their terminal speed to the terminal speed of a single filter?

10.53 A small balloon is inflated to a diameter of 20 cm and has a total mass of 0.40 g. When it is allowed to fall in air, the balloon has a drag force predominantly due to v^2, where v is its speed. Calculate the balloon's terminal speed given that the coefficient b equals 9.0×10^{-3} kg/m.

10.54 The speed of an automobile increases from 80 km/h (50 mi/h) to 115 km/h (71 mi/h). What is the ratio of the drag forces at the two speeds?

10.55● When the engines of a jet airliner develop 1.00×10^5 N of thrust (driving force), the jet reaches an air speed of 750 km/h. Calculate the thrust required for speeds of 800 km/h and 600 km/h. What does your result suggest about the relationship between fuel consumption and speed if the thrust is proportional to fuel consumption?

10.56● Two spherical objects have the same size and same surface roughness. One of them is heavier than the other. Show that if both objects are simultaneously released from rest from the same height, the heavier one strikes the ground first.

10.57● A geological sample from a river bed forms sediment at the rate of 1.0 g/day. How many revolutions per second would a centrifuge have to achieve to increase the sedimentation rate to 3.0 g/h? Assume that the sample is placed 5.0 cm from the axis of rotation of the centrifuge.

10.58● A baseball falling through the air experiences a drag force

$$F(\text{newtons}) = 8.06 \times 10^{-4} \, v^2,$$

where v is in m/s. What is the terminal speed for the ball if it falls from a great height? The ball has a mass of 0.145 kg.

10.59●● Show that the terminal speed of an object falling in air, such as a ball or parachute, can be estimated by

$$v_t = \sqrt{\frac{mg}{kA\rho}},$$

where m is the mass of the object, A is its cross-sectional area, ρ is the density of air, and k is a dimensionless constant whose value is 1 or less and depends on the shape of the object. (*Hint:* During its fall, the object "sweeps out" a vertical tube of air. Assume that the drag force is proportional to the rate at which the falling object transfers momentum to this tube of air, and that the rate of change of momentum is the product of the speed and the rate at which the mass of the air in the tube is displaced.)

*Section 10.9 Turbulent Flow

10.60 What is the order of magnitude of the lowest speed that a Ping-Pong ball can have in air at 20°C if the flow remains turbulent? The diameter of a Ping-Pong ball is 3.75 cm.

10.61 Show that the Reynolds number is dimensionless.

10.62● (a) Calculate the terminal speed for a steel ball of radius 0.50 cm and density 7.8×10^3 kg/m³ falling through water. Assume that Stokes's law applies. (b) Calculate the Reynolds number by setting L in Eq. (10.14) equal to the diameter of the ball. (c) Is the flow really laminar? What does this suggest about the value for the terminal speed? Assume a temperature of 20°C.

10.63● What is the minimum diameter of a pipe through which 1.00 m³ of glycerin at 20°C can be made to flow per hour if the flow is to be laminar?

10.64● If the flow of a liquid in a 2-cm-diameter pipe is just barely laminar, what size pipe would be needed to maintain laminar flow if the flow rate were to be twice as much as in the first pipe?

10.65● How much water per hour can be delivered by a $\frac{3}{4}$-in. pipe in which laminar flow is maintained? Assume a temperature of 20°C.

Additional Problems

10.66 What is the pressure at the bottom of a 2.0-km-deep oil well filled with oil of density 860 kg/m³?

10.67 A water tower provides pressure for a water supply system. What is the maximum water pressure available at the bottom of the tower, given that the water level is 33 m above the place where the pressure is to be measured?

10.68 A Rolex Sea-Diver wrist watch is guaranteed to be water resistant down to a depth of 4000 ft below sea level. A special version of this model ran after having been submerged to a depth of 35,000 ft in the Marianas Trench in the Pacific Ocean. To what pressures do these depths correspond? (Take the density of sea water to be 1.026×10^3 kg/m^3.)

10.69 Water is flowing in a horizontal pipe of varying cross section. At one point where the cross-sectional area is 1.0×10^{-2} m^2, the velocity of the water is 2.0 m/s and the pressure is 15 kPa. In another region of the pipe the velocity is 3.0 m/s. What is the cross-sectional area at the second position and what is the pressure there?

10.70 A glass marble of density 2.5×10^3 kg/m^3 and diameter 5.0 mm is dropped into a cylinder containing castor oil of density 900 kg/m^3. What is the terminal speed of the marble?

10.71● The deep research vessel Alvin can dive to depths of 4000 m. What is the force of the water on one of the submarine's 30.5-cm-diameter circular ports? The pressure inside the titanium hull is one atmosphere.

10.72● A garden hose has an interior diameter of 0.95 cm and a nozzle with a 0.40-cm-diameter opening. (a) If 30 liters per minute flow through the hose, what is the speed of the water emerging from the nozzle? (b) If the nozzle is held horizontally 1.1 m above level ground, how far will the stream go before it hits the ground?

10.73● A Venturi meter of the type shown in Fig. 10.21 has mercury in the U-shaped tube. The meter is used to measure the flow speed of water. At A the pipe diameter is 10.0 cm, at B it is 5.0 cm. What is the speed of the water at A if the differential height h is 5.0 cm?

10.74● On each heartbeat about 70 cm^3 of blood is forced from the heart at an average pressure of approximately 105 mm Hg. What is the average power output if the heart beats 70 times each minute?

10.75●● A glass disk G is held tightly against the end of a vertically held cylindrical tube C while the tube is lowered so that the bottom of the disk is 20 cm below the surface of the water (Fig. 10.38). What is the maximum thickness the disk can have and not fall away from the cylinder? The density of the glass is $\rho = 2.5 \times 10^3$ kg/m^3.

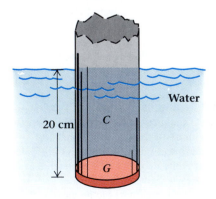

Figure 10.38
Problem 10.75.

10.76●● A hollow Ping-Pong ball is tethered to the bottom of a water-filled cylinder by a short string of length L. When the cylinder is mounted on a moveable arm and set into rotation, the ball is deflected toward the center of rotation (Fig. 10.39). Find an expression for the deflection angle θ in terms of the rotational frequency ω and the distance R.

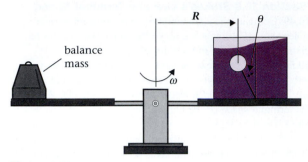

Figure 10.39
Problem 10.76.

10.77●● A cylinder of solid uranium weighs 9.34 kg in air, 8.84 kg in water, and 2.54 kg in another liquid. (a) What is the volume of the cylinder? (b) What is the density of the uranium cylinder? (c) What is the density of the liquid? (d) Identify the liquid.

10.78●● A block of oak wood floats at the interface between gasoline and water (Fig. 10.40). One-third of the volume of the block is in the water and two-thirds of the block is in the gasoline. What is the density of the block?

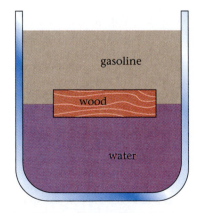

Figure 10.40
Problems 10.78 and 10.79.

10.79●● An ice cube floats at the interface between olive oil and water in the same manner as the block in Problem 10.78 (Fig. 10.40). If the olive oil has a specific gravity of 0.91, what fraction of the ice cube is submerged below the level of the oil-water interface?

10.80●● A dense liquid is poured into a 1-m-deep container and a less dense liquid carefully poured on top of it, so as to form two layers. After many days the liquids have become mixed, but not thoroughly so. At the bottom of the container the density is still that of the denser liquid and at the top the density is that of the lighter liquid. Tests show that the density is given by

$$\rho = (1 + 0.26x^2) \times 10^3 \text{ kg/m}^3,$$

where x is the distance below the surface in meters. What is the pressure in this liquid mixture at a depth of 0.5 m below the surface? (Try using the graphical technique of Chapters 2 and 6. Add up the weight of individual thin layers whose density is nearly constant.)

10.81•• If the internal volume of a hot-air balloon is 2180 m³, at what temperature must the air be to keep a 475-kg balloon and loaded gondola in the air when the outside temperature is 20°C? (See the footnote in Table 10.2.)

10.82•• Show that the problem posed in Example 10.3 can be solved without resorting to Pascal's principle by assuming that the output work is equal to the input work. (*Hint:* Remember that the volume of liquid remains constant.)

10.83•• A triangular prism of ice with uniform thickness floats in sea water with its base above the water (Fig. 10.41). (a) Show that the fractional volume of ice below the water is $V_{\text{below}}/V_{\text{total}} = \rho_{\text{ice}}/\rho_{\text{sea water}}$. (b) Show that the ratio of the depth of the peak below the water to the total height of the prism is $d/h = \sqrt{\rho_{\text{ice}}/\rho_{\text{sea water}}}$.

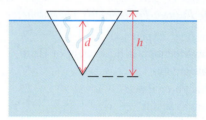

Figure 10.41
Problem 10.83.

Thermal Physics

The development of thermal physics was greatly influenced by the practical concerns and search for efficiency that characterized the Industrial Revolution in the eighteenth and nineteenth centuries. By that time, mechanics was relatively well developed, but electricity had not yet become of practical or commercial importance. As a consequence, the disciplines of mechanics, heat, and electricity all evolved along different paths. As recently as the late eighteenth century, the study of heat was not related to the study of mechanics. As a result, the definitions and units of measurement for temperature and heat were developed independently of the definitions and units for work and mechanical energy. Not until the mid-nineteenth century did James Prescott Joule quantitatively connect the unit of thermal energy with the unit of mechanical work, allowing us to see mechanics and thermal physics as parts of a greater whole.

Today we think of heat as a form of energy transfer. The effects of heat and temperature changes are a fundamental aspect of many physical situations, from the study of star formation to research in lasers. ■

11.1 Temperature and States of Matter

We begin our discussion of thermal physics by briefly describing what we mean by different states of matter. As an example, water can have the form of ice (solid), water (liquid), or steam (gas). Many other materials can also exist as solids, liquids, or gases. Such distinct forms, or states, of matter are called *phases* (Fig. 11.1). The change from one state, or phase, to another, such as the melting of ice, is usually caused by a transfer of thermal energy.

The molecules of a gas move about freely, except when they collide with other gas molecules or the walls of the container. The average separation between molecules is large compared with their own size, and as a result, a gas has no definite volume. Consequently, a gas may be compressed or expanded and will fill a container of any shape or size. In a liquid, the average separation between molecules is comparable to their own diameters. Individual molecules are free to move about, but because of the forces between them they move so that the average separation between near neighbors remains essentially constant. As a result, a liquid is virtually incompressible and has a definite volume, although its shape can change to match the shape of its container. In solids the separations are comparable to the separations in liquids, but the binding forces are so strong that the atoms in a solid are not free to move about. Instead, the atoms of a solid are confined to small oscillations about fixed positions. Thus, a solid has not only a definite volume but a definite shape as well.

Energy is associated with the motion of molecules in any state of matter. In fact, as we will see, changes in temperature and changes from one state to another are simply large-scale manifestations of changes in the energy of the random motions of the atoms and molecules that compose the material.

The concept of temperature originated in human sensory perception of the environment. When you touch an object, you say it is relatively "hot" or "cold." This response early led to attempts to describe the feeling in terms of the objects; for example, this rock is warmer than that rock. The desire to quantify and measure such differences in warmth culminated in the idea of **temperature:** the number assigned to an object as an indication of its warmth. A device used to measure temperature is called a **thermometer.** Thermometers are used not only as quantitative indicators to measure what the hand can feel, but also to extend the range of measurements far beyond the sense of touch. The range of thermometers extends from temperatures low enough to freeze the gases of the air to the enormous temperatures at the interiors of stars.

Our sensory perception allows us to define another useful term. Place an object A, which feels hot to your hand, in contact with an object B of the same material, which feels cold to your hand. After a period of time has passed, they will give the same sensation to your hand. Objects A and B are said to be in **thermal equilibrium,** and their temperatures are equal (Fig. 11.2). We can extend this idea to say that two objects that are not touching are in thermal equilibrium if, upon being placed in contact, their temperature would not change. This principle, which is discussed in greater detail in Chapter 13, is very useful in making a thermometer as described in the following section.

Figure 11.1

Matter that we commonly encounter exists in one of three phases: solid (glacier), liquid (water), and gas (air and water vapor).

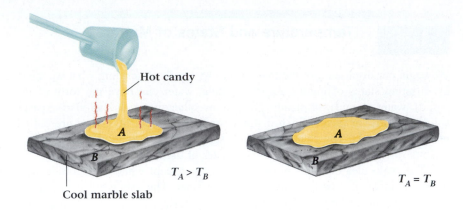

Figure 11.2
The hot molten candy and the cool marble slab reach thermal equilibrium after a period of time. At thermal equilibrium, their temperatures are equal.

Hot candy

A

B

$T_A > T_B$

Cool marble slab

A

B

$T_A = T_B$

11.2 Thermometry

The development and calibration of thermometers and the establishment of temperature scales are the essence of *thermometry,* the science of temperature and its measurement. The basis of thermometry is that some physical properties vary with temperature in a quantitative and repeatable fashion. Some of these thermometric properties are the volume of a gas or liquid, the length of a metallic strip, the electrical resistance of a conductor, and the light-transmitting properties of a crystal. Any physical system whose properties change with variations in temperature can be used as a thermometer. The choice of a particular thermometer depends primarily on the range of temperatures to be studied. We measure the change in some property, say the length of a column of liquid, and then associate a change in temperature with our measurement of the change in length.

The liquid-in-glass thermometer was invented about 1650 by the Grand Duke of Tuscany, Ferdinand II, one of Galileo's fellow countrymen and a patron of science. In this thermometer, a liquid indicator is sealed into a glass capillary tube having a bulb at one end. When the temperature increases, both the volume of the glass bulb and the volume of the liquid increase. If they both expanded at the same rate, we would observe no change. But because the liquid expands at a greater rate than the glass does, the liquid is forced to expand into the tube as the temperature increases. By using a relatively large bulb and a narrow tube, it is possible to make a thermometer that we can read easily from a scale scribed on the glass. The common fever thermometer is made this way.

We choose to discuss the liquid-in-glass thermometer here because of its simplicity and great familiarity. However, we emphasize that it is only one of many possible types of thermometers. For example, some thermometers determine temperature by measuring the electrical resistance of a platinum wire or a semiconductor crystal (Fig. 11.3). For a fixed applied voltage, the amount of current transmitted by the wire or crystal depends on the temperature and is reproducible. Thermometers of the type shown are used in medical applications. Similar thermometers are used in applications that require precision measurement over a wide range of temperatures.

Any thermometer, whether liquid-in-glass or one that depends on other thermal properties, must be calibrated to make it a useful instrument, capable of quantitative and reproducible measurements. For example, a liquid-in-glass ther-

Figure 11.3
A digital fever thermometer uses the temperature dependence of the electrical resistance of a semiconductor crystal to measure temperature.

mometer can be calibrated by marking on the glass the position of the liquid column at a set of reference temperatures or standards. Temperatures between the reference marks are interpreted as proportional to the length of the liquid column. In fact, this is just how the **Celsius temperature scale** is defined. The two fixed points of reference are the ice point and the steam point. The ice point is defined as the equilibrium temperature of a mixture of ice and water at a pressure of one atmosphere; the steam point is defined as the equilibrium temperature of water and steam at a pressure of one atmosphere. The numbers assigned to these two points in the Celsius scale are arbitrarily chosen as 0 for the ice point and 100 for the steam point.

Assuming that the cross section of the thermometer capillary is uniform and that the rate of expansion of the liquid with changes in temperature is constant, we then can mark the distance between the ice and steam points into 100 equal parts. We can easily compare the level of the liquid to the nearest mark, called a degree.* We thus measure temperature in units of degrees Celsius, abbreviated as °C. (This scale was originally known as the centigrade scale because it has one hundred divisions between the principal reference marks. The present name was adopted to honor the Swedish astronomer Anders Celsius, who popularized the scale in 1742.)

Figure 11.4 shows a Celsius scale thermometer at a temperature T_C, at which the liquid is extended a distance L beyond the zero position. We may calculate the temperature by

$$T_C = \left(\frac{L}{L_0}\right) \times 100,$$

where L_0 is the distance between the 0° and 100° marks. Here we have defined the temperature scale to be a linear function of the length L of the liquid column. (There is no fundamental reason for doing this; we could define other functions equally well. However, the linear relationship is the simplest.) When the temperature scale is defined this way, other important physical properties turn out to be approximately independent of temperature.

Although the Celsius temperature scale is widely used, there is nothing fundamental about choosing the ice point to be 0° and the steam point as 100°. The **Fahrenheit temperature scale** assigns a value of 32° to the ice point and 212° to the steam point, a difference of exactly 180°. It is easy to transform temperatures in one system into temperatures in the other system. For example, let us find the relationship that transforms temperature in °C to temperature in °F. Just remember that 0°C is equivalent to 32°F, and that a range of 180° on the Fahrenheit scale is 100° on the Celsius scale. Therefore, one Celsius degree is equivalent to $\frac{180}{100}$, or $\frac{9}{5}$, of one Fahrenheit degree. We can then write the Fahrenheit temperature T_F as

$$T_F = \frac{9}{5}T_C + 32. \tag{11.1}$$

You can convince yourself that this is correct by substituting the Celsius values for the freezing and boiling temperatures of water and seeing whether you get

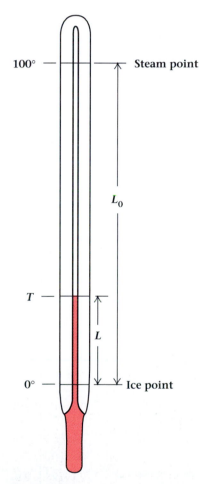

Figure 11.4
A liquid-in-glass thermometer with a Celsius scale. The temperature is proportional to the ratio of the distance L to the reference distance L_0.

---

*Some early thermometers were marked in 360 divisions, like the parts of a circle; thus, the term *degree* became applied to temperature.

the correct Fahrenheit values. You can use this same substitution to check whether you have remembered the formula correctly when you must recall it without the book.

Equation (11.1) gives the Fahrenheit temperature if the Celsius temperature is known. It can easily be rearranged to express the Celsius temperature in terms of Fahrenheit. Two examples of these transformations follow.

Example 11.1

What Fahrenheit temperature is equivalent to 37.0°C?

Solution Application of Eq. (11.1) yields

$$T_F = \tfrac{9}{5}T_C + 32 = \tfrac{9}{5}(37.0) + 32 = 66.6 + 32 = 98.6°F.$$

Example 11.2

On a day when the temperature is 86°F, what is the reading of a Celsius thermometer?

Solution From Eq. (11.1) we have

$$T_F = \tfrac{9}{5}T_C + 32,$$

which can be rearranged to give

$$T_C = \tfrac{5}{9}(T_F - 32),$$
$$T_C = \tfrac{5}{9}(86 - 32) = \tfrac{5}{9}(54) = 30°C.$$

In both the Fahrenheit and Celsius temperature scales, the assignment of the zero point is arbitrary. We can readily achieve temperatures below these zero points. However, one temperature scale has a more fundamental choice of zero. This scale was proposed in 1848 by William Thomson, Lord Kelvin (1824–1907), and arose from the study of gases. Kelvin's scale uses intervals equal to those of the Celsius degree, but with zero set at the lowest theoretical temperature that a gas can reach. The scale is based on the fact that a gas at 0°C will lose 1/273.15 of its volume for a 1°C drop in temperature. If this reduction in volume were to continue with decreasing temperature and if the gas did not liquefy, the volume would become zero at −273.15°C, a temperature called **absolute zero**. The temperature scale based on this zero is the **Kelvin temperature scale**. We will discuss the physical meaning of this observation in more detail in Chapter 12. In this chapter you can just consider the Kelvin scale to be a temperature scale with degree intervals of the same size as the Celsius degree, but with the zero point at −273.15°C. Thus a temperature *change* of 1°C is the same as a *change* of 1 K.

The conversion between Celsius and Kelvin temperatures is a simple one,

$$\boxed{T_K = T_C + 273.15.} \tag{11.2}$$

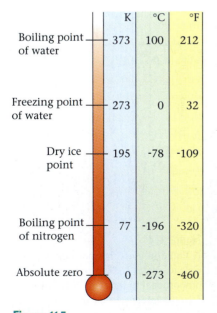

	K	°C	°F
Boiling point of water	373	100	212
Freezing point of water	273	0	32
Dry ice point	195	-78	-109
Boiling point of nitrogen	77	-196	-320
Absolute zero	0	-273	-460

Figure 11.5

A comparison of the Fahrenheit, Celsius, and Kelvin scales of temperature.

FAHRENHEIT'S THERMOMETER

Although the Celsius scale is becoming increasingly common in the United States, most people in the United States still think in terms of Fahrenheit temperatures when deciding what to wear outside. Have you ever wondered why the freezing temperature is 32°F? Why not 0° or 100°? What's so special about the numbers in 32°F, 212°F, or even 98.6°F?

At the beginning of the eighteenth century, the Danish astronomer Ole Roemer (famous for making the first measurements that showed that the velocity of light is finite) devised a temperature scale of his own for use with the alcohol-in-glass thermometers that he constructed. His thermometers attracted the attention of Gabriel Fahrenheit (1686–1736), a manufacturer of meteorological instruments in the Netherlands. In 1708, Fahrenheit traveled to Copenhagen to meet Roemer and see his thermometers, which were based on two reference points. For one reference Roemer used a mixture of ice, water, and salt to reach the lowest temperatures then attainable in the laboratory, which he called zero. His other reference was the boiling point of water, which he arbitrarily designated as 60 degrees.

Fahrenheit returned home to make thermometers like Roemer's. In 1714 he overcame technical difficulties with alcohol thermometers by substituting mercury as the expanding liquid. The use of mercury extended the range of temperature measurements from well below Roemer's zero to well above the boiling point of water. Furthermore, mercury expanded and contracted more uniformly than the other liquids then in use. As a result, Fahrenheit could mark his mercury thermometers more accurately and with finer divisions.

By 1724, Fahrenheit had adopted a new scale, similar to Roemer's but with much finer divisions. For the zero point he chose the same reference as Roemer. However, since his thermometer was intended for meteorological observations, he wanted a second reference point that would be nearer the maximum observed temperature for weather. He chose the normal temperature of the human body as the upper reference point, which he called 96°. Fahrenheit gave no reason for his choice of 96, but it may have been due to his desire for a finer scale and because 96 is evenly divisible by 2, 3, 4, 8, and 12.

Why didn't Fahrenheit choose the freezing point of water for his zero reference, as Newton had done before him and as Celsius did later on? Perhaps Fahrenheit was influenced by Roemer, or he may have wanted to avoid the inconvenience of repeatedly using negative temperatures during winter. Also, in the early 1700s it was widely believed that water did not always freeze at the same temperature. Soon, using his newly calibrated thermometers, Fahrenheit learned that water always froze at 32° on his scale. He immediately added this third reference point to his instruments.

A report of Fahrenheit's thermometers was published in the *Philosophical Transactions* in 1724. Almost at once his scale was adopted in Great Britain and the Netherlands and gained wide acceptance throughout the English-speaking countries.

The Fahrenheit scale in use today differs slightly from the original. The two fixed points are the ice point, assigned a value of 32°F, and the steam point, assigned a value of 212°F. On this scale the normal human body temperature is 98.6°F, slightly higher than the 96° originally chosen by Fahrenheit.

Today the Celsius scale and the Kelvin scale have replaced the Fahrenheit scale for scientific work. Also, the range of temperatures that can now be measured has been extended by many orders of magnitude since Fahrenheit's time. Modern thermometry uses many different physical properties to indicate temperatures, spanning a range from the extreme lows near 10^{-6} K to the surface temperature of the stars at about 10^4 K. The choice of thermometer depends on the temperature to be measured. For example, infrared pyrometers, which use the infrared radiation from hot matter to measure temperature, can measure temperatures ranging from $-30°C$ to $3000°C$. Steelworkers use pyrometers to find the temperature of molten steel. Parents quickly take their baby's temperature with special pyrometers that sense the radiation generated by the child's eardrum and surrounding tissue (Fig. B11.1).

Figure B11.1 A radiation thermometer (pyrometer) uses infrared to accurately measure a child's temperature in just one second.

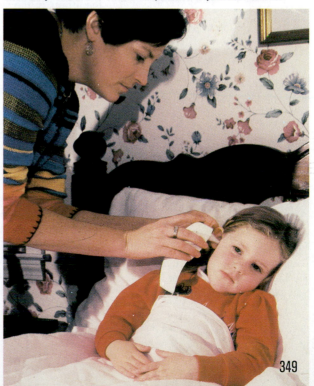

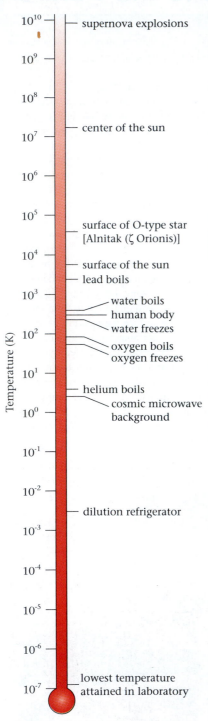

Temperature (K)

10^{10}	supernova explosions
10^{9}	
10^{8}	
10^{7}	center of the sun
10^{6}	
10^{5}	
10^{4}	surface of O-type star [Alnitak (ζ Orionis)]
10^{3}	surface of the sun / lead boils
10^{2}	water boils / human body / water freezes / oxygen boils / oxygen freezes
10^{1}	helium boils
10^{0}	cosmic microwave background
10^{-1}	
10^{-2}	
10^{-3}	dilution refrigerator
10^{-4}	
10^{-5}	
10^{-6}	
10^{-7}	lowest temperature attained in laboratory

Figure 11.6
Temperature range and corresponding physical situations that occur in nature. Note that the scale is logarithmic.

The unit of absolute temperature is the kelvin (K). It is not written with a degree sign. A temperature of 0°C is simply 273.15 K. A comparison of the various scales is shown in Fig. 11.5.

Today the range of temperatures that can be measured has been extended by many orders of magnitude since Fahrenheit's time. Modern thermometry uses many different physical properties to indicate temperatures, spanning a range from the extreme lows below 10^{-6} K to the surface temperature of the stars at about 10^{4} K. The particular choice of thermometer depends on the temperature to be measured. No single thermometer can span the enormous range of temperatures that occur in nature. Figure 11.6 shows the range of temperatures and the corresponding physical phenomena associated with them.

11.3   Thermal Expansion

When you loosen the metal cap on a glass bottle by holding it in a stream of hot water, you are making use of thermal expansion. The liquid-in-glass thermometer described in the preceding section works because of the difference in the rates of thermal expansion of the indicating liquid and of the glass envelope. We can calibrate such a thermometer without ever knowing the individual expansion rates. However, once a temperature scale is established, we can show that, to a good approximation, most solid objects change length in direct proportion to a change in temperature. Also, the change in length is proportional to the initial length of the object. Thus, for the same increase in temperature, a long copper bar expands more than a shorter one, but the ratios of change in length to initial length are the same for both bars (Fig. 11.7). We call this behavior **linear thermal expansion,** that is, expansion in one dimension. The reason for this expansion is that the increase in temperature causes greater amplitudes of vibration of the atoms in the solid, giving a greater average distance between them.

Let us describe linear thermal expansion mathematically. For an object of initial length L_0, the change in length ΔL due to a change in temperature ΔT can be expressed as

$$\Delta L = L_0 \alpha\, \Delta T. \tag{11.3}$$

The proportionality constant α, called the **linear thermal expansion coefficient,** has the dimension of inverse temperature, or °C^{-1}. Table 11.1 (p. 353) lists thermal expansion coefficients for a number of common materials. The coefficients themselves have some slight temperature dependence and are given here for 20°C, but you may take them as constant for the purposes of your study.

For an increase in temperature $\Delta T = T - T_0$, a rod of initial length L_0 expands to a new length $L = L_0 + \Delta L$. With the aid of Eq. (11.3), we can express the length L at the new temperature T in terms of the initial length L_0 at the initial temperature T_0:

$$L = L_0 (1 + \alpha \Delta T),$$

or

$$L = L_0[1 + \alpha(T - T_0)]. \tag{11.4}$$

Example 11.3

The Verrazano-Narrows Bridge between Brooklyn and Staten Island in New York City is one of the world's longest suspension bridges, with a center span of 1300 m (Fig. 11.8). Because the temperature variation over a year may be quite large, allowance must be made for thermal expansion and contraction of its materials. Assuming that the bridge is steel and, for safety, allowing for a temperature range of 120°C, how much thermal expansion must be allowed for in the center span? (This allowance is made by using expansion joints and components that can move with respect to each other.)

Solution We can obtain the change in length of the center span from our definition of thermal expansion:

$$\Delta L = L_0 \alpha\, \Delta T.$$

Upon inserting the numerical values, including α from Table 11.1,

$$\Delta L = (1300\ \text{m})(12 \times 10^{-6}\ °\text{C}^{-1})(120°\text{C}) = 1.9\ \text{m}.$$

Discussion The total allowance for expansion must be 1.9 m. This total expansion allowance is divided among a number of expansion joints, each allowing only a small amount of expansion.

Example 11.4

A copper hot-water pipe is 10.0 m long when cut and installed in a building on a day when the temperature is 10°C. How long is the pipe when it carries hot water at 60°C if the pipe is free to expand?

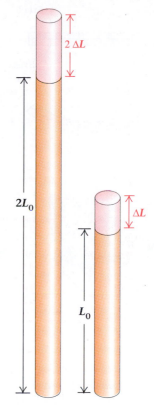

$$\Delta L = L_0\, \alpha\, \Delta T$$

Figure 11.7
A copper bar that is twice as long as a shorter copper bar undergoes twice as much expansion for the same temperature change.

Figure 11.8
Example 11.3: (a) The Verrazano-Narrows Bridge. Changes in temperature cause the bridge components to expand or contract. (b) A typical expansion joint used on bridges.

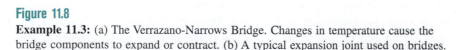

(a)

(b)

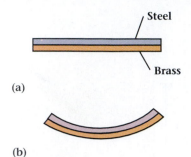

(a)

(b)

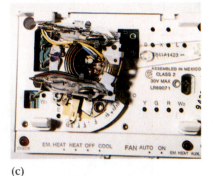

(c)

Figure 11.9

(a) A bimetallic strip of steel and brass at room temperature. (b) At a higher temperature, the brass expands more than the steel, causing the strip to bend. (c) A thermostat for controlling a home heat pump. The coil is a thermometer made from a bimetallic strip. The glass tubes containing drops of mercury are electrical switches.

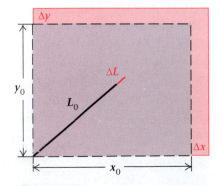

Figure 11.10

A rectangular plate expands in all directions with the same linear thermal expansion coefficient α.

Solution We can use Eq. (11.4):

$$L = L_0[1 + \alpha(T - T_0)]$$
$$L = (10.0 \text{ m})[1 + (17 \times 10^{-6} \text{ °C}^{-1})(60°C - 10°C)]$$
$$L = (10.0 \text{ m})[1.00085] = 10.0085 \text{ m} = 1000.85 \text{ cm}.$$

The pipe is 0.85 cm longer.

A common application of linear thermal expansion is the bimetallic strip, which is made by joining along their length two strips of metal with different thermal expansion coefficients (Fig. 11.9). Because of unequal expansion or contraction of the two metals with change in temperature, one side of the bimetallic strip becomes longer than the other, causing the strip to bend or curl. Bimetallic strips are frequently used to make thermometers and sensing elements in thermostats.

In some cases we do not observe the expansion anticipated with increasing temperature because the object is clamped or otherwise held fixed. This is the case with modern continuous-welded railroad track. The higher temperature produces a thermal stress in the rail, resulting in a force acting on the ties. Formerly, rail was laid in short sections containing gaps called expansion joints between them. The rails were free to slide back and forth on the ties as the temperature changed. Continuous-welded rail is laid in unbroken segments of any length, with a practical limit of 25 mi imposed by the need for electrically insulating breaks for signal purposes. Spikes clamp the rail firmly to the ties, which distribute the forces caused by temperature changes to the ballast (the small rocks and gravel packed around the railroad ties), and thereby to the earth. The large forces are distributed because the rails are clamped down at closely spaced intervals.

Because the linear dimensions of an object change with temperature, it follows that the area and the volume also change. For example, consider the expansion of a rectangular metal plate as its temperature changes by ΔT (Fig. 11.10). A straight line drawn on the plate in any direction would expand with the linear expansion coefficient of the metal. Along the horizontal (x) direction, the plate would expand with a coefficient α. If the material is homogeneous, the plate expands with the same α in the vertical (y) direction or, indeed, in any other direction. Thus, the plate enlarges horizontally by an amount $\Delta x = x_0 \alpha \, \Delta T$ and vertically by an amount $\Delta y = y_0 \alpha \, \Delta T$, giving an increase in area of approximately $2\alpha \, \Delta T$ (see Problem 11.19). Furthermore, if the plate contains a hole, the area of the hole increases by the same amount as would the portion of the plate that was removed to make the hole.

If we consider the thickness of the plate, it, too, increases with increasing temperature. If the temperature change is not too great, the change in volume ΔV of a homogeneous material is also proportional to the change in temperature ΔT and to the original volume V_0, so we have

$$\Delta V = \beta V_0 \, \Delta T, \qquad (11.5)$$

where β is the volume coefficient of thermal expansion. The units of β are also °C^{-1}. The volume coefficient of thermal expansion β is approximately three times the value of the linear coefficient of thermal expansion α: $\beta = 3\alpha$. Values for β are given in Table 11.1.

Table 11.1	Coefficients of Thermal Expansion at 20°C	
Material	**Linear Coefficient α (10^{-6} °C^{-1})**	**Volume Coefficient β (10^{-6} °C^{-1})**
Aluminum	24	72
Brass	19	57
Brick and concrete	10–12	30–36
Copper	17	51
Glass (ordinary)	9	27
Glass (Pyrex)	3	9
Invar	0.7	2.1
Iron and steel	12	36
Lead	29	87
Ice	51 (−20 to −1°C)	153 (−20 to −1°C)
Gasoline	—	950
Mercury	—	180
Water	—	210

The values given are approximate. They vary with the composition of alloys, glasses, and composite materials, and with temperature.

Master the Concept

Linear Expansion

Question: A circular copper plate of uniform thickness has a circular hole in its center. The plate expands when it is heated from room temperature to 500°C. Does the hole in the center expand or contract? Why?

Answer: The hole in the center of the plate expands because its diameter increases in the same proportion as does the diameter of the plate. To understand why, imagine that the hole is completely filled with a copper disk. As the plate is heated, the center disk expands at the same rate as the rest of the plate because it is made of the same material. Then, if the center disk is knocked out of the hot plate, the hole that remains is the size of the disk that is removed. Because the disk is bigger than it was when the plate was cold, the hole must also be bigger by the same amount.

Example 11.5

A 1.00-liter glass bottle is filled to the brim with water at a room temperature of 20°C. The temperature of the bottle and the water is then raised to 95°C. Does the water spill over, or does the level go down, and by how much? Because the volume coefficient of thermal expansion of water changes with temperature, use the average value of $\beta = 525 \times 10^{-6}$ °C for the range of 20°C to 95°C.

Strategy Think of the glass bottle as the "skin" of a solid piece of glass, all of which expands uniformly. Then the change in volume of the inside of the bottle is just the same as the change in volume of the solid interior. We may then compare this change in volume to the change in volume of the liquid.

Solution   We may write the change in volume ΔV_{glass} for the bottle as

$$\Delta V_{glass} = \beta V_0 \Delta T$$
$$\Delta V_{glass} = (27 \times 10^{-6} \text{ °C}^{-1})(1.00 \times 10^{-3} \text{ m}^3)(95°C - 20°C)$$
$$\Delta V_{glass} = 2.03 \times 10^{-6} \text{ m}^3 = 2.03 \text{ cm}^3.$$

For the water the change in volume ΔV_{water} is

$$\Delta V_{water} = \beta V_0 \Delta T$$
$$\Delta V_{water} = (525 \times 10^{-6} \text{ °C}^{-1})(1.00 \times 10^{-3} \text{ m}^3)(95°C - 20°C)$$
$$\Delta V_{water} = 39.4 \times 10^{-6} \text{ m}^3 = 39.4 \text{ cm}^3.$$

The expansion of the water is greater than the expansion of the bottle. The amount of water that will run over the edge is

$$\Delta V_{water} - \Delta V_{glass} = 39.4 \text{ cm}^3 - 2.03 \text{ cm}^3 = 37.4 \text{ cm}^3.$$

Most liquids expand smoothly with increasing temperature. Alternatively, we can say that they become less dense with increasing temperature. Water, however, is an exception (Fig. 11.11). Its density is greatest at 4°C and less for both higher and lower temperatures. Water also becomes less dense on freezing, again in contrast to most liquids. This effect has important consequences for aquatic life. When a body of water, such as a lake, cools, the cooler water at the top flows to the bottom because of its greater density. When the lake reaches 4°C, this flow stops because the colder top of the lake is less dense as the temperature drops below 4°C. As a consequence the top of the lake freezes first, while the lower depths remain at 4°C. So water freezes from the top down. If water behaved like other substances, lakes would freeze from the bottom up, and the continuous circulation of warmer water to the top would cause more efficient freezing. Under those conditions lakes would freeze solid more frequently than they actually do. But as it is, lakes do not frequently freeze solid, even in the coldest climates. This effect is aided by the fact that the ice layer acts as an insulating blanket over the water. Fish survive by staying on the bottom, where the temperature is at least 4°C.

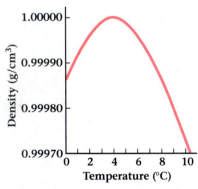

Figure 11.11
The density of water as a function of temperature.

11.4   The Mechanical Equivalent of Heat

Before the mid-eighteenth century, the distinction between temperature and heat was not clear and the two were often confused. At that time it was generally thought that heat was some kind of fluid, called *caloric,* which could be added to or taken away from a substance to make it hot or cold. We now know that **heat** is a form of energy transfer that occurs when there is a temperature difference between objects. An example of the distinction between heat and temperature is sometimes given by comparing a flaming candle and a warm radiator in the same cool room (Fig. 11.12). The candle flame is at a much higher temperature than the radiator, but you don't expect it to appreciably warm the room. On the other hand, although the radiator is at a lower temperature than the flame, enough heat flows from it to keep you warm. In both cases energy is transferred from an object at a higher temperature to surroundings at a lower temperature.

The first evidence for the connection between heat and energy transfer came when the American-born Benjamin Thompson, Count Rumford (1753–1814), was serving as minister of war in Bavaria. While supervising the boring of cannon, he became curious about the tremendous amount of heat generated. His interest led to some detailed experiments on the nature of heat and heat capacities. (See the definitions in Section 11.5.) He concluded that the increase in temperature was due to the work done in the boring process. Despite the implications of Rumford's work, the popular notion of heat as the fluid caloric still persisted, since that theory explained all the results in which people were generally interested.

In 1842, Julius Mayer (1814–1878) suggested that heat and mechanical work were equivalent and that one could be transformed into the other. He even went so far as to show that the temperature of water could be raised 1°C by mechanical agitation alone. However, he failed to determine the amount of work required for such a change.

(a)

The quantitative connection between heat flow and work was conclusively demonstrated a year later, in 1843, by James Prescott Joule (1818–1889). Joule devised an experiment in which the change of potential energy of falling weights was used to churn the water in an insulated container (Fig. 11.13). This famous apparatus contained paddles for stirring the water and stationary vanes to break up the flow, so that the water was not merely set into rotational motion (kinetic energy). The frictional drag of the water caused the weights to fall very slowly, so that their kinetic energy was quite small. The potential energy lost by the falling weights was imparted to the water and was detected as a change in temperature. In this way Joule showed that the temperature of one pound of water could be raised one degree Fahrenheit by the expenditure of 772 ft-lb of mechanical work.* He proved the direct conversion of mechanical energy into thermal energy (heat) and measured the numerical factor relating mechanical units to heat units.

As mentioned earlier, the definitions and units for mechanical energy developed independently of the definitions and units for heat, which grew out of the study of the properties of water. Two separate units for measuring heat were devised. In Britain, the primary unit was the British thermal unit (Btu), the amount of heat required to raise the temperature of one pound of water one degree Fahrenheit. In Europe, where a metric system was in use, the calorie (cal) was defined as the heat required to raise the temperature of one gram of water by one degree Celsius. The Calorie (spelled with a capital C) used in discussing diet and nutrition is a kilocalorie (10^3 cal).

In honor of Joule's contribution to science, his name was given to the common unit of energy. The joule equals one newton-meter and is roughly one-fourth the size of the calorie:

$$1 \text{ calorie} = 4.187 \text{ joules.}$$

This relation is called the *mechanical equivalent of heat*. The relationships between several energy units are given for reference in Table 11.2.

(b)

Figure 11.12
Heat and temperature are different physical quantities. (a) A candle has a high temperature but does not give off much heat; (b) a radiator can warm a room, but does not reach a very high temperature.

---

*Better measurements place this value as 778 ft-lb, less than 1% greater than Joule's carefully determined number.

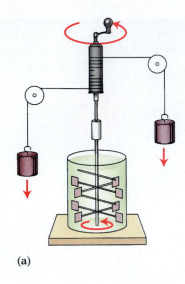

(a)

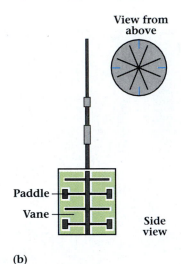

Paddle —

Vane —

Side view

(b)

Figure 11.13
(a) A sketch of Joule's apparatus. Falling weights turn a set of paddles in a water-filled container. (b) A cross-sectional view showing the paddles and the stationary vanes. Joule's apparatus measured the quantitative conversion of mechanical energy into thermal energy.

Table 11.2		Conversion Table for Some Common Energy Units				
		J	cal	kcal	Btu	kWh
1 J	=	1	0.239	2.39×10^{-4}	9.48×10^{-4}	2.78×10^{-7}
1 cal	=	4.187	1	10^{-3}	3.97×10^{-3}	1.16×10^{-6}
1 kcal	=	4187	1000	1	3.97	1.16×10^{-3}
1 Btu	=	1060	252	2.52×10^{-1}	1	2.93×10^{-4}
1 kWh	=	3.60×10^6	8.60×10^5	8.60×10^2	3.41×10^3	1

Neither the calorie nor the Btu is an SI unit. The appropriate SI unit for energy is the joule. However, the calorie is still used in many practical applications and in several fields of research and the Btu is common in engineering. We will use both the joule and the calorie in our examples. You should be able to work with either unit.

Example 11.6.

A 1500-W heater is submerged into one kilogram of water that is well below 100°C. At what rate, in °C/s, does the temperature rise when the heater is operating at its rated power?

Strategy Energy is supplied by the heater at a rate of 1500 W = 1500 J/s, which can be converted to units of cal/s. Then, since one calorie changes the temperature of one gram of water 1°C, we can find the rate of temperature increase by dividing the rate of energy input by the mass of water in units of grams.

Solution The rate of energy input is

$$1500 \text{ J/s} \times \frac{1 \text{ cal}}{4.187 \text{ J}} = 358.3 \text{ cal/s.}$$

The rate of energy input per gram is

$$\frac{358.3 \text{ cal/s}}{1000 \text{ g}} = 0.3583 \text{ cal/g} \cdot \text{s.}$$

Since each gram of water receives 0.3583 cal/s, the temperature of each gram of water, and therefore the entire volume of water, increases by 0.3583°C/s.

11.5 Calorimetry

The measurement of quantities of heat exchanged, a process known as *calorimetry,* was introduced in the 1790s. Chemists of the time found that when a hot object, such as a brass block, was immersed in a water bath, the resulting change in temperature of the water bath depended on both the mass and the initial temperature of the block. Further observations showed that, when two similar brass blocks at the same initial temperature were immersed in identical water baths, the more massive block caused a greater temperature change. Similarly, for two identical blocks at different temperatures, the hotter block gave rise to a greater temperature change in the bath. Finally, for blocks of the same mass and initial temperature but of different composition, the change in temperature was different for different materials.

We can synthesize these observations by describing the objects in terms of their **heat capacity,** which is the amount of heat required to change an object's temperature by 1°C. Blocks of the same material but of different masses have heat capacities proportional to their mass. Thus, we define an intrinsic quantity peculiar to each material, called its *specific heat capacity,* the ratio of the heat capacity to the mass. The specific heat capacity, or simply the **specific heat,** as it is usually called, is **the heat required per unit mass to change the temperature of a substance by one degree.** A material with a high specific heat, like water, requires a lot of heat to change its temperature, while a material with a low specific heat, like silver, requires little heat to change its temperature.

The amount of heat Q required to warm an object of mass m by raising its temperature ΔT is given by

$$Q = mc\,\Delta T, \qquad (11.6)$$

where c is the specific heat of the material from which the object is made. If the object cools, then the temperature change is negative and the heat Q is given off by the object. The units of specific heat are cal/g · °C, J/kg · °C, or Btu/lb · °F. A list of specific heats is given in Table 11.3.

Table 11.3	Specific Heat for Some Common Materials at 25°C	
Substance	Specific Heat (J/kg · °C)	Specific Heat (cal/g · °C) or (kcal/kg · °C)
Water (0°C –100°C)	4187	1.00
Ethyl alcohol	2430	0.581
Ethylene glycol	2390	0.571
Ice (−10°C–0°C)	2090	0.50
Steam (100°C)	2010	0.48
Wood	1700	0.4
Aluminum	900	0.215
Sodium chloride	871	0.208
Marble	860	0.21
Glass	840	0.200
Iron	448	0.107
Copper	390	0.0920
Zinc	386	0.0922
Silver	236	0.0564
Lead	128	0.0305

The specific heat of most materials varies slightly with temperature; however, you may take it to be constant.

We have said that heat is a form of energy transfer. Therefore, we can predict temperature changes when two or more substances are in thermal contact by applying the principle of conservation of energy: The heat (or energy) lost by the cooling objects must equal the heat (or energy) gained by the substances being warmed. We take the quantity of heat *added* to a body to be positive, and the heat *lost* by a body to be negative. Then we say that the sum of all the heat flows to all bodies in thermal contact is equal to zero. That is,

$$\boxed{\text{heat gained (positive)} + \text{heat lost (negative)} = 0.} \tag{11.7}$$

When we use this convention, the temperature change ΔT is *always* the difference between the final and initial temperatures, or $\Delta T = T_f - T_i$. A negative value of ΔT means that energy has left the body. Example 11.7 illustrates this principle using a Styrofoam cup as an insulating, low-heat-capacity container. You can easily do this experiment and compare your measured final temperatures with your calculations.

▼
Problem-Solving Strategy

Calorimetry

Calorimetry problems are applications of energy conservation. The total of the heat lost and the heat gained must be zero. Take care to consider the heat lost or gained by each object or item that is exchanging thermal energy. If you treat all temperature changes as $T_{\text{final}} - T_{\text{initial}}$, the signs will take care of themselves. Just sum all the heat exchanges and set the total to zero.

Be consistent with units. For safety, carry the units and cancel them as appropriate. Doing so is a good way to avoid errors due to improper units.

▼
Example 11.7

A Styrofoam cup of negligible heat capacity contains 150 g of water at 10°C. If you add 100 g of water at a temperature of 85°C, what is the final temperature of the mixture after it has been thoroughly mixed?

Strategy By the principle of conservation of energy, we expect that

$$\text{heat gained (positive)} + \text{heat lost (negative)} = 0.$$

We also expect that the final temperature T of the mixture will be between 10° and 85°C. We can calculate the two heat changes in order to determine T, and then check that our answer is in the expected range.

Solution The heat gained by the cooler water is

$$\text{heat gained} = m_1 c\ \Delta T_1 = m_1 c(T_f - T_{1i}) = (150\ \text{g})c(T_f - 10°\text{C}).$$

The heat lost by the hotter water is

$$\text{heat lost} = m_2 c \, \Delta T_2 = m_2 c (T_f - T_{2i}) = (100 \text{ g}) c (T_f - 85°C).$$

When the heat lost plus the heat gained is set equal to zero, the resulting expression determines a unique value for the final temperature T_f:

$$\text{heat lost} + \text{heat gained} = 150 c (T_f - 10°C) + 100 c (T_f - 85°C) = 0,$$
$$(150 + 100) T_f = (8500 + 1500)°C,$$
$$T_f = \frac{10,000°C}{250} = 40°C.$$

Discussion Our result falls within our expected temperature range. Note that in this particular problem there was no necessity for knowing the specific heat because both components of the mixture were of the same material, water, and had the same specific heat.

Example 11.8

A small metal block (mass of 74 g) is heated in an oven to 90°C. It is then taken from the oven and immediately placed in a calorimeter, a thermally insulated container designed for measurements of heat exchange. The calorimeter contains 300 g of water at 10°C. The heat capacity of the calorimeter is negligible, and the final temperature is 14°C. Identify the composition of the block from the following list: aluminum, iron, silver, or zinc.

Strategy Again we apply the rule of energy conservation: Heat lost plus heat gained equals zero. This will enable us to calculate the block's specific heat, which we can compare with the values in Table 11.3 in order to identify the metal.

Solution The heat lost by the block was

$$Q_1 = m_1 c_1 \, \Delta T_1 = (74 \times 10^{-3} \text{ kg}) c_1 (14°C - 90°C)$$
$$Q_1 = -5.62 \, c_1 \, (\text{kg} \cdot °C).$$

The heat gained by the water was

$$Q_2 = m_2 c_2 \, \Delta T_2 = (300 \times 10^{-3} \text{ kg})(4187 \text{ J/kg} \cdot °C)(14°C - 10°C)$$
$$Q_2 = 5024 \text{ J}.$$

Conservation of energy gives

$$\text{heat lost} + \text{heat gained} = 0,$$
$$-5.62 \, c_1 \, (\text{kg} \cdot °C) + 5024 \text{ J} = 0.$$

We rearrange to find

$$5.62 \, c_1 \, (\text{kg} \cdot °C) = 5024, \quad \text{or} \quad c_1 = \frac{5024 \text{ J}}{5.62 \text{ kg} \cdot °C} = 894 \text{ J/kg} \cdot °C.$$

By comparing this value for c with those in Table 11.3, we see that the metal block in this example is probably made of aluminum.

Discussion Notice that at one point in the computation of the heat gained by the water we multiplied by $(14°C - 10°C) = 4°C$, a number that has only one

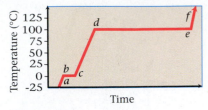

Figure 11.14

Temperature as a function of time for heat applied to water at a constant rate. The temperature remains constant during a change of phase. Ice is warmed to 0°C (*a* to *b*) and melts (*b* to *c*), the resulting water is heated (*c* to *d*) and boils (*d* to *e*). The final steam can obtain higher temperatures (*e* to *f*).

significant figure. Consequently, we give our result in a way that reflects this limitation: $c_1 = 9 \times 10^2$ J/kg · °C.

▼

11.6 Change of Phase

We know from experience that when heat is supplied to ice, it melts into water and that steam, when cooled, condenses into water. The transformation from one physical state to another (for instance, from solid to liquid or from gas to liquid) takes place *with no change in temperature* and is called a change of phase. If we perform a careful calorimetric measurement during a phase change, we find that our previous description of heat exchange is incomplete. In addition to the heat absorbed or released in proportion to changes in temperature, there is an amount of heat energy associated with a phase change (Fig. 11.14). This quantity is called the **heat of transformation** (sometimes the *latent heat of transformation*) L, defined as the ratio of the amount of heat Q absorbed (or released) to the mass m of material undergoing the phase change:

$$L \equiv \frac{Q}{m}.$$

We can express the heat absorbed (or released) in terms of L:

$$\boxed{Q = mL.} \tag{11.8}$$

The heat of transformation is expressed in units of J/kg or cal/g. The term *latent heat* goes back to the early days of the study of heat, when it referred to the absorption of heat without an accompanying change in temperature. In more recent times, people refer simply to the heat of transformation. If the phase change is from the solid to the liquid phase (or from the liquid to the solid), we refer to the **heat of fusion** (L_f); for the liquid-vapor phase change, we use the term **heat of vaporization** (L_v). The energy added (or removed) in the form of the heat of transformation goes into rearranging the internal structure of the substance. For example, when a solid becomes a liquid, energy is required to overcome the intermolecular forces that keep the material in the solid state.

Table 11.4 lists heats of transformation for several substances. We see from the table that the heat of fusion for ice is 3.34×10^5 J/kg (79.7 cal/g). This means that 3.34×10^5 J must be supplied to melt each kilogram of ice. Conversely, 3.34×10^5 J are removed from each kilogram of water that freezes.

A home ice cream churn provides a useful application of heat of fusion. The mixture to be turned into ice cream is placed in a metal can (with good heat-conducting properties) surrounded by ice (Fig. 11.15). Rock salt is then poured onto the ice. At the ice-salt interface, the ice melts because there is a chemical interaction with the salt. The resulting salt solution has a freezing point much lower than that of pure water. The solution can thus provide the energy for melting the rest of the ice. The melting ice absorbs heat energy from the salt solution, making the solution much colder than 0°C. Even though the temperature of the ice is also lowered, it continues to melt at its surface because of the effects

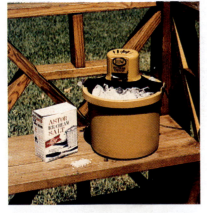

Figure 11.15

Ice cream is made in a churn by surrounding the can containing the ice cream mix with ice and salt. The surrounding salt water solution is much colder than 0°C, allowing the mix to freeze.

Table 11.4	Heat of Transformation of Various Substances at Atmospheric Pressure					
		solid → liquid			*liquid → vapor*	
		liquid → vapor				
Substance	Melting Point (K)	Heat of Fusion L_f (J/kg)	(cal/g)	Boiling Point (K)	Heat of Vaporization L_v (J/kg)	(cal/g)
Water	273	3.34×10^5	79.7	373	22.6×10^5	539
Mercury	234	0.115×10^5	2.74	630	2.97×10^5	71
Iron	1808	2.89×10^5	69.1	3023	63.4×10^5	1520
Lead	600	0.232×10^5	5.54	2023	5.69×10^5	205
Copper	1356	2.05×10^5	48.9	2840	48.0×10^5	1150
Oxygen	54.4	0.14×10^5	3.3	90.2	2.13×10^5	50.9
Nitrogen	63.3	0.26×10^5	6.1	77.3	2.01×10^5	48.0
Helium	—	—	—	4.2	21	5×10^{-3}

of the salt. The salt solution, in turn, absorbs heat energy from the ice cream mixture through the walls of the metal can, allowing the mixture to cool enough to become firm. Without the addition of salt to the ice, the temperature of the mixture would never become low enough to form ice cream.

Master the Concept

Heat of Vaporization: Steam Burns

Question: Why does exposure to steam at 100°C produce a more severe burn than exposure to the same amount of hot water at 100°C.

Answer: When hot water at 100°C touches your much cooler skin, the transfer of energy in the form of heat raises the temperature of the surrounding tissue thereby causing a burn. When steam at 100°C touches your skin it gives up its energy of vaporization as it condenses to water at 100°C. Because the heat of vaporization of water is so large, the energy transferred to the skin by condensing steam greatly exceeds the heat transferred by the hot water. Consequently, the burn caused by the steam is much worse than that due to the hot water.

Heat of Vaporization: Cooling by Perspiration

Question: Athlete's engaged in strenuous activities often sweat profusely. How does the perspiration help cool the athlete?

Answer: The athlete's body heats up as a result of the physical activity. In response, the body directs more blood to the surface where heat is lost by radiation, conduction, and the evaporation of sweat. When sweat evaporates from the skin, heat from the body provides the heat of vaporization. Under severe conditions, an athlete may sweat a liter or more of liquid (water) per hour. The evaporation of 1 liter of sweat can remove up to 2.26 MJ of thermal energy from the body. Thus perspiration cools the body by its evaporation.

Example 11.9

A 105-g copper calorimeter contains 307 g of water at room temperature ($T = 23°C$). If 52 g of ice at 0°C is added to the calorimeter, what is the final temperature of the system?

Strategy As before, we can solve this problem by applying the principle of conservation of energy:

$$\text{heat lost} + \text{heat gained} = 0.$$

However, in this case heat is gained by the ice in melting, as well as by the melted ice in being warmed from 0°C to the final temperature T.

Solution The heat gained by the ice is

$$Q(\text{gain}) = m_{\text{ice}}L_f + m_{\text{ice}}c_{\text{water}} \, \Delta T$$
$$Q(\text{gain}) = (52 \text{ g})(80 \text{ cal/g}) + (52 \text{ g})(1 \text{ cal/g °C})(T - 0°C)$$
$$Q(\text{gain}) = 4160 \text{ cal} + 52T \text{ cal/°C}.$$

Note that we have rounded off the heat of fusion of ice to 80 cal/g because the amount of ice and the temperatures are given to only two significant figures.

The heat lost is given up by the calorimeter (subscript c) and the original water (subscript w):

$$Q(\text{lost}) = m_{\text{w}}c_{\text{w}} \, \Delta T + m_{\text{c}}c_{\text{c}} \, \Delta T = (m_{\text{w}}c_{\text{w}} + m_{\text{c}}c_{\text{c}})(T - T_0)$$
$$= [(307 \text{ g})(1 \text{ cal/g} \cdot °C) + (105 \text{ g})(0.092 \text{ cal/g} \cdot °C)](T - 23°C)$$
$$Q(\text{lost}) = (317 \text{ cal/°C})(T - 23°C)$$
$$Q(\text{lost}) = 317T \text{ cal/°C} - 7290 \text{ cal}.$$

When the heat gained is added to the heat lost, we find

$$4160 \text{ cal} + 52T \text{ cal/°C} + 317T \text{ cal/°C} - 7290 \text{ cal} = 0.$$

Upon rearranging, we find

$$369T \text{ cal/°C} = 3130 \text{ cal},$$
$$T = 8.5°C.$$

Example 11.10

Repeat the calculations for Example 11.9, but this time use 95 g of ice. What happens now?

Solution The expression for heat lost is the same as before,

$$Q(\text{lost}) = 317T \text{ cal/°C} - 7290 \text{ cal}$$

and is a maximum of −7290 cal when the final temperature is 0°C. For the heat gained we find

$$Q(\text{gain}) = m_{\text{ice}}L_f + m_{\text{ice}}c_{\text{water}} \, \Delta T$$
$$Q(\text{gain}) = (95 \text{ g})(80 \text{ cal/g}) + (95 \text{ g})(1 \text{ cal/g °C})(T - 0°C)$$
$$Q(\text{gain}) = 7600 \text{ cal} + 95T \text{ cal/°C},$$

which is a minimum of 7600 cal at $T = 0°C$. When we apply the conservation equation, we arrive at a negative value for T, an obvious error. (If T were negative, all the water would turn to ice. In addition, we expect the final temperature to lie in the range between $0°C$ and $23°C$.) What went wrong?

We see that the maximum amount of heat available from the water and calorimeter is 7290 cal. Therefore *not all of the ice will melt,* since that would require 7600 cal, an amount that exceeds the heat available. If all of the available energy goes into melting ice, the amount melted is

$$\frac{7290 \text{ cal}}{80 \text{ cal/g}} = 91 \text{ g.}$$

The remaining mixture of water and 4 g of ice has a temperature of $0°C$. (In time, that ice will also melt as a result of the slow leakage of heat through the insulation of the calorimeter.)

11.7   Heat Transfer

Heat energy can be transferred in three ways: conduction, convection, and radiation (Fig. 11.16). When one end of a metal rod is heated, the other end gets warm. This is an example of **conduction,** in which thermal energy is transferred without any net movement of the material itself. Conduction is a relatively slow process. A more rapid process of heat transfer is accomplished through the mass motion or flow of some fluid, such as air or water, and is called **convection.** This transfer takes place when warm air flows about a room and when hot and cold liquids are poured together. A still more rapid transfer of thermal energy is accomplished by **radiation,** a process that requires neither contact nor mass flow. The energy from the sun comes to us by radiation. We also feel radiation from warm stoves, fires, and radiators. Although the processes of conduction, convection, and radiation may all take place at the same time, frequently one of them is dominant in a given situation. In this section, we will principally discuss some aspects of conduction and radiation. Because of the mathematical complexities, the details of convection will not be discussed.

An object's usefulness as a thermal conductor (or insulator) depends on a number of things, including its thickness and the nature of the material from which it is made. For example, no sensible person would try to use a good heat conductor like aluminum as a protective layer when picking up a hot pan. On the other hand, a padded cloth potholder is a good insulator and works quite well.

Suppose we imagine a wall of uniform material, such as plasterboard, that separates a warm room from a cold one. After a period of time, a steady temperature difference occurs across the wall and a steady flow of heat goes from the warmer room to the cooler one (Fig. 11.17). Experiments show that the time rate at which heat flows ($\Delta Q/\Delta t$) through the wall is proportional to its area A, proportional to the temperature difference ($T_2 - T_1$), and inversely proportional to the thickness L of the wall. This information is contained in the heat flow equation:

$$\frac{\Delta Q}{\Delta t} = KA \frac{(T_2 - T_1)}{L}.$$

(11.9)

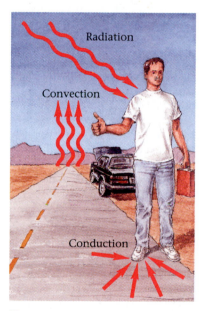

Figure 11.16
Thermal energy transfers by conduction, convection, and radiation.

Radiation

Convection

Conduction

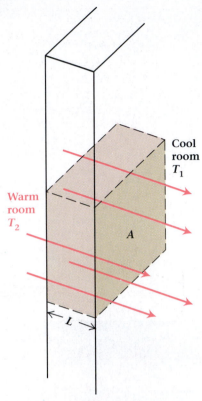

Figure 11.17
The rate of heat flow by conduction through a wall of area A depends on the temperature difference divided by the thickness, $(T_2 - T_1)/L$, as well as the thermal conductivity K of the wall's material.

The constant K is called the **thermal conductivity** and is characteristic of the material making up the wall. The SI units of heat flow are J/s or W, so the SI units of K are J/(s · m · °C) or W/m · °C.

A high thermal conductivity indicates a good heat conductor; a low thermal conductivity indicates a good heat insulator. Some representative values of K for common materials are tabulated in Table 11.5. In designing a good insulator, such as a potholder or the outside walls of a house, the first requirement is to choose a material with a small thermal conductivity, so that the heat flow in Eq. (11.9) is small. In addition, by minimizing the area of contact A and making the path length L as long as possible, we can further reduce the heat flow.

Example 11.11

A Styrofoam cooler has a surface area of 0.50 m² and an average thickness of 2.0 cm. How long will it take for 1.5 kg of ice to melt in the cooler if the outside temperature is 30°C? (The thermal conductivity of the Styrofoam used to make the cooler is 0.030 W/m · °C.)

Strategy First we compute the rate of energy transfer (heat) into the cooler due to conductivity through its walls, assuming an inside temperature of 0°C. Then we can compute the rate of melting of the ice and thus determine the time required to melt it.

Solution The rate of energy transfer is computed from

$$\frac{\Delta Q}{\Delta t} = KA\frac{(T_2 - T_1)}{L}$$

$$\frac{\Delta Q}{\Delta t} = (0.030 \text{ W/m} \cdot {}^\circ\text{C})(0.50 \text{ m}^2)\frac{30{}^\circ\text{C} - 0{}^\circ\text{C}}{0.020 \text{ m}} = 22.5 \text{ W}.$$

The heat of fusion of ice is $L_f = 3.34 \times 10^5$ J/kg. The mass of ice Δm melted in a time Δt is related to the rate of energy transfer through

$$\frac{\Delta Q}{\Delta t} = \frac{\Delta m}{\Delta t}L_f.$$

If we let Δm be the entire mass of ice, then Δt becomes the time required for it to melt,

$$\Delta t = \frac{\Delta m L_f}{\Delta Q/\Delta t} = \frac{(1.5 \text{ kg})(3.34 \times 10^5 \text{ J/kg})}{22.5 \text{ W}} = 2.23 \times 10^4 \text{ s}.$$

Upon dividing Δt by 3600 s/h, we find that it takes 6.2 h for all the ice to melt. ■

The effectiveness of insulation is rated by another quantity, called thermal resistance, or R value. The **R value** is the ratio of a material's thickness to its thermal conductivity:

$$R \equiv \frac{L}{K}.$$

(11.10)

In the United States, the R value is given in the British system units of $ft^2 \cdot h \cdot °F/Btu$. For the outside walls of your home you want a good heat insulator; that is, you want materials and insulation with a high R value. Some examples are found in Table 11.6. The R values are useful because you can simply add them to obtain the R value resulting from multiple layers of insulation (Problem 11.80). For example, a $6\frac{1}{4}$-in.-thick layer of fiberglass insulation has an R value of 19. Insulation made from two layers of $6\frac{1}{4}$-in. fiberglass has an R value of 38. For a given material, the R value is directly proportional to the thickness, which is represented by L in Eq. (11.10).

A hot object also loses energy by radiation.* This radiation, which is known as electromagnetic radiation, is similar to light (see Chapter 22) and can pass through empty space (a vacuum). The warmth you feel when you warm yourself by a fire is due to this radiation. If the object is hot enough, some of the radiation is visible and can indeed be seen. This emission is the blackbody radiation discussed in Section 1.1.

The rate at which an object radiates energy is proportional to its surface area A and to the fourth power of its absolute temperature T. The total energy radiated from an object per unit time (that is, its radiated power) is found experimentally to be

$$P = \sigma e A T^4, \tag{11.11}$$

where σ is the Stefan-Boltzmann constant, which has the value $\sigma = 5.67 \times 10^{-8}$ $W \cdot m^{-2} \cdot K^{-4}$, and e is a constant called the *emissivity*. The emissivity is a dimensionless number between 0 and 1 that describes the nature of the emitting surface. The emissivity is larger for dark, rough surfaces and smaller for smooth, shiny ones. For example, the emissivity of a black cast iron stove is near one

Table 11.5	*Thermal Conductivities of Some Common Materials at 27°C*
Material	**K (W/m · °C)**
Silver	427
Copper	398
Aluminum	237
Tungsten	178
Iron	80.3
Brick	0.4–0.8
Water	0.61
Asbestos	0.083
Glass	0.72–0.86
Air	0.026
Wood (pine)	0.11–0.14
Fiberglass	0.046
Polystyrene foam	0.033
Polyurethane foam	0.020

Table 11.6	*R Values of Some Common Building Materials*	
Material	**Thickness (in.)**	**R Value ($ft^2 \cdot h \cdot °F/Btu$)**
Gypsum board	$\frac{1}{2}$	0.45
Plywood	$\frac{1}{2}$	0.62
Brick	$3\frac{5}{8}$	0.6–1.2
Glass, single pane		1
Glass, double pane		2
Polystyrene foam	$\frac{3}{4}$	2.9
Fiberglass insulation	$3\frac{1}{2}$	11
Fiberglass insulation	$6\frac{1}{4}$	19

---

*This radiation is the blackbody radiation described mathematically by Planck (Chapter 1).

while the emissivity of the silver coating of a thermos bottle is near zero. Equation (11.11) is known as the **Stefan-Boltzmann law.**

According to the Stefan-Boltzmann law, all objects radiate energy, no matter what their temperature happens to be. Why then do they not lose all their thermal energy by radiation and cool down to 0 K? The answer is that they also absorb radiation from surrounding objects and eventually come to thermal equilibrium with their environment. The book in your hand is radiating, but it is also absorbing radiation from its surroundings. If the book (or other object) is at a temperature T and its surroundings are at a different temperature T_s, the net energy gained (or lost) per second by the book is given by

$$P_{net} = \sigma e A (T^4 - T_s^4), \qquad (11.12)$$

where A is the surface area of the book. Notice that we have used the same value for the emissivity for absorption as for radiation. This must be correct, because the net heat exchange must go to zero when $T = T_s$. Thus a good radiator is also a good absorber.

Because of the T^4 term in Eq. (11.11), the total power radiated grows rapidly as the temperature increases as seen in Fig. 1.3 (p. 4). For example, an object radiates 16 times more power at a temperature of 273°C (546 K) than it does at 0°C (273 K). The distribution of the radiation, which is composed of many different wavelengths, is also a function of temperature. It is the change in this distribution that accounts for the change of color of a glowing hot object as its temperature is raised. We will discuss these issues further when we describe blackbody radiation in Chapter 27 (p. 858).

Example 11.12

A patient waiting to be seen by his physician is asked to remove all his clothes in an examination room that is at 16°C. Calculate the rate of heat loss by radiation from the patient, given that his skin temperature is 34°C and his surface area is 1.6 m². Assume an emissivity of 0.80.

Solution   From Eq. (11.12), the rate of heat loss by radiation is

$$P_{net} = \sigma e A (T^4 - T_s^4),$$

where the temperatures are expressed in kelvins. Inserting the numbers, we get

$$P_{net} = (5.67 \times 10^{-8} \text{ W} \cdot \text{m}^{-2} \cdot \text{K}^{-4})(0.80)(1.6 \text{ m}^2)[(307 \text{ K})^4 - (289 \text{ K})^4]$$

$$P_{net} = 140 \text{ W}.$$

The problem of choosing good insulation is not always merely that of finding a poor thermal conductor. Air is a poor conductor, yet a hot object left exposed in the air cools rapidly as a result of convection currents in the air, which continually bring cool air in contact with the object. These convection currents are caused by the expansion of the air as it is warmed. The warmer air is lighter than the cooler surrounding air and rises as a result of buoyancy. Energy is also lost by direct radiation. To reduce these effects and still capitalize on the low conductivity of air, we may use an insulating material that contains many tiny pockets of air so that convection is reduced to nearly zero. Pockets of trapped air account for the good insulating qualities of Styrofoam, fiberglass, down, felt, and woolen clothing. In addition, the many surfaces of the insulating material help reduce the radiant loss by reflection and by radiation back toward the object.

The vacuum flask, used so effectively to keep hot foods hot and cold foods cold, has an inner glass container surrounded by an outer one (Fig. 11.18). The space between is evacuated and sealed. The vacuum between the containers offers little heat loss through either conduction or convection. Radiation losses are minimized by coating the wall of the evacuated space with a highly reflecting layer of silver. Thus, the principal means of heat leakage is through the plug at the mouth of the container and through the glass joining the inner and outer containers. This type of flask is called a *Dewar* flask after the Scottish scientist Sir James Dewar, who first used it in 1892 in his studies of liquid oxygen. It is also known by the trade name, Thermos bottle.

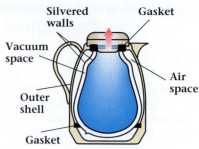

Figure 11.18

Cross-sectional view of a vacuum flask. The vacuum space reduces heat loss by conduction or convection, and the reflecting surfaces reduce heat loss by radiation.

Summary

Useful Concepts

■ Heat is a form of energy transfer between two objects; temperature is a measure of an object's warmth.

■ The relationship between Fahrenheit and Celsius temperatures is

$$T_C = \tfrac{5}{9}(T_F - 32).$$

■ The relationship between the Kelvin and Celsius temperature scales is

$$T_K = T_C + 273.15.$$

■ A body expands or contracts when the temperature changes. The change in length due to a change in temperature ΔT is

$$\Delta L = L_0 \alpha \, \Delta T,$$

where α is the linear expansion coefficient. The length at new temperature T in terms of the initial length L_0 at T_0 is

$$L = L_0[1 + \alpha(T - T_0)].$$

■ Joule showed experimentally that heat and mechanical work are equivalent, determining the relationship

$$1 \text{ calorie} = 4.187 \text{ joules.}$$

■ Calorimetry is based on the idea of conservation of thermal energy. In a thermally isolated system,

heat gained (positive) + heat lost (negative) = 0.

■ The heat Q required to change the temperature of a body of mass m is

$$Q = mc \, \Delta T,$$

where c is the specific heat capacity. The heat required to change the phase of a mass m of material with latent heat of transformation L is

$$Q = mL.$$

■ Heat of transformation is absorbed or given off during a substance's change of phase; however, the temperature remains constant during the phase change.

■ The heat conduction equation is

$$\frac{\Delta Q}{\Delta t} = \frac{KA(T_2 - T_1)}{L}.$$

■ The R value of insulation is the ratio of thickness to thermal conductivity:

$$R \equiv \frac{L}{K}.$$

■ Energy is radiated from all objects. The rate at which the energy is radiated is given by the Stefan-Boltzmann law:

$$P = \sigma e A T^4.$$

Important Terms

You should be able to write the definition or meaning of each of the following:

temperature	heat capacity
thermometer	specific heat
thermal equilibrium	heat of transformation
Celsius temperature scale	heat of fusion
Fahrenheit temperature scale	heat of vaporization
absolute zero	conduction
Kelvin temperature scale	convection
linear thermal expansion	radiation
linear thermal expansion	thermal conductivity
coefficient	R value
heat	Stefan-Boltzmann law

▼ Conceptual Questions

11.1 What differences are there between a fever thermometer that measures from 35 to 42°C and a laboratory thermometer that measures from −10 to 110°C?

11.2 Devise a thermometer that relies on some property other than thermal expansion to indicate changes in temperature.

11.3 How does thermal expansion affect the accuracy of a pendulum clock?

11.4 A square brass plate has a large circular hole cut in its center. If the plate is heated, it will expand. Will the diameter of the hole expand or contract? Explain your answer.

11.5 On a warm summer day you hold a thermometer by a thread while standing in front of an electric fan. When the fan is turned on, your hand feels cooler. Does the thermometer indicate a change in temperature? Explain what is happening.

11.6 Why are concrete highways made in short sections with tar-filled gaps between them?

11.7 A sports car going 80 km/h is braked to a stop without skidding. What happens to its kinetic energy?

11.8 Can you warm up a cup of coffee by stirring it vigorously?

11.9 Why is the climate of coastal cities milder than that of cities in the midst of large land areas?

11.10 What happens to the heat of transformation absorbed during a phase transition?

11.11 People in hot arid regions frequently store water in canvas bags through which some of the water can seep. What is the purpose of doing this?

11.12 How does the thickness of a pot or frying pan affect the way it cooks? What effect does the pot's composition (e.g., aluminum, steel, or ceramic) have on the way it cooks?

11.13 Why do some materials, such as glass and metal, usually feel cold and other materials, such as cloth, usually feel warm? (Refer to Tables 11.3 and 11.5.)

11.14 Do windows made of three panes of glass separated by two air spaces have any insulating advantage over those made of the equivalent total thickness of material in two panes of glass separated by one air space? Explain your reasoning.

▼ Problems

Section 11.2 Thermometry

Hints for Solving Problems

Be especially careful to use a consistent set of units, based on the same temperature scale.

11.1 What is generally regarded as the record high terrestrial temperature of 57.8°C occurred in Tripoli in Northern Africa in 1922. The record low of −89.2°C was recorded at the Soviet Antarctica station Vostok in 1983. Convert these temperatures to the Fahrenheit scale.

11.2 Human body temperature is about 98.6°F. Convert this to the Celsius scale and the Kelvin scale.

11.3 At atmospheric pressure, the boiling point of helium is 4.2 K. What is the boiling point of helium on the Celsius scale? On the Fahrenheit scale?

11.4 Express the following Kelvin temperatures in degrees Celsius and in degrees Fahrenheit: 77.3 K, 300 K, and 1356 K.

11.5 At atmospheric pressure, the boiling point of nitrogen is −195.8°C. What is the boiling point of nitrogen on the Kelvin scale? On the Fahrenheit scale?

11.6• Find a relationship expressing the temperature in degrees Fahrenheit in terms of the temperature in kelvins.

11.7• An approximate way of converting from the Fahrenheit scale to the Celsius scale is to subtract 32 from the Fahrenheit temperature and divide the result by two. How much error in degrees Celsius does this give for Fahrenheit temperatures of 80°, 40°, 10°, and −10°?

11.8• At what temperature do the Fahrenheit and Celsius scales have the same numerical value?

11.9•• A physicist defines a new temperature scale that has its zero at a room temperature of 70°F and its 10° mark at approximately body temperature (say 98°F). (a) Derive an expression for converting from the Fahrenheit scale to this new scale. (b) Derive an expression for converting from the Celsius scale to this new scale.

Section 11.3 Thermal Expansion

Hints for Solving Problems

A homogeneous material expands (or contracts) with the same coefficient of thermal expansion in any direction. A circular hole expands just as a solid circle would.

11.10 A tall steel flagpole was erected in Vancouver for Canadian Expo 86. If the flagpole gets 5.1 mm taller when the average ambient temperature increases by 5.0°C, approximately how tall is the flagpole?

11.11 A metal rod that is 100.0 cm long at 10°C is observed to be 100.2 cm long at a temperature of 80°C. What is its coefficient of linear expansion?

11.12 A copper rod is measured to be 3.00 m long at a temperature of 15°C. How much does it expand when heated to a temperature of 150°C?

11.13 A building with a steel framework is 50 m high. How much taller is it on a summer day when the temperature is 30°C than on a −5°C winter day?

11.14 In the days of horse-drawn wagons, iron tires used to be placed on wooden wagon wheels by heating the iron rims and slipping them over the wheels before they cooled. If an iron tire is made to fit tightly around a 1.50-m-diameter wheel at 15°C, what diameter will the tire have when it is heated to 800°C?

11.15 A motorist fills the 60-liter tank of his automobile with gasoline at 60°F. The automobile is left in the sun, and the temperature of the gasoline rises to 110°F. Approximately how much gasoline is lost to overflow caused by thermal expansion? Ignore the expansion of the tank.

11.16● A framework of rods is made as shown in Fig 11.19. How far and in what direction will the point A move when the temperature is increased by 100°C?

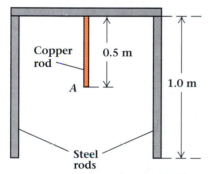

Figure 11.19
Problem 11.16.

11.17● An iron bar is exactly one meter long at 15°C. A brass bar is 0.5-mm shorter than the iron bar at 15°C. They are both heated in the same oven. At what temperature will they be the same length?

11.18● A steel plate with a circular hole of 2.005-cm radius and a copper ball with a radius of 1.998 cm are initially at 15°C. (a) If the temperature of the copper ball is raised to 275°C, will the ball still fit through the hole? (b) Will the ball fit through the hole if both the plate and the ball are at 275°C?

11.19● A flat plate with area A_0 at temperature T_0 is characterized by a linear expansion coefficient α. Show that to a good approximation the area A of the plate at temperature T is given by

$$A = A_0[1 + 2\alpha (T - T_0)].$$

11.20● A bottle made of ordinary glass has an internal volume of 1.00 liter at 20°C and contains 0.99 liter of water at the same temperature. At what temperature will water begin to overflow the bottle?

11.21● A motorist fills the 61-liter tank of her automobile

with gasoline at 15°C. The automobile is left in the sun, and the temperature of the gasoline rises to 44°C. Approximately how much gasoline is lost to overflow caused by thermal expansion? Assume the volume expansion coefficient of the tank is 70×10^{-6} °C^{-1}.

Section 11.4 The Mechanical Equivalent of Heat

11.22 (a) How many joules are required to raise 1.0 kg of water from room temperature of 22°C to its boiling point? (b) If this work were used to lift a 50-kg boy instead, how high could he be lifted?

11.23 When you do work against a frictional force, mechanical energy is transformed into thermal energy. Suppose you push a 2.4-kg textbook 0.85 m across a level table at a constant speed. If the coefficient of friction between the book and the table is 0.25, how much energy is dissipated? Give your answer in calories.

11.24 The food energy of a jelly donut is about 250 kcal. Show that the jelly donut could be used as a unit of energy where 1 jelly donut ≈ 1 MJ.

11.25 What is the kinetic energy of a 61-kg person running at 4.5 m/s? Express your answer in kcal.

11.26 What would be the maximum possible increase in temperature of water as it fell from the top to the bottom of a 23-m-high waterfall? Assume that the drop does not cool by evaporation as it falls.

11.27● What is the energy in calories of a 1780-kg car traveling at 20.4 km/h?

11.28● The Btu is defined to be the amount of thermal energy needed to raise the temperature of one pound of water by 1°F. Show that 1 Btu = 252 cal.

11.29●● A 1500-W heater is submerged into one kilogram of water that is well below 100°C. At what rate, in °C/s, does the temperature rise when the heater is operating at its rated power?

Section 11.5 Calorimetry

Hints for Solving Problems

For an object or a group of objects insulated from the surroundings, the heat lost plus the heat gained is zero, where heat gained is positive and heat lost is negative. Specific heat is the heat required per unit mass to change the temperature of a substance by one degree.

11.30 Two hundred grams of lead shot is placed in a 1.5-m-long cardboard tube, which is closed at both ends. If the tube is in a vertical position and then quickly inverted, the shot falls through the length of the tube. If this is done 50 times in succession, what is the maximum increase in temperature of the shot?

11.31 In an experiment similar to the one done by Joule (Fig. 11.13), a 10-kg weight is allowed to fall through a distance of 1.6 m. If there is 2.0 kg of water in the container, what will the temperature rise be?

11.32 Suppose that 250 g of water at 85°C is mixed with 95 g of water at 15°C in an insulated container of negligible heat capacity. What is the final temperature?

11.33 Assume that the specific heats of coffee and water are the same. How much cool water at 59°F must be added to an insulated cup that contains 145 g of coffee at 188°F to cool it to 135°F? Ignore the mass of the cup.

11.34 You pour 150 g of hot coffee at 85°C into a 210-g glass cup at 22°C. If they come to thermal equilibrium quickly, what is the final temperature? Assume no heat is lost to the surroundings.

11.35 A cook pours 400 g of hot water at 98°C into a 235-g aluminum pot initially at 15°C. If they come to thermal equilibrium quickly, what is the final temperature? Assume no heat is lost to the surroundings.

11.36 If 1.0 kg of lead shot at 150°C is poured into 1.0 kg of water at 23°C, what is the final temperature? Neglect the effect of the container and assume that no steam escapes and that no heat is lost to the surroundings.

11.37• A 2.0-cal/s heater is submerged in a 1.0-liter beaker full of water for 10 min. What is the temperature rise if all of the heat goes into raising the temperature of the water?

11.38• Two hundred fifty grams of water at 80°C is poured into a Styrofoam cup of negligible heat capacity containing 180 g of water at 10°C. After an additional 300 g of water is added to the cup, the mixture comes to an equilibrium temperature of 30°C. What was the temperature of the additional 300 g of water?

11.39• A thermometer with a mass of 0.055 kg and a specific heat of 0.20 kcal/kg · °C reads 15.0°C. The thermometer is then inserted into 0.300 kg of water and comes to a common temperature with the water of 44.4°C. What was the temperature of the water before the thermometer was inserted if other heat losses can be neglected?

11.40• A 175-g copper block at 90°C is dropped into an aluminum calorimeter cup initially at 20°C. The calorimeter cup has a mass of 400 g and contains 430 g of water, also at 20°C. What is the final temperature of the system?

Section 11.6 Change of Phase

Hints for Solving Problems

Heat exchanged includes $Q = mc \, \Delta T$, due to temperature change, and $Q = mL$, due to phase change. The final temperature of a mixture of hot and cold liquids will be intermediate between their initial temperatures.

11.41 How many joules are required to change one kilogram of ice at −15°C to water at 15°C?

11.42 From your experience, estimate the final temperature that will result if a single ice cube is placed in a cup of hot coffee that has just been boiled. Make reasonable estimates of the size of the ice cube and the volume of coffee.

11.43 How many calories are required to change 400 g of ice at −12°C to steam at 110°C?

11.44 How many joules are required to change 1.0 kg of solid lead at 23°C to a liquid at 2000°C? (Assume the specific heats of both solid and molten lead to be the same.)

11.45 If 20 g of steam at 100°C is mixed into 80 g of water at 20°C, what will be the final temperature if no thermal energy is lost and no steam escapes?

11.46 If the energy that goes into evaporating one kilogram of water from Lake Michigan at 15°C were used instead to raise that amount of water above the surface of the lake, how high would it be lifted?

11.47 A container of negligible heat capacity is filled with 5.0 kg of crushed ice and then placed on a hot plate that supplies heat to the ice at a rate of 30 W. What volume of water is produced per minute?

11.48• (a) If 120 g of hot coffee at 85°C is poured into an insulated cup containing 200 g of ice initially at 0°C, how many grams of liquid will there be when the system reaches thermal equilibrium? (b) How much ice will remain?

11.49• If you remove 1000 cal from 1.5 g of steam at 100°C, what will you have left?

11.50• A 50-W electrical heating element is placed in a well-insulated container into which has just been placed 500 g of water at 20°C and 300 g of ice at 0°C. How long will it take before all of the contents are evaporated?

11.51•• A 50-g piece of iron and a 40-g piece of copper at the same temperature of 80°C are put into an insulated container that has 400 g of water and 100 g of ice, all at 0°C. What is the equilibrium temperature?

Section 11.7 Heat Transfer

11.52 A wall is insulated with glass wool of thermal conductivity $K = 0.046$ W/m · °C. What is the rate of heat loss through an area 1.0 m wide by 1.8 m high, insulated with a layer of glass wool 15 cm thick, if the temperature difference across the layer is 20°C?

11.53 A large window of ordinary plate glass is 1 m wide, 1.5 m high, and 3.0 mm thick. Use the heat flow equation to estimate the rate of heat loss through the window on a day when the inside temperature is 22°C (72°F) and the outside temperature is 0°C. (In practice, the actual rate of heat loss through the window is much smaller than the value calculated on the basis of the heat flow equation with T_1 = inside temperature and T_2 = outside temperature. The layers of air on both sides of the glass act as additional insulation, so the temperature difference across the glass is substantially reduced.)

11.54 A small oven has a surface area of 0.20 m². The insulated walls are 1.5 cm thick with an average thermal conductivity of 4.0×10^{-2} W/m · °C. What is the rate of heat loss if the temperature inside the oven is maintained at 245°C and the outside temperature is 20°C?

11.55 What is the rate of energy radiated per unit area from a blackbody with emissivity = 1.00 at temperatures of 300 K, 1000 K, 3000 K, and 3200 K?

11.56 A lamp is designed to operate at a temperature of 3200 K. If the lamp is operated at a higher voltage that raises its temperature to 3400 K, what will be the fractional increase in radiant energy?

11.57● A recreational vehicle (RV) can be modeled as a rectangular box with exterior dimensions of 23′ × 7′10″ × 7′6″. The uninsulated exterior walls have an average R value of 4. The RV is heated with a furnace that delivers 19,000 Btu/h. Can the furnace keep the inside of the RV at 70°F on a day when the outside temperature is −10°F?

11.58● The bottom of an aluminum pot has an area of 177 cm² and a thickness of 3.25 mm. The pot is placed on a stove and heated until water boils away at a rate of 0.235 g/min. What is the temperature difference between the outside and the inside of the bottom of the pot? Be sure to state all assumptions needed to solve the problem.

11.59● The rate of radiation from the sun is determined to be 6.25 × 10⁷ W/m². Use this value to compute the effective temperature of the sun. Assume an emissivity of 1.00.

11.60● An outside wall of a room is 2.44 m (8 ft) high by 5.0 m (16.4 ft) wide. (a) Calculate the rate of heat loss through the wall, assuming there are no windows and the wall has an average R value of 15. Take the temperature difference across the wall to be 25°C. (b) What is the rate of heat loss through the wall if it has a single glass window with an R value of 1.0 and an area of 1 m² = 10.76 ft²? Give your answers in units of watts.

11.61● A Styrofoam cooler has dimensions of 0.20 m × 0.30 m × 0.35 m and an average thickness of 1.5 cm. How long will it take for 6.5 kg of ice to melt in the cooler if the outside temperature is 35°C? The thermal conductivity of the Styrofoam is 0.030 W/m · °C.

11.62● An iron bar 50 cm long is braised to a copper bar 50 cm long and of the same diameter. The free end of the iron bar is kept at 0°C while the free end of the copper bar is at 100°C. (a) What is the temperature of the junction point? (b) What is the heat flow down the rod if its cross-sectional area is 1.5 cm²?

11.63● The radiation from the sun is received from all parts of the sun's disk, including the less luminous outer edges. One estimate places the radiation rate at the center to be 16% greater than the average value of 6.25 × 10⁷ W/m². Use this estimate to determine the temperature of the sun near the center of the sun's disk. Assume an emissivity of 1.00.

11.64●● Derive an expression for heat flow in terms of the R value.

Additional Problems

11.65 A concrete highway has expansion joints at intervals of 18 m. How wide must the expansion joints be to allow for thermal changes over the temperature range from −10°C to 40°C? Use $\alpha = 10 \times 10^{-6}$ °C⁻¹.

11.66 What is the increase in volume of an aluminum sphere with a radius of 11.6 cm when it is heated from 0°C to 100°C?

11.67 What is the rate of heat flow along a copper bar 1.08 m long having a cross section of 1.64 cm² if one end of the bar is at 0.0°C and the other one is at 100.0°C?

11.68● Show that the density ρ of an object with respect to its density ρ_0 at some reference temperature is given approximately by $\rho = \rho_0(1 - \beta \Delta T)$, where β is the thermal coefficient of volume expansion and ΔT is the difference between the object's temperature and the reference temperature.

11.69● A 3.75-kg iron crucible holds 4.23 kg of silver at 20°C. How much thermal energy must be supplied to the system to melt the silver? Silver melts at 962°C with a heat of fusion of 1.11 × 10⁵ J/kg. Assume that the specific heats given in Table 11.3 are appropriate.

11.70● If a one-cubic-meter sealed container of air at 0°C and one atmosphere pressure is heated by supplying heat at the rate of 10.0 W for three minutes, by how much will the temperature rise? Assume the container to be perfectly insulated. Under these conditions, the density of air is 1.293 kg/m³ and the specific heat is 804 J · kg⁻¹ · K⁻¹.

11.71● An automobile having a mass of 1900 kg and traveling at a speed of 30 m/s is braked smoothly to a stop without skidding in 15 s. (a) How much energy is dissipated in the brakes? (b) What is the average power delivered to the brakes during stopping? (c) If the total heat capacity of the braking system (shoes, drums, etc.) is 0.75 kcal/°C, what is the temperature rise of the brakes during the stop?

11.72● What is the R value of a wall that loses heat at the rate of 10 W/m² when the temperature difference across the wall is 20°C? Express your answer in units of ft² · h · °F/Btu.

11.73● An iron bar is braised to a 50-cm long copper bar of the same diameter. The free end of the iron bar is kept at 0°C while the free end of the copper bar is at 100°C. How long is the iron bar if the temperature of the junction point is 50°C when the system reaches a steady state?

11.74● A person's metabolic rate can be measured using what is called a flow calorimeter. The person is placed in a large insulated container through which water can flow. The flowing water carries away the heat produced by the body. If a resting person is known to have a thermal power output of 85 W, what will the temperature difference between the intake and outflow water be when the flow rate is 1.0 liter each 5.0 min?

11.75● If you start with a container of 500 g of water at 20°C and add 300 g of ice, how many grams of steam at 100°C will you have to condense into the water, by bubbling it through a tube from the bottom, in order to return the mixture to its original temperature?

11.76● Two rods of equal length and diameter are connected end to end. The extreme ends of the rods are kept at different temperatures. The coefficient of thermal conductivity of one rod is 425 W/m°C and its free end is kept at 0°C. The coefficient of thermal conductivity of the other rod is 80 W/m°C and the free end of that rod is kept at 75°C. What is the temperature of the junction between the rods?

11.77•• A certain material has a linear expansion coefficient α. Show that, to a good approximation, the volume expansion coefficient is $\beta = 3\alpha$.

11.78•• A clock with a brass pendulum keeps correct time at 20°C. (a) What is the fractional change in pendulum length when the temperature rises to 38°C? The thermal expansion coefficient of brass is $\alpha = 18.5 \times 10^{-6}/°C$. (b) The period of a pendulum depends on the square root of its length L through $T = 2\pi\sqrt{L/g}$. By how many seconds will the clock be in error after running 24 h at 38°C? (*Hint:* You may find it helpful to use the binomial theorem.)

11.79•• A liquid of unknown specific heat at a temperature of 20°C was mixed with water at 80°C in a well-insulated container. The final temperature was measured to be 50°C, and the combined mass of the two liquids was measured to be 240 g. In a second experiment with both liquids at the same initial temperature, 20 g less of the liquid of unknown specific heat was poured into the same amount of water as before. This time the equilibrium temperature was found to be 52°C. Determine the specific heat of the liquid.

11.80•• Show that the effect of using two layers of insulation made of pieces of different material placed one after the other, with R values of R_1 and R_2, is to give an equivalent R value of $R_1 + R_2$.

11.81•• A 0.50-kg piece of glowing-hot iron at a temperature of 1000°C is put into an insulated container holding one kilogram of water at 23°C. How much water will be left after equilibrium has been reached? Assume that no steam escapes.

11.82•• A hollow insulating cube whose inside dimensions are 10 cm × 10 cm × 10 cm contains 300 g of water at 30°C. A 4.0-cm cube of aluminum at 95°C is lowered quickly into the water. Is it possible to reduce the temperature of the cube to 12°C by pouring cold water into the insulating container without allowing any water to spill over the sides of the container?

Gas Laws and Kinetic Theory

In this chapter we examine for the first time the concept of atoms. The atomic concept is that matter consists of enormous numbers of tiny components called atoms. The properties and interactions of atoms, or of the molecules into which they may combine, are responsible for the observed properties and behavior of matter in bulk form. Up to now we have dealt only with objects large enough to see and measure. Here we apply the ideas from mechanics to predict how a gas composed of a large number of atoms or molecules will behave.

Because we can neither see nor measure individual atoms directly, we must make some assumptions about what they are like. As we have discussed before, this process is part of making a theoretical model. In addition, because we clearly cannot follow the motions of every individual atom, we must decide how to evaluate average behavior. Our model will not be perfect—that is, it will not correspond to reality in every way—but the model will allow us to make predictions about the large-scale thermal behavior of gases. We can then test those predictions experimentally. The value of our model, called the kinetic theory of gases, depends on how close it comes to correctly describing the experimental observations.

Before considering the construction and use of a model of a gas, we briefly review the gas laws as they were determined over the years by observations and experiments. An understanding of the nature of air, and subsequently of other gases, could come only with the development of new devices and measuring instruments. The seventeenth century saw the discovery of such new instruments as the telescope, microscope, thermometer, barometer, pendulum clock, and air pump. These developments, as well as the founding of the first scientific societies, gave impetus to many kinds of scientific study, including the investigation of gases.

Today many researchers actively investigate the behavior of gases, using the latest computer simulation techniques and other modern methods of investigation. Much of their research has been spurred by environmental problems connected with the atmosphere, such as the depletion of the earth's ozone layer. Solutions to such problems depend on understanding the behavior of gases. ■

12.1 The Pressure of Air

Galileo was the first person to record a fact that others had surely observed: that a lift pump (Fig. 12.1) could raise water only as high as 10.4 m (34 ft). In 1643, Evangelista Torricelli (1608–1647), who was briefly associated with Galileo, provided an explanation for this fact. Previously it had been said that "nature abhorred a vacuum." People thought the lift pump worked because the water rose up into the chamber to avoid the formation of a vacuum at the top. But Torricelli believed that the water was forced up the pipe by the pressure of the atmosphere on the surface of the water at the bottom of the pump. He reasoned that since mercury is about 14 times as dense as water, if he was correct, atmospheric pressure should support a column of mercury approximately $\frac{1}{14}$ as high as the maximum water column:

$$(10.4 \text{ m water})\left(\frac{\text{density of water}}{\text{density of mercury}}\right) = (10.4 \text{ m})\left(\frac{1}{13.6}\right)$$
$$= 0.76 \text{ m mercury.}$$

Figure 12.2 shows a *Torricellian tube*, the forerunner of today's barometer. If a glass tube, closed at one end, is completely filled with mercury and then inverted into a bowl of mercury, the column of mercury in the tube drops until it reaches a height of about 76 cm above the lower surface, just as Torricelli predicted. In accord with Pascal's principle (Chapter 10), the pressure of the atmosphere on the surface of the mercury in the bowl is equal to the pressure due to the weight of the mercury in the tube. If this were not so, the mercury would flow because it would not be in static equilibrium. The space between the top of the liquid and the end of the tube contains no air and was named the *Torricellian vacuum.*

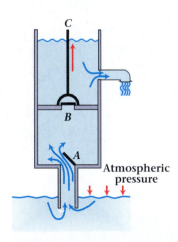

Figure 12.1

A lift pump. When rod *C* is lifted, valve *B* closes and valve *A* opens. Then water below the piston flows into the chamber and the water above flows out at the same time. When rod *C* is pushed down, *A* closes and *B* opens, allowing water to flow above the piston.

Example 12.1

Determine the pressure due to the atmosphere, using Torricelli's results.

Strategy We could make immediate use of the results contained in Chapter 10 on fluids, but it is instructive to consider the problem from a more fundamental standpoint. The weight of mercury in the tube in Fig. 12.2 is the product of its density ρ times the volume times the acceleration of gravity:

$$\text{weight} = \rho V g = \rho \cdot hA \cdot g,$$

where h is the height and A the cross-sectional area of the mercury column. For the fluid to be in static equilibrium, an upward force must be present. In this case it comes from the atmospheric pressure acting downward on the mercury in the bowl and then, according to Pascal's principle, being transmitted equally to all parts of the fluid. The upward force on the bottom of the tube is therefore

$$F = PA,$$

where P is the pressure due to the atmosphere.

Solution Because the fluid is in equilibrium, the magnitude of the upward force equals the magnitude of the downward force (the weight). Setting the two forces equal to each other gives

$$\rho \cdot hA \cdot g = PA,$$

or

$$P = \rho gh.$$

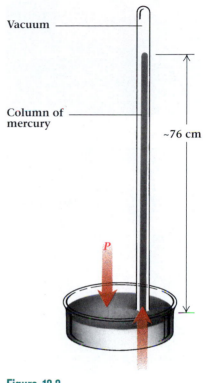

Figure 12.2
In a Torricellian tube, atmospheric pressure supports a column of mercury 76 cm tall.

This is the same expression we found earlier, in Section 10.1. Inserting the numbers for g, h, and the density of mercury, we have

$$P = (13.6 \times 10^3 \text{ kg/m}^3)(9.81 \text{ m/s}^2)(0.760 \text{ m})$$
$$P = 101 \times 10^3 \text{ N/m}^2$$
$$P = 1.01 \times 10^5 \text{ N/m}^2 = 1.01 \times 10^5 \text{ Pa.}$$

This pressure is atmospheric pressure. An atmosphere is a unit of pressure equal to the pressure of the earth's atmosphere at sea level. It is defined to be 1.01325×10^5 N/m^2.

Within a few years of Torricelli's experiments, Pascal suggested that the atmosphere is like an ocean of air, in which the pressure is greater at the bottom than at higher altitudes. This suggestion was soon confirmed when a Torricellian tube was carried from sea level to a mountain top and the mercury column was observed to be shorter at the higher elevations.

A dramatic example of atmospheric pressure was provided by Otto von Guericke (1602–1686), who used his own invention, the air pump, in the famous demonstration of the Magdeburg hemispheres (Fig. 12.3). Two hemispheres, each about 55 cm in diameter, were fitted together and most of the inside air removed. The hemispheres were held together by the pressure of the surrounding air. Two teams of eight horses each, when hitched to the hemispheres, could not pull them apart. Problem 12.12 asks you to find the force holding them together.

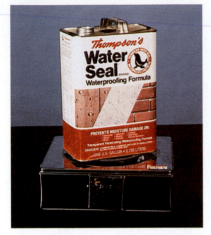

(a)

(b)

(c)

Figure 12.4

(a) A one-gallon metal can open to the air contains a small amount of boiling water. (b) The can is fitted with a tight-fitting stopper and the heat turned off. (c) As it cools the can is crushed by atmospheric pressure.

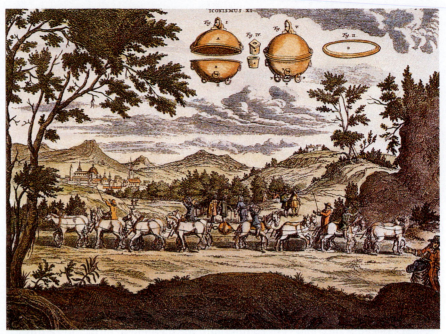

Figure 12.3

Otto von Guericke's experiment in Magdeburg, Germany, utilized an early vacuum pump. Two teams of eight horses could not separate the sealed and evacuated hemispheres against the force of atmospheric pressure.

Because we are always immersed in the atmosphere, we often overlook its effects. Its presence can be easily demonstrated, however. Consider the one-gallon can shown in Fig. 12.4(a). When a small amount of water was placed in the can and heated to the boiling point, the steam drove out the air and filled the can. While the water was still boiling, the can was closed with a rubber stopper and the heat source turned off. Immediately the can began to collapse as the steam condensed leaving a partial vacuum inside (Fig. 12.4b). With little internal pressure to resist, the pressure of the atmosphere outside crushed the can (Fig. 12.4c).

| 12.2 | **Boyle's Law** |

Robert Boyle* (1627–1691) advanced the study of gases using an air pump made by Robert Hooke, which was greatly improved over that of von Guericke. The result for which Boyle is best known today is the observed relationship between the pressure and the volume of an enclosed gas at a constant temperature.

---

*Boyle was one of the founders of the Royal Society, chartered in 1662, which published Newton's first papers and of which Newton was later president. Among the interests of this group were pneumatic experiments, which were often carried out as a diversion.

Boyle's law states that **the pressure exerted by a gas at constant temperature is inversely proportional to the volume in which it is enclosed.** Boyle's law is usually written

$$PV = \text{constant},$$

where P is the gas pressure, V its volume, and the value of the constant depends on the initial conditions. A complete statement of Boyle's law includes the condition that both the temperature and the amount of gas must be held constant.

Alternatively, Boyle's law may be written

$$P_1V_1 = P_2V_2, \qquad (12.1)$$

where the subscripts 1 and 2 refer to different physical states of the same sample of gas with the temperature held constant. Figure 12.5 illustrates Boyle's law.

We should point out that while Boyle's law is applicable over a wide range of pressures, it does not always apply. For example, if the temperature is low enough, a sample of gas will condense to a liquid at sufficiently high pressure. For carbon dioxide at 31°C, this pressure is about 7.38×10^6 N/m^2, or 72.9 atm.

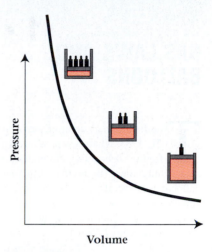

Figure 12.5
A graph of pressure versus volume for a gas enclosed in a cylinder at constant temperature. According to Boyle's law, PV is constant.

Example 12.2

A cylinder with a height of 0.20 m and a cross-sectional area of 0.040 m^2 has a close-fitting piston that may be moved to change the internal volume of the cylinder (Fig. 12.6). Air at atmospheric pressure (1.01×10^5 N/m^2) fills the cylinder. If the piston is pushed until it is within 0.12 m of the end of the cylinder, what is the new pressure of the air? Assume that the temperature of the gas remains constant and that the volume of gas in the gauge is small compared with the volume of the cylinder.

Solution Let subscripts 1 and 2 indicate the situation before and after the piston is pushed in. Then, according to Boyle's law, we can write

$$P_1V_1 = P_2V_2,$$

which can be written

$$P_2 = \frac{P_1V_1}{V_2}.$$

Inserting the numerical values, we find

$$P_2 = \frac{(1.01 \times 10^5 \text{ N/m}^2)(0.20 \text{ m} \times 0.040 \text{ m}^2)}{0.12 \text{ m} \times 0.040 \text{ m}^2}$$

$$P_2 = 1.7 \times 10^5 \text{ N/m}^2.$$

The new pressure is 1.7×10^5 N/m^2, or 1.7×10^5 Pa.

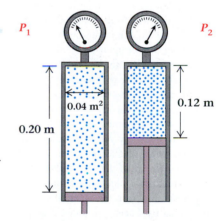

Figure 12.6
Example 12.2: Gas enclosed in a cylinder at constant temperature. The volume of gas in the gauge is small compared to the volume of the cylinder.

GAS LAWS AND BALLOONS

The first human flight occurred in Paris on November 21, 1783, when two passengers made a 25-minute flight in a hot-air balloon designed by Joseph and Etienne Montgolfier. The brothers had been experimenting with balloons for several years. Benjamin Franklin, then serving as ambassador to France, was an official observer of this first manned flight. When asked what was the use of flying he is said to have replied, "What use is a newborn baby?"

A few days after the Montgolfier flight, J. A. C. Charles, in a balloon of his own design, made an ascent with a companion. At the end of the flight his companion got out of the gondola and Charles alone soared to a height of 9000 ft in about ten minutes, making temperature and pressure measurements along the way.

According to Archimedes' principle, a balloon rises if its overall density is less than the density of the air it displaces. Thus, a balloon needs very-low-density gas to carry people through the air. Although the Montgolfiers knew about the discovery of hydrogen by Henry Cavendish in 1766, they elected to use hot air, for practical and economic reasons. Professor Charles, with encouragement from Franklin, selected hydrogen instead. (Hydrogen is the least dense of all gases.) For balloons of the same size, a hydrogen-filled balloon has several times more lifting force than a hot-air balloon. The first few years of ballooning were filled with controversy over the relative merits of hydrogen versus hot air, spurring studies of the laws of gas behavior by people who were often associated with ballooning. For instance, the first qualitative work on the thermal expansion of gases was probably done by Charles in about 1787, several years after his first hydrogen balloon flight.

Another balloonist, J. L. Gay-Lussac, was one of the first to make ascents for scientific purposes. He was an active chemist and made two important discoveries about gases. He independently studied the thermal expansion of gases and published his results in 1802, fifteen years after Charles's work. In his paper, Gay-Lussac referred to the earlier work of Charles, noting that Charles had obtained incorrect results for wet gases. The law of thermal expansion of gases is given the names of both Gay-Lussac and Charles.

During the next hundred years, ballooning evolved into sporting, military, and commercial applications, culminating in the invention of the rigid-frame airship by Count Ferdinand von Zeppelin in the last part of the nineteenth century. Zeppelin's airships became the luxury liners of the sky. The Zeppelin was made of a light, rigid metal framework covered with fabric. Inside were sealed bags of hydrogen gas and compartments

Figure B12.1 The Zepplin NT, a modern rigid-frame airship, on its maiden flight in 1997.

for passengers. The cabin that hung beneath the Zeppelin contained the bridge and navigation rooms. The *Hindenburg,* a Zeppelin built in 1936, was the largest flying machine ever made. It had space to carry 72 passengers at 80 mph for over 8000 mi. The *Hindenburg* made more than 50 successful flights, including 36 across the Atlantic, before it exploded on landing at Lakehurst, New Jersey, in 1937, ending the Zepplin era. A new technology airship was launched in 1997 using helium as the lifting gas (Fig. B12.1).

Today most ballooning takes place with colorful hot-air sports balloons (Fig. B12.2). From shortly after the time of the Montgolfiers until the 1950s, most balloons used hydrogen or helium, the two lightest gases. The expense of large quantities of these gases effectively kept private individuals from participating. On October 10, 1960, the age of the modern hot-air sports balloon was born when Ed Youst flew a hot-air balloon, using propane burners of his own design. These heat sources, which are at the heart of modern sports ballooning, are capable of delivering several miilion Btu per hour (10 million Btu/h = 2.9 MW), giving temperatures near 100°C inside the top of a rising balloon. Hot-air balloons must be relatively large because even very hot air is only slightly less dense than the atmosphere. A typical three- or four-person balloon has a volume of about 2200 m^3.

Figure B12.2 Modern hot-air balloons.

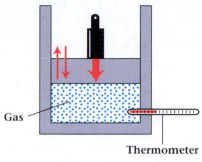

(a) Low temperature

12.3   The Law of Charles and Gay-Lussac

Boyle's law relates the pressure and the volume of a gas at constant temperature. However, we can also investigate the effect of temperature change on the volume of a gas at constant pressure, using the apparatus shown in Fig. 12.7(a,b). We introduce some particular gas, say gas A, into the cylinder and place a weight on the piston. This arrangement creates an enclosed variable volume in which the pressure (due to the weight) remains constant. Then we bring the gas to thermal equilibrium at several different temperatures. We record these temperatures and the corresponding volumes and graph the data (Fig. 12.7c, line A). If we repeat this procedure with another gas, say B, with the same initial volume and temperature, we get the same results (line B). However, if we use gas B with a different initial volume, we obtain data producing a new line, marked B' in Fig. 12.7(c). Whatever gas we use, the behavior is the same: The plotted data always lie along a straight line.

To see what this result means, we may write the equation of any of these straight lines in the form

$$V = V_0(1 + \beta T),$$

where V is the volume at temperature T and where V_0 and β are constants. The experimental result is that β is the same for all gases. If T is measured in degrees Celsius, β is found to have the value of $1/273.15°C$.

If a gas is brought to some temperature T_1, then its volume V_1 is given by

$$V_1 = V_0(1 + \beta T_1).$$

If the gas is brought to a different temperature T_2, its volume V_2 is

$$V_2 = V_0(1 + \beta T_2).$$

Dividing one of these equations by the other and writing in the numerical value for β gives

$$\frac{V_1}{V_2} = \frac{273.15 + T_1}{273.15 + T_2}.$$

It is evident that, at least so far as gases are concerned, it would be convenient to introduce a new temperature scale in which the temperature T is related to the Celsius temperature T_C by

$$T = 273.15 + T_C. \tag{12.2}$$

As we learned in Chapter 11, this temperature scale is called the Kelvin scale. The scale was introduced in 1848 by Lord Kelvin, who observed that if the lines for all gases in Fig. 12.7(c) were extrapolated to lower temperatures, they would intersect at $-273.15°C$ and zero volume. That does not mean that a gas would vanish at this temperature, even if it could be cooled so low before liquefying. It means, as we will see in more detail in Section 12.6, that the molecules of the gas would have a minimum amount of energy at that temperature. The scale is an absolute scale in the sense that its zero ($-273.15°C$) is the lower limit for temperatures as defined by macroscopic thermometers. *We will use the Kelvin scale throughout the remainder of this chapter.*

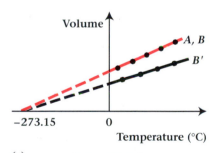

(b) Higher temperature

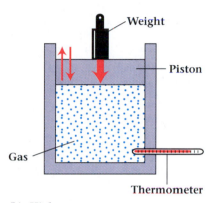

(c)

Figure 12.7

Apparatus for measuring the volume of a gas as a function of temperature at constant pressure shown at (a) lower temperature and (b) higher temperature. (c) Typical graphical representation. The volume is directly proportional to the temperature.

If the temperature is expressed in kelvins, we find that, when the pressure is held constant, the volume is proportional to the temperature. This statement is the **law of Charles and Gay-Lussac,** which can be expressed mathematically as

$$\frac{V_1}{V_2} = \frac{T_1}{T_2} \quad or \quad \frac{V}{T} = \text{constant.} \tag{12.3}$$

As with Boyle's law, the amount of gas also must be held constant for Eq. (12.3) to be valid.

Example 12.3

To what temperature would the air in a hot-air balloon have to be heated so that its mass would be 0.980 times that of an equal volume of air at a temperature of 25°C?

Strategy As the air in a hot-air balloon is heated, it expands. Some of the air escapes through the hole at the bottom of the balloon in order for the interior of the balloon to remain at constant atmospheric pressure. The mass of air remaining inside the constant volume of the bag is less than the mass of an equivalent volume of the surrounding cooler air. Thus, the density of the air inside the bag is reduced and the balloon floats, in accord with Archimedes' principle. The mass of cool gas m_1 at temperature T_1 that was originally inside the volume V_1 of the bag expands to a new volume V_2 upon heating to temperature T_2. Since the volume of the bag is constant, the amount of hot gas m_2 that remains inside is proportional to the original mass and to the ratio of V_1/V_2. Thus

$$m_2 = \frac{V_1}{V_2} m_1,$$

which can be rearranged as

$$\frac{V_2}{V_1} = \frac{m_1}{m_2}.$$

We can combine this equation with Eq. (12.3) to find the required temperature.

Solution The law of Charles and Gay-Lussac, Eq. (12.3), can be written as

$$T_2 = \frac{V_2}{V_1} T_1 = \frac{m_1}{m_2} T_1.$$

The ratio of m_1/m_2 is

$$\frac{m_1}{m_2} = \frac{1}{0.980}.$$

The temperature T_1 must be expressed in kelvins to be used in the law of Charles and Gay-Lussac,

$$T_1 = 25°C = (273 + 25) \text{ K} = 298 \text{ K.}$$

Inserting the values for T_1 and the ratio of the masses into the equation for T_2, we get

$$T_2 = \frac{1}{0.980} \; 298 \text{ K} = 304 \text{ K}.$$

Upon converting back to degrees Celsius, we get

$$T_2 = 31°\text{C}.$$

12.4 The Ideal Gas Law

Boyle's law and the law of Charles and Gay-Lussac are special cases of a more general expression called the **ideal gas law.*** It can be inferred from them and is usually written

$$PV = nRT. \tag{12.4}$$

Here, as before, P, V, and T stand for pressure, volume, and temperature, respectively; R is a constant that is the same for all gases and so is called the **universal gas constant.** If pressure is measured in the SI unit of pascals, volume in cubic meters, and temperature in kelvins, then R has the value 8.314 Joule/mole · K.

The quantity of gas, measured in moles, is given by n. A **mole** (abbreviated mol) is the amount of material whose mass in grams is numerically equal to the molecular mass of the substance. For example, the molecular mass of oxygen gas is 32; a mole of oxygen gas is 32 g. (Oxygen molecules are diatomic; that is, each molecule consists of two atoms, each with atomic mass 16.) Avogadro's principle, discussed in Chapter 26, leads to the conclusion that a mole of any gas contains the same number of molecules. This number is called **Avogadro's number,** N_A, and is $N_A = 6.02 \times 10^{23}$ molecules/mole. The modern definition of the mole encompasses the concept of isotopes, discussed in Chapter 29. *The mole is the amount of substance of a system that contains as many elementary entities as there are atoms in 0.012 kg of carbon-12.* Table 12.1 lists the molecular masses of a few common gases. The dependence of the ideal gas law on n, the number of moles, can be seen from the observation that doubling an amount of gas at constant temperature and volume will double its pressure. One mole of nitrogen gas at 273.15 K occupying a volume of 22.4 L has a pressure of 101 kPa; two moles of nitrogen gas at the same temperature and occupying the same volume have a pressure of 202 kPa.

An equation that links the pressure, volume, and temperature of a sample of matter is called an **equation of state.** Equation (12.4) is one such equation. It is the equation of state for an ideal gas, which is described in detail in the next section. Quantities that describe the condition or state of a system are called **state variables.**

---

*The definition of an ideal gas is given in Section 12.5. In brief, an ideal gas obeys Eq. (12.4). The behavior of real gases is closely approximated by Eq. (12.4).

Table 12.1	Approximate Molecular Masses of Some Common Gases	
Substance	**Symbol**	**Molecular Mass (g/mol)**
Molecular hydrogen	H_2	2
Helium	He	4
Water vapor	H_2O	18
Neon	Ne	20
Molecular nitrogen	N_2	28
Molecular oxygen	O_2	32
Argon	Ar	40
Carbon dioxide	CO_2	44

In ordinary situations, the ideal gas law predicts the behavior of many gases quite well. The agreement between the observed and predicted behavior of real gases is good near atmospheric pressure and ordinary temperatures. Deviations from the predictions are greatest when the pressure is very great or the temperature is very low. At the extremes of these conditions, the gases condense to liquids. For the examples and problems given in this chapter, the deviations from the ideal gas law are small.

Example 12.4

During a chemistry laboratory experiment, a sample of hydrogen gas is collected into a 0.50-liter flask at room temperature (23°C) and pressure (1.00 atm). It is cooled to 5.0°C and transferred to a container whose volume is 0.12 L. What pressure does the gas exert on the walls of the final container?

Solution Using the subscript 1 to indicate the initial conditions, we have

$$P_1V_1 = nRT_1.$$

Using the subscript 2 to denote conditions in the new container, we write

$$P_2V_2 = nRT_2.$$

Because the number of moles of gas is constant, we can combine the equations of state for the two situations to get

$$\frac{P_1V_1}{T_1} = \frac{P_2V_2}{T_2},$$

or

$$P_2 = \frac{V_1T_2P_1}{V_2T_1}.$$

Substituting the data gives

$$P_2 = \frac{(0.50 \text{ L})[(273 + 5.0) \text{ K}](1.00 \text{ atm})}{(0.12 \text{ L})[(273 + 23) \text{ K}]} = 3.9 \text{ atm}.$$

Discussion Notice that pressure and volume enter only as ratios and can be given in any system of units as long as they are the same. Temperatures, however, must always be given in kelvins.

Example 12.5

What is the density of carbon dioxide gas at a temperature of 23°C and atmospheric pressure?

Solution The mass m of a sample of matter is related to the number of moles by

$$m = nM,$$

where n is the number of moles and M is the gram molecular mass. For carbon dioxide, CO_2, the gram molecular mass is

$$M_{carbon} + 2M_{oxygen} = 12.0 + 2(16.0) = 44.0 \text{ g/mol}.$$

The density is given by

$$\rho = \frac{m}{V} = \frac{nM}{nRT/P} = \frac{MP}{RT}.$$

Substituting for the molecular mass of CO_2 in the units of kg/mol and for the other variables, we have

$$\rho = \frac{(44 \times 10^{-3} \text{ kg/mol})(1.01 \times 10^5 \text{ N/m}^2)}{(8.31 \text{ J/mol} \cdot \text{K})[(273 + 23) \text{ K}]}$$

$$\rho = 1.81 \text{ kg/m}^3.$$

This value compares well with the experimentally measured one of 1.84 kg/m^3.

Example 12.6

An automobile tire is filled to a gauge pressure of 240 kPa early in the morning when the temperature is 15°C. After the car is driven all day over hot roads, the tire temperature is 70°C. Estimate the new gauge pressure.

Strategy We assume the volume of the tire to be approximately constant. Then from the ideal gas law we get

$$\frac{P_1}{T_1} = \frac{P_2}{T_2}.$$

The appropriate pressure to use is the absolute pressure. Recall that the absolute pressure is gauge pressure plus atmospheric pressure.

Solution Rearrange the equation above to get the pressure P_2,

$$P_2 = P_1 \frac{T_2}{T_1}$$

$$P_2 = \frac{(240 \text{ kPa} + 101 \text{ kPa})(273 + 70) \text{ K}}{(273 + 15) \text{ K}} = \frac{(341 \text{ kPa})(343)}{288}$$

$$P_2 = 406 \text{ kPa}.$$

But this is the absolute pressure, so

$$\text{gauge pressure} = 406 \text{ kPa} - 101 \text{ kPa} = 305 \text{ kPa}.$$

12.5 The Kinetic Theory of Gases

The gas laws describe what happens to a gas under various conditions, but they say nothing about why gases act this way. Explaining the "why" behind observed behavior of matter has always been a fundamental motivation in physics. As early as the time of Boyle and Newton (that is, late seventeenth century), there were efforts to explain the observed behavior of gases on a fundamental basis. Newton proposed that a gas might consist of tiny particles called molecules, which exert a repulsive force on each other. Such an idea can lead to Boyle's law, though Newton and Boyle did not claim this to be the only possible explanation.

The development of a successful theory of gas behavior did not take place until the middle of the nineteenth century. Many people contributed to this development, and the theory has not been named after any particular scientist. However, since the theory assumes that a gas consists of particles in motion, it is usually called the **kinetic theory of gases.**

In the kinetic theory of gases, we assume that a gas consists of many particles. Here "many" means so numerous that we cannot hope to trace out their individual paths (Fig. 12.8). In fact, it may not be desirable to do this even if we could. The things we want to determine from a model of a gas are the things that determine its behavior—an equation of state and the thermal and mechanical properties of a gas—not the directions and speeds of a large number of individual particles. If we do not treat the molecules individually, then we must determine their average behavior. This implies that we need to develop a statistical theory that includes the rules of probability.

We will first construct a model of a gas by making detailed assumptions about its nature. Then we will calculate how this gas should behave, by using our knowledge of Newton's laws and of the conservation of momentum and mechanical energy and by taking the appropriate averages. For instance, we will derive a relationship between the pressure and volume of this model gas. One test of the validity of our assumptions about the gas model will be how closely our predictions agree with what we observe for a real gas.

This example of model making is typical of many theories in modern science. We cannot, in the ordinary sense, "see" individual molecules, atoms, or their constituent parts, so our explanations of what happens in many biological, chemical, and physical phenomena must be made in terms of an abstract model. Such models should not be taken as absolute replicas of reality. Furthermore, we will see that in many cases, as we examine smaller and smaller physical constituents of a system and therefore consider more of them, we are forced to calculate the observed behavior as statistical averages. This will be the case with a gas.

Our rules for formulating a model are as follows:

1. We will not make any assumptions that are not specifically needed in our derivation.

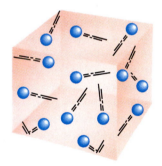

Figure 12.8
Model of a gas composed of a large number of identical molecules in random motion.

2. The assumptions made will be as simple as possible.

3. The assumptions will, when possible, have some basis in physical experience.

We will first list the *principal assumptions of the kinetic theory of gases* and then use them in deriving a relationship between the pressure and the volume of a gas. These assumptions constitute the definition of an ideal gas. Any gas that obeys the relationships derived from these assumptions at all temperatures and pressures is called an **ideal gas.**

1. *A sample of gas consists of many identical molecules. In this context "many" means so many that one could not hope to trace out their individual paths.*

2. *The molecules are very far apart in comparison to their size; that is, the total volume of the molecules is negligible when compared with the size of their container.*

3. *The direction of motion of any molecule is random; on the average, no direction is preferred above another and the molecules move with a variety of speeds.*

4. *The molecules are treated as if they were hard spheres. This assumption, which gives rise to the name "billiard ball model," means that there are no forces acting between molecules except when the molecules collide and that the collisions are elastic. In addition, we treat collisions with the walls of the container as elastic collisions.*

5. *The molecules obey Newton's laws of motion.*

These assumptions are to some extent based on everyday experience. For instance, we know that molecules of air cannot be seen by the unaided eye. This means they must be much smaller than spheres with a radius of about 0.1 mm. The assumption of random directions seems reasonable because if the molecules of a gas had a preferred direction, the gas would have a net flow in that direction.

The assumption that molecules of a gas are like hard spheres is not the only, or even the most realistic, assumption we could make. However, it is the simplest assumption and therefore, lacking specific information to the contrary, the most appropriate in this case. We have listed the use of Newtonian mechanics in order to emphasize that the results of our calculations with this model depend not only on the physical properties of the model, but also on the mathematical techniques used to obtain its predictions.

Now that we have stated the assumptions of the kinetic theory of gases, let us analyze their consequences and compare these with observable properties of a real gas. We will do this by deriving mathematical expressions for those physical quantities of importance in mechanics, such as momentum, force, and kinetic energy. We start by considering an ideal gas confined to a cubical box with sides of length L. The gas molecules move in a manner determined by Newton's three laws and the laws of conservation of energy and momentum. We assume that the number density, that is, the number of molecules per unit volume, is on the average the same throughout the box. For convenience we choose a coordinate system aligned with the edges of the box (Fig. 12.9).

Consider a molecule of mass m moving parallel to the x axis in the positive direction. For the moment, we neglect collisions between molecules and consider only collisions between the molecules and the walls of the box. If the molecule

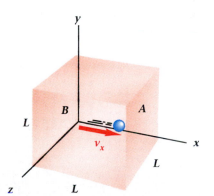

Figure 12.9
A model for calculating the pressure of a gas from **kinetic theory.** The gas molecule is confined to a box of length L as shown.

has a velocity v_x it will rebound from wall A with a velocity $-v_x$, because we assumed that the collisions are elastic, and the walls are essentially immobile. The molecule's change in velocity is $-2v_x$, and its change in momentum is $-2mv_x$. The momentum transferred to the wall is $+2mv_x$.

If the molecule encounters no other molecules but moves until it strikes wall B and rebounds again to strike wall A, the time between collisions with face A is

$$\Delta t = \frac{2L}{v_x}.$$

The average momentum change per unit time at face A is

$$\frac{\Delta p}{\Delta t} = \frac{2mv_x^2}{2L}.$$

We can interpret this expression as the average force on the wall due to any molecule that has an x component of velocity given by v_x. The total force due to N molecules is

$$F_{\text{total}} = N\frac{\overline{\Delta p}}{\Delta t} = \frac{Nm\overline{v_x^2}}{L},$$

where the bar indicates, as before, the average of that quantity.

By dividing both sides of the equation by the area of the face L^2, we have the pressure P,

$$P = \frac{F_{\text{total}}}{A} = \frac{F_{\text{total}}}{L^2} = \frac{Nm\overline{v_x^2}}{L^3} = \frac{Nm\overline{v_x^2}}{V},$$

or

$$PV = Nm\overline{v_x^2}, \tag{12.5}$$

where P is the pressure on the wall and $V = L^3$ is the volume of the box.

The square of the total velocity of a single molecule is the sum of the squares of the three component velocities:

$$v^2 = v_x^2 + v_y^2 + v_z^2.$$

For a large number of molecules, the assumption of no preferred direction (assumption 3) leads to an average squared velocity of

$$\overline{v^2} = \overline{v_x^2 + v_y^2 + v_z^2} = \overline{v_x^2} + \overline{v_y^2} + \overline{v_z^2}.$$

The average values of the components are identical because the x, y, and z directions are equivalent and independent. Therefore

$$\overline{v_x^2} = \overline{v_y^2} = \overline{v_z^2}$$

and

$$\overline{v^2} = 3\overline{v_x^2}.$$

Thus $\overline{v_x^2}$ in Eq. (12.5) can be replaced by $\overline{v^2}/3$ to give

$$PV = \tfrac{1}{3}Nm\overline{v^2}. \tag{12.6a}$$

Since $\overline{KE} = \frac{1}{2}m\overline{v^2}$, where $\overline{KE}$ is the average translational kinetic energy per molecule associated with random molecular motions, we have

$$PV = \frac{2}{3}N\overline{KE}. \qquad (12.6b)$$

The quantities on the left-hand side of Eq. (12.6) are macroscopic (large-scale) quantities. The quantities on the right-hand side are microscopic (molecular-scale) variables. We have derived an expression linking the microscopic, and generally unobservable, properties of molecules, such as their masses and speeds, with the easily observed large-scale properties, such as pressure and volume. For example, we can use Eq. (12.6a) to find the average of the speed squared or, as it is also called, the mean square speed. Then we can compute the quantity $\sqrt{\overline{v^2}}$, which is called the *root-mean-square speed* v_{rms}. The rms speed is a type of "average" speed used in describing the motion of a collection of particles.

In deriving Eq. (12.6), we assumed that there were no collisions between the particles. However, if we include elastic collisions, the result is unchanged. Because of the perfect exchange of velocities in elastic collisions between identical particles, the component of momentum mv_x of a particle leaving face B is still carried to face A of the box in Fig. 12.9 by some other particle. The time duration of elastic collisions is negligible compared with the time between collisions. So the absence of collisions in our derivation does not affect its validity.

Example 12.7

Estimate the rms speed of oxygen molecules at the standard temperature and pressure of 0°C and one atmosphere. Assume that oxygen can be treated as an ideal gas.

Strategy We can use Eq. (12.6a) to find the mean square speed. We can then compute the rms speed $v_{rms} = \sqrt{\overline{v^2}}$.

Solution Equation (12.6a) can be rewritten as

$$\overline{v^2} = \frac{3PV}{Nm}.$$

But since Nm is the total mass of the gas, we can write

$$\overline{v^2} = \frac{3P}{\rho},$$

where $\rho = Nm/V$ is the mass density.

The rms speed is computed from the square root of $\overline{v^2}$:

$$v_{rms} = \sqrt{\overline{v^2}} = \sqrt{\frac{3P}{\rho}}.$$

Inserting a pressure of one atmosphere and the density of oxygen at one atmosphere and 0°C (from Table 10.2), we get

$$v_{rms} = \sqrt{\frac{3(1.01 \times 10^5 \text{ N/m}^2)}{1.43 \text{ kg/m}^3}}$$

$$v_{rms} = 460 \text{ m/s} \approx 1700 \text{ km/h (1000 mi/h)}.$$

Discussion This result for the rms speed of oxygen molecules is of the same order of magnitude as the speed of sound (317 m/s) observed in oxygen at this temperature and pressure. This similarity is to be expected, since sound waves are transmitted by the motion of air molecules. According to our model, the molecules of a gas in a closed container are moving with great speeds.

12.6 The Kinetic-Theory Definition of Temperature

According to the ideal gas law (Eq. 12.4), the product of pressure and volume of a confined gas is proportional to the temperature,

$$PV = nRT.$$

We have also seen in Eq. (12.6b) that the product of pressure and volume is proportional to the average kinetic energy per molecule,

$$PV = \tfrac{2}{3}N\overline{KE}.$$

If our kinetic theory model is correct, the terms on the right-hand sides of these two equations must be equal.

$$\tfrac{2}{3}N\overline{KE} = nRT.$$

We can rearrange the equation to get

$$\overline{KE} = \tfrac{3}{2}\frac{n}{N}RT.$$

The number of molecules N is the number of moles n times Avogadro's number ($N = nN_A$). Thus

$$\overline{KE} = \tfrac{3}{2}\frac{RT}{N_A},$$

or

$$\boxed{\overline{KE} = \tfrac{3}{2}kT,} \tag{12.7a}$$

where the constant $k = R/N_A$ is called the **Boltzmann constant** and has a value of $1.3807 \times 10^{-23} \text{ J} \cdot \text{K}^{-1}$. The Boltzmann constant occurs frequently in the physics of gases and is sometimes called the gas constant per molecule. The Boltzmann constant can be used to write the ideal gas law as

$$PV = NkT.$$

Equation (12.7a) provides a new definition of temperature in terms of the microscopic mechanical properties of a gas. Specifically, temperature is a measure of the average random translational kinetic energy of the molecules of a gas (Fig. 12.10),

$$T = \frac{2}{3} \frac{\overline{KE}}{k}.$$ (12.7b)

Furthermore, we see that the average translational kinetic energy of a molecule in an ideal gas depends only on the temperature, not on the pressure or type of gas. Remember that the temperature in these equations is given in kelvins.

The connection between temperature and the average kinetic energy of the molecules is an important result. It tells us what is happening on a molecular scale. An increase in temperature corresponds to an increase in the average speed of the molecules; a decrease in temperature corresponds to a slowing down of the molecules. Neither phenomenon is cause or effect; instead, temperature is a large-scale manifestation of motion at the molecular level of gases, solids, and liquids. Knowing this helps us to better understand the relationships among temperature, heat flow, and energy that we discussed in Chapter 11.

Example 12.8

What is the rms speed of a nitrogen molecule at a temperature of 300 K? Assume that nitrogen behaves as an ideal gas. The mass of a nitrogen molecule is $m = 4.65 \times 10^{-26}$ kg (nitrogen is another diatomic molecule).

Solution From Eq. (12.7a) we have

$$\overline{KE} = \frac{3}{2} kT = \frac{1}{2} m\overline{v^2},$$

so

$$\overline{v^2} = \frac{3kT}{m},$$

$$v_{rms} = \sqrt{\overline{v^2}} = \sqrt{\frac{3kT}{m}}.$$

Upon inserting the values for k, T, and m, we find that

$$v_{rms} = \sqrt{\frac{3(1.38 \times 10^{-23} \text{ J} \cdot \text{K}^{-1})(300 \text{ K})}{4.65 \times 10^{-26} \text{ kg}}}$$

$$v_{rms} = 517 \text{ m/s}.$$

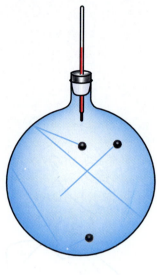

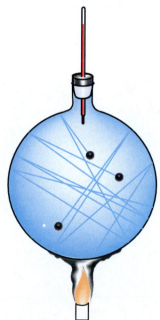

Figure 12.10
The average kinetic energy of a gas depends only on the temperature of the gas. The greater the temperature, the greater the average kinetic energy of the molecules.

12.7 Internal Energy of an Ideal Gas

We have seen that the product of pressure and volume of a confined ideal gas can be expressed in terms of the average translational kinetic energy per molecule, $\overline{KE}$:

$$PV = \frac{2}{3} N \overline{KE}.$$

The product $N\overline{KE}$ is the total translational kinetic energy of the gas due to the random thermal motion of its molecules. This kinetic energy represents the total **internal energy** U of an ideal monatomic gas. For that case there is no energy associated with rotations or internal vibrations of the molecules. Consequently we can write

$$PV = \tfrac{2}{3}U.$$

This equation is the same as Boyle's law if we assume that keeping the temperature constant and keeping the internal energy constant are the same thing.

By combining this equation with the ideal gas law (Eq. 12.4), we can relate the internal energy to the temperature through

$$nRT = \tfrac{2}{3}U.$$

We can rearrange to find

$$\boxed{U = \tfrac{3}{2}nRT.} \tag{12.8}$$

We will make use of the concept of internal energy when we discuss the laws of thermodynamics in the next chapter.

Equation (12.8) defines the internal energy in a monatomic gas such as helium or neon. Diatomic gases such as nitrogen and oxygen have additional internal energy associated with rotational and vibrational motions. Nevertheless their internal energy is still proportional to RT.

Figure 12.11
Example 12.9: Helium-filled balloons are popular items in the annual Macy's parade.

Example 12.9

A parade balloon contains 368 m³ of helium at a pressure of 115 kPa (Fig. 12.11). What is the internal energy of the helium in the balloon?

Strategy We need to find the internal energy from knowledge of the volume and pressure of the gas. The internal energy of the helium gas was given in Eq. (12.8) as $U = \tfrac{3}{2}nRT$. Although we are not given the temperature of the gas, we know from the ideal gas law that $nRT = PV$, so that the internal energy may be expressed as

$$U = \tfrac{3}{2}PV.$$

Solution The internal energy may be computed by substituting the numerical values for P and V to get

$$U = \tfrac{3}{2}(115 \times 10^3 \text{ Pa})(368 \text{ m}^3) = 6.35 \times 10^7 \text{ J}.$$

*12.8 The Barometric Formula and the Distribution of Molecular Speeds

So far in this chapter, we have considered only the average speeds of molecules. However, there are times when we need to know the distribution of molecular speeds in a gas. By distribution we mean a mathematical expression that tells us what fraction of the molecules have speeds in a given range.

The distribution of molecular speeds was first worked out by James Clerk Maxwell (1831–1879) in 1860. Maxwell's significant contribution was the introduction of statistical ideas into classical mechanics. To describe his conclusions, we first derive a formula for pressure as a function of height in the atmosphere as a way of introducing you to the exponential function. Then we will present Maxwell's results, which are expressed with this mathematical function. (We discuss the exponential function in detail in an appendix to this chapter.)

Consider a tall column of air, which might represent a cross section of the atmosphere (Fig. 12.12). We assume the temperature to be the same at all points in this atmosphere. Let us now examine the gas between two horizontal planes at altitudes z and $z + \Delta z$. The pressure at height z is greater than the pressure at height $z + \Delta z$ because of the weight of the air contained between z and $z + \Delta z$. Although the atmosphere does not have a well-defined height, we are able to write the difference in pressures because it depends on the difference in heights:

$$P_{z+\Delta z} - P_z = \Delta P = -\rho g\, \Delta z,$$

where ρ is the mass density of the gas. The minus sign indicates that the pressure decreases as the altitude increases. If we assume that the density of the air does not vary over a small change in height, we may use the ideal gas law to express this density. The total mass M of the gas is the number of moles n times the number of molecules in a mole N_A times the mass of one molecule m. The density is then

$$\rho = \frac{\text{mass}}{\text{volume}} = \frac{M}{V} = \frac{nN_A m}{nRT/P}$$

$$\rho = \frac{m}{kT} P,$$

where k is the Boltzmann constant. We can insert this value for ρ into the equation for ΔP to get

$$\frac{\Delta P}{\Delta z} = -\rho g = -\frac{mg}{kT} P. \tag{12.9}$$

This equation is a relationship between changes in pressure and changes in height. We wish to find an expression for pressure as a function of height alone. If we interpret Eq. (12.9) graphically, we see that we are looking for a function P whose slope $\Delta P/\Delta z$ is proportional to the value of P. The result is that

$$\boxed{P(z) = P_0 e^{-mgz/kT},} \tag{12.10}$$

where P_0 is the pressure at $z = 0$ and e is an irrational number with a value approximately equal to 2.718. (If you are not already familiar with the exponential function e^x, see the appendix to this chapter.) Although we would not expect full agreement between the prediction of Eq. (12.10) and the observed pressure in the earth's atmosphere because T is not constant, there is rather good general accord.

We may also express Eq. (12.10) in terms of the number density n, the number of molecules per unit volume. We can do so because at constant temperature, the pressure and density of an ideal gas are proportional. Thus

$$n = n_0 e^{-mgz/kT} \tag{12.11}$$

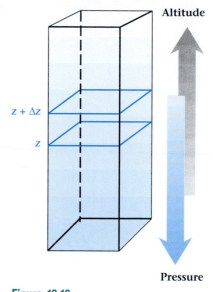

Figure 12.12
A simplified model of the atmosphere at constant temperature. Pressure is greater at lower altitudes.

This equation is called the *barometric formula*. It gives the number of molecules per unit volume as a function of height z in our idealized atmosphere.

Example 12.10

People who have been to high elevations on mountains know that it is more difficult to breathe at higher altitudes than at lower ones. This effect is due to the reduced air pressure or, equivalently, to the reduced number of air molecules (including oxygen) per unit volume. Estimate the number of molecules of air per unit volume on Pike's Peak (4300 m above sea level) relative to the number in Denver, Colorado (1610 m above sea level).

Solution   We need the average mass of the air molecules for m in Eq. (12.11). The atmosphere is about 21% oxygen (at 5.31×10^{-26} kg per molecule) and 79% nitrogen (at 4.67×10^{-26} kg per molecule). Thus, the average mass is

$$m = (0.21)(5.31 \times 10^{-26} \text{ kg}) + (0.79)(4.67 \times 10^{-26} \text{ kg}) = 4.80 \times 10^{-26} \text{ kg.}$$

The temperature is not uniform, but because we use the Kelvin scale, only a small error is introduced by assuming a constant value of $0°C = 273$ K. We can use Eq. (12.11) to find the ratio of the density of air molecules if we let n_0 represent the density at Denver and n represent the density at Pike's Peak. Then z is the difference in their altitudes, 2690 m. When we insert the numerical values into the equation we find

$$\frac{n}{n_0} = e^{-mgz/kT} = e^{-\frac{(4.80 \times 10^{-26} \text{ kg})(9.81 \text{ m/s}^2)(2690 \text{ m})}{(1.38 \times 10^{-23} \text{ J/K})(273 \text{ K})}} = 0.71.$$

Discussion   The difficulty of breathing on Pike's Peak is understandable because the number of molecules per unit volume is only 71% as great as in Denver. How does the density of air molecules on Pike's Peak compare with the density where you live?

Altitude sickness (or mountain sickness as it is also known) is the body's reaction to the reduced atmospheric pressure of high altitude. People whose travels include a change in altitude of about 8000 ft (2438 m) or more may experience symptoms that include shortness of breath, headache, and fatigue. Usually such symptoms subside after 2 or 3 days of acclimatization at the new altitude. Airlines pressurize the cabins of high-flying planes to the equivalent of a height of 6000 ft for the comfort of passengers and crew and to prevent altitude sickness.

By continuing to analyze gases in the same way and by using the results of the kinetic theory, we can extend the derivation of the barometric formula to get an expression for the fraction of particles in a gas within a range of speeds. This result was obtained (though by another method) by Maxwell, who showed that the distribution in speeds is given by

$$f(v)\Delta v = 4\pi \left(\frac{m}{2\pi kT}\right)^{3/2} v^2 e^{-mv^2/2kT} \Delta v, \tag{12.12}$$

where $f(v)\Delta v$ is the fraction of molecules that have speeds between v and $v + \Delta v$. Eq. (12.12) is called the **Maxwell-Boltzmann distribution** function. Figure 12.13 displays this function, computed for oxygen, neon, and helium gases at a temperature of 295 K. Figure 12.14 shows the distribution for oxygen at 295 K and 1000 K. Notice that the Maxwell-Boltzmann distribution function is not an exponential curve, although the equation does contain an exponential function. We do not expect you to memorize the details of this formula, but you should become familiar in a general way with the factors that determine the distribution and their significance. In particular, the shift of the curve with temperature and the factor in the exponent of kinetic energy $\frac{1}{2}mv^2$ divided by kT are worth noting.

The peak of each curve represents the most probable molecular speed for that temperature and gas. That is, the curve peaks at that speed exhibited by the largest fraction of molecules in the gas. The most probable speed depends on the temperature through

$$v_{\mathrm{mp}} = \sqrt{\frac{2kT}{m}}. \tag{12.13}$$

Although many molecules have speeds near v_{mp}, the overall range of speeds is large in every case. More massive molecules have lower most probable speeds, in agreement with the predictions of kinetic theory for a given temperature. For a given gas, the most probable molecular speed becomes greater with increasing temperature. In particular, more molecules have high speeds and fewer molecules have low speeds, throughout the range of speeds. This ties in with our earlier results from kinetic theory that higher temperatures mean higher molecular kinetic energies.

In 1955, Miller and Kusch conducted a high-precision experiment to compare a measured speed distribution with the Maxwell-Boltzmann equation. They placed a container A, filled with a known gas, and a molecule detector B at opposite ends of a cylinder C of length L (Fig. 12.15). A large number of helical grooves were cut into the surface of the cylinder, only one of which is shown in the diagram. The cylinder rotated about its axis at a constant angular velocity ω. As the cylinder rotated, it was bombarded by molecules coming out of an aperture in the heated container A. A molecule from A would reach detector B only if its velocity matched the motion of the slit as the cylinder turned. The velocity of those molecules that reached B was proportional to the cylinder's speed of rotation ω. Miller and Kusch recorded the number of molecules detected (beam

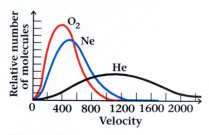

Figure 12.13

Maxwell-Boltzmann molecular-speed distribution for three different gases at $T = 295$ K. The differences are due to the different molecular masses.

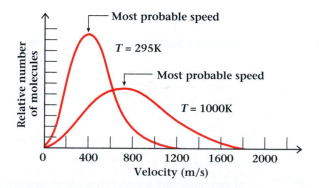

Figure 12.14

The distribution of molecular speeds for oxygen gas according to the Maxwell-Boltzmann distribution function at two different temperatures.

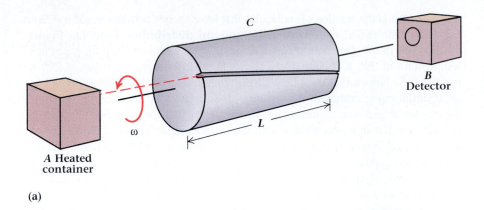

(a)

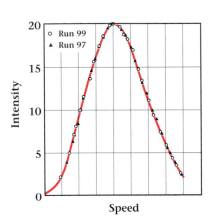

Figure 12.16

Distribution of molecular speeds in a gas. The points represent experimental observations; the solid line represents the Maxwell-Boltzmann distribution. (From R. C. Miller and P. Kusch, "Velocity Distributions in Potassium and Thallium Atomic Beams," *Physical Review,* vol. 99, August 15, 1955, p. 1314.)

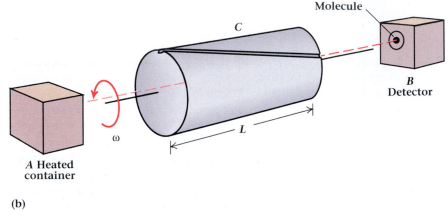

(b)

Figure 12.15

Apparatus for measuring the distribution of molecular speeds. As the grooved cylinder rotates, a molecule entering the groove at one end at just the right speed can continue in straight line motion and emerge from the far end. Molecules with other speeds will strike the edge of the groove and be scattered.

intensity) as a function of ω. This distribution of beam velocities was then used to get the distribution of molecular speeds in the container (Fig. 12.16). For other gases, the results were similar. In all cases, the distribution was consistent with the predictions of the Maxwell-Boltzmann distribution function.

Summary

Useful Concepts

■ Boyle's law states that if the temperature and the amount of gas are held constant, then

$$P_1V_1 = P_2V_2 \quad \text{or} \quad PV = \text{constant}.$$

■ The law of Charles and Gay-Lussac states that for a gas at constant pressure,

$$\frac{V}{T} = \text{constant}.$$

■ We can predict the behavior of gases using the ideal gas law,

$$PV = nRT.$$

The agreement between prediction and observation is best when temperatures and pressures are near ordinary values.

■ The kinetic theory of gases shows how it is possible to connect the macroscopic properties of a gas, such as pressure and volume, with its microscopic properties, such as average molecular kinetic energy. The kinetic theory gives, for an ideal gas,

$$PV = \tfrac{2}{3}N\overline{KE}.$$

■ The kinetic-theory definition of temperature is

$$\overline{KE} = \tfrac{3}{2}kT.$$

■ The internal energy in a monatomic gas is the total translational kinetic energy of the gas due to random thermal motion of its molecules. It is given by

$$U = \tfrac{3}{2}nRT.$$

■ The pressure at a given height in the atmosphere is given by

$$P(z) = P_0 e^{-mgz/kT}.$$

Important Terms

You should be able to write the definition or meaning of each of the following:

Boyle's law
law of Charles and
  Gay-Lussac
ideal gas law
universal gas constant
mole
Avogadro's number
equation of state

state variables
kinetic theory of gases
ideal gas
Boltzmann constant
internal energy
Maxwell-Boltzmann
  distribution

Conceptual Questions

12.1 A barometer of the type shown in Fig 12.2 is carried on an earth-orbiting satellite in which the cabin is pressurized to one atmosphere. Describe what an astronaut would observe if the apparatus were first laid on its side and then turned upside down.

12.2 If the atmosphere were made of only oxygen, instead of being primarily nitrogen, would barometers read higher or lower than they do now?

12.3 A car owner's manual says to measure the tire pressure before driving. What difference would it make if the pressure were measured after driving several miles at highway speeds?

12.4 Explain in words why the pressure of a gas increases when its volume is reduced.

12.5 Two gastight balloons are filled with helium at atmospheric pressure. One of the balloons is made from a latex material that stretches easily and the other is made from a material that does not stretch. Which balloon will rise higher?

12.6 Explain how the buoyancy of a submarine can be adjusted.

12.7 In the kinetic model of a gas we assumed elastic collisions between the molecules. How would the model change if inelastic collisions were allowed?

12.8 How would the results of our kinetic model be modified if we assumed that the size of the molecules is not negligible in comparison to the distance between them?

12.9 Give an explanation for the difference between heat and temperature, based on the kinetic theory.

12.10 At high altitude, the ratio of nitrogen to oxygen in the atmosphere increases above the ratio at sea level. Why?

12.11 Is the total kinetic energy of the molecules of air in a warm room greater than the total kinetic energy of the molecules of air in the same room when it is cool? Why? (*Hint:* The pressure is unchanged.)

12.12 An attempt is made to construct a barometer as indicated in Fig. 12.17. The barometer differs from the ordinary Torricellian tube only in that two liquids are employed as shown. (a) Can such a barometer be made if liquid A is mercury and liquid B is water? (b) Can such a barometer be made if liquid A is water and liquid B is mercury? Explain your reasoning in each case.

12.13 What is the difference between the average speed of the molecules in a gas and the rms speed?

12.14 How does the average velocity of the molecules in a gas differ from their average speed?

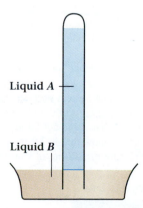

Figure 12.17
Question 12.12.

▼

Problems

Section 12.1 The Pressure of Air

Hints for Solving Problems

Remember that pressure is force per unit area. The gauge pressure due to the weight of a fluid of density ρ and depth h is $P = \rho gh$.

12.1 How high will the liquid stand in a Torricellian tube filled with methyl alcohol if the atmospheric pressure is 101 kPa and the temperature is 20°C? The density of methyl alcohol is 791 kg/m^3.

12.2 A Torricellian tube similar to the one in Fig. 12.2 supports a column of liquid 8.17 m high when the atmospheric pressure is 101 kPa. What is the liquid? (*Hint:* You may want to refer to the table of densities in Chapter 10.)

12.3 A mercury barometer is tilted at an angle of 45° from the vertical (Fig. 12.18). How far up the tube is the end of the mercury column, as measured along the tube? Assume that the temperature is 20°C and the atmospheric pressure is 101 kPa.

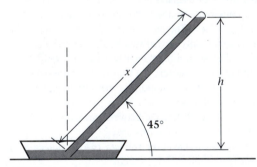

Figure 12.18
Problem 12.3.

12.4 What is the total downward atmospheric force on the top of a 0.21-m by 0.26-m book lying on a table? Why doesn't this force break the table?

12.5● Estimate the "total weight" of the atmosphere, using your knowledge of barometers and given the radius of the earth.

12.6● The air-conditioning system of a department store is designed to maintain a pressure of 0.20 in. of water above atmospheric pressure. What is the net force due to pressure on a rectangular display window that is 3.0 m × 4.0 m? (*Hint:* First convert the pressure of 0.20 in. of water to pascals.)

Section 12.2 Boyle's Law

Hints for Solving Problems

Remember to express temperatures in kelvins. It is often helpful to establish ratios of old and new state variables (P, V, T) when calculating changes in a gas.

12.7 If you can use your abdominal and chest muscles to decrease the volume of your lungs by 20%, what pressure can you develop by this method alone?

12.8 A cylinder with a cross-sectional area of 10.0 cm^2 is fitted with a movable piston. Air is introduced into the cylinder at atmospheric pressure and a temperature of 20°C. The temperature is held constant as the volume is compressed to half its initial volume. How much force must be applied to the piston to maintain it in its new position?

12.9● A 1.0-m long cylinder 6.0-cm in diameter is closed at one end and fitted with a moveable piston. The interior pressure is 1.0 atm when the piston is 0.50 m from the closed end. Make a graph of the pressure in the cylinder versus distance of the piston from the closed end for distances from 0.10 m to 0.80 m if the temperature of the cylinder and its contents is held constant.

12.10● The cylinder of a bicycle pump has an interior diameter of 2.0 cm and a length of 25 cm. It is used to put air into a tire where the pressure is already 240 kPa. How far down must you push the piston before air begins flowing into the tire?

12.11● A spherical bubble rises from the bottom of a lake. If the lake temperature is uniform and the bubble doubles its volume by the time it reaches the surface, how deep is the lake?

12.12● (a) How much force would the teams of horses have had to apply to pull the Magdeburg hemispheres apart? Their diameter is 55 cm. A rough estimate is that 90% of the air was removed from the sphere by the vacuum pump. (b) Is there anything surprising in this result?

Section 12.3 The Law of Charles and Gay-Lussac

Hints for Solving Problems

When working problems in this chapter, you may use the approximation $T_K = T_C + 273$ when converting temperatures.

12.13 A 1.00-L sample of an ideal gas at room temperature (23°C) is taken outside on a cool day (6°C) in a constant-pressure container similar to the one in Fig. 12.7(a). What is the new volume?

12.14 A one-meter-long glass tube is sealed at one end. A drop of mercury large enough to close off the tube is placed at the midpoint when the temperature is 0°C. Where will the mercury be when the closed end of the tube is immersed in boiling water?

12.15 We wish to double the volume of a gas held at constant pressure. To what temperature must we heat it if its original temperature was (a) 0°C, (b) 100°C, and (c) 1000°C?

12.16 (a) By what fraction of its initial volume does the volume of a gas decrease at constant pressure if the temperature is lowered from 100°C to 0°C? (b) What will be the fractional change if the temperature is lowered from 0°C to −100°C?

12.17 A column of dry air was sealed off from the atmosphere by closing one end of a glass tube and placing a drop of mercury in the tube 0.50 m from the closed end. This was done at some unknown temperature. The tube was then placed into a freezer where the temperature was known to be −10°C. After the tube reached the temperature of the freezer, the mercury slug was found to be 42 cm from the closed end. What was the original temperature?

Section 12.4 The Ideal Gas Law

Hints for Solving Problems

The value of the gas constant R in the ideal gas equation $PV = nRT$ is $R = 8.314$ J/mol · K.

12.18 The temperature of a 1.00-L sample of gas at atmospheric pressure is 33°C. How many moles of gas are in the sample?

12.19 Calculate the volume occupied by one mole of hydrogen gas at STP.

12.20 (a) How many molecules are there in one cubic meter of argon at STP? (b) How many molecules of radon are there in one cubic meter at STP? (STP means standard temperature and pressure: 0°C and 101 kPa.)

12.21 A cylinder closed at both ends has a piston in between that is free to slide. At 20°C, the piston is exactly in the center when hydrogen gas is in one end and helium in the other (Fig. 12.19). (a) What is the ratio of the number of hydrogen molecules to the number of helium molecules? (b) Where will the piston be if the temperature is raised to 57°C?

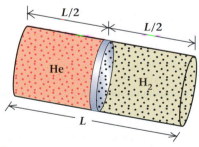

Figure 12.19
Problem 12.21.

12.22● Calculate the volume occupied by 50 g of carbon dioxide gas at a pressure of one atmosphere and a temperature of 25°C. Assume that the ideal gas law is obeyed.

12.23● A cylindrical propane tank used for portable cooking stoves has an interior length of 16.5 cm and a diameter of 6.5 cm. The mass of the gas inside is 400 g. Approximately how much pressure is exerted on the walls of the container when the temperature is 75°F? Propane has a molecular mass of 44 g/mol.

12.24● An unknown gas has a density of 1.784 kg/m³ at STP. (STP, for standard temperature and pressure, means a temperature of 0°C and a pressure of 101 kPa.) The gas does not

burn, support combustion, or react strongly with polished metal surfaces. (a) What is the gram molecular mass? (b) What is the most probable identity of the gas?

12.25● One cubic centimeter of water is converted to steam at atmospheric pressure and 100°C. What is the volume of the resulting steam?

12.26●● A spherical bubble rises from the bottom of a lake whose temperature is 7°C at the bottom and 27°C at the surface. If the bubble doubles its volume by the time it reaches the surface, how deep is the lake?

Sections 12.5 and 12.6 The Kinetic Theory of Gases and the Kinetic-Theory Definition of Temperature

12.27 Show, using the numbers 1, 3, 7, and 8, that the root-mean-square value and the average value are not the same.

12.28 What is the rms speed of hydrogen molecules at a temperature of 23°C? The mass of a hydrogen molecule is 3.34×10^{-27} kg.

12.29 A student plans to build a cubical box to hold a gas at a temperature of 17°C and a pressure of 101 kPa. He wants the number of gas molecules inside to equal the population of the United States (260 million). What length are the edges of the box?

12.30 (a) At what temperature is the rms speed of a molecule of hydrogen gas equal to 2200 m/s? (b) If you wished to reduce the rms speed of the molecules in hydrogen gas to 1100 m/s, what temperature would be required? The mass of a hydrogen molecule is 3.34×10^{-27} kg.

12.31 Find the average kinetic energy of helium atoms at a temperature of 500 K. Assume ideal gas behavior.

12.32 What is the average kinetic energy of oxygen molecules (O_2) at a temperature of 300 K? Assume oxygen to be an ideal gas.

12.33 What is the density of a gas, at a pressure of one atmosphere, in which the molecules have an rms speed of 450 m/s? (*Hint:* Study Example 12.7.)

12.34● Hydrogen (H_2) and helium (He) gases are mixed in a container. What is the ratio of the rms velocity of the heavier gas to that of the lighter gas? The mass of a helium molecule is approximately twice the mass of a hydrogen molecule.

12.35● The speed of sound in a gas is proportional to the rms speed of the molecules. What is the ratio of the speed of sound at 27°C to that at 0°C?

12.36● The speed of sound in a gas is proportional to the rms speed of the molecules. Show that the speed of sound in an ideal gas depends on the temperature and molecular mass, but not on the density. (See Example 12.8.)

12.37● (a) If the root-mean-square velocity of a molecule in a gas at 300 K were 100 m/s, what would be its mass? (b) Is this a realistic example?

12.38●● Starting with $v_{rms} = \sqrt{\dfrac{3P}{\rho}}$, where P is the pressure and ρ is the density of an ideal gas, show that the rms speed can be put in the form $v_{rms} = \sqrt{\dfrac{3kT}{m}}$, where k is Boltzmann's constant, T the temperature, and m the molecular mass.

Section 12.7 Internal Energy of an Ideal Gas

12.39 Calculate the total internal energy of an ideal gas confined to a volume of 10 L at a pressure of 2.0 atm and a temperature of 300 K.

12.40 Calculate the internal energy of one mole of helium gas at a temperature of 300 K. Assume ideal gas behavior.

12.41 Calculate the internal energy of 12 g of argon gas at a temperature of 300 K. Assume ideal gas behavior.

*Section 12.8 The Barometric Formula and the Distribution of Molecular Speeds

12.42 Plot a graph of the number density versus height for an atmosphere consisting of oxygen (O_2) only at a temperature of 23°C. Assume a unit density ($n = 1$) at $z = 0$. Plot your graph from $z = 0$ to $z = 15$ km.

12.43 Calculate the atmospheric pressure at a height of 5900 m for air with average molecular mass of 29 g/mol and uniform temperature of 290 K, given that the sea level pressure is P_0.

12.44● Use the barometric formula to estimate the ratio of the atmospheric pressure at Myrtle Beach, South Carolina (sea-level elevation), when the air temperature is 27°C, to the pressure at Denver, Colorado (1610 m above sea level), when the temperature is 10°C. (Use the mass given in Example 12.10.)

12.45● (a) At what height above sea level is the atmospheric pressure half the pressure at sea level? Assume that the temperature is a constant 0°C. (b) How high must you go for the pressure to drop to one-fourth the pressure at sea level? (Use the mass given in Example 12.10.)

12.46● Calculate the height at which atmospheric pressure is half the sea-level pressure P_0 for an atmosphere consisting solely of nitrogen (N_2) at a uniform temperature of 300 K.

12.47●● The maximum value of the Maxwell-Boltzmann distribution function occurs for a speed $v_{mp} = \sqrt{\dfrac{2kT}{m}}$. Find an expression for the magnitude of the distribution function at this speed.

Additional Problems

12.48 Air at a pressure of one atmosphere is confined to a volume V_0. If it is compressed at constant temperature to a volume of $\frac{1}{3}V_0$, what is the resulting pressure?

12.49 The velocity of sound in air is given by $v = \sqrt{\dfrac{1.4P}{\rho}}$, where P is the pressure and ρ the density of the air. What is the ratio of the rms velocity of the air molecules to the velocity of sound?

12.50● (a) Approximately how many molecules of air at STP (0°C and 101 kPa) are there in a room measuring 10 ft × 12 ft × 8 ft? (b) How much do the molecules weigh if the average molecular mass for air is 28.8 g/mol? (c) What is their average separation? (d) If they were all compacted so that the material had the same density as water, what volume would they occupy?

12.51● Nitrogen gas (N_2) is held in a 1.0-L container at a pressure of 15 atm and a temperature of 18°C. (a) How many moles of nitrogen are inside the container? (b) What is the total mass of nitrogen in the container?

12.52●● An upright cylinder, 1.00 m tall and closed at its lower end, is fitted with a light piston that is free to slide (Fig. 12.20). Initially the piston is in the center. A cuplike cavity is formed by the top of the piston and the upper cylinder walls. Water is poured into the cavity until it is full. At what fraction of the total height of the cylinder will the piston be when the cavity is full? Assume that the lower portion of the cylinder contains an ideal gas at constant temperature.

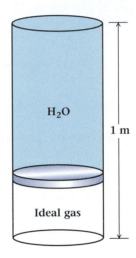

Figure 12.20
Problem 12.52.

12.53●● A bubble of 1.00-cm diameter is released at the bottom of a lake that is 30 m deep. The temperature at the bottom is 5°C, and near the surface it is 17°C. What is the diameter of the bubble when it reaches the surface?

12.54●● A 1.00-cm-diameter bubble is released at the bottom of a lake. The temperature at the bottom is 7°C, and near the surface it is 17°C. When the bubble reaches the surface its diameter is 1.80 cm. How deep is the lake?

12.55●● (a) How many molecules are there in a cubic meter of argon at a temperature of 23°C and a pressure of one atmosphere? (b) What is the total kinetic energy of these molecules? (c) How fast would someone have to throw a baseball (0.17 kg) in order for it to have a kinetic energy equal to the total kinetic energy in part (b)?

12.56●● A 1.0-L container of an ideal gas at 300 kPa and 20°C is connected through a small tube to a 10-L container of the same gas at 100 kPa and 25°C. After the gas has come to equilibrium, its temperature is 22°C. What is the pressure of the gas?

12.57●● (a) At what temperature is the rms speed of nitrogen gas equal to the escape velocity from the earth? (b) At what temperature is the rms speed of oxygen gas equal to the

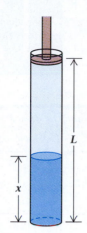

Figure 12.21
Problem 12.59.

escape velocity? (c) How would this result affect the composition of the air in the upper atmosphere, where the temperature is about 1000 K? The masses of oxygen and nitrogen molecules can be determined from Table 12.1.

12.58•• What is the ratio of the probability that a molecule of a gas has the most probable speed to the probability that it has a speed of twice the most probable speed?

12.59•• An instrument once used for measuring ocean depths from sailing ships consisted of a long heavy tube open at the bottom. When the tube was lowered into the water by a wire attached to the upper end (Fig. 12.21), the increased water pressure with depth compressed the air in the tube. When the tube was withdrawn from the water, the point to which the water had risen in the tube could be measured and used to determine the depth to which the tube had been lowered. Derive an expression for the approximate depth d in terms of the length of the tube L, the height x to which the water rises in the tube, atmospheric pressure P_0, the density of sea water ρ, and the acceleration of gravity g. For simplicity, assume that the temperature of the water is uniform and that the temperature of the gas in the tube is constant.

12.60•• Imagine making a proverbial lead balloon from a very thin lead foil. (a) Estimate the thickness of the foil for a lead balloon 1.00 m in diameter if it is to just float in air at STP when filled with hydrogen at STP. The average molecular mass for air is 28.8 g/mol. (b) Is this a realistic situation? Explain.

The Exponential Function

In many cases in the physical and biological sciences, the rate of change of a variable is proportional to the variable itself. The growth rate of a biological cell, the rate of growth of a population (whether people, plants, or bacteria), the rate of decay of a radioactive material, the rate of cooling of a hot object—all these examples represent rates of change that are proportional to the amount of material present. The function that reflects this characteristic is called the *exponential function*.

Let us examine some of the properties of the exponential function. We call our function, the dependent variable, y, and call the independent variable x. The variable x can be displacement, time, temperature, or any other quantity. In Fig. A12.1 we have drawn curve a whose slope $\Delta y/\Delta x$ increases as x increases. Curve b has a constant slope, and curve c has a slope that decreases with increasing x. The relationship between x and y in the exponential function has a graphical form that looks more like curve a than the other curves in Fig. A12.1. For the moment, we are considering only increasing rates of change.

Now let us look for a specific relationship between x and y that satisfies our requirements. Table A12.1 shows the powers-of-ten notation that we have frequently used earlier in this book. Powers of other numbers can be

Table A12.1	Several Base Numbers Raised to Various Powers	
$10^0 = 1$	$1.5^0 = 1$	$2^0 = 1$
$10^1 = 10$	$1.5^1 = 1.5$	$2^1 = 2$
$10^2 = 100$	$1.5^2 = 2.3$	$2^2 = 4$
$10^3 = 1000$	$1.5^3 = 3.4$	$2^3 = 8$
$10^4 = 10000$	$1.5^4 = 5.1$	$2^4 = 16$
$10^{4.5} = 31622$	$1.5^{4.5} = 6.2$	$2^{4.5} = 22.6$
$10^5 = 100000$	$1.5^5 = 7.6$	$2^5 = 32$

treated in the same way. The table gives several examples and shows that neither the base numbers nor the powers to which they are raised need to be integers. We display these relationships graphically in Figs. A12.2 and A12.3. Notice that the slope increases with increasing x for all of these curves. In Fig. A12.2, we have drawn a tangent to the curve at two points, A and B, and have determined the slope of the curve at these particular values of x. We can plot the slope $\Delta y/\Delta x$ of the curve as a function of x. The gray lines drawn in Fig. A12.3 are the slopes of the respective colored curves.

After studying Fig. A12.3, we can conclude that there is only one curve for which the slope and the original curve are identical. Such a curve corresponds to a base

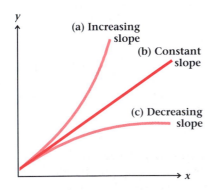

Figure A12.1
Curves with various slopes. The slope increases with increasing x for curve a. The slope is constant for curve b. The slope decreases with increasing x for curve c.

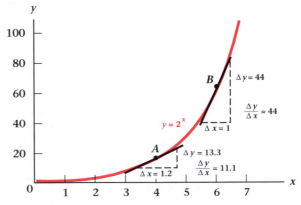

Figure A12.2
A graph of $y = 2^x$. The tangents at A and at B are the slopes of the curve at those points.

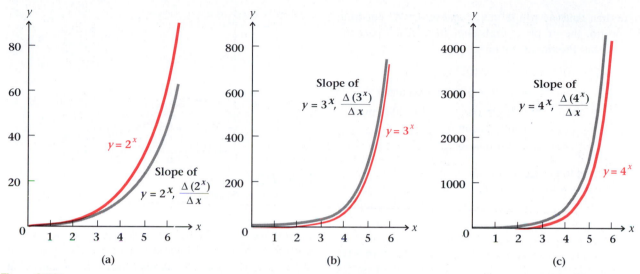

Figure A12.3

Graphs of the form $y = a^x$ and their slopes. The red lines give the values of y, the gray lines give the slopes of the red lines.

number we call e, whose magnitude is such that, for every point x,

$$\text{slope}(e^x) = \frac{\Delta(e^x)}{\Delta x} = e^x.$$

You should be able to guess from the figure that the number e must be between 2 and 3. In fact, it is an irrational number with a value of approximately 2.718. We see, then, that there is a function of x for which the slope of that function is just equal to the function itself. This function is the exponential function e^x. Table A12.2 gives values of e^x for some values of x.

In the equation for pressure in the atmosphere (Eq. 12.9), we required only that the slope of the curve be proportional to the curve, not equal to it. This requirement, too, is satisfied by an exponential function. For example, consider the number e raised to the power ax, where a is a constant. From the above equation we have

$$\text{slope}(e^{ax}) = \frac{\Delta(e^{ax})}{\Delta ax} = e^{ax},$$

where we have replaced x with ax. Since a is constant, $\Delta(ax) = a(\Delta x)$ and the equation becomes

$$\frac{\Delta(e^{ax})}{a\Delta x} = e^{ax}$$

$$\frac{\Delta(e^{ax})}{\Delta x} = ae^{ax}.$$

The exponential function e^x closely represents many processes observed in nature. It is characterized not only by its rapid increase with increasing x, but also by the increase in the rate of change with increasing x. To get a

Table A12.2	*Values of e^x*
x	**e^x**
0.0	1.000
0.5	1.649
1.0	2.718
1.5	4.482
2.0	7.389
2.5	12.182
3.0	20.09
3.5	33.12
4.0	54.60
4.5	90.02
5.0	148.4
5.5	244.7
6.0	403.4
6.5	665.1
7.0	1097
7.5	1808
8.0	2981

Note: $e^{-x} = 1/e^x$.

feeling for how rapidly the values of $y = e^x$ grow with increasing x, think of graphing the function on a large blackboard, with the axes scaled in centimeters. At $x = 1$ cm, the graph is $y = e^1 \approx 3$ cm above the x axis. At $x = 6$ cm, the graph is $y = e^6 \approx 403$ cm ≈ 4 m high (it is about to go through the ceiling if it hasn't done so already). At $x = 10$ cm, the graph is $e^{10} \approx 22,026$ cm ≈ 220 m high, higher than most buildings. At $x = 24$ cm, the graph is more than halfway to the moon, and at $x = 43$ cm from the origin, the graph is high enough to reach past the nearest star, Proxima Centauri:

$e^{43} \approx 4.7 \times 10^{18}$ cm $= 4.7 \times 10^{13}$ km

$\qquad\qquad = 1.57 \times 10^8$ light seconds

$\qquad\qquad$ (light travels at 300,000 km/s in a vacuum)

$\qquad\qquad = 5.0$ light-years.

The distance to Proxima Centauri is about 4.3 light-years. Thus, for $x = 43$ cm from the origin (less than 2 feet to the right of the y axis), the y component of the graph is nearly 5 light-years from the x axis.*

If we have a case in which the rate of change in y decreases in proportion to increasing x, then we have a result of the form $y = e^{-x}$ (Fig. A12.4), as is the case for the barometric formula of Eq. (12.11). Such a situation is called exponential decay. We will encounter equations of this form when we study radioactive decay in Chapter 26.

In the previous paragraphs, we have discussed the exponential function $y = e^x$. Table A12.2 gives values of y corresponding to a range of values of x. However, in some cases we may already know y but need to know x. We can find x by taking the logarithm of both sides of this equation. In general, if $N = a^b$, a is called the base. The logarithm of N with respect to the base a is the power to which the base must be raised to give N. In practice, both $e = 2.718 \ldots$ and 10 are commonly used as bases for

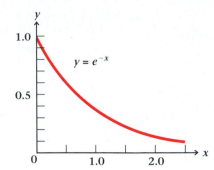

Figure A12.4
A graph of the function $y = e^{-x}$.

logarithms. When the base e is used, we refer to the logarithm as the natural logarithm and use the symbol ln. When base 10 is used, the symbol for the logarithm is log.

When we take the natural logarithm of both sides of the equation $y = e^x$, we get $\ln y = x$. We can find the natural logarithm with a pocket calculator or with the aid of a table. For example, suppose we know that $7.00 = e^x$ and want to find x. In this case, we find $x = \ln 7.00 = 1.95$. Be careful in your computations that you use the natural logarithm (ln) and not the logarithm to the base 10 (log).

The following relationships for logarithms are very useful:

$$\log (xy) = \log x + \log y,$$
$$\log (x/y) = \log x - \log y,$$
$$\log (x^n) = n \log x.$$

Although we have written these relations with log (base 10), they are valid for any base, including base e.

---

*From G. B. Thomas, Jr., and R. L. Finney, *Calculus and Analytic Geometry*, 8th ed. (Reading, Mass.: Addison-Wesley, 1992), p. 440.

Problems

A12.1 Plot the curve $y = e^{-0.1x}$ from $x = 0$ to $x = 10$. Determine the slope at $x = 5$ by calculation and by graphical techniques.

A12.2 Plot the curve $y = e^{-0.02x}$ from $x = 0$ to $x = 100$. Determine the slope at $x = 40$ by calculation and by graphical techniques.

A12.3 Plot a graph of $y = e^x$ from $x = -2$ to $x = +2$ in steps of $\Delta x = 0.2$.

A12.4• The intensity of a beam of light traveling in a glass fiber decreases with distance x according to

$$I = I_0 e^{-kx},$$

where I_0 is the intensity at $x = 0$ and k is the absorption coefficient. What is the absorption coefficient if the intensity decreases to $0.50\ I_0$ in a distance of 3.5 km?

A12.5•• The rate R_0 at which a sample of radioactive material emits radiation is measured to be 1200 particles per minute at time $t = 0$. The rate R at a later time t is given by $R = R_0 e^{-0.693t/T}$, where T equals 26 min and is called the half-life of the material. What is the rate of the radiation in particles per minute at $t = 2.0$ h?

A12.6•• In a biology experiment, the number of cells N in a particular population is given by $N = N_0 e^{at}$, where N_0 is the number of cells at time $t = 0$ and where $a = 0.30$/h. By what factor will the population increase in 24 h?

A12.7•• To what power does the base e have to be raised to give a number equal to the population of the United States (260×10^6)?

Thermodynamics

The area of physics concerned with the relationships between heat and work is called **thermodynamics.** Our recognition of heat as a form of energy transfer and our application of energy conservation in calorimetry provide introductory glimpses into this science. But the concerns of thermodynamics are much broader and more elegant than simply the measurement of heat. The underlying basis of thermodynamics is found in two general laws of nature abstracted from our universal observations and experience. The first law states that you cannot get more energy out of a system than you put into it, in all forms. While this sounds like a straightforward statement, its implications are quite profound and important. The second law says that the transfer of energy by heat flow has a direction; in other words, not all processes in nature are reversible. If a polar bear lies down in the snow, heat from its body will melt the snow; but the bear can't extract energy from the snow to warm itself. Thus, the flow of thermal energy has a direction—from hot to cold. By logical extension of these laws, we can correlate many measurable properties of matter with one another. Thermodynamic formulas predict many relationships between properties of matter and have the same general validity as the two laws on which they are based.

The development of thermodynamics grew out of a very practical concern with the operation of steam engines in the early nineteenth century. James Watt (1736–1819), a Scottish engineer and inventor, made steam power practical by markedly improving the efficiency of steam engines. His improvements, which resulted from his keen physical insight into thermal processes, were of such magnitude that he is often spoken of as the inventor of the steam engine.

The principles of thermodynamics that were developed in the eighteenth and nineteenth centuries made possible the tremendous advances in the power and efficiency of engines, from Watt's early steam engines to today's steam-driven electric power plants. The search for new and more efficient sources of energy continues today. But we recognize that all potential advances must still follow the laws of thermodynamics; no matter what new sources we tap, we can't get something for nothing. ■

13.1 Thermal Equilibrium

We saw in the previous chapter, that two objects in thermal contact can exchange heat as long as they are at different temperatures. The warmer object cools as the cooler one warms until they reach a common temperature at which no further changes take place. Two objects in this condition are said to be in a state of thermal equilibrium. In this state, a net flow of energy from one object to the other ceases, and they are at the same temperature throughout.

What happens if we introduce a third object? For example, suppose we take a bottle of milk *A* from the refrigerator and place it in an ideal insulating chest (Fig. 13.1a), where it exchanges heat with a can of orange juice *B* until they come to thermal equilibrium. (The chest insulates them from the outside world.) The objects remain in thermal equilibrium if they are separated. That is, if we place the juice in another insulated chest, it is still in equilibrium with the milk. Now, if a second can of juice *C* is also in thermal equilibrium with the milk *A* (Fig. 13.1b), we may ask the question: What is the relationship of *B* to *C*? Experiments show that *B* and *C* are in thermal equilibrium as well. This result may be stated as follows: **Two objects, each in thermal equilibrium with a third**

Figure 13.1

(a) Objects *A* and *B* at different temperatures come to thermal equilibrium when placed in thermal contact. (b) They remain in thermal equilibrium even when separated if no heat is exchanged with their environment. Then if object *C* is in thermal equilibrium with *A*, is it also in equilibrium with *B*?

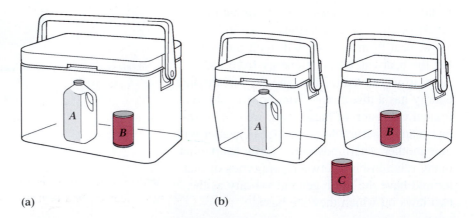

(a)   (b)

object, are in thermal equilibrium with each other. This rule is known as the **zeroth law of thermodynamics.** It is called the "zeroth" law because it logically precedes the statements of the first and second laws of thermodynamics, but was not recognized as an important and fundamental law of nature until after these other laws had been stated and named.

In the study of thermodynamics, we introduce the concept of a system and consider the transfer of energy into or out of the system by heat or work. The system could be any physical system, such as a machine (an automobile engine), a chemical system (a burning log), or a biological system (you). **A thermodynamic system** is any collection of objects considered together; the rest of the universe is the environment of the system. A thermodynamic system interacts with its surrounding environment by heat transfer and/or work. As a result of this energy exchange with the environment, the system's internal energy may change. By **internal energy,** we mean the total kinetic and potential energy associated with the internal state of the atoms composing the system. In addition, the system may also have kinetic and potential energies due to its collective motion and outside forces, such as the force of gravity.

Be sure you understand the differences among temperature, internal energy, and heat. Temperature is a measure of the warmth of an object; as we saw in Chapter 12, on an atomic level it is determined by the average random kinetic energy of the object's atoms. Internal energy is the sum of the kinetic and potential energies of the internal motion of all the atoms in the object. Heat is the transfer of energy to or from an object, either by changing the kinetic energy of the atoms (changing an object's temperature) or by changing the potential energy of the atoms (changing an object's phase).

13.2 The First Law of Thermodynamics

The first law of thermodynamics is based on the principle of the conservation of energy: that energy is neither created nor destroyed in any thermodynamic system. As is true of many other scientific laws, there is no absolute proof for the first law. Rather, it is an extrapolation of our experience and has no known exception. However, we must be sure we know all the forms in which energy can occur before we apply the first law to some specific system.

In the usual formulation of the first law, we consider the transfer of heat into a system, the work performed by the system, and the change in the system's internal energy. If we let Q be the net amount of heat flowing *into* a system during some process and W be the net work done *by* the system, then conservation of energy gives

$$Q = W + \Delta U,$$

where ΔU is the change in the system's internal energy. Upon rearranging, we find

$$\Delta U = Q - W. \tag{13.1}$$

The meaning of two of the terms in this equation should be clear from prior chapters: Work was encountered earlier in our study of mechanics; and heat, as a form of energy transfer, was treated in Section 11.4. A negative value of Q means that heat is being given out by the system instead of being added to the system. Similarly, a negative value of W means that work is being performed on the system rather than being done by the system. The internal energy U of the system can take a variety of forms and depends only on the temperature of the system. Equation (13.1) is the usual mathematical statement of the **first law of thermodynamics.** In words, **the change in internal energy of a system equals the difference between the heat taken in by the system and the work done by the system.** That is, when an amount of heat Q is added to a system, some of this added energy remains in the system increasing its internal energy by an amount ΔU while the rest of the added energy leaves the system as the system does work W.

Since the first law is a statement about energy, to apply it we need to specify, or measure, the energy or energy change of a system. For a purely mechanical system this is fairly easy to do because we can measure the masses and determine their velocities or positions. But with heat, things are different. There are no perfect thermal insulators to keep energy confined to a certain place. For that reason, among others, we must carefully identify the system under consideration and separate "the system" from "the rest of the universe" or what is usually called the *environment*.

A thermodynamic system, no matter what its composition, may undergo several special kinds of processes involving energy. If no heat enters or leaves the system during some process, then the system is said to be perfectly isolated from its environment and the process is called **adiabatic.** A good approximation to an adiabatic process is anything that happens so rapidly that heat does not have time to flow in or out of the system, as in the rapid compression of air in a tire pump. If $Q = 0$ in the first law (Eq. 13.1), we are left with

$$\Delta U = -W \qquad \text{(adiabatic process).}$$

NO AFFECT ON ENVIRONMENT

Thus, in an adiabatic process, the system does not exchange heat with its environment, but undergoes a change in internal energy that is the negative of the work done by the system. For example, when the system does work on the environment, W is positive and the resulting ΔU is negative.

If the temperature of a system does not change during a process, the process is said to be **isothermal.** An approximately isothermal process proceeds so slowly that the rate of change in temperature is negligible. Slow compression of air in a tire pump is an example of such a process, provided that air is not allowed to flow out of the cylinder. Because the kinetic energy of the air molecules is proportional to the temperature, the internal energy remains constant during an isothermal process that involves no change of phase or chemical change. Therefore, from the first law, the heat absorbed by the system must equal the work done by the system:

$$Q = W \qquad \text{(isothermal process).}$$

NO TEMP. CHANGE

Most processes in nature are neither strictly adiabatic nor strictly isothermal, but we can approximate many processes by treating them as one or the other, or as one followed by the other.

A process in which the volume of the system does not change is called isovolumetric or **isochoric.** Heating a gas in a rigid, tightly sealed container is an example of such a process. In a process that goes forward at constant volume, no displacement can take place, so the work done by the system is zero. (Remember, work is force times displacement.) Then, from the first law, we have only two terms,

$$Q = \Delta U \qquad \text{(isochoric process)}.$$

NO VOLUME CHANGE

That is, in an isochoric process, no work is done by the system, and any heat added to the system goes into increasing its internal energy.

If the pressure does not change during a process, the process is called **isobaric.** One example of an isobaric process is the boiling of water in an open container. Since the container is open, the process occurs at constant atmospheric pressure. At the boiling point, the temperature of the water no longer increases with the addition of heat; instead there is a change of phase from water to steam.

Many of the thermodynamic systems we will use to illustrate new principles consist of a fluid, often a gas. So we often find it convenient to express work in terms of pressure rather than force. In practice, a thermodynamic system might be the water and steam system in a steam engine or the system of gasoline and air in an internal combustion engine.

Figure 13.2 shows a cylinder of gas fitted with a piston. If the gas pushes the piston out an amount Δx, the increment of work done $F \Delta x$ can be written in terms of the pressure P and the change in volume ΔV:

$$\Delta W = F \, \Delta x = \frac{F}{A}(\Delta x \, A) = P \, \Delta V. \qquad (13.2)$$

The total work W done by the gas is the sum of these small increments, taking proper account of the relationship between pressure and volume in the case being investigated. A gas-filled cylinder fitted with a piston is the simplest form of what is called a *heat engine.* Such a heat engine is used to model more complicated systems, from automobile engines to biological systems.

As we noted, the change in a system's internal energy can take many different forms, including changes in temperature, changes of phase, and chemical changes. In the following examples we will consider a change in internal energy corresponding to a change in temperature. However, this is not the only possibility.

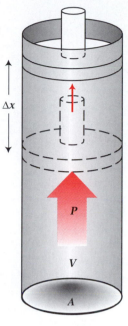

$$\Delta W = P \, \Delta V$$

Figure 13.2

The increment of work ΔW done by an expanding gas is $P \, \Delta V$.

Example 13.1

Show that in an isochoric process the change in temperature is proportional to the change in internal energy.

Solution   As we have seen, in an isochoric process, $Q = \Delta U$. From Chapter 11 we also know that the heat added to a mass m of a substance is related to the temperature change ΔT and the specific heat c by

$$Q = mc \, \Delta T.$$

(We assume no change in phase occurs in this process.) Upon equating these two expressions for Q, we obtain

$$\Delta U = mc \, \Delta T.$$

What we have shown in this special case is generally true: Temperature is associated with the internal energy of a system.

Example 13.2

Gas confined by a piston (Fig. 13.2) in a heat engine expands against a constant pressure of 100 kPa (nearly one atmosphere). When 2×10^4 J of heat are absorbed by the system, the volume of the gas expands from 0.15 m^3 to 0.25 m^3. (a) What is the work done by the system during this process? (b) What is the change in internal energy of the system?

Strategy   Here we have a gas kept at constant pressure by a moveable piston, so we can use the expression $P \, \Delta V$ to determine the work. Because the gas expands, work is done by the system and the work will be positive. We can then use this result and the first law to find the change in internal energy of the system, keeping in mind that because heat is added to the system, the heat is also positive.

Solution   (a) The work done at constant pressure is

$$W = P \, \Delta V = P(V_{\text{final}} - V_{\text{initial}})$$
$$W = (100 \times 10^3 \text{ N/m}^2)(0.25 \text{ m}^3 - 0.15 \text{ m}^3) = 1 \times 10^4 \text{ J}.$$

(b) The change in internal energy is obtained from the first law of thermodynamics:

$$\Delta U = Q - W.$$

In this case, Q is 2×10^4 J. The work W done by the gas was just found to be 1×10^4 J. Thus, the change in internal energy is

$$\Delta U = 2 \times 10^4 \text{ J} - 1 \times 10^4 \text{ J} = 1 \times 10^4 \text{ J}.$$

Discussion   In this case, in which there is no change of phase, the increase in internal energy shows up as an increase in temperature.

Example 13.3

A heat engine undergoes a process in which its internal energy decreases by 400 J while it is doing 250 J of work. What net heat is taken in (or given out) by the engine during this process?

Strategy   Work is done by the engine, so W is a positive quantity. The internal energy decreases, so ΔU is negative. We can insert the numerical values into the expression for the first law to find the value and sign of the thermal energy flow into the system.

Solution   The heat taken in (or given out) by the system can be found from the first law of thermodynamics, Eq. (13.1), which can be rewritten in the form

$$Q = \Delta U + W.$$

Here $\Delta U = -400$ J and $W = 250$ J. The heat Q is

$$Q = -400 \text{ J} + 250 \text{ J} = -150 \text{ J}.$$

Discussion   The negative sign for Q indicates that the net heat is given out by the system. The negative value of ΔU indicates that, in the absence of a phase change, the temperature decreases.

The first law of thermodynamics relates two measurable quantities that pertain to changes in a system: the heat added to or given out by the system and the work performed on or done by the system. We can measure the thermodynamic properties of systems in many ways: the length of a mercury column in a thermometer; the volume, pressure, or temperature of a sample of gas; or the resistance of a resistor. All of these properties are state variables or state coordinates. A state variable is a physical property that characterizes the state of a system independently of how that particular state is reached. For example, a cup of tea at room temperature has the same temperature whether it cools from boiling or is heated from freezing; temperature is a state variable. Under some conditions, the value of two state variables, such as pressure and volume, completely specify the thermodynamic state of a system. Then we can represent any possible state by a point on a two-dimensional plot, as in a *PV* diagram (Fig. 13.3). The amount of heat added to or released by a thermodynamic system is not a state variable, and although the work done by a system often does involve state variables, such as "*P* Δ*V* work" done by a gas, work itself is not a state variable. An interesting aspect of the first law is that the internal energy of a system is a state variable; that is, the energy difference between the heat into a system and the work done by that system does not depend on the details of how that heat was added or how that work was done. In this case, the difference between two quantities that are not state variables is itself a state variable.

An important goal of thermodynamic studies is to consider systems in different thermodynamic states and follow the changes in the state variables as the systems undergo change. However, thermodynamic state variables are meaningful only for systems in equilibrium. For example, when a large sample of gas undergoes a rapid expansion, the temperature and pressure may fluctuate from place to place within the gas. Thus, describing the temperature or pressure of this sample while it changes is not meaningful. To determine clearly defined values of state variables, we must consider reversible processes. A **reversible process** is one in which the system is very nearly in equilibrium all the time. For example, in the *PV* diagram of Fig. 13.4 for a system evolving from state A (P_1, V_1) to state B (P_2, V_2), a reversal of the controlling factors causes the system to exactly retrace its path in the opposite direction, back to its initial state. Alternatively, we can say that a reversal of the controlling factors causes a reversal of the energy transformation. There is no wasted energy.

Reversible processes are allowed by the first law of thermodynamics, and we will use them as examples of how to think about thermodynamic processes. However, a reversible process cannot take place if friction or any other form of energy loss from the system is present. The situation is somewhat like learning basic mechanics without including friction in everything from the start. We will see later that the second law of thermodynamics shows that completely reversible processes do not occur in nature. However, they are still useful models for analyzing changes in the state of a system.

Because many machines operate in approximately reversible cycles, it is important to examine the implications of the first law for a cyclic process in which a system begins with an internal energy U at an initial temperature T, exchanges heat and work with its surroundings, and returns to its initial state characterized by U and T. In any number of complete cycles, $\Delta U = 0$ so that $Q = W$. Thus, the first law tells us that it is impossible for a machine (or a system) in any number of complete cycles to put out more energy in the form of work than it takes in as heat. A machine that could do this would be called a perpetual motion

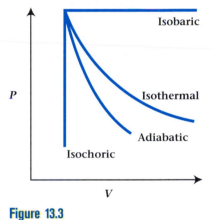

Figure 13.3

PV diagrams for isochoric, adiabatic, isothermal, and isobaric processes. The lines indicate the thermodynamic state of a system during the process, as measured by the two state variables pressure and volume.

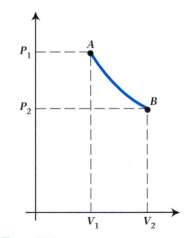

Figure 13.4

PV diagram for a system evolving from state A to state B along a reversible path.

machine of the first kind. The first law is sometimes stated in terms of such a machine as follows, *A perpetual motion machine of the first kind is impossible;* or, to put it more succinctly, *You can't win.*

13.3 The Carnot Cycle and the Efficiency of Engines

Let us now apply the first law of thermodynamics to analyzing the operation of a heat engine. Figure 13.5 shows an early steam engine built by the firm of Boulton and Watt. Steam came from the boiler (*B*) at the left and entered a condensing cylinder near the large upright cylinder (*C*). Here the steam was cooled and condensed to water, creating a partial vacuum in the large cylinder that caused a piston (*P*) to descend. This motion was transmitted to the vertical rods (*R*) on the right through the large reciprocating beam (*RB*) at the top. The upward motion of the right-hand rods pumped water from a mine. Rotary motion and the use of the pressure of expanding steam for the driving force did not come until later.

Naturally it was important to determine an engine's maximum power output. Watt accomplished this with a technique he developed and used privately for many years. In this method, Watt attached a device called an indicator to the engine. The mechanism included a card connected to a piston so that it could move back and forth as the piston moved within the cylinder (Fig. 13.6). The displacement of this main piston was proportional to the volume of the main cylinder. Another, smaller cylinder with a spring-loaded piston was connected to the main cylinder. The position of this small piston indicated the pressure. It also moved a lever with a pencil on its far end. As the card and small cylinder moved, the pencil traced out a curve of pressure versus volume. The resulting diagram was called an indicator diagram. This term is still used for measurements made on internal combustion engines. The indicator diagram is a cyclic *PV* diagram.

The area inside the curve of an indicator, or cyclic *PV*, diagram is proportional to the work done in each cycle. To see this, suppose that we divide the

Figure 13.5

An early steam engine built by the firm of Boulton and Watt. Its operation is described in the text.

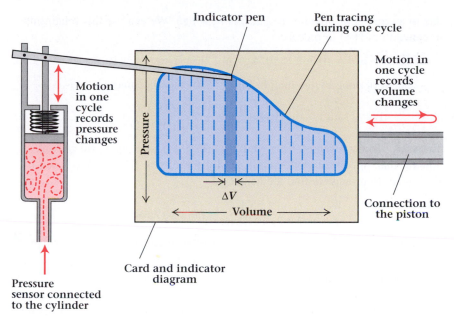

Figure 13.6
A schematic drawing of a steam engine indicator mechanism showing an indicator diagram. The area within the curve is proportional to the work done in each cycle.

area inside the indicator diagram into narrow vertical strips, just as we did in Chapter 6 for the area under the graph of force versus displacement. The width of each strip is proportional to the change in volume ΔV, and the height of each strip indicates the pressure change P. The small area of each strip $P \, \Delta V$ represents an increment of work. The sum of all these incremental values of work, which is the total work per cycle, corresponds to the area inside the curve. Because the time for each cycle is essentially constant for a given engine speed, the area inside an indicator diagram also measures the power of an engine.

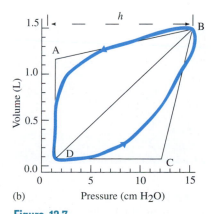

(a)

Example 13.4

Some of the lung's functions can be tested by breathing into a calibrated device (Fig. 13.7a). Lung pressure and lung volume are measured simultaneously and plotted during a breathing cycle. Approximately how much work is done in the breathing cycle shown in Fig. 13.7(b)?

Strategy Because the breathing curve is a plot of V against P, the area inside the curve is the work done, just as in the case of the indicator diagram on the steam engine. We can approximate the area inside the breathing curve by drawing two approximately equal triangles ABD and CBD. The work done is then given by twice the area inside one of the triangles (ABD).

Solution Let the side of the triangle ABD parallel to the volume (V) axis be the base, and let the distance parallel to the pressure (P) axis be the height, marked h in Fig. 13.7(b). The area of triangle ABD is $\frac{1}{2}$ base × height. The work done is twice that area, so $W = $ base × height. Reading directly from the graph we have

$$W = (AD)(h) = (1.17 \text{ L})(14.5 - 1.0) \text{ cm H}_2\text{O} = 15.8 \text{ L} \cdot \text{cm H}_2\text{O}$$

The measurements are not in SI units. The unit of cm H_2O represents the pressure due to a 1-cm column of H_2O. In Section 10.1 we saw that the pressure

(b)

Figure 13.7
(a) Measuring lung function. (b) Lung pressure and volume during a breathing cycle. The lower path from D to B is inspiration (breathing in) and the upper path from B to D is expiration (breathing out).

(a)

(b)

(c)

Figure 13.8
The motions of tornadoes, volcanoes, and jet engines are due to heat. Thus, their behavior is governed by the laws of thermodynamics.

due to a column of liquid of height h is $P = \rho gh$. We can use this relationship to convert cm H_2O to pascals:

$$1 \text{ cm } H_2O = \rho gh = (0.998 \times 10^3 \text{ kg} \cdot \text{m}^{-3})(9.81 \text{ m} \cdot \text{s}^{-2})(10^{-2} \text{ m}) = 97.9 \text{ Pa}.$$

The work done by the lung in Fig. 13.7(b) then becomes

$$W = (15.8 \text{ L} \cdot \text{cm } H_2O)\left(\frac{10^{-3} \text{ m}^3}{L}\right)\left(\frac{97.9 \text{ Pa}}{1 \text{ cm } H_2O}\right) = 1.55 \text{ J}.$$

Discussion   To get a sense of how much work 1.55 J represents, we can ask how far that amount of work can lift a 2.0-kg textbook. The work is given by $W = mgh$. Rearrange to find the height:

$$h = \frac{W}{mg} = \frac{1.55 \text{ J}}{(2.0 \text{ kg})(9.81 \text{ m/s}^2)} = 0.079 \text{ m} = 7.9 \text{ cm}.$$

Though steam engines were considerably improved by Watt and others, the basis for understanding the general principles of heat engines did not come until 1824, when the French engineer Sadi Carnot (1796–1832) published a treatise on this subject. In doing so, Carnot formulated the basic ideas of thermodynamics. He said that *all* movements were ultimately due to heat. It made no difference whether they occurred in natural phenomena, such as rain, storms, earthquakes, and volcanoes, or in mechanical devices such as steam engines (Fig. 13.8). In view of modern knowledge, Carnot's vision of nature was slightly simplified, but his understanding of thermal energy as the generator of motive power was essentially correct. His work was unappreciated except by his closest friends until, sixteen years after Carnot's death, Lord Kelvin pointed out its fundamental theoretical and practical importance.

Although we will discuss Carnot's ideas in terms of an ideal engine that cannot actually be built, the ideas have great practical importance even today. The ideal Carnot engine sets an upper limit on the efficiency of all real engines, including steam engines, Diesel and gasoline (Otto) engines, jet engines, and nuclear reactors. Furthermore, studies of the theoretical Carnot engine indicate some of the factors that affect the efficiency of real engines.

Carnot recognized that work could be done only when heat flowed from a higher temperature to a lower one. So Carnot proposed an *ideal* heat engine that operates cyclically and reversibly between two temperatures. This so-called Carnot engine is not 100% efficient, but is as efficient as any machine could be in transforming heat into work. Carnot analyzed the transformation of energy during one complete cycle of this engine's performance and determined the conditions for maximum efficiency. Only later was Watt's indicator diagram proposed as a basis for a mathematical discussion of Carnot's ideas.

In our example, the working substance of the engine is an ideal gas* confined within a cylinder by means of a frictionless piston (Fig. 13.9). We use an ideal gas for mathematical simplicity; however, the results would be the same for any working substance in the Carnot cycle.

Figure 13.9 shows diagrammatically the other components of a Carnot engine, in addition to the cylinder containing the working substance. A hot body

*Chapter 12 describes an ideal gas. Use of an ideal gas allows us to calculate the exact shape of the curves for the cyclic process shown in Fig. 13.10.

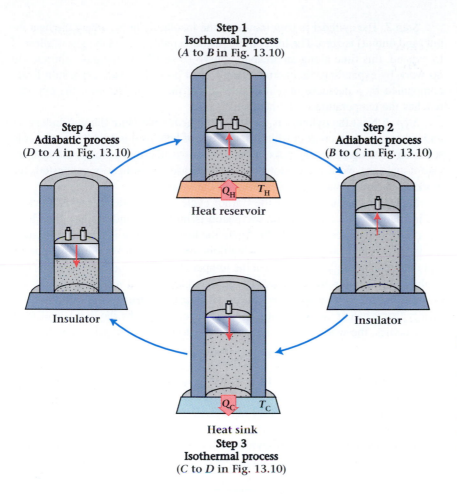

Figure 13.9
A diagrammatic version of a Carnot engine using an ideal gas as the working fluid. Heat is taken in isothermally at *A*, followed by adiabatic expansion at *B*, isothermal compression at *C* (where heat is expelled), and, finally, adiabatic compression at *D*.

of infinite thermal capacity, called a *heat reservoir,* supplies thermal energy without lowering its own temperature. (This reservoir can be approximated by any source of heat, such as the sun, that is much larger than the needs of the engine.) An insulating platform, together with the sides of the cylinder and piston, acts as a perfect insulator against the flow of heat. The cold body, called a *heat sink,* is also of infinite thermal capacity so that it can absorb heat without raising its own temperature. (This sink can be approximated by any large body, such as the ocean, that can absorb much more heat than the engine can generate.) Finally, there is a second insulating platform. Operation of the Carnot engine consists of moving the cylinder in a prescribed manner from one of these platforms to the other and then repeating the cycle.

The **Carnot cycle** consists of four reversible processes, two isothermal and two adiabatic:

Step 1. We start the cycle with the cylinder in contact with the heat reservoir, where the working substance (gas) takes in an amount of heat Q_H at a high temperature T_H. Because the system absorbs heat in a reversible process, its temperature is the same as the reservoir's; that is, this is an isothermal process. As the heat is absorbed, the gas expands and does work on the piston. This expansion is represented in Fig. 13.10 by going from *A* to *B* along an isothermal curve. During this isothermal process, the system's internal energy does not change, so according to the first law the work done by the system is equal to the heat input.

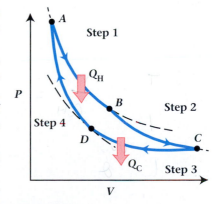

Figure 13.10
One cycle of a Carnot engine using an ideal gas as a working fluid. Curves *AB* and *CD* represent isothermal processes; curves *BC* and *DA* represent adiabatic processes. Compare with Fig. 13.9.

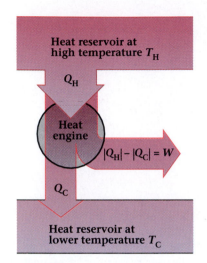

Figure 13.11

A schematic representation of a heat engine. The work done equals the net heat flow into the system, $|Q_H| - |Q_C|$.

Step 2. The cylinder is then moved to the insulating body, where the heat input (and output) is zero. The load on the piston is reduced, and the gas is allowed to expand, this time along an adiabatic curve (*B* to *C*). As the gas continues to do work by expanding, its internal energy must go down. This expansion is accompanied by a decrease in temperature along the curve *BC* until the cylinder reaches the temperature of the heat sink.

Step 3. Next the cylinder is moved to the heat sink. Here the gas undergoes an isothermal contraction in which an amount of heat $|Q_C|$ is expelled to the cold reservoir at temperature T_C.* As in the previous isothermal process, the heat intake equals the work done. However, in this case, since heat is exhausted, the work is negative; that is, work is done on the system.

Step 4. In the final step of the Carnot cycle, the cylinder is moved back to the insulating body. The load on the piston is increased, and the gas undergoes an adiabatic compression (*D* to *A*). Again the heat exchange is zero, and because the volume is decreasing (work is being done on the system), the internal energy and the temperature increase. When the temperature of the gas again reaches that of the heat reservoir, the cylinder is transferred to the heat reservoir and the cycle starts again. In this way the working substance returns to the same internal energy that it had at the start of the complete cycle. Thus, by the first law of thermodynamics, the work done must equal the net heat flow into the cylinder:

$$W = |Q_H| - |Q_C|,$$

where $|Q_H|$ and $|Q_C|$ are taken to be positive quantities. The process is shown schematically in Fig. 13.11.

You should go over Figs. 13.9 and 13.10 again to be sure you understand how they correspond.

We define the **thermal efficiency** of any system, such as a machine, to be the ratio of the work done to the heat input:

$$\text{thermal efficiency} = \frac{W}{Q_H}. \tag{13.3a}$$

Substituting for the work W, we get

$$\text{thermal efficiency} = \frac{|Q_H| - |Q_C|}{|Q_H|}. \tag{13.3b}$$

In Chapter 12 we showed that for an ideal gas the internal energy is proportional to the Kelvin temperature. From that fact and from a detailed examination of the Carnot cycle for an ideal gas, Kelvin showed that

$$\frac{|Q_C|}{|Q_H|} = \frac{T_C}{T_H},$$

where the temperatures are the absolute temperatures measured on the Kelvin scale. The thermal efficiency of an ideal engine is thus

$$\text{thermal efficiency} = 1 - \frac{T_C}{T_H} \quad \text{(ideal)}. \tag{13.4}$$

---

*Recall our sign convention that heat into the system is positive and heat out of the system is negative. Here, we consider the gas to be our system so that the heat Q_C expelled to the cold reservoir is negative.

Table 13.1	*Practical Efficiencies of Real Engines* ▼

Type of Engine	Efficiency (%)
Automobile engine (gasoline)	20–25
Diesel engine	26–38
Nuclear-powered steam turbine	35
Coal-fired steam turbine	40

It can be shown that *all* reversible engines operating in cycles between the same two heat reservoirs have the same efficiency, regardless of the operating fluid. Moreover, no heat engine of any kind, operating in cycles between the same two reservoirs, can have an efficiency greater than that of a reversible Carnot engine. Thus, even if there were no losses due to friction and heat leakage, the absolute maximum efficiency of a heat engine would be given by Eq. (13.4). The efficiency of any real engine is certain to be less than that of the ideal engine. Table 13.1 gives some examples of typical efficiencies.

▼

Problem-Solving Strategy

Comparing Quantities

1. When using the formulas for thermal efficiency remember that the temperatures are always expressed in kelvins. Take care to identify the system you are studying and its environment.
2. Remember that W is positive when the system expands and does work on the environment and negative when it is compressed. The value of Q is positive for heat coming into the system and negative for heat leaving the system.

▼

Example 13.5

What is the maximum possible thermal efficiency of a steam engine that takes in steam at 160°C and exhausts it at 100°C?

Solution We can calculate the efficiency using Eq. (13.4), after converting the temperatures to the Kelvin scale:

$$T_H = 160°C = 433 \text{ K},$$
$$T_C = 100°C = 373 \text{ K},$$
$$\text{efficiency} = 1 - \frac{T_C}{T_H} = 1 - \frac{373 \text{ K}}{433 \text{ K}} = 0.14.$$

The theoretical efficiency of this engine is only 14%. Note that this is true regardless of the details of the engine's operation.

Physics in Practice

GASOLINE ENGINES

A ds for new cars often stress the increased efficiency of the new models compared with what you're driving now. In fact, the last few years have seen real improvements in engine efficiency. But how far can this continue? Let's find out by analyzing a simple model of a typical car engine.

Internal combustion engines form a special class of heat engines that generate the input heat by the combustion of fuel within the engine itself. Examples of internal combustion engines include gasoline engines, Diesel engines, and gas turbines. Here we consider the gasoline engine as a representative example of internal combustion engines.

The operating cycle of the gasoline engine used in most cars is a four-stroke cycle (Fig. B13.1). In the *intake stroke* a mixture of air and gasoline vapor is drawn through the intake valve into the cylinder by the downward motion of the piston. The valve closes and the fuel-air mixture is compressed. At the top of this *compression stroke,* the gases are ignited by an electric spark from the spark plug, raising the temperature and pressure of the gases. The hot gases then expand against the piston in the *power stroke,* delivering energy to the crankshaft. The exhaust valve opens as the piston moves upward again, expelling the burned gases in the *exhaust stroke.* The exhaust valve closes, the intake valve opens, and the cycle is ready to repeat.

Analysis of an indicator diagram of a real gasoline engine is very difficult (Fig. B13.2a). For this reason, the gasoline engine is usually analyzed with a simplified model of the

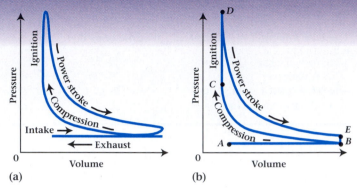

Figure B13.2 Indicator diagrams for (a) a real gasoline engine, and (b) an idealized Otto cycle.

cycle called the Otto cycle, after its developer, Nicholas Otto (1832–1891). The Otto cycle begins at point *A* on the *PV* diagram of Fig. B13.2(b). The volume expands at constant pressure to point *B* as the piston moves down during the intake stroke. During the compression stroke, the gases are compressed adiabatically to point *C*. Ignition of the gas by the spark causes an isochoric change to point *D* at higher temperature and pressure, which is followed by an adiabatic expansion to point *E* during the power stroke. Opening the exhaust valve causes the pressure to drop isochorically to point *B*, and this drop is followed by a decrease in volume at constant pressure as the piston moves through the exhaust stroke.

The work done by the gasoline engine is found from the area enclosed in the curve on the *PV* diagram, which for the idealized Otto cycle is the loop *B-C-D-E-B*. Comparison of the Otto cycle with a Carnot cycle operating between the same two temperatures shows that the efficiency of the Carnot cycle is more than that of the Otto cycle. The Otto cycle is, in turn, considerably more efficient than the actual gasoline engine cycle that it represents. Real gasoline engines achieve thermodynamic efficiencies of 20 to 25%, roughly half the value predicted from the simplified Otto model.

In recent years, manufacturers have been designing and building more efficient cars. Electronic sensors have been installed to monitor exhaust emissions, while computer control of air-fuel mixtures is now common. Other advances such as lean-burn engines, turbocharging, multiple valves, and cast-aluminum engine blocks have been used to make cars more fuel efficient. Future developments will undoubtedly include even more computer control of the combustion process and electronically controlled transmissions to provide the optimum gearing between the engine and the wheels for every situation.

Figure B13.1 The operating cycle of a four-stroke gasoline engine. The piston goes up and down twice during each cycle.

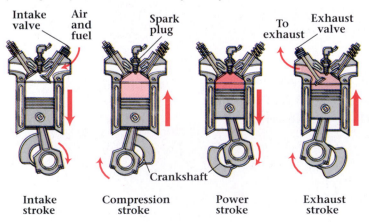

Intake valve | Air and fuel | Spark plug | To exhaust | Exhaust valve

Crankshaft

Intake stroke | Compression stroke | Power stroke | Exhaust stroke

416

Example 13.6

Calculate the maximum theoretical efficiency of a power plant that has a high-temperature reservoir at 500°C and a low-temperature exhaust (cold reservoir) at 50°C.

Solution Again we must convert to absolute temperatures:

$$T_H = 500°C = 773 \text{ K},$$
$$T_C = 50°C = 323 \text{ K},$$
$$\text{efficiency} = 1 - \frac{T_C}{T_H} = 1 - \frac{323 \text{ K}}{773 \text{ K}} = 0.58.$$

The maximum theoretical efficiency is 58%. The increase above that of Example 13.5 is due to the increase in T_H. Because real power plants do not reach ideal efficiency, a realistic value of the efficiency of a power plant is closer to 40%.

Example 13.7

An engine takes in 9220 J and does 1750 J of work each cycle while operating between 689°C and 397°C. (a) What is its actual efficiency? (b) What is its maximum theoretical efficiency?

Strategy We can calculate the actual efficiency as the ratio of the work done to the heat input. Then we can find the maximum theoretical efficiency from the efficiency of a Carnot engine operating between T_H and T_C as given in Eq. (13.4).

Solution (a) The actual efficiency of the heat engine is

$$\text{actual efficiency} = \frac{W}{Q_H} = \frac{1750 \text{ J}}{9220 \text{ J}} = 0.190.$$

(b) The maximum theoretical efficiency is that of a Carnot engine operating between $T_H = 689°C = 962 \text{ K}$ and $T_C = 397°C = 670 \text{ K}$.

$$\text{maximum efficiency} = 1 - \frac{T_C}{T_H} = 1 - \frac{670 \text{ K}}{962 \text{ K}} = 0.304.$$

Discussion Notice that the maximum theoretical efficiency (the Carnot efficiency) is larger than the actual efficiency. This is always the case, because Carnot engines do not take into account irreversible processes such as the loss of energy by friction. The processes in real engines are irreversible. Thus, the Carnot efficiency is an upper limit to what a real engine can accomplish.

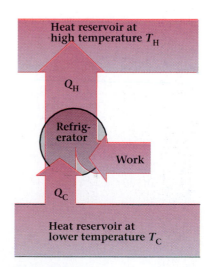

Figure 13.12
A schematic representation of a refrigerator. We have to put work into the system to transfer heat from lower temperature to higher temperature.

13.4 Refrigerators and Heat Pumps

By reversing the direction of the Carnot cycle, we can put work into the system and transfer heat from a low temperature to a higher one. A system operated in this manner is called a **refrigerator** (Fig. 13.12). In this case, the ratio of the heat extracted from the cold reservoir to the work supplied is similar to an efficiency and is called the **coefficient of performance** (c.p.):

$$c.p.(\text{refrigerator}) = \frac{Q_C}{W}. \tag{13.5a}$$

Substituting for W, we get

$$c.p.(\text{refrigerator}) = \frac{|Q_C|}{|Q_H| - |Q_C|}.$$

With ideal gas as the working substance, this becomes

$$\text{maximum c.p.}(\text{refrigerator}) = \frac{T_C}{T_H - T_C}. \tag{13.5b}$$

Because the coefficient of performance given in Eq. (13.5b) is that of a Carnot engine run in reverse, it represents the maximum c.p. of any real refrigerator. Notice that the amount of work required to run a refrigerator increases with the temperature difference between T_H and T_C. Also note that a refrigerator with a higher c.p. is a better refrigerator, that is, it extracts a given amount of heat for less work and, therefore, is less expensive to operate.

Figure 13.13(a) shows a diagram of the thermal part of a freon-cycle refrigerator, a type found in many homes. The typical refrigerator compresses a refrigerant gas (freon) to a pressure of several atmospheres. The resulting hot gas is forced through a heat exchanger (the condenser) external to or on the side walls of the refrigerator, where the gas is cooled to near room temperature and thereby condensed into a liquid. The cool liquid then flows at high pressure through a narrow tube to a much larger tube (the evaporator), which is a region of much lower pressure. In this region, the liquid evaporates because of the reduced pressure, absorbing heat from the contents of the refrigerator. The resulting cold gas is then drawn through the low-pressure tube back to the compressor, where the cycle starts over.

Compare the diagram of Fig. 13.13(a) with the schematic of Fig. 13.12. The work input is provided by the compressor; the heat input is absorbed by the gas at the evaporator at low temperature; and the heat exhausted is given up at higher temperature through the condenser. Thus, a refrigerator transfers heat energy from a cooler body to a warmer one at the expense of the work supplied.

An air conditioner (Fig. 13.13b) is a refrigerator that is designed to take heat from within a house and exhaust it to the outdoors. When such a system is reversed so that it cools the outdoors and delivers heat to the inside of the house, it is called a **heat pump.** The coefficient of performance of a heat pump is defined to be the ratio of the heat delivered inside the house (the high-temperature reservoir) to the work supplied:

$$c.p.(\text{heat pump}) = \frac{Q_H}{W}. \tag{13.6a}$$

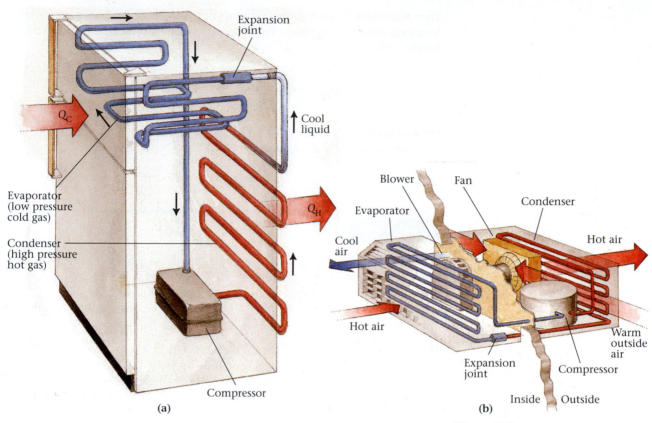

(a)

(b)

Figure 13.13
A diagram of (a) a common refrigerator and (b) a window air conditioner.

When we substitute for W, we get

$$c.p.(\text{heat pump}) = \frac{|Q_H|}{|Q_H| - |Q_C|}.$$

In terms of temperatures, the maximum coefficient of performance of the heat pump becomes

$$\text{maximum c.p.(heat pump)} = \frac{T_H}{T_H - T_C}. \qquad (13.6b)$$

Example 13.8

A household refrigerator has a coefficient of performance of 6.0. If the room temperature outside the refrigerator is 30°C, what is the lowest temperature that can be obtained inside the refrigerator?

Solution The maximum coefficient of performance, c.p., is

$$c.p.(\text{refrigerator}) = \frac{T_C}{T_H - T_C}.$$

We can rearrange this equation to give

$$T_C = \frac{(c.p.)T_H}{1 + (c.p.)}.$$

Then we obtain T_C by inserting the numerical values for the coefficient of performance and the temperature T_H. Remember that T_H must be measured on the Kelvin scale.

$$T_H = 30°C = (30 + 273)\ K = 303\ K,$$

$$T_C = \frac{6.0(303\ K)}{1 + 6.0} = 260\ K.$$

Upon converting T_C to the Celsius temperature, we see that the coldest temperature attainable inside the refrigerator is

$$T_C = -13°C.$$

Example 13.9

A household heat pump is used to maintain an inside temperature of 20°C on a day when the outside temperature is $-10°C$. (a) What is the theoretical maximum coefficient of performance for this heat pump? (b) If the heat pump delivers heat to the house at a rate 15 kW, how much power must be supplied to run the heat pump?

Strategy (a) We can calculate the maximum coefficient of performance from knowledge of the high and low temperatures by using Eq. (13.6b). The temperatures must be converted to Kelvin scale before using them in the equation.

(b) The power required may be found from the definition of the coefficient of performance (Eq. 13.6a) if we replace Q_H and W by the rate of heat delivered and the power input P.

Solution (a) We begin by converting the two temperatures from °C to kelvins.

$$T_H = 20°C = (20 + 273)\ K = 293\ K,$$
$$T_C = -10°C = (-10 + 273)\ K = 263\ K.$$

The coefficient of performance is given by

$$\text{c.p.(heat pump)} = \frac{T_H}{T_H - T_C} = \frac{293\ K}{293\ K - 263\ K} = \frac{293}{30}$$

$$\text{c.p.(heat pump)} = 9.8.$$

(b) The definition of the coefficient of performance of the heat pump is

$$\text{c.p.(heat pump)} = \frac{Q_H}{W} = \frac{\text{rate of heat}}{P},$$

$$P = \frac{\text{rate of heat}}{\text{c.p.}} = \frac{15\ kW}{9.8} \approx 1.5\ kW.$$

Discussion Note that we have computed the maximum coefficient of performance. A real heat pump operating between the same temperatures will have a smaller c.p. As a result, the power required to deliver that same heat to the house will be greater than the value computed here.

13.5 The Second Law of Thermodynamics

It might seem to you, as it did to those who followed immediately after Carnot, that there was either some outright contradiction or at least a lack of clarity in his ideas about heat. On the one hand, we have the concept of the mechanical equivalent of heat, which Joule's experiments had confirmed. On the other hand, we have Carnot's result that even the most efficient heat engine conceivable could not convert all of its heat input into mechanical output. Doesn't this result contradict the conservation of energy?

The key to resolving this problem lies in seeing that Carnot's results refer to the amount of work available for use as output, and not to whether the total energy is conserved. The situation is somewhat analogous to that of gravitational potential. In that case, work can be done only when a body goes from one height to a lower one. The greater the difference in heights, the greater the amount of work that can be done. If the body cannot fall to a lower height, then no work can be done, no matter how much potential energy the body originally had. Similarly, for a heat engine, no work can be done unless heat can be taken in at one temperature and exhausted at a lower temperature.

Out of this seeming contradiction came the formulation of the **second law of thermodynamics.** It was first expressed mathematically by the German theoretical physicist Rudolf Clausius (1822–1888) and shortly afterwards by Lord Kelvin. Although these two formulations appear to be different, both were an outgrowth of Carnot's ideas. The two statements can be shown to be equivalent.

1. **Clausius statement of the second law:** Heat cannot, by itself, pass from a colder to a warmer body.

2. **Kelvin-Planck* statement of the second law:** It is impossible for any system to undergo a cyclic process whose *sole* result is the absorption of heat from a single reservoir at a single temperature and the performance of an equivalent amount of work.

Let us look into some of the implications of these statements. Remember that they express in general ways the results of experimenting with and observing the behavior of heat. The Clausius statement of the second law of thermodynamics is consistent with our experience. If an ice cube and a cup of hot chocolate are placed in contact, heat will flow from the hot chocolate to the ice cube until they come to the same temperature. The principle of conservation of energy—the first law of thermodynamics—does not tell us anything about how this process proceeds. It would not be a violation of the first law of thermodynamics if heat were to flow from the ice cube to the hot chocolate.

The second law of thermodynamics is different from the laws of mechanics. It does not describe the interactions between individual particles, but instead describes the overall behavior of collections of many particles. The second law of

---

*It was Planck's thermodynamic studies that led to ideas that revolutionized our notions of thermal radiation (Chapter 1).

thermodynamics says something about the sequence, or order, in which events naturally take place. In mechanics, individual events are always reversible; we say that they are symmetric in time. If we make a movie of the collision between two air-track gliders and then look at the movie, the collision that we see satisfies all the laws of mechanics, regardless of whether we show the movie forward or backward. The collision has time-reversal symmetry. However, if we make a movie of an egg frying and then show the movie backward, the result violates all our previous experience. It is in this sense that the second law tells us which way is forward in time and which way is backward.

13.6   Entropy and the Second Law

We can gain additional insight into the meaning of the second law of thermodynamics by considering it from a standpoint first introduced by Clausius in 1850. He introduced a new thermodynamic state variable called entropy, which has two Greek roots and means much the same as "turning into." **Entropy** is a measure of how much energy or heat is unavailable for conversion into work.

When a system at Kelvin temperature T undergoes a *reversible* process by absorbing an amount of heat Q, its increase in entropy ΔS is

$$\Delta S \equiv \frac{Q}{T}.$$

absorb heat = $ (13.7)

Notice that we are defining entropy for a reversible process, which does not occur in nature. However, for our purposes, we can use this definition for processes that are approximately reversible. When calculating ΔS for an irreversible process we must do so by substituting a reversible process that has the same initial and final states as the irreversible process.

Before we examine the meaning of this state variable further, let's see some examples of calculating changes in entropy. We will find that, just as in problems involving potential energy or heat, the changes are the significant quantity. In Example 13.10, the temperatures are not constant. Ideally, we should use the techniques of calculus to add up the increments of entropy change over many small intervals of almost constant temperature. For this example and for the problems at the end of the chapter, the difference between the exact answer from calculus and the approximate answer obtained by using the average temperature for the process in Eq. (13.7) is less than 1% in all cases. However, you should keep in mind that this method is only approximate and is not always valid. Problems 13.62 and 13.65 are concerned with the difference between the exact and the approximate solutions.

Example 13.10

A student takes a 2.5-kg block of ice at 0°C, places it on a large rock outcropping, and watches the ice melt. (a) What is the entropy change of the ice (water)? (b) If the source of heat (the rock) is very massive and remains at a constant 21°C, what is the entropy change of the rock? (c) What is the total entropy change?

Solution (a) The block of ice melts at 0°C = 273 K. The energy required to melt the ice is

$$Q = mL = (2.5 \text{ kg})(3.34 \times 10^5 \text{ J/kg}) = 8.35 \times 10^5 \text{ J}.$$

We can use the defining equation for entropy, Eq. (13.7), directly to get the entropy change of the ice:

$$\Delta S_{ice} = \frac{Q}{T} = \frac{8.35 \times 10^5 \text{ J}}{273 \text{ K}} = 3060 \text{ J/K}.$$

(b) The entropy change of the rock is also found using Eq. (13.7), but in this case the heat flow is negative and the temperature is 21°C = 294 K.

$$\Delta S_{rock} = \frac{Q}{T} = \frac{-8.35 \times 10^5 \text{ J}}{294 \text{ K}} = -2840 \text{ J/K}.$$

(c) The total entropy change is

$$\Delta S = \Delta S_{ice} + \Delta S_{rock} = 3060 \text{ J/K} - 2840 \text{ J/K} = 220 \text{ J/K}.$$

Example 13.11

A student mixes 0.100 kg of water at 60°C (sample 1) with 0.200 kg of water at 40°C (sample 2). Determine the change in entropy of the system.

Strategy Using the methods of Chapter 11, we first determine the final temperature and the heat gained or lost by each sample of water. Then we compute the entropy separately for each sample, using Eq. (13.7) and the average temperature for each sample.

Solution We start with the equation

$$\text{heat lost} + \text{heat gained} = 0,$$

where the heat gained or lost equals $mc \, \Delta T$. The heat lost for sample 1 is

$$Q_1 = m_1 c \, \Delta T = (0.100 \text{ kg})(4187 \text{ J/kg} \cdot {}°\text{C})(T - 60°\text{C}).$$

The heat gained by sample 2 is

$$Q_2 = m_2 c \, \Delta T = (0.200 \text{ kg})(4187 \text{ J/kg} \cdot {}°\text{C})(T - 40°\text{C}).$$

When these are added and set equal to zero, we obtain a final temperature of 46.667°C. The heat gained or lost is found to be

$$Q_1 = -5.58 \text{ kJ},$$
$$Q_2 = +5.58 \text{ kJ}.$$

As expected, the heat lost by one sample is equal in magnitude to the heat gained by the other.

The average temperature T_{ave} of the first sample is 53.3°C, or 326 K. The average temperature of the second sample is 43.3°C, or 316 K. The change in entropy of the initially warm water (sample 1) is then approximately

$$\Delta S_1 = \frac{Q_1}{T_{ave \, 1}} = \frac{-5.58 \text{ kJ}}{326 \text{ K}} = -17.1 \text{ J/K}.$$

The change in the entropy of the initially cool water (sample 2) is

$$\Delta S_2 = \frac{Q_2}{T_{\text{ave 2}}} = \frac{+5.58 \text{ kJ}}{316 \text{ K}} = +17.7 \text{ J/K}.$$

The change in the entropy of the system is the sum of the changes in entropy of the component parts,

$$\Delta S = \Delta S_1 + \Delta S_2 = 0.6 \text{ J/K}.$$

Discussion Notice that the change in the entropy of the system is positive. Also notice that no matter what initial temperatures we start with and no matter what the mass of each sample we choose, the change *still* will be positive. You can see this from the fact that the heat energy gained always equals the heat energy lost and that the term for the initially warmer sample has both a minus sign and the larger denominator.

Example 13.12

Does the entropy of the universe change as a result of the operation of the power plant in Example 13.6?

Solution If we consider the power plant to be a reversible Carnot engine and if we assume that the surroundings interact reversibly, the entropy of the surroundings (the universe) is unchanged. This is so because the entropy lost by the high-temperature reservoir is exactly matched by the entropy gained by the low-temperature reservoir. The amount of entropy lost by the high-temperature reservoir when it gives up an amount of heat Q_{H} is

$$\Delta S_1 = \frac{-|Q_{\text{H}}|}{T_{\text{H}}}.$$

The entropy gained by the low-temperature reservoir when it absorbs an amount of heat Q_{C} is

$$\Delta S_2 = \frac{|Q_{\text{C}}|}{T_{\text{C}}}.$$

But, as we have seen, for a Carnot engine,

$$\frac{|Q_{\text{H}}|}{T_{\text{H}}} = \frac{|Q_{\text{C}}|}{T_{\text{C}}}.$$

Consequently, the net change in entropy of the surroundings is zero.

In a real power plant, however, the efficiency is always less than that of an ideal Carnot engine. As a result, the heat Q_{C} delivered to the low-temperature reservoir is *greater* than the amount that would be delivered by a Carnot engine. This means that the increase in entropy of the low-temperature reservoir is greater than the decrease in entropy of the high-temperature reservoir, so that there is a net increase in the entropy of the universe.

Examples 13.10 through 13.12 are special cases of a much broader principle, discovered by Clausius: *In any process the entropy of the universe increases or remains constant.* (This is another way of expressing the second law.) Entropy remains constant only in the case of reversible processes, which do not occur naturally. So the entropy principle predicts that the entropy of the natural universe always increases. It is extremely important to remember that this does *not* mean that the entropy of a local segment cannot decrease—the entropy did decrease for the warm water in Example 13.11—but the total entropy of a system and its surroundings always increases. All observations and calculations indicate that if entropy decreases in one place, it simultaneously increases by an equal or larger amount somewhere else. Thus, entropy is quite different from such concepts as energy, momentum, and angular momentum, which we have previously encountered, because entropy is not conserved. In fact, the opposite is true. Entropy can be created, and in natural processes entropy always increases if all systems taking part in the process are considered. We can examine the meaning of the entropy principle from two equivalent viewpoints, which we briefly outline in the remainder of this section.

If we have two bodies at different temperatures—say, a hot stove and a block of ice—we can connect a heat engine between them and extract useful work. If, instead, we place the two bodies in direct thermal contact, they will come to thermal equilibrium. In agreement with the first law of thermodynamics, the total energy content of the stove and the ice is the same before contact as that of the stove and water (melted ice) after they have been placed in contact. However, once they reach equilibrium, we cannot separate them again and expect to extract work from them with a heat engine. Something has changed, even though the total energy has not. What has changed is the availability of the energy to do work. An increase in entropy means a decrease in the energy available to do work, not a decrease in the total energy.

Another viewpoint connects entropy with probability and statistics. This insight is due to Ludwig Boltzmann (1844–1906), who showed that an increase in the entropy of a system or substance corresponds to an increased degree of disorder in the atoms or molecules composing the substance. The most probable—that is, the most statistically favored—arrangement of molecules is the one with the most molecular disorder. For example, suppose you have a box with a partition dividing it in two, with a gas on one side of the partition and the other side evacuated (Fig. 13.14). With all of the molecules of the gas in one side, you have a highly ordered situation. If the partition is removed, however, the gas molecules will soon distribute themselves throughout the box and be moving in random directions—a less ordered situation. We can calculate that this change of order corresponds to an increase in the entropy of the gas. The probability that the molecules will all return to their original corner position at the same time is vanishingly small. J. W. Gibbs (1839–1903), the first great theoretical physicist in the United States, once called entropy a measure of "mixed-upness."

The statistical view is represented by an example that originated with Sir Arthur S. Eddington (1882–1944). A new deck of 52 cards comes in a preestablished order. As you shuffle the deck and shuffle it again—an action corresponding to the occurrence of thermodynamic processes—the order of the cards becomes randomized (Fig. 13.15). No matter how many times you shuffle the deck, you do not expect it to return to its original order. Though this event is certainly a possibility, its *probability* is so low that it is not worth considering. Similarly, systems undergoing some physical process proceed from order to disorder because disorder is so much more probable. The first law does not say that systems will not, of their own accord, become ordered again; but the second law says that the probability of their doing so is, in practical terms, zero. The reason

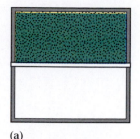

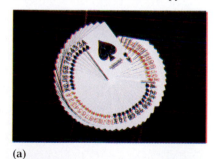

Figure 13.14
(a) A box in which all the gas molecules are confined to one side. (b) The same box after removing the dividing partition. The gas is now less ordered and has increased entropy.

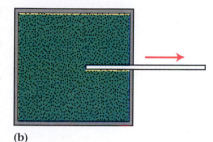

(a)

(b)

Figure 13.15
(a) The cards in a new deck are arranged in order. (b) After shuffling, the cards have a random arrangement. Continued shuffling will not bring the cards back to their original sequence.

is that any physical system is made up of so many molecules that the probability of its going back to its ordered state is infinitely smaller than the probability that a deck of 52 cards will be reordered upon repeated shufflings.

An advanced area of physics called *statistical mechanics* relates entropy to collections of large numbers of particles. Though we will not go into statistical mechanics in this text, it does provide a mathematical basis for calculating entropy as a measure of disorder, as well as for calculating other thermodynamic state variables and properties. The kinetic theory model described in Chapter 12 is a simple form of statistical mechanics.

The concept of increasing disorder applies to the entire universe as well as to systems here on earth. Current understanding of the early history of the universe is that it began as a highly compressed, hot "fireball," which has been expanding for something on the order of 15 billion years. This expansion corresponds to going from a more ordered to a more disordered state. During the expansion, the temperature of the universe decreases and its entropy increases. (Most of the universe, except for the stars and their associated planets, is quite cold. As we saw in Chapter 1, the average temperature is about 2.7 K.)

Just as in the case of the two liquids mixed in Example 13.11, the entropy and disorder of the universe increase as hot bodies cool and cold bodies warm. If the entropy principle is true throughout the universe, we can envision some time in the far distant future when everything in the universe will have reached a uniform temperature. No heat could flow, no work could be done, and no change in energy or motion could take place. Neither engines nor plants nor animals would be able to extract energy. This possible occurrence is often called the *heat death of the universe.*

The heat death of the universe may happen in billions of years, or it may not happen at all. If, at some point, the universe begins to contract again into a fireball, then heat death will not occur. At the present time, it is not known which of the two cases, contraction or endless expansion, will occur.

Clausius summarized the laws of thermodynamics by saying that (1) the energy content of the universe is constant and (2) the entropy of the universe always increases. However, because of the complexity of the problem, we cannot determine in advance at what point the entropy will have reached its maximum value and the available energy will have decreased to zero.

The first law expressed the idea of conservation of energy. We stated it succinctly as *You can't win.* The second law expresses the idea that all available energy cannot be converted into useful work. So not only can you not win, but *You can't break even.*

*13.7   Energy and Thermal Pollution

We obtain most of the necessities and comforts of life from the generation of power, in one way or another. The word generation should perhaps be replaced by the word conversion. The burning of fuel, as in a car engine, converts chemical potential energy in the fuel into heat energy; the use of water power, as in a hydroelectric power station, converts gravitational potential energy of the water into electrical energy; and so on. The conversion and utilization of energy from falling water, burning fossil fuels, nuclear reactors, solar converters, or any other source cannot take place without the cost of some heat discharge somewhere.

The conversion involved in producing and using energy is always accomplished by machines or organisms whose efficiency is less than that of a Carnot engine. Thus, an inevitable consequence of energy conversion is the discharge of heat. We have no option in this respect. The unwanted release of thermal energy into the environment is known as *thermal pollution*. It can be controlled and spread over a large area to minimize local changes in temperature, but it cannot be eliminated.

The rate of energy generation by humans on the earth is about 1×10^{13} W. It is possible that with conservation and other measures, this figure need not increase as rapidly as was once thought, but it is unlikely to decrease in the foreseeable future. Overall heat production from human devices is less than 6×10^{-5} of the rate of energy absorption from the sun (less than one-hundredth of one percent) and therefore would not have an appreciable influence if it were distributed uniformly over the earth. But, of course, it is not. In heavily populated regions, the heat production rate may be several percent of the rate of energy absorption from the sun.

Approximately 91% of the electrical power in the United States is generated in steam plants heated by fossil and nuclear fuels and by gas turbines that burn fossil fuels. Most of the resulting waste heat is deposited in streams and rivers. This concentrated dumping of energy can raise the temperature of the water by several degrees. Such a small change is insignificant to humans and most warm-blooded animals, but can drastically influence which type of fish will predominate in rivers and thereby modify the entire local food chain. Other effects can be even more unpleasant, since higher temperatures often provide a more hospitable environment for pathogenic organisms.

There are several approaches to solving, or at least lessening, the problem of thermal pollution. One approach in which active research is going on is the development of alternative energy sources. The techniques being studied include the harnessing of wind power, tidal power, solar power, and ocean temperature differences, as well as a host of other ingenious ideas (Fig. 13.16). A second more promising approach to reducing thermal pollution involves measures for saving energy, from better building insulation to smaller, more efficient cars. However, the fact remains that a large amount of energy will continue to be discharged into the environment.

Figure 13.16

Alternative energy sources. (a) A solar collector for generating power in France. (b) Some of the nearly 16,000 wind turbines operating in California. (c) Hoover Dam, one of the world's largest generators of hydroelectric power.

(a)

(b)

(c)

In response to governmental pollution abatement requirements, auto manufacturers have begun producing battery-operated electric cars. One such car is the EV-1, which General Motors has made available in California. These vehicles have been erroneously called zero-emission vehicles. While there is no direct emission of exhaust gases from the cars, there are emissions of both exhaust gases and waste heat from the power plants that provide the electricity for the cars. Moreover, there is the likelihood of increased emissions of lead and other exotic materials associated with the manufacture and disposal of the batteries needed to store the energy to propel these cars. Exhaust gas and thermal pollution are not eliminated, just relocated.

A possible remedy to thermal pollution is to extract further energy from heat exhausts. For example, exhaust heat can be used to warm buildings, resulting in lower direct-exhaust temperature and less need for additional fuel to heat the buildings. Another solution is to distribute heat exhaust more widely by depositing as much as possible into the atmosphere rather than into bodies of water. Figure 13.17 is a photograph of cooling towers that are designed for this purpose.

There is no single solution to the problems of thermal pollution, but perhaps the combination of alternative sources, conservation, and wide distribution of waste heat can make the problem manageable.

Figure 13.17
Cooling towers at the Paradise power plant.

Summary

Useful Concepts

■ Thermodynamics is the branch of physics concerned with the relationships between heat and work. These relationships govern the operation of all engines and devices that convert energy from one form to another, including processes that take place in living organisms.

■ The zeroth law of thermodynamics states that when two systems are each in thermal equilibrium with a third system, they are in thermal equilibrium with each other.

■ The first law of thermodynamics is a statement of conservation of energy,

$$\Delta U = Q - W,$$

where ΔU is the change in a system's internal energy, Q is the heat added to the system, and W is the work done by the system.

■ The thermal efficiency of a Carnot engine is

$$\text{thermal efficiency} = \frac{W}{Q_H} = 1 - \frac{T_C}{T_H}.$$

■ The coefficient of performance of an ideal refrigerator is

$$\text{c.p.(refrigerator)} = \frac{Q_C}{W} = \frac{T_C}{T_H - T_C}.$$

■ The coefficient of performance of an ideal heat pump is

$$\text{c.p.(heat pump)} = \frac{Q_H}{W} = \frac{T_H}{T_H - T_C}.$$

■ The second law of thermodynamics can be stated in several ways. Two of them follow:

Clausius statement of the second law: Heat cannot, by itself, pass from a colder to a warmer body.

Kelvin-Planck statement of the second law: It is impossible for any system to undergo a cyclic process whose *sole* result is the absorption of heat from a single reservoir at a single temperature and the performance of an equivalent amount of work.

■ When a system at temperature T absorbs an amount of heat Q in a reversible process, the change in entropy is

$$\Delta S \equiv \frac{Q}{T}.$$

■ Entropy is a measure of the disorder of a system. In any process, the entropy of the universe increases or remains constant.

■ Thermal pollution is the unwanted release of thermal energy into the environment.

Important Terms

You should be able to write the definition or meaning of each of the following:

thermodynamics
zeroth law of
   thermodynamics
thermodynamic system

internal energy
first law of
   thermodynamics
adiabatic process

isothermal process   reversible process   refrigerator   second law of
isochoric process   Carnot cycle   coefficient of performance   thermodynamics
isobaric process   thermal efficiency   heat pump   entropy

Conceptual Questions

13.1 Apple 1 is in thermal equilibrium with apple 2, apple 2 is in thermal equilibrium with apple 3, and so on to apple 6. Are apples 1 and 6 in thermal equilibrium? Can you prove your answer using the zeroth law of thermodynamics, or are you simply using your physical intuition (which is probably correct here)?

13.2 Assume that in Joule's paddle-wheel experiment (Chapter 11), the thermodynamic system is the water and it is well insulated from its surroundings. Has heat been added? Has work been done? Has the internal energy changed?

13.3 As you bend a wire back and forth, it becomes warm, then hot, and finally breaks. Discuss what is happening from the standpoint of the first law of thermodynamics.

13.4 According to the first law of thermodynamics, would you expect the temperature at which a substance melts or freezes to change if the substance is subject to a high pressure?

13.5 A weight is placed on top of a vertically held brass rod that has its lower end on a fixed support. Discuss what happens from an energy standpoint as the rod is heated.

13.6 If we place a hot brick in thermal contact with a cold brick, does the first law of thermodynamics allow the hot brick to become hotter and the cold brick to become colder? What is allowed by the second law?

13.7 Is there any change in the temperature of the air inside an automobile tire when you remove the valve and allow the air to escape rapidly?

13.8 A circular rod fits into a circular hole in a metal block. The interface is oiled to make it easier to turn the rod. When the rod is turned rapidly for a long time, the block and rod become warm. Extremely precise measurements show that the mechanical energy delivered in turning the rod is more than the heat energy developed. Discuss this result in terms of the first law of thermodynamics. (*Hint:* Can any changes occur in the molecular structure of oil?)

13.9 Can you cool a kitchen by leaving the refrigerator door open? What will happen?

13.10 A heat engine can extract energy from any system maintaining a temperature difference between two bodies. List as many cases as you can where this is actually done, starting with the steam engine. Give some cases in nature where a temperature difference is maintained but in which we do not, for economic or engineering reasons, extract energy.

13.11 Discuss and give specific examples of the effects that prevent real engines from reaching their Carnot efficiency.

13.12 Explain how a heat pump can add more energy to a house than is used by the electric motor driving the pump.

13.13 Assume that the exhaust from a conventional fossil fuel power plant is at a temperature of 50°C. (a) Plot the maximum theoretical efficiency as a function of the high, or input, temperature. (b) Does the shape of the curve suggest any practical considerations about the usefulness of increasing the input temperature?

13.14 A sealed container is divided into two equal parts by a removable partition. One side is filled with pure oxygen, and the other side is filled with air at the same pressure. (a) If the partition is removed, does the entropy of the system increase, remain constant, or decrease? (b) What would the answer be if both sides were originally filled with oxygen?

13.15 Discuss the statement that the eventual destiny of the physical universe is stagnation.

13.16 The earth is in approximate thermal equilibrium with its environment, absorbing radiant energy from the sun and reradiating energy at a lower temperature to its environment at the same rate. Recalling that processes on the earth increase its entropy, make an educated guess about the overall entropy change of the earth.

☆ STAR

Problems

Sections 13.1 and 13.2 Thermal Equilibrium and the First Law of Thermodynamics

Hints for Solving Problems

In using the first law of thermodynamics, you need to pay careful attention to the signs: Work done on a system is negative, work done by a system is positive, whereas heat into a system is positive, heat out of a system is negative.

13.1 One thousand joules of mechanical work are done by an insulated system that expands when 5.50 kcal of heat are added. What is the change in the internal energy of the system?

13.2 In a laboratory experiment, a heat engine takes in 200 J of heat and does 50 J of work by expanding in a reversible process. What is the change in internal energy of the engine during this process?

13.3 An insulated system takes in 6.50 kcal of heat and has

2000 J of work done on it. What is the change in internal energy of the system?

13.4 (a) What is the change in internal energy of 100 g of ice as the ice melts to water at 0°C if we assume the density of water and ice to be the same? (b) Would the answer be larger or smaller if you took into account the fact that ice is less dense than water at 0°C? Explain your answer.

13.5 An inventor tells you he has built a new type of electrical generator that produces 1000 W while using 75 Btu of fuel per minute. He mentions that an interesting side effect is that the apparatus does not heat up its environment. Is it possible for such a device to exist?

13.6 A 2.8-g piece of copper undergoes isochoric heating, which causes a temperature increase of 10°C. By how many joules does the internal energy change? (See Table 11.3.)

13.7 Gas confined by the piston in a particular heat engine expands against a constant pressure of 2.56×10^5 N/m². When 40.0 kJ of heat are added to the system, the volume expands from 0.105 m³ to 0.235 m³. What is the change in internal energy of the system?

13.8 Gas confined by the piston in a particular heat engine expands against a constant pressure of 3.0×10^5 N/m². When 10.0 kcal of heat are added to the system, the volume expands from 0.10 m³ to 0.20 m³. What is the change in internal energy of the system?

13.9 A rigid container is filled with 0.010 kg of carbon dioxide and sealed at atmospheric pressure and 20°C. It is then placed in a bath of ice water and allowed to come to thermal equilibrium. The container is then immersed in a water bath at 60°C and again allowed to come to thermal equilibrium. What is the difference in internal energy of the gas between the initial state, when the container was first filled, and the final state? The specific heat of carbon dioxide under these conditions is 638 J/kg · K.

13.10 A heat engine undergoes a process in which its internal energy increases by 275 J while it is doing 360 J of work. How much heat is taken in (or given out) by the engine during this process?

13.11 Show that the work done during an isobaric process is $W = P(V_2 - V_1)$.

13.12● (a) Consider a system that goes from a to b via the path acb on the PV diagram of Fig. 13.18. If the heat absorbed is 400 J and the system does 150 J of work, what is the change in internal energy of the system? (b) If the system does 50 J of work along path adb, how much heat energy must the system absorb?

13.13● When a system is taken from state a to b along the path acb in Fig. 13.18, 75 J of heat flow into the system while 25 J of work are done by the system. (a) How much heat flows into the system when it goes from a to b along the path adb if the work done is 10 J? (b) When the system returns to a along the curved path, the system performs −15 J of work. Does the system absorb or give out heat during that process? (c) How much heat is exchanged in part (b)?

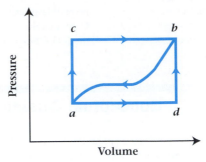

Figure 13.18
Problems 13.12 and 13.13.

Section 13.3 The Carnot Cycle and the Efficiency of Engines

> **Hints for Solving Problems**
>
> In formulas for thermal efficiency and coefficient of performance, the temperatures are always in kelvins. Be clear about what is the system you are studying and what are its surroundings. For simplicity in finding temperatures, take 0°C to be 273 K.

13.14 A Carnot heat engine takes in 100 J at a temperature of 100°C. At the end of the cycle, 50 J are exhausted. What is the temperature of the exhaust?

13.15 (a) What is the efficiency of an ideal engine operating between 275°C and 0°C? (b) What is the efficiency if the engine operates over the same temperature difference, but with the input and output temperature both raised by 100°C?

13.16 What is the temperature at which 45 J per cycle are added to a Carnot engine if 36 J are exhausted at a temperature of 100°C?

13.17 At the end of a cycle, 17 J are exhausted from a Carnot engine operating between 375°C and 150°C. (a) How many joules were taken in during each cycle? (b) How much work was done?

13.18 How much work is done per cycle by a Carnot engine operating between +10°C and −20°C if 12.0 J of heat are taken in at the higher temperature during each cycle?

13.19 In the summer, the temperature of water at the lower depths of Lake Geneva is 14°C. The surface temperature of the lake is about 20°C. What would the maximum efficiency be for any power-generating device that took advantage of this temperature difference?

13.20 Sketch a Carnot cycle using temperature for the vertical coordinate and volume for the horizontal coordinate.

13.21● The working fluid in a heat engine is initially at pressure P_0 and volume V_0. It is operated in a cycle shown in Fig. 13.19. What is the net work done by the engine in each cycle?

13.22● The gas in a heat engine undergoes the process shown in the PV diagram of Fig. 13.20. (a) What is the work done

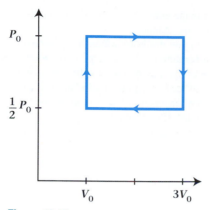

Figure 13.19
Problem 13.21.

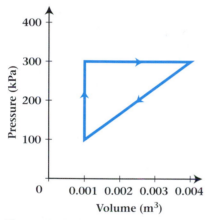

Figure 13.20
Problem 13.22.

during one complete cycle? (b) What is the net heat added to the system during one complete cycle?

13.23● A Carnot engine operating at 30% efficiency exhausts heat into a heat sink at 283 K. By how much would the temperature of the heat reservoir need to be raised to increase the efficiency of the engine to 45%?

13.24● A nuclear power plant in South Carolina produces electrical energy at the rate of 900 MW. The plant has a thermal efficiency of 30%. At what rate is energy released by the nuclear reactor? What is the rate of release of heat to the surrounding environment?

13.25● A hypothetical indicator diagram is traced out in which the pressure rises linearly with increasing volume to some value P_1 at volume V_1, then falls to zero immediately and remains there while the volume decreases to some volume V_0, at which point the cycle starts again. What is the work per cycle represented by this indicator diagram?

13.26● From the standpoint of efficiency, show whether it is better to increase the highest temperature of a Carnot engine

by ΔT or to decrease the lowest temperature by the same amount.

13.27● An engine does 2000 J of work and exhausts 2.25 kcal each cycle while operating between 700°C and 400°C. What is its actual efficiency? How does it compare with the maximum theoretical efficiency?

13.28● An engine has a maximum theoretical efficiency of 40%. If the temperature difference between the input and the output is 300°C, what is the exhaust temperature?

13.29● What is the maximum efficiency of an engine for which the ratio of the energy exhausted to the work done per cycle is 0.25?

13.30●● A steam-electric power plant delivers 900 MW of electric power. The surplus heat is exhausted into a river with a flow of 5.0×10^4 kg/s, causing a change in temperature of 12°C. (a) What is the efficiency of the power plant? (b) What is the rate of the thermal source?

13.31●● The ignition temperature of a typical automobile engine is 17,000°C, and the spent gas is exhausted from the engine at 120°C. (a) What is the maximum possible efficiency of such an engine? (b) The energy per unit volume released by combustion of gasoline is about 0.35×10^8 J/L. If the power required to overcome wind friction is 22 kW when the car is driven at 100 km/h and if the fuel consumption is 8.5 km/L, what is the actual efficiency when driving at 100 km/h? How does your answer compare with the maximum theoretical value?

13.32●● A fluid is heated so that its original volume V_0 is doubled. The pressure is observed to vary linearly with the volume according to $P = P_0(1 + bV)$, where b and P_0 are constants. Find an expression for the work done by the expanding fluid.

Section 13.4 Refrigerators and Heat Pumps

13.33 What is the maximum coefficient of performance of a refrigerator that maintains an inside temperature of 5.0°C in a room where the temperature is 25°C?

13.34 Show that the difference between the theoretical coefficient of performance for a heat pump and the same device running as a refrigerator is unity.

13.35 (a) Calculate the maximum coefficient of performance of a heat pump used to maintain an inside temperature of 21°C (70°F) on a day when the outside temperature is 7°C (45°F). (b) What is the coefficient of performance when the outside temperature drops to −12°C (10°F)?

13.36● What maximum temperature can be maintained inside a house with a heat pump whose coefficient of performance is 8 if the outside temperature is 0°C?

13.37● What is the minimum theoretical temperature that can be maintained in a refrigerator with a coefficient of performance of 5 when operated in a room where the temperature is 16°C?

13.38● Show that the efficiency e of a Carnot engine and the coefficient of performance c.p.$_{\text{refrig.}}$ of a Carnot refrigerator operating between the same temperatures are related by

$$e = \frac{1}{1 + \text{c.p.}_{\text{refrig.}}}.$$

13.39• (a) A Carnot engine operates between heat reservoirs at 420 K and 300 K. If it absorbs 500 J of heat from the hot reservoir each cycle, how much work is delivered by the engine each cycle? (b) If the engine is operated in reverse as a refrigerator acting between the two reservoirs, how much work input is required per cycle to remove 500 J of heat from the low-temperature reservoir?

13.40• Suppose that water is put into the refrigerator of Problem 13.33 at room temperature of 25°C. What is the cost of making 100 lb (45 kg) of ice if electricity costs $0.08 per kilowatt hour?

13.41•• Thermal insulation in modern refrigerators is not perfect. Consider a refrigerator with a heat leak of 2.2 kJ/min. The interior cooling coils are at −12°C and the external heat exchanger coils are in a room at 33°C. The refrigerator runs at 32% of the maximum theoretical coefficient of performance. What is the average power used to maintain the temperature inside the refrigerator?

Section 13.6 Entropy and the Second Law

Hints for Solving Problems

In calculating entropy changes, use the average temperature for a process when appropriate.

13.42 What is the change in entropy when a 230-g piece of ice melts to water at 0°C?

13.43 What is the change in entropy of the water when 500 g of water are converted to steam at 100°C?

13.44 An exhaust tube from a steam generator is submerged in a large container of cool water. When 10.0 g of steam are condensed into water at 100°C by bubbling the steam through the cold water, what is the change of entropy of the steam?

13.45 What is the change in entropy per second of the river water for the power plant in Problem 13.30? Assume the average water temperature to be 12°C.

13.46• What is the change in entropy when 500 g of water at 10.0°C are mixed with 500 g of water at 12.0°C?

13.47• What is the change in entropy when enough heat is supplied so that an 75-g piece of ice initially at 0°C melts, warms up to 100°C, and changes to steam at 100°C?

13.48• Determine the change in entropy when 50 g of cream at 14°C are put into 350 g of coffee at 80°C. Assume that the specific heat of both cream and coffee is essentially the same as that of water. Approximate by using average temperatures.

13.49• A child takes a 0.45-kg block of ice at 0°C, places it on a large marble slab, and watches the ice melt. (a) What is the entropy change of the ice (water)? (b) If the source of heat (the marble slab) is very massive and remains at a constant

20°C, what is the entropy change of the marble? (c) What is the total entropy change?

13.50• When you drink ice water, it is warmed by your body to body temperature (37°C). Energy is required to warm the water; therefore you could conceivably lose weight by consuming lots of ice water to help you burn off your food energy. (a) How much ice water at 0°C would it take to offset eating two jelly donuts? (See Problem 11.24.) (b) Estimate the change in entropy and state any assumptions made.

Additional Problems

13.51 What is the maximum efficiency of a heat engine that takes in heat at a temperature of 173°C and has an exhaust temperature of 75°C?

13.52 An inventor claims to have devised a steam turbine that is 75% efficient. The steam enters the turbine at a temperature of 300°C and exhausts at the temperature of the surrounding atmosphere. Do you believe his claim?

13.53 A 1.0-kg lead block slides slowly down an inclined plane and comes to rest at the bottom. If the initial height of the block was 0.45 m above its final resting place and if the temperature of the block, the table, and the surrounding air remains at 27°C, by how much does the entropy of the universe change as a result of this process?

13.54 Heat is added uniformly to a 1.0-kg sample of water at a constant rate of 35 W. The water is initially at a temperature of 10°C. (a) What is the change in entropy of the water over the temperature interval between 10°C and 11°C? (b) What is the change in entropy over the temperature interval from 98°C to 99°C?

13.55• A Carnot engine has an exhaust temperature of 120°C and exhausts 80 J per cycle. If the input temperature is 325°C, how much work does the engine do per cycle?

13.56• An ideal engine always takes in heat at the same input temperature T_H. When the exhaust temperature is 120°C, the efficiency is 30%. (a) What is the input temperature? (b) If the exhaust temperature is changed so that the efficiency rises to 37%, what is the new exhaust temperature?

13.57• A typical person consumes 2000 kcal of food per day and converts most of that energy into heat. (a) Under the assumption that all of the food energy is released as heat, calculate the rate of heat release in watts. (b) If all of this heat is transferred to the surrounding air, which remains at an essentially constant temperature of 24°C, what is the rate at which entropy is being delivered to the universe?

13.58• Heated metals begin to glow at temperatures near 600°C. If we take this temperature to be the temperature of the heat input, estimate the maximum efficiency of any mechanical device operated as a heat engine.

13.59•• Two similar coal-fired electric power plants generate electric energy at the same rate. One plant operates with an efficiency of 30%, while its newer companion operates with an efficiency of 40%. How much more waste heat is released to the environment by the first plant than by the newer one?

13.60•• Consider a large steam-electric plant that produces 1000 MW of electric power. (a) If the plant is 33.3% efficient, what must be the energy rate of the thermal source used? (b) What is the rate at which thermal energy is exhausted to the environment? (c) If this energy were dumped into a river whose flow was 6×10^4 kg/s, what would be the average rise in the temperature of the river?

13.61•• In order to increase the output of some heat engines, a two-step process is used. The input and output temperatures of the first engine are T_1 and T_2. The exhaust of the first engine provides the input to the second engine at temperature T_2 and with a final exhaust temperature of T_3. Show that the overall ideal efficiency is the same as that of a single engine operating between temperatures T_1 and T_3.

13.62•• (a) Using the technique of adding up the area under a curve, find the total change in entropy of a 1.0-kg sample of water as it is heated from 20°C to 40°C. You may want to divide the temperature range into 2° intervals. (b) Compute the entropy change as the total heat added, divided by the average temperature. (c) How do these two results compare?

13.63•• (a) Show that when a Carnot cycle is plotted as a temperature-versus-entropy diagram, the resulting graph is a rectangle. Calculate (b) the heat absorbed, (c) the heat exhausted, and (d) the work done by the system whose Carnot cycle is shown in Fig. 13.21.

13.64•• A copper rod 30 cm long having a cross-sectional area of 4.0 cm² is used to connect a boiler maintained at

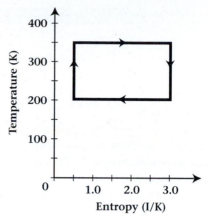

Figure 13.21
Problem 13.63.

100°C to a large block of ice at 0°C. (a) Once a steady state is reached, what is the rate of heat flow through the rod? What is the rate of change in entropy of (b) the boiler, (c) the rod, (d) the ice, and (e) the universe due to the heat flow through the bar?

13.65•• The exact expression for the entropy change of a material heated from $T_1 = 0$°C to $T_2 = 100$°C is $\Delta S = mc \ln(T_2/T_1)$, where m is the mass of the material and c is its specific heat. What percent error is made by calculating the entropy change by using the approximation $\Delta S = Q/T_{ave}$?

chapter 14

Periodic Motion

So far, we have used the basic laws of mechanics to study translational and rotational motions. Here we will use the same basic laws to introduce another important kind of motion: periodic motion. Such motion repeats itself, like a clock pendulum moving back and forth. It is also known as oscillatory, or vibratory, motion. We will find that oscillatory motion is also relevant to our study of electronic circuits, optics, and atomic structure.

Any behavior that repeats itself regularly is called **periodic.** Almost any physical object or system can be made to undergo periodic, or oscillatory, motion. Objects that display such motion abound in everyday life: a child's swing, a guitar string, a bell, or the quartz crystal in an electronic wristwatch. Even the earth itself undergoes oscillatory motion due to seismic activity. Many oscillating systems produce waves. For example, vibrating mechanical bodies generate sound waves, and electrical oscillations generate electromagnetic waves that make radio and television broadcasting possible. In Chapter 15, we will consider the details of wave motion.

For a mechanical system to be capable of self-sustaining oscillations, it must meet two requirements. The system must have inertia, or mass, and there must also be a force that acts to restore the system to its equilibrium, or lowest-energy, state. A stretched spring or a pendulum pulled to one side exhibits these properties. On the other hand, a light piece of thread held suspended from one end and a kicked pile of sand do not, and they also do not exhibit oscillatory behavior to any great extent.

There are many types of periodic motion, from the complicated, but repeating, pattern of the electrical impulses of a human heart (Fig. 14.1a) to the simple repeating pattern of the end of a vibrating tuning fork (Fig. 14.1b). The latter type of vibratory motion, which is both common and simple in its mathematical description, is called simple harmonic motion. In this motion, the displacement varies sinusoidally with time; that is, the graph of the displacement versus time is a sine curve, as shown in Fig. 14.1(b). Another example, which we will discuss in detail, is the oscillation of the pendulum of a grandfather clock. In this chapter, we restrict ourselves to simple harmonic motion. Even if the motion of a periodic system is more complex, its behavior has features in common with simple harmonic motion. ■

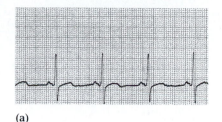

(a)

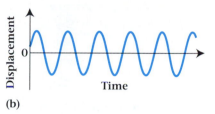

(b)

Figure 14.1
The repeating pattern of an electrocardiogram (a) is more complicated than the smooth, sinusoidal motion of the end of a tuning fork (b). However, both patterns are periodic.

<table>
<tr><td>14.1</td><td></td></tr>
</table>

Hooke's Law

One of the simplest forms of periodic motion is the up-and-down motion of a mass suspended from a spring. Because other types of oscillatory motion resemble the behavior of the mass on a spring, we can model complex systems after this behavior and apply what we learn about springs to other systems. Thus, we begin our study of periodic motion with a careful examination of the behavior of a mass moving at the end of a spring.

We saw in Chapter 6 that the force required to stretch a spring is, to a good approximation, proportional to the distance by which the spring is extended. This relationship between an applied force and the change in the length of a spring is known as **Hooke's law,** in honor of Robert Hooke,* who first enunciated it in 1678. The elastic behavior of many materials, including some woods, bone, and steel, can be described by Hooke's law if the materials are not stretched too far. Further stretching leads to a nonlinear relationship between force and extension and, finally, to breaking or rupture.

If we hang different weights from a spring that obeys Hooke's law, the elongation is proportional to the applied force (Fig. 14.2). When the spring reaches equilibrium, the gravitational force acting downward on the mass must be balanced by an upward force due to the spring. This spring force is called the **restoring force** because it acts in a direction opposite to the direction of the displacement of the end of the spring. Although we have illustrated the case of

---

*Hooke is also well known for his work in biology. His book *Micrographia,* published in 1665, contained the first drawings of tiny objects seen with the aid of a microscope; in fact, Hooke was the one who coined the word *cell* as it is used in biology.

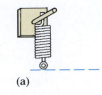

(a)

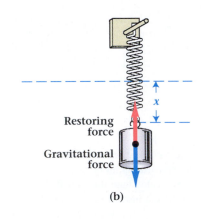

Restoring
force

Gravitational
force

(b)

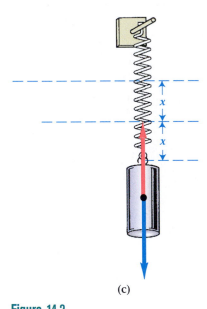

(c)

Figure 14.2

The extension of a spring is
proportional to the applied force.
(a) Spring with no load. (b) Spring
extended a distance x by a 0.5-N load.
(c) Spring extended a distance $2x$ by a
1.0-N load.

stretching a spring, the situation of compressing a spring is exactly the same. As
the spring is compressed beyond equilibrium, a restoring force acts to resist the
compression.

Figure 14.3 shows a block of mass m attached to a horizontal coil spring.
The block is at rest, but is free to move along a frictionless surface. We choose
this arrangement so that we can avoid the changes in gravitational potential energy that occur in a situation like the one in Fig. 14.2.

What happens when we pull the block a little to the right? Let's analyze this
situation mathematically. In Fig. 14.3(a), the block is at rest at the equilibrium
position $x = 0$. When the block is displaced a distance x to the right, the spring
exerts a restoring force F to the left (Fig. 14.3b). The relation between displacement and restoring force is

$$F = -kx. \tag{14.1}$$

If k is a constant, Eq. (14.1) is the mathematical form of Hooke's law. For many
materials, k is constant if the displacement x is not too large. This proportionality constant k is called the **spring constant.** It is also known as the stiffness constant. The negative sign in the Hooke's law equation indicates that the restoring
force due to the spring is in the opposite direction from the displacement. Since
we are considering only one-dimensional motion of the mass at the end of a
spring, we do not need to use vector notation in our equation.

Example 14.1

We would like to use a coil spring as a scale, as we discussed in defining force
in Chapter 4. To do so we must determine whether its extension is proportional
to the weight hung from it. Using a setup similar to the one in Fig. 14.2, we
measure the equilibrium length of the spring for different weights hung on one
end. The measurements are given in Table 14.1. Determine whether the spring
obeys Hooke's law, and if it does, find the spring constant.

Strategy The best way to understand the behavior of the spring is to make a
graph of spring length versus applied force (Fig. 14.4). Note that the measurements given are not the displacement of the spring from the unloaded equilibrium, but rather its total length. The graph is linear, which means that the spring
does obey Hooke's law over the range used. The slope of the line showing length
versus applied force is equal to the reciprocal of the spring constant. To see why
this is so, we write an equation for the straight line. If we let L represent the
length (the vertical coordinate) and F represent the applied force (the horizontal
coordinate)—the negative of the force in Eq. (14.1)—the appropriate equation is
$L = L_0 + mF$, where L_0 is the initial length and m is the slope. Here this equation can also be expressed as $L - L_0 = x = mF$, where x is the extension of the
spring. This last expression has the form of Hooke's law with $m = 1/k$.

Solution Draw the line that best fits the data and compute the slope of the line.
From the graph we see that

$$\text{slope} = m = \frac{\Delta L}{\Delta F} = \frac{\Delta x}{\Delta F} = \frac{1}{k},$$

Table 14.1	Measurements on a Spring	
Mass m Applied (kg)	Force = mg (N)	Length of Spring (cm)
0.00	0.00	16.2
0.050	0.49	21.4
0.100	0.98	26.5
0.150	1.47	31.7
0.200	1.96	36.9
0.250	2.45	42.0
0.300	2.94	47.2
0.350	3.43	52.4

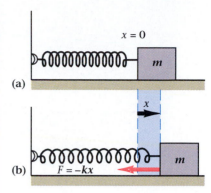

(a)

(b)

Figure 14.3
(a) A mass m attached to a spring rests on a frictionless surface. (b) When mass m is displaced by an amount x, the spring exerts a restoring force F in a direction opposite to x.

or

$$k = \frac{2.94 \text{ N} - 0.98 \text{ N}}{0.472 \text{ m} - 0.265 \text{ m}} = 9.47 \text{ N/m}.$$

Discussion Here we have drawn "by eye" a straight line that seems to best fit the data points. We have taken values of ΔF and ΔL from two of the data points that lie on that line. Numerical methods exist that take advantage of all of the data to determine the best fit of a line to a set of data points. One method, called the method of least squares, minimizes the square of the distances between the computed line and the data points and is used in many graphing programs. Calculating the line by such a standard method not only ensures that the resulting line is the best one to fit the data, but also allows others to arrive at the same value from the same data. A least-squares fit of the data here gives a value of $k = 9.48$ N/m. In this case, our estimate of the best line is quite close to the least-squares value. ■

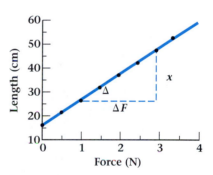

Figure 14.4
Example 14.1: A graph of spring length versus applied force.

If the block in Fig. 14.3 is displaced an amount x from its equilibrium position, an external force equal and opposite to the restoring force is required to keep it there. If this external force is suddenly released, the unbalanced spring force accelerates the block. The acceleration is found by combining Newton's second law with Hooke's law:

$$a = \frac{F}{m} = -\frac{k}{m}x. \qquad (14.2)$$

The acceleration is proportional to the displacement and in the opposite direction. We will see that this same proportionality between acceleration and displacement occurs for children's swings, clock pendulums, and many types of vibrating bodies, as well as for some quantities that describe the behavior of electrical circuits. In the next section, we will see that whenever this relationship occurs we get an especially simple type of oscillatory behavior.

Example 14.2

The spring in Example 14.1 and a block of mass 0.350 kg are placed horizontally, as in Fig. 14.3. The spring is compressed 6.0 cm and released from rest.

(a) What is the initial acceleration of the block? (b) What is the initial force on the block?

Solution (a) If we choose our coordinate axis so that x increases to the right, compressing the spring corresponds to $x = -6.0$ cm. The acceleration is then

$$a = -\frac{k}{m}x$$

$$a = -\frac{9.48 \text{ N/m}}{0.350 \text{ kg}}(-0.060 \text{ m}) = 1.625 \text{ m/s}^2 = 1.6 \text{ m/s}^2.$$

The acceleration is positive and thus to the right, a result that agrees with your intuition about which way the block will initially move.

(b) Because we know the mass and initial acceleration, the initial force can be found directly from Newton's second law,

$$F = ma = (0.350 \text{ kg})(1.625 \text{ m/s}^2) = 0.57 \text{ N}.$$

Discussion This force of 0.57 N is the restoring force that acts on the block and causes its initial acceleration in the positive x direction. This restoring force can also be found directly from Hooke's law as

$$F = -kx = -(9.48 \text{ N/m})(-0.060 \text{ m}) = 0.57 \text{ N}.$$

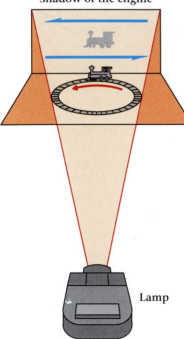

Shadow of the engine

Lamp

Figure 14.5
A simple arrangement illustrating the connection between uniform circular motion and simple harmonic motion. As the toy-train engine moves around the circular track, it casts a shadow on the wall that moves with simple harmonic motion.

14.2 The Simple Harmonic Oscillator

We have seen that for a spring that obeys Hooke's law, the acceleration a is proportional to the negative of the displacement x. We also noted that many oscillatory systems have motion similar to that of a spring. Now we want to find a general description of this motion. *Any system whose acceleration is proportional to the negative of the displacement undergoes* **simple harmonic motion.** As we will see, the projection (or shadow) of uniform circular motion onto a straight line is an example of simple harmonic motion.

We already know that uniform circular motion is periodic. When we view, from above, a toy-train engine going around a circular track, we see circular motion (Fig. 14.5). But if we look at the shadow of the train cast by a lamp edge-on to the track, the engine's shadow appears to oscillate back and forth. We will show this type of motion to be the same simple harmonic motion as the oscillations of a mass on a spring.

We begin by investigating the projection onto the x axis of the uniform circular motion of a point (Fig. 14.6a). If P is a point moving with constant speed v_0 in a circle of radius $R = x_0$, then the projection, or shadow, of P on the x axis oscillates back and forth between $+x_0$ and $-x_0$. The position of the point along the x axis is

$$x = x_0 \cos \theta.$$

Since $\cos \theta$ is always between -1 and $+1$, we see that x_0 is the maximum displacement of the projection of the point.

For uniform circular motion, the angle θ increases steadily with time; that is, θ is proportional to t. As we saw in Chapter 5, the time required to make one

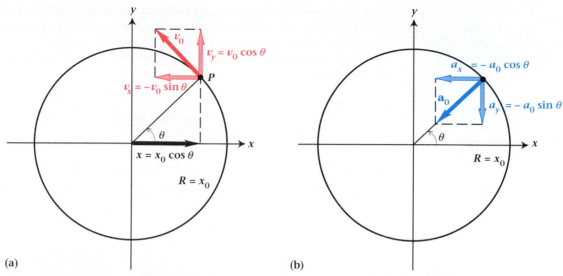

Figure 14.6
(a) A point P moving around a circular path with a constant tangential speed v_0. The projection along the x-axis of the radius vector to P is $x_0 \cos \theta$.
(b) The projection of the centripetal acceleration vector a_0 along the x-axis is $-a_0 \cos \theta$.

complete revolution ($\theta = 2\pi$) is the period T. Thus, the relationship between angle in radians and time is

$$\theta = 2\pi \frac{t}{T}.$$

Using the definition of frequency $f = 1/T$ and angular frequency $\omega = 2\pi f$ from Section 5.1, we can also write the angle as $\theta = 2\pi f t = \omega t$. The projection along the x axis of the circular motion (the x coordinate of the point in Fig. 14.6a) is therefore

$$x = x_0 \cos 2\pi f t. \tag{14.3}$$

The projection can also be expressed as $x = x_0 \cos 2\pi \frac{t}{T}$ or $x = x_0 \cos \omega t$. The quantity x_0 is the maximum displacement, the greatest distance from zero, and is called the **amplitude** of the displacement.

The velocity is also a function of time, and from Fig. 14.6(a) we find that v_x is proportional to $-\sin \theta$:

$$v_x = -v_0 \sin 2\pi f t. \tag{14.4}$$

The velocity can be expressed in terms of ω as $v_x = -v_0 \sin \omega t$. The quantity v_0 is the amplitude (or maximum value) of the velocity.

As we saw in Chapter 5, any object moving in a circle with constant speed is accelerated radially toward the center. The projection of this acceleration along the x axis (Fig. 14.6b) is

$$a_x = -a_0 \cos \theta,$$

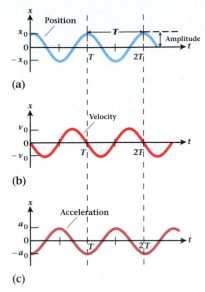

(a)

(b)

(c)

Figure 14.7

(a) The graph of x position as a function of time for the moving point of Fig. 14.6. This is also the graph of displacement as a function of time for the mass at the end of the spring of Fig. 14.3. The time interval between successive maxima is the period T. (b) Velocity along the x direction as a function of time. (c) Acceleration along the x direction as a function of time.

where a_0 is constant and positive and is the maximum acceleration of the point. The minus sign occurs because the acceleration vector is in the direction opposite to the position vector directed from the origin to the point P.

The acceleration in the x direction is a function of time. From Fig. 14.6(b) it is seen to be

$$a_x = -a_0 \cos 2\pi ft, \tag{14.5}$$

which can be expressed as $a_x = a_0 \cos 2\pi \dfrac{t}{T}$ or $a_x = a_0 \cos \omega t$.

Upon solving Eq. (14.3) for $\cos(2\pi ft)$ and substituting into Eq. (14.5), we have

$$a_x = -\frac{a_0}{x_0}x. \tag{14.6}$$

This equation says that for the oscillating projection of uniform circular motion on the x axis, the acceleration is proportional to the negative of the displacement. This relationship between acceleration and displacement is the same one that we had for Hooke's law. Therefore Eqs. (14.3) and (14.5) also describe the position and acceleration of a mass hung from a spring.

When the position of a harmonic oscillator is plotted as a function of time using Eq. (14.3), we obtain the graph in Fig. 14.7(a). The time required for the system to go through one complete oscillation is the period T. This is illustrated in the figure as the interval between two successive crests. More generally, the period is the interval between any two successive points along the curve that have the same magnitude and slope. In the earlier example of the toy-train engine, the time for the engine to make one complete revolution—the period—is the same as the time for one complete back-and-forth oscillation of the engine's shadow. The amplitude corresponds to the radius of the circle in circular motion, in this case the radius of the track.

If a block attached to a spring that obeys Hooke's law is displaced by an initial amount x_0 and released from rest, its motion is described by the curves in Fig. 14.7. We release it at time $t = 0$ with no initial velocity, but with an initial

Figure 14.8

Relationship between velocity and acceleration for a mass moving with simple harmonic motion at the end of a spring. The velocity is zero when the acceleration is maximum, and vice versa.

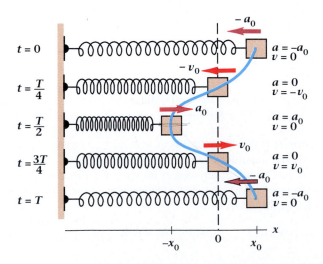

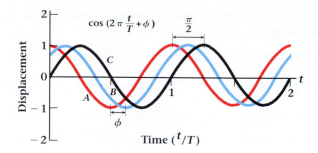

Figure 14.9

Displacement curves for a harmonic oscillator corresponding to different values of the phase angle ϕ: curve A, $\phi = 0$; curve B, $\phi = \phi$; and curve C, $\phi = \pi/2$. The amplitude $x_0 = 1$.

displacement and an initial acceleration opposite in direction to the displacement (Fig. 14.8). The block moves toward the equilibrium position $x = 0$, gaining speed as it moves. As it reaches the position of zero displacement, its momentum keeps it moving, even though the restoring force is zero at that point. The block is then displaced in the opposite direction from its initial displacement. A restoring force proportional to this new displacement gradually slows the block to a stop and then accelerates it back toward the initial position. In this way the block oscillates periodically with time. If there were no frictional forces to slow (damp) the motion, the block would oscillate indefinitely. We call this system a **simple harmonic oscillator,** and its motion is sinusoidal.

In the previous paragraphs, we described the motion by $x = x_0 \cos 2\pi ft$ and showed this motion in a number of graphs. In all of these cases we assumed that the displacement was x_0 at the time we called $t = 0$. However, the important general property of simple harmonic motion is not the displacement at time $t = 0$, but the shape of the curves that describe displacement, velocity, and acceleration. Notice that all of the following equations have the same behavior as a function of time but differ in magnitude at the time we choose to call $t = 0$ (Fig. 14.9):

$$x = x_0 \cos(2\pi ft + \phi)$$

$$x = x_0 \cos 2\pi ft \qquad\qquad (\phi = 0)$$

$$x = x_0 \cos\left(2\pi ft - \frac{\pi}{2}\right) = x_0 \sin 2\pi ft \qquad \left(\phi = -\frac{\pi}{2}\right).$$

The angle ϕ is called the **phase angle** and can be selected so that our choice of starting time and initial displacement are in agreement. The phase angle will be important in our later discussions of waves and electrical circuits.

Example 14.3

A metal block is hung from a spring that obeys Hooke's law. When the block is pulled down 12 cm from the equilibrium position and released from rest, it oscillates with a period of 0.75 s, passing through the equilibrium position with a speed of 1.3 m/s. (a) What is the displacement and (b) what is the speed of the block 0.28 s after it is released?

Strategy We must first choose a coordinate system. The initial displacement is downward, so we choose down for the positive values of x and up for negative values of x. Then a positive value for the velocity means motion in the positive, or downward, direction. Because the block starts from rest, the initial position is the maximum displacement. The motion corresponds to curve A in Fig. 14.9 with a phase angle of 0.

Solution (a) We can find the displacement by direct substitution into Eq. (14.3),

$$x = x_0 \cos 2\pi ft = x_0 \cos 2\pi \frac{t}{T} = (12 \text{ cm})\cos 2\pi \frac{0.28 \text{ s}}{0.75 \text{ s}} = (12 \text{ cm})\cos(2.35 \text{ rad})$$

$$x = (12 \text{ cm})(-0.700) = -8.4 \text{ cm}.$$

At $t = 0.28$ s, the block is 8.4 cm above the equilibrium position.
(b) Substitution into Eq. (14.4) gives the velocity,

$$v_x = -v_0 \sin 2\pi ft = -v_0 \sin 2\pi \frac{t}{T} = -(1.3 \text{ m/s})\sin 2\pi \frac{0.28 \text{ s}}{0.75 \text{ s}}$$

$$v_x = -(1.3 \text{ m/s}) \sin (2.35 \text{ rad}) = -(1.3 \text{ m/s})(0.711) = -0.92 \text{ m/s}.$$

The velocity is in the upward direction.

Discussion We could have chosen the coordinate system so that the initial displacement was negative. In that case, the signs of our answers would be different, but the physical description of what happens would be the same. For instance, we would still say that the displacement at 0.28 s is 8.4 cm above the equilibrium point.

From the preceding discussion it seems reasonable that a block on a spring oscillates in a sinusoidal manner. However, we want to know if any other type of motion is possible when the acceleration is proportional to the negative of the displacement. The answer is no, for the following reason. The velocity is the time rate of change of the displacement, and the acceleration is the time rate of change of the velocity. That is, the acceleration is given by the slope of the velocity curve, which in turn is given by the slope of the displacement curve. Curves of oscillations other than sinusoidal all give the result that the acceleration is not proportional to the displacement. We show this result explicitly for the waveform of Fig. 14.10. The solid curve in Fig. 14.10(a) departs only slightly from a cosine curve shown by the dashed line. The acceleration calculated from this displacement curve bears no resemblance to the initial curve. However, the acceleration curve of Fig. 14.7(c) is just the negative of the displacement curve of Fig. 14.7(a). Sinusoidal motion is indeed special.

14.3 Energy of a Harmonic Oscillator

We must do work to stretch a spring. As we saw in Chapter 6, the increase in the potential energy of a spring that obeys Hooke's law, due to extending the spring from $x = 0$ to $x = x_0$, is

$$PE = \tfrac{1}{2}kx_0^2. \tag{14.7}$$

Suppose a block of mass m is attached to the end of a spring of negligible mass. When the spring is extended to a distance x_0 and released from rest, its po-

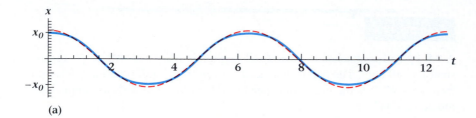

(a)

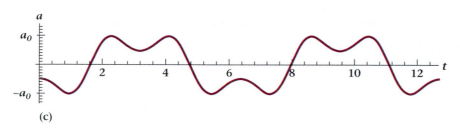

(b)

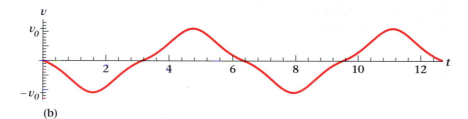

(c)

Figure 14.10
(a) The solid curve is a displacement that differs only slightly from a cosine curve (dashed line). (b) Velocity curve representing the slope of the displacement. (c) The acceleration curve is the slope of the velocity. Clearly the acceleration is not proportional to the displacement.

tential energy changes to kinetic energy as it moves toward the equilibrium position. As the block passes through $x = 0$, all its energy is kinetic and the magnitude of its velocity is a maximum. By applying the principle of energy conservation, we can find the relation between the maximum displacement x_0 and the maximum velocity v_0:

$$\tfrac{1}{2}mv_0^2 = \tfrac{1}{2}kx_0^2,$$

or

$$v_0 = x_0\sqrt{\frac{k}{m}}.$$

Note that the maximum velocity and maximum displacement do not occur at the same time or the same place. The velocity reaches a maximum v_0 when $x = 0$. Similarly, the displacement reaches a maximum x_0 when the velocity is zero.

The total energy is the sum of the kinetic and potential energies at any time and is a constant if we neglect dissipative (internal frictional) forces in the spring. We can write the total energy as

$$E = \tfrac{1}{2}mv^2 + \tfrac{1}{2}kx^2,$$

where the instantaneous values of displacement and velocity are given by Eqs. (14.3) and (14.6). We can use the energy relationship to show that the velocity at any displacement x is given by

$$v = \pm\sqrt{\frac{k}{m}(x_0^2 - x^2)}.$$

Example 14.4

A 3.0-kg ball is attached to a spring of negligible mass and with a spring constant $k = 40$ N/m. The ball is displaced 0.10 m from equilibrium and released from rest. What is the maximum speed of the ball as it undergoes simple harmonic motion?

Solution We can compute the maximum speed from the energy of the system. The maximum speed occurs at $x = 0$, when the kinetic energy is maximum and is equal to the initial potential energy. Then

$$\tfrac{1}{2}mv_0^2 = \tfrac{1}{2}kx_0^2.$$

So

$$v_0 = x_0\sqrt{\frac{k}{m}} = (0.10 \text{ m})\sqrt{\frac{40 \text{ N/m}}{3.0 \text{ kg}}} = 0.37 \text{ m/s.}$$

14.4 Period of a Harmonic Oscillator

We have seen that the velocity of a harmonic oscillator depends on the spring constant and on the mass. Therefore, the period, or the time required to complete one cycle of the motion, also depends on them. We can obtain the exact relationship by analyzing the circular motion of the particle in Fig. 14.6. The particle moves with constant speed v_0. In one period T, it traverses a circular path of length $2\pi x_0$; that is,

$$v_0 T = 2\pi x_0.$$

The period is then

$$T = \frac{2\pi x_0}{v_0}.$$

We have already seen that $v_0 = x_0\sqrt{\dfrac{k}{m}}$, so the period may be expressed as

$$T = 2\pi\sqrt{\frac{m}{k}}. \tag{14.8}$$

To show that this equation is consistent with our intuition, we have plotted the motion of several spring systems, each of which contains the same oscillator plus two others for which one of the parameters has been changed. Figure 14.11(a) shows the effect of displacing the block by different initial amounts. The amplitude of the oscillations is larger for larger initial displacements, but because m and k are unchanged, the period is the same. The period of the system is completely determined by the mass m and the spring constant k; how far it moves is influenced by what you do to it. In Fig. 14.11(b) we have used the same spring constant and the same initial displacement from the equilibrium position, but we have increased the mass for succeeding curves. As you would expect, the more massive systems are slower to respond and have longer periods. Figure 14.11(c) corresponds to springs of increasing k, or increasing stiffness. As expected, the stiffer the spring, the shorter the period. These curves emphasize that to change the period, you have to change the physical characteristics of the system—that is, m or k. You cannot vary the period by simply pulling the mass down farther before you release it. The period of a freely oscillating system depends on the properties of the system and is independent of the way in which the oscillation is initiated. This period is called the **natural period** of the oscillator.

As we saw in Chapter 5, the reciprocal of the period is the frequency f, the number of complete cycles per unit time:

$$f = \frac{1}{T}. \tag{14.9}$$

Recall that the dimension of frequency is inverse time and its unit is the hertz (Hz). The frequency associated with the natural period is the **natural frequency** of an oscillator. It is the frequency of an oscillator that has been set into motion and then left to oscillate freely. Examples include masses on springs, pendulums, and guitar strings. We will see in Section 14.7 that the natural frequency is the frequency at which energy is most easily fed into an oscillating system.

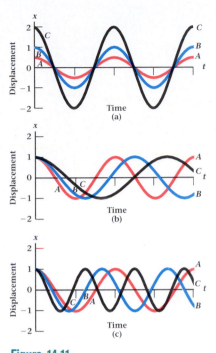

Figure 14.11
Displacement plotted against time for a simple harmonic oscillator. (a) The values of k and m are held constant while x_0 increases from curve A to B to C. (b) The values of k and x_0 are held constant while m increases from curve A to B to C. (c) The values of x_0 and m are held constant while k increases from curve A to B to C.

Master the Concept

Frequency of a Harmonic Oscillator

Question: A mass supported by a spring will oscillate when displaced from its equilibrium position and allowed to move freely. Why does a mass supported by a stiff spring oscillate with a higher frequency than does an equal mass supported by a weak spring? Ignore the mass of the springs.

Answer: When the masses on the two springs are displaced by the same amount, the stiffer spring exerts a greater force on the mass than does the weaker spring. Thus, the acceleration caused by the stiffer spring is greater than that caused by the weaker one. The mass on the stiffer spring moves more quickly between maximum and minimum displacements than does the mass on the weaker spring. As a result, the mass on the stiffer spring has the shorter period and the greater frequency.

Question: What happens if the initial amplitudes are different?

Answer: As long as we stay within the range of motions where the springs obey Hooke's law, the frequency (and the period) does not depend on the amplitude. So even if we start the two oscillators with different initial amplitudes, the mass on the stiffer spring will oscillate with the greater frequency.

Example 14.5

A block of 2.4 kg oscillates on a spring of constant $k = 26$ N/m with an amplitude of 17 cm. (a) What is the period of one complete oscillation? (b) What is the natural frequency of the oscillations? (c) What is the maximum speed of the oscillating block? (d) What is the speed of the block 0.21 s after it passes through its maximum position?

Strategy (a) Because we are given the spring constant and the mass, we can find the natural period from Eq. (14.8).

(b) The natural frequency is the reciprocal of the natural period found in part (a).

(c) The maximum speed v_0 may be found from the equation that relates maximum speed to maximum displacement: $v_0 = x_0\sqrt{k/m}$.

(d) Once we have both T and x_0, we can find the displacement at time t by using Eq. (14.3), remembering that the argument of the trigonometric function is in radians.

Solution (a) The period of oscillation is given by Eq. (14.8):

$$T = 2\pi\sqrt{\frac{m}{k}} = 2\pi\sqrt{\frac{2.4 \text{ kg}}{26 \text{ N/m}}} = 1.91 \text{ s} \approx 1.9 \text{ s}.$$

(b) The frequency is the reciprocal of the period:

$$f = \frac{1}{T} = \frac{1}{1.91 \text{ s}} = 0.52 \text{ Hz}.$$

(c) When we substitute the values for the amplitude x_0, the spring constant k, and the mass m, we have

$$v_0 = x_0\sqrt{\frac{k}{m}} = (0.17 \text{ m})\sqrt{\frac{26 \text{ N/m}}{2.4 \text{ kg}}} = 0.56 \text{ m/s}.$$

(d) We can now substitute the values found in parts (a) and (c) directly into Eq. (14.6) to give

$$|v_x| = \left|-v_0 \sin 2\pi\frac{t}{T}\right| = (0.56 \text{ m/s})\sin 2\pi\frac{0.21 \text{ s}}{1.91 \text{ s}} = 0.36 \text{ m/s}.$$

Discussion In part (c), we could have obtained the maximum speed from knowledge of the amplitude and the period. The maximum speed v_0 is related to the amplitude x_0 through the expression for uniform circular motion used in the beginning of this section,

$$v_0 = \frac{2\pi x_0}{T} = \frac{2\pi(0.17 \text{ m})}{1.91 \text{ s}} = 0.56 \text{ m/s}.$$

Notice that this result is the same as the result we found in part (c).

Example 14.6

The atomic force microscope consists of a sharp tip mounted on a soft cantilever spring (Fig. 14.12), a sensor that detects deflection of the spring, and a mechanical scanning system that moves the tip in a controlled path. If the

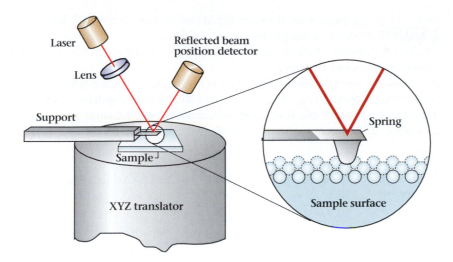

Figure 14.12
Example 14.6: Schematic of an atomic force microscope. A sharp tip is fastened to a cantilever spring having a very small spring constant. The microscope can be made to detect atomic-size displacements.

spring constant of the microscope spring is smaller than the equivalent spring constant between atoms in a solid, then the microscope can resolve positions comparable with atomic sizes without deforming the surface that is being examined. From knowledge that the vibrational frequencies of atoms in solids are of order 10^{12} Hz or higher and that the mass of an atom is of order 10^{-25} kg, calculate the effective interatomic spring constant.

Solution The relationship between the vibrational frequency and the mass is obtained from Eq. (14.8) as

$$f = \frac{1}{T} = \frac{1}{2\pi}\sqrt{\frac{k}{m}}.$$

Upon squaring both sides and rearranging, we find

$$k = (2\pi f)^2 m.$$

Inserting the approximate numerical values gives

$$k \approx (2\pi \times 10^{12}\ \text{s}^{-1})^2\ 10^{-25}\ \text{kg} \approx 4\ \text{N/m}.$$

Discussion The value we have just calculated is an upper limit to the spring constant that can be used in an atomic force microscope. Working microscopes typically use spring constants in the range of 0.01 to 1 N/m. For comparison, a spring constant of that size can be obtained from a piece of ordinary aluminum foil 1 mm wide and 5 mm long.

Atomic force microscopes are capable of giving three-dimensional profiles with nanometer resolution. Figure 14.13 is an atomic force micrograph showing the surface of a stamping die used in the manufacturing of compact disks.

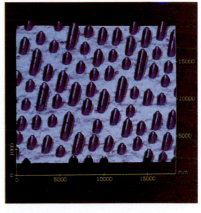

Figure 14.13
Example 14.6: Atomic force micrograph of a stamping die used to make compact disks. The audio information is encoded in bumps that are 60 nm high and arranged in tracks 1.6 μm apart.

14.5   The Simple Pendulum

A **simple pendulum** consists of a mass suspended by a light string of constant length L attached to a rigid support. When the mass, also called a bob, is pulled to one side and released, gravity causes it to swing down through the equilibrium

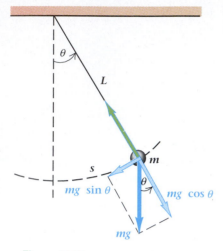

Figure 14.14

A simple pendulum consisting of a mass m at the end of a string of length L. The restoring force is $mg \sin \theta$.

point and up the other side. If the amplitude of its motion is small, the simple pendulum behaves as a simple harmonic oscillator. This is approximately the case in a grandfather clock, for example. Let us examine the pendulum and find out what determines its period.

When the bob of a simple pendulum is displaced to the side through an angle θ with the vertical (Fig. 14.14), the gravitational force $m\mathbf{g}$ has a component along the direction of the string and a component perpendicular to the string. This perpendicular component provides the restoring force F:

$$F = -mg \sin \theta.$$

When the angle θ is sufficiently small, $\sin \theta$ may be approximated by the angle θ in radians, so that*

$$F = -mg\theta.$$

The angle θ is related to the displacement s along the arc through which the pendulum swings by $s = L\theta$, giving

$$F = -\left(\frac{mg}{L}\right)s. \tag{14.10}$$

This equation has the same form as the spring equation (14.1) except that here the spring constant k is replaced by mg/L. Thus, the simple pendulum is a harmonic oscillator. The period of the pendulum is given by Eq. (14.8) with the substitution of mg/L for k. Thus,

$$T = 2\pi \sqrt{\frac{m}{k}} = 2\pi \sqrt{\frac{m}{mg/L}}$$

$$\boxed{T = 2\pi \sqrt{\frac{L}{g}}.} \tag{14.11}$$

This is an interesting result. The period of the simple pendulum is independent of the mass of the pendulum bob. The period is also independent of the amplitude of the motion (provided it is relatively small). However, the period of the pendulum does depend on its length.

The constancy of the period of a pendulum was discovered in the sixteenth century by Galileo. An often told story is that he made the discovery while observing that the periods of the chandeliers swinging in the Cathedral of Pisa were independent of their amplitudes. Although the story is instructive, it is almost certainly untrue. We do not know how Galileo made his discovery. Nevertheless, Galileo's conclusions led to the invention by Christiaan Huygens of the pendulum clock, which became the standard timekeeper for nearly 300 years. Pendulum clocks have a weight near the end of the pendulum that is used to adjust the period. Lowering the weight increases the effective length of the pendulum to make the period longer.

---

*How much error do we introduce by replacing the sine of an angle with the angle in radians for an angle of 10°? The sine of 10° is 0.1736. The angle measured in radians is 0.1745, just slightly larger. The relative error is $\dfrac{\sin \theta - \theta}{\sin \theta} = \dfrac{-0.0009}{0.1736} \approx -5 \times 10^{-3}$. That is, the error is about one-half of one percent for $\theta = 10°$. For smaller angles, the relative error is even less.

Notice that when the period and length of a pendulum are well known, Eq. (14.11) provides a method for measuring the gravitational acceleration *g*. It is also appropriate to note that this simple result for the pendulum is valid only for small-amplitude oscillations. At larger amplitudes, the period is no longer independent of the amplitude and the motion is no longer simple harmonic, though it is still periodic.

Master the Concept

Period of a Pendulum

Question: The pendulum on the clock that strikes the hours on the bell Big Ben in London was particularly noted for its accuracy. Formerly the period was regulated by placing a penny on top of the bob of the 3.98-m-long pendulum. Will placing a penny there make the clock run faster or slower?

Answer: We can approximate the clock pendulum as a simple pendulum with a bob centered at 3.98 m below the pivot axis. If a penny is added to the top of the bob, it does two things; it increases the mass of the bob and it raises the center of mass of the bob slightly. As we have seen, the period of the pendulum is independent of its mass, but it does depend on its length. So raising the center of mass has the effect of shortening the length of the pendulum and thus reducing its period. Hence, adding the penny makes the clock run faster.

Example 14.7

A grandfather clock has a pendulum 1.0 m long. What is its period?

Solution The period is given by Eq. (14.11):

$$T = 2\pi\sqrt{\frac{L}{g}} = 2\pi\sqrt{\frac{1.0 \text{ m}}{9.81 \text{ m/s}^2}} = 2.0 \text{ s}.$$

This length pendulum is frequently used in grandfather clocks. It is sometimes called a "seconds pendulum" because it makes a swing from one side to the other in one second, making the "tick" and the "tock" one second apart.

*14.6   Damped Harmonic Motion

Up to now, we have not considered the effects of friction on harmonic oscillators. Ignoring friction simplified the introduction to the topic and, as we have seen in both examples and problems, is justified because our results closely describe what actually happens when the frictional forces are weak or when we do not consider many oscillations. Yet friction is always present. Friction causes the amplitude of any real oscillating spring or pendulum to slowly decrease until the

WALKING AND RUNNING

To get some idea of the fundamental mechanical principles of walking and running, let us approximate the leg by a long thin rod of uniform cross section. We could more closely approximate a real leg with a more complicated model, but the nature of our conclusions would not change.*

Figure B14.1 shows a rod of length L supported at its upper end (point O) and free to swing. The period of a freely swinging rod supported at its upper end is

$$T = 2\pi\sqrt{\frac{2L}{3g}}.$$

This expression depends on the length of the rod and the acceleration of gravity in the same way as does the period of a simple pendulum. The factor of 2/3 is a result of the mass being distributed uniformly along the rod, rather than being concentrated at its lower end.

This expression gives an approximate value for the period of a freely swinging leg. You may wish to satisfy yourself that it is a reasonable approximation by doing the following simple experiment. Stand up and swing your leg back and forth as freely as you can. Do not use muscular effort to hold it back or to swing it rapidly. After you feel you can swing your leg freely, count the number of complete swings in ten seconds and calculate the period. Compare the period you observe with the prediction from the equation. To determine the length L, measure your leg length from the hip socket.

We can use this model to estimate a person's natural gait. Let us assume that a natural gait is the one involving the least muscular effort — the gait with the period found above. To a first approximation, we assume that the length of a stride is proportional to the length of the leg. The time for a single stride is one-half the period given above. The walking speed v thus depends on the leg length:

$$v_{walk} \propto \frac{L}{T/2} \propto \sqrt{L}.$$

This equation predicts that people with longer legs have a more rapid natural walking gait. The prediction is made on the basis of a model that assumes minimum energy expenditure and utilizes an oversimplified description of the leg. However, the prediction is borne out by common experience.

When a person runs, an important change must be made in our model. During running, the leg does not swing freely but is subject to a torque about O. The torque is a result of the force F applied by the muscles. This force is roughly proportional to the cross-sectional area of the muscles involved. If

we assume that, for people of different size, the relative proportions of the leg are the same, then the cross-sectional area, and therefore the force F, depends on the square of the length L. The torque is then proportional to the product of F with L:

$$\tau \propto FL \propto L^2 \cdot L \propto L^3.$$

The moment of inertia I is proportional to the mass times the square of the length (Chapter 9). Again we assume that all legs have essentially the same proportions; that is, width and thickness are proportional to length. Then the mass varies as the cube of the length:

$$I \propto mL^2 \propto L^5.$$

It can be shown that the period T of a long rod oscillating about one end and subject to a torque depends on the maximum torque and moment of inertia:

$$T \propto \sqrt{\frac{I}{\tau}}.$$

Upon substituting for I and τ, we find

$$T \propto \sqrt{\frac{L^5}{L^3}} \propto L.$$

The speed of running is proportional to the frequency of taking steps times the length of a step:

$$v_{run} \propto fL \propto \frac{L}{T} \propto \frac{L}{L} = 1.$$

So we have the prediction that, for animals with similarly shaped legs, the speed of running does not depend on the leg length. This prediction is not strictly true. However, the model does offer an explanation for the observation that the ordinary walking rate of people with long legs is greater than that of people with short legs, whereas the rate at which they can run is not always appreciably greater.

Figure B14.1 We can approximate the shape of a freely swinging leg by a uniform rod of length L that pivots about its upper end.

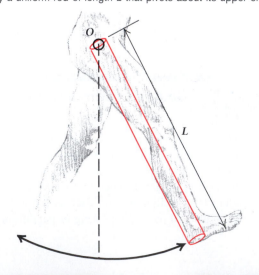

---

*This discussion of walking and running is adapted by permission of the author from P. Davidovits, *Physics in Biology and Medicine* (Englewood Cliffs, N.J.: Prentice-Hall, Inc., 1975).

oscillations eventually come to a stop. To be more complete, both Eq. (14.1) for the spring and Eq. (14.10) for the simple pendulum should have an additional term that describes the frictional force. This frictional force for an oscillator is often called **damping.**

Figure 14.15(a) shows an oscillator with damping provided by the resistance of the water in the jar to the motion of the vanes on the light rod attached to the bottom of the mass m. According to the principle of energy conservation, the amplitude of the oscillations of a damped system decreases as time goes by, unless energy is constantly supplied to replace the energy lost by friction. Thus, a spring oscillator that receives an initial displacement and is then left alone will gradually run down. For the particular system shown in Fig. 14.15, observations show that the damping is approximately proportional to the velocity of the vanes in the water. Including such a term in Eq. (14.1) leads to the result that if the mass is lifted up a distance y_0 and released, the displacement as a function of time is approximately

$$y \approx y_0 e^{-\gamma t} \cos(2\pi ft), \qquad (14.12)$$

where γ depends on the frictional force and the mass. If the damping is not too large, f is approximately the same as the natural frequency of the system.* The motion is shown by the blue curve in Fig. 14.15(b). We can look on this behavior as an oscillation at almost the natural frequency of the system but with an amplitude that decreases exponentially with time. This is shown by the gray lines in Fig. 14.15(b), which graphs the factor $e^{-\gamma t}$. Recall that the natural frequency is the frequency of the freely oscillating system.

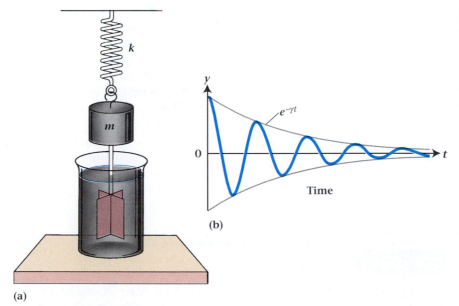

Figure 14.15
(a) A simple harmonic oscillator with damping provided by viscous (frictional) forces. (b) Displacement of an underdamped harmonic oscillator as a function of time. The amplitude of the motion decreases exponentially with time.

*The frequency f is related to the natural frequency f_0 by $f^2 = f_0^2 - (\gamma/2\pi)^2$. The complete expression for the displacement under the conditions just described is
$y = y_0 e^{-\gamma t}[\cos 2\pi ft + (\gamma/\omega) \sin 2\pi ft]$. For lightly damped systems this last term is small.

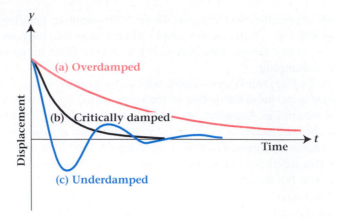

Figure 14.16

The displacement of (a) an overdamped harmonic oscillator, (b) a critically damped oscillator, and (c) an underdamped oscillator as a function of time.

Though not all harmonic oscillators have damping that is proportional to the velocity, Eq. (14.12) closely represents what happens in many real physical situations. As an extreme example, if the friction is constant but independent of velocity, a graph of the motion will be similar to Fig. 14.15(b), but the curves that form the upper and lower boundaries of the oscillatory motion will be replaced by straight lines.

If the harmonic oscillator is a pendulum, rather than a spring, Eq. (14.12) still describes the motion, but the linear displacement x is now replaced by the angular displacement of the pendulum. The same basic mathematical form holds for other damped oscillatory systems, whether they are mechanical, electrical, or acoustical systems.

Equation (14.12) and Fig. 14.15 describe the case of relatively weak damping and correspond to what is called *underdamped* motion. If the frictional force is too strong, the oscillator will not move at all. For damping that is strong, but yet allows the system to move, a related equation describes the motion that is called *overdamped,* shown by curve (a) in Fig. 14.16. A particularly important case lies between the underdamped and overdamped cases. This case is called *critical damping* and corresponds to the situation in which the damping coefficient γ is equal to the natural frequency of the system. In this case, a damped spring that is displaced and released returns as quickly as possible to the equilibrium position without overshooting (Fig. 14.16b). If we apply a constant force to a critically damped system, it moves to a new equilibrium position in the minimum time with no overshoot. The design of pointers on meters, shock absorbers, hydraulic and pneumatic door closers, and other practical devices is based on the proper design of damped harmonic oscillators.

Example 14.8

A swing with 2.50-m ropes is pulled aside and let go. After 20 oscillations, the maximum amplitude is one-half of its original value. Assume that the swing behaves like the damped harmonic oscillator described in Eq. (14.12). (a) What is the value of γ? (b) Approximately what amplitude, relative to its original amplitude, will the swing have after 35 oscillations?

Solution (a) We assume that the swing oscillates at its natural frequency, so that the period T is

$$T = 2\pi\sqrt{\frac{L}{g}} = 2\pi\sqrt{\frac{2.50 \text{ m}}{9.81 \text{ m/s}^2}} = 3.172 \text{ s}.$$

Twenty oscillations will then take

$$20T = 63.44 \text{ s}.$$

Because the amplitude decreases exponentially to 0.5 of the original amplitude in time $t = 20T$, we can write

$$0.5 = e^{-\gamma t} = e^{-\gamma(63.44 \text{ s})}.$$

We may now determine γ by taking the natural logarithm of both sides of the equation. You can find the logarithm of 0.5 with a calculator:

$$\ln(0.5) = -0.6931.$$

The logarithm of the exponential is just the value of the exponent:

$$\ln(e^{-\gamma\, 63.44 \text{ s}}) = -\gamma\,(63.44 \text{ s}).$$

Thus,

$$0.6931 = \gamma\,(63.44 \text{ s}), \quad \text{and} \quad \gamma = 0.0109/\text{s}.$$

(b) Thirty-five oscillations will take

$$35T = 111.0 \text{ s},$$

so the amplitude, as a fraction f of the original amplitude, is

$$f = e^{-\gamma t} = e^{-(0.0109/\text{s})(111.0 \text{ s})} = 0.297.$$

After 35 oscillations, the swing will have about one-third of its original amplitude.

*14.7   Forced Harmonic Motion and Resonance

Before continuing, try the following simple experiment. Make an oscillator by joining several rubber bands together and hanging an unopened soft drink can from the end (Fig. 14.17). You will find it easy to attach the can to the rubber band by looping the bottom rubber band under the pull-up tab. You may try some other combination of elastic elements and mass; the idea is to make an oscillator with a natural period of about one-half to one second so that you can easily observe what happens. Get a rough idea of the natural frequency of the system by looping one end of the rubber band around the tip of a finger, pulling the can down, and observing what happens when you release it.

Next, suspend the system from your finger and move your finger up and down slowly. You will find that the can moves up and down in phase with your finger, and with about the same amplitude. Increase the rate of your finger's motion and observe the change that occurs as the frequency of your finger comes closer to the natural frequency of the system. The amplitude of the can's motion increases greatly, and at the natural frequency of the system, a very small motion of your finger makes the can move up and down with a large amplitude. At

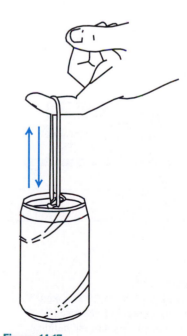

Figure 14.17
A soft drink can suspended from rubber bands can be used to demonstrate resonance.

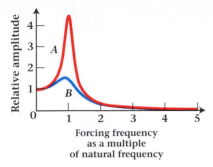

Figure 14.18
The response of a damped harmonic oscillator plotted as a function of the frequency of the driving force. Curve *B* corresponds to greater damping than does curve *A*.

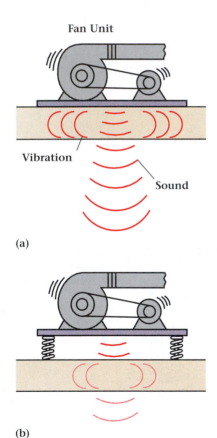

(a)

(b)

Figure 14.19
Example 14.9: (a) A fan bolted directly to the roof generates vibrations throughout the building. (b) Shock-mounting the fan on springs reduces the vibrations transmitted to the building, provided we choose the proper spring constants.

this point, energy is being fed into the system at the natural frequency of the system, a condition called **resonance.** The natural frequency is also known as the **resonant frequency.** As we will see in the next chapter, natural resonant systems usually have more than one resonant frequency.

As you continue to increase the frequency with which you move your finger up and down, you will find that the amplitude of the can's motion decreases. Eventually you will reach a frequency at which the can stays relatively motionless, even though you move your finger through quite a large distance.

What you have done is to observe the response of a lightly damped harmonic oscillator (the rubber bands and can) to an applied force as a function of the frequency of that force. Actually, rubber bands do not obey Hooke's law strictly, but the behavior of the system is close enough to a harmonic oscillator for us to study the general principles.

A graph of the amplitude of a forced damped harmonic oscillator as a function of the forcing frequency shows the behavior we have just described. In Fig. 14.18, two curves have been plotted, corresponding to different amounts of damping. As you would expect, large damping in the system causes the maximum amplitude at resonance to be lower than otherwise.

The idea of resonance is extremely important for mechanical and electrical systems because energy is most effectively transferred when it is supplied at the natural frequency of the system. When you push a swing to make it go higher, you push at its natural frequency. When you tune a radio or TV set, you are adjusting the natural, or resonant, frequency of the electrical circuit to match the frequency of the radio or TV station you want to receive. We will make additional use of the idea of resonance in Chapter 15 on waves.

Resonance effects can also have undesirable or even destructive effects. The rattle in a car's body or an annoying buzz in a stereo speaker is often due to resonance. Most people have heard that a powerful singer can shatter a glass by singing at the right frequency. Equally famous is the warning that a group of people should not march in step across a bridge, lest the frequency of the footsteps match some natural frequency of the bridge. These are all examples of resonance.

Frequently we need a system that does not transfer energy efficiently. An example is a mechanism for isolating sensitive apparatus from vibrations. A common solution to a problem of vibration is to fasten the source of vibration on an elastic mounting to cushion and absorb the shock. What may not be obvious is that the selection of the wrong elastic mounting can make the problem worse. Isolation is accomplished by decreasing the natural frequency of the system relative to the frequency of the vibration source. The reason this technique works can be seen from Fig. 14.18 and from recalling the experiment with the rubber band and the can. The least energy is transferred when the frequency of the driving force is much higher than the natural frequency of the system.

We have based our discussion of resonance, and the response of a system to forcing, entirely on the behavior of a mass attached to a spring that obeys Hooke's law. However, the same general principles and results apply to other oscillating systems, whether they be mechanical, electrical, or other.

Example 14.9

The fan unit for a building's heating and air-conditioning system is rigidly mounted on the roof and runs continuously (Fig. 14.19a). The vibrations are transmitted throughout the structure of the building and generate unacceptable vibration levels. To reduce the vibration felt below, the system is to be attached to a spring-mounted slab. The fan shaft turns at 1800 rpm (revolutions per minute), and the combined mass of the unit and the mounting slab (Fig. 14.19b) is 576 kg. What is the proper stiffness constant for the springs used to support the slab? Assume four springs, one at each corner.

Strategy The oscillation system in this case consists of the motor, the fan, the mounting platform, and the springs. One rule of thumb sometimes used is that the driving, or disturbing, frequency should be at least 3 times the natural frequency of the system. For many cases, a factor of 5 is adequate, and for critical conditions, a factor of 12 or more is appropriate. We can achieve these factors by lowering the natural frequency of the system. If we choose a one-to-five ratio, which corresponds to a reduction in the force of the vibrations to the building of about 96%, the desired natural frequency of the system is

$$\frac{1}{5}(1800 \text{ rpm})\left(\frac{1}{60 \text{ s/min}}\right) = 6.0 \text{ Hz.}$$

Solution The proper springs can be selected by using

$$f = \frac{1}{T} = \frac{1}{2\pi}\sqrt{\frac{k}{m}}.$$

Solving for the spring constant k gives us

$$k = m(2\pi f)^2 = (576 \text{ kg})[2\pi(6.0/\text{s})]^2 = 8.19 \times 10^5 \text{ N/m.}$$

This would be the largest desirable spring constant if all of the mass were supported by one spring. Since there are four springs, one at each corner of the mounting slab, each one of these four springs will have a spring constant, or stiffness, of $\frac{1}{4}(8.19 \times 10^5 \text{ N/m}) = 2.05 \times 10^5 \text{ N/m.}$

Summary

Useful Concepts

■ Many phenomena can be described by an equation such as Hooke's law,

$$F = -kx.$$

These phenomena are examples of simple harmonic motion and occur in electrical as well as mechanical systems.

■ For simple harmonic motion,

$$x = x_0 \cos 2\pi ft,$$

$$v = -v_0 \sin 2\pi ft = -\sqrt{\frac{k}{m}} x_0 \sin 2\pi ft,$$

$$a = -a_0 \cos 2\pi ft = -\frac{k}{m} x_0 \cos 2\pi ft.$$

■ The frequency is the reciprocal of the period,

$$f = \frac{1}{T}.$$

■ The potential energy of a stretched spring is

$$\text{PE} = \tfrac{1}{2}kx_0^2.$$

■ The period of a mass oscillating on a spring is

$$T = 2\pi\sqrt{\frac{m}{k}}.$$

■ The period of a simple pendulum is

$$T = 2\pi\sqrt{\frac{L}{g}}.$$

■ The displacement of a damped harmonic oscillator is the product of an exponentially decreasing term and an oscillating term:

$$y = y_0 e^{-\gamma t} \cos(2\pi ft).$$

■ Energy is transferred most effectively at the natural frequency of the system, a condition known as resonance.

Important Terms

You should be able to write the definition or meaning of each of the following:

periodic	phase angle
Hooke's law	natural period
restoring force	natural frequency
spring constant	simple pendulum
simple harmonic motion	damping
amplitude	resonance
simple harmonic oscillator	resonant frequency

▼
Conceptual Questions

14.1 Determine the effective spring constant of a bathroom scale by estimating the amount the scale compresses when you stand on it.

14.2 A glider on a horizontal air track is connected to the end of the track by a spring and placed on an elevator. What is the effect on the period of its oscillations when the elevator is (a) accelerated upward, (b) accelerated downward, (c) moving with a constant speed?

14.3 A hollow plastic sphere is held well below the surface of a swimming pool by a spring attached to the bottom. Do the vertical oscillations of the sphere show damped simple harmonic motion?

14.4 A lump of clay is dropped onto an upright spring. What determines the maximum compression of the spring? (*Hint:* Consider conservation of energy.)

14.5 Name as many examples as you can of harmonic motion found in nature.

14.6 Two metal blocks of different mass are suspended inside a large closed box by identical springs attached to the interior top of the box. When the box is at rest, the spring supporting the more massive block is longer than the other spring. Discuss what happens to the relative positions of the masses if the box is dropped from a great height.

14.7 A simple pendulum suspended from the ceiling of an elevator has a period T_0 when the elevator is not moving. What is the effect on the period when the elevator is (a) accelerated upward, (b) accelerated downward, (c) moving with a constant speed?

14.8 Consider the motion of a meterstick swinging from one end and that of a one-meter-long simple pendulum of the same mass. Are their periods the same? Give a physical reason for your answer.

14.9 You can consider an automobile suspension to be a forced damped harmonic oscillator, consisting of the springs and shock absorbers, which is forced by the wheels' up-and-down motion over bumps and ruts. What are some of the important design criteria? Should there be much or little damping?

14.10 If the outer case of an air conditioner rattles when the unit runs, what can you do that will probably make it stop?

14.11 Why is it that sometimes you hear rattles when you drive a car at one speed, but they go away when you drive faster or slower?

14.12 A cord and a spring with an equilibrium length equal to the length of the cord are hung from a frame inside a rapidly moving train. Equal masses are attached to the lower end of the cord and the spring. Which makes a greater angle with respect to the vertical when the train accelerates? Explain your answer.

▼
Problems

Hints for Solving Problems

Hooke's law relates the force to the displacement for springs and systems that behave linearly. In some cases, we are interested in the applied or external force for which the force and displacement are in the same direction. In other cases, we are interested in the restoring force due to the spring. The restoring force is opposite in direction to the displacement. You should draw a diagram for each situation carefully indicating the direction of the forces and the displacements.

Section 14.1 Hooke's Law

14.1 A 15,000-N Pontiac Grand Prix is supported by four coil springs, each with a spring constant of 7.00×10^4 N/m. By how much is each spring compressed under the weight of the car if the weight is evenly distributed?

14.2 Determine whether the following two sets of data were taken from springs that obey Hooke's law. In each case, objects of increasing mass were hung from a suspended spring and the length of the spring measured. If Hooke's law is obeyed, determine the spring constant. If not, decide whether there is any part of the range over which Hooke's law is obeyed, and determine the spring constant for that part.

Mass (kg)	Length of Spring A (cm)	Length of Spring B (cm)
0.0	15.7	8.4
1.0	16.5	15.6
2.0	17.8	20.5
3.0	19.3	21.4
4.0	20.4	21.6
5.0	21.3	22.1
6.0	22.8	22.5
7.0	24.1	22.6
8.0	25.0	23.1
9.0	26.6	25.8
10.0	27.6	28.2

14.3 A coil spring has a spring constant of 54 N/m. If the full length of the spring is 35 cm when a 1.0-kg mass is hung from it, what is the equilibrium length of the spring when the 1.0-kg mass is removed?

14.4 A coil spring is extended by 2.50 cm when it supports a 1.00-N weight. If an identical spring is joined to the end of the first one, what is the extension of the combined spring when it supports the same 1.00-N weight? Assume the spring masses to be negligible.

14.5● A fisherman's spring scale is extended to a total length of 0.18 m when a 6.12-kg fish is suspended from it. If the spring constant is 1000 N/m, what is the total length of the spring when an 11.4-kg fish is suspended from it?

14.6● Two identical coil springs are mounted side by side so that they jointly support a weight hanger of 2.00 N. When an additional 4.00 N is added to the hanger, it is displaced downward by 5.00 cm. What is the force constant of the individual springs?

14.7● Two springs have the same spring constant. One of them is 0.400 m long, and the other 0.250 m long. If they are connected end-to-end between rigid supports 1.300 m apart, where will the connection point of the two springs lie relative to the supported end of the short spring?

14.8● A spring gun consists of a horizontal spring with $k = 50$ N/m that is compressed 17 cm when the gun is cocked. The gun fires a 150-g projectile. What is the initial acceleration of the projectile?

14.9●● A block hung from the ceiling on a vertical spring stretches the spring by one-fourth of its equilibrium length. The spring is taken down and the free end placed over a peg on a horizontal frictionless surface. The block is set in motion in a circular orbit about the peg with a linear speed of 5.00 m/s. The motion of the block causes the spring to extend by exactly the same amount it did when the spring was in the vertical position. Is the equilibrium length of the spring longer or shorter than your arm?

14.10●● A spring has an equilibrium length of 25.6 cm and a spring constant of 52.9 N/m. The spring is connected to the underside of the roof of a car and a 0.264 kg block suspended

from it. (a) How long is the spring when the car is at rest? (b) How long is the spring if the car is accelerating horizontally at 1.73 m/s²?

Section 14.2 The Simple Harmonic Oscillator

Hints for Solving Problems

Simple harmonic motion occurs when the acceleration is proportional to the negative of the displacement. In simple harmonic motion, maximum acceleration occurs at maximum displacement and zero velocity.

14.11 The position of a harmonic oscillator is described by $x = x_0 \cos 2\pi t/T$, where the displacement amplitude is $x_0 = 10$ cm and the period T is 0.25 s. Calculate the displacement at $t = 0.75$ s, at $t = 2.0$ s, and at $t = 2.3$ s.

14.12 A block moving with simple harmonic motion has a period of 2.0 s. The maximum displacement from equilibrium in any direction is 5.0 cm. (a) Write an equation to describe the displacement, given that at time $t = 0$ the displacement is zero and the velocity is positive. (b) Write an equation for the displacement, given that at time $t = 0$ the displacement is a maximum and the velocity is zero.

14.13 A harmonic oscillator with an amplitude of 30 cm has a displacement of 30 cm at $t = 0$. At $t = 0.20$ s, it has a displacement of 27 cm, without having passed through zero displacement. What is the period of the oscillator's motion?

14.14● A can of beans hung on a vertical spring is pulled down from its equilibrium position and released. It travels up and then back down to the position from which it was released in 0.75 s, traveling a total distance of 0.75 m. How far from its release position was the can 0.12 s after it was released?

14.15● A harmonic oscillator has a period of 0.314 s and an amplitude of 7.0 cm. At $t = 0$ it is at $x = 0$. (a) How far does it travel between $t = 0$ and $t = 0.063$ s? (b) How far does it travel between $t = 0.283$ s and $t = 0.345$ s?

14.16● Show that for a harmonic oscillator $v_0 = 2\pi x_0/T$, where x_0 is the amplitude of the displacement, v_0 is the amplitude of the velocity, and T is the period. (*Hint:* Use Fig. 14.6.)

14.17● A 0.50-kg air-track glider is attached to the end of the track by a horizontal coil spring of $k = 20.0$ N/m. The glider is displaced 15.0 cm from its equilibrium position and released, so that it oscillates back and forth on the track. (a) What is the maximum acceleration of the glider? (b) What is its acceleration at a time equal to one-eighth of the oscillator's period? (c) What is its position at a time equal to one-eighth of the oscillator's period?

14.18●● Show that for a harmonic oscillator the displacement x and velocity v are related by

$$x = x_0\sqrt{1 - \left(\frac{v}{v_0}\right)^2},$$

where x_0 is the maximum displacement and v_0 is the maximum velocity.

Section 14.3 Energy of a Harmonic Oscillator

Hints for Solving Problems

Assume that all the potential energy of a harmonic oscillator at maximum displacement is converted to kinetic energy at the equilibrium position.

14.19 The spring described in Example 14.1 is stretched to 5.0 cm beyond its equilibrium length. How much work is required to stretch the spring?

14.20 The potential energy stored in the compressed spring of a toy dart gun is 0.37 J. The spring constant is 29 N/m. By how much is the spring compressed?

14.21 Two joules are required to extend a spring by 0.25 m. What is the spring constant?

14.22 A 4.0-kg block oscillates with an amplitude of 0.080 m on a light spring of constant $k = 15$ N/m. (a) What is the maximum value of the block's velocity? (b) What is its maximum acceleration? $a = |A\omega^2|$

14.23• A 0.50-kg plush toy oscillates at the end of a long light spring of constant $k = 10.0$ N/m and negligible mass. The maximum speed of the toy as it oscillates is 0.30 m/s. (a) What is the amplitude of the oscillation? (b) How much energy is stored in the spring at its greatest extension?

14.24• A physics student wants to build a spring gun that when aimed horizontally will give a 20-g ball an initial acceleration of g by releasing a spring that has been compressed 15 cm. (a) What spring constant is necessary? (b) What is the velocity of the ball as it emerges from the gun?

14.25•• By substituting for v and x as functions of time in the expression for the total energy of a harmonic oscillator, show that the total energy is conserved; that is, show that the total energy is independent of time.

Section 14.4 Period of a Harmonic Oscillator

Hints for Solving Problems

The period of a simple harmonic oscillator made with a spring depends only on the mass and the spring constant or, for a simple pendulum, on the length and the acceleration of gravity.

14.26 A 0.400-kg brass block is attached to a spring of negligible mass. What is the value of the spring constant if the vibrational frequency is 0.550 Hz?

14.27 A 0.500-kg puppet oscillates at the end of a light spring of constant $k = 2.00$ N/m. (a) What is the frequency of the oscillations? (b) What is the period?

14.28 A 2.0-kg fish is attached to a spring of constant $k = 56$ N/m and of negligible mass. The spring is stretched 8.0 cm from equilibrium and released from rest. (a) What is the maximum speed of the fish as it oscillates? (b) What is the frequency of the oscillation?

14.29• A circus performer bobs up and down at the end of a long elastic rope at a rate of once every two seconds. By how much is the rope extended beyond its unloaded length when the performer hangs at rest?

14.30• The prongs of a 440-Hz tuning fork vibrate with simple harmonic motion. A scratch on the end of one of the prongs travels through a total distance of 1.00 mm as the prong moves from one extreme position to the other. (a) What is the maximum speed of the scratch? (b) What is the maximum acceleration of the scratch?

14.31• An oscillator with a 0.46-s period is made from a mass suspended from a spring. The mass is placed on a frictionless surface that makes a 45° angle with the horizontal, and the spring is attached at the top of the incline (Fig. 14.20). What is the new period of the oscillator?

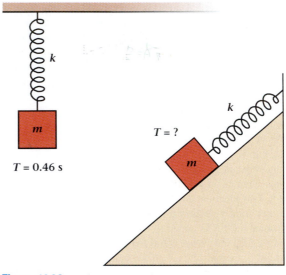

Figure 14.20
Problem 14.31.

14.32• One end of a light hacksaw blade is clamped in a vise with the long axis of the blade horizontal and with the sides vertical. A 0.845-kg mass is attached to the free end. When a steady sideways force of 20.5 N is applied to the mass, it moves aside 14.3 cm from its equilibrium position. When released, the blade moves back and forth with harmonic motion. Approximately how many complete oscillations will the blade make in 20 s when the force is removed?

14.33• A harmonic oscillator of frequency $f = 0.500$ Hz has an amplitude of 12.0 cm. (a) What is the maximum speed of its motion? (b) What is the maximum acceleration?

14.34• A 20.5 kg mass suspended on a spring is pulled down and released. The mass oscillates with a period of 0.750 s and an amplitude of 3.50 cm. What energy was initially stored in the spring before it was released?

14.35•• Plot on the same graph the potential energy, the kinetic energy, and the total energy of a harmonic oscillator as a function of time if $m = 0.10$ kg, $k = 20$ N/m, and $x_0 = 5.0$ cm. Plot for t between 0 and 0.6 s.

14.36•• A long 3.0-cm-diameter cylinder of ice is weighted at the bottom so that it floats upright in a container of water at 0°C. The mass of the cylinder with the weight is 300 g. When the cylinder is pushed down and released, it bobs up and down. Show that this motion is simple harmonic motion and find its frequency. (*Hint:* You will need to know the density of water at 0°C.)

14.37•• Two masses A and B freely slide in a cylinder (Fig. 14.21). The lower mass B is supported by a spring. When both A and B are in the cylinder, the period of the systems oscillation is 0.850 s. When the 2.00-kg mass A is removed, the period of the system is 0.350 s. What is the mass of B?

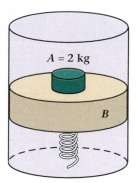

Figure 14.21
Problem 14.37.

Section 14.5 The Simple Pendulum $T = 2\pi\sqrt{L/g}$

14.38 A 21-kg child swings on a playground swing 3.4 m long. (a) What is the period of her motion? (b) If her older brother, who weighs twice as much as she does, rides in the swing instead, what is the period of the motion? Assume the center of mass to be at the position of the seat in both cases.

14.39 A 30-kg child is playing on a swing of negligible mass. The child swings so high that at the peak of the motion her center of mass is 1.0 m above where it is at the bottom of the swing. How fast is the child moving when she passes through the minimum point of her swing?

14.40 A simple pendulum makes 83 complete swings in one minute. (a) What is its frequency? (b) What is its period of oscillation?

14.41 A child swings on a playground swing attached to chains 4.0 m long. (a) Calculate the period of the swing for small-amplitude oscillations. (b) What would be the new period if the seat height were raised 1.0 m? Assume the center of mass to be at the position of the seat.

14.42 Until 1987, a 22.25-m-long pendulum hung in the Smithsonian Institute, National Museum of American History. (a) Calculate the period of the pendulum for small-amplitude oscillations. (b) What would be the new period if the pendulum were reinstalled with a length one-third of its original length?

14.43 An 800-kg wrecking ball hangs from the end of a crane by a 30.0-m cable. If the crane operator quickly moves the end of the crane 3.00 m to the left, how many seconds pass before the wrecking ball passes below the end of the crane?

14.44 Students in an elementary physics laboratory measured the period of a simple pendulum to be 2.475 s. Careful measurement showed the length of the pendulum to be 151.8 cm. What was the acceleration of gravity in their laboratory?

14.45• Tarzan is across the river from Jane, who is in danger of being blown up by a time bomb set for 6.9 s. Tarzan can save Jane only by swinging across the river on a 21-m vine hung directly over the center of the river. Because Tarzan is clinging to another vine, he can swing over only with the force of gravity and is unable to push off. (a) Will he arrive in time to save Jane? If so, by what margin? (b) Where are Tarzan and Jane when the bomb goes off?

14.46• The lengths of pendulums with periods of two seconds are given for several locations. Find the distance through which an object will fall in a half second at these locations, neglecting air resistance: (a) St. Thomas, 99.11 cm; (b) New York, 99.32 cm; (c) London, 99.41 cm; (d) Spitzbergen, 99.61 cm.

14.47• A rule sometimes used by clock makers is that a pendulum oscillating with V oscillations per minute has a length L in inches given by $L = (187.6/V)^2$. Verify this expression.

14.48•• A pendulum makes 135 complete oscillations in 3.00 min. When carried on a train rounding a curve at 77.4 km/h, the same pendulum makes 136 complete oscillations in 3.00 min. What is the radius of curvature of the track?

*Section 14.6 Damped Harmonic Motion

14.49• A 94-cm simple pendulum is pulled aside a small distance and released. After 120 oscillations the amplitude is one-half of its starting value. The damping is proportional to the speed of the pendulum bob. (a) What is the value of the damping coefficient γ if the behavior is described by Eq. (14.12)? (b) What fraction of the original amplitude will remain after 15 min?

14.50•• Graph the motion of a damped harmonic oscillator described by Eq. (14.12) for which $\gamma = 0.30$/s and $2\pi f = 2.0$/s. Extend your graph out to $t = 12$ s. You may want to draw the exponential curves first. Next mark the maximum, minimum, and zeros of the cosine function and then sketch the displacement curve.

14.51•• A 1.26-m pendulum is pulled aside and released. Air resistance gives a damping coefficient of $\gamma = 0.02$/s. The support wire makes an angle of 3.30° with the vertical direction 1.50 s after the release. What was the angle from the vertical of the original displacement?

*Section 14.7 Forced Harmonic Motion and Resonance

14.52 Repeat the experiment with the rubber bands and drink can described in Section 14.7. This time determine the percent difference between your observed frequency of finger motion at the resonant frequency and the frequency of the system calculated from the spring constant and the mass of the can. Because of the assumptions, you should expect only approximate agreement. Determine the spring constant by measuring the extension of the rubber bands when the can is hung from them. (Assume that Hooke's law holds, neglect the mass of the metal can compared with the mass of the liquid, and, if necessary, assume the density of soft drinks to be the same as the density of water.)

14.53• What should be the length of the pendulum in Fig. 14.22 to give the 1.0-kg air-track glider the maximum amplitude if the spring constant is $k = 120$ N/m?

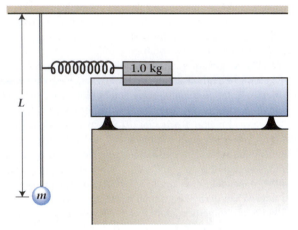

Figure 14.22
Problem 14.53.

14.54• A motor that turns at 12,000 rpm is suspended from above by a single spring to isolate the building from its vibration. If the natural frequency of the system is to be one-fourth of the vibration frequency, by how many centimeters should the spring extend beyond its unloaded length when the motor is hung on it?

14.55•• A 1550-kg Pontiac Grand Prix is supported by four coil springs, each with a spring constant of 7.00×10^4 N/m. (a) What is the natural oscillation frequency of this system? (b) The car bumps as it rolls over the tar strips when driven along a concrete highway. If the tar strips are 18.5 m apart, how fast is the car moving when the frequency of the bumps is resonant with the natural frequency?

Additional Problems

14.56 A 0.500-kg lead weight is attached to a spring of negligible mass. When the weight is set into motion, it oscillates with a period of 2.00 s. (a) What is the spring constant k? (b) What force is required to stretch the spring by 2.00 cm?

14.57 A clock pendulum has a period of 0.750 s. (a) How long is the pendulum? (b) How long must it be to have a period of 2.00 s?

14.58• Two 35-cm springs have different spring constants. If the springs are connected end to end and the outer ends are connected to supports that are 100 cm apart, the connection point is 45 cm from the nearest support. What is the ratio of the spring constants?

14.59• When a physics textbook is hung from a spring attached to the ceiling, the spring stretches 4.0 cm. The book is next hung from three springs exactly like the first one (Fig. 14.23). What is the total extension of the spring system?

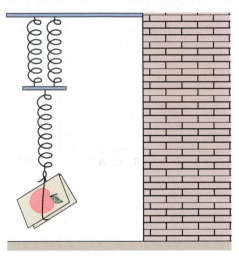

Figure 14.23
Problem 14.59.

14.60• A child's pail is filled with sand and hung from a vertical spring, extending the spring 14 cm. If the pail is pulled down an additional 28 cm and released from rest, what is the pail's initial upward acceleration?

14.61• When a box of cookies is hung on the end of a spring, the spring stretches by 5.90 cm. What is the oscillation frequency of the system if the box is pulled down and released?

14.62• A solid plastic block lies on a smooth floor. It is connected to the wall by a long horizontal coil spring with a spring constant of 86 N/m. If the block is pulled aside 37 cm and released, how much heat will have been generated before the block comes to rest?

14.63•• When a 3.00-kg cube of metal is hung from a spring, the spring stretches by 7.50 cm. The same block is placed on a frictionless horizontal surface and the spring connected to a support at the same level as the block. The block is pulled 3.00 cm away from its equilibrium position and released. What is the initial acceleration of the block?

14.64•• A person whirls a 0.250-kg stone in a 1.00-m-radius horizontal circle at the end of a string. The horizontal component of the force with which the person pulls on the string is 5.00 N. Write expressions for the projections of the displacement, velocity, and acceleration on a diameter of the circle.

14.65•• It is claimed that if a hole could be bored straight through the earth and if an object were dropped into the hole, the object would oscillate back and forth along this diameter with simple harmonic motion. Test this claim by assuming that the earth is of uniform density, getting an expression for the force on the object, and comparing the expression with Eq. (14.1). (*Hint:* A small object at a distance R from the center of a mass with spherical symmetry will experience a gravitational force toward the center that depends only on the mass inside an imaginary sphere of radius R. This is true whether the object is outside or inside the spherical mass.)

14.66•• A mass m is attached to a vertical spring and released. The mass oscillates but eventually comes to rest a distance h below its initial position when the spring was unstretched. Find expressions for the change in gravitational potential energy of the mass and for the energy stored in the spring as a function of m, h, and the acceleration of gravity. Explain why the two expressions are or are not the same.

14.67•• A simple pendulum is in an upward accelerating elevator. A student says that the 0.500-m-long pendulum has a period of 1.40 s. Another student in the same elevator finds that an object falls from rest through 1.00 m in 0.450 s. Given that both experimental results are subject to ± 2.00% accuracy, are the students' results in agreement?

14.68•• A clock pendulum that "ticks seconds" (has a period of two seconds) on earth is set up on the moon. What is the pendulum's period on the moon?

14.69•• A pendulum bob on a string of length L is arranged as shown in Fig. 14.24. The point marked P is a smooth peg and is at a distance $L/2$ below the support. Write an expression for the period for small oscillations of this system.

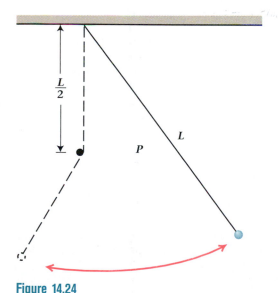

Figure 14.24
Problem 14.69.

14.70•• A mass is suspended from a fixed point by a light cord of length L. The mass is set in motion in a horizontal cir-

cle of radius R (Fig. 14.25). Such an arrangement is called a conical pendulum because the moving cord sweeps out the surface of an inverted cone. (a) Show that the frequency f of a conical pendulum is given by

$$f = \frac{\sqrt{g}}{2\pi(L^2 - R^2)^{1/4}}.$$

(b) Show that for L much greater than R the period of a simple pendulum is approximately the same as that of a conical pendulum of the same length.

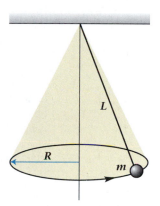

Figure 14.25
Problem 14.70.

14.71•• A U-shaped tube is partially filled with water. (a) Show that if the water is displaced from equilibrium, it executes simple harmonic motion. (b) Find an expression for the period of the oscillations. (Let the diameter of the tube be d and the total length of the water column be L.)

14.72•• Assume that the length of the water column in a very long upright U-shaped tube is 50 cm. If the damping coefficient γ is 0.30/s, how many oscillations must occur for the amplitude to decrease to one-fourth of its original value?

14.73•• A swing is hung from the limb of a tree. When a young child gets into the swing, it settles 2.0 cm. Because of the flexibility of the tree limb, the swing bobs up and down as well as swinging back and forth. When the swing is gently bobbed up and down with no swinging motion, the frequency is 10 times the frequency of the swing when it is gently swung so as not to excite bobbing motion. How long is the swing?

14.74•• The apparatus shown in Fig. 14.26 consists of a mass m attached to the end of a light rod (negligible mass) of length L and hinged at one end. A spring of spring constant k is joined to the rod a distance a from the hinged end. Show that the frequency of oscillation of this apparatus is

$$f = \frac{a}{2\pi L}\sqrt{\frac{k}{m}}$$

by first showing that it is equivalent to a mass hanging from a spring of constant $k(a/L)^2$.

14.75•• A simple pendulum consists of a bob of mass m supported by a thin metal wire with a linear thermal expansion coefficient α. At temperature T, the period for small amplitude oscillations of the pendulum is τ_0. Show that if the temperature increases by an amount ΔT, the period of the pendulum changes by $\Delta \tau = \frac{1}{2}\alpha\tau_0\,\Delta T$. You may use the approximation that $(1 + x)^{1/2} \approx 1 + \frac{1}{2}x$ for $x \ll 1$.

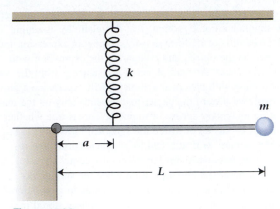

Figure 14.26
Problem 14.74.

Waves and Sound

T he study of periodic vibrations and waves is one of the oldest scientific studies. The Pythagoreans first established the connection between musical sounds and mathematics in the fourth century B.C. This study continued through the middle ages and was later developed by Huygens and Galileo. In the eighteenth century, an interest in musical instruments prompted the mathematical study of vibrating bodies and of the propagation of sound through air. By the nineteenth century, the study of sound had become an integral part of physics and was known as *acoustics.*

Studies of wave motion also advanced through research into the behavior of light, the discipline known as *optics.* Robert Hooke and Christiaan Huygens, contemporaries of Newton, believed light to be a form of wave motion, and Huygens developed a mathematical model for light waves. However, not until the early 1800s did Thomas Young firmly establish the wave nature of light. By the end of that century, Heinrich Hertz had established that what we now call radio waves were of the same nature as light waves. In the early twentieth century, wave concepts were used to develop the theory of quantum mechanics, which is sometimes called wave mechanics.

In our daily lives we see many examples of waves and oscillations: Radio and television waves reach around the world; musical sounds are created and modified by computers, as well as by traditional instruments; and ultrasound and x rays are commonly used in medical diagnosis and treatment.

There are similarities between vibratory motion and wave motion. We use mechanical concepts such as displacement, velocity, acceleration, and energy in describing them both. However, you must take care to distinguish between these two phenomena. As we develop the wave concept, we will point out both the similarities and the differences. ■

15.1   Pulses on a Rope

The simplest way to begin the study of waves is to think of the propagation of a pulse along a rope. If one end of a tautly stretched rope is suddenly snapped up and back down to its initial position, the action generates a wave pulse that travels along the rope to the other end (Fig. 15.1). As you can see from the figure, the displacement of the rope due to this wave pulse is at right angles to the direction in which the pulse travels. For this reason we call this type of motion a **transverse wave pulse.** If we watch any particular point of the rope very closely, we see that the motion of that point along the direction of the rope is very slight compared with its motion perpendicular to the rope. Each section of the rope makes a single oscillation up and down. But the oscillations are not independent, for each section of the rope is connected to the adjacent sections. Thus, the propagation of a transverse wave pulse along the rope is a collective motion of the whole rope, not simply the isolated behavior of any one section.

In addition to distinguishing between the motion of a pulse, or wave, and the motion of the medium (here, the rope), we must distinguish the kind of motions involved. For example, the velocity of a wave on a rope may be constant, while the transverse velocity of a point on the rope may be sinusoidal in time. In this case, the particles of the rope oscillate back and forth about an equilibrium position, while the wave propagates along the rope at its own speed. Waves on a rope, sound waves, and water waves all require a medium for their propagation. Electromagnetic waves, such as light and radio waves, are different because they do not require a medium for their propagation, as we will see in Chapter 20.

The wave pulse we have just described is called a **traveling wave.** The pulse occurs at one place at one time and at another place at a later time. The distance the pulse travels is proportional to the elapsed time. We have here a very general description of a wave: A **wave** *is a disturbance that transfers energy from one point to another without imparting net motion to the medium through which it propagates.* This description is quite general and, as you can see, does not require the wave to be repetitive.

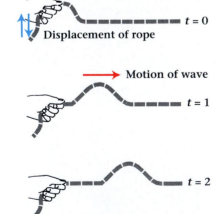

Figure 15.1
A wave pulse traveling along a rope. The displacement of the rope is vertical, whereas the motion of the wave pulse is horizontal.

15.2 Harmonic Waves

Let us now consider another special one-dimensional wave: a wave generated by the simple harmonic motion of one end of a long extended rope. In such a **harmonic wave,** the distance between successive maxima, or wave crests, is the **wavelength** λ (Fig. 15.2). As the end of the rope moves up and down with harmonic motion, it generates a sinusoidal wave. (Remember, a sinusoidal wave has the shape of a sine curve.) Note that while any given point on the *rope* moves up and down, the point does not undergo any displacement along the x axis. By contrast, any given point on the wave—a crest, say—moves one wavelength λ along the x axis in one period.

The displacement of a harmonic wave is a function of both position and time. At the instant of time $t = 0$ (Fig. 15.2), the displacement of the wave $y(x, t)$ is described by a sinusoidal function of position x,

$$y(x, t)\big|_{t=0} = y_0 \sin \frac{2\pi x}{\lambda},$$

where λ is the wavelength of the wave and y_0 is its amplitude. (That is, y_0 is the maximum vertical displacement of the rope from its equilibrium position.) However, if we look only at the position $x = 0$, the motion of the rope as a function of time is described by

$$y(x, t)\big|_{x=0} = -y_0 \sin \frac{2\pi t}{T},$$

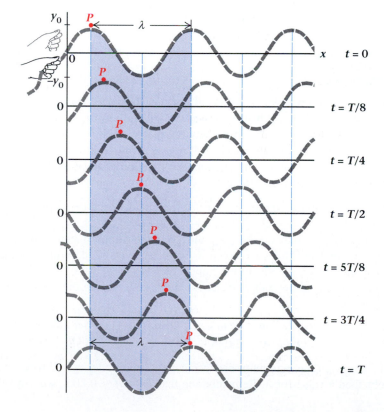

Figure 15.2

Harmonic waves traveling along a rope are generated by a sinusoidal motion of one end. A point on the crest of the wave, such as P, moves one wavelength λ in one period T.

where T is the period of the wave. (Compare this with Section 14.2.) The general mathematical form of a wave moving in the positive x direction, including both position and time, is given by

$$y(x, t) = y_0 \sin 2\pi\left(\frac{x}{\lambda} - \frac{t}{T}\right). \tag{15.1}$$

The function $y(x, t)$ describes the displacement of the rope at any position x and any time t. Equation (15.1) is the mathematical description of a *traveling harmonic wave* moving in the positive x direction. We could use a cosine function for this equation just as well, but we choose the sine function so that $y = 0$ when $x = 0$ and $t = 0$.

If we follow the motion of a single wave crest, we see that in one complete period it travels one full wavelength. The wave, therefore, is traveling with a speed v given by

$$v = \frac{\lambda}{T}.$$

Remembering that the frequency is the reciprocal of the period, we see that

$$v = \lambda f. \tag{15.2}$$

This relation between a wave's speed, frequency, and wavelength is very important. It is valid for all waves, mechanical or otherwise, and we shall frequently make use of it throughout this book.

Example 15.1

Figure 15.3 represents two snapshots of a wave on a rope. The snapshots were taken $\frac{1}{10}$ s apart. We know that the wave was traveling to the right and that it moved by less than one wavelength between pictures. Find its (a) wavelength, (b) wave speed, and (c) frequency. (d) Write an expression for the rope's displacement from its equilibrium position as a function of position and time. The maximum displacement is 3.0 cm.

Solution (a) Examining the figure, we see that the distance between two successive crests, or the wavelength, is $\lambda = 2.0$ m.
(b) During the $\frac{1}{10}$-s interval the wave moved to the right a distance of half a wavelength, or 1 m. The wave speed is the distance traveled divided by the time interval, or $v = 10$ m/s to the right.
(c) Since we now know both wavelength and speed, we can obtain the frequency from

$$f = \frac{v}{\lambda} = \frac{10 \text{ m/s}}{2.0 \text{ m}} = 5.0 \text{ s}^{-1} = 5.0 \text{ Hz}.$$

(d) We are given the maximum displacement or amplitude, $y_0 = 3.0$ cm, and just obtained a value for λ. Remembering that $T = 1/f = 0.20$ s, we can substi-

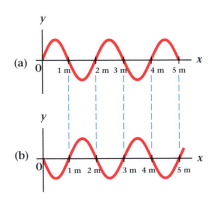

Figure 15.3
Example 15.1: A diagram of two snapshots of a wave on a rope. Picture (b) was taken $\frac{1}{10}$ s after picture (a). The scale is shown in meters. The amplitude of the wave is exaggerated to make it easier to see.

tute into Eq. (15.1) to get

$$y(x, t) = (3.0 \text{ cm}) \sin 2\pi\left(\frac{x}{2.0} - \frac{t}{0.20}\right) = (3.0 \text{ cm}) \sin (3.1x - 31t),$$

where x is given in meters and t in seconds.

*15.3 Energy and Information Transfer by Waves

We experience the energy transferred by waves in many situations: We feel the force of an ocean wave, our skin is warmed by the light waves from the sun, we hear sound waves. Furthermore, most of the information that we receive comes to us by waves. Speech and music are transmitted by sound waves, radio and television by electromagnetic waves. The reflected light by which you read this page is a wave. How does the energy (and, hence, the information) transmitted by the wave depend on the properties of the wave? We answer this question by first considering how energy is transferred by a single pulse. Then we extend our results to get an expression for the energy of a harmonic wave.

At time $t = 0$, a small segment of the rope around point P in Fig. 15.4, with mass Δm and length Δl, is at rest and has no kinetic energy. The up-and-down hand motion provides the energy required to start the pulse along the rope. As the leading edge of the pulse reaches P, the segment Δl begins to move upward. As the wave crest passes the segment Δl, the segment moves to its highest position and then starts down again, possessing kinetic energy while it is in motion. When the entire pulse has passed P, the segment Δl returns to rest and again has no kinetic energy. The progress of the pulse along the rope corresponds to the flow of energy along the rope. Any other type of pulse, including a pulse traveling through the air, would transfer energy along the direction of propagation in a similar manner.

How much energy has been transferred past P during a time t? For a traveling harmonic wave on a rope, each point moves with simple harmonic motion in the transverse (y) direction. As we saw in Section 14.3, Eq. (14.7), in the absence of damping, the total energy of a harmonic oscillator is equal to its potential energy at maximum displacement y_0, that is, $\frac{1}{2}ky_0^2$. Also in Chapter 14, the relationship between mass, spring constant, and frequency was found to be $f = (1/2\pi)\sqrt{k/m}$. If we treat the segment of the rope as a harmonic oscillator with mass Δm moving at frequency f, we can rearrange the equation to find an effective spring constant $k = (2\pi f)^2 \Delta m$. The energy associated with the motion of this segment of the rope is then

$$\Delta E = \tfrac{1}{2}ky_0^2 = \tfrac{1}{2}(2\pi f)^2 \Delta m\, y_0^2$$

$$\Delta E = 2\pi^2 \Delta m\, f^2 y_0^2.$$

We now have an important result: The energy of a wave depends on the square of the amplitude of the wave. Thus, a wave with twice the amplitude of an otherwise equivalent wave (same frequency, same medium) will have four times as much energy.

To find the rate of energy flow, or power, we observe that Δm can be written as $\rho A\, \Delta l$, where ρ is the density, A the cross-sectional area, and Δl the length

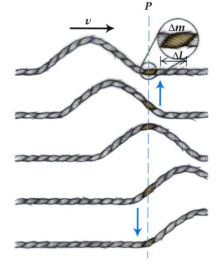

Figure 15.4

An element of mass Δm at point P is given kinetic energy as the wave pulse passes by with a speed v.

of the rope segment. In a time Δt, the wave with speed v passes a length $\Delta l = v \, \Delta t$, so that we can substitute $\Delta m = \rho A v \, \Delta t$ into the equation for ΔE. We obtain an expression for the energy transported in time Δt:

$$\Delta E = 2\pi^2 A \rho v f^2 y_0^2 \, \Delta t.$$

The rate at which energy propagates along the rope is the power P,

$$P = \frac{\Delta E}{\Delta t} = 2\pi^2 A \rho v f^2 y_0^2.$$

The more generally useful parameter is the **intensity** I, defined to be the power flowing through unit area. For the case at hand, the intensity in watts per square meter is

$$I = \frac{P}{A} = 2\pi^2 \rho v f^2 y_0^2. \qquad (15.3)$$

Although we have derived this result for the specific case of waves on a rope, it does give the correct dependence of the intensity on the density of the medium, the wave velocity, the frequency, and the amplitude appropriate for any traveling harmonic wave.

Example 15.2

Waves of the same frequency and velocity travel along two identical ropes. The power transmitted down rope 1 is 0.30 mW. If the waves in rope 2 have an amplitude 1.6 times the amplitude of the waves in the first rope, how much power is transmitted along rope 2?

Solution We may write the ratio of the transmitted powers as

$$\frac{P_2}{P_1} = \frac{2\pi^2 A \rho v f^2 y_{02}^2}{2\pi^2 A \rho v f^2 y_{01}^2},$$

which becomes

$$\frac{P_2}{P_1} = \frac{y_{02}^2}{y_{01}^2}.$$

Thus

$$P_2 = P_1 \frac{y_{02}^2}{y_{01}^2} = 0.30 \text{ mW} \left(\frac{1.6^2}{1.0^2}\right) = 0.77 \text{ mW}.$$

15.4 Sound Waves

So far, we have been discussing **transverse waves,** such as a wave pulse along a rope, in which the particles of the rope move at right angles to the direction of propagation of the wave. However, waves also occur in which the particles of the wave medium move back and forth along the direction of propagation. Such waves are known as **longitudinal waves.** For example, if a stretched spring, such as a slinky, is alternately expanded and compressed, a compressional oscillation travels along the spring (Fig. 15.5). The sections of the spring where the coils

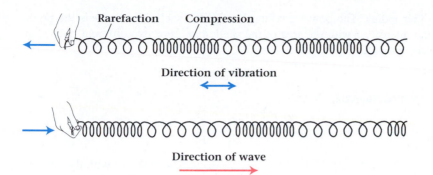

Figure 15.5

Example of a longitudinal wave in a spring. The direction of vibration is parallel to the direction of the wave.

are tightly packed are called *compressions,* and the sections where the coils are farther apart are called *rarefactions.* The terms compression and rarefaction are also used to describe relative density for other types of longitudinal waves.

Sound waves in air are longitudinal waves. The vibrations of a drumhead or loudspeaker exert varying pressure on the air. Increasing the pressure crowds the gas molecules together and pushes them against adjacent molecules. These molecules, in turn, strike their neighbors. The resulting pulse of compressed air moves away from the pressure source (Fig. 15.6). As the compression passes, the individual gas molecules move back to their original positions. Thus, while the wave travels in a longitudinal fashion, the molecules themselves only vibrate back and forth along the line of propagation.

Sound waves travel out in all directions from an isolated source and are therefore three-dimensional waves. The outward-moving compressions (or rarefactions) correspond to expanding spherical shells called **wavefronts** (Fig. 15.7). The radius of any shell increases at the speed of sound. If we are far away from the source, a small piece of the spherical wavefront will be approximately flat, or planar. We call these plane wavefronts or simply **plane waves.**

As the spherical wave expands, energy is conserved if the damping is negligible, as it is over short ranges in air. Therefore, the power passing through a spherical shell of one radius is the same as that passing through a shell of any

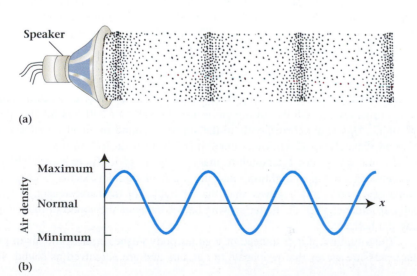

Figure 15.6

(a) Representation of a sound wave in air. The wave is longitudinal. The more closely spaced dots indicate regions of higher density. (b) A graph of the density as a function of position.

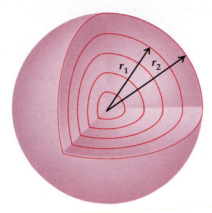

Figure 15.7
Waves expanding from a point source have spherical wavefronts.

other radius. The power passing through the shell of radius r is the product of the intensity I with the area of the shell, $4\pi r^2$. For two different radii r_1 and r_2, we can write

$$I_1 4\pi r_1^2 = I_2 4\pi r_2^2.$$

Upon rearranging, we get

$$\frac{I_2}{I_1} = \frac{r_1^2}{r_2^2}. \tag{15.4}$$

Thus, the intensity of a spherical wave decreases inversely with the square of its distance from the source.

Example 15.3

A loudspeaker on a tall pole standing in a field of tall grass generates a high-frequency sound at an intensity of 1.0×10^{-5} W/m^2 at the position of the ears of a person standing 8.0 m directly below it. If the person walks away from the pole so that she is 24 m from the loudspeaker, what is the sound intensity at the new position of her ears?

Solution The grass will almost completely absorb the high-frequency sound that reaches it, so the expanding sound wave reaching the listener's ears will be a section of a spherical wave with no additional effects from reflected waves. We can thus compare the intensities at the two positions:

$$\frac{I_2}{I_1} = \frac{r_1^2}{r_2^2},$$

$$I_2 = \frac{r_1^2}{r_2^2}I_1 = \frac{(8\ \text{m})^2}{(24\ \text{m})^2}(1.0 \times 10^{-5}\ \text{W/m}^2)$$

$$I_2 = 0.11 \times 10^{-5}\ \text{W/m}^2.$$

The speed with which sound waves propagate in air depends on atmospheric pressure, temperature, and humidity. At standard sea-level pressure and 0°C, the speed of sound in dry air is 331.5 m/s. The speed of sound at other temperatures is adequately represented by the expression

$$v(T) = (331.5 + 0.6T)\ \text{m/s},$$

where the temperature T is in degrees Celsius. For the purpose of working problems in this text, it will be convenient to use the value of 340 m/s for the speed of sound. This value corresponds to the speed of sound in air at a temperature of approximately 15°C. (In other units, it is 1100 ft/s or 760 mi/h.)

Sound travels slowly enough to make us aware of its finite speed. The delay between a flash of lightning and the crash of thunder occurs because the speed of sound is much slower than that of light. For the same reason, a baseball outfielder can see the batter's swing before he hears the sound of the bat hitting the ball.

The vibration of a drumhead or a guitar body pushes against the air to produce pressure waves that propagate to our ears and are perceived as sound. The

Table 15.1	Sound Speeds in Some Representative Materials at 15°C ▼
Material	**Speed (m/s)**
Air	340
Polyethylene	920
Helium	977
Water	1500
Marble	3810
Wood (oak) along the fiber	3850
Aluminum	5000
Iron	5120

response of human ears is limited to a range of frequencies from about 20 Hz to about 20,000 Hz. In general, as we grow older the upper limit of audible frequencies drops. It is not uncommon to find people with adequate hearing whose range of audible frequencies extends to only 10,000 or 15,000 Hz. Those frequencies between 20 Hz and 20,000 Hz are usually referred to as *audio,* or *sonic,* frequencies. Vibrational frequencies above 20,000 Hz are beyond our hearing and are referred to as *ultrasonic* frequencies. Similarly, extremely low frequencies are called *infrasonic* frequencies. Ultrasonic vibrations generate soundlike waves, but we do not hear them because of the limitations of our ears. It is well known that many animals, including dogs, can hear frequencies well above those audible to people.

The *pitch* of a sound or musical note is a subjective judgment of its highness or lowness. It is determined primarily by its frequency: a high pitch corresponds to a high frequency. However, the loudness of a sound at a given frequency can influence its apparent pitch.

Sound waves can propagate in solids and liquids, as well as in gases. The velocity of sound is much greater in denser solids and liquids than in air. The sound waves in liquids are longitudinal, just as in air. However, vibrations can propagate in solids as both longitudinal and transverse waves, and the corresponding wave speeds may be different. A list of representative sound speeds is given in Table 15.1.

There are many applications of ultrasonic waves. Because of their relatively short wavelengths, they can be focused into narrow beams and directed more easily than the longer waves of audible sound. (We will see this effect again in the discussion of wave optics in Chapter 24.) Figure 15.8 illustrates the use of ultrasonic waves to measure distances. A pulse of waves of frequency between 25 and 40 kHz is emitted from the ranging device, which automatically measures the time for the echo of the pulse to return. The distance to the wall or other object is computed from the travel time of the pulse and the speed of sound in air, taking into account the dependence of the speed on temperature and humidity. The computation is made by the microcircuits within the ranging device, and the result is displayed on a numerical panel. These instruments can measure with a precision of about 0.6 cm. Similar devices are used on some automatically focusing cameras to determine the proper position for the lens.

When the ultrasonic waves strike the wall in Fig. 15.8, they are partly reflected, partly transmitted, and partly absorbed. The reflections occur whenever

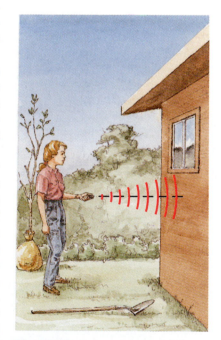

Figure 15.8
Ranging devices use ultrasonic waves to measure distances with a precision of 0.6 cm.

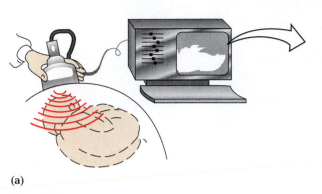

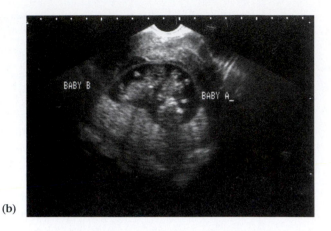

(a)   (b)

Figure 15.9

(a) Physicians can "see" a baby inside its mother's body by using the hand-held external probe of an ultrasonic imaging system. (b) An ultrasonic image of 10-week old human twins inside the mother's body.

the waves encounter a difference in density—here, the boundary between the air and wall. The amount of the reflection depends on the relative speeds of sound on each side of the boundary. (This behavior is a general property of waves, not just a property of ultrasonic waves.) The returning reflected signals from each boundary can be used to generate an image of the objects encountered by the waves. Within your body, incident ultrasonic waves are reflected from boundaries between tissues, bones, and fluids of different densities. Ultrasonic imaging is routinely used by physicians to "look" inside the human body (Fig. 15.9). The visual image is reconstructed from the information contained in the amplitude and phase of the reflected waves. Ultrasonic imaging is generally considered to be a safer technique than x-ray imaging.

Example 15.4

What is the wavelength of the sound waves when someone sings a "standard A" ($f = 440$ Hz)?

Solution   Assume the room temperature to be about 15°C, so that the speed of sound is 340 m/s. We may use Eq. (15.2) to get

$$\lambda = \frac{v}{f} = \frac{340 \text{ m/s}}{440 \text{ s}^{-1}} = 0.77 \text{ m}.$$

*15.5   Measuring Sound Levels

The human ear is an extremely sensitive detector, capable of hearing sounds over an extremely large range of intensities. For example, a passing train generates sound intensities that may be 10^4 to 10^6 times greater than the sound intensity due to a buzzing mosquito, yet we can hear both sounds clearly. Such a range of pressure responsiveness is remarkable. Although our ears are sensitive to this enormous range of sound intensities, our subjective judgment of loudness does not directly correspond to the magnitude of the sound intensity. Experiments have shown that, to a good approximation, people perceive a sound to be about twice as loud as a reference sound when its intensity is ten times as large as that of the reference. A sound perceived to be four times as loud as a reference requires an increase in sound intensity by a factor of 100. This relationship is approximately logarithmic; that is, loudness is proportional to the logarithm of the sound intensity. Thus, it is natural to employ a scale of measurement that is also logarithmic.

The unit of sound intensity measurement is the **decibel,** dB. The decibel unit is one-tenth the size of the bel (a unit named in honor of the inventor of the telephone, Alexander Graham Bell). The **intensity level** L_I in decibels is defined to be ten times the logarithm of the ratio of two intensities I and I_0; that is,

$$L_I \text{ (in decibels)} = 10 \log_{10} \frac{I}{I_0}, \tag{15.5}$$

where I_0 is the reference intensity. For example, if the intensity I exceeds the reference intensity I_0 by a factor of 4, the intensity level of I is 6 decibels (6 dB) above I_0: $L_I = 10 \log 4 = 10(0.6) = 6$ dB.

Since the measure of sound intensity level in decibels is really a comparison of two intensities, a given sound level cannot be stated in units of decibels unless the reference level is known. The standard reference level of sound intensity is 10^{-12} W/m², which is approximately the intensity that can just barely be heard by a person with good hearing. This intensity is called the threshold of hearing. The intensity level of a very quiet recording studio is about 20 dB, corresponding to a sound intensity 100 times greater than the quietest sound you can hear. Figure 15.10 shows a comparison of sound intensity levels due to various sources.

The ear does not respond equally well to all frequencies in the audio range. It is considerably more sensitive to frequencies between 2000 and 5000 Hz than to either higher or lower frequencies. For this reason, sound level meters are often designed to have a frequency response matching that of the human ear. A meter that does so is said to be A-weighted, and measurements are often given in A-weighted decibels, or dBA.

Example 15.5

At a party, a stereo tape deck is playing at maximum volume. Suddenly a dancer trips over a wire, and one speaker stops playing. What is the reduction in sound intensity level, measured in dB?

Strategy Let's assume that the two speakers are balanced—that is, that an equal sound level is produced by each. When one fails, the resulting sound intensity is one-half of the initial intensity. This change can be expressed in dB by using Eq. (15.5).

Solution If we let the initial intensity be I_0 and the intensity with only one speaker be $I_0/2$, we find the

$$\text{intensity change in dB} = 10 \log \frac{I_0}{2I_0} = 10 \log \frac{1}{2} = -3 \text{ dB.}$$

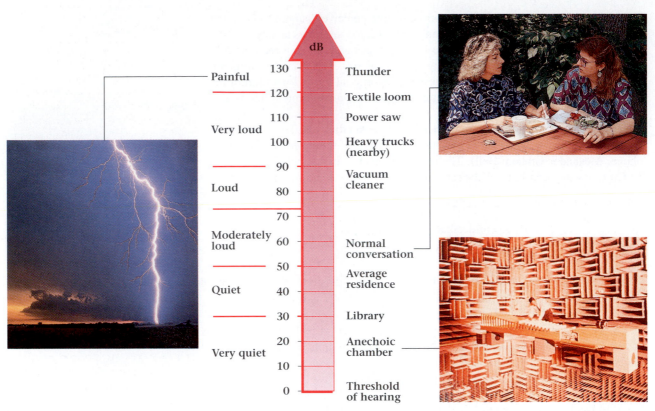

Figure 15.10

Approximate sound intensity levels for an assortment of familiar sound sources.

Discussion The reduction in sound level due to failure of one-half of the sound system is only 3 dB. Thus, a 3-dB reduction in sound level corresponds to reducing the intensity by a factor of one-half. Conversely, a 3-dB increase in sound level corresponds to twice as much sound intensity incident on the ear. In either case, it represents a small change in sound level in comparison with the range of hearing.

15.6   The Doppler Effect

Most of us are familiar with the rise and subsequent drop in pitch of an automobile horn as the car approaches and then passes. As the moving car approaches a stationary listener, the sound waves crowd together, causing an increase in the frequency of the sound heard. After the car has passed and is moving away from the listener, the waves spread out and the observed frequency is lower. This change in frequency associated with the relative motion between a sound source and a listener (observer) is called the **Doppler effect,** after Christian Doppler (1803–1853), the Austrian physicist who first explained the phenomenon.

We can understand the cause of the Doppler effect with the aid of Fig. 15.11, which shows a bob oscillating up and down on the surface of a container of water. Its motion causes circular water waves to spread out from the point of contact (Fig. 15.11a). When the bob is moved to the right, each wave crest moves

ROOM ACOUSTICS

Have you noticed that the same music group sounds better in one auditorium than in another, or that the same lecturer is easier to understand in one room than in another? What you hear depends not only on the source of the sound but also on the acoustical properties of the room.

About 1895, Professor Wallace Sabin of Harvard University discovered that one of the more important acoustical properties of a room is its reverberation time, which is the time for the sound pressure level to decrease by 60 dB. The reverberation time in a large stone-walled cathedral is several seconds, while a small, heavily carpeted and draped room has a much shorter reverberation time. Experience has shown that the optimum reverberation times depend on the size of the room and its intended use (Fig. B15.1).

The reverberation time in a given room is determined by the amount and type of sound-absorbing material present. The sound absorption a of a surface is the product of the actual surface area and the sound-absorption coefficient of the surface material (Table B15.1). The absorption unit is named the *metric sabin*. The total sound absorption in a room is the sum of the absorptions for the individual surfaces.

Sabin developed an empirical relationship between the reverberation time T_{60}, the volume V of the room, and the sound absorption a in the room. When distances are measured in meters, the reverberation time in seconds is

$$T_{60} = 0.16 \frac{V}{a}.$$

For example, a room 9 m wide $\times$ 12 m long $\times$ 4 m high has a volume of 432 m³. The total area of the walls is 168 m², and the areas of the ceiling and of the floor are 108 m² each. If the walls are plywood paneling, the floor is wood, and the

ceiling is gypsum board (dry wall), the total sound absorption at 500 Hz is 44.8 metric sabins. Therefore, the predicted reverberation time of the empty room is 1.5 s. This value is somewhat longer than is optimum for general use in a room of this size (Fig. B15.1). However, with seats and people in the room, the reverberation time is shortened. As an example, if the floor were three-quarters covered with students in tablet-arm chairs, approximately 30 metric sabins would be added. The new reverberation time would be less than a second, which is in the acceptable range. Such calculations are not expected to give the precise value of the actual reverberation time, but they do indicate the approximate value expected under specific conditions.

The overall acoustical quality of a room is influenced by the reverberation time at all frequencies, not just 500 Hz. Subjective descriptions of an auditorium's sound, such as *warm, live,* and *brilliant,* depend not only on the length but also on the frequency dependence of the reverberation time. For example, a *warm* sound is the result of a longer reverberation time in the lower, or bass, frequencies.

Modern auditorium design also includes computer simulation and scale models. The most successful designs result from the work of a skilled and experienced person using the best of technology, research, and artistry. Even then, the desired result is not always achieved for large halls. Electronic reinforcement, with amplification and delays, can help adjust an auditorium's acoustical properties, but it cannot turn a bad auditorium into a good one.

Figure B15.1 Optimum reverberation times at 500 Hz for rooms of different size and use.

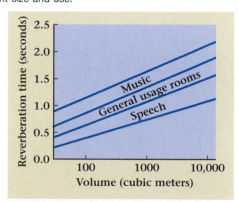

Table B15.1	*Typical Sound Absorption Coefficients*		
Material	**Absorption Coefficients at**		
	125 Hz	**500 Hz**	**2000 Hz**
Acoustical ceiling tile (3/4 in.)	0.76	0.83	0.99
Brick, unglazed	0.02	0.03	0.05
Carpet, heavy, on concrete	0.02	0.14	0.60
Concrete block, painted	0.10	0.06	0.09
Heavy drapes (to half area)	0.14	0.55	0.70
Concrete or terrazzo	0.01	0.02	0.02
Linoleum, or asphalt tile	0.02	0.03	0.03
Wood flooring	0.15	0.10	0.06
Gypsum board, 1/2 in.	0.29	0.05	0.07
Ordinary window glass	0.35	0.18	0.07
Plaster on lath	0.14	0.06	0.04
Plywood paneling, 3/8 in.	0.28	0.17	0.10
Person in an upholstered chair	0.39	0.80	0.92
Students in tablet-arm chairs	0.30	0.49	0.87

Source: From M. David Egan, *Architectural Acoustics* (New York: McGraw-Hill, 1988).

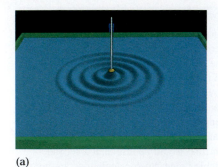

(a)

(b)

Figure 15.11
(a) A computer-generated image of circular waves spreading out from the point of contact of an oscillating bob in contact with the surface of water in a shallow tank. Notice that the waves spread out uniformly in all directions on the surface. (b) Image of the waves when the oscillating bob is moved to the right with constant speed $v_s = 0.4v_{wave}$. Notice how the waves ahead of the bob are crowded together, while those behind are spread apart.

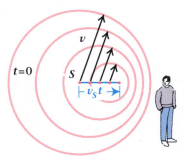

Figure 15.12
The Doppler effect: As a sound source moves toward an observer, the wavelength decreases, causing the observer to hear a higher frequency. During a time t, the wave advances a distance vt while the source advances a distance $v_s t$.

out as an expanding circle, but since the source is moving, it emits each successive wave at a different location. As a result, the waves moving in the same direction as the source are crowded together, while those moving in the opposite direction are spread farther apart (Fig. 15.11b). The wave speed is constant whether the source moves or not. Thus, where the wavelength is shortened, the frequency is increased, and where the wavelength is lengthened, the frequency is reduced.

Figure 15.12 also shows a source of sound waves moving to the right radiating spherical waves, shown here as circles. For this case of a source of frequency f moving toward an observer, we can readily determine the new frequency f' that the observer hears. During one period T of the source, the wave moves out a distance vT, where v represents the speed of sound. During this same time interval, the source, moving at a speed v_s, moves a distance $v_s T$. The difference between these two distances is the new wavelength λ':

$$\lambda' = (v - v_s)T.$$

The frequency is obtained from the wavelength in the usual way:

$$f' = \frac{v}{\lambda'} = \frac{v}{(v - v_s)T},$$

or, using the fact that $f = 1/T$,

$$f' = \frac{f}{(1 - v_s/v)} \qquad \text{(source approaching)}.$$

If the source is moving away from the observer, the wavelength reaching the observer is increased instead of shortened. Using the same argument as above, we find the new frequency by the same expression with a plus sign in front of the v_s/v term. The two cases can be written as one:

$$f' = \frac{f}{(1 \mp v_s/v)}, \qquad \text{(moving source, stationary observer)} \qquad (15.6)$$

where the upper sign means that the source is approaching the observer (higher frequency) and the lower sign means that the source is moving away (lower frequency).

▼

Example 15.6

A police car horn emits a 250-Hz tone when sitting still. What frequency does a stationary observer hear if the police car sounds its horn while approaching at a speed of 27.0 m/s (60 mi/h)? What frequency is heard if the horn is sounded as the car is leaving at 27.0 m/s?

Solution The apparent frequency for the approaching car is found from Eq. (15.6), where we use the negative sign and the speed of sound is taken as 340 m/s:

$$f' = \frac{250 \text{ Hz}}{1 - \dfrac{27 \text{ m/s}}{340 \text{ m/s}}} = 272 \text{ Hz}.$$

When the car passes, the frequency is lowered according to Eq. (15.6) with the positive sign, and f' becomes

$$f' = \frac{250 \text{ Hz}}{1 + \dfrac{27 \text{ m/s}}{340 \text{ m/s}}} = 232 \text{ Hz}.$$

For waves transmitted by a medium, such as sound waves in air, a slightly different Doppler formula results when the observer, rather than the source, is in motion. If a stationary observer hears a frequency f, an observer O moving with a speed v_o toward the source would hear a higher frequency f', because he would encounter wavefronts more rapidly as a result of his motion toward the source. In a time t, the stationary observer receives ft wave crests. In the same time t, the moving observer travels a distance $v_o t$, which corresponds to $v_o t/\lambda$ wavelengths. The total number of wave crests encountered is number of wave crests $= ft + v_o t/\lambda$. The resulting frequency is the number of waves divided by the time,

$$f' = f + \frac{v_o}{\lambda} = f + \frac{v_o f}{v} = f\left(1 + \frac{v_o}{v}\right) \qquad \text{(observer approaching).}$$

If the observer were moving away from the source, fewer waves would reach him per second. The general result for a moving observer is

$$\boxed{f' = f\left(1 \pm \frac{v_o}{v}\right)} \qquad \text{(moving observer, fixed source)} \qquad (15.7)$$

where the plus sign corresponds to an observer approaching the sound source (higher frequency) and the minus sign corresponds to the observer moving away (lower frequency). Although this is similar to Eq. (15.6), the two equations are not quite the same. However, when v_s and v_o are much less than v, the two sets of formulas give essentially the same results.

Physicians detect blood flow and measure its speed by reflecting ultrasonic waves from blood moving toward the ultrasonic source (Fig. 15.13a). The reflected waves are Doppler-shifted to a higher frequency. When the reflected waves are combined with the incident waves, they produce a new wave at a frequency that is the difference between the reflected and incident frequencies. For a 2-MHz source and normal blood flow, the difference frequency is in the audio range.

The Doppler effect is not limited to sound waves alone but applies to other kinds of waves as well. It even applies to electromagnetic waves such as light and radar, although the formulas derived here for sound waves differ from the formulas that apply to light and radio waves. The correct equations to use for light are found in Chapter 25. In astronomy, the measured Doppler shifts in the light received from stars have been the principal sources of information on stellar motions. In more familiar situations, radar waves bounced off a moving target are Doppler-shifted as a result of the motion of the reflecting object. By measuring the shift in frequency of a radar wave beamed at a moving object, we can measure the speed of the object precisely. This technique has been widely adapted for measuring the speed of cars, planes, thunderstorms, and even baseballs (Fig. 15.13b).

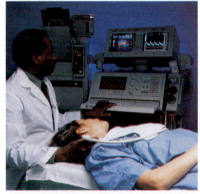
(a)

(b)

Figure 15.13
(a) Detecting blood flow in the carotid artery with ultrasonic waves.
(b) Measuring the speed of a baseball with radar.

Example 15.7

If the police car in Example 15.6 were sounding its horn while stationary, what frequency would be heard by an observer who was approaching it at a speed of 27.0 m/s (60 mi/h)?

Solution Here we have the case of an observer moving toward a stationary source. Equation (15.7) with the plus sign is the appropriate expression:

$$f' = f\left(1 + \frac{v_o}{v}\right)$$

$$f' = 250 \text{ Hz}\left(1 + \frac{27 \text{ m/s}}{340 \text{ m/s}}\right) = 270 \text{ Hz}.$$

Notice that this is not the same answer as in Example 15.6.

*15.7 Formation of a Shock Wave

We have seen in the Doppler effect that the wavefronts produced by a moving source of sound are crowded together in the direction toward which the source is traveling. As the speed of the source increases, the crowding becomes more pronounced. What happens when the speed of the source becomes greater than the wave speed? In this case, the source moves faster than the waves and the arguments used to describe the Doppler effect no longer apply. Instead, the spherical waves expanding from the source at subsequent positions along the path of the source all combine, forming a single conical wavefront known as a **shock wave** (Fig. 15.14). Because the shock wave is composed of many wavefronts acting together, it has a large amplitude.

At time $t = 0$ the source emits a wave from point O. At a later time t, the wavefront has expanded to a radius $r = vt$ and the source has traveled a distance $v_s t$ to reach point S. Subsequent wavefronts also expand as indicated in Fig. 15.14, so that at time t they just reach the tangent line drawn from S to the wavefront centered at O. The resulting envelope of wavefronts forms a cone of half-angle θ given by

$$\sin \theta = \frac{v}{v_s}. \tag{15.8}$$

The ratio v_s/v, called the Mach number, is often used to give the speed in terms of the speed of sound. Thus, a speed 1.5 times the speed of sound is referred to as Mach 1.5.

When the shock wave is produced by an airplane moving at a speed greater than the speed of sound—that is, at supersonic speed—the shock wave is known as a sonic boom. Figure 15.15(b) shows the shock wave produced in air by a supersonic aircraft moving at Mach 1.1. Notice that in addition to the shock wave produced at the front end, lesser shock waves appear at the rear of the plane.

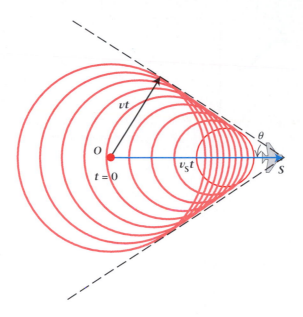

Figure 15.14
Shock waves are generated when the source of sound moves faster than the speed of sound waves in the medium. The waves combine to produce a conical wavefront.

High-speed aircraft often produce two or more shock waves, which are associated with the nose, the tail, and other projections on the aircraft.

15.8   Reflection of a Wave Pulse

One of the most interesting aspects of acoustics is the physics of music—how a violin or clarinet makes sound and why the sounds are as they are. To investigate these topics, we first need to know some other general properties of waves, including reflection and interference. We introduce these concepts by returning to our model of mechanical wave pulses traveling along a rope. These wave properties introduced here apply to all waves, and you will encounter them again when we discuss light waves.

When a wave traveling along a string reaches the end of the string, it is reflected. The exact way in which it is reflected depends on whether the end of the string is fixed or free to move. Let us examine the two cases separately.

When a wave pulse reaches the far end of a string that is fixed to a wall at that end, the wave does not suddenly stop, but is reflected. If no energy is dissipated at the far end of the string, the reflected wave has a magnitude equal to that of the incident wave; however, the direction of the displacement will be reversed (Fig. 15.16a). This reversal happens because as the pulse encounters the wall, the upward force of the pulse on the end of the string pulls upward on the wall. As a result, according to Newton's third law, the wall pulls downward on the string. This reaction force causes the string to snap downward, initiating a reflected pulse that moves off with an inverted (or negative) amplitude.

What happens, now, if the string is free to move at its far end? Again, a wave pulse traveling along the string is reflected when it reaches that end (Fig. 15.16b). But in this case we see that the reflected wave has the same direction of displacement as the incident wave. As the pulse reaches the end of the string, the string moves up in response to the pulse. As the end of the string begins to

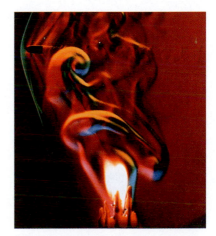

(a)

(b)

Figure 15.15
A shock wave generated by (a) a bullet and (b) an aircraft traveling at supersonic speeds.

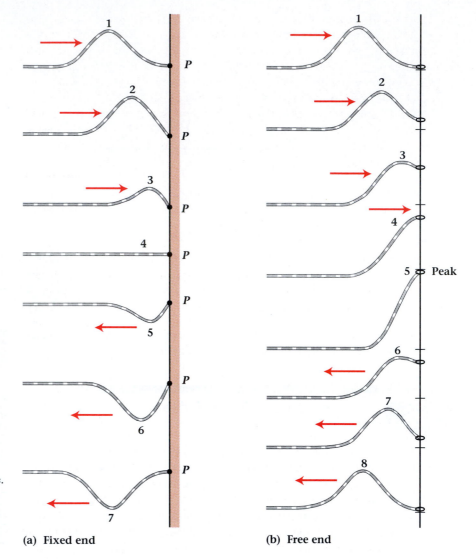

Figure 15.16
(a) Reflection of a wave pulse from a fixed end point. The direction of displacement of the reflected wave is inverted. (b) Reflection of a wave pulse from an end that is free to move. The direction of displacement of the reflected wave is unchanged. (Numbers indicate the sequence of motion.)

(a) Fixed end

(b) Free end

return to its initial position, it starts a pulse back along the string, just as if the end motion were due to some outside force. The result is a pulse exactly like the incident wave pulse, except that its direction of travel is reversed.

15.9   Standing Waves on a String

If two wave pulses are started along a rope from opposite ends, the waves will meet, pass through each other, and continue their motion as though nothing had happened. The resultant wave observed during the time the two individual waves overlap is the algebraic sum of the individual amplitudes (Fig. 15.17). The same behavior is displayed by sound waves: When two or three people are talking in the same room, you can distinguish the individual voices even if they speak simultaneously. The sound of each person's voice is not disturbed by the others.

The loudness of the combined sound does vary, however, depending on how many voices are heard simultaneously. This effect is an example of super-position. The **principle of superposition** says that at any instant, the resultant combination wave is the algebraic sum of all the component waves. The principle of superposition applies to many types of wave motion and so has enormous impact—it allows us to analyze combinations of waves.

Let us consider the vibrational behavior of a taut, but flexible, string held fixed at each end. A guitar string is a good example. If the string is pulled aside at some point and then released, it begins to vibrate. We can obtain a description of its motion by analyzing it in terms of oppositely directed wave pulses origi-nating at the point where the initial displacement occurred. These waves travel back and forth along the string, undergoing reflection each time they reach the ends of the string. The resulting motion of the string is then the superposition of the amplitudes of all these pulses.

Imagine that the guitar string oscillates with harmonic motion. For a given string, the speed of waves along the string depends on the tension in the string. If the tension is constant, the speed of the waves is also constant. Now, suppose a wave propagates along the string and back and reaches the origination point at just the right time so that the displacement of the reflected wave is in the same direction as the displacement of the next wave. Then the two displacements add together, and the resulting wave has a larger amplitude. The two waves are said to be *in phase,* and the resulting build-up of amplitude is called **constructive in-terference.** At many other frequencies (and wavelengths), the displacements of the waves are in opposing directions. These waves are said to be *out of phase,* and their amplitudes tend to cancel. This effect is called **destructive interfer-ence.** Thus, waves of these frequencies are suppressed.

If the string is driven periodically, it responds by vibrating at the frequency of the driving force. However, the amplitude of the vibration will be much greater if the string is driven at a resonance frequency. On the other hand, if the string is struck sharply and thereafter allowed to vibrate on its own, only the res-onant frequencies will persist.

The lowest resonant frequency of vibration of the string (or other object) is called its **fundamental frequency.** Resonant frequencies that are integer multi-ples of the fundamental frequency are called **harmonic frequencies.** For our string, the fundamental frequency is the first harmonic; the frequency that is dou-ble this value is the second harmonic, and so on. All resonant frequencies higher than the fundamental, whether they are integer multiples of the fundamental or not, are called **overtones.** For example, the overtones of an idealized drum head are not harmonic but stand in the frequency ratio of $1.0 : 1.6 : 2.1 : 2.3$.

Each resonant frequency corresponds to an oscillation of the entire string. For the present case, the ends of the string are fixed. The vibrational motion with lowest frequency (first harmonic) is shown in Fig. 15.18(a); this vibration corre-sponds to a sinusoidal motion of the entire string moving up and down between the supports. At any instant the displacement of the string is described by a sine curve; but this curve does not travel along the string, as we have seen for trav-eling waves. Instead, the whole string moves up and down. We have a **standing wave.** Such a wave can be represented as the product of two sine functions: one that describes the shape of the waveform and another that gives the time depen-dence. In Problem 15.82 you are asked to use the principle of superposition to find a mathematical expression for this standing wave.

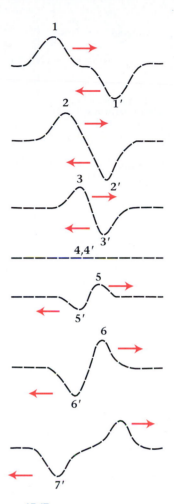

Figure 15.17

Two wave pulses moving in opposite directions pass through each other. The resultant wave is the algebraic sum of the individual waves. (Numbers indicate the sequence of motion for each wave pulse.)

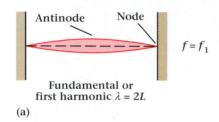

Fundamental or
first harmonic $\lambda = 2L$

(a)

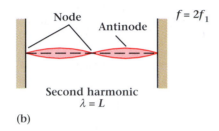

Second harmonic
$\lambda = L$

(b)

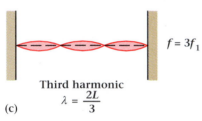

Third harmonic
$\lambda = \dfrac{2L}{3}$

(c)

Figure 15.18
Standing waves on a string. The points labeled as nodes do not move. The points that vibrate with maximum amplitude are called antinodes.

Since each end of the string does not move, the end points must be special. They are **nodes**—points of zero vibrational amplitude—and the distance between adjacent nodes is always one-half of the wavelength. Halfway between each pair of nodes is an **antinode,** a point on the string that vibrates with the greatest amplitude. Thus, the fundamental vibration of a string with fixed end points has a node at each end and a single antinode in between. The length of the string is half a wavelength. The fundamental frequency of the string depends on the speed of waves along the string and on the length of the string. The lowest frequency of vibration of the string is

$$f = \frac{v}{\lambda} = \frac{v}{2L},$$

where L is the length of the string and v is the speed of the wave along the string. Other resonant frequencies also occur. Since the two end points are nodes, the length of the string must be an integral number of half wavelengths. Thus, the distance between the ends can be $\lambda/2$, $2(\lambda/2)$, $3(\lambda/2)$, etc. That is, for a string of length L, the resonant wavelengths must be consistent with

$$L = n\frac{\lambda}{2} \qquad \text{where } n = 1, 2, 3, \ldots.$$

For fixed L, there is a wavelength λ_n associated with each integer n,

$$\lambda_n = \frac{2L}{n}.$$

The frequency is obtained from Eq. (15.2),

$$f_n = \frac{v}{\lambda_n} = n\frac{v}{2L}. \tag{15.9}$$

For flexible strings, the transverse wave speed is given in meters per second by

$$v = \sqrt{\frac{T}{\mu}}, \tag{15.10}$$

where T is the tension in the string in newtons and μ is the linear mass density in kilograms per meter. Standing waves on a stretched flexible string have resonant frequencies given by

$$f_n = \frac{n}{2L}\sqrt{\frac{T}{\mu}}, \qquad n = 1, 2, 3, \ldots. \tag{15.11}$$

Increasing the tension in the string raises the speed of waves along it and thus raises the natural vibrational frequencies. When you hear a stringed instrument being tuned, the musician is adjusting the tension in the strings. In addition, by pressing the string down at different points on the fingerboard, it is possible to shorten the vibrational length of the string—and thus increase the fundamental frequency—by varying amounts. That is how a musician is able to play many different notes with one string (Fig. 15.19).

Figure 15.19
A guitar player changes the length of the vibrating string by pressing the string against the neck of the instrument with her fingers.

Example 15.8

(a) What is the wave speed of a guitar string whose fundamental frequency is 330 Hz if the length of string free to vibrate is 0.651 m? (b) What is the tension if the string's linear mass density is 0.441 g/m?

Strategy   (a) The wave speed may be determined from Eq. (15.2), which relates wave speed, frequency, and wavelength. The wavelength of the fundamental frequency is twice the free length of the string.
(b) The tension in the string may be obtained from the wave speed and the linear density of the string through the relationship of Eq. (15.10).

Solution   (a) The wave speed is found from the product of the wavelength and the frequency.

$$v = f\lambda = f(2L) = (330 \text{ Hz})(2)(0.651 \text{ m}) = 430 \text{ m/s}.$$

(b) The wave speed is related to the tension by

$$v = \sqrt{\frac{T}{\mu}}.$$

Squaring both sides and rearranging gives

$$T = \mu v^2 = (0.441 \text{ g/m})(10^{-3} \text{ kg/g})(430 \text{ m/s})^2$$
$$T = 81.5 \text{ N}.$$

■

Recall from our earlier discussion that the harmonic frequencies of a vibrating string are integer multiples of the fundamental frequency. A frequency of twice the fundamental frequency is called the second harmonic or first overtone; one of three times the fundamental is the third harmonic or second overtone; and so on. Not all harmonics are present in all vibrating systems.

A plucked string vibrates in a complicated way, with many of its harmonics contributing to its motion. In fact, it is this combination of harmonics that gives a vibrating string its distinctive sound. What's more, the particular combination of harmonics present in motion of the string depends on where it is plucked. A string plucked in the center has only odd-numbered harmonics (Fig. 15.20). The

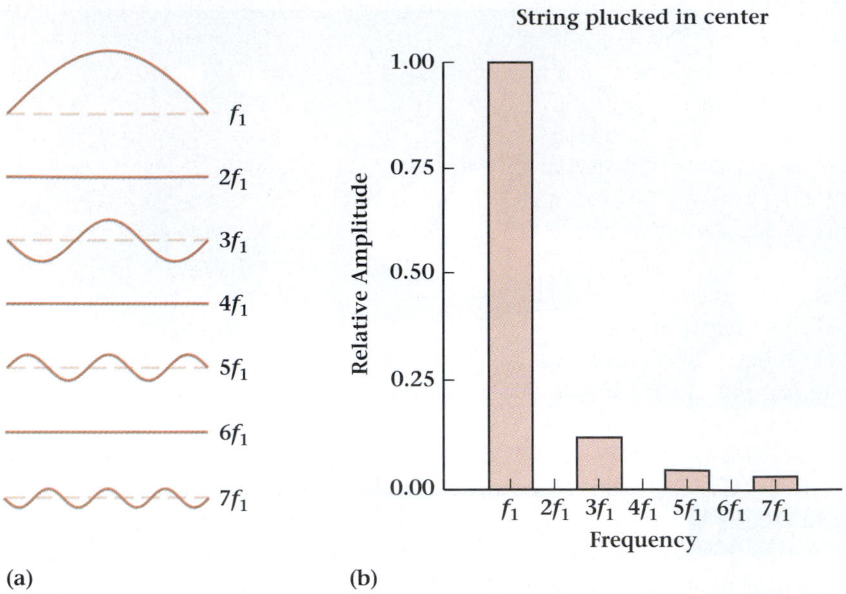

Figure 15.20

(a) A string plucked in the center has only odd-numbered harmonics. The sum of these harmonics with appropriate amplitudes and phases gives the proper shape of the string. (b) The bar chart shows the correct relative amplitude of each harmonic. The amplitudes in the wave diagram have been exaggerated for clarity.

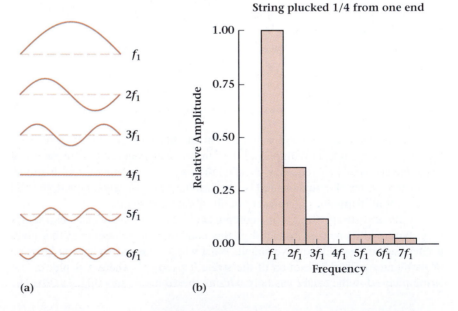

same string plucked at a point 1/4 of its length from one end has an entirely different set of harmonics (Fig. 15.21). The resulting sounds are different.

The distinctive sounds of different musical instruments depend on the harmonic content of the sound waveforms produced by the instruments and on how the waves are excited. For example, is the string excited by being plucked, struck, or bowed? Similarly, the mixture of harmonics makes a human voice distinctive. You recognize someone's voice not just because of its pitch and the way that person enunciates, but also because of that person's particular combination of overtones.

Figure 15.21

(a) A string plucked 1/4 of its length from one end lacks all harmonics that are integer multiples of $4f_1$. The sum of the allowed harmonics with appropriate amplitudes and phases gives the proper shape of the string. (b) The bar chart shows the correct relative amplitude of each harmonic. The amplitudes in the wave diagram have been exaggerated for clarity.

The sound from many musical instruments, such as pianos, guitars, and violins, originates in a set of strings, each vibrating with a specific fundamental frequency. In music, the term *frequency* is not usually employed, but other terms are used to describe essentially the same thing. When the frequency of two tones is in the ratio of $1:2$, they are said to be one *octave* apart. Between these two tones, each octave contains intermediate tones, arranged in a sequence that is repeated in every octave. Specifically, in Western music the octave is divided into twelve semitones, of which seven are selected to form a major or minor scale. A scale is just a regular series of tones arranged in order of frequency. A scale using the full sequence of twelve semitones is called a chromatic scale.

To Western ears, pleasing combinations of tones are those with frequencies that stand in whole-number ratios, such as $1:2$, $4:5$, and even $4:5:6$. Without going into detail, we will say that difficulties arise when this idea is used to fill an octave with twelve tones. Therefore, several scales have been developed that almost satisfy the requirements. The scale to which most musical instruments are tuned today is called the *equal-tempered scale*. In this scale each of the twelve tones in an octave has a frequency $\sqrt[12]{2}$ times the frequency of the previous one. As you can see in Table 15.2, the frequency ratios are not exactly whole numbers, although they are fairly close.

Although we refer to an acoustic guitar or a violin as a "stringed instrument," the sound does not come directly from the string. The string has too little surface area and insufficient amplitude to set much air in motion. Instead, the sound you hear from a guitar or violin comes from the body of the instrument, which is made to vibrate by the oscillating string. The frequency that is heard depends on the length of the particular string plucked or bowed. As we noted earlier, this length is determined by the performer, who chooses where to press the string to the fingerboard.

The quality of the tone you hear, say, from a violin, results primarily from the way the body of the violin is made. The vibrating string forces the body to vibrate and give off sound. The construction of the instrument determines the

Table 15.2	*Frequencies of the Notes in an Octave Beginning with Middle C*		
Name	Equal-tempered Frequency	Ratio to C_4	Nearest Whole-number Ratio to C_4
C_4	261.63	1.0000	1.0000
$C_4\#$	277.18	1.0594	
D_4	293.66	1.1224	1.1250 (9/8)
$D_4\#$	311.13	1.1892	
E_4	329.63	1.2599	1.2656 (81/64)
F_4	349.23	1.3348	1.3333 (4/3)
$F_4\#$	369.99	1.4142	
G_4	392.00	1.4983	1.5000 (3/2)
$G_4\#$	415.30	1.5874	
A_4	440.00	1.6818	1.6875 (27/16)
$A_4\#$	466.16	1.7818	
B_4	493.88	1.8877	1.8984 (243/128)
C_5	523.25	2.0000	2.0000

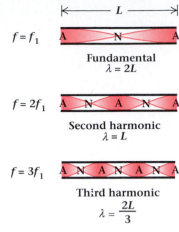

Figure 15.22
Standing waves in an open pipe. The ends of the pipe are antinodes. The fundamental resonant wavelength is twice the length of the pipe.

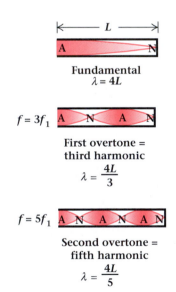

Figure 15.23
Standing waves in a pipe closed at one end. The fundamental resonant wavelength is four times the length of the pipe.

relative strength of the overtones for each tone that the body radiates. The maker must choose exactly which pieces of wood to use, their thickness at every point, the way in which the parts are joined, the finishing process, and many other details of the construction. Although it is now fairly well known what acoustical properties a violin must have to give it a good tone over its entire range, to actually build a good instrument requires knowledge, skill, and practice.

15.10 Waves in a Vibrating Column of Air

Hollow pipes have long been used for making musical sounds. Flutes, organ pipes, and children's whistles produce sounds in similar ways. To understand how they work, let us examine the behavior of air in a hollow pipe that is open at both ends. If you blow air across one end, the disturbance due to the moving air at that end propagates along the pipe to the far end. When it reaches the far end, part of the wave is reflected, much as a wave is reflected along a string whose end point is free to move. Since there is motion of the air at that end, the end point is an antinode with respect to the flow of air (Fig. 15.22). All harmonic frequencies are possible, just as for a string, because the pipe open at both ends has the same symmetry as the string fixed at both ends. And, as with the string, the fundamental wavelength is twice the length of the pipe.

If one end of the pipe is closed off, the air is not free to move any further in that direction and the closed end becomes a node. The resonant behavior of the pipe is completely changed. Since one end is a node and the other is an antinode, the lowest frequency (longest wavelength) vibration has no other nodes or antinodes between the ends. Thus, the fundamental wavelength is four times the length of the pipe (Fig. 15.23). The fundamental frequency is therefore a factor of 2 lower than for a pipe of the same length but open on both ends.

Above the fundamental frequency, the next resonant vibrational mode in our closed pipe has both an antinode and a node between the ends. Since this mode represents the next higher frequency, it is called the first overtone. But it is not the first harmonic. Its wavelength is one-third of the fundamental wavelength and its frequency is three times the fundamental, so it is the third harmonic. Similarly, the next higher frequency, the second overtone, is the fifth harmonic. Thus, for a pipe closed at one end, not all harmonics are possible; only odd-numbered harmonics can be excited.

What harmonics are excited also depends on the initiating disturbance. For example, if you blow gently across the top of a pop bottle, it resonates softly at its fundamental frequency. But if you blow much harder, you hear the higher pitch of an overtone because the faster airstream creates higher frequencies in the exciting disturbance. This same effect can also be achieved by increasing the air pressure to an organ pipe.

Example 15.9

The second overtone of standing waves in a pipe closed at one end is 512 Hz. How long is the pipe?

Strategy The fundamental resonant wavelength of a pipe closed at one end is four times the length of the pipe. For such a pipe, only odd harmonics, such as the first, third, fifth, seventh, . . . , occur. The second overtone in such a series is the fifth harmonic, which has a wavelength

$$\lambda = \frac{4L}{5},$$

as shown in Fig. 15.23, where L is the length of the pipe.

Solution The frequency of the second overtone is

$$f = \frac{v}{\lambda} = \frac{5v}{4L}.$$

We can rearrange this equation to solve for L. Upon inserting the numerical values, we get

$$L = \frac{5v}{4f} = \frac{5(340 \text{ m/s})}{4(512 \text{ s}^{-1})} = 0.83 \text{ m}.$$

*15.11 Beats

When two sound sources that have almost the same frequency are sounded together, an interesting effect occurs. You hear a sound with a frequency that is the average of the two. However, the loudness of this sound repeatedly grows and then decays, rather than being constant. Such repeated variations in amplitude are called **beats,** and the occurrence of beats is a general characteristic of waves.

If the frequency of one of the wave sources is changed, there is a corresponding change in the rate at which the amplitude varies. This rate is called the beat frequency. As the frequencies come closer together, the beat frequency becomes slower. Thus, a musician can tune a guitar to another sound source by listening for the beats while increasing or decreasing the tension in each string. Eventually the beats become so slow that they effectively vanish, and the two sources are then in tune.

Beats are easily explained by considering two sinusoidal waves y_1 and y_2 of the same amplitude y_0, but of different frequencies f_1 and f_2. The superposition principle says the combined amplitude y is the algebraic sum of the individual amplitudes:

$$y = y_1 + y_2$$
$$y = y_0 \sin 2\pi f_1 t + y_0 \sin 2\pi f_2 t$$
$$y = y_0(\sin 2\pi f_1 t + \sin 2\pi f_2 t).$$

Using the trigonometric identity for the sum of the sines of two angles, we have*

$$y = \left(2y_0 \cos 2\pi\left(\frac{f_1 - f_2}{2}\right)t\right) \sin 2\pi\left(\frac{f_1 + f_2}{2}\right)t. \tag{15.12}$$

---

*The following relationship follows from the definition of the trigonometric functions:
$$\sin A + \sin B = 2 \cos \frac{A - B}{2} \cdot \sin \frac{A + B}{2}.$$

Physics in Practice

HEARING AND THE EAR

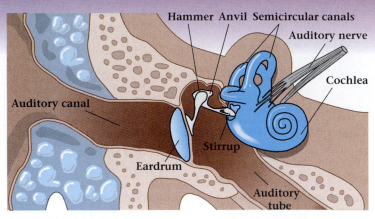

Figure B15.3 Schematic diagram of the human ear.

How good are human ears as detectors of sound? We can hear sounds ranging from the faint buzz of a mosquito's wings to the roar of a jet engine. In each case, air is set in vibration. Eventually the vibrations of the air reach your ear and act on your eardrum, letting you hear the sound. The human ear is an exceptional detector because it works over such a wide range of intensities and frequencies.

Your ears are not equally sensitive to all frequencies. In Fig. B15.2 we plot the intensity of the softest sound heard at different frequencies. The lowest region of the curve corresponds to the range of frequencies where the ear is most sensitive, which is in the range of 3–4 kHz. This is the frequency range of the upper notes on a piano.

We can see just how sensitive the ear is by using the relationship between vibrational amplitude and intensity given in Eq. (15.3). To find the amplitude of a sound that can just be heard by a person with good hearing, we use the value for the minimum detectable intensity, 10^{-12} W/m^2, and a frequency of 4000 Hz corresponding to the most sensitive frequency region. On using these values, along with representative values for the other parameters, we find a vibrational amplitude of about 3×10^{-12} m. This truly is a small displacement; it is approximately 1/100 of the diameter of the oxygen molecules

Figure B15.2 Frequency response of the human ear.

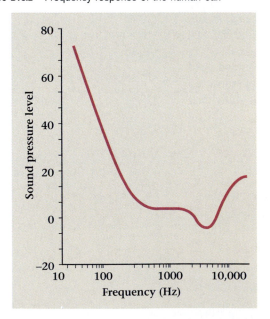

that make up the air. Your eardrum is unquestionably very sensitive to respond to such small fluctuations.

On the other hand, a sound loud enough to cause pain, say 120 dB or louder, has an amplitude of about 3×10^{-6} m. Though this value is quite a bit larger than the amplitude of the softest sound, it is still quite small. If a sheet of paper were of this thickness, a stack of 400 sheets would be about as thick as a dime.

Figure B15.3 shows a schematic diagram of the human ear. The auditory canal of the outer ear acts in an interesting way as an amplifier for certain frequencies. Consider the canal to be a pipe closed at one end by the eardrum. In Section 15.10, we discuss the resonant frequencies of such pipes. The average length of the auditory canal is about 2.5 cm, with a resonant frequency of 3400 Hz. Fig B15.2 shows an increase in sensitivity in the region of this resonance. It is interesting to note that this is also the frequency range of a baby's cry. This is, however, somewhat higher than the range of adult human speech, which lies primarily in the 500–2000 Hz region. Passive hearing aids that do not require batteries have been designed that effectively shift the resonance region downward toward the range of adult speech.

Beginning in the midteens, a gradual decrease in the sensitivity of hearing begins—both in frequency range and threshold of hearing. A young child may be able to hear sounds with frequencies as high as 40 kHz. By the teens, this upper limit has dropped to about 20 kHz, and from then on a relatively steady decrease of about 160 Hz per year is observed. For people 50 years old, an upper limit of 10–15 kHz is typical.

Temporary loss of hearing often follows exposure to a single, short, loud noise. For the most part, the ear recovers from such short duration overloads. More damaging are extended and repeated loud noises. People who work in loud environments are known to suffer permanent and irreversible hearing loss. For this reason they often wear special headgear or earplugs to reduce the sound's intensity level at their ears. Musicians who play loud music for long periods also frequently wear hearing protectors.

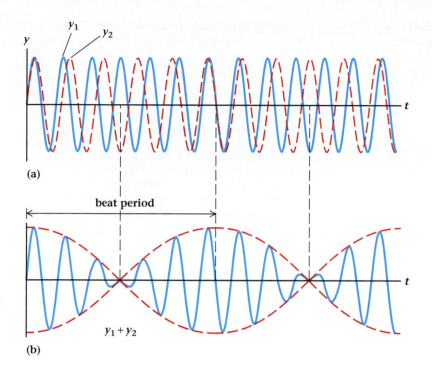

(a)

beat period

$y_1 + y_2$

(b)

Figure 15.24
(a) Two waves of slightly different frequencies add to produce a wave (b) that oscillates with the average of their frequencies. The resulting amplitude also oscillates, with a beat frequency equal to the difference between the source frequencies.

This equation represents the amplitude of a sound wave with a frequency that is the average of the two source frequencies (the sine term) and with a slowly varying amplitude (the cosine term) (Fig. 15.24). Our sensation of loudness depends on the intensity, which is proportional to the square of the amplitude. The rate at which the loudness maxima occur is the beat frequency. Its magnitude is the difference in the two source frequencies, $|f_1 - f_2|$. Beats can easily be heard up to frequencies of about 10 Hz. Beyond that, they are hard to distinguish.

Summary

$V_s \rightarrow$ Velocity of sound $= 340$ m/s

Useful Concepts

■ The general expression for a one-dimensional traveling wave is

$$y(x, t) = y_0 \sin 2\pi\left(\frac{x}{\lambda} - \frac{t}{T}\right).$$

■ The relationship between the speed, wavelength, and frequency of a wave is

$$v = \lambda f.$$

■ For a spherically expanding wave, the intensity decreases with the square of the distance from the source.

■ The intensity level L_I of a sound is measured in decibels relative to a reference sound intensity I_0:

$$L_I \text{ (in dB)} = 10 \log_{10} \frac{I}{I_0}. \qquad I_0 = 1 \times 10^{-12}$$

■ The observed frequency change due to motion of the source or observer is called the Doppler shift. The frequency heard is given by

$$f' = \frac{f}{(1 \mp v_s/v)} \qquad \text{moving source,}$$

$$f' = f(1 \pm v_o/v) \qquad \text{moving observer.}$$

The upper sign in these equations corresponds to approaching motion and the lower sign corresponds to leaving.

■ If the speed of the source exceeds the speed of the wave, a shock wave is formed. The shock wavefront makes an angle θ with the direction of motion of the source, given by

$$\sin \theta = \frac{v}{v_s}.$$

■ The transverse wave speed v in a flexible string or wire is given by

$$v = \sqrt{\frac{T}{\mu}},$$

where T is the tension in the string and μ is the mass per unit length of the string.

■ The frequency of standing waves in a string is

$$f_n = \frac{n}{2L}\sqrt{\frac{T}{\mu}}, \qquad n = 1, 2, 3, \dots.$$

■ When two sounds of almost the same frequency are heard at the same time, the sensation is of a single tone whose amplitude is rising and falling. These variations in amplitude are called beats. The rate at which they occur, called the beat frequency, is

$$|f_1 - f_2|.$$

Important Terms

You should be able to write the definition or meaning of each of the following:

transverse wave pulse	Doppler effect
traveling wave	shock wave
wave	principle of superposition
harmonic wave	constructive interference
wavelength	destructive interference
intensity	fundamental frequency
transverse waves	harmonic frequencies
longitudinal waves	overtones
wavefronts	standing wave
plane waves	nodes
decibel	antinode
intensity level	beats

Conceptual Questions

15.1 List as many transverse and longitudinal waves as you can.

15.2 You can estimate the distance to a lightning flash by counting the seconds that elapse between the time you see the flash and when you hear the accompanying thunder. (a) The approximate distance in miles is obtained by dividing the time in seconds by 5. Explain why this is so. (b) Estimate the distance to a lightning flash that you see 10 s before you hear the thunder.

15.3 The speed of sound in helium is much faster than in air at the same temperature and pressure. Use this fact to explain the observation that if you fill your lungs with helium and try to talk, your voice will be higher pitched than normal.

15.4 A stereo amplifier is said to have a channel separation of 40 dB. Explain what this means.

15.5 If an organ pipe is placed in a large chamber in which the air pressure is reduced to slightly below atmospheric pressure, will the frequency of the sounded note change?

15.6 If the whine of a bullet were 1000 Hz as it passed through the air, what would you hear if the bullet were traveling away from you with the speed of sound?

15.7 Describe what you will hear in the following situation: You swing a constant-frequency source attached to a string around your head in a 1-m-diameter horizontal circle. You are standing near a brick wall that reflects the sound.

15.8 Discuss why the lower-pitched strings in most musical instruments are wrapped with a thin coil of wire.

15.9 If you moisten your finger and rub it around the rim of a thin-stemmed glass, you often produce a sound. How is this tone produced? On what does it depend? Why must you wet your finger? What changes will take place if the glass is half filled with water?

15.10 How could you get a stretched string to vibrate in only its second harmonic mode?

15.11 What does it mean to say that musical scales are logarithmic?

Problems

Section 15.2 Harmonic Waves

Hints for Solving Problems

For harmonic waves, assume that $y = 0$ at $x = 0$ when $t = 0$ (corresponding to a sine wave). Remember the fundamental relationship $v = \lambda f$.

15.1 Write the general expression for a traveling harmonic wave of amplitude y_0 moving in the negative x direction with wavelength λ and period T.

15.2 A harmonic wave moving in the positive x direction has an amplitude of 3.0 cm, a speed of 40 cm/s, and a wavelength of 40 cm. Calculate the displacement due to the wave at (a) $x = 0.0$ cm, $t = 2.0$ s and (b) $x = 10$ cm, $t = 20$ s.

15.3 Harmonic waves are sent along a rope at a speed of 10 m/s. What is the frequency of the wave if successive wave crests are 2.5 m apart?

15.4 Radio waves travel with the speed of light, approximately 3.00×10^8 m/s in both vacuum and air. What are the wavelengths of the radio signals from (a) an FM station broadcasting at 97.5 MHz, (b) a standard broadcast AM station operating at 560 kHz, and (c) a shortwave station operating at 11,900 kHz?

15.5 The human eye is sensitive to light of wavelengths from about 400 to 700 nm. What range of frequencies does this range correspond to? Take the speed of light to be 3.00×10^8 m/s.

15.6 A shortwave receiver has "90-m band" written on one end of the dial and "11-m band" on the other. What range of frequencies does this radio receive? Take the speed of radio waves to be 3.00×10^8 m/s.

15.7● At $x = 15.0$ cm and $t = 2.00$ s, the displacement of a traveling wave is 8.66 cm. The amplitude of the wave is 10.0 cm, and its wavelength is 8.00 cm. What is its period? Assume the smallest positive phase angle.

15.8● A cart moves with a constant speed of 0.30 m/s in the x direction. A 2.3-kg lead brick is hung from a support on the cart with a spring of spring constant $k = 20$ N/m. The brick is pulled down 6.0 cm below the equilibrium point and released at $t = 0$. Write an expression for the y coordinate of a red dot painted on the center of the brick as a function of x and t. Assume that $x = 0$ when $t = 0$ and call the equilibrium position $y = 0$.

15.9● A strip of paper is pulled along at 20 cm/s at a right angle to the plane of a 0.49-m-long pendulum. The pendulum has a small, continuously running ink-jet device attached to the bob, which is arranged so that it marks the position of the pendulum on the paper (Fig. 15.25). The bob is held 8.0 cm to one side and released at $t = 0$. Write the equation for the wave drawn on the paper.

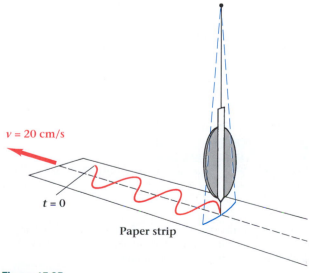

$v = 20$ cm/s

$t = 0$

Paper strip

Figure 15.25
Problem 15.9.

15.10● Rewrite the general expression for a traveling harmonic wave moving in the positive x direction in terms of the angular frequency ω and the wave number $k = 2\pi/\lambda$.

*Section 15.3 Energy and Information Transfer by Waves

15.11 A child sends traveling waves along a rope that passes through the narrow slits in a picket fence. The waves are transmitting 0.30 W. Another child puts some boards on the fence to restrict the amplitude of the waves that can pass, so they become only one-third of the original amplitude. Assuming that the speed of the waves is constant, what must the first child do to maintain the same level of power transmission?

15.12 A 6.0-m-long string has a mass of 8.5 g. Transverse waves propagate along the string with a speed of 189 m/s. One end of the string is forced to oscillate at 120 Hz with an amplitude of 0.53 cm. What power is transmitted along the string? You may want to use the expression for the power given in Problem 15.13.

15.13● Show that the expression for the power transmitted by a harmonic wave on a rope can also be written as

$$P = \tfrac{1}{2}\,\mu v\,\omega^2\, y_0^2,$$

where μ is the mass per unit length of the rope and $\omega = 2\pi f$.

Section 15.4 Sound Waves

Hints for Solving Problems

Assume the speed of sound in air to be 340 m/s, corresponding to a temperature of about 15°C, and ignore variations due to temperature unless specifically stated otherwise in the problem. The speed of sound in other materials is found in Table 15.1.

15.14 What are the shortest and longest wavelengths produced in air by a stereo system that has a useful range of 50 to 17,000 Hz?

15.15 What is the wavelength of a 2.00-kHz sound wave (a) in air, (b) in helium, and (c) in water?

15.16 (a) Calculate the frequencies of the pressure waves in air having the following wavelengths: 3.4 mm, 3.4 cm, 34 cm, 340 cm, and 34 m. (b) Which of these frequencies can be heard by human ears?

15.17 What is the wavelength of the waves produced in air by an ultrasonic distance-measuring device if the frequency is 40 kHz?

15.18 Calculate the speed of sound in dry air at a temperature of 28°C. $V = 331.5 + 0.6T$

15.19 When sounded, two identical car horns produce sound intensities at your ear in the ratio of 6:1. If the nearer car is 8.0 m away from you, how far away is the other car? (*Hint:* Assume spherical wavefronts.)

15.20 You are 10 m from a loudspeaker, where you hear a sound intensity I_0. How far must you be from the speaker to

Handwritten at top:

a) $f' = f_0 \div (1 - v_s/v)$

b) $\Delta f = f_0 \left(\frac{1}{1 - v_s/v} - \frac{1}{1 + v_s/v} \right)$

reduce the sound intensity that you hear to $I_0/3$? (*Hint:* Assume spherical waves.)

15.21• Show that for spherical harmonic traveling waves the amplitudes y_i at distances r_i from the source are related through

$$\frac{y_1}{y_2} = \frac{r_2}{r_1},$$

where the subscripts 1 and 2 refer to different locations.

15.22• Reference is sometimes made to a singer's "hitting high C." What is the wavelength of this sound at 12°C? (The frequency of high C, more precisely called C_6, is 1047 Hz.)

15.23• A loudspeaker sounds a 600-Hz tone. What is the difference in the wavelength of the sound waves produced in air at −3°C and at 38°C?

15.24•• The ultrasonic distance-measuring device shown in Fig. 15.8 emits waves at a specific frequency. The time between sending and receiving the pulse is measured to within $\pm T$, where T is the wave period. Show that the measured distance is accurate to within $\pm \lambda/2$, where λ is the wavelength.

*Section 15.5 Measuring Sound Levels

15.25 The sound in a classroom is 63,000 times as intense as the minimum threshold of hearing. What is the intensity level of this sound in decibels?

15.26 What is the ratio of the sound intensity near a textile loom to that in a library?

15.27 An audio amplifier has a power gain of 120,000. Express this gain in units of decibels.

15.28 In a good FM radio receiver, the radio signal detected may be as much as 65 dB greater than the noise signal. What is the ratio of signal intensity to noise intensity?

15.29 The upper limit of the intensity level of sound waves is about 194 dB. This intensity corresponds to 100% modulation of the air; that is, the amplitude of the pressure oscillations corresponds to that of atmospheric pressure. What intensity (in watts per square meter) would be required to generate this level of sound?

15.30 When you are listening to music at 78 dB, what is the intensity of the sound at your ears? Express your answer in watts per square meter.

15.31• By how many decibels do you reduce the sound intensity level due to a source of sound if you quadruple your distance from it? Assume that the waves expand spherically.

15.32•• Five identical looms operating in a textile mill produce a sound intensity level of 85 dB. If two additional looms are put into operation in the same room, what is the new sound intensity level?

15.33•• The sound intensity level in a factory is 93 dB when seven identical metal stamping machines are all in operation. If four of the machines are shut down for maintenance, what is the sound intensity level due to the remaining machines?

Handwritten at left:
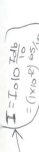
$I = I_0 10^{\frac{I_{db}}{10}}$
$= (1 \times 10^{-12}) 10^{65/10}$

Section 15.6 The Doppler Effect

Hints for Solving Problems

Remember to use the correct sign of v_s or v_o when calculating Doppler shifts.

15.34 A train approaching a station at a speed of 34 m/s sounds a 2000-Hz whistle. (a) What is the apparent frequency heard by an observer standing at the station? (b) What is the change in frequency heard as the train passes by?

15.35 A train going 40 m/s approaches a crossing bell whose frequency is 820 Hz. (a) What frequency is heard by passengers on the approaching train? (b) What frequency is heard by passengers after the train passes the bell?

15.36• You are standing by the railroad track when a train sounding a 750-Hz whistle passes you at 80 km/h. What is the difference in the frequencies you hear as the train approaches and departs?

15.37• A 440-Hz source has been sounding in air for a long time. (a) What frequency will you hear if you move away from it at 0.90 times the speed of sound? (b) What frequency will you hear if you move away from it at the speed of sound?

15.38• A car traveling 30 m/s overtakes another car going only 25 m/s. When the faster car is still behind the slower one, it sounds a horn of frequency 1500 Hz. What is the frequency heard by the driver of the slower car?

15.39• If you move at 18 m/s toward a 2300-Hz source that is moving toward you with a ground speed of 30 m/s, what frequency do you hear? $v_s = \left(\sqrt{(\Delta f / f)^2 + 1} - 1 \right) / \frac{\Delta f}{f}$

15.40•• If you detect a frequency shift of 30 Hz in the 6000-Hz bell of a bicycle as it approaches and then leaves you, how fast is the bicycle going? [*Hint:* You may find it helpful to use the approximation that for $x << 1$, $1/(1 \pm x) \approx 1 \mp x$.]

15.41•• A trailer truck traveling east at 28 m/s sounds a 1000-Hz horn. (a) What frequency is heard by an approaching driver headed west at 38 m/s? (b) What frequency is heard if the driver is headed east away from the truck at 38 m/s?

15.42•• (a) Derive an expression for the frequency as a function of time that you would hear if you dropped a source of frequency f_0 from a tall tower. (b) Derive an expression for the frequency heard by an observer on the ground.

*Section 15.7 Formation of a Shock Wave

Hints for Solving Problems

The speed of sound in several materials is found in Table 15.1.

15.43 What is the speed of the air flow around a model airplane in a wind tunnel if the half-angle of the shock wave is 42°? Give your answer as a Mach number.

15.44 A rocket model in a wind tunnel generates a shock wave with a half-angle of 65° when the gas is flowing at 400 m/s. What is the speed of sound in the gas?

15.45 What is the half-angle for the shock wave generated by the Concorde aircraft when it is flying at Mach 2.1?

15.46 A projectile moving at a speed of 418 m/s through an atmosphere of pure nitrogen creates a shock wave with a half-angle of 53°. What is the speed of sound in nitrogen?

15.47• A bullet traveling in helium at 15°C creates a shock wave with a half-angle of 45.0°. What is the half-angle of the shock wave if the bullet travels with the same velocity through air at the same temperature?

15.48• You hear the sonic boom of a high-speed jet plane exactly 3.0 s after it passes directly overhead in level flight. At the time you hear the boom, you see the plane at an angle of 20° above the horizon. (a) How fast is the plane traveling? (b) What is the altitude of the plane? Assume that the speed of sound at the altitude of the plane is 325 m/s.

Section 15.9 Standing Waves on a String

> **Hints for Solving Problems**
>
> Be sure to determine whether a vibrating string or column of air has nodes or antinodes at its ends. Remember the distinction between overtones and harmonics.

15.49 What is the wave speed in a guitar string stretched between supports 0.65 m apart if the fundamental frequency of the string is 392 Hz?

15.50 A nylon string is stretched between supports 1.20 m apart. Given that the speed of transverse waves in the string is 800 m/s, find the frequency of the fundamental vibration and the first two overtones. $f = V/2L$ ① ⇒ ×2 ② ⇒ ×3

15.51 An experiment shows that a pulse propagates with a speed of 260 m/s along a nylon cord subject to a tension of 68.0 N. What is the mass per unit length of the cord?

15.52 Two adjacent strings on an unusual stringed instrument have the same mass per unit length and are subjected to the same tension. One of the strings is 1.44 times as long as the other. What is the ratio of the fundamental frequency of the longer string to the second harmonic of the other one?

15.53 A wire stretched between supports 50 cm apart has a fundamental frequency of 400 Hz. Can it be resonantly excited by an object vibrating at a frequency of (a) 2000 Hz and (b) 3000 Hz?

15.54• (a) How far does a transverse pulse travel in 1.23 ms on a string of density 5.47×10^{-3} kg/m under tension of 47.8 N? (b) How far will the pulse travel in the same time if the tension is doubled?

15.55• A steel wire is stretched taut between supports one meter apart. (a) What is the fundamental wavelength of vibration of the wire? (b) What is the fundamental frequency if the

speed of sound in the wire is 2050 m/s? (c) What is the wavelength of the fundamental frequency in air?

15.56• The vibrations from an 800-Hz tuning fork set up standing waves in a string clamped at both ends. The wave speed in the string is known to be 400 m/s for the tension used. The standing wave is observed to have four antinodes and an amplitude of 2.0 mm. How long is the string?

15.57• A 5.00-kg mass is suspended from the ceiling by a 20.0-g wire 1.60 m long. What is the fundamental frequency of the wire?

15.58• What is the ratio of frequencies of a higher note to a lower note that is 16 semitones below it on the equal-tempered scale?

15.59• A string stretched between two supports sets up standing waves with two nodes between the ends when driven at a frequency of 230 Hz. (a) What order harmonic is such a wave? (b) What is the frequency of the fundamental? (c) At what frequency will the wave have three nodes?

15.60•• A uniform string of linear density 10 g/m is tied to the ceiling of an elevator. A 5.0-kg mass is hung from the other end of the string at a point 1.00 m from the ceiling. If the elevator accelerates upward at 0.10 g, what is the fundamental frequency of the string?

Section 15.10 Waves in a Vibrating Column of Air

15.61 Calculate the fundamental frequency and the first three overtones of a hollow pipe 36 cm long and open at both ends.

15.62 (a) What is the fundamental frequency of a hollow tube 50 cm long and open at both ends? (b) What would be the frequency if the tube were closed at one end? ⓐ $f = V/2L$ ⓑ $f = V/4L$

15.63• Calculate the fundamental and first three overtones of a hollow pipe 25 cm long and closed at one end.

15.64• Low-frequency standing waves can sometimes be generated in tall cylindrical structures, such as unused smokestacks. (a) What would be the frequency of the fundamental acoustical vibration in a 50-m-tall smokestack if it were closed at the bottom? (b) What would be the frequency of the first three overtones?

15.65• Singing in the shower is a habit with some people, even those who do not normally sing elsewhere. To get some insight into the reason for this, calculate the first four harmonics associated with each of the linear dimensions of a 1.65 m $\times$ 2.30 m $\times$ 2.43 m room and put them in ascending order. Assume that the speed of sound is $v = 340$ m/s.

*Section 15.11 Beats

15.66 Two tuning forks are sounded simultaneously, and a beat frequency of 2.0 Hz is heard. If the frequency of the higher-pitched fork is 262 Hz, what is the frequency of the other one?

15.67 How many beats are heard when two organ pipes, each open at both ends, are sounded together if one pipe is 60 cm long and the other is 61 cm long?

15.68• Two identical guitar strings are stretched with the same tension between supports that are not the same distance apart. The fundamental frequency of the higher-pitched string is 400 Hz, and the speed of transverse waves in both wires is 150 m/s. How much longer is the lower-pitched string if the beat frequency is 3.0 Hz?

15.69• Piano tuners use overtones and beats as an aid in tuning a piano. If a tuner hears a beat frequency of 1.0 Hz between the second harmonic of a string whose fundamental frequency is 440 Hz and the third harmonic of another note, what is the other note? Assume that the tuner has already determined that the second harmonic of 440 Hz is the higher of the two frequencies.

15.70•• Two strings are each 1.00 m long with a mass of 0.10 g. Both are subjected to the same tension and are located in an elevator. One string is stretched between fixed supports. One end of the other string is fixed to the ceiling of the elevator while the other end supports a 6.00-kg mass. When the elevator is at rest, both strings vibrate with the same fundamental frequency. When the elevator accelerates upward, the notes are different and a beat frequency of 2 Hz is heard. What is the acceleration of the elevator?

Additional Problems

15.71 If one sound is 25 dB greater than another sound, what is the ratio of their intensities?

15.72 A wire of mass 0.030 kg is stretched between fixed supports a distance 1.50 m apart. (a) If the tension in the wire is 850 N, what is the speed of a wave along the wire? (b) How does the wave speed compare with the speed of sound in air?

15.73• Rewrite the expression for the speed of sound as a function of temperature so that the speed is in feet per second and the temperature in degrees Fahrenheit.

15.74• A car moving with a speed of 20 m/s sounds a horn of frequency 1000 Hz. (a) What is the frequency heard by a stationary observer positioned ahead of the car? (b) What frequency does the observer hear if the wind is blowing at a speed of 5 m/s from the car toward the observer?

15.75• An automobile passes by on the street blowing its horn. A music student standing on the sidewalk observes that the pitch of the horn drops a musical third; that is, the frequency heard after it passes is only 27/32 of the frequency heard when it was approaching. (a) Calculate the speed of the car in meters per second. (b) Also give your answer in units of kilometers per hour.

15.76• The six strings of a classical guitar are all 65.5 cm long and are tuned to frequencies of 82, 110, 147, 196, 247, and 330 Hz. The mass per unit length of the set of strings is, starting with the one of lowest frequency, 5.05×10^{-3}, 3.71×10^{-3}, 2.21×10^{-3}, 1.01×10^{-3}, 0.58×10^{-3}, and 0.44×10^{-3} kg/m. (a) What is the tension in each string? (b) What is the total string force?

15.77•• A student whirls an electronic buzzer around her head. The buzzer, which makes an 800-Hz tone when at rest, is swung on a string in a 0.75-m-radius horizontal circle at three revolutions per second. Derive an expression for the frequency as a function of time of the sound heard by a distant observer.

15.78•• The text following Eq. (15.7) claims that when v_s and v_o are much less than v, the Doppler formulas for moving source and for moving observer give essentially the same results. (a) Show that the claim is true, and (b) determine the limiting value of $v_o = v_s$ for which the two formulas agree to within 1.0%. (c) Make a graph of the % difference as a function of v_s/v.

15.79•• A traveling wave has a wavelength of 0.25 m and a period of 0.10 s. (a) If you stand in one place, how many crests pass you per second? (b) If you travel in the same direction as the wave at a speed of 1.0 m/s, how many crests pass you in one minute? (c) If you travel in the same direction as the wave at a speed of 3.0 m/s, how many crests do you intercept in one minute? (d) If you go in the opposite direction from the direction of the wave motion at 0.50 m/s and start from a crest at $t = 0$, how far away from you will that same crest be at $t = 32$ s?

15.80•• Show that the change in wavelength $\Delta\lambda$ of the sound from a source due to a change in the temperature of the air ΔT is approximately

$$\Delta\lambda \approx (1.8 \times 10^{-3})\lambda \, \Delta T,$$

where $\Delta\lambda$ and λ are in meters and ΔT is in degrees Celsius.

15.81•• In Chapter 12 we stated that the speed of sound in a gas is proportional to v_{rms} of the gas molecules and therefore is proportional to the square root of the temperature in kelvins. In this chapter we asserted that the speed of sound is well represented by a constant plus a term that is proportional to the temperature in degrees Celsius. Compare these two statements by plotting both predictions on the same graph from 0°C to 100°C and determining the maximum percentage difference between the two predictions in the specified temperature range. (*Hint:* The expression from Chapter 12 may be written in the form $v = 331.5\sqrt{T/273.15}$, where T is in kelvins.)

15.82•• Show that the expression for a standing wave,

$$y = 2y_0 \cos\left(2\pi \frac{x}{\lambda}\right) \cdot \sin\left(2\pi \frac{t}{T}\right),$$

may be derived by adding together two traveling waves that are of the same amplitude y_0 and same frequency but that travel in opposite directions.

15.83•• A copper and a steel pipe are of the same diameter and length at 8°C. The fundamental frequency of the air resonance of both is 180 Hz at this temperature when open at both ends. How many beats per second will be heard at 27°C?

Physics in Practice: Dipoles and Microwave Ovens

Electric Charge and Electric Field

Y ou are already familiar with many electrical phenomena. Lightning is one; another is the shock you get on touching a doorknob after walking across a rug on a dry day. In each case you see a brief spark, but the effect does not persist. Such events are due to what we call static electricity.

You are also familiar with the pull of a magnet on the refrigerator. Though seemingly unrelated, electricity and magnetism are different aspects of a single fundamental force of nature, called electromagnetism. We begin our study of electromagnetism by considering the forces that arise from static electricity. As you will see, the law describing the electrostatic force has the same mathematical form as the law of universal gravitation. We then define the electric field in a manner similar to the way we defined the gravitational field. Because the field concept plays a prominent role in the theory of electricity, you should give particular attention to the field and its applications in order to understand the chapters that follow. The laws of electricity given here apply both to large-scale events and to atomic and nuclear interactions, as you will see in later chapters.

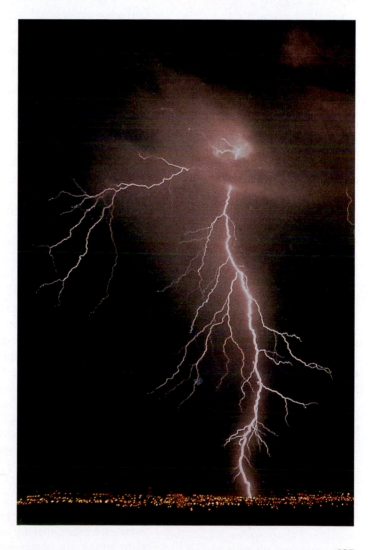

In the modern world, virtually no human activity takes place without being touched somehow by electricity. The use of electricity for energy, illumination, and the transfer of information is one of the facets of life in the twentieth century that most sharply separates it from life in previous centuries. Today, the rate of change continues at an even greater pace with rapid developments in communications technology and computers. Electricity has revolutionized not only the technological side of our lives, but also the cultural, political, and economic areas.

As we study electricity, we still use the concepts of mechanics, especially the ideas you already learned about the conservation of energy and the relationship between force and acceleration. In fact, the nongravitational forces that we studied in mechanics, such as friction and Hooke's law, are actually due to electrical forces between molecules and atoms. The common contact forces of daily life are electrical in origin. Consequently, understanding the electrical nature of matter helps us understand mechanics at the molecular level. ■

Figure 16.1
When rubbed briskly with a cloth, amber attracts bits of paper.

16.1 Electric Charge

Humans have known about electrical effects for thousands of years. As far back as the ninth century B.C., people searched the seashores for a yellowish-brown fossil resin called amber, which they carved and polished into beads and jewelry. By 600 B.C., the Greeks had discovered that when amber was rubbed briskly with a cloth, it would attract and pick up small, light objects such as bits of feathers or straw, or even thin scraps of metal (Fig. 16.1). The Greek word for amber is *elektron,* and it is from this root word that we get our word *electricity.* Lightning (seen in the chapter opener photo) is another example of an electrical effect in nature. However, the relationship between lightning and the behavior of amber went unrecognized for centuries.

As we will see, the basic element of electricity is electric charge, which is a fundamental property of matter. The Englishman William Gilbert (1544–1603) was an early investigator of what we now call **electrostatics,** the study of electric charges at rest. Gilbert made a sensitive test instrument from a lightweight stick pivoted about its center on a needle point. Gilbert found that if he rubbed a piece of amber and held it near the end of the stick, the stick moved noticeably about its pivot, indicating an attraction by the amber. He tried rubbing other objects and found that many of them produced similar effects. In this way Gilbert found a force that was present only after the amber or other substance was rubbed. This force was due not to gravitation, but to a different source: electricity.

The realization that rubbing two materials together could produce electricity led to the development of frictional electric machines. The first true frictional electric machine was developed about 1705 by Francis Hauksbee. Hauksbee's machine consisted of a glass sphere turned by a crank (Fig. 16.2). When rubbed, the sphere was capable of making sparks and causing light threads to stand out from each other, thereby demonstrating the ability of electricity to repel as well as to attract. An object such as Hauksbee's globe is said to become "charged" when rubbed, meaning that it possesses a net electric charge. It is this net **elec-**

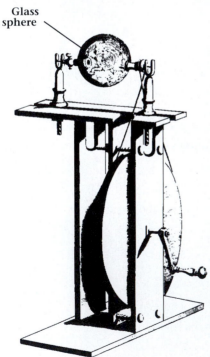

Glass
sphere

Figure 16.2
Hauksbee's electrostatic machine. The glass sphere was turned by a crank, becoming charged by friction.

tric charge that is the source of the electric force of attraction or repulsion. Charge, like mass, is a fundamental property of matter.

Through a series of experiments in 1731, Stephen Gray found that he could conduct electric charge over a great distance through a metal wire supported by silk cords. This observation led to the classification of materials into two types, either conductors or insulators. A **conductor** is a material through which charge may flow easily; an **insulator** is a material through which charge flows poorly or not at all.* Most metals are good conductors, while silk and cotton threads, wood, rubber, and many plastics are good insulators.

Two years later, in 1733, Charles-François du Fay, superintendent of gardens to the king of France, discovered that rubbing various objects together could generate two different kinds of electricity. Du Fay found that while two pieces of rubbed glass repelled each other and two pieces of rubbed amber repelled each other, the rubbed amber was strongly attracted to rubbed glass (Fig. 16.3). His experiments provided evidence for two kinds of electric charge and for a basic rule describing their behavior: **like charges repel** and **unlike charges attract.**

To explain this observed behavior of electricity, Benjamin Franklin (1706–1790) proposed a "single fluid" theory of electricity. According to his theory, only one kind of charge is mobile. The two types of electric behavior—that found on rubbed amber and that found on rubbed glass—are due to a deficiency or an excess of the more mobile kind of charge, transferred during the rubbing. Franklin arbitrarily designated the effect produced by rubbing glass as positive electricity and that produced by rubbing amber as negative electricity. We still use these terms today.

In solids, the mobile charges are negative **electrons,** which are discussed in detail in later chapters. When electrons are removed from an object, positive charges remain. These positive charges are the nuclei of the atoms that make up the material. In solids, the positive charges are fixed and do not move around. However, in liquids and gases, both positive and negative charges are free to move, so both types contribute to the conduction of charge.

Macroscopic objects are nearly always electrically neutral; that is, they have neither excess negative nor excess positive charge. However, we can give them a negative charge by adding extra electrons. Conversely, we can make them positive by removing electrons, so as to leave them deficient in negative charge (and thus positive). It is important to note that no charge is removed from any object except by being transferred to another object. For example, in rubbing glass with silk (Fig. 16.4), electrons are removed from the glass, leaving it positive. At the same time, the transfer of electrons to the cloth makes the cloth negative. The excess positive charge left on the glass is equal in amount (but opposite in type) to the excess negative charge transferred to the cloth. This equality illustrates a basic natural law known as the **law of conservation of charge,** which states that the *total amount of electric charge in the universe remains constant.* That is, *single charges can be neither created nor destroyed.* This principle is one of the fundamental observations of nature, equal in importance to the conservation laws for energy, momentum, and angular momentum. No violations of this principle have ever been observed. Charges can be created (and destroyed) only in pairs

---

*Modern electronic devices are made from semiconductors, which are materials that are neither good conductors nor good insulators, but show a behavior intermediate between the two. The nature of semiconductors is described in Chapter 31.

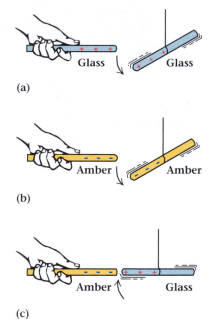

Figure 16.3

(a) Two rubbed glass rods repel each other. (b) Two rubbed amber rods repel each other. (c) A rubbed glass rod and a rubbed piece of amber attract each other.

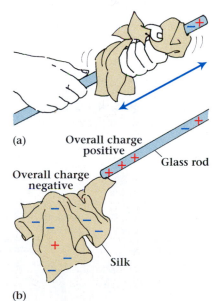

Figure 16.4

(a) Rubbing a glass rod with silk transfers electric charge. (b) The glass becomes positively charged and the silk becomes negatively charged.

of equal magnitude and opposite sign. Thus, the electrostatic effects that you have observed when running a comb through your hair or rubbing a balloon on your sweater are the result of charges separated by rubbing.

The presence of excess electric charge can be detected by an *electroscope* (Fig. 16.5). The one shown consists of a conducting metal ball and rod, *A,* with a pair of flexible metallic leaves, *B,* attached to its lower part. The rod is insulated from the case, *C.* When excess charge is present on the assembly of rod and leaves, the leaves swing out because of repulsion between the like charges on each leaf. If the leaves are made very light, the electroscope can be made quite sensitive to small amounts of charge.

We can illustrate the nature of electric charge by charging an electroscope, as shown in Fig. 16.5. We induce a charge on the electroscope without ever touching it with a charged object. This technique is known as **charging by induction.** Step 1: The electroscope is initially uncharged; that is, it contains equal amounts of positive and negative charge. Step 2: When we bring a negatively charged rod near (but not touching) the metal knob on top of rod *A,* negative charges in the electroscope rod are repelled to the lower part, leaving the knob positively charged. The negative charges go to the two leaves, which move apart because of the repulsive force between the negative charges. Step 3: When we touch the knob with a finger, the negative charges, still repelled by the negatively charged rod, flow out of the electroscope through our body to the earth. This procedure is called *grounding.* The leaves on the electroscope then collapse. Step 4: When we remove the finger, the position of the leaves does not change. Step 5: When the negatively charged rod is removed, the entire electroscope is left with a deficiency of negative charge—that is, with an excess of positive charge. The leaves again diverge, this time from the repulsive force between two positive charges.

We can explain what happens during the charging process in terms of moving positive and negative charges or in terms of moving negative charges only. Notice that the charge remaining on the electroscope is opposite to that of the charging rod.

We can also charge the electroscope by touching it with the charged rod. If we touched the negatively charged rod to the knob in Step 2 of Fig. 16.5, some

Figure 16.5

Charging an electroscope by induction. Step 1: An electroscope without any excess charge. Steps 2 through 5 use a negatively charged rod to induce a positive charge on the electroscope.

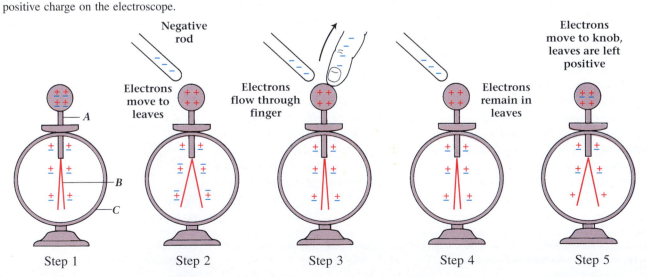

negative charge would be left behind when we moved the rod away. In this case, the charge remaining on the electroscope is the same as that of the charging rod.

16.2 Coulomb's Law

The French engineer and physicist Charles Coulomb (1736–1806) attacked the problem of quantifying the attractive and repulsive forces in static electricity with little concern for the nature of charge itself. Coulomb said that the important thing was to get a mathematical expression for the electrostatic force in terms of the magnitudes of the charges and their separation.

To measure the relatively small electrostatic forces, Coulomb used a torsional balance (Fig. 16.6), similar to the balance Cavendish used later to measure the gravitational force between masses (discussed in Chapter 5). In his experiments on electrical forces, Coulomb charged two small pith balls (pith is a light, spongy material from the center of plant stems) and measured the force between them. Coulomb found that "the repulsive force between two small spheres electrified with the same type of electricity is inversely proportional to the square of the distance between the centers of the two spheres."* He also verified the inverse-square distance law for the attractive force between unlike charges.

The electrostatic force also depends on the product of the charges. Thus, the magnitude of the force between two point charges is

$$F = k\frac{|q_1||q_2|}{r^2}, \tag{16.1}$$

where q_1 and q_2 are the values of the two charges, r is the separation between them, and k is a proportionality constant. The vertical bars indicate the magnitude, or absolute value, of the charges. The relationship of Eq. (16.1) is known as **Coulomb's law.** The force is directed along the line joining the two charges. Notice that there are two forces, one acting on one charge and another acting on the other charge. These two forces form an action-reaction pair, in accord with Newton's third law. Thus, the force $\mathbf{F}_{1,2}$ exerted on charge q_1 by charge q_2 is equal in magnitude and opposite in direction to the force $\mathbf{F}_{2,1}$ exerted on charge q_2 by charge q_1 (Fig. 16.7). If both q_1 and q_2 are positive, the force is repulsive. Similarly, if both charges are negative, the force is again repulsive. However, if one charge is positive and the other negative, then the force is attractive. Coulomb's law has the same mathematical form as Newton's law of universal gravitation, except that in the electric case the force can be either attractive or repulsive. Also, like the gravitational force, the electrostatic force is a conservative force.

The magnitude of the proportionality constant k in Eq. (16.1) depends on the system of units we choose for measuring the charges and their separation. If we choose k to be a dimensionless constant, then Eq. (16.1) could be the defining equation for electrical charge. In that case, the dimension of charge would be length times the square root of force. An entire system of electrical units can be built up on the basis of this choice. However, present practice favors the SI

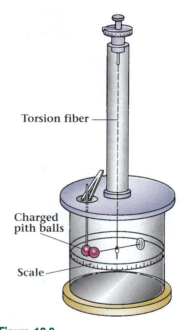

Figure 16.6
Coulomb's apparatus. The torsion balance measured the electric force on the charged spheres in the same way that Cavendish's balance measured the gravitational force.

Labels on figure: Torsion fiber; Charged pith balls; Scale

---

*C. Coulomb, *Mémoires de l'Académie des Sciences* (1785, p. 569 ff.), translated in *Great Experiments in Physics,* edited by M. H. Shamos (New York: Holt, Rinehart and Winston, 1959), p. 63.

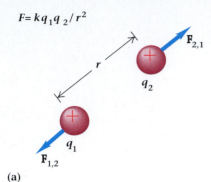

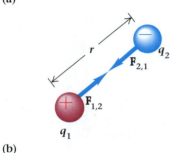

$F = kq_1q_2/r^2$

(a)

(b)

Figure 16.7
The force $\mathbf{F}_{1,2}$ exerted on charge q_1 by charge q_2 is equal in magnitude and opposite in direction to the force $\mathbf{F}_{2,1}$ exerted on charge q_2 by charge q_1. The direction of the force depends on whether the objects have the same sign (a) or opposite signs (b).

system, in which the base unit is the unit for electric current, the ampere (defined in Chapter 18). The SI unit for charge, the **coulomb** (C), is derived from the ampere. If we measure the amount of charge in coulombs and the separation in meters, then Eq. (16.1) correctly gives the force in newtons if

$$k = 8.988 \times 10^9 \text{ N} \cdot \text{m}^2/\text{C}^2 \approx 9.0 \times 10^9 \text{ N} \cdot \text{m}^2/\text{C}^2.$$

It is often convenient to define the constant in Coulomb's law to be

$$k = \frac{1}{4\pi\epsilon_0}.$$

While this appears to unnecessarily complicate the force equation, it does eliminate a factor of 4π from other equations we shall encounter later. The quantity ϵ_0 is known as the **permittivity of free space** and has the value

$$\epsilon_0 = \frac{1}{4\pi k} = 8.854 \times 10^{-12} \frac{\text{C}^2}{\text{N} \cdot \text{m}^2}.$$

When the permittivity constant is used, Coulomb's law takes the form

$$F = \frac{1}{4\pi\epsilon_0} \frac{|q_1||q_2|}{r^2}. \tag{16.2}$$

Although the coulomb is the SI unit of charge, it is not the most fundamental unit. The smallest and most basic unit of charge found in nature is the charge on an electron (negative) or a proton (positive). The magnitude of this quantity is denoted by e, the **elementary charge,** which has the value

$$e = 1.602 \times 10^{-19} \text{ C}.$$

Experimentally, it has been found that electric charge always occurs in multiples of the elementary charge e. For this reason we say that charge is *quantized;* that is, it occurs only in integer multiples of e.*

Problem-Solving Strategy

Forces Between Electric Charges

When working problems that involve the electrostatic force, remember that the force between two electric charges is directed along the line joining the charges and that the force between like charges is repulsive and the force between unlike charges is attractive. The magnitude of the force can be found from Coulomb's law.

---

*Current models of nuclear particles, such as the proton, postulate the existence of component particles called quarks with fractional values of the electron charge. Most interpretations of these models indicate that isolated quarks, and hence their fractional charges, are not observable in the ordinary sense. So far, no isolated quarks have been detected. (See Chapter 32.)

Table 16.1	Approximate Charge on Some Objects	
Object		**Charge (C)**
Single electron		$\approx 10^{-19}$
Aerosols and ink-jet printer drops		$\approx 10^{-15}$
Person on insulating stand		$\approx 10^{-6}$
Charge transferred between storm cloud and earth		$\approx 10^2 - 10^4$

Source: Joseph M. Crowley, *Fundamentals of Applied Electrostatics* (New York: John Wiley, 1986).

It takes about 6.24×10^{18} electrons to make one coulomb of negative charge. The electrostatic force between two charges of one coulomb each is so great that separate charges of coulomb magnitude are rarely encountered. A more typical range of charge encountered in electrostatic experiments is from 10^{-9} to 10^{-6} C (nC to μC). Table 16.1 lists the charge on several objects. Two examples of Coulomb's law and the magnitude of electrostatic forces are given below.

Example 16.1

Two tiny Styrofoam balls each with a mass of 152 mg are coated with conducting paint and suspended from a common point by light insulating threads 20.0 cm long. A static electric charge is given to one of the balls. When they are touched together, the charge is shared equally between them as the charge moves on their conducting surfaces. Afterward, they repel each other and swing apart until each makes an angle of 15° with the vertical (Fig. 16.8a). What is the charge on each ball?

Strategy It is helpful to draw a free-body diagram like Fig. 16.8(b). From Newton's third law we know that the force on one ball must be equal in magnitude and opposite in direction to the force on the other ball. Each ball is also acted on by gravity and by the thread. When the balls are at rest, the net force acting on each is zero and the system is in equilibrium. From the figure, we see that in the equilibrium position the electric and gravitational force components are related by

$$\tan 15° = \frac{F_{\text{electric}}}{F_{\text{gravity}}} = \frac{k\dfrac{q^2}{r^2}}{mg}.$$

The separation r between the balls can be found using trigonometry from knowledge of the length of a thread and the angle between the thread and the vertical direction.

Solution We see from Fig. 16.8(b) that the value of r is $2L \sin 15°$, where L is the length of the thread. Thus, we get

$$\tan 15° = \frac{kq^2}{mg(2L \sin 15°)^2},$$

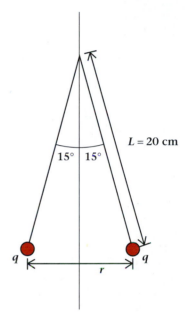

(a)

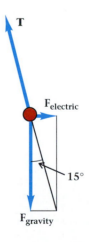

(b)

Figure 16.8

Example 16.1: (a) Two identically charged Styrofoam balls are separated by their repulsive forces on each other. (b) A free-body diagram for one of the charged balls.

which can be rearranged to give

$$q = \pm 2L \sin 15° \sqrt{\frac{mg \tan 15°}{k}}.$$

We can insert the numerical values into the equation and calculate the charge directly. The result is

$$q = \pm 2(0.200 \text{ m}) \sin 15° \sqrt{\frac{(152 \times 10^{-6} \text{ kg})(9.81 \text{ m/s}^2) \tan 15°}{9.0 \times 10^9 \text{ N} \cdot \text{m}^2/\text{C}^2}}$$

$$q = \pm 2.2 \times 10^{-8} \text{ C}.$$

Discussion The $\pm$ sign in the calculation of the charge arises naturally when we compute the square root. Without some other information we have no way of knowing whether the charge given to the Styrofoam balls is positive or negative. Their behavior is exactly the same for either case. However, the charges must be either both positive or both negative.

Example 16.2

Compare the electrostatic repulsive force between two electrons with the attractive force due to gravitation by finding the ratio of the magnitudes of the forces. (The properties of the electron are described in later chapters; however, some of its properties are listed in the table in the back endpaper of this book.)

Solution The electrostatic force is given by Coulomb's law, and the gravitational force is given by Newton's law of gravitation. Both of these forces depend on $1/r^2$. The ratio of the magnitude of these forces is then independent of the separation r between the electrons.

The magnitude of the electrical force is

$$F_e = k\frac{q^2}{r^2} \qquad \text{(repulsive)}$$

and the magnitude of the gravitational force is

$$F_g = G\frac{m^2}{r^2}. \qquad \text{(attractive)}$$

The ratio of these two forces is

$$\frac{F_e}{F_g} = \frac{k}{G}\frac{q^2}{m^2}.$$

By substituting in the values of the charge of the electron, -1.6×10^{-19} C, and its mass, 9.1×10^{-31} kg, along with the values for k and G, we get a ratio of the forces of

$$\frac{F_e}{F_g} = \frac{9.0 \times 10^9 \text{ N} \cdot \text{m}^2/\text{C}^2}{6.67 \times 10^{-11} \text{ N} \cdot \text{m}^2/\text{kg}^2} \frac{(1.6 \times 10^{-19} \text{ C})^2}{(9.1 \times 10^{-31} \text{ kg})^2}$$

$$\frac{F_e}{F_g} = 4.2 \times 10^{42}.$$

Discussion The electrostatic force between electrons is immensely greater than the gravitational force between them. Thus, when describing the behavior of electrons, it is usually sufficient to consider only the electrical forces and neglect the gravitational effects.

The magnitude of the force between electric charges is, in large part, the reason why doing static electric experiments is so difficult. Because the force between charges is so large and insulators are not perfectly insulating, it is hard to separate a large number of charges and keep them separated.

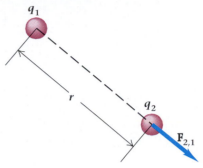

Figure 16.9
The direction of the force $\mathbf{F}_{2,1}$ on a charge q_2 due to another charge q_1 is along the line between the charges and depends on the signs of both charges.

16.3 Superposition of Electric Forces

Up to now we have dealt with the problem of finding the electrostatic force on a charge q_2 due to a charge q_1 (Fig. 16.9). Often it is necessary to find the net force on a charge when more than two charges are present. It is helpful to approach this type of problem in parts. We can determine the net force on any one charge by summing up the individual contributions to the force due to each of the other charges. This independence of the forces is known as the **principle of superposition.** For example, suppose you have the three charges shown in Fig. 16.10 and you wish to find the repulsive force on the positive charge q_3 due to the positive charges q_1 and q_2. First find the magnitude and direction of the force exerted on q_3 by q_1. Then determine the force exerted on q_3 by q_2. The net force on q_3 is the *vector sum* of these two individual forces. Notice that the force $\mathbf{F}_{3,1}$ exerted on q_3 by the charge q_1 is independent of the presence of q_2. Similarly, the force $\mathbf{F}_{3,2}$ exerted on q_3 by q_2 is independent of the presence of q_1. The examples that follow illustrate this procedure.

Superposition is not a self-evident principle for all types of forces; each case must be examined separately. However, superposition is valid for electrical forces.

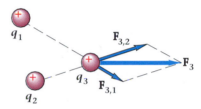

Figure 16.10
The force $\mathbf{F}_3$ on charge q_3 is the vector sum of the forces due to q_1 and q_2, considered independently.

Example 16.3

Calculate the force on the charge q_3 due to the other two charges located as shown in Fig. 16.11(a). Take the magnitudes of the charges to be $q_1 = -4.2 \ \mu C$, $q_2 = +1.3 \ \mu C$, and $q_3 = +1.1 \ \mu C$.

Strategy Before making any calculations, we observe from the signs of the charges that the force between q_2 and q_3 is repulsive while that between q_1 and q_3 is attractive. Thus, knowing that like charges repel and opposite charges attract, we can draw a free-body diagram (Fig. 16.11b). Notice that the force $\mathbf{F}_{3,2}$ is directed to the right and $\mathbf{F}_{3,1}$ is directed to the left. The total force on q_3 is the difference between the magnitudes of $\mathbf{F}_{3,2}$ and $\mathbf{F}_{3,1}$. Thus

$$F_3 = k\frac{|q_2||q_3|}{(r_{2,3})^2} - k\frac{|q_1||q_3|}{(r_{1,3})^2},$$

and is directed to the right if positive and to the left if negative.

Solution The expression for F_3 can be manipulated to give

$$F_3 = k|q_3|\left(\frac{|q_2|}{(r_{2,3})^2} - \frac{|q_1|}{(r_{1,3})^2}\right).$$

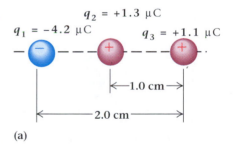

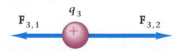

Figure 16.11
Example 16.3: (a) Three charges arranged along a line. (b) Free body diagram for charge q_3. The two forces $\mathbf{F}_{3,1}$ and $\mathbf{F}_{3,2}$ are oppositely directed. The magnitude of each force is calculated independently of the other.

Inserting the values for k, the charges, and the distances gives

$$F_3 = (9.0 \times 10^9 \text{ N} \cdot \text{m}^2/\text{C}^2)|1.1 \times 10^{-6} \text{ C}|\left(\frac{|1.3 \times 10^{-6} \text{ C}|}{(1.0 \times 10^{-2} \text{ m})^2} - \frac{|-4.2 \times 10^{-6} \text{ C}|}{(2.0 \times 10^{-2} \text{ m})^2}\right)$$

$$F_3 = (9.0 \times 10^9 \text{ N} \cdot \text{m}^2/\text{C}^2)(1.1 \times 10^{-6} \text{ C})\left(\frac{1.3 \times 10^{-6} \text{ C}}{(1.0 \times 10^{-2} \text{ m})^2} - \frac{4.2 \times 10^{-6} \text{ C}}{(2.0 \times 10^{-2} \text{ m})^2}\right)$$

$$F_3 = +25 \text{ N}.$$

Because the value is positive, the net force on the charge q_3 is 25 N to the right.

Discussion Notice that the magnitude of force $\mathbf{F}_{3,1}$ did not change because charge q_2 is located between charges q_1 and q_3. This illustrates the power and beauty of the principle of superposition of electrostatic forces due to individual charges. The force on one charge is found by considering the effects of other charges one at a time and then summing them. Of course, we have to remember that the sum is a vector sum because force is a vector.

Example 16.4

Three charges $q_1 = +3.7$ μC, $q_2 = -3.7$ μC, and $q_3 = +4.8$ μC are fixed at the corners of an equilateral triangle 3.0×10^{-2} m on a side. Find the magnitude and direction of the net force on charge q_3 due to the other charges.

Strategy We can use the principle of superposition. First make a drawing of the situation, as in Fig. 16.12(a). Then calculate the force $\mathbf{F}_{3,1}$ on q_3 due to q_1, ignoring the presence of charge q_2. Next, calculate the force $\mathbf{F}_{3,2}$ due to charge q_2, ignoring the presence of q_1. Finally, add the two force vectors to get the net force $\mathbf{F}$ acting on charge q_3.

Solution Using Coulomb's law, we find the magnitude of the force $\mathbf{F}_{3,1}$:

$$F_{3,1} = k\frac{|q_1||q_3|}{(r_{1,3})^2} = 9.0 \times 10^9 \frac{\text{N} \cdot \text{m}^2}{\text{C}^2} \frac{(3.7 \times 10^{-6} \text{ C})(4.8 \times 10^{-6} \text{ C})}{(3.0 \times 10^{-2} \text{ m})^2}$$

$$F_{3,1} = 178 \text{ N}.$$

Since both q_1 and q_3 are positive, the force $\mathbf{F}_{3,1}$ is repulsive and is directed as shown in Fig. 16.12(b). Note that because the charges and the distance are known to only two significant figures, we should round off our answer to two significant figures. However, we will wait until we reach the final result to do the rounding off.

We compute the force $\mathbf{F}_{3,2}$ in the same way that we found $\mathbf{F}_{3,1}$. Its magnitude is

$$F_{3,2} = k\frac{|q_2||q_3|}{(r_{2,3})^2} = 9.0 \times 10^9 \frac{\text{N} \cdot \text{m}^2}{\text{C}^2} \frac{|-3.7 \times 10^{-6} \text{ C}||4.8 \times 10^{-6} \text{ C}|}{(3.0 \times 10^{-2} \text{ m})^2}$$

$$F_{3,2} = 178 \text{ N}.$$

Because the sign of charge q_2 is opposite to the sign of q_3, the force $\mathbf{F}_{3,2}$ is attractive and is directed toward q_2 as shown in the figure.

The two forces we have just computed are equal in magnitude and make equal angles of 60° with the x axis. Consequently, their y components are

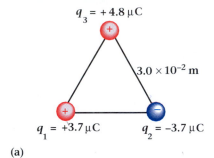

$q_3 = +4.8$ μC

3.0×10^{-2} m

$q_1 = +3.7$ μC $q_2 = -3.7$ μC

(a)

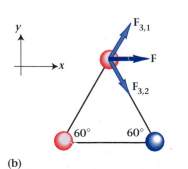

(b)

Figure 16.12
Example 16.4: (a) Three charges are arranged in an equilateral triangle 3.0 cm on a side. (b) The repulsive force $\mathbf{F}_{3,1}$ and the attractive force $\mathbf{F}_{3,2}$ have the same magnitude and make equal angles of 60° about the x axis. The resultant force $\mathbf{F}$ lies along the positive x axis.

equal and opposite and thus sum to zero. However, their x components add to give a net force with magnitude

$$F = 2|\mathbf{F}_{3,1}| \cos 60° = 2(178 \text{ N})\frac{1}{2} = 178 \text{ N} = 1.8 \times 10^2 \text{ N},$$

where the result is rounded to two significant figures. The direction of the net force is to the right along the x axis as shown in Fig. 16.12(b).

16.4   The Electric Field

We have seen how Coulomb's law can be used to find the force on one charge due to its interaction with other charges. Now we want to consider the effects of charges in terms of a concept introduced by Michael Faraday: the electric field. We introduced the concept of a field when describing gravitation in Chapter 5. In the gravitational case, field strength is the gravitational force per unit mass acting on a test mass. The sources of that force are other masses, but we can completely describe the behavior of the test mass by its interaction with the *field*, a physical quantity defined at all points in space. We can also predict the behavior of any other known mass, just from a knowledge of the field in the region where the mass is located. In the same way, we can describe the behavior of a test charge by its interaction with the electric field. Once we know the electric field at a point in space, we can readily compute the force on any electric charge placed at that point. Since we can determine the force, we can calculate the motion of the charge.

We define the **electric field E** at any point in space as the electric force **F** per unit charge exerted on a small positive test charge q_0 placed at that point,*

$$\mathbf{E} = \frac{\mathbf{F}}{q_0}. \tag{16.3}$$

This field is caused by other electric charges distributed about the test charge. Thus, Eq. (16.3) defines the field due to this distribution of charge, not the field caused by the test charge. We choose the test charge q_0 to be very small so that its presence does not distort the field being measured.

The electric field is a vector that points in the direction of the force on a positive test charge. Its magnitude is determined by dividing the magnitude of the force exerted on a test charge by that charge. Thus, the units of electric field are newtons per coulomb, N/C.

To find the field of a single point charge Q that is isolated from other charges, we first locate Q at the origin. Then we use Coulomb's law to determine the force **F** on a positive test charge q_0 at a displacement **r** from the origin. The magnitude of **F** is

$$F = k\frac{Qq_0}{r^2}.$$

---

*Compare the definition of the electric field with the definition of the gravitational field given in Eq. (5.12) on p. 166.

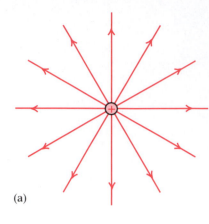

(a)

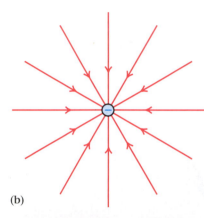

(b)

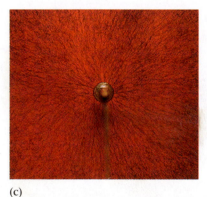

(c)

Figure 16.13

A pictorial representation of the electric field about (a) a positive charge and (b) a negative charge. The field extends out from the charges equally in all directions, that is, the field is spherically symmetric. (c) Field lines are indicated by the alignment of small threads along the direction of the electric field of a small circular conductor when the threads are suspended in oil.

Therefore, the magnitude of the electric field at point r is

$$E = \frac{F}{q_0} = k\frac{Q}{r^2}. \tag{16.4a}$$

In terms of ϵ_0, the magnitude of the electric field of a point charge Q is

$$E = \frac{1}{4\pi\epsilon_0}\frac{Q}{r^2}. \tag{16.4b}$$

If Q is a positive charge, then the field is directed radially away from it. If Q is negative, the direction of the field is toward it.

The direction of the electric field of an isolated point charge is illustrated in Fig. 16.13. Here we have used *lines of force* (also called *field lines*) to represent the electric field, just as we did earlier for the gravitational field. (Remember that the relative number of lines of force is proportional to the magnitude of the force and hence to the strength of the field. You may want to review the representations of gravitational fields shown in Figs. 5.20 and 5.21.) In an electrostatic field, lines of force always begin at positive charges and end at negative charges. Near a positive charge, the field is radially outward (Fig. 16.13a); and near a negative charge, the field lines are radially inward (Fig. 16.13b). The field lines are closer together where the field is stronger and farther apart where the field is weaker. The patterns of lines of force for gravitational and electrostatic fields are similar, which is not surprising because both fields have a $1/r^2$ behavior; that is, they decrease inversely as the square of the distance from the source.

Example 16.5

(a) What are the magnitude and direction of the electric field 1.5 cm from a fixed point charge of $+1.2 \times 10^{-10}$ C? (b) How far away from the point charge does the field drop to 0.10% of the field 1.5 cm from the charge?

Strategy (a) We can compute the magnitude of the electric field directly from Eq. (16.4):

$$E = \frac{1}{4\pi\epsilon_0}\frac{Q}{r^2}.$$

(b) Once we know the value of the field 1.5 cm from the charge, we can use the same relationship to compute the distance from the charge that the field becomes 0.0010 the value of the field at 1.5 cm.

Solution (a) The magnitude of the electric field is computed from

$$E = \frac{1}{4\pi\epsilon_0}\frac{Q}{r^2} = 9.0 \times 10^9 \text{ N} \cdot \text{m}^2/\text{C}^2 \frac{1.2 \times 10^{-10} \text{ C}}{(0.015 \text{ m})^2} = 4.8 \times 10^3 \text{ N/C}.$$

Notice that r was expressed in the SI unit of meters before inserting into the equation. The direction of the field is radially outward from the point charge because the charge is positive.

(b) We can use the same mathematical relationship to find the distance at which the field drops to 0.10% of 4.8×10^3 N/C. We rearrange Eq. (16.4) to get

$$r^2 = \frac{1}{4\pi\epsilon_0}\frac{Q}{E}.$$

We insert the same numerical value for Q. For E we use the value of 0.10% of 4.8×10^3 N/C or 4.8 N/C. Then the distance r becomes

$$r = \sqrt{\frac{1}{4\pi\epsilon_0}\frac{Q}{E}} = \sqrt{\frac{(9.0 \times 10^9 \text{ N} \cdot \text{m}^2/\text{C}^2)(1.2 \times 10^{-10} \text{ C})}{4.8 \text{ N/C}}} = 0.47 \text{ m} = 47 \text{ cm}.$$

Discussion   Although we used a straightforward technique to obtain the distance for part (b), we could have found the same result without having a numerical value of the field E. Since k and Q are constants for this problem, we only have to ask what value of r makes $1/r^2$ equal to 0.0010 times $1/(1.5 \text{ cm})^2$. That is,

$$\frac{1}{r^2} = \frac{0.0010}{(1.5 \text{ cm})^2},$$

or

$$r^2 = \frac{(1.5 \text{ cm})^2}{0.0010}.$$

On taking the square root, we get

$$r = \frac{1.5 \text{ cm}}{\sqrt{0.0010}} = 47 \text{ cm}.$$

As you can see, we obtain the same result for r without having to know the value of the field at that point.

Example 16.6

An electron (charge $e = -1.6 \times 10^{-19}$ C and mass 9.1×10^{-31} kg) is injected into a region of uniform electric field of magnitude $E = 1.0 \times 10^5$ N/C (Fig. 16.14). What is the initial acceleration of the electron?

Strategy   From Newton's second law, the magnitude of the acceleration is given by F/m. To find the acceleration, we must first find the magnitude of the force on the electron. We can do this by multiplying the field strength by the electron charge.

Solution   The magnitude of the force on the electron is

$$F = |\mathbf{F}| = |qE| = |-e||E|.$$

The acceleration is the force divided by the mass,

$$a = \frac{F}{m} = \frac{|qE|}{m} = \frac{(1.6 \times 10^{-19} \text{ C})(1.0 \times 10^5 \text{ N/C})}{9.1 \times 10^{-31} \text{ kg}} = 1.8 \times 10^{16} \text{ m/s}^2.$$

Discussion   The direction of the acceleration is along the direction of the force on the electron. Because the charge on the electron is negative, the force is

$$\mathbf{F} = -e\mathbf{E},$$

which is *opposite* to the direction of the electric field. Note that Newton's laws apply to electric forces just as they do to mechanical forces.

   The actual force is relatively small compared to forces we encounter in everyday life. For example, the gravitational force (weight) on a dime is more than 10^{12} times as great as the electrostatic force found here. However, because the electron's mass is so small, it is given a very large acceleration.

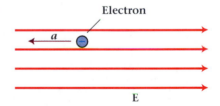

Figure 16.14
Example 16.6: An electron of charge -1.6×10^{-19} C is injected into a region of uniform electric field.

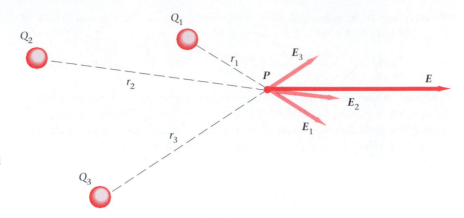

Figure 16.15
Three charges produce an electric field in space. The field measured at point P is the vector sum of the individual fields.

16.5   Superposition of Electric Fields

We have just examined the electric field due to a single isolated point charge. What is the field due to several source charges? For example, suppose we have three point charges, arranged as shown in Fig. 16.15. The force on a test charge q_0 at point P is the vector sum of the forces due to each source charge calculated separately. As we described earlier, this is nothing more than a superposition of the individual forces. Thus

$$\mathbf{F} = \mathbf{F}_1 + \mathbf{F}_2 + \mathbf{F}_3.$$

The electric field at P is found by dividing the force $\mathbf{F}$ by q_0, obtaining

$$\mathbf{E} = \frac{\mathbf{F}}{q_0} = \frac{\mathbf{F}_1}{q_0} + \frac{\mathbf{F}_2}{q_0} + \frac{\mathbf{F}_3}{q_0},$$

Figure 16.16
(a) Lines of electric field near two identical charges. The field is weakest directly between the charges, as indicated by the absence of field lines. (b) Field lines are indicated by small threads suspended in oil.

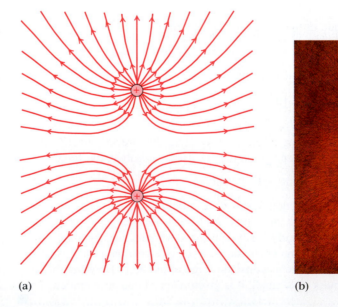

(a)   (b)

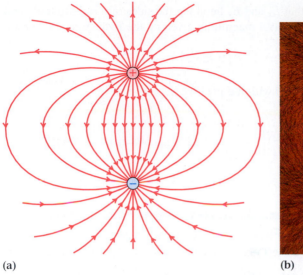

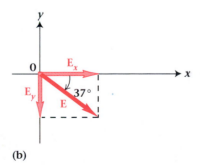

Figure 16.17
(a) Lines of electric field near two equal but opposite charges. The field is strongest directly between the charges, as indicated by the high concentration of field lines. (b) Field lines are indicated by small threads suspended in oil.

or

$$\mathbf{E} = \mathbf{E}_1 + \mathbf{E}_2 + \mathbf{E}_3.$$

In other words, the electric field resulting from several point charges is just the superposition of their individual fields. Thus, we have arrived at a rule for superposition of fields from the rule for superposition of forces.

Figures 16.16 and 16.17 illustrate the fields of two equal positive charges and of two equal but unlike charges. Although these figures show only the field in a plane containing the two charges, the field is actually three-dimensional. It is symmetric about the axis joining the two charges, so that if you imagined rotating the charges about that axis, the shape of the field would be unchanged.

(a)

Example 16.7

A tiny Styrofoam ball with charge $-q$ is placed on the x axis a distance a from the origin. A second Styrofoam ball with charge $+3q$ is placed on the y axis a distance $2a$ from the origin. What is the electric field at the origin?

Strategy It is helpful to draw a diagram such as Fig. 16.18(a) to illustrate the situation. To solve the problem, we need to find the magnitude and direction of the field at the origin O due to each of the two charges. Then we can add them vectorially to find the resultant field.

Solution The field at O due to the charge $-q$ is directed along the positive x axis with magnitude

$$E_x = k\frac{q}{a^2}.$$

The field at O due to the charge $+3q$ is directed along the negative y axis with magnitude

$$E_y = k\frac{3q}{(2a)^2} = 0.75k\frac{q}{a^2}.$$

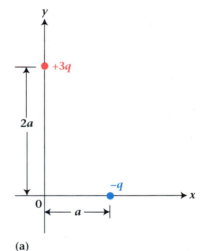

(b)

Figure 16.18
Example 16.7: (a) Position of the charges. (b) Calculated x and y components and resultant electric field at the origin.

In Fig. 16.18(b), we see that E_x and E_y lie along the sides of a right triangle and the resulting field vector **E** has a magnitude given by the Pythagorean theorem

$$E = \sqrt{\left(k\frac{q}{a^2}\right)^2 + \left(0.75\frac{q}{a^2}\right)^2} = 1.3k\frac{q}{a^2}.$$

The field **E** makes an angle θ with the positive x axis given by

$$\theta = \tan^{-1}\left(\frac{E_y}{E_x}\right) = \tan^{-1}\left(\frac{-0.75k\frac{q}{a^2}}{k\frac{q}{a^2}}\right) = \tan^{-1}(-0.75) = -37°.$$

16.6 Electric Flux and Gauss's Law

We now wish to examine a law of electrostatics that relates the electric field to the charge that generates it. In principle, we can always calculate an electric field using Coulomb's law, the definition of the field, and the principle of superposition. However, an alternative way of looking at the same physical situation leads us to a new law called Gauss's law, which is more powerful than Coulomb's law and which provides more insight into the connection between field and charge. Our first step is to define a quantity, related to the electric field, that we call the electric flux. Then we show that the net amount of electric flux that passes through a closed surface* is directly related to the amount of electric charge inside that surface. This relationship is Gauss's law. In the following section we will analyze this relationship mathematically and show that it is true for any electrostatic field. We will find Gauss's law to be especially useful in situations with a high degree of symmetry—for example, in the case of a sphere or a cylinder.

First we need to define flux. In general, flux is the rate at which something passes through a surface. For example, the rate at which water moves through a cross section of a pipe may be described by its flux. In this case, the flux would be the amount of water passing through the pipe per unit time.

We define electric flux in a similar manner. To begin, consider a uniform electric field passing through a small area A of a surface, as shown in Fig. 16.19(a). Equal spaces between the lines of force in the figure indicate that the field is uniform. Because the magnitude of the electric field is proportional to the number of lines of force in a given area, we define the **electric flux** to be the number of field lines that pass through a given surface.** If we have a larger surface (Fig. 16.19b), more field lines pass through, corresponding to a larger flux. If the electric field were stronger, we would indicate it by drawing the field lines closer together, as described in Section 16.4. In that case more lines would pass through our surface, corresponding to an even larger flux (Fig. 16.19c). Notice that when the surface area is turned so that it is no longer perpendicular to the electric field, fewer electric field lines pass through (Fig. 16.19d). In this case, the flux is reduced, even though the area is the same. In fact, when the surface

---

*The surface of a basketball is closed but the surface of a sheet of writing paper is not closed. The surface of the ball has no edges like those found on the sheet of paper.
**Although the lines of electric field extend through the surface, the field is not flowing.

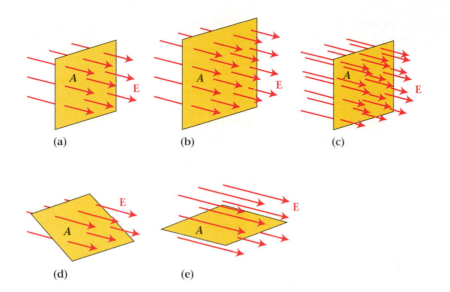

(a)   (b)   (c)

(d)   (e)

Figure 16.19

(a) A small surface area oriented perpendicular to a uniform electric field. The number of lines of force that pass through the surface is proportional to the electric flux. (b) In the same field, more lines pass through the larger surface than through the surface in (a). (c) More field lines pass through the same surface in a stronger field than in a weaker one. (d) When the surface is not perpendicular to the direction of the field, fewer lines pass through. (e) If the surface is parallel to the field direction, no lines pass through it and the flux is zero.

area is turned parallel to the direction of the field, no field lines pass through at all and the flux is zero (Fig. 16.19e).

We now use our definition of flux and our knowledge of point charges to arrive at Gauss's law. Imagine that we surround an isolated point charge $+q$ with an imaginary closed spherical surface of radius r centered about the point charge (Fig. 16.20). The electric flux due to the point charge that passes through this closed surface is represented by the number of field lines that cross the surface. The agreed-upon convention is that if the field lines are directed outward through the surface, the flux is positive; if the field lines are directed inward, the flux is negative. The total number of field lines passing through our closed spherical surface is proportional to the magnitude of the charge and is independent of the radius r. Thus, the electric flux through the surface is directly proportional to the charge enclosed within the surface. If we replace the charge in Fig. 16.20 with $-q$, then the direction of field lines is reversed and the flux becomes negative. If we combine several charges at the central point of the sphere, the sign of the net charge will determine the sign of the flux. We are led to **Gauss's law for electrostatics,** which says that **the net electric flux through any (real or imaginary) closed surface is directly proportional to the net electric charge enclosed within that surface.**

We have not proved Gauss's law here. Instead we have described a special case, but one that clearly conforms to Gauss's law. However, Gauss's law is more general. It is valid regardless of the number of charges involved, their positions inside or outside the closed surface, or the shape of the surface itself. Gauss's law stands as one of the fundamental laws of electricity. The (usually imaginary) surface used for application of Gauss's law is often called a **Gaussian surface.**

We can qualitatively verify Gauss's law by examining Fig. 16.21. Look at the various Gaussian surfaces drawn in cross section in the figure and think about the net flux through each one. The field lines that emerge through surface A surrounding the positive charge q are all directed outward and correspond to a net positive flux. The field lines that enter surface B surrounding the negative charge $(-q)$ are all directed inward and correspond to a net negative flux. The field lines that enter one part of surface C, which surrounds no charge, emerge from the

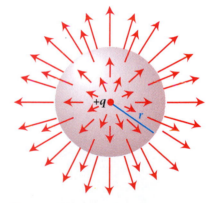

Figure 16.20

The lines of electric field emerging from a positive point charge $+q$ pass through a spherical surface centered about the charge. The number of field lines passing through the surface is independent of the radius of the surface.

Figure 16.21

Four Gaussian surfaces *A, B, C, D* in the region of two point charges of the same magnitude but opposite sign are shown in cross section. Surface *A* surrounds a positive charge and has a net positive flux. Surface *B* surrounds the negative charge and has a net negative flux. Surfaces *C* and *D* surround zero net charge and thus have zero flux.

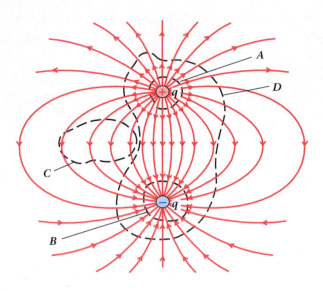

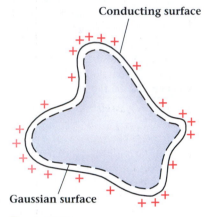

Conducting surface

Gaussian surface

Figure 16.22

The dashed line represents a Gaussian surface just inside the surface of a charged conductor, shown here in cross section. We can use Gauss's law to show that when a charged conductor has reached equilibrium, any excess charge on the conductor must reside on its outer surface.

Figure 16.23

A man inside a Faraday cage is shielded from an electric discharge.

other side. Thus, the net flux is zero. Finally, the large surface *D* surrounding both charges contains no net charge. The lines of force that originate on $+q$ pass out of surface *D* at one end and then enter the surface on the other end. The net flux is zero. Each of these situations is consistent with Gauss's law.

We can use Gauss's law to find out what happens to excess charge placed on a conductor. There are charges within a conductor that are free to move about. We say that the conductor is in *equilibrium* when there is no net motion of these charges either within the conductor or on its surface. **When a conductor is in equilibrium, the electric field everywhere inside the conductor is zero.** (If there is an electric field inside, the mobile charges within the conductor move about until the internal field reduces to zero.) Because the internal electric field inside a conductor in equilibrium is zero, Gauss's law tells us that the flux through an imaginary surface just inside the actual surface of the conductor is also zero. Such a Gaussian surface is indicated by the dashed line of Fig. 16.22.

Now, if excess charge is placed on an isolated conductor of any arbitrary shape (Fig. 16.22), the charge sets up a field within the conductor. Again, this field causes the mobile charges within the conductor to move about until the internal field reduces to zero. Application of Gauss's law tells us that there is no net or excess charge inside the conductor. Because the excess charge does not reside inside the conductor, it must necessarily reside on the outer surface. So, **at equilibrium, any excess charge placed on a conductor resides entirely on its outer surface.** This effect was observed experimentally by Benjamin Franklin long before Gauss's law was known.

If excess charge resides on the surface of a conductor, then even a thin covering of conducting material can be used to surround a volume and shield it from a field of external static charges. Such an arrangement is often called a Faraday cage (Fig. 16.23), after Michael Faraday, who made extensive experiments on shielding. The volume within the Faraday cage is shielded from the effects of external electric fields. Thus, the occupants of a room surrounded by conducting material are safe from external electrical effects as long as they remain inside the room. For this reason you are safe inside a car or an airplane during a lightning storm.

*16.7

A Quantitative Approach to Gauss's Law

In the previous section, when we discussed electric flux and Gauss's law, we reached the important conclusion that for static cases, any excess electric charge on a conductor resides on its surface. In this section, we take a more quantitative approach to Gauss's law and show how we can use it to find the value of an electric field.

First let's consider a uniform electric field passing through a small area ΔA of an arbitrary surface (Fig. 16.24). We wish to describe the electric flux in terms of the electric field $\mathbf{E}$ that exists in the region of this small area. In the previous section, we defined the flux as the number of field lines that pass through the surface. Since we expressed the magnitude of the electric field E in terms of the relative number of field lines, we can define the electric flux as the amount of electric field that passes through a given surface area multiplied by the area. As we have just seen, when a surface is perpendicular to the direction of the field, the maximum number of field lines can pass through and we have the maximum flux. As the surface tips so that the angle between the plane of the surface and the field gets smaller, the effective area becomes smaller and fewer field lines can pass through. For this reason, our mathematical formula for flux must include the effect of the angle between the surface and the field lines. In practice, we usually define the orientation of a surface as the direction of a line perpendicular to the surface (the normal). So, in defining flux we consider the angle between the field direction and the normal.

The electric flux $\Delta\phi_E$ through a small portion of surface area ΔA is the product of the magnitude of the electric field $|\mathbf{E}|$, the magnitude of the surface area ΔA, and the cosine of the angle θ between the direction of the field and the direction of the normal to the surface (Fig. 16.24). The electric flux $\Delta\phi_E$ is then

$$\Delta\phi_E = E\Delta A \cos\theta.$$

Now let's consider the surface to be a sphere of radius r centered about a source charge Q (Fig. 16.25). With the charge in the center, the field is perpendicular to the surface at all points and has magnitude

$$E = k\frac{Q}{r^2}$$

everywhere. Since $\mathbf{E}$ is perpendicular to the surface, the angle θ between the field and the normal at any point on the surface is zero and $\cos\theta = \cos 0 = 1$. The flux through an area ΔA then becomes

$$\Delta\phi_E = E\Delta A,$$

or

$$\Delta\phi_E = k\frac{Q\Delta A}{r^2}.$$

The total flux ϕ_E through the entire surface is the sum of all contributions $\Delta\phi_E$ over the surface of the sphere:

$$\phi_E = (\Delta\phi_E)_1 + (\Delta\phi_E)_2 + (\Delta\phi_E)_3 + \ldots$$

$$= k\frac{Q}{r^2}\Delta A_1 + k\frac{Q}{r^2}\Delta A_2 + k\frac{Q}{r^2}\Delta A_3 + \ldots.$$

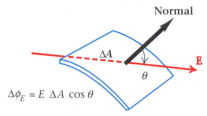

$$\Delta\phi_E = E\,\Delta A\cos\theta$$

Figure 16.24
Electric field passing through a small surface of area ΔA. The normal to the surface makes an angle θ with the direction of the electric field. The electric flux $\Delta\phi_E$ is equal to the product of the field strength E, the area ΔA, and the cosine of the angle between the direction of the field and the surface normal.

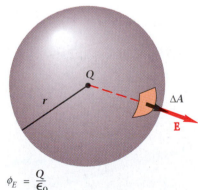

$$\phi_E = \frac{Q}{\epsilon_0}$$

Figure 16.25
A spherical surface symmetrically surrounding a point charge Q. The surface normal is parallel to the electric field at all points of the surface.

We can remove the factor $k\dfrac{Q}{r^2}$ from the sum because it has the same constant value everywhere on the surface of the sphere:

$$\phi_E = k\frac{Q}{r^2}(\Delta A_1 + \Delta A_2 + \Delta A_3 + \dots).$$

We are left with the sum of all the individual areas of surface, which is just the total surface area of the sphere, $4\pi r^2$. The total flux thus becomes

$$\phi_E = k\frac{Q}{r^2} \times 4\pi r^2 = 4\pi k Q.$$

If we replace k by $1/4\pi\epsilon_0$, we find that

$$\boxed{\phi_E = \frac{Q}{\epsilon_0}.} \tag{16.5}$$

Equation (16.5) is the mathematical statement of Gauss's law. This is one place where introducing the permittivity ϵ_0 results in a simpler equation. Notice that the total flux through the surface depends only on the amount of charge Q contained *within* the surface. (The amount of charge *outside* the surface has no net effect on the flux.) If Q is positive, then the flux is positive, meaning that the field is directed outward from the charge; if Q is negative, the flux is negative, so the field is directed inward toward the charge; and if Q is zero, the flux is also zero, and there is no net field through the surface.

Although we have derived Gauss's law for a spherical surface, it is valid regardless of the shape of the surface involved. Because of this, we can choose a Gaussian surface of any shape around a charge. Proper choice of this surface can often simplify the flux calculation. Gauss's law holds regardless of where the charge is located within the surface. It even holds for the case where the charge is outside the closed surface. In that case, the charge inside is zero, so the net flux is also zero.

In a more general case, as shown in Fig. 16.21, the net electric flux through any (real or imaginary) closed surface is directly proportional to the net electric charge enclosed within that surface, regardless of the number of charges involved, their positions inside or outside the surface, or the shape of the surface itself. Two examples of using Gauss's law follow. Note that in both of these examples, the calculation of the flux is made possible by their symmetry.

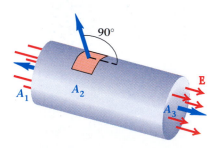

Figure 16.26

Example 16.8: A cylindrical Gaussian surface in a uniform electric field. The net charge enclosed by the surface is zero, so the net flux is zero. (As many field lines pass into the surface as pass out of the surface.)

Example 16.8

Verify that Gauss's law correctly describes the flux through a closed cylindrical surface in a region of uniform electric field **E** in which there are no charges.

Strategy To simplify flux calculations, we orient the cylindrical surface so that its axis is parallel to the direction of the electric field (Fig. 16.26). Then we designate the area of the cylinder's left-end surface as A_1, the area of the side of the cylinder as A_2, and that of the right-end surface as A_3. Since the total flux is the sum of all contributions $\Delta\phi_E$ over the surface and $\Delta\phi_E = E\,\Delta A\cos\theta$, the flux is

$$\phi_E = EA_1\cos\theta_1 + EA_2\cos\theta_2 + EA_3\cos\theta_3.$$

Solution Since the field lines are directed perpendicularly inward through the surface A_1, $\theta_1 = 180°$ and $\cos\theta_1 = -1$. The surface A_2 is everywhere at right angles to **E**, so $\theta_2 = 90°$ and $\cos\theta_2 = 0$. The field lines pass perpendicularly outward through surface A_3, so $\theta_3 = 0°$ and $\cos\theta_3 = +1$. Thus

$$\phi_E = EA_1(-1) + EA_2(0) + EA_3(1).$$

But A_1 and A_3 are equal, so

$$\phi_E = E(-A_1 + A_3) = 0.$$

Discussion This result is exactly what we expect from Gauss's law because no charges are located within the cylindrical surface. Notice that there is electric field within the cylinder, but the net flux is zero because the inward flux and the outward flux are equal. If the cylinder is rotated so that its axis is no longer parallel to the direction of the field, Gauss's law still applies and the net flux through the cylinder is still zero. However, for that asymmetric orientation, calculation of the flux would be much more difficult.

Example 16.9

Figure 16.27 shows a section of an infinitely long charged plastic rod. The rod is uniformly charged with a constant linear charge density λ (the charge per unit length, $\lambda = Q/l$). Use Gauss's law to show that the electric field due to this uniform line of charge is proportional to $1/r$, where r is the distance from the line.

Strategy Notice that the electric field near the uniformly charged rod must be radially directed because of the symmetry of the situation. The field must have cylindrical symmetry because the appearance of the charged rod is unchanged by rotating the rod about its axis. The field must also be independent of position along the rod because the distance to either end is infinite, so any part of the rod is exactly the same as any other part. Knowing these facts, we can apply Gauss's law to an imaginary cylindrical surface centered about the line of charge (Fig. 16.27).

Solution Again, we choose the shape of the surface to simplify the calculations through symmetry considerations. The flux is

$$\phi_E = \sum E\,\Delta A\cos\theta = \sum_{\text{ends}} E\,\Delta A\cos\theta_1 + \sum_{\text{cylinder}} E\,\Delta A\cos\theta_2.$$

The field is parallel to the plane of the end surfaces, so $\theta_1 = 90°$ and

$$\sum_{\text{ends}} E\,\Delta A\cos\theta_1 = 0.$$

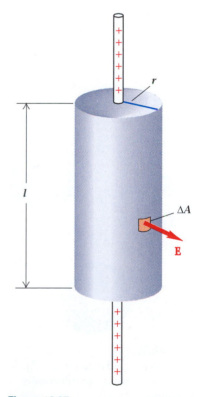

Figure 16.27
Example 16.9: A cylindrical Gaussian surface of radius r and length l centered about an infinite line of charge. This situation approximates the field around a uniformly charged long thin rod.

Since the cylinder is symmetrically positioned about the line, the magnitude of the field, E, is constant over the cylindrical surface and the angle θ_2 between $\mathbf{E}$ and the normal is zero. Thus

$$\theta_E = \sum_{\text{cylinder}} E\,\Delta A\,\cos 0 = E \sum \Delta A,$$

where the summation is over the cylindrical surface. The surface area of a cylinder of radius r and length l is $2\pi rl$, so

$$\phi_E = 2\pi rlE.$$

By Gauss's law, the flux must be

$$\phi_E = \frac{Q}{\epsilon_0} = \frac{\lambda l}{\epsilon_0},$$

since λl is the charge enclosed by the surface. Upon combining these two expressions for the flux, we see that

$$E = \frac{\lambda}{2\pi\epsilon_0 r}.$$

The magnitude of the field is proportional to $1/r$.

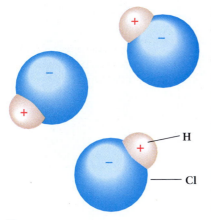

Figure 16.28

The HCl molecule is an electric dipole because its charge is not spherically symmetric.

| 16.8 | **The Electric Dipole** |

Many common electrical situations involve two equal but opposite charges separated by a fixed distance. This combination of charges is called an **electric dipole.*** For example, the hydrogen chloride (HCl) molecule is negative at one end, the Cl end, and positive at the other end, the H end (Fig. 16.28). Many other molecules also have some asymmetry of charge and act as electric dipoles, including water (H_2O), sulfur dioxide (SO_2), and chloroform ($CHCl_3$). Spherically symmetric molecules such as methane (CH_4) and carbon tetrachloride (CCl_4) do not act as dipoles.

A dipole placed in a uniform electric field experiences no net force because the force on the positive charge is equal and opposite to the force on the negative charge. As a result, there is no net translational acceleration. However, this does not imply that nothing happens. On the contrary, a net *torque* arises that causes the dipole to rotate so as to align along the direction of the field. The situation is illustrated in Fig. 16.29, where the clockwise torque generated by F_1 about the center of the dipole is

$$\tau_1 = F_1 \frac{d}{2}\sin\theta.$$

Here θ is the angle between the dipole axis and the electric field and d is the distance between the charges. Similarly, the torque resulting from the force on the

*The term *dipole* implies that there are two poles, or charges.

negative charge is also clockwise and is given by

$$\tau_2 = F_2 \frac{d}{2} \sin \theta.$$

The magnitudes of the forces are equal because the magnitudes of the charges are identical. So the total torque, which is the sum of the individual torques, is

$$\tau = \tau_1 + \tau_2 = Fd \sin \theta.$$

The force F is given by qE. This leads to

$$\tau = (qE)d \sin \theta = qdE \sin \theta.$$

The torque depends on the magnitude of the field (E), the charge (q), the separation between the charges (d), and the angle θ.

The product qd is usually given the name **dipole moment** and for electric dipoles is often represented by the letter p. If we define the dipole moment as a vector of magnitude p directed from the negative to the positive charge (Fig. 16.30), then the torque is the product of the dipole moment $\mathbf{p}$, the field $\mathbf{E}$, and the sine of the angle between them:

$$\boxed{\tau = pE \sin \theta.} \tag{16.6}$$

The direction of the rotation produced by the torque rotates $\mathbf{p}$ into alignment with $\mathbf{E}$.

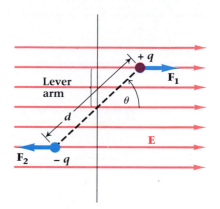

Figure 16.29
The forces acting on an electric dipole in a uniform electric field cause it to rotate.

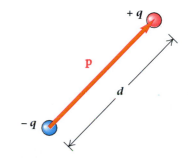

Figure 16.30
Direction of the dipole moment $\mathbf{p}$ of two opposite charges of magnitude q separated by a distance d.

Example 16.10

What is the maximum angular acceleration of a molecule of HCl (dipole moment $p = 3.6 \times 10^{-30}$ C · m and moment of inertia of 2.7×10^{-47} kg · m²) in a region of electric field $E = 1.7 \times 10^4$ N/C?

Strategy As we saw in Section 9.6, the angular acceleration is proportional to the applied torque. Thus the maximum angular acceleration occurs when the torque is a maximum. The maximum torque occurs when the direction of the dipole moment vector is at right angles to the direction of the electric field vector.

Solution When the dipole moment vector is at right angles to the electric field vector the angle between them is 90°. In that case, the torque is maximum and is given by

$$\tau = pE \sin 90° = pE.$$

The angular acceleration α is related to the torque through $\alpha = \tau/I$, where I is the moment of inertia. Thus, the angular acceleration is

$$\alpha = \frac{\tau}{I} = \frac{pE}{I} = \frac{(3.6 \times 10^{-30} \text{ C} \cdot \text{m})(1.7 \times 10^4 \text{ N/C})}{2.7 \times 10^{-47} \text{ kg} \cdot \text{m}^2} = 2.3 \times 10^{21} \text{ rad/s}^2.$$

Discussion Although the torque is quite small, the moment of inertia is also quite small, so the molecule accelerates very rapidly.

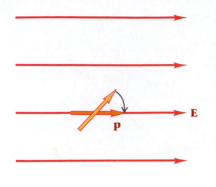

Figure 16.31
When dipole **p**, free to rotate, is placed in a region of uniform field **E**, the effect of the field is to rotate the dipole to bring its dipole moment in line with the field.

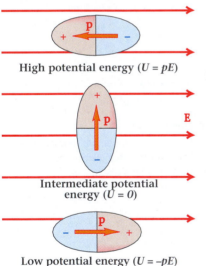

High potential energy ($U = pE$)

Intermediate potential energy ($U = 0$)

Low potential energy ($U = -pE$)

Figure 16.32
A dipole in an electric field has a potential energy that depends on the orientation of the dipole moment vector relative to the direction of the field.

We have shown that when a dipole is placed in an external electric field, a torque acts to align it with the field. For example, consider the behavior of a free dipole **p** interacting with a uniform field that is fixed in space (Fig. 16.31). The free dipole rotates into alignment with the field. It follows that we must do work to rotate the dipole to some position other than alignment along the field direction. We can, therefore, associate a potential energy with the orientation of the dipole in the field. If we consider the position in which the dipole is perpendicular to the field as the reference position, then the potential energy decreases when the dipole rotates into the direction of the field and increases when the dipole rotates in the opposite direction (Fig. 16.32). The potential energy is

$$U = -pE \cos \theta. \tag{16.7}$$

Although it is helpful to visualize the motion of a dipole in terms of the torques acting on it, frequently the physically important quantity is the potential energy associated with the dipole's orientation. We shall examine the magnetic dipole in Chapter 19. Then in later chapters we will show how the electric and magnetic dipole energies enter into the theory of atoms and how we may use them to explain the magnetic behavior of solids.

Example 16.11

A water molecule has an electric dipole moment $p = 6.2 \times 10^{-30}$ C · m. If the molecule is in an electric field of 1.0×10^3 N/C, how much energy does it take to rotate the dipole from parallel alignment to antiparallel alignment with the field?

Solution The energy required is equal to the change in potential energy:

$$\Delta U = U_{antiparallel} - U_{parallel} = -pE \cos 180° - (-pE \cos 0°)$$
$$\Delta U = pE + pE = 2pE = 2(6.2 \times 10^{-30} \text{ C} \cdot \text{m})(1.0 \times 10^3 \text{ N/C})$$
$$\Delta U = 1.2 \times 10^{-26} \text{ J}.$$

DIPOLES AND MICROWAVE OVENS

Although microwave ovens are a relatively recent invention, they have become widespread around the world. Many people don't give it a thought when they pop leftover meals into the microwave oven to heat them, or when they use the oven for fast and easy cooking. But why is microwave cooking so fast? How does it work and how safe is it?

An electrical force exists between two polar molecules, such as the water molecule in Fig. B16.1, even though they are both uncharged. Some of the bonding between molecules is due to this dipole-dipole interaction. In a uniform electric field, a polar molecule experiences a torque that tends to align its dipole moment with the field. If we reverse the direction of the field, a new torque arises that tends to rotate the molecule through 180°. If the molecule is isolated, it can rotate freely. However, if the molecule is bound to other molecules in a substance, it encounters retarding "friction" to the applied torque. This friction is due to the disruption of the bonds between molecules. If the direction of the field is changed rapidly, the energy used against friction appears as heat in the substance itself.

Microwaves, which are high-frequency radio waves, can provide such a rapidly changing electric field. In North America, microwave ovens designed for home use operate at a frequency of 2450 MHz (wavelength of 12.2 cm). Microwaves are produced and confined within the small volume of the oven (Fig. B16.2). These waves, like sound waves or light waves, may be transmitted, reflected, or absorbed. Microwaves are easily transmitted through air, glass, paper, and many types of plastics. Microwaves are reflected from metal, but absorbed by water, fat, and sugar. In most foods, the microwaves penetrate one to two inches. As they penetrate, they are absorbed by the food, which generates thermal energy primarily as a result of the disruption of the intermolecular bonds of the water molecules as they rapidly change direction.

The rapidity with which microwave ovens cook is due not so much to the total power delivered into the oven as to the fact that the microwaves penetrate the outer layers of the

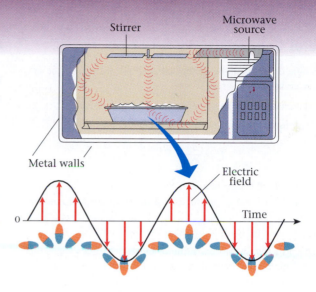

Figure B16.2 A typical microwave oven. Microwaves from the source are directed into the oven cavity, where they are contained by metal walls. The metal stirrer helps distribute the microwave energy evenly throughout the oven.

food and deliver their energy within the food itself. In a conventional oven, energy is transferred to the surface of the food by relatively inefficient convection through the low-thermal-conductivity air. In both conventional and microwave ovens, food cooks from the outside in.

Because microwaves reflect from the inside walls of the oven, it is possible to set up standing waves within the oven, similar to the way standing sound waves are set up in a pipe. The nodes and antinodes correspond to cold and hot spots, which lead to uneven cooking. This effect is reduced by rotating metal fans (stirrers) intended to break up standing waves and by a rotating cooking platform. Food enclosed within metal containers does not cook because the metal shields the food from the microwaves.

Caution is appropriate when using microwave ovens. The microwave power output of typical household ovens is about 500 W, a level high enough to cause burns. However, a level of microwave power that is too low for cooking may cause serious damage to humans. Therefore, standards have been set to limit how much radiation may escape from the oven. The 1985 U. S. federal standard is that the microwave power two inches from the door shall not exceed 1 mW/cm^2 from a new oven. Studies have indicated that levels of from 0.1 to 1.0 mW/cm^2 have no apparent harmful effect over an eight-hour period. For safety, a conducting plastic seal ensures that microwave energy does not leak out around the door. For additional safety, an electrical interlock shuts off the oven when you open the door. To allow you to see inside the oven with the door shut, most microwave ovens have windows with a metal screen or mesh. This mesh is opaque to the long wavelengths of the microwaves but easily passes the shorter waves of visible light.

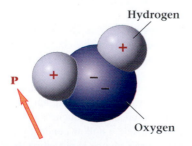

Figure B16.1 Water molecules have a permanent dipole moment directed from the negative (oxygen) end to the positive (hydrogen) end.

Summary

Useful Concepts

■ Single charges can be neither created nor destroyed (the law of conservation of charge).

■ Like charges repel and unlike charges attract.

■ Coulomb's law for the force between charges is

$$F = \frac{1}{4\pi\epsilon_0} \frac{|q_1||q_2|}{r^2},$$

where r is the distance between charges q_1 and q_2. The force is given in N if q_1 and q_2 are measured in coulombs and r in meters and $\epsilon_0 = 8.85 \times 10^{-12}$ $C^2 \cdot N^{-1} \cdot m^{-2}$. The direction of the force is along the line between the two charges.

■ The principle of superposition of electrical forces states that the combined effect of many forces acting simultaneously is the same as the vector sum of the individual forces.

■ The electric field is defined as the electrical force per unit positive charge:

$$\mathbf{E} = \frac{\mathbf{F}}{q_0}.$$

■ The principle of superposition of electrical fields states that the combined effect of many fields acting simultaneously is the same as the vector sum of the individual fields.

■ Gauss's law states that the net electric flux through any closed surface is directly proportional to the net electric charge enclosed within that surface. The mathematical statement of Gauss's law is

$$\phi_E = \frac{Q}{\epsilon_0}.$$

■ When a conductor is in equilibrium, the electric field everywhere inside the conductor is zero.

■ The excess charge on a conductor lies on its exterior surface.

■ The torque on a dipole in an electric field is

$$\tau = pE \sin\theta,$$

where p is the magnitude of the dipole moment and θ is the angle between the direction of the field and the direction of the dipole moment vector.

■ The potential energy of a dipole of moment $\mathbf{p}$ in an electric field $\mathbf{E}$ is

$$U = -pE \cos\theta.$$

Important Terms

You should be able to write the definition or meaning of each of the following:

electrostatics	elementary charge
electric charge	principle of superposition
conductor	electric field
insulator	electric flux
electrons	Gauss's law for
law of conservation of charge	electrostatics
charging by induction	Gaussian surface
Coulomb's law	electric dipole
coulomb	dipole moment
permittivity of free space	

Conceptual Questions

16.1 It is common to observe a static spark when you touch the doorknob after shuffling across a rug in the winter. Why is it uncommon to get a similar spark in the summer?

16.2 Gasoline tank trucks can become charged as they travel. How can this happen and how can it be prevented?

16.3 Why does your hair tend to stand on end when combed on a cold dry day?

16.4 Where does the charge come from when an object is charged by friction?

16.5 When tiny scraps of paper are placed between two charged plates, they bounce back and forth between the plates. Explain this occurrence.

16.6 When charging an electroscope by induction, why is it necessary to remove your finger before removing the charging rod?

16.7 What electrostatic experiment can you do to tell whether two charges have the same magnitude?

16.8 Discuss the similarities and differences between the gravitational force between masses and the electric force between charges.

16.9 Tall buildings made of concrete and steel are frequently struck by lightning without any apparent damage. Can you explain why?

16.10 Two equal negative charges are each placed at a corner of an equilateral triangle. A positive charge of equal magnitude is placed at the third corner. Sketch the field lines for this situation.

16.11 Are the Coulomb forces between electrical charges conservative or nonconservative forces? Explain your answer.

16.12 What are the practical consequences of the observation that the electric field inside a hollow conductor is zero regardless of how much electric charge is placed on its outer surface? What does this imply regarding the safety of a person inside an automobile in a thunderstorm?

Problems

Section 16.2 Coulomb's Law

16.1 Calculate the magnitude of the repulsive force between a pair of like charges, each of one microcoulomb, separated by a distance of one centimeter.

16.2 A point charge of $+6.3$ μC is located 0.15 m from a second point charge of -4.8 μC. What are the magnitude and direction of the force on each charge?

16.3 Two equal charges of 3.7 μC are placed a distance x apart. What must be the value of x if the force between the charges is 4.0×10^{-8} N?

16.4 Two charges of equal magnitude exert an attractive force of 4.0×10^{-4} N on each other. If the magnitude of each charge is 2.0 μC, how far apart are the charges?

16.5 Two equal point charges are located one centimeter apart. If the electrostatic force between them is 1.6×10^{-3} N, what must be the magnitude of the charges?

16.6 A charge of $+3.0$ μC exerts a force of 3.2×10^{-2} N on a charge of -1.6 μC. How far apart are the charges?

16.7 Two charged Styrofoam balls are moved so that the force between them becomes twelve times greater than it was originally. What is the ratio of their new separation to their original separation?

16.8 Modern particle accelerators can be used to produce particles that are not usually observed by other means. One such particle is the antiproton, a particle with the same mass as a proton and with charge of the same magnitude but opposite sign. When a proton and an antiproton are simultaneously produced, what is the Coulomb force between them when they are separated by a distance comparable to atomic dimensions, say 2.50×10^{-10} m? The charge on the proton is $+1.60 \times 10^{-19}$ C.

16.9 The nucleus of an atom contains protons of charge $+1.60 \times 10^{-19}$ C and neutrons, which are uncharged. (a) What is the ratio of the electrical force to the gravitational force between two protons? (b) Can the attractive force of gravity be responsible for holding the nucleus together? Proton mass $= 1.67 \times 10^{-27}$ kg.

16.10 A model of the hydrogen atom consists of a proton at the center with an electron in a surrounding orbit. The charges of the proton and the electron have opposite signs and equal magnitude of $q = 1.6 \times 10^{-19}$ C. The average separation between the electron and the proton is approximately 5.3×10^{-11} m. What is the Coulomb force between them?

16.11• Two conducting spheres of the same size have charges $q_1 = -4.0 \times 10^{-6}$ C and $q_2 = +8.0 \times 10^{-6}$ C, and their centers are separated by 3.0 cm. (a) What is the force between them? (b) What would be the force if the spheres were touched together and again separated by the same distance?

16.12• Suppose that the attraction between the moon and the earth were due to Coulomb forces rather than gravitational force. What would be the magnitude of the charge required if equal but opposite charges resided on both earth and moon? Mass of earth $= 5.98 \times 10^{24}$ kg; mass of moon $= 7.35 \times 10^{22}$ kg; earth-moon distance $= 3.84 \times 10^8$ m.

16.13•• Two small, positively charged spheres experience a mutual repulsive force of 1.52 N when their centers are 0.200 m apart. The sum of the charges on the two spheres is 6.00 μC. What is the charge on each sphere?

16.14•• Two equally charged insulating balls each weigh 0.10 g and hang from a common point by identical threads 30 cm long. The balls repel each other so that the separation between their centers is 6.4 cm. What is the magnitude of the charge on each ball?

16.15•• Two insulating balls each weigh 0.10 g and hang from a common point by identical threads 30 cm long. Each ball carries a charge of $+8.0$ nC and repels the other ball. If, after two hours, the balls are found to be only 4.0 cm apart, what is the rate at which charge has leaked off? Assume that the charge leaks off both balls at the same rate.

Section 16.3 Superposition of Electric Forces

16.16 Two point charges of magnitude $+q$ are separated by a distance r. A third charge of equal magnitude and opposite sign is placed midway between the two positive charges. (a) What is the net electrostatic force on the end charges? (b) What is the net electrostatic force on the middle charge?

16.17 Four charges are spaced along the positive x axis at 0.50-m intervals, beginning at the origin. Starting with the charge at the origin, the charges have values of $+5.0$ μC, -3.0 μC, -4.0 μC, and $+5.0$ μC. What is the net force on the -3.0 μC charge?

16.18• Four charges are arranged on the corners of a square whose edge length is a. Two positive charges $+Q$ are placed on diagonally opposite corners, and two negative charges $-Q$ are placed on the other corners. (a) What is the force on a test charge q_0 placed at the center of the square? (b) What are the magnitude and the direction of the force on the test charge q_0 if it is placed at the midpoint of one of the edges?

16.19• Four equal point charges $+q$ are placed at the corners of a square of edge length L. Find the force on any one of the charges.

16.20• Three equal charges are placed at the corners of an equilateral triangle 0.50 m on a side. What are the magnitude and the direction of the force on each charge if the charges are each -3.7 nC?

16.21• Three equal charges, each of $+6.0 \ \mu C$, are spaced along a line. The end charges are each 0.23 m from the central charge. What are the magnitude and the direction of the force on each charge?

Section 16.4 The Electric Field

> **Hints for Solving Problems**
>
> Remember that the electric field is directed away from a positive source charge and toward a negative source charge. The force on a charge in a field is the product of the electric field at that point and the magnitude of the charge.

16.22 What is the electric field strength E at a point 0.200 m from a point charge of $+1.75 \ \mu C$?

16.23 A proton of charge $q = 1.60 \times 10^{-19}$ C is placed in a region of uniform electric field with field strength $E = 1.46 \times 10^5$ N/C. What is the electric force on the proton?

16.24 Calculate the electric field 10.0 cm from a point charge of 1.35 μC.

16.25 When placed in a region of electric field, a test charge of 1.25 μC experiences a force of 1.00 N. What is the magnitude of the field?

16.26 A point charge q produces an electric field of magnitude 90.5 N/C at a distance of 1.68 m. If the field is directed toward the charge, what is the value of q?

16.27 A spherical insulating shell has a charge q uniformly spread over its surface. What is the electric field at the center of the shell?

16.28 A positive charge of 1.00 μC is located in a uniform field of 1.00×10^5 N/C. A negative charge of $-0.100 \ \mu C$ is brought near enough to the positive charge that the attractive force between the charges just equals the force on the positive charge due to the field. How close are the two charges?

16.29 (a) Sketch the pattern of the electric field around an isolated positive charge Q. (b) Sketch the pattern of the electric field around an isolated negative charge of magnitude $2Q$.

16.30 Sketch the pattern of the electric field around two otherwise isolated charges separated by a small distance d. Assume that both charges are positive and that the magnitude of one of the charges is twice that of the other.

16.31• A proton accelerates from rest to 3.00×10^6 m/s in 1.00×10^{-6} s in a uniform electric field E. What is the magnitude of the electric field? (*Hint:* The proton has a mass $m_p = 1.67 \times 10^{-27}$ kg and a charge $+e = 1.60 \times 10^{-19}$ C.)

16.32• A small droplet of oil with mass of 2.00×10^{-15} kg is held suspended in a region of uniform electric field directed upward with a magnitude of 6125 N/C. (a) Is the excess charge on the droplet positive or negative? (b) How many excess elementary charges reside on the droplet? (The elementary charge is $e = 1.60 \times 10^{-19}$ C.)

16.33• An electric charge $Q = 1.50 \ \mu C$ is in a region of electric field with components $E_x = 4000$ N/C, $E_y = 3000$ N/C,

and $E_z = 0$. What are the magnitude and the direction of the force on the charge Q?

16.34• A tiny ball of mass 0.500 g is suspended by a thread of negligible mass in a horizontal electric field of 400 N/C. If the ball has a charge $q = 9.00 \ \mu C$, what will be the deflection angle of the string measured from the vertical (Fig. 16.33)?

16.35• A tiny Styrofoam ball of mass 0.500 g is suspended by a light thread of negligible mass. The ball is electrically charged. Then the ball is placed in a uniform horizontal electric field of 400 N/C. What is the charge q on the ball when it is deflected by 15° (Fig. 16.33)?

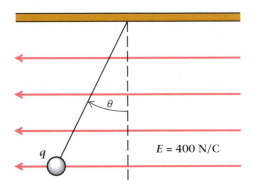

Figure 16.33
Problems 16.34 and 16.35.

16.36• Show that the kinetic energy acquired in a time t by a particle of mass m and charge q released from rest in a uniform electric field E is given by

$$KE = \frac{1}{2m}(Eqt)^2.$$

Section 16.5 Superposition of Electric Fields

> **Hints for Solving Problems**
>
> Remember that electric fields are vectors, so when combining electric fields be sure to treat them as vectors.

16.37 Three equal positive point charges are located at the corners of an equilateral triangle. Sketch the pattern of the electric field in the plane of the triangle.

16.38 An electric field of 11.6 N/C directed due north is superposed on an existing electric field of 16.9 N/C directed due east. What are the magnitude and the direction of the resultant field?

16.39• A charge of $+3.0 \times 10^{-6}$ C is located at the point $x = +0.10$ m, $y = 0$. A second charge of -3.0×10^{-6} C is located at $x = -0.10$ m, $y = 0$. What are the magnitude and the direction of the electric field at (a) the point $x = 0$, $y = 0$ and (b) the point $x = 0$, $y = +0.30$ m?

16.40● Three equal positive point charges of magnitude Q are located at three corners of a square of edge length d. A negative charge $-3Q$ is placed on the fourth corner. (a) Sketch the pattern of the electric field in the plane of the square. (b) At the position of the negative charge, what is the magnitude of the electric field due to the three positive charges? (c) What is the force on the negative charge?

16.41●● Four equal point charges of magnitude Q are arranged on the corners of a square of edge length 1.0 cm. Evaluate the electric field at (a) the center and (b) the midpoint of one edge, given that $Q = 1.0 \times 10^{-9}$ C.

*Sections 16.6 and 16.7 Electric Flux and Gauss's Law and a Quantitative Approach to Gauss's Law

Hints for Solving Problems

Remember that Gauss's law applies to the net flux and to the net charge enclosed within a Gaussian surface.

16.42 Use Gauss's law to determine the electric field inside a hollow conducting sphere that carries a total charge Q. Extend your answer to a conductor of any shape.

16.43 Figure 16.34 shows a closed surface surrounding some electric charges. (a) What is the net electric flux through the surface? (b) Is the electric flux directed inward or outward from the surface?

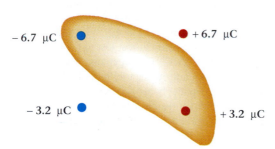

Figure 16.34
Problem 16.43.

16.44 Figure 16.35 shows three surfaces enclosing some electric charges. (a) Which surface has the greatest net electric flux emerging from it? (b) Which surface has the greatest net inward electric flux? (c) Does any surface have zero net flux through it?

16.45 A plane rectangular surface 0.100 m on an edge is placed in a region of uniform electric field of magnitude 24.6 N/C. The normal to the surface makes an angle of 33° with the direction of the electric field. What is the electric flux through the surface?

16.46 A point charge Q is at the center of a conducting spherical shell of radius R. The total charge of the shell is $-Q$. (a) What is the field in the region between the point charge and the shell? (b) What is the field outside the shell?

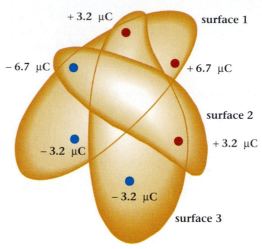

Figure 16.35
Problem 16.44.

16.47● A hollow spherical shell is uniformly charged with a total charge Q. Show that the electric field outside the shell is everywhere the same as the field due to a point charge Q located at the center of the shell.

16.48● A large insulating sheet is given a uniform static charge of surface charge density σ. Show that the electric field near the sheet is given by $E = \dfrac{\sigma}{2\epsilon_0}$. (The surface charge density is the charge per unit area of the surface.)

16.49●● A narrow hole is drilled along the diameter of a solid sphere. The sphere carries a positive electric charge distributed uniformly throughout with a charge per unit volume ρ. A negative charge q is placed within the narrow hole. (a) Show that the charge q will be attracted to the center of the sphere with a force proportional to the distance from q to the center of the sphere. (b) Find an expression for the motion of the charged particle as a function of time.

16.50●● A tiny plastic ball of mass m carries an electric charge q. The ball is attached to a charged nonconducting sheet with uniform surface charge density σ by means of a nonconducting string of negligible mass (Fig. 16.36). Derive a formula for the angle θ between the string and the vertically oriented sheet in terms of m, q, σ, and g.

Section 16.8 The Electric Dipole

16.51 A water molecule of dipole moment $p = 6.1 \times 10^{-30}$ C · m is in a region of electric field $E = 1.4 \times 10^3$ N/C. What is the maximum torque on the dipole?

16.52 What is the maximum torque on a molecule of HCl (dipole moment $p = 3.6 \times 10^{-30}$ C · m) in a region of electric field $E = 1.0 \times 10^4$ N/C?

16.53 A dipole composed of two charges of magnitude 3.25×10^{-6} C experiences a maximum torque of 1.34×10^{-5} N · m when placed in a region of electric field of

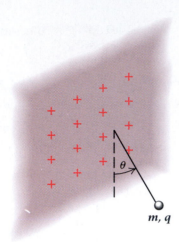

Figure 16.36
Problem 16.50.

1.06×10^3 N/C. What is the separation between the two charges that make up the dipole?

16.54 An electric dipole $p = 3.2 \times 10^{-30}$ C · m is in a region of uniform electric field of 2.5×10^3 N/C. Calculate the energy required to rotate the dipole from parallel alignment to antiparallel alignment with the field.

16.55 A sulfur dioxide molecule has an electric dipole moment $p = 5.3 \times 10^{-30}$ C · m. If the molecule is in an electric field of 1.0×10^4 N/C, what is the change in potential energy if it rotates from antiparallel to parallel alignment of its electric dipole moment with the direction of the electric field?

16.56 A water molecule, $p = 6.2 \times 10^{-30}$ C · m, is oriented along the direction of an electric field **E**. If the potential energy of the dipole in the field is -6.2×10^{-28} J, what is the magnitude of the electric field?

16.57● A dipole of $+q$ and $-q$ separated by distance $2a$ is surrounded by an imaginary Gaussian surface. (a) What is the electric flux through that surface? (b) Can Gauss's law be used to determine the electric field at the surface? Explain.

16.58● An experiment is done with a substance whose molecular dipole moment is not known. If the substance is subjected to a field of 1.50×10^4 N/C, each molecule releases an average energy of 2.28×10^{-25} J when rotating from antiparallel to parallel to the applied field. What is the molecular dipole moment of this material?

16.59● A molecule of HCl (dipole moment $p = 3.6 \times 10^{-30}$ C · m and moment of inertia of 2.7×10^{-47} kg · m²) is suddenly subjected to an electric field **E** at right angles to the direction of the molecule's dipole moment. If the initial angular acceleration of the molecule is 3.33×10^{21} rad/s², what is the magnitude of the electric field?

16.60●● Show that the electric field along the axis of a dipole of dipole moment p is given by $E = p/2\pi\epsilon_0 r^3$ when the distance r from the center of the dipole is large compared with the separation between the charges. (*Hint:* Use the binomial theorem to evaluate the field.)

16.61●● An electric dipole consists of two equal charges with opposite signs separated by a distance a. The charge $+q$ is located at $a/2$ along the y axis and the charge $-q$ is located at $y = -a/2$. (a) What is the direction of the electric field at a point P on the x axis? (b) Calculate the magnitude of the electric field due to the dipole at a point P on the x axis. Assume that $x >> a$.

Additional Problems

16.62 Two small charged pieces of amber experience a force when they are separated by a distance R. If the charge on each of the bodies is tripled and the separation between them is halved, what is the ratio of the new force to the original force between them?

16.63● How far apart must two protons be if the Coulomb force between them is equal to the weight of a single proton at the earth's surface? Assume that the two protons are far removed from any other charges. The charge of the proton is 1.60×10^{-19} C, and its mass is 1.67×10^{-27} kg.

16.64● A small nonconducting sphere of mass m falls a distance d from one metal plate to another directly beneath it. If the sphere possesses a charge q, show that it will return to its original position in the same time that it took to fall when a uniform electric field $E = 2mg/q$ is present between the plates.

16.65● Four point charges with magnitude 5.0 μC are placed at the corners of a square that is 30 cm on a side. Two charges, diagonally opposite each other, are positive, and the other two are negative. What are the magnitude and the direction of the force on one of the charges?

16.66● An electron (mass = 9.1×10^{-31} kg and charge = -1.60×10^{-19} C) is initially at rest just to the right of the leftmost plate in Fig. 16.37. The separation between the two plates is 4.00 cm. A uniform electric field of 5000 N/C directed to the left is suddenly applied between the two plates. (a) What is the force on the electron? (b) What kinetic energy will the electron have when it strikes the other plate?

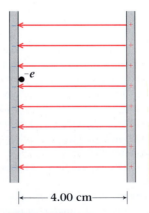

Figure 16.37
Problem 16.66.

16.67•• Imagine that the effects of gravity have been "turned off" and that you want to place a 50-kg satellite into a synchronous circular orbit with the same radius that it would obtain if gravity were "turned on." How much electric charge would be required if the satellite were held in orbit by Coulomb forces? For simplicity, assume that the magnitudes of the net charge on the satellite and on the earth are the same and that the charge on the earth is uniformly spread over the earth's surface.

16.68•• The hydrogen atom may be described as a positively charged proton surrounded by an orbiting, negatively charged electron. In a simple model, the electron is in a circular orbit of radius $r = 5.29 \times 10^{-11}$ m about a stationary proton. The mass of the proton is $m_p = 1.67 \times 10^{-27}$ kg, and the mass of the electron is $m_e = 9.11 \times 10^{-31}$ kg. The charge on the proton is $+e$ and the charge on the electron is $-e$, where $e = 1.60 \times 10^{-19}$ C. (a) What is the Coulomb force between the proton and the electron? (b) What is the linear speed of the electron in its orbit? (c) What is the orbital frequency of the electron about the proton?

16.69•• A spherical shell with inner radius a and outer radius b has a total charge Q distributed uniformly throughout its volume with a charge density ρ. Calculate the resulting electric field for $r < a$, $a < r < b$, and $b < r$.

16.70•• A solid insulating sphere of radius R has an electric charge uniformly distributed throughout its volume. The charge per unit volume is ρ. (a) Show that outside the sphere the electric field is given by $E = \dfrac{\rho R^3}{3\epsilon_0 r^2}$, where r is the distance from the center of the sphere. (b) Show that the field inside the sphere is given by $\dfrac{\rho r}{3\epsilon_0}$. (c) Draw a graph of the field from $r = 0$ to $r = 3R$. (*Hint:* Use Gauss's law and remember that it is the total charge within the Gaussian surface that contributes to the electric field at that surface.)

16.71•• Show that Coulomb's law can be obtained from Gauss's law.

16.72•• Four equal charges of $+3.0 \times 10^{-7}$ C are placed on the corners of one face of a cube of edge length 10.0 cm. A charge of -3.0×10^{-7} C is placed at the center of the cube. What are the magnitude and the direction of the force on the charge at the center of the cube?

16.73•• Two identical electric dipoles of magnitude p are separated by a distance r that is large compared to the separation of the charges within either of the dipoles. Dipole 1 has its direction fixed perpendicular to the line joining the dipoles (Fig. 16.38). At a distance r along the x axis, the magnitude of the field due to dipole 1 is $E = p/4\pi\epsilon_0 r^3$. (a) In which direction will dipole 2 point if it is free to rotate? (b) Calculate the energy required to flip the dipole 2 by 180° from its position in part (a).

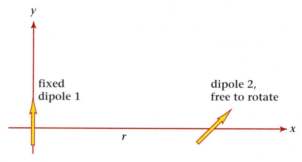

Figure 16.38
Problem 16.73.

16.74•• An electron is injected horizontally into a vertical electric field. Show that the path of the electron is parabolic. (*Hint:* Neglect the effects of gravity and compare this case with the case of projectile motion in a gravitational field in Chapter 3.)

Electric Potential and Capacitance

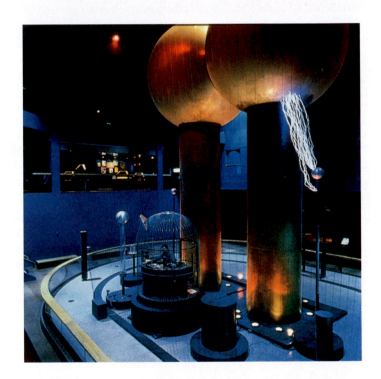

Having studied the electrostatic field of a charged body, we now turn our attention to electric potential, a concept that arises from the idea of electric potential energy. The advantage of using the potential will be evident in the simpler calculations in this chapter compared with those in the preceding chapter.

What causes a flow of electric charge when a spark jumps across a small distance in air? This question about the behavior of electric charge is comparable to the question in mechanics of why a ball rolls down a hill. In either case, we could say that the ball (or charge) moves in response to an applied force. However, in the case of the ball, we found it useful to introduce the idea of potential energy. Then we saw that the motion of the ball could be explained in terms of changes in its gravitational potential energy. In studying electricity, we define an analogous quantity called electric potential energy. In addition, we introduce the concept of an electric potential, which will allow us to describe many physical situations in simpler ways than if we used only the field concepts from the previous chapter. The idea of electric potential—or voltage, as it is often called—is one of the key concepts in studying electric circuits and electronic devices.

The basic ideas of work and energy are the same when applied to electrical forces as when applied to gravitational or any other forces. We develop the idea of electric potential by analyzing the work that is done and the change in energy that occurs when a charge is moved in an electric field. Then, to illustrate what a useful idea potential can be, we discuss an important application of electrostatics: the Van de Graaff electrostatic generator. We go on to generalize the idea of potential, which gives us a background for understanding a simple electric circuit device, the capacitor. Capacitors, which evolved from early devices for storing charge, are part of most common electronic circuits, from the tuning circuit in your radio to the ignition system in your car. Their operation can be understood by building on the ideas of electric charge and electric potential. ■

17.1 Electric Potential

We learned in Chapter 16 that according to Coulomb's law, a positive test charge q_0 located a distance r_A away from a positive charge Q experiences an electrostatic force in the direction away from Q (Fig. 17.1a). If q_0 is moved closer to Q, say to a point r_B, as in Fig. 17.1(b), it then experiences a greater force, still directed away from Q. To move charge q_0 from r_A to r_B, we must apply a force in a direction opposite to the electrostatic force. Using the techniques of calculus, we find that the work required is

$$W = kQq_0\left(\frac{1}{r_B} - \frac{1}{r_A}\right),\tag{17.1}$$

where k is the constant in Coulomb's law. This equation is similar to Eq. (6.8) for the work done in moving a mass m in the gravitational field of another mass M.

If the test charge is released from rest at r_B, it will be in motion as it passes r_A and will therefore have kinetic energy. From conservation of energy, we know that this kinetic energy must come from somewhere. In other words, the work given by Eq. (17.1) corresponds to an increase in potential energy of the test charge. Since the interaction is electrical, we call this *electric potential energy.**

Equation (17.1) gives us the *difference* in electric potential energy between two points in space due to a point charge. In establishing a zero of potential energy, we are free to pick any absolute reference potential that suits our convenience. Traditionally, we choose the zero of potential to correspond to the case in which the charges are infinitely separated—that is, for $r_A = \infty$. With this choice, the electric potential energy of the test charge q_0 located a finite distance r away from a source charge Q is

$$PE_{elec} = W = k\frac{Qq_0}{r}.\tag{17.2}$$

---

*The electrostatic force is a conservative force, just like the gravitational force, so the work done does not depend on the path of the charge, but only on its starting and ending positions.

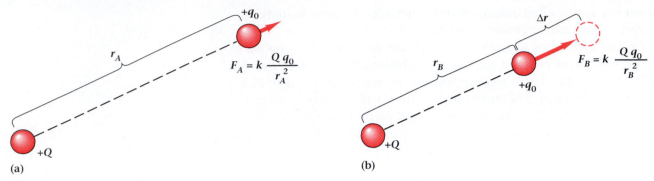

(a)   (b)

Figure 17.1

(a) A test charge q_0 located a distance r_A from a source charge Q. (b) The test charge is moved a distance Δr closer to source charge Q, which results in a change in the potential energy.

There is a strong similarity between the equation for electric potential energy and the equation for gravitational potential energy, given in Chapter 6 as

$$PE_{grav} = -G\frac{Mm}{r}.$$

In both cases we take the zero reference point to be at $r = \infty$. In the gravitational case, the force is attractive and the potential is always negative. For a positive charge Q, the electrical force on a positive test charge q_0 is repulsive and the potential energy is positive. However, if Q is negative, resulting in an attractive force between Q and q_0, the electric potential energy is negative. We can draw a potential-energy diagram for q_0 in the electric field of Q similar to the gravitational potential-energy diagrams shown in Chapter 6. However, in this case, when the charges have the same sign the potential energy is positive and when the charges have unlike signs the potential energy is negative (Fig. 17.2).

We find it helpful to introduce a related quantity called the **electric potential** V at a point in an electric field, defined to be the electric potential energy divided by the magnitude of the test charge q_0:

$$V \equiv \frac{PE_{elec}}{q_0}. \tag{17.3}$$

The electric potential due to a point charge Q is obtained from the potential energy of Eq. (17.2) by dividing by q_0,

$$V = k\frac{Q}{r} = \frac{Q}{4\pi\epsilon_0 r}. \tag{17.4}$$

If the charge Q is positive, the potential is positive, and if Q is negative, the potential is negative. The unit of electric potential is the **volt**:

1 volt = 1 joule/coulomb.

The electric potential at any point in space is defined to be the work per unit charge required to bring a charge from infinity to that point. The electric potential due to two or more point charges is obtained from the principle of superposition of forces. The work done in bringing a test charge from infinity can be

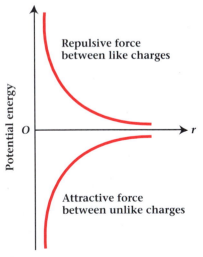

Figure 17.2

The electrostatic force between two charges corresponds to either of two potential energy curves, depending on whether the force is attractive or repulsive.

separated into parts associated with the individual forces due to each single charge. The total work is the sum of these individual contributions. Because electric potential is a scalar, not a vector quantity, the potential at a given point in space is just the algebraic sum of the potentials due to the separate source charges.

Frequently, just as for the case of gravitation, the quantity of interest is not the absolute potential, but the potential difference between two points. The **electric potential difference** is the ratio of the work done by an external force in moving a charge from one point to another to the magnitude of the charge. The potential difference between points A and B is

$$V_{AB} = V_B - V_A = \frac{W_{AB}}{q}. \tag{17.5}$$

Since the potential is measured in volts, the potential difference is also measured in volts. For this reason we often refer to a potential difference as a voltage. Sources of electrical energy, such as batteries and generators, are used to maintain a fixed electric potential difference. For example, during operation of a flashlight the batteries maintain a constant electric potential difference across the lamp and thus provide the energy necessary for producing the light.

Many times, the potential difference is taken with reference to the *ground* (earth), which we generally choose as the zero potential. We do this for the same reason that we often chose the ground to be the reference when describing gravitational potential energy. There we could calculate a change in potential energy mgh, where h was the height above the ground. Here we want the change in electric potential from the ground to some other point. Table 17.1 lists potential differences for several cases.

In electrical problems, as in other areas of physics, we often need to know the amount of work required to make something happen or the work that is done by a particular device. The key to finding such work in electrical problems is to use the idea of the potential difference. We can rearrange Eq. (17.5) as

$$W_{AB} = qV_{AB}, \tag{17.6}$$

in order to emphasize that the work to move a charge q from A to B is just the product of the charge and the potential difference from A to B.

Table 17.1	*Electric Potential Differences Found in Common Situations* ▼	
Situation	**Potential Difference (volts)**
pn-junction in a transistor	10^{-1}
Biological cell	10^{-1}
Single cell battery	1
Household electric outlet	10^2
TV sets	10^4
Thunderclouds	10^8

Example 17.1

A proton has a positive electric charge $e = 1.60 \times 10^{-19}$ C. What is the electric potential energy of an electron (charge $-e$) at a point 5.29×10^{-11} m from the proton (the average distance of an electron from the proton in a hydrogen atom)? For simplicity in computation in this and other examples, you may use $k = 9.00 \times 10^9$ N · m²/C².

Strategy We can compute the electric potential energy from Eq. (17.2) if we let Q be the charge on the proton and q_0 be the charge on the electron.

Solution The potential energy is

$$\text{PE}_{\text{elec}} = k\frac{Qq_0}{r} = k\frac{(e)(-e)}{r}$$

$$= 9.00 \times 10^9 \text{ N} \cdot \text{m}^2/\text{C}^2\left(\frac{(1.60 \times 10^{-19} \text{ C})(-1.60 \times 10^{-19} \text{ C})}{5.29 \times 10^{-11} \text{ m}}\right)$$

$$\text{PE}_{\text{elec}} = -4.36 \times 10^{-18} \text{ N} \cdot \text{m} = -4.36 \times 10^{-18} \text{ J}.$$

Discussion When the negatively charged electron is placed near the positively charged proton, the resulting potential energy is negative, indicating that the electron is attracted to the proton.

Example 17.2

What is the electric potential at the center of the square shown in Fig. 17.3? The values of the charges are $q_1 = 1.0$ nC, $q_2 = -2.0$ nC, $q_3 = +3.0$ nC, and $q_4 = -4.0$ nC. Assume the square to have an edge length $d = 1.0$ m.

Strategy Because *electric potential* is a scalar, not a vector quantity, we can find the net potential due to several point charges by just adding together the contribution to the potential due to each source. We can do this without considering the relative directions of the charges. When finding the *electric field* from a collection of charges, we must find the magnitude and direction of the field due to each charge and then add the fields together using the rules for vector addition. You may want to compare this example with Example 16.7, which asked you to find the electric field.

Solution The center of the square is equidistant from all four charges, a distance of $d/\sqrt{2}$. Let's call this distance r. Then we can calculate the potential V by adding up the potentials due to the individual charges,

$$V = V_1 + V_2 + V_3 + V_4 = \frac{kq_1}{r} + \frac{kq_2}{r} + \frac{kq_3}{r} + \frac{kq_4}{r}$$

$$V = \frac{k}{r}(q_1 + q_2 + q_3 + q_4)$$

$$V = \frac{9.00 \times 10^9 \text{ N} \cdot \text{m}^2/\text{C}^2(1.0 - 2.0 + 3.0 - 4.0) \times 10^{-9} \text{ C}}{1 \text{ m}/\sqrt{2}}$$

$$V = -25 \text{ N m/C} = -25 \text{ V}.$$

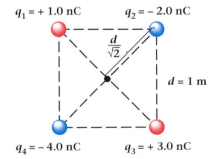

$q_1 = + 1.0$ nC $\quad q_2 = - 2.0$ nC

$\frac{d}{\sqrt{2}}$

$d = 1$ m

$q_4 = - 4.0$ nC $\quad q_3 = + 3.0$ nC

Figure 17.3
Example 17.2: Four point charges arranged in a square of edge d.

Discussion The electric potential at the center of the square is 25 V below the zero level reference potential, which corresponds to the case in which charges are infinitely far away. This means you would have to do work on a positive test charge to move it from the center of the square to infinity.

Example 17.3

A small metal sphere, carrying a net electric charge of $+6.50$ μC, is held fixed on an insulating stand (Fig. 17.4). An alpha particle of charge $+3.20 \times 10^{-19}$ C is projected along a radial path toward the sphere. When the alpha particle is 1.00 m from the center of the sphere, it has a speed of 5.40×10^5 m/s. How close to the sphere does the particle get?

Strategy The charge on the sphere is spherically symmetric, so it behaves as if all of its charge is concentrated at its center. Because the alpha particle and the sphere are both positive, the force between them is repulsive and the potential energy increases as the alpha gets nearer to the sphere. This increase in potential energy comes at the expense of the kinetic energy. As the alpha gets closer, its kinetic energy decreases and the potential energy increases. We can apply the law of conservation of energy to find the distance of closest approach, which will be the distance at which the kinetic energy goes to zero.

Solution We can express the law of conservation of energy as

$$KE_i + PE_i = KE_f + PE_f,$$

where KE_i is the initial kinetic energy and $KE_f = 0$. Using Eq. (17.2) for the potential energy we get

$$\tfrac{1}{2}m_\alpha v_\alpha^2 + \frac{1}{4\pi\epsilon_0}\frac{q_\alpha q_{sphere}}{r_i} = \frac{1}{4\pi\epsilon_0}\frac{q_\alpha q_{sphere}}{r_f}.$$

Upon rearranging we get

$$\frac{1}{r_f} = \frac{1}{r_i} + \frac{\tfrac{1}{2}m_\alpha v_\alpha^2}{\frac{1}{4\pi\epsilon_0}q_\alpha q_{sphere}}$$

$$\frac{1}{r_f} = \frac{1}{1.00\ \text{m}} + \frac{\tfrac{1}{2}(6.62 \times 10^{-27}\ \text{kg})(5.40 \times 10^5\ \text{m/s})^2}{(9.00 \times 10^9\ \text{N} \cdot \text{m}^2/\text{C}^2)(3.20 \times 10^{-19}\ \text{C})(6.50 \times 10^{-6}\ \text{C})}$$

$$\frac{1}{r_f} = 1.00\ \text{m}^{-1} + 0.0516\ \text{m}^{-1} = 1.052\ \text{m}^{-1}.$$

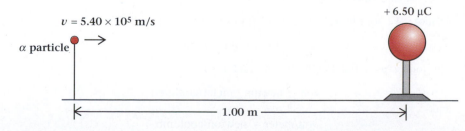

$v = 5.40 \times 10^5$ m/s

α **particle**

$+ 6.50$ μC

1.00 m

Figure 17.4
Example 17.3: An alpha particle is projected toward a positively charged metal sphere.

The distance of closest approach is

$$r_f = 0.951 \text{ m}.$$

Discussion Ernest Rutherford used an analysis like that used here to determine the size of the atomic nucleus of gold. (See Chapter 26.) In that case, he was trying to explain the experimental results obtained by bombarding a gold foil with alpha particles. He estimated the charge on the gold nucleus to be $+100e$. Because the mass of the gold was so large compared to that of the alpha, he treated it as if it remained at rest during the collision. ■

We can gain further insight into the connection between the electric potential and the electric field by considering the behavior of a positive test charge placed in a uniform electric field. The field exerts a force on the test charge given by $\mathbf{F} = q_0\mathbf{E}$, in the direction of the field (Fig. 17.5). The work done by the field on the charge is

$$W_{AB} = q_0 Ed,$$

where d is the distance from A to B. From the work-energy theorem we know that the work done is equal to the change in kinetic energy. Applying conservation of energy, we find that the change in potential energy is just the negative of the change in kinetic energy, so

$$\Delta PE = -q_0 Ed.$$

As the charge moves from point A to point B under the influence of this force, it moves from a higher potential to a lower potential. The corresponding change in potential energy is

$$\Delta PE = q_0 V_{AB}.$$

Upon comparing these two equations for ΔPE, we see that

$$V_{AB} = -Ed. \tag{17.7}$$

The negative sign indicates that if a positive charge moves through a displacement d in the direction of the electric field, then its potential energy decreases.

It is apparent from Eq. (17.7) that the dimensions of the electric field are the same as those of electric potential divided by distance. In SI units, the electric field is given in volts per meter. Indeed, the most common way to measure E is in units of potential divided by a convenient length unit, such as V/m, V/mm, or kV/m.

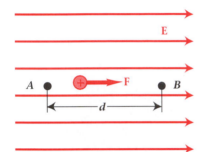

Figure 17.5

A uniform electric field **E** exerts a force **F** on a positive test charge q_0 and does work in moving the charge from A to B.

Example 17.4

Show that the units of newtons per coulomb, which we first used for electric field, are equivalent to volts per meter.

Solution We start by recalling the definition of the volt:

$$\text{volt} = \text{joule/coulomb}.$$

The joule was defined as a newton · meter, so

$$\text{volt} = \text{newton} \cdot \text{meter/coulomb}$$

and

$$\text{volt/meter} = \text{newton/coulomb}.$$

17.2   The Van de Graaff Electrostatic Generator

The electrostatic generator is a machine for producing very large electric potential differences, of the order of millions of volts (Fig. 17.6a). The modern form of the electrostatic generator was developed in 1931 by the American physicist Robert J. Van de Graaff (1901–1967). Since the chief application of these electrostatic generators is to accelerate charged particles to very high kinetic energies, they are often referred to as accelerators.

The operation of a **Van de Graaff generator** can be explained in terms of the principles described in the preceding sections. These same principles apply to research accelerators as well as to the demonstration static generators. In the self-excited, frictional Van de Graaff generator frequently seen in the classroom, a motor-driven roller drives a belt of insulating material within a column of the machine (Fig. 17.6b). Charge is separated by the rubbing of the belt against the driver roller while the belt is moving. Positive charge collects on the roller, and an equal amount of negative charge collects on the inner side of the belt. This negative charge is less dense because it is spread over the entire length of the belt. Thus, in the region of the lower roller the net charge is positive. At the base of the column, where the belt passes the sharp points of a charged metal "comb," this positive charge attracts negative charges (electrons) from the ground, leaving a high negative-charge density on the belt. The charged belt moves upward into a large, hollow, conducting dome. Here a conducting wire, connected to a comb of sharp points near the top of the belt, allows the negative electrons from the charged belt to flow to the dome, making the charge on the dome negative.

Figure 17.6

(a) Student with a demonstration Van de Graaff generator. (b) Schematic representation of a classroom Van de Graaff generator. A moving belt carries electrons to the hollow dome, giving the dome a large negative charge.

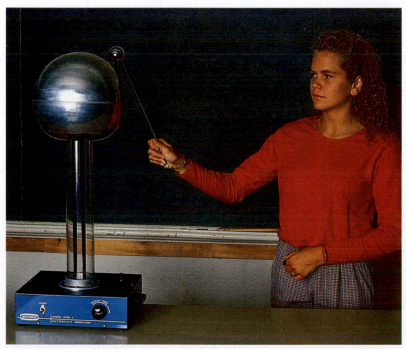

(a)

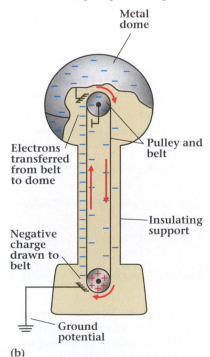

Metal dome

Electrons transferred from belt to dome

Pulley and belt

Negative charge drawn to belt

Insulating support

Ground potential

(b)

Since the dome is a conductor, its excess charge resides only on the outside surface, as was shown in Section 16.6. By Gauss's law, there is no field inside the dome due to these outer charges. However, there is an electric field inside the dome due to the negatively charged belt. This field drives negative charges to the dome, regardless of the dome's potential relative to the ground. As a result, a large negative charge builds up on the dome, accompanied by a large increase in voltage, which is proportional to the total charge.

There are limits to the potential difference that can be maintained between the dome and surrounding objects. Eventually the potential difference becomes so great that the air can no longer act as an insulator, and the dome discharges with a spark of miniature lightning. The motion of ions in the air can also cause the dome to lose charge. In addition, if the generator is operated in air when the humidity is high, charge leaks off because of moisture on the insulating surfaces and on the belt. Consequently, the maximum potential is reduced. For this reason, the electrostatic generators used in research are usually enclosed in a sealed container, where they can be surrounded by a controlled, dry atmosphere of air, nitrogen, or other gas. Figure 17.7 shows a research electrostatic generator with the outer enclosure removed. The domes of such research accelerators can be charged positively, instead of negatively, by charging the upward moving belt positively.

If the dome is charged positively to several million volts above ground potential, positive charges can be accelerated away from the dome. By placing a source of charged particles inside the dome, we can generate a stream of highly energetic particles. The energy of these particles is determined by the potential of the dome. In practice, electrostatic accelerators are used to produce beams of high-energy particles for use in nuclear physics, biomedicine, and semiconductor manufacturing.

Figure 17.7
A research electrostatic generator used for producing beams of energetic particles.

Example 17.5

In the manufacture of some integrated circuit chips (Fig. 17.8), which are the hearts of computers, watches, and other electronic equipment, boron ions (atoms with one or more electrons added or removed) are implanted in the upper surfaces of silicon to control its electrical behavior. The boron ions, which acquire a high velocity in passing through the potential difference of an electrostatic accelerator, bury themselves in the silicon when they hit the surface. What is the speed of a singly charged boron ion that has been accelerated from rest by a potential difference of 86.0 kV?

Strategy We can find the speed by using the principle of conservation of energy. In moving through the given potential difference, a boron ion of mass m_B gains a kinetic energy $\frac{1}{2}m_Bv^2$, where v is the final velocity. This kinetic energy is a result of the work done on the ion by the electric field as the ion moves through the electric potential difference. We can equate the two energies and solve for v.

Solution As we saw in the last section, the work done is $W = qV$, where q is the charge on the ion and V is the potential difference. Then, from the work-energy theorem, we get

$$\tfrac{1}{2} m_B v^2 = qV,$$
$$v = \sqrt{\frac{2qV}{m_B}}.$$

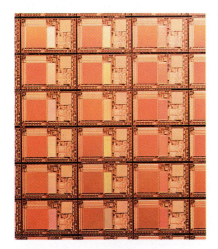

Figure 17.8
Example 17.5: A magnified view of an array of silicon integrated circuits. The characteristics of the silicon can be modified by ion implantation.

The charge q is equal to the magnitude of the electron charge, 1.60×10^{-19} C, and the mass of the boron ion is 1.83×10^{-26} kg. Inserting these values into the equation for v, we find

$$v = \sqrt{\frac{2(1.60 \times 10^{-19} \text{ C})(86.0 \times 10^3 \text{ V})}{1.83 \times 10^{-26} \text{ kg}}}$$

$$v = 1.23 \times 10^6 \text{ m/s.}$$

17.3 The Electron Volt

There are many occasions on which the SI unit of energy, the joule, is very large compared to the energy being considered. Example 17.1 is a good illustration of where we could use a unit of energy more suitable to atomic-scale interactions. As we have seen, whenever a charge q is moved through an electric potential difference V, an amount of work $W = qV$ is required. If the charge q is the magnitude of the charge on an electron, e, and the potential difference is one volt, then the work done is

$$W = (e)(1 \text{ V}) = (1.602 \times 10^{-19} \text{ C})(1 \text{ V}) = 1.602 \times 10^{-19} \text{ J.}$$

This amount of work or energy is given the special name **electron volt,** abbreviated eV:

$$1 \text{ eV} = 1.602 \times 10^{-19} \text{ J.} \tag{17.8}$$

The electron volt is a practical unit for measuring energy on a molecular or smaller level. It is a natural unit to use for both the potential and the kinetic energies of particles such as the electron (charge $-e$), the proton (charge $+e$), and the helium nucleus (charge $+2e$).

   The electron volt is a valid unit of energy, regardless of how the energy is acquired. In the defining physical situation, the electron gains kinetic energy as it moves through an electric potential. It could also gain or lose kinetic energy by collision with other electrons or atoms. The kinetic energy is still conveniently expressed in eV, although it was not attained by actually moving through a particular potential difference.

   Even though the electron volt is a very useful unit of energy, it is not an SI unit. For this reason, when you make a calculation involving energy, you should be sure to use the unit of joule, even if it requires you to convert the value you have in eV into joules before continuing the calculation, as shown in the following example.

Example 17.6

What is the speed of an electron with a kinetic energy of 500 eV? The mass of the electron is 9.11×10^{-31} kg.

Solution   This is a straightforward calculation using the definition of kinetic energy. However, in making the calculation, we must convert the energy from electron volts to the SI unit of joules.

$$KE = 500 \text{ eV} \left(\frac{1.602 \times 10^{-19} \text{ J}}{1 \text{ eV}} \right) = 8.01 \times 10^{-17} \text{ J.}$$

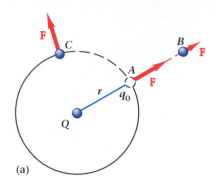

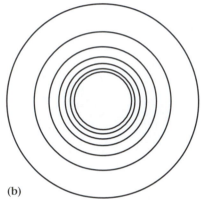

(a)

(b)

Figure 17.9

(a) A test charge q_0 changes potential energy when moving along a radial path from a fixed charge Q, but not when moving in a circular path about Q. The circular path is an equipotential line. (b) Equipotential lines in a plane around a point charge. The potential difference from one line to the next is the same for all lines.

Figure 17.10

The equipotential surfaces around a point charge have the form of a nested set of spherical shells. The surface closest to the charge has the largest value of potential. The potential difference from one shell to the next is the same for all shells.

The definition of kinetic energy is

$$KE = \frac{1}{2} mv^2.$$

Upon rearranging, we get

$$v = \sqrt{\frac{2\ KE}{m}} = \sqrt{\frac{2(8.01 \times 10^{-17}\ J)}{9.11 \times 10^{-31}\ kg}} = 1.33 \times 10^7\ m/s.$$

17.4 Equipotential Surfaces

We have seen pictorial representations of electric fields in Figs. 16.14, 16.17, and 16.18. These representations were based on the concept of electrostatic force, a vector. We now wish to introduce a pictorial representation of electric potential based on the concept of energy, a scalar. You may find it helpful to review the ideas used in Chapter 6 for making potential-energy diagrams (see p. 193).

Figure 17.9(a) shows a positive test charge q_0 initially located a distance r from a charge Q whose position is fixed. If the test charge q_0 moves from point A to point B along a radial path, the force due to the electric field of Q does work on the test charge. As a result, the potential energy of q_0 is changed as it moves. On the other hand, if the charge moves from point A to point C along the circular path of constant radius, its motion is perpendicular to the electric force and no work is done by (or against) the field. Thus, the potential energy of q_0 is the same at point A as at point C. As long as the charge moves at right angles to the electric field (and hence to the electric force), its potential energy remains unchanged. In this case, q_0 could move completely around Q along the circular path without any work being done by the electric field. The potential energy of the test charge is the same everywhere on the circle, so the electric potential must also be constant along the circle. For this reason, we call any circle about the source charge Q an *equipotential line.* Figure 17.9(b) shows equipotential lines in a plane about the source charge. The potential difference between successive rings is constant.

As we noted in Chapter 16, lines of force about an isolated point charge extend symmetrically in three dimensions. Similarly, the circular two-dimensional equipotential line around a point charge is actually part of a three-dimensional spherical surface. Any surface for which all points are at the same potential is called an **equipotential surface.** Figure 17.10 depicts the equipotential surfaces around a point charge as a set of nested spherical shells. Each equipotential surface represents a different potential, decreasing in magnitude with increasing distance from the source charge. The choice of specific potentials is arbitrary; an equipotential surface can be drawn through any point in the field.

Equipotential surfaces may be drawn corresponding to any given configuration of electric field. We construct the contours of the equipotential surfaces by making them everywhere perpendicular to the electric field vector (or lines of force). If the equipotential surface was not perpendicular to the field, motion of a charge along the surface would correspond to a change in potential energy due to the work done by the field on the charge. Figure 17.11 shows a cross section of the equipotential surfaces for (a) a pair of like charges, (b) a pair of unlike charges. In both cases, the cross section is taken in the plane containing the two charges. We can get a better appreciation for the potentials by shifting our point

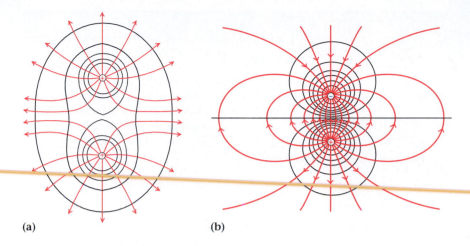

(a) (b)

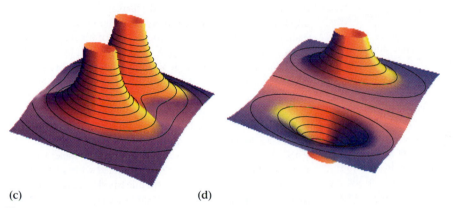

(c) (d)

Figure 17.11

Cross sections of equipotential surfaces around (a) a pair of like charges with equal magnitudes, (b) a pair of unlike charges with equal magnitudes. The black lines represent the equipotential surfaces, and the red lines indicate the electric field. (c) The potential in the plane containing the like charges and (d) the unlike charges.

of view from directly overhead to observe the plane from an angle and then showing the potential in that plane as the height above (or below) the $V = 0$ plane as in parts (c) and (d) of the figure. The potential rises near the positive charges and sinks near the negative charges. Equipotential lines are drawn on the surface.

In Chapter 16 we found that any excess charge placed on a conductor must reside on its surface. If the charge is at rest, whatever electric field is present must be perpendicular to the surface, since any field parallel to the surface would cause the charges to move. Because the field is perpendicular to the surface, all points along the surface must be at the same potential. We conclude that **when all charges have come to rest, the surface of a conductor is always an equipotential surface.** Also, as we saw in Chapter 16, at equilibrium there is no field inside a conductor, so all interior points have the same potential as the surface. Thus, it takes no work to move a charge around anywhere within a conductor.

THE LEYDEN JAR AND FRANKLIN'S KITE

Early experimenters in electrostatics faced a very difficult problem. The magnitude of the electric force made it difficult to separate and store large amounts of charge. Even with Hauksbee's electrostatic machine, isolating and storing charge was not easy.

In 1746 at the University of Leyden, the Dutch scientist Pieter van Musschenbroek (1692–1761) was trying to build up and store large amounts of electric charge. The result was a device that became widely known as the *Leyden jar* (Fig. B17.1), and was the ancestor of the modern capacitor. The Leyden jar is a glass jar covered inside and out with metal foil. It has a wooden lid with a metal rod running through it. A metal chain hangs down from the rod to touch the metal foil lining the inside of the jar.

To charge a Leyden jar, first the outer metal coating of the jar is grounded, say by a metal wire running into the ground. Then the center conductor is brought near an electrostatic machine, which, for illustration, we consider to be positively charged. When the center conductor is brought into contact with the machine, the negative charges (electrons) in the center conductor flow off onto the machine. As a result, a positive charge is left on the inner foil liner. This inner charge attracts negative charge to the outer foil, the negative charge coming from the earth by means of the ground wire.

The charging process can be repeated several times to build up an extremely large charge. After the charging is completed, the ground is disconnected. Because the inner and outer conductors are well insulated from each other by the glass, the jar is able to store a surprisingly large amount of charge. If it were not for the leakage of charge into the air and across the insulating surfaces of jar and lid, the charge would remain indefinitely. As it is, the charge can remain for long periods of time if the jar is clean and dry and the air is dry.

If a conductor attached to the outer coating of a charged Leyden jar is brought near the center conductor, the jar will discharge with a large spark. In Musschenbroek's laboratory, people acting as experimental subjects were knocked about and stunned by the terrifying shock they received from the jar. If one person held the jar while another person touched the insulated conductor, both felt a shock when they joined their free hands together. In some experiments, several people holding hands made a human chain through which the charge could pass. The end persons primarily felt the shock if the others

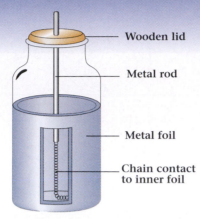

Figure B17.1 The Leyden jar, consisting of a glass insulator between two metal foils.

Wooden lid

Metal rod

Metal foil

Chain contact to inner foil

clasped their hands firmly. Human chains exceeding 200 persons in length were formed as the experimenters demonstrated their new tool.

Benjamin Franklin also experimented with Leyden jars and wondered if electricity might be the same as lightning. In 1752 he charged a Leyden jar with the electricity obtained from a kite that he flew during a thunderstorm (Fig. B17.2). For this proof that lightning and electricity were one and the same, Franklin was made a member of England's scientific Royal Society.

Franklin's experiment was extremely hazardous; he was fortunate to carry it out unscathed. Indeed, several people who tried to duplicate his experiment were struck by lightning and killed.

Franklin's work in electricity made him one of the leading scientists of his time. Today, one of his best known scientific works is his invention of the lightning rod. Franklin noted that he could discharge a charged object by bringing a metal sphere close to it. However, if instead, he brought a sharply pointed metal rod near the charged object, it would discharge when the rod was considerably farther away. (In modern terms, the electric field is greater near a sharp point than near a rounded surface, and the field causes the discharge of the object.) Franklin reasoned that a pointed metal rod placed on the roof of a house and grounded with a heavy wire could discharge thunderclouds harmlessly. In addition, this "lightning rod" would provide a safe electrical path to ground for any lightning strike, protecting the house from damage. Lightning rods are still in common use in the United States, especially on tall buildings and in rural areas.

Figure B17.2 Benjamin Franklin's experiment.

Capacitors

Any system of two conductors placed near each other but separated by an insulator, such as the metal foils on a Leyden jar (see p. 538), can store more charge than can an isolated conductor. It was said that the Leyden jar concentrated, or condensed, the charge, and the name *condenser* has persisted to the present day. However, since the important measure of these devices is their capacity to store charge, their preferred name is capacitors. A **capacitor** is a device for storing charge.

Capacitors come in many shapes and sizes, made from a variety of materials. The only thing they all have in common is the close placement of two or more conducting sheets or plates separated by a thin layer of insulation. In some capacitors, the conductors are metal foil and the insulator is paper; other capacitors contain alternating layers of metal plates and mica. Ceramic capacitors used on electronic circuit boards can be as small as a grain of rice. Still others built into integrated circuits consist of thin strips of silicon oxide deposited on silicon and covered with a layer of metal. Figure 17.12 shows some typical capacitors.

(a)

Capacitors play an important role in many electrical and electronic circuits. For example, in computers and television sets, the insulating layers of capacitors can be used to isolate portions of a circuit from unwanted electric potentials. Capacitors are also used to store charge, and hence energy, to be given up when needed. In this sense, a charged capacitor is similar to a compressed spring, for both store energy that can be released when needed.

If a potential difference, or voltage, is applied across a capacitor, charge is transferred, making one plate positive and the other negative. If the potential difference between the plates is increased, the charge stored on each of the capacitor's plates is proportionally increased. We find that the charge q stored on each plate is proportional to the potential difference V between the plates: $q \propto V$. We define the proportionality constant as the **capacitance** C, which is the ratio of the quantity of charge q stored on each plate to the potential difference V between the conductors,

(b)

Figure 17.12
(a) Some typical aluminum oxide capacitors. (b) Ceramic capacitors are smaller than grains of rice.

$$C \equiv \frac{q}{V}. \tag{17.9}$$

The magnitude of the capacitance is a constant for each particular capacitor and depends only on the geometrical shape of the capacitor and on the nature of the insulating material between the conductors. When the charge is measured in coulombs and the potential in volts, the capacitance is measured in **farads** (F):

1 farad = 1 coulomb/volt.

The unit for capacitance is named in honor of the English physicist and chemist Michael Faraday (1791–1867), who was renowned for his work in electrochemistry and electricity. The farad is a large unit; many practical capacitors have values of microfarads (μF), nanofarads (nF), or even picofarads (pF).

Master the Concept

How Capacitors Store Charge

Question: Capacitors are known for their ability to store electric charge. What keeps the charge on the capacitor?

Answer: When a potential difference is applied across the plates of a capacitor, charges flow, making one plate positively charged and the other plate negatively charged. Because the plates are separated by an insulating layer, no charge flows between them. When the potential source is removed, the attractive electrostatic force between the positive and negative charges holds them in place, enabling the capacitor to store the charge until a conductor is connected across the two plates.

Example 17.7

A 25-μF capacitor is charged to a potential of 18 V. (a) How much charge is stored on the capacitor? (b) The potential is then changed so that the charge becomes 5.9×10^{-4} C. What is the change in the potential?

Solution (a) We can find the charge by rearranging Eq. (17.9) in the form

$$q = CV.$$

Inserting the numerical quantities gives

$$q = (25 \times 10^{-6} \text{ F})(18 \text{ V}) = 4.5 \times 10^{-4} \text{ C}.$$

(b) The change in potential ΔV is

$$\Delta V = V_2 - V_1,$$

where the subscript 2 indicates the new situation and the subscript 1 indicates the original situation. The capacitance is a fixed property of the capacitor and does not change. Using the definition of capacitance, the expression for ΔV becomes

$$\Delta V = \frac{q_2}{C} - \frac{q_1}{C} = \frac{1}{C}(q_2 - q_1)$$

$$\Delta V = \frac{1}{25 \times 10^{-6} \text{ F}}(5.9 \times 10^{-4} \text{ C} - 4.5 \times 10^{-4} \text{ C}) = \frac{1.4 \times 10^{-4} \text{ C}}{25 \times 10^{-6} \text{ F}}$$

$$\Delta V = 5.6 \text{ V}.$$

17.6 The Parallel-Plate Capacitor

In its simplest form, a capacitor consists of a pair of parallel metal plates separated by a small distance d (Fig. 17.13). The air between the plates is the insulator. When a charge $+q$ is stored on one plate and a charge $-q$ on the other, an electric potential difference exists between the plates. For a given separation between the plates, we expect that larger plates can hold more charge and thus contribute to a larger capacitance. Similarly, for a fixed area we expect the capacitance to decrease as the separation increases because the influence of each plate on the other is reduced; that is, the force of attraction between the plates is reduced. The capacitance of two parallel plates of area A that is large compared with their separation d is

$$C = \frac{\epsilon_0 A}{d}, \qquad (17.10)$$

where ϵ_0 is the permittivity of free space given in Section 16.2. We will derive this equation formally in the next section. Equation (17.10) applies when the plates are in a vacuum. When the plates are in air, the capacitance is changed only slightly, as we shall see in Section 17.8.

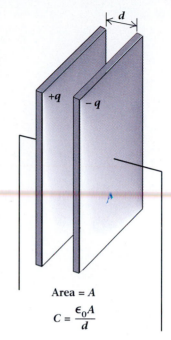

Area = A

$C = \frac{\epsilon_0 A}{d}$

Figure 17.13
Two oppositely charged parallel plates form a capacitor. The capacitance depends on the area A of the plates and inversely on the distance d between them.

Example 17.8

A parallel-plate capacitor is designed to have a capacitance of 1.00 F when the plates are separated by 1.00 mm in a vacuum. What must be the area of the plates?

Solution   We can use Eq. (17.10), which relates capacitance, area, and separation:

$$C = \frac{\epsilon_0 A}{d}.$$

Rearranging to put the unknown quantity A on one side and the known quantities on the other, we find

$$A = \frac{Cd}{\epsilon_0}.$$

We insert the values for C, d, and ϵ_0 to get

$$A = \frac{(1.00 \text{ F})(1.00 \times 10^{-3} \text{ m})}{8.85 \times 10^{-12} \text{ C}^2/\text{N} \cdot \text{m}^2} = 1.13 \times 10^8 \text{ F} \cdot \text{N} \cdot \text{m}^3/\text{C}^2.$$

Discussion   We expect that the area should be in m². Let's examine the units to be sure. Remember that

$$1 \text{ farad} = 1 \text{ coulomb/volt},$$
$$1 \text{ joule} = 1 \text{ newton} \cdot \text{meter},$$

and

$$1 \text{ volt} = 1 \text{ joule/coulomb}.$$

So

$$\text{F} \cdot \text{N} \cdot \text{m}^3/\text{C}^2 = \frac{\text{C}}{\text{V}} \frac{\text{J}}{\text{m}} \frac{\text{m}^3}{\text{C}^2} = \frac{1}{\text{V}} \frac{\text{J}}{\text{C}} \text{m}^2 = \text{m}^2.$$

The area of a 1.00-F parallel-plate capacitor with a 1.00-mm separation is 1.13×10^8 m²—almost the size of the city of San Francisco!

Example 17.9

Two parallel-plate capacitors are made from circular plates. One capacitor has plates with radii $r_1 = 12.3$ cm and a spacing between the plates of 2.34 mm. The second capacitor has plates with radii $r_2 = 14.8$ cm. If the two capacitors have the same capacitance, what is the spacing between the plates in the second capacitor?

Solution Because the capacitors have the same value, we may write

$$C_1 = C_2,$$

$$\frac{\epsilon_0 A_1}{d_1} = \frac{\epsilon_0 A_2}{d_2}.$$

Using $A = \pi r^2$, and solving the previous equation for d_2, we have

$$d_2 = d_1 \left(\frac{r_2}{r_1}\right)^2 = 2.34 \text{ mm}\left(\frac{14.8 \text{ cm}}{12.3 \text{ cm}}\right)^2$$

$$d_2 = 3.39 \text{ mm}.$$

In this case it was not necessary to convert distances to meters because we are taking ratios. However, as a general rule, you should always convert all variables to their SI values.

Example 17.10

Show that the units of the permittivity constant ϵ_0 can be expressed as farads per meter.

Solution The units of ϵ_0 were given as $C^2/(N \cdot m^2)$. From Eq. (17.10) we see that ϵ_0 must have dimensions of farads per meter. We can rearrange the units to put them in the desired form as follows:

$$\frac{C^2}{N \cdot m^2} = \frac{C^2}{J \cdot m} = \frac{C}{\frac{J}{C} \cdot m},$$

where we use the fact that one joule = 1 newton · meter. We further observe that joule/coulomb = volt and coulomb/volt = farad, so

$$C^2/N \cdot m^2 = F/m.$$

We will usually find it convenient to express ϵ_0 in these units of F/m.

17.7 Electric Field of a Parallel-Plate Capacitor

In the previous sections, we discussed capacitors in terms of the charge on their plates and the potential difference between them. We now want to examine the field within a capacitor and use what we have learned to find the capacitance of

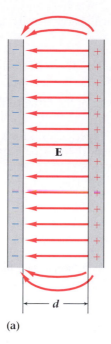

(a)　　　(b)

Figure 17.14
The electric field due to a pair of oppositely charged parallel plates is uniform except near the edges. (a) Schematic representation of the field. (b) Field lines as indicated by small threads suspended in oil between oppositely charged plates.

a parallel-plate capacitor. Suppose we place two large identical flat conducting plates parallel to each other and close together to form a parallel-plate capacitor. Then we charge one of the plates uniformly with positive charge and give the other an identical amount of negative charge. From the symmetry, the resulting field between the plates will be very nearly uniform and directed from the positive plate to the negative plate (Fig. 17.14). Some fringing of the field occurs at the edges, where the field lines become distorted and the magnitude rapidly diminishes, but that can be made arbitrarily small by increasing the area of the plates and reducing their separation. Thus, except for positions near the edges, the field is constant everywhere between the plates and by symmetry must be perpendicular to their surfaces (Fig. 17.15).

We can use Gauss's law to find the magnitude of the electric field between the plates. Imagine a Gaussian surface, a cylinder with one end inside one of the metal plates and the other end in the space between the plates (Fig. 17.16). Under the conditions we have described, the surface of the plate is an equipotential and **E** must be perpendicular to it, as shown in Fig. 17.16. According to Gauss's law, the flux emerging from the Gaussian surface is proportional to the charge q_1 enclosed. For this alignment, there is no flux through the curved side of the cylinder. Neither is there any flux through the end of the cylinder that is inside the metal plate because the electric field is zero within a conductor. Thus, all the flux must emerge through the end of area A_1. The statement of Gauss's law becomes

$$\phi_{\text{net}} = EA_1 = \frac{q_1}{\epsilon_0},$$

where A_1 is the area of the end of the cylinder, q_1 is the charge within, and ϵ_0 is the proportionality constant. For a flat plate we expect the charge to be uniformly distributed over the surface. In that case the ratio of charge to area is constant and can be taken as equal to the total charge q on the plate divided by its total

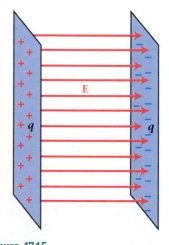

Figure 17.15
The electric field between two charged parallel plates is uniform and is directed from the positive plate to the negative plate.

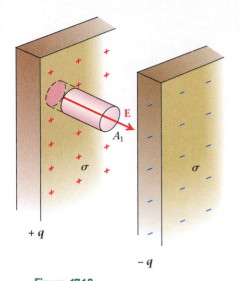

Figure 17.16
A cylindrical Gaussian surface oriented with its axis perpendicular to two parallel charged plates. One end of the Gaussian surface is inside the metal plate.

surface area A. The ratio of q/A is called the surface charge density, usually denoted by σ. Thus, the magnitude of the field is

$$E = \frac{q_1}{\epsilon_0 A_1} = \frac{q}{\epsilon_0 A} = \frac{\sigma}{\epsilon_0}. \tag{17.11}$$

The field between the plates is constant, independent of the distance from either plate.

A test charge q_0 placed in this constant field experiences a force $\mathbf{F} = q_0\mathbf{E}$. If the charge moves through the distance d from the positive to the negative plate, the work done by the field is $W = Fd = q_0Ed$. This amount of work is equal to the loss of electric potential energy by the test charge. The electric potential difference between capacitor plates is then given by

$$V = \frac{W}{q_0} = \frac{q_0Ed}{q_0} = Ed. \tag{17.12}$$

We can now formally calculate the capacitance of the parallel-plate capacitor. We start with the definition of capacitance,

$$C = \frac{q}{V}.$$

We replace the potential V with Ed, according to Eq. (17.12). Thus

$$C = \frac{q}{Ed}.$$

We then substitute for E using Eq. (17.11) to get

$$C = \frac{q}{qd/\epsilon_0 A},$$

or

$$C = \frac{\epsilon_0 A}{d}.$$

This equation is the same expression for the capacitance of a parallel-plate capacitor that was given earlier as Eq. (17.10).

Example 17.11

A parallel-plate capacitor whose plates are 2.5 cm apart is charged to a potential difference of 100 V. What would be the force on a test charge of 1.0 μC placed between the plates?

Strategy The force on a charge q is

$$F = qE.$$

We need to find E in terms of the potential difference V and the separation between the plates d. Then we can find the force F.

Solution Rewriting Eq. (17.12), we find that

$$E = \frac{V}{d}.$$

Substituting this expression for E in the equation for the force, we get

$$F = \frac{qV}{d} = \frac{(1.0 \times 10^{-6} \text{ C})(100 \text{ V})}{2.5 \times 10^{-2} \text{ m}}$$
$$F = 4.0 \times 10^{-3} \text{ N}.$$

Discussion The 1.0-μC charge could be placed on a small Styrofoam ball having a mass of 0.022 g, as might be done in an electrostatics experiment. If no other forces were present, the electric force would give the ball an acceleration of 180 m/s^2.

17.8 Dielectrics

The capacitance given by Eq. (17.10) for a parallel-plate capacitor is appropriate only when there is a vacuum between the capacitor plates. When a nonconducting (electrically insulating) material, known as a **dielectric,** is inserted between the plates, the capacitance increases, even though the area and separation remain constant. The ratio of the new capacitance to the capacitance in a vacuum is called the **dielectric constant** κ (Greek letter kappa):

$$\kappa \equiv \frac{C(\text{dielectric layer})}{C(\text{vacuum layer})}. \tag{17.13}$$

Table 17.2 lists the dielectric constants for several materials commonly used in capacitors.

Table 17.2	*Representative Dielectric Constants and Breakdown Strengths*	
Material	**Dielectric Constant** κ	**Dielectric Strength** (10^6 V/m)
Vacuum	1.0000	∞
Air (dry at 1 atm)	1.0006	3
Teflon	2.1	60
Polyethylene	2.6	25
Mylar	3.1	240
Paper	3.5	14
Mica	3–6	160
Glass	5–10	13
Aluminum oxide (amorphous)	8–9	700–900
Polyvinyldienedifluoride (PVDF)	11	200
Tantalum pentoxide	25	526

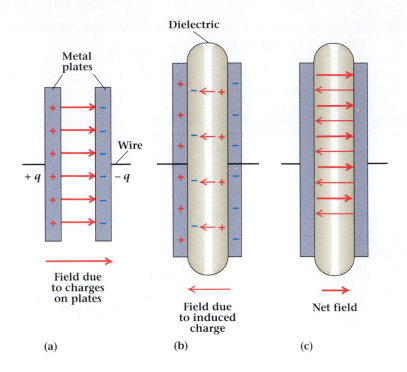

Figure 17.17
(a) The field of a parallel-plate capacitor without any material between the plates. (b) An induced charge on the surfaces of an insulating material placed between the plates generates a field within the material opposite in direction to the original field. (c) The net field between the plates is reduced, leading to an increased capacitance.

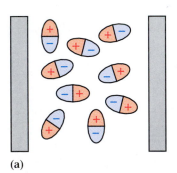

(a)

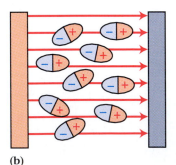

(b)

Figure 17.18
(a) Polar molecules are randomly oriented in the absence of an electric field. (b) In the presence of an electric field, the dipoles become partially aligned along the direction of the field.

For a parallel-plate capacitor, adding an insulating layer of dielectric constant κ that fills the space between the plates modifies the capacitance so that it becomes

$$C = \frac{\kappa \epsilon_0 A}{d} \qquad \text{(parallel-plate capacitor with dielectric).} \qquad (17.14)$$

Thus, the capacitance depends on the area of the plates, the dielectric constant of the insulating material, and the thickness of the dielectric layer.

We can gain some insight into the behavior of dielectric materials by considering a piece of dielectric placed between the plates of a capacitor (Fig. 17.17). When the plates are charged, an electrical force acts on the dielectric to pull any negative charges toward the positively charged plate and push any positive charges toward the negatively charged plate. Because the dielectric material is nonconducting, the charges within the dielectric are not mobile. Still, a slight reorientation takes place, resulting in an induced charge that appears on the surface of the dielectric. The field created by the induced charge is directed opposite to the original field (Fig. 17.17b). Thus, the net field between the plates is reduced (Fig. 17.17c).

In some cases, the dielectric consists of polar molecules whose dipole moments become oriented by the external electric field (Fig. 17.18). Within the material these dipolar charges tend to cancel each other, but at the surfaces they induce a charge. The induced charge is bound and cannot be removed from the dielectric. Because the polarity of the induced charge is opposite to the charge on the plates, its presence serves to reduce the electric field in the region between the plates. When the field is reduced, the potential difference between the plates is also reduced according to the relation $V = Ed$. Thus, for the same charge residing on the plates, the presence of the dielectric reduces the potential difference and thereby increases the capacitance.

Suppose we charge a parallel-plate capacitor when there is a vacuum in the gap between the plates. Then we isolate the capacitor so the charge does not leak off and fill the gap with an insulator of dielectric constant κ. Let V_0 be the potential difference across the capacitor when there is a vacuum in the gap and V be the potential when the dielectric is present. Because the charge is the same in both situations, Eq. (17.13) becomes

$$\kappa = \frac{C}{C_0} = \frac{q/V}{q/V_0} = \frac{V_0}{V}. \tag{17.15}$$

Thus, for the same charge, the voltage on the capacitor with a vacuum in the gap is greater than the voltage on the capacitor filled with the dielectric. The ratio of these two voltages is given by the dielectric constant.

An electronic stud finder (Fig. 17.19) uses a variation in dielectric constant to locate the studs hidden behind a wall. Two conducting plates on the back of the stud finder form a capacitor as though the plates of a parallel-plate capacitor were opened up like the pages of a book. The field between the plates then extends several centimeters into the space behind the stud finder. When the stud finder is placed against a wall, its field is modified by the presence of the dielectric material in the wall, resulting in a reference capacitance. As the stud finder is moved near a stud, the capacitance changes because of the stud's dielectric material. It is this change in capacitance that is measured and used to find the location of the stud.

Another important characteristic of a dielectric is its **dielectric strength,** which is the maximum applied potential per unit thickness that the material can withstand before breaking down, or losing its ability to insulate. The dielectric strength is the maximum electric field that the dielectric can support. If the dielectric strength is exceeded, a spark occurs, along with a catastrophic failure of the material. The electric sparks you encounter from shuffling across the rug on a dry day result from a breakdown of the insulating layer of air. Lightning bolts are also sparks that occur when the dielectric strength of the air is exceeded. Table 17.2 includes the dielectric strengths of the materials listed.

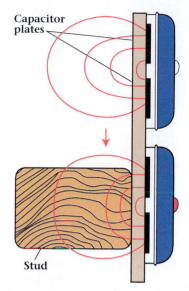

Capacitor plates

Stud

Figure 17.19

An electronic stud finder uses a change in capacitance to locate the studs hidden behind a wall. When the stud finder moves over the stud, the capacitance increases because of the dielectric constant of the wood.

Example 17.12

A capacitor is made with a dielectric layer of Mylar film that is 12 μm thick. The effective area of the film and conducting plates is 0.1 m^2. What is the capacitance of the capacitor and what voltage can it withstand?

Solution We can compute the capacitance from Eq. (17.14),

$$C = \frac{\kappa \epsilon_0 A}{d}.$$

From Table 17.2, we find the dielectric constant κ for Mylar to be 3.1. Upon inserting the values, we get

$$C = \frac{3.1(8.85 \times 10^{-12} \text{ F} \cdot \text{m}^{-1})(0.1 \text{ m}^2)}{12 \times 10^{-6} \text{ m}}$$

$$C = 0.23 \times 10^{-6} \text{ F} = 0.23 \ \mu\text{F}.$$

The dielectric strength of Mylar is also given in Table 17.2 as 240×10^6 V/m. If we multiply the dielectric strength by the film thickness, we get the maximum voltage that the capacitor can withstand:

$$V_{max} = (240 \times 10^6 \text{ V/m})(12 \times 10^{-6} \text{ m}) = 2900 \text{ V}.$$

17.9   Energy Storage in a Capacitor

We said earlier that a major application of capacitors is their use as energy storage devices. When charge is added to the plates, work is required to overcome the potential between the plates. As more charge is accumulated, the potential increases so the additional work needed for each increment of charge also increases. The work done in charging the capacitor goes into electric potential energy.

Let us now examine the energy-storing capability of a capacitor. We start by considering an initially uncharged capacitor. When a charge q' is transferred from one plate to the other, the potential difference V between the plates due to this transfer is q'/C. Then, if a small additional charge Δq is transferred, the amount of work required is

$$\Delta W = V \Delta q.$$

As more charge is transferred, the potential difference increases with the total charge q according to

$$V = \frac{q}{C}.$$

The total work done in charging the capacitor from zero to a voltage V is the sum of the work needed to transfer each amount of charge Δq:

$$W = \Sigma \, \Delta W = \Sigma \, V \Delta q.$$

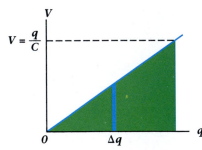

Figure 17.20

Graph of voltage versus charge for a capacitor.

We can compute this sum from the area under the voltage-charge curve (Fig. 17.20). The area of this triangular region is $Vq/2$. Thus, the energy required to charge an initially uncharged capacitor to a potential difference V and charge q is

$$W = \frac{1}{2}Vq.$$

This is also the energy stored on the capacitor when it becomes charged. If we use the relation $V = q/C$, we can express this energy as

$$W = \frac{1}{2}\frac{q^2}{C},$$

or

$$\boxed{W = \frac{1}{2}CV^2.} \tag{17.16}$$

The energy stored on the capacitor may be considered as an electrical potential energy, which can be released when the capacitor is discharged by connecting a conducting path between its plates. We can think of capacitors as devices for storing energy or as devices for storing electric charge.

We have seen that, when fringing can be ignored, the electric field within a parallel-plate capacitor is uniform; that is, it is the same at all points between the plates. Thus, the *energy density u,* the stored energy per unit volume, should also be uniform. We can determine the energy density from the ratio of the energy stored to the volume occupied by the field,

$$u = \frac{\text{energy stored}}{\text{volume}}$$

$$u = \frac{W}{Ad} = \frac{\frac{1}{2}CV^2}{Ad},$$

where Ad is the volume between the plates. If we substitute the relationship $C = \kappa\epsilon_0 A/d$, the energy density becomes

$$u = \frac{1}{2}\kappa\epsilon_0\left(\frac{V}{d}\right)^2.$$

However, V/d is the electric field within the capacitor, so

$$u = \frac{1}{2}\kappa\epsilon_0 E^2. \tag{17.17}$$

Although we have derived Eq. (17.17) for the special case of the parallel-plate capacitor, it is true in general. That is, if an electric field $\mathbf{E}$ is present at any point in space, then we can think of that point as having a stored energy density given by Eq. (17.17).

Master the Concept

Energy Storage, Dielectric Breakdown, and Capacitors

Question: Commercial capacitors with a large energy storage ability (large C and large V) are always physically larger than capacitors with smaller energy storage ability. For example, a 1.0 F, 300 V capacitor is much larger than a 300 μF, 25 V capacitor. Why is this so?

Answer: We have seen that the energy stored on a capacitor is the product of its capacity C with the square of the voltage V. We have also seen that the capacity is proportional to the ratio of the area A of the plates to thickness d of the dielectric material separating them. The maximum voltage that can be maintained on the capacitor is limited by the dielectric strength of the insulating material through $V_{max} = E_{max}d$, where E_{max} represents the dielectric strength. Consequently, the energy stored is proportional to $\left(\frac{A}{d}\right)d^2 = Ad$, which is the volume of the insulating material. Thus, a capacitor that can store more energy than another must necessarily be physically larger.

Figure 17.21
Example 17.13: The light from a photoflash is generated from the energy released in the sudden discharge of a capacitor.

Example 17.13

A 370-μF capacitor in a photoflash unit (Fig. 17.21) is charged to a potential difference of 330 V. (a) How much charge is stored on the capacitor? (b) How much energy is stored?

Solution The charge stored on the capacitor is computed from the definition of capacitance as

$$q = CV = (370 \times 10^{-6} \text{ F})(330 \text{ V}) = 0.122 \text{ C}.$$

The energy stored may be calculated from

$$W = \tfrac{1}{2}CV^2 = \tfrac{1}{2}(370 \times 10^{-6} \text{ F})(330 \text{ V})^2$$
$$W = 20.1 \text{ J}.$$

Example 17.14

A capacitor made of polyethylene film 25 μm thick has an effective surface area of 0.10 m². How much energy can be stored on the capacitor? The dielectric constant is found in Table 17.2.

Strategy This is a three-step problem. To find the maximum energy, we need to know the maximum voltage we can apply and the capacitance of the capacitor. The capacitance can be found directly from Eq. (17.14), since we know all of the necessary quantities. We can then calculate the maximum voltage we can put on the capacitor from the relationship

$$V_{\max} = \text{dielectric strength} \times \text{thickness}.$$

After obtaining these values, we can use Eq. (17.16) to find W.

Solution First we compute the capacitance, using $\kappa = 2.6$.

$$C = \frac{\kappa \epsilon_0 A}{d} = \frac{2.6(8.85 \times 10^{-12} \text{ F} \cdot \text{m}^{-1})(0.10 \text{ m}^2)}{25 \times 10^{-6} \text{ m}}$$
$$C = 9.2 \times 10^{-8} \text{ F}.$$

The maximum voltage is obtained from the product of the film's dielectric strength and its thickness:

$$V_{\max} = (25 \times 10^6 \text{ V} \cdot \text{m}^{-1})(25 \times 10^{-6} \text{ m}) = 625 \text{ V}.$$

The energy stored is then

$$W = \tfrac{1}{2}CV^2 = \tfrac{1}{2}(9.2 \times 10^{-8} \text{ F})(625 \text{ V})^2$$
$$W = 0.018 \text{ J} = 18 \text{ mJ}.$$

Summary

Useful Concepts

■ Electric potential is the potential energy per unit positive charge:

$$V = \frac{PE_{elec}}{q_0}.$$

■ The electric potential a distance r from a single point charge Q is

$$V = \frac{Q}{4\pi\epsilon_0 r}.$$

■ The electric potential difference between points two A and B is

$$V_{AB} = V_B - V_A = \frac{W_{AB}}{q},$$

where W_{AB} is the work done by an external force in moving the charge from A to B.

■ A surface for which all points are at the same potential is called an equipotential surface. When all the charges have come to rest, the surface of a conductor is always an equipotential surface.

■ A capacitor is a device for storing electric charge. The capacitance is defined to be

$$C \equiv \frac{q}{V}.$$

■ The dielectric constant of a material is

$$\kappa = \frac{C(\text{dielectric layer})}{C(\text{vacuum layer})}.$$

■ The capacitance of a parallel-plate capacitor filled with material of dielectric constant κ is

$$C = \frac{\kappa\epsilon_0 A}{d}.$$

■ The energy stored in a capacitor is

$$W = \tfrac{1}{2}CV^2.$$

Important Terms

You should be able to write the definition or meaning of each of the following:

electric potential	capacitor
volt	capacitance
electric potential difference	farads
Van de Graaff generator	dielectric
electron volt	dielectric constant
equipotential surface	dielectric strength

Conceptual Questions

17.1 Sketch the equipotential lines about a charged sphere.

17.2 Sketch the equipotential lines about a charged cube located a great distance from any other conductors. How do the equipotentials look close to the cube? Far away from the cube?

17.3 Why can a Leyden jar not be charged very well if the outer coating is insulated during the charging?

17.4 How can an electric field have a direction in space although the electric potential has no direction?

17.5 Would you be safe on top of the dome of a large Van de Graaff generator when it was operating? Assume that you would be in contact with the dome only. Discuss the conditions under which this behavior would be safe or unsafe.

17.6 Would you be safe inside the dome of a large Van de Graaff generator when it was operating?

17.7 A hollow metal surface supported on an insulating stand has a large radius on one end and a small radius on the other (Fig. 17.22). What can you say about the electric potential, the electric field, and the electric charge density on the surface when it contains a net charge Q? Make a sketch of the equipotential lines about the object.

Figure 17.22
Question 17.7.

17.8 Work is required to pull apart two charged capacitor plates. What happens to the energy expended in pulling them apart?

17.9 Explain how a charged comb can attract uncharged bits of paper.

17.10 How does the breakdown strength of the dielectric affect the physical size of practical capacitors?

17.11 What is the purpose of the short strap frequently seen dragging along behind a gasoline tanker truck?

17.12 Explain the following advice that is often given for protection from lightning. During an electrical storm, if your hair begins to stand on end, lightning is likely to strike nearby. For safety, drop to your knees and bend forward, placing your hands on your knees, being careful not to place your hands on the ground.

17.13 A parallel plate capacitor has a glass plate between the metal plates. The capacitor is charged to a potential V_0 and the connecting wires are removed so that the capacitor is isolated. How does the voltage on the plates change if the glass plate is now removed?

17.14 Sketch the lines of force corresponding to the equipotential lines about two like charges of unequal magnitude (Fig. 17.23).

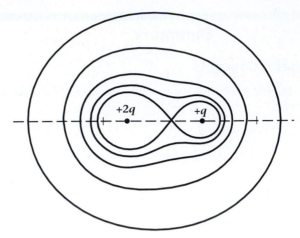

Figure 17.23
Question 17.14.

Problems

Sections 17.1 and 17.2 Electric Potential and the Van de Graaff Electrostatic Generator

Hints for Solving Problems

Remember that electric potential is a scalar and has no direction, unlike the electric field, which is a vector. A positive charge moved opposite to the direction of a field will have an increase in 'its electric potential energy.

17.1 What is the potential energy of two alpha particles separated by a distance of 1.50×10^{-10} m? The charge of an alpha particle is 3.20×10^{-19} C.

17.2 What is the electric potential at a point 0.45 m away from a point charge of 2.5 mC?

17.3 The potential 0.010 m from a point charge is 0.10 V. What is the magnitude of the point charge?

17.4 If 3.75×10^{-4} J of work are required to move 13.7 μC of charge from one point to another, what is the electric potential difference between the two points?

17.5 To move a charged particle through an electric potential difference of 10^{-3} V requires 2×10^{-6} J of work. What is the magnitude of the charge?

17.6 The electric potential difference between the parallel deflection plates in an oscilloscope is 300 V. If the potential drops uniformly when going from one plate to the other and if the distance between the plates is 0.75 cm, what is the magnitude of the electric field between them and in which direction does it point?

17.7 (a) What is the electric potential a distance 0.25 m away from a point charge of 1.0×10^{-9} C? (b) How much work is required to bring an identical charge from very far away up to this distance of 0.25 m from the first charge?

17.8● A Van de Graaff generator has a dome 50 cm in diameter. Calculate the maximum electric potential it can reach before the surrounding air breaks down because of the electric field near the surface of the dome. The maximum field that the air can withstand is 3×10^6 V/m.

17.9● A proton of mass $m_p = 1.67 \times 10^{-27}$ kg and charge $+e = 1.60 \times 10^{-19}$ C is accelerated from rest through an electric potential of 300 kV. What is its final velocity?

17.10● Two point charges q_1 and q_2 are arranged as shown in Fig. 17.24. What is the electric potential difference $V_B - V_A$ between the two points labeled A and B?

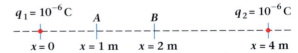

Figure 17.24
Problem 17.10.

17.11● A charge $q_1 = 0.244$ μC is initially positioned 3.00 cm directly above a fixed charge $q_2 = 21.6$ μC. The charge q_1 is moved 6.25 cm straight upward and then moved 7.50 cm horizontally away from the charge q_2. What is the change in the electrical potential energy of the movable charge q_1?

17.12● The dielectric strength of air, $E = 3.0 \times 10^6$ V/m, is the maximum field that air can withstand before it breaks down and becomes conducting. (a) How much charge can be placed on a spherical conductor with a 10-cm radius before the field at its surface exceeds the breakdown strength of the air? (b) What would be the electric potential at the surface of this conductor?

17.13● A demonstration Van de Graaff generator has a charge of 3.0 μC stored uniformly on its 30-cm-diameter dome.

(a) What is the value of the electric field at the surface of the dome? (b) What is the electric field 0.50 m from the surface of the dome? (c) What is the electric potential of the dome?

17.14● A demonstration Van de Graaff generator has a 30-cm-diameter dome. The generator is capable of producing a potential of 150,000 V. (a) Calculate the charge required to raise the dome to this potential. (b) Find the value of the electric field near the dome.

17.15● Four point charges $q = +3.6$ μC are arranged in a square of edge length 2.0 cm. (a) What is the electric potential at the center of the square? (b) What is the electric potential at the midpoint of one edge?

17.16●● A positive point charge of 1.0 nC is located at position $x = 0.00$. A second point charge $q = -1.0$ nC is located at $x = 0.10$ m. (a) What is the electric potential on the x axis as a function of x? (b) What is the value of the electric potential at $x = -0.050$ m?

17.17●● (a) What is the electric potential energy of three equal positive charges of 1.25 μC when they are arranged to form a triangle 2.75×10^{-3} m on a side? (b) What is the potential energy if the charges are all negative?

Section 17.3 The Electron Volt

Hints for Solving Problems

For problems 17.18–17.21, use the following values for mass and charge: an electron has mass $m_e = 9.11 \times 10^{-31}$ kg and charge $-e$, a proton has mass $m_p = 1.67 \times 10^{-27}$ kg and charge $+e$, an alpha particle has mass $m_\alpha = 6.65 \times 10^{-27}$ kg and charge $+2e$, where $e = 1.602 \times 10^{-19}$ C.

17.18 An electron is released from rest in a vacuum between two flat, parallel metal plates that are 10.0 cm apart and are maintained at a constant electric potential difference of 750 V. (a) If the electron is released at the negative plate, what is its speed just before it strikes the positive plate? (b) What is the kinetic energy of the electron in units of eV?

17.19 (a) What is the speed of a 300-eV proton? (b) What is the speed of a 300-eV electron?

17.20 The speed of an alpha particle is determined to be 1.38×10^6 m/s. If all of its kinetic energy is acquired by passing through an electric potential, what is the magnitude of that potential?

17.21● Protons in a research Van de Graaff accelerator are accelerated from rest through a potential difference of 250,000 V. (a) After they pass through the potential difference, what is the kinetic energy of the protons in units of eV? (b) What is their energy in units of joules? (c) What is the speed of the protons?

Section 17.4 Equipotential Surfaces

Hints for Solving Problems

Remember that electric field lines are everywhere perpendicular to the equipotential surfaces.

17.22 From your knowledge of the behavior of the electric field and potential near a point charge and of the fact that the surface of a conductor is an equipotential, sketch the lines of electric field and the equipotentials for a point charge q located a distance x from a large conducting plane.

17.23 An isolated metal sphere of radius R carries a positive charge Q. An identical isolated metal sphere, initially uncharged, is brought into contact with the first sphere, and the two are then separated. How much charge remains on the first sphere?

17.24● A thick, conducting, spherical shell of inner radius a and outer radius b is uncharged. The shell surrounds a point charge $+Q_0$ at its center (Fig. 17.25). (a) What is the surface charge residing on the inner surface of the shell? (b) What charge resides on the outer surface? (c) What is the electric field in the region $0 < r < a$? (d) What is the electric field in the region $a < r < b$? (e) What is the electric field in the region $r > b$?

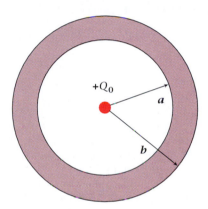

Figure 17.25
Problem 17.24.

17.25* Two isolated metal spheres are charged to the same potential V. The first sphere has a radius that is twice as large as the radius of the second sphere. (a) Which sphere carries the greater electric charge? (b) Which sphere has the greater electric field at its surface?

17.26● An isolated metal sphere of radius R carries a positive charge Q. Another isolated metal sphere, initially uncharged and of radius $R/2$, is brought into contact with the first sphere, and the two are then separated. How much charge remains on the first sphere?

17.27● Two hollow metal spheres are mounted on insulating stands. Sphere 1 has a radius r_1, and sphere 2 has a radius $r_2 = 2r_1$. A charge Q is placed on sphere 1 to give it an electric potential of $+180$ V. The other sphere is uncharged at a potential of 0 V. If the two spheres are placed in contact and then separated, what will be their new potential?

Section 17.5 Capacitors

17.28 When a charge of 180 μC is applied to the plates of a capacitor, the potential difference between the plates is 15 V. What is the capacitance?

17.29 A 25-μF capacitor is charged to 15 V. How much charge is stored on the capacitor?

17.30 When the plates of a radio capacitor are charged with 18×10^{-5} C, the potential difference between them is 9.0 V. What is the capacitance?

17.31 A charge of 3.0×10^{-3} C is stored on a 25-μF capacitor. What is the electric potential difference across the capacitor?

17.32 How much charge is stored on each plate of a 5000-μF capacitor whose plates are held at a potential difference of 350 V?

17.33● The charge on a capacitor increases by 20.0 μC when the potential difference between the plates is increased from 15.0 V to 30.0 V. What is the capacitance of the capacitor?

Section 17.6 The Parallel-Plate Capacitor

> **Hints for Solving Problems**
>
> For Problems 17.34–17.38, assume that the plates are separated by air.

17.34 The plates of a capacitor are separated by 0.10 mm. If the area of each plate is 100 cm^2, what is the value of the capacitance?

17.35 A parallel-plate capacitor is made of metal plates separated by 0.12 mm. What is its capacitance if the area of the plates is 113 cm^2?

17.36 Calculate the capacitance of two flat metal pie pans 23 cm in diameter separated by 2.0 mm.

17.37● A parallel-plate capacitor is made of two 50.0-cm-diameter circular plates spaced 1.50 mm apart. How much charge is stored on the capacitor when it is charged to 750 V?

17.38● A parallel-plate capacitor is designed so that the plates can be pulled apart. The capacitor is initially charged to a potential difference of 50 V when the plates are 1.0 mm apart. The plates are insulated so that the charge cannot leak off. What is the potential difference between the plates when they are pulled to a new separation 2.0 mm apart?

Section 17.7 Electric Field of a Parallel-Plate Capacitor

17.39 How much work is required to move an electron 0.015 m through a uniform field of 7.65 V/m?

17.40 A parallel-plate capacitor has plates with an area of 250 cm^2 separated by 1.0 mm. What is the electric field between the plates when they are charged with 0.50 μC?

17.41 A parallel-plate capacitor whose plates are 0.650 mm apart is charged to a potential difference of 125 V. What would be the force on a test charge of 1.00 μC placed between the plates?

17.42 A homemade capacitor is made from two metal pie pans 18.5 cm in diameter separated by 2.00 mm. The capacitor is charged to a potential of 15.7 V. How much charge is stored on each plate of the capacitor?

17.43● A pair of charged metal plates are separated by 2.0 mm. An electric field of 1.0×10^5 V/m exists between the plates. An electron with mass $m_e = 9.1 \times 10^{-31}$ kg and charge $e = -1.6 \times 10^{-19}$ C is released from the negative plate and is accelerated by the field toward the positive plate. (a) What is the change in kinetic energy of the electron? (b) What is the energy change in units of electron volts? (c) If the electron starts from rest, what is its velocity just before it strikes the positive plate?

17.44●● Use Gauss's law to show that the field inside a parallel plate capacitor is constant everywhere between the plates.

Section 17.8 Dielectrics

17.45 When reaching for the door handle after sliding across a car seat on a dry winter day, you get a spark when your fingertip is 5.0 mm away from the handle. What was the potential difference between you and the door handle just before the spark?

17.46 During a demonstration of electricity, a professor charges a Leyden jar from a Van de Graaff generator. Then he takes a wire connected to the outside conductor and brings it near the metal rod that sticks through the insulating top (Fig. B17.1). What is the magnitude of the potential difference if a spark occurs when the wire is 12 mm from the rod?

17.47 The parallel plates of a parallel-plate capacitor are separated by 1.0 mm. The gap between the plates is filled with polyethylene, and the plates are charged to a potential of 80 V and insulated to keep the charge from leaking off. Then the polyethylene layer is pulled out. What is the potential difference between the plates of the capacitor?

17.48 A manufacturer plans to make a capacitor with tantalum pentoxide as the dielectric. What must be the minimum thickness of the dielectric layer if the capacitor must withstand 50 V?

17.49 What must be the effective surface area of a 10-μF aluminum oxide capacitor if the average film thickness is 2.0×10^{-8} m? Use $\kappa = 8$.

17.50 Calculate the maximum potential difference that can be maintained across a 25-μm layer of polyethylene.

17.51 A parallel-plate capacitor is made of two flat metal plates pressed against a thin slab of dielectric material. The capacitor is connected to a power supply, and a potential difference of 90 V is applied to the plates. With the power supply disconnected, the dielectric material is removed and the potential difference between the plates is measured to be 500 V. What is the dielectric constant of the material that was initially used to fill the gap between the plates?

17.52● A capacitor is made from a Teflon film that is 25 μm thick. (a) What is the effective area of the plates of this capacitor if it has a capacitance of 0.50 μF? (b) What is the maximum working voltage of the capacitor—that is, how much voltage can it stand before it breaks down?

17.53● Show that for a given plate area, the product of capacitance and the maximum voltage of a capacitor is fixed by the type of material used for the dielectric.

17.54• A capacitor is made from a PVDF film that is 25 μm thick. (a) If the effective area of the film and the conducting plates is 0.60 m², what is the capacitance of the capacitor? (b) How much voltage can the capacitor withstand without breaking down?

17.55•• A capacitor is made from two parallel plates, each with an area of 146 cm². The plates are separated by a distance of 0.58 mm. Half of the area within the plates is filled with paper and the other half is filled with air (Fig. 17.26). Calculate the capacitance of this capacitor.

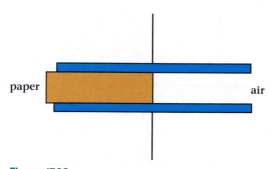

paper   air

Figure 17.26
Problem 17.55.

Section 17.9 Energy Storage in a Capacitor

Hints for Solving Problems

The work done in charging a capacitor is stored as electric potential energy that can be released when the capacitor is discharged.

17.56 How much energy is stored on a 1000-μF capacitor that is charged to a potential difference of 500 V?

17.57 A capacitor of 0.010 μF is charged to 5.0 V. (a) What is the charge and (b) what is the energy stored on the capacitor?

17.58 An electrolytic capacitor is rated at 25 μF at 450 V. (a) How much charge can be stored on that capacitor? (b) How much energy can it store?

17.59• A parallel-plate capacitor is made of square metal plates 30 cm × 30 cm. The plates are pressed against a glass plate 3.0 mm thick of dielectric constant $\kappa = 6.3$ that exactly fills the space between them. The capacitor is charged to a potential difference of 500 V and then insulated from the surroundings. (a) How much energy is stored on the capacitor? (b) If the piece of glass is removed, how much energy will be stored on the capacitor?

17.60•• The stored electrical energy of a 4000-μF capacitor charged to 500 V is converted to thermal energy by discharging the capacitor through a heating element submerged in 200 g of water in an insulating cup. What is the increase in temperature of the water if the heat capacity of the heater can

be ignored? Assume that all of the energy stored on the capacitor goes into heating the water.

Additional Problems

17.61 A pair of parallel plates are separated by 1.5 mm of air. An electric field of 3.0×10^4 V/m exists between the plates. What is the potential difference between the plates?

17.62 A pair of parallel plates is separated by 1.0 mm of air. If the plates are charged to a potential difference of 75 V, what is the electric field between the plates?

17.63 An alpha particle is accelerated through a potential difference of 2500 V. What is the change in potential energy of the alpha? Give your answer in both eV and J.

17.64 Two identical point charges of $q = +1.00 \times 10^{-8}$ C are separated by a distance of 1.00 m. How much work is required to move them closer together so that they are only 0.50 m apart?

17.65• Three point charges are arranged in an equilateral triangle of edge length 1.0 cm. What is the electric potential at the midpoint of one edge if each charge is 1.0 μC?

17.66• An isolated metal sphere of radius R_1 carries an electrical charge Q. A second sphere of radius R_2 and initially uncharged is brought into contact with the first sphere. Charge is exchanged between the two and they are separated. How much charge remains on the first sphere?

17.67• A parallel-plate capacitor has a surface charge density $\sigma = 8.5 \times 10^{-5}$ C/m² and an electric field $E = 2.5 \times 10^6$ V/m in the material between the plates. (a) Express the dielectric constant κ in terms of E and σ. (b) What is the value of the dielectric constant for the material between the plates?

17.68• A capacitor is made from two parallel conducting plates, each with an area of 860 cm². The plates are separated by a dielectric film 0.250 mm thick. Measurements show that the potential difference between the plates is 185 V when the charge on the plates is 1.75 μC. (a) What is the capacitance of the capacitor? (b) What is the most likely material used for the dielectric?

17.69• Derive an expression for the work done in pulling out the polyethylene layer in Problem 17.47. Give your answer in terms of C_0 and V_0, the initial capacitance and voltage.

17.70•• A large insulating sheet has a uniform static charge of surface-charge density $\sigma = 4.57 \times 10^{-8}$ C/m². (a) Calculate the electric field near the sheet. (b) Calculate the separation of equipotential surfaces spaced 100 V apart.

17.71•• A +30-μC charge is placed 1.00 m from a charge of −30 μC. (a) How much work is required to bring a +15-μC charge from very far away to a point midway between them? (b) How much work would be required to move the test charge from the midway point to a point 25 cm from the positive charge along the line joining the charges?

17.72•• A 100-μF parallel-plate capacitor is charged to 500 V. It is then disconnected from the charging source and insulated. (a) How much energy is stored on the capacitor? (b) If the plates are pulled apart to a separation that is twice their initial separation, what energy is stored on the capacitor? (c) If the answers are not identical, account for their difference.

17.73•• A capacitor is made from two parallel metal plates, each with an area of 127 cm^2 (Fig. 17.27). The plates are separated by a distance of 0.46 mm. Half of the area between the plates is filled with Mylar and the other half is filled with polyethylene. What is the capacitance of this capacitor?

17.74•• A parallel-plate capacitor made of circular plates of radius 25 cm separated by 0.20 cm is charged to a potential difference of 1000 V by a battery. Then a sheet of polyethylene is pushed between the plates, completely filling the gap between them. How much additional charge flows from the battery to one of the plates when the polyethylene is inserted?

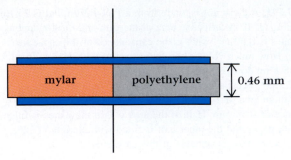

Figure 17.27
Problem 17.73.

Electric Current and Resistance

Today, electric circuits are a commonplace part of our daily lives. In this chapter we use the concepts of charge, field, and potential to introduce the practical subject of charges in motion, or current electricity. In our ordinary experience, electric currents move along closed conducting paths called circuits. To understand how circuits work, we will use the laws of conservation of charge and conservation of energy, fundamental laws that apply to all electric circuits. We introduce several new terms, such as current, resistance, and electromotive force, that are necessary for describing these circuits. In the rest of the chapter, we develop several other new concepts as we apply this knowledge to other circuit combinations. A familiarity with fundamental ideas of circuit operation will help you understand practical aspects of electricity, from home power distribution to basic safety from electrical shock.

In 1791, the Italian Luigi Galvani (1737–1798) found a way to produce electric current. Galvani noticed that the muscles of a dissected

557

Figure 18.1
Count Alessandro Volta (1745–1827).

Figure 18.2
A voltaic pile given to Michael Faraday by Alessandro Volta. It consists of a stack of disks, alternately silver and zinc, each pair sandwiched between moistened paper.

frog's leg twitched when a nearby electrostatic generator was in operation. Further experiments showed that the electrostatic generator was not the cause of the twitching, but that the twitching occurred when the frog's leg was attached to a brass hook and placed on an iron plate against which the hook was pressed. He concluded, incorrectly, that the nerves and muscles of the frog produced electricity, which he had detected. Galvani's work was soon overshadowed by that of Count Alessandro Volta (Fig. 18.1), who maintained that the electricity detected by the frog's muscles was due to contact between the dissimilar metals brass and iron. Volta was a careful investigator who was already famous for other electrical discoveries. He invented the first battery, which became known as the voltaic pile. This pile consisted of a repeating stack of disks of copper, zinc, and cardboard moistened with salt water (Fig. 18.2). A stack ("battery") of twenty such units gave a strong shock and could produce sparks.

It was Volta who first recognized the crucial difference between the electricity from a battery and that from an electrostatic device such as a Leyden jar: The electricity from a battery flowed continuously. This discovery opened up opportunities for recognizing effects not available from static electricity. These effects include electrochemistry, electric heating, electromagnetism, and all the applications that follow from them, including investigations into the properties of matter itself. ■

18.1 Electric Current and Electromotive Force

We know from common experience that a battery is a source of energy because it can turn a motor that does mechanical work. In doing so, the battery converts stored chemical energy into electrical energy, which, in turn, is converted to mechanical work. One simple source of electrical energy is a single dry cell, commonly used in flashlights and portable radios. Several cells connected together form a battery, which is essentially the same as Volta's original pile of cells. Although the word **battery** literally means an array of cells, this word often refers to single cells as well. For example, common flashlight batteries are really single cells. Because of their combined convenience, reliability, and portability, batteries are used in a wide variety of applications, ranging from portable video cameras to watches and hearing aids. Several common types of batteries are shown in Fig. 18.3.

When we use a conducting wire to provide a continuous path from one terminal of a battery to the other, a current of electric charge passes through the wire. We define this **electric current** to be the rate at which electric charge passes through a conductor. If we use the conventional symbol I for electric current, we have

$$I \equiv \frac{\Delta q}{\Delta t},$$

(18.1)

where Δq is the amount of charge that passes a given point during the time interval Δt (see Fig. 18.4a). For the present, we consider only the case in which the charge always flows in the same direction and at the same rate. This is called **direct current** or **dc.** In Chapter 21 we will discuss the case in which the rate and direction change. The dimensions of electric current are charge per unit of time, and the SI unit for current is the **ampere** (abbreviated A).* The unit for charge is the coulomb, so

$$1 \text{ ampere} = \frac{1 \text{ coulomb}}{\text{second}}.$$

The source of this current is the battery. In particular, the battery provides a potential difference, or voltage V, between its terminals. The corresponding electric field causes charges to move within the wire, thus generating the electric current. (The existence of a field within a conducting wire does not contradict our statement in Chapter 17 that a field cannot exist within a conductor. There we were discussing charges at rest; here we are talking about charges in motion.) Energy is released from the battery when an amount of charge q moves through this potential difference: $W = qV$. Electrical energy can be released from a battery (or other source of electrical energy) only when the current has a complete conducting path available from one side of the potential difference to the other. We call this conducting path a complete **electric circuit.**

Although current is a *scalar* quantity, it has a sign associated with it. By convention, we indicate the sign of the current with an arrow pointing in the direction in which a *positive* charge would move in the electric field provided by the battery (Fig. 18.4a,b). Thus, the current in the external circuit is said to be directed from the positive terminal to the negative terminal of the battery. In solids, such as wires, the mobile charges that are actually free to flow, and therefore make up the current, are the negatively charged electrons. Consequently, the motion of the electrons is opposite to the conventional direction assigned to the current. However, since a flow of positive charges in one direction is equivalent to a flow of negative charges in the opposite direction, we use the convention of positive current in almost all cases. Although the current consists of the motion of discrete charges, the number of them is so large that we may ignore this granularity and consider the current to be a smooth, continuous flow of charge.

As we will see in more detail later, electrons do not flow through a wire unimpeded. If they did, the field would eventually accelerate them to very great speeds. Instead, the moving electrons interact with the relatively stationary positive ions in the metal. The result of the applied field is to impose a net velocity on the random motions of the electrons. In this respect they behave much like the particles of a gas, having large random velocities, but a small net velocity, as they drift with a fairly constant average speed through the ions of the conductor. This average speed of translation through the conductor is called the *drift velocity* and ranges from a few hundredths of a millimeter per second to several centimeters per second for copper wires in ordinary situations (Fig. 18.5). The drift velocity is the rate at which electrons themselves move through a wire; it is not the rate at which electric signals move. The speed of electric signals is very high, close to the speed of light. Charge in a conductor is somewhat analogous to water

Figure 18.3
Common batteries are available in an assortment of sizes. The batteries shown here are not rechargeable.

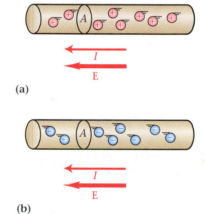

(a)

(b)

Figure 18.4
(a) Current is defined to be the rate of charge passing a given cross section in a conductor. The direction of electric current is the direction in which a positive charge moves. (b) The motion of (negatively charged) electrons is opposite to the direction associated with the current.

---

*The ampere is specified more completely in Section 19.6.

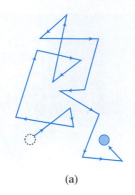

(a)

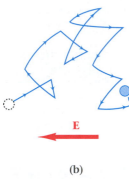

E

(b)

Figure 18.5
(a) Random motion of an electron through a metal when no electric field is present. After many collisions, the electron remains close to its original position. (b) Motion of the electron in a metal when an electric field is present. Notice the net drift of the electron in the direction opposite to the electric field.

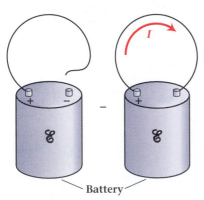

I

Battery

Figure 18.6
A source of emf $\mathscr{E}$ causes a current in a conductor when the circuit is completed.

in a pipe. When water begins flowing into one end of a full pipe, water begins to flow out the other end almost immediately. Similarly, when a difference of electric potential is applied across a wire, electrons throughout the whole wire move in response to the resulting field.

The potential difference that appears between the terminals of a battery when no current is present is called the **electromotive force** or **emf.** (The term *emf* has historical origins, but it is *not* a force, despite its name.) The symbol for emf is $\mathscr{E}$. The emf is measured in volts, like any other potential difference.

Any device that can maintain a potential difference and supply current to an external circuit is a source of emf. Examples include batteries, solar cells, and generators. If the emf of a battery is zero, there is no current when a wire is connected across its terminals. In this case, there is no potential difference to drive the charges. But if the emf is nonzero, a current is present when the terminals are connected with a conducting wire to form a complete circuit (Fig. 18.6). The greater the emf, the greater the current in the circuit.

An alkaline flashlight cell has an emf of 1.5 V; lithium cells sometimes used in watches and cameras have emfs of 3 V; and a newly charged nickel-cadmium rechargeable cell (Nicad cell) has an emf of about 1.2 V. The low emf of Nicad cells, in comparison to alkaline cells, explains why they cannot be used to replace alkaline cells for some applications. A typical multicell lead-acid automobile battery has an emf of approximately 12 V, corresponding to six 2-V cells. The principal characteristic of the lead-acid battery is its ability to deliver very high currents.

Example 18.1

When you press one of the buttons on a pocket calculator, the battery provides a current of 300 μA for 10 ms. (a) How much charge flows during that time? (b) How many electrons flow in that time?

Solution (a) From the definition of current we get

$$\Delta q = I\,\Delta t.$$

The charge that flows in 10 ms is then

$$\Delta q = (300 \times 10^{-6}\ \text{A})(10 \times 10^{-3}\ \text{s}) = 3.0 \times 10^{-6}\ \text{C}.$$

(b) The magnitude of the charge of an electron is 1.60×10^{-19} C, so that

$$\text{number of electrons} = \frac{\text{total charge in 0.010 s}}{\text{charge on one electron}}$$

$$\text{number of electrons} = \frac{3.0 \times 10^{-6}\ \text{C}}{1.60 \times 10^{-19}\ \text{C}} = 1.9 \times 10^{13}.$$

Discussion Even though the current of 300 μA is relatively small, nearly 20 trillion electrons move past a given point in the circuit in a time interval of only 10 ms. Because such large numbers are involved, we don't usually see effects of individual charges. Instead we think of the current as a smooth, continuous flow of charge.

18.2 Electric Resistance and Ohm's Law

As we mentioned earlier, if you maintain an electric potential difference, or voltage V, across any conductor, an electric current occurs (Fig. 18.7). In general, the magnitude of the current depends on the potential difference. For any particular material, the ratio of the applied voltage to the resulting current is defined to be the **resistance** R:

$$R \equiv \frac{V}{I}. \tag{18.2}$$

You can think of resistance as the ability of a material to resist the flow of charge when it is subject to a given potential difference. A material that is considered to be a useful insulator passes only a small current for a particular voltage V and thus has a large resistance. By contrast, a good conductor carries a large current at the same voltage V and has a small resistance. If V is measured in volts and I in amperes, then resistance has the unit of volt per ampere, which is called an **ohm** (abbreviated Ω). We emphasize that Eq. (18.2) is the *definition* of resistance and does not imply that the resistance measured under one set of conditions is equal to the resistance measured under different conditions. In particular, the definition of resistance does not imply a proportional relationship between V and I.

If you change the potential difference applied *across* a conductor, the resulting current *through it* also changes. Each combination of V and I determines a particular value of resistance, as given by Eq. (18.2). In general, the ratio of V to I changes as the applied voltage changes. For example, measurements show that for an incandescent lamp bulb the relationship between current and voltage is not linear (Fig. 18.8).

In some important special cases, the change in the current is proportional to the change in the voltage so the ratio of V to I is constant. For these cases, the resistance remains constant as the potential difference changes, and the I-V graph is a straight line. Using symbols, we say that V and I are proportional ($V \propto I$), and the constant of proportionality is the resistance R. This relationship may be expressed as

$$V = IR, \qquad \text{when } R \text{ is a constant.} \tag{18.3}$$

Equation (18.3) is known as **Ohm's law** in honor of Georg Simon Ohm (1787–1854), the German physicist who first demonstrated this relationship between current and potential difference in 1827. Materials whose resistance is constant over a wide range of voltages are said to obey Ohm's law. In these materials, current is proportional to the applied potential difference V and inversely proportional to the resistance R. Notice that Ohm's law is not a law of nature in the sense that conservation of momentum or the universal law of gravitation is a law of nature; rather, it is an experimental observation about the behavior of *some* materials under a *limited range of circumstances.*

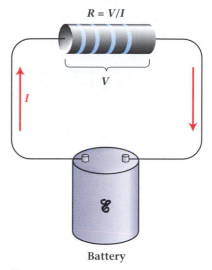

Figure 18.7
An electric potential across a conductor causes a current. The conductor has a resistance defined by $R = V/I$.

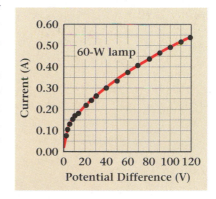

Figure 18.8
Current-voltage characteristic of a 60-W incandescent lamp.

Figure 18.9
(a) Current through a typical silicon *pn*-junction diode as a function of the applied voltage. The relationship between current and voltage is not linear. (b) Current through a piece of copper wire kept at constant temperature. The relationship between current and voltage is linear.

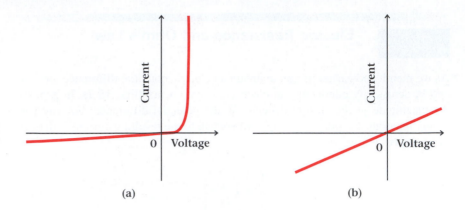

(a)   (b)

Note the distinction between the definition of resistance, Eq. (18.2), and Ohm's law, Eq. (18.3). We can measure the electric resistance of a conductor by making simultaneous measurements of current and voltage. Figure 18.9(a) shows the current through a typical silicon *pn*-junction diode as a function of the applied voltage. The current increases with increased voltage, but the relationship between them is not linear, so the resistance is not constant. Such a material is said to be nonohmic because the resistance is not constant. Figure 18.9(b) shows the current through a piece of copper wire kept at constant temperature. Here the relationship between current and voltage is linear, and the copper is said to obey Ohm's law over this range. Indeed, Ohm's law is often used in describing electric circuits. Nevertheless, you should remember that it may not always be applicable.

As an aid to understanding the relationship between current, potential difference, and resistance, an analogy is sometimes used. The rate at which water flows in a closed pipe is analogous to the electric current in a wire; a pump that delivers a specific pressure is analogous to a battery of a specific emf; and the electric resistance is analogous to the pipe's resistance to the flow of the water. This is an instructive analogy when you first encounter the idea of current, but it is only an analogy, and you should try to think in terms of electrical, rather than mechanical, quantities.

Example 18.2

(a) The current-voltage characteristic of a 60-W incandescent (tungsten filament) lamp bulb is shown in Fig. 18.8. What is its resistance when operated at 18 V? (b) What is its resistance when operated at 120 V? (c) The filament of Thomas Edison's first practical electric light conducted a current of about 0.30 A when a potential of approximately 18 V was applied. What was the approximate resistance of the filament under these conditions?

Strategy The current through the lamp can be found for any voltage from the graph of Fig. 18.8. Then the resistance can be found easily from the definition of resistance in Eq. (18.2).

Solution (a) From the graph we find that when the potential difference is 18 V, the current is approximately 0.20 A, for a resistance of

$$R = \frac{V}{I} = \frac{18 \text{ V}}{0.20 \text{ A}} = 90 \ \Omega.$$

(b) Similarly, at 120 V the current is 0.54 A, giving a resistance of

$$R = \frac{V}{I} = \frac{120 \text{ V}}{0.54 \text{ A}} = 220 \ \Omega.$$

(c) The resistance of Edison's lamp when operated at 18 V is

$$R = \frac{V}{I} = \frac{18 \text{ V}}{0.30 \text{ A}} = 60 \ \Omega.$$

Discussion When the modern light bulb is operated at 18 V, its resistance is 1.5 times greater than the resistance of Edison's lamp. At the higher potential (120 V), the lamp's resistance is more than double its low-voltage resistance. This difference is due to its increased temperature. From the graph, we see that for the lamp the relationship between current and voltage is definitely nonlinear.

■

Electric components manufactured especially for their resistance are called **resistors.** They are commonly available in a wide range of values from a few ohms to millions of ohms. These resistors are introduced into circuits to provide resistances that are large compared with the resistance of the wires and connectors joining together the other circuit components.

One of the most common resistor types is the carbon composition resistor (Fig. 18.10). Composition resistors are normally molded into a cylindrical shape from carbon granules compacted together with a binding resin. Resistors made this way are relatively inexpensive. They come in various sizes according to their power-handling ability, with power ratings of 1/8, 1/4, 1/2, 1, and 2 watts being common. When higher power ratings are required, wire-wound resistors are often used. These wire-wound resistors are commonly made from the alloys manganin and constantan, which have only small variations in resistance with changes in temperature.

Figure 18.10
Hot-molded carbon composition resistors. The black carbon grains of the resistor are encased in an insulating phenolic jacket. Colored stripes are used to indicate the value of the resistance.

▼

*18.3   Resistivity

By working with wires of different thicknesses and lengths, Ohm found that the amount of current transmitted by a wire for a given potential difference was directly proportional to the cross-sectional area of the wire and inversely proportional to its length. Ohm's observation makes sense intuitively when we consider that current is carried through a conductor by the motion of electrons. We pointed out before (Section 18.1) that electrons have a drift velocity in a current-carrying conductor, interacting with the ions of the conductor and carrying charge along in this fashion. A conductor with a larger cross-sectional area allows more charges to flow easily, so a larger current results. However, for a longer conductor and the same applied potential, the electric field is reduced, resulting in a reduced electron flow.

Ohm had already shown that the current was inversely proportional to the resistance (Ohm's law). When these observations are combined, we find that the

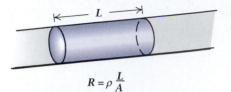

$$R = \rho \frac{L}{A}$$

Figure 18.11

The resistance of a piece of wire or other conducting material is proportional to its length L and inversely proportional to its cross-sectional area A.

resistance of a wire must be proportional to its length L and inversely proportional to its cross-sectional area A (Fig. 18.11):

$$R = \rho \frac{L}{A}. \tag{18.4}$$

The constant of proportionality ρ is the electric **resistivity.** In SI units, resistivity is given in ohm-meters ($\Omega \cdot$ m). The resistivities of a selected list of metals are found in Table 18.1. For most metals the resistivity increases with increasing temperature (Fig. 18.12). For some materials, over narrow ranges, the change in resistivity is approximately proportional to the change in temperature.

Example 18.3

What is the electric resistance of an iron wire 0.50 m long with a diameter of 1.3 mm if the resistivity of iron is $9.7 \times 10^{-8}\ \Omega \cdot$ m?

Solution We can calculate resistance from Eq. (18.4) once we know the cross-sectional area of the wire. If d represents the diameter of the wire, then the cross-sectional area is

$$A = \frac{\pi d^2}{4} = \frac{\pi(1.3 \times 10^{-3}\ \text{m})^2}{4} = 1.33 \times 10^{-6}\ \text{m}^2.$$

Figure 18.12

The resistivity of several metals as a function of temperature. Notice that the resistivities of all these metals are nonlinear over the temperature range shown.

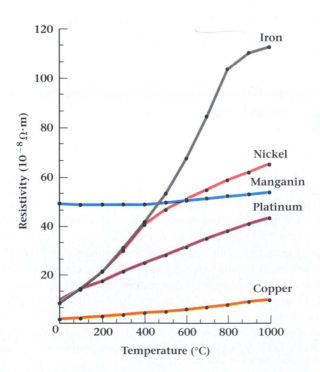

The resistance is then

$$R = \frac{\rho L}{A} = \frac{(9.7 \times 10^{-8} \; \Omega \cdot m)(0.50 \; m)}{1.3 \times 10^{-6} \; m^2} = 0.037 \; \Omega.$$

Table 18.1	Electric Resistivity of Some Metals	
Material		**Resistivity at 20°C ($\Omega \cdot m$)**
Conductors		
Aluminum		2.65×10^{-8}
Copper		1.72×10^{-8}
Gold		2.24×10^{-8}
Iron		9.71×10^{-8}
Nichrome		100×10^{-8}
Platinum		10.6×10^{-8}
Silver		1.59×10^{-8}
Tungsten		5.65×10^{-8}
Semiconductors		
Carbon (graphite)		1.5×10^{-5}
Germanium (pure)		5×10^{-1}
Silicon (pure)		3×10^{3}
Insulators		
Glass		$10^7 - 10^{10}$
Quartz		7.5×10^{17}

Example 18.4

A piece of copper wire has a cross section of 4.0 mm² and a length of 2.0 m. (a) What is the electric resistance of the wire at 20°C? (b) What is the potential difference across the wire when it carries a current of 10 A?

Strategy (a) We can calculate the resistance using the relationship of Eq. (18.4) that $R = \rho l/A$. We are given the values for l and A. The resistivity of copper is found in Table 18.1 to be $1.72 \times 10^{-8} \; \Omega \cdot m$.
(b) The potential difference across the wire is found from the definition of resistance $R = V/I$.

Solution (a) Before substituting the numerical values into the equation, we convert the cross-sectional area from mm² to the SI unit of m². Then we get

$$R = \frac{\rho L}{A} = \frac{(1.72 \times 10^{-8} \; \Omega \cdot m)(2.0 \; m)}{4.0 \times 10^{-6} \; m^2} = 8.6 \times 10^{-3} \; \Omega.$$

(b) When the wire carries a current of 10 A, the voltage is determined from

$$V = IR = (10 \; A)(8.6 \times 10^{-3} \; \Omega)$$
$$V = 8.6 \times 10^{-2} \; V = 0.086 \; V.$$

Power and Energy in Electric Circuits

18.4

If we connect a lamp to a pair of batteries, using good conductors whose resistance is small compared with that of the lamp itself, we can consider the lamp as the only resistive component in the circuit. We can then treat the connecting wires as ideal conductors with no resistance. Figure 18.13(a) illustrates this physical setup, with a carbon resistor substituted for the lamp. The same situation

(a)

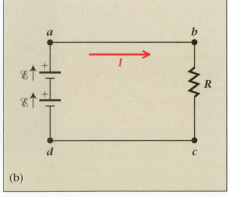

(b)

Figure 18.13
A simple resistive circuit. (a) The physical circuit. (b) Schematic diagram of the circuit, using standard symbols for the circuit components. The black arrows show the direction of increasing potential for the sources of emf.

SUPERCONDUCTIVITY

We have seen that normal materials have electric resistance, which leads to power loss and heating. Many electrical applications require materials with very low resistance—the lower the better. Power transmission lines, electromagnets, computer chips—all would be revolutionized by resistanceless materials. In fact, under special conditions, materials with zero resistance do exist; but making practical use of them continues to be a difficult problem.

In 1908 the Dutch physicist H. Kamerlingh Onnes (1853–1926) succeeded in liquefying helium, for which he received the 1913 Nobel Prize in physics. At atmospheric pressure, helium liquefies at 4.2 K. Moreover, when the pressure above liquid helium is reduced, its temperature can be lowered to below 1 K.

It was well known at that time that the electric resistance of metals decreases with decreasing temperature, approaching a limiting, or residual, value as the temperature approaches zero. However, in 1911, when Kamerlingh Onnes cooled mercury to temperatures below 4 K, its resistance suddenly dropped to zero at a particular transition temperature T_c (Fig. B18.1). This loss of resistance occurred even when the mercury was impure. Kamerlingh Onnes realized that the mercury had undergone a phase transition to a new state, and had become a *superconductor*. In this new superconducting state, the resistance of the mercury was truly zero.

In the years that followed, many other materials were identified as superconductors, including aluminum ($T_c = 1.2$ K), lead ($T_c = 7.2$ K), niobium ($T_c = 9.3$ K), and a number of intermetallic compounds such as niobium-tin (Nb_3Sn, $T_c = 18$ K). Materials such as niobium-tin have been used for the current-carrying windings of superconducting magnets for research and medicine. Magnets of this type are capable of very large magnetic fields. However, to keep them cold (so that they are below their transition temperature), they must be well insulated and cooled with liquid helium.

Because of the expense and difficulty of using liquid helium as a coolant, many researchers sought superconductors with higher transition temperatures. In particular, they sought superconductors with transition temperatures above 77 K, which is the boiling point of liquid nitrogen, a relatively inexpensive coolant. For many years, the highest known superconducting transition temperature was 23 K, the transition temperature of niobium-germanium (Nb_3Ge). However, in 1986

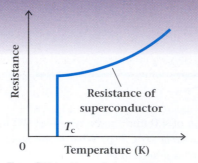

Figure B18.1 The resistance of a superconductor suddenly drops to zero as its temperature is lowered through the superconducting transition temperature, T_c.

J. G. Bednorz and K. A. Müller of the IBM Zurich laboratory found an oxide compound of barium, lanthanum, and copper that became superconducting at 35 K. This discovery earned them the 1987 Nobel Prize in physics, and it set off a wave of activity in laboratories around the world by investigators searching for materials with even higher transition temperatures. By 1998, superconductors with transition temperatures as high as 135 K had been reported (Fig. B18.2). In spite of intensive efforts, only slightly higher transition temperatures have been observed by early 1998.

Because the new high-temperature superconductors require relatively inexpensive coolants, they hold promise for a wide range of applications. Possibilities include electric power transmission without resistive losses, new magnets, and perhaps even magnetically levitated vehicles. However, these applications require improved materials because the available high-temperature superconductors lose their superconductivity when carrying large currents.

Figure B18.2 Two small magnets, one above and one below, are held in place by a high temperature superconductor cooled below its superconducting transition temperature.

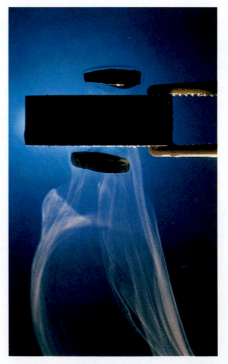

is represented by the circuit diagram in Fig. 18.13(b). In drawing electric circuit diagrams, special symbols are used to indicate resistors, batteries, capacitors, and other devices. A number of common circuit symbols is shown in Fig. 18.14.

Since the resistance of the wires is negligible, there is virtually no potential difference across them even for large currents, so the full potential difference of the batteries appears across the resistor. The potential difference (voltage) from points b to c in Fig. 18.13(b) is the same as the voltage from a to d because the voltage differences from a to b and from c to d are both zero. The magnitude of the electric current is constant in this circuit, the condition known as direct current or dc.

As the current passes through the lamp, charges are moving from a higher potential to a lower one. Energy is being lost from the battery and converted in the filament of the lamp into heat and light. The amount of energy released by a charge q as it falls through the potential V across the lamp is $W = qV$. Since the potential is constant, the rate at which energy is released, or the power P, is

$$P = \frac{\Delta W}{\Delta t} = \frac{\Delta(qV)}{\Delta t} = V\frac{\Delta q}{\Delta t}.$$

But $\Delta q/\Delta t$ is just the current I, so we have

$$P = IV. \tag{18.5}$$

When the potential difference is measured in volts and the current in amperes, their product gives the power in watts. We can see this from the definitions of the units:

$$(\text{ampere})(\text{volt}) = \left(\frac{\text{coulomb}}{\text{second}}\right)\left(\frac{\text{joule}}{\text{coulomb}}\right) = \frac{\text{joule}}{\text{second}} = \text{watt}.$$

In a resistor, the energy dissipated appears as thermal energy, a thermodynamically irreversible process in the sense introduced in Chapter 13. This effect is used in appliances such as electric stoves, hair dryers, and heaters (Fig. 18.15). In an incandescent lamp, the energy delivered to the filament raises its temperature so high that light is emitted. In other circuit elements, the energy may take on different forms. For example, the energy may appear as mechanical work done by a motor, as sound from a loudspeaker, or as stored chemical energy in a battery when the battery is being recharged. Conversion from electrical to mechanical energy is never 100% efficient. The difference appears as heat. Anyone who has felt an operating electric motor is well aware of this, especially if the motor was drawing a lot of current.

▼

Problem-Solving Strategy

Power and Energy in Electric Circuits

In working problems, you should use SI units of ampere, volt, ohm, and watt. Although the kilowatt-hour is often used for electrical energy, the SI unit of energy is the joule.

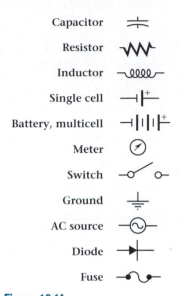

These are a few common circuit symbols used in this book:

Capacitor

Resistor

Inductor

Single cell

Battery, multicell

Meter

Switch

Ground

AC source

Diode

Fuse

Figure 18.14
Standard symbols used in electric circuit diagrams.

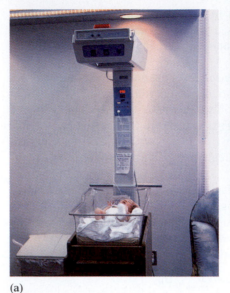

(b)

Figure 18.15
(a) A baby warmer used in a hospital nursery converts electrical energy into thermal energy. (b) Close-up of the radiant heater.

(a)

Example 18.5

A typical electric hair dryer (Fig. 18.16) designed to operate on a 120-V household circuit is rated at 1500 W. What is the resistance of the dryer?

Strategy Since we know the voltage, we can determine the resistance from its definition if we can first find the current. The current can be calculated from the power relation of Eq. (18.5).*

Solution Solving Eq. (18.5) for the current, we have

$$I = \frac{P}{V} = \frac{1500 \text{ W}}{120 \text{ V}} = 12.5 \text{ A}.$$

When this value for the current is inserted into the definition of resistance, we obtain

$$R = \frac{V}{I} = \frac{120 \text{ V}}{12.5 \text{ A}} = 9.60 \ \Omega.$$

Example 18.6

What is the power consumption of an electric iron if its resistance is 13.1 Ω and it operates on a household circuit? Assume an effective household voltage of 120 V.

---

*The voltage in household circuits oscillates with time, a condition commonly called alternating current (ac). However, the value of the voltage is usually expressed as the dc equivalent. The effective values of oscillating currents and voltages are discussed in Chapter 21.

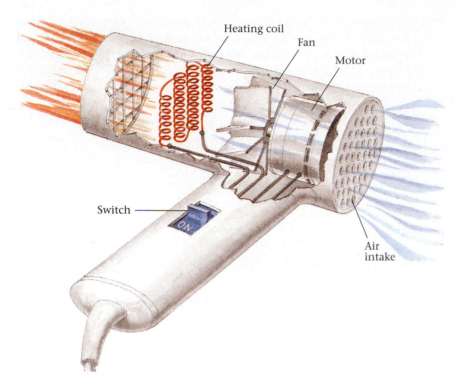

Heating coil

Fan

Motor

Switch

ON

Air
intake

Figure 18.16
Example 18.5: An electric hair dryer.

Strategy This problem is similar to Example 18.5. Here we are given the resistance (13.1 Ω) and the voltage and are asked to find the power. Using voltage and resistance, we can easily find the current. We can then combine current and voltage to get the power.

Solution The current is obtained from Eq. (18.2):

$$I = \frac{V}{R} = \frac{120\ \text{V}}{13.1\ \Omega} = 9.16\ \text{A}.$$

Using this value for I, we can compute the power:

$$P = IV = (9.16\ \text{A})(120\ \text{V})$$
$$P = 1100\ \text{W}.$$

In the two preceding examples, we used two equations sequentially to obtain the answer. We could have combined these equations at the outset and then substituted the values only once. For example, if we use Eq. (18.2) to eliminate the current in Eq. (18.5), we get a new equation relating power to the resistance and voltage across an electric device:

$$P = IV = \left(\frac{V}{R}\right)V$$

$$P = \frac{V^2}{R}. \qquad\qquad (18.6)$$

This last equation could have been used from the start in both examples to calculate the answer in one step. In both cases we were given two of the three quantities resistance, power, and voltage, and the third could have been easily calculated from Eq. (18.6). Similarly, we can eliminate the voltage from Eqs. (18.2) and (18.5) to get a relation between the current through a device, resistance, and power:

$$P = IV = I(IR)$$

$$\boxed{P = I^2R.} \tag{18.7}$$

As we explained earlier in this section, when an electric current passes through a resistor, electrical energy is irreversibly transformed to thermal energy. Equations (18.6) and (18.7) describe the rate of transfer of electrical energy to heat energy in a resistor. Both of these equations are known as **Joule's law.**

Example 18.7

A piece of wire has a resistance of 30 Ω. How much power is dissipated in the wire if it carries a current of 0.50 A?

Solution   We may use Eq. (18.7) to calculate the power directly,

$$P = I^2R = (0.50 \text{ A})^2 (30 \text{ Ω}) = 7.5 \text{ A}^2 \cdot \text{Ω}.$$

Discussion   Since we have consistently used SI units, the power is in watts. Let's check the units to be sure. The ohm is one volt per ampere, so

$$A^2 \cdot \Omega = (\text{ampere})^2(\text{ohm}) = (\text{ampere})^2\left(\frac{\text{volt}}{\text{ampere}}\right)$$

$$= (\text{ampere})(\text{volt}) = \left(\frac{\text{coulomb}}{\text{second}}\right)\left(\frac{\text{joule}}{\text{coulomb}}\right)$$

$$= \frac{\text{joule}}{\text{second}} = \text{watt}.$$

The power is indeed given in watts and is $P = 7.5$ W.

Example 18.8

In the stairwell of a ten-story building, there are two continuously burning 75-W safety lamps for each floor. (a) What is the total energy (in kilowatt-hours) used in one year? (b) What will it cost to use the lamps for a year if the cost of electricity is $0.078/kWh?

Strategy   (a) The total power used is the product of the number of lamps with the power per lamp. The total energy used is the product of the total power with the time.
(b) The cost can be computed as the product of the total energy with the cost of electricity per kWh.

Solution (a) The total power use is

$$2 \text{ lamps/floor} \times 10 \text{ floors} \times 75 \text{ W/lamp} = 1.5 \text{ kW}.$$

The energy, or work, is then

$$W = Pt$$

$$W = 1.5 \text{ kW}(1 \text{ year})\left(\frac{365 \text{ days}}{1 \text{ year}}\right)\left(\frac{24 \text{ hours}}{\text{day}}\right) = 1.3 \times 10^4 \text{ kWh}.$$

(b) The cost for a year of operation is

$$(1.3 \times 10^4 \text{ kWh})(\$0.078/\text{kWh}) \approx \$1000.$$

Master the Concept

Electric Energy and Power

Question: Although we often speak of electric utilities as power companies, what is the product that we buy?

Answer: Power is the rate at which energy is used, so the energy used in our homes is the product of power and time. The SI unit of energy is the joule or watt-second. But this unit is inconveniently small because in our homes we commonly use energy at the rate of kilowatts and do so for extended periods of hours or days. As a result, a common unit for electrical energy is the kilowatt-hour (kWh). One kilowatt-hour is the energy consumed in one hour (3.6×10^3 s) by operating at a constant power of 1 kW (10^3 W),

$$1 \text{ kWh} = 3.6 \times 10^6 \text{ J}.$$

Our bill from the utility company is for the number of kilowatt-hours of electricity used. The product that we buy is energy.

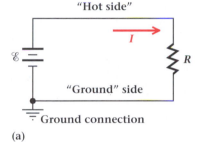

(a)

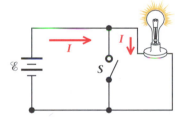

(b)

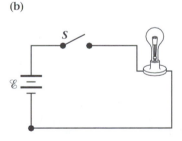

(c)

Figure 18.17
(a) A circuit with one side grounded. The symbol for ground is shown.
(b) Conditions for a short circuit. When the switch S is closed, the current passes through the switch instead of going through the lamp.
(c) An open circuit. With the switch S open, there is no current through the lamp.

18.5 Short Circuits and Open Circuits

A battery produces a voltage between its terminals. Either terminal can be chosen as the reference potential. In household circuits we choose the reference potential to be that of the earth or ground. That is, one side of the circuit is maintained with a zero potential difference with respect to the earth and is called the **ground potential** (Fig. 18.17a). The other side of the circuit is called the "hot" side. A current will pass through any conducting object that simultaneously contacts the hot side of the circuit and a conducting path to ground. Birds and small animals can sit or run along high-voltage wires without harm because their bodies do not make contact with the hot side of the circuit and ground at the same time.

Occasionally, electrical equipment experiences a type of failure in which an alternative unwanted conducting path occurs, allowing large currents to pass.

Such a path is usually called a **short circuit** because most of the current is not flowing through the desired circuit but instead passes through a parallel "shorter" path of lower resistance (Fig. 18.17b). Since we usually refer one side of a circuit to ground potential, we also describe the condition of a short circuit as being grounded out, or simply *grounded*. The result of a short circuit is frequently a high current, since for a given potential difference, a lower resistance leads to a higher current.

Another common problem in electric circuits is the interrupted or **open circuit.** If the conducting path is interrupted or broken at any point, no current can occur in the circuit. It is just as if a switch had been opened. Open circuits most often occur when the conducting path is mechanically broken or burns out from an overload. The classic example of this is the failure of a string of decorative Christmas lights when a bulb is removed. All of the lights go out when one bulb is removed because no current can flow through the open circuit that results.

18.6 Kirchhoff's Rules and Simple Resistive Circuits

In analyzing resistive circuits, we use two fundamental ideas: conservation of energy and conservation of charge. When applied to electric circuits, these laws are known as **Kirchhoff's rules.** The first rule is the law of conservation of energy, which says that *the algebraic sum of the potential differences around a closed conducting loop must be zero.* This means that if a charge is taken from some point in the circuit and moved completely around the loop to its starting point, the net work done on the charge is zero. This must be so because the charge has returned to the same energy level. This statement is called **Kirchhoff's voltage rule,** sometimes referred to as the *loop rule.*

The second fundamental law that applies to circuit analysis is the conservation of charge at a *junction,* a place in the circuit where two or more conductors join. When expressed in terms of current, this means that the sum of the currents flowing into a junction must equal the sum of the currents leaving the junction. If this were not so, then the junction would be either creating or destroying charge. Thus, a single-loop circuit must have the same current in all parts of the loop. For a multiloop circuit, the *net current entering into any junction must be zero.* This last statement is known as **Kirchhoff's current rule,** sometimes called the *junction rule.*

Let us return to our example of the simple circuit consisting of two batteries and a resistor (Fig. 18.18a). For simplicity, we consider the batteries to be ideal with no internal resistance and with emf $\mathscr{E}$. We treat the resistor and the conducting wires as ideal so that we consider all the resistance to reside in the resistor. The potential difference of each battery is 1.5 V. When the two batteries are joined in series (that is, the positive side of one battery connected to the negative side of the other), their emfs are in the same direction in the sense of the Kirchhoff voltage rule. Thus, the voltages add to provide a potential difference of 3.0 V. When a second resistor is added in series with the first (Fig. 18.18b), so that current passes through one and then the other, the total potential difference across them must add up to the three volts available from the batteries. Because there is only one source of current and only one conduction path, the current I is the same throughout the entire circuit.

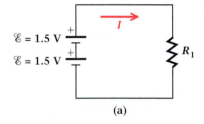

(a)

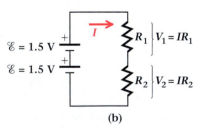

(b)

Figure 18.18

(a) A simple resistive circuit with one resistor connected across the potential of two batteries. (b) Two resistors connected in series across the same two batteries. The current is the same in each resistor connected in series.

The potential across the first resistor is

$$V_1 = IR_1,$$

and across the second is

$$V_2 = IR_2.$$

Kirchhoff's voltage rule says that the algebraic sum of the potential differences across the battery and the resistors is zero. If we imagine traversing the loop in a clockwise manner, then the potential increases when we pass through the batteries and decreases when we pass through the resistors:

$$\mathscr{E} + \mathscr{E} - IR_1 - IR_2 = 0.$$

Thus, the magnitude of the total potential across the resistors is equal to the magnitude of the total battery voltage V,

$$V = \mathscr{E} + \mathscr{E} = IR_1 + IR_2,$$

or

$$V = I(R_1 + R_2) = IR_s.$$

From this equation we see that two resistors in series can be considered equivalent to a single resistor R_s, whose resistance is equal to the sum of the individual resistances. This reasoning is true not only for the two resistors in series in Fig. 18.18(b), but for any number of resistors joined in series. Thus, **the equivalent resistance of resistors connected in series** is

$$R_s = R_1 + R_2 + R_3 + \cdots. \qquad (18.8)$$

Example 18.9

Three 10-Ω resistors and one 15-Ω resistor are connected in series across a 9.0 V battery (Fig. 18.19). Find the voltage drop across the 15-Ω resistor.

Strategy For resistors in series we can use Eq. (18.8) to find the equivalent resistance of the four resistors. Then we can find the current from the ratio of the potential difference to the resistance. Finally, knowing the current, we can compute the voltage drop across a single resistor.

Solution The equivalent resistance of the four series resistors is $R_s = 45\ \Omega$, the sum of their individual resistances. The current in the circuit is

$$I = \frac{V}{R_s},$$

where V is the battery voltage. Thus

$$I = \frac{9.0\ \text{V}}{45\ \Omega} = 0.20\ \text{A}.$$

The voltage across the 15-Ω resistor is

$$V_3 = IR_3 = (0.20\ \text{A})(15\ \Omega) = 3.0\ \text{V}.$$

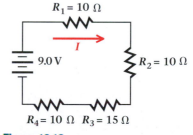

Figure 18.19
Example 18.9: Four resistors joined in series.

Discussion Note that the voltage drop across the 15-Ω resistor is the same regardless of the sequence in which the resistors are arranged, as long as they remain connected in series across the same applied voltage.

■

Two resistors can also be joined together in parallel across the same battery (Fig. 18.20). This parallel connection allows current to flow independently through the parallel branches of the circuit, while the voltage across each branch is equal. In this case, the voltage across each branch is the full voltage of the battery. The current I_1 through resistor R_1 is

$$I_1 = \frac{V}{R_1},$$

and the current through resistor R_2 is

$$I_2 = \frac{V}{R_2}.$$

From the conservation of charge (Kirchhoff's current rule), the total current provided by the battery is the sum of the two branch currents I_1 and I_2,

$$I = I_1 + I_2.$$

Therefore

$$I = \frac{V}{R_1} + \frac{V}{R_2},$$

or

$$I = V\left(\frac{1}{R_1} + \frac{1}{R_2}\right).$$

The total current is the potential divided by the equivalent resistance,

$$I = \frac{V}{R_p}.$$

By comparing these last two expressions for the current, we see that the equivalent resistance R_p of two resistors connected in parallel is

$$\frac{1}{R_p} = \frac{1}{R_1} + \frac{1}{R_2}.$$

This argument can be extended to show that **the equivalent resistance of resistors connected in parallel** is

$$\boxed{\frac{1}{R_p} = \frac{1}{R_1} + \frac{1}{R_2} + \frac{1}{R_3} + \cdots.} \tag{18.9}$$

For the special case of only two resistors, the equivalent parallel resistance may be written as

$$R_p = \frac{R_1 R_2}{R_1 + R_2}.$$

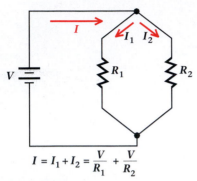

$$I = I_1 + I_2 = \frac{V}{R_1} + \frac{V}{R_2}$$

Figure 18.20
Two resistors joined in parallel. The voltage is the same across each resistor.

From this last equation we observe that the equivalent resistance of two resistors joined in parallel is always less than the magnitude of the smaller resistor. To see this, suppose that $R_1 < R_2$. Then

$$R_p = R_1 \left(\frac{R_2}{R_1 + R_2} \right) < R_1.$$

In the special case of two identical resistors, $R_1 = R_2 = R$, and the equivalent parallel resistance becomes $R/2$.

An important point to remember concerning series and parallel connection of resistors is that *for series connection the current through the resistors is the same,* and *for parallel connection the voltage across the resistors is the same.*

Example 18.10

Three resistors joined in parallel have values of 330 Ω, 100 Ω, and 220 Ω (Fig. 18.21). (a) What is the equivalent resistance of the combination? (b) When the combination is connected across a battery, the current in the 100-Ω resistor is 0.12 A. What power is expended in the 330-Ω resistor?

Strategy (a) To find the equivalent resistance of three resistors in parallel, we can directly substitute into Eq. (18.9).
(b) In order to find the power in the 330-Ω resistor, we need to know the potential across it. As we have seen, the potential difference across resistors in parallel is the same for each of them. Thus, the potential difference across the 330-Ω resistor is the potential across the 100-Ω resistor. Because we know both R and I, we can use Ohm's law to find the potential difference V across the 100-Ω resistor. Then we use this value of V to calculate the power expended in the 330-Ω resistor from $P = V^2/R$.

Solution (a) The combined resistance is found directly from Eq. (18.9):

$$\frac{1}{R_p} = \frac{1}{330 \ \Omega} + \frac{1}{100 \ \Omega} + \frac{1}{220 \ \Omega} = \frac{116}{6600 \ \Omega}.$$

The equivalent resistance is

$$R_p = \frac{6600 \ \Omega}{116} = 56.9 \ \Omega.$$

(b) The potential drop across the 100-Ω resistor is

$$V = IR = (100 \ \Omega)(0.12 \ A) = 12 \ V.$$

This 12 V is also the potential across the 330-W resistor. The power expended by the 330-Ω resistor is then

$$P = \frac{V^2}{R} = \frac{(12 \ V)^2}{330 \ \Omega} = 0.44 \ W.$$

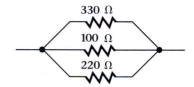

Figure 18.21
Example 18.10: Three resistors joined in parallel.

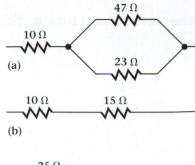

Figure 18.22

Example 18.11: (a) Two resistors connected in parallel are connected in series to a third resistor. (b) The equivalent resistance of the parallel resistors in series with the third resistor. (c) The single resistor equivalent to the three resistors of part (a).

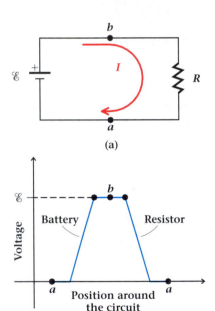

(a)

(b)

Figure 18.23

(a) A current exists in the circuit when a resistance is connected to a battery. (b) According to Kirchhoff's voltage rule, the sum of the potential changes around the loop is zero. In the diagram of voltage versus position around the circuit, sources of emf increase potential and resistors (traversed in the direction of the current) decrease the potential.

Example 18.11

A 10-Ω resistor is joined in series with the parallel combination of a 47-Ω resistor and a 23-Ω resistor (Fig. 18.22a). What is the equivalent resistance of this combination?

Strategy We can determine the resistance of this arrangement by first reducing the parallel combination to a single equivalent resistor and then combining that with the 10-Ω resistor, using the rule for series combination. This general procedure can be used to treat more complicated combinations than the one given here.

Solution The equivalent resistance of the two resistors connected in parallel is

$$\frac{1}{R_p} = \frac{1}{R_1} + \frac{1}{R_2} = \frac{1}{47\ \Omega} + \frac{1}{23\ \Omega}.$$

Solving for R_p gives

$$R_p = 15\ \Omega.$$

This, in turn, is combined in series (Fig. 18.22b) to give

$$R = 15\ \Omega + 10\ \Omega = 25\ \Omega.$$

This is the value of the equivalent single resistance shown in Fig. 18.22(c).

Discussion The procedure followed here for stepwise reducing a complicated network to a simple one is a very useful and practical method for treating circuits with resistor combinations. However, some circuits do not yield to this analysis and can only be solved using the full power of Kirchhoff's rules.

*18.7 Application of Kirchhoff's Rules

Now that we have used Kirchhoff's rules with simple circuits in order to learn how to combine resistors in series or parallel, we want to see how the rules can be used with more complicated networks. In order to apply the loop rule, we must carefully determine the sign of each potential difference. As a start, consider the potential difference across a battery or other source of emf. The positive direction of the emf is defined to be from the terminal of lower potential to the terminal of higher potential. Frequently the arrow and the plus and minus signs are understood and are not explicitly given on the symbol for a battery. However, the direction of the emf is always understood to be the same as that indicated earlier in Fig. 18.13(b).

When a resistor is connected to the terminals of a battery (Fig. 18.23a), charge flows through the circuit. If we use the conventional direction for the current (the direction in which a positive charge would move), then the direction of current through the resistor is in the direction of *decreasing* electric potential (Fig. 18.23b). We can combine this observation with the definition of an emf to list the convention for finding the potential differences for the Kirchhoff voltage rule. For an imaginary traversal of a loop, we use the following convention:

1. The potential difference is $+\mathcal{E}$ when a source of emf is traversed in the forward direction of the emf.

2. The potential difference is $-\mathcal{E}$ when a source of emf is traversed in the backward direction.

3. The potential difference is $-IR$ when a resistor is traversed along the direction of the current.

4. The potential difference is $+IR$ when a resistor is traversed in the direction opposite to that of the current.

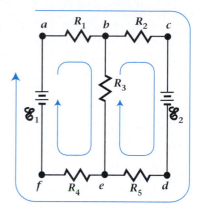

Figure 18.24
The sum of the voltages around a loop is zero. This holds for loop *a-b-e-f-a*, loop *a-b-c-d-e-f-a*, and loop *b-c-d-e-b*.

The Kirchhoff voltage rule holds equally well for the loop of Fig. 18.23 or any of the loops in Fig. 18.24. The algebraic sum of the potential differences is zero around the loop *a-b-e-f-a* in Fig. 18.24. If the potential difference between points *a* and *b* is written as V_{ab}, then we have

$$V_{ab} + V_{be} + V_{ef} + V_{fa} = 0.$$

Had we chosen a different loop, such as *b-c-d-e-b*, the result would still be the same:

$$V_{bc} + V_{cd} + V_{de} + V_{eb} = 0.$$

In Fig. 18.25 we may consider currents I_1 and I_2 to be positive since their directions are into the junction. The other current is taken as negative because it is directed away from the junction. Kirchhoff's current rule for this junction can be written as

$$I_1 + I_2 - I_3 = 0.$$

Example 18.12

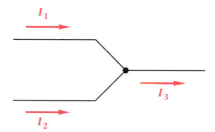

Figure 18.25
Kirchhoff's current rule: The algebraic sum of the currents into the junction is zero.

Find the currents I_1, I_2, and I_3 in the circuit shown in Fig. 18.26.

Strategy We can find the currents by applying Kirchhoff's rules. In doing so we will need to find three equations that can be solved simultaneously for the three unknown quantities I_1, I_2, and I_3. We start by writing the junction equation for the junction *b*. Then we find the loop equations appropriate for the two loops *a-b-e-f-a* and *b-e-d-c-b*.

Solution Applying Kirchhoff's junction rule to the junction at *b*, we see that currents I_1 and I_2 are entering the junction and are positive, while current I_3 is negative because it is leaving the junction.

(1) $$I_1 + I_2 - I_3 = 0.$$

Starting at position *a* and going clockwise, we apply the loop rule to the loop *a-b-e-f-a* to get

(2) $$-(27 \ \Omega)I_1 - (33 \ \Omega)I_3 + 9.0 \ \text{V} = 0.$$

Starting at position *b* and going counterclockwise (in the direction of I_3 and I_2), we apply the loop rule to loop *b-e-d-c-b* to get

(3) $$-(33 \ \Omega)I_3 - (66 \ \Omega)I_2 + 6.0 \ \text{V} = 0.$$

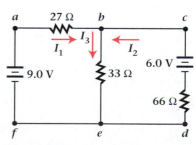

Figure 18.26
Example 18.12: Use Kirchhoff's rules to find the currents in the circuit.

We now have three independent equations that must be solved simultaneously for the three currents. We begin by rearranging Eq. (2) to get

$$(4) \qquad 9.0 \text{ V} = (27 \ \Omega)I_1 + (33 \ \Omega)I_3.$$

Next we rearrange Eq. (3), using the junction equation (Eq. 1) to eliminate I_2.

$$6.0 \text{ V} = (66 \ \Omega)(I_3 - I_1) + (33 \ \Omega)I_3$$
$$6.0 \text{ V} = -(66 \ \Omega)I_1 + (99 \ \Omega)I_3.$$

Next we divide the last equation by 3 to get

$$(5) \qquad 2.0 \text{ V} = -(22 \ \Omega)I_1 + (33 \ \Omega)I_3.$$

Upon comparing Eq. (5) with Eq. (4), we see that both equations contain the term $(33 \ \Omega)I_3$. Thus, we can eliminate I_3 by subtracting Eq. (5) from Eq. (4). The result is

$$7.0 \text{ V} = (49 \ \Omega)I_1,$$

$$I_1 = \frac{7.0 \text{ V}}{49 \ \Omega} = 0.14 \text{ A}.$$

Now that we have I_1, we can obtain I_3. Using Eq. (4) we get

$$9.0 \text{ V} = (27 \ \Omega)(0.14 \text{ A}) + (33 \ \Omega)I_3,$$

$$I_3 = \frac{9.0 \text{ V} - 3.8 \text{ V}}{33 \ \Omega} = 0.16 \text{ A}.$$

Then from Eq. (1), we find the current I_2,

$$I_2 = I_3 - I_1 = 0.16 \text{ A} - 0.14 \text{ A} = 0.02 \text{ A}.$$

Discussion In this particular example, all of the currents came out positive. That means that the current directions assigned at the beginning correspond to the actual directions of the currents. For the same kind of circuit but with different values of components, it is possible for one or more of the currents to come out negative. A negative result for the current in such a computation is not an error but merely means that the direction of the current is opposite to the direction chosen for solving the circuit equations.

18.8 Capacitors in Combination

So far in this chapter, the only circuit components we have discussed are current sources and resistors. Now that we have learned to analyze these resistive circuits, let's add in the circuit component we examined in the previous chapter: the capacitor. There we learned that capacitance is proportional to the area of the plates of a parallel-plate capacitor. Thus, if we could double the area, we would double the capacitance. We can effectively do just this by joining two identical capacitors in parallel. Then, when we apply a potential across them, charge flows to each capacitor, resulting in a total stored charge that is twice the charge on each individually. Hence, the capacitance of the combination is twice that of one capacitor alone.

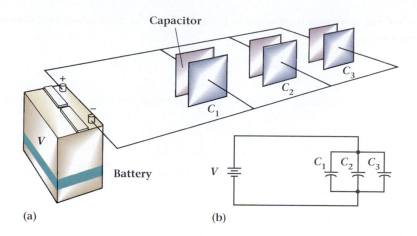

Capacitor

C_1 C_2 C_3

Battery

(a)   **(b)**

Figure 18.27
(a) Three capacitors connected in
parallel. The voltage is the same
across each capacitor. (b) Schematic
representation of three capacitors
connected in parallel.

We can use this same reasoning to calculate the capacitance of any number of capacitors, of any size, joined together in parallel. Consider, for example, three capacitors in parallel (Fig. 18.27). The voltage V supplied by the battery appears across each capacitor. The charge on each capacitor depends on this voltage, and the capacitance of each is determined from the equation $C = q/V$. Thus, the charge q_1 on capacitor 1 is

$$q_1 = C_1 V.$$

Similarly, the charges on the other two capacitors are

$$q_2 = C_2 V \quad \text{and} \quad q_3 = C_3 V.$$

The equivalent capacitance C_p of the parallel combination of capacitors is the ratio of the total charge to the applied voltage. Since the total charge is just the sum of q_1, q_2 and q_3, we have

$$C_p = \frac{q_1 + q_2 + q_3}{V} = \frac{q_1}{V} + \frac{q_2}{V} + \frac{q_3}{V} = C_1 + C_2 + C_3.$$

We thus arrive at an addition rule for capacitors: **The equivalent capacitance C_p of capacitors connected in parallel is**

$$C_p = C_1 + C_2 + C_3 + \cdots. \tag{18.10}$$

Suppose instead that we join the capacitors end-to-end in series across the battery (Fig. 18.28). For this situation, as the first capacitor becomes charged, it induces an equal charge in the second capacitor, which in turn induces an equal charge in the third capacitor. Since the capacitors permit no flow of charge between their plates, there is no net charge within the region marked by the colored rectangle in Fig. 18.28. Therefore, all of the capacitors must have the same charge q:

$$q_1 = q_2 = q_3 = q.$$

Now, conservation of energy (Kirchhoff's voltage rule) requires that the sum of the individual capacitor voltages be equal to the total voltage supplied by the battery. That is,

$$V = V_1 + V_2 + V_3.$$

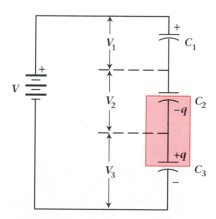

Figure 18.28
Capacitors connected in series. The
magnitude of the charge is the same
on each plate of each capacitor. (That
is, there is no net charge in the region
marked by the colored rectangle.)

The voltage across each capacitor is given by the ratio of the charge on it to its capacitance,

$$V_1 = \frac{q_1}{C_1}, \qquad V_2 = \frac{q_2}{C_2}, \qquad V_3 = \frac{q_3}{C_3}.$$

Replacing the individual voltages by their equivalents in terms of q/C, we find that

$$V = q\left(\frac{1}{C_1} + \frac{1}{C_2} + \frac{1}{C_3}\right).$$

This last equation has the form $V = q/C$ if we identify the equivalent series capacitance C_s as

$$\frac{1}{C_s} = \frac{1}{C_1} + \frac{1}{C_2} + \frac{1}{C_3}.$$

Thus, we arrive at a second addition rule for capacitors: **The equivalent capacitance C_s of capacitors connected in series is**

$$\frac{1}{C_s} = \frac{1}{C_1} + \frac{1}{C_2} + \frac{1}{C_3} + \cdots. \tag{18.11}$$

Note that for capacitors connected in parallel, the voltage is the same across each capacitor, while for capacitors connected in series, the charge is the same on each capacitor.

Example 18.13

Calculate the capacitance of the network shown in Fig. 18.29.

Strategy We can approach this problem in a manner similar to that used in Example 18.11. We begin by finding the equivalent capacitance of the three capacitors joined in parallel. Then we combine that capacitance in series with the remaining capacitance to obtain the single equivalent capacitance.

Solution The equivalent capacitance of the three parallel capacitors is just the sum of their individual capacitances,

$$C_p = C_1 + C_2 + C_3 = 5 \ \mu F + 5 \ \mu F + 10 \ \mu F = 20 \ \mu F.$$

Next we add their equivalent capacitance of 20 μF to the 10 μF, according to Eq. (18.11) for adding capacitors in series. We get

$$\frac{1}{C_s} = \frac{1}{20 \ \mu F} + \frac{1}{10 \ \mu F} = \frac{3}{20 \ \mu F}.$$

Upon solving this equation for C_s, we have the capacitance of the network,

$$C = \frac{20}{3} \ \mu F = 6.7 \ \mu F.$$

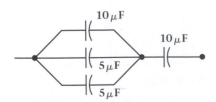

Figure 18.29
Example 18.13: A capacitor network.

Example 18.14

Calculate the voltage across the 10-μF capacitor in the circuit of Fig. 18.30(a).

Strategy We first find the capacitance of the parallel combination of the 10-μF capacitor with the two 20-μF capacitors. Then we find the potential across the combination, which will be the same as the potential across the 10-μF capacitor.

Solution The two 20-μF capacitors in series have an equivalent capacitance given by

$$\frac{1}{C_1} = \frac{1}{20 \ \mu F} + \frac{1}{20 \ \mu F} = \frac{2}{20 \ \mu F},$$
$$C_1 = 10 \ \mu F.$$

The capacitance C_1 is in parallel with the 10-μF capacitor (Fig. 18.30b). Their combination is

$$C_2 = 10 \ \mu F + 10 \ \mu F = 20 \ \mu F.$$

This capacitance is equal to the capacitance of the other remaining capacitor (Fig. 18.30c). Thus, half the voltage appears across the 20-μF capacitor and half across the combination C_2. This means that the voltage across the 10-μF capacitor is

$$V = \frac{9.0 \ V}{2} = 4.5 \ V.$$

Discussion In this example, we were able to get quickly to the answer by simplifying the circuit diagram and by recognizing the symmetries involved. We could have plugged away with equations for charge and capacitance and ultimately obtained the same result. Although you want to make use of your mathematical skills where needed, it is always wise to carefully look at the physical situation and use your understanding and the symmetries present rather than to just look for formulas and substitute values into them. As was the case in mechanics, you should look at each step of a problem and try to understand what you have.

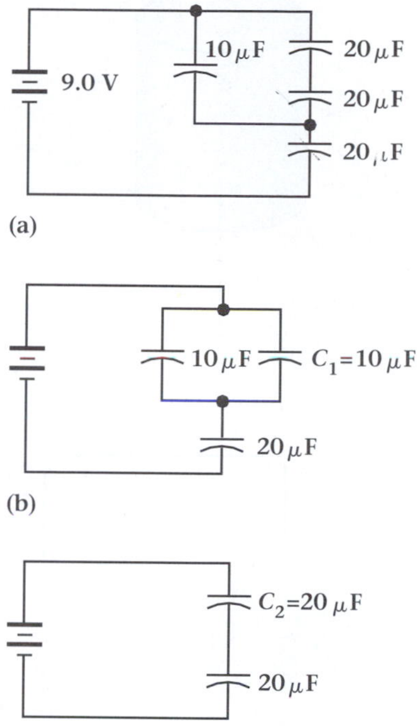

(a)

(b)

(c)

Figure 18.30
Example 18.14: (a) A capacitor network joined to a battery.
(b) Network resulting from combining two of the 20-μF capacitors.
(c) Network resulting from combining the parallel capacitors.

18.9   Internal Resistance of a Battery

We described batteries as devices that maintain a fixed electric potential difference between two points. However, when a real battery is used to provide electrical energy, the external voltage across the terminals is less than the emf. This reduction in voltage is due to the potential drop occurring across the internal resistance of the battery itself. As more current is drawn from a battery, a greater voltage drop occurs across its internal resistance. This effect is most easily visualized by considering a real battery to consist of an ideal emf (resistanceless battery) in series with a resistance (Fig. 18.31a). The resistance r is the internal resistance of the battery.

If the battery is connected to an external resistance R (called a load resistor), the circuit could be drawn as shown in Fig. 18.31(b), which explicitly

(a)

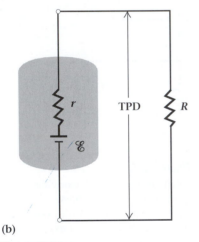

(b)

Figure 18.31

(a) A real battery can be represented as an ideal emf in series with an internal resistance. (b) A real battery in a circuit. The terminal potential difference (TPD) is less than the emf.

includes the internal resistance. The current through the circuit depends on the emf $\mathscr{E}$ and the total resistance,

$$I = \frac{\mathscr{E}}{r + R}.$$

The potential difference across the terminals of the battery is the **terminal potential difference,** abbreviated TPD.* It is the emf reduced by the voltage drop across the internal resistance r. Thus, the TPD has a value

$$\text{TPD} = \mathscr{E} - Ir = \mathscr{E} - \frac{\mathscr{E}r}{r + R},$$

which reduces to

$$\text{TPD} = \frac{R}{r + R}\mathscr{E}. \tag{18.12}$$

According to this equation, when the load resistance R is small, the terminal voltage is appreciably less than the emf. However, when the load resistance is large compared with the internal resistance of the battery, the terminal voltage approximately equals the emf.

Carbon-zinc dry cells are notorious for their relatively large internal resistance, which becomes particularly noticeable as the cells age. Consequently, simply checking the output voltage of a dry cell battery with a voltmeter is no real indication of the battery's condition. The reason that this simple test is meaningless is that, as we will see later, a voltmeter has a high resistance. When a voltmeter is connected alone across a battery, it gives a reading very nearly that of the emf. But if the battery is connected first across a low resistance, say 200 Ω, the TPD measured by the voltmeter is much less than the emf. As an illustration of this, you might measure the TPD of the battery in a radio while it is operating and see what reading you get.

The decrease in TPD associated with increased current is not limited to batteries. Any network composed of sources of emf and resistors can be represented

---

Master the Concept

Charging and Discharging Batteries

Question: When a rechargeable battery is supplying current to a load, the TPD is less than the emf. What happens to the TPD when the same battery is being charged?

Answer: In order to charge the battery, an external potential difference is applied in a direction that causes charge to flow through the battery in the backward direction. The potential difference across the internal resistance of the battery is thus in the same direction as the battery's emf. As a result, the TPD is larger than the emf. This means that it takes a higher voltage to charge a battery than the battery can deliver.

---

*The TPD is also known as the *terminal voltage*.

by an equivalent series combination of a single battery and a single resistor. (This statement is known as Thévenin's theorem.) Such a circuit is fully equivalent in the way in which it influences an external (load) circuit. The output behavior of any battery, battery network, or power supply can be described by this equivalent circuit. Thus, the general behavior of any voltage source is that the terminal voltage decreases as the current drawn by the load increases.

Example 18.15

A transistor radio battery has an emf of 9.0 V. When a short copper wire is connected directly across the battery terminals, a current of 4.0 A passes through the wire. What is the internal resistance of the battery? What is the terminal potential difference across a 10-Ω load?

Solution Because the resistance of the wire is negligibly small, the current is limited by the internal resistance of the battery. Thus

$$r = \frac{\mathscr{E}}{I} = \frac{9.0 \text{ V}}{4.0 \text{ A}} = 2.25 \ \Omega.$$

When the battery is placed across a 10-Ω load, the terminal potential difference is less than the emf. From Eq. (18.12),

$$\text{TPD} = \frac{R}{r + R}\mathscr{E}.$$

Here $R = 10 \ \Omega$, $r = 2.25 \ \Omega$, and $\mathscr{E} = 9.0$ V, so

$$\text{TPD} = \frac{10}{2.25 + 10}9.0 \text{ V} = 7.3 \text{ V}.$$

*18.10   Home Power Distribution

One very important characteristic of parallel circuits is that the current can be interrupted in one branch without interrupting the current in the other branches. This is especially significant in a power distribution system, where many separate users are simultaneously provided with electricity from the same network. Each user would like to operate his or her own lamps and appliances independent of the others. For this reason, the power distribution circuits that are so familiar to us in our homes are indeed parallel circuits.

In wiring a house or building, usually several separate circuits are connected in parallel to the main incoming power line. In turn, each of these circuits carries several outlets, appliances, or lamps, also in a parallel configuration (Fig. 18.32). Because of the parallel circuitry, each outlet and each lamp can be used independently.

The wiring that carries current through the house has a low resistance, but it is not zero. Consequently, the temperature of the wires increases when the current gets large. If the amount of current is not limited somehow, the wires may overheat and become a fire hazard. This danger can be avoided by installing a safety

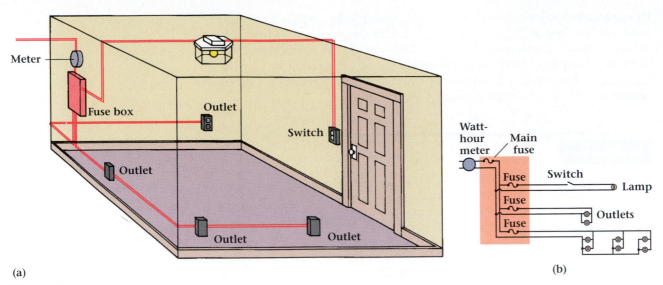

(a)

(b)

Figure 18.32

(a) Wiring scheme for a house. (b) Schematic diagram of the same house. In practice, each separately fused circuit would have several outlets and lamps.

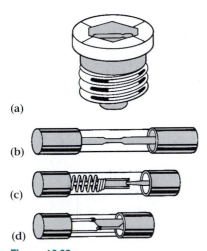

Figure 18.33

(a) A fuse of the type used in older buildings. Typical fuses of the type found in cars, amplifiers, and tape decks: (b) standard fuse, (c) slow-blowing type, and (d) fast-acting fuse.

device that interrupts the circuit and cuts off the current whenever it gets too large to be safe. The most common safety devices are fuses and circuit breakers.

A fuse consists of a metal wire or ribbon that melts at a particular current level (Fig. 18.33). For small currents, the fuse merely acts as a conductor. But when the current exceeds the rated capacity of the fuse, it overheats, melts, and leaves an air gap in the circuit. Once a fuse "blows," it has done its job of protecting the circuit and is no longer useful. It must then be replaced with a new fuse before the circuit can operate again.

In newer buildings, the role of the fuse is handled by a circuit breaker (Fig. 18.34). This device mechanically interrupts the circuit in response to an overload current. The most widely used circuit breakers are thermal-magnetic devices. They use a combination of bimetallic strips and electromagnets to provide both thermal and magnetic protection from excessive currents. Unlike a fuse, a circuit breaker can be reset and used again. Each time the current exceeds the overload value, the circuit breaker trips and remains off until it is reset. Whenever a circuit breaker or fuse is blown repeatedly, the circuit should be carefully checked to find the cause of the overload. Then the source of the overload should be removed before resetting the breaker or replacing the fuse.

Household circuits are normally fused at 15 or 20 A. Exceptions are the special high-current circuits used for water heaters, ranges, and other large appliances. These circuits may carry up to 60 A and will not usually have other outlets on the same circuit. Such circuits also have larger-diameter wires to reduce their resistance.

Appliances such as toasters, frying pans, and radios are normally rated according to their power consumption in watts. Lamps, too, are commonly identified by their power rating. The result is that while the circuits are rated by current, the loads are rated by power. The connection is Eq. (18.5), which states that $P = IV$. The house voltage in the United States is nominally 120 V. Thus, a 20-A circuit can deliver 2400 W at 120 V. This means that if a 1350-W frying pan is operated on the same 20-A circuit as a 1500-W toaster, the circuit breaker will trip and shut off the current to that circuit. Similar occurrences are familiar to everyone.

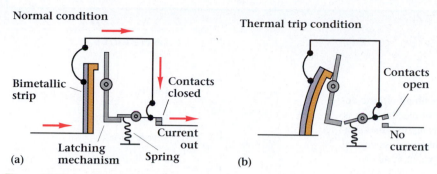

Figure 18.34

Schematic diagram of a circuit breaker showing only the thermal trip mechanism. (a) Normally the current path is closed. (b) When overheated, the bimetal strip bends, releasing the trip mechanism and allowing the contacts to open. Additional protection is provided by a magnetically activated trip mechanism that responds more quickly to overload than does the thermal trip.

Many electrical appliances are wired with three conductors and use a three-pronged plug on their power cord (Fig. 18.35). Homes and offices are wired with three conductors and have outlets to match the three-pronged plugs. The third conductor is added for safety. In normal operation, two of the wires provide the path for the current. The third wire, which is grounded directly to the earth, is usually connected to the case of the appliance, thereby ensuring that should a short circuit occur, the case remains at ground potential. Since the case remains at ground potential, a person touching the case will not receive a shock.

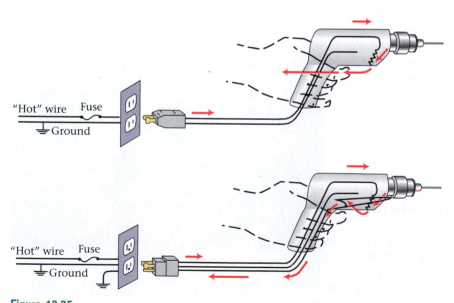

Figure 18.35

(a) If a short circuit develops in a metal-cased drill, the user can get a dangerous electric shock if the drill is wired with only a two-pronged plug. (b) If the case is grounded by the third wire of a three-pronged plug, the user is protected from harm when a short circuit develops.

ELECTRIC SHOCK

Electric shock is a hazard associated with all electric appliances and equipment. The danger of electrocution is not just for the high voltages of transmission lines. Unfortunately, people have been killed by ordinary house current at 120 V or by contact with industrial equipment at 40 or 50 V.

The important measure of shock intensity is not the voltage, but the amount of current that passes through the body (Fig. B18.3). Thus, any electric device using ordinary household voltages can potentially supply a fatal current. The amount of current can be determined from Ohm's law, but since the body's electric resistance varies enormously, it is not possible to give precise statements of safe or dangerous voltages. For example, the body's effective resistance depends largely upon the area of contact and the condition of the skin. But skin resistance may vary from about 500,000 Ω dry to as little as 500 Ω when wet.

The hazards of electric shock depend not only on the amount of current involved, but also on the path of the current. A current passing through your arm from fingertip to elbow may produce a painful and unpleasant shock, but the same current passing from one hand to the other through your chest may be fatal.

Electric current can damage the body in three distinct ways: (1) it may subject the body to intense heat and cause burns; (2) it may disrupt the proper functioning of the nervous system and heart; and (3) it may cause the muscles to twitch uncontrollably. Currents as low as 20 mA may cause difficulty in breathing, and at 75 mA breathing may stop completely. Currents between about 100 and 200 mA result in ventricular fibrillation of the heart, which means an uncoordinated and uncontrolled twitching of the heart muscles. The resulting loss of pumping action is fatal. At still greater currents, the heart may stop completely without going into fibrillation. Under such conditions, the chance of survival may actually be improved, since the heartbeat can be more easily restored from being stopped than from fibrillation. The defibrillators used in medical emergencies (Fig. B18.4) apply a large momentary voltage to the body to stop the heart and facilitate the restoration of the normal heart rhythm.

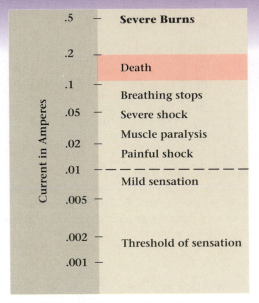

Figure B18.3 Electrical shock hazard at various values of current.

The best medicine for electric shock is prevention. Have respect for electricity at all voltages. Be cautious and follow normal safety procedures when working with electrical equipment.

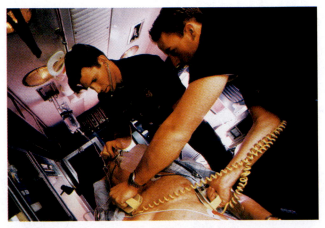

Figure B18.4 Paramedics working inside an ambulance attempt to revive a man by applying a jolt from a defibrillator.

Summary

Useful Concepts

■ Electric current is the rate at which electric charge flows and is defined by

$$I \equiv \frac{\Delta q}{\Delta t}.$$

■ Electric resistance is defined by

$$R \equiv \frac{V}{I}.$$

■ A resistive material is said to obey Ohm's law if current is proportional to the potential difference across a sample of that material:

$$V = IR.$$

■ The resistance of a specific piece of wire is

$$R = \rho \frac{L}{A}.$$

■ The power dissipated in an electric circuit is

$$P = IV = \frac{V^2}{R} = I^2 R.$$

■ Kirchhoff's voltage rule states that the algebraic sum of the potential differences around a closed conducting loop must be zero.

■ Kirchhoff's current rule states that the net current entering into any junction must be zero.

■ The equivalent resistance R_s of a number of resistors connected in series is

$$R_s = R_1 + R_2 + R_3 + \cdots$$

■ The equivalent resistance R_p of a number of resistors connected in parallel is

$$\frac{1}{R_p} = \frac{1}{R_1} + \frac{1}{R_2} + \frac{1}{R_3} + \cdots$$

■ Capacitors connected in parallel have an equivalent capacitance of

$$C_p = C_1 + C_2 + C_3 + \cdots$$

■ Capacitors connected in series have an equivalent capacitance given by

$$\frac{1}{C_s} = \frac{1}{C_1} + \frac{1}{C_2} + \frac{1}{C_3} + \cdots$$

■ The terminal potential difference (TPD) of a battery with internal resistance r connected across an external resistance R is

$$\text{TPD} = \frac{R}{r + R}\mathscr{E}.$$

Important Terms

You should be able to write the definition or meaning of each of the following:

battery	resistors
electric current	resistivity
direct current	Joule's law
ampere	ground potential
electric circuit	short circuit
emf	open circuit
resistance	Kirchhoff's rules
ohm	terminal potential difference
Ohm's law	

Conceptual Questions

18.1 Why are the 1.5-V D cells commonly used in flashlights so much larger than the 9-V batteries commonly used in small radios?

18.2 What advantage is there to arranging cells in series? In parallel?

18.3 What is the difference between resistance and resistivity?

18.4 Can you show that the resistance of two resistors joined in parallel is always less than the resistance of the smaller of the two resistors?

18.5 Compare electric resistivity with thermal resistivity. (The thermal resistivity is the reciprocal of the thermal conductivity described in Chapter 11.) Are good electric conductors also

good thermal conductors? Are good thermal conductors always good electric conductors? Give some examples.

18.6 Incandescent lamps usually burn out just as they are turned on. Can you explain why this is so?

18.7 How does a fuse or circuit breaker act to protect electrical equipment?

18.8 Some electrical appliances, notably TV sets, come with a two-pronged, polarized electric plug, which fits into an electric outlet in only one way. How do the plugs accomplish this polarization, and what is the purpose of having them do so?

18.9 Three-way lamps allow you to control the light level by means of a switch. A typical three-way light bulb is rated

50-100-150 W. How is such a bulb constructed? How many filaments must there be and how are they connected?

18.10 Figure 18.12 shows the temperature dependence of the resistivity of several metals. Explain why each of these metals is or is not a suitable choice of material for the temperature sensor in a resistance thermometer.

18.11 Explain why it is dangerous to reach into the back of a TV set even when it has been turned off and unplugged.

Problems

Section 18.1 Electric Current and Electromotive Force

18.1 A steady electric current of 0.50 A flows through a wire. (a) How many coulombs of charge pass through the wire per second? (b) How many per minute?

18.2 A charge of 1020 C was passed through a wire in 5.50 min. What was the average electric current during that interval?

18.3 A charge of 833 C was passed through a wire in 1.00 min. What was the average electric current during that interval?

18.4 How long does it take for 67 C to pass a given point in a wire that carries an electric current of 0.80 A?

18.5 A charge corresponding to one electron for every person in the United States passes a point in 10.0 ms. What is the resulting current? (Take the population to be 260 million people. The charge on an electron is 1.60×10^{-19} C.)

Section 18.2 Electric Resistance and Ohm's Law

Hints for Solving Problems

Remember that potential is measured *across* a resistor or other circuit component, but current passes *through* the element. Assume resistors obey Ohm's law unless stated otherwise.

18.6 What is the potential difference across a 220-Ω resistor when a current of 3.50 A flows through it?

18.7 What is the current drawn by a 470-Ω resistor when a potential difference of 25.0 V is maintained across it?

18.8 A 40-W electric lamp draws a current of 0.33 A when operated with a potential difference of 120 V. What is the resistance of the lamp?

18.9 A three-cell flashlight draws a current of 0.60 A. What is the operating resistance of the light bulb if each cell provides a potential of 1.0 V when delivering this current?

18.10 A piece of Nichrome wire passes a current of 0.853 A when a potential of 1.64 V is applied. What is the resistance of the wire?

18.11 What is the resistance of a resistor through which 8.0×10^4 C flow in one hour if the potential difference across it is 12 V?

18.12 The current through an electronic device is measured for several different voltages applied across the device. When the potential difference is 0 V, 0.50 V, and 0.75 V, the current is 0 A, 0.010 A, and 0.015 A, respectively. Does the device obey Ohm's law? (*Hint:* Make a graph of the data and examine it.)

18.13 The current through an electronic device is measured for several different voltages applied across the device. When the potential difference is 1.7 V, 7.6 V, 20 V, and 45 V, the current is 0.075 A, 0.15 A, 0.22 A, and 0.35 A, respectively. Does the device obey Ohm's law? (*Hint:* Make a graph of the data and examine it.)

18.14• Compute the current drawn by the 60-W lamp of Fig. 18.36 at 20-V intervals. Draw a graph of I versus V.

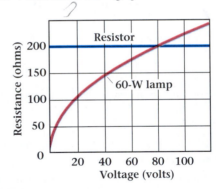

Figure 18.36

Problems 18.14 and 18.15: The resistance of a 60-W lamp is shown as a function of applied voltage. The horizontal line is the resistance of a 200-W resistor.

18.15• Compute the current drawn by the 200-Ω resistor of Fig. 18.36 at 20-V intervals and draw a graph of I versus V.

*Section 18.3 Resistivity

Hints for Solving Problems

Resistivity is a property of a material itself; resistance is a property of the material and its dimensions.

18.16 Calculate the resistance of a piece of 20-gauge copper wire 2.00 m long. The cross-sectional area of 20-gauge wire is 0.5176 mm^2.

18.17 A piece of 20-gauge wire one meter long has an electric resistance of 0.19 Ω. Calculate its resistivity and identify the composition of the wire from the values given in Table 18.1. The cross-sectional area of 20-gauge wire is 0.5176 mm².

18.18 Calculate the electric resistance of an iron rod 2.00 m long, assuming that its cross-sectional area is 0.90 mm².

18.19 Calculate the resistance of a Nichrome wire 0.50 mm in diameter and 1.25 m long.

18.20● The light-duty extension cords that are routinely sold for household use are made from 18-gauge copper wire. Cords for higher current usage are often made from 16-gauge wire. Compute the resistances of wires 5.00 m long. The cross-sectional area of 18-gauge wire is 0.8231 mm², and that of 16-gauge wire is 1.309 mm².

18.21● When aluminum is used for electrical wire, it is common to use a larger-diameter wire than would be necessary for copper. Compare the resistance of 16-gauge aluminum wire with that of 18-gauge copper wire of equal length by finding the ratio of their resistances. The cross-sectional area of 18-gauge wire is 0.8231 mm², and that of 16-gauge wire is 1.309 mm².

18.22● Two cylindrical bars, each with diameter of 2.30 cm, are welded together end to end. One of the original bars is copper and is 0.370 m long. The other bar is iron and is 0.185 m long. What is the resistance between the ends of the welded bar at 20°C?

Section 18.4 Power and Energy in Electric Circuits

Hints for Solving Problems

Remember to use SI units in your calculations. The SI unit of current is the ampere (A), of resistance the ohm (Ω), of power the watt (W), and of energy the joule (J). The kilowatt-hour (kWh) is often used as a unit for energy, but it is not an SI unit.

18.23 (a) How much current is drawn by a 150-W lamp operating at its rated voltage of 120 V? (b) What is the resistance of the lamp when operating?

18.24 A flashlight lamp connected to a battery that provides 1.4 V draws a current of 0.10 A. What electric power is used by the lamp?

18.25 The label on a toaster reads 800 W at 120 V. How much current does it draw?

18.26 A drilling machine operates with an electric power consumption of 840 W. How much does it cost to operate the machine continuously for eight hours if the electricity costs $0.080/kWh?

18.27 A refrigerator is equipped with a motor that draws 100 W but operates only 25% of the time. What is the cost of operating the refrigerator for 30 days if electricity costs $0.080/kWh?

18.28 A 150-W street lamp is operated for 12 hours a day. How much energy does it take to operate the lamp for 30 days? Express your answer in kilowatt-hours and in joules.

18.29 How much does it cost to operate a 100-W lamp 8 hours a day for 30 days if electricity costs $0.075/kWh?

18.30● A 100-Ω resistor is rated at 1.00 W maximum power capacity. (a) What is the maximum voltage that can be applied across the resistor without exceeding its maximum power rating? (b) What is the current at this voltage?

18.31● (a) What is the operating resistance of a lamp rated 40 W at 130 V? (b) How does that compare with the "cold" resistance of 31 Ω measured at very low voltage? (c) What does this calculation tell you about the relationship between resistance and temperature for the material of the filament?

18.32● A slide projector operating on a 120-V circuit has a 75-W lamp and a 1/256-hp motor fan. How long can the projector and fan run before using one cent's worth of electrical energy if electricity costs $0.076/kWh?

Sections 18.5 and 18.6 and *18.7 Short Circuits and Open Circuits; Kirchhoff's Rules and Simple Resistive Circuits; Application of Kirchhoff's Rules

Hints for Solving Problems

Remember to draw a circuit diagram for each problem when you begin working it. The voltage across resistors in parallel is the same; the current through resistors in series is the same.

18.33 Two identical 100-Ω resistors are joined together as shown in Fig. 18.37. (a) What is the current through each when the switch S is open? (b) What is it when S is closed?

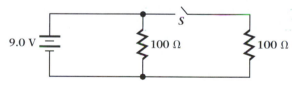

Figure 18.37
Problem 18.33.

18.34 A 100-Ω resistor is joined in parallel with a 47-Ω resistor. What is the equivalent resistance?

18.35 Three resistors of 100, 47, and 33 Ω are joined together in parallel. What is the equivalent resistance?

18.36 Two identical 100-Ω resistors are joined together as shown in Fig. 18.38. (a) What is the current through each resistor when the switch S is open? (b) When S is closed?

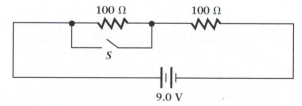

Figure 18.38
Problem 18.36.

18.37 Find the resistance of the network shown in Fig. 18.39.

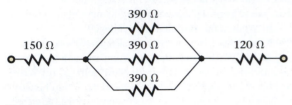

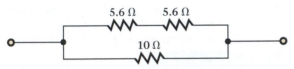

Figure 18.39
Problem 18.37.

18.38 Find the resistance of the network shown in Fig. 18.40.

Figure 18.40
Problem 18.38.

18.39• Three 47-Ω resistors and one 15-Ω resistor are joined in series across a 9.0-V battery. (a) What is the current through the resistor? (b) Find the voltage drop across the 15-Ω resistor.

18.40• Explain the operation of the common household circuit shown in Fig. 18.41. The resistor symbol represents a lamp. Consider all combinations of switches S_1 and S_2. Both switches have two positions A and B. For each switch, position A is connected to the "hot" side of the household voltage; position B is connected to the "ground." How is such a circuit used?

Figure 18.41
Problem 18.40.

18.41• A 100-Ω resistor is joined in series with a 33-Ω resistor. (a) If the circuit has a current of 340 mA, what is the voltage applied? (b) What voltage drop appears across the 100-Ω resistor?

18.42• A Wheatstone's bridge (Fig. 18.42) is used to measure resistance. When the bridge circuit is balanced, there is no current through the meter G and the voltage from point A to point C is zero. Verify that the current through the meter vanishes when $R_x = R_2R_3/R_1$.

18.43• A Wheatstone's bridge similar to Fig. 18.42 is balanced for an unknown resistor R_x. What is R_x if $R_1 = 100.0$ Ω, $R_2 = 20.00$ Ω, and $R_3 = 18.31$ Ω? (*Hint:* Use the results of Problem 18.42.)

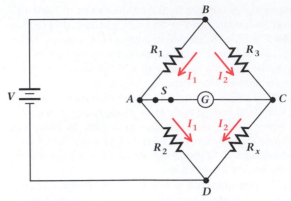

Figure 18.42
Problems 18.42, 18.43, and 18.72.

18.44•• Two resistors are connected in parallel across an ideal 12.0-V battery. Resistor A has a value of 24.0 Ω and resistor B carries a current of 0.250 A. (a) What is the potential difference across each resistor? (b) What is the current in resistor A? (c) What is the resistance of B? (d) What is the total power dissipated in the circuit?

Section 18.8 Capacitors in Combination

Hints for Solving Problems

Capacitors added in series have the same charge; capacitors added in parallel have the same voltage across them.

18.45 Three capacitors of 1.5, 2.0, and 5.0 μF are joined together in parallel. What is their equivalent capacitance?

18.46 Three capacitors of 2.0, 5.0, and 20 μF are joined together in parallel. What is their equivalent capacitance?

18.47 Capacitors of 10, 20, and 50 μF are joined together in series. What is their equivalent capacitance?

18.48 Three capacitors of 1.0, 1.5, and 5.0 μF are joined together in series. What is their equivalent capacitance?

18.49• A radio capacitor is made of metal plates, each with an area of 6.0 cm^2. The plates are arranged to make 15 parallel-plate capacitors wired in a parallel configuration. (a) What is the capacitance of the capacitor if the separation between plates is 0.50 mm? (b) What is the capacitance if the plates are rotated so that the effective area of overlap is 4.0 cm^2.

18.50• Calculate the effective capacitance of the capacitor network shown in Fig. 18.43.

18.51• Calculate the effective capacitance of the network shown in Fig. 18.44.

18.52• Calculate the voltage across the 20-μF capacitor in the circuit of Fig. 18.45. The battery voltage is 9.0 V.

18.53• A student has three capacitors of 15 μF each. How many different combinations of capacitance can he make and what are their values?

18.54• A 15-μF capacitor is joined in series with a 25-μF capacitor and a 12-V battery. (a) What is the potential difference

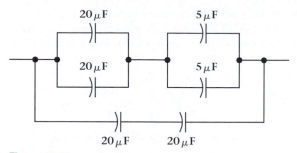

Figure 18.43
Problem 18.50.

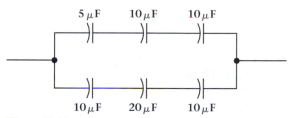

Figure 18.44
Problem 18.51.

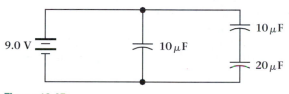

Figure 18.45
Problem 18.52.

across each capacitor? (b) What is the charge on each capacitor? (c) How much energy is stored by each capacitor?
18.55• Suppose the two capacitors of Problem 18.54 were joined in parallel across the battery. (a) What would be the potential difference across each? (b) What would be the charge on each capacitor? (c) What would be the energy stored on each?
18.56• Calculate the voltage across the 10-μF capacitor in the circuit of Fig. 18.46. The battery voltage is 12.0 V.

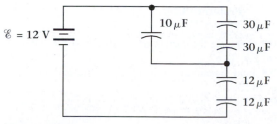

Figure 18.46
Problem 18.56.

Section 18.9 Internal Resistance of a Battery

18.57 A flashlight operates with three D cell batteries, each with an emf of 1.5 V. When operating, the flashlight draws a current of 0.60 A and the terminal voltage of the three-cell combination drops to 2.8 V. What is the internal resistance of each cell when delivering this current?
18.58• (a) Find the terminal potential difference of the battery in Fig. 18.47, given that the internal resistance r is 375 Ω and the load resistance R_L is 1500 Ω. (b) What is the TPD if the load becomes 750 Ω? (c) How much power is delivered to the 750-Ω load?

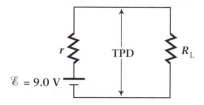

Figure 18.47
Problem 18.58.

18.59• A 9.0-V battery delivers 170 mA to a 47-Ω load. (a) What is the internal resistance of the battery? (b) What is its terminal potential difference when joined to this load?
18.60•• A certain battery has a terminal potential difference of 9.60 V when connected across a 200-Ω load and a terminal potential difference of 10.30 V when connected across a 300-Ω load. What is the emf of this battery?

*Section 18.10 Home Power Distribution

18.61 A particular household circuit is fused for 15 A at 120 V. (a) Can the circuit carry a 1200-W blow dryer? (b) Can it carry two dryers at the same time?
18.62 A single household circuit operates at 120 V and is protected by a 15-A circuit breaker. How many 60-W lamps can be operated simultaneously without tripping the circuit breaker?
18.63• A household circuit is wired for 120 V with a 20-A circuit breaker. (a) Can a 1500-W toaster, a 2200-W griddle, a 960-W iron, and a 300-W lamp be operated at the same time on this circuit? (b) Can any of them be operated simultaneously? If so, which ones?
18.64• An 1100-W iron, a 1200-W frying pan, and a lamp are all connected to a 120-V household circuit fused for 20 A. (a) What is the maximum wattage lamp that can be used simultaneously with the iron and the frying pan? (b) What would happen if a 60-W lamp were used? (c) A 200-W lamp?

Additional Problems

18.65• Two conducting wires of the same material are to have the same resistance. One wire is 20 m long and 0.40 mm in diameter. If the other wire is 0.30 mm in diameter, how long should it be?

18.66• Two cylindrical bars, each with diameter of 2.30 cm, are welded together end to end. One of the original bars is copper and is 0.470 m long. The other bar is iron and is 0.125 m long. What is the resistance between the ends of the welded bar at 20°C?

18.67• An elevator in a 20-story building is used to raise a 7200-N load 50 m. How much will it cost to raise the load if electricity costs $0.078/kWh? Assume that the elevator system is 50% efficient—that is, that the energy expended in raising the load is half of the total energy consumed.

18.68• A 1500-Ω resistor is rated at 2.0 W maximum power capacity. (a) What is the maximum voltage that can be applied across the resistor without exceeding its maximum power rating? (b) What is the maximum current?

18.69• A 120-V motor draws a current of 2.0 A while lifting a load at a speed of 0.65 m/s. If the efficiency for transforming electrical energy into mechanical energy is 62%, what mass is being lifted?

18.70• What is the current through (a) the 10-Ω resistor of Fig. 18.48 and (b) the 20-Ω resistor?

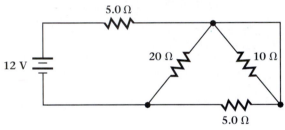

Figure 18.48
Problem 18.70.

18.71• Find the equivalent resistance of the network shown in Fig. 18.49.

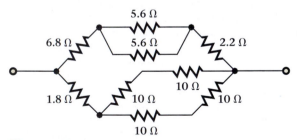

Figure 18.49
Problem 18.71.

18.72• The meter of a Wheatstone's bridge (Fig. 18.42) is disconnected by opening switch *S*. If the resistor values are $R_1 = 10.0\ \Omega$, $R_2 = 5.00\ \Omega$, $R_3 = 20.0\ \Omega$, and $R_x = 9.00$, what is the potential difference between point *A* and point *C* if the battery potential is 12.0 V?

18.73• Identical 100-Ω resistors are joined together as shown in Fig. 18.50. What is the current through each resistor (a) when the switch *S* is open and (b) when it is closed?

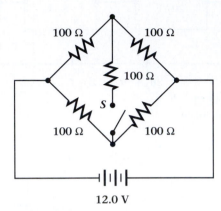

Figure 18.50
Problem 18.73.

18.74•• Two resistors are connected in series across an ideal 12.0-V battery. Resistor *A* has a value of 24.0 Ω and the potential difference across resistor *B* is 3.60 V. (a) What is the potential difference across *A*? (b) What is the current in the resistors? (c) What is the resistance of *B*? (d) What is the total power dissipated in the circuit?

18.75•• (a) Calculate the potential difference between points *A* and *B* in Fig. 18.51. (b) Calculate the power delivered to the 15-Ω resistor.

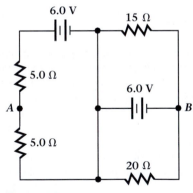

Figure 18.51
Problem 18.75.

18.76•• Calculate the current in the 10-Ω resistor in the network of Fig. 18.52.

18.77•• Find the current in the 68-Ω resistor in the network of Fig. 18.53.

18.78•• Find the currents I_1, I_2, and I_3 in the network of Fig. 18.54.

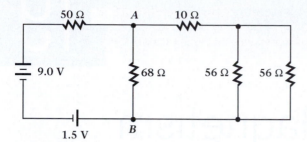

Figure 18.52
Problem 18.76.

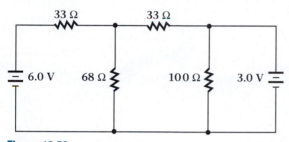

Figure 18.53
Problem 18.77.

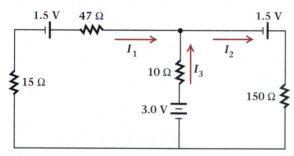

Figure 18.54
Problem 18.78.

18.79•• Three resistors are joined together across a 24.0 V battery (Fig. 18.55). The voltage drop across resistor R_1 is 8.0 V, the current I_2 through resistor R_2 is 0.20 A, and the power dissipated in resistor R_3 is 2.56 W. (a) What is the value of each resistor? (b) What is the current through each resistor? (c) What is the power dissipated in each resistor? Find (d) the total resistance of the network, (e) the current drawn from the battery, and (f) the total power used by the circuit.

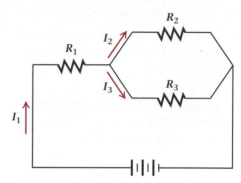

Figure 18.55
Problem 18.79.

Magnetism

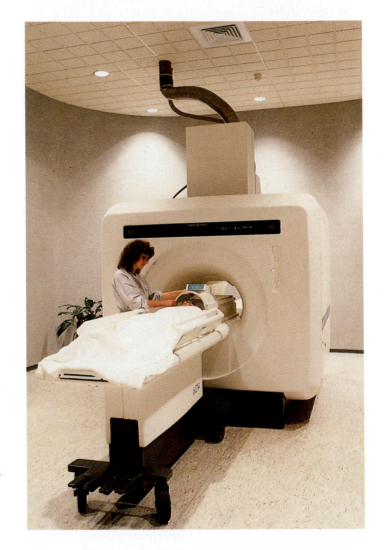

Magnetism has been known for thousands of years, dating back to the discovery recorded by the ancient Greeks that the naturally occurring black mineral called lodestone could attract iron objects.* Very little else was recorded until 1269, when Petrus Peregrinus de Maricourt, writing to a friend, gave a clear description of the magnetic compass, which had been developed during the eleventh and twelfth centuries. De Maricourt observed that spherically shaped lodestones have two special points called "poles," which he located by laying bits of iron wire on the stone. The lines drawn along these wires intersected at two points that he named the north pole and the south pole in analogy with the celestial sphere. He

---

*Lodestone is found in most parts of the world and is usually in the form of magnetite (Fe_3O_4). Large deposits were found in Asia Minor, near an ancient town named Magnesia. These rocks became known as magnesia rocks and later as magnets.

also observed the attraction of unlike poles and repulsion of like poles. These ideas were incorporated in the work of William Gilbert, who published the results of his own experiments in 1600. Gilbert was the first to realize that the earth itself acts as a magnet, and he clearly stated the differences and similarities known at that time between electricity and magnetism.

By the beginning of the nineteenth century, a large amount of experimental information about the nature of electricity and magnetism had been accumulated. The discoveries and ideas of Gilbert, Franklin, Coulomb, Volta, and many others were well known. The similarities between electrical and magnetic attraction, the knowledge that compass needles on ships struck by lightning sometimes changed polarity, and experiments like Franklin's magnetization of needles by passing an electric discharge through them—all hinted at a possible connection between electric and magnetic behavior. But magnetism and electricity were considered to be two independent phenomena until early in 1820. In that year, Hans Christian Oersted, a professor of physics at the University of Copenhagen, discovered that an electric current could indeed affect a magnet.

Today, most people are familiar with magnetic compasses and permanent magnets. Electromagnets, which are perhaps less familiar, are found in such places as motors, tape recorders, and power plants. We now recognize the connection between electricity and magnetism as so fundamental that we no longer refer to the two forces of electricity and magnetism, but instead speak of a single force of *electromagnetism*. ■

19.1   Magnets and Magnetic Fields

A compass needle is simply a slender magnet that is supported at its center so that it can rotate freely (Fig. 19.1). If there are no other magnets nearby, the needle lines up in an approximately north-south direction. The end of the needle that points toward the geographic north is the north-seeking pole, or simply the *north pole*. The other end of the needle is the *south pole*. Larger bar magnets exhibit the same behavior. For this reason, the ends of bar magnets are often marked as north and south poles.

If two bar magnets are brought near each other, they exert forces on each other. If the north pole of one is brought near the north pole of the other, the force between them is repulsive. Similarly, if the south pole of one is brought near the south pole of the other, the force is also repulsive. But if the north pole of one is brought near the south pole of the other, the force between them is attractive. Thus, we say that *like poles repel* but *unlike poles attract*.

In showing both attractive and repulsive behavior, magnetic poles are somewhat like electric charges. But don't take the analogy too far, because isolated magnetic poles do not exist. If a bar magnet is broken in two, you do not get isolated north and south poles; instead, you get two new magnets, each having a north and a south pole (Fig. 19.2). If these pieces are in turn broken in two, each

Figure 19.1

A compass needle is a small magnetic dipole supported at its center so that it can rotate.

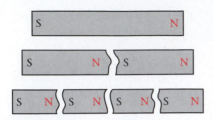

Figure 19.2
When a magnet is broken in two, each piece contains both a north and a south pole. Isolated magnetic poles (monopoles) are not observed to exist.

new piece contains both a north and a south pole. Magnets of this type are called **magnetic dipoles** because they have two poles.* Single magnetic poles, called monopoles, have not been observed. Although some theories suggest that isolated magnetic monopoles exist, no convincing experiments have verified their existence.

When a magnet is brought near a compass, the compass needle deflects. The amount by which the needle deflects depends on the location of the magnet relative to the compass. We explain this behavior by saying that a **magnetic field** exists in the region surrounding the magnet. This magnetic field exerts a torque on the compass needle, causing it to rotate into the direction of the field. (As we shall see in the next section, we can determine the magnitude of the field from the magnitude of the torque acting on the magnetic dipole.) Thus, the direction of the magnetic field vector **B** at any given point in space is defined as the direction indicated by the north pole of a compass needle when placed at that point. We can map out the lines of magnetic field around a bar magnet (or any other magnet) by placing many small compass needles in the neighborhood of the magnet and observing their orientations (Fig. 19.3a). We can then sketch the magnetic field by drawing lines tangent to the directions of the needles. Note that the field is directed away from a north pole and toward a south pole (Fig. 19.3b).

Some objects, such as iron paper clips and nails, do not usually attract each other. But when they are brought near a magnet, they become strongly attracted to it and to each other. We say they are ferromagnetic† (from the Latin word *ferrum,* meaning iron). In the presence of the magnetic field, they become magnetically polarized; that is, they behave as tiny magnets themselves. For this reason, when iron filings are sprinkled on a sheet of paper held close to a magnet, they orient themselves and clump together along the lines of magnetic field, revealing the shape of the field (Figs. 19.3c and 19.4).

As Gilbert first pointed out, the earth itself is a magnetic dipole. It is the earth's magnetic field that causes compass needles to line up along the approximately north-south direction. However, since the north pole of a compass is attracted toward the geographic north, the magnetic pole that is located in that

Figure 19.3
(a) The direction of the magnetic field near a magnet is shown by the directions of small compass needles. (b) Representation of the field lines around a bar magnet. (c) Iron filings reveal the shape of the field near a bar magnet.

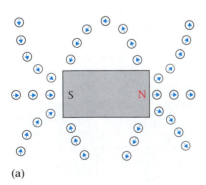

(a)

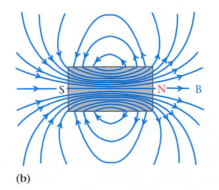

(b)

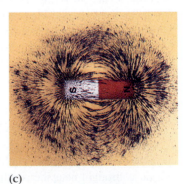

(c)

---

*Dipole magnets have the simplest configuration, but other arrangements of poles are possible, including quadrupole, octapole, and so on, all being multiples of dipoles.
†Ferromagnetism is discussed in greater detail in Section 19.10.

(a)

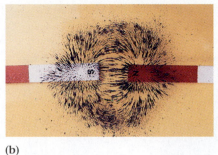

(b)

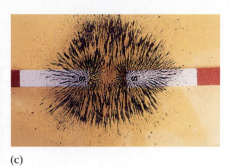

(c)

Figure 19.4

Iron filings reveal the shape of the magnetic field around (a) a horseshoe magnet and (b) two bar magnets (c).

region is in reality a south magnetic pole (Fig. 19.5). Note also that the earth's magnetic poles do not coincide exactly with its geographic poles. For this reason, the magnetic north indicated by a compass is not quite the same as the true geographic north.

Oersted's Discovery: Electric Current Produces Magnetism

Oersted's discovery that an electric current produces magnetism is thought to be the only case of a major scientific discovery being made during a lecture to students. While demonstrating an electrical experiment as part of a lecture at the University of Copenhagen, Oersted positioned a wire near to and parallel with a compass needle. When he passed a current through the wire, the compass acted as if a magnet had been brought nearby.

Using an improved experimental setup, similar to that shown in Fig. 19.6(b), Oersted conducted an extensive series of experiments. He connected a battery of some 20 copper-zinc cells to a straight section of the wire, which was held horizontally above the compass needle and parallel to it. When Oersted closed the circuit, the compass needle rotated so that the end nearest the negative side of the battery deflected to the west. The effect decreased as he moved the wire away from the compass. However, no noticeable change in the effect occurred when glass, metal, wood, water, or stone was placed between the wire and the compass. In July 1820, Oersted announced his discovery in a four-page pamphlet.

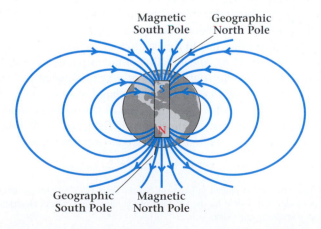

Figure 19.5

The earth behaves like a giant magnetic dipole with a south magnetic pole near the north geographic pole. The magnetic dipole is tilted approximately 11° from the earth's rotation axis.

Figure 19.6

(a) Hans Christian Oersted (1777–1851). (b) Experimental setup for Oersted's experiment. A current through the wire caused the compass needle to deflect.

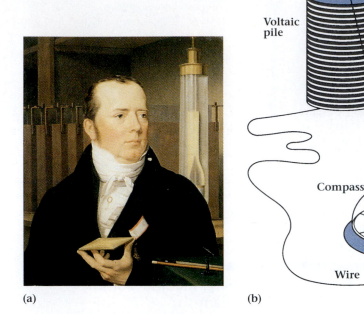

(a)   (b)

In Oersted's experiment, the presence of an electric current in the wire produces a torque, causing the magnetic compass needle to rotate. Since the compass needle is a magnetic dipole (that is, two magnetic poles separated by a fixed distance), we can describe it with a **magnetic dipole moment μ** similar to the electric dipole moment **p** defined in Chapter 16. The deflection of the compass needle is due to the interaction of the magnetic moment with the magnetic field **B** (Fig. 19.7). The behavior of a magnetic dipole in a magnetic field can be described in exactly the same way that we described the behavior of an electric dipole in an electric field. The torque on the compass needle is proportional to the product of its magnetic dipole moment **μ** and the magnetic field **B**,

$$\tau = \mu B \sin\,\theta, \tag{19.1}$$

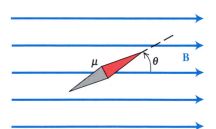

Figure 19.7

A magnetic field **B** exerts a torque on a magnetic dipole μ given by $\tau = \mu B \sin\,\theta$.

where θ is the angle between **μ** and **B**, in analogy with Eq. (16.6). The SI unit for magnetic field B is the **tesla,** abbreviated as T,* and the unit of magnetic moment is $A \cdot m^2$. The origin of these units will be explained later in this chapter.

The magnetic field lines due to a current-carrying wire form concentric circles about the wire (Fig. 19.8). The direction of the magnetic field about the wire is given by a right-hand rule (Fig. 19.9) that was introduced by André-Marie Ampère (1775–1836): *If the thumb of the right hand is taken as the direction of current, the curled fingers point in the direction of the magnetic field about that current.* If we trace out the lines of the magnetic field, we see that they are continuous and do not terminate on magnetic charges or poles the way electric field

---

*The unit for magnetic field is named in honor of Nikola Tesla (1856–1943), an engineer and inventor who was largely responsible for the adoption of the alternating current electrical distribution system that is in worldwide use today.

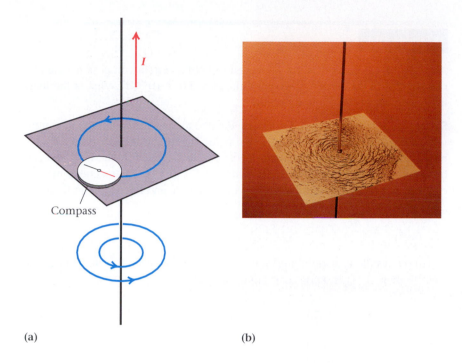

(a)   (b)

Figure 19.8
(a) Magnetic field lines around a long straight wire carrying an electric current form concentric circles about the wire. (b) Iron filings align with the magnetic field, indicating its circular symmetry.

lines terminate on electric charges. This observation implies that isolated magnetic charges or poles (monopoles) do not exist independent of other poles in the same way that we have isolated positive and negative electric charges. It also leads us to a statement for magnetic flux similar to Gauss's law for electrostatics.

We define the **magnetic flux** $\Delta\phi_m$ through a small element of surface as the product of the magnitude of the magnetic field B, the magnitude of surface area ΔA, and the cosine of the angle between the direction of the field and the normal to the surface (Fig. 19.10). Thus, the element of flux is

$$\Delta\phi_m = B\ \Delta A \cos\theta. \tag{19.2}$$

The net magnetic flux is the sum over the entire surface of all elements $\Delta\phi_m$. This magnetic flux is analogous to the electric flux defined in Chapter 16.

If the magnetic field lines do not start or stop at any point in space, but form closed loops, then any magnetic field line entering a closed surface must also leave that surface, so the net flux is zero. Thus, we can make a statement for magnetic fields similar to Gauss's law for electrostatics: **The net magnetic flux through any (real or imaginary) closed surface is zero.** This statement is known as **Gauss's law for magnetism.**

As mentioned in Section 19.1, some theories of modern physics predict the existence of magnetic monopoles. Because these theories have been successful in many other areas, it is important to test them in as much detail as possible. Accordingly, several ingenious, highly sensitive experiments have been carried out in the search for magnetic monopoles. To date, no reproducible experiments have been conducted that verify the existence of monopoles. Thus, the present consensus is that isolated magnetic poles do not exist.

Figure 19.9
The direction of the magnetic field is given by the right-hand rule: When the thumb of the right hand points in the direction of the conventional (positive) current, the fingers curl around the wire in the direction of the magnetic field.

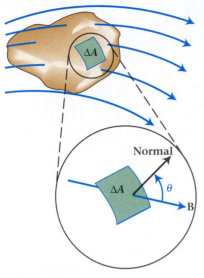

Figure 19.10
The magnetic flux through an element of surface ΔA is the product of the magnitude of magnetic field B, the area ΔA, and the cosine of the angle between the surface normal and the direction of the field. The total magnetic flux is the sum of these products over the entire surface.

Example 19.1

A torque of 4.3×10^{-3} N · m is needed to hold a certain compass needle at right angles to the earth's magnetic field, $B = 5.0 \times 10^{-5}$ T. What is the magnetic moment of the compass needle?

Strategy The torque on the compass needle due to the earth's field is maximum when the needle is at right angles to the field. The torque needed to hold the needle in that position is equal and opposite to that applied by the field and can be computed from the defining relationship of Eq. (19.1). The magnetic moment can then be computed from knowledge of the torque and the magnetic field.

Solution The magnitude of the torque is

$$\tau = \mu B \sin \theta.$$

When the needle is at right angles to the field $\mathbf{B}$, the angle θ is 90° and the sine becomes 1. Then upon rearranging we get

$$\mu = \frac{\tau}{B} = \frac{4.3 \times 10^{-3} \text{ N} \cdot \text{m}}{5.0 \times 10^{-5} \text{ T}} = 86 \text{ A} \cdot \text{m}^2.$$

Discussion In computing the magnetic moment, we used the SI unit for torque (N · m) and the SI unit for magnetic field (T). The magnetic moment is then given in SI units, which are A · m^2.

Example 19.2

A rectangular sheet of paper 21.5 cm by 28.0 cm rests on a flat horizontal tabletop. Calculate the magnetic flux through the sheet of paper due to the earth's magnetic field at a location where the field has a magnitude of 5.31×10^{-5} T and is directed downward at an angle of 37° from the horizontal (Fig. 19.11).

Strategy The magnetic field may be considered uniform over the sheet of paper. We may choose the direction of the surface area of the paper to be downward so that the angle θ between the direction of $\mathbf{B}$ and the direction of $\mathbf{A}$ is 53°.

Solution The magnetic flux through the paper may be found using the defining equation

$$\Delta \phi_m = B \, \Delta A \cos \theta.$$

Because the field is uniform and the paper is flat, the total flux through the paper is just

$$\phi_m = BA \cos \theta.$$

The area of the paper is the product of its length and width. Thus, the flux becomes

$$\phi_m = (5.31 \times 10^{-5} \text{ T})(0.215 \text{ m} \times 0.280 \text{ m})(\cos 53°)$$
$$= 1.92 \times 10^{-6} \text{ T} \cdot \text{m}^2.$$

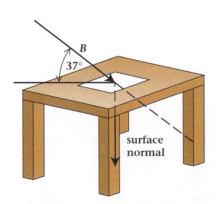

Figure 19.11
Example 19.2: The earth's magnetic field generates a magnetic flux through a horizontal sheet of paper.

Discussion The SI unit for magnetic flux of T · m^2 is given the name *weber* (abbreviated Wb). Thus, the flux computed here could be expressed as 1.92×10^{-6} Wb = 1.92 μWb.

MAGNETIC RESONANCE IMAGING

In the last two decades, a revolution has taken place in diagnostic medical imaging. New methods have been developed that provide a detailed look at structures deep within the human body without surgical invasion or ionizing radiation. One of the best of these methods is magnetic resonance imaging (MRI).

The nuclei of hydrogen atoms (protons) have magnetic dipole moments. When these protons are placed in a uniform magnetic field, most of their magnetic moments will align along the direction of the field since that alignment corresponds to the lowest energy state. If the protons are then exposed to high frequency radio waves at the resonant frequency, some will be excited to the higher energy state that corresponds to alignment opposite to the field direction. When that happens, power is absorbed from the incident radio beam. It is this resonance absorption that gives MRI its name.

When a subject containing hydrogen atoms is placed in a magnetic field, the resonance frequency of each hydrogen nucleus depends on the field that it experiences. If the field is not constant across the subject, but instead has a smoothly changing magnitude (a gradient), then the resonant frequency will depend on the position of the nucleus in the field. Thus, the magnetic resonance signal will give information about the position of the nuclear magnetic moments. By carefully controlling the gradient, a picture of the hydrogen concentration in the subject can be formed.

If the alignment of the nuclear magnetic moments is disturbed by a pulsed magnetic field, they return to their previous state in a characteristic time known as the relaxation time. The hydrogen nuclei in different tissues in the body have different relaxation times. By using the effects of these relaxation times on the resonance signal when generating the images, MRI can distinguish between different types of tissue even when the densities of hydrogen atoms are the same (Fig. B19.1). In some situations this provides a distinct advantage over x-ray methods that give poor images of soft tissues.

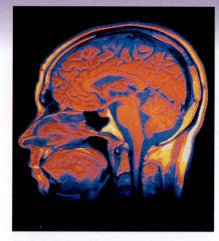

Figure B19.1 MRI image.

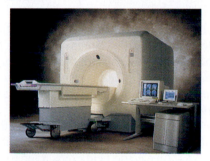

Figure B19.2 MRI magnet.

For the patient, the most visible part of the MRI system is the magnet that produces the large magnetic field (Fig. B19.2). Typically the magnet is made from several solenoidal coils wound about a common axis with a bore of about one meter. These coils are wound with superconducting wire and encased in a large insulating container filled with liquid helium, which acts as the refrigerant to keep the coils cooled below their superconducting transition temperature. These magnets are capable of fields up to 1.5 T with a homogeneity of one part in 10^6 over a volume of one cubic meter.

Additional coils are used to establish the field gradients in the directions as desired. By proper choice of gradients, a two dimensional slice of the subject can be imaged. These images can be taken in the sagittal, transverse, and coronal orientations (Fig. B19.3). Separate antenna coils are used for detecting the signals. The proper coil can improve the image resolution. For example, special coils are routinely used in order to get the best images of the head.

In a typical applied magnetic field of about 1.0 T, the resonant frequency of the radio waves that interact with the hydrogen nuclei in the human body is 42.6 MHz. Because the energy of these radio waves is so small, MRI is much safer than x radiation which can break up the chemical bonds holding molecules together within the body.

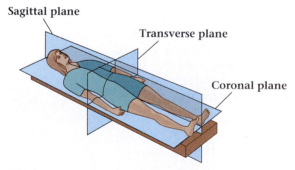

Figure B19.3 Three orthogonal views for the MRI image of a patient.

19.3 Magnetic Forces on Electric Currents

Oersted's experiment showed that a wire carrying an electric current exerts a force on a magnet. It follows from Newton's third law that a magnet must also exert a force on the current-carrying wire. For example, consider a current-carrying wire in a region of uniform magnetic field (Fig. 19.12). We observe that the wire experiences a magnetic force that is always perpendicular to the magnetic field and also perpendicular to the length of the wire in the field. Thus, the force is perpendicular to the motion of the electric charges in the wire—that is, to the current. If we change the direction of the wire relative to the direction of the field, we find that the force is maximum when the wire is perpendicular to the field and that the force is zero when the wire is parallel to the field.

The magnitude of the force on a section of wire of length l carrying a current I in the presence of a magnetic field $\mathbf{B}$ is given by

$$F = IlB \sin \theta, \qquad (19.3)$$

where θ is the angle between the direction of the current and the direction of the field. We emphasize that the force vector is perpendicular to both the current and the field. The direction of the force is that given by another right-hand rule: *When the right hand is held so that the fingers can be curled from the direction of the current into the direction of the magnetic field, the vector $\mathbf{F}$ points in the direction of the thumb* (Fig. 19.13).

Notice that the effect of the magnetic field on a current-carrying wire is quite different from the effect of an electric field on a charge, or a gravitational field on a mass. An electric field exerts a force on a charged particle that is parallel to the direction of the field (or opposite, if the charge is negative); a gravitational field exerts a force on a mass that is parallel to the direction of the gravitational field. But a magnetic field exerts a force on a current-carrying wire that is *perpendicular* to both the direction of the magnetic field and the direction of the current.

We can use Eq. (19.3) to define the units of the magnetic field. If the current is measured in amperes, length in meters, and force in newtons, then the

Figure 19.12

The magnetic force on a current-carrying wire in a magnetic field is perpendicular to the field **B** and to the length of the wire in the field. The force is maximum when the wire is perpendicular to the field.

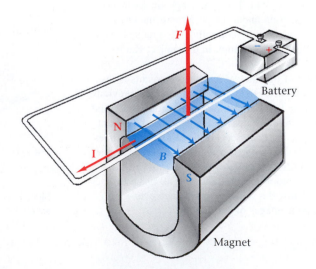

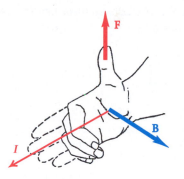

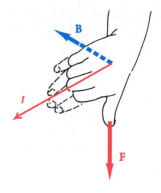

Figure 19.13
The direction of the magnetic force on a current-carrying wire is perpendicular to both the current *I* and the magnetic field **B**. When the right hand is held so that the fingers can be curled from the direction of *I* into the direction of **B**, the thumb points in the direction of the force.

magnetic field must have units of newtons per ampere-meter (N/A · m). This particular combination of units is given the name *tesla*. A field of one tesla is a relatively large magnetic field. For example, the magnitude of the earth's magnetic field near the earth's surface is $\approx 5 \times 10^{-5}$ T. Even the field near the end of a common bar magnet is only about 10^{-2} T, while the largest fields available from superconducting magnets are on the order of 20 T.

Example 19.3

A straight wire is placed between the poles of a laboratory magnet that produces a uniform field of 1.00 T over a region that is 0.250 m wide (Fig. 19.14). The wire is perpendicular to the direction of the field and, when connected to a source of electricity, carries a current. What must the current be if the field exerts a force of 9.81 N on the wire? (A force of 9.81 N is the weight of a 1.00-kg mass.)

Solution Equation (19.3) gives the force on the wire as

$$F = IlB \sin \theta.$$

Because we want to obtain the current required for a given force, we rearrange the equation to get

$$I = \frac{F}{lB \sin \theta}.$$

When we insert the numerical values of $F = 9.81$ N, $l = 0.250$ m, $B = 1.00$ T, and $\sin \theta = \sin 90° = 1$, we find the current:

$$I = \frac{9.81 \text{ N}}{(0.250 \text{ m})(1.00 \text{ T})(1.00)} = 39.2 \text{ A}.$$

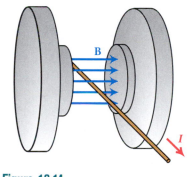

Figure 19.14
Example 19.3: A current-carrying wire in a magnetic field.

19.4   Magnetic Forces on Moving Charged Particles

As you know, a current in a wire consists of moving charges. Since a current-carrying wire may experience a force when placed in a magnetic field, it is not surprising that a moving charge that is not confined within a wire may also experience a force due to a magnetic field. Experiments confirm that this force does occur. The magnitude of this force is given by

$$F = qvB \sin \theta, \tag{19.4}$$

where q is the magnitude of the moving charge, v is its speed, B is the magnitude of the magnetic field, and θ is the angle between the velocity of the moving charge and the direction of the field. The factor $\sin \theta$ means that the force is proportional to the component of **v** perpendicular to **B**—that is, $v\perp = v \sin \theta$ (Fig. 19.15). The component of **v** parallel to **B** does not contribute to a force on the charge. The direction of the force is always perpendicular to both **v** and **B**. For a positive charge, the direction of the force is given by the same right-hand rule we used to find the force on a current-carrying wire because the direction of the velocity is the same as the direction of the current. However, if the charge is negative, the force is in the direction opposite to that given by the right-hand rule. Thus, the force exerted on a moving negative charge by a magnetic field is equal in magnitude but opposite in direction to the force exerted on a positive charge of the same magnitude moving with the same velocity. Note that Eq. (19.4) may also be used as the defining equation for the magnetic field, because F, q, v, and θ may be measured by independent means.

We can show that Eq. (19.3), for the force on a wire segment of length l carrying a current I perpendicular to a magnetic field **B**, is equivalent to Eq. (19.4), for the force on a single charge q moving with speed v perpendicular to a magnetic field **B**. Consider a wire segment of length l that contains a total moving charge Q (Fig. 19.16). The current due to the charge Q passing a given point in the wire is

$$I = \frac{Q}{t},$$

where t is the time for an individual charge to go a distance l. This time t depends on the speed v of the charge and is $t = l/v$. Consequently, the current is

$$I = \frac{Qv}{l}.$$

From Eq. (19.3), the force on a length l of wire with current I is

$$F_w = IlB = \frac{Qv}{l}lB = QvB,$$

Figure 19.15

The magnitude of the force on a particle of charge q moving with a velocity **v** in a magnetic field **B** is $F = qvB \sin \theta$. The force **F** is perpendicular to the plane containing **v** and **B** and is directed according to the right-hand rule. If the charge is negative, the particle experiences a force of the same magnitude in the opposite direction.

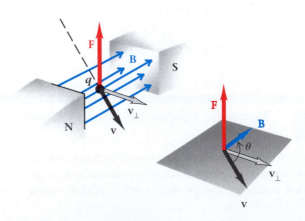

where Q is the total charge moving in that segment of wire at any instant. The total charge Q is the sum of all the individual moving charges q. Thus, the force on the wire is the sum of the forces on all of the individual charges, and the force on each charge is $F = qvB$, in accord with Eq. (19.4).

Suppose we inject a positively charged particle with mass m and charge q into a region of uniform magnetic field $\mathbf{B}$. If the particle's velocity $\mathbf{v}$ is perpendicular to $\mathbf{B}$, what is the particle's subsequent motion? According to Eq. (19.4), the particle experiences a force

$$F = qvB \sin \theta = qvB \sin 90° = qvB$$

because $\mathbf{v}$ is perpendicular to $\mathbf{B}$. The particle moves in accordance with Newton's laws of motion under the influence of this force. Thus, the particle undergoes an acceleration of magnitude

$$a = \frac{F}{m} = \frac{qvB}{m}.$$

Because the force is always perpendicular to both $\mathbf{v}$ and $\mathbf{B}$, the acceleration is also perpendicular to $\mathbf{v}$ and $\mathbf{B}$. Since the acceleration is always perpendicular to the velocity, the speed of the particle is unchanged and the path of the particle is curved, as we learned in Chapter 5. In fact, since the magnitude of the acceleration is constant and $\mathbf{a}$ is perpendicular to the instantaneous velocity, the particle's path is a circle (Fig. 19.17).

The centripetal acceleration required to keep the particle in a circular orbit is provided by the interaction of the moving charge with the magnetic field. If we equate the expression for centripetal acceleration with the acceleration due to the magnetic field, we get

$$a_c = \frac{v^2}{r} = \frac{qvB}{m},$$

where r is the radius of the particle's circular path. This equation can be rearranged to give

$$\boxed{r = \frac{mv}{qB}.} \tag{19.5}$$

Thus, the particle moves in a circular path whose radius r depends on the momentum mv of the particle, its charge, and the magnitude of the magnetic field.

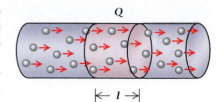

Figure 19.16
A wire segment of length l contains a total charge Q. The motion of those charges constitutes a current.

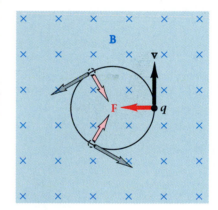

Figure 19.17
A magnetic field $\mathbf{B}$ causes a particle of charge q moving with a velocity $\mathbf{v}$ in a plane perpendicular to the field to move in a circular orbit. The crosses represent the tail-ends of magnetic field lines directed into the page.

▼

Problem-Solving Strategy

Magnetic Forces

1. The magnetic force on a moving charge (or a current) depends on the angle θ between the direction of the magnetic field and the direction of the velocity of the charge. In working problems, be sure to identify these two directions carefully so that you can determine θ.
2. Remember that the sign of the charge is important. The magnetic force on a moving negative charge is oppositely directed to the force on a positive charge moving in the same direction in the same field.

Example 19.4

A proton of mass 1.67×10^{-27} kg and electric charge 1.60×10^{-19} C is injected into a magnetic field $B = 18.5$ mT with a velocity of 1.55×10^5 m/s in a direction at right angles to the field. Compute the magnetic force on the proton and compare it with the weight of the proton.

Solution The magnitude of the force on the proton is given by Eq. (19.4),

$$F = qvB \sin \theta,$$

where q is the charge, v is the speed, B is the magnetic field, and θ is the angle between the velocity and the field. The direction of the velocity is perpendicular to the direction of the field, so $\theta = 90°$ and $\sin \theta = 1$. Upon inserting the numerical values, we find the force to be

$$F = (1.60 \times 10^{-19} \text{ C})(1.55 \times 10^5 \text{ m/s})(18.5 \times 10^{-3} \text{ T})(1)$$
$$= 4.59 \times 10^{-16} \text{ N}.$$

The gravitational force on the proton is

$$F = mg = (1.67 \times 10^{-27} \text{ kg})(9.81 \text{ m/s}^2) = 1.64 \times 10^{-26} \text{ N}.$$

The gravitational force is negligible compared with the magnetic force.

Example 19.5

A deuteron of mass 3.34×10^{-27} kg and charge 1.60×10^{-19} C is injected into a magnetic field of 5.50×10^{-2} T with a velocity of 1.84×10^5 m/s at an angle of $65°$ from the field direction. What is the subsequent motion of the deuteron?

Strategy The initial velocity of the deuteron can be resolved into components parallel and perpendicular to the magnetic field. The motion along the direction of the field is unchanged by the field. However, there is a force on the particle due to the velocity component perpendicular to the field. This force is at right angles to that component and to the field and would cause the particle to travel in a circle if the velocity component parallel to the field were zero. As a result, the actual path of the particle is a helix with a definite radius and a definite pitch, the distance traveled along the field during one complete rotation about the field.

Solution The velocity component parallel to the field direction is

$$v_\parallel = v \cos \theta = v \cos 65°.$$

The velocity component perpendicular to the field is

$$v_\perp = v \sin \theta = v \sin 65°.$$

The projection of the deuteron's motion perpendicular to the field is a circle of radius

$$r = \frac{mv_\perp}{qB} = \frac{(3.34 \times 10^{-27} \text{ kg})(1.84 \times 10^5 \text{ m/s})(\sin 65°)}{(1.60 \times 10^{-19} \text{ C})(5.50 \times 10^{-2} \text{ T})}$$

$$r = 6.33 \times 10^{-2} \text{ m} = 6.33 \text{ cm}.$$

The particle moves along the direction of the field with a constant speed $v_\parallel$. The distance traveled along the field during the time of one complete rotation perpendicular to the field is the pitch. The time for one rotation is the distance around a circle of radius r divided by the speed $v_\perp$:

$$T = \frac{2\pi r}{v_\perp} = \frac{2\pi m}{qB}.$$

Thus, the pitch becomes

$$\text{pitch} = v_\parallel T = (v \cos 65°)\left(\frac{2\pi m}{qB}\right)$$

$$= (1.84 \times 10^5 \text{ m/s})(\cos 65°)\frac{(2\pi)(3.34 \times 10^{-27} \text{ kg})}{(1.60 \times 10^{-19} \text{ C})(5.50 \times 10^{-2} \text{ T})}$$

$$= 0.185 \text{ m} = 18.5 \text{ cm}.$$

Discussion The deuteron moves along a helical path in the magnetic field (Fig. 19.18). The radius of the helix is 6.33 cm, and its pitch is 18.5 cm. Helical motion of this type is routinely observed for subatomic particles moving in the field of large magnets in particle detectors used in particle physics research. Beyond the laboratory, we find charged particles traveling in helical paths around lines of the earth's magnetic field at high altitudes, where they form the Van Allen radiation belts. Near the poles, these particles collide with air molecules in the upper atmosphere to cause the Aurora Borealis in the northern hemisphere and the Aurora Australis in the southern hemisphere.

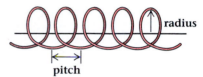

Figure 19.18
Example 19.5: A deuteron moves in a helical path in a magnetic field. In this instance the helix has a radius of 6.33 cm and a pitch of 18.5 cm.

*19.5 The Cyclotron

Electrostatic accelerators, such as the Van de Graaff accelerator described in Chapter 17, can generate beams of charged particles with very high velocities. These beams of energetic charged particles have many uses in research, industry, and medicine. Once such a beam emerges from an accelerator, we can change its direction most easily by sending it through a region of intense magnetic field. By adjusting the magnitude of this field, we can control the amount of deflection of the beam. Thus, beam-steering magnets may be used to direct the beam to different targets.

Magnetic fields also play a direct role in the operation of many accelerators, including the **cyclotron,** invented by Ernest O. Lawrence (1901–1958). In 1930, Lawrence and M. S. Livingston (1905–1986) constructed an accelerator that used a magnetic field to bend the paths of the particles into nearly circular orbits. A simplified diagram of a cyclotron (Fig. 19.19) includes two semicircular cavities (called "dees" because of their shape), an ion source S for producing the charged particles, and an oscillating-voltage source. The magnets that produce the magnetic field are not shown.

A properly timed oscillating voltage source makes the electric potential difference between the dees change direction each time the charged particles pass through the gap between them, so that the particles encounter an accelerating voltage each time they pass through the gap. A particle with speed v moves in a circular path of radius r within one of the dees. After being accelerated across

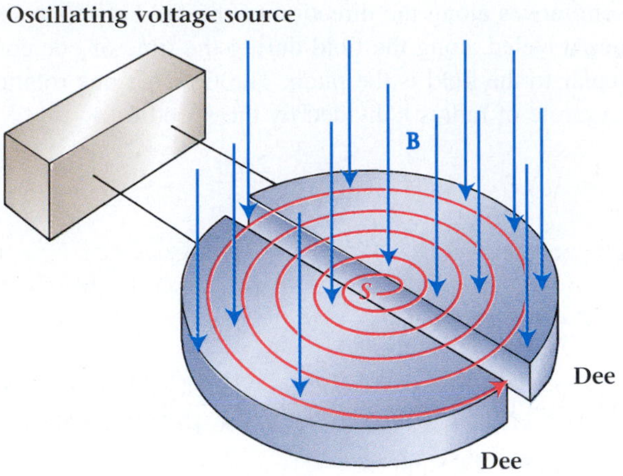

Figure 19.19

A simplified diagram of a cyclotron. The ion source S is located within the dees, and the magnetic field is perpendicular to the plane of the dees.

the gap, it moves with a greater speed and a greater radius of curvature. However, we can show that the time required to complete a revolution in the field is constant.

The time T required to complete one orbit (the period) is the path length divided by the speed:

$$T = \frac{2\pi r}{v}.$$

Solving Eq. (19.5) for v and substituting gives us

$$T = \frac{2\pi r}{qBr/m} = \frac{2\pi m}{qB}.$$

This tells us that the period needed by the charged particles to complete an orbit is constant, regardless of the values of r and v.* A constant period means that the frequency is constant, since the frequency is the reciprocal of the period. If the electric potential between the dees oscillates at this same frequency, then the charged particles accelerate each time they cross the gap. When subjected to many of these accelerations, the particles can attain very high velocities and thus possess large kinetic energies.

Other charged-particle accelerators have been invented and developed since the first cyclotron. All of them use magnetic fields to guide the accelerated particles in the desired paths. New acceleration techniques have been perfected, but the circular shape of most large accelerators reflects Lawrence's original idea. Further discussion of accelerators appears in Chapter 32.

Example 19.6

A cyclotron operating at a magnetic field of 1.6 T is used to accelerate protons (mass = 1.67×10^{-27} kg and charge = 1.6×10^{-19} C). What must be the frequency of oscillation of the accelerating voltage?

─────────────

*At speeds approaching the speed of light, the period changes because of effects discussed in Chapter 25.

Solution The frequency is the reciprocal of the period. Using the equation for T just given, we get

$$f = \frac{1}{T} = \frac{qB}{2\pi m} = \frac{(1.6 \times 10^{-19} \text{ C})(1.6 \text{ T})}{(2\pi)(1.67 \times 10^{-27} \text{ kg})}$$

$$f = 2.4 \times 10^7 \text{ Hz} = 24 \text{ MHz}.$$

Discussion A frequency of 24 MHz lies in the range of broadcast frequencies for short-wave radio (1.3-30 MHz). Such a frequency is easily obtained with conventional electronic components.

19.6   Magnetic Field Due to a Current-Carrying Wire

In the earlier sections of this chapter, you have seen that an electric current produces a magnetic field. How strong is this field for a given current? The answer to this question was found experimentally by two French scientists, Jean-Baptiste Biot (1774–1862) and Felix Savart (1791–1841), shortly after Oersted's discovery of the effect of a current-carrying wire on a compass needle. Biot and Savart found an expression for the magnetic field due to a long, straight, current-carrying wire. Figure 19.20 shows a point P located a distance d from a long straight wire carrying a constant current I. The magnitude of the field B at point P is

$$B = k'\frac{2I}{d}, \tag{19.6}$$

where k' is a constant with the value in SI units of exactly 10^{-7} N/A^2. This relation for the field due to a long wire is often known as the Biot-Savart law.

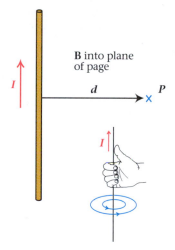

Figure 19.20
The magnetic field at point P a distance d from a long current-carrying wire is directed into the plane of the page as given by the right-hand rule. The field is perpendicular to both I and d and has the magnitude $B = k'\frac{2I}{d}$.

Example 19.7

What is the magnitude of the magnetic field at a distance of 3.0 m from a long straight wire that carries a direct current of 15 A?

Solution We can calculate the magnitude of the field from Eq. (19.6), where $I = 15$ A, $d = 3.0$ m, and $k' = 10^{-7}$ N/A^2:

$$B = k'\frac{2I}{d} = 10^{-7} \text{ N/A}^2 \frac{(2)(15 \text{ A})}{3.0 \text{ m}}$$

$$B = 1.0 \times 10^{-6} \text{ N/A} \cdot \text{m}.$$

Thus

$$B = 1.0 \times 10^{-6} \text{ T},$$

where we have recognized that the combination of units N/A · m is the tesla. The field calculated here is about 2% as large as the earth's magnetic field at its surface.

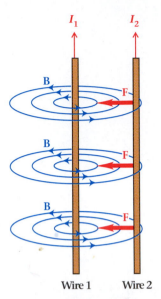

Wire 1   Wire 2

Figure 19.21

Two wires carrying parallel currents attract each other. The current in each wire produces a magnetic field that interacts with the current in the other wire, resulting in an attractive force.

If two parallel wires each carry a current, each wire exerts a force on the other. To understand what is happening, consider one wire as setting up a magnetic field that interacts with the current of the second wire. Figure 19.21 shows two parallel wires carrying currents I_1 and I_2 in the same direction. The direction of the magnetic field about the first wire is given by the right-hand rule (Fig. 19.9). Since the two wires are parallel, the magnetic field due to the first wire is perpendicular to the second wire and thus to current I_2. The resulting force on the second wire is directed toward the first, as shown in the figure. This agrees with the right-hand rule, since for rotation of the fingers from the direction of current I_2 toward the direction of the field **B**, the thumb points toward the first wire. Thus, the force between the wires is attractive.

By Newton's third law, the first wire experiences an attractive force of the same magnitude toward the second wire. The wires are thus drawn toward each other. We would reach this same conclusion by considering the interaction of the current I_1 in the field due to current I_2. If the two currents had been directed opposite to each other (that is, antiparallel), then the force between the two wires would be repulsive and they would be pushed apart.

The force between two current-carrying wires is used to define the ampere in terms of mechanical units. Once the ampere has been established, then the coulomb (ampere-second) and the volt (joule/coulomb) are set in accordance with the size of the ampere. The unit for magnetic field, the tesla, also depends on the size of the ampere.

To define the ampere, first consider a long wire carrying a current I_1 that generates a magnetic field **B**. If another wire of length l parallel to the first wire carries a current I_2 in the same direction, as in Fig. 19.21, the attractive force on that wire is

$$F = I_2 l B,$$

in accord with Eq. (19.3). If the two wires are a distance d apart, then according to Eq. (19.6) the magnitude of the magnetic field of the first wire at the position of the second wire is

$$B = 2k' \frac{I_1}{d},$$

and the force per unit length on the second wire becomes

$$\frac{F}{l} = k' \frac{2I_1 I_2}{d}.$$

If the two currents are equal, this equation becomes

$$\frac{F}{l} = k' \frac{2I^2}{d}.$$

The force per unit length, which is an easy quantity to measure, depends on the separation d, the current I, and the proportionality constant k'. In the SI system, we choose $k' = 10^{-7}$ N/A^2 so that the equation we just derived becomes the defining equation for the ampere. That is, **when the attractive force per unit length of two long parallel wires placed one meter apart and carrying the same current is 2×10^{-7} N/m, that current is defined to be exactly one ampere.**

The values of k' and of the ampere are not independent. This particular value of k' is chosen to make the ampere as just defined equal to the ampere that was in use prior to the adoption of the SI standard units. Current balances based on the principles described here provide sensitive and practical instruments for measuring currents and calibrating other current meters. The choice of the ampere for the base unit of electricity in the SI system is an operational choice based on the relative ease of measurement and observation of the forces involved and on the precision with which they can be determined.

For reasons similar to those that led us to replace the constant k in Coulomb's law by $1/4\pi\epsilon_0$, we introduce a constant called the magnetic permeability of free space μ_0, related to k' through the equation

$$\mu_0 = 4\pi k' = 4\pi \times 10^{-7} \text{ N/A}^2. \tag{19.7}$$

Hence,

$$k' = \frac{\mu_0}{4\pi}.$$

Soon after discovering the expression for the magnetic field of a long, straight, current-carrying wire, Biot and Savart found a more general expression that allows us to calculate the magnetic field arising from a wire of any shape and current distribution. For the special case of a circular loop (Fig. 19.22), the results are that the field at the center of the loop is

$$B = \frac{\mu_0 I}{2r}. \tag{19.8}$$

The direction of B is perpendicular to the plane of the loop, in accord with the right-hand rule.

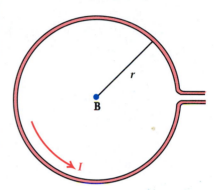

Figure 19.22
A current loop of radius R carrying a current I. The magnetic field at the center of the loop is directed out of the plane of the loop, in accord with the right-hand rule.

Master the Concept

Electric Currents and Magnetic Fields

Question: If the magnetic field of the earth is due to a circulating current below the earth's crust, does it flow from west to east or east to west?

Answer: The magnetic field of the earth is a dipolar field oriented so that the direction of the field along the polar axis is from the north geographic pole to the south geographic pole (Fig. 19.5). Let us model the problem as a large current-carrying loop aligned with the equator. This current would generate a dipolar magnetic field aligned along the polar axis. If the current is directed from east to west, the right hand rule tells us that the field is directed from the geographic north to the south in alignment with the earth's known field. Thus, a source of the earth's magnetic field could be current beneath the surface, circulating in an east to west direction.

Example 19.8

A circular loop of copper wire is connected to a 6.3-V storage battery. What must be the radius of the loop if the magnetic field at its center due to the current is about the same as the earth's magnetic field, say 5.0×10^{-5} T? The resistance per unit length of the wire is 0.068 Ω/m. Assume that the battery has very low internal resistance and maintains the full 6.3 V across the wire.

Strategy We can calculate the magnetic field of a loop using Eq. (19.8). To do so, we need to know the current in the loop. The current can be determined from Ohm's law by using our knowledge of the potential difference V and the total resistance of the loop.

Solution For a loop of radius r (Fig. 19.22), the resistance R is the product of the length of the wire $2\pi r$ with the resistance per unit length. The current through the loop of wire is

$$I = \frac{V}{R} = \frac{V}{(2\pi r)(\text{resistance/unit length})}.$$

Substituting this expression for I into the equation for the field at the center of a loop gives

$$B = \frac{\mu_0 I}{2r} = \frac{\mu_0 V}{(4\pi r^2)(\text{resistance/unit length})}.$$

Solving for r, we get

$$r = \sqrt{\frac{\mu_0 V}{(4\pi B)(\text{resistance/unit length})}}$$

$$r = \sqrt{\frac{(4\pi \times 10^{-7}\ \text{N/A}^2)(6.3\ \text{V})}{(4\pi)(5.0 \times 10^{-5}\ \text{T})(0.068\ \Omega/\text{m})}} = 0.43\ \text{m}.$$

Discussion A storage battery with a potential difference of 6.3 V is sufficient to produce a magnetic field at the center of a single, 43-cm-radius loop of wire that is comparable to the earth's magnetic field. A larger loop will produce a smaller field for two reasons. First, the longer wire will have a greater resistance and thus carry a smaller current. Second, the field is reduced because, according to Eq. (19.8), even if the current is constant, B decreases as r increases.

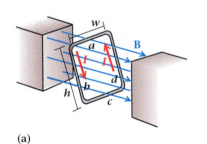

(a)

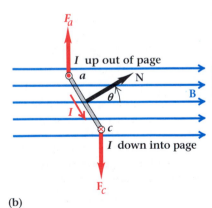

(b)

Figure 19.23
(a) A rectangular current loop in a magnetic field. (b) View of the same loop from the side, showing the direction of forces along the edges a and c. The forces on edges b and d are equal and opposite. They lie on the same line and thus generate no torque.

19.7 **Torque on a Current Loop**

We have already discussed the effect of a magnetic field on a straight current-carrying wire. However, if the wire is not straight, but forms a loop instead, we must consider the force acting on each part of the loop. A knowledge of the behavior of current loops and the shapes of their magnetic fields is important for understanding electric generators, motors, and transformers, among many other electrical applications.

The simplest situation to analyze is a rectangular loop of wire in a region of uniform magnetic field **B** (Fig. 19.23a). For simplicity we have aligned the loop so that two of its edges (a and c) are perpendicular to **B**. Figure 19.23(b) shows a view of the same loop from the side. Note that edges b and d are not perpendicular to **B**. The normal (perpendicular) to the plane of the loop is represented by the vector **N**, which makes an angle θ with the magnetic field.

The force on the loop's near edge (b) of length h is directed outward (away from the center of the loop), perpendicular to **B** (Fig. 19.23a), as you can see by applying the right-hand rule (rotate I into B). Similarly, the force on the far edge (d) is also directed away from the center of the loop, perpendicular to **B**. Because these two forces are equal and opposite and lie along the same line, there is no net torque on the loop due to sides b and d.

The forces on the other two edges (a and c) are also equal to each other and opposite in direction, but they are not directed along a common line, as seen in Fig. 19.23(b). The net effect is to exert a torque about a horizontal axis through the loop. The direction of the torque (clockwise in Fig. 19.23b) tends to rotate the loop so that its normal aligns along **B**.

From Eq. (19.3), the force on edge a is

$$F_a = IwB$$

and is directed upward. The force on edge c is

$$F_c = IwB$$

and is directed downward. The torque due to $\mathbf{F}_a$ about a horizontal axis through the middle of the loop is

$$\tau_a = \frac{h}{2}IwB \sin \theta,$$

while that due to $\mathbf{F}_c$ is

$$\tau_c = \frac{h}{2}IwB \sin \theta.$$

Since both torques are in the same direction, the total torque is the sum of the individual torques,

$$\tau = IwhB \sin \theta,$$

or

$$\tau = IAB \sin \theta, \tag{19.9}$$

where A is the area of the loop and is equal to wh.

Equation (19.9) has the same form as Eq. (19.1), which gave the torque on a magnetic dipole due to an applied magnetic field. Thus, the current loop is a magnetic dipole with magnetic moment

$$\mu = IA.$$

The direction of the magnetic moment is along the direction of **N**, the normal to the plane of the current loop. It is given by another right-hand rule (Fig. 19.24): *The direction of μ is the same as the direction in which your thumb points if you curl the fingers of your right hand in the direction of positive current around the loop.*

A current loop can consist of more than one single loop of wire. In this case, the magnetic moment is proportional to the total number of loops or turns around a given area A. For example, if the wire makes five turns around an area A, then the magnetic moment is $5IA$. For N loops, the moment becomes

$$\mu = NIA. \tag{19.10}$$

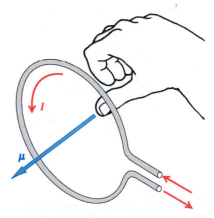

Figure 19.24
Orientation of the magnetic dipole moment of a current loop. The direction of μ is the same as the direction in which your thumb points if you curl the fingers of your right hand in the direction of positive current.

That is, the magnetic moment depends on the number of turns of wire, the current in the wire, and the area of the loop. We chose a rectangular loop for its obvious simplicity, but our result is valid for any plane loop, no matter what its shape. Then, from Eq. (19.1), the torque on a current-carrying loop is $\tau = \mu B \sin \theta$, where the magnetic moment μ is given by Eq. (19.10).

Example 19.9

A flat coil is made by wrapping wire around a 2-liter bottle and then gently slipping it off. The resulting coil of 25 turns of wire has a radius of 5.5 cm. (a) What is the magnetic moment of the coil when it carries a current of 1.5 A? (b) What is the maximum torque exerted on the coil by the earth's magnetic field of 5.0×10^{-5} T when the coil conducts a current of 1.5 A?

Solution (a) The magnetic moment may be computed from Eq. (19.10) as

$$\mu = NIA.$$

In this case $N = 25$, $I = 1.5$ A, and the area $A = \pi(5.5 \times 10^{-2} \text{ m})^2$. So,

$$\mu = 25(1.5 \text{ A})[\pi(5.5 \times 10^{-2} \text{ m})^2] = 0.36 \text{ A} \cdot \text{m}^2.$$

(b) The torque is found from Eq. (19.1) as

$$\tau = \mu B \sin \theta,$$

where μ is the magnetic moment that we just computed. The torque is a maximum when $\boldsymbol{\mu}$ is perpendicular to **B** so that $\sin \theta = 1$. It is

$$\tau = \mu B = (0.36 \text{ A} \cdot \text{m}^2)(5 \times 10^{-5} \text{ T}) = 1.8 \times 10^{-5} \text{ m} \cdot \text{N}.$$

The magnetic field around a circular current loop is shown in Fig. 19.25. It is similar to the magnetic dipole field near a bar magnet (Fig. 19.3). The direction of the field is consistent with the right-hand rule given in Fig. 19.9. Note that the magnetic field lines are all continuous loops having neither beginning nor end. The magnetic field lines associated with a bar magnet also form continuous loops. In this respect the magnetic dipole is different from the electric dipole, in which the field lines terminate on the electric charges.

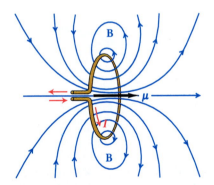

Figure 19.25
Magnetic field due to a current loop. The magnetic moment μ points in the direction of a north magnetic pole.

19.8 Galvanometers, Ammeters, and Voltmeters

A practical instrument whose operation depends on the torque developed on a current-carrying coil is the **galvanometer.** Its most direct application is for sensitive measurement of electric current. Galvanometers are the basic component of many electrical meters. A particularly reliable galvanometer was built in 1880 by the French physicist Jacques Arsène d'Arsonval, who surrounded a cylindrical iron core with a movable coil placed between the poles of a magnet. When a current was passed through the coil, it produced a magnetic dipole whose interaction with the field of the permanent magnet created a torque on the coil. The coil turned through an angle that depended on the magnitude of the current causing the field. Precise measurements of the angle permitted precise measurement of the current I.

In 1888, Edward Weston used d'Arsonval's idea to construct the first portable current-measuring instrument that could be read directly from its scale. His design became standard and is still in use. Weston's design uses a horseshoe-shaped permanent magnet with iron pole pieces shaped to a cylindrical gap (Fig. 19.26). A cylindrical iron core, only slightly smaller than the gap, is fixed in the center between the pole pieces. A coil placed in the gap between the core and the pole pieces is supported on jeweled bearings. A tiny coil spring holds the coil in the zero position when no current is applied. Current is carried to the coil through the spring and the pivots. The movement is rugged enough for portable use while at the same time being quite sensitive.

A current-measuring instrument is usually called an **ammeter.** It measures the current passing through it. Although the heart of the ammeter is a galvanometer, the ammeter may be constructed and calibrated so that it can measure currents larger than the maximum current for full-scale deflection of the galvanometer movement. The galvanometer movement itself is normally very sensitive, so the current through the movement must be limited. Thus, to make an ammeter, we connect a resistor in parallel with the galvanometer movement so that some of the current bypasses the galvanometer (Fig. 19.27a). By choosing the proper bypass resistor, we can construct an ammeter of any desired range. Meters of this sort are often called by such names as milliammeter or microammeter, depending on their range.

An instrument for measuring potential difference, or voltage, is usually called a **voltmeter.** We can construct a voltmeter from a galvanometer movement by connecting a resistor in series with it to limit the current (Fig. 19.27b). Ideally, the voltmeter would measure the potential difference without allowing any current to pass through the meter. However, some current must pass in order to move the galvanometer coil. The resistor limits that current and determines the range of the meter. The maximum range of the voltmeter is the product of the current required for full-scale deflection of the meter and the combined resistance of the galvanometer and the series resistor.

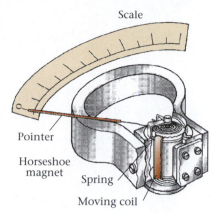

Figure 19.26

Operation of a d'Arsonval-Weston meter movement. The field of the horseshoe magnet exerts a torque on the moveable coil when a current is present. The deflection of the coil is proportional to the current.

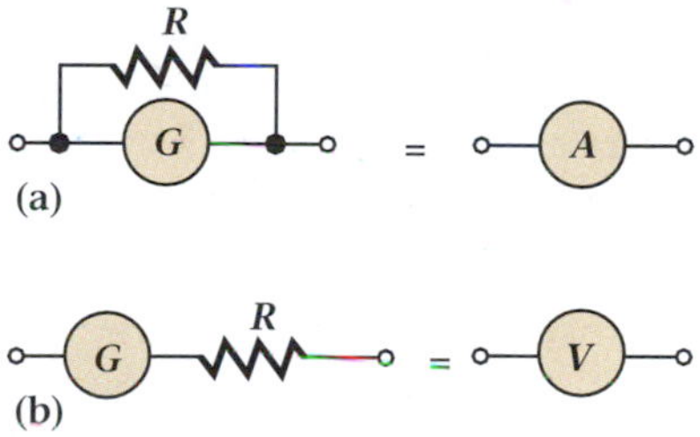

Figure 19.27

(a) A parallel connection of a resistor (R) to a galvanometer (G) to make an ammeter (A). (b) A series connection of a resistor to a galvanometer to form a voltmeter (V). Remember that the galvanometer itself also has a resistance.

Example 19.10

A student needs to make a voltmeter that measures 100 V full scale. The only galvanometer movement available has full-scale deflection of the pointer at a current of 100 μA, and its resistance is 10 kΩ. What additional resistance is needed and how should it be connected to the galvanometer?

Strategy To make a voltmeter from the galvanometer movement, the student needs to connect a resistor in series with the movement. We are told that for full-scale deflection at 100 V, the current through this particular meter must be 100 μA. So, we need to calculate the total resistance that would limit the current to 100 μA when the applied voltage is 100 V. Then we can determine how much series resistance must be added to the resistance of the movement to give the necessary total resistance.

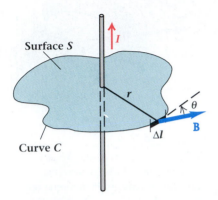

Figure 19.28

Illustration of Ampère's law. A closed curve C bounds a surface S. Ampère's law states that the sum of the contributions $B \, \Delta l \cos \theta$ around the curve is proportional to the net current I crossing the surface S.

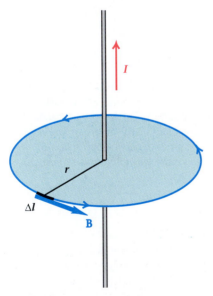

Figure 19.29

A circular closed loop located a distance r about a long straight wire. Since the magnitude of the magnetic field **B** is constant around the loop and the field is directed parallel to Δl at each point, $\theta = 0$ and $\cos \theta = 1$.

Solution The resistance of the voltmeter is

$$R_m = \frac{V}{I} = \frac{100 \text{ V}}{100 \times 10^{-6} \text{ A}} = 10^6 \ \Omega$$

$$R_m = 1000 \text{ k}\Omega.$$

The required voltmeter resistance is the sum of the series resistance R and the resistance of the galvanometer movement R_G,

$$R_m = R + R_G.$$

The value of the series resistor is then

$$R = R_m - R_G = 1000 \text{ k}\Omega - 10 \text{ k}\Omega$$

$$R = 990 \text{ k}\Omega.$$

Thus, the student can make a 100-V meter by connecting a 990-kΩ resistor in series with the galvanometer movement.

*19.9   Ampère's Law

Soon after the discoveries of Oersted and of Biot and Savart, Ampère found a useful relation between electric currents and magnetic fields. This relation may be applied in highly symmetric situations to find the magnetic field more easily than by computing with the Biot-Savart law. In either case, the result is the same. In this respect, using Ampère's law is somewhat like finding the electric field by using Gauss's law rather than Coulomb's law. In cases that lack the proper symmetry, Ampère's law is not easily applied. It is always valid, even though the Biot-Savart law is needed sometimes for detailed calculation of the magnetic field.

According to Ampère's law, for an arbitrary closed curve around a current-carrying conductor (Fig. 19.28), the component of magnetic field tangent to the curve is related to the net current passing through the (imaginary) surface bounded by the curve. In particular, for each element Δl of the curve, we can calculate a contribution $B \, \Delta l \cos \theta$ to the magnetic field, where θ is the angle between **B** and $\Delta \mathbf{l}$. **Ampère's law** states that the sum of these contributions around the entire loop is proportional to the net current I through the loop. That is,

$$\boxed{\Sigma \, B \, \Delta l \cos \theta = \mu_0 I.} \tag{19.11}$$

We can easily apply Ampère's law for the case of the magnetic field about a long straight wire. We know already that the field is always perpendicular to the wire, because the lines of B make circles centered about the wire. Thus, if we choose a circular path with the wire at its center (Fig. 19.29), then **B** is in the same direction as $\Delta \mathbf{l}$ and $\cos \theta = 1$ everywhere around the loop. Furthermore, the magnitude of the field B is constant around the loop because the loop is everywhere equidistant from the wire. Thus

$$\Sigma \, B \, \Delta l \cos \theta = B \, \Sigma \, \Delta l = B(2\pi r) = \mu_0 I,$$

where $\Sigma \, \Delta l$ is $2\pi r$, the circumference of the loop. The field due to the current in the wire is

$$B = \frac{\mu_0 I}{2\pi r}, \qquad (19.12)$$

which is equivalent to the expression found by Biot and Savart (Eq. 19.6) when $\mu_0/4\pi$ is substituted for k'.

Another important current configuration is that of a **solenoid,** which is a helical winding of wire that may carry a current I. The magnetic field of the solenoid is the vector sum of the fields due to each nearly circular loop comprising the solenoid. Figure 19.30 depicts the field around a loosely wound solenoid. Its behavior is similar to that of a magnetic dipole. When the coil is tightly wound with adjacent wires nearly touching, the solenoid resembles a cylindrically flowing current. As the solenoid is wound tighter and made longer, the field just outside the central part of the solenoid becomes smaller. Eventually, as the length of the solenoid greatly exceeds its diameter, the field just outside the central part of the solenoid becomes negligibly small.

We can apply Ampère's law to determine the magnitude of the field inside a very long solenoid. Figure 19.31(a) illustrates a section of an infinitely long, tightly wound solenoid. In the cutaway diagram in Fig. 19.31(b), the closed loop for computing with Ampère's law is the rectangle $abcd$. From the symmetry we expect that **B** should be perpendicular to the sides ab and cd. Outside, along da, the field is zero. Inside, the field along bc is constant. For this case, Ampère's law (Eq. 19.11) gives

$$\Sigma \, B \, \Delta l \cos \theta = 0_{(ab)} + Bl_{(bc)} + 0_{(cd)} + 0_{(da)} = Bl = \mu_0 n l I,$$

where n is the number of turns per unit length of the coil. Thus, nl gives the number of turns with current I that pass through the surface bounded by $abcd$. The resulting field is parallel to the axis of the solenoid, with a magnitude given by

$$B = \mu_0 n I, \qquad (19.13)$$

and is independent of position within the solenoid. That is, near the center of a long solenoid the magnetic field is uniform.

Solenoids are used to generate magnetic fields in many practical devices. Their applications in cars include the solenoids that engage the starter gear and those that control the electric door locks. In the home they are found in door bells and in speakers for TVs and tape players. Solenoids are often used to generate regions of controlled, uniform magnetic field. In hospitals, they are used for generating the fields needed for magnetic resonance imaging (Fig. 19.32).

In door bells or starter solenoids in cars, the function of the solenoid is to generate a large field that acts on an iron rod to magnetize it and pull it into the middle of the coil. Normally, when there is no current in the solenoid, there is no magnetic field and a spring keeps the rod pushed to one side of the solenoid. When a switch is closed, allowing current to pass through the coil, the resulting field quickly draws the rod into the solenoid. In the door bell, the rod moves far enough to strike a metal bar, generating the sound. In the car, the rod moves far enough to engage a gear that allows the starter motor to turn the engine.

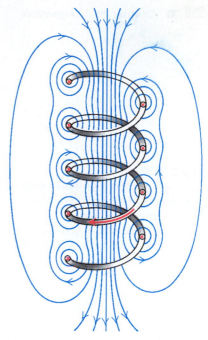

Figure 19.30

The magnetic field around a loosely wound solenoid. Some field lines leak out between the coils.

(a)

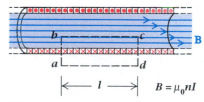

(b)

Figure 19.31

(a) A section of a long solenoid. If the coils are tightly wound, we can neglect any field leakage through the sides. (b) Cutaway diagram of a section of a long solenoid. The dots represent current out of the page, and the crosses indicate current into the page. Applying Ampère's law to the closed path $abcda$, we can determine the magnitude of the magnetic field in a tightly wound solenoid.

Figure 19.32
(a) A superconducting solenoid magnet is suspended from the lid of a Dewar prior to testing. The solenoid is contained in the hollow aluminum cylinder. (b) A 0.5-T superconducting magnet enclosed within its Dewar.

(a)   **(b)**

Example 19.11

Compute the magnetic field near the center of a long solenoid having 20 turns per centimeter and carrying a current of 5.0 A.

Solution   The field is given by Eq. (19.13), with $n = 2.0 \times 10^3$/m:

$$B = \mu_0 n I = (4\pi \times 10^{-7} \text{ N/A}^2)(2 \times 10^3\text{/m})(5.0 \text{ A})$$
$$B = 4\pi \times 10^{-3} \text{ N/A} \cdot \text{m} = 0.013 \text{ T}.$$

This field is about 250 times greater than the magnitude of the earth's field in most parts of the United States (about 5×10^{-5} T).

---

*19.10   Magnetic Materials

In general, whenever we place an object in a magnetic field, the object becomes polarized; that is, it develops a magnetic dipole moment. This polarization may be measured and used to characterize the object. We define the magnetization **M** to be the dipole moment per unit volume, a vector quantity whose direction is that of the magnetic dipole moment.

   The total magnetic field inside the object is the sum of the magnetic field arising from the magnetization and the magnetic field due to electric currents inside and outside the material. The ratio of the magnetization to the external magnetic field **B** that induces it is the **magnetic susceptibility** per unit volume, χ:

$$\chi \equiv \mu_0 \frac{M}{B}. \tag{19.14}$$

Here, B is the external magnetic field without the sample object present. When the field and the magnetization are in SI units, the susceptibility is dimensionless.

Table 19.1	Commonly Observed Types of Magnetic Behavior	
Type	**Comments**	**Examples**
Diamagnetism	Small, negative χ; independent of temperature	Benzene Silicon
Paramagnetism	Positive χ; $\chi \propto 1/T$	Aluminum
Ferromagnetism	Positive χ; complex temperature dependence below the transition temperature T_C; $\chi \propto 1/(T - T_C)$ above the transition temperature	Iron (T_C = 1043 K) Nickel (T_C = 631 K) Gadolinium (T_C = 289 K)

The magnitude of the magnetic susceptibility depends on the inherent properties of the material. In fact, it depends on the very nature of the electronic structure of the ions that make up the material. Therefore, we can use susceptibility measurements to categorize magnetic materials. Table 19.1 lists several types of common magnetic behavior. The weakest effect is *diamagnetism,* characterized by a small, negative susceptibility that is usually independent of temperature and applied field. The negative susceptibility indicates that the direction of the induced moment is opposite to the direction of the applied field. The origin of diamagnetism is a change in the atomic electron orbits about the direction of the field, resulting in a magnetic field opposite to the applied field. The diamagnetic susceptibility is small and is usually less interesting than the other types of magnetic behavior. However, diamagnetism is always present, even when masked by much larger, positive effects. Note that the polarization of individual ions disappears whenever the external field is removed; that is, diamagnetic materials have no permanent dipoles.

Superconductors belong to a special class of diamagnetic materials. When you place a superconducting material in a magnetic field, electric currents are generated at its surface that screen out the external field so there is zero magnetic field within the material. This behavior is known as *perfect diamagnetism.* If a magnet is brought near a superconductor, the diamagnetic polarization of the superconductor causes a repulsive force between them. This effect may be used to float a magnet above a superconductor (Fig. 19.33). A practical application of this effect on a larger scale may one day lead to magnetically levitated vehicles.

Another common magnetic behavior is *paramagnetism,* which is described by a positive susceptibility that depends inversely on the absolute temperature:

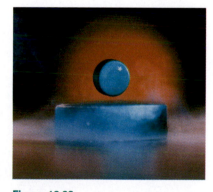

Figure 19.33
A small magnet floats above a high-temperature superconductor. The presence of the magnet induces in the superconductor a polarization that repels the magnet.

$$\chi = \frac{C}{T}.$$

(19.15)

This equation is known as the **Curie law,** and C is the Curie constant. Each particular paramagnetic material has a characteristic Curie constant.

Paramagnetism arises from atoms or ions with permanent magnetic dipole moments that exist independent of any applied field. The orientation of each moment is independent of the orientation of its neighbors. If no field is applied, these moments are randomly directed and there is no net magnetization of the material. However, when a magnetic field is present, the moments align preferentially along the direction of the external field. The sum of these moments causes the magnetization.

Example 19.12

Copper sulfate is paramagnetic with a susceptibility of 1.68×10^{-4} at 293 K. What is the susceptibility of copper sulfate at the temperature of liquid nitrogen (77.4 K) if it follows the Curie law?

Solution According to the Curie law (Eq. 19.15), the susceptibility depends inversely on the temperature:

$$\chi = \frac{C}{T}.$$

By dividing the susceptibility at one temperature by the susceptibility at another temperature, we get the ratio

$$\frac{\chi(T_1)}{\chi(T_2)} = \frac{T_2}{T_1}.$$

Substituting the known values, we find

$$\chi(77.4 \text{ K}) = \frac{293}{77.4}\chi(293 \text{ K})$$

$$\chi(77.4 \text{ K}) = 3.786(1.68 \times 10^{-4})$$

$$\chi(77.4 \text{ K}) = 6.36 \times 10^{-4}.$$

Discussion At the temperature of liquid nitrogen, the susceptibility of paramagnetic copper sulfate is almost four times larger than it is at room temperature. At the very low temperature of liquid helium (4.2 K), it would be almost 70 times larger. Because of the strong temperature dependence of the magnetic behavior, experiments on magnetic materials are often performed at low temperatures.

Other magnetic effects, distinct from diamagnetism and paramagnetism, result from the cooperative interactions of individual ionic moments. By "cooperative" we mean that the individual moments are not independent, but interact strongly. The most widely known of these magnetic phenomena is **ferromagnetism**. Ferromagnetic behavior is what the ancient Greeks observed in their lodestones and is what you observe when a magnet attracts a nail or other object. Thus, materials that we loosely call "magnetic" because they are noticeably attracted by a magnet are really ferromagnets.

Ferromagnetic solids are characterized by small regions in which all of the ionic moments are aligned in the same direction. These tiny regions are called

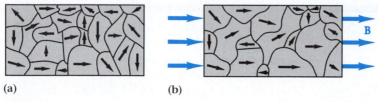

Figure 19.34

(a) Magnetic domains in an unpolarized ferromagnetic solid are random because of the random orientation of the grains within the material. (b) Under the influence of an external field, the polarization direction of the individual domains and their size may change irreversibly, creating a permanent magnet.

magnetic domains. A given ferromagnetic object (such as an iron nail) usually consists of many small crystals, or grains, and each grain contains several magnetic domains whose polarizations are in different directions (Fig. 19.34a). Thus, the net magnetization of the nail is zero. When an external field is applied, the boundaries between the domains move (Fig. 19.34b). Those domains aligned close to the field direction increase in size, while those in other directions decrease. The result is a net magnetization that is very much larger than that of a paramagnet. For a small polarizing field (say 10^{-2} T), the magnetic field resulting from the magnetization of a ferromagnet is typically 10,000 to 100,000 times larger than the polarizing field. Furthermore, if the external field is large enough, the domain structure may be permanently affected, so that a net magnetization remains even when the applied field is removed. Consequently, the resulting material is a permanent magnet.

The alignment of the ionic moments in the domains is characteristic of ferromagnets. This alignment is the result of a quantum-mechanical effect known as exchange, which is beyond the scope of the present discussion. However, it is often useful to think of ferromagnetism in terms of an internal field (the exchange field) that tends to align the individual ionic moments. The effect of the internal field is counteracted by the randomizing effect of thermal agitation, which tends to disorient these moments. The result is a critical temperature above which the material ceases to be ferromagnetic and becomes paramagnetic instead. The temperature that marks the transition from ferromagnetism to paramagnetism is known as the *Curie temperature* T_C. Below T_C the individual moments are all aligned in domains. Above T_C the individual moments behave independently, as in a paramagnet, and the material no longer has a permanent magnetic moment. The susceptibility above T_C follows a modified Curie law known as the *Curie-Weiss law:*

$$\chi = \frac{C}{T - T_C}.\qquad(19.16)$$

In addition to ferromagnetism, there are other types of magnetic behavior corresponding to a definite alignment of magnetic moments. One example is ferrimagnetism, in which the moments do not align in the same direction but still have a net permanent magnetization. Many ferrimagnets are insulators. They are widely used in refrigerator magnets.

Summary

Useful Concepts

■ Like magnetic poles repel; unlike poles attract.

■ The simplest type of magnet is the magnetic dipole. It can be characterized by its magnetic dipole moment $\boldsymbol{\mu}$. The torque on a magnetic dipole in a magnetic field is

$$\tau = \mu B \sin \theta.$$

■ The magnetic flux through an element of surface ΔA is

$$\Delta \phi_m = B \, \Delta A \cos \theta,$$

where B is the magnitude of the field and θ is the angle between the direction of the field and the direction of the surface normal.

■ Gauss's law for magnetism states that the net magnetic flux through any closed surface is zero.

■ The force on a current-carrying wire in a magnetic field is

$$F = IlB \sin \theta.$$

■ A charged particle moving in a magnetic field is subject to a force given by

$$F = qvB \sin \theta.$$

■ A charged particle moving perpendicular to a magnetic field travels in a circle of radius

$$r = \frac{mv}{qB}.$$

■ The magnetic field a distance d from a long straight wire carrying a current I is

$$B = k' \frac{2I}{d} = \frac{\mu_0 I}{2\pi d}.$$

■ Definition of the ampere: When the attractive force per unit length between two long parallel wires placed one meter apart and carrying the same electric current is 2×10^{-7} N/m, that current is defined to be exactly one ampere.

■ A current loop has a magnetic moment given by

$$\mu = NIA.$$

■ Ampère's law can be written

$$\Sigma B \, \Delta l \cos \theta = \mu_0 I,$$

where the sum is carried out around a closed loop.

■ The magnetic susceptibility per unit volume is

$$\chi = \mu_0 \frac{M}{B}.$$

■ The Curie law for paramagnetism is

$$\chi = \frac{C}{T}.$$

Right-Hand Rules

■ The direction of the magnetic field about a wire is given by a right-hand rule: If the thumb of the right hand is taken as the direction of current, the curled fingers point in the direction of the magnetic field about that current.

■ The direction of the force on a current-carrying wire in a magnetic field is also given by a right-hand rule: When the right hand is held so that the fingers can be curled from the direction of the current into the direction of the magnetic field, the force lies in the direction of the thumb.

■ The direction of the magnetic moment of a current loop is given by another right-hand rule: The magnetic moment is in the direction that your thumb points when you curl the fingers of your right hand in the direction of positive current.

Important Terms

You should be able to write the definition or meaning of each of the following:

magnetic dipoles	ammeter
magnetic field	voltmeter
magnetic dipole moment	Ampère's law
tesla	solenoid
Gauss's law for magnetism	magnetic susceptibility
magnetic flux	Curie law
cyclotron	ferromagnetism
galvanometer	magnetic domains

Conceptual Questions

19.1 The current-carrying wire shown in Fig. 19.8(b) is turned to lie in a horizontal plane, and a horizontal card is placed on each side of the wire. When iron filings are sprinkled on the card, what pattern will form? Sketch your answer.

19.2 Sketch the pattern of iron filings that will form on the card in Fig. 19.8(b) if two closely spaced current-carrying wires pass through the card. Sketch patterns for the two cases in which the currents are in the same or opposite directions.

19.3 Why does Gauss's law for magnetism imply that there are no magnetic monopoles?

19.4 A current-carrying wire passing through a certain region of space does not experience any force acting on it. Can we conclude that there is no magnetic field in that region?

19.5 A positively charged proton is projected toward you in a uniform magnetic field that is directed straight upward. In what direction is the proton deflected?

19.6 Equation (19.6) for the magnitude of the magnetic field near a long, straight, current-carrying wire becomes infinitely large when the distance d becomes zero. Does this mean that you can actually have an infinitely large magnetic field?

19.7 Some older textbooks describe the behavior of a current-carrying loop in a magnetic field in terms somewhat like this: The loop aligns itself so as to enclose as many of the lines of force of the magnetic field as possible. Explain this statement in terms of the discussion given in this text.

19.8 Suppose you make an electromagnet by winding a coil on each leg of a U-shaped iron bar and connecting the coils in series with each other across a battery. If you look directly at the ends of the pole pieces and conclude that the current flows clockwise in one coil, in what direction should it flow in the other coil?

19.9 Discuss how you would go about constructing an ammeter and a voltmeter from a given galvanometer and whatever resistors and wires you need. What would be the effects of using large or small resistance for each?

19.10 How can a piece of iron (such as a nail) be ferromagnetic and yet not have a net magnetic dipole moment?

19.11 How many ways can you think of to give an iron rod a net permanent magnetic moment?

Problems

Section 19.1 Magnets and Magnetic Fields

Hints for Solving Problems

The direction of the magnetic field points toward a south magnetic pole and away from a north magnetic pole.

19.1 Sketch the magnetic field due to two identical bar magnets (dipoles) placed together (Fig. 19.35).

Figure 19.35
Problem 19.1.

19.2 Sketch the magnetic field due to two identical bar magnets (dipoles) placed together (Fig. 19.36).

Figure 19.36
Problem 19.2.

19.3 Sketch the magnetic field due to four identical bar magnets (dipoles) placed together, as in Fig. 19.37, with all north poles close together.

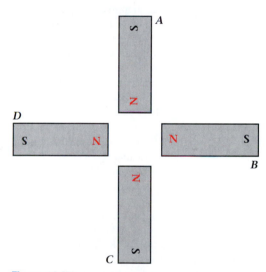

Figure 19.37
Problems 19.3 and 19.4.

19.4 Sketch the magnetic field due to four identical bar magnets (dipoles) placed together as in Fig. 19.37 but with magnets A and C reversed so that their south poles are close together.

Section 19.2 Oersted's Discovery: Electric Current Produces Magnetism

Hints for Solving Problems

A magnetic field exerts a torque on a magnetic dipole to rotate it into the direction of the field.

19.5 The units of magnetic field are N/A · m. Use this fact and Eq. (19.1) to show that the units of magnetic moment are A · m².

19.6 Imagine that you are standing directly beneath a current-carrying wire and facing in the direction of the current. There is a magnetic field passing through your body due to the current in the wire. In which direction does the field point?

19.7 A torque of 4.0×10^{-3} N · m is needed to hold a certain compass needle at right angles to the earth's magnetic field, $B = 5.0 \times 10^{-5}$ T. What is the magnetic moment of the compass needle?

19.8 A compass needle in a uniform magnetic field of 0.30 T experiences a torque of 0.050 N · m when it points in a direction 45° away from the direction of the field. What is the magnitude of the needle's magnetic moment?

19.9 A magnet with a magnetic moment of 6.5×10^6 A · m^2 experiences a torque of 8.7×10^{-3} N · m when the axis of the magnet is 27° away from the direction of a uniform magnetic field. What is the magnitude of the field?

19.10 A small bar magnet has an effective magnetic moment $\mu = 7.50 \times 10^6$ A · m^2. What is the maximum possible torque on the magnet due to the earth's field, $B = 5.00 \times 10^{-5}$ T?

19.11• A circular loop of wire of radius of 4.62 cm is oriented so that the plane of the loop makes an angle of 27° with a uniform magnetic field of 3.64×10^{-2} T. What is the magnetic flux through the loop?

19.12•• A slender, 150-g bar magnet 15.0 cm long has a magnetic moment of 5.30×10^6 A · m^2. It is held at an angle of 45° to the direction of a uniform magnetic field. When released, it begins to rotate about its center with an initial angular acceleration of 5.09×10^{-3} rad/s^2. What is the magnitude of the magnetic field?

19.13•• In a region of uniform magnetic field B, how much work is required to rotate a bar magnet of dipole moment μ from parallel alignment to antiparallel alignment along the magnetic field? (*Hint:* This is similar to the situation of an electric dipole seen in Chapter 16.)

19.14•• (a) A small bar magnet has an effective magnetic moment $\mu = 8.55 \times 10^6$ A · m^2. What is the torque required to hold the magnet at an angle of 48° from the direction of an applied magnetic field of 8.40×10^{-2} T? (b) If the magnet is released, how much work is done by the field in rotating the magnet into alignment with the field?

Section 19.3 Magnetic Forces on Electric Currents

Hints for Solving Problems

A magnetic field exerts a force on a current-carrying wire that is at right angles to both the direction of the current and the direction of the field.

19.15 A 0.50-m-long wire of mass $m = 1.0$ g lies on a horizontal platform in a region of uniform magnetic field $B = 1.0 \times 10^{-3}$ T. The field is perpendicular to the wire so that an upward force is exerted on the wire when a current passes through it. What is the minimum current required to lift the wire against its weight?

19.16 A laboratory magnet (Fig. 19.14) with a field $B = 0.20$ T over an effective area 10 cm in diameter exerts a force on a current-carrying wire that crosses between the pole faces at right angles to the field. What is the force on the wire if the current is 10 A?

19.17 An electric trolleycar cable carries a direct current of 380 A in a region where the vertical component of the earth's magnetic field is 5.00×10^{-6} T. What is the horizontal force on a 12.0-m section of the wire due to magnetic effects?

19.18 A wire 0.250 m long carries a current of 0.750 A in a region of uniform magnetic field. The wire is subject to a force of 2.50 N when held perpendicular to the direction of the field. What is the magnitude of the magnetic field?

19.19 A wire 0.540 m long carries a current of 6.08 A and makes an angle of 45.0° with a uniform magnetic field. If the force on the wire is 0.106 N, what is the magnitude of the magnetic field?

19.20• A horizontal wire 30.0 cm long makes an angle of 45.0° with a magnetic field $B = 0.0500$ T horizontally directed to the north. What are the magnitude and the direction of the force on the wire when it carries a current of 1.00 A in the direction from southwest to northeast?

19.21•• Show that the total force on a current-carrying straight wire perpendicular to a uniform magnetic field is independent of the length of the wire if the potential difference across the wire is held constant. (*Hint:* Assume Ohm's law to be valid for the wire.)

Section 19.4 Magnetic Forces on Moving Charged Particles

Hints for Solving Problems

A magnetic field exerts a force on a moving charge that is at right angles to both the velocity of the charge and the direction of the field.

19.22 Alpha particles from a particular radioactive source have a speed of 1.85×10^7 m/s. How large a magnetic field is required to bend the path of the particles into a circle of 0.580-m radius? An alpha particle has a mass of 6.64×10^{-27} kg and a charge of 3.20×10^{-19} C.

19.23 A proton of charge 1.60×10^{-19} C and mass 1.67×10^{-27} kg is projected into a region of magnetic field $B = 0.140$ T. What is the acceleration of the proton if it is initially traveling perpendicular to the field with a speed $v = 1.05 \times 10^6$ m/s?

19.24 A proton of charge $+1.60 \times 10^{-19}$ C and mass 1.67×10^{-27} kg is introduced into a region of $B = 1.15$ T with an initial velocity of 1.25×10^6 m/s perpendicular to B. What is the radius of the proton's path?

19.25• Suppose the initial velocity of the proton in Problem 19.24 makes an angle of 45° with the direction of the magnetic field B. What is the subsequent motion of the proton?

19.26• Suppose the initial velocity of the proton in Problem

19.24 makes an angle of 50° with the direction of the magnetic field B. What is the pitch of the helix that describes the path of the proton? The pitch is the distance the proton travels along the axis of the helix in the time required for the proton to make one complete loop around the axis.

19.27• A charged particle moving through a bubble chamber travels in a circular arc of radius 0.53 m. If the magnetic field perpendicular to the plane of the arc is 0.10 T, what is the momentum of the particle? Assume that the charge of the particle is that of an electron, $e = -1.6 \times 10^{-19}$ C.

19.28• A particle having an electric charge $q = 3.2 \times 10^{-19}$ C is injected into a magnetic field $B = 0.10$ T with a speed of 3.0×10^{6} m/s. (a) If the velocity of the particle is perpendicular to the direction of the magnetic field, what is the force on the particle? (b) How great an electric field would be required to exert a force with the same magnitude on the particle?

19.29• A proton moving at right angles to a magnetic field of 0.100 T travels in a circular orbit of radius 5.00 cm. The proton has an electric charge $q = 1.60 \times 10^{-19}$ C and a mass of 1.67×10^{-27} kg. (a) What is the speed of the proton? (b) What is its kinetic energy?

19.30•• In a high-energy physics experiment, a subnuclear particle moves in a circular arc of 0.27-m radius perpendicular to a magnetic field of 2.7×10^{-2} T. The kinetic energy of the particle is determined to be 4.1×10^{-16} J. Identify the particle from its mass. The masses of the electron, pion, and proton are 9.1×10^{-31} kg, 2.5×10^{-28} kg, and 1.67×10^{-27} kg, respectively. Assume that the particle is known to have a positive charge equal to the magnitude of the electron charge.

*Section 19.5 The Cyclotron

19.31 A laboratory cyclotron operates in a magnetic field of 1.50 T. If it is used to accelerate deuterons (mass 3.34×10^{-27} kg and charge 1.60×10^{-19} C), what must be the frequency of oscillation of the accelerating voltage?

19.32 A cyclotron used to accelerate deuterons is operated at a frequency of 9.16 MHz. What is the operating magnetic field of the cyclotron? The deuteron mass = 3.34×10^{-27} kg; charge = 1.60×10^{-19} C.

19.33 What is the cyclotron frequency of an electron of mass 9.1×10^{-31} kg and charge -1.6×10^{-19} C in a region of field $B = 1.2$ T?

19.34 What would be the operating frequency of the cyclotron of Example 19.6 if it had a magnetic field of only 0.50 T?

19.35• Derive the expression

$$KE = \frac{q^2 B^2 r^2}{2m}$$

for the kinetic energy of a particle of charge q and mass m moving at a radius r in a cyclotron with uniform magnetic field B.

19.36• At what radius must deuterons be extracted from a cyclotron with a 1.00-T magnetic field if they are to have an energy of 2.00×10^{-12} J each? (Try using the results of Problem 19.35 and the deuteron information in Problem 19.31.)

Section 19.6 Magnetic Field Due to a Current-Carrying Wire

19.37 A long straight wire carries a current of 35 A. At what distance from the wire will the magnetic field become comparable to the earth's field of 5.0×10^{-5} T?

19.38 A long straight wire carries a current of 100 mA. What is the magnitude of the magnetic field 10.0 cm from the wire?

19.39 Two parallel wires spaced 0.50 m apart in a horizontal plane carry oppositely directed currents of 200 A. (a) What is the force per unit length on the wires? (b) Are the wires attracted or repelled?

19.40 Two parallel wires spaced a distance d apart experience an attractive force per unit length of 6.00×10^{-4} N/m. If the currents have a magnitude of 10.0 A each, what is the distance d?

19.41 Two parallel wires repel each other with a force per length of 1.00×10^{-3} N/m when spaced a distance of 2.50 cm apart. If one wire has a current of 100 A, what is the current in the other wire?

19.42 Show that the magnetic field at the center of a circular coil of N turns of radius r carrying a current I is given by

$$B = \frac{\mu_0 N I}{2r}.$$

19.43 What is the current in a 10-cm-radius wire loop if the magnetic field generated at the center of the loop is equal to the magnetic field of the earth at its surface? Take the earth's field to be 5.0×10^{-5} T.

19.44• A rectangular loop of wire carries a current of 6.3 A. Nearby in the plane of the loop is a long straight wire carrying a current of 5.4 A (Fig. 19.38). What are the magnitude and the direction of the force acting on the loop if it has the dimensions given in the figure?

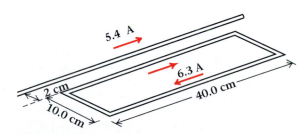

Figure 19.38
Problem 19.44.

19.45• Fifty turns of wire are wrapped into a coil 0.32 m in diameter and connected to a 12-V battery. If the total resistance of the coil is 0.90 Ω, what is the magnetic field at its center?

Section 19.7 Torque on a Current Loop

Hints for Solving Problems

A current loop has a magnetic moment $\mu = NIA$.

19.46 A coil of 32 turns of wire is wound about a circular form. The average diameter of the individual loops is 28.5 cm. Calculate the magnetic moment of the coil when it carries a current of 1.75 A.

19.47 What is the magnetic moment of a 12.5-cm-diameter coil of 150 turns when it carries a current of 75 mA?

19.48 A small coil has a cross section of 3.0 cm² and consists of 8 turns of wire. If the coil carries a current of 100 mA, what is the torque required to hold the coil at right angles to a field of $B = 0.010$ T?

19.49 A rectangular coil 4.3 cm wide by 5.8 cm long is made from 6 turns of wire. The coil carries a current of 200 mA in a magnetic field $B = 0.055$ T. How much torque is required to hold the coil so that it makes an angle of 45° with the magnetic field?

19.50● A rectangular coil 4.0 cm wide by 5.0 cm long is made from 10 turns of wire carrying a current of 50 mA. The coil is aligned with its magnetic moment along the direction of a uniform applied field $B = 0.50$ T. How much work is required to rotate the coil 180°?

Section 19.8 Galvanometers, Ammeters, and Voltmeters

19.51 A technician is building a voltmeter from a galvanometer movement that has a resistance of 10 kΩ and full-scale deflection at a current of 100 μA. If she places a 490-kΩ resistor in series with the galvanometer, what will be the full-scale voltage reading of the resulting meter?

19.52 A lab technician needs to make an ammeter that measures 15 A full scale. The only galvanometer available deflects full scale at a current of 100 μA, and its resistance is 10 kΩ. What resistance is needed and how should it be connected to the galvanometer?

19.53● The resistance of a galvanometer movement is 500 Ω, and the current required for full-scale deflection is 250 μA. (a) What resistance is needed to convert the galvanometer to an ammeter that reads 5.0 A full scale and how should it be connected? (b) Draw a circuit diagram for the resulting meter.

19.54● A student needs to make a voltmeter that measures 200 V full scale. The only galvanometer available deflects full scale at a current of 10.0 mA, and its resistance is 1.00 kΩ. (a) What resistance is needed and how should it be connected to the galvanometer? (b) Draw a circuit diagram for the resulting meter.

19.55●● A galvanometer movement (Fig. 19.26) is made with a rectangular coil of cross section A having N turns of wire and moving in a magnetic field B. A coil spring with spring constant κ restricts the rotational motion of the moving coil. When the torque due to the magnetic field equals the opposing torque of the spring, the pointer comes to rest. The spring follows a rotational version of Hooke's law; the torque is given by $\tau = \kappa\theta$. Derive an expression for the angular deflection of the moving coil in terms of the area and number of turns of the coil, the magnetic field in which it moves, the spring constant κ, and the current in the coil.

*Section 19.9 Ampère's Law

19.56 A 0.53-m-long solenoid consisting of 1000 turns is used to generate a magnetic field 0.15 T. How much current is required?

19.57 A solenoid 0.425 m long has 950 turns of wire. What is the magnetic field in the center of the solenoid when it carries a current of 2.75 A?

19.58 A large solenoid used in experiments with subnuclear particles produces an axial magnetic field of 0.800 T. The 2.00-m-long solenoid has 320 turns of wire. What current is required to produce the required field?

19.59● Figure 19.39 shows a cross-sectional view of a coaxial cable. The center conductor is surrounded by an insulating layer surrounded by an outer conductor. The current in the outer conductor is equal and opposite to the current in the inner conductor. Show that the magnetic field outside the coaxial cable due to its current is zero.

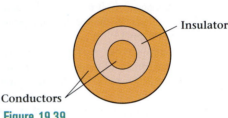

Figure 19.39
Problem 19.59.

19.60● A cylindrical conductor of radius a carries a uniformly distributed current I (Fig. 19.40). Show that the magnetic field within the conductor measured at a distance r from the center is given by the following equation:

$$B = \frac{\mu_0 I r}{2\pi a^2}.$$

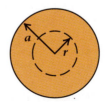

Figure 19.40
Problem 19.60.

19.61●● A toroid is a coil wrapped around a doughnut-shaped core (Fig. 19.41). (a) Use Ampère's law to calculate the field inside the toroid along the circle of radius r. (b) How is the field different from that of a solenoid? (c) What is the field outside of the toroid?

Additional Problems

19.62● A charged particle is observed to move in a circle of 0.0823-m radius in a plane perpendicular to a uniform mag-

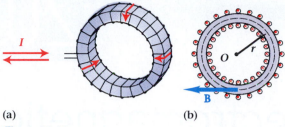

Figure 19.41

Problem 19.61: (a) A toroidal winding. (b) Cross section of a toroid.

netic field of 7.69×10^{-3} T. From other measurements, the momentum of the particle is known to be 2.03×10^{-22} kg · m/s. What is the electric charge of the particle?

19.63• A coil of area $A = 120$ cm^2 is made from 40 turns of wire. When a current of 2.5 A flows through the coil, it aligns with a uniform external field **B**. If it takes 0.65 J of energy to rotate the coil so that it is aligned antiparallel to **B**, what is the magnitude of the magnetic field?

19.64• A rectangular loop of wire oriented in the xy plane carries an electric current I. Show that for a uniform magnetic field directed along the z axis, the net magnetic force on the loop is zero.

19.65• Positive and negative particles are created in an experiment in which subnuclear particles are made to collide at high velocities in the center of a large empty solenoid. The design of the experimental apparatus calls for protons with speeds of 2.00×10^6 m/s perpendicular to the magnetic field to be deflected in a circular path of radius 2.00 m. If a current

of 1000 A can be supplied to the solenoid, how many turns per meter should it have?

19.66• A magnetic dipole with $\mu = 5.07 \times 10^{-3}$ A · m^2 is placed inside a long solenoid so that the magnetic moment makes an angle of 34.6° with the axis of the solenoid. The 50.3-cm-long solenoid is wound with 120 turns of wire and carries a current of 2.75 A. What is the torque on the dipole?

19.67•• Suppose you wish to make a magnetic dipole from a piece of wire of length l that can support a current I. How should you wind the coil to get the maximum dipole moment? Should you wind many small loops, thereby increasing N, or one large loop, increasing A? Reach your answer by deriving an expression for the magnetic moment as a function of I, l, and N.

19.68•• A freely turning compass needle in a uniform magnetic field oscillates with simple harmonic motion when displaced from equilibrium by a small angle. Find the equation for the oscillation frequency in terms of the moment of inertia I, the magnetic dipole moment μ, and the magnetic field strength B. (Refer to Sections 9.6, 14.1, and 14.5.)

19.69•• A deuteron with mass of 3.34×10^{-27} kg and charge of 1.60×10^{-19} C moves in a magnetic field of 5.70×10^{-2} T. What is the speed of the deuteron if it moves in a helical path of radius 0.0512 m and pitch 0.167 m?

19.70•• A galvanometer movement (Fig. 19.26) is made with a rectangular coil of cross section A having N turns of wire and moving in a magnetic field B. The spring that restricts the rotational motion of the moving coil follows a rotational version of Hooke's law; the torque is given by $\tau = \kappa\theta$. When the movement carries a current I, the coil is deflected through an angle θ. Find an expression for the magnetic field in which the coil moves.

Electromagnetic Induction

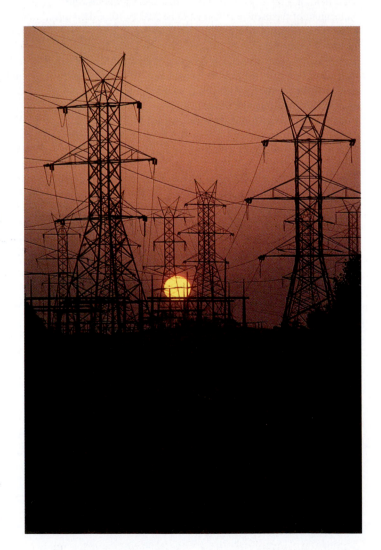

As we saw in the previous chapter, it was Oersted who first showed that an electric current could produce a magnetic field. Shortly afterward, Biot, Savart, and Ampère extended Oersted's discovery and put their results into a mathematical form. The converse question of whether a magnetic field could produce an electric current was a natural one, and it was a topic of much interest to Michael Faraday (Fig. 20.1). Faraday began to experiment with magnets and current-carrying wires at the Royal Institution in London. In 1831 he discovered that electric currents could be induced by magnetic fields. The same discovery was independently made at about the same time by the American Joseph Henry (1797–1878) and by the Russian Heinrich Lenz (1804–1865).

When Henry published his results in 1832, he also explained that a changing electric current in a coil can induce another current in the *same* coil. As a result, the current in a coil consists of two components, the initial current plus an induced current. This effect is known as self-inductance or, more simply, inductance.

(In 1834, Faraday also discovered self-inductance independent of Henry. This time, however, Henry received the credit for being first.) Inductance, like resistance and capacitance, plays an important role in the behavior of electrical circuits.

During the middle of the nineteenth century, James Clerk Maxwell (1831–1879) summed up what was then known about electromagnetism in a set of only four equations. These four equations, now known as Maxwell's equations, are one of the great unifying achievements in physics. They form a complete theory and hold a central position in the theory of electromagnetism, analogous to the central position of Newton's laws in the theory of mechanics. Maxwell's equations sum up the concepts developed in this and the previous four chapters and often serve as the starting point for advanced texts in electromagnetism.

Applications of electromagnetism surround us in such forms as electric generating plants and their network of power lines, and appliances that depend on motors for their operation. Our communication and information networks are all based on electromagnetism, and we can trace their roots to the work of Maxwell and his predecessors. ■

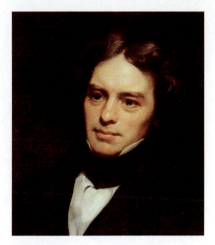

Figure 20.1
Michael Faraday (1791–1867) discovered magnetic induction.

20.1   Faraday's Law

Faraday made an important discovery while he was investigating the effect of a current-carrying solenoid on a second solenoid wound over the first one (Fig. 20.2a). One coil was joined to a battery, while the other was connected to a galvanometer. Even a large current flowing through the first coil caused no detectable current in the other. However, Faraday did notice an effect when he connected or disconnected the circuit of the first coil. At the moment when the first circuit was closed, a sudden and slight movement occurred in the galvanometer in the second circuit. When the circuit was opened, again a momentary deflection of the galvanometer occurred, but this time in the opposite direction. Further experiments showed that the current in the second (or secondary) coil depended not on the primary current in the first coil, but on the *rate of change* of the current. This effect was greatly enhanced by winding the coils around an iron ring (Fig. 20.2b).

Faraday continued his research and observed that if he thrust a magnet into a coil of wire (Fig. 20.3a), a current was induced in the coil *while the magnet was moving* relative to the coil. Moving the magnet away from the coil caused the galvanometer to deflect in the opposite way (Fig. 20.3b). Thus, the motion of the north pole of a bar magnet away from the coil induced a current whose direction was opposite to that caused by the motion of the north pole toward the coil. Faraday also noted that moving the magnet toward the coil had the same effect as moving the coil toward the magnet; only the relative motion was important. Furthermore, he observed that bringing the south pole of a bar magnet toward the coil caused the galvanometer to deflect opposite to the way it deflected when the north pole of the magnet was brought toward the coil. In other words, moving the south pole of a bar magnet toward the coil produced the same effect as moving the north pole away from the coil.

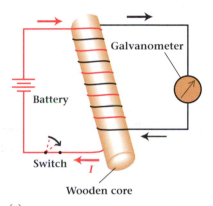

(a)

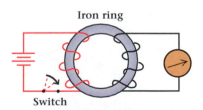

(b)

Figure 20.2
(a) A simplified drawing of Faraday's experiment. Two separate coils are shown, one connected to a battery and the other connected to a galvanometer. A change in the current in the first coil produced a current in the second coil. (b) Faraday wound the coils about a ring of soft iron to enhance the effect.

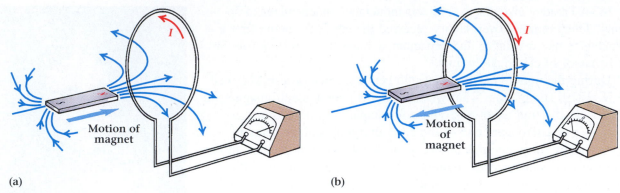

(a)   (b)

Figure 20.3

(a) Moving a magnet toward a coil generates an electric current in the loop, detected by the galvanometer. (b) Moving the same end of the magnet away from the coil generates a current in the opposite direction.

All these observations were finally explained by Faraday's realization that the currents observed in the secondary coil were entirely due to the changing magnetic flux within the coil. We can summarize Faraday's observations by saying that **the emf induced in a loop of wire is proportional to the rate of change of magnetic flux through the coil.** This statement is known as **Faraday's law.** We often call the emf induced by a changing magnetic flux an **induced emf,** and the current it produces is called an **induced current** or an induction current.

Remember that the amount of magnetic field that passes through a given area, such as the area enclosed by a current-carrying loop of wire, contributes to the magnetic flux, which was introduced in Section 19.2. The maximum number of field lines pass through a wire loop when its area is perpendicular to the direction of the magnetic field. As the loop turns so that the angle between its plane and the field gets smaller, the effective area becomes smaller and fewer field lines can pass through it. For this reason, the definition of magnetic flux includes the effect of the angle between the area and the field lines (Fig. 20.4). As before, we define the orientation of the area as the direction of its normal (a line perpendicular to the area). In defining the flux, the angle θ is the angle between the field lines and the normal line.

For a region of uniform magnetic field, the magnetic flux ϕ_{m} through an area A is the product of the magnitude of the magnetic field $|\mathbf{B}|$, the area A, and the cosine of the angle θ between the direction of the field and the direction of the normal to the area:

$$\phi_{\mathrm{m}} = BA \cos \theta. \tag{20.1}$$

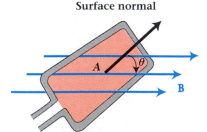

Figure 20.4

Magnetic field passing through an area A. The normal to the area makes an angle θ with the direction of the magnetic field. The magnetic flux depends on the orientation of the field to the area through the factor $\cos \theta$.

The SI unit for the magnetic flux is the weber (Wb). The relationship between the weber and the unit for the magnetic field, the tesla, is $1 \text{ Wb} = 1 \text{ T} \cdot \text{m}^2$.

Shortly after Faraday's discovery, Heinrich Lenz, working in Russia, extended our understanding of induction currents by stating a related principle. Although Faraday knew how to determine the direction of an induced current, Lenz expressed this relationship more directly. **Lenz's law** states that **the direction of an induced current is such that its own magnetic field opposes the original change in magnetic flux that induced the current.** This law is illustrated in Fig. 20.5, which shows the south pole of a bar magnet moving toward a conducting loop. The direction of the magnetic field is indicated by the small arrows. As the magnet approaches the loop, the amount of field passing through the loop increases, thus increasing the magnetic flux through the loop. The increasing flux induces an emf in the loop, and, since the loop is closed, the emf induces a current I. Now, the direction of this induced current is such as to generate a magnetic field in the direction that opposes the change in the flux. In this case, the initial field is opposite to the direction of motion of the magnet (Fig. 20.5a), so the induced field must be in the direction of motion of the magnet (Fig. 20.5b). According to the right-hand rule, the current is clockwise as seen from the position of the magnet. If the magnet were moved away from the loop, the flux through the loop would decrease and a current would be induced in the opposite direction. The field due to the induced current would try to maintain the field through the loop constant. That is, the induced field opposes the change in external flux.

We can write Faraday's law of induction mathematically with the equation

$$\mathcal{E} = -\frac{\Delta\phi_m}{\Delta t}. \tag{20.2}$$

Here $\mathcal{E}$ is the induced emf in one loop of wire, $\Delta\phi_m$ is the change in magnetic flux, and Δt is the time interval over which the change takes place. The minus sign indicates the direction of the emf and is in agreement with Lenz's law. If we apply this equation to a coil containing N loops or turns of wire, then

$$\mathcal{E} = -N\frac{\Delta\phi_m}{\Delta t}. \tag{20.3}$$

Faraday's and Lenz's laws are in accord with the law of conservation of energy. When a magnet is moved closer to a coil (or the coil is moved toward the magnet), by Faraday's law an emf is created that can generate an electric current if the coil is a closed circuit. By Lenz's law, the direction of this induced current is such that the resulting magnetic field exerts an opposing force on the moving

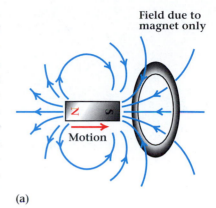

(a)

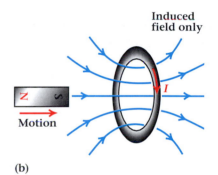

(b)

Figure 20.5
(a) The motion of a magnet toward a coil induces a current in the coil. As the S pole gets closer to the loop, the flux through the loop increases. (b) By Lenz's law, the direction of the induced current produces a magnetic field that opposes the increase in the flux.

Master the Concept

Lenz's Law

Question: Why does the changing magnetic flux in a conducting loop always induce a current that generates a field opposing that change in flux?

Answer: Suppose the changing flux is due to the motion of a bar magnet moving along the axis of the loop. If, contrary to Lenz's law, the induced field assisted the change in flux, we could give the magnet a slight push toward the loop and the resulting induced field would attract the magnet into the loop. Because the accelerating magnet would be simultaneously gaining kinetic energy and dissipating electrical energy in the loop, such behavior would violate the principle of conservation of energy. Thus the behavior of the induced current must oppose the change in flux as Lenz's law requires in order to be consistent with the conservation of energy.

magnet. In other words, if we push the magnet toward the coil, the induced current generates a magnetic field that opposes the push. If we pull the magnet away from the coil, the induced magnetic field reverses direction and opposes that, too. Thus we always encounter a resisting force and are required to do work to induce a current. The work done on the magnet-coil system appears as energy expended in the form of I^2R heating losses in the coil.

Problem-Solving Strategy

Induced Emf and Induced Current

The emf induced in a loop of wire is proportional to the rate of change of the magnetic flux through the loop. For a fixed loop, the flux changes if the magnetic field changes magnitude or direction. Even when the field is fixed, the flux will change if the area of the loop is changed or if the loop orientation is changed relative to the direction of the field. A current due to the induced emf occurs when the loop is closed or is part of a closed circuit.

Example 20.1

A square coil 10 cm on a side has 20 turns of wire. Initially, it is at rest between the poles of a magnet whose field is 0.25 T, with the plane of the coil perpendicular to the field (Fig. 20.6). What average voltage (emf) is produced on the coil if it is withdrawn completely from the field in 0.10 s?

Strategy To find the emf, we must first find the change in magnetic flux $\Delta\phi_m$. Then from knowledge of the time interval Δt, we can use Eq. (20.3) to compute $\mathcal{E}$. From Eq. (20.1), we know that $\phi_m = BA\cos\theta$. Because the field is perpendicular to the plane of the coil, $\cos\theta = 1$, and the initial flux is just the product of the field B with the area of the coil. For a square coil of side length l, the area is l^2. Thus the initial flux is

$$\phi_i = Bl^2.$$

After the coil is withdrawn from the magnet, $\phi = 0$. Thus

$$\Delta\phi_m = 0 - \phi_i = -\phi_i.$$

Solution The induced emf is

$$\mathcal{E} = -N\frac{\Delta\phi_m}{\Delta t} = \frac{NBl^2}{\Delta t} = \frac{20(0.25 \text{ T})(0.10 \text{ m})^2}{0.10 \text{ s}}$$

$$\mathcal{E} = 0.50 \text{ V}.$$

Discussion Let us also check to be sure that the units are correct. The units above are, as expected,

$$\frac{\text{T} \cdot \text{m}^2}{\text{s}} = \frac{(\text{N/A} \cdot \text{m}) \cdot \text{m}^2}{\text{s}} = \frac{\text{N} \cdot \text{m}}{\text{A} \cdot \text{s}} = \frac{\text{J}}{\text{C}} = \text{V}.$$

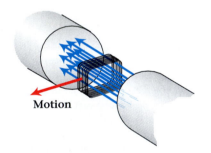

Figure 20.6

Example 20.1: A coil is quickly pulled out of the field.

Motion

Example 20.2

Compute the current through a 37-Ω resistor connected to a 5-turn circular loop 10 cm in diameter (Fig. 20.7), assuming that the magnetic field through the loop is increasing at a rate of 0.050 T/s. What is the current in the loop and what is its direction?

Strategy We can find the current through the resistor if we know the voltage across it. That voltage is the emf developed in the coil by the changing magnetic flux. Faraday's law (Eq. 20.3) gives the emf in terms of the changing flux. In turn, the flux is found from Eq. (20.1). In this case the flux changes because the magnetic field B changes.

Solution If the resistance of the wire is small compared with that of the resistor, the current through the loop depends only on the emf and the 37-Ω resistance. According to Ohm's law:

$$I = \frac{\mathscr{E}}{R}.$$

The emf is obtained from Faraday's law,

$$\mathscr{E} = -N\frac{\Delta\phi_m}{\Delta t}.$$

From the definition of the magnetic flux we see that

$$\Delta\phi_m = \Delta BA = \Delta B\pi\left(\frac{D}{2}\right)^2,$$

where D is the diameter of the loop. Thus, the emf becomes

$$\mathscr{E} = -N\pi\left(\frac{D}{2}\right)^2\frac{\Delta B}{\Delta t}.$$

The current is then found to be

$$I = \frac{\mathscr{E}}{R} = -\frac{N\pi D^2\frac{\Delta B}{\Delta t}}{4R}$$

$$I = -\frac{5\pi(0.10\text{ m})^2(0.050\text{ T/s})}{(4)(37\text{ }\Omega)} = -53\mu A.$$

The minus sign reminds us that for a particular case, we need to determine the direction of the current so that the induced magnetic field opposes the given change in flux. Since the field is increasing into the loop as seen in Fig. 20.7, the direction of the current corresponds to generating a magnetic field out of the page. Thus, the current is counterclockwise.

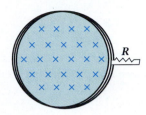

Figure 20.7
Example 20.2: The magnetic field within the coil is increasing. The crosses indicate that the magnetic field is directed into the plane of the page.

20.2   Motional Emf

In our discussion of Faraday's law, we examined the changing flux in a loop of wire due to the relative motion of the field and the loop. Now we extend our study to include the effects of a magnetic field on a straight piece of wire or other

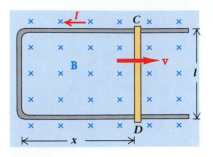

Figure 20.8

A conducting bar sliding along a stationary U-shaped conductor that is perpendicular to a magnetic field **B**. The area of the conducting loop changes, inducing an emf. The crosses indicate a magnetic field directed into the plane of the loop.

conductor that moves relative to the field. We will see that an emf is developed between the ends of the wire.

To investigate the induction of emf on a moving conductor, let's consider a conducting bar of length l sliding along a stationary U-shaped conductor that is perpendicular to a uniform magnetic field **B** (Fig. 20.8). The bar moves with a constant speed v in the x direction. As the bar moves, the area enclosed by the loop consisting of the bar and the U-shaped conductor increases. Consequently, the magnetic flux through the loop increases. The emf developed around the loop is obtained from Faraday's law:

$$\mathcal{E} = -\frac{\Delta \phi_m}{\Delta t} = -\frac{\Delta(BA)}{\Delta t} = -B\frac{\Delta A}{\Delta t} = -B\frac{(l\,\Delta x)}{\Delta t} = -Bl\frac{\Delta x}{\Delta t},$$

where A represents the area enclosed by the loop. Since $\Delta x/\Delta t = v$, we have

$$\mathcal{E} = -Blv.$$

According to Lenz's law, the induced current yields a magnetic field to oppose the change in flux. This field is directed out of the page and corresponds to a counterclockwise current in the loop shown in Fig. 20.8.

We should also be able to understand this result by considering the force due to the magnetic field acting on the charges inside the moving bar. The development of an emf between the ends of the moving bar CD depends on the motion of the bar perpendicular to the magnetic field. The emf generated in this manner is called a *motional emf.*

We can explain motional emf as follows. In Chapter 19 we saw that a charge q moving with speed v at right angles to a magnetic field B experiences a force $F = qvB$. The direction of this force is perpendicular to both the direction of motion and the magnetic field. When the bar moves in the x direction (Fig. 20.8), the magnetic field exerts a force on the mobile charges along the direction of the bar. The work needed to move a positive charge q from D to C is*

$$W = Fl = qvBl.$$

The emf is an electric potential difference. Thus, emf is the work per unit charge to move a charge q from D to C:

$$\mathcal{E} = \frac{W}{q} = Blv. \tag{20.4}$$

The emf of Eq. (20.4) corresponds to the potential difference $VC - VD$. The direction of this emf would drive a counterclockwise current if contact were made between the bar and the U-shaped conductor.

Motional emf is due to the force of a magnetic field on moving charges and cannot be used to deduce Faraday's law. Faraday's law predicts an induced emf in a loop when we move the wire loop *or* when we hold the loop at rest and change the field. However, as we have seen, when the bar slides on the U-shaped conductor (a closed loop), the emf obtained from Faraday's law for the loop is the same as the motional emf for the bar.

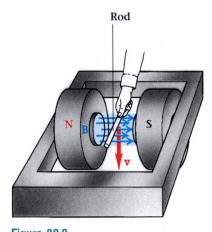

Figure 20.9

Example 20.3: An aluminum rod moves through the magnetic field of a laboratory magnet.

---

*As we mentioned earlier, the mobile charges in a metal rod are the negatively charged electrons, which would move in the opposite direction (from C to D). However, the resulting emf is unchanged.

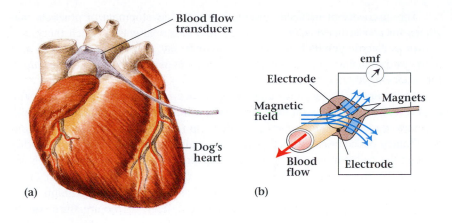

Blood flow transducer

Electrode

Magnetic field

Magnets

emf

Dog's heart

Blood flow

Electrode

(a)

(b)

(c)

Flow signal

Time →

Figure 20.10
(a) A magnetic blood flow transducer can be used to make a blood-flow meter. (b) Schematic diagram of the transducer that generates and emf that depends on the blood flow. (c) Amplitude of the signal as a function of time.

Example 20.3

A 0.27-m aluminum rod is rapidly moved between the poles of a laboratory magnet that generates a magnetic field of 0.89 T (Fig. 20.9). With what speed must the rod be moved so that the emf developed between the ends is 1.5 V—about the same as the emf of a single dry-cell battery? The rod and its motion are perpendicular to the direction of the magnetic field.

Solution   We may use Eq. (20.4) and solve for the speed v.

$$\mathscr{E} = Blv,$$

$$v = \frac{\mathscr{E}}{Bl} = \frac{1.5 \text{ V}}{(0.89 \text{ T})(0.27 \text{ m})} = 6.2 \text{ m/s}.$$

Electrodynamic flow meters are made possible by the behavior of charges moving in a magnetic field. When a conducting fluid, such as blood, flows in an insulating tube between the poles of a small magnet, ions* in the fluid experience a force at right angles to both the direction of flow and the magnetic field. The motion of these ions transverse to the direction of flow generates an electric potential that can be measured externally by electrodes inserted into the tube on opposite ends of a diameter. In the case of blood flow through intact blood vessels, the walls are sufficiently conductive to permit use of a flow meter applied as a cuff around the vessel (Fig. 20.10). The magnitude of the potential difference is given by Eq. (20.4), where B is the magnetic field, l is the distance between the electrodes, and v is the flow velocity.

20.3   Generators and Motors

Faraday's results had very practical consequences. They led directly to the invention of the transformer, the alternator, and the generator. These inventions, together with that of the electric motor, brought about the many applications of electric energy that are so widespread today.

*Ions are atoms or molecules that have a net electric charge.

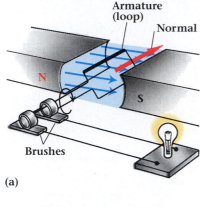

(a)

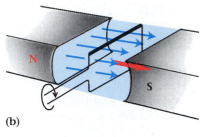

(b)

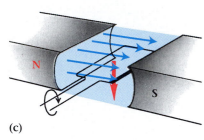

(c)

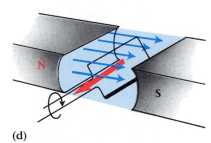

(d)

Figure 20.11

Operation of an ac generator, called an alternator. (a) A loop of wire (the armature) rotates in a magnetic field. The induced alternating emf enters the external circuit through metal brushes that rub against rotating metal rings, called slip rings, attached to the armature. As the armature rotates clockwise to (b), the magnetic flux increases to a maximum. When the armature rotates to (c), the flux is zero. At position (d), the flux is negative relative to positions (a) and (b).

The discovery of induced currents led to the development of practical machines for converting mechanical energy into electrical energy. Such machines, known as electric **generators,** may be classified into two main types: *dc generators,* for producing unidirectional current, and ac generators, called *alternators,* for producing alternating current (ac).

The simplest ac generator consists of a loop of wire turning in a magnetic field (Fig. 20.11a). The wire loop (called the armature in this application) is connected to a pair of rotating contacts called slip rings, which rub against stationary contacts called brushes. As the armature rotates in the magnetic field, the flux through the coil continuously changes. This induces a changing emf, which appears at the brush contacts. As the armature rotates clockwise from part (a) to part (b) of Fig. 20.11, the magnetic flux increases, reaching a maximum when the normal to the loop is parallel to the magnetic field. As the armature continues to rotate past the position of part (b), the flux decreases, reaching zero when the normal to the loop is 90° from the direction of the field (part c). When the loop rotates past the 90° position (part d), the direction of the normal is opposite to the direction of the field, so the flux through the armature is negative with respect to the flux in parts (a) and (b).

If the field B is uniform and the loop rotates with a constant angular speed ω, the magnetic flux through the single loop of area A may be expressed as

$$\phi_m = BA \cos \theta = BA \cos \omega t.$$

This flux is shown as a function of time in Fig. 20.12(a). From Faraday's law, we calculate the emf to be

$$\mathcal{E} = -\frac{\Delta \phi_m}{\Delta t} = -\frac{BA \; \Delta(\cos \omega t)}{\Delta t}.$$

Now, the change of cos ωt with the change in time is just the slope of the curve of cos ωt versus t. In Chapter 14 we saw that the rate of change of the cosine function was proportional to the sine function. Using this result in the last equation, we get

$$\mathcal{E} = \omega BA \sin \omega t. \tag{20.5}$$

Thus, the output of an alternator is a sinusoidal emf (Fig. 20.12b). When an alternator is connected to a closed circuit, it produces a sinusoidally **alternating current.** In the next chapter we will examine alternating-current circuits, which make up the majority of electric circuits you are familiar with. When we mention ac sources there, we will generally be referring to alternators.

A dc (unidirectional) generator can be made from the same rotating armature if we replace the slip rings by a commutator (Fig. 20.13a). The commutator is simply a slip ring split across the diameter, so that as the loop rotates, it changes the direction of its connection to the output circuit. During one half of a rotation, $\Delta\phi_m/\Delta t$ is positive and the induced current goes toward brush 1. During the other half of the rotation, the change in flux is negative and the induced current goes in the other direction; however, because the commutator switches its contact with the brushes, the current still goes toward brush 1. If the brush contacts are properly arranged with respect to the commutator and the position of the loop, the polarity of the emf can remain constant (Fig. 20.13b).

Practical generators are similar in principle to the single-loop generator just described. In order to increase the emf in a practical generator, many turns of wire are wrapped about a rotor or armature, which turns in a magnetic field. The magnetic field may be due to permanent magnets or to electromagnets surrounding the rotor. If there are N loops in the armature, then the right-hand side of Eq. (20.5) is multiplied by N.

As long as the external circuit is open, no current flows through the armature and it turns freely. But when the terminals of the generator are joined to a closed circuit (Fig. 20.14) so that current flows through the loop, a torque results that resists the motion of the loop. Mechanical energy must be supplied to the generator to overcome this countertorque. As more current is drawn through the loop, more external torque is required to turn the generator. In other words, mechanical energy must be supplied as electrical energy is used in the external circuit. The mechanical energy needed to turn the generators of Fig. 20.15 comes from the water flowing through the Grand Coulee Dam.

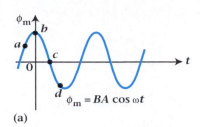

(a)

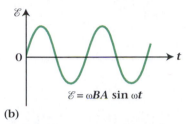

(b)

Figure 20.12
(a) Graph of the magnetic flux as a function of time for an armature rotating at constant angular velocity. The points on the curve refer to the positions of the armature in parts (a)–(d) of Fig. 20.11. (b) Graph of the voltage output of an alternator versus time.

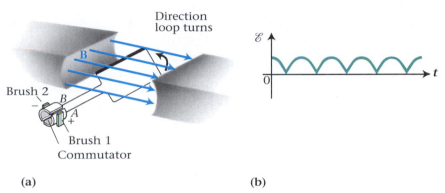

(a)

(b)

Figure 20.13
A dc generator. (a) As the loop rotates in the magnetic field, an induced emf is produced between contacts A and B. As the loop rotates to a position where the polarity of the emf between the commutator contacts reverses, the commutator moves to a new position so that contact A touches brush 2 and contact B touches brush 1. Thus, the polarity of the output potential from brush 1 to 2 is maintained. (b) Graph of the voltage output of the dc generator versus time.

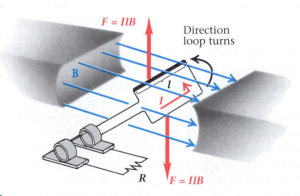

Figure 20.14
As the loop rotates counterclockwise, an emf is induced, which, according to Lenz's law, causes a current through the loop in the direction shown by the arrows. The force F of the field on the current-carrying wire is $F = IlB$, which causes a torque opposing the motion of the loop.

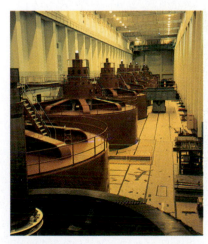

Figure 20.15
A row of ac generators at the Grand Coulee Dam hydroelectric plant.

Example 20.4

A coil made of 25 loops, each of area 0.010 m², is rotated about an axis perpendicular to the earth's magnetic field (about 5.0×10^{-5} T) at a frequency of 60 Hz. What is the peak emf generated by the coil?

Solution The peak emf in the coil is given by Eq. (20.5), modified by including the number of loops N and with sin $\omega t = 1$:

$$\mathscr{E} = N\,\omega BA.$$

For 25 loops, the total emf is $\mathscr{E} = 25\,\omega BA$. The angular frequency $\omega = 2\pi f = 2\pi(60 \text{ Hz}) = 377$ rad/s. The emf becomes

$$\mathscr{E} = (25)(377 \text{ rad/s})(5.0 \times 10^{-5} \text{ T})(0.010 \text{ m}^2)$$
$$\mathscr{E} = 4.7 \times 10^{-3} \text{ V} = 4.7 \text{ mV}.$$

An **electric motor** is similar in operation and design to a generator, but works in a converse manner. That is, the motor turns electrical energy into mechanical energy. In the simplest case, a current is passed through a single-loop armature in a magnetic field, causing the loop to rotate. For example, if we apply an external dc potential to the brushes of the elementary dc generator of Fig. 20.13, producing a current in the loop, the magnetic field will exert a torque on the armature, causing it to rotate. As the armature rotates through the position where the plane of the loop is perpendicular to the field, the commutator reverses the contacts, so a torque in the same direction continues to turn the armature. Here the input is electrical energy (from an external source) and the output is mechanical energy (a torque acting through an angle).

Although small motors are often built with permanent magnets, most motors are built so that electromagnets provide the field for the armature. In the simplest type of motor, the armature is wired in series with this field magnet. In such a series-wound motor, the magnetic field increases proportionally with increases in armature current. Since the torque depends on both the field strength and the current, we can control the speed of the motor by regulating its current. Motors of this kind are often used in power tools. In a shunt-wound motor, the field magnet is wired in parallel with the armature winding. In this case, the motor runs with a speed that is nearly independent of the load on the motor. Both series-wound and shunt-wound motors are used primarily in dc applications.

Most of the motors you are likely to use are induction motors, which have no contacts to the armature. Instead, the armature windings are wrapped around an iron core. The motor is designed so that an alternating current applied to the field electromagnet produces a rotating magnetic field, which induces a current in the armature windings. The torque on the armature due to the rotating field causes the armature to rotate. Induction motors are widely used in fans, refrigerators, and other common appliances. Induction motors can be designed to rotate synchronously with the field, which means that their rotation frequency is directly related to the frequency of the alternating current. Motors of this type are widely used in phonograph turntables and electric clocks.

We have already mentioned the presence of countertorque in generators. A related effect occurs in motors. As the armature moves in the field, the motion induces an emf in the armature. This emf is directed in the opposite sense to the applied voltage and is thus called the **back emf.** When a motor is first switched on, a large current surges through the armature wire. As the motor gets up to speed, the back emf opposes the applied potential. The net effect is a reduced potential across the armature and thus a smaller armature current.

You often see the effect of back emf in household circuits when you turn on a large appliance motor. For example, a momentary dimming of the lights often accompanies the start-up of an air conditioner or refrigerator motor. The dimming occurs because the terminal potential difference of the house circuit drops as the motor draws a large start-up current. As the motor comes up to speed, its current demand decreases, along with the corresponding I^2R losses in the supply circuit. When this happens, the household circuit potential rises back to its initial value.

Because the back emf reduces the current, a motor that can be safely operated under normal conditions may become overheated and burn out if its rotation is hindered or stopped. For example, an electric mixer that is unable to turn because the batter to be mixed is too thick or because a spoon is caught in its blades will draw more current than it is designed to carry. Consequently, it will overheat, and may even burn out, if it is not switched off or disconnected from the power line.

In recent years, much effort has gone into developing extremely small motors known as microelectromechanical devices (Fig. 20.16). These devices operate on the same basic physical principles we have discussed. Because of their revolutionary small size, they have potential applications in many different fields including medicine.

(a)

(b)

Figure 20.16
(a) The world's smallest electric car. (b) The micro motor that propels the micro car.

Example 20.5

When the motor in a window air conditioner is first turned on, it momentarily draws 40 A. Then the current quickly drops to a steady value of 13.8 A. If the motor is operated on a 120-V power source, what is the back emf generated while the motor is running?

Solution For the purpose of this problem, we will approximate a motor by a series connection of an inductive source of emf and a resistance. If we apply Kirchhoff's voltage rule (Section 18.6) to the circuit, we obtain

$$V - \mathcal{E} = IR,$$

where V is the 120 V of the power line, $\mathcal{E}$ is the back emf of the motor, I is the current in the circuit, and R is the resistance of the motor. When the air conditioner is first turned on, the motor is at rest, there is no back emf ($\mathcal{E} = 0$) and the current is 40 A. The resistance can then be determined:

$$R = \frac{V}{I} = \frac{120\,\text{V}}{40\,\text{A}} = 3.0\ \Omega.$$

When the motor is running at normal speed, the current is 13.8 A, so the back emf must be

$$\mathcal{E} = V - IR = 120\,\text{V} - (13.8\,\text{A})(3.0\ \Omega) = (120 - 41)\,\text{V} = 79\,\text{V}.$$

The Transformer

An extremely important application of electromagnetic induction is the **transformer,** which is used to increase or decrease an ac voltage without appreciable loss of power. A transformer has two multiturn coils of wire wound on the same iron core (Fig. 20.17). The common ferromagnetic core increases the magnetic flux due to the input current and maximizes the magnetic coupling between the coils. When an ac voltage V_1 is applied across the primary coil (input) of the transformer, the resulting current induces a changing magnetic flux through the iron core that is proportional to the number of turns N_1 in the primary coil. This changing flux travels through the core and intercepts the secondary winding, where it induces a voltage V_2 proportional to the number of turns N_2. This voltage is the output voltage of the transformer. The ratio of input to output voltages is

$$\frac{V_1}{V_2} = \frac{N_1}{N_2}.$$

(20.6)

By a suitable choice of N_1 and N_2, we can make the transformer step up (increase) the voltage output or step down (decrease) the output with respect to the input.

If we connect a resistive load across the secondary coil of the transformer, a current I_2 will pass through the load. This current is due to the potential difference V_2 induced by the changing flux. If we neglect any energy losses within the transformer, the law of conservation of energy requires that the electrical energy output of the transformer equal the electrical energy input. Since power is the rate of change of energy, this is equivalent to saying that the input and output powers are equal. The power is $P = IV$, so we can express the conservation rule as

$$I_1 V_1 = I_2 V_2.$$

(20.7)

The above argument assumes no energy loss due to Joule heating or flux leakage. In practice, these effects make the output power less than the input power. However, in a well-designed transformer the loss is a small fraction of the total power. Typical transformers have losses of about 4–8%.

The primary application of the transformer is in raising or lowering an ac voltage to a desired level. But, as Eq. (20.7) shows, this change cannot be

Figure 20.17

(a) Sketch of a transformer. We can choose the number of turns on the input coil N_1 and the output coil N_2 to obtain the desired ratio of input voltage to output voltage. (b) The circuit symbol for a transformer.

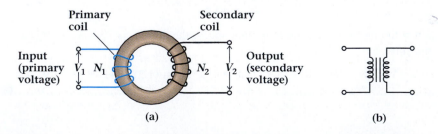

(a)

(b)

Figure 20.18
(a) Commercial transformers at the Hoover Dam hydroelectric power station. You can judge the size of the transformers from the truck parked near them. (b) Transmission lines carry electric power from the generators to the customers.

achieved without affecting the available current. That is, for a given input voltage and current, you cannot increase the output voltage without decreasing the available output current. This is a consequence of energy conservation.

The transmission of electrical power from a generating station to the consumer often takes place over many miles of wire. In order to minimize the I^2R losses in the transmission lines, it is standard practice to use transformers to step up the voltage to very high levels (Fig. 20.18a). Typical transmission-line voltage is 138 kV, but in some instances voltages as high as 1 MV are used. Because power is the product of current and voltage, the current required for a given level of power is reduced when the transmission lines are operated at high voltage. The lower current reduces the I^2R power loss. At the end points of the distribution system, the voltage levels are stepped down to more modest levels. These features—the ability to shift voltage levels up and down and to minimize power losses during transmission—are important reasons for the use of ac instead of dc electricity in power distribution networks.

Example 20.6

A 20-W high-intensity bulb in a desk lamp has a resistance of 7.2 Ω when burning. The lamp's power comes from the secondary of a small transformer, the primary of which is connected to a 120-V circuit. (a) What is the ratio of the number of turns on the primary winding to the number of turns on the secondary winding? (b) What is the minimum current through the primary when the lamp is on?

Strategy To find the relative number of turns, we first need to find the secondary voltage. We can calculate the secondary voltage from knowledge of the lamp's resistance and the power expended. Then we can use the primary and secondary voltages to find the ratio of turns. We will assume that the power input and the power output are equal and use Eq. (20.7) to find the primary current.

Solution   (a) The secondary voltage V_2 is found from

$$P_2 = I_2 V_2 = \frac{V_2^2}{R_2},$$

$$V_2 = \sqrt{P_2 R_2} = \sqrt{(20 \text{ W})(7.2 \text{ } \Omega)} = 12 \text{ V}.$$

According to Eq. (20.6), the ratio of the number of turns is equal to the ratio of the voltages. Setting $V_1 = 120$ V, we have

$$\frac{N_1}{N_2} = \frac{V_1}{V_2}$$

$$\frac{N_1}{N_2} = \frac{120 \text{ V}}{12 \text{ V}} = 10.$$

That is, the primary coil has 10 turns of wire for each turn of wire in the secondary coil.

(b) We now equate the input power to the output power (Eq. 20.7) to find the primary current:

$$I_1 V_1 = I_2 V_2 = 20 \text{ W},$$

$$I_1 = \frac{20 \text{ W}}{120 \text{ V}} = 0.17 \text{ A}.$$

Discussion   The current in the primary will actually be larger than the calculated value of 0.17 A. You can understand why this is so by feeling such a transformer when it is operating. It will be warm, indicating that some of the electrical energy is being converted to thermal energy. The power in the primary circuit must include the power delivered to the secondary circuit plus the power that produces the heating. The actual input current for the high-intensity lamp was measured at 0.25 A. Thus the lamp consumes 30 W input, delivers 20 W to the bulb, and dissipates 10 W as waste heat in the transformer.

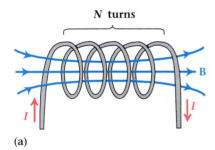

N turns

(a)

(b)

Figure 20.19
(a) A current I generates a flux ϕ passing through N turns. When the current changes the flux changes, causing an induced emf to appear that is proportional to the change in the current. The proportionality constant is the inductance of the coil. (b) The circuit symbol for an inductor.

20.5	**Inductance**

We now want to examine how a magnetic field builds up when a current is passed through a wire and what emf is induced by this changing magnetic field. Let's begin by considering a coil of N turns of wire carrying a current I. The current generates a magnetic field that passes through each loop of the coil (Fig. 20.19a). The total magnetic flux passing through the coil, which is the product of the number of turns and the flux through each turn, is proportional to the current carried by the coil. We can therefore write

$$N\phi_\text{m} = LI. \tag{20.8}$$

The proportionality constant L is called the **inductance** of the coil and depends only on the shape and size of the coil. Sometimes L is referred to as *self-inductance*, to distinguish it from the *mutual inductance* between two coils.

If the current in a coil changes, as in the case of alternating current, a corresponding change in the flux occurs, given by

$$\Delta(N\phi_m) = L\, \Delta I.$$

According to Faraday's law, the changing flux induces an emf given by

$$\mathcal{E} = -\frac{\Delta(N\phi_m)}{\Delta t} = -L\frac{\Delta I}{\Delta t} \qquad (20.9)$$

Thus, a changing current in a coil induces an emf opposing that change, and this induced emf is proportional to the inductance of the coil. It is important to remember that this induced emf is present only while the current is changing.

The units of inductance are volt-second per ampere. This combination of units has been given the name **henry** (abbreviated H):

$$1 \text{ henry} = 1 \text{ volt-second/ampere.}$$

A circuit component that primarily exhibits inductance, like a solenoid or other wire coil, is called an **inductor.** In a circuit, an inductor may be regarded as a source of emf of $-L\,(\Delta I/\Delta t)$. As such, an inductor acts to resist changes in current. Along with resistors and capacitors, inductors are found as circuit elements in many applications including radios, television sets, and tape recorders. The circuit symbol used to represent the inductor is a stylized coil (Fig. 20.19b).

Inductors are usually made from many turns of wire wound about a supporting core. The greater the number of turns of wire, the greater the total magnetic flux for a given current, and that flux passes through more turns. So the inductance is larger for a larger number of turns. Inductors are commercially available in sizes ranging from a few microhenries to several hundred henries. Inductors made with only a few turns of wire about a nonmagnetic support have small inductances. Large inductances are made with many windings about a magnetic core. The magnetic material increases the magnetic flux passing through the coils of wire and thus increases the value of the inductance.

Inductors have resistance in addition to their inductance because the wire used in the windings necessarily has resistance. In a circuit, a real inductor contributes both inductance and resistance.

Example 20.7

Calculate the self-inductance L of a solenoid containing N turns of wire in a coil of length l and cross-sectional area A. Then find the value of L given that $N = 100$ turns, $l = 3.0$ cm, and $A = 2.0$ cm^2.

Strategy We can find the self-inductance L from Eq. (20.8), $L = N\phi_m/I$, but first we must find the flux ϕ_m. The field of a solenoid was calculated in Chapter 19 to be

$$B = \mu_0 n I = \mu_0\left(\frac{N}{l}\right)I.$$

Since the field is uniform inside the solenoid, the corresponding flux is

$$\phi_m = BA = \frac{\mu_0 NIA}{l}.$$

Solution Substituting for ϕ_m in Eq. (20.8), we find

$$L = \left(\frac{N}{I}\right)\left(\frac{\mu_0 NIA}{l}\right) = \frac{\mu_0 N^2 A}{l}.$$

We can compute the numerical size of L from this last equation if we insert the values for N, l, A, and μ_0. Thus, the inductance becomes

$$L = \frac{(4\pi \times 10^{-7}\ \text{N/A}^2)(100)^2(2.0 \times 10^{-4}\ \text{m}^2)}{3.0 \times 10^{-2}\ \text{m}}$$

$$L = 8.4 \times 10^{-5}\ \text{H} = 84\ \mu\text{H}.$$

Discussion Note that we calculated the inductance from purely geometrical considerations; the value of the inductance does not depend on the current.

*20.6 Energy Storage in a Magnetic Field

In Chapter 18 we found that the power delivered to a resistive load in an electric circuit was the product of current and electric potential difference across the resistor. When the load is an inductance of magnitude L, the power required to produce a current I changing at a rate $\Delta I/\Delta t$ is

$$P = \mathscr{E}I = LI\frac{\Delta I}{\Delta t}. \tag{20.10}$$

In a small time interval Δt, the circuit's source of emf delivers an amount of energy ΔW to the inductor, given by

$$\Delta W = P\ \Delta t = LI\ \Delta I.$$

The total energy required to build up the current from zero to some value I_0 is given by the sum

$$W = \Sigma\ \Delta W = \Sigma\ LI\ \Delta I.$$

This sum may be computed from the area under the graph LI versus I, in the same way that we calculated the energy stored on a capacitor in Section 17.9. Thus, the total energy is found to be

$$W = \tfrac{1}{2}LI_0^2. \tag{20.11}$$

This is the energy required to build up a current I_0 in an inductor.

If the current is held constant at I_0, no back emf is induced across the inductor and no further energy is required to sustain the current. (We are not now concerned with the energy loss due to the resistance present in a real inductor.) The energy of Eq. (20.11) has gone into creating a magnetic field surrounding the inductor. If the circuit is altered to reduce the current, the magnetic field decreases. The energy that was stored in the field provides the energy for the new emf of the inductor as it attempts to sustain the current and oppose the change.

We saw in Chapter 17 that when a capacitor was charged, energy was stored in the electric field within the capacitor. In the present case, we find that when an inductor carries a current, energy is stored in the magnetic field surrounding the inductor. In both cases, the stored energy can be released and used to do work.

We can compute the energy density in the magnetic field by considering the case of a very long solenoid. We begin by writing the inductance of the solenoid in the form given in Example 20.7. There we showed that $L = \mu_0 N^2 A/l$. If we let $N/l = n$, the number of turns per unit length, then the inductance becomes $L = \mu_0 n^2 Al$, which can be substituted into Eq. (20.11) to give

$$W = \tfrac{1}{2}LI^2 = \tfrac{1}{2}\mu_0 n^2 All^2.$$

We saw earlier that the field within a long solenoid is $B = \mu_0 nI$. Using this relationship for B, we can approximate the energy as

$$W = \frac{1}{2\mu_0}B^2 Al.$$

The quantity Al is the volume of the solenoid where the field is present. Thus, the energy per unit volume is

$$uB = \frac{W}{Al} = \frac{1}{2}\frac{B^2}{\mu_0}. \tag{20.12}$$

Although we have derived the energy density in the magnetic field for a special case and have approximated the field as uniform within the solenoid, the result is true in general. That is, if a magnetic field $\mathbf{B}$ is present at any region of space, then we can think of that region as having a stored energy density given by Eq. (20.12).

20.7 The Experimental Laws of Electromagnetism

Over the last few chapters, we have studied the contributions to electromagnetic theory embodied in the laws of Coulomb, Gauss, Faraday, Ampère, and others. Each of these laws was discovered independently, and each extended our understanding of electricity and magnetism in a distinct area. Let us now show the unity of the laws of electricity and magnetism by gathering together the experimental laws that we have already discussed and see how they combine to form a single theory, electromagnetism.

We began the study of electricity by considering the force between two point charges. The mathematical statement of this electric force is Coulomb's law. Later we introduced the electric field and Gauss's law for electrostatics that relates the net electric flux through a closed surface to the net electric charge enclosed within the surface. Although they seem quite different at first glance, Coulomb's law and Gauss's law are just different mathematical expressions of the same experimental observation. Some calculations are easier when you use Coulomb's law and others are easier using Gauss's law. Gauss's law emphasizes aspects of the field, rather than the force and its use makes it easier for us to see the symmetry and unity of electromagnetism. Even though the mathematical form of Gauss's law may seem complex, remember that this law—and the other laws of electromagnetism—has its origin in simple experiments. Gauss's law is number 1 in Table 20.1.

In studying magnetism, we found another Gauss's law that is analogous to Gauss's law for electrostatics. Gauss's law for magnetism arises from the experimental observation that the lines of magnetic field close on themselves, forming closed loops. Therefore, the net magnetic flux through any arbitrary closed surface is zero and there are no net magnetic "charges" enclosed within the surface. Gauss's law for magnetism is number 2 in Table 20.1.

Table 20.1   *Observational Laws of Electromagnetism*

Physical Situation	Observation	Name of Law
	Electric field lines begin and end on electric charges.	Gauss's law for electrostatics
	Magnetic field lines form closed loops.	Gauss's law for magnetism
	Magnetic fields are produced by electric currents.	Ampère's law
	Changing magnetic flux induces an electric field and,	Faraday's law

The discovery that magnetic fields are caused by electric currents was made by Oersted in 1820. A mathematical description of the magnetic field caused by a constant electric current was soon found by Biot and Savart and, later in a different form, by Ampère. (Remember that Ampère's law holds only for steady currents.) Ampère's law is number 3 in Table 20.1.

Faraday discovered that an emf is generated in a loop when the magnetic flux enclosed by the loop changes with time. If the loop is closed, the induced emf produces an electric current. Faraday's law of induction is the only one of the basic experimentally discovered laws of electromagnetism that explicitly depends on the time. It is listed as number 4 in Table 20.1.

Although these four basic laws contain the basis for all of electromagnetism, we need one more idea to link electromagnetism with mechanics. The important linking relationship is the force on a charge moving in the presence of both electric and magnetic fields. The force on the charge is the vector sum of two components. One component is proportional to the electric field and is along (or opposite to) the direction of the field. The other component is proportional to the product of the velocity of the charge, the magnetic field, and the sine of the angle between them and is perpendicular to the direction of both **v** and **B**.

The laws found in Table 20.1 are based on experiments that can be carried out under relatively simple conditions. Each can be tested individually. The laws represent static—or constant current—conditions using equipment of ordinary size. Everything that can be said about the behavior of quasi-static electromagnetism is summed up in these equations. There are, of course, other important and practical phenomena not explicitly described by these laws. For example, conductivity depends on properties of the material. The four experimental laws listed here deal explicitly with electric charges, electromagnetic fields, and their interactions. Nevertheless, there is something missing.

20.8 Maxwell's Equations

In 1864, at about the time that the American Civil War was ending, the great Scottish physicist James Clerk Maxwell (Fig. 20.20) combined all of the ideas and discoveries of electricity and magnetism in one unified picture of electromagnetism. Maxwell showed that a complete description of electromagnetic effects could be based on a set of only four equations. These equations are mathematical statements of the four experimental laws reviewed in the last section and listed in Table 20.1. There is, however, one important difference. Maxwell added another term to Ampère's law. This additional term, which contributes to the magnetic field in the same way that a current does and is called the displacement current, depends on the time rate of change of electric flux. With this term included, Ampère's law then depends on the time rate of change of the electric flux in a manner similar to the way Faraday's law depends on the time rate of change of the magnetic flux. Adding the extra term makes the equations for the electric field E and the magnetic field B look more symmetric, especially when expressed in modern mathematical notation.

Maxwell emphasized the unity of electromagnetism and worked out some of the predictions of the unified theory—which included his additional displacement current term in the Ampère law. Today, these four equations are

Figure 20.20
James Clerk Maxwell became the first Cavendish Professor of Experimental Physics at Cambridge University in 1871.

known as *Maxwell's electromagnetic equations,* even though they did not originate entirely with him. Yet it was Maxwell who showed conclusively that these four equations could be used to interpret and explain an impressive array of electromagnetic phenomena. Moreover, this set of equations is believed to be complete. That is, any correct additional equation relating electromagnetic fields can be derived from this set.

We shall not list the mathematical statements of Maxwell's equations, since their formal complexity is beyond the scope of this text. However, we can summarize them descriptively as follows:

Maxwell's Electromagnetic Equations

1. **Gauss's law for electricity.** This law describes the electric field due to electric charges and can be used to derive Coulomb's law.

2. **Gauss's law for magnetism.** According to this law, magnetic field lines are continuous and without end, and thus there are no isolated magnetic poles.

3. **Ampère-Maxwell's law.** Maxwell extended this law to describe the production of magnetic fields not only by electric currents but by changing electric fields as well.

4. **Faraday's law of induction.** This law describes the production of electric fields as a result of changing magnetic fields.

It was Faraday who introduced the concepts of fields and of lines of force to describe the interaction of objects separated in space. This formulation still remains a powerful way to describe electric, magnetic, and other fields. But Maxwell's precise mathematical description of these fields—and their reality, as demonstrated by many experiments—remains an important part of our understanding of electromagnetism. In addition, Maxwell showed that electric and magnetic fields radiate in the form of waves from an oscillating electric charge. The experimental confirmation of these **electromagnetic waves** did not come in Maxwell's lifetime, but was achieved by Heinrich Hertz (1857–1894) in 1887, eight years after Maxwell's death.* This discovery opened the way to the wireless telegraph, radio, television, microwaves, and radar. In fact, today Maxwell's equations are considered the foundation of our understanding of electromagnetism, equal in importance to Newton's laws of motion in mechanics.

By the late nineteenth century, careful study of Maxwell's theory showed it to be in conflict with the formulation of mechanics by Galileo and Newton. This created considerable consternation for the physicists of the time. Maxwell's theory gave a unique value to the speed of light, but Newtonian mechanics did not allow this uniqueness and was unable to account satisfactorily for the behavior of objects whose speeds approached the speed of light. It became painfully obvious that one of these hallowed theories had to be incorrect. The problem was resolved in 1905 when Albert Einstein described his theory of special relativity. According to relativity, Maxwell's theory of electromagnetism was correct, but Newtonian mechanics was incomplete. In Chapter 25 we will discuss Einstein's resolution of this problem and some of his unexpected results.

---

*As early as 1844, Joseph Henry recognized that electromagnetic disturbances could propagate through space and discussed their existence with his classes.

LINEAR ACCELERATORS FOR RADIATION THERAPY

Radiation therapy is a common medical tool, especially for the treatment of cancer patients. One common source of radiation is a precisely controlled beam of electrons produced with a linear accelerator, a device that accelerates charged particles from lower to higher velocities. These electron beams can be used directly or they can be used to generate highly energetic x rays. (Production of x rays is discussed in Chapters 26 and 27.)

We can construct a linear accelerator in which charged particles are accelerated in a straight line by using a series of conducting tubes connected to an alternating voltage supply (Fig. B20.1). Positively charged particles traveling along the axis of the tubes are accelerated from left to right across the gap between the tubes if the electric field between them is directed from left to right. For example, if the first tube is positive and the second tube is negative, a positively charged particle is accelerated across the gap between them. If the voltage changes to make the second tube positive and the third tube negative by the time the particle has reached the gap between them, the particle is again accelerated across the gap. If we adjust the period of the alternating voltage supply to match the transit time of the particles through each pair of adjacent tubes, the particles accelerate as they cross each gap between the tubes.

When electrons are accelerated in a linear accelerator of this type, they quickly reach very high speeds because of their small mass. Even if the voltage oscillates at radio frequencies ($\approx 10^6$ Hz), the tube lengths become excessively long due to the electron velocity. If the source is made to oscillate at the even higher frequencies of microwave generators ($\approx 10^9$–10^{12} Hz), the wavelengths of the microwave oscillations are comparable to the dimensions of the tubes, and a system like the one of Fig. B20.1 is no longer practical.

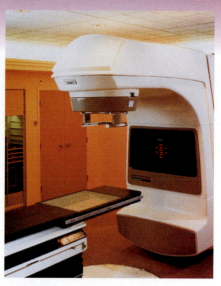

Figure B20.2 A linear accelerator for radiation therapy.

In the contemporary linear accelerator, the accelerator tubes are replaced by a waveguide, a hollow metal conductor through which high-frequency microwaves are propagated. This waveguide can be used to accelerate electrons, which are injected into the waveguide and travel in the same direction as the waves. An electron moving down the symmetry axis will be accelerated, depending on its position in the field. If the wave is propagating to the right, an electron moving with the same velocity and initially placed at a position in the field where it experiences a force to the right, will be continuously accelerated. In this manner, the electrons are moved along with the wave in much the same way that a surfer rides along on an ocean wave.

The electrons move along the waveguide with an increasing speed that approaches the speed of light. They gain energy from the electromagnetic field as they go. Electron beams with energies equal to that of electrons accelerated through a potential of up to 24 MV can be generated in waveguide systems less than one meter long (Fig. B20.2). Linear accelerators are reliable sources of intense beams of high-energy electrons. Precise energy control is obtained by careful control of the microwave frequency. In many cases, the electron beams are directed into a metal target where they produce x rays on impact. These intense beams of highly penetrating x rays are then directed at cancerous tumors to destroy their cells.

Figure B20.1 A simple linear accelerator, consisting of conducting tubes alternately connected to opposite sides of an alternating voltage source.

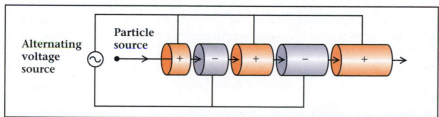

Electromagnetic Waves

Electromagnetic waves can be generated by oscillating electric charges. When a sinusoidally varying current source is attached to a simple antenna composed of two conducting rods (Fig. 20.21), a sinusoidally oscillating electromagnetic wave radiates from the antenna. Consequently, at some distance away from the antenna an electric field arises that oscillates sinusoidally with time as the wave passes.

We can understand the nature of the waves from a simple model. Imagine that the antenna of Fig. 20.21 is statically charged with a positive charge Q on the upper end and a negative charge $-Q$ on the lower end. This dipolar charge distribution would generate an electric field as shown in Fig. 20.22(a). As long as the charge is maintained constant, the field remains as shown. If at some later time the charge is reversed so that the upper end is negative and the lower end is positive, then the field must also reverse. However, it will take a finite time for the reversed field to reach an observer at a point P some distance away from the antenna (Fig. 20.22). If the charge on the antenna is made to oscillate periodically, then the field at point P will also oscillate periodically as the field radiates out from the antenna (Fig. 20.22c). The direction of the electric field vector in the radiated wave is parallel to the orientation of the antenna. An electric current will exist in the antenna whenever the charge on the antenna changes. This current will also generate a magnetic field. If the charge is oscillating periodically, then the current will also oscillate periodically. Thus, the electric field makes up only part of the electromagnetic wave, for there is also a magnetic field that accompanies the electric field. If the charge on the antenna is made to oscillate sinusoidally, then the complete wave will have a sinusoidally oscillating magnetic field coupled to a sinusoidally oscillating electric field. It is this combination that radiates out from the antenna.

At distances far from the antenna, the wave is essentially a plane wave; that is, at a given instant of time the electric and magnetic fields are uniform over a plane perpendicular to the direction of propagation. The electric and magnetic fields are perpendicular to the direction of propagation and to each other. These fields vary sinusoidally with space and time (Fig. 20.23). The electromagnetic

Figure 20.21

(a) A microwave antenna.
(b) A sinusoidally oscillating electromagnetic wave radiates from an antenna driven by a sinusoidally varying current.

(a)

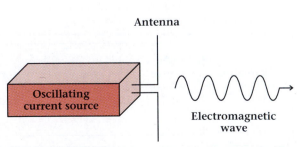

(b)

wave is a transverse wave because the wave disturbance (the field) is transverse to the direction of propagation.

The electric field of broadcast television waves is horizontal because of the orientation of the transmitting antennas. This is the reason that television receiving antennas are made as they are. It is also the reason that one turns an antenna to "aim" it at the station. The maximum signal, and hence the best reception, occurs when the elements of the antenna are aligned parallel to the direction of the electric field in the radiated wave.

Analysis of Maxwell's equations not only predicts the existence of electromagnetic waves, but also predicts the speed of propagation of the waves. The value predicted for the speed depends on the value of the constant ϵ_0 found in Coulomb's law and on the value of μ_0 found in Ampère's law. The speed of an electromagnetic wave in free space depends on these constants alone and is given by

$$c = \frac{1}{\sqrt{\mu_0 \epsilon_0}}. \qquad (20.13)$$

By carefully making the appropriate electrical measurements to determine the value of these constants, the speed of electromagnetic waves was calculated. Such measurements, made largely between the time of Maxwell's first work and the end of the nineteenth century, gave results in agreement with the value for the velocity of light. The inference, later shown to be correct, was that light is an electromagnetic wave. The most meaningful test of Maxwell's equations (including the displacement current) came when Hertz produced electromagnetic waves and measured their speed to be approximately that of light.

The properties and behavior of light, discussed in Chapters 22 through 24, can all be deduced from Maxwell's equations. These include not only wavelike behavior of the type just discussed, but the straight-line propagation that we normally observe and even the bending of the light path as it passes through a lens or prism.

Over the years, many determinations of the speed of light have been made. (Some of these methods are described in Chapter 22.) By the early 1980s, the best value for the speed of light in vacuum was $c = 299{,}792{,}458 \pm 1.2$ m/s. The primary limitation of the measurement was the precision with which the length of the meter could be established. In 1983, the Seventeenth General Conference on Weights and Measures adopted a new definition of the meter,

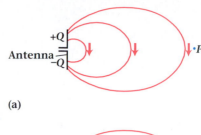

(a)

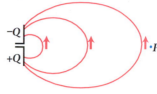

(b)

(c)

Figure 20.22
(a) In the static case for the antenna charged as shown, the electric field in space is the field due to a dipole. Along the symmetry plane, the electric field is directed downward as shown and decreases with distance from the antenna. (b) If the position of the charges is reversed, the direction of the field is reversed. (c) If the charges are made to oscillate with time, the field oscillates in time as it radiates away from the antenna. At an instant of time, the field will oscillate in space as shown.

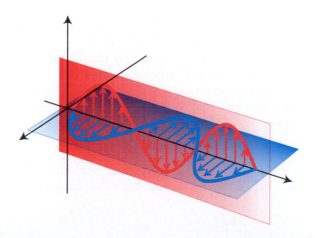

Figure 20.23
The appearance of the electric and magnetic fields in a plane electromagnetic wave. Each field varies sinusoidally with space and time. The speed of the wave is c, the speed of light.

based on the best value for the speed of light: **The meter is the length of path traveled by light in vacuum during a time interval of 1/299,792,458 of a second.**

When we warm ourselves by sunlight or the radiation from a heat lamp, we receive energy from the sun or the lamp that is brought to us by the radiation. Thus, the electromagnetic waves carry energy as they move through space. Where is the energy and how do we account for it?

We found earlier in Chapter 17 that in free space an electric field has an energy density given by

$$uE = \tfrac{1}{2}\epsilon_0 E^2.$$

In this chapter, we found that a magnetic field has an energy density given by

$$uB = \frac{1}{2}\frac{B^2}{\mu_0}.$$

The total energy density u in a region of space that contains both electric and magnetic fields is then

$$u = uE + uB = \tfrac{1}{2}\epsilon_0 E^2 + \frac{1}{2}\frac{B^2}{\mu_0}. \tag{20.14}$$

The relationship between E and B for electromagnetic waves is found from the Maxwell equations to be

$$\boxed{B = \sqrt{\epsilon_0\mu_0}\,E = \frac{E}{c}.} \tag{20.15}$$

By combining Eq. (20.15) with Eq. (20.14), the energy density in an electromagnetic wave can be written in terms of only the electric field or only the magnetic field. The result is

$$u = \epsilon_0 E^2 \quad \text{or} \quad u = \frac{B^2}{\mu_0}. \tag{20.16}$$

Notice that the result we have just obtained implies that for a traveling electromagnetic wave the energy density of the electric field is equal to the energy density of the magnetic field. Even so, Eq. (20.15) shows that when expressed in SI units the magnitude of the magnetic field is smaller than the magnitude of the electric field by a factor equal to the speed of light.

We are usually more interested in the energy transmitted by a wave rather than in the energy density in the field at some location. Specifically, we are interested in the rate of energy transmitted by the wave which is known as the *irradiance* ($\mathscr{I}$). The irradiance is given by the product of the energy density and the speed of the wave.

$$\mathscr{I} = \text{rate of energy flow per unit area} = cu. \tag{20.17}$$

This product has the dimensions of power per unit area and in SI units is measured in watts per square meter. If we insert one of the values from Eq. (20.16) into Eq. (20.17) and take the time average*, we get

$$\mathcal{I} = \frac{c\epsilon_0 E_0^2}{2} \quad \text{or} \quad \mathcal{I} = \frac{cB_0^2}{2\mu_0}, \tag{20.18}$$

for the irradiance (or intensity) of the electromagnetic wave. Here E_0 is the amplitude of the electric field, and B_0 is the amplitude of the magnetic field.

Example 20.8

What are the approximate amplitudes of the electric and magnetic fields 1.0 m away from an operating 60-W incandescent lamp? You may assume that 85% of the input electrical power is given off as electromagnetic radiation, the remainder goes into heating the socket and the air.

Strategy The power per unit area is the irradiance. Once we have a value for the irradiance, we can calculate both the electric field E and the magnetic field B. As an approximation we assume the radiation from the lamp to be uniform in all directions and to be of a single frequency.

Solution The irradiance at 1.0 m is the total power radiated spherically divided by the area of a sphere with a radius of one meter.

$$\mathcal{I} = \frac{\text{Power}}{\text{Area}} = \frac{P}{4\pi r^2} = \frac{60 \text{ W} \times 0.85}{4\pi(1.0 \text{ m})^2} = 4.06 \text{ W/m}^2.$$

The electric field can be obtained from Eq. (20.18) when 4.06 W/m² is inserted for $\mathcal{I}$. Upon rearranging we have

$$E_0 = \sqrt{\frac{2\mathcal{I}}{c\epsilon_0}} = \sqrt{\frac{2(4.06 \text{ W/m}^2)}{(3.0 \times 10^8 \text{ m/s})(8.85 \times 10^{-12} \text{ F/m})}}$$

$$E_0 = \sqrt{3.06 \times 10^3 \text{ V}^2/\text{m}^2} = 55 \text{ V/m}.$$

We can find B from Eq. (20.15) as

$$B_0 = \frac{E_0}{c} = \frac{55 \text{ V/m}}{3.0 \times 10^8 \text{ m/s}} = 1.8 \times 10^{-7} \text{ T}.$$

Discussion These fields are small compared to the natural electric and magnetic fields near the earth's surface of 150 V/m and 5×10^{-5} T.

■

---

*For a sinusoidal traveling wave, the electric and magnetic fields can be represented as an $E = E_0 \sin \frac{2\pi}{\lambda}(x - ct)$ or $B = B_0 \sin \frac{2\pi}{\lambda}(x - ct)$, where E_0 and B_0 are constant amplitudes. The time average of E and B is then just the amplitude multiplied by the time average of the sine function given by $\sin \frac{2\pi}{\lambda}(x - ct) = \frac{1}{2}$. Thus, the 2 that appears in the denominator of Eq. (20.18) is a result of the time average of the oscillating field.

▼
Summary

Useful Concepts

■ Magnetic flux is defined as

$$\phi_m = BA \cos \theta.$$

■ Faraday's law of induction gives the emf in a coil of N turns due to changing magnetic flux,

$$\mathcal{E} = -\frac{\Delta \phi_m}{\Delta t}.$$

■ Lenz's law states that induced current occurs in a direction that opposes the change in flux. Thus, the minus sign in Faraday's law is an expression of Lenz's law.

■ The ratio of input to output voltages of a transformer is equal to the ratio of turns in the primary coil to turns in the secondary coil:

$$\frac{V_1}{V_2} = \frac{N_1}{N_2}.$$

■ The voltages and currents in a transformer's primary and secondary coils are related by

$$V_1 I_1 = V_2 I_2.$$

■ In a coil of N turns carrying a current I and through which a magnetic flux ϕ_m passes, the inductance is

$$L = \frac{N \phi_m}{I}.$$

■ The emf induced in an inductor by a changing current is

$$\mathcal{E} = -L \frac{\Delta I}{\Delta t}.$$

■ The energy stored in an inductor carrying current I_0 is

$$W = \frac{1}{2} L I_0^2.$$

■ The laws of electromagnetism are summed up in Maxwell's equations.

■ The speed of light is

$$c = 299,792,458 \text{ m/s}, \quad \text{or} \quad c \approx 3 \times 10^8 \text{ m/s}.$$

■ An electromagnetic wave has both electric and magnetic field components. Both fields are transverse to the direction of propagation of the wave and are at right angles to each other. In free space, the magnetic field component of the wave is related to the electric component by

$$B = \frac{E}{c}.$$

Important Terms

You should be able to write the definition or meaning of each of the following:

Faraday's law	transformer
induced emf	inductance
induced current	henry
Lenz's law	inductor
generators	Maxwell's electromagnetic
alternating current	equations
electric motor	electromagnetic waves
back emf	

▼
Conceptual Questions

20.1 Discuss several different ways of producing an emf in a wire.

20.2 A flexible wire is held in the shape of a loop perpendicular to a magnetic field. The ends of the wire are suddenly jerked so that the wire is pulled straight. Is an emf generated in the wire? Explain.

20.3 Most automobiles use alternators to provide the emf for charging the battery. How can this be done, since the battery requires a unidirectional (dc) current for charging?

20.4 The belt on a belt-driven generator breaks while the generator is being used to charge a storage battery. You observe that the generator continues to turn. What is the explanation of this observation?

20.5 A bar magnet is moved toward a circular loop of wire as shown in Fig. 20.24. In which direction does the induced current flow? In which direction will it flow if the magnet is moved away from the loop?

20.6 How does the inductance of a coil change when an iron core is put into it? Why?

20.7 Two coils connected in series are originally so far apart that the magnetic field of each does not overlap the other coil. If they are brought together so that there is flux in each coil due to current in the other, can the total inductance increase? Can they be brought together in such a way that the total inductance is decreased? Explain your answers.

20.8 An airplane is flying from west to east in the northern

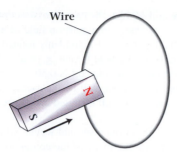

Figure 20.24

Question 20.5: A magnet with its north pole nearest a loop of wire is moved closer to the loop.

hemisphere. Which wing tip (left or right) acquires a positive electric potential relative to the other tip as a result of the motional emf induced by the earth's magnetic field?

20.9 Explain why a spinning metal top slows down when a magnet is brought near.

20.10 Hospitals have emergency generators to provide electricity when the incoming electric service is interrupted. Usually, the emergency generators are designed to provide three times as much power as the hospital normally uses. Why is it necessary to have generators with so much capacity?

20.11 A coil of many turns, connected to a battery, carries a steady current. Is energy stored in this circumstance? Where? How can you test the correctness of your answer?

20.12 In the United States, the standard frequency for electric power distribution is 120 V at 60 Hz, while in many other parts of the world it is 240 V at 50 Hz. If a transformer is available to provide the proper voltage level, what kinds of electrical appliances or equipment can be used on either system and what kinds cannot be used?

20.13 Most AM radios have antennas that consists of a coil of wire surrounding a bar of magnetic material. Such antennas respond to the magnetic field of the traveling radio wave. How should such an antenna be oriented relative to the transmitting station for the most efficiency?

Problems

Section 20.1 Faraday's Law

Hints for Solving Problems

An emf is induced in a conducting loop when the magnetic flux through the loop changes. If the loop makes a closed circuit, an induced current is generated in a direction to create a magnetic field that opposes the change in flux.

20.1 A coil contains 100 turns of wire in a loop 15 cm in diameter. The loop is placed between the poles of a large electromagnet, $B = 1.0$ T, with the plane of the loop perpendicular to the field. If the magnetic field is steadily reduced from 1.0 T to 0 in 16 s, what is the average emf in the coil while the field is changing?

20.2 The flux in a single-loop coil of area 37 cm^2 steadily changes from 6.5×10^{-3} T to 9.3×10^{-3} T in 0.50 s. What emf is induced in the coil?

20.3 A bar magnet is rotated at a steady rate about its center (Fig. 20.25). Describe the effect this will have on a loop of wire located nearby. Sketch a graph of the voltage output of the loop as a function of the position of the magnet and as a function of time.

20.4 The plane of a square loop of wire with edge length of 8.0 cm is perpendicular to a 0.017-T magnetic field (Fig. 20.26a). What is the average emf between the points E_1 and E_2 when the corner D is quickly folded about the diagonal AC so as to lie on top of B (Fig. 20.26b) if it takes 0.13 s to make the fold?

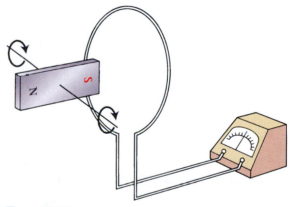

Figure 20.25
Problem 20.3.

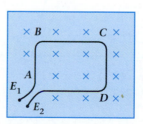
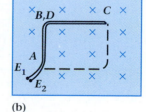

(a)   (b)

Figure 20.26
Problem 20.4.

20.5 A 10-turn square coil with 12-cm-long sides is held in a uniform magnetic field of 0.090 T. The normal to the plane of the coil makes an angle of 43° with the direction of the field. The coil is withdrawn to a region of zero field in 0.17 s. What is the average emf developed in the coil as it is withdrawn from the field?

20.6 A 2.8-cm-radius 50-turn coil of 12.5-Ω resistance rotates about a diameter in a uniform magnetic field of 0.090 T. The coil starts with the normal to its plane parallel to the direction of the field and ends with the normal perpendicular to the direction of the field. If an instrument to measure electric charge is connected to the coil, how much charge would it record when the coil is flipped?

20.7 A "flip coil" of 50-mm radius and 50 turns is oriented perpendicular to a magnetic field. The coil is quickly flipped through 180° about its diameter in 0.070 s. The average emf in the coil is measured to be 2.0 V. What is the strength of the magnetic field in the region of the coil?

20.8● A single thin coil of very high resistance wire is used as a flip coil. (See Problem 20.7.) When the coil is quickly rotated about its diameter in a uniform magnetic field B, the average current in the coil is I. A new coil is wound on the same form to have the same shape but with N turns of identical wire. The new coil is rotated about its diameter in the same way in the same field as the first coil. Compare the average current from the new coil to that from the first coil.

20.9● A copper wire (*ab*) is wound around a wooden rod (Fig. 20.27). Another wire (*cd*) is wound on top of it and connected to a resistor R. What is the direction of current in R if (a) a current flowing from a to b is increased, (b) a current flowing from a to b is decreased, (c) a current flowing from b to a is increased, and (d) a current flowing from b to a is decreased?

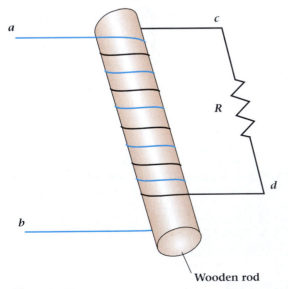

Figure 20.27
Problem 20.9.

20.10● A wire coil of 25 turns has a cross-sectional area of 10 cm². The coil is placed in a uniform field of a large magnet with $B = 0.20$ T. The coil is suddenly rotated 90° from an orientation parallel to the field to one perpendicular to the field. (a) If the time to flip the coil is 0.50 s, what is the average emf produced? (b) What will be the average current produced if the circuit of the coil is closed and the resistance is 75 Ω?

20.11● A coil 17 cm in diameter and wound with three turns of wire is placed with the plane of the coil at right angles to a magnetic field of 3.8×10^{-2} T. What emf is induced in the coil if (a) the field is halved in 0.27 s, (b) the field is reversed in 0.27 s, (c) the coil is rotated through an angle of 90° about its diameter in 0.27 s, or (d) the coil is rotated through an angle of 180° about its diameter in 0.27 s?

20.12● Two loops made from a single wire lie in a plane (Fig. 20.28). A uniform magnetic field is perpendicular to the plane of the coils and is directed outward from the plane of the figure. (a) What is the magnitude of the emf between points A and B when the magnetic field perpendicular to the plane of the loops increases at a rate of 0.25 T/s? The radius $r_1 = 0.20$ m and the radius $r_2 = 0.15$ m. (b) Is the potential difference $VA - VB$ positive or negative?

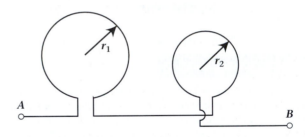

Figure 20.28
Problem 20.12.

Section 20.2 Motional Emf

20.13 A flat-bed truck travels due south at 120 km/h in a location where the vertical component of the earth's magnetic field is 5×10^{-6} T. What is the emf induced in a 1.0-m-long copper bar held horizontally above the bed of the truck and perpendicular to its direction of motion?

20.14 What emf is developed between the tips of the wings of a Boeing 767 jet liner in level flight at 850 km/h in a location where the vertical component of the earth's magnetic field is 1.75×10^{-5} T? The distance between the wing tips of the 767 is 47.65 m.

20.15● A conducting rod of resistance R is moved along horizontal rails joined at one end to make a loop similar to the one in Fig. 20.8. A uniform magnetic field B is perpendicular to the plane of the loop and extends over the region in which the rod moves. Assume that the rod moves with constant speed v and that the rails have negligible resistance and friction. (a) Find an expression for the emf around the loop. (b) Find an expression for the emf across the rod. (c) What is the current in the circuit? (d) What force is required to keep the rod moving with constant velocity?

20.16● A 0.57-m-long rod lies in a plane containing a magnetic field $B = 7.0 \times 10^{-3}$ T. The axis of the rod makes an angle of 60° with the field as shown (Fig. 20.29). The rod moves perpendicular to the plane at a speed of 1.03 m/s in the direction shown. (a) What is the emf between the two ends of the rod? (b) Which end is at the higher potential?

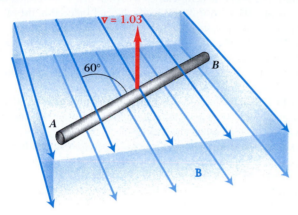

Figure 20.29
Problem 20.16.

20.17● A 1.0-m-long aluminum bar is held horizontally in the east-west direction and dropped from a height of 10 m at a place where the horizontal component of the earth's magnetic field is 3.8×10^{-5} T. What is the emf between the ends of the bar just before it strikes the ground?

20.18● (a) Show that the product Blv is dimensionally equivalent to an emf. (b) Show that when SI units are used, the unit of Blv is the volt.

20.19●● What is the emf between the ends of a 1.0-m-long light metal chain if you swing it in a horizontal circle about your head at 2.0 revolutions per second at a location where the vertical component of the earth's magnetic field is 2.5×10^{-5} T?

20.20●● A rod of length l and mass m slides down parallel conducting rails making an angle θ with the horizontal (Fig. 20.30). The rails have negligible resistance and are joined at the bottom to form a loop. A uniform magnetic field **B** is directed vertically downward in the region of the rails. What is the terminal velocity of the rod if it has an electrical resistance R? Ignore friction between rod and rails.

Section 20.3 Generators and Motors

20.21 A rectangular loop with an area of 120 cm² is placed in a magnetic field of 1.3×10^{-2} T and rotated about its long axis at 5.0 Hz. What is the instantaneous emf generated at the instant when the normal to the loop makes an angle $\theta = 0.27°$ with the magnetic field?

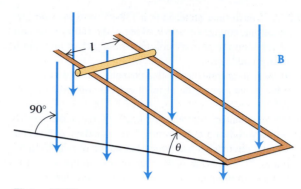

Figure 20.30
Problem 20.20.

20.22 A 50-turn loop of wire has a cross-sectional area of 0.010 m². It is rotated about an axis perpendicular to the earth's field of 5.0×10^{-5} T. What is the peak value of the emf generated if the loop rotates at 3600 rotations per minute?

20.23 What would be the peak emf of the generating coil of Problem 20.22 if it were turned in a magnetic field of 0.100 T?

20.24● A single loop of area 0.173 m² is placed perpendicular to a magnetic field of 0.354 T. The loop is rotated about an axis lying in the plane of the loop and passing through its center. An instantaneous emf of 8.41 V is generated when the plane of the loop makes an angle of 45.0° with the field direction. What is the angular speed of the loop?

Section 20.4 The Transformer

Hints for Solving Problems

Assume, unless stated otherwise, that there are no energy losses in a transformer, so the input power and output power are equal.

20.25 A transformer is made by winding a primary coil of 300 turns around an iron core. A secondary winding of 750 turns is made about the same core. If the primary voltage is 120 V, what is the output voltage on the secondary?

20.26 A transformer is made by winding a primary coil of 250 turns around an iron core. A secondary winding of 50 turns is made around the same core. If the primary voltage is 120 V, what is the output voltage on the secondary?

20.27 A doorbell transformer has an output voltage of 12 V when connected to a 120-V household supply. What is the ratio of the number of turns on the primary coil to the number of turns on the secondary coil?

20.28 A step-down transformer that converts 120 V ac to 10 V ac is rated for 1.5 A output current on the secondary winding. What is the input current when the output is 1.5 A?

20.29 A transformer attached to a 138-kV transmission line has an output voltage of 38 kV. What is the ratio of the number of turns on the primary coil to the number of turns on the secondary coil?

20.30 A step-up transformer that changes 115 V ac to 230 V ac. What is the magnitude of input current when the output is 2.0 A?

20.31• A potential difference of 120 V is maintained across the primary coil of a transformer. The secondary coil is connected to a circuit that has a total resistance of 6.42 Ω and dissipates 22.5 W. What is the ratio of the number of turns on the primary to the number of turns on the secondary?

20.32• Two transformers are connected as shown in Fig. 20.31, where N_1 and N_2 are the numbers of primary and secondary turns on the transformer T_1 and N_3 and N_4 are the numbers of primary and secondary turns on the transformer T_2. Show that

$$\frac{V_4}{V_1} = \frac{N_4}{N_3}\frac{N_2}{N_1}.$$

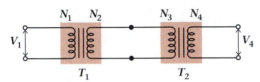

Figure 20.31
Problem 20.32.

20.33• A low-voltage outdoor lighting system uses a transformer that steps the 120-V household voltage down to 24 V for safety. The equivalent resistance of all of the low-voltage lamps is 9.6 Ω. (a) What is the current in the secondary coil? (b) What is the current in the primary coil? (c) How much power is used, neglecting losses in the transformer and line?

20.34• A resistor R is connected across the secondary coil of a transformer with N_1 primary turns and N_2 secondary turns. Show that the relationship between the primary current I_1 and voltage V_1 is

$$I_1 = \left(\frac{N_2}{N_1}\right)^2 \frac{V_1}{R}.$$

20.35•• Two separate 1000-turn secondary coils are wound on the same transformer, which has a 500-turn primary coil. The two secondary coils are each connected to identical resistors. If the input to the primary coil is 0.15A at 120 V, what is the current in each secondary coil? (*Hint:* Use conservation of energy.)

20.36•• A transformer near your house steps down higher voltage to 240 V for use by heavy appliances in your home. The transformer loses 4.0% of the input electrical energy by Joule heating and other effects. The ratio of primary to secondary turns is 52. What is the current in the primary if the only load on the secondary is a 1500-W electric stove?

Section 20.5 Inductance

Hints for Solving Problems

An inductor behaves as a source of emf when the current in it is changing. The direction of the inductor emf is such that it opposes the change in current.

20.37 A 25-mH inductor carries a current that is increasing at a rate of 1.05×10^{-2} A/s. What is the emf induced in the inductor and what is its polarity?

20.38 A 0.85-H inductor carries a current that decreases at a rate of 0.12 A/s. What is the induced emf?

20.39 At what rate does the current change in a 35-mH inductor when an emf of 0.019 V develops across it?

20.40 Calculate the self-inductance of a small solenoid 5.0 cm long made of 10 turns of wire, each loop enclosing 1.0 cm². Assume that end effects are negligible.

20.41 An inductor is wound about a paper core 5.0 cm long and 12 cm² in cross section. How many turns of wire are required to make an inductance of 1.3 mH?

20.42• Show that the units of μ_0 are henries per meter.

20.43•• Show that the self-inductance of an air-core toroid of cross-sectional area A and mean circumferential length l wound with N turns of wire is given by

$$L = \frac{\mu_0 N^2 A}{l}.$$

To determine the self-inductance of the toroid, use the approximation that the magnetic flux across the cross section is just the product of the area times the value of B at the mean radius.

20.44• Find the inductance of an air-core toroid with a cross-sectional area of 6.0 cm² and a mean circumferential length of 30 cm, wound with 200 turns of wire. (Use the results of Problem 20.43.)

*Section 20.6 Energy Storage in a Magnetic Field

20.45 A 50-mH inductor carries a current of 48 mA. How much energy is stored in the field of the inductor?

20.46 A 38-mH inductor stores 4.0×10^{-5} J when carrying a dc current. What is the magnitude of that current?

20.47• A 0.80 H-inductor carries a current of 0.53 A. (a) How much energy is stored in the field of the inductor? (b) What is the rate of energy loss in the inductor if it has a resistance of 0.87 Ω?

20.48•• Use the results of Problem 20.43 to show that the energy density in the magnetic field of the toroid is given by $u = \frac{B^2}{2\mu_0}$. (*Hint:* The magnetic field is confined within the toroid. Calculate the energy stored in the toroidal inductor and relate it to the magnetic field of the toroid. The energy density is the energy divided by the volume.)

20.49•• Calculate the energy density in a region of space where the magnetic field magnitude is 0.0045 T and the electric field is 360 kV/m.

Section 20.9 Electromagnetic Waves

20.50 Calculate the speed of light using the relationship $c = 1/\sqrt{\mu_0 \epsilon_0}$.

20.51 A computer is designed to perform calculations in nanoseconds (ns). What is the maximum separation between any two elements in the computer if an electrical signal is to go from one to the other and back again in 1.0 ns? (*Hint:* The maximum velocity of the propagation of an electrical signal is the velocity of light.)

20.52 What is the peak value of the electric field in the beam of a 2.5 W argon laser if the beam has a diameter of 0.85 mm. (*Hint:* For simplicity, consider the beam to be uniformly spread over its diameter.)

20.53• Imagine that there exists another universe with the same particles (electrons, protons, and so on) and the same fundamental charge as in our universe. In that universe, the same laws of physics that we know apply except that the force between charged particles identical to the ones in our universe is only 1/4000 as much as it is in our universe. What is the speed of light in this new universe?

20.54• A circular laser beam has a diameter of 1.5 mm and a power of 3.0 mW. (a) Assuming the light is uniformly spread over the circle, what is the energy density in the beam? (b) What are the values of the electric and magnetic field amplitudes in the beam?

20.55• On a bright day, the sunlight reaching the ground has an irradiance of 1.0 kW/m². (a) What are the values of the electric and magnetic field amplitudes in this light? (b) How does the magnetic field amplitude compare with the earth's magnetic field of about 5.0×10^{-5} T?

Additional Problems

20.56 A flat coil of 300 turns is wound into a square loop 5.0 cm on a side. The coil is designed so that it can be rotated by 90° in 0.25 s. The coil is placed so that the magnetic flux is zero, and it is rotated so that the flux is a maximum. If the average voltage on the coil due to the induced emf is 0.30 mV, how large is the magnetic field?

20.57 A transformer attached to a 10-kV transmission line has 25 times as many primary turns as it does secondary turns. What is the output voltage?

20.58• Two loops made from a single wire have their planes perpendicular to a uniform magnetic field **B** (Fig. 20.32). (a) What is the magnitude of the emf between points A and C when the magnetic field decreases at the rate of 0.025 T/s? The radius $r_1 = 0.14$ m and the radius $r_2 = 0.16$ m. (b) Is the potential difference $VA - VC$ positive or negative?

20.59•• A copper rod 25 cm long rotates about one end in a plane perpendicular to a 3.0×10^{-2}-T uniform magnetic field. The rod rotates at a rate of 2.5 revolutions per second. The outer end makes contact with a stationary conducting ring

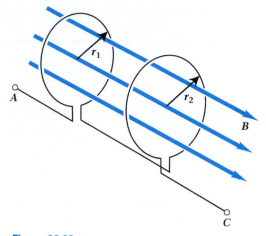

Figure 20.32
Problem 20.58.

(Fig. 20.33). (a) What is the emf between the conducting ring R and the center of rotation A? (b) If a 2.3-Ω resistor is connected between A and R, how much power is required to keep the copper rod turning at the same rate? Neglect friction and neglect the resistance of the bar and the ring.

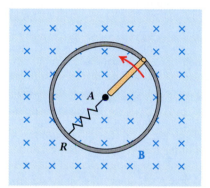

Figure 20.33
Problem 20.59.

20.60•• A conducting rod lies on two parallel, horizontal rails 0.10 m apart that are connected by a resistor of 200 Ω in a region of uniform vertical magnetic field of 0.15 T (Fig. 20.34). How much work is required to move the bar along the rails at constant speed for a distance of 1.0 m in 0.25 s? Neglect friction between the bar and the rails and assume that the bar and the rails have no electric resistance.

20.61•• The direct transmission lines from a generating station to a manufacturing plant are 20 km long and have a total resistance of 6.2 Ω. The generators produce electric power at 13,800 V. (a) Assuming that no step-up transformer is used at the power station, determine the rate of energy loss in the transmission lines if the potential difference at the plant is 12,000 V. (b) Suppose that transformers are to be used to step

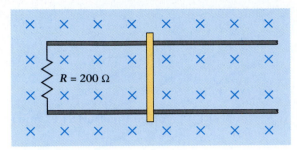

Figure 20.34
Problem 20.60.

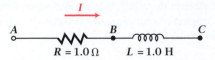

Figure 20.35
Problems 20.65 and 20.66.

up the voltage output from the generators and to step it down again at the plant. If the power loss in the line is to be cut to 10^{-3} of the loss found in part (a), what should be the output voltage of the transformer at the generating plant? (c) If electricity cost $0.08/kWh, how much money is saved in one year by using the higher voltage for transmission?

20.62•• A transformer has a primary voltage of 3750 V and a primary current of 20 A. The secondary voltage is 1200 V, and the secondary current is 60.25 A. If all of the losses in the transformer are resistive losses, what will be the emergent temperature of cooling oil which flows through the transformer at 5.0 kg/min if the initial temperature of the oil is 15°C? Take the specific heat of the oil to be 3.0 kJ/kg · °C.

20.63•• A 500-μF capacitor is charged to 450 V. The capacitor discharges when connected across a 1.3-H inductor. What is the maximum current through the inductor? (*Hint:* Use conservation of energy and assume that at some instant of time all of the energy stored in the capacitor is transferred to the inductor.)

20.64•• A circular loop of wire with a diameter of 5.0 cm and a total resistance of 1.6 Ω is placed inside a long solenoid with the axis of the loop parallel to the axis of the solenoid. The solenoid is 10.0 cm in diameter and has 2.0×10^4 turns per meter. At time $t = 0$, the current in the coil is turned on and increases linearly with time at the rate of 0.020 A/s for 3.0 s, after which it remains constant. How much heat (in joules) is given off by the circular wire loop during the time that current is changing in the solenoid?

20.65•• At the instant of time $t = 0$, the current from A to C in Fig. 20.35 is 30 A and is decreasing at the rate of 10 A/s. What is the electric potential difference between A and C at $t = 0$?

20.66•• For the circuit shown in Fig. 20.35, is it possible to pick values of R, L, initial current from A to C, and $\Delta I/\Delta t$ so that the potential at A is the same as the potential at C?

20.67•• Show that the equivalent inductance, L_s, of two inductors added in series is

$$L_s = L_1 + L_2.$$

(*Note:* The total potential difference across the two inductors is the sum of the individual emfs of each inductor, and the rate of change of current with time is the same for both inductors.)

20.68•• Show that the equivalent inductance, L_p, of two inductors added in parallel is

$$\frac{1}{L_p} = \frac{1}{L_1} + \frac{1}{L_2}.$$

(Recall that Kirchhoff's law requires that the two emfs be equal and that the total current be the sum of the individual currents.)

20.69 A 480-mH inductor is connected in series with a 370-mH inductor. What is the total inductance of the combination? (See Problem 20.67.)

20.70 A 0.27-H inductor is connected in series with a second inductor to give a total inductance of 0.87 H. What is the inductance of the second inductor? (See Problem 20.67.)

20.71 A 0.83-H inductor is joined in parallel to a 0.52-H inductor. What is the inductance of the combination? (See Problem 20.68.)

20.72 Three 0.25-H inductors are to be joined together to give a maximum inductance. What is the maximum value of the combination and how should they be joined together? (See Problems 20.67 and 20.68.)

20.73• Two inductors connected in parallel have an effective inductance of 0.037 H. If the inductance of one inductor is twice that of the other, what is the inductance of each inductor? (See Problem 20.68.)

Alternating-Current Circuits

In the preceding chapter we introduced inductance, a third major component of electric circuits. However, since the effects of inductance are observed only with changing currents, we have waited until now to complete our study of circuits. Many common circuits in electrical and electronic equipment use alternating current, which produces a different response in circuit components than does direct current. We examine this behavior now, seeing how we can employ different combinations of resistance, inductance, and capacitance in useful ways. Compared with the basic circuits we will examine, the circuits in today's handheld calculators, television sets, and compact-disk players are miracles of complexity and manufacturing technology. However, all of these devices rely for their operation on the same fundamental principles, which we present in this chapter.

The development of our knowledge of electromagnetism, from bits of rubbed amber and lodestone in the time of the ancient Greeks to today's world filled with complex electronic equipment, is a fascinating story. We will continue to use our knowledge of electromagnetism as we study other areas of physics. In fact, much of what we have learned about physics in this century has come from experiments in which electronic instruments enable us to observe the world far beyond the reach of our normal senses. Understanding how these instruments work is an essential part of planning and interpreting experiments. Our experimental verification of contemporary physical theories rests in part on our confidence in modern electronics. ■

21.1 The *RL* Circuit

Many electric circuits contain resistors, capacitors, and inductors in combination. One type of circuit, known as an *RL* circuit, contains just a resistor (R) and an inductor (L). Another type of circuit, called an *RC* circuit, contains only a resistor (R) and a capacitor (C). In this section and the next, we will discuss how the current and voltage in these circuits vary with time when a source of constant emf is applied to the circuit. This will help us understand what happens in a circuit when a source of varying emf is applied to it.

We have already seen that an inductor reacts to a change in current with an induced emf that opposes that change. We can analyze the behavior of an inductor in a circuit by using the Kirchhoff voltage rule. Consider a circuit containing an inductor L, a resistor R, and a battery of constant emf V_0 (Fig. 21.1). The resistance R represents the total resistance in the circuit, including the resistance of the inductor. When we close the switch S, a current starts through the circuit. The inductance resists the sudden increase in current and delays its buildup to its ultimate value.

From Kirchhoff's voltage rule, the equation for potential difference around the loop is

$$V_0 + V_R + V_L = 0.$$

We have seen that, following our sign convention, the potential drop across the resistor is $-IR$ and the potential difference across the inductor is $-L(\Delta I/\Delta t)$. Thus

$$V_0 - IR - L\frac{\Delta I}{\Delta t} = 0.$$

Upon rearranging, this equation becomes

$$\frac{\Delta I}{\Delta t} = \frac{V_0 - IR}{L} = -\frac{R}{L}\left(I - \frac{V_0}{R}\right).$$

Because V_0/R is constant, the change in $I - V_0/R$ is the same as the change in I. Thus, we can write

$$\frac{\Delta(I - V_0/R)}{\Delta t} = -\frac{R}{L}(I - V_0/R).$$

This equation has the form of the equations we encountered in Chapter 12. In both cases, we have a rate of change of a quantity that is proportional to the value of the quantity. Thus, the quantity $I - V_0/R$ has the form of an exponential function of time,

$$I - V_0/R = I_0 e^{-Rt/L},$$

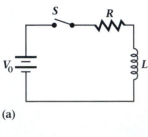

(a)

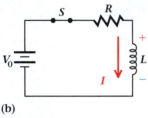

(b)

Figure 21.1

An *RL* series circuit. (a) A resistance R, a battery V_0, and an inductance L are joined in series with a switch S. (b) When the switch is closed, a current exists in the circuit.

which may be rearranged to give

$$I = V_0/R + I_0 e^{-Rt/L}.$$

We can evaluate the proportionality constant I_0 by considering what happens at time $t = 0$, when the switch is closed. There was no current prior to closing the switch, and because the inductor inhibits any change in current, the current is zero *immediately* after the switch is closed. Thus, at $t = 0$, $I = 0$ and the equation for the current becomes

$$0 = V_0/R + I_0.$$

Consequently,

$$I_0 = -V_0/R.$$

We can then express the current as a function of time by

$$I = \frac{V_0}{R}(1 - e^{-Rt/L}). \tag{21.1}$$

A graph of the current in this circuit is shown in Fig. 21.2(a). Note that for long times ($t \gg L/R$), the current in the circuit is determined by the battery and the resistance alone; that is, the current approaches $I = V_0/R$.

We can also obtain the potential difference across the inductor (Fig. 21.2b) from the loop equation as

$$V_0 - IR + V_L = 0,$$

$$V_L = IR - V_0 = -V_0 e^{-Rt/L}. \tag{21.2}$$

The inductor reacts to the closing of the switch by producing an emf opposite in direction to the battery voltage. This emf inhibits the current initially, but the current gradually builds up as the inductor voltage decays. The time for the current to build up to within $1/e$ (37%) of its final value is

$$\tau_L = \frac{L}{R} \tag{21.3}$$

and is called the **inductive time constant.** Circuits with small inductive time constants attain their steady-state value of current quickly; circuits with large time constants take a longer time to attain maximum current.

If the switch in Fig. 21.1 is suddenly opened, the rate of change of current $\Delta I/\Delta t$ may be quite large. Since $\mathcal{E} = -L(\Delta I/\Delta t)$, the inductive voltage will also be large. This voltage, called an inductive kick, may be large enough to cause a dielectric breakdown of the air and produce arcing in the switch. When the inductance in a circuit is large, the inductive kick can present serious problems.

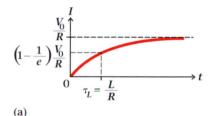

(a)

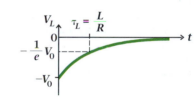

(b)

Figure 21.2
(a) Graph showing the buildup of current in the *RL* circuit of Fig. 21.1. (b) Graph of inductor voltage for the circuit of Fig. 21.1.

Example 21.1

An 8.0-H inductor of negligible resistance is joined in series with a 5.0-Ω resistor. What is the time constant of this combination? If the inductor and resistor are connected in a circuit like that of Fig. 21.1 along with a 3.0-V battery, what is the current in the loop at $t = \tau_L$ if the switch S is closed at time $t = 0$? What is the voltage across the inductor at $t = \tau_L$?

Strategy We can easily find the time constant τ_L by using the values given for R and L in the definition of the time constant, Eq. (21.3). Then setting t equal to τ_L, we can use Eqs. (21.1) and (21.2) to find I and V_L.

Solution The time constant of the LR combination is readily computed to be

$$\tau_L = \frac{L}{R} = \frac{8.0 \text{ H}}{5.0 \text{ }\Omega} = 1.6 \text{ s}.$$

At time $t = \tau_L$, the current may be evaluated from Eq. (21.1) as

$$I = \frac{V_0}{R}(1 - e^{-Rt/L}) = \frac{V_0}{R}(1 - e^{-1}) = \frac{V_0}{R}(0.63),$$

where V_0 is the battery voltage and R is the resistance. Thus

$$I = \frac{3.0 \text{ V}}{5.0 \text{ }\Omega}(0.63) = 0.38 \text{ A}.$$

At $t = \tau_L$, the current has risen to 63% of its final value. As t becomes very large compared with τ_L, the current rises to its final value of $V_0/R = 0.60$ A.

The voltage across the inductor is given by Eq. (21.2) as

$$V_L = -V_0 e^{-Rt/L}.$$

When $t = \tau_L$, the inductor voltage becomes

$$V_L = -V_0 e^{-1} = -0.37 \text{ } V_0 = -1.1 \text{ V}.$$

Thus, in one time constant, the potential across the inductor falls to 37% of its initial value.

21.2 The *RC* Circuit

In Chapter 17 we found that a capacitor could be used to store charge. Because the dielectric material of the capacitor is nonconducting, a capacitor does not allow a steady dc current to pass. However, when a potential difference is first applied across a capacitor, negative charge is added to one plate, and equal numbers of negative charges are removed from the other plate, making it positive. The effect is as if a time-varying current passed through the capacitor for a short time.

A simple circuit for charging a capacitor consists of a battery of potential V_0 connected in series with a resistor R and a capacitor C through a switch S (Fig. 21.3). Initially there is no charge on the capacitor and the switch is open. At time $t = 0$, the switch is closed.

We apply Kirchhoff's voltage law to the circuit to find

$$V_0 - IR - q/C = 0,$$

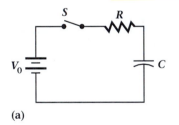

(a)

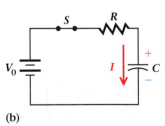

(b)

Figure 21.3

An *RC* circuit for charging a capacitor. (a) A resistance *R*, a battery V_0, and a capacitance *C* are joined in series with a switch *S*. (b) When the switch is closed, charge flows from the battery to the capacitor.

where q/C is the voltage across the capacitor. We have included the negative sign because the potential difference built up across the capacitor is opposite to that of the battery.

The current is defined to be the time rate of change of electric charge. If we substitute $\Delta q/\Delta t$ for the current in the Kirchhoff equation above, we find

$$\frac{\Delta q}{\Delta t} = -\frac{1}{RC}(q - V_0 C). \qquad (21.4)$$

Notice the similarity between this equation and the equation for the *RL* circuit in the preceding section. Although the combination of constants is different, the equations have the same form. The solution also has the same form:

$$q = V_0 C + q_0 e^{-t/RC}.$$

We can evaluate q_0 from the initial conditions. At $t = 0$, $q = 0$, and $q_0 = -V_0 C$. Inserting this value of q_0 in the equation for q, we find that the charge on the capacitor grows with time according to

$$q = V_0 C(1 - e^{-t/RC}). \qquad (21.5)$$

The potential difference V_C across the capacitor is $V_C = q/C$. Upon substituting Eq. (21.5) for the charge, we find the voltage to be

$$\boxed{V_C = V_0(1 - e^{-t/RC}),} \qquad (21.6)$$

where V_0 is the battery potential.

The current in the circuit is $\Delta q/\Delta t$, which may be obtained from Eq. (21.4) when Eq. (21.5) is inserted for the charge q:

$$\boxed{I = \frac{V_0}{R}e^{-t/RC}.} \qquad (21.7)$$

Figure 21.4 shows graphs of the charging current and the voltage on the capacitor as a function of time. The current in the *RC* circuit is at a maximum the instant the switch is closed and decreases exponentially with time. The characteristic time constant for the current to decay to $1/e$ (37%) of its initial value is

$$\tau_C = RC, \qquad (21.8)$$

called the **capacitive time constant.** A circuit with a large capacitive time constant charges and discharges slowly; a circuit with a small time constant charges and discharges quickly.

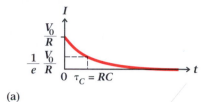

(a)

(b)

Figure 21.4

Graphs of (a) the charging current and (b) the voltage on the capacitor of Fig. 21.3. Time $t = 0$ corresponds to the closing of the switch.

Example 21.2

How many time constants must elapse before the voltage across a capacitor in a defibrillator reaches 98% of its final value? Give your answer in terms of the time constant τ_C.

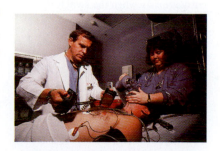

A physician waits for the defibrillator capacitor to charge before applying the paddles during CPR.

Solution The capacitor voltage is expressed in terms of the time by Eq. (21.6) with $\tau_C = RC$. Upon rearrangement, this equation is

$$e^{-t/\tau_C} = \frac{V_0 - V_C}{V_0}.$$

Next we insert the value of the capacitor voltage at time t, $V_C = 0.98V_0$, to get

$$e^{-t/\tau_C} = \frac{1.00 - 0.98}{1.00} = 0.02.$$

Taking the natural logarithm of both sides of this equation and using a calculator to evaluate the logarithm, we find that

$$\frac{-t}{\tau_C} = \ln 0.02 = -3.9 \quad \text{or} \quad t = 3.9\,\tau_C.$$

We conclude that a charging capacitor reaches 98% of its final voltage in 3.9 time constants.

■

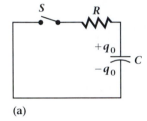

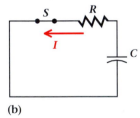

Figure 21.5
An *RC* circuit for discharging a capacitor. (a) The capacitor is charged and the switch is open. (b) A discharge current exists when the switch is closed.

We can discharge a charged capacitor by completing a closed circuit across its terminals (Fig. 21.5). The resistor R could be a large resistance, or it could be merely the effective resistance of the conducting wire. When the switch S is closed, Kirchhoff's voltage law gives

$$RI + \frac{q}{C} = 0.$$

If we replace I by $\Delta q/\Delta t$, we get

$$\frac{\Delta q}{\Delta t} = -\frac{q}{RC}.$$

Again we have an equation that has an exponential function for a solution. The charge on the capacitor decreases with time according to

$$q = q_0 e^{-t/RC},$$

where q_0 is the initial charge on the capacitor at time $t = 0$. The capacitor voltage is again given by the ratio of q/C:

$$V_C = \frac{q_0}{C} e^{-t/RC} \quad \text{or} \quad V_C = V_0 e^{-t/RC}.$$

The current is given by

$$\boxed{I = \frac{V_0}{R} e^{-t/RC}.} \tag{21.9}$$

The *RL* and *RC* circuits are similar in that both of them respond to an increase in applied voltage with a current that changes exponentially with time. However, they respond in different ways. Because of the induced emf, the *RL* circuit inhibits the current initially, but allows it to grow with time as the induced emf decays. The *RC* circuit permits an initial current, but as the capacitor becomes charged, that current approaches zero. As we will see later in this chapter, the different behavior of these devices leads to dramatically different responses to ac signals.

21.3 Effective Values of Alternating Current

In discussing household circuits in Chapter 18, we mentioned that the instantaneous electric potential (the voltage) is not constant, but we assumed that it had a dc equivalent. In practice, the voltage in household circuits changes continuously, smoothly oscillating back and forth from positive to negative (Fig. 21.6). As a result, the currents also oscillate in a sinusoidal manner, with the same frequency as the voltage. As we discussed in Chapter 20, this alternating current, or ac, is generated naturally by a conducting loop rotating in a magnetic field. In North America, household current alternates at a rate of 60 complete oscillations per second, or 60 Hz. Much of the rest of the world uses 50-Hz electrical power.

The curve of Fig. 21.6 describes a sinusoidally oscillating source voltage. The instantaneous potential difference v between the two terminals of the source oscillates with time according to

$$v = V_m \sin 2\pi ft,$$

where f is the frequency of the oscillation, t is the time, and V_m is the amplitude (the maximum or peak value) of the oscillation. We can describe the sinusoidally varying voltage (or current) in terms of its amplitude and frequency. However, in dealing with ac voltages and currents, it is more common to use their dc equivalent values than their amplitudes.

To see how we can determine dc equivalent values for an ac circuit, consider the example of a current passing through a resistor R. For a direct current I, the power dissipated is I^2R. Because dc is constant, the instantaneous dc power is also the average power. Now if we apply a time-varying current of instantaneous magnitude i to the circuit,* the instantaneous power dissipation is still

$$p = i^2R,$$

as before. However, the average power is

$$\bar{p} = \overline{i^2R} = \overline{i^2}R$$

if R is constant.

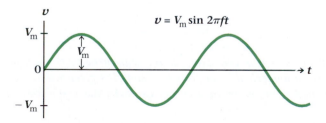

Figure 21.6
Graph of a sinusoidal voltage plotted as a function of time. The amplitude is V_m, the peak value of the signal.

---

*We represent the instantaneous value of a sinusoidal quantity with a lowercase letter and the amplitude, or maximum value, of the quantity with an uppercase letter.

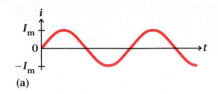

(a)

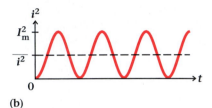

(b)

Figure 21.7

(a) Graph of a sinusoidally oscillating current i. The average of the current over a complete oscillation is zero. (b) Graph of i^2, the square of the current of part (a). The dashed line shows the average value of i^2.

The average power depends on the average of the square of the current, $\overline{i^2}$. A sinusoidal current reverses direction periodically, and its average over one period is zero; however, the square of the current is always positive or zero (Fig. 21.7). As a result, the time average of i^2 over one period is nonzero. The square root of $\overline{i^2}$ is called the **root-mean-square current** or the **rms current,** I_{rms}. The rms value of the current is also called the **effective value** of the current. We can then express the average power as

$$\overline{p} = I^2{}_{\text{rms}}R, \tag{21.10}$$

where I_{rms} is the root-mean-square current,

$$I_{\text{rms}} = \sqrt{\overline{i^2}}. \tag{21.11}$$

Had we chosen to start with instantaneous power given by $p = iv$, then the average power dissipated in the resistor would be*

$$\overline{p} = I_{\text{rms}}V_{\text{rms}}, \tag{21.12}$$

where V_{rms} is the root-mean-square value of the oscillating voltage. The **rms voltage** is

$$V_{\text{rms}} = \sqrt{\overline{v^2}}. \tag{21.13}$$

We can compute the rms value of current from the average of the square of the sinusoidal current over one period. The average value of the square of the sine function over one period is $\frac{1}{2}$. So we have

$$I_{\text{rms}} = \frac{I_{\text{m}}}{\sqrt{2}}, \tag{21.14}$$

where I_{m} is the amplitude of the oscillating current. Note that the relationship of Eq. (21.14) is appropriate only for sinusoidally oscillating currents. The rms value of a sinusoidally oscillating voltage has the same form as that of the current and is given by

$$V_{\text{rms}} = \frac{V_{\text{m}}}{\sqrt{2}}. \tag{21.15}$$

The effective or rms values of voltage and current are so commonly used that whenever a numerical value for an ac signal is given without qualification, it is understood to be the rms value. For example, the 120-V outlets ordinarily

---

*Eq. (21.12) is appropriate for a circuit containing only an oscillating source of emf and resistors. When the circuit also includes capacitors and inductors, as discussed in the next two sections, the current and voltage may no longer be in phase. In that case, the average power becomes $\overline{p} = I_{\text{rms}}V_{\text{rms}} \cos \phi$, where ϕ is the phase angle between current and voltage.

ELECTROCARDIOGRAPHY

As we have seen in the last few chapters, the applications of electronics to industry, communications, computers, and entertainment have been spectacular. However, electronics has also had tremendous impact in medicine. Laboratory workers routinely use automated electronic devices to analyze blood samples and use sophisticated computer controlled instruments for many diagnostic purposes.

One instructive example of medical electronics is the electrocardiograph, an instrument that has become standard for diagnosis. Basically the *electrocardiograph* consists of an electronic amplifier coupled to a chart recorder that displays the electric potential associated with heart beats as a function of time.

In its normal resting state, muscle fiber, such as in the heart, maintains a negative charge within its membrane and a positive charge outside. In this state the fiber is said to be polarized. As long as the fiber remains undisturbed, the potential difference across the membrane remains at about 85 mV, which is called the resting potential (Fig. B21.1a).

When a nerve impulse excites the muscle, a rapid exchange of charge occurs through the membrane, resulting in a rapid depolarization. In fact, the exchange of ions that carry this charge across the membrane usually results in a small reverse polarization (Fig. B21.1b). Immediately after the depolarization, the muscle returns to its initial state and is repolarized. This sequence of changes is known as the action potential. Within milliseconds of the spread of the action potential through the fiber, the muscle begins to contract. The potential difference associated with an action potential wave (Fig. B21.2) is about 100 mV.

The result of all this activity is that an electrical pulse occurs with each heartbeat, as well as with the contraction of

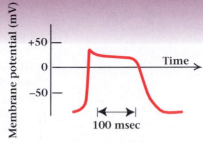

Figure B21.2 A graph of the action potential versus time for a heart muscle.

any other muscle. The electric potentials associated with the pulsing heart muscle can be detected externally (Fig. B21.3a). The resulting measurements as recorded on a chart recorder can be seen in Fig. B21.3(b). The external potentials are typically on the order of 1 mV. Since the contraction of any muscle generates similar pulses, electrocardiograms are normally made with the patient lying still.

Action potentials also occur in nerve activity. Consequently, the electrical activity of the brain can be detected in essentially the same manner as that used to monitor the heart. These signals are due to nerve activity within the brain. At the surface of the scalp, these signals typically have an amplitude of only 50 μV. Thus, to measure brain waves, muscular activity must be suppressed, since the signals detected from the muscles could be up to twenty times larger than those from the brain. A recording of the brain waves is known as an *electroencephalogram.*

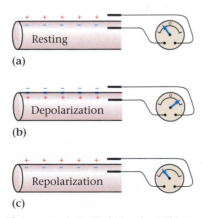

Figure B21.1 Sequence of events during the action potential, showing (a) the normal resting potential, (b) development of a reverse potential during depolarization, and (c) reestablishment of the normal resting potential during repolarization.

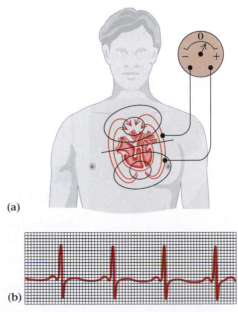

Figure B21.3 (a) Electric field lines (red) and equipotential lines (black) in the chest around a partially depolarized heart. (b) Electrocardiogram recording of a normal heart.

found in homes supply a constant 120 V rms. Voltmeters and ammeters used in ac circuits are usually calibrated to read rms values. In fact, rms usage is so common that subscripts are usually not included; that is, instead of V_{rms} and I_{rms} we simply use V and I.

Note that the average or effective value of the power given by Eq. (21.10) or (21.12) is the same as that produced by dc current and voltage of magnitude equal to I and V. In other words, one ampere rms ac produces exactly the same heating of a resistor as does one ampere dc.

When we discussed transformers in Chapter 20, we talked about ac voltages without giving the details we have just considered here about peak voltages and rms voltages. Normally the specifications of a transformer are given in terms of the rms voltages on input and output. Note that transformers are necessarily ac devices; there are no dc transformers.

Example 21.3

The output of a step-down transformer is measured at 12.6 V ac when connected to a 12-W light bulb. (a) What is the peak value of the output voltage? (b) What is the peak value of the output current?

Solution   (a) The measured output voltage of the transformer is the effective, or rms, value. The relation between the rms value and the peak value of the voltage is given by Eq. (21.15):

$$V_{rms} = \frac{V_m}{\sqrt{2}}.$$

So the peak value of the voltage output is

$$V_m = \sqrt{2}V_{rms} = \sqrt{2}(12.6 \text{ V}) = 17.8 \text{ V}.$$

(b) The rms current can be calculated from Eq. (21.12) as

$$I_{rms} = \frac{\bar{p}}{V_{rms}} = \frac{12 \text{ W}}{12.6 \text{ V}} = 0.95 \text{ A}.$$

The peak current is then

$$I_m = \sqrt{2}I_m = \sqrt{2}(0.95 \text{ A}) = 1.3 \text{ A}.$$

21.4   Reactance

We now consider the response of capacitors and inductors subjected to a sinusoidal potential or current. We know that a capacitor does not permit a direct current to pass through it, but does respond to changes in potential by allowing a charging or discharging current to pass. We expect that a capacitor subject to a rapidly alternating potential difference will allow some alternating current to pass. However, when the potential difference alternates more slowly, less current will pass. Thus, the response of a capacitor to an oscillating potential depends on the frequency of the oscillation.

For example, suppose we place an oscillating potential of frequency f and amplitude V_m across a capacitor C (Fig. 21.8). We may express the potential as

$$v = V_m \sin 2\pi ft.$$

The charge on the capacitor is proportional to its voltage through

$$q = Cv = CV_m \sin 2\pi ft. \qquad (21.16)$$

The current is obtained from the rate of change of the charge with time,

$$i = \frac{\Delta q}{\Delta t}.$$

The rate of change of a sine function is proportional to a cosine function:*

$$\frac{\Delta(\sin 2\pi ft)}{\Delta t} = 2\pi f \cos 2\pi ft.$$

Thus, the current is described by

$$i = 2\pi f CV_m \cos 2\pi ft = I_m \cos 2\pi ft. \qquad (21.17)$$

These equations for current and voltage tell us that *the current and the voltage do not reach their maxima at the same time.* We can see this behavior with the aid of Fig. 21.9(a). The current reaches its maximum one-quarter of a cycle before the voltage reaches its maximum. Thus, we say that the current *leads* the

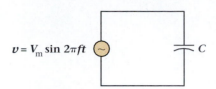

$v = V_m \sin 2\pi ft$ C

Figure 21.8
An alternating voltage source connected across a capacitor.

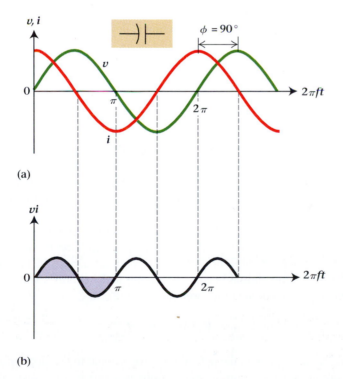

(a)

(b)

Figure 21.9
(a) The current and voltage across a capacitor do not reach their maxima at the same time. The current *leads* the voltage by one-quarter of a cycle, corresponding to a phase shift of 90°. (b) The power delivered to the capacitor, given by the product vi, oscillates in time. The time average of the power is zero.

---

*In studying the harmonic oscillator in Chapter 14, we examined the relationships between displacement, velocity, and acceleration. In the case of the sinusoidally varying velocity, the rate of change (the acceleration) was proportional to the cosine.

voltage across a *capacitor* by 90°, or $\pi/2$ radians. In fact, the current can be expressed as

$$i = 2\pi fCV_{\mathrm{m}} \sin\left(2\pi ft + \frac{\pi}{2}\right).$$

The difference in time between the occurrences of the maximum current and the maximum voltage is usually given in terms of a phase angle ϕ. One full cycle or period corresponds to a phase angle of 360°, or 2π radians; thus, a shift of one-quarter of a cycle is equivalent to a shift of 90°, or $\pi/2$ radians. For the capacitor, we say that the current is shifted 90° ahead of the voltage. We could also say that the voltage lags the current by 90°.

The instantaneous power delivered to the capacitor is vi. When this product is summed over an entire cycle, the result is zero because of the 90° phase shift between v and i. Figure 21.9(b) shows the product of the v and i curves of Fig. 21.9(a). The sum over one cycle is represented by the shaded area of the curve. The area above the time axis is positive, and the area below it is negative. From the symmetry of the curve we see that the average power delivered to the capacitor is zero.

The maximum value of the current varies linearly with the maximum value of the applied voltage for a capacitor, in analogy with Ohm's law for resistors. From Eq. (21.17), the maximum value of the current is

$$I_{\mathrm{m}} = 2\pi fCV_{\mathrm{m}}.$$

If we compare this relationship with the current-voltage relationship for resistors, $R = V/I$, we get a proportionality factor analogous to resistance. This proportionality factor is the **capacitive reactance,** defined to be

$$X_C \equiv \frac{V_{\mathrm{m}}}{I_{\mathrm{m}}} = \frac{1}{2\pi fC}. \qquad (21.18)$$

Notice that the reactance is frequency-dependent, as we expected, and becomes infinitely large as the frequency f approaches zero. This corresponds to the fact that capacitors do not pass a constant (dc) current. At high frequencies, the reactance becomes quite small and the capacitor does little to impede the current. Thus, a capacitor that serves to isolate one dc voltage from another may readily transmit ac.

An inductor also displays a linear relationship between current and voltage. As in the case of the capacitor, this relationship depends on frequency. For example, if we place a source of oscillating potential across an inductor (Fig. 21.10), an oscillating current occurs. According to Kirchhoff's voltage law, the sum of the potentials around the loop must be zero, so

$$v = V_{\mathrm{m}} \sin 2\pi ft = L\frac{\Delta i}{\Delta t}.$$

According to this equation, the potential is proportional to the slope of the curve for the current. Since the current whose slope is a sine is a negative cosine, we can write the current as

$$i = -I_{\mathrm{m}} \cos 2\pi ft = I_{\mathrm{m}} \sin\left(2\pi ft - \frac{\pi}{2}\right).$$

Thus, the current *lags* the voltage across an *inductor* by 90° (Fig. 21.11).

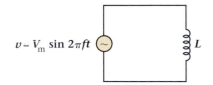

$v = V_{\mathrm{m}} \sin 2\pi ft$ L

Figure 21.10

An alternating voltage source connected across an inductor.

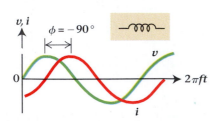

Figure 21.11

Graphs of current and voltage across an inductor. The current *lags* the voltage by one quarter of a cycle.

The instantaneous power delivered to the inductor is iv. When this product is summed over a complete cycle, the result is zero because the current and voltage are 90° apart. Thus, over a complete cycle, the inductor takes no power from the circuit. In this respect, it behaves like the capacitor.

We define the term **inductive reactance** analogously to the way we defined capacitive reactance. If we calculate the rate of change of the current using the techniques of Chapter 14, we find

$$\frac{\Delta i}{\Delta t} = 2\pi f I_m \sin 2\pi ft.$$

The proportionality between peak current and peak voltage in an inductor is the inductive reactance X_L. We obtain the value of the reactance by inserting the expression for $\Delta i/\Delta t$ into the equation for the voltage. The result is

$$X_L \equiv \frac{V_m}{I_m} = 2\pi fL. \tag{21.19}$$

At high frequencies, the inductive reactance is large and the inductance impedes the current. This behavior corresponds to the fact that an inductor opposes changes in current. At frequencies approaching zero, the reactance is small and the inductance offers little hindrance to the current.

Example 21.4

What is the reactance of a 25-μF capacitor at a frequency of 15 kHz? What is the voltage across the capacitor if the current is 5.8 mA?

Solution The reactance can be found from Eq. (21.18)

$$X_C = \frac{1}{2\pi fC} = \frac{1}{(2\pi)(15 \times 10^3 \text{ Hz})(25 \times 10^{-6} \text{ F})}$$

$$X_C = 0.424 \ \Omega = 0.42 \ \Omega.$$

(Problem 21.34 asks you to show that the unit of reactance is the ohm.)

The voltage across the capacitor is the product of the current with the reactance:

$$V_C = IX_C = (5.8 \times 10^{-3} \text{ A})(0.424 \ \Omega) = 2.46 \times 10^{-3} \text{ V} = 2.5 \text{ mV}.$$

21.5 The *RLC* Series Circuit

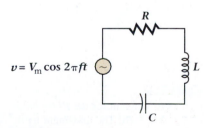

$v = V_m \cos 2\pi ft$

Figure 21.12
A series circuit containing resistance R, inductance L, and capacitance C. The circuit is driven by an oscillating voltage source.

Let us now tie all of these ideas together by examining the behavior of a series circuit containing resistance, inductance, and capacitance when the circuit is driven by an oscillating voltage source. This so-called *RLC* circuit (Fig. 21.12) is a fundamental circuit underlying many kinds of electronic equipment. We can analyze the behavior of the circuit by considering the voltages around the loop. According to Kirchhoff's voltage law, at any instant the voltage provided by the

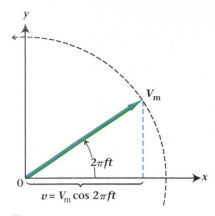

Figure 21.13

Phasor representation of an oscillating voltage $v = V_m \cos 2\pi ft$.

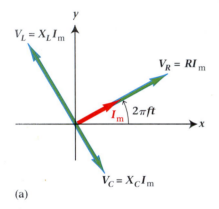

(a)

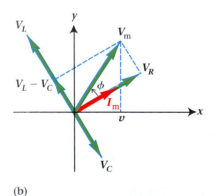

(b)

Figure 21.14

(a) Phasor diagram for the *RLC* circuit of Fig. 21.12. (b) The maximum net voltage is V_m, the phasor sum of V_L, V_C, and V_R. At any instant of time, the net voltage v is the projection of V_m along the *x* axis.

source must equal the sum of the potential drops across the three elements *R*, *L*, and *C*. That is,

$$v = v_R + v_L + v_C.$$

Each of these potentials oscillates with the frequency of the source. However, each voltage has a difference phase relationship with the current in the circuit. The voltage v_R across the resistor is in phase ($\phi = 0$) with the current. The voltage v_L across the inductor leads the current by a phase angle of 90°. The capacitive voltage v_C lags the current by 90°. As a consequence of these phase differences, we cannot simply add the voltages by summing either their peak values or their rms values.

To handle this situation, we return to the relationship we studied in Chapter 14 between simple harmonic motion and circular motion. Specifically, we represent an oscillating voltage (or current) by a **phasor,** which is a two-dimensional mathematical quantity that rotates with a constant angular frequency $2\pi f$. The magnitude (length) of the phasor represents the amplitude of the voltage (or current). Phasors are added head to tail, like vectors. The direction of the voltage phasor represents the phase angle relative to a reference phase and is not related to any spatial direction of the potential difference. For example, we represent a voltage $v = V_m \cos 2\pi ft$ by the projection along the *x* axis of a phasor of magnitude V_m rotating about the origin (Fig. 21.13).

The phasor diagram for the *RLC* circuit of Fig. 21.12 is drawn in Fig. 21.14(a). We choose the reference phasor to be the current in the circuit, shown as a phasor of magnitude I_m. The phasor V_R describing the voltage across the resistor lies along the same direction as the current phasor. We see them both drawn at a particular instant of time *t*. The phasor representing the inductor voltage leads the current by 90°; its magnitude is $V_L = X_L I_m$. Similarly, the phasor for the capacitor voltage has a magnitude $V_C = X_C I_m$ and lags the current by 90°. The net voltage is the sum of these individual phasors, which we can obtain as if we were adding vectors (Fig. 21.14b). Note that this sum gives us the maximum voltage V_m; the net voltage v at any instant is the projection of V_m along the *x* axis.

The voltage across the *RLC* combination has a maximum value V_m and oscillates with frequency *f*. However, it is phase-shifted by an amount ϕ relative to the current. This phase angle can be evaluated with the aid of Fig. 21.14(b), which shows ϕ as the angle between phasor I_m and phasor V_m. The tangent of ϕ is given by

$$\tan \phi = \frac{V_L - V_C}{V_R},$$

or, dividing each voltage by the current I_m,

$$\phi = \tan^{-1}\left(\frac{X_L - X_C}{R}\right). \tag{21.20}$$

We can also calculate the peak or maximum value of the voltage from the figure by applying the Pythagorean theorem:

$$V_m = \sqrt{V_R^2 + (V_L - V_C)^2}.$$

When the individual voltages are expressed in terms of the current I_m, the last equation becomes

$$V_m = I_m \sqrt{R^2 + (X_L - X_C)^2}.$$

The factor that relates the peak voltage to the peak current is called the **impedance** Z,

$$Z = \sqrt{R^2 + (X_L - X_C)^2}, \qquad (21.21a)$$

or

$$Z = \sqrt{R^2 + \left(2\pi fL - \frac{1}{2\pi fC}\right)^2}. \qquad (21.21b)$$

Because the rms voltage and current are directly proportional to the peak values of voltage and current, the impedance is also the ratio of the rms voltage to the rms current. Thus, we can use it to express a generalized form of Ohm's law, applicable to ac circuits:

$$V_m = I_m Z, \qquad (21.22)$$

or

$$V_{rms} = I_{rms} Z. \qquad (21.23)$$

Note that impedance depends not only on the resistance, capacitance, and inductance of the *RLC* circuit, but also on the frequency of the applied voltage (because reactance is frequency-dependent).

Example 21.5

A series circuit has $R = 30\ \Omega$, $L = 0.10$ H, and $C = 100\ \mu$F. What is the peak voltage across each of these elements if a 60-Hz sinusoidal current with a peak value of 1.0 A flows in the circuit? What is the peak voltage across the combination of these elements? What is the phase angle between voltage and current?

Strategy We will use phasor addition to determine the combined voltage in the circuit, so the first step is to calculate the peak voltage for each individual component. We will do this by applying the generalized versions of Ohm's law for resistance, inductive reactance, and capacitive reactance. We will carry out the phasor addition algebraically, but you may want to draw your own phasor diagram to help you see the relationships involved in this type of circuit analysis.

Solution We calculate the peak voltages one at a time. The resistor voltage is

$$V_R = RI_m = (30\ \Omega)(1.0\ \text{A}) = 30\ \text{V}.$$

The peak voltage across the inductor is

$$V_L = X_L I_m = 2\pi fL I_m = (2\pi)(60\ \text{Hz})(0.10\ \text{H})(1.0\ \text{A})$$
$$V_L = 37.7\ \text{V}.$$

The peak voltage across the capacitor is

$$V_C = X_C I_m = \frac{I_m}{2\pi f C} = \frac{1 \text{ A}}{(2\pi)(60 \text{ Hz})(10^{-4} \text{ F})},$$

$$V_C = 26.5 \text{ V}.$$

The voltage across the combination is found from the phasor sum of the three voltages just calculated. It is

$$V_m = \sqrt{V_R^2 + (V_L - V_C)^2}$$

$$V_m = \sqrt{(30 \text{ V})^2 + (37.7 \text{ V} - 26.5 \text{ V})^2} = 32 \text{ V}.$$

Notice that while the peak voltage across the inductor is 37.7 V, the peak voltage across all three elements is only 32 V because of the phase relationships.

The phase angle between voltage and current is

$$\phi = \tan^{-1}\left(\frac{V_L - V_C}{V_R}\right) = \tan^{-1}\left(\frac{37.7 \text{ V} - 26.5 \text{ V}}{30 \text{ V}}\right)$$

$$\phi = 20°.$$

The voltage leads the current by 20°.

Example 21.6

A power supply that generates 20 V rms output at a frequency of 1000 Hz is connected to a series *RLC* circuit containing a resistor $R = 100 \ \Omega$, an inductor $L = 0.015$ H, and a capacitor $C = 0.40 \ \mu$F. What is the rms current in the circuit?

Solution We may combine Eqs. (21.21b) and (21.23) to give

$$I_{rms} = \frac{V_{rms}}{Z} = \frac{V_{rms}}{\sqrt{R^2 + \left(2\pi f L - \dfrac{1}{2\pi f C}\right)^2}}.$$

Upon substituting the given values we have

$$I_{rms} = \frac{20 \text{ V}}{\sqrt{(100 \ \Omega)^2 + \left((2\pi)(1000 \text{ Hz})(0.015 \text{ H}) - \dfrac{1}{(2\pi)(1000 \text{ Hz})(0.40 \times 10^{-6} \text{ F})}\right)^2}}$$

$$I_{rms} = 0.063 \text{ A} = 63 \text{ mA}.$$

21.6 Resonant Circuits

Electrical systems, like mechanical systems, can display resonance. Resonant circuits have long been used in the tuning circuits of radios and TVs. To understand electrical resonance, remember that the response of a series *RLC* circuit to an applied voltage $v = V_m \cos 2\pi f t$ depends on the frequency. We may express the rms current as

$$I_{rms} = \frac{V_{rms}}{Z}.$$

Using Eq. (21.21b) for the impedance, we can write this as

$$I_{rms} = \frac{V_{rms}}{\sqrt{R^2 + \left(2\pi f L - \frac{1}{2\pi f C}\right)^2}},$$

which shows the frequency dependence explicitly.

When the frequency is zero, the term $1/2\pi f C$ is infinite and the current is zero. (The capacitor blocks dc and offers a large resistance to very low frequencies.) At very high frequencies, the term $2\pi f L$ becomes large and again the current approaches zero. (The inductor blocks high frequencies.) At intermediate frequencies, the current is appreciable and reaches a maximum when $2\pi f L = 1/2\pi f C$. This condition of the circuit is called *resonance,* and the frequency at which the maximum current occurs is the *natural* or *resonant frequency* f_0, given by

$$f_0 = \frac{1}{2\pi\sqrt{LC}}. \qquad (21.24)$$

When the resistance is small, the resonance is very sharp, leading to an appreciable current over only a narrow range of frequencies. However, when the resistance is large, the resonance becomes broad. The effect of the resistance is seen in Fig. 21.15, which shows the current as a function of frequency in an RLC series circuit for two different values of R. The lower curve corresponds to a resistance two times larger than that of the upper curve. Notice the similarities between electrical resonance and mechanical resonance (Section 14.7).

Resonance also occurs for other circuit combinations of R, L, and C. In particular, it occurs for the parallel combination of inductance and capacitance (Fig. 21.16). For this case, resonance is characterized by a maximum impedance, rather than a minimum. At low frequencies, the reactance of the inductor is small and a large current passes through the branch containing the inductor. At high frequencies, the capacitive reactance is low, allowing a large current in the branch containing the capacitor. At some intermediate frequency, the magnitudes of the impedance in the two branches are equal and the total current and voltage are in phase. If the resistance is small, the resonant frequency for this parallel combination of L and C is also given by Eq. (21.24).

Frequency-selective circuits (resonant circuits) are commonplace in radio, television, and other electronic applications. The tuning circuit in a radio often consists of an inductor in parallel with a variable capacitor (Fig. 21.17). You tune the resonant frequency of the circuit to the frequency of the desired station by adjusting the value of capacitance. The radio's antenna brings many radio signals of different frequencies to the resonant circuit. Signals not at the resonant frequency of the tuning circuit pass to the ground through the low impedance of the circuit for those frequencies. Consequently, the signal voltage reaching the amplifier is very small. At the resonant frequency, the impedance of the tuning circuit is high so that an appreciable signal voltage appears across the circuit and is delivered to the input of the amplifier.

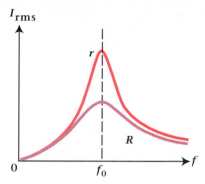

Figure 21.15

Graph of current versus frequency for a series RLC circuit. The lower curve corresponds to a resistance value two times larger than that of the upper curve.

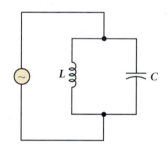

Figure 21.16

An oscillating voltage source drives a parallel combination of inductance and capacitance in a resonant circuit.

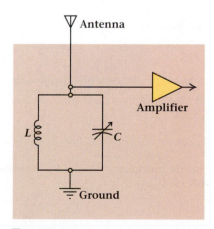

Figure 21.17

A parallel LC combination forms the tuning circuit for a radio receiver.

Example 21.7

A radio tuning circuit is made from a 150-μH inductor connected in parallel with a variable capacitor. If the capacitor is adjusted so that its capacitance is 169 pF, what is the resonant frequency of the circuit?

Solution The resonant frequency is given by Eq. (21.24):

$$f_0 = \frac{1}{2\pi\sqrt{LC}} = \frac{1}{2\pi\sqrt{(150 \times 10^{-6}\ \text{H})(169 \times 10^{-12}\ \text{F})}}$$

$$f_0 = 1000\ \text{kHz}.$$

Discussion The range of standard AM radio broadcast frequencies is from 530 kHz to 1610 kHz. The frequency we have just calculated is near the center of the frequency band allowed for AM radio.

Summary

Useful Concepts

■ In an ac circuit, an inductor opposes the change in current. The current in a closed RL circuit containing a voltage source V_0 is

$$I = \frac{V_0}{R}(1 - e^{-Rt/L}),$$

where L/R is the inductive time constant. The potential difference across the inductor is

$$V_L = -V_0 e^{-Rt/L}.$$

■ The potential difference across the capacitor in an RC circuit containing a voltage source V_0 is

$$V_C = V_0(1 - e^{-t/RC}),$$

and the current is

$$I = \frac{V_0}{R}e^{-t/RC},$$

where RC is the capacitive time constant.

■ The rms value of a sinusoidally oscillating current with amplitude I_m is

$$I_{\text{rms}} = \frac{I_m}{\sqrt{2}},$$

and the rms value of a sinusoidally oscillating voltage with amplitude V_m is

$$V_{\text{rms}} = \frac{V_m}{\sqrt{2}}.$$

■ The average power in a circuit with ac current and voltage is given by

$$\bar{p} = I_{\text{rms}}V_{\text{rms}} = I^2{}_{\text{rms}}R.$$

■ The capacitive reactance is

$$X_C = \frac{1}{2\pi fC}.$$

■ The inductive reactance is

$$X_L = 2\pi fL.$$

■ A phasor represents an oscillating voltage or current. Its length is proportional to the magnitude of the voltage or current and rotates with an angular frequency $2\pi f$. Phasors can represent circuit combinations of alternating voltages or currents.

■ In an RLC series circuit, the impedance is

$$Z = \sqrt{R^2 + (X_L - X_C)^2}.$$

■ The phase angle between the voltage and the current is

$$\phi = \tan^{-1}\left(\frac{X_L - X_C}{R}\right).$$

■ In an RLC series circuit, the resonance frequency is given by

$$f_0 = \frac{1}{2\pi\sqrt{LC}}.$$

Important Terms

You should be able to write the definition or meaning of each of the following:

inductive time constant
capacitive time constant
rms current
effective value
rms voltage

capacitive reactance
inductive reactance
phasor
impedance

Conceptual Questions

21.1 How can an ac distribution system transmit power if the average of the current over every cycle is zero?

21.2 What is the difference between the square root of the average value squared and the square root of the average of the squared value of a sinusoidal voltage?

21.3 What would be the advantages or disadvantages of changing the frequency of household electricity to 1200 Hz rather than 60 Hz? What would they be if it were changed to 10 Hz?

21.4 Is it possible to adjust R, L, and C in some circuit so that the impedance is zero, thereby giving you free electrical energy?

21.5 What does it mean to say that the voltage across an inductor leads the current by $\pi/2$?

21.6 How does the resonant tuning circuit in a radio determine what station you hear?

21.7 Explain the difference between peak value, rms value, and average value for a sinusoidally oscillating current.

21.8 Meters calibrated for ac give the correct rms values for sinusoidal current and voltage. If such meters are used to measure voltages with triangular waveforms or square waveforms, would they give the correct rms values? Explain your reasoning.

Problems

Section 21.1 The *RL* Circuit

> **Hints for Solving Problems**
>
> Remember that the time constant for an *RL* circuit is $\tau_L = L/R$. The current in an *RL* circuit containing a voltage source V_0 increases with time after the circuit is closed.

21.1 Show that the dimension of henries per ohm is time.

21.2 A series *RL* circuit containing a 0.17-H inductor has a time constant of 0.13 s. What is the resistance in the circuit?

21.3 (a) What is the inductive time constant of an *RL* circuit of $L = 0.80$ H and $R = 27$ Ω? (b) What is the ratio of the current in the circuit at time $t = L/R$ to the current after a very long time ($t \gg L/R$)?

21.4• What is the inductive time constant for the circuit shown in Fig. 21.18? Note that inductors combine according to the same rules used for combining resistors.

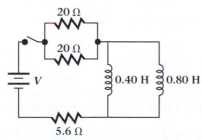

Figure 21.18
Problem 21.4.

21.5• A resistance of 100 Ω and an inductance of 50 mH are connected in series across a 9.0-V battery. How long will it take after the circuit is completed for the current to reach 80 mA?

21.6• A 1000-Ω resistor is joined in series with a 400-mH inductance and a 12-V battery. (a) What is the time constant of the circuit? (b) How long will it take after the circuit is completed for the voltage across the resistor to reach 3.0 V?

21.7• The current in an *RL* circuit rises to one-half of its maximum value in 7.0 s. (a) What is the inductive time constant for the circuit? (b) If $L = 0.80$ H, what is the value of the resistance?

21.8• (a) What is the inductive time constant of an *RL* circuit of $L = 0.80$ H and $R = 5.6$ Ω? (b) How long does it take after the switch is closed for the current in the circuit to reach 99% of its ultimate value?

Section 21.2 The *RC* Circuit

> **Hints for Solving Problems**
>
> Remember that the time constant for an *RC* circuit is $\tau_C = RC$. The current in an *RC* circuit decreases with time after the circuit is closed.

21.9 Show that the SI unit of resistance times capacitance is the second.

21.10 A 250-μF capacitor is discharged through a 4.7-MΩ resistance. What is the time constant?

21.11 What is the time constant of the circuit in Fig. 21.19?

21.12• Two 5000-μF capacitors are joined in parallel and charged to 48.6 V. The capacitors discharge through a resistance of 130 kΩ. (a) What is the time constant of the circuit? (b) How much voltage will remain across the capacitors after one hour has elapsed?

21.13• Two 470-Ω resistors are joined in series with two capacitors. One of the capacitors has a value of 150 μF. If the time constant of this series combination of resistors and capacitors is 29.0 ms, what is the capacitance of the second capacitor?

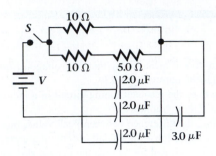

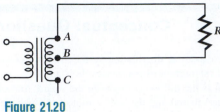

Figure 21.19
Problem 21.11.

Figure 21.20
Problem 21.22.

21.14• A 5000-Ω resistor is joined in series with a 100-μF capacitor. (a) What is the capacitive time constant of this combination? (b) If a 9.0-V battery is suddenly connected across the RC combination, how long will it take for the capacitor voltage to reach 8.0 V?

21.15•• A resistance of 100,000 Ω is used to discharge a 0.47-μF capacitor that is charged to 1500 V. (a) How many time constants must elapse before the capacitor voltage falls to 10 V? (b) How much time must elapse?

Section 21.3 Effective Values of Alternating Current

Hints for Solving Problems

The effective value of a sinusoidal current or voltage is the rms value, equal to $1/\sqrt{2}$ of the peak value. Whenever a numerical value for an ac signal is given without qualification, it is understood to be the rms value.

21.16 An electric signal oscillates sinusoidally at a frequency of 60 Hz. What is the period of this oscillation?

21.17 The period of the alternating voltage in many European countries is 20.0 ms. What is the frequency of such oscillations?

21.18 A radio circuit carries a sinusoidal current with a peak of 15.1 mA. What is the effective value of the current?

21.19 What is the peak voltage of a household voltage outlet of 120 V ac?

21.20 Use the fact that $\overline{\sin^2 x}$ is $\frac{1}{2}$ over one cycle to derive the relation between the rms value of a sinusoidal oscillation and its amplitude.

21.21 (a) Graph the function $y = \sin^2 x$ over an interval from $x = 0$ to $360°$. (b) Verify graphically that the average value of $\sin^2 x$ over that interval is $\frac{1}{2}$.

21.22• (a) Calculate the rms current through the 330-Ω load resistor in the circuit of Fig. 21.20. The transformer output is 12.6 V rms from A to C, and the connection B is halfway between A and C. (b) What is the peak value of the voltage across the resistor? (c) What is the peak value of the current through the resistor?

21.23• A transformer is designed to have an output of 12.6 V

rms when connected to an input of 110 V rms. If the input is raised to 120 V, what is the peak value of the output voltage?

Section 21.4 Reactance

21.24 What is the reactance of a 1.0-μF capacitor at a frequency of (a) 60 Hz, (b) 500 Hz, (c) 1000 Hz, and (d) 10 kHz?

21.25 A 0.10-μF capacitor is driven by a voltage oscillating at a frequency of 10 kHz. (a) What is the reactance of the capacitor? (b) If the peak voltage on the capacitor is 9.0 V, what is the maximum current through it?

21.26 At what frequency will a 100-μF capacitor have a reactance of 4.0 Ω?

21.27 A 4.7-μF capacitor is driven by a voltage oscillating at 490 Hz. (a) What is the reactance of the capacitor? (b) What is the peak current if the peak voltage across the capacitor is 5.0 V?

21.28 At what frequency will a 25-mH inductor have a reactance of 100 Ω?

21.29 Calculate the reactance of a 525-mH inductance operated at 1850 Hz.

21.30 An inductor has an inductance of 25 mH. What is the reactance of this inductor at frequencies of 60, 1000, and 10,000 Hz?

21.31 A 0.43-H inductor is placed across a 60-Hz ac voltage source that maintains a steady rms voltage V. When the inductor is replaced by an unknown capacitor, the current doubles. What is the value of the capacitor?

21.32• A 50-mH inductance is driven by a voltage oscillating at 120 Hz. (a) What is the peak value of the current if the maximum voltage is 10 V? (b) What is the rms value of the current?

21.33• A 1.5-μF capacitor is connected to an 75-V, 60-Hz power supply. (a) What is the effective current through the capacitor? (b) What is the maximum value of the current?

21.34• Show that the dimensions of inductive and capacitive reactance are the same as the dimension of resistance.

Section 21.5 The RLC Series Circuit

Hints for Solving Problems

Remember that phasors add like vectors. Voltage across a capacitor lags the current by 90°; voltage across an inductor leads the current by 90°.

21.35 Calculate the amplitude of the current for a series circuit containing R, L, and C driven by a voltage source $v = V_m \cos 2\pi ft$ if the source frequency is $f = 2000$ Hz. Let $R = 50\ \Omega$, $L = 10$ mH, $C = 1\ \mu F$, and $V_m = 10$ V.

21.36• Construct the phasor diagram for the situation of Problem 21.35.

21.37• A 0.80-H inductor is connected in series with a 330-Ω resistor across a 60-Hz voltage source. An ac voltmeter across the source indicates 120 V_{rms}. (a) What rms current flows in the circuit? (b) What is the phase angle between current and voltage? (c) What is the power dissipation in the circuit?

21.38• The current to a loudspeaker is 350 mA (rms), and the terminal voltage is 2.8 V_{rms}. The phase angle between current and voltage is 42°. Find the power input and the effective impedance.

21.39• A series RC circuit is driven by an alternating voltage source at a frequency of 100 Hz. (a) If $R = 330\ \Omega$ and $C = 100\ \mu F$, what is the phase angle between current and voltage? (b) What is the phase angle between capacitor voltage and resistor voltage? (c) What is the phase angle between capacitor current and resistor current?

21.40•• A series circuit has $R = 33\ \Omega$, $L = 0.15$ H, and $C = 110\ \mu F$ across a 60-Hz sinusoidal voltage source with a peak value of 18-V. (a) What is the peak voltage across each of these elements? (b) What is the peak voltage across the combination of these elements? (c) What is the current in the circuit? (d) What is the phase angle between voltage and current?

21.41•• An RLC series circuit operating at 1000 Hz has a 2.38-Ω resistor and a 23-mH inductor. What is the value of the capacitor that when added to the circuit will make the impedance of the entire circuit twice as large as the magnitude of the resistance?

Section 21.6 Resonant Circuits

> **Hints for Solving Problems**
>
> In a series RLC circuit, the impedance is a minimum at the resonant frequency.

21.42 A capacitor $C = 4.7 \times 10^{-9}$ F is connected across a 2.5-mH inductor. What is the natural oscillation frequency of the combination?

21.43 A 0.010-μF capacitor is connected across a 2.5-mH inductor. What is the natural oscillation frequency of the combination?

21.44 A parallel circuit like that of Fig. 21.17 forms the tuning circuit for an AM radio. If the inductor has a value of 1.90 μH, what must be the maximum and minimum values of the variable capacitor if the radio receives frequencies from 530 kHz to 1610 kHz?

21.45• A series circuit of $R = 1500\ \Omega$, $L = 350$ mH, and $C = 0.15\ \mu F$ is driven with an oscillating source of variable frequency. (a) What is the resonant frequency of the circuit? (b) What is the impedance at the resonant frequency?

21.46• A 270-μF capacitor and a 25-mH inductor are in series with a 330-Ω resistor. (a) What is the resonant frequency? (b) What is the impedance at the resonant frequency?

21.47• A series circuit containing R, L, and C is driven by a voltage source $v = V_m \cos 2\pi ft$. If $R = 51\ \Omega$, $L = 10$ mH, and $C = 1.0\ \mu F$, what is the current at the resonant frequency? Let $V_m = 12$ V.

21.48• Construct the phasor diagram for Problem 21.47.

21.49• A 10-μF capacitor is charged to 30 V and then connected across a 0.45-H inductor. (a) What is the maximum value of the current in the circuit? (b) What is the resonant frequency of the circuit? (*Hint:* Use conservation of energy.)

21.50• A 15-μF capacitor is connected across a 0.50-H inductor. The capacitor is initially charged to 30 V. (a) What is the peak current in the circuit? (b) What is the frequency of the oscillations?

21.51• A series circuit of $R = 470\ \Omega$, $L = 225$ mH, and $C = 1.50\ \mu F$ is driven by a voltage source oscillating at a frequency of 365 Hz. (a) Calculate the impedance at 365 Hz. (b) What is the resonant frequency? (c) What is the impedance at the resonant frequency?

Additional Problems

21.52 A series circuit of a resistor R, an inductor L, and a battery of emf V_0 has a time constant τ_L. The circuit is closed at time $t = 0$. (a) What is the value of the current in the circuit when $t = \tau_L/2$? (b) When $t = 2\tau_L$? (c) When $t = 4\tau_L$?

21.53• A resistor R, an inductor L, and a battery are joined in series with a switch (Fig. 21.1). After the switch is closed, how much time does it take for the current in the circuit to reach 50% of its final value? Give your answer as a fraction or multiple of the inductive time constant.

21.54• A 1000-μF capacitor is discharged across a 500-kΩ resistor. (a) What is the time constant of this discharge? (b) If the initial voltage on the capacitor is 100 V, what is the voltage at $t = 120$ s after the circuit is closed?

21.55• The voltage from a square-wave oscillator varies with time as shown in Fig. 21.21. Calculate (a) the average and (b) the rms voltage for such a wave in terms of the peak voltage V_m.

Figure 21.21
Problem 21.55.

21.56• Show that the average power delivered over one cycle to a resistive load R is

$$\bar{p} = \frac{V^2_{rms}}{R}.$$

21.57• Find the current in the circuit of Fig. 21.22 at (a) very low frequencies and (b) very high frequencies. Assume that the voltage source has a constant amplitude independent of the frequency and that the inductor has zero resistance.

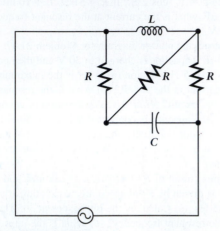

Figure 21.22
Problem 21.57.

21.58•• A series combination of $R = 1000\ \Omega$ and $C = 500\ \mu F$ is connected to a 12-V battery. How long after the circuit is completed will it take for the capacitor voltage to reach 10 V?

21.59•• Show that for an inductor and a resistor connected in a series ac circuit the power dissipated in the resistance is given by $P = V_{rms}I_{rms} \cos \phi$, where ϕ is the phase difference between current and voltage.

21.60•• The voltage source in Fig. 21.23 has a constant amplitude at all frequencies. At very high frequencies the current in the circuit is three times the value of the current at very low frequencies. What is the ratio of the resistors R_2/R_1?

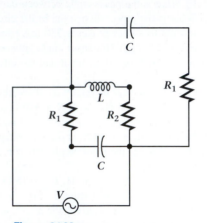

Figure 21.23
Problem 21.60.

Geometrical Optics

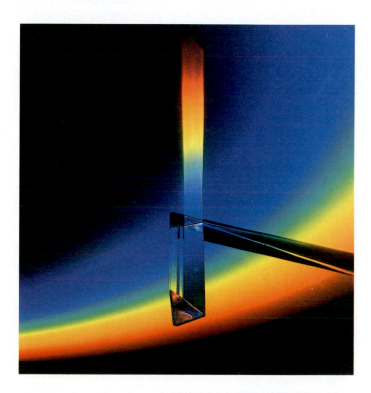

W e saw at the end of Chapter 20 how Maxwell's equations unified electricity and magnetism into one comprehensive theory. The outstanding achievement of this theory was the prediction of electromagnetic radiation, culminating in the realization that light is a form of this radiation. Thus, the study of electromagnetism leads directly to optics—the study of light.

Because of the importance of vision, people have been interested in the nature and behavior of light for many centuries. Although we now know that light behaves like an electromagnetic wave, as described by Maxwell's equations, most of the fundamental laws of optics were discovered before the time of Maxwell. This is particularly true for the material in this chapter. For example, the apparent bending of a stick partially immersed at an angle into water, an effect due to the phenomenon of optical refraction, was described nearly 2000 years ago by the Greek astronomer and geographer Ptolemy (ca. 100 A.D.). A related effect, the ability to see the sun after it has gone below the horizon, was known to Tycho Brahe, Johannes Kepler, and others in the sixteenth century.

By the end of the seventeenth century, many of the now-familiar laws and relationships of optics had been stated. However, the nature of light was still not agreed upon. Some scientists, such as Robert Hooke and Christiaan Huygens, believed that light was a wave, while Isaac Newton proposed that light consisted of particles, or corpuscles as he called them. Newton's reputation was so great that the corpuscular theory predominated for some time.

In the early 1800s, the wave theory of light began to attract renewed attention as a result of experiments by Thomas Young and theoretical work by Augustin Jean Fresnel. By the last quarter of the century, Maxwell had predicted the existence of electromagnetic waves with a speed equal to the speed of light. Maxwell's waves were verified experimentally by Heinrich Hertz at the close of the century.

What is light in our contemporary view of physics—waves or particles? There is no simple answer to this question. In fact, as we will see in Chapter 28, quantum mechanics has shown that the difference between a wave and a particle depends to a large extent on your point of view. Perhaps the best way to think of it is that waves and particles are both simplified models of reality, and that light is a complicated phenomenon that doesn't quite fit either model alone. Both models are used extensively. Generally, we adopt whichever model provides an explanation of the optical behavior under study.

In this chapter, we will examine optical behavior that does not depend on the nature of light, but only on the path it travels. This study is called geometrical optics and includes reflection and refraction. ■

22.1 Models of Light: Rays and Waves

Although we usually consider light to be a wave phenomenon, that is not the only possible view. The idea that light is composed of a stream of particles is also consistent with the observation that a beam of light from a flashlight or a laser seems to travel in straight lines except when it encounters an interface between two optical media (Fig. 22.1). We call this straight-line path a **ray** of light. The ray model allows us to explain in simplest terms the formation of images by lenses and mirrors.

However, as we will see in Chapter 27, light displays particlelike characteristics when it interacts with matter, as it does, for example, when sunlight falls on a leaf and photosynthesis takes place. The apparent conflict over this wave-particle duality cannot be understood with just the theories of Newton or Maxwell, but is resolved by the twentieth-century theory of quantum mechanics discussed in Chapter 28.

The realization that light was the same as the electromagnetic radiation predicted by Maxwell's equations united the studies of electromagnetism and optics. The range of wavelengths that comprise visible light makes up only a small portion of the entire electromagnetic spectrum (Fig. 22.2), which spans many orders of magnitude. There are no sharp dividing lines separating the various regions; there is just a continuous blending from one region to the next.

Figure 22.1
A ray of light from a laser. The light seems to travel in straight lines unless it encounters an interface between two optical media.

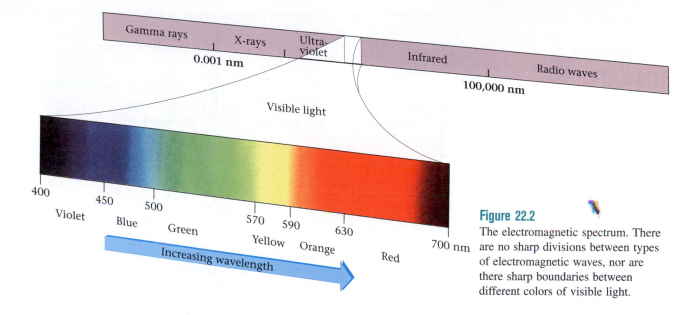

Figure 22.2

The electromagnetic spectrum. There are no sharp divisions between types of electromagnetic waves, nor are there sharp boundaries between different colors of visible light.

When sunlight is spread into a spectrum, we see the characteristic band of visible colors from red to violet. Beyond the violet edge of the visible spectrum are frequencies of radiation that exceed that of the violet. We use the name **ultraviolet** to describe this invisible extension of the spectrum. Beyond the red end of the visible spectrum lie frequencies below those we can see. These wavelengths make up the **infrared** region of the spectrum.

The fundamental behavior of all components of the electromagnetic spectrum is the same. They differ only in their wavelengths and frequencies and in the kinds of devices that can be used to generate and detect them. The behavior of all electromagnetic waves can be predicted from Maxwell's equations and a knowledge of the composition and shape of the lenses, reflectors, and other components involved. For example, the design of microwave antennas (Fig. 22.3) follows the same underlying principles as does the design of telescope mirrors.

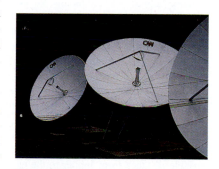

Figure 22.3

Microwave antennas are curved reflectors used to focus microwaves.

22.2   Reflection and Refraction

Suppose we shine a beam of light at a mirror (Fig. 22.4). The light strikes the mirror at a point P and is then reflected. What is the direction of the reflected beam? You may know the answer even if you have not formulated it mathematically: The incident and reflected beams make equal angles with the mirror.

In optics it is an established convention to measure angles with respect to the normal, which is a line perpendicular to the surface, as indicated in Fig. 22.4. The angle between the incoming ray and the normal is called the **angle of incidence,** θ_i. The angle between the outgoing ray and the normal is the **angle of reflection,** θ_r. Both angles are measured positive from the normal. When these angles are measured, the angle of reflection is always found to equal the angle of

THE SPEED OF LIGHT

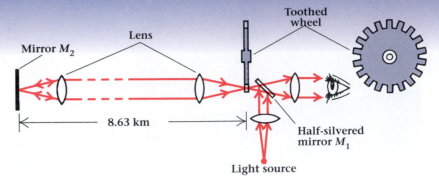

Figure B22.2 The experimental arrangement for Fizeau's experiment to measure the speed of light.

Galileo was perhaps the first person to suggest a method for measuring the speed of light. He proposed an experiment with two observers stationed on mountaintops some distance apart, each having a lantern enclosed in an opaque housing except for a hole covered by a shutter. Both observers were to begin with closed shutters. At the start, one observer would open his shutter and begin to mark time. When the second observer saw the light from the first lantern, he was to open his own shutter. When the light from the second lantern reached the first observer, he was to note the elapsed time since his shutter was opened. The speed of light could then be determined from knowledge of the elapsed time and the distance between the mountains. We do not know whether the experiment was ever performed. As it happens, the speed of light is far too great and human reactions are much too slow for this experiment to succeed. However, the principle is sound and forms a basis for a number of successful determinations of the speed of light.

In 1676, the Danish astronomer Ole Roemer (1644–1710) announced his discovery of systematic variations in the time intervals, as observed on the earth, between successive disappearances of Jupiter's moons as they moved into the planet's shadow. These variations were associated with the variable distance from Jupiter to the earth (Fig. B22.1). He

Figure B22.1 Orbits of Earth, Jupiter, and one of Jupiter's moons at two times, six months apart. Intervals between the moon's eclipses as measured on the Earth depend on the relative positions of Earth and Jupiter.

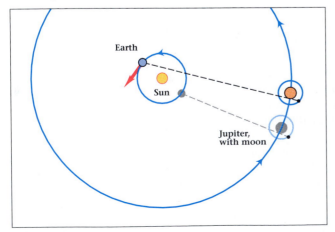

found a time difference of about 22 minutes over six months, corresponding to the time needed for light to cross the diameter of the earth's orbit. Roemer is thus credited with discovering that the speed of light is finite. In 1678, Huygens combined Roemer's time measurement with the estimated distance traveled to obtain a value for the speed of light equivalent to about 2.3×10^8 m/s.

An experiment similar in concept to that of Galileo was performed by Armand Hippolyte Louis Fizeau (1819–1896) in 1849. His experiment provided the first measurement of the speed of light performed on the earth. In order to avoid measuring extremely short intervals of time, he used a rotating toothed wheel and a mirror (Fig. B22.2). Light was focused on the rim of the wheel after reflection from a half-silvered mirror (M_1 in the figure). The light then passed through one of the gaps between the 720 teeth at the edge of the wheel and continued to a second mirror (M_2), where it reflected back along the same path to the wheel, passed through a gap, and reached the mirror M_1. Some of the light reaching the mirror was transmitted to an observer.

When the wheel was rotated at certain speeds, the light passing through a gap as it proceeded to the mirror M_2 struck a tooth when it returned and the observer saw a decreased intensity. At other speeds of rotation, the light would encounter a gap on the return trip and be seen. From knowledge of the number of gaps, the rotational speed of the wheel, and the path length of the light, Fizeau calculated a value for the speed of light equivalent to 3.15×10^8 m/s.

More recently, measurements of the speed of light have utilized its wave behavior and the knowledge that the velocity of a wave is the product of its frequency and wavelength. The wavelength of a highly stabilized laser was determined by high-precision techniques, while its frequency was measured by comparison with cesium-beam atomic clocks. From the values obtained, the best value for the speed of light is $299,792,458 \pm 1.2$ m/s. Because the chief limitation in these measurements was the uncertainty in the length of the meter based on the previous standards, this value of the speed of light has been adopted and the meter redefined in terms of the speed of light (Section 20.9).

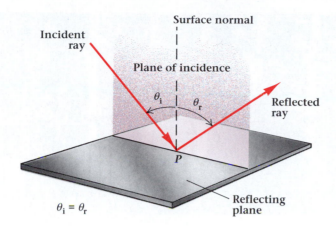

θ_i = θ_r

Figure 22.4
Reflection of light from a flat mirror. The incident and reflected beams make equal angles with the normal. The plane of incidence is defined by the incident and reflected rays.

incidence. Thus, the observed **law of reflection** is that **the angle of reflection is equal to the angle of incidence,** or

$$\theta_r = \theta_i. \qquad (22.1)$$

The normal, the incident ray, and the reflected ray all lie in a plane perpendicular to the reflecting surface, known as the **plane of incidence.** The light ray does not "turn" out of the plane of incidence as it is reflected.

The law of reflection applies to both flat and curved surfaces. For a curved surface, the angle of reflection is determined by the angle of incidence between the incident ray and the surface normal at the point where the incident ray strikes the surface.

Reflection from a smooth mirrorlike surface is called *specular reflection* (Fig. 22.5a). When you look at a mirror, you do not usually see the mirror surface itself; you see the specularly reflected image of other objects instead. For example, you may look at a mirror and see an image of yourself or someone else. What happens when light is reflected from an object whose surface is not perfectly smooth? In that case, the light is *diffusely reflected,* with different parts of the incident light beam scattered in different directions according to the law of reflection (Fig. 22.5b). The light reflected at each little region of the surface is reflected at an angle equal to the local angle of incidence. Most objects that we see are made visible by the diffuse reflection of light from their surfaces. The paper in this book reflects light diffusely so that you can see it from any angle. To see a light reflected in a mirror, however, requires that you be in just the right place so that the specularly reflected light can reach your eye. This effect is used at Boston's Logan Airport to create interesting patterns of multiple images formed by a wall of plane mirrors intentionally set at slightly different angles (Fig. 22.6).

Sometimes both diffuse and specular reflection occur simultaneously from the same surface. For example, when sunlight shines on a car, you can see the car from any direction around it. That is diffuse reflection. If the car is highly polished, you may also see the image of distant objects reflected from its surface. That is specular reflection.

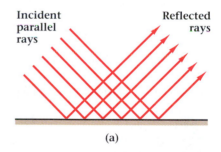

(a)

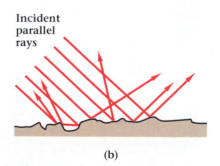

(b)

Figure 22.5
(a) Specular reflection from a smooth surface. (b) Diffuse reflection from a rough surface.

Figure 22.6
A wall of plane mirrors set at slightly different angles reflects multiple images of the authors.

Master the Concept

Reflection

Question: How do the day/night rearview mirrors commonly found in automobiles work?

Answer: The day/night mirror is made from glass that is wedge shaped and silvered on the back side. The mirror is mounted so that the base of the wedge is at the top. When light strikes the mirror, a small portion (4%) is specularly reflected from the air-glass interface and the rest is transmitted through the glass to the back surface where it is strongly reflected. Because the front and back surfaces are not parallel, the light reflected from the back surface emerges in a direction that is several degrees above the front surface reflection. In normal daytime use, the mirror is positioned so that the strong reflection of light from the rear window is reflected to the driver's eyes. At night the mirror is tilted slightly upward so that the light reflected from the silver backing is directed over the drivers head. In that position only the dimmer light from the front surface reflection reaches the driver's eyes.

Example 22.1

How far behind a mirror does your reflection appear to be as you view yourself?

Solution Figure 22.7 is a schematic diagram of a person looking directly into a mirror. Consider any ray of light that leaves any part of the person, is reflected in the mirror, and arrives at the person's eye. All such rays are reflected in accordance with the law of reflection. We have selected and drawn a ray in the figure from the foot to the eye. Since the brain interprets light striking the eye as having traveled in a straight-line path, the dashed line represents the path of the light as it appears to the viewer. The intersection of the dashed line with the horizontal line representing the floor locates the apparent position of the

Figure 22.7
Example 22.1: A person looking into a plane mirror sees her image behind the mirror. The person and the image are both equidistant from the mirror.

Object   Mirror   Image

θ

Normal

θ

θ   θ

D   *B*   *A*

image. From the geometry, we see that triangles ABC and DBC are similar and have a common side. Thus, $AB = BD$. Therefore, the apparent distance of the image behind the mirror is the same as the real distance of the person in front of the mirror.

■

If light strikes a *transparent* body, we observe both reflection and transmission. (A transparent body contrasts with an *opaque* body, which does not permit transmission of light through it.) The transmitted beam is bent, or *refracted,* as it crosses the surface between one medium (such as air or even a vacuum) and another (water, for example). Ptolemy described the observation that a beam of light entering water is bent toward the normal (Fig. 22.8a). He also knew that the amount of bending depends on the angle of incidence. Later investigation showed that the amount of bending also depends on the particular materials found on either side of the interface (Fig. 22.8b). However, not until 1621 did the Dutch mathematician Willebrod Snell (1591–1626) empirically discover the exact law of refraction. We can now derive this relationship from Maxwell's equations or from a more elementary wave picture of light (see Chapter 24). The **law of refraction,** also called **Snell's law,** can be written

(a)

$$n_1 \sin \theta_1 = n_2 \sin \theta_2, \qquad (22.2)$$

where n_1 depends only on the optical properties of medium 1 and n_2 depends only on the optical properties of medium 2. The constant n is called the **index of refraction** of the medium. We define the index of refraction n of a medium to be the ratio of the speed of light in vacuum c to the speed of light in that medium v,

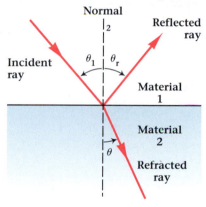

(b)

$$n \equiv \frac{c}{v}. \qquad (22.3)$$

Figure 22.8
(a) The apparent bending of a pencil at the water surface due to the refraction of light. (b) Refraction of light at a boundary between two different materials. The direction of a single ray is changed when it crosses the boundary between materials 1 and 2. Some of the light is also reflected at the surface.

We can verify Snell's law for a large number of materials and use it to determine their indices of refraction. In fact, we can do this without any knowledge of the nature of light; we assume only that light travels in straight lines unless it encounters the boundary between two materials. Table 22.1 lists the indices of refraction for a number of common transparent materials. Usually the index of refraction of air can be taken as unity without introducing any significant error. In most materials, the speed of light depends upon its wavelength. Generally, the index of refraction decreases slowly with increasing wavelength.

If the refracted ray in Fig. 22.8(b) strikes a mirror so that it is reflected back along the same path, it will be refracted at the interface and emerge along the same path as the incident ray. We say that light rays are reversible. That is, if a light ray traveling in some direction takes a particular path, a light ray going in the opposite direction will also follow exactly the same path.

When light goes from a material with a smaller index of refraction to one with a larger index of refraction, we say that the light has gone from a less dense to a more dense optical material. We can use this terminology to state a qualitative version of Snell's law: *When light travels from a less dense to a more dense*

Table 22.1	Index of Refraction of Some Common Materials ($\lambda = 589$ nm)	
Material		**Index of Refraction (n)**
Gases (at atmospheric pressure and 0°C)		
Hydrogen		1.0001
Air		1.0003
Carbon dioxide (CO_2)		1.0005
Liquids (at 20°C)		
Water		1.333
Ethyl alcohol		1.362
Glycerine		1.473
Solids (at room temperature)		
Ice (0°C)		1.31
Acrylic (polymethylmethacrylate)		1.49
Polystyrene		1.59
Crown glass		1.50–1.62
Flint glass		1.57–1.75
Diamond		2.417

optical material, the rays are bent toward the normal, as shown in Fig. 22.8(b). Because of the reversibility of light rays, we can also say, *When a light ray travels from a more dense to a less dense optical material, the rays are bent away from the normal.* The bending occurs at the boundary, or interface, between the two materials.

If a ray of light goes through more than one medium, Eq. (22.2) holds for each interface. For example, light entering a glass of water passes first from air to glass and then from glass to water. Snell's law may be applied at each interface to determine the direction of the path of the light. The following example considers such a case.

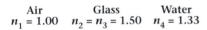

Air   Glass   Water
$n_1 = 1.00$ $n_2 = n_3 = 1.50$ $n_4 = 1.33$

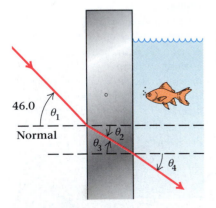

Figure 22.9
Example 22.2: A beam of light in air is refracted as it enters a slab of glass. The beam is refracted again as it passes from the glass into water.

Example 22.2

The side of a fish tank is made of a thick parallel-sided piece of glass with an index of refraction of 1.50 (Fig. 22.9). A beam of light strikes the surface of the glass at an angle of 46.0° with respect to the normal. What is the direction of the beam in the water?

Strategy We solve this problem by first using Snell's law to find the angle of refraction at the air-glass interface. Then we use Snell's law a second time to determine the angle of refraction at the glass-water interface. Our calculations thus involve four angles (see Fig. 22.9); the angle θ_4 is the one we are looking for.

Solution At the first surface, let θ_1 and θ_2 be the angles of incidence and refraction, respectively. The indices of refraction are n_1 for the air and n_2 for the glass. We can insert the values of these quantities into Snell's law to find θ_2:

$$n_1 \sin \theta_1 = n_2 \sin \theta_2,$$

$$\sin \theta_2 = \left(\frac{n_1}{n_2}\right) \sin \theta_1 = \left(\frac{1.00}{1.50}\right) \sin 46.0°$$

$$\sin \theta_2 = 0.480,$$

$$\theta_2 = 28.7°.$$

Thus, in the glass the refracted ray makes an angle of 28.7° with the normal.

At the second surface, we use Snell's law again to get

$$\sin \theta_4 = \left(\frac{n_3}{n_4}\right) \sin \theta_3,$$

where θ_3 is the angle of incidence of the light within the glass and θ_4 is the angle of refraction in the water. Since the faces of the glass are parallel, $\theta_3 = \theta_2$ and $\sin \theta_3 = \sin \theta_2$. The index of refraction of the glass is $n_3 = n_2 = 1.50$ and of the water is $n_4 = 1.33$ (Table 22.1). Upon inserting these values and the value of $\sin \theta_2$ from our first calculation, we find that

$$\sin \theta_4 = \left(\frac{1.50}{1.33}\right) 0.480 = 0.541,$$

$$\theta_4 = 32.7°.$$

22.3   Total Internal Reflection

A particularly interesting situation arises when the light incident on an interface comes from within the optically denser medium—that is, from the material with the greater index of refraction. In this case, n_1 is greater than n_2. Consequently, in accord with Snell's law, θ_2 is larger than θ_1. As θ_1 increases, θ_2 must also increase, but it reaches a limit at 90° (Fig. 22.10a). At this point, the refracted ray runs along the interface. The incidence angle corresponding to a 90° angle of

Figure 22.10
(a) When the angle of incidence exceeds the critical angle θ_c, light is totally internally reflected. (b) The image of a ruler is totally reflected from the back surface of a triangular prism.

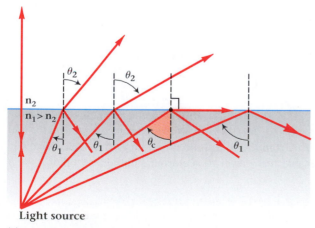

(a)

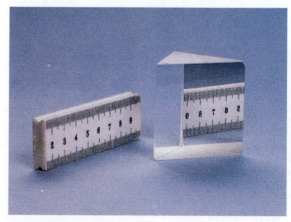

(b)

refraction is called the **critical angle** θ_c and is found from Snell's law, where

$$n_1 \sin \theta_c = n_2 \sin 90°,$$

$$\sin \theta_c = \frac{n_2}{n_1}. \tag{22.4}$$

When the angle of incidence is smaller than θ_c, the light is transmitted at some θ_2 between 0 and 90°. But for angles of incidence *greater* than θ_c, Snell's law is no longer valid and experiments show that no light penetrates into the less-dense medium. Instead, when the angle of incidence exceeds the critical angle, *all* the incident light is reflected back into the denser medium by the interface between two normally transparent materials. We call this phenomenon **total internal reflection.** When light is reflected in by total internal reflection, no light is lost at the reflecting surface. By comparison, some light is lost in reflection from even the best silvered mirrors, which typically reflect only about 90% of the incident light. The light from the ruler in Fig. (22.10b) is totally reflected from the back of the prism to create the mirror image.

Example 22.3

Show that a 45°-45°-90° glass prism can deflect a light beam 90° by total internal reflection. The index of refraction of the glass is 1.52.

Strategy   If the incident ray strikes the short side of the prism normally (that is, perpendicularly; Fig. 22.11), there is no refraction at the interface I_1 and the transmitted beam hits the back surface I_2, making an angle of incidence of 45°. Thus, if the critical angle is less than 45°, all of the light is reflected from the back of the prism. So we need to find the critical angle θ_c and compare it with the angle of incidence of 45°.

Solution   To calculate θ_c, we use Eq. (22.4) with $n_1 = 1.52$ for the glass and $n_2 = 1.00$ for air. The critical angle is

$$\theta_c = \sin^{-1}\left(\frac{n_2}{n_1}\right) = \sin^{-1}\left(\frac{1.00}{1.52}\right)$$

$$\theta_c = 41.1°.$$

The angle of incidence of 45° exceeds the critical angle. Thus, the light striking I_2 is indeed totally reflected. It strikes the third interface, I_3, normally and emerges at an angle of 90° from the initial direction. Periscopes make use of two such prisms to enable the viewer to see around corners or other obstacles.

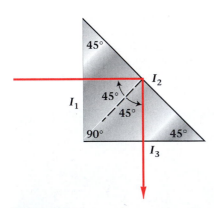

Figure 22.11

Example 22.3: Light incident normally on a 45°-45°-90° prism is totally internally reflected and emerges from the other face.

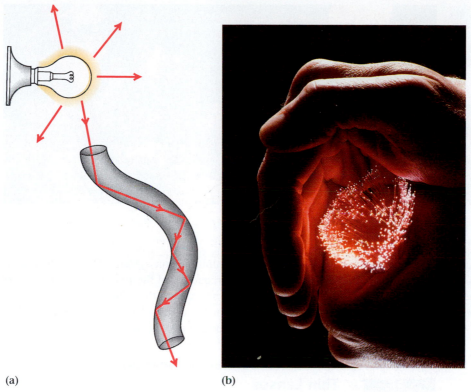

(a)   (b)

Figure 22.12

(a) Light rays trapped by total internal reflection are channeled along the length of a light pipe. (b) Light can be seen emerging from the ends of tiny optical fibers.

22.4   Fiber Optics

Total internal reflection can work for us in *light pipes.* If we direct light into one end of a long rod of glass or plastic, the light is totally internally reflected by the walls, bouncing back and forth until it emerges at the far end (Fig. 22.12a). If we bend the light pipe into a particular shape, the light follows the shape of the pipe and emerges only at the end.

Sometimes light pipes are made of very thin *optical fibers,* which may be grouped into a bundle (Fig. 22.12b). If the fiber ends are polished and their spatial arrangement is the same at both ends, an image may be transmitted from one end of the fiber bundle to the other (Fig. 22.13a). A fiber bundle that transmits an image is called a *coherent bundle,* or an *image conduit.* Bundles of fibers that do not have exactly the same alignment at both ends transmit light but not images. They are known as *incoherent bundles* or *light guides.* Because their fibers are so flexible, fiber optics light guides and coherent bundles are used in instruments designed to permit direct visual observation of otherwise inaccessible objects. Prime examples are endoscopes (Fig. 22.13b), which are used by physicians to examine the interior of a patient's body.

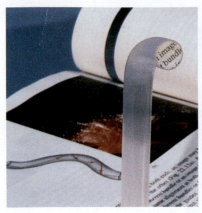

(a)

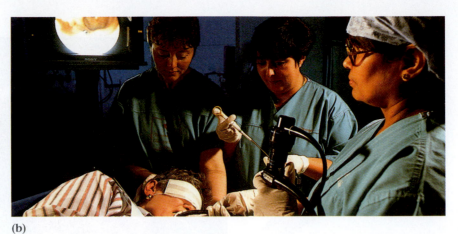

(b)

Figure 22.13

(a) An image conduit composed of a coherent bundle of optical fibers transmits an image from one end to the other. (b) Physicians using a fiber-optic endoscope examine the inside of a patient's body by viewing a television image.

Optical fibers are also used in communication systems to transmit modulated light beams. Because of the high frequency of the light waves, these fibers can carry more information in the same space than the metal wires that they replace. The fibers are less expensive to produce than are copper wires, and they are resistant to noise caused by stray electromagnetic signals.

22.5 Thin Lenses

An optical lens is a piece of glass or other transparent material used to direct or control rays of light. Usually the lens surfaces are spherical, although other shapes (parabolic, cylindrical, etc.) are not uncommon. The refraction of light at the surface of a lens depends on its shape, its index of refraction, and the nature of the medium surrounding it (usually air), in accordance with Snell's law. Because lenses can be used to produce images of objects, they form the basis of most optical instruments, from cameras and projectors to microscopes and telescopes. In fact, the eye itself contains a lens and functions as an optical system.

There are only a few distinct ways of combining flat, convex, and concave surfaces to form a single lens (see Fig. 22.14). If you hold a single lens like one of those in the figure at a moderate distance in front of your eye and look at something through it, you can make several observations. Consider first, lenses that are thicker in the center than at their edges, such as those in Fig. 22.14(a). These are called positive or **converging lenses,** because they refract incident parallel rays so that they converge on a focal point located on the opposite side of the lens. A distant object viewed through such a lens appears smaller and inverted if the lens is held some distance from the eye (Fig. 22.14b). An object held close to the same lens appears erect and enlarged (Fig. 22.14c). Now look at lenses that are thinner in the center than at their edges, such as those in Fig. 22.14(d). These are called negative or **diverging lenses,** because they refract incident parallel rays so that they appear to diverge from a focal point located on the incident side of the lens. Any object viewed through such a lens always appears erect and smaller than when viewed with the unaided eye. The image remains erect no matter how far the object is from the lens or how far the lens is from the eye (Fig. 22.14e).

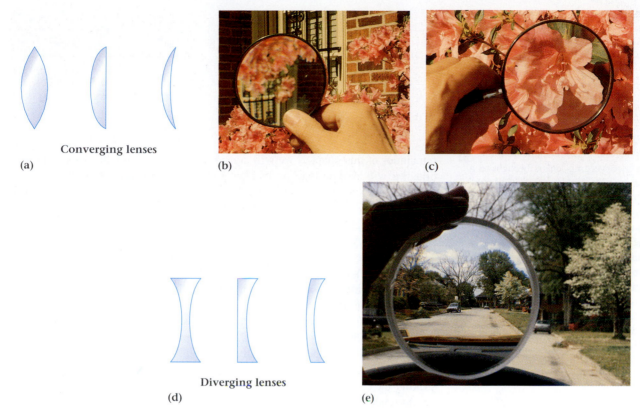

Converging lenses

(a)

(b)

(c)

Diverging lenses

(d)

(e)

Figure 22.14
(a) Cross-sectional views of the most common types of thin converging lenses having spherical surfaces. (b) View through a converging lens of a distant object. (c) View through a converging lens of an object close to the lens. (d) Cross-sectional views of thin diverging lenses. (e) View through a diverging lens.

Let's take a converging lens and allow light from the sun to fall on it. Because the sun is so far away, the rays striking the lens are essentially parallel to one another. When these light rays strike the lens parallel to its axis of symmetry, the rays converge to a point called the *focal point* of the lens (Fig. 22.15). Thus, we can say that any ray incident on a converging lens and parallel to its *optical axis* (that is, the symmetry axis of the lens) passes through the focal point upon leaving the lens. A lens whose thickness is small in comparison with its focal length is called a **thin lens.** The distance from the center of the thin lens to the focal point is the **focal length** of the lens. Each lens has its own particular focal length determined by the curvature of its surfaces and the index of refraction of the material from which it is made.

There is a symmetry to the passage of light through a thin lens. Any light ray entering the converging lens from the left parallel to the optical axis (Fig. 22.15a) passes through the focal point F on the right after it leaves the lens. Conversely, any ray coming from the focal point on the right and striking the lens emerges parallel to the axis (Fig. 22.15b). Parallel light entering the same lens from the right converges to a point F' on the left of the lens (Fig. 22.15c). For a thin lens, the distances f and f' from the center of the lens to F and F' are the same, even though the lens surfaces are not identical. Thus, we consider a thin lens to have two identical focal lengths, one on either side of the lens.* In the

---

*For a thick lens—that is, one for which the thickness is not small in comparison with the focal length—it is possible for f and f' to be different.

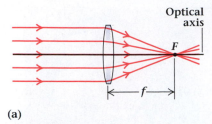

(a)

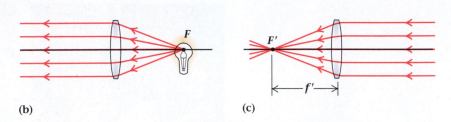

(b)   **(c)**

Figure 22.15

(a) Parallel light incident from the left on a converging lens is brought to a focus at the focal point F. (b) Light emerging from this focal point passes through the lens and emerges parallel to the optical axis. (c) Parallel light incident from the right converges to a focal point F' on the left side of the lens. For a thin lens, these two focal points are equidistant from the lens.

remainder of this chapter, we will consider all lenses to be thin lenses. In addition, we will assume that the incident light rays are nearly parallel to the lens axis (the paraxial approximation). Later, in Section 22.9, we discuss some of the ways in which real lenses differ from our idealized thin lens.

Parallel light incident on a concave lens parallel to its optical axis diverges (Fig. 22.16a). After passing through the lens, the light behaves as if it came from a point source located at F. Thus f is the focal length for this lens. Figure 22.16(b) shows that the same lens refracts light directed toward the focal point on the other side so that the rays emerge parallel to the optical axis after passing through the lens.

An ordinary piece of flat glass, such as a window with parallel faces, passes light without changing its direction. Although the direction of the ray does not change, the ray is displaced laterally (Fig. 22.17a). At the center of a lens, the front and back surfaces are parallel, so to a very good approximation, a light ray striking the center of the lens goes through the lens undeviated. If the incoming ray does not make too large an angle with the axis and if the lens is thin, the offset in the ray is negligible (Fig. 22.17b).

| 22.6 | **Locating Images by Ray Tracing** |

The major usefulness of lenses is their ability to form images of an object. The object may be self-luminous, giving off its own light (like the sun or a light bulb), or it may reflect the light that falls on it (like an apple or a page of this

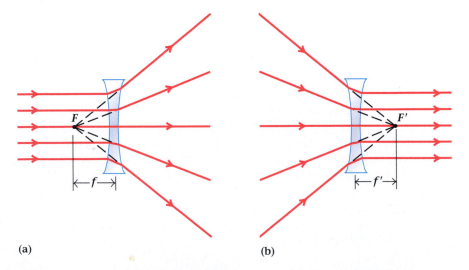

Figure 22.16

(a) Parallel light incident on a concave lens diverges as if it came from the point F. (b) Light converging toward the point F' diverges so that it emerges parallel after passing through the concave lens.

(a)   **(b)**

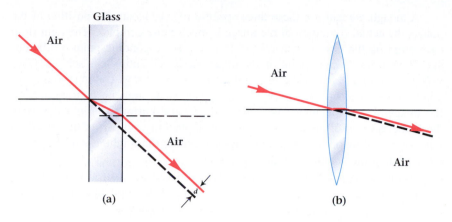

Figure 22.17
(a) A ray of light incident on a parallel-sided slab of glass is displaced an amount *d*, proportional to the thickness of the glass. The direction of the emerging beam is parallel to the incident beam. (b) A ray of light incident on the center of a thin lens is essentially undeviated if the angle of incidence is small.

book). In either case, an image of the object is formed where light rays that come from points on the object intersect or at the points from which the rays appear to originate. When the light rays actually intersect at the image, we call this a **real image.** A screen placed at a real image point would show the image in the same way that pictures appear on a movie screen. A **virtual image** is formed at a point where the light appears to converge, or from which it appears to come. If you place a screen at the position of a virtual image, no image is observed on the screen, for the light rays do not actually intersect there. An example of images in a mirror may help to explain virtual images. If you stand one meter in front of a mirror, your image appears to lie one meter behind the mirror. However, you will not find an image on a screen placed at the image position one meter behind the mirror.

An object, such as a pencil, reflects light rays in all directions from all points of the pencil. We illustrate this in Fig. 22.18 by using arrows from the tip of the pencil to represent light rays. However, to locate the image of the pencil formed by a thin lens, you don't need to trace the light rays from all over the object. Instead, you can locate the image by tracing three particular rays from a point on the object. Figure 22.18(a) shows these rays refracted by a converging lens; Fig. 22.18(b) shows them for a diverging lens. In each case the focal length of the lens and the distance from the object to the center of the lens (the **object distance**) determine the distance of the image from the lens (the **image distance**).

Figure 22.18
Graphical technique for locating the image position with (a) a converging lens and (b) a diverging lens. Dashed lines indicate where light appears to come from (or go to).

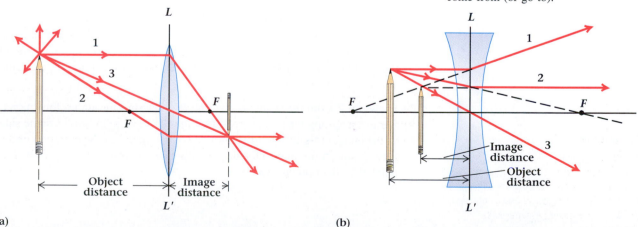

(a)
(b)

Although we can use these three specific rays to locate the position of the image, the actual formation of the image is much more complex. The light from each point on the object converges to form the corresponding point on the image. Rays from a single point on the object strike the entire lens surface. Conversely, any small area of the lens surface is struck by rays coming from every point on the object. All of the rays striking each small area are focused to form the image. Thus, the image is composed of contributions of light that pass through every part of the lens. For this reason, blocking out part of the lens does not cast a shadow on the image, but simply reduces the brightness of the image by restricting the amount of light that gets to it.

When the rays are close to and nearly parallel to the axis (the paraxial approximation), a small planar object perpendicular to the optical axis is imaged by a thin lens into a planar region that is also perpendicular to that axis. Thus, all points in the object plane are represented as points in an image plane. When the object is three-dimensional, like the pencil in Fig. 22.18(a), points located in different object planes do not focus in a single image plane. For instructional purposes we restrict our discussion to planar objects and their planar images. Thus, we normally represent objects and images with simple arrows perpendicular to the optical axis.

Let us state the graphical procedure to follow for locating the image from a given lens when you know the object's position. In the next section, we will describe how to calculate the image height and location algebraically from this graphical ray diagram. You should refer to both parts of Fig. 22.18 as you read these steps.

A. Select an appropriate scale (meters, centimeters, etc.) and mark the position of the lens on the optical axis (also referred to as the principal axis).

B. Draw a line LL' through the center of the lens and perpendicular to the optical axis, as shown. This line will replace the actual lens surfaces that refract the light.

C. Mark the focal points F of the lens on the optical axis and locate the object on the axis, all to the same scale. (In problems of this sort, you will ordinarily be given the focal length f of the lens.)

D. Draw the following rays. Any two of these rays are sufficient to locate the image, but you should always draw the third ray as a check.

    1. Draw ray 1 parallel to the optical axis from the object to the line LL' (the lens). For a converging lens, extend this ray from LL' through the focal point on the side opposite the incident light. For a diverging lens, extend the ray from LL' as though the ray came from the focal point on the same side as the incident light.

    2. Draw ray 2 from the object to the line LL', passing through a focal point. For a converging lens, draw ray 2 through the focal point on the same side as the incident light; for a diverging lens, draw ray 2 in the direction of the focal point on the opposite side. In both cases, continue the ray from LL' parallel to the optical axis.

3. Draw ray 3 from the object to line LL' at the center of the lens and continue it undeviated.

E. If the rays converge on the side opposite the incident light, the image is real and is located at the intersection of these three rays. If the rays diverge, the image is virtual and is located where these three rays appear to originate. You can measure the position and image height directly from the scale drawing.

You could make a ray diagram for each point on the object to fully locate the image. However, if all of the object is at approximately the same distance from the lens, then you need only consider the image of one point of the object and the rest of the object is thereby located. Again, note that for a given object distance, the size and position of the image are completely determined by the focal length of the lens. Once you have determined the height of the image, you can determine the **lateral** or **linear magnification,** the ratio of the image height to the object height.

Example 22.4

A thin converging lens of focal length 6.0 cm is used to image a 35-mm slide on a screen. If the slide is 10.0 cm from the center of the lens, where should you locate the screen so that it coincides with the image? What is the linear magnification of the image?

Strategy The answer can be found by making a scale drawing and measuring the distance from the lens to the image in the drawing. The distance so determined can then be scaled to give the actual distance. It is especially necessary to use care when making the scale diagram. You will find it helps to draw the diagram to a scale as large as your paper will allow in order to minimize the errors due to the finite width of the lines.

Solution This situation is drawn to scale in Fig. 22.19. You can determine the distance from the lens to the image position directly from the figure as 15 cm. Also, upon comparing the image height to the object height, we see that the image is 1.5 times taller than the object. This ratio of image height to object height is the lateral magnification.

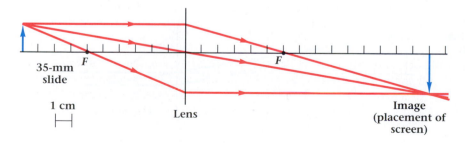

Figure 22.19
Example 22.4: Scale drawing for locating the image of a 35-mm slide.

35-mm slide

1 cm

Lens

Image (placement of screen)

▼

Problem-Solving Strategy

Ray Tracing

Images can be located by carefully drawing three chief rays through a lens. Use a ruler to locate the position of the lens, object, and focal points. Then use the ruler to draw the following straight lines.

1. From one point on the object, draw an incident ray parallel to the optical axis of a positive lens to the line representing the lens. Then for the ray emerging from the lens draw a straight line that passes through the focal point of the lens. (Red line in Fig. 22.20).
2. From the same point on the object, draw a second ray passing through the center of the lens that continues in a straight line. (Blue line in Fig. 22.20.)
3. From the same point on the object, draw a third ray passing through the focal point before striking the lens. Then draw it emerging as a line parallel to the optical axis. (Green line in Fig. 22.20.)

Figure 22.20

Locating an image by ray tracing.

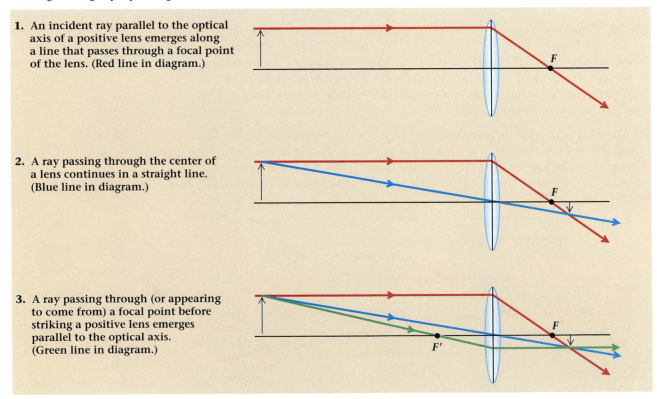

1. An incident ray parallel to the optical axis of a positive lens emerges along a line that passes through a focal point of the lens. (Red line in diagram.)

2. A ray passing through the center of a lens continues in a straight line. (Blue line in diagram.)

3. A ray passing through (or appearing to come from) a focal point before striking a positive lens emerges parallel to the optical axis. (Green line in diagram.)

The Thin-Lens Equation

We now derive an algebraic expression relating the object distance, the image distance, and the focal length of the lens, using a geometrical ray diagram for a general case. The resulting mathematical expression will allow you to calculate distances and focal lengths more precisely than can be done by ray tracing. However, applying this equation requires careful treatment of positive and negative distances from the lens, which we will discuss after presenting the basic derivation. You should always sketch a ray diagram when solving problems, even when you are using the algebraic method.

A general ray diagram for a convex lens is drawn in Fig. 22.21. We designate the object distance by o, the image distance by i, and the focal length by f. The focal points are labeled F and F'. The height of the object is labeled h_o and that of the image h_i. We use the geometry of the similar triangles APH and GPI to get

$$\frac{HA}{GI} = \frac{AP}{PG}.$$

From the similar triangles CPF and IGF, we have

$$\frac{CP}{GI} = \frac{PF}{GF}.$$

But $HA = CP$, so

$$\frac{AP}{PG} = \frac{PF}{GF}.$$

From the figure we see that $AP = o$, $PG = i$, $PF = f$, and $GF = i - f$, so that this equation becomes

$$\frac{o}{i} = \frac{f}{i - f},$$

which, upon rearranging, becomes

$$\boxed{\frac{1}{i} + \frac{1}{o} = \frac{1}{f}.}$$

(22.5)

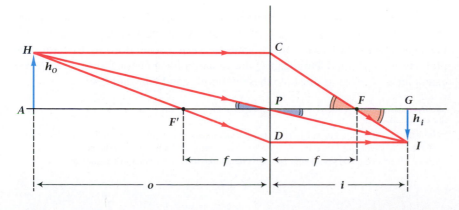

Figure 22.21

Geometry for derivation of the thin-lens formula.

This equation is called the **thin-lens equation** in Gaussian form.* Recall that by a thin lens we mean one for which the focal length is much greater than the thickness of the lens. For such a lens, Eq. (22.5) is sufficient to specify the position of the image.

The lateral or linear magnification of an object is given by the ratio of the image height to the height of the original object:

$$m = \frac{h_i}{h_o}. \tag{22.6}$$

From the similar triangles *HAP* and *IGP* in Fig. 22.21, we see that

$$\frac{h_i}{h_o} = \frac{i}{o},$$

so the lateral magnification becomes

$$\boxed{m = -\frac{i}{o}.} \tag{22.7}$$

The minus sign is included because, as we will see, it is convenient to have a positive magnification for an erect image and a negative magnification for an inverted (upside-down) image.

In the preceding derivation, we examined the special case of a converging lens with the object placed outside the focal point. However, the thin-lens equation holds for both converging and diverging thin lenses and for any object distance, provided the following sign conventions are maintained:

1. The sign of f is positive for a converging lens and negative for a diverging lens.

2. The sign of o is positive if the object is on the same side of the lens as the incident light (real object), negative if the object is on the other side (virtual object).†

3. The sign of i is positive (real) if the image is on the side of the lens opposite the incident light, negative (virtual) if the image is on the same side.

4. Object and image heights are positive if above the optical axis, negative if below it.

These conventions are consistent with the results of geometrical ray tracing. In fact, the use of Eq. (22.5) and the use of ray tracing techniques always give the same results. In general, it is helpful to make a scale drawing to clarify the problem in your mind and then use the equation for numerical precision. A few examples will serve to illustrate these principles and conventions.

---

*The behavior of a thin lens is described using other characteristic distances in the Newtonian form of the thin-lens equation. (See Problem 22.59.) The Gaussian form is more common.
†Virtual objects occur in compound optical systems like those described in Chapter 23.

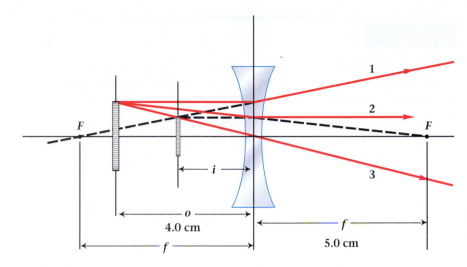

Figure 22.22
Example 22.5: Ray diagram for a diverging lens.

Example 22.5

A diverging lens has a focal length of -5.0 cm. If you look through the lens at a dime held 4.0 cm from the lens, what are the location and the magnification of the image?

Strategy Figure 22.22 diagrams the situation. As we work through the algebraic solution, note how we pay careful attention to the sign of each distance so that we can interpret the image location correctly.

Solution We start by solving Eq. (22.5) for the image distance:

$$i = \frac{fo}{o - f}.$$

The numerical values of f and o with their proper signs are $f = -5.0$ cm and $o = +4.0$ cm. The image distance is

$$i = \frac{(-5.0 \text{ cm})(+4.0 \text{ cm})}{4.0 \text{ cm} - (-5.0 \text{ cm})}$$

$$i = -2.2 \text{ cm}.$$

The negative image distance tells us that the image is virtual, that is, the rays appear to diverge from that position. If your eye were on the axis somewhere to the right of the lens in Fig. 22.22 and looking toward the lens, the coin would appear to be located 2.2 cm on the other side of the lens.

The magnification is given by

$$m = -\frac{i}{o} = -\left(\frac{-2.2 \text{ cm}}{4.0 \text{ cm}}\right) = 0.55.$$

A magnification of less than one means that the image is smaller than the object. The positive value for m indicates that the image is erect, as shown in the figure.

Example 22.6

A penny is placed 4.0 cm in front of a converging lens whose focal length is 15 cm. Where is the image and what is its nature? That is, is the image real or virtual, erect or inverted? What is its magnification?

Solution The following calculation corresponds to the diagram in Fig. 22.23. Solving Eq. (22.5) for the image distance, we have

$$i = \frac{fo}{o - f}.$$

Inserting the numerical values of o and f, we get

$$i = \frac{(15 \text{ cm})(4.0 \text{ cm})}{4.0 \text{ cm} - 15 \text{ cm}} = -5.5 \text{ cm}.$$

The negative image distance tells us that the image is virtual. The linear magnification is

$$m = -\frac{i}{o} = -\frac{-5.5 \text{ cm}}{4.0 \text{ cm}} = 1.4.$$

The positive value of m indicates an erect image, as shown in the figure.

We can use the thin-lens equation to reach some basic conclusions about the images obtained from lenses. For a converging lens, a real object outside the focal length produces a real, inverted image; a real object within the focal length produces a virtual, erect image. For a diverging lens, the image of a real object is always virtual and erect, regardless of the object's distance from the lens. You should confirm these statements for yourself by using the thin-lens equation and the sign conventions given above.

When lenses are used in combination, as in a compound instrument such as a microscope, the image of one lens becomes the object for a second lens. When that happens, the first image may lie on the negative side of the second lens, acting as a virtual object. For a converging lens, a virtual object always forms a real image. For a diverging lens, a virtual object within the focal length forms a real image; a virtual object beyond the focal length forms a virtual image. We will see some of this imaging in Chapter 23 on optical instruments.

Figure 22.23
Example 22.6: Ray diagram for a converging lens.

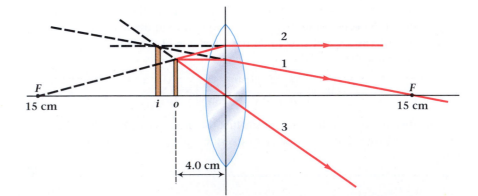

Example 22.7

Two thin lenses, each with +8.0-cm focal length, are placed 12.0 cm apart along their axes. Locate the image of an object placed on the axis 16.0 cm from the nearest lens.

Strategy We will use the thin-lens equation to locate the image of the original object formed by the first lens alone. Then we will use that image as the object for the second lens in order to find the image formed by that lens. In doing so, we will find it very helpful to diagram the situation so that we can determine the proper signs to use with the quantities that appear in the lens equations.

Solution We start by solving the thin-lens equation for the distance of the first image from the first lens:

$$i_1 = \frac{f_1 o_1}{o_1 - f_1}.$$

We insert the numerical values of $f_1 = +8.0$ cm and $o_1 = 16.0$ cm to get

$$i_1 = \frac{(8.0 \text{ cm})(16.0 \text{ cm})}{16.0 \text{ cm} - 8.0 \text{ cm}} = +16.0 \text{ cm}.$$

The image due to the first lens is a real image that lies 16 cm beyond the lens as shown in Fig. 22.24(a).

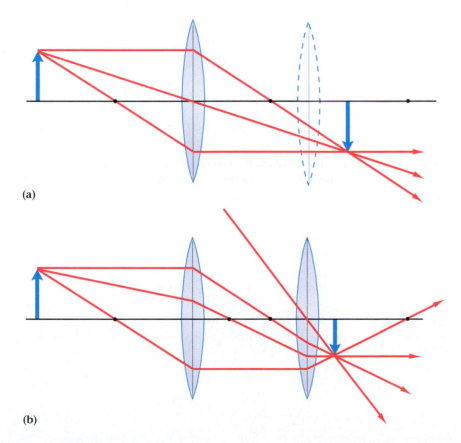

(a)

(b)

Figure 22.24
Example 22.7: Ray diagram for a combination of lenses.

Next we treat the image of the first lens as the object for the second lens. From the diagram, we see that the new object is a virtual object because it lies on the side of the second lens that is away from the incident light (Fig. 22.24b). The new object distance o_2 is the distance between the lenses minus i_1.

$$o_2 = 12.0 \text{ cm} - i_1 = 12.0 \text{ cm} - 16.0 \text{ cm} = -4.0 \text{ cm}.$$

Again using the thin lens equation, we get

$$i_2 = \frac{f_2 o_2}{o_2 - f_2} = \frac{(8.0 \text{ cm})(-4.0 \text{ cm})}{-4.0 \text{ cm} - 8.0 \text{ cm}} = +2.7 \text{ cm}.$$

The image of the two-lens combination lies 2.7 cm beyond the second lens. It is a real image.

Example 22.8

Show that the effective focal length of two thin lenses of focal lengths f_1 and f_2 placed close together is

$$\frac{1}{f} = \frac{1}{f_1} + \frac{1}{f_2},$$

when the distance between the lenses approaches zero.

Strategy As before in Example 22.7, we can let the image formed by the first lens be the object for the second lens. In this case we let the lenses be in contact so that the distance between them is zero. Then using the thin-lens approximation, the image of the first lens is the object for the second lens. Because it lies on the side away from the incident light, the object for the second lens is virtual and the object distance is negative.

Solution We begin with the thin-lens equations for each lens.

$$\frac{1}{o_1} + \frac{1}{i_1} = \frac{1}{f_1} \quad \text{and} \quad \frac{1}{o_2} + \frac{1}{i_2} = \frac{1}{f_2}.$$

When the two lenses are in contact, we can make the approximation that $o_2 = -i_1$. If we substitute this relationship into the equation for the first lens, we get

$$\frac{1}{o_1} + \frac{1}{i_1} = \frac{1}{o_1} + \frac{1}{-o_2} = \frac{1}{f_1}.$$

Next we can use the second thin-lens equation to eliminate o_2.

$$\frac{1}{o_1} + \frac{1}{-o_2} = \frac{1}{o_1} + \left(\frac{1}{i_2} - \frac{1}{f_2}\right) = \frac{1}{f_1}.$$

Now upon rearranging the last two terms, we get

$$\frac{1}{o_1} + \frac{1}{i_2} = \frac{1}{f_1} + \frac{1}{f_2}.$$

This last equation has the form

$$\frac{1}{o} + \frac{1}{i} = \frac{1}{f},$$

where

$$\frac{1}{f} = \frac{1}{f_1} + \frac{1}{f_2}.$$

Discussion When two positive thin lenses are held close together, the combination behaves like a single positive thin lens with a new focal length that is shorter than the focal length of either lens by itself. This relationship for combining thin lenses also holds even if one or both of the thin lenses are negative lenses.

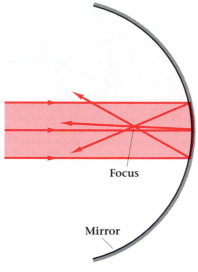

 Spherical Mirrors

Many optical instruments use curved mirrors as image-forming devices. Curved mirrors, like lenses, are used to focus light and create images. However, mirrors work by reflection of light, rather than by refraction, so their practical applications are different.

Curved mirrors are often spherical in shape. A concave mirror reflects light from the inner surface of a sphere; a convex mirror reflects light from the outer surface. Figure 22.25 depicts a beam of parallel light reflecting from a concave spherical mirror and converging to a single point, called the focus of the mirror. However, if the diameter of the incident beam is a large fraction of the diameter of the mirror (Fig. 22.26), the reflected rays do not all converge to the same point, and the image is somewhat blurred. This effect, called spherical aberration, is discussed in more detail in Section 22.9. For this reason we limit our discussion of spherical mirrors to those cases where the diameter of the incident light beam is small compared with the radius of curvature of the surface.

Figure 22.27 shows a concave mirror illuminated by a beam of light parallel to the optical axis. A single ray parallel to the axis strikes the mirror surface

Figure 22.25
The reflection of light rays from a curved surface obeys the law of reflection at each point. When parallel light rays strike near the center of a concave spherical mirror, they are reflected to a common point, called the focus.

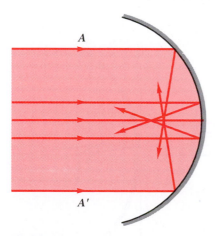

Figure 22.26
Parallel light rays A and A' strike the spherical mirror at so great a distance from the central axis that they are not reflected to the common focus of the rays incident close to the axis.

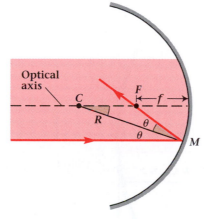

Figure 22.27
Geometrical construction showing that rays near the axis of a spherical mirror converge through a point F located a distance R/2 from the mirror.

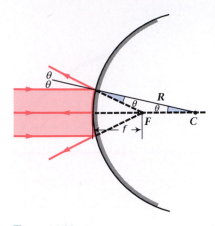

Figure 22.28

Parallel light rays diverge upon reflection from a convex spherical mirror. The diverging rays appear to come from a virtual source point located at $F = R/2$ inside the mirror.

at M and is reflected through the axis at the focal point F. A line drawn from the center of curvature C to the mirror at M forms the base of an isosceles triangle CMF, with equal sides CF and FM. In the limit of small angle θ, the distances CF and FM are equal to one-half the radius R of the surface. The focal length f of the concave spherical mirror is then

$$f = \frac{R}{2},$$

(22.8)

where R is the radius of curvature of the mirror.

A concave mirror focuses a parallel beam of light in a manner analogous to the way a converging lens does. On the other hand, if the mirror surface is convex, the reflected light diverges, in a manner analogous to the spreading of a beam by a diverging lens. A convex mirror makes the reflected beams appear to come from a focal point F inside the mirror surface (Fig. 22.28). For a small angle θ, this focal point is halfway between the reflecting surface and the center of curvature, so again $f = R/2$.

The geometrical procedure for locating the image produced by a spherical mirror is similar to that used for lenses. As you read these steps, follow along with the aid of Fig. 22.29.

A. Select an appropriate scale and mark the positions of the object and the mirror on the optical axis, as shown.

B. Draw a line MM' through the center of the mirror and perpendicular to the optical axis.

C. Mark the focal point F of the mirror at $f = R/2$ on the optical axis and locate the object on the axis, all to the same scale.

D. Draw the following rays to locate the image. As with lenses, any two of these reflected rays are sufficient to locate the image position at their intersection, but you should always draw the third ray as a check.

Figure 22.29

(a) Graphical technique for locating the image position for a concave mirror. (b) Graphical technique for locating the image position for a convex mirror.

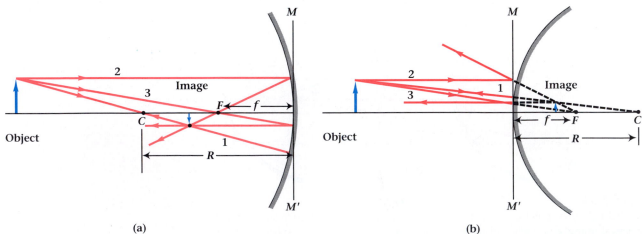

(a)

(b)

1. Ray 1 is drawn from the object through the center of curvature. It reflects back along itself upon striking the mirror.

2. Ray 2 is drawn parallel to the optical axis from the object to the mirror (*MM'*), where it is reflected *through* the focal point of a concave mirror or *from* the focal point of a convex mirror.

3. Ray 3 is drawn from the object to the mirror (*MM'*) along a path through or toward the focal point and is reflected parallel to the optical axis.

Example 22.9

An object is placed 2.0 cm in front of a concave spherical mirror whose radius of curvature is 8.0 cm. Locate the position of the image by ray tracing. Is the image erect or inverted?

Strategy This is an example of ray tracing for a spherical mirror, just as Example 22.4 is an example of ray tracing for a thin lens. The same general procedures are used for mirrors as for lenses and the same precautions should be followed. You should diagram your own solution and compare it with the answer in the text. Remember to make your diagram as large as your paper will allow.

Solution The situation is drawn to scale in Fig. 22.30. Since the mirror's radius is 8.0 cm, the focal point *f* is 4.0 cm from the reflecting surface. Ray 1, drawn through the center of curvature, reflects back along itself. The extension of this reflected ray is drawn as a dashed line on the other side of the mirror. Ray 2, drawn parallel to the optical axis, reflects back through the focal point. The extension of ray 2, also shown as a dashed line, intersects the extension of the first ray at a measured distance 4.0 cm behind the mirror. Ray 3, which appears to come from the focal point, reflects back parallel to the axis. Its extension also intersects the other rays. Thus, the image is located 4.0 cm behind the mirror. The image is erect and measures twice as high as the object. ■

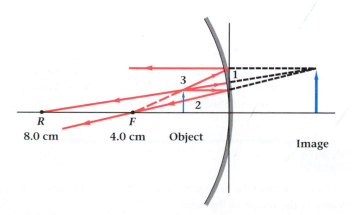

Figure 22.30
Example 22.9: Graphical ray diagram. The dashed lines indicate where light appears to come from.

We can compute image positions for mirrors by using the same equation we used for thin lenses,

$$\frac{1}{i} + \frac{1}{o} = \frac{1}{f},$$

where o is the distance from the object to the mirror, i is the distance from the image to the mirror, and $f = R/2$ is the focal length. However, we require a new set of sign conventions for mirror problems:

1. The sign of f is positive if the focal point is on the same side of the mirror as the incident light, negative otherwise. (In other words, f is positive for a concave mirror and negative for a convex mirror.)

2. Both o and i are positive in sign if they lie on the same side of the mirror as the incident light and negative if they lie on the opposite side.

The use of the mirror equation and its sign conventions is illustrated in the following two examples. Note that the definition of magnification given earlier for lenses applies to mirrors as well.

▼

Example 22.10

An object is placed 2.0 cm in front of a concave spherical mirror whose radius of curvature is 8.0 cm. Use the mirror equation to locate the position of the image and its size. (This problem is the same as Example 22.9 except that we now find the image position and height by computation instead of by ray tracing.)

Solution First note that the focal length f is positive, according to rule 1. It is $f = R/2 = 4.0$ cm. According to rule 2, the object distance is positive. The image distance can then be computed by rearranging the mirror equation,

$$i = \frac{fo}{o - f} = \frac{(4.0\ \text{cm})(2.0\ \text{cm})}{2.0\ \text{cm} - 4.0\ \text{cm}} = -\frac{8.0\ \text{cm}}{2.0} = -4.0\ \text{cm}.$$

The image distance is negative, indicating that the image is virtual and lies on the opposite side of the mirror. Its magnification is

$$m = -\frac{i}{o} = -\frac{-4.0\ \text{cm}}{2.0\ \text{cm}} = +2.0.$$

The image is erect and is twice as tall as the object. Notice that the answer here is the same as we obtained in Example 22.9, even though the method used to obtain the answer is entirely different.

▼

Example 22.11

A flea is located 3.0 cm from a convex spherical mirror of radius 10 cm (Fig. 22.31). Where is the image of the flea?

Solution In this case, the mirror surface is convex, so the center of curvature is behind the mirror, leading to a negative focal length $f = -5.0$ cm. Again we obtain the image distance from the mirror equation,

$$i = \frac{fo}{o - f} = \frac{(-5.0 \text{ cm})(3.0 \text{ cm})}{3.0 \text{ cm} - (-5.0 \text{ cm})} = -1.9 \text{ cm}.$$

The image is virtual and lies 1.9 cm within the mirror.

■

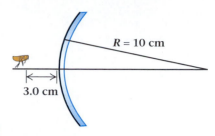

Figure 22.31
Example 22.11: Where is the image of the flea?

Note that a convex mirror always produces a virtual image of a real object. The image always lies within the mirror and is always erect and reduced. Notice is taken of this property of convex mirrors in the message on automobile passenger-side rearview mirrors: "Objects in mirror are closer than they appear." For a concave mirror, a real object placed outside the focal length is reflected to form a real, inverted image. An object placed within the focal length is reflected to form a virtual, erect image. In this case, the image is enlarged. Shaving and makeup mirrors are convex mirrors with focal lengths larger than the normal distance from the mirror to your eye.

The most common nonspherical mirrors are concave parabolic reflectors. A parallel beam incident along the optical axis of a parabolic mirror is imaged to a point, without the complications of spherical aberration mentioned earlier. Similarly, light from a small source can be reflected into a parallel beam, as is done in searchlights. The dish antennas that are used for microwave communication, satellite TV links, and radio telescopes are parabolic reflectors.

*22.9 Lens Aberrations

Up to now, our discussion of lenses has not taken into account some of the optical imperfections that are inherent in single lenses made of uniform material with spherical surfaces. These failures of a lens to give a perfect image are known as **aberrations.** They have nothing to do with the composition of a lens or the smoothness of its surfaces, but arise naturally from the geometry of image formation. They may be reduced to an acceptable level, but, in general, the reduction of one type of aberration tends to increase another.

If we carefully reexamine our original observation that the sun's rays all converge to a focal point after passing through a simple lens, we see that this is only approximately so. Figure 22.32(a) shows incident rays parallel to the optical axis of a converging lens. Careful measurement of the refraction angles shows that incident rays farther from the optical axis intersect slightly closer to the lens than the focal point. This effect is called *spherical aberration.* Figure 22.32(b) shows incident rays at an angle to the optical axis. Again, careful measurement shows that the refracted rays do not meet at one point, but create a trailing blur away from the optical axis. This effect is called *coma.* We see that what we said earlier about thin lenses is strictly true only in the so-called paraxial approximation—that is, only for incoming rays that are not far from the axis and that are parallel to the axis. These limitations explain why the center of an image is often relatively clear while the edges are considerably more distorted.

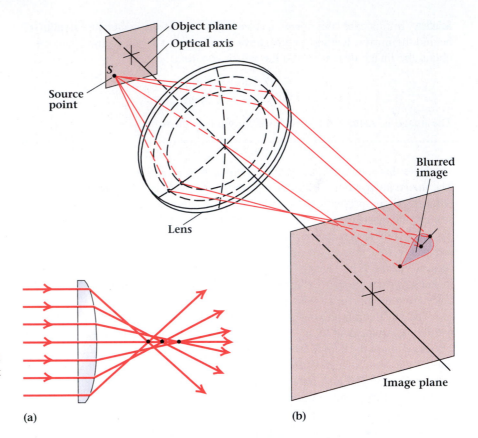

Figure 22.32
(a) Rays coming in parallel to the optical axis do not all intersect at the same point on the axis. Incident rays farther from the optical axis intersect closer to the lens. This effect is called spherical aberration. (b) Rays incident at an angle to the optical axis of the lens fail to intersect at a point, an effect called coma.

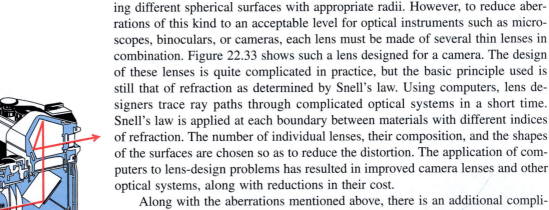

(a)   **(b)**

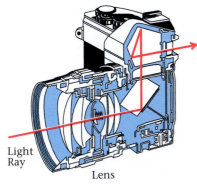

Figure 22.33
Cross section of a single-lens reflex camera with a modern photographic lens. The combination of several lenses helps correct for aberrations.

Aberrations can be reduced, but not removed entirely, by suitably combining different spherical surfaces with appropriate radii. However, to reduce aberrations of this kind to an acceptable level for optical instruments such as microscopes, binoculars, or cameras, each lens must be made of several thin lenses in combination. Figure 22.33 shows such a lens designed for a camera. The design of these lenses is quite complicated in practice, but the basic principle used is still that of refraction as determined by Snell's law. Using computers, lens designers trace ray paths through complicated optical systems in a short time. Snell's law is applied at each boundary between materials with different indices of refraction. The number of individual lenses, their composition, and the shapes of the surfaces are chosen so as to reduce the distortion. The application of computers to lens-design problems has resulted in improved camera lenses and other optical systems, along with reductions in their cost.

Along with the aberrations mentioned above, there is an additional complication: The index of refraction varies slightly for light of different colors—that is, different wavelengths. This variation is most easily seen in the action of a prism (Fig. 22.34). Incoming white light is a mixture of all colors. When white light strikes the prism surface, it is refracted according to Snell's law, but since the indices of refraction are slightly different for the various colors, each color refracts through a different angle. This spreading of light according to wavelength is called **dispersion.** A similar thing happens at the second surface, resulting in a spectrum of colors. Because the surfaces of a single lens are not parallel, it behaves like a prism. Figure 22.35 shows this behavior in somewhat

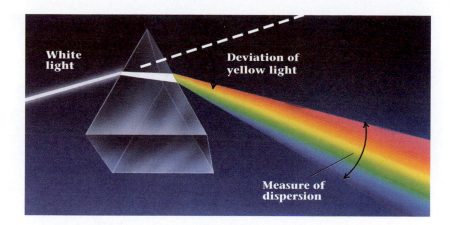

Figure 22.34
Dispersion of light into a spectrum by a prism. The angular spread of the emerging beam has been exaggerated. In a typical glass prism, it is much less than shown here.

exaggerated form, in an effect called *chromatic aberration.* It cannot be removed by geometrical shaping of a single lens. The variation of the index of refraction is described further in Section 24.8.

Chromatic aberration can be troublesome in any optical instrument that uses lenses. To overcome this problem in telescopes, Isaac Newton invented the reflecting telescope, which uses a focusing mirror instead of an objective lens. No dispersion occurs when light is reflected from the mirror, so chromatic aberration is eliminated from that element.

As we have seen, modern camera lenses are compound lenses made with groups of simple lenses used together. By choosing materials with indices of refraction that depend in different ways on the color of the light, lens designers can ensure that positive and negative lens elements compensate for each other by having equal but opposite chromatic effects. Such lenses are said to be *achromatic,* because they exhibit only minimal chromatic aberration. The first achromatic lens was made by John Dollond, a London optician, who succeeded in building an achromatic telescope in 1758.

We can take a complicated grouping of lenses like that in Fig. 22.33 and consider the whole as if it were a single lens with a single focal point on either side. Clearly, such a lens is not literally a thin lens. Often, the lens is almost as thick as its stated focal length. However, the reason for all the aberration corrections is to make the lens behave just as we said an ideal simple lens does. We can therefore deal with such compound lenses as single lenses and use the formulas developed for thin lenses to a good approximation.

Figure 22.36 illustrates lens aberrations and their correction. It shows the same scene photographed through a simple uncorrected lens and through a well-corrected lens. The blurring in the first photograph (Fig. 22.36a) is primarily the result of spherical aberration, and the inaccuracy in the color is due to chromatic aberration. A color photograph taken through the well-corrected lens (Fig. 22.36b) shows no visible color errors.

In recent years, lensmakers have begun to depart from the traditional manufacturing techniques of grinding lenses to spherical surfaces. Newer methods allow the production of lenses with one or both surfaces neither planar nor spherical. The resulting lenses, known as *aspherics,* have less aberration than an equivalent spherical lens. Alternatively, they can be made with shorter focal lengths than those possible with spherical lenses of equal diameter and equal

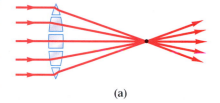

(a)

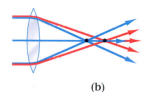

(b)

Figure 22.35
(a) A lens can be considered to be made of sections of prisms. (b) Dispersion of the light by a lens causes a spreading of the focus, with blue light converging nearer to the lens than red light.

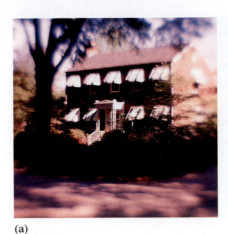

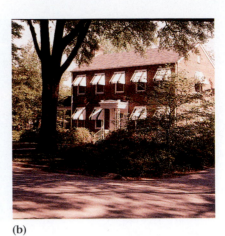

Figure 22.36
(a) A photograph taken using a single thin lens, a plano-convex 50-mm lens. (b) The same scene photographed through a well-corrected camera lens of the same focal length (50 mm).

(a) **(b)**

spherical aberration. Lenses of this type are now being used in cameras and as condenser lenses in projectors. (Cameras and projectors are discussed in Section 23.3.)

▼ Summary

Useful Concepts

■ The law of reflection is

$$\theta_r = \theta_i.$$

■ Snell's law of refraction is

$$n_1 \sin \theta_1 = n_2 \sin \theta_2,$$

where n_1 is the index of refraction of material 1 and n_2 is the index of refraction of material 2. The index of refraction of a medium is the ratio of the speed of light in vacuum c to the speed of light in the medium v,

$$n = \frac{c}{v}.$$

■ The critical angle for total internal reflection within an optically denser medium of refractive index n_1 is

$$\sin \theta_c = \frac{n_2}{n_1}.$$

■ The thin-lens (and spherical mirror) equation is

$$\frac{1}{i} + \frac{1}{o} = \frac{1}{f}.$$

■ The lateral magnification due to a thin lens is given by

$$m = -\frac{i}{o}.$$

■ The focal length of a spherical mirror is

$$f = \frac{R}{2}.$$

■ The location of the image of a point formed by a lens can be found from a scale diagram by drawing at least two of the following rays:

1. A ray parallel to the optical axis from the object to the lens. For a converging lens, extend the ray through the focal point on the opposite side. For a diverging lens, extend the ray from the lens as though the ray came from the focal point on the same side.

2. A ray from the object to the lens passing through a focal point. For a converging lens, draw the ray through the focal point on the same side as the incident light; for a diverging lens, draw the ray in the direction of the focal point on the opposite side. Continue the ray parallel to the optical axis after passing through the lens.

3. A ray from the object through the center of the lens, which continues undeviated.

■ The intersection of these rays (or their extensions) locates the image of the object point.

■ The location of the image of a point formed by a mirror can be found from a scale diagram by drawing at least two of the following rays:

1. A ray from the object through the center of curvature that is reflected back along itself upon striking the mirror.

2. A ray parallel to the optical axis from the object to the mirror, where it is reflected through the focal point of a concave mirror or from the focal point of a convex mirror.

3. A ray from the object to the mirror along a path passing through or toward the focal point that is reflected parallel to the optical axis.

■ The intersection of these rays (or their extensions) locates the image of the object point.

Important Terms

You should be able to write the definition or meaning of each of the following:

ray	diverging lens
ultraviolet	thin lens
infrared	focal length
angle of incidence	real image
angle of reflection	virtual image
plane of incidence	object distance
law of reflection	image distance
law of refraction	lateral magnification
index of refraction	thin-lens equation
critical angle	aberrations
total internal reflection	dispersion
converging lens	

Conceptual Questions

22.1 If you use an ordinary electric lamp with a frosted bulb to cast a shadow of your hand on the wall, are all parts of the shadow equally dark? Why?

22.2 Explain what, if anything, is wrong with the following statement: When you look in a mirror, it reverses left and right; it should therefore also reverse up and down.

22.3 Could you see the surface of a mirror that was in all ways perfectly reflecting?

22.4 Explain the operation of one-way mirrors. Under what conditions do they work best? Are they really one-way?

22.5 Under what conditions is it possible to immerse a transparent object in a clear liquid and not be able to see the object?

22.6 If you stand on a river bank trying to spear a fish swimming in the water, where should you aim the spear?

22.7 If someone hands you a lens, how can you tell by looking through it whether it is a positive or negative lens?

22.8 What happens to the focal length of a glass lens immersed in water? Consider both positive and negative lenses.

22.9 The right-hand side-view mirror on most automobiles carries a message that "Objects in the mirror are closer than they appear." What is the meaning of this statement. What kind of mirror is used? Why is such a mirror used?

22.10 The index of refraction of the atmosphere is slightly greater than unity. Do you see the rising sun sooner or later than you would if the earth had no atmosphere? Draw a diagram illustrating your answer.

22.11 Describe the images you would see upon looking into the bowl of a silver spoon. How does the image change if you move the spoon closer to your eye? What happens if you turn it over and look at the back side? What is the relationship of the image size for one side of the spoon to the image size for the other side?

22.12 What happens to the image formed on a screen by a lens if you cover the central portion of the lens with a small piece of opaque paper? What happens if you cover the upper half of the lens?

22.13 The following statement is found in a handbook for scuba divers: "Underwater objects seen through a flat face mask appear about 25% larger or closer." Using diagrams, explain the reasons behind this statement.

22.14 A hollow transparent container is shaped like a lens, being thicker in the center and thinner at the edges. Describe the optical behavior of the container if it is sealed and then immersed in a fish tank filled with water. How would your answer change if the container were thinner in the center and thicker at the edges?

Problems

Section 22.2 Reflection and Refraction

Hints for Solving Problems

The two basic rules needed to solve problems of reflection and refraction are the law of reflection ($\theta_r = \theta_i$) and Snell's law ($n_1 \sin \theta_1 = n_2 \sin \theta_2$). Remember to measure all angles from the normal to the surface.

22.1 A ray of light strikes a mirror and is reflected so that the angle between the incident and the reflected beam is 30°. (a) If the mirror is rotated to increase the angle of incidence by 1°, what will be the new angle between the incident and the reflected beam? (b) If, instead, the mirror is moved to decrease the angle of incidence by 1°, what will be the new angle between the incident and the reflected beam?

22.2 A woman 168 cm tall stands in front of a plane mirror attached to a vertical wall. What is the minimum length that the mirror must have for the woman to see her entire height?

22.3 A light ray is directed at the inside of a highly polished silver hemispherical bowl and is incident parallel to the symmetry axis of the bowl. If it strikes the bowl so that the angle between the incident ray and a radius of the hemisphere is 45°, in what direction will it emerge from the bowl?

22.4 A highly polished silver hemisphere of radius R rests with its base on a level table. A parallel beam of light with circular cross section and diameter D shines vertically down on the hemisphere. The central axis of the beam coincides with the symmetry axis of the hemisphere. What must be the diameter of the beam if light from the outer edges of the beam is reflected horizontally?

22.5 What is the speed of light in water, ice, and diamond? Use the information in Table 22.1.

22.6 If you run toward a plane mirror at a speed of 2.7 m/s, how fast do you approach your image?

22.7• Two flat mirrors held together at one edge make an angle θ. A bee sits near the intersection of the mirrors along the line bisecting the angle θ (Fig. 22.37). A student at A looking into the mirrors sees five bees, the real one plus four images, equally spaced around the vertical axis at C. What is the value of angle θ?

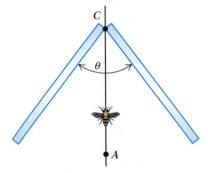

Figure 22.37
Problem 22.7.

22.8• A person whose eyes are 1.6 m above ground stands on one shore of a still lake and observes the reflection of a 1.4-m-tall fence on the opposite shore. His line of sight to the reflection of the top of the fence is 6.0° below the horizontal. How wide is the lake?

22.9• A student standing near an empty open barrel can see half-way down the far side of the barrel. The barrel is a cylinder with a diameter that is half its height. While the student is standing there, the barrel is completely filled with a transparent oil with an index of refraction $n = 1.37$. Without moving, can the student now see to the bottom of the barrel?

22.10• Water in a fish tank is 38 cm deep, and a coin rests on the bottom. How far below the surface does the coin appear to be when viewed from above? (Assume a small angle of incidence.)

22.11• A box of fish food is held 23 cm above the water surface in the fish tank of Problem 22.10. How far above the surface does the box appear to a fish in the tank? (Assume a small angle of incidence.)

22.12• A beam of light makes an angle of 30° with the normal of a mirror made of 2.0-mm-thick glass silvered on the back. If the index of refraction of the glass is 1.5, how far is the point at which the beam leaves the glass surface (after being reflected from the silver backing) from the point at which the beam entered the glass?

22.13•• A light ray is normally incident on the top face of a thin transparent prism of index $n = 1.48$. The bottom face of the prism makes an angle of 4.3° with the top face. The point at which the incident ray strikes the bottom surface is 0.73 cm below the top face. This ray is partially reflected from the bottom surface and then leaves the prism at the top surface.
(a) At what angle does the ray emerge from the top surface?
(b) How far is the point at which the beam exits the prism from the entry point?

22.14•• A beam of light is directed at a glass ($n = 1.50$) cylinder in a plane perpendicular to the cylinder axis. It strikes the cylinder at an angle of 30° with respect to the normal at the point of incidence. In what direction, with respect to the original beam direction, does it leave the cylinder?

22.15•• A rectangular glass block has opposite sides that are exactly parallel. (a) Show that light incident on one side emerges from the other side parallel to its original direction. (b) Show also that the displacement of the two beams depends linearly on the thickness of the block.

22.16•• Two prisms with the same angle but different indices of refraction are put together to form a parallel-sided block of glass (Fig. 22.38). The index of the first prism is $n_1 = 1.52$ and that of the second prism is $n_2 = 1.71$. A laser beam is normally incident on the first prism. What angle will the emerging beam make with the incident beam?

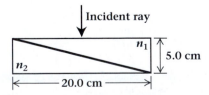

Figure 22.38
Problem 22.16.

Section 22.3 Total Internal Reflection

22.17 What is the critical angle for total internal reflection in plastic (polymethylmethacrylate) at an interface with air?

22.18 What is the critical angle for a glass-to-water interface? Assume a refractive index of 1.50 for the glass.

22.19 A person lying on the bottom of a pool notices that she can see all of the above-water surroundings. The view is distorted and compressed, but everything is visible that could be seen from the surface of the water. What is the angular diameter of this view?

22.20● A fish tank filled with water is made of flat glass walls of $n = 1.50$. What is the maximum angle of incidence for a light ray within the water to strike the glass wall and still emerge to the outside air?

22.21● A glass cube 5.00 cm on an edge has an index of refraction $n = 1.52$. An opaque disk of radius R is pasted in the center of each face of the cube. (a) What is the minimum value of R to completely hide the center of the cube? (b) What fraction of the surface area of the cube is covered by the disks for R found in part (a)?

22.22● A beam of light in air is incident at an angle of 45° on the surface of a transparent solid. The light is deviated by 18° toward the normal upon entering the solid. What is the critical angle for total internal reflection from within this solid?

22.23●● Light is incident normally upon the long face of a symmetric prism (Fig. 22.39). What is the range of values of α that will permit total reflection from both rear faces for glass of $n = 1.55$?

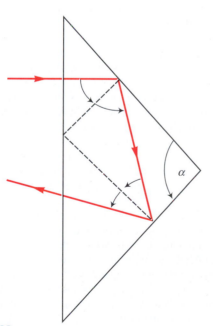

Figure 22.39
Problem 22.23.

22.24●● A lamp that gives off light in all directions is placed 2.00 m below the surface of the water in the middle of a large swimming pool. What fraction of the light can escape directly from the water? (Consider the fraction of light emerging through part of the surface of a sphere centered on the lamp. The surface area of a sphere of radius R is $4\pi R^2$. The surface area of the portion of the sphere cut off by a plane that intersects the sphere at a radial distance h from the surface is $2\pi Rh$. Assume no light is reflected from the sides or bottom of the pool.)

Section 22.6 Locating Images by Ray Tracing

Hints for Solving Problems

Begin by selecting and recording the scale factor you will use. Draw the diagram to a scale as large as your paper will allow. Next draw an optical axis and mark the locations of the given optical components and positions. Then draw the rays described in Section 22.6. Measure the appropriate distances on your diagram and use your scale factor to convert to the answer.

22.25 A diverging lens with focal length of 25 cm is used to view a postage stamp 50 cm from the lens. (a) How far from the lens does the stamp appear? (b) What is the magnification of the image? Work this problem by ray tracing.

22.26 An overhead projector with a lens of 35.6-cm focal length is used to throw an image on a screen 1.80 m away. If you consider the lens to be a thin lens, how far is the slide from the lens when the image is in focus? What is the magnification in this setup? Work this problem by ray tracing.

22.27 Prove by geometrical (ray-diagram) techniques that if an object is at a distance of twice the focal length from a positive lens, the image is the same size as the object.

22.28 An incandescent light bulb is marked "60 W" on its upper end. A lens of 25-cm focal length, when held above the light bulb, forms an image of the "60 W" on the ceiling. If the distance from the lens to the ceiling is 1.30 m, how much larger are the letters and numbers in the image on the ceiling than on the object light bulb? Work this problem by ray tracing.

22.29 A diverging lens of 0.75-m focal length is held so that the image of a chair appears to be 0.40 m from the lens. How far from the lens is the chair? Work this problem by ray tracing.

22.30● A hollow cube 6.0 cm on an edge is aligned with its center on the optical axis of a converging lens of focal length 20 cm. The cube is turned so that its sides are parallel to the axis of the lens. The center of the cube is 30 cm from the lens. (a) Determine the position and size of the images of the front and rear faces of the cube by ray tracing. (b) Sketch the image of the cube.

22.31• A small light bulb, a lens, and a piece of white cardboard are mounted on a meterstick so that an inverted image of the bulb is formed on the cardboard. The image is three times as large as the actual size of the bulb. The bulb is located 10 cm from one end of the meterstick, and the cardboard is 10 cm from the other end. (a) What is the location of the lens? (b) What is the focal length of the lens? Work this problem by ray tracing.

Section 22.7 The Thin-Lens Equation

Hints for Solving Problems

Be careful to observe the sign conventions for image and object distances. A negative image distance corresponds to a virtual image; a negative object distance corresponds to a virtual object. A real object and a real image have positive object and image distances. If the image formation results from the light passing through two or more lenses, treat the lenses one at a time. For example, compute the image position for the first lens and then use it to obtain the object position for the second lens. Then find the image position for the second lens.

22.32 Prove that if an object is at a distance of twice the focal length from a lens, the image is the same size as the object. Show this using an algebraic method.

22.33 A thin lens forms an image of an overhead electric lamp on a piece of paper. The distance to the lamp from the lens is 148 cm; the distance to the paper from the lens is 16.0 cm. What is the focal length of the lens?

22.34 Work Problem 22.25 using the thin-lens equation.

22.35 Work Problem 22.26 using the thin-lens equation.

22.36 A screen is placed exactly one meter away from an illuminated test pattern. A lens placed between them forms a clear (focused) image of the pattern on the screen when the center of the lens is 180 mm from the test pattern. What is the focal length of the lens?

22.37 An image of a flower is produced on a screen by a lens 1.60 m from the screen. The image of the flower has a linear magnification of -2.5. What is the focal length of the lens?

22.38• A luminous object is placed 3.00 m from a screen. When a lens is interposed between the object and the screen at a distance of 0.50 m from the screen, an in-focus image of the object forms on the screen. It is claimed that if the object is moved 1.70 m farther from the screen and the lens moved 0.30 m farther from the screen, the image will again be in focus. Is this claim true?

22.39• A small hand-held camera has a lens with a focal length of 50 mm. The camera is arranged so that by moving the lens in or out, objects from 0.50 m away to "infinity" can be brought to an exact focus on the film. How many centimeters will the lens have to be moved as the focus is changed from the closest to the farthest object?

22.40• Show that the thin-lens formula for a converging lens can be written as $i = (1 - m)f$.

22.41•• A converging lens held near a flashlight images the bulb on a wall 1.60 m from the lens. With the flashlight fixed, the lens is moved closer to the wall and a sharp image is again formed when the lens is 15 cm from the wall. (a) What is the focal length of the lens? (b) How far is the flashlight bulb from the wall?

22.42•• Two thin lenses, each with $+12$-cm focal length, are placed 60 cm apart. An object 18 cm to the left of the first lens is imaged a distance x to the right of the second lens. Determine x.

22.43•• Two thin lenses, one with $+12$-cm focal length and the other with -6.0-cm focal length, are placed 18 cm apart. Locate the final image of an object placed on the axis 18 cm from the nearest lens, which is the positive lens. Give your answer with respect to the position of the diverging lens. For simplicity, assume light propagating from left to right.

Section 22.8 Spherical Mirrors

22.44 A 3.0-cm-tall domino is located 100 cm from a concave spherical mirror of $R = 30$ cm. Locate the position of the image and find the magnification.

22.45 A dentist holds a 1.5-cm focal length dental mirror 1.0 cm from a patient's tooth. What is the magnification of the tooth? (*Hint:* The mirror surface is concave.)

22.46 A pencil is held perpendicular to the optical axis of a concave spherical mirror. The pencil is 87 cm from the center of the mirror, and its image is found 18 cm from the mirror. (a) Find the focal length of the mirror. (b) What is the radius of curvature of the mirror? (c) What is the magnification of the pencil's image?

22.47 The radius of a concave spherical mirror is 20 cm. A chess piece 4.2 cm high is located at distances of (a) 12 cm and (b) 6.0 cm in front of the mirror. Find the image positions for both of these object distances. Describe the images.

22.48 A coin 2.0 cm in diameter is held 20 cm from a concave spherical mirror of 30 cm radius of curvature. Locate the image of the coin and its height.

22.49 The radius of a convex spherical mirror is 20.0 cm. A sugar cube 1.50 cm high is located at distances of (a) 20.0 cm, (b) 12.0 cm, and (c) 6.00 cm in front of the mirror. Locate the image position for each of these object distances.

22.50 A shiny, spherical Christmas tree ornament is 95 mm in diameter. (a) Where is the image of a child standing 2.0 m away from the ornament? (b) What is the magnification of the image? (c) Is the image inverted or erect?

22.51 A wide-angle mirror in a grocery store is made from a section of a spherical surface whose radius of curvature is 0.75 m. How far behind the mirror is the image of a person standing 2.4 m in front of it?

22.52• When you hold a concave mirror of radius R at just the right distance, you see a double-size image of your eye. What is the distance between your eye and the mirror?

22.53•• Derive the mirror equation using a geometrical diagram. Employ a method similar to the one used in the text to derive the equation for a thin lens.

Additional Problems

22.54 What is the critical angle for light inside a diamond having an index of refraction $n = 2.417$?

22.55 A light beam in air enters a transparent material at an angle of $47°$ with respect to the normal. After refraction, it makes an angle of $32.5°$ with respect to the normal. Identify the material on the basis of the information in Table 22.1.

22.56 Suppose that the prism of Example 22.3 were made of ice ($n = 1.31$). (a) Would a ray normally incident be totally reflected? (See Fig. 22.11.) (b) Design a symmetric ice prism that would totally reflect the light incident on one of the smaller faces.

22.57• Two plane mirrors are joined at right angles to each other. A small statue of Newton is placed along the diagonal between the mirrors at a small distance from the vertex. The statue is viewed from a point not on the diagonal. Determine, by the use of appropriate drawings, the total number of images of Newton that can be seen in the mirrors.

22.58• A converging lens of 15-cm focal length is placed on an optical bench that is 120 cm long. At one end a self-luminous object is fixed; at the other end is a fixed screen. (a) Measuring from the object end of the bench, determine the *two* positions for which an image will be in focus on the screen. (b) What is the magnification in each case? Compare your results with a geometrical construction in one case.

22.59• Show that you may write the thin-lens equation in the form

$$f^2 = xx',$$

where x and x' are the distances of the object and the image from the focal points. This equation is known as the Newtonian form of the thin-lens equation. (*Hint:* Make a drawing like Fig. 22.21 with $x = o - f$ and $x' = i - f$.)

22.60• Show that the lateral magnification of a thin lens may be expressed in the form

$$m = -\frac{x'}{f},$$

where x' is the distance from the focal point to the image point, $x' = i - f$. This expression is known as the Newtonian formula for lateral magnification. (*Hint:* Make a drawing similar to Fig. 22.21 with $x' = i - f$.)

22.61• Show that the effective focal length of three thin lenses of focal length $f_1, f_2,$ and f_3 placed close together is

$$\frac{1}{f} = \frac{1}{f_1} + \frac{1}{f_2} + \frac{1}{f_3},$$

when the distance between the lenses approaches zero.

22.62•• A 10-cm-long pencil embedded in a block of transparent ice is parallel to and 20 cm from the surface. Assume that you place your eye 60 cm from the surface along a line that passes through the center of the pencil. (a) What angle would the pencil subtend at your eye if the ice were not present? (b) What angle does the pencil in the ice subtend at your eye? Use the small-angle approximation given in Chapter 14, that is, $\sin \theta \approx \tan \theta \approx \theta$.

22.63•• A spherical glass ball of index 1.5 and radius R is used as a burning lens. By direct calculation using Snell's law, find the approximate point to which rays converge. To do this, find the intersection of a ray directed parallel to the optical axis with a ray incident on the ball a distance $R/2$ from the optical axis. It will help to make a scale drawing.

22.64•• A converging lens of 16-cm focal length is placed at the center of curvature of a concave spherical mirror whose radius is 32 cm. Locate the final image of an object that is very far away.

22.65•• The mathematical expression that relates the shape of a lens, its index of refraction, and its focal length is called the *lensmaker's equation*. For a lens in air, it is

$$\frac{1}{f} = (n - 1)\left(\frac{1}{R_1} - \frac{1}{R_2}\right),$$

where R_1 and R_2 are the radii of curvature of the first and second surfaces, respectively (Fig. 22.40). The radius is positive if the surface encountered by the light beam is convex, negative if it is concave. What is the focal length of a double convex lens made from glass with index of refraction $n = 1.52$ if $R_1 = 0.35$ m and $R_2 = -0.35$ m?

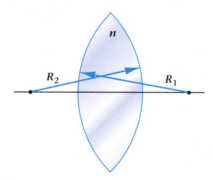

Figure 22.40
Problem 22.65.

22.66•• A capillary tube with an internal diameter d is made of glass of refractive index n. What is the apparent internal diameter of the capillary when viewed in air from a distance that is large compared with the diameter of the tube?

22.67•• Derive an expression for determining the speed of light using a Fizeau apparatus in terms of the number of teeth N, the angular velocity ω of the wheel, and the distance D between the mirrors. (*Hint:* Refer to Fig. B22.2.)

22.68•• The lateral magnification of a thin lens is $m = -i/o$. Show that the longitudinal magnification—that is, the magnification in the direction of the optical axis—is given by m^2.

22.69•• Two thin lenses, one with $+12$-cm focal length and the other with -6.0-cm focal length, are placed 12 cm apart. Locate the final image of an object placed on the axis 18 cm from the nearest lens, which is the positive lens. Give your answer with respect to the position of the diverging lens. For simplicity, assume light propagating from left to right.

22.70•• Three thin lenses are aligned along their optical axis. The first lens has a $+24$-cm focal length. The second lens has a -18-cm focal length and is placed 18 cm from the first lens. The third lens has a $+12$-cm focal length and is placed 9.0 cm beyond the second lens. Find the position of the image of an object located 36 cm from the first lens.

22.71•• A lens forms an image of a candle flame on a wall 2.50 m away from the flame. The linear size of the image is 3.5 times the size of the flame. (a) Using the techniques of ray tracing, find the focal length of the lens. (b) Check your results by using the thin-lens equation to find the focal length.

Optical Instruments

Back to the Future: Development of the Telescope

The human eye is extremely sensitive and functions effectively over a wide range of distances and light levels. However, we can greatly expand the boundaries of our vision through the use of optical instruments that extend our vision in new ways and to new dimensions. Optical microscopes let us see objects as small as 0.1 μm in diameter. Telescopes show us galaxies so far away that their light has taken billions of years to reach us. Film and video cameras record images from visible and infrared light. Lenses that gather more light and display fewer aberrations are being designed and produced continually.

The ideas of optics and optical instruments have led to new concepts and designs for instruments that extend the range of our visual senses far beyond the limits of visible light. For example, beginning with the development of the electron microscope in the 1930s, improvements, innovations, and new methods have reached the point where we can now image individual atoms. In hospitals, acoustic waves are routinely used to generate ultrasonic images of the interior of the human body. However, before we describe the principles of optical instruments, we first examine the most fascinating and interesting of all optical instruments, the human eye. ■

The Eye

The human eye is one of the most familiar optical instruments, yet it is also one of the most complex. The overall mechanism of vision is extremely complicated and even now is not fully understood. Yet we can understand some of the important principles of vision by considering the optical properties of the human eye.

Figure 23.1 is a schematic drawing of a human eye, seen in cross section from above. An image focused on the retina by the optical elements of the eye is the stimulus for nerve signals that are ultimately interpreted by the brain. The principal optical elements are the cornea (a transparent sheath across the front of the eye), the pupil (an opening), the iris (which adjusts the size of the pupil), a fluid called the aqueous humor in front of the lens, the lens, the ciliary muscle (which controls the shape of the lens), the gel-like vitreous humor, and the retina (the light-sensitive layer that lines the back of the eye). However, when we consider the way in which the incident light is refracted by the eye, we cannot treat it as a thin lens because there are more than two refracting boundaries and the object and image are in different optical media. The greatest relative change in the index of refraction takes place between the air ($n \approx 1.000$) and the cornea ($n \approx 1.376$). Consequently, the greatest refraction takes place at that surface also. The eye lens is actually a gradient-index lens with a larger index ($n \approx 1.406$) near the center than near the edges ($n \approx 1.386$). Because the aqueous and vitreous humors that surround the lens have indices of $n \approx 1.336$, the refraction at the lens surfaces is small compared with the refraction at the cornea.

The normal relaxed eye presents distant objects in focus on the retina (Fig. 23.2a). For close objects to be in focus on the retina, some optical parameters of the eye must change. If a distant object moves toward the eye, the eye changes to keep the image in focus on the retina. This effect is called **accommodation** and is accomplished primarily by contraction of the ciliary muscles, which changes the shape of the lens. When the eye focuses on nearby objects, the lens becomes thicker and the surfaces more curved (Fig. 23.2b). Because the lens has a higher index of refraction than the surrounding medium, the net effect is to shorten the focal length of the lens.

Figure 23.1

Cross section of a human eye viewed from above, showing the main optical elements. The function of each part is described in the text.

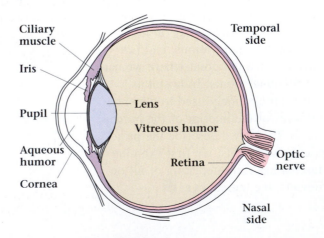

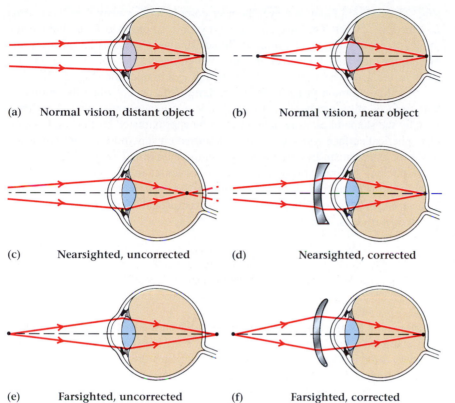

(a)   Normal vision, distant object

(b)   Normal vision, near object

(c)   Nearsighted, uncorrected

(d)   Nearsighted, corrected

(e)   Farsighted, uncorrected

(f)   Farsighted, corrected

Figure 23.2
(a) The relaxed, normal eye forms an image of distant objects on the retina. (b) For close objects, the lens of the normal eye changes shape to focus the image on the retina. (c) In a nearsighted eye, the image of a distant object forms in front of the retina. (d) A diverging lens corrects for nearsightedness. (e) In a farsighted eye, the image of a nearby object forms beyond the retina. (f) A converging lens corrects for farsightedness.

The ability of the eye to accommodate is limited. We may not move an object any closer than a distance called the near point and still view it clearly. The **near point** is the distance from the unaided eye that produces the largest retinal image without blurring. Objects closer than the near point cannot be brought into focus. The average value for the near point is about 25 cm, although there is considerable individual variation, even for people with so-called normal vision. By convention, the standard near point is taken to be 25 cm. The **far point** is the greatest distance from the unaided eye that produces a distinct image. The far point of the normal eye is at infinity.

Most people do not have ideal vision. The range of visual ability that is considered normal is quite broad. Two common types of vision defects are simply related to the optical properties of the eye and can be readily corrected. They are called nearsightedness (*myopia*) and farsightedness (*hyperopia*).* A nearsighted eye is unable to accommodate over the normal range from 25 cm to infinity. Instead, there is a far point beyond which vision is not distinct. The image of a more distant object comes to a focus in front of the retina, so only a blurred image forms on the retina (Fig. 23.2c). Clear vision of distant objects is restored by placing a diverging lens in front of the eye. This diverging lens forms an image of the distant object within the accommodation range of that particular eye

---

*Farsightedness commonly occurs in people after middle age because of thickening and lack of flexibility in the lens. For this reason it is also known as *presbyopia,* from the Greek words *presbys* for old and *ops* for eye.

(Fig. 23.2d). For a farsighted eye, the near point is farther away than the normal 25 cm (Fig. 23.2e). The image of a nearby object comes to focus behind the retina, so again the image is blurred. To see things as close as 25 cm from the eye requires the use of a converging lens, which images the object at least as far away as the actual near point of the particular eye in question (Fig. 23.2f).

Another common visual problem, **astigmatism,** occurs when the cornea or lens surfaces are not spherical. An astigmatic eye images point objects as lines. Usually the shape of an astigmatic eye can be approximated by the combination of a spherical surface with a cylindrical deformation superposed. Corrections are made for astigmatism by using a compensating cylindrical eyeglass lens.

Opticians generally express the strength of lenses used to correct visual defects in terms other than of their focal length. The common unit is the **diopter. The strength of a lens in diopters is the reciprocal of the focal length expressed in meters.** That is,

$$\mathscr{D} = \frac{1}{f},$$

(23.1)

where f is the focal length in meters and $\mathscr{D}$ is the strength of the lens in diopters. For example, a lens with a focal length of $+0.25$ m has a strength of $+4.0$ diopters. Shorter focal lengths correspond to greater dioptic strengths. The use of lens strength is especially convenient when several thin lenses are used in close proximity. Then the strength of the combination is just the sum of the strengths of the individual lenses. (See Example 22.8.) Selecting a lens of proper strength for correcting a visual defect of the kind shown in Fig. 23.2 is usually accomplished by trying various combinations of lenses in front of the eye until the clearest vision is obtained. Then a single lens is chosen that has the same strength as the combination of trial lenses.

For people whose eyes are unable to accommodate fully at both ends of the range, lenses are available with two distinct regions of different dioptic strength. Such lenses are known as *bifocals.* They have an upper region of dioptic strength appropriate for distant vision and a lower region designed for close vision. In some cases, *trifocal* lenses are used to provide better vision at intermediate distances. The numerical example below illustrates the use of a corrective lens.

Example 23.1

Your nearsighted friend has a far point of 2.00 m. Objects farther away than the far point are not imaged clearly. What lens strength will provide clear vision for very distant objects?

Strategy   Your friend needs a lens that forms an image of a distant object at a distance from the eye equal to her far point. For a distant object ($o \approx \infty$), the lens must bring the image to a distance of 2.00 m from her eye. Because the image will be on the same side of the lens as the incident light, the image distance is negative and is $i = -2.00$ m, the far point. We need only apply the thin-lens equation to solve this problem.

Solution   From the thin-lens equation we get

$$\frac{1}{f} = \frac{1}{o} + \frac{1}{i} = \frac{1}{\infty} + \frac{1}{-2.00 \text{ m}},$$

or

$$f = -2.00 \text{ m}.$$

The proper lens has a focal length $f = -2.00$ m. It is a diverging lens. The strength of the lens in diopters is

$$\text{strength of lens} = \mathcal{D} = \frac{1}{f} = \frac{1}{-2.00 \text{ m}} = -0.50 \text{ diopter.}$$

Example 23.2

A farsighted classmate is unable to clearly focus on objects closer than 125 cm from his eyes. What strength corrective lens is needed to allow him to read books held 30 cm from his eyes?

Strategy   We must select a lens that will place an image of the book at a distance at which your classmate's eyes can focus—that is, at his near point of 125 cm. The object distance is $o = 30$ cm $= 0.30$ m, and the image distance is $i = -125$ cm $= -1.25$ m, the negative sign indicating that object and image are on the same side of the lens. (Alternatively, we could say the sign is negative because the image is virtual and erect.)

Solution   From the thin-lens equation we get

$$\frac{1}{f} = \frac{1}{o} + \frac{1}{i} = \frac{1}{0.30 \text{ m}} + \frac{1}{-1.25 \text{ m}} = \frac{2.5}{\text{m}}.$$

Thus, the dioptic strength of the lens is

$$\mathcal{D} = \frac{1}{f} = 2.5 \text{ diopters.}$$

Discussion   The farsighted classmate needs a positive lens to allow a book to be held comfortably close and still be seen clearly. In this case, the lens required has a strength of 2.5 diopters, which corresponds to a focal length of 0.40 m, or 40 cm.

23.2   The Magnifying Glass

Many optical devices are used to produce magnified images of objects. The simplest magnifying instrument of all is the **magnifying glass,** or simple microscope. The magnifying glass is a single converging lens that, when held near the eye, gives an image whose size on the retina is larger than that observed by the unaided eye. By adjusting the distances at which the lens and object are held from the eye, the viewer can obtain maximum magnification without undue eye strain or blurring of the retinal image.

Magnifying glasses are not only used singly to enlarge such things as fine print and small objects, they are also used to enlarge the images formed by other lenses. When they are utilized in this way, they are referred to as eyepieces. The primary function of magnifying glasses and eyepieces is to increase the angular size of the image and to allow viewing with a relaxed eye.

As an object is moved closer, it subtends* a larger angle at the eye, so that the image produced covers a larger part of the retina (Fig. 23.3). However, we cannot bring the object any closer than the near point and still see it clearly. A magnifying glass allows us to increase the visual angle that an object subtends at the eye—that is, to form a larger image on the retina—without requiring our eye to focus closer than the near point. For example, Fig. 23.4(a) shows an object of height h placed at the standard near point (25 cm), where it subtends an angle θ. If the angle θ is small, we can write θ in radians as

$$\theta = \frac{h}{25},$$

where both h and the near point are expressed in centimeters. In Fig. 23.4(b), we have placed a magnifying glass (converging lens) next to the eye and moved the object within the focal length of the lens. The positions of the lens and the object have been adjusted so that the enlarged virtual image of the object falls at the eye's near point. Since the image is on the same side of the lens as the original object, the image distance is negative.

What magnification have we achieved? To find out, we first calculate the object distance, using the thin-lens formula. If distances are measured in centimeters, we get the result

$$o = \frac{25f}{f + 25}.$$

Figure 23.3

(a) The image of an object seen with the unaided eye forms on the retina. (b) As the object comes closer, its image covers a larger part of the retina.

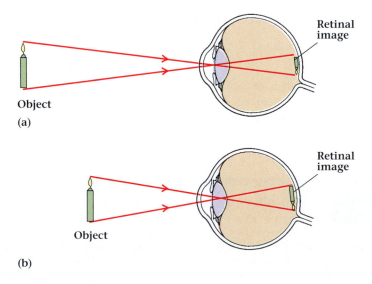

Retinal image

Object

(a)

Retinal image

Object

(b)

---

*Recall that in a triangle, a line *subtends* the angle opposite to it; that is, the line extends across the entire angle. The object in Fig. 23.3 forms one side of a triangle whose opposite vertex is at the observer's eye. Thus, the object subtends an angle at the eye.

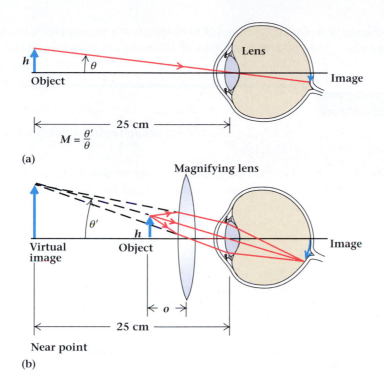

Figure 23.4
(a) An object placed at the near point of an unaided eye. (b) The same object viewed through a magnifier with the virtual image at the near point. The angular magnification M is θ'/θ.

From the figure we see that the angle θ' subtended by the virtual image is approximately

$$\tan \theta' \approx \theta' = \frac{h}{o} = \frac{h(f + 25)}{25f}.$$

The magnification of interest for a magnifying glass is the angular magnification. In general, the **angular magnification** is defined by

$$M \equiv \frac{\theta'}{\theta}, \tag{23.2}$$

which for this case becomes

$$M = \frac{25 \text{ cm}}{f} + 1 \qquad \text{(image at near point).} \tag{23.3}$$

Remember that f is measured in centimeters in this equation.

If the object is held at or just inside the focal point of the lens, the image forms very far away, essentially at infinity, rather than at the near point. This corresponds to the most comfortable viewing distance because the eye is relaxed. (See Problem 23.18.) In this case, $\theta' = h/f$, so the magnification is given by

$$M = \frac{25 \text{ cm}}{f} \qquad \text{(image at infinity).} \tag{23.4}$$

Manufacturers commonly use Eq. (23.4) to specify the magnification of a magnifying glass. Thus, a magnifying glass of 10-cm focal length is marked 2.5×.

Example 23.3

A photographer has an 8× magnifier for examining negatives. What is the focal length of the magnifier lens?

Solution The magnification of the lens is specified for producing an image at infinity. Therefore, we can obtain the focal length of the magnifier lens from Eq. (23.4),

$$M = \frac{25 \text{ cm}}{f} = 8.$$

Upon rearranging, we get

$$f = \frac{25 \text{ cm}}{8} = 3.1 \text{ cm.}$$

Example 23.4

A biology student wishes to use a 6-cm focal-length lens as a magnifier.
(a) What is the magnification of the lens when used with a relaxed eye?
(b) What is the maximum magnification of the lens?

Solution (a) For the relaxed eye, the image will be at infinity, so the magnification can be obtained with Eq. (23.4):

$$M = \frac{25 \text{ cm}}{f} = \frac{25 \text{ cm}}{6 \text{ cm}} = 4\times.$$

(b) The maximum magnification occurs when the image is at the near point. In this case, the magnification is obtained from Eq. (23.3):

$$M = \frac{25 \text{ cm}}{f} + 1 = \frac{25 \text{ cm}}{6 \text{ cm}} + 1 = 5\times.$$

23.3   Cameras and Projectors

In its simplest form, a photographic camera is a light-tight box with a lens set into one side and a light-sensitive material (film) placed on the side opposite the lens (Fig. 23.5). Normally, light is prevented from entering the camera by means of a shutter, either at the lens position or just in front of the film. Unlike the eye, the camera lens has a fixed focal length. Therefore, you focus a camera by moving the lens closer to or farther from the film, depending on the object's distance from the lens. To take a photograph, you first adjust the lens position so that a real, inverted image of the object is in focus in the plane of the film. Then, when you press a button, the shutter momentarily opens and an image forms on the

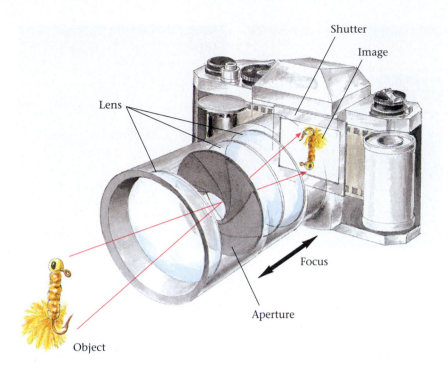

Shutter

Image

Lens

Focus

Aperture

Object

Figure 23.5
A camera is a light-tight box for holding the film and lens in proper position. It also includes a shutter to control the time during which light reaches the film.

film. This image is stored in the light-sensitive material for later chemical processing, yielding a reproduction of the scene as its reflected light originally fell on the film.

Modern cameras, such as the 35-mm camera of Fig. 22.33, are basically the same as the simple camera. They have well-corrected lenses of the type discussed in Section 22.9. Many have automatic exposure control and automatic focusing, features that are made possible through the use of integrated circuit electronics. Computer-aided design of lenses and computer-aided manufacturing allow us to have inexpensive lenses of high quality that were not available for any price only twenty years ago. Many cameras take interchangeable lenses that allow photographers control over the composition of their pictures (Fig. 23.6).

Other cameras, such as digital still cameras, motion picture cameras, and television cameras, use the same basic principle. In a digital camera, the light is imaged on a semiconductor detector rather than photographic film. The information in the image is stored in the same manner as computer data and may be displayed on a computer monitor or printed with a computer printer. A motion picture camera takes a rapid sequence of still pictures, which are eventually projected at the same rate at which they were taken. The apparent smoothness of the motion is due to the fact that the rate at which the pictures are taken and projected, ordinarily 24 frames per second, is higher than the rate at which we can distinguish between individual images. The persistence of vision gives us the appearance of smooth continuous motion. In a television camera, although the image is detected, transmitted, and reproduced by electronic means, the optical principles are the same as for a movie or still camera.

With modern photographic films and with adequate lighting, the shutter need be open only a small fraction of a second to successfully record an image on film. Many cameras have a range of available shutter speeds—that is, lengths of time during which the shutter is open. Shutter speeds of 1/30, 1/60, 1/125,

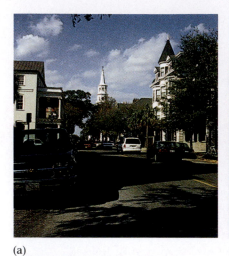

(a)

(b)

(c)

Figure 23.6

These three photographs were taken from the same place, using three different lenses. The focal length used were (a) 28 mm, (b) 55 mm, and (c) 200 mm. The longer the focal length of the lens, the greater the angular magnification of the image.

1/250, 1/500, 1/1000, and 1/2000 second are standard. Note that each of these is approximately half as long (or, in photographic terminology, "twice as fast") as the preceding one. The faster the shutter speed, the faster the object can be moving and still produce a sharp image.

Getting the "correct exposure" in a photograph corresponds to allowing the appropriate amount of light to strike the film.* This amount is different for different types of film. The amount of light that strikes the film is determined not only by how long the shutter stays open, but also by how large the effective lens opening is. Thus, it is analogous to the amount of water flowing from a faucet, which depends on both the length of time the faucet stays open and the cross-sectional area of the opening. The size of the lens opening, or aperture, is often measured by what is called the *f*-value or **f-number.** This is defined to be

$$f\text{-number} \equiv \frac{\text{focal length of lens}}{\text{diameter of lens}} = \frac{f}{d}. \quad (23.5)$$

For example, a lens with a diameter one-half its focal length has an *f*-number of 2, which is written *f*/2. A variable-diameter opening in the camera, called an iris diaphragm (after the iris in your eye), can be adjusted to decrease the effective lens opening and therefore increase the *f*-number. By being able to adjust both shutter speed and *f*-number, the photographer can give the proper exposure to the film while still having the option of using a particular shutter speed or a particular lens opening.

Table 23.1 lists the standard *f*-number intervals. These numbers are often referred to as *f*-stops or simply stops. Notice that the values of successive *f*-stops differ by a factor of approximately $\sqrt{2}$. Thus, if you change the *f*-number of a lens from *f*/4 to *f*/5.6, you reduce the diameter of the aperture by a factor of $1/\sqrt{2}$. The area of the aperture decreases by the square of this factor. So, if the shutter speed remains constant, the amount of light passing through the lens opening is cut in half. A change in aperture equivalent to going from one *f*-stop

Table 23.1	Standard Full-Stop *f*-numbers	
f-number	**(f-number)²**
0.7	0.49
1.0	1
1.4	1.96
2.0	4
2.8	7.84
4	16
5.6	31.4
8	64
11	121
16	256
22	484

*By "amount of light," we mean the energy deposited on the film by the light.

to the next successive one is known as a change of one full stop. It increases or decreases the light passed through the lens by a factor of 2.

A useful consequence of the *f*-number method of classifying lens openings is that for the same shutter speed, lenses of different focal lengths give proper exposure at the same *f*-numbers. This result is due to two compensating factors. First, the amount of light that passes through the aperture is proportional to its area, and hence to the square of its diameter, d^2. Second, the light per unit area that reaches the film depends inversely on the area of the image. For the usual situation, in which the object distance is large compared with the focal length, the linear magnification is proportional to the focal length *f* of the lens, so the area of the image is proportional to f^2. The rate at which a photographic image is formed, or the *speed* of the lens, is then

$$\text{speed of lens} \propto \frac{d^2}{f^2} = \frac{1}{(f\text{-number})^2}. \qquad (23.6)$$

For most purposes, the speed of the lens depends only on the *f*-number and is independent of the particular focal length.

Example 23.5

The light meter in your camera indicates that your film is properly exposed at a shutter speed of 1/50 s and an aperture setting of *f*/16. However, to photograph a bird flying across the park, you must shorten the exposure time to 1/200 s to avoid blurring. What should the new aperture setting be?

Solution Shortening the exposure time by a factor of 4 requires that the aperture be increased to four times its original area. This corresponds to opening the lens two full stops. Thus, the lens opening should be *f*/8.

The film camera has an optical inverse—the slide projector. The basic components of a slide projector are shown in Fig. 23.7. The transparent slide with the likeness of the photographed subject, the lens, and the image projected on the screen are the inverse of the camera used to make the slide. By choosing the proper focal-length lens, you can make the image of the slide completely fill the screen at the chosen projector-screen distance. The need to accommodate different sizes of screens and rooms has led to the use of variable-focal-length, or zoom, lenses for home projectors.

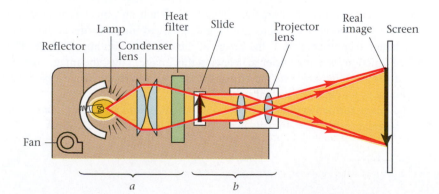

Figure 23.7
The basic optical components of a slide projector. The illuminating system is indicated by bracket *a* and the image-forming system by bracket *b*. Most projectors also include a fan to cool the slide and prevent damage due to overheating.

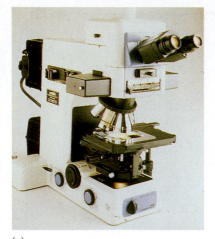

(a)

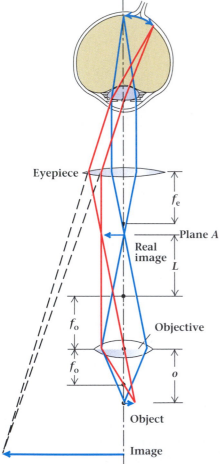

Eyepiece

f_e

Plane *A*

Real
image

L

f_o

Objective

f_o

o

Object

Image

(b)

Figure 23.8

(a) A modern microscope. (b) Ray
diagram of a compound microscope.

The illuminating system of a projector is equally as important as the image-forming system. It usually consists of a lamp and reflector, a condensing lens, and a piece of heat-absorbing glass to protect the slide from the heat of the lamp. The reflector simply directs more light in the direction of the slide. The condenser lens is a strongly converging lens, ensuring that light passing through the edges of the slide passes through the projector lens. Thus, the condenser not only increases the amount of light passing through the slide, but also increases the amount of light reaching the screen. If the condenser lens is removed, the entire slide will still be illuminated, but the image from the edges of the slide will not appear on the screen, and only the central part of the picture will be seen.

The optical principles of a movie projector are the same as those of a slide projector. However, a movie projector also contains a mechanical means for moving the film and for interrupting the light when the film is moving. The result is that the viewer sees a series of individual still pictures. As mentioned earlier, the pictures are presented at a rate of 24 per second, and the brain blends the sequence of individual images into a smoothly flowing scene. Ordinarily, a three-bladed shutter is used to block the light when the film is moving. Because the shutter makes one rotation for each frame, it also interrupts the light during the presentation of each individual picture. The shutter interrupts the light 72 times per second, a rate so fast that the eye cannot see the flicker.

23.4 Compound Microscopes

The microscope and the telescope were developed at about the same time in the early 1600s. Early naturalists soon utilized microscopes to make important discoveries. Marcello Malpighi's discovery of capillaries, Anton van Leeuwenhoek's discovery of protozoa, and Robert Hooke's beautiful drawings of magnified cells were important contributions to the understanding of biological processes. These advances would have been impossible without the microscope. The microscope and telescope have perhaps played a greater role than any other scientific instruments in establishing our current understanding of natural laws.

A typical **compound microscope** consists of a tube with a converging lens at each end (Fig. 23.8a). Though in modern microscopes each lens may actually consist of a group of lenses to minimize distortion from lens aberrations, we can understand the optical principles by treating each group as a single lens. The lens close to the object being viewed is called the **objective.** The lens through which one looks is called the **eyepiece** or **ocular.**

Figure 23.8(b) is a ray diagram of a compound microscope. The objective is a lens of comparatively short focal length f_o. When an object is placed just outside the focal point of the objective, a real, enlarged image forms at the plane *A*. If a screen were placed at *A*, the image of the object would appear there. The eyepiece functions as a magnifier for viewing this image. Thus, the image formed by the objective lens serves as the object for the eyepiece lens. The eyepiece, in turn, produces an enlarged virtual image.

The image formed at plane *A* by the objective has a *linear magnification* m_o. The Newtonian expression for this magnification (see Problem 22.60) is

$$m_o \approx -\frac{L}{f_o},$$

where L is the distance from the focal point to the image plane. This image is then viewed through the eyepiece with an *angular magnification*

$$M_e = \frac{25 \text{ cm}}{f_e}.$$

The overall magnification M of the microscope is the product of the linear magnification of the objective and the angular magnification of the eyepiece, resulting in a magnification

$$M = m_o M_e = \left(-\frac{L}{f_o}\right)\left(\frac{25 \text{ cm}}{f_e}\right). \tag{23.7}$$

For practical microscopes, f_o is much less than L and f_e is less than 25 cm, resulting in a large value for the magnification. The negative sign in Eq. (23.7) indicates that the image is inverted.

Microscope objectives and eyepieces are commonly labeled according to their effective magnifications when used with a standard separation between them. Most (but not all) manufacturers design their microscopes so that the distance L, between the focal point of the objective and that of the eyepiece, is 16.0 cm. The magnification of an objective lens can then be expressed as $m_o = 16.0 \text{ cm}/f_o$. If we know the magnifications of the eyepiece and the objective lens, we can use Eq. (23.7) to find the overall magnification. A 10× eyepiece has an angular magnification of 10 times, and a 10× objective has a linear magnification of 10 times. Used together, they give an overall magnification of 100×.

Example 23.6

A laboratory microscope has a 20× objective and a 10× eyepiece. Determine (a) the overall magnification, (b) the focal length of the objective lens, and (c) the focal length of the eyepiece.

Solution (a) The overall magnification is the product of the two magnifications:

$$M = m_o M_e = 20 \times 10 = 200\times.$$

(b) The focal length of the objective lens may be obtained from its magnifying power,

$$m_o = \frac{16.0 \text{ cm}}{f_o}.$$

Upon rearranging, we get

$$f_o = \frac{16.0 \text{ cm}}{20} = 0.80 \text{ cm}.$$

(c) The focal length of the eyepiece is obtained from its magnification,

$$M_e = \frac{25 \text{ cm}}{f_e}.$$

Rearranging, we find

$$f_e = \frac{25 \text{ cm}}{10} = 2.5 \text{ cm}.$$

You might expect that you could use stronger and stronger lenses to make an optical microscope as powerful as you desire, but magnification alone is not the only criterion to consider. What is really important is not simply the size of the image, but the ability to distinguish, or resolve, two object points that are very close together. As we will see in the next chapter, the minimum distance between two object points that can be resolved in an image depends on the wavelength of the illumination and on the diameter of the lens. In optical systems, we reach fundamental, unavoidable limits resulting from the wave nature of light.

23.5 Telescopes

Though different in purpose, the telescope has a great deal in common with the microscope as an optical instrument. A telescope also consists of a long tube with an objective lens toward the object and an eyepiece lens toward the viewer. Several types of telescopes exist. Figure 23.9(a) shows a diagram of a **refracting astronomical,** or **inverting, telescope.** In this type of telescope, the objective lens has a relatively long focal length and the object being viewed is far away compared with this focal length. As a result, rays from the object come in nearly parallel and a real image forms near the focal point of the objective. This image would show up clearly on a screen placed there, just as in the case of the microscope. Note, however, that in this case the image is smaller than the physical size of the object. Again we use an eyepiece to view the inverted image that has been brought to a focus by the objective lens. The angular size of the viewed image is larger than the angular size of the object when viewed without the telescope, and so the initial object appears closer.

We define the magnification M of a telescope as the ratio of the angle θ' subtended by the object when viewed through the telescope to the angle θ subtended when the object is viewed with the unaided eye. The angle subtended at the eye is essentially the same as that subtended at the objective by the object. If we let h be the height of the image (the blue arrow) formed by the objective lens, we can show from Fig. 23.9 that $\theta \cong h/f_o$ and $\theta' = h/f_e$, so

$$M = \frac{\theta'}{\theta} \cong -\frac{f_o}{f_e}. \tag{23.8}$$

If we replace the converging eyepiece with a diverging lens, we get an erect image. Telescopes of this type are called **Galilean telescopes** after Galileo (Fig. 23.9b). If the eyepiece were not present, incoming rays from a distant object would come to a focus essentially at the focal point of the objective lens. A real inverted image would be formed on a screen placed at this point. However, when the diverging eyepiece lens is placed within the focal length of the objective, so that the rays striking the eyepiece emerge from it parallel to each other, a virtual image is produced. From the diagram, we see that the magnification is again given by Eq. (23.8). Note that when f_e is negative, as it is here, the magnification is positive, indicating an erect image.

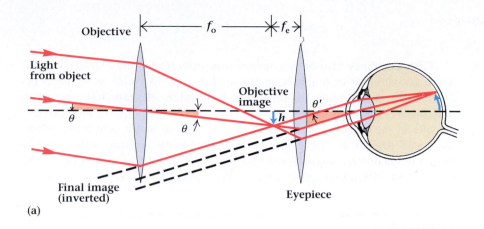

Objective

f_o f_e

Light from object

Objective image

θ'

θ

θ ↑ h

Final image (inverted)

Eyepiece

(a)

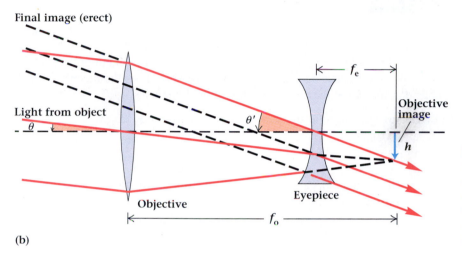

Final image (erect)

Light from object

θ

θ'

f_e

Objective image

h

Eyepiece

Objective

f_o

(b)

Figure 23.9
(a) Principle of operation of a refracting astronomical (or inverting) telescope. (b) Ray diagram of a Galilean telescope. The diverging eyepiece produces an erect, virtual image.

Binoculars are essentially twin refracting telescopes mounted side by side (Fig. 23.10). The prisms, which give most binoculars their characteristic shape, are used to erect the image that would otherwise be inverted. (The light is totally internally reflected in the prisms.) Because the index of refraction of the prisms is greater than that of air and because they fold the light path, the tubes are shorter than the tubes of a simple telescope of the same magnification. Opera glasses differ from binoculars in that they consist of a pair of side-by-side Galilean telescopes. Because their images are already erect, no prisms are needed.

Figure 23.10
Prisms are used to erect the image in a binocular.

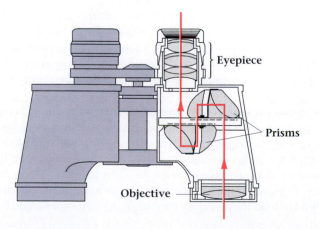

Eyepiece

Prisms

Objective

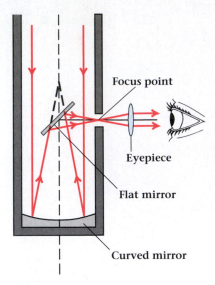

Figure 23.11
Ray diagram of a Newtonian telescope, which uses a mirror instead of a lens to focus light.

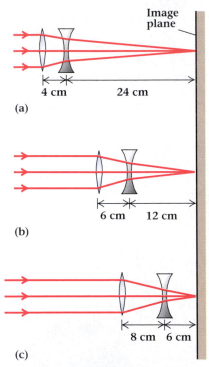

Figure 23.12
A varifocal lens made from two components of equal but opposite focal length (12 cm). The distance between the diverging lens and the effective focal point depends on the separation of the lenses: (a) 4 cm, (b) 6 cm, and (c) 8 cm.

Telescopes may also be constructed with mirrors for the objective. Isaac Newton constructed such a telescope in 1668 in order to avoid the problems of chromatic aberration found in lenses. The **Newtonian telescope** has a converging (concave) mirror as its first element (Fig. 23.11). A flat mirror mounted diagonally in the middle of the tube reflects the light so that it comes to a focus just outside the tube. An eyepiece is used to view the resulting image. Other types of reflecting telescopes differ mainly in where the image is brought to focus. Modern telescopes used in astronomy are reflecting telescopes.

Example 23.7

An astronomical telescope is used to study the moon, which subtends an angle of approximately 0.5° at the earth's surface. The objective lens of the telescope has a focal length of 0.75 m, and the eyepiece has a focal length of 0.10 m. (a) What is the angular magnification? (b) What angle does the moon's image subtend at the eye of the person looking through the telescope?

Solution (a) The magnification is found directly from Eq. (23.8),

$$M = -\frac{f_o}{f_e} = -\frac{0.75 \text{ m}}{0.10 \text{ m}} = -7.5\times.$$

(b) The angular appearance of the moon's image can also be found from Eq. (23.8). Since $M = \theta'/\theta$, the image angle θ' is

$$\theta' = M\theta = (7.5)(0.5°) \approx 4°.$$

*23.6 Other Lenses

One of the marvels of modern technology is the zoom lens, widely used in photography and television. The zoom lens can be changed quickly from a wide-angle lens (short focal length) to a telephoto lens (long focal length). A true **zoom lens** maintains focus throughout the entire zoom range at any focusing distance, provided you have focused sharply on an object. A **varifocal lens** must be refocused whenever you change its focal length.

Modern zoom lenses used with television cameras are available with focal-length changes of as much as 20 to 1. Lenses used on 35-mm cameras come in a range of zoom ratios from the limited 2:1 ratio of a 35-70-mm lens to the 7.5:1 ratio of a 28-210-mm lens. Thus, a single lens can be adjusted to obtain the focal length needed to provide the desired composition and magnification (Fig. 23.6).

To understand the operation of a varifocal lens, consider the behavior of two thin lenses of equal and opposite focal length as the separation between them changes (Fig. 23.12). Suppose the lenses have focal lengths of +12 cm and −12 cm. When the lenses are separated by 4 cm, light incident on the converging lens parallel to its axis is ultimately brought to focus 24 cm from the diverging lens. If the separation of the lenses is increased to 8 cm, the light converges only 6 cm from the diverging lens, resulting in a wider-angle field of view.

DEVELOPMENT OF THE TELESCOPE

Combinations of lenses were used to make telescopes in Holland by about the year 1600. Shortly afterward, Galileo made a telescope and trained it on the heavens. Though he was not the first to use a telescope, Galileo was the first to make a systematic study of the heavens with a telescope. He published his observations in 1610 in a work called "The Starry Messenger." In it Galileo describes his observations of the moon, giving the first evidence that the surface of the moon is not smooth and featureless, as was previously thought and as it appears to the unaided eye. He also discovered four moons circling Jupiter and made the first observations of sunspots. These observations were in dramatic conflict with the teachings of the established church concerning the "perfect, unblemished heavens" and the earth as the center of the cosmos.

The primary purpose of Galileo's telescope was to magnify, so he could better see the details of the planets. For modern astronomers, one of the important properties of telescopes is their ability to gather large amounts of light and thereby let us see fainter and more distant objects. Telescopes that gather more light have larger diameters. During the first half of the twentieth century, a succession of increasingly larger reflecting telescopes were built, culminating with the completion in 1947 of the 5-m Hale Telescope at Mt. Palomar. The magnitude of the effort required for construction of the Hale raised doubts that significantly larger telescopes would ever be built due to the technological difficulties.

A new technology emerged with multiple mirror telescopes. The first large example was the Multiple Mirror Telescope (MMT) in Arizona, completed in 1979 (Fig. B23.1). Lessons learned from the MMT paved the way for the design of the 10-m-diameter Keck I Telescope in Hawaii. The Keck I Telescope is made of 36 hexagonal segments. Each hexagonal segment can be individually moved for fine adjustments as little as 4 nm by computer-controlled motors. Keck I became operational in 1990 with only nine of its mirrors in place.

More recently, steps have been taken to reduce the blurring due to atmospheric turbulence. At least two approaches have been successful: sending the telescope into orbit above the atmosphere, as with the Hubble Space Telescope, and actually flexing or reshaping a mirror to compensate for atmospheric turbulence. This latter procedure, called adaptive optics, is especially effective in improving ground-based telescopes.

With adaptive optics an optical sensor detects the effects of atmospheric turbulence on light from a known "guide star." Information is sent to a computer that controls actuators that move a deformable mirror inserted in the optical train of the telescope. The computer calculations and the mirror deformations must be done quite rapidly (within several microseconds) or the atmo-

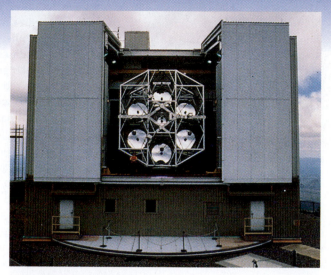

Figure B23.1 The Multiple Mirror Telescope broke with tradition by using six separate mirrors to form a single image.

sphere will change too much for the corrections to be useful. When a suitable guide star is not within the telescope's field of view, artificial guide stars are used. The light from a laser that is directed upward and focused on a point many kilometers above the earth's surface is back-scattered from the atmosphere and thus forms an artificial guide star for reference in correcting for the atmospheric turbulence. Figure B23.2 shows the improved resolution obtained using adaptive optics.

Keck II, an instrument similar to Keck I, will have adaptive optics. After Keck II's completion, Keck I will be retrofitted with adaptive optics. Then by using the two telescopes together, astronomers can examine the sky in even finer detail than with either telescope alone.

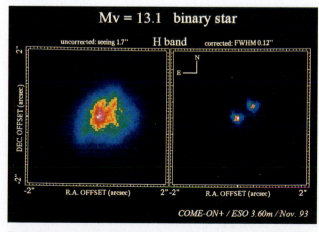

Figure B23.2 Images of a binary star without adaptive optics (left) and with adaptive optics (right) made with the 3.6-m telescope at the European Southern Observatory in La Silla, Chile.

Figure 23.13
A typical zoom lens for a 35-mm camera has 12 or more individual lenses arranged in four groups. The Canon 28-105 mm lens has 15 elements.

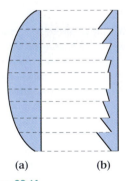

(a)   (b)

Figure 23.14
Cross sections of (a) an ordinary plano-convex lens and (b) the corresponding Fresnel lens. Note that the surface contours are the same for the corresponding rings.

Commercial zoom lenses are much more complicated than a simple combination of two thin lenses. A typical zoom lens has 12 or more separate glass lenses arranged in four groups (Fig. 23.13). Two or more of these groups move relative to the others in order to vary the effective focal length, all the while maintaining a constant image position so that focus is maintained. The numerous lenses in each group are needed to minimize aberrations and produce a lens that is not restricted to paraxial rays.

Because a zoom lens has so many elements, it is extremely important that each surface be coated with an antireflection layer. (See Section 24.4.) These coatings not only increase the amount of light passing through the system, but also increase the contrast and reduce the flare that results from multiple reflections from the surfaces. Some reduction in the number of elements is achieved by using aspheric lenses. In the future, the incorporation of gradient-index lenses may further reduce the number of elements, making zoom lenses lighter and cheaper.

Another special lens was developed by Augustin Jean Fresnel early in the nineteenth century to solve the problem of how to make large lenses for lighthouses. Fresnel recognized that refraction occurs only at the lens surfaces. By designing a lens that eliminated some of the material between the surfaces, he was able to produce a thin lens that was equivalent to a much thicker one. Thus, a **Fresnel lens** has concentric rings with the surface contour of the equivalent ring of a thick lens but with each successive ring stepped back to eliminate the unnecessary material between the front and back surfaces (Fig. 23.14).

Fresnel lenses do not produce high quality images, but they can be made very thin and light. Originally used in lighthouses (Fig. 23.15), they are now used in other ways, such as condenser lenses in overhead projectors, as light collectors for solar cells, and as flat pocket magnifiers (Fig. 23.16).

Figure 23.15
A Fresnel lighthouse lens.

One of the more recent advances in lens technology is the development of the **gradient-index lens** (GRIN), in which the index of refraction decreases as a function of the radius. Light rays incident near the edge of the lens are not refracted as much as those near the center of the lens. Consequently, a gradient-index lens can be used to correct for spherical aberration (Fig. 23.17). Lenses of this type may soon be used in cameras, but at present they are limited to small-diameter rods with parallel, flat faces. The rod axis is the optical axis of the lens. Because of their nonuniform index of refraction, they can be used for coupling light sources to optical fibers, even though both faces are flat.

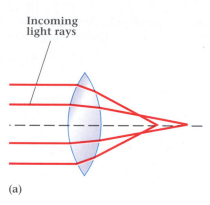

(a)

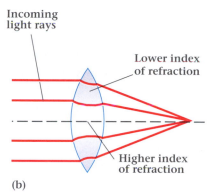

(b)

Figure 23.17
(a) An ordinary lens bends light more at the edges than near the center.
(b) A gradient-index lens brings all the light to the same focus.

Figure 23.16
Fresnel lenses are often used as magnifiers.

Summary

Useful Concepts

▪ The strength of a lens in diopters is the reciprocal of the focal length when the focal length is expressed in meters,

$$\mathcal{D} = \frac{1}{f}.$$

▪ A converging lens may be used as a magnifying glass. The important magnification is the angular magnification,

$$M \equiv \frac{\theta'}{\theta}.$$

▪ If the lens is held close to the eye, the angular magnification for the relaxed eye is

$$M = \frac{25 \text{ cm}}{f},$$

where 25 cm is used for the standard near point and corresponds to having the image at infinity. This equation is normally used to specify the magnification of a magnifying glass.

▪ The *f*-number of a lens is given by

$$f\text{-number} \equiv \frac{\text{focal length}}{\text{diameter}}.$$

■ The overall magnification of a compound microscope is the product of m_o, the linear magnification of the objective lens given by $m_o = -\dfrac{L}{f_o}$, and M_e, the angular magnification of the eyepiece.

■ The magnification of a telescope is the ratio of the focal length of the objective lens to the focal length of the eyepiece:

$$M = -\frac{f_o}{f_e}.$$

■ Zoom and varifocal lenses have multiple elements that move relative to one another to allow the effective focal length of the combination to be changed.

Important Terms

You should be able to write the definition or meaning of each of the following:

accommodation
near point
far point
astigmatism
diopter
magnifying glass
angular magnification
f-number
compound microscope
objective lens

eyepiece
ocular
refracting astronomical
    telescope
Galilean telescope
Newtonian telescope
zoom lens
varifocal lens
Fresnel lens
gradient-index lens

Conceptual Questions

23.1 Explain the operation of a simple periscope. What would happen to the image if the mirror at the top could be rotated around so that you were looking behind you instead of in front?

23.2 A magnifying glass held at arms length is about a meter away from a wall-mounted mirror. The magnifying glass is held so that you can look through it and see the reflection of your eye. Is the image of your eye erect or inverted? Is the image magnified or reduced? Explain. Test yourself by trying this demonstration for yourself.

23.3 Explain the optical function and practical use of bifocal glasses.

23.4 A nearsighted person wishes to buy both glasses and contact lenses. The dioptic power of the prescription contact lenses is different from the power of the lenses used in the glasses. Why?

23.5 Many simple cameras have fixed-focus lenses; that is, the lens does not move with respect to the film position. How is it possible to take pictures in which both distant objects and nearby objects seem to be in focus simultaneously?

23.6 You can make a pinhole camera by taking a light-tight box, making a small hole in the center of one face, and placing a sheet of film on the opposite face inside the box. A flap serves as a shutter. For such a camera, is the image erect or inverted? What is the magnification? Use diagrams to explain your answer.

23.7 Explain the operation of an overhead projector.

23.8 What are the advantages of constructing and operating large reflecting telescopes rather than large refracting telescopes?

23.9 What is the advantage of having large-diameter objective lenses on telescopes or binoculars?

23.10 Given an additional converging lens, how can you use it to modify an inverting telescope so that the final image will be erect? You need not keep the tube length the same.

23.11 Does the diagonal mirror placed on the central axis of a Newtonian telescope cause a dark place in the middle of the image (Fig. 23.11)? Explain your answer.

23.12 Archaeologists frequently make records of their excavations with photographs taken from a ladder or other high support directly above the site. The following rule is often used to estimate the support height and lens focal length needed: The ratio of the lens focal length to the linear size of the negative should equal the ratio of the camera height above the ground to the linear size of the excavation area. Show that this rule gives the optimum image for a lens of a given focal length.

23.13 What are some of the reasons your vision improves in bright light? What role does the narrowing of the pupils play?

23.14 Pick reasonable values for the dimensions of a 35-mm slide projector and determine appropriate values for the focal lengths and f-numbers of both the condenser and projection lenses. Assume that the projector is to be used at home, rather than in an auditorium. The dimensions of the image on the film are 24.5 mm × 36.3 mm.

Problems

Section 23.1 The Eye

23.1 What is the strength in diopters of a camera lens of 35-mm focal length?

23.2 An eyeglass lens has a strength of +5.50 diopters. What is the focal length of the lens?

23.3 What is the equivalent focal length of a +5-diopter lens and −2-diopter lens held together? Use the result of Example 22.8.

23.4 A farsighted eye has a near point of 150 cm. Objects closer than 150 cm are not seen clearly. A converging lens is used to permit clear vision of a book placed 25 cm in front of the eye. Find the focal length of the lens and express its strength in diopters.

23.5 A nearsighted person has a far point of one meter. Objects beyond one meter are not sharply focused. What lens should be used to obtain clear vision for objects at infinity? Express your answer in diopters.

23.6 A myopic person wears eyeglasses with a lens strength of −2.5 diopters. Where is the far point for that person's eyes?

23.7 A farsighted person wears eyeglasses with a lens strength of +1.5 diopters. Where is the near point for that person's eyes?

23.8• Show that to find the equivalent strength, in diopters, of two thin lenses held close together, you add their individual dioptic powers. (*Hint:* See Example 22.8.)

23.9• A person who has been wearing −2.75-diopter lenses is informed that her far point has moved inward by 20% of its former value. What is the correct strength of a new lens that will correct her vision?

23.10• The prescription for a student's eyeglasses is changed from 0.25 diopter to 1.25 diopters. By what distance had the student's near point shifted?

Section 23.2 The Magnifying Glass

23.11 What is the power in diopters of a 3× magnifying glass?

23.12 A magnifying glass enlarges an object by an angular factor of 4. What is its approximate focal length?

23.13 A dime (diameter about 1.8 cm) is viewed through a 5× magnifying glass. Approximately what angle does it appear to subtend at the eye?

23.14 By what constant do you multiply the strength of a lens in diopters to obtain its magnification?

23.15• From top to bottom, the letters on a coin subtend an angle of 2.5° when viewed through an 8× magnifying glass by a person with normal vision. What is the height of the letters on the coin?

23.16• An 8× magnifying glass is placed 4.0 cm away from a postage stamp. (a) Where is the image of the stamp? (b) Is it real or virtual?

23.17• A magnifier has two lenses of focal length 0.10 m and 0.16 m, which may be used singly or in combination. What are the possible magnifying powers? (*Hint:* See Example 22.8.)

23.18• Show that Eq. (23.4) gives the correct magnification when the image is at infinity.

23.19• Suppose you want to examine your own eye with a magnifying glass by holding it close to your eye and looking into a mirror. (a) If you use a 3× magnifier, how far is the lens from the mirror when the image is in focus for relaxed vision. (b) What is the angular magnification of the image?

23.20•• When expressions for the magnification of a simple magnifier, Eqs. (23.3) and (23.4), were derived the distance between the lens and the eye was neglected. Redraw Fig. 23.4 with a nonzero spacing d between the eye and lens and show that

$$M = \frac{25}{f} + \frac{25}{s} - \frac{25d}{sf},$$

where s is the distance of the magnified image from the eye and f is the focal length of the magnifier. (Distances are to be measured in centimeters.)

Section 23.3 Cameras and Projectors

23.21 A movie camera lens has a focal length of 17 cm and a diameter of 40 mm. What is its f-number?

23.22 You are taking photos with a shutter speed of 1/500 s and a lens aperture of $f/1.4$. If you then set the lens to $f/2.8$, what should be the shutter speed to give the same exposure to the film?

23.23 What is the focal length of a 3.2-cm-diameter $f/2.8$ lens?

23.24 The 50-mm focal-length lens of a video surveillance camera is rated $f/1.8$. It is normally operated at maximum aperture, that is, at $f/1.8$. In order to obtain better magnification, the old lens is replaced with a lens of 100-mm focal length. What is the minimum diameter of the new lens if the illumination is unchanged?

23.25 The Hale telescope on Mt. Palomar in California, which is 5.0 m (200 in.) in diameter, has a focal length of about 17 m. What is its *f*-number?

23.26 Each of two lenses is used to form the image of a distant window on a screen. One has a diameter of 4.0 cm and a focal length of 10 cm; the second has a diameter of 6.0 cm and a focal length of 24 cm. Which lens produces a brighter image of the object?

23.27 In darkness, the iris of your eye opens to a diameter of about 8 mm. It is known that in darkness the eye has an aperture of about *f*/2.8. What is the approximate equivalent focal length of a thin lens in air that has corresponding values?

23.28● The lens of a projection TV system has a maximum aperture of 5.8 cm and is rated *f*/4.3. In operation, the lens projects an image on a screen 4.5 m away. The diagonal size of the image is 150 cm. What is the approximate diagonal size of the object element in the projection system?

23.29● A camera has a lens with a focal length of 55 mm and an effective maximum lens diameter of 1.5 cm. A correct exposure of a particular scene is obtained for *f*/5.6 at 1/50 s. (a) Can the correct exposure still be obtained if the shutter is set at 1/500 s to avoid blurring? (b) If not, what maximum shutter speed could be used?

23.30● The Nikon N8008 camera has a shutter speed of 1/8000 s. (a) If the correct exposure is obtained at a full aperture of *f*/1.8 and a speed of 1/8000 s, what is the correct aperture (*f*-number) for an exposure of 1/500 s using the same lens? (b) How far will a car traveling 130 km/h go during a 1/8000-s exposure?

23.31● Given that the size of the negative in a 35-mm camera is 24.5 mm × 36.3 mm, determine the approximate angular field of view for lenses of 24-, 55-, and 250-mm focal length.

23.32● A "normal" lens for a particular camera is one with a focal length approximately equal to the diagonal dimension of the film used. Thus, for a 35-mm camera with a negative size of 24.5 mm × 36.3 mm, the "normal" lens has a focal length of about 43.8. (In reality most normal 35-mm camera lenses have focal lengths nearer 50 mm.) Show that for a camera with a "normal" lens, a given scene photographed from the same place will form an image that occupies the same relative proportion of the film regardless of the size of the camera. (*Hint:* Assume the object to be very far away.)

23.33● A special effects photograph is to be made by double exposure of the same piece of film. The correct normal exposure for the scene is 1/125 s at *f*/4. The first exposure is made of the background alone at 1/125 s. The second exposure, which includes a new object in the scene as well as the background, is to be made at 1/30 s. What should be the *f*-value for the aperture for each exposure if the background is to receive a total exposure that is correct. Assume the individual exposures contribute equally to the background exposure.

23.34● A 17.9-cm focal-length lens is used to project 35-mm slides onto a screen 4.27 m from the projector lens. What must be the size of the screen if the image of the slide is to just fill the screen? The dimensions of the image on the slide are 23.5 mm × 34.3 mm.

23.35● A camcorder with a 65-mm focal-length lens is used with a +3 diopter close-up lens attachment. (a) If the camera is set to focus at infinity without the attachment, what should be the distance between the camera and the subject when the close-up attachment is used? (b) What is the effective focal length of this lens combination?

23.36●● An object of height *h* is photographed with a camera a distance *D* away. The camera lens has a focal length f_c. The resulting transparent slide is projected onto a screen a distance *L* from the projection lens, which has a focal length of f_p. Show that the relationship between the height h_i measured on the screen and the actual height *h* is given by

$$\frac{h_i}{h} = \left(\frac{L - f_p}{D - f_c}\right)\left(\frac{f_c}{f_p}\right).$$

Section 23.4 Compound Microscopes

Hints for Solving Problems

Remember that we treat optical instruments as combinations of thin lenses, so that they can be understood by successive application of the thin-lens equation. In a compound microscope, the standard distance *L* between the focal point of the objective and that of the eyepiece is 16.0 cm.

23.37 The maximum useful magnification for ordinary visible-light microscopes is about 2000×. What should be the focal length of the objective used in combination with a 20× eyepiece to give this magnification in a standard microscope?

23.38 An eyepiece of 2.5-cm focal length and an objective of 0.32-cm focal length are used in a standard microscope. (a) What power should be marked on each element? (b) What is the overall magnification of the combination?

23.39 In a laboratory microscope, the first image of a specimen is formed inside the microscope 15 cm from the objective lens. If the specimen is 3.0 mm from the objective when the image is in focus, what is the focal length of the objective?

23.40● A 20× objective and a 5× eyepiece from a standard-length (16.0-cm) microscope are placed in a microscope with a 16.5-cm spacing. What is the ratio of the overall magnification of the combination to that of the standard microscope?

23.41● Given a 180.0-mm-long tube with a lens of 2.00-mm focal length at one end and a lens of 30.0-mm focal length at the other, where should an object be placed to use the tube and lenses as a microscope with the maximum magnification?

23.42●● A specimen is viewed in sharp focus with a standard microscope having a 10× eyepiece and a 20× objective. The final image position is at infinity. The specimen is then moved 0.010 mm toward the objective. How much will the eyepiece have to be moved to restore the focus of the image at infinity?

Section 23.5 Telescopes

23.43 An astronomical telescope is used to view the moon. If the objective has a focal length of 50 cm and the eyepiece a focal length of 3.5 cm, what is the angular magnification of the moon?

23.44 An astronomical telescope is designed with an overall magnification of 35×. (a) If the objective has a focal length of 100 cm, what should be the focal length of the eyepiece? (b) How far should the eyepiece be from the objective?

23.45 The distance between the objective and eyepiece lenses of an inverting telescope with a 5× eyepiece is 55 cm. What is the telescope's overall magnification? Assume a very distant object.

23.46 Binoculars denoted as 7 × 50 have a total magnification of 7 times and objective lenses of 50-mm diameter. What is the focal length of the eyepiece if the objective focal length is 21 cm?

23.47 A 1.90-m-tall football player is 50 m from you. If you look at him with a pair of 7× binoculars, what angle in radians does he subtend at your eye?

23.48 What is the magnifying power of a Galilean telescope with an objective of 24-cm focal length and an eyepiece of −4.0-cm focal length?

23.49• An upright meterstick 53.8 m away is viewed through an inverting telescope with a 5× eyepiece. The observed image subtends an angle of 0.151 radian. (a) What is the angular magnification of the telescope? (b) What is the focal length of the objective lens?

23.50• A Galilean telescope with two lenses spaced 30 cm apart has an objective of 50-cm focal length. (a) What is the focal length of the eyepiece? (b) What is the magnification of the telescope? Assume the object to be very far away. (c) What must be the separation between the two lenses when the subject being viewed is 30 m away? Assume the viewing is done with a relaxed eye.

23.51• Opera glasses are usually made from two Galilean telescopes side by side. One pair is made with objective lenses of 20-cm focal length and eyepieces of −5.0-cm focal length. What should be the separation between the two lenses in these glasses if you want to view a soprano 30 m away? Refer to Fig. 23.9(b).

23.52•• An inverting telescope with a 5× eyepiece is focused on a distant car. As the car approaches, the eyepiece is moved to keep the car in focus. When the car is 30 m away, the eyepiece has been moved 0.437 cm from its original position. (a) What is the focal length of the objective lens? (b) What is the overall magnification of the telescope?

*Section 23.6 Other Lenses

23.53 A simple varifocal lens is made from a converging lens with a 10-cm focal length followed by a diverging lens with a −10-cm focal length. (a) Calculate the distance between the diverging lens and the image of a very distant object when the lenses are separated by 5.0 cm. (b) Calculate the distance when the lenses are separated by 8.0 cm.

23.54 A varifocal lens is made from a converging lens with a 12-cm focal length followed by a diverging lens with a −10-cm focal length. (a) Calculate the distance between the diverging lens and the image of a very distant object when the

lenses are separated by 4.0 cm. (b) Calculate the distance when the lenses are separated by 8.0 cm.

Additional Problems

23.55 An advertisement in an astronomy magazine offers a 20-cm-diameter f/10 telescope mirror. What is the mirror's focal length?

23.56 How much more energy reaches the screen from a projector with an f/2.0 lens than from an identical projector with an f/3.5 lens of the same focal length?

23.57 One lens in a pair of eyeglasses is used to form the image of an overhead light fixture on a tabletop. The light fixture is 1.42 m above the table, and the eyeglasses are 0.16 m above the table. What is the strength of the lens in diopters?

23.58 A magnifying lens focuses the sun's rays to a point 15 cm away from the lens. What power magnifying glass will this lens make?

23.59 What is the diameter of a 2.3-diopter lens with an f-number of f/5.6?

23.60 A new microscope comes with 5× and 10× eyepieces and a revolving "nosepiece" that contains 10×, 20× and 40× objectives. (a) List all the overall magnifications. (b) List all the magnifications that would be possible if the 5× eyepiece were replaced by a 15× eyepiece.

23.61• An ordinary camera has a lens that can be moved closer or farther from the film plane. The minimum and maximum lens to film distances are 6.0 cm and 10.0 cm. (a) If the camera is in focus for objects at infinity when the lens is in the 6.0 cm position, what is the closest camera-to-subject distances for the image to be in focus? (b) What is the magnification at the distance found in (a)?

23.62• A slide projector with a 12.7-cm focal-length lens is focused on a screen that is 25.5 m from the projector. The screen is then brought to a distance of 1.50 m from the lens. How far and in what direction must the projector lens be moved in order to refocus the image?

23.63• On some lenses, the widest aperture does not correspond to one of the standard full-stop f-numbers. A half-stop is an f-number corresponding to a lens opening with an area half way between the full-stop values given in Table 23.1. What is the f-number of the half-stop between f/1.4 and f/2.0?

23.64• A magnifying glass held 13 cm in front of a television screen projects an image on a wall 3.37 m from the television screen. What magnifying power is marked on the lens?

23.65• Use the definition of angular magnification and simple geometry to derive Eq. (23.8) for the magnification of a refracting astronomical telescope.

23.66• A 30-cm focal-length lens is placed over a rectangular 1.5 cm × 1.0 cm opening in the center of one side of a closed cubical box. A shutter and film are placed so that the box serves as a camera. For the purpose of exposing the film, what is the equivalent f-number of the lens in the rectangular aperture?

23.67• An 8× astronomical telescope has a 30-cm focal-length objective lens. After looking at stars, an astronomer moves the eyepiece 1.0 cm farther away from the objective to focus on nearer objects. What is the distance to the nearer objects?

23.68•• A close-up lens is mounted onto a camera lens to allow focusing on closer-than-normal objects. A 55-mm focal-length lens on a particular 35-mm camera can be focused from infinity to 1.0 m from the lens. If it is desirable to have the nearest focus only 15 cm from the lens, approximately what should be the strength of the close-up attachment lens in diopters?

23.69•• A camera with a 55-mm focal-length lens cannot be focused on objects closer than 0.54 m. When a close-up lens is mounted onto the camera, the camera can be moved to 0.27 m from the object. (a) What is the focal length of the close-up lens? (b) What is the power of the close-up lens in diopters? (c) What is the maximum camera to object distance when the close-up lens is in place?

23.70•• The numerical aperture, N.A., of a microscope objective is a measure of its light-gathering power. In air, the N.A. is defined to be the sine of half the apex angle α of the cone of light received by the lens (Fig. 23.18). A large N.A. corresponds to a large light-gathering power. (a) What is the N.A. of a 0.92-cm-diameter objective with a working distance s of 1.79 cm? (b) What is the approximate f-number of the

lens? (c) Show that the light energy entering the lens is proportional to the square of the N.A.

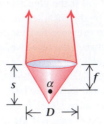

Figure 23.18
Problem 23.70.

23.71•• A telephoto lens is made with a front element of positive focal length f, followed by a second element with a negative focal length $-f$. The spacing between the lenses is $0.5f$. (a) Show that the image of a distant object comes to focus at a distance $1.5f$ from the front element. (b) Show that the image is the same size as one produced by a single element of focal length $2f$. (c) What is the ratio of the equivalent focal length to the distance from the front element to the film?

Wave Optics

I n the previous two chapters we saw that geometrical optics provides a lot of practical and useful information. The assumption that light travels in a straight line except at the interface between two media satisfactorily explains most of what we see. It is sufficient to allow us to design and understand the majority of optical instruments, including cameras, projectors, telescopes, and microscopes.

However, in considering geometrical optics, we did not need to inquire into the nature of light itself. In this chapter, we see that when the optical components, such as apertures and lenses, become sufficiently small, wavelike properties of light become more noticeable. Additional evidence for the wave nature of light comes from such apparently diverse phenomena as the behavior of polarized sunglasses, the blue of the sky, and the red of the setting sun.

On the other hand, we will see later, in Chapter 27, that situations occur in which light clearly displays particlelike properties. As we have said before, this apparent duality between the wave and particle nature of light cannot be explained in terms of the classical mechanics and electromagnetism discussed so far in this text. A different type of theory, called quantum mechanics, is needed. It is especially useful in describing phenomena on the molecular, atomic, and smaller scales, as mentioned earlier in Section 4.10 and shown in Fig. 4.29.

As we will see later, in Chapter 30, wave optics and quantum mechanics play key roles in the operation of lasers. Furthermore, these ideas are central to the development of recent cosmological theories on the creation of the universe and its possible future. ■

24.1 Huygens' Principle

The first thing necessary to discuss a wave theory of light is a technique for describing wave motion. A simple technique was developed by the Dutch scientist Christiaan Huygens to explain reflection and refraction. We can understand his principle in terms of simple water waves. Suppose we toss a rock into a pond. When the rock hits the water, it generates circular waves that spread out from the point of impact (Fig. 24.1). As we watch the wavefront expand, we are actually following a series of points of constant phase. Every point along the wave crest all around the circle has the same phase, and as the wave expands, this circle of constant phase expands. If the wave encounters a barrier, the wave is reflected; but if the barrier has an opening, part of the wave passes through.

Huygens recognized that it is possible to determine how the wave crest advances by considering each point along the wave crest to be a source point for tiny, expanding, circular wavelets, which expand with the speed of the wave. The contour of the advancing wave is the envelope tangent to these wavelets. In turn, this envelope generates source points for determining a later position of the wave. Figure 24.2 illustrates Huygens' constructions for describing circular waves and plane waves. The radius of each wavelet, equal to the distance between successive wavefronts, is taken to be one wavelength. This technique for describing the motion of waves, called **Huygens' principle,** is valid for all types of waves.

Figure 24.1
Circular ripples spread from a disturbance at the surface of a pond.

Figure 24.2
Huygens' construction of (a) a circular wave and (b) a plane wave. The radius of each wavelet, equal to the distance between successive wavefronts, is usually taken to be one wavelength.

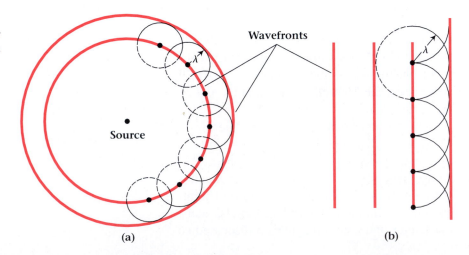

Wavefronts

Source

(a)

(b)

We can extend Huygens' principle beyond two-dimensional waves to include three-dimensional waves, such as sound and light. The points of constant phase define a surface called the **wave surface** or **wavefront.** We can determine the shape of the wave surface as the wave advances by applying Huygens' principle as just described. Because the wavefront at any point is perpendicular to the direction in which the wave is advancing, the wavefront is perpendicular to the ray that describes the path of the light.

24.2   Reflection and Refraction of Light Waves

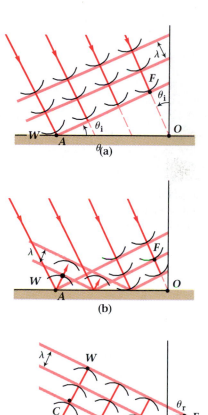

We can use Huygens' principle to derive the laws of reflection and refraction. Recall that the law of reflection states that the angle of incidence and the angle of reflection are equal. This rule comes straight from everyday observations, as we discussed in Chapter 22. However, we can derive it geometrically with Huygens' principle. Imagine a beam of light incident at an angle θ_i on a smooth surface (Fig. 24.3a). The wavefront WF is perpendicular to the path of the beam. The angle OAF is equal to θ_i, since OA is perpendicular to the normal and AF is perpendicular to the incident ray.

According to Huygens' principle, we can locate the next successive wavefront by considering the expansion of spherical wavelets from the wavefront WF. In Fig. 24.3, wavelets along WF strike the surface and reflect upward. The wavelet from W reaches the surface first. As it reflects up by one wavelength, the wavelet from F extends one wavelength toward the surface (Fig. 24.3b). Then, one by one, other wavelets along WF strike the surface and reflect up. Now, the right triangle AOC in Fig. 24.3(c) is congruent to the right triangle OAF of Fig. 24.3(a) because first, they share the common side AO, and second, $AC = FO$, since the wave speed remains the same. This means that the angle AOC is the same as OAF (part a). Hence, the *angle of reflection* θ_r *is identical to the angle of incidence* θ_i. This statement is the law of reflection given earlier in Chapter 22:

$$\theta_i = \theta_r. \qquad (24.1)$$

We can also derive the law of refraction from Huygens' principle. In Chapter 22, we discussed Snell's law, the relationship between the indices of refraction and the angles of a light ray as it passes from a medium of one index to that of another. Let's now look at a wave derivation of Snell's law. Consider a Huygens wavefront incident upon an interface between medium 1 (say air) with index of refraction n_1 and medium 2 (perhaps glass) with a larger index of refraction n_2. The resulting behavior is like that shown in Fig. 24.4. The wave speed in air is v_i (incident light), and the wave speed in the glass is v_t (transmitted light). Since we assumed that n_2 is greater than n_1, v_i is greater than v_t. As the wave enters into medium 2 at point A, its speed is reduced, and thus its wavelength is shortened (Fig. 24.4b). The wavefront transmitted in the lower medium consequently travels at a different angle with respect to the normal.

We observe that the two triangles AOF (Fig. 24.4a) and AOC (Fig. 24.4d) have one side (AO) in common. Further, we see that

$$\sin \theta_i = \frac{OF}{AO} \quad \text{and} \quad \sin \theta_t = \frac{AC}{AO}.$$

Figure 24.3
Reflection of a plane wave by a smooth (mirror) surface as described with Huygens' waves. Illustrations (a), (b), and (c) show the progress of the wavefronts with time.

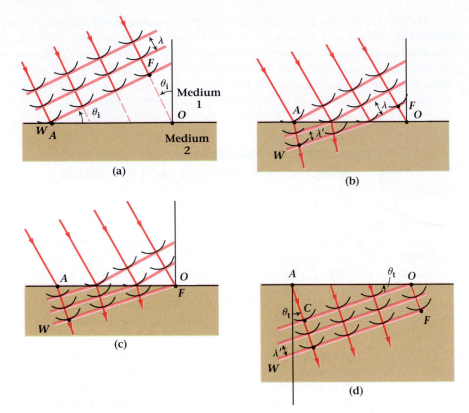

Figure 24.4
Refraction of a plane wave at a flat interface between two transparent media, with n_2 greater than n_1. Illustrations (a), (b), (c), and (d) show the progress of the wavefronts with time.

Upon dividing, we find

$$\frac{\sin \theta_i}{\sin \theta_t} = \frac{OF}{AC}.$$

However, OF and AC are proportional to the speeds v_i and v_t of the wave. In a time t while the incident wave moves a distance $v_i t = OF$ (Fig. 24.4a), the transmitted wave moves a distance $v_t t = AC$ (Fig. 24.4d). So we obtain

$$\frac{\sin \theta_i}{\sin \theta_t} = \frac{v_i t}{v_t t} = \frac{v_i}{v_n}.$$

Then using the definition of the index of refraction, we get

$$\frac{\sin \theta_i}{\sin \theta_t} = \frac{c/n_i}{c/n_t} = \frac{n_t}{n_i}.$$

This equation is the law of refraction given in Chapter 22 as

$$n_i \sin \theta_i = n_t \sin \theta_t. \tag{24.2}$$

Thus, we have shown here that the observed laws of geometrical optics—the law of reflection and Snell's law—follow naturally from the assumption that light is a wave.

We have already seen that the velocity of light in a medium is not the same as the free-space velocity c, but is given by $v = c/n$. When a light wave passes from one material into another, the frequency remains constant.* The wavelength inside a material of index of refraction n is smaller than the free-space wavelength because, as we saw in Chapter 15, the velocity of a wave is the product of its wavelength and frequency, $v = f\lambda$. Thus if the velocity changes, the wavelength changes proportionately. The new wavelength inside the material of index n is

$$\lambda' = \lambda/n. \tag{24.3}$$

We will see the effects of the wavelength λ' when we consider interference in thin films.

24.3 Interference of Light

In 1807, Thomas Young (1773–1829) published his *Lectures on Natural Philosophy,* containing the description of an optical experiment now referred to as Young's double-slit experiment. This demonstration of interference effects firmly established the wave theory of light on experimental grounds and provided a straightforward means for measuring the wavelengths.

To understand why Thomas Young's double-slit experiment was crucial to a wave theory of light, we first need to examine the interference effects of two in-phase wave sources. Then we will show that these same effects were produced by Young's experimental setup, implying that light is a wave. Finally, we will follow Young's calculations in determining the wavelength of visible light.

In Chapter 15, we observed that a wave disturbance due to two or more sources can usually be taken as the algebraic sum of the individual waves. If the individual sources vibrate with different frequencies, there is nothing special about the resulting disturbance. But if two or more sources vibrate at the same frequency and with a constant relative phase, interesting interference effects occur.

Consider a little bob vibrating up and down on the surface of a body of water. Its motion causes circular water waves to spread out from the bob (Fig. 24.5).

---

*In this respect, light waves are like sound waves. Remember, from Section 15.9, that the frequency of sound waves from a guitar or violin is determined by the vibrational frequency of the string. As the waves pass into the air, their frequency does not change, even though the speed of sound waves in air is different from the speed of waves in the string.

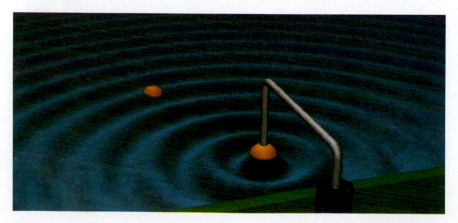

Figure 24.5
A computer-generated image of circular waves spreading out from the point of contact of an oscillating bob. A small yellow sphere is seen floating on the surface.

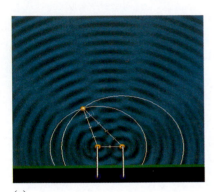

Figure 24.6
A pattern of constructive and destructive interference, produced by two in-phase sets of circular water waves.

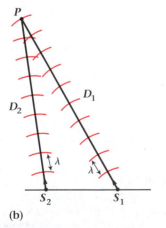

(a)

(b)

Figure 24.7
(a) Waves from the two sources reach the floating yellow sphere in phase when their path lengths differ by an integer multiple of the wavelength.
(b) Waves from sources S_1 and S_2 reach point P in phase when $D_1 - D_2 = m\lambda$.

Now if two bobs are made to vibrate with the same frequency, each of them causes circular water waves to spread out from the point of contact. The waves from these two sources interfere with one another. In some directions, the waves combine constructively, making waves of larger amplitude. In other directions, they combine destructively, so that there is little or no wave amplitude. A pattern develops, as seen in Fig. 24.6. The wedge-shaped areas of sharp contrast indicate the crests and troughs of strongly reinforced waves. However, in some radial directions, waves from the two sources arrive exactly out of phase. Since the resulting amplitude is zero, we see no contrasting lines representing the crests and troughs; instead we see a smooth region indicating no wave motion.

The condition for **constructive interference** is that waves from two sources with the same frequency arrive at the same point together with the same phase. The result is an amplitude that is greater than the amplitude of either wave alone. We can see from Fig. 24.7(a) that the resultant wave at the position of the floating sphere has maximum amplitude because the two contributing wave crests (and subsequently the troughs) arrive at the same time. If the two sources oscillate with exactly the same phase, then the condition for maximum constructive interference is that the path lengths of the two waves must be identical or else differ by an integer multiple of the wavelength (Fig. 24.7b); that is,

$$D_1 - D_2 = \Delta D = m\lambda, \qquad m = 0, 1, 2, 3, \ldots \qquad \text{(constructive interference)},$$

where D_1 and D_2 are the path lengths of the waves from their source to the point P.

When the two waves arrive exactly out of phase, **destructive interference** occurs and the resulting wave is diminished.* If the two sources have the same amplitude, destructive interference results in zero net amplitude. This occurs when the path lengths differ by half a wavelength or by any odd half-integer mul-

---

*The conditions for constructive and destructive interference depend on the relative phase of the waves. They apply, in general, to all types of waves, no matter what the source of the phase difference.

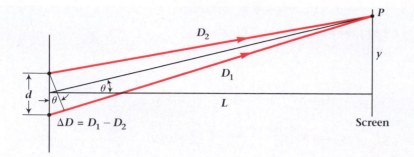

Figure 24.8
Geometrical construction for describing the interference pattern of two sources when the path length is much greater than the distance between the sources.

tiple of a wavelength; that is,

$$\Delta D = (m + \tfrac{1}{2})\lambda, \qquad m = 0, 1, 2, 3, \ldots \qquad \text{(destructive interference)}.$$

If the distances D_1 and D_2 are quite large compared with the separation d between the sources, we can express the conditions for constructive and destructive interference in terms of the angle θ shown in Fig. 24.8. When the distance L between the sources and the plane containing the observation point is much greater than d, the two paths D_1 and D_2 are nearly parallel. We can approximate the difference between them by $\Delta D \cong d \sin \theta$. This result leads to two new equations for determining the maxima and minima of the resultant wave at P:

$$m\lambda = d \sin \theta, \qquad m = 0, 1, 2, 3, \ldots \qquad \text{(maxima)}, \qquad (24.4a)$$

$$(m + \tfrac{1}{2})\lambda = d \sin \theta, \qquad m = 0, 1, 2, 3, \ldots \qquad \text{(minima)}. \qquad (24.4b)$$

Equation (24.4a,b) enable us to relate the directions of the maxima and minima with the wavelength and the source separation d. These equations are true for all waves, not just water waves. In particular, they accurately describe the interference of two sound waves and, as we shall see, two light waves.

Notice that when the separation d gets larger in Eqs. (24.4a,b), the angles θ corresponding to the respective maxima and minima get smaller, as seen in Fig. 24.9(a). Similarly, when the separation between the wave sources gets smaller, the angles of the corresponding maxima and minima increase (Fig. 24.9b).

Figure 24.9
The angle between successive maxima (a) decreases when the separation between the bobs is increased and (b) increases when the separation is decreased.

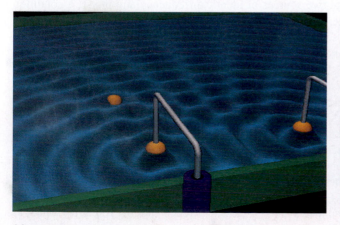

(a)

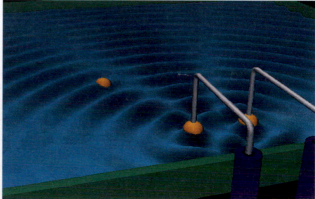

(b)

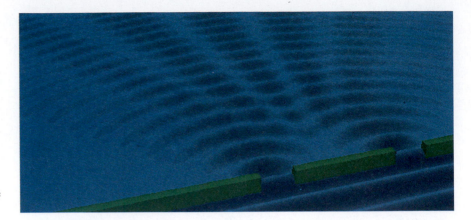

Figure 24.10
Interference pattern of water waves caused when a plane wave (bottom) passes through a pair of slits. Note the similarity with Fig. 24.6.

In Fig. 24.8, the maximum (or minimum) occurs at a point P above the center line. When D_2 is greater than D_1, the maximum (or minimum) lies below the center line. Thus, the pattern of maxima and minima is symmetric about the center line. (See Fig. 24.6.) A central maximum occurs along the center line for $m = 0$ in Eq. (24.4a). For this reason, it is also referred to as the zero-order maximum. The next maxima on either side of the central beam are called the first-order maxima and correspond to $m = \pm 1$. Similarly, the other peaks are labeled by their order. For example, the second-order maximum is the second maximum away from the center line and corresponds to $m = \pm 2$ in Eq. (24.4a). Similarly, the minima given by Eq. (24.4b) are labeled in order of their position from the central maximum. A first minimum occurs to either side of the central maximum. A second minimum occurs beyond the first, and so on. Note, however, that the first minimum corresponds to $m = 0$ in Eq. (24.4b).

The interference pattern just described also arises when a single wave strikes a barrier pierced by two narrow slits. The wave passes through the slits, which act as if they were new sources of waves (Fig. 24.10). The waves emerging from the two slits must have the same frequency because they were generated from the same initial wave. Because the incident wave crests strike both slits at the same time, the waves passing through the two slits also have the same phase. The slits thus act like two sources of identical frequency and phase, a condition known as **coherence.** The emerging waves produce an interference pattern exactly like the one described for two sources. This type of interference occurs for all kinds of waves and is known as double-slit interference.

Thomas Young's experimental setup (Fig. 24.11a) allowed sunlight emerging from a small aperture to strike two very small slits. When light from the two slits fell upon a screen, dark stripes appeared, dividing the area into regularly spaced light and dark portions. Figure 24.11(b) shows the appearance of a typical double-slit pattern, made with the red light of a helium-neon laser.

Young recognized that for interference to occur, the light falling on the two slits must be coherent. The purpose of the first aperture was to ensure that the only light striking the two narrow slits came from the same source and was thus coherent. Figure 24.12 shows the wave diagram drawn by Young to explain the origin of the light and dark bands of the interference pattern. Young reasoned, as we did above, that the maxima (bright bands) occurred when the path lengths

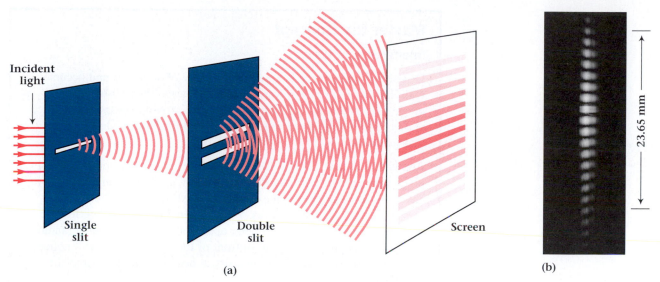

(a)

(b)

Figure 24.11

(a) The arrangement for Young's double-slit experiment. Sunlight passing through the first slit is coherent and falls on two slits close to each other. Light passing beyond the two slits produces an interference pattern on a screen. (b) The double-slit pattern of a small-diameter beam of light from a helium-neon laser. The center-to-center separation of the two slits was 0.113 mm. The film was located 29.1 cm from the slits. The reference marks seen along the edge of the film were spaced 23.65 mm apart. From these data, try to compute the wavelength of the laser light.

from the two slits to the screen differed by integral multiples of the wavelength, and that the minima (dark bands) occurred when the paths differed by an odd number of half-wavelengths. The situation is exactly the same as that described in Fig. 24.8.

Once Young had worked out this explanation of the double-slit interference pattern, he realized that by measuring the spacing between the slits, the distance L to the screen, and the positions y of the maxima and minima, he could determine the wavelength of light. Young used an analysis similar to our use of Eq. (24.4) to find that the wavelengths of light range from about 400 nm in the extreme violet to about 700 nm in the extreme red. (Young's actual values were given as one 60-thousandth of an inch to one 36-thousandth of an inch.)

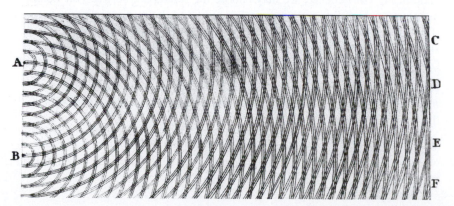

Figure 24.12

Diagram of the interference of light waves emerging from two points A and B as drawn by Thomas Young. Points C, D, E, and F indicate regions of destructive interference.

Master the Concept

Interference of Waves

Question: Interference is a general property of all waves and occurs with light waves, water waves, and sound waves. When listening to music on a stereo system with two speakers, why do you not experience places in the room where you hear no sound (because of destructive interference)?

Answer: You have already seen interference in standing sound waves in pipes, although the term *interference* was not used (Section 15.10). The equations for interference due to two in-phase sources of the same frequency hold for sound waves. You can demonstrate this interference with two speakers held a half-meter apart and a common source (audio oscillator). As you move about in front of the speakers, you will notice strong variations in the sound intensity in accord with the double-slit equations.

The music that you hear from the stereo contains many wavelengths, not just one. Thus, a location for destructive interference of one wavelength will not be destructive for other wavelengths. This observation is also true for points of constructive interference. Even though you can detect interference effects with two speakers and a single frequency, you do not normally observe them when listening to music. In addition, waves reflected from the walls and scattered by objects within the room contribute to what you hear.

Example 24.1

A narrow beam of coherent light from a laser passes through a pair of narrow slits and strikes a screen placed 1.00 m beyond the slits. The slits are spaced 0.050 mm apart. The resulting interference pattern on the screen has maxima that are 12.7 mm apart. What is the wavelength of the laser light?

Strategy We may use Eq. (24.4a) to find the wavelength, once we know the value for m. The maxima are evenly spaced near the center of the double-slit pattern. For convenience, we may choose the separation between the central maximum, which lies on the symmetry axis, and the maximum nearest it (first-order maximum) as the distance y in Fig. 24.8. This distance corresponds to a path difference of one wavelength—that is, to $m = 1$ in Eq. (24.4a).

Solution The angle θ is so small that we may approximate $\sin \theta$ as $\sin \theta \approx \tan \theta \approx y/L$. The condition for an interference maximum, Eq. (24.4a), then becomes

$$\lambda = \frac{dy}{L},$$

where we have substituted y/L for $\sin\theta$ and set $m = 1$. Inserting the numerical values, we compute the wavelength:

$$\lambda = \frac{(5.0 \times 10^{-5}\text{ m})(1.27 \times 10^{-2}\text{ m})}{1.00\text{ m}}$$

$$\lambda = 6.4 \times 10^{-7}\text{ m} = 640 \times 10^{-9}\text{ m} = 640\text{ nm}.$$

24.4 Interference in Thin Films

We can observe the interference of light in ways other than the double-slit experiment. For example, interference also occurs when light is reflected from or transmitted by a thin film of transparent material. This interference is responsible for the colors of soap bubbles and oil slicks. As we will show, it is also the basis for the nonreflecting coatings commonly used on binoculars and photographic lenses.

(i) A Thin Film of Index n Surrounded by a Medium of Lower Index

Figure 24.13 shows a diagram of monochromatic light incident from above on a thin transparent dielectric film of thickness t and index of refraction n that is greater than that of the surrounding medium. At the first interface, the light is partly transmitted (refracted) and partly reflected. At the second interface, a portion of the transmitted light is reflected and follows the path indicated in the figure. Thus, an incident light beam produces two coherent reflecting beams, one from each surface of the thin film.

If light strikes the film at nearly normal incidence, the two reflected beams may interfere constructively or destructively, depending on whether they are in phase or out of phase. The path length of the beam reflected from the second surface is $2t$ greater than that of the beam reflected from the first surface. If there are no other effects, the beams will interfere constructively when this path difference equals a whole number of wavelengths, for reasons similar to those discussed in connection with the double-slit experiment. However, there is another effect at work here. When light of wavelength λ is transmitted from a medium of lesser index of refraction to one of greater index, a phase change of 180°, corresponding to $\frac{1}{2}\lambda$, takes place upon reflection. The situation is analogous to the case of pulses reflecting back from the boundary between two ropes of different density (Fig. 24.14). A pulse from a lighter rope to a heavier rope reverses its phase upon reflection, whereas a pulse from a heavier rope to a lighter one does not change phase on reflection. Similarly, light going from a medium of smaller index of refraction (air) to one of larger index (oil or water) changes phase upon reflection, while light going in the opposite direction does not change phase.

Applying these ideas of phase changes to our thin film, we see that the light ray reflected from the top surface of the film undergoes a phase change of 180°. The transmitted ray does not experience any phase change during refraction at the upper surface of the film, nor does it undergo any phase change as it reflects from the bottom surface of the film. There is, however, a phase change associated with the path traveled through the film. The total effect taking place here,

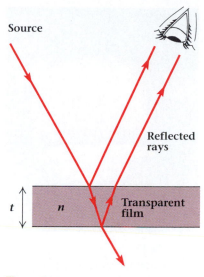

Figure 24.13
Monochromatic light incident on a thin transparent film is reflected from both the top and bottom surfaces.

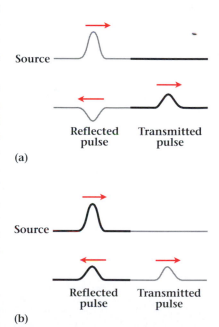

Figure 24.14
Reflection of a pulse wave at the boundary between ropes of different linear density. (a) Pulse incident in the less-dense rope changes phase upon reflection. (b) Pulse incident in the denser rope does not change phase.

Figure 24.15
Interference pattern of a soap film suspended in a loop of wire.

combining the phase change upon reflection with the path difference through the film, introduces an extra difference of $\lambda/2$ into our previous conditions for constructive and destructive interference.

Inside the film, where the index of refraction is n, the wavelength λ' is smaller than the incident wavelength λ by a factor of $1/n$. For constructive interference of light at normal incidence reflected from a thin film, the path length in the film must be an odd half-integer multiple of the wavelength λ'. Thus, the condition for constructive interference is

$$(m + \tfrac{1}{2})\lambda = 2nt, \qquad m = 0, 1, 2, 3, \ldots \qquad \text{(maxima).} \qquad (24.5a)$$

Similarly, minima in the reflected intensity occur for

$$m\lambda = 2nt, \qquad m = 0, 1, 2, 3, \ldots \qquad \text{(minima).} \qquad (24.5b)$$

An interesting case occurs when the thickness of the film changes along its length, giving rise to alternating regions of constructive and destructive interference. For example, for a soap film suspended in a loop (Fig. 24.15), the upper portion of the film is thinner than the lower portion. When the light striking the soap film of varying thickness is white, the various wavelengths of light constructively interfere at different places in the film, leading to a separation of the colors of white light. This thin-film phenomenon is also responsible for the rainbow colors visible on oil slicks. Some fish have scales with thin-film coatings that produce colors in the same way.

Problem-Solving Strategy

Reflection from a Thin Film Surrounded by a Medium of Different Index

There will be constructive interference on reflection when the optical path length of light in the film is an odd number of half wavelengths. The optical path is twice the thickness of the film times the index of refraction. The interference will be destructive when the optical path is an integer number of wavelengths. These same results hold regardless of whether the film index is larger or smaller than that of the surrounding medium.

Example 24.2

A thin soap film with an index of refraction of 1.4 is made by dipping a wire loop into a solution of soapy water. When viewed in sunlight, a large portion of the film reflects green light. Estimate the minimum thickness of that portion of the film.

Strategy For a thin film in air, the reflected intensity is greatest when the light reflected from the front surface is out of phase with the light from the back surface. This condition is expressed mathematically by Eq. (24.5a). Next we need to estimate the wavelength of the light. Since the reflected color looks green, the wavelength for constructive interference lies somewhere in the green part of the spectrum. (See Fig. 22.2.) We can estimate the wavelength to be approximately 540 nm and use this number to compute the approximate thickness of the film.

Solution   The film thickness is given by Eq. (24.5a) to be

$$t = \frac{(m + \frac{1}{2})\lambda}{2n}.$$

The minimum thickness corresponds to $m = 0$. Substituting in the numerical values gives

$$t = \frac{\frac{1}{2}(540 \text{ nm})}{2(1.4)} = 96 \text{ nm}.$$

(ii) A Thin Film of Index n Surrounded by a Medium of Higher Index

The same considerations of interference also hold for light reflected from a thin film of air (of thickness t) between two media of higher index, such as two glass plates (Fig. 24.16). Light reflected from the first air-glass interface does not have a change in phase (higher refractive index to lower). However, light reflected from the second air-glass interface does undergo a phase change (lower refractive index to higher). The resulting equations are the same as Eqs. (24.5a,b), where n is the index of refraction of air.

We can see this effect when a piece of glass with a slightly curved bottom (such as a lens with a large radius of curvature) is placed on a flat glass plate and illuminated from overhead by a point source of light. When you look from above, you see a series of concentric rings. These rings, known as Newton's rings, arise from the interference between light reflected at the curved surface and light reflected from the underlying flat surface. Their appearance can be used to judge the flatness of the plate or the sphericity of the lens surface. If the source is not monochromatic but provides white light instead, the rings will be colored.

(iii) A Thin Film of Index n Surrounded by a Medium of Lower Index on One Side and a Medium of Higher Index on the Other Side

Interference of reflected light occurs in many thin-film situations involving different indices of refraction. For example, a thin film of oil ($n = 1.36$) on a glass plate ($n = 1.55$) involves two phase changes, one for reflection from each interface. The conditions for constructive and destructive interference are the same as Eqs. (24.5), only now the locations of maxima and minima are reversed. Rather than try to remember which combination of equations fits which thin-film situation, simply remember to count up the phase changes that occur for light reflected from each interface between media of different refractive indices. Then write down the conditions for constructive interference (path difference equals integer number of wavelengths) and destructive interference (path difference equals odd half-integer number of wavelengths) and add a factor of one half-wavelength for each difference in phase change between reflected rays.

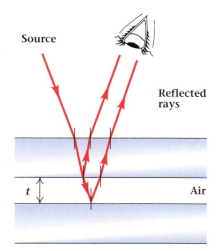

Figure 24.16
Reflection of light by a thin film of air between two glass plates.

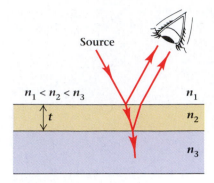

Figure 24.17
Reflection of light from a thin transparent coating on a material of higher index of refraction.

One of the most important applications of interference in thin films occurs when we place a thin film in contact with a third medium of still-greater index of refraction (Fig. 24.17). As we have already explained, a phase change occurs upon reflection at both surfaces, since both surfaces are low-index to high-index boundaries. As a result, the condition for constructive interference in the reflected beams is

$$m\lambda = 2n_2 t, \qquad m = 1, 2, 3, \ldots \quad \text{(maxima)}, \qquad (24.6a)$$

where n_2 is the index of refraction of the thin film.

When monochromatic light strikes the surface at normal incidence, a phase difference occurs between the two reflected beams as a result of the difference in their paths. In this case, destructive interference occurs for

$$(m + \tfrac{1}{2})\lambda = 2n_2 t, \qquad m = 0, 1, 2, 3, \ldots \quad \text{(minima)}. \qquad (24.6b)$$

When the thickness of the film is $\lambda/4n_2$, corresponding to $m = 0$, there is no reflection at that wavelength. So, if we coat a piece of glass, such as a lens, with a thin film that has the right thickness and an index of refraction intermediate between those of air and glass, we can minimize reflection from the glass. Such a film is called an **antireflection coating.** Because reflections are reduced, more light is transmitted through lenses that have antireflection coatings.

Problem-Solving Strategy

Reflection from a Thin Film Surrounded by a Medium of Lower Index on One Side and a Medium of Higher Index on the Other

There will be constructive interference on reflection when the optical path length of light in the film is an integer number of wavelengths. The interference on reflection will be destructive when the optical path is an odd number of half-wavelengths. Consequently, an antireflection coating has a thickness that is an odd multiple of $\lambda/4n_2$.

Example 24.3

For a photographic lens, we want to design a simple antireflection coating that will be most effective for green light of $\lambda = 550$ nm. The coating material is magnesium fluoride, which has an index of refraction of 1.38. How thick should the coating be?

Solution We want maximum transmission and minimum reflection of the green light. Thus, we want a film of the correct thickness to create maximum destructive interference for reflected light. This condition is given above in Eq. (24.6b). The minimum thickness of coating that will accomplish the purpose corresponds to $m = 0$, so we get

$$\tfrac{1}{2}\lambda = 2nt, \quad \text{or} \quad t = \frac{\lambda}{4n}.$$

(a)

(b)

Figure 24.18
(a) Plane waves incident on a wide single slit pass through with little spreading into the geometric shadow. (b) When the slit width approaches the wavelength, the waves spread into a diverging beam that expands into the region of the geometric shadow.

Substituting the numbers into the expression gives

$$t = \frac{550 \times 10^{-9} \text{ m}}{4(1.38)} = 99.6 \text{ nm.}$$

Discussion The film thickness chosen here, one-fourth of the wavelength of light in the coating, will provide good antireflection behavior for the green light specified. However, this coating will not be as effective at either end of the visible spectrum, since the wavelengths there are quite different. Better antireflection coatings, effective over the whole visible spectrum, are built up from several layers of thin films of alternating low and high indices of refraction.

24.5 Diffraction by a Single Slit

We are all familiar with shadows thrown on a wall by an object, such as a hand, that blocks part of a beam of light, and we know that the shadow has approximately the same geometric shape as the object. Young's double-slit experiment shows that light does not travel past objects in simple straight lines, but instead spreads out in wavefronts that can interfere with each other. This spreading out of light passing through a small aperture or around a sharp edge is called **diffraction.** Diffractive spreading is exactly what we would expect from Huygens' principle and, as we saw, was used by Young to illuminate his two slits.

Light always spreads out as it travels, but diffractive effects become noticeable only when light travels through a small enough aperture or past a sharp edge. Figure 24.18 shows a plane wave passing through (a) a wide slit and (b) a narrower slit. You can see that once the slit width approaches the dimension of the wavelength, the waves spread out as if from a point source. Figure 24.19 shows another example: the shadow of a ball bearing. Since the ball bearing has circular symmetry, light diffracted around its edge interferes constructively at the very center of the shadow. This bright spot in the center of a shadow was first predicted in 1819 by the French physicist Siméon Poisson as a necessary consequence of Augustin Fresnel's wave theory of light. Poisson believed the

Figure 24.19
Shadow of a ball bearing illuminated with laser light.

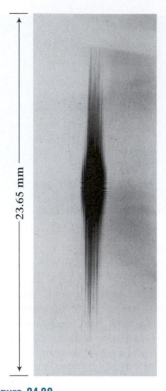

23.65 mm

Figure 24.20
Diffraction pattern of a single horizontal slit 0.05 mm wide. The photographic film was placed 29.1 cm from the slit. The reference marks along the edge of the film were 23.65 mm apart. Diffraction makes the vertical image taller than the slit width.

Figure 24.21
Geometry of the single-slit diffraction pattern.

prediction—and the theory—ridiculous; but in fact, François Arago showed experimentally that the bright spot did exist, thus supporting rather than disproving the wave theory.

The pattern produced by a plane light wave illuminating a single slit depends on the size of the slit relative to the wavelength λ. When the slit is very wide compared to λ, the pattern closely resembles the geometrical shadow of the slit. As in the case of the water waves in Fig. 24.18, the pattern spreads out as the width of the slit is narrowed (Fig. 24.20).

The explanation for the pattern of single-slit diffraction is similar to the explanation for the double-slit pattern. Figure 24.21 shows the geometry for the diffraction of light by a single slit of width b. When the paths from the slit to the screen for light beams passing across the upper and lower edges of the slit differ by an integral multiple of λ, a dark region appears on the screen. This happens because light from the center of the slit is out of phase with light from the edges. (That is, the path difference is one half-wavelength.) Thus, we have a condition for minima in the single-slit pattern,

$$m\lambda = b \sin \theta, \qquad m = 1, 2, 3, \ldots \qquad \text{(single-slit minima).} \qquad (24.7)$$

Between each pair of minima is a maximum. The brightest maximum occurs right in the center, and the other maxima get successively dimmer (Fig. 24.22). We can calculate the intensities and positions of these maxima, but the process requires techniques beyond the scope of this book.

Example 24.4

A single narrow slit is illuminated with red light of wavelength 632.8 nm. A screen placed 1.60 m from the slit shows a typical single-slit diffraction pattern. The separation between the two first minima is 4.0 mm. What is the width of the slit?

Strategy Refer to Fig. 24.21, which illustrates the geometry of single-slit diffraction. The angle θ is given by

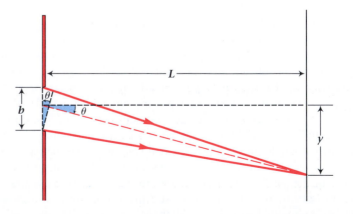

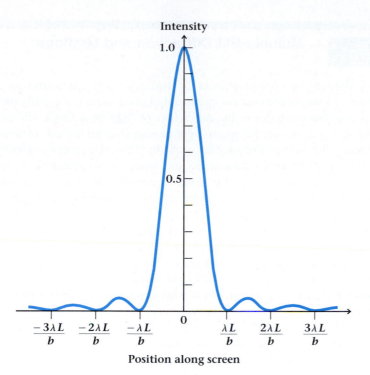

Figure 24.22
Light intensity along the screen for diffraction by a single slit of width b located a distance L from the screen. By far the brightest spot occurs at the center line of the slit.

$$\tan \theta = \frac{y}{L},$$

where y, measured from the symmetry axis, is the distance from the middle of the central maximum to the first minimum. We are given that the total distance between the two first minima, one on either side of the center, is 4.0 mm. Thus, the quantity y is half of this, or 2.0 mm. The slit-to-screen distance L is 1.60 m and is many times greater than y. We are thus able to approximate the angle by $\theta \approx \sin \theta \approx \tan \theta = y/L$.

Solution   Using the approximation for $\sin \theta$, we express Eq. (24.7), the condition for the single-slit minimum, in the form

$$\lambda = \frac{by}{L},$$

where we have set m equal to one. Upon rearranging, we find that

$$b = \frac{L\lambda}{y}.$$

Substituting the appropriate values into the equation, we find that the width of the slit is

$$b = \frac{(1.60 \text{ m})(632.8 \times 10^{-9} \text{ m})}{2.0 \text{ mm} \times \dfrac{10^{-3} \text{ m}}{\text{mm}}} = 5.1 \times 10^{-4} \text{ m} = 0.51 \text{ mm}.$$

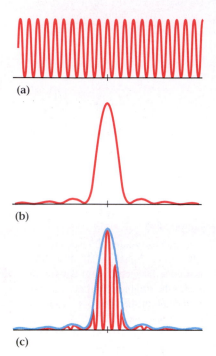

Figure 24.23
(a) Intensity pattern expected from double-slit interference without considering effects due to the width of the slits. (b) Single-slit diffraction intensity pattern. (c) Double-slit interference pattern showing the effects of the single-slit diffraction envelope. This figure is drawn for the slit separation to be three times the slit width.

24.6 Multiple-Slit Diffraction and Gratings

By now you may be wondering about the distinction between diffraction and interference. These phenomena are inseparably linked and are not really different. The pattern that results from the diffraction of light by a single slit could be thought of as the self-interference of light passing through the slit. In double-slit interference, the light is diffracted by each slit. Thus, the patterns describe both interference and diffraction at the same time. However, we generally use the term *diffraction* to describe the effect of a wave encountering an obstacle, while we use the term *interference* to describe the effects of combining multiple sources or parts of a wave.

In our initial treatment of the double-slit experiment, we were not concerned with the effect of the finite slit width. However, the slit width does determine the overall extent of the double-slit pattern, in that we find the interference pattern only in the region where light is diffracted by each single slit. Figure 24.23 illustrates the interference pattern of two slits superposed on the intensity pattern of a single slit. For this case, we have chosen the slit width to be small compared with the spacing between the slits. Consequently, since the spreading of the patterns is inversely related to slit spacing and to slit width, the two first minima of the single-slit pattern are spread farther apart than are the double-slit minima. The first minimum of the single-slit pattern occurs at an angle θ given by

$$\sin \theta = \frac{\lambda}{b} \qquad \text{(first minimum)}. \qquad (24.8)$$

The double-slit maxima are separated by an angle θ' given by Eq. (24.4a):

$$\sin \theta' = \frac{\lambda}{d}.$$

Thus, illuminating two narrow slits separated by a distance d several times their width b produces a broadly spread diffraction pattern with closely spaced interference fringes.

Example 24.5

An interference pattern is generated by a pair of 0.30-mm-wide slits. The fifth-order maximum of the double-slit pattern is not observed because it occurs at the position of the first-order minimum of the single-slit pattern. What is the separation of the slits?

Solution The first minimum of the single-slit pattern occurs at an angle θ given by

$$\sin \theta = \frac{\lambda}{b},$$

where b is the width of the slit and λ is the wavelength of the light.

The fifth-order maximum of the double-slit pattern occurs at angle θ' given by Eq. (24.4a) for $m = 5$,

$$\sin \theta' = \frac{5\lambda}{d}.$$

Since the single-slit minimum occurs at the same position as the fifth-order maximum for the double slit, $\theta = \theta'$ and

$$\frac{\lambda}{b} = \frac{5\lambda}{d}.$$

So

$$d = 5b = 5(0.30 \text{ mm}) = 1.5 \text{ mm}.$$

The separation of the slits is 1.5 mm, five times their width.

.... ■

What happens to the resulting pattern if we increase the number of slits? If we use three equally spaced slits instead of two, the pattern looks like Fig. 24.24(a). The intensity peaks still occur at the same positions as in the double-slit pattern, since these are the same positions for which the light from all three slits arrives in phase. However, the peaks are narrower, and subsidiary maxima appear between the principal maxima (Fig. 24.24b). For two slits, no light appears at the angle where the two waves are exactly out of phase. But when three waves are present, one wave remains when the other two exactly cancel. Thus, the smaller maxima in the three-slit pattern occur at the positions of the minima in the two-slit pattern.

When we add a fourth slit, another subsidiary peak occurs between the principal maxima (Fig. 24.24c). The ratio of the intensities of the smaller maxima to that of the principal maxima is even smaller in the four-slit pattern than for the three-slit pattern. In fact, as more and more slits are used, the general behavior in the resulting interference pattern is that the subsidiary peaks are suppressed and the principal peaks become narrower. If a large number of slits are used, the principal maxima can become quite sharp. An array of a large number of parallel, equally spaced slits is called a **diffraction grating.** We can analyze its resulting pattern of light intensity as due to the interference of many slits.

When white light from the sun passes through a diffraction grating, the light forms a spectrum, a rainbowlike pattern of colors. Spectra can also be produced by reflection from a grating. For example, you can see colors in the light reflected from the ridged surface of a black phonograph record. The grooves of the record act as a diffraction grating. Similar color patterns can be seen on a compact disk (CD), these being due to diffraction from the spiraling line of dimples pressed into the disk. Brilliant colors are also produced from embossed plastic film gratings with reflective backings (Fig. 24.25).

A spectrum is produced because white light consists of a mixture of colors, each with its own wavelength, as we saw in Chapter 22. When white light passes through a grating, each wavelength is diffracted through its own characteristic angle. Thus, the light disperses into a spectrum of component colors. Diffraction gratings are routinely used for spreading light into a spectrum (Section 24.9).

The angle of diffraction of each wavelength of light passed through a grating is given by the diffraction equation,

$$m\lambda = d \sin \theta, \qquad m = 0, 1, 2, 3, \ldots \quad \text{(grating maximum)} \qquad (24.9)$$

where d is the spacing between the centers of the slits in the grating, and we have assumed the incident light to be normal to the plane of the grating. This equation

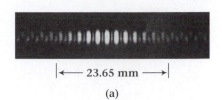

|← 23.65 mm →|

(a)

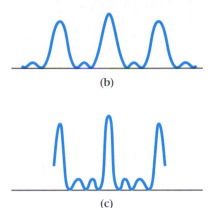

(b)

(c)

Figure 24.24
(a) Photo of a three-slit diffraction pattern. The conditions were the same as in Fig. 24.11. (b) Sketch of the intensity profile due to three slits. (c) Sketch of the intensity profile due to four slits.

Figure 24.25
One of the authors (RLC) holds a sphere covered with diffraction foil. White light striking the foil is spread into many colors.

is the same one we used to describe the maxima of double-slit interference, and it occurs here for exactly the same reasons.

In diffraction, larger wavelengths are deflected through larger angles. Thus, diffraction spectra are formed with violet closer to the normal and red farther away. When the adjacent path lengths differ by two or more wavelengths, corresponding to $m = 2$ or more, a secondary spectrum occurs at an even greater angle of deflection. In general, light falling on a grating is diffracted into several principal maxima or orders simultaneously, in addition to the zeroth-order beam that goes straight through.

Example 24.6

White light is normally incident on a diffraction grating of 5000 lines/cm. What is the angular separation between red light ($\lambda = 650$ nm) and blue light (450 nm) in the first-order spectrum?

Solution   Since the problem involves a beam incident normally upon the grating, we can use Eq. (24.9). We need to determine θ for the two wavelengths given for $m = 1$. The spacing d of the grating is the reciprocal of the number of lines per centimeter, that is,

$$d = \frac{1}{5000 \text{ lines/cm}} = 2 \times 10^{-4} \text{ cm/line} = 2 \times 10^{-6} \text{ m.}$$

For the red light, the angle is obtained from Eq. (24.9),

$$\sin \theta_r = \frac{\lambda_r}{d} = \frac{650 \times 10^{-9} \text{ m}}{2 \times 10^{-6} \text{ m}} = 0.325,$$

$$\theta_r = 19°.$$

At the blue end, the angle is given by

$$\sin \theta_b = \frac{\lambda_b}{d} = \frac{450 \times 10^{-9} \text{ m}}{2 \times 10^{-6} \text{ m}} = 0.225,$$

$$\theta_b = 13°.$$

The angular separation between the two beams is $\Delta \theta = \theta_r - \theta_b = 6°$.

24.7   Resolution and the Rayleigh Criterion

An important prediction of the wave theory of light is that the ability of an optical instrument to produce distinct images of objects that are very close together is limited. The *resolving power,* or *resolution* as it is often called, is a measure of this ability to produce sharp images. No matter how carefully a lens is designed and made, there is a limit to its resolution. This limit is determined by the diffraction pattern of the lens.

In Section 24.5, we saw that when parallel light passes through a single slit of width b, the light spreads out by diffraction. The resulting diffraction pattern consists of a bright central region bordered by alternating dark and bright bands. The angle θ between the center of the central maxima and the first minima

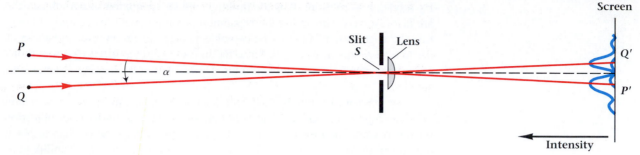

Figure 24.26
The light from two point sources P and Q is imaged on a screen after passing through a slit S. The images P' and Q' are not points, but are spread out by diffraction.

(Fig. 24.21) is obtained from Eq. (24.7):

$$\sin \theta = \frac{\lambda}{b},$$

where λ is the wavelength of the light. For small values of the angle θ, we can approximate $\sin \theta$ by θ so that the equation becomes

$$\theta = \frac{\lambda}{b}.$$

As we will see, this angle determines the minimum separation that must exist between objects in order for them to be imaged individually.

Let's start by considering two point sources of light, P and Q, that subtend an angle α at the slit S (see Fig. 24.26). The light passing through the slit is imaged on a screen by a converging lens. A ray drawn through the center of the lens from each source point determines the positions on the screen of their images P' and Q'. However, the images are not simply points, but are spread into a larger area as indicated. If the angle α is larger than the angular spread θ of each image due to diffraction, the two sources are imaged distinctly (Fig. 24.27a,b). We say the images are well resolved. If the angle α is smaller than θ, the two images overlap and cannot be resolved into two images (Fig. 24.27d). When $\alpha = \theta$, the variation in intensity across the pattern is discernible to the human eye (Fig. 24.27c). For this separation, $\alpha = \theta$, the first minimum of one diffraction image falls on the central maximum of the other image. Lord Rayleigh* chose this separation as the minimum angle that could be resolved. His choice is often referred to as **Rayleigh's criterion.**

When the diffracting aperture is a circle rather than a slit, the angle of the first minimum in the diffraction pattern is somewhat larger than that given above. The angular position of the first minimum in the diffraction pattern of a circular aperture of diameter D occurs at an angle

$$\theta_m = 1.22\frac{\lambda}{D}. \qquad (24.10)$$

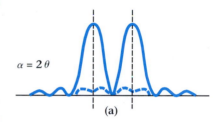

$\alpha = 2\,\theta$

(a)

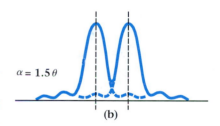

$\alpha = 1.5\,\theta$

(b)

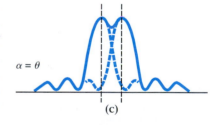

$\alpha = \theta$

(c)

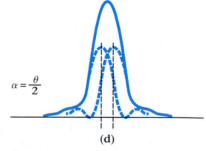

$\alpha = \frac{\theta}{2}$

(d)

Figure 24.27
The diffracted image of two point sources at different angular separations α. Rayleigh's criterion for just resolving the two images is shown in (c).

---

*John William Strutt, Baron Rayleigh (1842–1919) is famous for his study of waves. He was awarded the 1904 Nobel Prize in physics for his discovery of argon gas.

By Rayleigh's criterion, θ_m is the limiting angular separation of two objects that are just resolved. This minimum separation is a theoretical lower limit. In practice, the angular separation of the two objects may have to be even greater in order to achieve distinct images. Since the limiting angle is inversely proportional to the diameter D, we can obtain improved resolution by making D larger or λ smaller.

The diameter of the objective lens (or mirror) sets a limit on the resolution of a telescope. Astronomical telescopes are made with large-diameter objectives for this reason—and for increased light-gathering ability—rather than to obtain high magnification. No matter what the magnification, a telescope's ability to resolve two nearby stars is limited by the diffraction pattern of its objective. In addition, atmospheric turbulence sets a practical limitation on the resolution of ground-based telescopes. For this reason, telescopes are now being placed in earth-orbiting satellites to put them above the image-degrading atmosphere.

The Hubble Space Telescope (Fig. 24.28) was launched into earth orbit from the space shuttle in 1990. Traveling above the atmospheric turbulence and absorption, it can detect radiation from wavelengths of 105 nm in the ultraviolet through the visible to infrared wavelengths of 1100 nm. The Hubble's 2.4-m-diameter mirror was designed to have an angular resolution of about 0.08 arcsecond at a wavelength of 632.8 nm. However, shortly after launch, it was discovered that although the mirror was polished to an exceptional degree of smoothness, its curvature corresponds to the wrong shape, thus creating spherical aberrations that seriously degrade the images. Because the mirror surface is so good, it was possible to sharpen up the direct images using computer image enhancement techniques. As a result, images have been obtained with resolution of about 0.05 arcsecond (Fig. 24.29).

Although the algorithms used on the Hubble pictures significantly sharpened the images, they did not allow the recovery of faint objects. In December 1993, corrective optics were installed to compensate for the spherical aberration in the primary mirror. The improvement in resolution was dramatic. Images taken with the Wide Field Planetary Camera before and after the corrections were made

Figure 24.28

The Hubble Space Telescope as it was launched from the Space Shuttle Discovery on April 25, 1990.

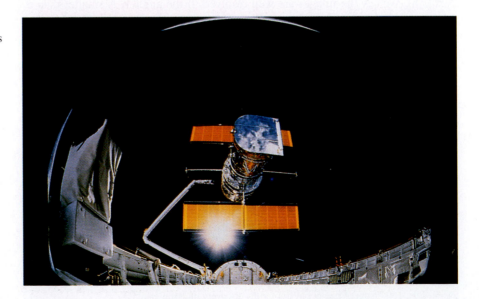

Figure 24.29
Orion Nebula as seen with the ground-based telescope at the Anglo-Australian Southern Observatory (upper right) and with the improved resolution of the Hubble Space Telescope's Wide Field Planetary Camera (lower left).

show the difference. Not only are the images sharpened but faint objects can be seen that were previously invisible (Fig. 24.30).

Example 24.7

A photo printed on a 720 dots/inch ink-jet printer is held 25 cm from your eye. Can you resolve adjacent dots in the photo when your pupil diameter is 3.0 mm and $\lambda = 550$ nm?

Solution The eye's minimum angle of resolution is determined from Eq. (24.10),

$$\theta_m = 1.22 \frac{\lambda}{D} = 1.22 \frac{550 \times 10^{-9} \text{ m}}{3.0 \times 10^{-3} \text{ m}} = 2.2 \times 10^{-4} \text{ radian.}$$

The angular separation of adjacent dots is

$$\theta = \left(\frac{1 \text{ in.}}{720}\right)\left(\frac{2.54 \text{ cm}}{1 \text{ in.}}\right)/(25 \text{ cm}) = 1.4 \times 10^{-4} \text{ radians.}$$

The angle made by the dots is smaller than your eye can resolve.

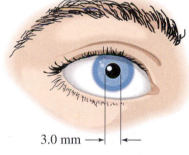

3.0 mm →| |←

Example 24.7: In bright light the pupil narrows to 3.0-mm diameter.

Figure 24.30
Images of the core of the galaxy M100 taken with the Hubble Space Telescope's Wide Field Planetary Camera (left) before and (right) after the installation of corrective optics. The improved camera resolves objects as small as 30 light-years across in a galaxy that is tens of millions of light-years away.

Figure 24.31
A scanning electron micrograph showing the skeleton of a *Radiolaria* plankton.

The resolving power of microscopes is also limited by diffraction. It is the resolution that limits the useful magnification. Increasing the magnification by the use of stronger lenses only enlarges an already-blurred image. Increasing the diameter of the objective lens is not practical for microscopes, as it is for telescopes. The inherent limitation of optical microscopes is the wavelength of visible light, from about 400 nm to 700 nm. Any improvement in resolution can be made only by using shorter wavelengths.

After the discovery of the wave properties of electrons (see Chapter 28), it was soon realized that electrons could easily be accelerated to high velocities at which their wavelengths were a thousand times shorter than the wavelengths of visible light. Furthermore, electron beams can be controlled by electrostatic and magnetic fields acting as lenses. Thus, the idea of the *electron microscope* was born. The first electron microscope was built by Ernst Ruska in Germany in 1932, and improved versions became commercially available in 1938. Professor Ruska's efforts were belatedly recognized in 1986 when he received the Nobel Prize in physics.

Modern electron microscopes are commonplace in laboratories throughout the world, where they are used to form images of extremely tiny objects. The resolution of these microscopes routinely allows the observation of objects as small as 0.5 nm, and some of the most powerful microscopes permit the observation of detail as small as 0.05 nm. Thus, large molecules such as DNA can be seen with electron microscopes. Figure 24.31 was taken with an electron microscope. The colors seen in the figure are due to computer enhancement and are completely arbitrary. We discuss the principle of electron microscopes in more detail in Chapter 28.

24.8   Dispersion

White light can be spread into its component colors by a glass prism as well as by diffraction. Isaac Newton showed that one prism could separate white light into a spectrum and another prism turned the opposite way could combine that light back into a white beam again. He also showed that a small portion of the spectrum could not be spread into any other colors by passing the light through a second prism. However, the spreading of white light into colors is not just a trick you can do with a prism; it is also responsible for chromatic aberration (see Section 22.8), for the attractiveness of diamonds and cut glass, and even for the beauty of the rainbow.

The production of a continuous spectrum by a prism is due to the variation of wave velocity with wavelength that commonly occurs in transparent media. As stated in Chapter 22, the dependence of the index of refraction upon the wavelength (or color) of light is called *dispersion*. Water, glass, transparent plastics, and quartz are all dispersive materials. Generally, shorter wavelengths travel with slightly smaller wave velocities than do longer wavelengths. This means that the prism's index of refraction is not constant across the visible spectrum, but decreases continuously as the wavelengths increase from violet to red. Since the index of refraction is greater for violet than for red, violet light deflects through a greater angle than does red. (See Fig. 22.33.) This effect of a prism is just the opposite of the dispersive effect of a diffraction grating, which deflects red light more than violet. The graph in Fig. 24.32 shows the typical dispersive behavior of decreasing refractive index with increasing wavelength for a transparent plastic.

Diamonds are highly valued as gems because they are rare and beautiful. Their beauty is in part due to their large index of refraction and their large dispersion. Gems are cut and polished to refract and reflect the light incident on them. The brilliant colors you see in an otherwise colorless diamond are due to the dispersion of the light as it is refracted and reflected from within. Slightly rotating a properly faceted diamond causes the colors to change and flash from the different facets as the angle of incidence of the light changes.

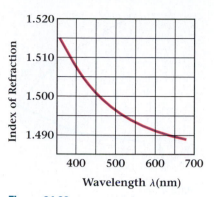

Figure 24.32
Index of refraction as a function of wavelength for polymethylmethacrylate, a transparent plastic used in lenses and other optical components.

Example 24.8

Light from a hydrogen source is normally incident on a 30° crown glass prism (Fig. 24.33). Find the angular separation of the emerging rays of light, assuming $\lambda_{red} = 656$ nm and $\lambda_{violet} = 434$ nm. The refractive index of the glass at these two wavelengths is 1.514 (red) and 1.528 (violet).

Strategy Because the light enters the prism at normal incidence, no change in direction of the ray occurs at the first interface. We see that the ray then strikes the back surface with an angle of incidence of 30°. This is where the change in direction of the rays will occur. We can calculate the angles of refracted (transmitted) rays relative to the surface normal.

Solution For red we get

$$n_i \sin \theta_i = n_t \sin \theta_t,$$

$$\sin \theta_t(\text{red}) = \frac{n_i \sin \theta_i}{n_t} = \frac{1.514 \sin 30°}{1.000} = 0.7570,$$

$$\theta_t(\text{red}) = 49.20°.$$

For violet,

$$\sin \theta_t(\text{violet}) = \frac{1.528 \sin 30°}{1.000} = 0.7640,$$

$$\theta_t(\text{violet}) = 49.82°.$$

The angular separation is θ_t (violet) − θ_t (red) = 0.62°. Notice that the violet light is deflected through the larger angle because the index of refraction is greater for violet than it is for red.

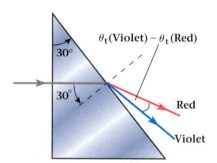

Figure 24.33
Example 24.8: Light from a hydrogen source is normally incident on a 30° glass prism. The light is dispersed as it emerges.

The formation of a rainbow by drops of water in the atmosphere is another example of dispersion. In a rainbow, water drops disperse reflected sunlight, revealing the colors of the spectrum across the sky.

To see a rainbow, you must have the sun to your back while you look toward a cloud of water drops. (These drops may be in a natural cloud or in a spray or mist that you could make with a garden hose.) For you to see the primary rainbow (there is also a secondary bow), the angle between the direction from the sun to you and your line of sight to the cloud of mist must be about 42°. Normally, we see rainbows when the sun is low in the sky and its light refracts and

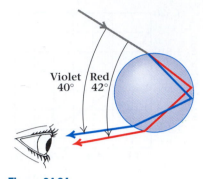

Figure 24.34

Refraction and dispersion of light by a spherical drop of water.

reflects from many drops of water in the air. The complete rainbow is a circle and can be seen as such from an airplane. Usually we see only the top portion of the circular arc, as the rest lies below the horizon, where it is not easily seen against the background and where the number of reflecting drops in our line of sight is small. For this same reason, we do not see rainbows when the sun is higher than 42° above the horizon.

Figure 24.34 illustrates the refraction of light by a spherical drop of water, showing those rays responsible for the rainbow. The effect of refraction and reflection bunches the rays so that more leave the drop near the angles indicated than in other directions. As a result, the reflected and refracted light is enhanced at the angles shown. Dispersion by the water drop causes light from the sun to spread out into a spectrum. If there are many drops (Fig. 24.35), the eye receives the different colors from drops at different heights. The result is that the top of the rainbow appears red and the inner arc appears violet.

24.9 Spectroscopes and Spectra

If you look through a prism or a diffraction grating at a light source, you will see a band of overlapping images, each in a different color. If the source is an ordinary incandescent lamp with a frosted bulb, the images overlap so that most of the resulting broadened image appears white, with a violet border on one side and a red one on the other. If you place a narrow slit in front of the lamp so that you can see only a thin strip of the source through the prism, the band of images has very little overlap and you see bright spectral colors across the band.

A **spectrometer** is an optical instrument designed to enhance this effect and permit analysis of spectra. A *spectroscope* (Fig. 24.36) is a spectrometer that per-

Figure 24.35

The eye sees the different colors of the rainbow in the light reflected from drops at different heights.

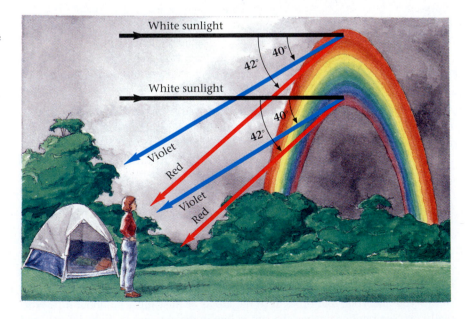

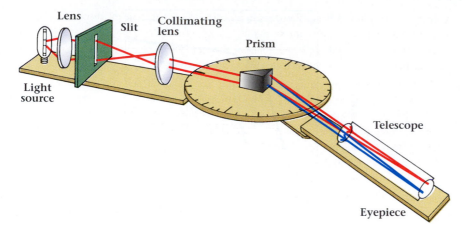

Figure 24.36
A prism spectroscope. The image of the slit formed by the red light is deviated less than the image formed by the violet light. The angle of deviation can be measured and used to calculate the wavelengths of the light.

mits direct observation of spectra. Light from the source we wish to analyze passes through a narrow slit and is focused into a parallel beam by a *collimating lens.* (A collimating lens is a converging lens that focuses a diverging light beam into a parallel beam.) After this beam passes through a prism, parallel beams emerge, each at a different angle according to its wavelength. A telescope focuses the parallel beams and allows an observer to see an image of the slit. Most spectroscopes also have a calibrated circular scale. This scale enables the observer to measure the angle of the emerging light for each image, which is then used to determine the wavelengths of the light in the spectrum. Each image of the slit is called a spectral line. For an incandescent source, the lines merge together into a continuum. Other sources, however, give characteristic lines. As we will see in Chapter 27, knowing what wavelengths are emitted by a light source gives us important information about the composition of the source.

Most modern spectrometers are made with a diffraction grating in place of a prism because the dispersion of a grating can be made much greater than that of a prism. Other instruments called *spectrographs,* which are very similar to spectroscopes, permit recording of the spectra photographically. *Spectrophotometers* are devices that convert the resultant light intensity to an electrical signal and then use a chart recorder or computer terminal to display the spectra graphically. Figure 24.37 shows the spectrum taken with a spectrophotometer of light transmitted by a green-colored glass filter. The spectrum is displayed as the percentage of light passed at each wavelength. As you can see, the filter transmits green light (540 nm) but does not pass red or blue light.

In recent years, the public has become more aware of the harm to the eyes caused by ultraviolet radiation. This concern has been heightened by the reduction of the protective layer of ultraviolet-absorbing ozone in the upper atmosphere. This loss of ozone, thought to be due to pollutants such as chlorofluorocarbons, means that the intensity of ultraviolet radiation at ground level is increased. For this reason, many people now have their sunglasses and their regular glasses treated with a protective coating that is transparent in the visible range but opaque to the ultraviolet range. The effectiveness of such coatings can be seen in Fig. 24.38, which compares the transmission spectra of treated and untreated lenses.

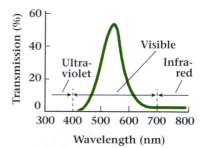

Figure 24.37
Intensity spectrum of the light transmitted by a green glass filter (Hoya #G-533).

Figure 24.38

Transmission spectra of eyeglass lenses showing the effect of treatment for blocking ultraviolet (UV) radiation.

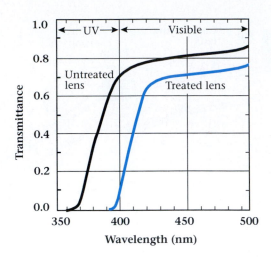

Figure 24.39

(a) Light from an ordinary source contains a mixture of waves oscillating in different directions. (b) A plane-polarized light beam contains only one direction of oscillation.

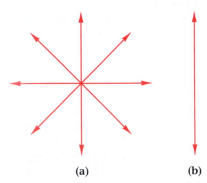

(a)   (b)

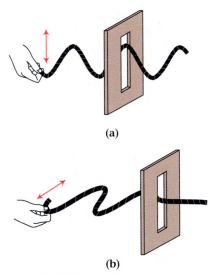

(a)

(b)

Figure 24.40

(a) Shaking a rope up and down produces a vertically polarized wave that passes through a vertical slit. (b) Shaking a rope from side to side produces a horizontally polarized wave that will not pass through a vertical slit.

24.10

Polarization

So far in this chapter, we have discussed the optical properties of light that are due to interference and diffraction, showing how these characteristics support a wave theory of light. However, another important property of light, called polarization, is due not just to light being a wave, but to light being a transverse wave. Longitudinal waves, such as sound in air, do not exhibit polarization, but light, as well as other electromagnetic radiation, can be polarized.

Recall from Chapter 15 that the distinguishing characteristic of transverse waves is that their oscillating motion occurs in planes perpendicular to the direction in which the wave itself is moving. As we discussed in Chapter 20, electromagnetic waves, including visible light, are transverse waves, consisting of sinusoidally oscillating electric and magnetic fields perpendicular to each other (see Fig. 20.23).

Electromagnetic waves are usually generated by oscillating electric charges, such as the oscillating current in a radio or TV antenna or an electron in an atom. The direction of the oscillation determines the orientation of the wave's electric field. In most common light sources, such as a candle flame, the sun, or an incandescent lamp, the many oscillating atoms are randomly oriented. Although the electric field of each wave produced by a particular atom lies in a single plane, the overall beam of light contains electric fields oscillating in all planes (Fig. 24.39a). However, if the oscillating charges are confined to move in only one plane, as is the oscillating current in an antenna, then the entire electric field of the resulting beam oscillates in only that direction (Fig. 24.39b). Such a wave is said to be polarized, and the orientation of the electric field vector is taken as the direction of **polarization.** The complete wave has a sinusoidally oscillating magnetic field coupled to the electric field, but traditionally we refer to only the electric field to indicate the direction of polarization.

A useful mechanical analogy for polarized light is the motion of transverse waves along a rope. If you shake a rope up and down, the wave is vertically plane-polarized. Such a wave can travel through a vertical slit (Fig. 24.40a), but

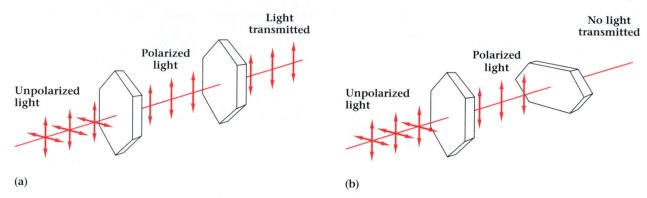

(a)   (b)

Figure 24.41
Polarization of light by tourmaline crystals. (a) When the polarization axes of the crystals are parallel, plane-polarized light passes through. (b) When the axes are perpendicular, no light passes through.

not through a horizontal slit. Similarly, if you shake a rope side to side, the wave is confined to a horizontal plane; it can pass through a horizontal slit, but not a vertical one (Fig. 24.40b).

It is possible to produce polarized light by filtering ordinary light through a polarizer, which is a material that transmits only those waves that oscillate in a single plane. Several materials have this property, including the naturally occurring mineral tourmaline. When an ordinary beam of light strikes a tourmaline crystal, part of the light is absorbed and part is transmitted (Fig. 24.41a). If a second tourmaline crystal is placed behind the first and aligned in the same orientation, light is transmitted. If the second crystal is rotated by 90° about the axis of the light beam, no light passes through (Fig. 24.41b). We can explain this observation in terms of polarization. The initial light beam is unpolarized. We think of it as containing a mixture of random polarizations. When this beam strikes the first crystal, the tourmaline passes only light polarized along the polarization axis of the crystal. The light emerging from the first crystal is completely polarized along the direction shown. When this light falls on the second tourmaline crystal, all the light is absorbed, because the second crystal has been turned so that its polarization axis is 90° away from that of the first crystal.

Crystal polarizers like tourmaline are not used much now because of the development of inexpensive plastic polarizers called *Polaroids*. Synthetic sheet polarizers were invented in 1928 by Edwin H. Land and improved by him ten years later. When viewed individually, these transparent sheets have a neutral grey appearance. However, when two polarizers overlap with their polarization directions at right angles, the region of overlap looks black (Fig. 24.42).

If the angle between the two Polaroids is made less than 90°, the area of overlap is no longer opaque. It gradually gets more and more transparent as the angle decreases to 0°. The intensity of light that passes through two polarizers that are aligned with an angle θ between their polarization directions is

$$I = I_{\mathrm{m}} \cos^2 \theta, \tag{24.11}$$

where I_{m} is the maximum amount of light transmitted when the two polarizers are aligned along the same direction of polarization. Equation (24.11) is known as **Malus's law** after its discoverer, who first observed this effect experimentally in 1809.

(a)

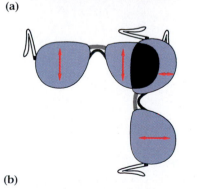

(b)

Figure 24.42
(a) One of the authors (ERJ) holding two large polarizing filters with their polarization directions at a right angle. (b) You can demonstrate the same effect yourself with two pairs of polarizing sunglasses.

An ideal polarizer passes 100% of the incident light that is plane-polarized along its polarization direction and 0% of the light that is polarized at 90° to its polarization direction. Since unpolarized light is a mix of all possible polarizations, an ideal polarizer transmits only 50% of the incident unpolarized light intensity.

Example 24.9

Two ideal polarizers are aligned with a 30° angle between their polarization directions. If unpolarized light of intensity I_0 is incident upon them, what is the intensity of the transmitted light?

Solution The light passing through the first polarizer is half the incident intensity, so

$$I_1 = \tfrac{1}{2}I_0.$$

The quantity I_1 becomes the I_m of Malus's law, Eq. (24.11). The intensity of the light emerging from the second polarizer is

$$I_2 = I_1 \cos^2 \theta = \tfrac{1}{2} I_0 \cos^2 30° = 0.375I_0.$$

Malus also discovered that light becomes polarized upon reflection from glass windows and surfaces of water. In 1814, some six years after this discovery, David Brewster* (1781–1868) found that at a particular angle of incidence, now called the Brewster angle, polarization of the reflected light is complete, with polarization perpendicular to the plane of incidence (Fig. 24.43). At other angles, the reflected light is only partially polarized, with more light polarized in this direction than in any other.

Brewster found that maximum polarization occurs when the reflected ray and the refracted ray are at right angles to each other. By combining Brewster's observation with Snell's law, we can find a relation between a material's index of refraction and its Brewster angle. Maximum polarization occurs for $\theta_r + \theta_t = 90°$, where θ_r is the angle of reflection and θ_t is the angle of transmission (refraction). From Snell's law we can show that

$$n_i \sin \theta_i = n_t \sin \theta_t = n_t \sin (90° - \theta_r) = n_t \cos \theta_r.$$

This equation reduces to

$$n_i \sin \theta_i = n_t \cos \theta_i$$

when we use the fact that $\theta_i = \theta_r$. At this condition of maximum polarization, we can denote the angle of incidence as the Brewster angle θ_B and rewrite this equation as

$$\tan \theta_B = \frac{n_t}{n_i}. \tag{24.12}$$

---

*Brewster is also known for his invention of the kaleidoscope and for his improvements on the stereoscope invented by Sir Charles Wheatstone.

This last equation is known as **Brewster's law.** Note that the Brewster angle depends on the index of refraction of the materials on both sides of the reflecting dielectric surface.

Example 24.10

At what angle of incidence is the light reflected from the surface of water completely polarized? Assume that the light is incident in air.

Solution The index of refraction of water is 1.33. Inserting this value into Brewster's law, we find that

$$\tan \theta_B = \frac{n_t}{n_i} = \frac{1.33}{1.00} = 1.33,$$

$$\theta_B = 53.1°.$$

■

Sunglasses made from Polaroid sheet are widely used because of their ability to block the glare of specularly reflected sunlight from water surfaces or from surfaces of other smooth reflecting objects such as automobile windows. The glasses are made with a vertical axis of polarization, because, as can be seen from Fig. 24.43(a), glare resulting from the sun is usually horizontally polarized. You can see this by rotating such glasses by 90°. In this new position, the glare light passes through because the transmission axis of the glasses is aligned with the polarization of the light. Normally, of course, the polarization direction of the glasses is crossed with that of the glare light so that the glare is extinguished.

*24.11   Scattering

We conclude this chapter by examining one more optical phenomenon that is best explained by a wave theory of light. This time, we answer an interesting question about our physical world—why is the sky blue?

In a perfectly transparent, homogeneous, and structureless medium, a beam of parallel light proceeds without broadening or lateral spreading. A beam of light in a vacuum behaves this way. But all real materials have some molecular or crystal structure and are not perfectly transparent to visible light. Furthermore, most materials contain impurities; even air has water vapor, dust, smoke, and other minute particles suspended in it. When a beam of light, such as a searchlight beam, passes through fog or dust-laden air, you can see the beam itself from the side. We say that the light has been scattered; molecules of air and its impurities have absorbed some light from the beam and then reradiated it in other directions, a process we call **scattering.** Most of the light travels in the initial beam direction, but some is scattered to the sides and even backwards.

The details of the scattering process depend on the relative size of the scattering particles compared with the wavelength of light. When the particles that cause the scattering are smaller than the wavelength of the incident light, we have what is called **Rayleigh scattering.** This type of scattering was first

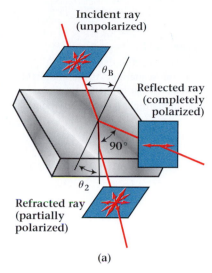

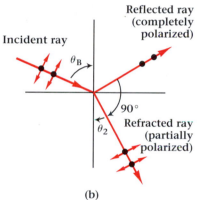

(b)

Figure 24.43
(a) Polarization of light by reflection at the Brewster angle. (b) The reflected light is completely polarized in a direction perpendicular to the plane of incidence. The refracted light, which is partially polarized parallel to the plane of incidence, makes an angle of 90° with the reflected ray.

Figure 24.44

An experiment to demonstrate Rayleigh scattering. As the blue light is scattered out of the beam, the transmitted light appears red-orange.

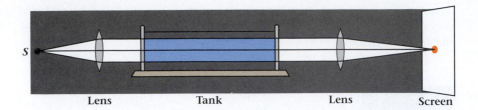

Lens   Tank   Lens   Screen

explained by Lord Rayleigh, who showed that the scattering was proportional to the fourth power of the frequency of the light.

We can see what this means by examining a simplified version of an experiment for viewing Rayleigh scattering (Fig. 24.44). Light from a bright source of white light passes through a glass tank and strikes a screen. The collimating lens makes the beam parallel in the tank; the second lens focuses an image of the end of the tank on the screen. The tank is filled with water, into which we have mixed fine microscopic particles. (Such a mixture can be made by adding a few drops of milk to a tank of water.) From the side, the path of the beam is visible and has a bluish tint. Blue light has a higher frequency than red light and, according to Rayleigh's f^4 law, is scattered more. In fact, it is scattered in all directions, as can be observed by walking around the tank. Because the blue end of the spectrum is scattered out of the beam, less blue and relatively more red appears as the beam proceeds. The light emerging from the right-hand end of the tank appears somewhat reddish-orange. This is not due to any change in the scattering, but happens because this end of the tank is illuminated by light that is now deficient in blue.

Light from the sun incident on the earth's atmosphere is scattered most strongly in the violet and blue regions of the spectrum. In fact, the sky would appear violet except that our eyes are not very sensitive to violet. The combination of Rayleigh scattering and the sensitivity of our eyes explains the blue of the daytime sky. When the sun is on the horizon, as at sunrise or sunset, the optical path through the atmosphere is much longer than when the sun is directly overhead, so more blue is scattered out of the direct beam. Thus, both the rising and the setting sun appear red (Fig. 24.45). These effects are due to Rayleigh scattering of the molecules of the air itself, which have dimensions of the order of a fraction of a nanometer, and not to suspended dust or vapor in the atmosphere.

Smoke rising from the end of a lighted cigarette or from around the edges of a pile of burning leaves appears blue because of Rayleigh scattering. On the other hand, exhaled smoke appears white because it contains larger water droplets. When the scattering particles are many times larger than the wavelength of light, the light is reflected and refracted as in geometrical optics, not absorbed and reradiated, so the scattered light is white. This type of scattering is called Tyndall scattering and is responsible for the appearance of fog, clouds, powders, and ground glass.

Suppose we look through a polarizing filter at the light scattered out the sides of the tank in Fig. 24.44, looking at right angles to the incident beam. We find that the intensity of the light grows and diminishes as we rotate the filter. This demonstrates that the scattered light is polarized, an effect that was also predicted by Rayleigh. Because light is a transverse wave, the plane of polarization is perpendicular to the plane containing the incident light beam and the perpen-

Figure 24.45

The setting sun looks red because the atmosphere scatters the blue light from the rays that reach our eyes.

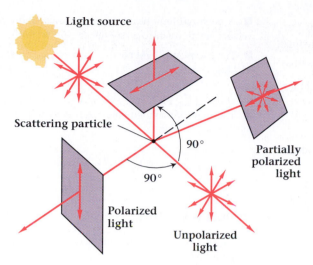

Figure 24.46
Light scattered at right angles to the initial beam direction is polarized.

dicular line of sight of the viewer (Fig. 24.46). When the light is viewed at other angles, the polarization is not complete.

If a polarizer is placed in the path of the light entering the tank in Fig. 24.44, the scattered light is a maximum in the direction perpendicular to the direction of polarization and zero along the direction of polarization. Thus, when one is looking horizontally into the tank, the scattered light will appear brightest if the incident beam is polarized vertically and weakest if the beam is polarized horizontally.

If you look at the sky through a polarizing filter in a direction perpendicular to the direction of the sun's rays, you will see the sky change from bright to dark as you rotate the filter. The polarization of the light is not 100%, but may be as much as 70% to 80% when the air is clear. Some insects, especially bees, are able to find their way over long distances by using the direction of polarization as a sort of compass.

Some materials cause the direction of polarization to rotate as the light passes through, an effect known as **optical activity.** As the light moves through the medium, the direction of polarization traces out a helical (screwlike) path. This effect is due to the geometric shape of the molecules in the material. In concentrated solutions of ordinary sugar (dextrose) the pitch of the helix changes with wavelength (color) because of dispersion. Consequently, passing polarized white light through a column of corn syrup creates a multicolored "barber pole" (Fig. 24.47).

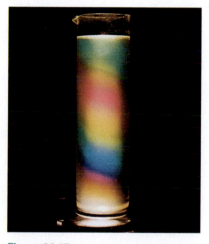

Figure 24.47
A multicolored "barber pole" produced by polarized white light entering a column of corn syrup from below.

Summary

Useful Concepts

■ The laws of geometrical optics, such as the law of reflection and Snell's law, can be derived from a wave interpretation of light using Huygens' principle. This principle considers each point on a wavefront to be a point source for expanding wavelets. The resulting wave is the envelope of these wavelets and, in turn, can be considered as a source for finding a still later position of the wave.

■ For Young's double-slit experiment, the angular locations of the maxima and minima are given by

$$m\lambda = d \sin \theta, \quad m = 0, 1, 2, 3, \ldots \quad \text{(double-slit maxima)},$$
$$(m + \tfrac{1}{2})\lambda = d \sin \theta, \quad m = 0, 1, 2, 3, \ldots \quad \text{(double-slit minima)},$$

where d is the separation between the slits.

■ When light is normally incident in air on a thin film of index of refraction n, the interference between the front and back surface reflections gives maxima and minima when

$$(m + \tfrac{1}{2})\lambda = 2nt, \quad m = 0, 1, 2, 3, \ldots \quad \text{(reflection maxima)},$$

$$m\lambda = 2nt, \quad m = 0, 1, 2, 3, \ldots \quad \text{(reflection minima)},$$

where t is the thickness of the film and λ is the wavelength of the light in free space.

■ Antireflection coatings can be made by coating a surface with a thin film of thickness t and index of refraction n that is intermediate between the index of the surface and the index of the surrounding medium (usually air) when

$$(m + \tfrac{1}{2})\lambda = 2nt, \quad m = 0, 1, 2, 3, \ldots \quad \text{(reflection minima)}.$$

■ The spreading out of light passing through a small aperture or around a sharp edge is called diffraction. For diffraction by a single slit of width b, the angular locations of the minima are given by

$$m\lambda = b \sin \theta, \quad m = 1, 2, 3, \ldots \quad \text{(single-slit minima)}.$$

■ For a diffraction grating, the angle of diffraction of each wavelength is given by

$$m\lambda = d \sin \theta, \quad m = 0, 1, 2, 3, \ldots \quad \text{(grating maxima)},$$

where d is the spacing between slits.

■ For a circular aperture of diameter D, the minimum angle of resolution θ_m given by the Rayleigh criterion is

$$\theta_m = 1.22 \frac{\lambda}{D}.$$

■ A polarized light beam contains waves whose electric field vectors are all oriented along the same direction.

■ The intensity of light passing through two polarizers is given by Malus's law:

$$I = I_m \cos^2 \theta.$$

■ The angle of maximum polarization by reflection θ_B is given by Brewster's law:

$$\tan \theta_B = \frac{n_t}{n_i}.$$

Important Terms

You should be able to write the definition or meaning of each of the following:

Huygens' principle	Rayleigh's criterion
wavefront (or wave surface)	spectrometer
	polarization
constructive interference	Malus's law
destructive interference	Brewster's law
coherence	scattering
antireflection coating	Rayleigh scattering
diffraction	optical activity
diffraction grating	

Conceptual Questions

24.1 Describe one method of using sound waves to carry out an experiment analogous to Young's double-slit experiment.

24.2 Can a Young's double-slit experiment be done for the case of microwaves ($\lambda \approx 1$ cm)? How?

24.3 The source of illumination for a Young's double-slit setup is white light. Suppose you cover one slit with a filter that transmits only blue light and the other slit with a filter that transmits only red light. What would you expect to see on the screen?

24.4 Suppose you have a Young's double-slit arrangement that produces an interference pattern of light on a screen. If you submerge the entire apparatus in water, what effect will that have on the pattern?

24.5 A very high-quality telescope mirror is said to be "diffraction limited." What does this mean?

24.6 A card is placed between a point source and a screen, but relatively close to the screen. Do you expect the card to cast a sharp shadow? If not, what do you expect and why?

24.7 Monochromatic light falls on a single slit and forms a pattern on the screen. What do you observe when the slit width is equal to the wavelength of the light? What will you observe as the slit width is increased to many times the wavelength of the light?

24.8 Can you arrange two glass plates so that a narrow beam of light reflected from the first one and incident on the second is not reflected by the second one? Explain your answer.

24.9 Maxwell showed that electromagnetic waves would not penetrate very far into conductors. Therefore, conductors should not be transparent to light. Mention several cases and discuss whether Maxwell's prediction is consistent with your observations.

24.10 Is there any optical basis for the old saying regarding weather, "Red light at night, sailors delight; red light in morning, sailors take warning"?

24.11 How is the resolution of a camera lens affected by the f-number of the lens?

24.12 If a salesperson claims to be selling Polaroid sunglasses, how can you check to see if this is true?

24.13 What is the approximate size of the smallest things that can be successfully investigated with a microscope using visible light?

24.14 When a drop of oil is placed on the surface of a bowl of water, the surface becomes less reflecting. What can you say about the index of refraction and any other properties of the oil?

24.15 If you were given a photograph of a diffraction pattern and told that it was made by one or more slits with monochromatic light, how much could you determine about the slit arrangement that made the pattern? (This is the underlying principle of holography, discussed in detail in Chapter 30.)

Problems

Sections 24.1 and 24.2 Huygens' Principle and Reflection and Refraction of Light Waves

Hints for Solving Problems

The wavelength of light is proportional to its velocity and inversely proportional to its frequency.

24.1 Show by Huygens' geometric construction that the image of an object located a distance D from a plane mirror is located a distance D on the other side of the mirror.

24.2 Light with a wavelength of 486 nm in air is passed through glass with an index of refraction of 1.54. (a) What is the wavelength of the light in the glass? What is the frequency of the light (b) in air and (c) in the glass?

24.3 Light from a helium-neon laser with $\lambda = 633$ nm in air passes through dense flint glass with an index of refraction of 1.65. (a) What is the speed of the light in the glass? (b) What is the wavelength of the light in the glass?

24.4 Using Huygens' principle (rather than Snell's law), find the angle of refraction when light is incident on the interface between air ($n = 1.00$) and glass ($n = 1.50$) at an angle of 45°.

24.5 A beam of light has a wavelength $\lambda = 435$ nm in air. (a) What is its wavelength in diamond? (b) What is it in ethyl alcohol? (Refer to Table 22.1.)

24.6 Light from a sodium lamp has a wavelength $\lambda = 589$ nm in empty space. (a) What is the wavelength of this light in water? (b) What is it in polymethylmethacrylate? (Refer to Table 22.1.)

24.7• Using Huygens' principle, find the point at which plane waves incident on a concave spherical mirror converge. Use only that part of the wavefront closest to the axis.

Section 24.3 Interference of Light

Hints for Solving Problems

Constructive interference occurs when two waves are in phase. In the double-slit experiment, the waves are in phase when their path difference is an integer multiple of the wavelength ($m\lambda$). They are out of phase (destructive interference) when the path lengths differ by an odd half-integer multiple of the wavelength ($m + \frac{1}{2}\lambda$).

24.8 A pair of narrow slits is illuminated with green light of wavelength $\lambda = 546.1$ nm. The resulting interference maxima are found to be separated by 1.10 mm on a screen 1.00 m from the slits. What is the separation of the slits?

24.9 Two small loudspeakers sounding in phase at 2000 Hz are placed side by side and 20 cm apart in a large grassy field where sound is not strongly reflected from the ground. At what minimum angle, with respect to the perpendicular bisector of the line joining the two speakers, can you stand and hear no sound? Assume $v_{sound} = 340$ m/s.

24.10 At what angle with respect to the optical axis will the second-order maximum lie when a pair of slits with a separation of 0.50 mm is illuminated with green light of wavelength 550 nm?

24.11 Monochromatic light is incident on two narrow slits whose separation is 0.500 mm. An interference pattern forms on a screen 90.0 cm away. If adjacent interference maxima are separated by 1.06 mm, what is the wavelength of the light?

24.12 Red light of wavelength $\lambda = 632.8$ nm is incident on a pair of slits whose separation is 0.75 mm. What is the separation of neighboring interference maxima on a screen 180 cm from the slits?

24.13 A pair of narrow slits is illuminated with red light of wavelength $\lambda = 633$ nm. The slits are separated by 0.085 mm center to center. (a) What is the angular separation of the interference maxima near the center of the pattern? (b) How far apart are neighboring maxima if they are observed on a wall 6.25 m away from the slits?

24.14 Red light of wavelength $\lambda = 633$ nm is incident on a pair of slits whose center-to-center separation is 0.120 mm. What is the linear separation of the maxima corresponding to $m = +7$ and $m = -7$ if they are observed on a wall 5.75 m away from the slits?

24.15 What ratio of λ/d produces a first-order interference maximum at angle $\theta = 0.010°$?

24.16• Double slits of 0.105-mm separation are placed at the front of a lecture room that is 21.3 m long. The interference pattern is cast on a screen at the front by reflecting the pattern from a mirror on the back wall of the room. If the light source is a helium-neon laser ($\lambda = 633$ nm), what is the separation between the first two bright fringes?

24.17• A collimated (parallel) beam of blue light of wavelength 440 nm falls normally on a pair of double slits and forms an interference pattern on a distant screen. If the light source is changed, what wavelength of visible light can be used to give an interference minimum at the same place that the first-order maximum occurred for the blue light?

24.18• The first-order maximum formed by a double slit illuminated with red light of $\lambda = 650$ nm lies at a given point on a screen. (a) If the slits are illuminated with light of another wavelength, is it possible for the second-order maximum of the new wavelength to lie at the same point? (b) Is this wavelength visible?

24.19• A pair of narrow slits is illuminated with blue light of wavelength $\lambda = 434$ nm. The resulting interference maxima are separated by 1.00 mm on a screen 1.00 m from the slits. What would be the separation of the maxima if the illuminating light were red light of $\lambda = 656$ nm?

Section 24.4 Interference in Thin Films

Hints for Solving Problems

In computing interference from the reflection from thin films, you must also include the phase change (equivalent to a path interval of $\lambda/2$) associated with reflection from a denser medium.

24.20 How thick should a sheet of mica ($n = 1.58$) be if it is to be as thin as possible and still give rise to destructive interference for reflection, given that the incident light is in the red ($\lambda = 650$ nm) part of the spectrum?

24.21 How thick should be the coating of a material with index of refraction 1.25 on a piece of flat glass if we wish for normally incident light of wavelength 450 nm to be transmitted with maximum efficiency?

24.22 How thick should be the coating of a material with index of refraction 1.32 on a piece of flat glass if we wish normally incident light of wavelength 560 nm to be reflected with maximum efficiency?

24.23 Determine (a) the thickness that gives the maximum transmission and (b) the thickness that gives the minimum transmission for light of wavelength 570 nm when a material with index of refraction 2.00 is deposited on glass that has an index of refraction of 1.50.

24.24• The index of refraction of ice ($n = 1.31$) is slightly less than that of water ($n = 1.33$). Is it theoretically possible to have a sheet of ice floating on water that does not reflect the green (550 nm) part of the spectrum from the sun when it is directly overhead? If so, what is the minimum thickness of the sheet?

24.25• If you are below the surface of a still pond on which a 0.001-mm thickness of oil ($n = 1.25$) is floating, what, if any, colors will be enhanced in the sunlight from directly overhead?

24.26•• Two glass microscope slides of 9.0-cm length are placed in contact. A 0.50-mm-thick spacer is placed between them at one end, forming a wedge of air. When they are illuminated from above with monochromatic light of $\lambda = 650$ nm, an alternating pattern of bright and dark bands is seen. What

is the distance between the center of one dark band and the next?

Section 24.5 Diffraction by a Single Slit

24.27 Calculate the width of the central maximum in the single-slit diffraction pattern of yellow light of $\lambda = 589.0$ nm by a slit 0.250 mm wide viewed on a screen 2.50 m away. (Try determining the separation of the two first-order minima.)

24.28 A single-slit diffraction pattern is formed when light of $\lambda = 633$ nm is passed through a narrow slit. The pattern is viewed on a screen placed one meter from the slit. What is the width of the slit if the width of the central maximum is 2.53 cm?

24.29 A single-slit pattern is formed by light passing through a narrow slit 0.050 mm wide. If the width of the central maximum is 3.8 cm on a screen 1.5 m away from the slit, what is the wavelength of the light?

24.30• Laser light $\lambda = 633$ nm is normally incident on a 0.100-mm-wide slit in a thin opaque plastic sheet floating on the surface of a swimming pool. Calculate the width of the central maximum at the bottom of the pool 4.50 m below the surface.

24.31•• Loudspeaker arrays are often required to spread the sound out in a large angle in the horizontal plane. Such speaker arrays are frequently made up of a vertical line of individual round speakers spaced close together and wired so as to move in phase. Suppose that it is desired that the sound be spread out over an angle of $90°$. Estimate the size of the individual speakers, given that the program material has an average frequency of 4000 Hz. Consider the speakers to form a vertical slit. Assume $v_{sound} = 340$ m/s.

Section 24.6 Multiple-Slit Diffraction and Gratings

Hints for Solving Problems

For a diffraction grating, the number of lines per centimeter is the reciprocal of the spacing between slits in centimeters.

24.32 Repeat Example 24.6 for second-order diffraction.

24.33• A pair of narrow slits is used to generate an interference pattern. The third-order maxima of the double-slit pattern are missing because they occur at the position of the first minima of the single-slit pattern. What is the ratio of slit separation to slit width?

24.34• A pair of narrow slits 0.50 mm in width is used to generate an interference pattern. The fourth-order maxima of the double-slit pattern occur at the positions of the first minima in the single-slit pattern. What is the separation of the slits?

24.35• Light of wavelength $\lambda = 632.8$ nm is normally incident upon a grating of 5000 lines/cm. How many different diffraction orders can be seen in transmission?

24.36• Small loudspeakers are evenly spaced in a long row. The loudspeakers are connected to move in phase when are fed a 8000-Hz signal. A person walking parallel to the row of speakers and 5.0 m away hears a decrease in the intensity every 0.50 m. How far apart are the speakers? Assume $v_{sound} = 340$ m/s.

24.37• In a double-slit experiment, the sixth-order maxima are missing. The slits are illuminated with green light of $\lambda = 546$ nm and observed on a screen 4.05 m away. If the separation between maxima is 4.25 mm, what is the width of the slits?

24.38•• Prove that for light incident on a diffraction grating at an angle ϕ from the normal to the plane of the grating (Fig. 24.48), the equation for diffraction maxima becomes $m\lambda = d(\sin \phi + \sin \theta)$, where m is a positive or negative integer, λ is the wavelength of the light, and d is the grating spacing.

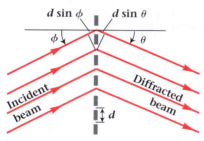

Figure 24.48
Problem 24.38.

Section 24.7 Resolution and the Rayleigh Criterion

Hints for Solving Problems

Unless specified otherwise, take the wavelength of light to be 550 nm.

24.39 What is the minimum angle of resolution of an eye with a pupil diameter of 2.0 mm? This diameter corresponds to the value for a typical human eye in bright light.

24.40 Calculate the minimum theoretical angle of resolution of a telescope 5.08 m (200 in.) in diameter for light of $\lambda = 550$ nm.

24.41 Two ducks are flying close together, with one leading the other by 0.60 m. Their motion is transverse to a duck hunter who sees them through binoculars with 35-mm-diameter objective lenses. How far away from the hunter are the ducks, according to the Rayleigh criterion, if they are just resolved as a pair?

24.42 Imagine an airplane passenger looking down on two automobiles side by side on the ground. According to the Rayleigh criterion, at what maximum altitude will the passenger just resolve them as two if their center-to-center separation is 2.3 m? Assume that the passenger has good eyesight with pupils narrowed to 2.0 mm because of bright light.

24.43 An amateur astronomer has a reflecting telescope of 0.20-m (8-in.) diameter. What is the maximum distance between objects on the moon that can be resolved, according to Rayleigh's criterion? Take the wavelength of light to be 550 nm and the earth-moon distance to be 3.8×10^8 m.

24.44• Two small street lights 2.00 m apart are photographed by a camera 30 m away. The camera's 50-mm focal-length lens has an aperture of $f/1.8$. Can the two lamps be distinguished on the photograph?

24.45• A sensitive galvanometer uses a telescope to read a scale reflected off a 3.0-mm-diameter mirror mounted on a moving coil. The scale is a circular arc with a radius of 1.50 m centered about the mirror. What is the smallest division on the scale that can be read with the telescope regardless of the magnification? Assume $\lambda = 550$ nm.

24.46• A camera with a 135-mm focal-length lens is to be used to obtain a sharp image of a dot pattern on a wall 200 m away. The dots are small compared with the spacing between them and are regularly spaced 3.3 mm apart. What is the largest f-number that can be used if the lens performance is limited by diffraction alone?

24.47•• A weather satellite in stationary orbit has a camera lens 75 mm in diameter. What is the minimum-size object that it can resolve on the earth's surface? (*Hint:* First find the altitude of the satellite above the earth's surface. Then assume that the object diameter will subtend an angle θ_m at the camera lens. Take $\lambda = 550$ nm.)

24.48•• Imagine that you are an astronaut on the moon photographing the earth with your Hasselblad camera whose $f/2.8$ lens has an 80-mm focal length. Is is possible (even in principle) for you to see the Great Wall of China (width 7.6 m) on the photograph? If not, then what is the size of the smallest object that can be seen?

Section 24.8 Dispersion

24.49 Using the data below, make a graph of index of refraction versus wavelength for high dispersive crown glass and heavy flint glass. Compare your curves with Fig. 24.32.

Wavelength (nm)	434	486	589	656	768
(Types of glass)		Index of refraction			
High dispersive crown	1.546	1.533	1.527	1.520	1.517
Heavy flint	1.675	1.664	1.650	1.644	1.638

24.50 Light is normally incident on one face of a 30° flint-glass prism. Calculate the angular separation of red light ($\lambda = 650$ nm) and violet light ($\lambda = 450$ nm) emerging from the back face. Use $n_{red} = 1.644$ and $n_{violet} = 1.675$. (See Fig. 24.33.)

Section 24.10 Polarization

> **Hints for Solving Problems**
>
> Remember to measure angles of reflection and refraction (including Brewster's angle) from the normal to the surface.

24.51 Two polarizers transmit light of intensity I_m when their polarization directions are aligned. How much light is passed when they are oriented at an angle of 20° with respect to each other?

24.52 Two ideal polarizers transmit light of intensity I_m when their polarization directions are aligned. What is the angle between their directions of polarization when the transmitted intensity is 21% of I_m?

24.53 What is the angle of incidence for maximum polarization for light reflected from the surface of glass of index 1.52?

24.54 Find the Brewster angle for light reflected at a water-glass interface. Use $n = 1.52$ for the glass. Assume incident light in the water.

24.55 The Brewster angle for light reflected from a piece of glass is 58.9°. (a) What is the index of refraction of the glass? (b) What kind of glass is this? (Refer to Table 22.1.) Assume the incident light is in air.

24.56 Find the Brewster angle for maximum polarization for light reflecting off a water-air interface; that is, the incident light comes from within the water. Compare this result with the result of Example 24.10.

24.57• An ideal polarizer passes 50% of the incident light intensity when the incident light is unpolarized. Unpolarized light from a particular lamp has an intensity measured with a light meter. (a) What is the intensity reading on the meter when a single ideal polarizer is inserted in the beam? (b) What is the intensity reading if a second ideal polarizer is inserted in the beam with its polarization direction aligned parallel to that of the first polarizer? (c) What is the intensity when the second polarizer is rotated by 45°?

24.58•• Polarizers are oriented at an angle of 90° with each other, blocking the passage of light. When a third polarizer is added between them, light is transmitted. (a) Explain how this can occur. (b) Assuming ideal polarizers, find the angle of the third polarizer for maximum light transmission through the system.

24.59•• A commercially available Polaroid sheet passes only 32% of the incident unpolarized light. (a) If two of these Polaroids are aligned so that their polarization directions coincide, what percentage of an incident unpolarized beam will be transmitted by the pair of Polaroids? (b) What percentage of the incident light will be transmitted if their polarization directions differ by 32°?

*Section 24.11 Scattering

24.60 Make a graph of the relative intensity of Rayleigh scattering in the visible region against wavelength. Use any arbitrary value for the intensity of the scattering at the violet end of the spectrum and calculate for other wavelengths on a relative basis.

Additional Problems

24.61 An interference pattern is produced by illuminating double slits of center-to-center separation 0.23 mm with monochromatic light. What is the wavelength of the light if the third-order maximum beam makes an angle of 0.51° with respect to the optical axis?

24.62 Two narrow slits 0.20 mm apart are illuminated with monochromatic light. The third-order maximum lies at a certain point on a screen some distance away. What would be the separation of the slits required to give a fifth-order maximum at the same point on the screen if the wavelength and the distance to the screen are unchanged?

24.63 A single-slit pattern is formed by light passing through a narrow slit 0.050 mm wide. If the central maximum has a width of 4.7 cm on a screen 2.0 m from the slit, what is the wavelength of the light?

24.64 The first-order diffracted beam of a normally incident well-collimated 546-nm light beam emerges from a grating at an angle of 5° with respect to the normal. How many lines per centimeter does the grating have?

24.65 What wavelength in the visible region will be best reflected at near normal incidence from a glass sheet that is 400 nm thick? Take the index of refraction of the glass to be 1.50.

24.66• White light is normally incident on a diffraction grating that has 6000 lines/cm. What is the angular separation between red light ($\lambda = 650$ nm) and blue light ($\lambda = 450$ nm) in the second order?

24.67• White light is normally incident on a 4000-lines/cm diffraction grating. (a) Do the first- and second-order visible spectra overlap? (b) Do the second- and third-order visible spectra overlap? Take the visible spectrum to range from 400 nm for violet to 700 nm for red.

24.68• White light is normally incident on one face of a 30° prism of polymethylmethacrylate. Find the approximate angular separation of red light of $\lambda = 650$ nm and blue light of $\lambda = 450$ nm as they emerge from the second face. Refer to Fig. 24.32 for the index of refraction.

24.69•• Laser light of wavelength $\lambda = 633$ nm is incident on a 5000-lines/cm diffraction grating. If the incident light makes an angle $\phi = 30°$ with the normal to the grating, what are the angles of the two first-order diffracted beams? (*Hint:* See Problem 24.38.)

24.70•• Light of $\lambda = 633$ nm is incident upon a transmission grating of 2000 lines/cm at an angle of 15° from the normal. (a) Find the angles of emergence of the two first-order diffracted beams. (b) Draw a diagram showing what occurs. (*Hint:* See Problem 24.38.)

24.71•• A radio station broadcasting on a frequency of 1500 kHz generates a directional beam by using an array of four antennas that are all driven in phase. The antennas are arranged along an east-west line so that each antenna is 50 m from the next. (a) In what direction will the radiated signal be the greatest? (b) How much signal will radiate along the east-west line?

24.72•• If a rainbow were made by drops of liquid with an index of refraction greater than that of water, say $n = 1.47$, would the bow appear higher or lower in the sky? Justify your answer and draw a diagram of the angles involved.

24.73•• The resolution of photographic lenses is often determined by photographing a chart with groups of parallel lines. The developed film is examined under magnification, and the most closely spaced group of lines that can be resolved is noted to give the resolution of the lens in lines per millimeter. Typical resolutions for the central image of good-quality 35-mm camera lenses are in the range of from 50–80 lines/mm, depending on the aperture. Because effects other than diffraction also limit resolution, the theoretical resolution is much better than commercial camera lenses usually achieve. The best lenses approach the "diffraction limit." Determine the diffraction limit, in lines per millimeter, of the resolution of an $f/1.8$, 50-mm focal-length lens. (The resolution given is for the lines on the film, not at the object position. Assume a thin lens and let the object be so far away that the lens-film distance is equal to the focal length.)

chapter 25

Relativity

Toward the end of the nineteenth century, the once-separate disciplines of mechanics, optics, acoustics, thermodynamics, and electromagnetism had become interrelated through the concepts of energy and energy conservation. Physicists had developed these various subfields of physics and used them to explain many natural phenomena. In fact, physicists of that day were familiar with most of the ideas we have discussed so far in this book—ideas that came to be known as classical physics. Chief among the classical theories was Newtonian mechanics, which, more than any other science, had shaped people's view of the world. Scientists relied on mechanistic models to explain almost everything in and out of physics; thus, Newton's ideas transformed society as well as physics.

Into this setting came Maxwell's extremely successful theory of electromagnetism.

Its explanation of the wave nature of light contributed to the unification of physics. However, Maxwell's theory was not entirely consistent with Newtonian mechanics. Maxwell predicted that an electromagnetic wave should propagate in a vacuum with a unique speed $c = 3 \times 10^8$ m/s. This speed became a point of conflict because classical mechanics had no explanation for why light should always travel at the same characteristic speed.

In 1905, Albert Einstein devised a theory that reconciled the discrepancies between mechanics and electromagnetism. This theory of relativity challenged the Newtonian view and has affected the way people look at the world. The effects of the theory of relativity on contemporary society have been so great that "relativity" and "Albert Einstein" are among the outstanding words and images of modern science. ■

25.1 Principle of Relativity

The concept of relative motion was introduced in Chapter 3 when we considered relative velocities. There we pointed out that before we can measure an object's velocity, we must first specify the coordinate system or reference frame within which we make our measurements. Both Galileo and Newton recognized the basic ideas of relative motion. Newton conceived of space as absolute, and he defined absolute motion as the translation of a body from one absolute place to another. However, he also recognized that all motion is necessarily measured relative to the position from which it is observed.

We saw in Chapter 4 that Newton's laws apply in any nonaccelerating frame of reference, which we called an *inertial reference frame* because in such a frame the law of inertia (Newton's first law) holds. If the laws of mechanics are valid in one inertial frame, such as a railway station, they must also be valid in any other inertial frame moving with constant velocity relative to the first one, such as a train moving past. In fact, there is no way to prove that any given inertial frame is absolutely at rest. It is reasonable to say that the train is at rest and the earth is moving beneath it.

The statement that the laws of mechanics are valid in all inertial frames is known as the principle of **Galilean relativity.** It is so named in honor of Galileo, who first discussed the motion of an object in one coordinate system as seen from another. Galilean relativity is part of our everyday experience. For example, if a passenger riding in a car moving with constant velocity pours a soft drink into a paper cup, the results are the same as if she had performed this task in her home. The same laws of gravity and motion apply in each inertial reference frame.

Applying Galilean relativity to Maxwell's theory of electromagnetic waves led to contradictory results. The dominance of Newton's mechanistic ideas of motion led physicists to think that electromagnetic waves must propagate in some mechanical medium, which they called the ether. The absolute reference frame in which Maxwell's equations were valid and light would travel with speed c was thought to be a frame that was stationary or fixed with respect to the ether. Since the earth rotates about its axis and revolves about the sun, the earth must surely be moving through the ether—if indeed the ether does exist.

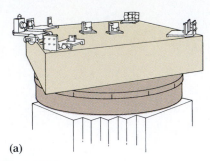

(a)

Figure 25.1

In the Michelson-Morley experiment, a light beam is divided in two by a half-silvered mirror. The two beams travel along different paths, are reflected several times, and then are recombined. The resulting interference pattern is extremely sensitive to changes in path length or wave speed in either branch. (a) The apparatus, called an interferometer. (b) Diagram of the paths of light.

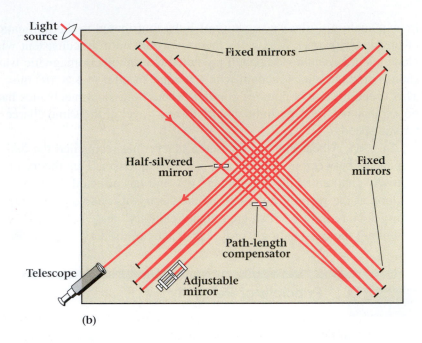

(b)

Galilean theory predicted that the speed of light would not be the same if viewed from different frames of reference moving with constant velocity relative to each other. Therefore, the speed of light should be different depending on whether it is measured along a direction parallel to the earth's motion through the ether or along a direction at right angles to its motion.

A crucial experiment, designed to measure how the earth's motion through the ether affected the speed of light, was first performed by the American physicist A. A. Michelson (1852–1931) in 1881. The experiment was later repeated with greater precision many more times by Michelson and E. W. Morley (1838–1923). In the Michelson-Morley experiment, a beam of light was split into two parts traveling at right angles to each other (Fig. 25.1). The beams were then reflected and recombined to form an interference pattern. Changes in the interference pattern could be used to detect very minute changes in the speed of light. Thus, by rotating the apparatus, they could tell whether the velocity of light changed with direction. Michelson's experiments gave a negative result; no changes were observed in the velocity of light. Thus, there was no preferred frame of reference and therefore no ether.

| 25.2 | **Einstein's Postulates of Special Relativity** |

Numerous attempts were made to reconcile the discrepancies between Galilean relativity and Maxwell's equations. However, they were all unsuccessful until Albert Einstein proposed his special theory of relativity. Einstein's work resolved the problem raised by the Michelson-Morley experiments, although his ideas were not motivated by their experimental results. In his famous paper "On the Electrodynamics of Moving Bodies," which appeared in 1905, Einstein suggested that absolute motion had no meaning in electrodynamics and that the ether concept was superfluous. His **special theory of relativity** was based on two simple postulates:

I. The laws of physics are the same in all inertial reference frames.

II. The speed of light in free space is independent of the motion of its source and of the motion of the observer.

From these two postulates Einstein formed a theory that brought mechanics and electromagnetism together, but that demanded changes in the formulation of mechanics for speeds approaching the speed of light. Special relativity is no longer considered a theory to be tested; like Newton's laws or Maxwell's equations, it is a physicist's everyday working tool in its appropriate domain.

The simplicity of the postulates is perhaps a little surprising, given the reputation of special relativity for producing unusual predictions. The first postulate appears to be a reasonable requirement of any theory. After all, it is natural to expect that the laws of nature do not change according to the situation. In particular, the laws of nature should not depend on relative constant motion. Though simple, the second postulate is not intuitively apparent—in fact, it stands in direct contradiction to what we have assumed in everyday life to be the correct way of adding velocities as well as to the mechanics we have described so far. However, it is possible to test this statement, and we find that it does agree with experiments, as we show in this chapter.

What was so unexpected was that two simple postulates, when followed to their logical conclusion, gave such unexpected results, but results that are nevertheless confirmed by experiments. As we shall see, these results revolutionized our conceptions of time and space. Even the seemingly simple matter of how we add velocities must be reconsidered in light of these postulates.

25.3 Velocity Addition

When you run forward and throw a ball in the direction in which you are running, the ball moves faster than it would if you were standing still. For example, if you can throw a ball 30 m/s when standing still, the same throw when you are running at 3 m/s gives the ball a speed of 33 m/s. You can summarize your ordinary experience in adding velocities as

$$u = u' + v,$$

where v is your velocity with respect to the ground, u' is the velocity of the ball with respect to you, and u is the velocity of the ball with respect to the ground. This expression is called the Galilean, or classical, velocity addition formula.

Though the Galilean velocity addition formula seems to agree with our everyday experience, it violates the second postulate of special relativity, which, as we have already indicated, agrees with experimental observations. This difficulty is resolved by use of the correct velocity addition formula, which was first given by Einstein. It is

$$u = \frac{u' + v}{1 + \dfrac{u'v}{c^2}}, \tag{25.1}$$

ALBERT EINSTEIN

Albert Einstein burst into the world news in 1919 in a way no other scientist ever has. That year saw a dramatic confirmation of his prediction for the amount of bending of a light ray by a gravitational field; as a result, he became *the* scientist in the public eye. Most readers did not know what the theory of relativity was. Nevertheless, its confirmation by British astronomers during a total solar eclipse in 1919, the intriguing name of the theory, and the personality of the author, all combined to create unprecedented interest in a theory with no apparent practical applications. Einstein quickly became, and still remains, one of the most widely recognized public images (Fig. B25.1). His usually rumpled appearance made him popular with photographers and his unaffected and open nature made him popular with any audience. He was so popular that the London Palladium once tried to engage him for a three-week appearance.

As a young man, the German-born Swiss citizen spent the years 1902–1908 at the Swiss Patent Office in Bern. There he had the freedom to concentrate and work alone on basic questions, a way of working that characterized his entire career. In 1905, he published three unsually important papers. The first was on special relativity. The second gave an explanation of Brownian motion and helped establish the reality of the molecular view of matter (Chapter 26). The third paper, about the nature of light and its interaction with matter, was a part of the foundation of quantum mechanics (Chapter 28). Later in the same year, he published a fourth paper, which contained the well-known equation $E = mc^2$. Einstein went on to develop the general theory of relativity, publishing his first results in 1915. In 1921, after the confirmation of the general theory of relativity, he received the Nobel Prize in physics for his contribution to theoretical physics and his explanation of the interaction of light with matter.

The widely told story that young Einstein was a poor student is untrue. He made above-average marks, but did not like formal schooling, preferring to follow his own train of thought without regard for authority. He worked in his own way while at the patent office, his most productive period and one that he remembered as the happiest in his life.

At least as important as Einstein's inventive mind and deep knowledge of physics was his ability to concentrate on

Figure B25.1 Albert Einstein (1879–1955), shown at his desk at the Swiss Patent Office around the time of his 1905 papers.

problems for as long as it took to solve them. The ideas of general relativity took eight years to develop. He so thoroughly followed his own train of thought that when he arrived at such ideas as doing away with the conventional concepts of space and time, they seemed entirely natural and unavoidable to him. Anyone with such a disposition is likely to have little regard for authority and convention, either in science or in any other area.

Einstein is invariably compared with Newton with respect to the importance of their theories and the intensity with which they worked on problems. But they had little else in common. We earlier mentioned Newton's pride, intellectual vanity, and combative personality. Einstein was just the opposite. Comments from two scientists who knew him convey something of his character. According to Otto Frisch, "The quality that dominated his personality was a very great and genuine modesty. When anybody contradicted him he thought it over and if he found he was wrong he was delighted because he felt that he had escaped from an error and that now he knew better than before."

Robert Oppenheimer wrote,

Einstein is also, and I think rightly, known as a man of great good will and humanity. Indeed, if I had to think of a single word for his attitude toward human problems, I would pick the Sanskrit word *Ahinsa,* not to hurt, harmlessness. He had a deep distrust of power; he did not have that convenient and natural converse with statesmen and men of power that was quite appropriate to Rutherford and to Bohr, perhaps the two physicists of this century who most nearly rivaled him in eminence . . .

After what you have heard, I need not say how luminous was his intelligence. He was almost wholly without sophistication and wholly without worldliness. I think that in England people would have said that he did not have much "background", and in America that he lacked "education". This may throw some light on how these words are used. . . . There was always with him a wonderful purity at once childlike and profoundly stubborn.*

---

*From *Science and Synthesis* (Paris: Unesco Publications, 1967), with permission.

where the symbols u, u', and v have the same meaning as before and c is the velocity of light. Equation (25.1) is in agreement with both our everyday experience and the second postulate of relativity. When the velocities involved are small compared with the velocity of light, the relativistic velocity addition rule gives the same result as would be expected from classical mechanics. (See Problem 25.11.) When the velocities are not small compared with the speed of light, the results of the velocity addition rule are correct, but no longer so intuitive.

Note that the velocities in Eq. (25.1) all lie in the same direction. The formula can also be used when one velocity is directed opposite to the other, if you choose one direction positive and the other negative. An appropriate generalization of Eq. (25.1) is used for motion that is not colinear.

We find that we must modify our ideas of velocity addition to bring them into agreement with the postulates of special relativity. Because velocity is the displacement divided by the time, we need to reevaluate our concepts of both time and space in light of the postulates of special relativity. Before looking more closely at space and time, we first need to ask a question that does not even occur in classical mechanics: "What do we really mean when we say that things happen simultaneously?"

Example 25.1

Show that the velocity addition rule is consistent with Einstein's postulate that the speed of light is the same to all observers.

Strategy Suppose one observer moves with velocity v relative to another observer. The first observer shines a light beam directly ahead of him at a velocity c measured in his frame. The speed of light as seen by the second observer can be found from the velocity addition rule of Eq. (25.1).

Solution We set $u' = c$. Then the second observer measures the speed of the light beam as

$$u = \frac{c + v}{1 + \dfrac{cv}{c^2}} = \frac{c + v}{1 + \dfrac{v}{c}} = \frac{c(c + v)}{c + v} = c.$$

Discussion The velocity addition rule does give us the desired result, as indeed it must since it was derived from Einstein's postulates.

Example 25.2

A spaceship is moving away from the earth at a speed of $0.5c$ when it launches a probe in the same direction with a speed of $0.5c$, as seen from the ship. What is the probe's speed as seen from earth?

Strategy We can solve this problem with straightforward application of the velocity addition formula (Eq. 25.1). We are free to choose the direction of motion of the spaceship to be positive. Then the direction of the probe's motion is also positive.

Solution The spaceship moves with velocity $v = 0.50c$, and the probe moves with velocity $u' = 0.5c$. The resulting speed of the probe seen from the earth is

$$u = \frac{0.5c + 0.5c}{1 + \dfrac{(0.5c)(0.5c)}{c^2}} = \frac{c}{1 + 0.25} = 0.8c.$$

Discussion The probe moves away from the earth with a speed of $0.8c$ as measured by an observer on earth—less than the speed of c predicted by the classical addition of relative velocities.

Example 25.3

A spaceship exploring a strange planet moves away from the planet with a speed of $0.48c$ and fires a probe back toward the planet at a speed of $0.63c$ as seen from the spaceship (Fig. 25.2). What is the speed of the probe as seen from the planet?

Strategy Again we apply the relativistic velocity addition formula, choosing the direction away from the planet as positive. Since the probe is fired back toward the planet, its velocity is negative in the frame of the spaceship.

Solution The velocity of the spaceship is $v = +0.48c$, and the velocity of the probe measured in the frame of the spaceship is $u' = -0.63c$. The velocity of the probe as seen from the planet is

$$u = \frac{u' + v}{1 + \dfrac{u'v}{c^2}} = \frac{-0.63c + 0.48c}{1 + \dfrac{(-0.63c)(0.48c)}{c^2}} = \frac{-0.15c}{1 - 0.30} = -0.22c.$$

Discussion To an observer on the planet, the probe approaches at a speed of $0.22c$. Notice that the probe's velocity as seen from the planet is not simply the algebraic sum of the two velocities u' and v. ■

Figure 25.2

Example 25.3: A spaceship, moving away at speed $0.48c$ from a strange planet, fires a probe back toward the planet at speed $0.63c$. Seen from the planet, the probe's speed is greater than the difference between these two speeds.

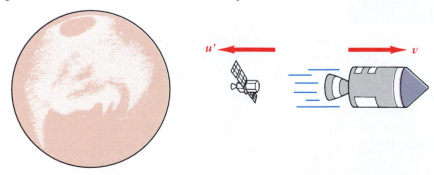

25.4 Simultaneity

Relativity theory—whether special relativity or classical Galilean relativity—compares measurements made by two observers in relative motion to each other. Without any loss of generality, we can pick one observer to be stationary and the other to be in motion with constant velocity relative to the first observer. For example, suppose the "stationary" observer is on the ground and the "moving" observer is on a train. To the observer seated in the railway car, the train seems at rest and the world seems to be rushing past. You may have noticed the same thing if you've ever sat at a train window as another train opposite you started to move. It is often difficult to know at first whether you are moving and the other train is stationary, or you are at rest and the other train is moving.

We can describe the situation using inertial reference frames (Fig. 25.3), showing the coordinates of the outside stationary observer and the passenger on the train. Reference frame S corresponds to the observer on the ground; frame S' corresponds to the train passenger, moving with velocity v in the x direction. The distance x' is a distance measured by the passenger in the frame S' within the train. For example, it may be the distance from the passenger to the empty seat in front of her. The distance x represents the x component of the distance from the outside observer to the empty seat and is measured from the outside observer's position on the ground in the frame S. The observer sees this distance to be increasing with time as the train moves away with speed v.

If we want to discuss motion in the two reference frames moving relative to each other, then we need to compare time measurements in two frames. Thus, we must establish a common reference between them. That is, we need to synchronize the clocks at some particular time. How can we do this? First, consider two observers in the same reference frame, say frame S (Fig. 25.4), measuring time with separate but identical clocks. Imagine one observer to be at the origin and the other at a distant point x. They can synchronize their clocks in the following way. A light pulse flashed by the observer at the origin at time $t = 0$ requires a time interval x/c to reach the second observer at x. When the observer

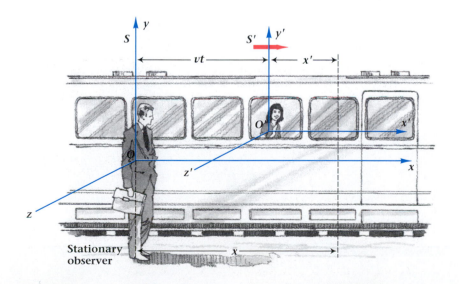

Figure 25.3

An observer on the ground watches a traveler passing by on a train. The passenger's frame of reference S' moves with constant speed v along the positive x direction relative to the observer's stationary frame S.

Stationary observer

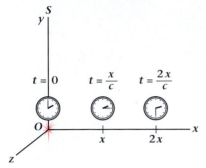

Figure 25.4
A clock at a distance x from the origin may be synchronized to a clock at the origin. When a light pulse sent from the origin at time $t = 0$ is received at x, the clock at x is set to $t = x/c$. Other clocks may be set in the same manner, where the time is related to their position divided by the speed of light.

sees the light flash, he sets his clock to time $t = x/c$. Thus, the two clocks are synchronized. Other observers in S can set their clocks in a similar manner.

Observers in an inertial frame S' moving with constant velocity relative to the frame S (as in Fig. 25.3) can also measure time with identical clocks. Using the same procedure, they can synchronize their clocks with a clock at the origin of S'. Thus, a clock at any point x' measured in the frame S' is set to $t' = x'/c$ when a light pulse emitted from the origin of S' at $t' = 0$ reaches the point x'.

At the instant when the origins of the two reference frames coincide, we can synchronize the clocks at the origins of both reference frames. Then when an observer in S records an event occurring at point x and time t, an observer in frame S' records the same event at a position measured as x' and time t'. However, time measured in frame S' depends not only on the time t of a clock in frame S, but also on the *position* of that clock. The greater the separation of the clock in S from the origin of S, the greater the disagreement between t and t'.

Through his insistence on the constancy of the speed of light as a necessary part of a correct relativistic relationship, Einstein revealed an unexpected phenomenon. Time itself is relative. This was a startling revelation to people accustomed to Newton's intuitive view that time is absolute and completely separate from space. In his *Principia,* Newton had written that "absolute, true and mathematical time, of itself, and from its own nature flows evenly without regard to anything external." But time and space are inseparably linked.

The fact that an observer's measurements of the space and time intervals between two events depend on the motion of the observer's reference frame forces us to reexamine our notions of simultaneity. For example, consider an observer at rest in a coordinate system S. He perceives two events—say two lightning flashes—as being **simultaneous** if light waves from the events reach his position at the same time (Fig. 25.5). However, since the transit time of a light signal depends on the distance from source to observer, these same two events will not appear simultaneous to a second observer stationed a distance x away from the first. Thus, two clocks synchronized in the manner described above appear to show the same time only to an observer located equidistant between them. Everyone else sees the two clocks showing different times. (Of course, in most everyday situations, these differences are extremely small and go unnoticed.)

Now let's consider two observers in relative motion. Suppose an observer in stationary reference frame S sees two clocks to be synchronized in his reference frame. An observer in moving reference frame S' will not see these same two clocks as synchronized. Figure 25.6 illustrates this loss of simultaneity with a thought experiment due to Einstein. We suppose that two lightning bolts strike opposite ends of a boxcar that is moving to the right with velocity v relative to

Figure 25.5
If an observer at A sees two lightning flashes to be simultaneous, the observer at B does not observe them to be simultaneous.

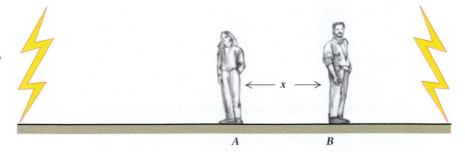

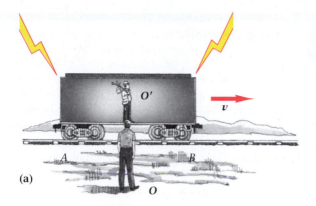

(a)

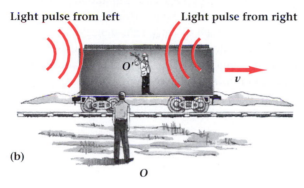

Light pulse from left   **Light pulse from right**

(b)

(c)

Figure 25.6
Observer O, on the ground, sees two lightning flashes strike both ends of the railroad car simultaneously. Observer O', riding in the car, sees the right-hand flash first. The sequence (a), (b), and (c) shows the position of the railroad car and the light pulses at successive times.

an observer O standing near the tracks halfway between A and B. A passenger O' riding in the boxcar observes the same two bolts. However, since the observer O' is moving toward the right, the light pulse from the right reaches him before the light pulse from the left. He concludes that lightning struck the front of the boxcar first. However, light pulses from left and right reach observer O at the same time. He concludes that the two events are simultaneous.

Thus, two events that one observer sees as simultaneous may be sequential to someone else, and may perhaps even be in reverse sequence to a third viewer. However, causal relations are always maintained. That is, if event A causes event B to happen, no observer, regardless of his or her motion, will ever be able to see these two events occur in the reverse order of B before A. No observer can ever see an effect precede the event that causes it to take place.

25.5 Time Dilation

One of the most intriguing effects of special relativity is the phenomenon of time dilation. Because of this effect, a mechanical (or biological) clock in an inertial reference frame in motion relative to you runs more slowly than an identical clock at rest in your reference frame. As Einstein himself observed—when we move we not only change our position in space, we also change the rate at which we advance into the future. This observation is in disagreement with the Newtonian concept of absolute, evenly flowing time.

We can derive the mathematical expression of time dilation from the following thought experiment. Imagine a spaceship moving to the right with a constant speed v. A flashlamp emits a light pulse that travels upward until it hits a mirror fixed to the ceiling of the spaceship a distance d above the lamp. The mirror reflects the pulse back down to a light detector mounted beside the lamp. An observer in the moving reference frame S', at rest with respect to the spaceship, uses an electronic clock to measure the time interval Δt_0 for the light pulse to travel from the lamp to the mirror and back to the detector (Fig. 25.7a). This time interval is given by the path length traveled by the pulse divided by the speed of light,

$$\Delta t_0 = \frac{2d}{c}.$$

However, since the spaceship moves with a speed v relative to the reference frame S, an observer in S sees the light pulse travel a longer path than $2d$ in the time Δt measured in his frame (Fig. 25.7b). Because the speed of light is the same in both reference frames (Postulate I), the time interval measured in the frame S must be longer than the time interval measured in the moving frame. It is given by

$$\Delta t = \frac{2L}{c},$$

where $2L$ is the total path traveled by the light. Applying the Pythagorean theorem to Fig. 25.7(b), we obtain

$$\left(\frac{c\Delta t}{2}\right)^2 = d^2 + \left(\frac{v\Delta t}{2}\right)^2.$$

Figure 25.7

(a) The time of flight of a light pulse as measured with a clock stationary in the spaceship (frame S') is $2d/c$.
(b) An observer in S who sees the spaceship moving with speed v sees the light beam travel a longer path and thus measures a longer time of flight.

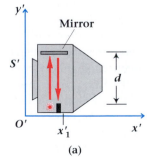

(a)

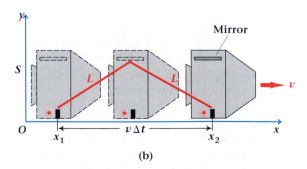

(b)

Since $2d/c$ is Δt_0 in the reference frame S', we get

$$(\Delta t)^2 = (\Delta t_0)^2 + \frac{v^2}{c^2}(\Delta t)^2.$$

Upon rearranging, we find that

$$\Delta t = \frac{\Delta t_0}{\sqrt{1 - \dfrac{v^2}{c^2}}}. \tag{25.2}$$

Thus, an observer moving with respect to the clock observes the time interval indicated by that clock to be stretched out. For this reason we call this effect **time dilation.**

The time interval Δt_0 measured in a reference frame in which the clock is at rest is called the **proper time.** The time interval measured in any frame moving with speed v relative to the clock is greater than the proper time according to Eq. (25.2). Consider two observers with identical clocks, each at rest in its own frame of reference. The frames move relative to each other with speed v. Each observer measures the other's clock to be running slow. This must be so because there is no preferred reference frame and all that matters is their relative motion.

Example 25.4

The starship *Enterprise* makes a long trip into space, traveling at a speed $v = 0.80c$ relative to the earth. The trip takes 20 years, according to a shipboard clock. How many years elapse on earth during the voyage?

Solution We use Eq. (25.2) to determine the elapsed time. The proper time Δt_0 of the shipboard clock is 20 years. To an earthbound observer, the elapsed time is

$$\Delta t = \frac{\Delta t_0}{\sqrt{1 - \dfrac{v^2}{c^2}}} = \frac{20 \text{ years}}{\sqrt{1 - (0.80)^2}} = 33 \text{ years.}$$

The first measurement of time dilation for radioactive particles moving at high speed relative to the earth was made in 1941 by Bruno Rossi and D. B. Hall. We will describe a similar measurement made in 1963 by D. H. Frisch and J. H. Smith.* The experiment involves the detection of muons, which are subatomic particles that are produced in the upper atmosphere and that rain down toward the earth with a speed close to the speed of light. As they travel downward, some of them spontaneously decay in flight. Consequently, the number arriving at a medium altitude—say on top of a mountain—is greater than the number that

---

*D. H. Frisch and J. H. Smith, "Measurement of the Relativistic Time Dilation Using μ-Mesons," *American Journal of Physics,* May 1963, p. 342.

survive to reach sea level. In their experiment, Frisch and Smith counted the number of muons having a narrow range of speeds near the peak of Mount Washington, New Hampshire. Then they went down to Cambridge, Massachusetts, and counted the muons that survived the trip down to sea level. The probability that a muon will decay, and thus its mean life, is determined only by forces within the muon itself. Therefore, any dependence of their mean life on their speed relative to us is due only to relativity.

Let us examine Frisch and Smith's findings in detail. The Mount Washington apparatus was set up to detect and stop muons with speeds of $0.995c$. The time intervals between the arrival of a muon in the detecting apparatus and its subsequent decay were measured. Measurements from one run (Fig. 25.8) corresponded to a mean life of 2.2 μs. The average number of muons per hour arriving at this detector was 563. The time required for the muons to travel the 1907 m from the elevation of the mountaintop to the elevation of the laboratory in Cambridge, measured with a clock stationary with respect to the laboratory, is

$$\Delta t = \frac{d}{v} = \frac{1907 \text{ m}}{(0.995)(3 \times 10^8 \text{ m/s})} = 6.4 \ \mu\text{s}.$$

From Fig. 25.8 we see that after 6.4 μs elapsed, only about 27 muons per hour would be expected to survive the trip down. But when detectors were set up in Cambridge to count muons of initial speed $0.995c$, they observed an average rate of 408 muons per hour. We conclude that the muons decay more slowly when they are moving rapidly relative to us than when they are at rest relative to us.

How does the observed count of 408 muons per hour correspond to relativity predictions? On the graph we see that 408 muons/h corresponds to an elapsed time of only 0.7 μs. This means that time runs slower in the frame of reference of the moving muons than in the laboratory by a factor of 0.7/6.4. We can check these results against the prediction of Eq. (25.2) by rearranging that equation for v and substituting for $\Delta t_0/\Delta t$:

$$v = c\sqrt{1 - (\Delta t_0/\Delta t)^2} = c\sqrt{1 - (0.7/6.4)^2} = 0.994c.$$

This calculated value for the speed of the muon in the laboratory frame is consistent with the value previously established for the experiment.

This experiment confirms time dilation and supports the special theory of relativity. Furthermore, it shows that relativistic effects can become large when the relative speeds approach the speed of light. The effects observed here are general effects of relativity and are not limited to the special case of radioactive decay.

Figure 25.8

Surviving muons as a function of time. (From Frisch and Smith, with permission.)

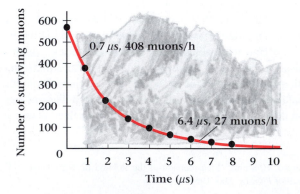

Example 25.5

The mean life of muons at rest is 2.2 μs. What is the mean life of a beam of muons moving with a speed of 0.95c?

Solution We use the time dilation equation with the proper time $\Delta t_0 = 2.2$ μs and $v/c = 0.95$. The result is

$$\Delta t = \frac{2.2 \times 10^{-6} \text{ s}}{\sqrt{1 - (0.95)^2}} = 7.0 \times 10^{-6} \text{ s} = 7.0 \text{ } \mu\text{s}.$$

The apparent lifetime of the muons increases by about a factor of 3 over their proper mean life.

25.6 Length Contraction

We have seen that observers at rest relative to one another occupy the same reference frame and thus agree on their measurements of space and time. However, if they are moving relative to one another, their measurements of space and time do not agree. These differences are significant only at speeds approaching the speed of light, but though negligible, they exist at everyday speeds as well. With these concepts in mind, let's examine in detail the problems of measuring the length of a moving object.

Consider an observer in a frame S' moving with speed v relative to frame S (Fig. 25.9). This observer is at rest with respect to a rod moving with the system. The observer in frame S' measures the rod to have a length l_0 along the direction of motion, $l_0 = x_2' - x_1'$. This is called its **proper length:** the length measured by the observer at rest with respect to the rod. The observer also sees a marker at point P in frame S approach and then recede at speed v, covering a distance equal to the length of the rod in a time $\Delta t = l_0/v$. This time interval Δt is not a proper time, for the events that define it occur at different places (the

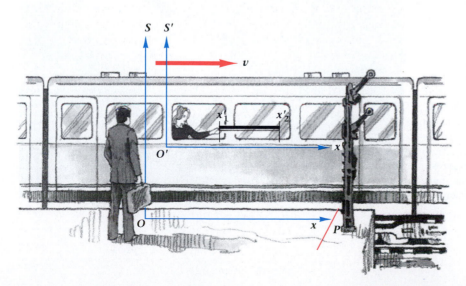

Figure 25.9

A rod at rest in system S' moves with speed v relative to the frame S. The length of the rod as measured in S' is l_0, the proper length of the rod.

THE TWIN PARADOX

A fascinating aspect of the theory of relativity is what is often called the *twin paradox*. The problem statement is quite simple. Imagine a pair of identical twins, one a scientist and the other an astronaut. The astronaut departs from earth in a rocket ship to visit a nearby star. He travels at a speed of 0.99c relative to the earth on a round trip that lasts 35 years, according to clocks on earth.

Time passes more slowly for the moving space traveler than for his earthbound brother, who occupies a different reference frame. When the astronaut returns, he will have aged less than the scientist who stayed home:

$$\Delta t_0 = \Delta t \sqrt{1 - \frac{v^2}{c^2}} \approx 0.14\ \Delta t.$$

Here Δt_0 is the proper time interval of the astronaut and Δt is the time interval of the brother on earth. The age difference between the twins depends on the duration of the trip and the speed of the spaceship. In this case, the traveler ages 5 years while his earthbound brother ages 35 years.

If we view everything from the reference frame of the astronaut, he perceives his twin brother on earth moving at 0.99c relative to the rocket ship. Therefore, it would seem that the scientist would be the one who ages more slowly. This presents the seeming paradox: Each twin thinks he is 30 years older than his brother.

We can resolve this dilemma by recalling the postulates on which the theory is based. The first postulate states that the laws of physics are the same in all inertial frames of reference. But in the case of the twins, the two frames of reference are clearly not equivalent. The astronaut twin does not travel at constant velocity—first he must be accelerated to enormous speed, then be accelerated again as he turns around the star, and then be accelerated a third time as he returns to earth. No matter how the astronaut's trip is made, he must be accelerated if he is to leave the earth and then return to it. Thus, the astronaut's frame of reference is not an inertial frame and the two views of time are not equivalent.

Because in this case the earth may be considered an inertial frame, the earthbound twin has the correct idea: the traveling twin does not age as much as he does. There is no real paradox.

The effect of time dilation is very real. The traveling twin really would return only 5 years older than when he left, while the earthbound brother would have aged 35 years. However, neither twin would have sensed anything unusual about the passage of time during those years. Everything would have seemed normal to each. Only when they came back together to the same inertial frame would the differences become apparent.

Figure B25.2 The galaxy Centaurus A is at a distance from the earth estimated as from 10 to 28 million light-years. If you could travel a distance of 10 million light-years at a speed of 0.99c, the voyage would seem to you to take 1.4 million years. If you could travel at a speed of 0.999999999999c, the trip would only take 14 years.

marker passing first one end of the rod and then the other). Moreover, the observer must use synchronized clocks to measure Δt.

An observer at rest in the frame S sees the rod move past the marker in a time $\Delta t_0 = l/v$, where l is the length of the rod as observed from S. The time measurement is made at one point with a clock at rest, so it is a proper time.

The length measured in frame S is $l = v\Delta t_0$, and the length measured in frame S' is $l_0 = v\Delta t$. These two measurements of length may be compared. The result is

$$\frac{l}{l_0} = \frac{v\Delta t_0}{v\Delta t} = \sqrt{1 - v^2/c^2},$$

where we have used Eq. (25.2) for $\Delta t_0/\Delta t$.

The observer in motion with speed v relative to the rod sees a length l given by

$$l = l_0\sqrt{1 - v^2/c^2}. \qquad (25.3)$$

That is, this observer measures the rod to be shorter than does the observer at rest with the rod. This effect is known as **length contraction.** It is also called Lorentz contraction.

The earlier discussion of the muon decay experiment can be reformulated in terms of observations from the muon's rest frame. From that point of view, the height of the mountain appears shortened (by length contraction) so that the time to travel from top to bottom is reduced. When the calculations are made, we get the same result for the number of muons reaching sea level that we found before. (See Problem 25.32.)

Example 25.6

A spaceship with a proper length of 100 m is moving with a speed $v = 0.6c$ relative to an earthbound observer. What is the apparent length of the spaceship as determined by the observer on earth?

Solution   The length is found from the length contraction formula (Eq. 25.3), where the proper length is $l_0 = 100$ m and $v/c = 0.60$:

$$l = l_0\sqrt{1 - v^2/c^2} = (100 \text{ m})\sqrt{1 - (0.60)^2} = 80 \text{ m}.$$

25.7   Mass and Energy

Einstein showed that the basic physical quantities of time, distance, and velocity are not absolute, as previously thought, but are all relative to the motion of the observer. What, then, can we say about the hallmarks of classical physics, the laws of conservation of energy and momentum? We shall see that in order to maintain their validity, we have to modify our understanding of mass, energy, and momentum. First, we will discuss the relationship Einstein derived between mass and energy; then, in the next sections, we will show how the postulates of

THE APPEARANCE OF MOVING OBJECTS

We have talked rather freely about the effects of time dilation and length contraction associated with clocks and other objects moving at great speed relative to an observer. In describing the muon experiment, we explicitly calculated the effects of time dilation. However, observing systems traveling at speeds near c is no simple matter.

What would we see if we could look at an object passing by at a relativistic speed? For more than 50 years the accepted notion was that you could readily see or photograph the length contraction of a rapidly moving object. For example, it was thought that a passenger in a high-speed rocket seeing another rocket passing in the opposite direction would see the rocket shortened in the direction of motion. But in 1959, J. Terrell critically analyzed the situation and showed that length contraction would not be directly observed with the eye. Because of the finite speed of light, the light coming from different parts of the rocket does not strike the eye at the same time. Therefore, as the rocket approaches, some of the light coming from the front of the rocket reaches the eye at the same time as light that left other parts of the rocket slightly earlier. The view of the rocket becomes distorted. Its exact appearance depends on the angle between the line of sight and the direction of motion, as well as upon its speed. An interesting account of this effect has been published by G. D. Scott and H. J. van Driel in the *American Journal of Physics,* August 1970, p. 971 (Fig. B25.3).

Suppose we could devise a camera with a very-large-diameter lens and a properly designed shutter. If the camera

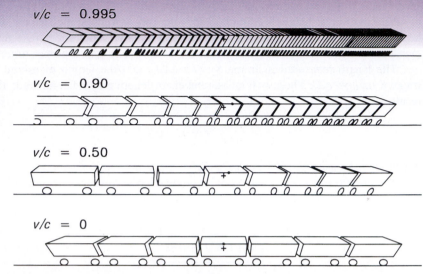

Figure B25.3 Views of a train of boxcars (each of $1 \times 1 \times 3$ units) by an observer 10 units away at an angle 30° above the plane of the tracks. The axis of the observer's camera is directed at the center of the middle car, shown by a cross (+). The dot (·) marks the center of the near edge of the middle car in each case. (After Scott and van Driel, with permission.)

were large enough—that is, if it ran the length of the rocket—and if it received only light coming perpendicularly to the path of the passing rocket, then we could photograph the Lorentz contraction.

An object coming straight toward you at very high speeds would also look distorted. For example, computer modeling of a cubic array of balls joined by straight rods and viewed parallel to one of its edges shows a curving of the rods as the array approaches at speeds near the speed of light (Fig. B25.4). The faster the relative motion, the greater the distortion.

All of this serves as a warning to be careful in the interpretation of relativistic effects and the equations that describe them. The effects of relative motion on length and time are real enough, but how they are observed is not always obvious.

Figure B25.4 This computer model shows a cubic array of balls connected by straight rods, viewed parallel to one of the edges (left). If the array approaches you at a speed of $0.90c$ (center), you see light at any one time that left the array at different times, depending on the distance from you. The rods appear to curve and the edges of the array appear to recede. At a speed of $0.99c$ (right), the distortion is more pronounced.

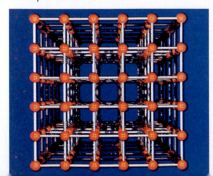
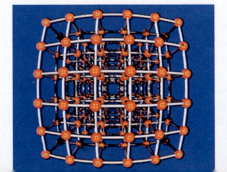
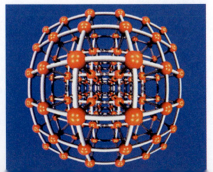

special relativity affect our understanding of momentum and energy at high speeds.

One result of Maxwell's electromagnetic theory is the prediction that when light strikes an object, it exerts a pressure. The presence of this pressure implies that light waves carry momentum. Moreover, the theory requires that this momentum be proportional to the energy carried by the light wave. Experiments support the theory, having verified this effect in several ways. The result is that a wave carrying an amount of energy E has a momentum given by

$$p = \frac{E}{c},$$

where c is the speed of light. We shall use this result and the principle of conservation of momentum to derive a relation between mass and energy.

Let us consider a thought experiment due to Einstein. Imagine a closed box of length L and mass M. The box is initially at rest with respect to us, but is suspended so that it is completely free to move without friction. Now suppose that at some instant of time, a bank of flashbulbs at one end of the box emits a flash of light (Fig. 25.10a). Since the light carries momentum, the box must recoil with an equal but opposite momentum, of magnitude

$$p_{box} = Mv = p_{light} = \frac{E}{c}, \tag{25.4}$$

where v is the recoil velocity of the box and E is the energy carried by the flash of light.

The light requires a time t to reach the other end of the box, where it is completely absorbed. During this time t, given approximately by $t \approx L/c$, the box moves a distance $-x$ as it recoils (Fig. 25.10b). Because the speed of light is so great, the box moves only a very tiny distance while the light travels the length of the box. When the light is absorbed at the other end of the box, the system returns to rest, in agreement with conservation of momentum.

Since no outside forces act on the box, its center of mass must not move during this event, even though the position of the box obviously changes. The only way this can happen is for a shift of mass to occur within the box as it moves, maintaining the position of the center of mass. Consequently, there must be an equivalent mass m associated with the light, of such magnitude that while the box of mass M moves a distance $-x$, its movement about the center of mass is compensated by the light moving a distance L:

$$mL = Mx. \tag{25.5}$$

Since we know the momentum of the box, and thus its velocity, we can use this information along with the time of flight of the light to evaluate the distance x that the box moves by multiplying the speed of the box by the elapsed time. Equation (25.4) can be solved for the speed v to give

$$v = \frac{E}{Mc}.$$

The elapsed time t is the time it takes for the light to travel the distance L. Thus, we have

$$x = vt = \left(\frac{E}{Mc}\right)\left(\frac{L}{c}\right).$$

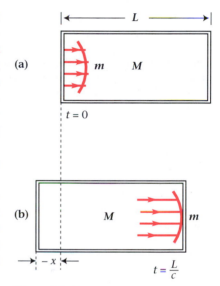

Figure 25.10
(a) At time $t = 0$ the box at rest emits a flash of light. (b) At time $t \approx L/c$ the light strikes the other end of the box and is absorbed. During this time interval, the box moves a distance $-x$.

Using this value for x in Eq. (25.5) and solving for the energy E, we get

$$E = mc^2. \tag{25.6}$$

This result is Einstein's famous **mass-energy relation.** For the example at hand, it says that when radiation is emitted by one end of the box, the box loses an amount of mass given by E/c^2. Similarly, when the light is absorbed at the other end, the box must gain an amount of mass given by E/c^2.

The significance of the mass-energy relation lies in its generalization of the law of conservation of energy to include mass. It tells us that any change in the energy of an object necessarily involves a change in the object's mass. Conversely, a change in an object's mass is accompanied by the absorption or emission of energy. The most dramatic example of energy released by a change in an object's mass is the release of energy in nuclear reactions, especially in fission and fusion. These topics will be discussed in detail in Chapter 29.

Example 25.7

A free neutron is unstable and spontaneously decays into a proton and an electron. The mass of the neutron is $m_n = 1.67495 \times 10^{-27}$ kg, the mass of the proton is $m_p = 1.67265 \times 10^{-27}$ kg, and the mass of the electron is $m_e = 9.1095 \times 10^{-31}$ kg. How much energy is released when the neutron decays?

Solution   The sum of the masses of the electron and the proton is less than the mass of the neutron. That is, the total mass of the end products is less than the mass of the original particle. This difference in mass corresponds to the energy released in the decay of the neutron:

$$\Delta E = \Delta mc^2 = (m_n - m_p - m_e)c^2$$
$$\Delta E = (1.39 \times 10^{-30} \text{ kg})(3.00 \times 10^8 \text{ m/s})^2,$$
$$\Delta E = 1.25 \times 10^{-13} \text{ J}.$$

Discussion   The mass that disappears in the decay of the neutron is converted to kinetic energy. Some of this kinetic energy goes into the resulting motion of the proton and electron; the rest is carried off by a massless particle called the neutrino (see Chapter 29).

25.8   Relativistic Momentum

One of the most basic physical laws is the law of conservation of momentum. It is fundamental to Newtonian mechanics and is consistent with our observations. Is it also valid from the relativistic point of view?

To answer this question, let us imagine an elastic collision involving two identical balls. When measured in a frame in which the two balls are initially at rest, their masses are identical. Suppose the two balls move toward each other with sufficient speed so that relativistic effects will be obvious. Now, let the two balls collide with a glancing blow that sends each ball recoiling to some extent perpendicular to the direction of their initial relative motion. (In such a collision, neither mass receives any appreciable velocity in its own reference frame along the direction of relative motion.) We can describe the collision in each of two frames in which one of the balls is at rest.

To an observer in the frame S, containing ball B, ball A at rest in the moving frame S' approaches from the negative x direction with a speed v (Fig. 25.11a). After the collision takes place, ball B in the frame S has a small velocity in the positive z direction (Fig. 25.11b). As viewed from S, the moving ball A is deflected by a very small angle so that it has a small component of velocity in the negative z direction.

Figure 25.11
(a) Ball A in frame S' approaches ball B in S at a relative speed v along the x direction. The two balls are situated on opposite sides of the x axis so that the collision is a glancing one. (b) From the frame S, the ball B moves along the positive z axis after the collision, while ball A is deflected by a small angle. (c) From the viewpoint of frame S', ball B is initially moving and ball A is stationary. (d) In frame S', after the collision, the ball A moves along the negative z' axis, while ball B moves off with a small component of velocity along the positive z' axis.

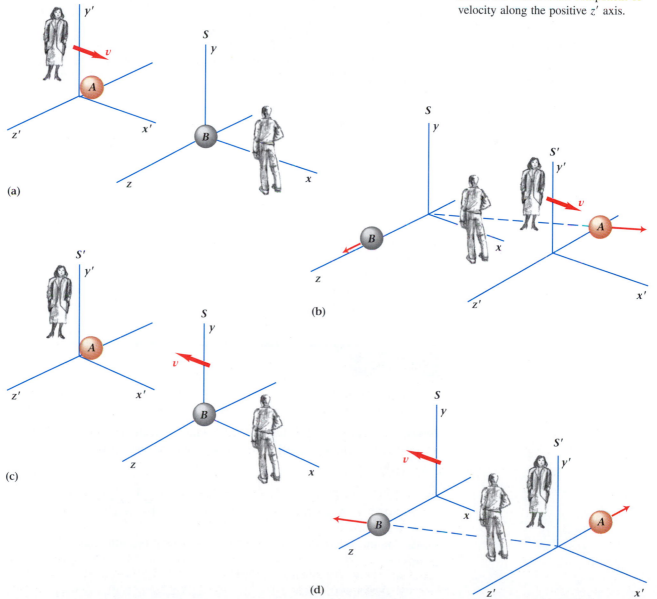

An observer in the frame S' sees the same collision differently. From her viewpoint, ball A, at rest in S', is struck by ball B moving fast in the negative x' direction (Fig. 25.11c). After the glancing collision, ball A recoils slowly along the negative z' axis, while ball B speeds off in the negative x' direction, with a small component of velocity in the positive z' direction (Fig. 25.11d). Both observers determine the velocity of the mass in their reference frame and report the result to the other.

Suppose both observers watch each other make their measurements of time and distance traveled by the two balls after the collision. They will agree on the measurement of distance because there is no length contraction in the positive z direction, perpendicular to their relative motion. However, because of time dilation, the observer in S concludes that the time required for the ball to travel one meter in S' is longer than the value reported by the observer in S' by a factor of $1/\sqrt{1 - v^2/c^2}$. (Remember that the lifetime of the moving muons was greater than their proper lifetime.) Thus, he concludes that the velocity of the ball is smaller than the value reported by the observer in S'.

Similarly, the observer in S' can observe the motions of the balls and compare her results with the results of the observer in S. The observer in S' concludes that the motion of the ball in S as seen by the observer in S is improperly reported, because of time dilation. Thus, the observer in S' concludes that the transverse velocity of ball B is too slow.

The result is that the law of conservation of momentum holds for both observers only if we redefine momentum to be

$$p = \frac{m_0 v}{\sqrt{1 - v^2/c^2}}, \tag{25.7}$$

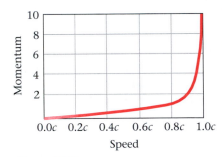

Figure 25.12

Momentum as a function of speed for an object moving at speeds near the speed of light.

where m_0 is the mass of the ball. This equation is identical to our original, classical definition of momentum except for the factor $\sqrt{1 - v^2/c^2}$. When the speed v is small compared to the speed of light, the denominator in Eq. (25.7) is essentially 1 and the momentum is linear with the speed (Fig. 25.12). As the speed of a particle approaches the speed of light, its momentum increases without limit.

Example 25.8

Compare the momentum of a muon traveling at a speed of $0.990c$ to the value computed from the nonrelativistic definition of momentum by finding their ratio.

Strategy The correct computation of momentum is made with Eq. (25.7). The nonrelativistic definition gives $p = m_0 v$. The ratio of these two momenta is

$$\frac{p_{\text{relativistic}}}{p_{\text{nonrelativistic}}} = \frac{\dfrac{m_0 v}{\sqrt{1 - v^2/c^2}}}{m_0 v} = \frac{1}{\sqrt{1 - v^2/c^2}}.$$

Solution In this case, $v/c = 0.990$ and $v^2/c^2 = 0.980$. Inserting these values, we find

$$\frac{p_{\text{relativistic}}}{p_{\text{nonrelativistic}}} = \frac{1}{\sqrt{1 - v^2/c^2}} = \frac{1}{\sqrt{1 - 0.980}} = \frac{1}{0.141} = 7.09.$$

Discussion Note that when v/c is close to 1, the momentum is strongly dependent on v. For example, if we had used $v/c = 0.995$ instead of 0.990, the ratio of momenta would have been 10.

| 25.9 | ## Relativistic Kinetic Energy |

In classical mechanics, we defined the quantity $\frac{1}{2}mv^2$ to be the kinetic energy of an object of mass m moving with speed v. Now let's examine the modifications of this expression imposed by special relativity.

We begin with the relativistic expression for momentum found in the last section. It is possible to interpret Eq. (25.7) as meaning that an object's mass is velocity-dependent according to the mass-velocity relationship of $m = m_0/\sqrt{1 - v^2/c^2}$. When the velocity v is zero, the apparent mass m is m_0, which we call the **rest mass** or proper mass of the object. In the limit of small velocities, $v/c \ll 1$, we can expand the square root term in a binomial series so that to a very good approximation, the mass becomes

$$m \approx m_0(1 + \tfrac{1}{2}v^2/c^2).*$$

If we multiply through by c^2, we get

$$mc^2 \approx m_0c^2 + \tfrac{1}{2}m_0v^2.$$

The left-hand term is the total energy of the object. The right-hand term consists of two parts: a term related to the rest mass of the object (m_0c^2) and a term equal to the classical kinetic energy ($\frac{1}{2}m_0v^2$). At higher speeds, the equation may not look so simple, but we can still separate the total energy into a rest-mass energy and a kinetic energy. Now, suppose we *redefine* the object's kinetic energy to be the difference between its total energy and its rest mass energy. Then the kinetic energy is equal to the work required to give the object a velocity v. Thus

$$KE = mc^2 - m_0c^2. \tag{25.8}$$

When we insert the value of the relativistic mass into this equation, we get

$$KE = m_0c^2\left[\frac{1}{\sqrt{1 - v^2/c^2}} - 1\right]. \tag{25.9}$$

---

*The binomial series that represents $1/\sqrt{1-x}$ is

$$\frac{1}{\sqrt{1-x}} = 1 + \frac{1}{2}x + \frac{3}{8}x^2 + \frac{15}{48}x^3 + \ldots,$$

where $x < 1$. When x is very small, the series is closely approximated by the first two terms:

$$\frac{1}{\sqrt{1-x}} \approx 1 + \frac{1}{2}x, \qquad x \ll 1.$$

When the velocity is small, this relativistic expression for kinetic energy goes smoothly to the classical value. (You can check this for yourself by applying the binomial theorem to the first term.) So long as an object's velocity is much less than the speed of light, the classical formulas are accurate enough. But when v approaches c, the classical formula for kinetic energy is no longer adequate and we must use Eq. (25.9) instead.

In some cases, it is helpful to express an object's total energy in terms of its rest mass and momentum. If we square the relation $E = mc^2$, we get

$$E^2 = m^2c^4 = m^2c^2(c^2 + v^2 - v^2).$$

Observing that $p = mv$ and that $m = m_0/\sqrt{1 - v^2/c^2}$, we find after some algebra that

$$E^2 = p^2c^2 + m_0^2c^4. \tag{25.10}$$

This expression gives us the *total* energy E of a particle of mass m_0 moving with momentum p.

One of the effects mentioned in the previous section is that it becomes harder to accelerate an object as its speed approaches c. We see from the kinetic energy equation that a large amount of energy is required to accelerate an object to a speed close to c. The object behaves as if it gets more and more massive and requires more and more kinetic energy for each small increase in speed. Therefore, an infinite amount of energy would be required to accelerate an object with rest mass m_0 to the speed of light. Since an infinite amount of energy is not available, the conclusion is inescapable that particles with rest mass cannot attain the speed of light.

Example 25.9

An electron is accelerated to a kinetic energy twice its rest-mass energy. How fast is it traveling?

Solution We evaluate the speed v using the relativistic expression for kinetic energy (Eq. 25.9) with KE $= 2m_0c^2$:

$$\text{KE} = 2m_0c^2 = m_0c^2\left[\frac{1}{\sqrt{1 - v^2/c^2}} - 1\right].$$

Upon factoring out the m_0c^2 term, we get

$$2 = \frac{1}{\sqrt{1 - v^2/c^2}} - 1.$$

This can be rearranged to give

$$\sqrt{1 - v^2/c^2} = \tfrac{1}{3}.$$

Now square both sides and solve for v to give

$$v = 0.94c.$$

■

The Relativistic Doppler Effect

*25.10

In Chapter 15 we discussed the Doppler effect, which describes the apparent change in frequency of a sound wave from a source in motion relative to an observer. Now let's examine the effect of relative motion on light waves, making sure that our description is consistent with the special theory of relativity.

Imagine a starship moving with velocity v toward an observer on earth. As the starship approaches, it emits a light signal of frequency f_0 measured in the reference frame in which the rocket is at rest. Because of time dilation, the period T of the signal appears longer to the observer on earth than to the starship's captain, who measures a period T_0. Moreover, as the ship approaches the observer, the wavefronts of light are compressed from their normal wavelength, measured with a stationary source, by an amount equal to the distance traveled by the starship during one period (Fig. 25.13). The resulting wavelength is

$$\lambda = cT - vT = (c - v)T = \frac{(c - v)T_0}{\sqrt{1 - v^2/c^2}},$$

where T_0 is the proper period. Consequently, the frequency f measured by the earthbound observer is

$$f = \frac{c}{\lambda}$$

or, in terms of the proper frequency of the source, $f_0 = 1/T_0$,

$$f = \frac{f_0\sqrt{1 - v^2/c^2}}{1 - v/c}.$$

We can rearrange this equation to give

$$f = \frac{f_0\sqrt{(1 - v/c)(1 + v/c)}}{\sqrt{(1 - v/c)(1 - v/c)}},$$

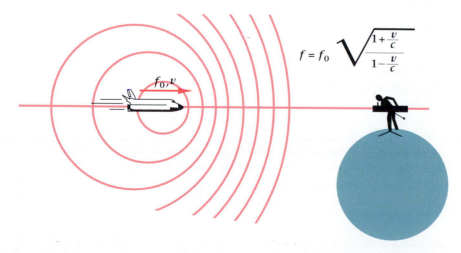

$$f = f_0\sqrt{\frac{1 + \dfrac{v}{c}}{1 - \dfrac{v}{c}}}$$

Figure 25.13

A starship approaching the earth sends out a radio signal with frequency f_0. The wavefronts are compressed by the motion of the starship so that an observer on earth detects them at a higher frequency f.

which may be simplified to

$$f = f_0 \sqrt{\frac{1 + v/c}{1 - v/c}} \quad \text{(approaching).} \qquad (25.11)$$

The observed frequency increases as a result of the relative motion of source and observer. If the source and observer were moving apart, then v in Eq. (25.11) would become negative and the frequency would decrease. Notice that in the relativistic case, the only velocity that matters is the *relative* velocity of the source and observer. This was not true for sound (see Chapter 15).

The relativistic Doppler effect applies to any electromagnetic wave. If the waves are reflected from a moving object back toward the source, the shift in frequency can be used to determine the velocity of that object relative to the source. A practical application of this effect is the use of radar to measure the speed of cars and aircraft.

Detailed examination of the light from distant stars and galaxies shows that, in general, the light is shifted to lower frequencies. Equivalently, we say that the light is red-shifted; that is, it is shifted to longer wavelengths. This red shift arises from the relativistic Doppler effect associated with the motion of the stars away from us. Everywhere we look in the sky, the galaxies are moving away. When we analyze their motion using the Doppler formula, we find that the farther away the galaxies are from us, the faster they are receding (Fig. 25.14). Thus, we live in an expanding universe.

Example 25.10

A spacecraft exploring a distant galaxy is moving very rapidly toward a strange star that emits a yellow light of frequency 5.0×10^{14} Hz, measured in the rest frame of the planet. What frequency does the spacecraft crew observe for the light if their velocity relative to the star is $0.095c$?

Solution A straightforward application of Eq. (25.11) determines the frequency of the light as

$$f = 5.0 \times 10^{14} \text{ Hz} \sqrt{\frac{1 + 0.095}{1 - 0.095}} = 5.5 \times 10^{14} \text{ Hz.}$$

The frequency seen by the crew is $f = 5.5 \times 10^{14}$ Hz. Since we more often refer to light in terms of wavelength, we convert this to

$$\lambda = \frac{c}{f} = \frac{3.0 \times 10^8 \text{ m/s}}{5.5 \times 10^{14} \text{ Hz}} = 550 \text{ nm.}$$

This wavelength corresponds to green light. The yellow light is shifted to green for the observers moving toward the source at the speed $v = 0.095c$.

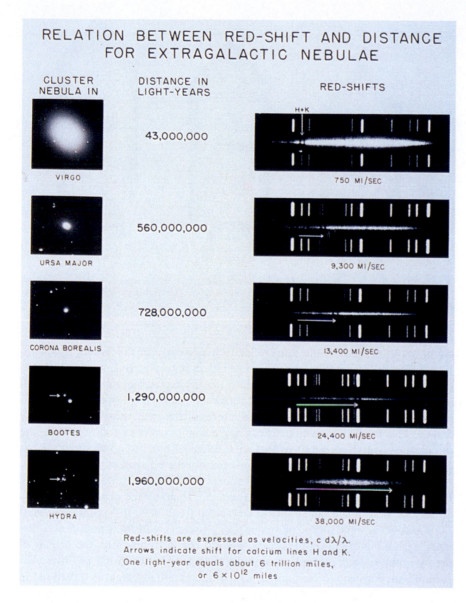

RELATION BETWEEN RED-SHIFT AND DISTANCE
FOR EXTRAGALACTIC NEBULAE

CLUSTER NEBULA IN	DISTANCE IN LIGHT-YEARS	RED-SHIFTS
VIRGO	43,000,000	750 MI/SEC
URSA MAJOR	560,000,000	9,300 MI/SEC
CORONA BOREALIS	728,000,000	13,400 MI/SEC
BOOTES	1,290,000,000	24,400 MI/SEC
HYDRA	1,960,000,000	38,000 MI/SEC

Red-shifts are expressed as velocities, c dλ/λ.
Arrows indicate shift for calcium lines H and K.
One light-year equals about 6 trillion miles,
or 6×10^{12} miles

Figure 25.14
Light from more distant galaxies is red-shifted more. The spectra of light from the galaxies are compared with spectra from the laboratory. (Line spectra are discussed in Chapter 27.)

*25.11   The Principle of Equivalence

A few years after his first work in special relativity, Einstein advanced a new relationship between accelerated motion and gravitational force. This relation, called the **principle of equivalence,** states that *experiments performed in a uniformly accelerating reference frame, having an acceleration* a *with respect to an inertial frame, give exactly the same results as identical experiments carried out in an inertial frame containing a uniform gravitational field* −a.

   To illustrate the principle of equivalence, imagine a spaceship accelerating with a constant acceleration equal to the acceleration of gravity *g* (Fig. 25.15). An astronaut in the spaceship releases a ball. Because of its inertia, the ball

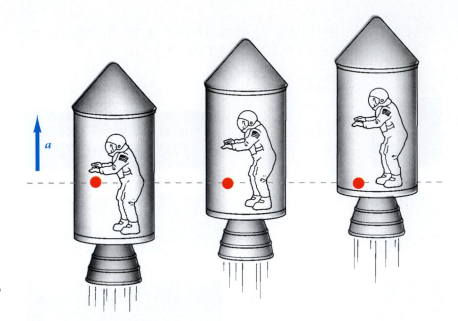

Figure 25.15

An astronaut standing in an accelerating spaceship releases a ball, which falls in the direction opposite to the rocket's acceleration.

continues to move with the same velocity it possessed at the instant it was released. However, since the spaceship is accelerating, it overtakes the ball. From the astronaut's point of view, the ball falls toward the back of the spaceship with an acceleration $-g$. That is, from the reference frame of the spaceship, the astronaut cannot distinguish between the effects of the spaceship's acceleration g and the effects of a gravitational field in the ship of acceleration $-g$. Thus, the principle of equivalence tells us that an accelerated motion in one direction has exactly the same effect as a gravitational field in the opposite direction.

Let us now examine the Doppler effect of a light in a stationary reference frame S as seen from an accelerating frame S'. Then, by the principle of equivalence, we shall relate our results to an inertial frame with a gravitational field.

Consider a light source of frequency f_0 stationary in the frame S at a distance l from the origin on the x axis (Fig. 25.16). At time $t = 0$, the origins of the two frames coincide and the frame S' begins to move in the positive x direction with constant acceleration a. Also at time $t = 0$, the light is turned on. In

Figure 25.16

At time 0 the light at position l flashes and the system S' acquires a uniform acceleration a.

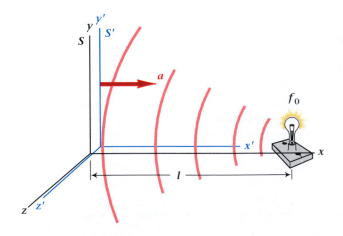

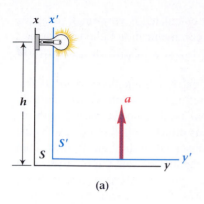

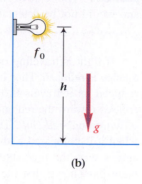

Figure 25.17
(a) An accelerating frame S' moving toward the light source. (b) The light source viewed in the equivalent stationary frame with a gravitational acceleration $-g$.

a time approximately equal to l/c, a light pulse reaches the origin of the frame S', just as it attains a velocity $v = at = al/c$ relative to S. From the relativistic Doppler equation, the frequency of the light observed at the origin of S' is

$$f = f_0 \sqrt{\frac{1 + v/c}{1 - v/c}} = f_0 \frac{\sqrt{1 + v/c}}{\sqrt{1 - v/c}}.$$

Using the binomial theorem to evaluate both square roots for $v << c$, we get

$$f \approx f_0 \left(1 + \frac{v}{c}\right) = f_0 \left(1 + \frac{al}{c^2}\right). \qquad (25.12)$$

According to the principle of equivalence, we should expect to get the identical frequency in a system that is at rest relative to S but that has a gravitational field of magnitude a in the negative x direction. We can relate this to the earth and its gravitational field by considering the x direction of the frame S' to be directed vertically upward from the earth's surface (Fig. 25.17). Here the gravitational field is directed downward as required and has magnitude g near the earth's surface. We can then make the following interpretation: A transmitter of electromagnetic waves located a height h above the ground and radiating a signal at frequency f_0 is detected on the ground at a frequency f given by

$$f = f_0 \left(1 + \frac{gh}{c^2}\right). \qquad (25.13)$$

Example 25.11

A radio signal is broadcast from a plane at an altitude of 10,000 m and received on the ground. What is the relative shift in the frequency due to gravitational effects?

Solution The relative shift in frequency is given by

$$\frac{\Delta f}{f_0} \approx \frac{f - f_0}{f_0}.$$

Using Eq. (25.13) for f, we find that

$$\frac{\Delta f}{f} = \frac{gh}{c^2} = \frac{(9.81 \text{ m/s}^2)(10^4 \text{ m})}{(3.0 \times 10^8 \text{ m/s})^2} \approx 10^{-12}.$$

Discussion The relative shift that we have just calculated is very small indeed and would not be detectable by broadcast on communication radio equipment.

··■

If light is traveling opposite to the direction of a gravitational field, its frequency is reduced. Thus light traveling from the sun to the earth has its frequency reduced as it travels in the sun's gravitational field. Since its wavelength increases toward the red end of the spectrum, this effect is often referred to as the *gravitational red shift.* This effect has been observed in the light from the sun when it is compared with light generated in a laboratory, and the observed shift agrees with the predicted shift to within a few percent. More recently, experiment has shown that the measured frequency shift agrees with the value predicted by the theory to within 0.1%.

An additional experiment involving the principle of equivalence and kinematic time dilation was conducted in 1971 when J. C. Hafele and R. E. Keating flew four cesium beam atomic clocks around the world to test the theory of relativity. They flew the clocks on regular jet flights once around the world eastward and once around westward and, at the end of the trips, compared the time on their clocks with that on reference clocks at the U.S. Naval Observatory. According to the theory of relativity, the airborne clocks should have lost about 40 ns during the eastward trip and gained about 275 ns during the westward trip. Their measured values are given in Table 25.1. You can see that they agree with the predicted values within the limits of the uncertainties.

Let's see where these effects come from. We have seen that a clock in motion with speed v relative to an observer runs slowly compared with a clock at rest with the observer. Now suppose a clock is flown along the equator in an airplane with ground speed v. Because of time dilation, time on the plane will run at a different rate than time on the earth's surface. To an observer at rest with respect to the fixed stars, a clock on the earth is moving because of the earth's rotation. If the airborne clock is flown eastward, he observes it moving faster than a clock on the earth. Thus, we expect that time should run slower on the plane. If the clock is flown westward, the fixed observer sees it moving slower than a clock on the earth. Thus, time runs faster on the plane than on the earth. But there is more. We need to include the gravitational effect. We have seen that a source of radiation with frequency f_0 at a height h is detected on the ground with a higher frequency. There is a corresponding difference in the periods of the radiation and in the time measured as a number of periods. The result is, time runs faster on the plane as it flies higher. This effect is due to gravitation and is independent of the plane's velocity. The rate of the clock is determined by both gravitational and kinematic effects.

▼

Table 25.1	*Measured and Predicted Time Differences in the Experiment of Hafele and Keating*	
	Eastward	**Westward**
Measured	-59 ± 10	273 ± 7
Predicted	-40 ± 23	275 ± 21

Source: J. C. Hafele and R. E. Keating, *Science*, Vol. 177, 14 July 1972, pp. 166–170. Copyright ©1972 by the AAAS.

The significance of the Hafele-Keating experiment lies in its confirmation that the time differences predicted by the theory of relativity are real—not only for muons, but also for real macroscopic objects, such as clocks. The time differences involved were small, but were well within the capability of the atomic clocks to resolve. Thus, merely by traveling eastward around the earth, the experimenters aged a few nanoseconds less than their earthbound colleagues. If the speeds were greater the relative effect would be greater also. This is the result of the experiment: Time is indeed relative and is not the same for all observers.

*25.12 General Relativity

In proposing his special theory of relativity, Einstein postulated that the laws of physics were the same to all observers in all inertial frames of reference. The acceptance of this theory and the validity of the postulates are supported by a wide range of observations and experiments. The "special" of special relativity is the restriction to inertial frames of reference. To describe the laws of physics in accelerating systems, we need a more general theory. Such a theory was proposed by Einstein in 1916. This **general theory of relativity** creates a theoretical framework applicable to all systems, inertial or noninertial. That is, it includes accelerating systems.

The basic postulate of general relativity is that **all physical laws can be formulated so as to be valid for any observer, regardless of the observer's motion.** When this postulate is combined with the principle of equivalence, which treats gravitational fields as equivalent to accelerations, we see that the general theory becomes more than a description of accelerated systems. It is a theory of gravitation.

The special relationship between gravitation and acceleration has intrigued physicists since the time of Isaac Newton. In Newton's second law, $F = ma$, the acceleration that an object acquires from the application of an external force depends on its mass m. Moreover, according to Newton's law of universal gravitation, the gravitational attraction of that object by another object also depends on the mass. These two entirely separate effects depend on the same quantity, the mass. Other forces do not show this behavior. For example, the electromagnetic force on a charged particle depends on the magnitude of its charge, while its acceleration depends on its mass. There seems to be something special about gravitational forces.

By using the principle of equivalence, we can transform our viewpoint to a reference frame with just the right acceleration to eliminate the gravitational field at any particular point in space. But in general, the acceleration required to eliminate the gravitational field at an arbitrary second point in space will be different. For example, consider the gravitational field of the earth. Above the North Pole, the field is directed toward the center of the earth along its axis of rotation. But at some other point above the earth, say above the equator, the gravitational field is radial to the center of the earth and perpendicular to the axis of the earth's rotation. Thus, while a transformation of coordinates may appear to remove a local gravitational field at one point, it cannot eliminate all gravitational effects.

Observations such as these suggested to Einstein that the usual conception of space itself should be reexamined. Einstein proposed that instead of being

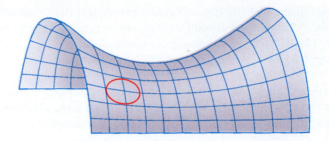

Figure 25.18
A two-dimensional surface representing a curved two-dimensional space. We draw the figure as if the distortion existed in the third dimension, but that is only to help visualize the curvature. An observer on the surface can tell that the surface is not flat because the ratio of circumference to diameter of a finite circle drawn on the surface does not equal π.

always the rectilinear Euclidean space of Newton's laws, space itself might be curved (Fig. 25.18). In such a space, the motion of an object could be described in terms of the geometry of the space rather than in terms of external forces.

For example, in a Euclidean inertial reference frame, a ray of light moves in a straight line. However, when this event is viewed from an accelerating frame of reference, the path of the light ray appears curved. By the principle of equivalence, we should expect the light ray to follow a curved path in a gravitational field. Since light has no mass, we cannot state the force involved in terms of Newton's laws. However, from the perspective of general relativity, it is *space itself* that is viewed as being curved or distorted by the presence of the gravitational field (that is, by the presence of mass). In such a curved space, light merely travels along a *geodesic,* which is the curved path that represents the minimum distance between two points in that space. Such behavior has an analogy in travel on the earth's surface. The shortest route between cities is not a straight line, but a curved path along a great circle around the earth (that is, along a circumferential circle, Fig. 25.19). Such a path is the geodesic on the spherical surface of the earth. This is a natural consequence of the fact that the earth is spherical and not flat.

We do not have to explain the deflection of light in terms of Newtonian forces. Instead, we say that light always follows a geodesic through space, which may itself be curved. Thus, where space is Euclidean (that is, far away from any mass), light travels in a straight line. Where space is curved (that is, near a significant mass, like a star), light travels along the geodesic in that space.

There are only a few experiments that can distinguish the predictions of the general theory of relativity from those of Newtonian mechanics. Such experiments are known as *tests* of the theory. These experiments include, among others, the deflection of light in a gravitational field; the exact path of Mercury's orbit around the sun; and the time delay of radar signals that pass near the sun. The agreement between the observations and the predictions shows that the general theory of relativity provides a suitable explanation. However, these tests do not prove the theory to be correct. The tests are examined briefly in the following paragraphs. Note that there are experiments, such as those mentioned in the previous section, that test the principle of equivalence but do not really test general relativity.

When light from a star passes close to the sun, it is deflected by the sun's gravitational field. This deflection causes the star to appear slightly displaced (Fig. 25.20). The displacement has been measured by photographing the apparent positions of stars during a solar eclipse and comparing these positions with those observed in the night sky six months later. Apparent shifts of less than 2 seconds of arc have been measured this way, in close agreement with the theoretical prediction of 1.75 seconds of arc.

Figure 25.19
The shortest path between New York and Moscow is the great circle path (geodesic) shown.

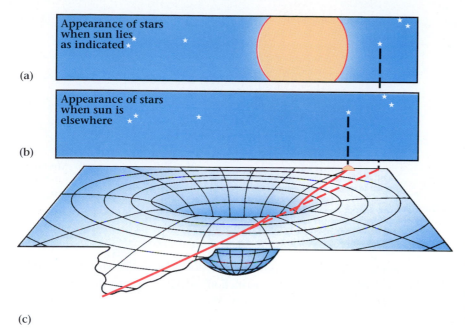

(a) Appearance of stars when sun lies as indicated

(b) Appearance of stars when sun is elsewhere

(c)

Figure 25.20
Deflection of a light beam results from curvature of space near a massive object, such as the sun. The appearance of stars when the sun lies as indicated (a) is different from their appearance when the sun lies elsewhere (b). (c) The path of the light from the star to the observer's eye follows a geodesic in the curved space, represented here by the curved surface.

When a massive object lies squarely between the earth and a distant light source, the object acts as a gravitational lens and bends the path of the light so that it may be seen from the earth. If the massive object has circular symmetry, the effect will be to create a ring of light. If the massive object is a galaxy with nonuniformly distributed mass, the ring will be broken into distinct patches such as the four patches seen in Fig. 25.21(a). If the alignment is imperfect, the cross pattern will not be symmetric. Figure 25.21(b) is a detailed image of such a gravitational lens taken with the Hubble space telescope. The light from a source 8 billion light-years away is deflected by the gravitational field of a galaxy 400 million light-years from earth.

As the planets travel about the sun in elliptical orbits with the sun at one focus, they affect one another so that their orbits are not perfect ellipses. In the

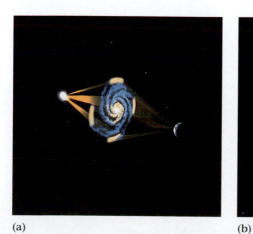

Gravitational Lens G2237+0305

(a)

(b)

Figure 25.21
(a) Artist's view of a gravitational lens creating a cross-like pattern of light from a distant source. (b) Hubble space telescope photo of gravitational lens G2237 + 0305.

Figure 25.22
(a) Elliptical orbit of a planet about the sun. The closest point of approach is the perihelion. (b) Exaggerated drawing of the advancing orbit of Mercury, showing the motion of the perihelion from point P_1 to P_2 to P_3.

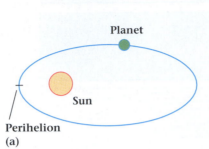

(a)

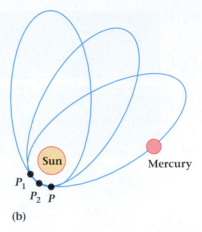

(b)

case of the planet Mercury, its orbit moves slowly about the sun (Fig. 25.22). The point of closest approach (the perihelion) advances around the sun at a rate of nearly 500 seconds of arc per century. When all of the complex calculations of Newtonian forces are made, including the gravitational effects of Venus, the earth, Jupiter, and others, a discrepancy remains of about 43 seconds of arc per century between observation and prediction. If we apply special relativity to account for effects of mass change and time dilations, the resulting discrepancy between experiment and theory is still about 21 seconds of arc per century. But if we apply the theory of general relativity, the prediction is for an additional 43 seconds of arc per century, in agreement with the observations.

These tests cannot be cited as absolute proof of the general theory of relativity, since other theories could conceivably explain these results as well. However, contemporary physicists generally agree that the principle of equivalence is valid and that Einstein's general theory is consistent with experimental observations. Various details of the general theory have been modified over the years by the insights of other theoretical physicists, but the overall structure stands as one of the great general concepts of twentieth-century physics.

Summary

Useful Concepts

■ Special relativity is based on two postulates:

I. The laws of physics are the same in all inertial reference frames.

II. The speed of light in free space is independent of the motion of its source and of the motion of the observer.

■ Velocities are added according to

$$u = \frac{u' + v}{1 + \dfrac{u'v}{c^2}},$$

where u' is a velocity measured in a reference frame moving with velocity v relative to the frame in which u is measured.

■ An observer at rest with respect to a clock records a proper time interval Δt_0. An observer moving with respect to the clock sees the time interval indicated by the clock stretched out to Δt:

$$\Delta t = \frac{\Delta t_0}{\sqrt{1 - \dfrac{v^2}{c^2}}}.$$

This effect is called time dilation.

■ An observer in motion with speed v relative to a rod of proper length l_0 sees a length l given by

$$l = l_0\sqrt{1 - v^2/c^2}.$$

This effect is known as length contraction.

■ The mass-energy relationship is

$$E = mc^2.$$

■ The momentum of an object is

$$p = \frac{m_0 v}{\sqrt{1 - v^2/c^2}}.$$

■ A particle of rest mass m_0 has kinetic energy

$$KE = mc^2 - m_0 c^2 = m_0 c^2 \left[\frac{1}{\sqrt{1 - v^2/c^2}} - 1 \right].$$

■ The total energy of a particle of rest mass m_0 is

$$E^2 = p^2 c^2 + m_0^2 c^4.$$

■ The Doppler effect for light is

$$f = f_0 \sqrt{\frac{1 + v/c}{1 - v/c}} \qquad \text{(approaching).}$$

■ The principle of equivalence states that experiments performed in a uniformly accelerating frame of reference, having an acceleration a with respect to an inertial frame, give exactly the same results as identical experiments carried out in an inertial frame having a uniform gravitational field $-a$.

■ The general theory of relativity creates a theoretical framework applicable to both inertial and noninertial systems. The basic postulate of general relativity is that all physical laws can be formulated so as to be valid for any observer, regardless of the observer's motion.

Important Terms

You should be able to write the definition or meaning of each of the following:

Galilean relativity	length contraction
special theory of relativity	mass-energy relation
simultaneous	rest mass
time dilation	principle of equivalence
proper time	general theory of relativity
proper length	

Conceptual Questions

25.1 Passengers in an accelerating railroad car have with them a ball and a ruler but no clock. The windows are painted over, so they may not see outside. Can they, by observing what happens when they drop the ball, tell whether they are in an inertial reference frame? Could they answer the question if they were in an accelerating elevator?

25.2 Is it reasonable to expect the physical theories describing nature to be in accord with "common sense"? What does the term really mean, and how important is it likely to be in formulating and judging theories for cases in which speeds are near the speed of light?

25.3 Is it correct to say that Galilean relativity is wrong?

25.4 What would be the effect on our daily lives if the velocity of light were 30 m/s? How would the world differ from the way it seems now?

25.5 Is the twin paradox aptly named? Before you answer, look up the meaning of "paradox" in a dictionary large enough to give several definitions for the word.

25.6 Is it true that "matter can be neither created nor destroyed"? If not, what is the correct statement? How might this bear on the conservation of mass when balancing chemical equations?

25.7 Is the mass of a red-hot piece of metal greater than its mass when it is at room temperature? Explain your answer.

25.8 The special theory of relativity establishes an upper limit to the speed of a material body. Does this theory set an upper limit to the momentum and kinetic energy of the same body?

25.9 Twins separate, with one going to live on top of a very high mountain and the other living in a cave deep underground. Which twin ages faster?

25.10 Suppose you set your clock using time signals from the radio transmitter at the National Bureau of Standards in Fort Collins, Colorado. Approximately how long does it take for the radio signal to travel to your receiver?

25.11 What does it mean to say that when you look at the stars at night, you are really looking into the past?

Problems

Section 25.1 Principle of Relativity

Hints for Solving Problems

It is important in working relativity problems to be certain which reference frame you are in.

25.1 A train moves slowly past a station at 6 km/h. A passenger on the train runs 8 km/h down the aisle in the direction opposite to the train's velocity. (a) What is the passenger's speed relative to the station? (b) What would it be if the passenger reversed direction?

25.2 Two airplanes are flying side by side at the same

altitude. Plane A is slowly overtaking plane B at 4 km/h. A stewardess in plane A is walking toward the rear of the plane at a speed of 2 km/h, and a passenger is walking toward the front at 2 km/h. What are their speeds relative to a passenger watching them from plane B?

25.3• A ferry boat headed directly away from the river bank is traveling at 3.00 km/h with respect to the water in a river flowing at 4.00 km/h. A passenger walks diagonally toward the stern of the boat with a speed of 1.39 m/s. The passenger's path makes an angle of 53.1° with the length of the boat. Her path is directed toward the upstream direction. What is the passenger's speed relative to the bank?

25.4• A stunt diver jumps from the top of a 20.0-m-high building. After falling for 1.50 s, the diver flips a coin upward with a relative speed of 14.7 m/s. What is the initial speed of the coin with respect to the ground as it leaves the diver's hand?

Section 25.3 Velocity Addition

Hints for Solving Problems

In the velocity addition formula, remember that u' is a velocity in a frame moving with speed v relative to the frame in which u is measured.

25.5 A space station moving away from the earth at $0.55c$ launches a rocket toward earth with a velocity $0.75c$. What is the rocket's speed as seen from the earth?

25.6 A spaceship hurtling toward Mars at a speed of $0.80c$ launches a probe toward the planet at a speed $v = 0.50c$ relative to the rocketship. How fast does the probe approach Mars as measured by an observer on Mars?

25.7 A spaceship racing toward Jupiter at $0.70c$ launches a probe toward the planet at a relative speed of $0.50c$. How fast does the probe approach Jupiter as measured by an observer on Jupiter?

25.8 In a colliding-beam experiment, particles traveling at $0.950c$ relative to the laboratory met in a head-on collision with other particles traveling in the opposite direction at $0.950c$ relative to the lab. What is the relative speed of approach of the particles as seen in the rest frame of one of them?

25.9 A spaceship approaching the moon at a speed $0.50c$ sends a light signal to an observer stationed on the lunar surface. (a) What is the speed of the light signal as measured by the pilot of the spaceship? (b) What is the speed of the light signal seen by the lunar observer?

25.10 Two space stations approach each other at a relative speed of $0.50c$. A rocket is sent from station A to station B with a speed $0.40c$ as measured from station A. What is the approach speed as seen from station B?

25.11 Show that for speeds that are very fast but still small compared to the speed of light that the velocity addition formula of Eq. (25.1) reduces to the expected classical formula.

25.12• A spaceship approaching a planet at a speed of $0.23c$ fires a probe toward a planet. Observers on the planet see the probe approaching at a speed of $0.67c$. What is the speed of the probe as measured from the spaceship?

25.13• A giant spaceship traveling at a speed of $0.50c$ toward a distant planet launches a smaller spaceship in the same direction with a speed of $0.50c$ relative to the giant ship. The smaller spaceship then launches another still smaller spaceship (number two) with a speed of $0.50c$ relative to itself. This process is continued through successive generations of spaceships until the speed of the last ship exceeds $0.95c$ as measured in the frame of the giant spaceship. How many spaceships, not counting the original giant ship, must be launched before this speed is reached?

25.14•• The spaceship *Explorer* moving away from Mars at speed $v = 0.60c$ relative to an observer there is pursued by an enemy ship, traveling at $v_e = 0.70c$ relative to Mars. The pursuit ship launches a missile with relative velocity $v_m = 0.30c$ toward the *Explorer*. How fast does the missile appear to move as seen from the spaceship *Explorer*?

Section 25.4 Simultaneity

25.15 An astronaut on the moon wants to synchronize his clock to a time signal on earth. He receives a radio message saying that the time at the tone will be exactly 6:00. To what time should his clock be set at the instant of the tone?

25.16 An astronaut on the Martian surface, 8.96×10^{10} m from home, receives a time signal from earth. The message says that at the sound of the tone it will be exactly 5:00. When the tone sounds, she sets her clock to 5:00. When she returns home, she compares her clock to the standard clock. What does she find? Neglect any effect due to the trip home.

25.17 How long does it take for a radio signal to go from the North Pole to the South Pole? For simplicity, assume that the radio waves follow a path that approximates a semicircular arc with a radius equal to the earth's radius.

Section 25.5 Time Dilation

Hints for Solving Problems

Take care to discern which time is the proper time.

25.18 An astronaut journeying to the nearby star Lacaille 9352 at a speed of $0.850c$ relative to earth requires 20.0 years as measured on earth to complete his trip. How many years does he age during his journey?

25.19 An astronaut journeying to the nearby star Alpha Centauri at a speed of $0.99c$ relative to the earth requires 8.57 years as measured on earth to complete her trip. How many years does the astronaut age during this journey?

25.20 Captain Picard travels at high speed for two years according to his clocks and calendar. Upon returning to earth, he finds that six years have elapsed. What was his average speed?

25.21 A cosmonaut journeys at high speed for three years according to his clocks and calendar. Upon returning, he finds that eight years have elapsed on earth. What was his average speed?

25.22 The mean life of muons is 2.2 μs. What is the mean life of a beam of muons moving at a speed of 0.9995c relative to you?

25.23 The charged pion, a particle of importance in subnuclear physics, has a mean life of 2.6×10^{-8} s. What is the mean life of a beam of charged pions moving with a speed of 0.99c relative to you?

25.24● Imagine another universe with the same natural laws as ours, but in which the values of the physical constants (such as the speed of light) are different. On a planet in that universe, a physicist on a railroad platform watches a train passing by at a speed of 157 km/h. The physicist determines that a clock on the train shows an elapsed time of 60.0 s when a clock on the platform shows an elapsed time of only 59.5 s. What is the speed of light in that universe?

Section 25.6 Length Contraction

Hints for Solving Problems

Take care to discern which length is the proper length.

25.25 A spaceship with a proper length of 300 m passes near a space platform at a relative speed of 0.86c. What is the length of the spaceship when measured in the frame of the space platform?

25.26 A spaceship moves with a speed $v = 0.90c$ relative to a space platform that has a landing strip 5000 m long. What is the apparent length of the landing strip as measured in the frame of the spaceship?

25.27 A meterstick is accelerated to such a velocity that to an interested observer it appears only 80 cm long. How fast must the stick be moving in the observer's frame of reference?

25.28 A meterstick is accelerated to a velocity such that it is contracted to only 50 cm relative to an interested observer. How fast must the meterstick be moving in the observer's frame of reference?

25.29 An electron travels along a straight 10-m section of a particle accelerator at a speed of 0.99c. If you could ride along with the electron, how long would that straight section appear to you?

25.30● By how much is a 100-m-long train shortened as observed by someone at rest relative to the track if the train is traveling 320 km/h? (*Hint:* Use the approximation that for $x << 1$, $\sqrt{1 - x} \approx 1 - x/2$.)

25.31● An observer on earth measures the length of a receding spaceship and finds it to be 0.86 of the length measured by the passengers on board the ship. Without directly calculating the relative speed, determine how much time will elapse on the earthbound observer's clock for each 60 minutes that elapse on the shipboard clock as seen by the passengers on the

spaceship. (*Hint:* Find a relationship between lengths and times in the two systems.)

25.32● In the muon experiment described in the text, the muons appeared to have an abnormally long lifetime as a result of the time dilation associated with their velocity. (a) From the rest frame of the muons, calculate the contracted distance from the mountaintop to sea level, 1907 m in the rest frame of the earth. (b) How long does it take the muon to travel this distance at a relative speed of 0.994c? (c) What fraction of the muons should remain after the time found in (b)? (d) Does this result agree with the experimental results?

25.33● The Concorde airplane is 62.1 m long. (a) What difference will an observer on the ground find between the Concorde's length when at rest on the runway and during flight at 1700 km/h? (b) How fast would the Concorde have to fly to increase this difference by a factor of five? (*Hint:* Use the approximation that for $x << 1$, $\sqrt{1 - x} \approx 1 - x/2$.)

Section 25.7 Mass and Energy

25.34 Suppose that all of the matter in a 125-mg grain of sand was entirely converted to energy. How much energy would be released?

25.35 An electron and a positron (a positive electron) can come together and annihilate; that is, they disappear, producing a flash of electromagnetic radiation. If each of these particles has a mass of 9.11×10^{-31} kg, what is the total energy of the radiation?

25.36● A certain nuclear power plant is capable of producing 1.0×10^9 W of electric power. During operation of the reactor, mass is converted to energy. How much mass is converted per hour if the efficiency of the plant is 30%?

25.37● Suppose that one milligram of matter was converted to energy and that all of the energy was used (with no waste) to operate a 200-W lamp. How long could the lamp be operated?

Section 25.8 Relativistic Momentum

25.38 With what speed must a muon travel so that its correct (relativistic) momentum is 6.0% greater than the value that would be computed nonrelativistically? Give your answer in terms of the speed of light.

25.39 A muon ($m = 1.88 \times 10^{-28}$ kg) travels with a speed of 1.87×10^8 m/s as measured in the laboratory. What is the momentum of the muon in the frame of the laboratory?

25.40 How fast must a particle travel so that correct relativistic momentum would be six times the value expected from a non-relativistic calculation?

25.41● An alpha particle ($m = 6.64 \times 10^{-27}$ kg) has a momentum of 3.75×10^{-19} kg · m/s as measured in the laboratory. What is the speed of the alpha in the frame of the laboratory?

25.42● What must be the speed of an alpha particle that has the same momentum as a neutron traveling with a speed $v = 0.90c$? The mass of the alpha is $m_\alpha = 6.64 \times 10^{-27}$ kg, and the mass of the neutron is $m_n = 1.67 \times 10^{-27}$ kg.

Section 25.9 Relativistic Kinetic Energy

Hints for Solving Problems

Relativistic energy consists of two parts: energy due to rest mass and energy due to motion.

25.43 A muon travels with a kinetic energy equal to its rest-mass energy. How fast is it traveling?

25.44 A particle of mass m_0 moves with a speed of 2.5×10^8 m/s relative to an observer. What is the kinetic energy of the particle in the rest frame of the observer?

25.45 How fast must a particle travel relative to an observer for its total energy to be double its rest-mass energy?

25.46● A particle of mass m_0 is given a kinetic energy equal to one-half its rest-mass energy. How fast must the particle be traveling?

25.47● Calculate the velocity of a particle whose kinetic energy is 10 times its rest-mass energy.

25.48● A particle's total energy is five times its rest-mass energy. How fast is the particle traveling?

25.49● A particle of mass m_0 travels at a speed of $0.75c$ relative to the laboratory. (a) What is its kinetic energy? (b) What is its total energy?

25.50● Show that the relativistic expression for kinetic energy goes smoothly to the classical expression when the velocity is small compared with the speed of light. (*Hint:* Use the binomial series to evaluate the relativistic expression.)

*Section 25.10 The Relativistic Doppler Effect

25.51 Radio signals from a distant planet are received at a frequency of 106 MHz on a spaceship headed directly toward the planet at a speed of $0.30c$. What is the frequency of the radio signal as measured on that planet?

25.52 A radio station broadcasting on a frequency of 106 MHz is received by an astronaut headed toward the station at a speed 6.0×10^7 m/s. To what frequency must the astronaut's radio be tuned?

25.53 A spaceship traveling at $v = 0.35c$ is moving transverse to the line of sight of an earthbound observer. (a) If the ship's radio has a proper frequency of 105 MHz, at what frequency will it be detected by the observer? (b) What frequency would be observed if the ship were headed directly toward the earth? For transverse motion, only the shift due to time dilation holds.

25.54 The red light emitted by hydrogen has a wavelength of 656 nm measured in the laboratory. An astronomer studying a distant star observes light at a longer wavelength, which she believes is hydrogen light that has been Doppler-shifted because of the star's motion. (a) Is the star moving toward or away from her? (b) If the star is moving at a speed of 40,000 km/s, what is the observed wavelength of the light?

25.55● The red light emitted by hydrogen has a wavelength of 656 nm measured in the laboratory. An astronomer studying a distant star observes light at a longer wavelength, which he believes is hydrogen light that has been Doppler-shifted because of the star's motion. (a) If the observed wavelength is 733 nm, is the star moving toward or away from him? (b) What is the speed of the star relative to the earth?

25.56● An astronaut moves rapidly past a galactic traffic signal that emits yellow light of frequency 5.17×10^{14} Hz measured in the rest frame of the signal. The astronaut looks at the signal in her rearview mirror. (a) What frequency does the astronaut observe if her velocity is $0.095c$ relative to the signal? (b) To what color does this frequency correspond?

*Section 25.11 The Principle of Equivalence

25.57 A radio signal of frequency f_0 is broadcast by a station on a mountaintop 3500 m above sea level. What is the shift in frequency of that station as measured by an observer at sea level?

25.58 A radio signal of frequency f_0 is broadcast by a station at sea level. What is the shift in frequency of that station as measured by an observer on a mountaintop 2000 m above sea level?

25.59 A satellite at an altitude of 100 km broadcasts a radio signal at 108 MHz. What is the shift in frequency at sea level due to gravitational effects? Is the shift positive or negative? Take g as constant.

*Section 25.12 General Relativity

25.60● The separation between the planet Mercury and the sun is 46×10^9 m at the perihelion. The period of revolution of Mercury about the sun is 88 days, and its perihelion advances at the rate of 573 seconds of arc per century. Calculate the distance through which the perihelion of Mercury advances in one period of the planet's motion about the sun.

Additional Problems

25.61● A beam of particles traveling at $0.75c$ is observed in the laboratory. In a time equal to one mean life in the frame of the particles, the beam has traveled 6.12 m in the laboratory. What is the mean life of these particles?

25.62● A cosmonaut orbits the earth at an altitude of 300 km above the earth's surface. (a) Calculate the tangential velocity of the spaceship. (b) Calculate the period of one complete orbit. (c) If the cosmonaut remains in orbit for 14 days, how much younger will he be upon his return than his twin brother who stayed behind? Use the approximation that for $x << 1$, $\sqrt{1 - x} \approx 1 - x/2$. Ignore the effects of altitude.

25.63●● A spaceship traveling away from the earth at a speed of $0.50c$ launches a smaller ship with a relative speed of $0.50c$ in the same direction. The smaller ship then fires a missile back toward the earth with a speed of $0.50c$ relative to the smaller ship. What is the velocity of the missile relative to the earth?

25.64●● An electron travels at high speed through a 10.0-m-long section of an accelerator. In the reference frame of the moving electron, the length of the accelerator section is contracted to 7.83 m. What is the momentum of the electron?

25.65•• In a hypothetical experiment, a plane flies eastward around the earth at an average supersonic speed of 600 m/s relative to the ground and an average altitude of 12 km. The flying time is 18 hours, and the flight path is equatorial. Calculate the time difference that should result when a clock that was on the plane is compared with a clock that was stationary on the ground at the equator. (*Hint:* To find the difference due to special relativity, choose an imaginary clock that is stationary with respect to the earth's center of mass. Then find the time on a clock moving with the earth's surface as the earth rotates about its axis, and the time on a plane moving with speed v relative to the earth. Then add the differences due to gravitation to the difference due to special relativity.)

25.66•• In a hypothetical experiment, a plane flies westward around the earth at an average supersonic speed of 600 m/s relative to the ground and an average altitude of 12 km. The flying time is 18 hours, and the flight path is equatorial. Calculate the time difference that should result when a clock that was on the plane is compared with a clock that was stationary on the ground. (*Hint:* See the hint for Problem 25.65. Notice that the direction of the plane's velocity is negative with respect to the direction in that problem.)

25.67•• An astronaut orbits the earth in an equatorial orbit at an altitude of 350 km above the mean radius of the earth. Her twin sister remains on earth. The astronaut remains in orbit for an interval of 150 days. (a) Is there a difference in their ages and if so, how much is it? (b) When she returns to earth will the astronaut be younger or older than her twin sister who stayed behind? (*Hint:* See the hint for Problem 25.65. Take the direction of motion of the spacecraft to be in the same direction as the motion of the earth.)

The Discovery of Atomic Structure

The atomic concept is fundamental to the modern view of physics and chemistry—but it has not always been so. Only in the twentieth century did the existence of atoms become universally accepted. The size of individual atoms is so small that we are unable to see them directly, and therefore we find evidence for them difficult to comprehend. It is not unusual to meet people who accept atoms as the objects that chemists and physicists talk about, but who cannot cite any experimental justification for believing in the existence of atoms. For these reasons, we turn our attention to the development of the concept of atomism and to the many observations that have established its validity.

Atomic theory has proved successful because it explains observations that could not otherwise be understood. As we saw in Chapter 12, an elementary atomic theory helped explain the behavior of gases. As more details of atomic structure have become known, older theories of matter and its behavior have given way to newer and more comprehensive ideas. Knowledge of atomic structure has deepened our understanding of other areas of physics. For example, greater understanding of atomic structure led to better understanding of solids, which in turn led to the development of new kinds of materials, including new semiconductors, magnets, and superconductors.

In this chapter, we discuss the development of atomic theory up to the early twentieth century, describing the basic facts upon which later ideas were built and which remain fundamental to contemporary understanding. In addition, we present the principal observations about natural radioactivity and the law of radioactive decay and then show how Rutherford's scattering experiment led to our present concept of the nuclear atom. In the next chapter, we describe how classical physics was insufficient to explain observations of atomic structure and how the concepts introduced by Planck, Einstein, and Bohr led to a new understanding of matter. ■

26.1 Evidence of Atoms from Solids and Gases

An atomic theory of matter was first proposed by the Greek philosopher Leucippus about 500 B.C. and developed by his pupil Democritus. They believed that matter was made up of particles so small that they could not be divided. These ultimate particles were called atoms, from the Greek word *atomos,* which means indivisible. Democritus proposed that the different properties of various substances were due to differences in the nature of their atoms. However, his atomic theory was rejected by the dominant philosophers of the next generation. In particular, Aristotle argued that matter was continuous and therefore infinitely divisible. Thus, there was no need for atoms. This rejection by Aristotle doomed the atomic concept until the seventeenth century, when the ideas of Galileo and Newton undermined Aristotle's authority and atomism was revived.

An important stimulus to the acceptance of atomism came from studies of crystals made by many different people. One of the most remarkable properties of crystals is that many possess a regular geometric shape. For example, quartz crystals are hexagonal (Fig. 26.1a). Other crystals have their own particular shapes (Fig. 26.1b). The shapes of small crystals intrigued the English scientist

Figure 26.1

(a) Crystals of quartz have hexagonal shapes. (b) Pyrite (FeS_2) occurs in three characteristic shapes: cubic, octahedral, and pyritohedral. (c,d) Drawings of crystals made by Robert Hooke.

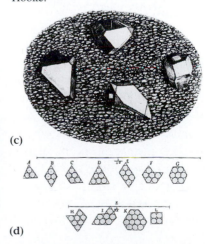

(a)   (b)   (c)

(d)

1895 Wilhelm Roentgen discovers x rays, demonstrating they can penetrate matter.

1896 Henri Becquerel is the first to observe radioactive decay; three kinds of radioactive emissions are eventually found.

1897 J. J. Thompson discovers the electron, measures its mass-to-charge ratio, and establishes that electrons are basic constituents of all atoms.

(a)

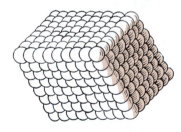

(b)

Figure 26.2

(a) Calcite crystals. (b) A model of calcite composed of small rounded constituent particles. (After Huygens.)

Robert Hooke when he examined them with his microscope. He assumed that the symmetry of the crystals reflected the regular arrangement of the tiny particles that composed the solid. He illustrated his ideas with several drawings published in 1665 (Fig. 26.1c,d). In 1690, the Dutch scientist Christiaan Huygens showed that the geometrical shape of Iceland spar (calcite) could be explained if the mineral were composed of small rounded bodies arranged as shown in Fig. 26.2.

The ability of some crystals to cleave—that is, to split along definite planes, leaving smooth surfaces—led the French abbé René Just Haüy to a deeper understanding of crystal structure. In 1781, Haüy accidentally dropped a calcite cluster, which broke apart. Haüy noticed that the broken crystal had a single fracture along one edge. When he tried to break it in other directions, it broke into rhombohedral pieces. He later cleaved calcite crystals of other initial shapes, always finding that they could be broken to reveal the rhombohedral form.

After similar experience with other crystals, Haüy proposed that continued cleavage of crystals into smaller and smaller pieces ultimately reduces the crystal to the smallest possible unit or building block. Today we call this smallest shape the **unit cell.** A whole crystal can be formed by stacking these unit cells side by side.

Haüy showed that his model of unit cells could also be used to explain the angles between other natural faces that occur when the crystal layers are successively smaller and recede from the edge in a regular manner, like the steps of a large building (Fig. 26.3). Thus, the crystallographers of the early nineteenth century became convinced that the shapes of crystals were a direct result of their construction from invisibly small building blocks in a regular fashion. Eventually, links were established between the shapes and properties of crystals and the physical and chemical properties of their constituent atoms.

At the same time that crystallographers were discovering Haüy's ideas, chemistry was beginning to provide independent evidence for atoms. By 1800 the law of definite proportions had been demonstrated by the French chemist J. L. Proust, though it was by no means universally accepted. According to this rule, the proportions (by mass) of the elements in a chemical compound are constant. If water, consisting of the elements hydrogen and oxygen, is decomposed, the mass of the oxygen released is *always* eight times greater than the mass of the hydrogen. Similarly, when oxygen and hydrogen combine to form water, they *always* do so in the same eight to one proportion by mass.

1908 Jean Perrin determines the value of Avogadro's number, thereby demonstrating the existence of atoms.

1909 Ernest Rutherford proposes the nuclear model of the atom, based on his analysis of particle scattering experiments.

1911 W. L. Bragg (shown) and his father, W. H. Bragg, measure the wavelength of x rays by using the structure of crystals.

Inspired in part by this rule, the English chemist John Dalton about 1803 developed a quantitative atomic theory of chemistry. In addition to ideas similar to the atomic theory of Democritus, the principal postulates of Dalton's theory were that chemical reactions only separate or join atoms, and that when different atoms combine to form a particular compound, it always contains the same relative number of atoms. That is, Dalton's theory incorporated the observed law of definite proportions.

Dalton and others applied these ideas to understanding chemical combinations. Tables of relative atomic masses were compiled, which compared the mass of an element with that of hydrogen. Still, no one knew either the mass or the size of a single atom.

Soon after the publication of Dalton's atomic theory, J. L. Gay-Lussac found an empirical rule governing the behavior of gases. He rediscovered an observation by Cavendish that hydrogen and oxygen gases combined in volume proportions of two to one to form water. He further showed that other gases reacted in volumes whose ratios were small whole numbers. This rule, known as the law of volumes, says that gases unite in simple and definite proportions by volume.

The law of definite proportions, Dalton's atomic theory, and the law of volumes were brought together by the Italian count Amadeo Avogadro (1776–1856). Avogadro suggested that equal volumes of gases (at the same temperature and pressure) contain equal numbers of molecules. This hypothesis clearly differentiates between molecules and atoms: The molecule is a combination of atoms. Avogadro had no way of knowing how many molecules were present in a given volume of gas, but he knew the number must be large.

Using his hypothesis of equal numbers of molecules in equal volumes of gases and Dalton's theory, Avogadro satisfactorily explained the law of definite proportions and the law of volumes. In his view, gases consisted of molecules, and molecules consisted of atoms. In addition, equal volumes of gases contained equal numbers of molecules, and gases reacted by exchanging atoms, thus changing the ratios of numbers of molecules (and hence volumes) by small whole numbers. Avogadro applied his hypothesis to explain Cavendish's experimental observation, correctly concluding that a molecule of water was formed from half a molecule of oxygen and two half-molecules of hydrogen.

Unfortunately, Avogadro's hypothesis was ignored by his contemporaries. Had it been accepted, chemists would have been spared half a century of confusion. However, we should remember that few facts were available in Avogadro's time that could support his hypothesis, and even these "facts" were not without controversy.

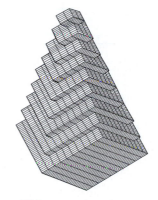

Figure 26.3
Haüy's interpretation of crystal facets arising from stepped layers.

Electrolysis and the Quantization of Charge

Because atoms are very small, it follows that vast numbers of them are required to make any appreciable amount of matter, such as one gram. How many atoms are in one gram of matter? This question was central to the development of atomic theory, as well as to the theory of chemical reactions. The next step in answering this question came from Faraday's experiments with electricity.

In the early 1800s, experiments showed that chemical compounds could be separated into their constituent components by passing an electric current through them, a process called **electrolysis.** Michael Faraday discovered that the amount of substance decomposed in electrolysis was proportional to the magnitude of the electric current and to the elapsed time (Fig. 26.4). Faraday concluded that the mass of material released or deposited on the electrodes is proportional to the electric charge that passes through the system.

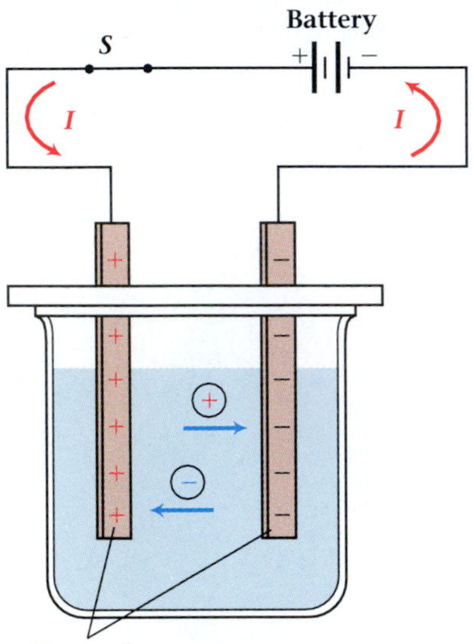

Faraday also found that the mass of material deposited was proportional to its chemical equivalent mass, which is the atomic mass divided by the valence, or most common combining ratio. For example, electrolysis of sodium chloride (table salt) released amounts of sodium and chlorine in direct proportion to their atomic masses, as given by Dalton. However, electrolysis of copper chloride sufficient to release an amount of chlorine equal to its atomic mass would only deposit an amount of copper equal to half its atomic mass. Faraday would have said that copper has a valence of 2, so in each case, electrolysis released the equivalent mass of the substance. For many elements these equivalents were the same as Dalton's atomic masses (corresponding to a valence of 1). This correlation was explained by assuming that charge is *quantized* (existing in only discrete amounts), with a unit of charge e. Each single-valent ion (the particle collected by the electrolysis) was thought to carry one unit of charge. Faraday carefully measured the total amount of charge required to deposit one gram atomic mass (one mole) of a single-valence element. (One gram atomic mass is the amount of substance, in grams, equal to its atomic mass.) This amount of charge became known as the *faraday,* and its presently accepted value is 96,485 coulombs.

According to Avogadro's hypothesis, a gram atomic mass of a substance would contain a definite number N_A of molecules. The number N_A is called **Avogadro's number.** Thus the faraday, F, must be the product of e and the number N_A. That is,

$$F = N_A e. \tag{26.1}$$

Since the faraday constant is known with great precision, Eq. (26.1) is important because it can be used to determine either Avogadro's number N_A or the elementary charge e if the other is known. Faraday recognized the significance of this relationship, but was unable to measure either N_A or e independently.

Faraday did not prove that atoms exist; he could not determine their size or mass, for example. However, he used the assumption of atomism to explain successfully the quantitative aspects of electrolysis, and he extended this assumption to include the concept of quantization of electric charge. In doing so, he gave still more credibility to the idea that atoms do, indeed, exist.

Figure 26.4

An electrolytic cell. The amount of substance that decomposes is proportional to the applied current and the elapsed time.

Example 26.1

In an electrolysis experiment, a student plated 0.503 g of silver onto an electrode. During the plating, a constant current of 0.500 A was maintained for 15.0 min. Assuming singly charged ions, determine the atomic mass of the silver.

Strategy For singly charged ions, the ratio of the atomic mass to the mass deposited during electrolysis is equal to the ratio of the faraday to the charge q delivered. Once we find the charge delivered, we can find the atomic mass from knowledge of the mass deposited, the faraday, and the charge.

Solution We start by computing the amount of charge delivered in the electrolysis. This charge is the product of current with time,

$$q = It = (0.500 \text{ A})(15.0 \text{ min})(60 \text{ s/min}) = 450 \text{ C}.$$

If we let M be the atomic mass, then the ratio of M to the mass deposited is the ratio of F to q:

$$\frac{M}{0.503 \text{ g}} = \frac{96485 \text{ C/mol}}{450 \text{ C}},$$

$$M = 108 \text{ g/mol}.$$

26.3 **Avogadro's Number and the Periodic Table**

By the middle of the nineteenth century, chemistry was in a state of confusion. Independent application of the law of definite proportions and the law of volumes without the use of Avogadro's hypothesis had led to different interpretations of reactions and different tables of atomic masses. To clear up some of this confusion, an International Chemical Congress was convened in Karlsruhe, Germany, in 1860. At this congress, Stanislao Cannizzaro recommended acceptance of Avogadro's hypothesis. Among the participants at the congress who were influenced by Cannizzaro were the German chemist Lothar Meyer (1830–1895) and the Russian chemist Dmitri Mendeleev (1834–1907).

Subsequently, Meyer and Mendeleev independently produced tables of the elements. Mendeleev's version was published first (1869), and Meyer's followed soon after. Figure 26.5 shows Mendeleev's table of the elements known at that time, as it appeared in the German translation of his work. He arranged the elements in order of increasing atomic mass, noting the periodic occurrence of elements with similar physical and chemical properties. This arrangement became known as the *periodic table* of the elements. Mendeleev left gaps in the table to fit the elements into the proper columns according to their properties, and he proposed that the gaps represented elements not yet discovered. Subsequent discovery of these elements, with properties matching Mendeleev's predictions, helped win acceptance of his table among chemists. The absolute masses of atoms were still unknown, but chemists were now in agreement about their relative masses.

					Ti = 50	Zr = 90	? = 180
					V = 51	Nb = 94	Ta = 182
					Gr = 52	Mo = 96	= 186
					Mn = 55	Rh = 104.4	Pt = 197.4
					Fe = 56	Ru = 104.4	Ir = 198
					Ni = Co = 59	Pd = 106.6	Os = 199
H = 1					Cu = 63.4	Ag = 108	Hg = 200
	Be = 9.4	Mg = 24			Zn = 65.2	Cd = 112	
	B = 11	Al = 27.4			? = 68	Ur = 116	Au = 197?
	C = 12	Si = 28		W	? = 70	Sn = 118	
	N = 14	P = 31			As = 75	Sb = 122	Bi = 210?
	O = 16	S = 32			Se = 79.4	Te = 128?	
	F = 19	Cl = 35.5			Br = 80	J = 127	
Li = 7	Na = 23	K = 39			Rb = 85.4	Cs = 133	Tl = 204
		Ca = 40			Sr = 87.6	Ba = 137	Pb = 207
		? = 45			Ce = 92		
		?Er = 56			La = 94		
		?Yt = 60			Di = 95		
		?In = 75.6			Th = 118?		

Figure 26.5

Mendeleev's periodic table of the elements as he arranged them in 1869. His predicted elements are marked in color.

Avogadro's number was still unmeasured at the beginning of the twentieth century. The first reliable measurements of this quantity were published in 1908 by Jean Perrin. Perrin made quantitative measurements of **Brownian motion**, which is the continual irregular movement of minute particles suspended in a liquid. This motion was first observed in 1827 by the Scottish botanist Robert Brown. The cause of Brownian motion is molecular motion, as molecules in the liquid incessantly move and collide with one another. For a large object, the net effect of collisions due to molecules of the surrounding fluids is zero because, on average, all the forces are balanced. But when the size of the object is sufficiently small, the instantaneous force is not zero. Thus, a tiny particle suspended in a fluid moves wildly about with random motions. Figure 26.6 shows a drawing, similar to those made by Perrin, of a particle moving in a fluid. The line segments join the consecutive positions of the same particle at intervals of 30 s.

Figure 26.6

Drawings by Perrin of Brownian motion, made by connecting the consecutive positions of the same granules at intervals of 30 s.

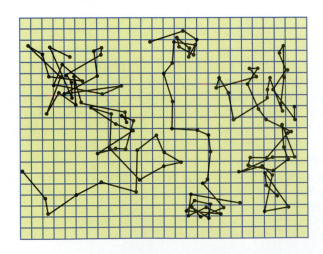

In addition, Perrin found that the suspended particles distribute themselves in the same way as do the particles of a gas. That is, the density of particles decreases exponentially with height in the same way that the density of the atmosphere decreases with height (Section 12.7). A pressure called osmotic pressure is associated with the motion of these particles. It is proportional to the mean kinetic energy $\overline{KE}$ of the particles. At a height h where the number of particles per unit volume is n, the pressure is

$$P(h) = \tfrac{2}{3}n\overline{KE}.$$

At greater height $h + \Delta h$, the number of particles changes to $n + \Delta n$, so

$$P(h + \Delta h) = \tfrac{2}{3}(n + \Delta n)\overline{KE}.$$

Because the system is in thermal equilibrium and because the average kinetic energy of the particles is proportional to the temperature, the average kinetic energy $\overline{KE}$ must be the same regardless of height.

Since the particles are suspended in a liquid of different density, a buoyant force acts on them. The forces on the particles due to osmotic pressure, gravitation, and buoyancy are balanced. Let the volume of each particle be taken as V, its density as ρ, and the density of the surrounding liquid as ρ'. The number of particles in the space between h and $h + \Delta h$ in a column of cross section S (Fig. 26.7) is given by $nS\,\Delta h$. The total force on the particles is the force per particle times the number of particles in the volume. The resultant of the gravitational force and the buoyant force (Section 10.3) is a downward force:

$$\text{downward force} = (nS\,\Delta h)g(\rho - \rho')V.$$

The upward force due to the osmotic pressure is

$$\text{upward force} = [P(h) - P(h + \Delta h)]S = -\tfrac{2}{3}\,\overline{KE}S\,\Delta n.$$

Upon equating these two forces, we see that

$$\frac{\Delta n}{\Delta h} = \frac{-g(\rho - \rho')Vn}{\tfrac{2}{3}\,\overline{KE}}.$$

The change with height in the number of particles per unit volume is proportional to the number of particles per unit volume. As we showed earlier (Chapter 12), equations of this type can be described with an exponential function,

$$n(h) = n(0)e^{-3g(\rho - \rho')Vh/2\overline{KE}}. \tag{26.2}$$

Recall that in Chapter 12 we found that the average kinetic energy $\overline{KE}$ can be related to the temperature through

$$\overline{KE} = \frac{3RT}{2N_A},$$

where R is the molar gas constant, T is the absolute temperature, and N_A is Avogadro's number. These two equations above can be combined to give

$$N_A = \frac{RT}{g(\rho' - \rho)Vh}\ln\!\left[\frac{n(0)}{n(h)}\right]. \tag{26.3}$$

We now have an expression for Avogadro's number N_A in terms of measurable quantities.

In 1908, Perrin prepared a suspension of tiny particles of a yellow pigment used in watercolors. He measured the density of particles and their diameter,

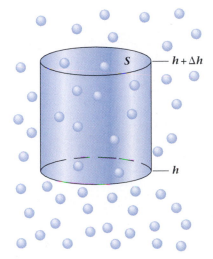

Figure 26.7

Particles with individual volume V suspended in a fluid move with random motion. The number of particles between h and $h + \Delta h$ in a column of cross section S is $nS\,\Delta h$, where n is the number of particles per unit volume.

from which he computed their volume. He then counted the number of particles at different heights in the solution. From these measurements he showed that, indeed, the number of particles per unit volume decreased exponentially with height as predicted, verifying Eq. (26.2). Perrin then used his measurements to determine the value of Avogadro's number. Perrin's initial result was a little larger than the value

$$N_A = 6.02 \times 10^{23} \text{ molecules/mole}$$

which we use today.

The atomic concept existed before Perrin's time. The kinetic-theory model of a gas, for example, had been successfully used to account for many observed gas properties. However, the success of the atomic model in kinetic theory and in chemistry was not proof of the reality of atoms and molecules. The concept might be only a convenient fiction useful for calculations. With Einstein's 1905 paper on Brownian motion, another way of subjecting the atomic hypothesis to a qualitative test was found. Perrin performed several other independent experiments on Brownian motion in addition to the one discussed here, all of them yielding similar values for Avogadro's number. Some of these experiments also tested the details of Einstein's predictions.

The agreement between the values found for Avogadro's number in independent experiments analyzed under the assumption of the reality of molecules, along with the agreement with Einstein's prediction for Brownian motion, firmly established the reality of molecules and atoms. Because of Perrin's work, the distinguished chemist Wilhelm Ostwald, long an opponent of the atomic theory, wrote, "The atomic hypothesis is thus raised to the position of a scientifically well-founded theory."

26.4 The Size of Atoms

By making use of what we have just learned, especially the value of Avogadro's number, we can estimate the size of atoms. We are now in a position to determine the number of atoms that occupy a given volume. From there it is just a short step to determine the volume occupied by a single atom—that is, its size.

Example 26.2

How many copper atoms are there in a solid cube of copper that is 1.00 cm on each side?

Strategy We can use Avogadro's number to calculate the number of atoms if we know the number of moles of copper that we have. We can calculate the number of moles as the mass divided by the atomic mass, and we can calculate the mass from the given volume and the density of copper. So once we look up the values of atomic mass and density for copper, we can determine the number of copper atoms present.

Solution Chemical and physical measurements establish that copper has an atomic mass of 63.6 g/mol and a density of 8.96 g/cm³. The mass of copper in the cube is

$$\text{mass} = \text{density} \times \text{volume} = (8.96 \text{ g/cm}^3)(1.00 \text{ cm}^3) = 8.96 \text{ g}.$$

The number of moles present is given by the mass divided by the atomic mass:

$$\text{number of moles} = \frac{\text{mass}}{\text{atomic mass}} = \frac{8.96 \text{ g}}{63.6 \text{ g/mol}} = 0.141 \text{ mol}.$$

Finally, multiplying the number of moles by Avogadro's number, we get the number of atoms:

$$\text{number of atoms} = \text{number of moles} \times N_A$$
$$\text{number of atoms} = (0.141 \text{ mol})(6.02 \times 10^{23} \text{ atoms/mol})$$
$$\text{number of atoms} = 8.49 \times 10^{22} \text{ atoms}.$$

Example 26.3

Use the result of Example 26.2 to estimate the size of an atom of copper.

Solution   We see in the preceding example that one cubic centimeter of copper contains 8.49×10^{22} atoms, a very large number indeed. We can find the volume occupied by one atom by dividing the total volume (1 cm^3) by the number of atoms, which gives an average volume per atom of 1.18×10^{-23} cm^3, or 1.18×10^{-29} m^3. If, for simplicity, we assign each atom a small cubic volume of that size, then the cube edges have a length equal to the cube root of the volume, or 2.28×10^{-10} m. Without making any assumption about the shape of atoms or how they are arranged, we may conclude that the linear dimensions of an atom are of this order of magnitude; that is,

$$\text{atomic size} \approx 10^{-10} \text{ m}.$$

Master the Concept

The Size of Molecules

Question: When a small drop of oleic acid, an inorganic substance that is insoluble in water, is placed on the surface of a large bowl of water, it spreads out forming a layer only one molecule thick. The long thin molecules align with one end toward the water and the other end up. How can this effect be used to estimate the size of the oleic acid molecules?

Answer: When the oleic acid is not hemmed in by the edge of the container, it spreads out to form a large circular area. The volume of this circular layer is the same as the volume of the initial drop. We can find the thickness of the layer by dividing the volume of the drop by the area of the circular layer. If the layer is indeed only one molecule thick, then the thickness of the layer is the length of the molecules that make up the layer.

▼
26.5 Crystals and X-Ray Diffraction

Perrin's experiments proved the existence of atoms and molecules and determined the value of Avogadro's number. However, other questions remained. Are atoms really indivisible? Do they have any internal structure? Today we know the answers to these questions, but the discoveries that led to these answers did not come all at once. We first discuss the discovery of x rays, which, as we will see, served as a primary tool for examining details of atomic structure.

During the late nineteenth century, improvements by Sir William Crookes in the production of vacuums led to pressures that were thousands of times smaller than those previously attainable. To investigate the behavior of electric discharges in such a vacuum, researchers sealed electrodes into tubes and removed the air inside with a vacuum pump. Tubes of this sort became commonly known as Crookes tubes (Fig. 26.8). When a high voltage was applied across the electrodes, conduction took place within the tube and a green glow became visible in the glass at the positive end. Sometimes, fluorescent minerals were placed in the tube and were observed to glow when the current was present. The fluorescence was due to rays emanating from the cathode (the negative electrode), and these were called **cathode rays.** We will see in Section 26.6 that attempts to understand the nature of cathode rays led to the discovery of the first subatomic particle: the electron.

Among the experimenters who studied the behavior of Crookes tubes was the German scientist Wilhelm Konrad Roentgen (1845–1923). Late in 1895, Roentgen noticed that operating the tube caused a card coated with fluorescent material to glow, even though the card was outside the tube. Even more remarkable was Roentgen's observation that the card's fluorescence was still visible even if he completely surrounded the Crookes tube with black cardboard. In fact, the fluorescence was independent of whether the coated side or the plain side of the card faced the tube.

Figure 26.8

(a) A type of Crookes tube known as a Maltese Cross tube. The shadow of the cross seen on the face of the tube implies the presence of rays that emerge from the cathode and travel in straight lines. (b) The glow from the electric discharge and the shadow of the cross are visible.

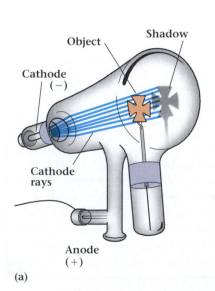

(a)

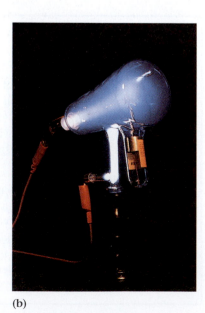

(b)

Roentgen concluded from these observations that some kind of unknown radiation, distinct from cathode rays, was emanating from the Crookes tube and was penetrating objects that were opaque to both visible and ultraviolet light. He named this radiation **x rays** and tested other materials to see if they were at least partially transparent to these rays. Roentgen observed that when a hand was held between a fluorescent screen and the Crookes tube, a shadow image of the bones could be seen within the lighter shadow of the hand itself (Fig. 26.9). This observation was seized immediately by the medical profession, and within weeks of Roentgen's announcement many medical researchers were experimenting with this important new discovery. In a very brief time, x rays became a routine diagnostic tool in hospitals all over the world. As a result, Roentgen became extremely famous and many people referred to x rays as Roentgen rays. In 1901, he was awarded the first Nobel Prize in physics for his discovery.

Roentgen's x rays were so intriguing that scientists began intense investigation and speculation about what they really were. The rays were not deflected by electric or magnetic fields, so they could not be streams of charged particles. They were not diffracted by diffraction gratings, so they were not waves with wavelengths close to those of light. However, there was a possibility that they could be electromagnetic waves with wavelengths very much shorter than those of visible light.

In 1912, Max von Laue (1879–1960) proposed using a crystal as a diffraction grating for x rays. As we have seen, a crystal may be thought of as layer upon layer of atoms, all spaced apart at regular intervals. These intervals are of the order of 10^{-10} m, as shown in the previous section. Thus, if x rays are waves of very short wavelength, a beam of x rays striking a crystal might be diffracted, similar to the way light is diffracted by a grating. The experiment was carried out by Laue's colleagues W. Friedrich and P. Knipping, who bombarded a crystal of zinc sulfide with x rays and recorded the diffraction pattern on a photographic plate. Figure 26.10 shows a similar pattern of x rays diffracted from sodium chloride.

The diffraction of x rays was an historic event. First of all, it established the wave behavior of x rays and provided a means for measuring their wavelengths. Second, it confirmed the atomic structure of crystals and provided a means of determining the actual three-dimensional arrangement of the atoms within crystals.

Within a few months of Laue's discovery, William Lawrence Bragg (1890–1971), while still a student at Cambridge, worked out a simple relationship between the observed angles of x-ray diffraction and the spacing between planes of atoms in the crystal. Together with his father, William Henry Bragg (1862–1942), he determined the atomic crystal structure of a number of crystals.

Bragg reasoned that since x rays are very penetrating, they are only partially reflected from each layer of atoms as they pass through the crystal. Suppose the incident waves are specularly reflected from parallel planes of atoms, with each plane reflecting only a small fraction of the incident radiation. As with ordinary optical diffraction, for some angles of incidence the reflected beams are all in phase and a diffracted beam is observed (Fig. 26.11a). At other angles of incidence, the waves from successive layers are not in phase and no diffraction is seen (Fig. 26.11b). For maximum intensity of the reflected beam, the waves from successive layers must be exactly in phase. Consequently, the path difference between the rays labeled 1 and 2 in Fig. 26.12 must be an integral number of

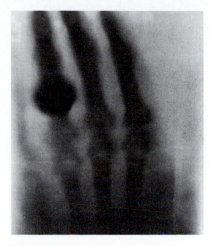

Figure 26.9

An x-ray photograph made by Roentgen at his first public lecture on his new discovery on January 23, 1896.

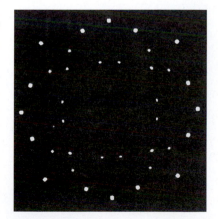

Figure 26.10

A diffraction pattern made by passing a beam of x rays through a small sodium chloride crystal. The symmetry of the pattern is a result of the cubic symmetry of the crystal.

Figure 26.11

Reflection of x rays from successive layers of atoms in a crystal. (a) The wavelets are reflected exactly in phase and reinforce each other. (b) The wavelets are reflected out of phase and cancel each other.

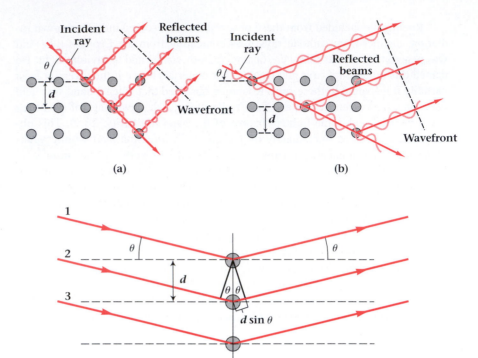

(a) (b)

Figure 26.12

In Bragg scattering of x rays from a crystal, scattering maxima occur at angles where $2d \sin \theta$ is an integral multiple of wavelengths.

wavelengths. This path difference depends on the angle θ and the separation between layers, d, and equals $2d \sin \theta$. Thus

$$n\lambda = 2d \sin \theta, \qquad (26.4)$$

where n is an integer and λ is the x-ray wavelength. This statement is known as **Bragg's law.** The angle θ in the Bragg equation is often called the Bragg angle. The scattering angle, which is the angle between the incident beam and the scattered beam, is 2θ.

Using Avogadro's number, we can compute the number of atoms in a crystal of known size, as in Example 26.2. From the number of atoms per unit volume, we can calculate the average spacing between atoms. This information, together with knowledge of crystal symmetry, means that we can calculate the spacing between layers, d, and then use Bragg's law to determine the wavelength of x rays. The smallest spacing between atoms in a crystal is typically about 2×10^{-10} m. From the Bragg equation, we see that diffraction is not observed unless λ is smaller than $2d$. Thus, the x rays used for diffraction must have wavelengths comparable to or smaller than the interatomic spacing in the crystal.

Example 26.4

In a silver bromide crystal, the atoms are arranged in parallel planes 2.88×10^{-10} m apart. If diffraction is observed from these planes at a scattering angle $2\theta = 31°$, what is the wavelength of the x rays? Assume first-order diffraction, $n = 1$.

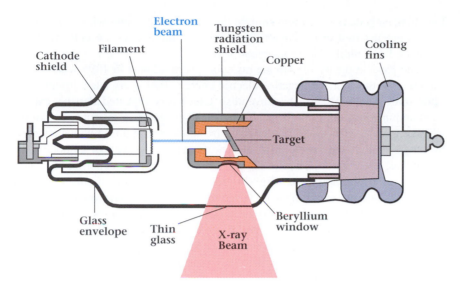

Figure 26.13
Schematic diagram of an x-ray tube with a hooded anode to reduce stray radiation. (After a design by Philips laboratories.) Electrons leave the cathode filament and strike the target anode at high energies, producing the beam of x rays.

Solution Since the scattering angle is $2\theta = 31°$, the Bragg angle θ is 15.5°. From Bragg's law we have

$$\lambda = 2d \sin \theta = 2(2.88 \times 10^{-10} \text{ m}) \sin 15.5°$$
$$\lambda = 1.54 \times 10^{-10} \text{ m}.$$

We have discussed some of the properties of x rays, but not the details of how they are produced. According to Roentgen, x rays were "produced by the cathode rays at the glass wall of the tube." He also found that x rays were produced when other materials were struck by cathode rays. In modern terminology, x rays are produced when high-speed electrons strike matter. In a modern x-ray tube, the electrons are "boiled off" a hot metal cathode and accelerated by an electric potential of 10^4 to 10^6 volts in an evacuated tube (Fig. 26.13). When the electrons strike the positive target, or anode, their rapid deceleration produces x rays. Further details of x-ray production will be given in Chapter 27 after we have developed the necessary concepts.

X rays differ from light in their frequency, which affects the way they interact with matter. The higher-frequency (and shorter-wavelength) x rays are more penetrating. Also, the penetrating power of the x rays is greater for higher-accelerating potentials. Their relative transmission by matter depends on the material's composition. This range of penetrating power allows us to use x rays in a tremendous variety of applications, including medical diagnosis and treatment, solid-state research, archeology, industry, and many other applications where it is desirable to "see beneath the surface" (Fig. 26.14).

X-ray images are widely used in medical diagnosis. Here a woman is being positioned for a chest x ray.

26.6   Discovery of the Electron

During the last decade of the nineteenth century, controversy abounded as to the nature of cathode rays. Were they a wave phenomenon similar to light? Or were they corpuscular radiation, consisting of tiny bits of matter? As we saw in

(a)

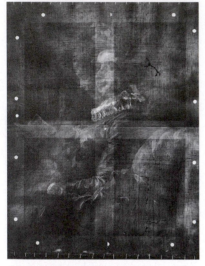

(b)

Figure 26.14

(a) *A Young Girl Reading,* painted by Jean-Honoré Fragonard about the year 1776. (National Gallery of Art, gift of Mrs. Mellon Bruce.) (b) X rays reveal an image of a young man beneath Fragonard's painting.

Fig. 26.8, cathode rays travel in straight lines and cast shadows when small metal objects are inserted in their path. In addition, cathode rays are deflected by a magnetic field, unlike electromagnetic waves.

The particle theory of cathode rays was given a boost in 1895 when Perrin showed that they carried negative electric charge. Two years later, Joseph John Thomson (1856–1940) of the Cavendish Laboratory repeated Perrin's experiment and unequivocally confirmed his result. Thomson then set out to determine whether cathode rays could be deflected by an electric field. Previous attempts by others had shown no effect of electric fields on the path of cathode rays, which was a serious objection to the particle theory. For his experiment, Thomson built a special apparatus (Fig. 26.15a), which was more highly evacuated than those of earlier experimenters. When the air was removed from the tube and a potential difference applied across the electrodes, cathode rays emerged from the negatively charged cathode C and passed through slits in the grounded anodes A and B. The rays then traveled between the two parallel plates D and F and struck the end of the tube, where they produced a small, well-defined fluorescent spot.

When the two plates were connected to a battery and plate D was made negative with respect to F, the rays were deflected downward (Fig. 26.15b). When D was made positive with respect to F, the deflection was upward. Also, the magnitude of the deflection was proportional to the potential difference (and hence the electric field) between the plates. This result lent support to the idea that cathode rays were really charged particles.

Let's follow Thomson's analysis of his experiment. Assume, as Thomson did, that the cathode rays are particles of mass m and electric charge e. A particle traveling between the plates will be transversely deflected because the electric field $\mathbf{E}$ exerts a force $e\mathbf{E}$ transverse to the particle's initial direction. The magnitude of the transverse acceleration due to the electric field is

$$a_E = \frac{F}{m} = \frac{eE}{m}.$$

If the particle has an initial speed v_0, it will travel the length l of the plates in a time $t = l/v_0$. During that time, the particle acquires a transverse velocity v_E given by

$$v_E = a_E t = \frac{eE}{m} \frac{l}{v_0}.$$

As the deflected particle leaves the region between the deflection plates and enters the field-free region, it makes an angle θ_E with the undeflected beam,

$$\theta_E \approx \tan \theta_E = \frac{v_E}{v_0} = \frac{eEl}{mv_0^2}. \tag{26.5}$$

Then Thomson deflected the cathode rays in the same tube with a magnetic field, produced in the space between the electric deflection plates by a matched pair of current-carrying coils positioned outside the glass tube. Thomson shaped the coils to produce a nearly uniform magnetic field at right angles to the deflection plates and extending over the distance l. The acceleration of the particles due to just the magnetic field is

$$a_B = \frac{F}{m} = \frac{ev_0 B}{m}.$$

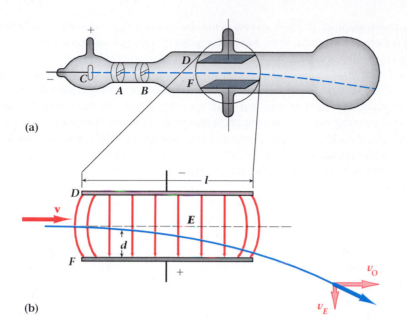

(a)

(b)

Figure 26.15
(a) Schematic diagram of Thomson's apparatus for determining the ratio e/m of an electron. (b) Motion of the cathode ray particles as they pass through the charged plates. In addition to their horizontal velocity v_0, which is constant, they acquire a transverse velocity v_E.

The transverse velocity imparted to the cathode rays by the magnetic field is

$$v_B = a_B t = \frac{ev_0 B}{m} \frac{l}{v_0} = \frac{eBl}{m}.$$

Thus, the angular deflection θ_B due to the magnetic field is

$$\theta_B \approx \tan \theta_B = \frac{eBl}{mv_0}. \tag{26.6}$$

By first deflecting the particles with the electric field alone, Thomson measured θ_E. Then by turning on and adjusting the magnetic field, he brought the beam of particles back to the undeflected position, thus making θ_B equal and opposite to θ_E. Under this condition, Eqs. (26.5) and (26.6) can be set equal to each other and we find that $v_0 = E/B$. Then, using this evaluation of v_0 in Eq. (26.6), we can determine the ratio of e/m to be

$$\frac{e}{m} = \frac{E\theta_E}{B^2 l}. \tag{26.7}$$

All the quantities on the right-hand side of this equation are measured, so in this way Thomson was able to determine the ratio of the electric charge to the mass of the cathode rays.* Today the accepted value is

$$\frac{e}{m} = 1.7588 \times 10^{11} \text{ C/kg.}$$

Thomson conducted his experiments with several different gases at low pressure in the evacuated tube. Each time the result was the same and was consistent with the idea that the cathode rays were charged particles. Upon considering

---

*Thomson actually published the ratio m/e in his 1897 paper. It has become conventional in contemporary physics to use the reciprocal ratio e/m.

the implication of all these experiments, he reached the following conclusions: (1) Atoms are not indivisible, because negatively charged particles had been torn from normally neutral atoms. (2) All cathode ray particles have the same charge and the same mass and are a part of all atoms. (3) If the charge e was the same size as the smallest unit of charge observed in electrochemistry, then the mass of the cathode ray particle was less than one thousandth the mass of a hydrogen atom—the smallest mass known up to that time. Thomson first called these particles "corpuscles," but they soon became known as **electrons,** a name G. J. Stoney had coined several years earlier for the elementary electric charge found in Faraday's electrolysis experiments. Thus, Thomson is considered the discoverer of the electron.

Following Perrin's measurement of Avogadro's number in 1908, the electric charge e was determined from N_A and the Faraday constant F by using Eq. (26.1). Shortly afterward, in a series of experiments begun about 1909, the American physicist Robert A. Millikan (1865–1953) succeeded in measuring the charge of the electron with greater precision than anyone had done before him. Moreover, his work conclusively proved what Thomson had assumed, that the charge on each electron was exactly the same. In his experiment, Millikan examined the motion of single tiny drops of oil that picked up static charge from ions in the air. He suspended the charged drops in air by applying an electric field and watched them through a short-focus telescope (Fig. 26.16).

Millikan derived an expression for the charge q on the drop in terms of large-scale measurable quantities. In a sequence of experiments, he observed many drops and obtained a series of values for q. In practice, q was found to be an integral multiple of the elementary charge e. The magnitude of the elementary charge is

$$e = 1.6022 \times 10^{-19} \text{ C.}$$

This is the charge of one electron. Using the best available value for e/m, we find that the electron's mass is

$$m_e = 9.1094 \times 10^{-31} \text{ kg.}$$

Thus, the mass of the electron is approximately 1/1837 the mass of an atom of hydrogen.

Figure 26.16

Schematic view of Millikan's oil drop experiment. Tiny drops of oil enter the space between the conducting plates through the hole in the top plate. Their motion under the simultaneous influence of the electric field between the two charged plates and the gravitational field is observed through the short-focus telescope.

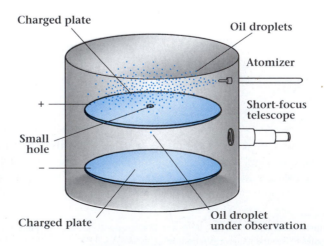

Charged plate

Oil droplets

Atomizer

Short-focus telescope

+

Small hole

−

Charged plate

Oil droplet under observation

SEEING ATOMS

Atoms are extremely small. We have already seen that a reasonable estimate of atomic diameter is about 2×10^{-10} m. Thus, atoms are thousands of times smaller than the wavelengths of visible light, and we cannot see atoms with our naked eyes. Even the use of an optical microscope is no help, since the wavelengths of light are so much greater than the size of the objects we wish to see.

To be able to observe the shapes of objects smaller than optical wavelengths we must use extraordinary techniques and instruments. One early instrument for doing so is the field ion microscope, developed in 1951 by Professor Erwin W. Müller. Essentially, the ion microscope consists of an extremely sharp needle point raised to a high electric potential above a fluorescent screen in an evacuated enclosure (Fig. B26.1). Within the evacuated chamber is a small amount of helium gas. The helium atoms become polarized by the large applied electric field and are drawn toward the positively charged needle point. As they get very close to the tip, the electric field becomes very large, increasing the likelihood of ionization. Once an atom becomes ionized, it is accelerated away from the tip by the electric field. Its motion lies along the direction of the field, that is, along the field lines. In this manner, ions produced at a particular point on the tip are accelerated to a unique point on the screen. Thus, the pattern on the screen is an image of the tip of the needle.

If the rate of ionization is great enough, a steady pattern of light is observed on the screen from the impinging ions (Fig. B26.2a). If the tip radius is small enough, we can obtain

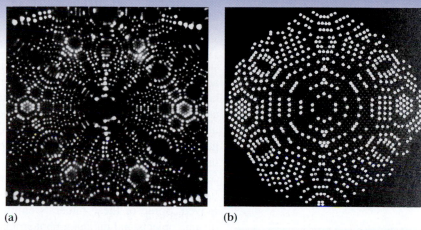

(a)   **(b)**

Figure B26.2   (a) Field ion micrograph of a tungsten tip with a radius of 25 nm. (b) A ball model of a 15-nm radius tungsten tip with the protruding atoms enhanced. These are the atoms imaged by the field ion microscope. The similarity between the model and the micrograph is apparent.

an image of the tip in which the bright spots of the pattern correspond to individual atoms on the tip's surface. Those atoms lying on the step edges of the atomic planes contribute a greater local field and thus are seen as bright spots on the screen. Atoms sitting on top of an otherwise smooth atomic plane are also imaged brightly. Thus, the ion microscope can directly reveal the atomic order in a metallic needle.

If a model of the needle tip, constructed from balls, is painted so that the balls on the edges can be easily seen, it appears as in Fig. B26.2(b). The similarity between the ball model and the microscope picture is striking.

More recently, a new microscope capable of examining surfaces on an atomic scale was developed by Gerd Bennig and Heinrich Rohrer in 1984. The scanning tunneling microscope, as it is called, can resolve features even smaller than the size of an atom and does not require the high electric fields of the field ion microscope. A representative micrograph is shown in Fig. B26.3. A further description of the microscope and how it operates is found in Chapter 31. Although these microscopes are limited in the kinds of surfaces and materials they can image, they have produced easily understandable proof of the atomic theory: We can actually see pictures of the individual atoms.

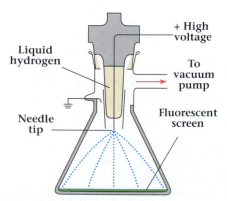

Figure B26.1   Schematic view of a field ion microscope. Dashed lines show the shape of the electric field between the needle point (object) and the fluorescent screen (image plane).

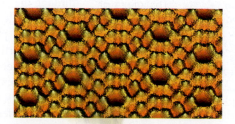

Figure B26.3   The atomic structure of a silicon crystal surface obtained from a scanning tunneling microscope.

(a)

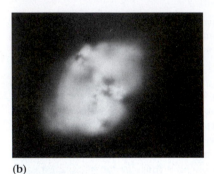

(b)

Figure 26.17

(a) A rock containing the uranium-bearing mineral uranophane. (b) An autoradiograph made by placing the rock on top of a photographic plate. The exposure is due to the radiation emitted by the mineral.

Table 26.1	Thickness of Aluminum Required to Reduce Radioactive Intensity by One-half
Radiation from Radium	**Thickness of Aluminum**
Alpha rays	0.0005 cm
Beta rays	0.05 cm
Gamma rays	8 cm

Source: E. Rutherford, *Philosophical Magazine,* Vol. 26, 1903, p. 177.

26.7 Radioactivity

As the atomic theory became firmly established, modifications of some of its basic postulates became necessary. The discovery of the electron and of positive and negative ions showed that atoms were not indivisible. If atoms could be broken apart, what was their internal structure? How did electrons fit inside an atom? If atoms were not indivisible, then just how permanent were they? The next discovery that affected these issues came in 1896, only one year after Roentgen's announcement of x rays. This event was Henri Becquerel's discovery of radioactivity.

Becquerel was studying phosphorescent minerals that continued to emit light (fluoresce) after the cause of their excitation (usually sunlight) was removed. Some of these minerals emitted a radiation that would pass through opaque objects and darken a photographic plate. It was presumed that this radiation was associated with the phosphorescence. At one point, Becquerel developed a photographic plate that had been wrapped in heavy black paper and then left in the dark for several days beneath a piece of uranium sulfate. To his great surprise, he found that the mineral had produced an intense image on the plate (Fig. 26.17). Becquerel soon demonstrated that the radiation was due to the presence of the uranium itself and was independent of any phosphorescent effects. This emission of a penetrating natural radiation is called **radioactivity.**

After Becquerel's discovery, chemists searched for other radioactive materials. At first, of the many compounds tested, only those of uranium and thorium showed the property of radioactivity. However, in 1898 Marie Curie and her husband, Pierre, reported the isolation of two new radioactive elements, which they named polonium and radium. Radium turned out to have much higher radioactivity than either uranium or polonium.

Shortly after the discovery of radioactivity, Ernest Rutherford (1871–1937) showed that the radiation from uranium had two distinct components. He found that one component of the radiation was readily stopped by a single layer of thin aluminum foil. Rutherford named this easily absorbed component **alpha (α) rays.** The second, more penetrating component he called **beta (β) rays.** Further experiments by the French physicist Paul Villard showed that the radiation from uranium contained a third component that was even more penetrating than beta rays. This third type of radiation is called **gamma (γ) radiation.** Table 26.1 gives the relative penetrations of the three radiations from radium through aluminum absorbers. Table 26.2 lists the masses and charges of these radiations.

Over the course of several years, Rutherford gradually explained the nature of alpha rays. The absorption of alpha rays was proportional to the density of the absorbing material, and the rate of absorption increased with distance traveled through the absorber. On the basis of this indirect evidence, Rutherford proposed that alpha rays were particles. Because they strongly ionized any gas through which they passed, Rutherford suggested they must possess electric charge. Charged particles should be deflected by a magnetic field, but such deflection was not observed initially for alpha rays. However, deflection was observed later. Rutherford explained the initial failure to notice deflection by suggesting that alpha particles were very massive. Subsequent measurements established that the alpha particles have an *e/m* less than one-thousandth that of electrons.

Rutherford and his associate Hans Geiger made accurate measurements of the electric charge carried by alpha rays, finding the charge of the alpha particle to be exactly twice the charge carried by a hydrogen ion. Using this information and the ratio of e/m, they showed that the mass of the alpha particle was the same as that of a helium atom. Separate experiments confirmed that alpha particles really are doubly charged helium ions.

Many properties of beta rays were first observed by Becquerel. Using the experimental setup illustrated in Fig. 26.18, he found that beta rays were deflected by a magnetic field. Becquerel placed the radioactive material in a small lead container, open at the top. This container was then placed on top of a photographic plate, which was positioned in a magnetic field directed parallel to the plane of the plate. The beta rays emitted by the radioactive source were deflected by the magnetic field in a direction indicating that they carried a negative electric charge.

We showed earlier that a particle of charge e and mass m, moving with speed v perpendicular to a magnetic field B, travels in a circular path of radius

$$r = \frac{mv}{eB}. \qquad (26.8)$$

A beta particle initially directed upward from the lead box in Fig. 26.18 would be deflected by the magnetic field so that it would strike the plate a distance $2r$ from the source. With this arrangement, Becquerel found that the rays struck the plate over a range of distances, indicating a wide variation in the speeds of the beta rays. By placing thin absorbing foils over the plate, he discovered that the most easily deflected rays were also the most easily absorbed rays.

Becquerel also investigated the deflection of beta rays by electrostatic fields. By comparing the electrostatic deflection with the magnetic deflection, he was able to show that beta rays had the same ratio of e/m as cathode rays. In a series of experiments, Becquerel established that beta rays are electrons, with velocities greater than those of cathode rays. These large velocities indicated that the radiation was not due to ordinary chemical behavior.

Gamma rays were found to have no charge and no mass. In this respect they are similar to x rays. Gamma rays are short-wavelength electromagnetic waves.

These three kinds of radioactive emissions represent a behavior so different from normal chemical behavior that their origins must also be different. As we will see, the discovery of the atomic nucleus (Section 26.9), which led to a better understanding of atomic structure, revealed the source of radioactivity and completes our outline of the experimental evidence for our atomic model.

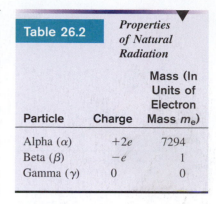

Table 26.2		Properties of Natural Radiation
Particle	Charge	Mass (In Units of Electron Mass m_e)
Alpha (α)	$+2e$	7294
Beta (β)	$-e$	1
Gamma (γ)	0	0

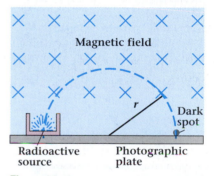

Figure 26.18
Experimental arrangement for showing the deflection of beta rays by a magnetic field. The direction of the field is into the plane of the figure.

26.8　Radioactive Decay

The intensity of the radiation from uranium (or any other radioactive element) is independent of temperature, pressure, and chemical combination. It depends only on the quantity of uranium actually present. When the amount of uranium is increased, the amount of emitted radiation increases proportionally. The measure of radioactivity is the number of radioactive disintegrations per second, called the **activity.** The SI unit of activity is the **becquerel** (Bq), which is one disintegration per second.

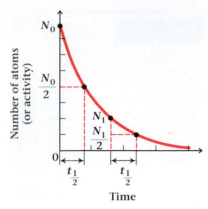

Figure 26.19
The number of atoms of a radioactive species decreases exponentially with time as the atoms decay to other species. The activity decreases in the same way because it is proportional to the number of atoms. The time interval for N_0 atoms to decrease to $N_0/2$ atoms is the same as the time for N_1 atoms to decrease to $N_1/2$ atoms. The time is the half-life $t_{\frac{1}{2}}$.

Table 26.3	Effect of Time on Number of Unstable Atoms Remaining	
Time (in Half-Lives)		**Number of Unstable Atoms Remaining**
0		N_0
1		$N_0/2$
2		$N_0/4$
3		$N_0/8$
4		$N_0/16$
5		$N_0/32$

Each radioactive atom is unstable and has a probability of decaying spontaneously into another, lighter atom with the emission of radiation. It does so independently and without any influence from its neighbors. However, when more atoms are present, the activity is proportionally increased. Mathematically, this means that the rate of decay, or activity A, is given by

$$A = -\frac{\Delta N}{\Delta t} = \lambda N, \tag{26.9}$$

where N is the number of radioactive atoms, ΔN is the number of decays in the time Δt, and λ is a proportionality constant (called the decay constant). The negative sign in Eq. (26.9) expresses the fact that the number of radioactive atoms decreases. This equation has the same form that we saw in Chapter 12 and leads to an exponential relation between the number of unstable atoms and the time. If N_0 unstable atoms are present at time zero, then the number remaining at some time t is

$$N = N_0 e^{-\lambda t}. \tag{26.10}$$

A graph of Eq. (26.10) is shown in Fig. 26.19. Since the activity is proportional to the number of radioactive atoms present, the activity satisfies an exponential decay law also.

After a sufficient time has elapsed, the number of unstable atoms remaining is only one-half of the initial number N_0. This time interval is known as the **half-life.** It is different for each type of radioactive atom. In each successive half-life, the number of radioactive atoms remaining is one-half the number present at the beginning of the interval. Table 26.3 shows the effect of radioactive decay on an initial sample of N_0 atoms.

The half-life is closely related to the decay constant λ. When the time interval is equal to one half-life, $t_{1/2}$, the number of atoms remaining is $N = N_0/2$ and Eq. (26.10) becomes $\frac{1}{2} = e^{-\lambda t_{1/2}}$. We can use this relation to obtain the half-life in terms of λ. Taking logarithms of both sides gives

$$\ln \tfrac{1}{2} = -0.693 = -\lambda t_{1/2}, \quad \text{or} \quad t_{1/2} = \frac{0.693}{\lambda}.$$

Equation (26.10) can then be written

$$\boxed{N = N_0 e^{-0.693 t/t_{1/2}}.} \tag{26.11}$$

Notice that if the decay constant is large, the half-life is correspondingly short. Similarly, a small decay constant means a long half-life.

Example 26.5

Radioactive cobalt-60 has a half-life of 5.26 years. A certain sample of cobalt-60 has an initial activity of A_0. What will its activity be 10.52 years from now?

Solution The passage of 10.52 years is equivalent to two half-lives of cobalt-60. After one half-life, the activity falls to $A_0/2$. After the second half-life, it will be reduced by another factor of $\frac{1}{2}$ to $\frac{1}{2}(A_0/2) = A_0/4$.

Example 26.6

Cancers inside the head may be treated nonsurgically with gamma radiation from cobalt-60 that penetrates the skull. A "gamma knife" directs many precisely aimed gamma-ray beams that converge on the cancer simultaneously. If the cobalt-60 sources have an initial activity A_0, what will be their activity after three years have passed?

Solution   We can use the exponential decay equation (Eq. 26.11), recognizing that both sides are proportional to the activity:

$$A = A_0 e^{-0.693t/t_{1/2}}.$$

When we insert the value for the half-life and elapsed time, we have

$$A = A_0 e^{-0.693(3.0/5.26)} = A_0 e^{-0.395} = 0.67\ A_0.$$

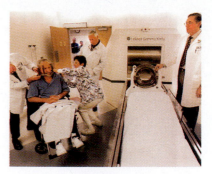

A doctor and a nurse carefully fit a steel frame to a patient's head in preparation for "gamma knife" treatment. The frame fits into the collimator seen at the head of the moveable bed. When the doors to the radiation source are opened, the bed moves the collimator into position where its 201 holes direct the cobalt-60 gamma radiation to a precise location inside the patient's skull.

Example 26.7

A particular sample of radioactive material has an initial activity of 12,876 disintegrations/second. After 36.2 hours, the activity has fallen to 8146 disintegrations/second. What is the half-life of this sample?

Solution   Again we can use the exponential decay equation (Eq. 26.11), recognizing that both sides are proportional to the activity:

$$A = A_0 e^{-0.693t/t_{1/2}}.$$

Rearrange the equation and then take the logarithm of both sides:

$$e^{-0.693t/t_{1/2}} = A/A_0,$$

$$\frac{-0.693t}{t_{1/2}} = \ln(A/A_0).$$

Solving for the half-life, we get

$$t_{1/2} = \frac{-0.693t}{\ln(A/A_0)}.$$

Using the values given, we find

$$t_{1/2} = \frac{(-0.693)(36.2\text{ h})}{\ln(8146/12{,}876)} = 54.8\text{ h}.$$

Two facts combine to make it possible to use the known half-life of radioactive carbon-14 as a kind of clock or calendar for determining the age of archaeological objects, in a process known as radiocarbon dating. The first fact is that living plants take in carbon dioxide (CO_2) from the air. The second fact is that a small proportion of the carbon in the atmosphere is radioactive. The ratio of the amount of radioactive carbon-14 to nonradioactive carbon-12 is almost constant and is about 1.3×10^{-12}. (The distinction between the two types of carbon is not important here, but will be discussed in Chapter 29.) Radioactive carbon-14 is constantly decaying with a half-life of about 5700 years. It is constantly replenished from atmospheric nitrogen, which is converted into radioactive carbon-14 after bombardment by cosmic rays that reach the earth's atmosphere from all directions from outer space. The rates of decay and production are approximately equal.

As long as a plant is alive and growing, it takes in both radioactive and nonradioactive carbon in the form of carbon dioxide. But once the plant is cut or is eaten by an animal or person, the intake of carbon stops and the amount of radioactive carbon begins to decrease as it decays. By determining the relative amounts of radioactive to nonradioactive carbon, we can determine how long ago the object stopped living. For instance, if the relative amount of radioactive carbon in a piece of wood is one-half as much as it originally was, then one half-life of carbon-14, or about 5700 years, must have passed since the tree was cut down.

Some uncertainty in the time estimate of radiocarbon dating occurs because the relative fraction of carbon-14 in the atmosphere changes with time over long periods. This difficulty is partly overcome by comparing radiocarbon dates with dates established by other techniques, such as tree-ring dating. Radiocarbon dating is useful until the carbon-14 remaining has become too small to measure reliably. This puts a practical limit on radiocarbon dating of about 40,000 years.

Other radioactive elements can be used to determine the age of some archaeological objects and geological formations. For instance, the amounts of several different radioactive isotopes found in rocks independently indicate that the age of the oldest rocks in the earth's crust is about 5×10^9 years.

26.9 Discovery of the Atomic Nucleus

By 1909, physicists could piece together a description of the atom from assorted experimental evidence, yielding the following model: Atoms are about 10^{-10} m in diameter and are electrically neutral. They are composed of negatively charged electrons of small mass and positively charged ions, which contain most of the atomic mass. However, as the first decade of the century drew to a close, the arrangement of these atomic components was still unknown and, in the absence of any direct evidence, was the subject of much speculation.

Because electrons represent such a small part of the mass of the atom, one model of the atom consisted of a heavy positive charge spread uniformly over the entire atomic volume (Fig. 26.20). The electrons were imagined to be stuck in the positive charge (like plums in a pudding) in just enough numbers to exactly neutralize the positive charge. This model, advocated by J. J. Thomson, became known as the plum-pudding model.

By 1909, it was known that certain atoms, such as uranium, give off alpha and beta radiations, which penetrate thin layers of matter. Because of their enormous speeds, these particles must pass near or through any atoms that lie in their path. The deflection of an alpha particle, say, from its straight-line path by its encounter with an atom would depend on both the magnitude and the distribution of the electric charge within the atom. Thus, a powerful tool was available for obtaining information on the internal structure of the atom.

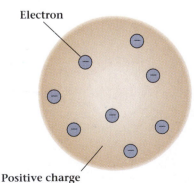

Electron

Positive charge

Figure 26.20
The plum-pudding model of the atom.

It was already established both theoretically and experimentally that a well-defined narrow beam of alpha rays was broadened by passage through air or very thin metal foil. This broadening was due to one or more collisions with atoms, which produced small-angle scattering, a well-understood phenomenon. In 1909, Hans Geiger and E. Marsden, while working with Rutherford, observed a striking new effect. A very small fraction of the alpha rays incident on a metal foil were scattered through such a large angle that they emerged from the same side of the foil from which they entered. This effect was very surprising. Because the alpha particles were comparatively heavy and were moving with high velocities, it was inconceivable that they should be scattered backward as a result of successive small-angle scatterings. Also, since the alpha particles mostly penetrated the foil, it seemed surprising that there was anything to hit that could send them recoiling backward.

Geiger and Marsden made their observations with the experimental arrangement shown schematically in Fig. 26.21. The alpha particles were directed toward a thin metal foil. A fluorescent screen was placed near the foil on the same side as the source of radiation. A thick sheet of lead was placed between source and screen to prevent the alpha particles from striking the screen directly. A microscope was brought close to the screen. Individual alpha particles that were scattered back toward the screen could be seen as tiny flashes of light where they struck the screen.

Geiger and Marsden studied the effect of using different materials for the scattering foil and found that materials of greater atomic mass scattered a greater number of alpha particles into the backward direction. Even with heavy elements for the foil, however, the backward scattering was very small. When the target foil was platinum, only about one alpha was scattered backward for every 8000 alphas incident on the foil.

Two years later, Rutherford published an explanation for the strong scattering of the alpha rays. He proposed that all of the positive charge of the atom was concentrated in a tiny **nucleus** at its center and that the compensating negative

Figure 26.21

The Geiger and Marsden experiment. (a) A collimated beam of alpha particles bombards a thin gold foil; some alpha particles are scattered through a large angle and are observed by the flashes produced on a fluorescent screen. (b) The large-angle scattering occurs when the alpha particle passes very close to a nucleus.

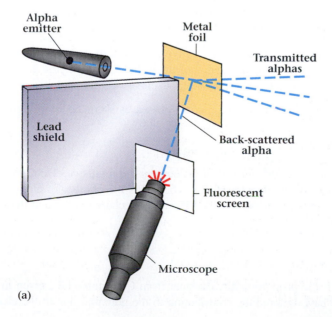

(a)

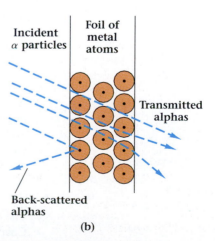

(b)

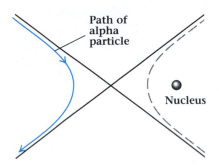

Figure 26.22
Rutherford scattering of an alpha particle by a nucleus. The colored line is the hyperbolic path of the alpha. If the force between the alpha and the nucleus were attractive, the trajectory would still be hyperbolic and would correspond to the dashed line.

charge (due to the electrons) was uniformly distributed over a sphere of radius comparable with the observed radius of an atom. He also found it necessary to assume that most of the mass of the atom was associated with the positive charge. For this model, the repulsive Coulomb force between the central positive charge and a deeply penetrating alpha particle could be many times greater than the force on the alpha due to the diffuse outer charge. When the distance of the positive alpha particle from the positive nucleus of the atom becomes small compared with the atomic radius, the repulsive force on the alpha becomes great enough to cause the large-angle scattering.

Rutherford calculated the behavior of an alpha scattered by a nucleus so heavy that it could be considered to remain at rest during the collision. Using that assumption, together with the Coulomb force law, he showed that the general trajectory for the alpha was one branch of a hyperbola, with the nucleus of the atom as a focus (Fig. 26.22). (See page A-3 of Appendix A for information on hyperbolas.) Experimental verification of Rutherford's scattering formula led to the acceptance of the nuclear model of the atom.

The heart of the technique used by Rutherford—that is, the measurement of subatomic properties by the scattering of incident particles—is still used today. Much of our knowledge of the nucleus (Chapter 29) and of its component parts (Chapter 32) comes from such experiments. It can even be said that most of our information about the physical world comes from scattering data, because to look at an object is to observe the scattering of light waves from it. We will find in Chapter 28 that light waves have particlelike properties also, so the connection between scattering experiments is closer than it first appears.

Example 26.8

Estimate the size of an atomic nucleus by considering a head-on collision of an alpha particle with an atom of gold, assuming that the Coulomb force law applies in this microscopic region. Use the information that was available to Rutherford: that the speed of the alpha particles was about 1.8×10^7 m/s, that the atomic mass of gold was 197, and that the electric charge on the gold nucleus was about half its atomic mass, say $100e$.

Strategy We can solve this problem by applying the law of conservation of energy. In doing so, we use our knowledge of the initial kinetic energy and the electrostatic potential energy when the alpha is closest to the nucleus.

Solution Initially, an incoming alpha particle possesses a kinetic energy given by

$$\text{KE} = \tfrac{1}{2}m_\alpha v_\alpha^2.$$

As the alpha approaches the nucleus, it encounters the repulsive electric force due to the nuclear charge, denoted by Ze, where Z is the number of elementary electric charges e present in the nucleus. The alpha's charge is $2e$, and the potential energy is

$$\text{PE} = k\frac{2Ze^2}{r}.$$

The constant k is the proportionality constant from Coulomb's law, given in Chapter 16. We have ignored the contribution to the potential due to the electrons because we assume that the electrons are spread over a region of space that is large compared to the radius of the nucleus. Consequently, the contribution to the overall potential by the electrons is small.

At the distance of closest approach b, the alpha particle's motion is reversed (Fig. 26.23). At this point, its kinetic energy is zero. From the law of conservation of mechanical energy, we know that the potential energy of the alpha at $r = b$ must equal its initial kinetic energy:

$$\tfrac{1}{2}m_\alpha v_\alpha^2 = \frac{k2Ze^2}{b}.$$

This equation thus sets an upper limit on the value of b and hence gives an estimate of the size of the nucleus. Upon rearranging, we find

$$b = \frac{4kZe^2}{m_\alpha v_\alpha^2}.$$

The mass of the alpha was known to be about 6.62×10^{-27} kg. When this number is substituted into the equation above along with the values of the other quantities, we get

$$b = \frac{4(9 \times 10^9 \text{ N} \cdot \text{m}^2/\text{C}^2)100(1.6 \times 10^{-19} \text{ C})^2}{(6.62 \times 10^{-27} \text{ kg})(1.8 \times 10^7 \text{ m/s})^2}$$

$$b = 4.3 \times 10^{-14} \text{ m}.$$

Discussion This value of b gives an upper limit to the nuclear radius. Since it is many times smaller than the value of 10^{-10} m that is typical for the atomic radius, our assumption that we could neglect the electronic contribution to the potential is justified.

At the time of Rutherford's theoretical predictions, there was no independent method of determining the charge on the nucleus. However, the general success of his predictions confirmed the validity of the theory. Acceptance of the nuclear theory in turn led to successful descriptions of the electronic structure of the atom, as we shall see in the next chapter. In turn, these descriptions led to a way to carefully determine the charge on the nucleus.

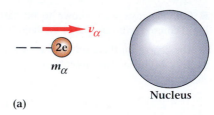

(a)

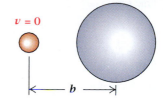

(b)

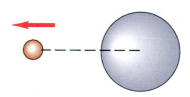

(c)

Figure 26.23
(a) Schematic diagram of an alpha particle approaching a heavy nucleus head-on. (b) At some point, the alpha particle has zero velocity as it reverses direction. This point marks the distance of closest approach b, which gives us an upper limit for the radius of the nucleus. (c) Alpha particle rebounding from the nucleus.

Summary

Useful Concepts

■ The observation that crystals have regular geometries and can be cleaved along smooth planes can be explained by saying that they are made up of submicroscopic components or building blocks, called atoms.

■ The law of definite proportions states that the proportions (by mass) of the elements in a chemical compound are constant. The law of volumes states that gases unite by volume in simple and definite proportions. These two statements led to Avogadro's law: Equal volumes of gases at the same temperature and pressure contain equal numbers of molecules.

■ The experiments of Perrin on Brownian motion established the reality of atoms and led to the measurement of the charge on the electron.

■ The size of atoms is of the order of 10^{-10} m.

■ X rays are produced when electrons are suddenly stopped upon striking a target. The wave character of x rays is shown by the fact that they undergo diffraction. The Bragg law for the diffraction of x rays by crystalline atomic layers separated by a distance d is

$$n\lambda = 2d \sin\theta.$$

■ Cathode rays were shown to be particles (electrons) by J. J. Thomson, who measured the ratio e/m.

■ The charge on the electron was measured in an experiment by R. A. Millikan. The presently accepted value for the electron charge is

$$e = 1.6022 \times 10^{-19} \text{ C}.$$

■ In radioactive decay, the number of unstable atoms left after time t is

$$N = N_0 e^{-0.693t/t_{1/2}},$$

where N_0 is the number of atoms present at time 0 and $t_{1/2}$ is the half-life.

■ Our present model of the atom is due to Rutherford, who explained the strong backward scattering of alpha particles by proposing a small heavy positive nucleus surrounded by a diffuse sphere of electrons.

Important Terms

You should be able to write the definition or meaning of each of the following:

unit cell	radioactivity
electrolysis	alpha (α) rays
Avogadro's number	beta (β) rays
Brownian motion	gamma (γ) radiation
cathode rays	activity
x rays	becquerel
Bragg's law	half-life
electrons	nucleus

Conceptual Questions

26.1 What reasons might Hooke and Huygens have had for considering the smallest components of crystals to be spherical rather than, say, cubical?

26.2 Explain why the law of definite proportions is a direct result of the atomicity of matter.

26.3 Is Avogadro's number a fundamental constant of nature, in the same way as the speed of light and as the charge on the electron, or is its value just happenstance?

26.4 Name some large-scale phenomena that would be observably different if Avogadro's number were 10^{20} times as large as its actual value.

26.5 What factors govern whether Brownian motion can be observed for a particle suspended in a fluid? Is it a matter of size only?

26.6 When one metal is electroplated onto another, the conducting solution used normally contains ions of the metal being deposited. Describe what happens at the anode and the cathode and in the solution.

26.7 Assume you have a radiation detector that is sensitive to alpha, beta, and gamma radiations. Describe a simple method for determining which radiations are present in a particular radioactive source.

26.8 In view of the Rutherford model of the atom, is the relative penetrating power of natural radioactivity to be expected? (See Table 26.1.) Explain.

26.9 Imagine that you could roll small hard balls at an obstacle on a smooth table. The obstacle is hidden from view, but you can see the directions of the outgoing balls. How much can you determine about the obstacle? Give several specific examples.

26.10 Which of the experiments discussed in this chapter best support the claim that all electrons have exactly the same charge?

26.11 Examine the periodic table of the elements and note those elements that, from your personal knowledge, have physical or chemical properties in common. Do these elements belong to particular rows or columns in the table?

26.12 What type of materials can be examined with the field ion microscope? What, if any, restrictions hold for the materials to be examined with the scanning tunneling microscope?

Problems

Section 26.1 Evidence of Atoms from Solids and Gases

26.1 Place several pennies on a flat surface with their edges touching. Show by diagrams that if they are arranged to form a square array, they take up more surface area than when they are arranged in a hexagonal array.

26.2 When water is decomposed, the weight of the oxygen released is 8 times that of the hydrogen released. Oxygen and hydrogen combine in volume proportions of 1 to 2 to make water. Use this information, along with Avogadro's hypothesis, to find the ratio of the mass of a molecule of oxygen to that of a molecule of hydrogen.

26.3 (a) A chemist reports an experiment in which 1.37 liters of gas combine with 2.15 liters of another gas to make a third gas. Do you believe his results? (b) In a separate report, 1.74 liters of gas are combined with 2.61 liters of a second gas to form a third gas. Do these results seem likely?

26.4 Nitrogen may be combined with oxygen in proportion by mass of 1.75 g nitrogen to 1.00 g oxygen to form the gas nitrous oxide, N_2O. The same elements may be combined in the ratio of 0.438 g nitrogen to 1.00 g oxygen. What is the atomic composition of the second compound?

Section 26.2 Electrolysis and the Quantization of Charge

Hints for Solving Problems

Remember that in electrolysis the transfer of one faraday of electric charge corresponds to the transfer of one mole of electric charge. The amount of charge transferred is the product of current and time. The atomic masses of the elements are found in the periodic table in the endsheets.

26.5 In an electrolysis experiment, a current of one ampere is maintained for 38.0 min during which copper is deposited on the negative electrode. Copper has an atomic mass of 63.6, and its ions are divalent; that is, each ion carries two units of electric charge e. Calculate how much copper (mass) was deposited on the electrode.

26.6 In an electrolysis experiment, a student plated silver onto an electrode, maintaining a current of 0.800 A for 46.0 min. What mass of silver was deposited?

26.7 In an electrolysis experiment, a student plated copper from a solution of copper ions. The student did not know whether the ions were singly charged or doubly charged. After a current of 0.25 A was passed for 40 min, the electrode was removed, dried, and weighed. The mass of copper deposited was 0.396 g. Were the copper ions singly or doubly charged?

26.8 Calculate the volume in liters occupied by one mole of a gas at standard temperature and pressure. (Remember the ideal gas law.)

26.9• Hydrogen and oxygen gas are generated by passing an electric current through water. What volume of oxygen gas (O_2) can be liberated in one hour in a cell that carries a current of 10 A? (Assume atmospheric pressure and a temperature of 0°C.)

26.10• Hydrogen gas and oxygen gas are generated by passing an electric current through water. What volume of hydrogen (H_2) can be liberated in one hour in a cell that carries a current of 6.0 A? (Assume atmospheric pressure and a temperature of 0°C.)

Section 26.3 Avogadro's Number and the Periodic Table

26.11 In his study of Brownian motion, Perrin measured the numbers of particles at four distinct heights: 5 μm, 35 μm, 65 μm, and 95 μm. The concentrations at those heights were

proportional to the numbers 100, 47, 22.6, and 12. Graph these numbers on semilogarithmic paper to see if the concentration depends exponentially on the height.

26.12 In another study, Perrin measured the concentration of particles at four levels, with each successive level 10 μm above the preceding one. The concentrations in a given area were 100, 43, 22, and 10. Graph this information on semilogarithmic paper. At what height above the first layer will the concentration drop to 50?

26.13 The density of particles n in a suspension is $n = n_0 e^{-\lambda h}$, where λ is a constant and h is the height. Show that the density falls to $n_0/2$ at the height $h = 0.693/\lambda$.

26.14• The density of particles n in a suspension is found to go as $n = n_0 e^{-\lambda h}$, where λ is a constant and h is the height. What is the value of λ if the density falls to $n_0/2$ when $h = 10$ μm?

26.15•• Derive Eq. (26.3) from the two equations immediately preceding it.

Section 26.4 The Size of Atoms

26.16 Silicon has a density of 2.23 g/cm^3 and an atomic mass of 28.09 g/mol. Calculate the number of atoms in a cubic centimeter of silicon.

26.17 Silver has a density of 10.5 g/cm^3 and an atomic mass of 107.9 g/mol. Calculate the number of atoms in a cubic centimeter of silver.

26.18• Aluminum has an atomic weight of 26.98 g/mol and a density of 2.70 g/cm^3. (a) How many atoms are there in a cube of aluminum 1.00 cm on an edge? (b) Calculate the average volume per atom. (c) If you treated the volume per atom as a cube, what would be the length of the cube edge associated with a single atom?

Section 26.5 Crystals and X-Ray Diffraction

Hints for Solving Problems

Remember that in Bragg diffraction, the angle θ is measured from the plane of atoms, not from the normal to that plane.

26.19 First-order diffraction of x rays of wavelength $\lambda = 0.709 \times 10^{-10}$ m occurs for a certain set of planes in a nickel crystal. If the scattering angle $2\theta = 23.2°$, what must be the spacing of the planes responsible for the diffraction?

26.20 X rays of $\lambda = 0.709 \times 10^{-10}$ m are diffracted in second order ($n = 2$) from a silver bromide crystal. If the scattering angle is $2\theta = 28.50°$, what is the spacing between the atomic planes?

26.21 X rays are Bragg-diffracted from a cobalt crystal with interplanar spacing of 4.07×10^{-10} m. In first order, the scattering angle 2θ is 24.0°. What is the wavelength of the x rays?

26.22 X rays are Bragg-diffracted from a cesium chloride crystal with interplanar spacing of 4.11×10^{-10} m. In first or-der, the scattering angle 2θ is $7.59°$. What is the wavelength of the x rays?

26.23• X rays of wavelength $\lambda = 1.54 \times 10^{-10}$ m are aimed at a crystal with interplanar spacing of 2.88×10^{-10} m. Beams diffracted from these planes emerge for more than one angle of incidence. Determine the allowed values of 2θ.

26.24• X rays of $\lambda = 0.709 \times 10^{-10}$ m are diffracted from a set of crystal planes separated by $d = 1.44 \times 10^{-10}$ m. Calcu-late the allowed values of 2θ.

Section 26.6 Discovery of the Electron

26.25 Calculate the radius of the circular path of an electron accelerated through an electric potential of 200 V and injected into a region of magnetic field $B = 5.00 \times 10^{-3}$ T.

26.26 Calculate the angular deflection of an electron beam traveling at 1.0×10^7 m/s through a region of magnetic field $B = 1.0 \times 10^{-3}$ T that is 1.5 cm long.

26.27• Electrons in a cathode ray tube are accelerated through a potential of 19.0 kV. The electron beam then passes through a pair of deflection plates 2.00 cm long separated by 4.00 mm. What potential must be applied to the plates to achieve a 3.00-cm deflection of the beam on a fluorescent screen 30.0 cm beyond the deflection plates?

26.28• Electrons moving with a speed of 1.00×10^7 m/s pass through a pair of parallel plates 1.00 cm long separated by 8.00 mm (Fig. 26.15b). (a) If a potential of 100 V is applied to the plates, what is the angle of deflection of the electron beam? (b) What is the linear deflection on a fluorescent screen placed 20.0 cm beyond the deflection plates?

26.29• An electron of mass m_e and charge e, moving with speed v perpendicular to a magnetic field B, travels in a circle of radius r. Show that the ratio of charge to mass of the elec-tron is

$$\frac{e}{m_e} = \frac{v}{Br}.$$

26.30•• Electrons in a cathode ray tube pass through a pair of deflection plates 2.00 cm long separated by 4.00 mm. What electric potential must be applied to offset an angular deflec-tion of $15°$ (= 0.26 radian) caused by a magnetic field $B = 1.00 \times 10^{-2}$ T?

Section 26.7 Radioactivity

Hints for Solving Problems

Refer to Table 26.2 for the charges and masses of alpha and beta particles.

26.31 Calculate the radius of curvature of the path of an alpha particle of mass 6.6×10^{-27} kg and speed 2.4×10^7 m/sec perpendicular to a magnetic field of 0.17 T.

26.32 Calculate the radius of curvature of the path of a beta particle of mass 9.1×10^{-31} kg and speed 2.0×10^7 m/sec perpendicular to a magnetic field of 0.50 T.

26.33 Alpha particles are deflected by a magnetic field of 6.40×10^{-2} T into a curved path with a radius of 4.06 m. What is the speed of the alpha particles?

26.34• Charged particles are deflected in a circular path of 6.60-m radius in a magnetic field of 5.00×10^{-2} T. The speed of the particles is known to be 1.60×10^7 m/s. What is their ratio of charge to mass? Could the particles be alpha particles?

26.35• The hydrogen ion has a mass of 1.67×10^{-27} kg and a charge of $+e$. Compare the radius of curvature of the path of a hydrogen ion to that of an alpha particle of mass 6.64×10^{-27} kg of the same speed v in the same magnetic field B.

26.36• Calculate the radius of curvature of the path of an al-pha particle of mass 6.6×10^{-27} kg moving with a kinetic en-ergy of 8.7×10^{-14} J in a magnetic field of 1.00 T.

Section 26.8 Radioactive Decay

Hints for Solving Problems

In radioactive decay, the number of radioactive atoms (and thus the activity) decreases exponentially with time. If you know the number present at any time, say t, the number re-maining at time Δt later is $e^{-\lambda \Delta t}$ times the number present at t, regardless of what time t you begin counting from.

26.37 Radioactive bismuth (^{210}Bi) undergoes beta decay with a 5.0-day half-life. How long will it take a sample of initial activity A to decrease to one-eighth its initial activity?

26.38 A sample of polonium has a half-life of 3.0 min and an initial activity of 10×10^6 Bq. What is its activity after 30 min have elapsed?

26.39• One gram of uranium (^{238}U) contains 2.5×10^{21} atoms. Its half-life is 4.5×10^9 years. What is the activity of 1.0 g of uranium?

26.40• The ratio of radioactive to nonradioactive carbon in a sample of wood is found to be one-third that found in a re-cently cut piece of wood. Approximately how long ago was the first sample cut? The half-life of carbon-14 is 5700 years.

26.41• The activity of a certain sample was measured at in-tervals over a 12-hour period. The data are given on the fol-lowing page. (a) Plot the decay curve on semilogarithmic graph paper, with activity along the logarithmic coordinate and time along the linear coordinate. (b) Determine the disin-tegration constant and the half-life. (c) What would you expect the activity to have been at $t = 6$ hours?

Time (h)	Activity (Disintegrations/Min)
0	8550
1	7015
2	5750
3	4720
4	3890
5	3165
7	2140
8	1740
9	1439
10	1182
11	970
12	795

26.42• The penetration of gamma rays through copper was studied by measuring the number of gamma rays that pass through successive copper layers. The data recorded are given below. (a) Plot the data on semilogarithmic graph paper to verify that the transmitted intensity is $I = I_0 e^{-mx}$, where m is a linear absorption coefficient and x is the thickness of the copper. Plot intensity along the logarithmic coordinate and thickness along the linear coordinate. (b) Also determine the half-thickness—that is, the thickness at which an incident beam of gamma rays is diminished by one-half.

Activity (Events/Minute)	Thickness (cm)
21700	0
20600	0.1
18300	0.3
14500	0.6
11900	0.95
9560	1.3
6450	1.9
5050	2.2
3470	2.9

26.43•• Actinium (^{227}Ac) has a half-life of 22 years. One gram of actinium contains 2.65×10^{21} atoms. (a) What is the activity of one gram of actinium? (b) What is the activity in curies? (One curie is 3.7×10^{10} disintegrations per second and is the activity of one gram of pure radium.)

Section 26.9 Discovery of the Atomic Nucleus

26.44 What is the distance of closest approach of an alpha particle of speed $v = 1.8 \times 10^7$ m/s to a lead nucleus of $Z = 82$? Assume that the lead nucleus remains at rest.

26.45 The electric charge on a uranium nucleus is $+92e$, where e is the elementary charge. What is the distance of closest approach b of an alpha particle with initial velocity of 2.0×10^7 m/s directed straight at the uranium nucleus? Assume that the uranium nucleus remains at rest.

26.46• What is the distance of closest approach of an alpha particle of speed $v = 1.70 \times 10^7$ m/s to a gold nucleus of $Z = 79$? How close would the alpha get to a copper nucleus of $Z = 29$? Assume that the gold and copper nuclei remain at rest.

Additional Problems

26.47 X rays are Bragg-diffracted in third order ($n = 3$) from a cesium chloride crystal. If the wavelength of the x rays is $\lambda = 0.709 \times 10^{-10}$ m and the scattering angle is $2\theta = 29.99°$, what is the interplanar spacing?

26.48• Radioactive ^{57}Co has a half-life of 270 days. How long will it take for a source of activity A to decrease its activity to $A/10$?

26.49• A small radioactive source of beta rays is monitored by a detector placed 4.0 cm from the source. The detector records 5280 counts/min. What is the count rate when the detector is moved back to a position 9.0 cm from the source? (*Hint:* Assume that the radiation is emitted uniformly in all directions and that the size of the detector is fixed.)

26.50• A small radioactive source of gamma rays is monitored by a detector placed 5.0 cm from the source. The detector records an average count rate of 4320 counts/min. What would be the count rate if the detector were moved back to a position 10 cm from the source? (*Hint:* Assume that the radiation is emitted uniformly in all directions and that the size of the detector is fixed.)

26.51•• A particle of mass m and initial speed v_0 collides head-on with a free stationary atom of mass M. Using conservation of kinetic energy and momentum, find the recoil velocity and energy of the incident particle and of the atom when $M = 10m$.

26.52•• A particle of mass m and initial speed v_0 collides head-on with a free stationary atom of mass M. (a) Using conservation of kinetic energy and momentum, find the recoil velocity and energy of the incident particle and of the atom when $M = 50m$. (b) What can you say about Rutherford's approximation that the struck nucleus was so heavy that it did not move during the collision?

Origins of the Quantum Theory

During the early twentieth century, several seemingly unrelated developments in physics converged to provide the basis for a surprising new model of atomic structure. One contribution was the collective observations from the study of optical spectra. Another was Planck's explanation of the distribution of the energy of light emitted from an incandescent object. Planck's proposal that the energy of the light depended on its frequency was in turn used by Einstein to explain an observation known as the photoelectric effect.

These diverse ideas, along with Rutherford's discovery of the atomic nucleus in 1911, led Niels Bohr in 1913 to conceive a radically new atomic model. Bohr's theory successfully explained many of the puzzles that had arisen in the studies of optical spectra; it also provided an explanation of x-ray spectra. His ideas led directly to the development of modern quantum theory and its applications, including lasers and semiconductors.

Bohr's theory was highly successful, but had its limitations as well. Today we have replaced Bohr's original theory with the quantum mechanical approach to atomic structure, described in the next chapter. However, many concepts introduced by Bohr are essentially correct and helped point the way to contemporary quantum theory. His atomic model is still useful as a tool for visualizing the quantum description of atoms. This is another example of the evolution of ideas in physics, in which one successful explanation becomes modified by later ideas, yet some of the original assumptions are accepted in later theories.

Bohr's theory was a key step in another major development in physics: the overthrow of the Newtonian world view. Einstein's special theory of relativity showed that Newtonian mechanics was insufficient to explain the behavior of objects moving at speeds approaching the speed of light. Bohr's theory of the atom showed that classical physics also failed to explain the behavior of objects on the extremely small scale of atomic dimensions. Thus, this chapter documents the progress of the second great revolution of modern physics, the early development of quantum mechanics. ■

27.1   Spectroscopy

The beginning of a new understanding of atomic structure was based on observational evidence, much of it from the spectra of light emitted or absorbed by the various chemical elements. These observations were important in the development of new theories because the theoretical models had to be consistent with the observations. Initially, **spectroscopy** was the study of optical spectra, which result from dispersing light into its component colors—that is, spreading the light according to its wavelength (Section 24.9). Eventually, this idea was extended to other regions of the electromagnetic spectrum, so that today spectroscopy includes the study of spectra from microwaves, infrared rays, ultraviolet rays, and x rays, as well as visible light. Spectroscopy in all its forms plays a major role in determining the chemical composition of substances; much of modern chemistry depends on spectroscopic analysis of materials.

Since the Middle Ages, people have known that introducing various substances into a flame changes the color of the flame. For example, copper produces a green flame and potassium produces a violet one. By the end of the nineteenth century, it was known that if an electric current is passed through a low-pressure gas, the gas emits light whose color is characteristic of that particular gas. This effect finds commercial use today in the ubiquitous neon sign (Fig. 27.1).

As we saw in Chapter 24, Newton used a prism to disperse sunlight into a spectrum of visible colors. Then, about 1750, the Scotsman Thomas Melville began investigating the spectra emitted by various substances held in a flame. He found that the distribution of wavelengths (colors) emitted was not the same, but varied from one element to another. Eventually, as the quality of prisms improved, scientists were able to catalog spectra, associating particular sets of emitted wavelengths with particular chemical elements. It became clear that since each kind of atom produces a different spectrum, the cause of spectra—whatever it is—must be connected with the nature of atomic structure.

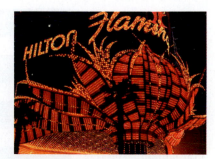

Figure 27.1
The characteristic red glow of neon lights up Las Vegas after dark.

1814 Joseph Fraunhofer discovers absorption spectra in sunlight. He is the first to use diffraction gratings to measure spectral wavelengths.

1895 Johann Balmer finds a formula for the visible series of spectroscopic lines of hydrogen.

1900 Max Planck proposes quantization of energy to explain the shape of the blackbody radiation curve.

Today we think of spectra as the fingerprints of matter. Every atom or molecule emits its own characteristic spectrum of light—no two spectra are alike. We may produce spectra by heating a gas at low pressure in an electronic discharge tube. *Emission spectra* appear as a series of bright lines, each representing a particular wavelength of light (Fig. 27.2). (Each line is actually the image of a slit, as described in Section 24.9.) *Absorption spectra* appear as a series of dark lines, not always equivalent to the bright lines of emission spectra. Each line represents a particular wavelength of radiation absorbed from an otherwise continuous spectrum of transmitted light. The cataloging of the different line patterns was an important step in being able to identify the chemical composition of substances isolated in chemical and biological research. However, the nineteenth-century theories of physics could not explain why the spectral patterns differed.

Some of the most important work in cataloging spectra was done by the lensmaker and physicist Joseph Fraunhofer (1787–1826). He discovered a series of dark lines (an absorption spectrum) present in the otherwise continuous solar

Figure 27.2

(a) Continuous spectrum of the sun, showing the absorption lines. Emission spectrum of (b) hydrogen, (c) barium, (d) mercury, and (e) sodium.

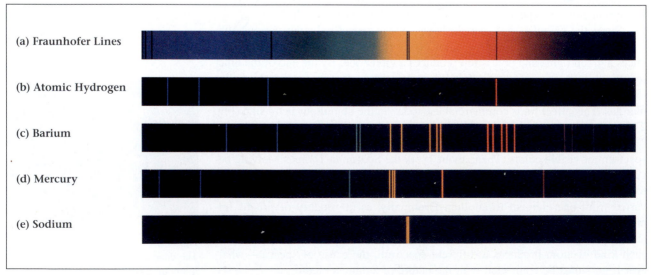

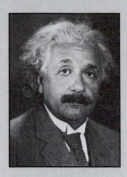

1905 Albert Einstein uses Planck's quantum idea to explain the photoelectric effect, introducing the concept of photons of light.

1913 Niels Bohr makes the first quantum model of the hydrogen atom, which successfully explains the wavelength of the spectral lines.

1914 Henry Moseley uses x-ray spectroscopy to relate atomic number to the charge of the nucleus.

spectrum. These Fraunhofer lines established the presence of individual chemical elements in the sun. But the explanation for the particular patterns of atomic spectra was still unknown.

27.2 Balmer's Series

Fraunhofer's work spurred a great interest in spectroscopy, leading to the development of better techniques·and instruments. By the late nineteenth century, spectroscopy had become a well-developed field of physics. The spectra of most elements had been carefully measured, and detailed tables of wavelengths were available. But still, the reasons for the existence of spectral lines were not understood.

In 1885, a Swiss schoolteacher, Johann Jacob Balmer, found a simple mathematical formula that related the wavelengths of the prominent lines in the visible and near-ultraviolet spectrum of hydrogen gas. (Hydrogen exhibits one of the simplest atomic spectra.) *Balmer's formula* for the wavelength λ of the hydrogen lines is

$$\lambda = 364.56 \text{ nm } \frac{n^2}{n^2 - 2^2}, \tag{27.1}$$

where n is an integer that takes on the values $n = 3, 4, 5, 6, \ldots$. The corresponding lines observed in the visible spectrum of hydrogen are called the **Balmer series** (Fig. 27.3).

Using his formula, Balmer calculated the wavelengths of the nine lines (four visible and five ultraviolet) that were then known to exist in the spectrum of hydrogen. Balmer's formula was strictly empirical. That is, it was not derived from any physical model or theory of physical behavior; instead, Balmer offered his formula simply as a mathematical relationship that was consistent with observations. There was no apparent reason why it should work. Nevertheless, it provided amazingly precise computation of the wavelengths in the hydrogen spectrum. Even for the worst case, which occurred for $n = 11$, Balmer's calculated wavelength was within 0.1% of the measured value.

FRAUNHOFER AND THE SOLAR SPECTRUM

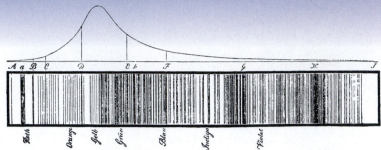

Figure B27.1 Fraunhofer's drawing of the dark lines observed in the solar spectrum.

What is the sun made of? Does it consist of the same chemical elements we have on earth, or is it formed from its own kind of matter? The first clue to the answers to these questions came in 1814 from an analysis of the spectrum of sunlight by Joseph Fraunhofer. Fraunhofer was a skilled lensmaker as well as a physicist. He discovered a method of accurately shaping lenses and learned how to make larger pieces of optical quality glass than his predecessors could make.

Fraunhofer wanted to determine the index of refraction of different kinds of glass for different colors of light, so that he might design more accurate achromatic lenses. During this work, he discovered a pair of bright yellow lines in the spectrum of the light from an oil lamp. He found that similar lines were also present in other kinds of firelight. These bright lines always occurred at the same place in the spectrum, and so would be useful in determining refractive indices. We now know that these lines are due to the presence of the element sodium. Modern sodium vapor lamps, easily recognized by their distinctive orange color, are often used for outdoor lighting because of their efficiency.

To determine accurately the wavelengths of the yellow lines, Fraunhofer constructed a spectroscope (see Chapter 24) consisting of a narrow slit placed some distance from a flint glass prism. He observed the spectrum of the lamp through a telescope directed at the prism. He next studied sunlight to see whether the bright lines were there, too. To his surprise, he found a large number of dark lines in the otherwise bright solar spectrum. At the exact position of the yellow lines seen in the firelight, he saw dark lines rather than bright lines. Fraunhofer's sketch of what he saw is shown in Fig. B27.1.

Fraunhofer measured the wavelengths of the more prominent lines, labeling them with the letters from A to K. The lines coinciding with the bright lines in the firelight were called the D lines. Because of his contribution to the discovery of the dark lines in the solar spectrum, these lines have become known as the Fraunhofer lines.

Fraunhofer was also the first person to use diffraction gratings to observe spectra. He formed his own gratings of fine wire, closely spaced, and used them to make measurement of optical wavelengths.

By the late nineteenth century, the identification of spectra with chemical elements on the earth led to the realization that Fraunhofer's lines indicated the presence of chemical elements in the sun. For example, the fact that the D line is part of the Fraunhofer series proves the presence of sodium in the sun. Other prominent parts of the Fraunhofer series indicate the presence of hydrogen and helium, which are known to be the major constituents of the sun. (Helium was discovered in the analysis of solar spectra before it was discovered on earth. The word *helium* comes from the Greek *helios,* the sun.) Later study of spectra from starlight showed that the sun and the stars are made of the same material. In short, the sun is a star.

The study of spectra has continued to be a major source of information about astronomical objects. For example, ionized gases have a slightly different spectral pattern from nonionized gases, so by analysis of spectral intensities, we can estimate the percentage of ionization in a star. This, in turn, enables us to estimate the temperature of stars or interstellar dust clouds. The Doppler shift of spectral lines, described in Chapter 25, provides information on the speeds of objects in outer space. Our picture of the universe, on the enormous scale of galaxies as well as on the tiny scale of atoms, is built up in large part from spectroscopic observation and analysis (Fig. B27.2).

Figure B27.2 The Horsehead Nebula. The red light in the photograph is due to hydrogen emission; the blue light is starlight scattered from dust clouds.

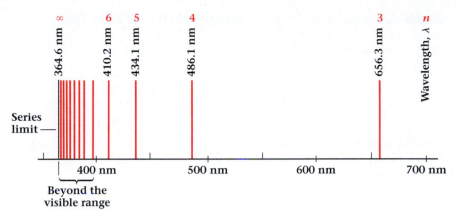

Figure 27.3
Sketch of the Balmer lines in the spectrum of atomic hydrogen.

In announcing his formula, Balmer suggested it might possibly be a special case of some more general formula that would apply to other series of lines in other elements. The Swedish spectroscopist Johannes Robert Rydberg sought to find such a formula. From the large amount of data available in 1889, Rydberg found several spectral series that fit an empirical formula that he showed was equivalent to the Balmer formula. The **Rydberg formula** can be written to give the reciprocal of the wavelength of the emitted light as

$$\frac{1}{\lambda} = R_\infty \left(\frac{1}{n_1^2} - \frac{1}{n_2^2} \right), \quad n_1 < n_2, \tag{27.2}$$

where R_∞ is the *Rydberg constant,* 10,973,731.534 m^{-1}, and n_1 and n_2 are integers. For the Balmer series, $n_1 = 2$ and n_2 takes on values of 3, 4, 5, 6, ... As n_2 becomes very large, the lines converge toward a *series limit.* Later observations by other spectroscopists confirmed additional spectral series in the infrared and ultraviolet for hydrogen, corresponding to other values of n_1 including $n_1 = 1, 3, 4,$ and 5.

By the year 1900, mathematical formulas were known that could provide extremely accurate calculations of the spectral lines in hydrogen. And yet, as far as atomic structure was concerned, no one had devised any model that could account for the existence of the observed spectra, nor could anyone explain why Rydberg's formula worked as well as it did.

Example 27.1

Calculate the wavelength of the line in the Balmer series with the longest wavelength.

Strategy To give the largest wavelength λ, the right-hand side of the Rydberg formula (Eq. 27.2) must be as small as possible. This means that n_1 and n_2 should differ by unity; or when $n_1 = 2$, as it does for the Balmer series, then $n_2 = 3$.

Solution Using the values of $n_1 = 2$ and $n_2 = 3$ in Eq. (27.2), we find

$$\frac{1}{\lambda} = R_\infty\left(\frac{1}{2^2} - \frac{1}{3^2}\right)$$

$$\frac{1}{\lambda} = (1.097 \times 10^7 \text{ m}^{-1})\left(\frac{1}{2^2} - \frac{1}{3^2}\right) = 1.5236 \times 10^6 \text{ m}^{-1},$$

$$\lambda = 656 \text{ nm.}$$

This line is in the red, or long-wavelength, part of the visible spectrum. You can see where this line falls in the spectrum by looking back at the diagram of the visible spectrum in Fig. 22.2.

27.3 Blackbody Radiation

About the same time that spectroscopists were wrestling with the problems of spectral series, other scientists were studying the continuous spectrum of light emitted from a hot object. The interest was in part theoretical, but it was also a practical interest growing out of the needs of the growing public—and home—illumination industry. The eventual explanation for the shape of this spectrum departed from classical physics and pointed the way to a key aspect of the new model of the atom.

When a solid object is heated above several hundred degrees Celsius, it becomes incandescent. The radiation emitted by an incandescent object forms a continuous range of wavelengths, part of which lies in the visible range. As the object's temperature increases, the relative intensities of the light across the visible spectrum change. This causes a perceptible shift in the observed color, which can be used to estimate the temperature (see Table 27.1).

An ideal object for producing incandescent light is called a **blackbody.** A blackbody is a surface or object that perfectly absorbs all radiation falling on it. A blackbody is also a perfect radiator. Such an object might be an enclosed oven or kiln with only a tiny opening for light to get out. When the oven is cold, light from outside that passes into the opening is not reflected back (Fig. 27.4a) and so the hole looks completely black. When the oven is very hot, light emitted inside passes out through the hole. Figure 27.4(b) shows the radiation emitted by a kiln operated at a temperature of 1400 K. The light depends only on the temperature of the oven and not upon the oven's composition.

Figure 27.5 shows the radiant energy distribution for a blackbody at several temperatures. The total radiant energy is proportional to the area under the curve, which grows rapidly with increasing temperature. (According to the Stefan-Boltzmann law, Section 11.7, the radiated energy varies as T^4.) The radiation appears as a continuous spectrum, having neither the bright lines seen in flame spectra and in gas discharge spectra nor the dark lines seen in the light from the sun. Instead, all wavelengths are present over a wide range. As the temperature increases, the wavelength of maximum intensity λ_m shifts to smaller values. (Note that the curve for 6000 K corresponds to the approximate temperature of the sun and that λ_m occurs in the visible part of the spectrum.)

Table 27.1 *Color Scale of Temperatures*

Color of Glowing Object	Approximate Temperature (°C)
Incipient red	500–550
Dark red	650–750
Bright red	850–950
Yellowish red	1050–1150
Incipient white	1250–1350
White	1450–1550

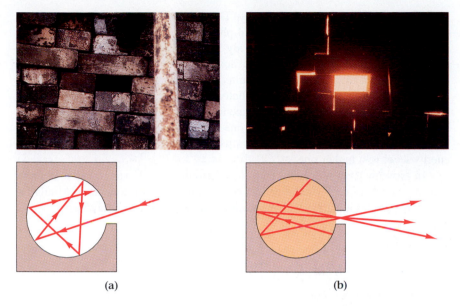

Figure 27.4
A small kiln used to fire ceramics behaves nearly like an ideal blackbody cavity. (a) At a temperature of approximately 300 K, the kiln is very dark inside. Light entering from outside is absorbed. (b) When the kiln is heated to 1400 K, the interior glows brightly and the light emitted inside the cavity passes out through the opening.

Measurements of blackbody spectra show that the wavelength of maximum intensity λ_m decreases with increasing temperature according to the rule

$$\lambda_m T = 2.90 \times 10^{-3}\ \text{m} \cdot \text{K}. \tag{27.3}$$

This rule is known as the **Wien displacement law.**

Even though the Wien displacement law successfully predicted the position of the maxima in the blackbody radiation curve, this law gave no information about the detailed shape of the curve. Numerous attempts were made to explain the shape of the blackbody spectrum. Models based on the well-established laws of classical electromagnetism and thermodynamics failed to reproduce the

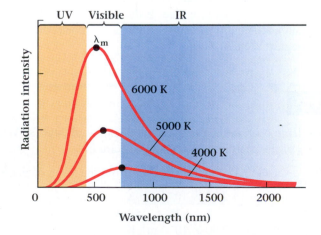

Figure 27.5
Blackbody spectra for several different temperatures. The wavelength λ_m of the peak of the curve shifts to shorter wavelengths at higher temperatures. The total amount of energy radiated (given by the area under the curve) is proportional to T^4.

experimental measurements adequately. The explanation of blackbody radiation was simply not to be found in classical physics.

The detailed shape of the blackbody radiation curve was first explained in 1900 by Max Planck as a result of a novel, *ad hoc* hypothesis (Section 1.1). Planck proposed that the radiation resulted from the behavior of a large number of identical oscillators. When the oscillators exchange energy, they do so by the emission and absorption of electromagnetic radiation. Radiation is emitted when an oscillator makes a transition from one energy level to a lower one. Absorption of radiation is an inverse process in which the oscillator jumps from a lower energy level to a higher one.

In arriving at his successful empirical formula, Planck was forced to assume that the energy of each oscillator was proportional to its frequency f. Similarly, the energy of the radiation was proportional to the frequency; that is,

$$E = hf, \tag{27.4}$$

where the constant of proportionality h is a universal constant. Thus, Planck proposed a model in which energy is *quantized*—that is, limited to certain discrete quantities. Planck himself determined the value of h from previous experimental measurements of blackbody radiation. The value of h, which today is called **Planck's constant,** is

$$h = 6.626 \times 10^{-34} \text{ J} \cdot \text{s}.$$

We should remind you here that in formulating his theory of blackbody radiation, Planck did not draw upon any direct evidence of energy quantization, nor did his results come from extension of the existing classical theories. Instead, he introduced the **quantum** concept as a modification of classical ideas that brought his theory into agreement with experimental observations. (The word *quantum* means the smallest possible unit of energy.)

The replacement of the traditional view, that energy flowed like a smooth unbroken stream of water, by a new one, in which energy needed to be thought of as coming in little packets, marked the beginning of quantum mechanics and the end of a time in which all physical explanations were in terms of continuous flows or motions. However, like many revolutionary ideas, Planck's idea had little influence when it was first proposed. It did not gain credence until Einstein used it to explain a seemingly unrelated effect, the photoelectric effect, which is discussed in the next section.

Example 27.2

A photoflood lamp operates at a filament temperature of 3400 K. What is the wavelength λ_m of the peak in its blackbody spectrum?

Solution The wavelength λ_m may be obtained from the Wien displacement law:

$$\lambda_m = \frac{2.90 \times 10^{-3} \text{ m} \cdot \text{K}}{T}$$

$$\lambda_m = \frac{2.90 \times 10^{-3} \text{ m} \cdot \text{K}}{3.40 \times 10^3 \text{ K}} = 853 \text{ nm}.$$

The spectral peak lies in the near-infrared region, just beyond the visible spectrum, and most of the energy radiated is not visible.

Example 27.3

What is the wavelength of a quantum of radiation whose energy is 3.05×10^{-19} J?

Solution   From the Planck relation, the energy of the radiation is

$$E = hf.$$

Applying the relationship between frequency and wavelength, $c = f\lambda$, we find that the energy is

$$E = \frac{hc}{\lambda}.$$

The wavelength is obtained from

$$\lambda = \frac{hc}{E} = \frac{(6.626 \times 10^{-34} \text{ J} \cdot \text{s})(3.00 \times 10^8 \text{ m/s})}{(3.05 \times 10^{-19} \text{ J})}$$

$$\lambda = 652 \text{ nm}.$$

The radiation lies in the red portion of the visible spectrum.

27.4   The Photoelectric Effect

By the year 1905, many areas of physics had undergone a transition from confidence to confusion. The existence of line spectra was still unexplained. Blackbody radiation had been successfully explained, but only by an assumption that contradicted classical physics. X rays and radioactivity had been discovered and also were unexplained. However, the next ten years saw the beginning of understanding for all these areas of physics, including an explanation of the photoelectric effect. Einstein's photoelectric theory, proposed in 1905 (the same year as his special theory of relativity), shed new light on Planck's idea of the quantization of energy. Along with Rutherford's proposal in 1911 that atoms have nuclei, it led to Bohr's 1913 theory of atomic structure.

In 1887, Heinrich Hertz was studying the generation of electromagnetic waves with a spark gap. He found that electric discharges between the electrodes in the gap were enhanced when ultraviolet light was allowed to shine on those electrodes. This enhancement was an unexpected behavior that could not be readily explained. Later investigators showed that a freshly polished zinc plate, when negatively charged, would lose its charge when exposed to ultraviolet light, while a positively charged plate showed no such effect. The conclusion was drawn that a negatively charged plate emitted negatively charged particles when illuminated with ultraviolet light. This emission of negative charges from material as a result of light falling on it is called the **photoelectric effect.**

The nature of the negatively charged particles was unknown at first, but was resolved by J. J. Thomson's discovery of the electron in 1897. The negatively

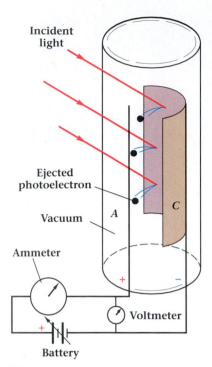

Figure 27.6

Schematic view of an apparatus for observing the photoelectric effect. When light strikes the plate *C*, it ejects electrons that are collected at the wire *A*, causing a current.

charged particles emitted in the photoelectric effect were found to be electrons. Still, the phenomenon could not be explained on the basis of classical physics.

To examine the photoelectric effect quantitatively, consider an evacuated quartz tube containing two electrodes (Fig. 27.6). In normal operation, the thin wire *A* is held at a positive potential with respect to the curved metal plate *C*. When light of a given frequency strikes plate *C*, it ejects electrons from its surface. These electrons (sometimes called photoelectrons) are attracted to the wire by the positive electric potential and give rise to a measurable current.

If we gradually reduce the potential applied to the wire until it becomes negative with respect to the plate, some of the electrons ejected from the plate will not have enough kinetic energy to reach the wire. They will be repelled back to the plate, reducing the current. As the potential of the wire is made still more negative, the current decreases until it becomes zero at a potential called the *stopping potential* V_s. The maximum kinetic energy of the photoelectron is the energy equivalent of the stopping potential.

Experiments with this apparatus show that for light of a given frequency, the magnitude of the current produced (that is, the number of photoelectrons per unit time) depends on the intensity, or brightness, of the light (Fig. 27.7). However, the stopping potential is independent of the intensity of the light and is different for different materials (Fig. 27.8a, b). Thus, the maximum kinetic energy of the photoelectrons is also independent of the light intensity. Instead, the electron's maximum kinetic energy *does* depend on the *frequency* of the light falling on the cathode (Fig. 27.9).

In addition to determining the photoelectron's maximum kinetic energy, the frequency of the light enters into the photoelectric effect in another way. Not all frequencies of light cause photoelectric emission from a given substance. In-

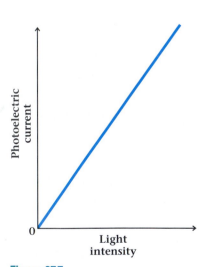

Figure 27.7

Photoelectric current plotted against the intensity of the incident light for a case in which photoelectrons are emitted.

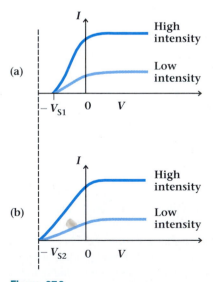

Figure 27.8

Photoelectric current plotted against the retarding potential for two different intensities of light. The stopping potentials (V_s) are different for (a) material 1 and (b) material 2.

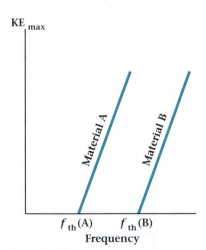

Figure 27.9

The maximum kinetic energy of photoelectrons as a function of the frequency of the incident light for two materials. Each material exhibits a threshold frequency f_{th}.

stead, each cathode material exhibits a **threshold frequency** f_{th}. Illumination with light of frequency less than f_{th} will not cause the ejection of photoelectrons, no matter how great the intensity of the radiation. But illumination with a frequency greater than f_{th} will produce photoelectrons instantaneously, even if the light intensity is very small.

We may analyze these experimental observations by noting that they fall into two classes: those effects that can be explained by classical physics and those effects that cannot be so explained. One effect that is understandable on the basis of classical ideas is the increase in current with an increase in the intensity of light. The principal observations that have no classical physics explanation are as follows:

1. *Electrons are emitted only when the frequency of the light is above some threshold value, no matter how intense the light.* The classical expectation is that you should be able to expel electrons if you provide enough energy in the form of sufficiently intense light.

2. *The maximum kinetic energy of the emitted electrons depends on the frequency of the light.* According to classical physics, you could increase the electron's kinetic energy with any frequency of light simply by making the light more intense.

3. *The photoelectrons are emitted almost at once when the light strikes.* The expectation of classical physics is that the photoelectrons will absorb energy over a period of time as the light continues to shine, eventually gaining enough energy to escape the material.

The explanation of the photoelectric effect was given in 1905 by Albert Einstein. His theory was beautiful in its simplicity and accounted for all of the experimental observations. Einstein hypothesized that incident light consists of streams of Planck's energy quanta, called **photons.** When monochromatic light shines on the cathode in the photoelectric experiment, the photons penetrate the surface and give up their energy to the electrons of the cathode. The simplest process envisioned by Einstein was one in which the entire energy of the photon is given up to a single electron. The energetic electron then makes its way through the surface and escapes from the material. A certain minimum amount of energy must be acquired by the electron before it can escape from the material of the cathode. This energy is called the **work function** ϕ of the particular material used and is the basis of the explanation of the threshold frequency. Table 27.2 lists the photoelectric work functions for selected metals.

Einstein's idea incorporated Planck's quantum hypothesis into a statement of energy conservation. The energy hf of the incident photon must equal the energy needed to free the electron plus the electron's kinetic energy:

$$hf = \text{energy to free the electron} + \text{KE}.$$

The maximum kinetic energy of the photoelectrons is then the difference between the energy of the incident photon and the minimum energy ϕ necessary to free the electron from the material. That is,

$$\text{KE}_{max} = hf - \phi. \qquad (27.5)$$

Table 27.2	Photoelectric Work Functions of Selected Metals
Metal	**Work Function ϕ (eV)***
Cesium	2.14
Potassium	2.30
Sodium	2.75
Silver	4.74
Copper	4.94
Gold	5.31
Platinum	5.65

*1 eV = 1.6×10^{-19} J.

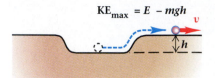

$$\text{KE}_{\text{max}} = E - mgh$$

Figure 27.10
A mechanical analogy of the photoelectric effect. If a ball in a ditch is given kinetic energy E greater than mgh, it will escape from the ditch. The maximum KE after escape is $\text{KE}_{\text{max}} = E - mgh$.

This equation is known as the **Einstein photoelectric equation.** Figure 27.10 presents a mechanical analog of the photoelectric effect.

The minimum (threshold) energy required to remove an electron from the surface with no kinetic energy is given by Eq. (27.5) with KE = 0. Thus, the threshold frequency is defined in terms of the work function ϕ as

$$hf_{\text{th}} = \phi. \qquad (27.6)$$

For lower frequencies, where hf is less than ϕ, no emission of photoelectrons occurs. At frequencies for which hf exceeds the work function, photoelectrons are emitted in numbers proportional to the number of incident photons (that is, the light intensity) and with kinetic energy maxima given by the Einstein equation. Subsequent experiments by Millikan verified Einstein's equation and provided an independent measurement of the Planck constant.

Einstein's successful explanation of the photoelectric effect furthered acceptance of the idea of the quantization of radiation, even though the particlelike behavior of the photon seemed contradictory. On one hand, the light is treated like a wave having a frequency f. On the other hand, a single photon transfers all of its energy in an encounter with a single electron, a particlelike behavior. This dual behavior is characteristic of all waves and particles on the atomic scale.

In discussing photoelectrons, and afterward as well, we find it convenient to use a unit of energy suitable to atomic-scale interactions. As we showed in Chapter 17, whenever a charge q is moved through an electric potential difference V, an amount of work $W = qV$ is required. If the amount of charge is the electron charge e and the potential difference is 1 V, then the work done is

$$W = (e)(1 \text{ V}) = (1.602 \times 10^{-19} \text{ C})(1 \text{ V}) = 1.602 \times 10^{-19} \text{ J}.$$

This amount of work or energy is given the special name **electron volt,** abbreviated eV:

$$1 \text{ eV} = 1.602 \times 10^{-19} \text{ J}. \qquad (27.7)$$

The electron volt is a practical unit for measuring energy on a molecular or smaller level. It is a natural unit to use in the photoelectric effect for both the work function and the electron kinetic energy.

The electron volt is a valid unit of energy regardless of how the energy is acquired. In the case at hand, the electrons gain energy from an encounter with a photon. Upon escaping from the material, the electrons have a kinetic energy that is conveniently expressed in electron volts, though it was not attained by actually moving through a particular potential difference.

Example 27.4

What is the energy in electron volts of a photon of green light of wavelength 546 nm?

Solution The energy of the photon is given by

$$E = hf = \frac{hc}{\lambda}$$

$$E = \frac{(6.626 \times 10^{-34} \text{ J} \cdot \text{s})(3.00 \times 10^{8} \text{ m/s})}{(546 \times 10^{-9} \text{ m})(1.60 \times 10^{-19} \text{ J/eV})} = 2.28 \text{ eV}.$$

Example 27.5

The photoelectric work function for cesium is 2.14 eV. (a) What is the maximum kinetic energy of electrons ejected from the surface of cesium by light of wavelength $\lambda = 546$ nm? (b) What is the maximum speed of the electrons?

Solution   (a) The maximum kinetic energy is given by the Einstein equation as

$$KE = hf - \phi = \frac{hc}{\lambda} - \phi,$$

which is

$$KE = \frac{(6.626 \times 10^{-34} \text{ J} \cdot \text{s})(3.00 \times 10^8 \text{ m/s})}{(546 \times 10^{-9} \text{ m})(1.60 \times 10^{-19} \text{ J/eV})} - 2.14 \text{ eV}$$

$$KE = 2.28 \text{ eV} - 2.14 \text{ eV} = 0.14 \text{ eV}.$$

(b) Because the kinetic energy is small, we may express it as

$$KE = \tfrac{1}{2}mv^2,$$

$$v = \sqrt{\frac{2KE}{m}}$$

$$v = \sqrt{\frac{2(0.14 \text{ eV})(1.60 \times 10^{-19} \text{ J/eV})}{9.11 \times 10^{-31} \text{ kg}}}$$

$$v = 2.2 \times 10^5 \text{ m/s}.$$

Discussion   Note that the speed is small compared with the speed of light ($v/c = 7.3 \times 10^{-4}$). If the speed were closer to the speed of light, we would need to use the relativistic expression relating kinetic energy and speed (Eq. 25.9). In cases where you are not sure, it is a good idea to use the relativistic expression, which is correct for all speeds.

Example 27.6

Find an expression for the energy of a photon in electron volts when the wavelength of the photon is given in nanometers.

Solution   We begin with the Planck formula,

$$E = hf = \frac{hc}{\lambda},$$

and insert the values of h, c, and the conversion factors:

$$E = \frac{(6.626 \times 10^{-34} \text{ J} \cdot \text{s})(3.00 \times 10^8 \text{ m/s})}{\lambda(10^{-9} \text{ m/nm})} \frac{1 \text{ eV}}{1.60 \times 10^{-19} \text{ J}}$$

$$E = \frac{1240}{\lambda} \text{ eV} \cdot \text{nm}.$$

This is a useful rule for people who work with light and x rays.

PHOTONS AND VISION

What is the weakest flash of light that you can see? Put another way, what is the minimum number of photons required on the retina for you to detect a flash? We discussed the optical properties of the eye in Chapter 23, showing how incident light is focused to an image on the retina by the cornea and the lens. Later, in Section 27.4, we saw how Einstein explained the interaction of light with matter by utilizing the idea that light rays consist of large numbers of individual photons. Einstein's theory often forms the basis for our interpretation of how light causes physical, chemical, or biological changes.

You see a flash of light when the photosensitive receptors in the retina are stimulated. There are two types of receptors—rods and cones—each with distinct types of light-sensitive molecules called visual pigments (Fig. B27.3). The cones are responsible for color vision and lie primarily in the fovea, the area of most acute vision. The more numerous rods lie mainly outside the fovea. They convey no color information, but are much more sensitive to light than are the cones. Figure B27.4 shows the spectral absorption curves of the visual pigments in the rods and

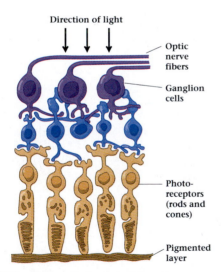

Figure B27.3 Schematic drawing of the retina.

cones. Notice that the human retina is, in some sense, wrong-way out. The rods and cones are on the back surface of the retina, away from the direction of the incoming light. In order for light to reach the photoreceptors, it must pass through most of the retina, which consists of relatively transparent cells.

The answer to our initial question was provided by experiments first done by Hecht, Shlaer, and Pirene in the 1940s. The first part of an experiment consisted of determining the minimum number of photons that must enter the cornea in order to cause the sensation of light. The second part was to determine how many of these photons actually reach the visual receptors.

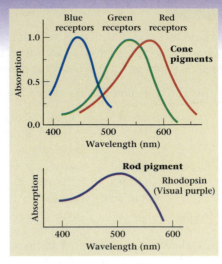

Figure B27.4 Absorption curves of the visual pigments.

The experimental apparatus gave a light flash of 0.1 s duration that fell on an area of the retina corresponding to about 500 rods. The light had its maximum intensity at 510 nm, corresponding to the greatest sensitivity of the rods. Experimental subjects were exposed to random flashes and asked to indicate when a flash was seen. The intensity of the source was lowered in steps until no flashes were seen, thereby establishing a threshold. By knowing the details of the apparatus, the experimenters determined that the threshold for a visual response was about 100 photons incident on the cornea. This is truly a small amount of light: a continuous output of 100 photons every 0.1 s of 510-nm light corresponds to less than 10^{-15} watts.

However, not all of the photons arriving at the outside of the cornea reach the rods in the retina. Some of the photons are reflected by the cornea and the lens. In addition, the fluids in the eye absorb or scatter about half of the light so that only about 45 of the original 100 photons actually reach the front surface of the retina. Most of these photons are absorbed in the retina before they can reach the rods. Further experiments have shown that about 5 photons must fall on a 500-rod area of the retina in order to produce a flash.

The probability that any one of the 500 rods will be struck by more than one of the 5 photons is extremely small. Therefore, we conclude that a single photon will activate a single rod. But excitation of one rod is not sufficient—five or more must be stimulated simultaneously for us to see a flash.

How do we explain that a single photon will activate a single rod, but that about 5 photons are needed to elicit a visual response? Photoreceptor rods are activated when they are supplied with enough energy. Sometimes this energy is provided by a photon of energy $E = hf$. However, a rod may also be excited by the thermal energy present in its environment. Calculations indicate that random thermal excitations occur at about a rate that is the same as that of the 5 photons for a visual signal. Thus, it seems that one does not see a flash unless the signal produced is comparable to or greater than that due to the random thermal excitation of the rods.

27.5 Bohr's Theory of the Hydrogen Atom

The experimental evidence of spectral lines, blackbody radiation, and the photoelectric effect all provided clues to a deeper understanding of the nature and behavior of atoms, but these clues were subtle and not readily interpreted. In 1911, Rutherford added a vital contribution when he proposed that the atom had a tiny, positively charged nucleus that held most of the atom's mass.

Rutherford's description of a positively charged nucleus surrounded by electrons conflicted with classical electromagnetic theory. If the electrons were assumed to be at rest, no possible stable configuration could be found; the electrons would just fall in toward the nucleus. On the other hand, if they revolved in a circular orbit about the nucleus, they would undergo constant acceleration toward the center. Maxwell's equations predict that the accelerating charges would radiate away their energy and spiral into the nucleus. What is it, then, that keeps the atom from collapsing?

In 1913, the Danish physicist Niels Bohr published his answer to the question. Bohr had worked a year with J. J. Thomson and a year with Rutherford. Upon returning home to Copenhagen, he combined Rutherford's concept of the nucleus with the quantum hypothesis of Planck and Einstein to develop a radically new theory of the atom. Bohr evaded the classical difficulties by observing that the existence of stable atoms implied that the laws of classical electrodynamics were not appropriate for the description of atoms and therefore must be altered.

Bohr began by considering the simplest case, that of a positively charged nucleus with a single electron traveling in a closed orbit around it (Fig. 27.11). This model corresponds to the hydrogen atom. He further assumed that the mass of the electron was negligibly small compared with the mass of the nucleus. Then he postulated that

1. **The electron revolves about the nucleus in stationary, or stable, nonradiating orbits corresponding to fixed energy states.** Even though this postulate is not consistent with classical physics, Bohr assumed that while the electrons are in stationary orbits they may be treated with classical dynamics, including Newton's laws, Coulomb's law, and the ordinary conservation rules.

2. **The atom emits light only when the electron makes a sudden change from one energy state to another lower energy state.** Bohr assumed that when radiation does take place, the relationship between the energy radiated and the frequency is given by Planck's equation.

Bohr applied conservation of energy by assuming that the energy of a photon emitted by an atom is exactly equal to the energy difference between two energy levels (Fig. 27.12). Thus, when an electron in an atomic energy level E_2 makes a transition to a lower energy level E_1, a photon is emitted with an energy

$$hf = E_2 - E_1. \tag{27.8}$$

This equation is known as the *Bohr frequency condition.*

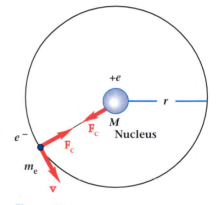

Figure 27.11
A positively charged nucleus with a single electron traveling in a closed orbit around it. The centripetal force on the electron is given by Coulomb's law.

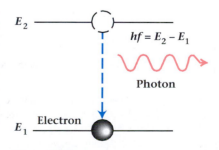

Figure 27.12
The energy of an emitted photon is equal to the energy difference between two energy levels, or states.

By using these postulates, Bohr was able to calculate the energy levels of the hydrogen atom and thereby account for the frequencies of the Balmer lines observed in the hydrogen spectra. The derivation we present here is not the one given in Bohr's original paper. In that paper he showed that, as a result of his assumptions, it follows that angular momentum (first discussed in Chapter 9) is also quantized for the motion of electrons in the hydrogen atom. The quantization of angular momentum is an extremely important concept for atomic and smaller-scale phenomena, and we will use it in succeeding chapters. By employing the quantization of angular momentum from the outset, we can considerably shorten the derivation of the consequences of Bohr's assumptions.

We begin by assuming an electron of mass m traveling with speed v in a circular orbit of radius r_n about a stationary nucleus. For such an orbit, Bohr showed that the angular momentum L_n of an electron is

$$L_n = mvr_n = n\frac{h}{2\pi}, \quad n = 1, 2, 3, 4, \ldots, \tag{27.9}$$

where h is Planck's constant and n is an integer. This equation expresses the requirement that the orbital angular momentum of an electron in a stationary state must be an integer multiple of $h/2\pi$. Thus, we have a quantization rule for angular momentum: **Angular momentum is quantized in units of $h/2\pi$.**

In the Bohr model of the hydrogen atom, the electron travels in a circular orbit about a stationary nucleus of charge $+e$. For the electron to travel in a stationary, or stable, orbit, the electrostatic attraction due to the nucleus must provide the force required for circular motion. Thus, the centripetal force mv^2/r_n is equal to the electrostatic force, given by

$$F_e = \frac{e^2}{4\pi\epsilon_0 r_n^2},$$

where ϵ_0 is the permittivity constant introduced in Chapter 16. When we equate the centripetal force to the electrostatic force and use the angular momentum of Eq. (27.9) to eliminate the speed, we get

$$r_n = \frac{\epsilon_0 n^2 h^2}{\pi m e^2}, \quad n = 1, 2, 3, 4, \ldots, \tag{27.10}$$

Equation (27.10) gives the radii of the orbits in the Bohr model of the atom. Only those orbits corresponding to integral values of n are allowed (Fig. 27.13). The integer n is called the **principal quantum number,** for it establishes an electron's orbit and, as we will see, the allowed energy for the electron.

You should not equate the Bohr radii r_n with atomic radii. The idea of electron orbits as circles with sharply defined radii contradicts experimental evidence. However, for hydrogen, good agreement exists between the predictions of Bohr's theory and experimental observations of spectra and energy levels.

Example 27.7

Calculate the radius of the smallest orbit of hydrogen.

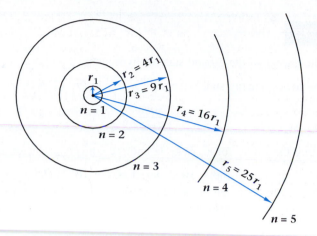

Figure 27.13
Sketch of the allowed electron orbits in Bohr's model of hydrogen. The radii increase with the square of n.

Solution The smallest orbit corresponds to the situation in which the principal quantum number takes its smallest value, $n = 1$. From Eq. (27.10) we have

$$r_1 = \frac{1^2 h^2 \epsilon_0}{\pi m e^2},$$

where m is the mass and e the charge of an electron. So

$$r_1 = \frac{(6.626 \times 10^{-34} \text{ J} \cdot \text{s})^2 (8.854 \times 10^{-12} \text{ C}^2/\text{J} \cdot \text{m})}{\pi (9.109 \times 10^{-31} \text{ kg})(1.602 \times 10^{-19} \text{ C})^2}$$

$$r_1 = 5.29 \times 10^{-11} \text{ m} = 0.0529 \text{ nm}.$$

This number is called the *first Bohr radius of the hydrogen atom* and is comparable to the atomic radius of hydrogen as determined by other means.

■

The total energy E_n of an electron orbiting around the nucleus includes both kinetic and potential terms. The potential energy is due to the electric interaction and is negative (Chapter 17). The total energy is

$$E_n = \text{KE} + \text{PE} = \tfrac{1}{2} m v^2 - \frac{e^2}{4\pi\epsilon_0 r_n},$$

where only certain orbits, with radius r_n, are allowed. By substituting the value of v from the angular momentum quantization rule (Eq. 27.9) into the KE term, we get an expression for E_n as a function of r_n. Then by substituting in both terms for r_n from Eq. (27.10), we get

$$E_n = \frac{-m e^4}{8\epsilon_0^2 h^2 n^2}, \qquad n = 1, 2, 3, 4, \ldots. \tag{27.11}$$

The total energy is negative, which means that energy from outside is required to remove the electron from its orbit around the nucleus. Note that the smallest value of n and, therefore, the smallest value of r correspond to the lowest, or most tightly bound, energy state of hydrogen. The state of lowest energy of a system is called its **ground state.**

Example 27.8

Calculate the energy of the ground state of hydrogen.

Solution   The lowest energy state occurs for $n = 1$ in Eq. (27.11),

$$E = \frac{-me^4}{8\epsilon_0^2 h^2}.$$

Upon inserting the values for m, e, ϵ_0, h, and the conversion factor, we get

$$E = \frac{-(9.109 \times 10^{-31} \text{ kg})(1.602 \times 10^{-19} \text{ C})^4}{8(8.854 \times 10^{-12} \text{ C}^2/\text{J} \cdot \text{m})^2(6.626 \times 10^{-34} \text{ J} \cdot \text{s})^2(1.602 \times 10^{-19} \text{ J/eV})}$$

$$E = -13.6 \text{ eV}.$$

This value of 13.6 eV is in excellent agreement with the experimentally observed value for the ionization energy of hydrogen—that is, the energy required to remove the single electron from the lowest energy state of the hydrogen atom.

■

The light emitted when an electron makes a transition from one energy level to another carries away energy equal to the difference between the energy levels, according to Eq. (27.8). If these two levels are characterized by two integers n_1 and n_2, we get

$$hf = E_2 - E_1 = \frac{-me^4}{8\epsilon_0^2 n_2^2} - \frac{-me^4}{8\epsilon_0^2 h^2 n_1^2},$$

where we have used Eq. (27.11) for the energy. Applying the relation between frequency and wavelength, $c = f\lambda$, we have

$$\frac{1}{\lambda} = \frac{me^4}{8\epsilon_0^2 h^3 c}\left(\frac{1}{n_1^2} - \frac{1}{n_2^2}\right). \tag{27.12}$$

We have just developed the Rydberg equation given earlier as Eq. (27.2). If we calculate the quantity in front of the parentheses from the fundamental constants, we find a value of 1.097×10^7 m^{-1}, the same value determined spectroscopically for the Rydberg constant R_∞.

Bohr's value for the combination of constants in front of the parentheses differed from the spectroscopically measured value by only 6%, an error due to the poor precision in the values for e, m, and h available to him. The difference in the two values was within the uncertainty caused by the errors in the values of the constants. His success was truly remarkable.

27.6   Successes of the Bohr Theory

Starting with the simple model of a single electron making circular orbits about a positive charge, Bohr explained the Rydberg formula. The essential assumptions he made were that the electron could exist in a stationary orbit and that radiation was emitted only when the electron made a transition from one stationary orbit to another. In addition, he used Planck's constant to relate the frequency of the radiation to the energy difference between the orbits. A consequence of the assumptions was that angular momentum, as well as energy, was quantized.

Bohr recognized that Eq. (27.12) gave the wavelengths of the Balmer series when n_1 was set equal to 2 and n_2 was allowed to be 3, 4, 5, ... Similarly, if $n_1 = 3$ and $n_2 = 4, 5, 6, . . .$, his formula gave the wavelengths of the infrared series observed for hydrogen by Friedrich Paschen in 1908. Bohr then predicted that other spectral series not yet observed would be found, corresponding to $n_1 = 1$ and to n_1 greater than 3. Indeed, a series of ultraviolet lines in hydrogen was observed by Theodore Lyman in 1914, corresponding to $n_1 = 1$. Other series corresponding to $n_1 = 4$ and $n_1 = 5$ were discovered in the 1920s.

The interpretations of Bohr's equations can be understood with the aid of Fig. 27.14(a), which represents the electron orbits in space as given by Bohr's model. As we have seen, the radii of the orbits increase as the square of the quantum number n (Eq. 27.10). Figure 27.14(b) shows a representation of the allowed orbits in terms of their energy E. A figure like this is called an **energy-level diagram**. Vertical arrows drawn between the energy levels represent electronic transitions from one energy level to another. A downward transition corresponds to emission of light. The energy of the emitted photon is given by the energy difference between the levels. Transitions from higher energy states to the $n = 2$ level correspond to light emitted at the frequencies seen in the Balmer series.

Absorption spectra can be understood as upward transitions in Fig. 27.14. Notice that absorption is a resonant phenomenon; that is, not all frequencies of light can be absorbed. There are two requirements for absorption of light by an atom. First, only those frequencies can be absorbed that exactly correspond (by Planck's equation) to the difference between two distinct energy states. Second, the absorption can happen only if an electron is present in a lower energy level to absorb the energy and if there is a vacancy in the allowed level at an energy hf above it, into which the electron can jump. Resonant absorption accounts for

Figure 27.14

(a) Electron orbits of Bohr's model of the hydrogen atom, showing the transitions corresponding to the frequencies of spectroscopic lines in hydrogen. (b) Energy-level representation of the hydrogen atom. The downward arrows represent electronic transitions from higher energy states to lower energy states, corresponding to light emission. Upward arrows would correspond to resonant absorption of light. The lengths of the arrows are proportional to the energy of the photons emitted (or absorbed).

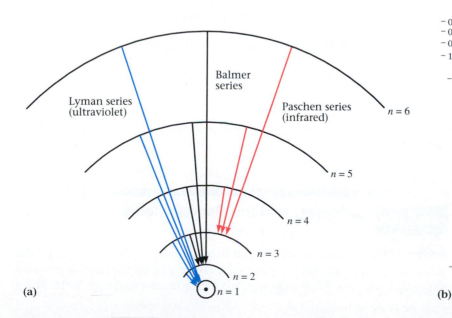

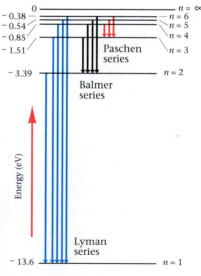

the dark lines observed by Fraunhofer in the solar spectrum. These absorption lines, which are observed superimposed on the continuous spectrum of the sun, are due to absorption in the outer (cooler) layers of the sun. Atoms in these cooler layers have more electrons in lower energy levels and a larger number of higher energy levels available than do atoms in the sun's hotter interior.

The absorption spectra of cool hydrogen do not show lines corresponding to the Balmer series observed in emission. In emission, the transitions of the Balmer series correspond to electrons making transitions from higher states to the $n = 2$ state, which is 10.2 eV above the ground ($n = 1$) state. In cool hydrogen gas, the $n = 2$ state is essentially empty and the electrons are primarily in the ground state. Because there are so few atoms with electrons in the $n = 2$ state, light at the Balmer frequencies cannot be resonantly absorbed, but simply passes through the gas.

Example 27.9

What are the shortest and longest wavelengths in the Paschen series of hydrogen?

Solution The Paschen series corresponds to $n_1 = 3$ in the Bohr formula for radiation from atomic hydrogen. The extreme wavelengths correspond to transitions from $n_2 = 4$ and from $n_2 = \infty$. For the Paschen series, we rewrite Eq. (27.12) as

$$\frac{1}{\lambda} = \frac{me^4}{8\epsilon_0^2 h^3 c}\left(\frac{1}{3^2} - \frac{1}{n_2^2}\right).$$

Remember that the calculated value of the term outside the parentheses on the right-hand side is the same as the Rydberg constant, 1.097×10^7 m^{-1}. The longest wavelength occurs for $n_2 = 4$. It is

$$\frac{1}{\lambda} = (1.097 \times 10^7 \text{ m}^{-1})\left(\frac{1}{3^2} - \frac{1}{4^2}\right),$$

$$\lambda = 1.88 \times 10^{-6} \text{ m} = 1880 \text{ nm}.$$

The shortest wavelength occurs for $n_2 = \infty$ and is given by

$$\frac{1}{\lambda} = (1.097 \times 10^7 \text{ m}^{-1})\left(\frac{1}{3^2} - \frac{1}{\infty}\right) = (1.097 \times 10^7 \text{ m}^{-1})\left(\frac{1}{3^2} - 0\right),$$

$$\lambda = 8.20 \times 10^{-7} \text{ m} = 820 \text{ nm}.$$

Note that the shortest wavelength of the Paschen series is still outside of the visible spectrum.

*27.7 Moseley and the Periodic Table

Another successful correlation between Bohr's atomic model and experimental observations came through the study of x-ray spectra by the English physicist Henry G. J. Moseley. While Bohr was developing his atomic theory, W. H. Bragg and W. L. Bragg were carrying out experiments with their newly constructed x-ray spectrometer (Section 26.5). In 1913, they showed that an x-ray spectrum consists of two parts: a continuous part superimposed on a line spectrum similar to the emission spectrum of a hot gas (Fig. 27.15). The continuous part is due to an electromagnetic effect: Electrons radiate when decelerating as they are stopped by the target of the x-ray tube. The lines are characteristic of the material in the target and have no classical explanation. The most intense line is called the K_α line.

Just a few months after Bohr's theory was published, Moseley reported a study of the x-ray spectra of the elements of atomic mass between calcium and zinc. He showed that the frequencies of all of the K_α lines were given by a single formula,

$$f = \tfrac{3}{4}cR_\infty(Z-1)^2, \qquad (27.13)$$

where c is the velocity of light and R_∞ is the Rydberg constant (Fig. 27.16). The parameter Z is an integer that was found to be the charge on the nucleus in units of the electron's charge. This equation can be rearranged and written in terms of the wavelength as

$$\frac{1}{\lambda} = \frac{f}{c} = R_\infty(Z-1)^2\left(\frac{1}{1^2} - \frac{1}{2^2}\right).$$

In other words, the K_α line has a frequency described by an equation similar to the Rydberg equation, but with an additional factor that depends on the nuclear charge.

The interpretation of Moseley's equation is straightforward in light of the Bohr theory. The equation gives the frequency of the x ray when an electron goes from the $n = 2$ state to the $n = 1$ state of an atom of nuclear charge Ze. The simple Bohr theory would have predicted a frequency proportional to Z^2 for an atom with a single electron. The origin of the $(Z-1)^2$ factor lies in the fact that elements heavier than hydrogen contain two electrons in the lowest energy level. When one electron is removed from the lowest level, there is still one negative charge near the nucleus. This negative charge reduces the effective positive nuclear charge to $(Z-1)e$.

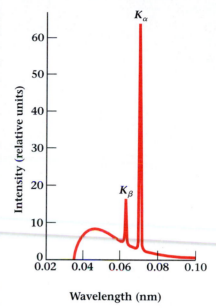

Figure 27.15

The x-ray spectrum of a molybdenum target. (Line widths are not to scale.)

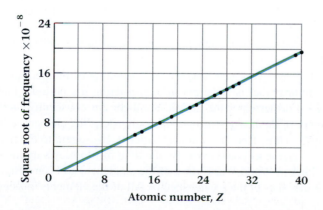

Figure 27.16

A graph of Moseley's data showing the linear relation between the square root of the frequencies of the x-ray lines and the atomic number.

Example 27.10

Calculate the wavelength and energy of the K_α x rays from chromium ($Z = 24$).

Solution First we calculate the wavelength, using Moseley's equation:

$$\frac{1}{\lambda} = \frac{f}{c} = \frac{3}{4}R_\infty(Z - 1)^2,$$

$$\lambda = \frac{4}{3R_\infty(Z - 1)^2} = \frac{4}{3(1.097 \times 10^7 \text{ m}^{-1})(23)^2} = 0.2298 \text{ nm}.$$

Then we determine the energy from the wavelength:

$$E = \frac{hc}{\lambda} = \frac{(6.626 \times 10^{-34} \text{ J} \cdot \text{s})(3.00 \times 10^8 \text{ m/s})}{(0.2298 \times 10^{-9} \text{ m})}$$

$$E = (8.650 \times 10^{-16} \text{ J})\left(\frac{1 \text{ eV}}{1.602 \times 10^{-19} \text{ J}}\right) = 5.40 \text{ keV}.$$

Example 27.11

Find an expression for the energy in electron volts of the K_α line of an element of atomic number Z.

Solution The energy may be determined from $E = hf$, which becomes

$$E = hf = \frac{3}{4}hcR_\infty(Z - 1)^2.$$

When we use the values of h, c, R_∞, and the conversion constant, we have

$$E = \frac{3}{4}(6.626 \times 10^{-34} \text{ J} \cdot \text{s})(3.00 \times 10^8 \text{ m/s})(1.097 \times 10^7 \text{ m}^{-1})(Z - 1)^2$$

$$E = \left(1.636 \times 10^{-18}(Z - 1)^2 \text{ J}\right)\frac{1 \text{ eV}}{1.602 \times 10^{-19} \text{ J}}$$

$$E = 10.2 \ (Z - 1)^2 \text{ eV}.$$

The real significance of Moseley's work lies in his characterization of the elements. His observation that the square root of an element's K_α line frequency is linearly related to its nuclear charge, Eq. (27.13), has become known as **Moseley's law.** It provides a direct way to number the elements according to their nuclear charge. The number of nuclear charges in an atom is its atomic number, which gives the sequence of elements as they occur in the periodic table (see the inside back cover of this book).

A result of this work was a better understanding of the periodic table. It is the atomic number that is important in constructing the periodic table, not the atomic mass. Furthermore, the atomic numbers are consecutive integers. The atomic number of the heaviest known naturally occurring element, uranium, was found to be 92. Thus, from the simplest element, hydrogen, to the heaviest, uranium, only 92 elements could exist. By the time of Moseley's untimely death as a British soldier in World War I, most of the elements had been assigned atomic numbers on the basis of their x-ray spectra. All of the elements missing in Moseley's time were subsequently found. Elements beyond uranium having atomic numbers up to 112 have been created in the laboratory.

Moseley's work was of primary importance because it provided an unambiguous method of numbering the elements. Furthermore, Moseley's results reinforced the new models of Rutherford and Bohr. Together these ideas form the basis of our modern view of atoms and how one atom differs from another.

Summary

Useful Concepts

■ The Rydberg formula for the spectral lines of hydrogen is

$$\frac{1}{\lambda} = R_\infty \left(\frac{1}{n_1^2} - \frac{1}{n_2^2} \right), \qquad n_1 < n_2.$$

For the Balmer series, $n_1 = 2$ and $n_2 = 3, 4, 5, \ldots$.

■ For blackbody radiation, the temperature of the body and the maximum wavelength are related through the Wien displacement law,

$$\lambda_m T = 2.90 \times 10^{-3} \text{ m} \cdot \text{K}.$$

■ According to Planck's hypothesis, the energy of a quantum of radiation is determined by its frequency,

$$E = hf,$$

where Planck's constant $h = 6.626 \times 10^{-34}$ J $\cdot$ s.

■ The Einstein photoelectric equation gives the maximum energy of a photoelectron as

$$\text{KE}_{max} = hf - \phi.$$

■ Bohr postulated that atomic electrons exist in definite energy levels, or orbits, and do not radiate (or absorb) energy except when they make a transition to another energy level or orbit. The Bohr theory of the hydrogen atom gives the Rydberg and

Balmer formulas through the use of

$$hf = E_2 - E_1.$$

■ The orbital angular momentum of an electron in the hydrogen atom is

$$L = n\frac{h}{2\pi}, \quad \text{where } n = 1, 2, 3, \ldots.$$

■ The frequency of the K_α line of an element is given by Moseley's law,

$$f = \tfrac{3}{4}cR_\infty(Z - 1)^2.$$

Important Terms

You should be able to write the definition or meaning of each of the following:

spectroscopy	photons
Balmer series	work function
Rydberg formula	Einstein photoelectric equation
blackbody	electron volt
Wien displacement law	principal quantum number
Planck's constant	ground state
quantum	energy-level diagram
photoelectric effect	Moseley's law
threshold frequency	

Conceptual Questions

27.1 Stars appear to have distinct colors. Some stars look red, some yellow, and others blue. What is a possible explanation for this?

27.2 How does the gravitational red shift affect the wavelengths of the Fraunhofer lines measured in the solar spectrum, compared with those of the same lines measured in the laboratory?

27.3 When the interior of an operating kiln reaches a uniform temperature, it is virtually impossible to see the objects that are being heated. Explain why this is so.

27.4 Some theories propose that Planck's constant is not really constant but changes with time. What would be some of

the consequences of an increasing value of h if all other fundamental constants remained unchanged?

27.5 Do all electrons emitted in the photoelectric effect have the same kinetic energy?

27.6 What difference would it make if the electrons in an atom obeyed classical mechanics?

27.7 Explain the differences between the emission of light by an incandescent lamp, a neon sign, and a fluorescent lamp.

27.8 Can x rays be emitted by hydrogen?

27.9 How are x rays produced? Explain the origin of the line spectra and the continuous spectra. What limits the minimum size of x-ray wavelengths?

27.10 You are examining the spectrum of a particular gas that is excited in a discharge tube. You are viewing the discharge through a transparent box that contains the same gas. Under what conditions would you expect to see dark lines in the spectrum?

27.11 Why is it desirable for suntan lotions to block out the ultraviolet rays of the sun?

27.12 What kinds of materials have the lowest work functions? A study of Table 27.2 and the periodic table should help you answer this question.

▼ Problems

Section 27.2 Balmer's Series

27.1 Calculate the wavelengths of the first four lines in the Balmer series. These lines are in the visible spectrum.

27.2 (a) What are the wavelength and color of the line in the Balmer spectrum for $n_2 = 4$? (b) What are the wavelength and color of the line for $n_2 = 5$?

27.3 Calculate the wavelength of the Balmer line corresponding to $n_2 = 7$. Why is this line not seen visually?

27.4• Show that the Balmer formula and the Rydberg formula are equivalent. Find the relationship between the numerical constant in Balmer's formula (call it b) and the Rydberg constant.

Section 27.3 Blackbody Radiation

Hints for Solving Problems

The wavelength for maximum intensity in blackbody radiation is related to the temperature of the radiating body through the Wien law. Remember that photon energy and frequency are related through $E = hf$ and that for all electromagnetic waves in free space, $\lambda f = c$.

27.5 A certain star has a peak intensity in its blackbody spectrum at 380 nm. What is the surface temperature of the star?

27.6 The peak intensity of the solar spectrum occurs for $\lambda_m = 475$ nm. What is the temperature of the surface of the sun?

27.7 A star has an effective surface temperature of 5100 K. What is the wavelength λ_m of the peak in its blackbody spectrum?

27.8 What is the wavelength λ_m of the most intense radiation from a blackbody at room temperature (20°C)?

27.9 What is the energy in joules of a quantum of infrared light whose wavelength is 955 nm?

27.10 What is the energy in joules of a quantum of blue light of wavelength 432 nm?

27.11 What is the wavelength of a quantum of light having an energy of 4.82×10^{-19} J?

27.12 What is the wavelength of a quantum of light whose energy is 3.50×10^{-19} J?

27.13• A photoflood lamp operates at a temperature of 3200 K. (a) What is the wavelength of the peak in its blackbody spectrum? (b) What is the energy of a photon with the wavelength found in (a)?

27.14• Write the Wien displacement law in terms of the temperature and the frequency.

27.15•• (a) Find an expression for the momentum of a photon in terms of its frequency. (b) A beam of light is directed upward at a small, horizontally held piece of aluminum foil. How many photons per second of blue light (475 nm) must strike the 0.50-g, 3.0 cm × 3.0 cm piece of foil from below in order to make it float? (c) What is the power of such a beam, and (d) is the experiment a practical one? Why? (*Hint:* Use the relationship given in Chapter 25 that the momentum and energy of light are related by $p = E/c$. Recall that the force is the rate of change of the momentum. Assume that the photons are perfectly reflected from the aluminum foil and recall what this meant for the molecules striking the wall in the kinetic theory model of a gas.)

Section 27.4 The Photoelectric Effect

Hints for Solving Problems

The energy of the incident photon in the photoelectric effect equals the work function plus the maximum kinetic energy of the emergent electron. Work functions of several materials can be found in Table 27.2.

27.16 Express the Planck constant in terms of eV · s.

27.17 Calculate the wavelength of a photon whose energy is 1.80 eV.

27.18 A gamma ray from a nuclear decay has an energy of 59 keV. What is its wavelength?

27.19 X rays are produced when energetic electrons collide with a material target. The radiation consists of a line spectrum plus a broad-banded continuous spectrum, which extends up to the maximum energy of the electrons. Find the shortest x-ray wavelength due to electrons accelerated through 35.7 kV.

27.20 What is the minimum voltage required in an x-ray tube to produce photons whose wavelength is 0.10 nm?

27.21 The mean wavelength of the D lines in sodium is 589.3 nm. What is the mean photon energy of the D lines? Give your answer in electron volts.

27.22 The eye is most sensitive to light of wavelength 550 nm. What is the energy of photons with that wavelength? Give your answer in electron volts.

27.23 What is the maximum kinetic energy of photoelectrons ejected from the surface of sodium by blue light of wavelength 434 nm?

27.24 What is the threshold wavelength for photoelectric emission from silver?

27.25● (a) Calculate the wavelength of an x-ray photon whose energy is 90.0 keV. (b) What is the frequency of the photon?

27.26● A material has a photoelectric work function of 3.66 eV. What is the maximum speed of photoelectrons ejected from it by light of wavelength 250 nm?

27.27● Light from the 253.7-nm UV line in the mercury spectrum ejects electrons from the surface of metallic sodium. (a) What is the maximum kinetic energy of these photoelectrons? (b) Can photoelectrons be ejected from sodium by the green light ($\lambda = 546$ nm) in the mercury spectrum?

27.28● (a) Calculate the work function of a surface from which the green light from mercury ($\lambda = 546.1$ nm) ejects photoelectrons with maximum kinetic energy of 0.13 eV. (b) What is the material of the surface?

27.29● A photoflood lamp operates at a temperature of 3400 K. (a) What is the wavelength of the peak in its blackbody spectrum? (b) What is the energy of a photon with the wavelength found in (a)? Give your answer in units of eV.

27.30● The human eye is quite a sensitive detector. Only a few photons are needed to trigger a visual stimulus. Assume that 10 photons of red (600 nm) light cause a visible flash in your eye. (a) What is the total energy deposited in your eye? (b) If a thousand times as much energy is used to raise a 2.0-μg speck of pollen, how high will it be lifted?

Section 27.5 Bohr's Theory of the Hydrogen Atom

27.31 Calculate the third Bohr radius of the hydrogen atom—that is, the radius for $n = 3$.

27.32 Calculate the second Bohr radius of the hydrogen atom—that is, the radius for $n = 2$.

27.33 As the integer n_2 in the Rydberg equation (Eq. 27.2) approaches infinity, the spectral lines converge to a minimum wavelength known as the series limit. Calculate the series limit for the Balmer lines.

27.34 If you twirl a small 5.00-g wooden ball in a circle at 3.0 revolutions per second on a light 10-cm string, approximately how many units of angular momentum does it have?

27.35● (a) What is the energy of the photon that will raise a hydrogen atom from its ground state to the $n = 4$ excited state? (b) What is the wavelength of this photon?

Section 27.6 Successes of the Bohr Theory

Hints for Solving Problems

Emission spectra correspond to electrons making a transition from a higher to a lower energy state; for absorption spectra, the converse is true.

27.36 What is the longest-wavelength photon that can ionize a neutral hydrogen atom initially in the ground, or lowest, energy state?

27.37 A free electron combines with a free hydrogen nucleus to form a hydrogen atom in the $n = 4$ level. What is the wavelength of the photon emitted during this process?

27.38 Calculate the series limit for the infrared series for which $n_1 = 5$.

27.39 As the integer n_2 in the Rydberg equation (Eq. 27.2) approaches infinity, the spectral lines converge to a minimum wavelength known as the series limit. Calculate the series limit for the Lyman series of lines.

27.40 What is the angular momentum of an electron in the third Bohr orbit?

27.41 Evaluate the Rydberg constant from its definition, using the values for m, c, e, h, and ϵ_0.

27.42● Show that the quantity $h/2\pi$ has the proper dimensions of an angular momentum.

27.43● (a) What is the energy of the least energetic photon that can be absorbed by hydrogen gas at room temperature? (b) What is the wavelength of such a photon? Assume the electrons are in the lowest state at the start.

27.44● Derive the equation for the first Bohr radius of singly ionized helium. (Helium has a nuclear charge of $Z = +2e$.)

27.45● (a) Calculate the speed of an electron in the second Bohr orbit. (b) How does it compare with the speed of light? What does this suggest?

27.46● The spectral series corresponding to $n_1 = 4$ in hydrogen is called the Brackett series in honor of its discoverer. What are the longest and shortest wavelengths in the Brackett series?

*Section 27.7 Moseley and the Periodic Table

Hints for Solving Problems

You may need the periodic table found on the inside back cover of this book.

27.47 On the basis of Moseley's law, calculate the energy of the K_α line from (a) gadolinium ($Z = 64$) and (b) thorium ($Z = 90$).

27.48 On the basis of Moseley's law, calculate the energy of the K_α line from (a) aluminum ($Z = 13$) and (b) lead ($Z = 82$).

27.49 What are the wavelengths of the K_α line from (a) nickel ($Z = 28$) and (b) iron ($Z = 26$)?

27.50 What are the wavelengths of the K_α lines from (a) tungsten ($Z = 74$) and (b) molybdenum ($Z = 42$)?

27.51● What is the minimum possible voltage for an x-ray tube that produces x rays of wavelength 0.048 nm?

27.52● In an x-ray tube, electrons are accelerated through a potential of 100 kV and allowed to strike a metal target.

(a) What is the maximum energy of the x rays produced?

(b) What is the wavelength of these x rays?

27.53• The energy of the K_α line in an unknown sample is 8.07 keV. What is the element producing these x rays?

27.54• K_α x rays from a certain target are found to have a wavelength of 0.079 nm. What is the element of the target?

27.55• K_α x rays from a certain target are found to have a wavelength of 0.143 nm. What is the element of the target?

Additional Problems

27.56• A 40-W incandescent lamp operates at a temperature of 2650 K. (a) What is the wavelength of the peak in its blackbody spectrum? (b) What is the energy in eV of a photon with wavelength found in (a)?

27.57• Calculate the longest and shortest wavelengths in the Lyman series, corresponding to $n_1 = 1$ in hydrogen.

27.58• (a) What is the energy of the photon that will raise a hydrogen atom from its first excited state ($n = 2$) to the $n = 5$ excited state? (b) What is the wavelength of this photon?

27.59• What is the ratio of the wavelength emitted by the $n = 3$ to $n = 2$ transition in the hydrogen atom to the Bohr radius of the $n = 2$ orbit?

27.60• The wavelength of a 0.500-mW He-He laser is 633 nm. How many photons per second pass along the beam?

27.61• (a) What energy is required to completely remove an electron from the $n = 3$ level of hydrogen? (b) What is the electron's energy when removed? (c) What frequency of radiation will be emitted if the electron falls back to its former level?

27.62• While swinging a one-kilogram mass in a one-meter-radius circle, you want to increase its angular momentum by one quantum ($h/2\pi$). By how much must you increase the speed of the mass if you maintain the radius constant? Give your answer in units of m/s.

27.63•• At what temperature will hydrogen gas have an average kinetic energy per molecule equal to the energy required to remove the electron from the hydrogen?

27.64•• Work out in detail each step of the derivation of Bohr's results between Eq. (27.9) and Eq. (27.12).

Quantum Mechanics

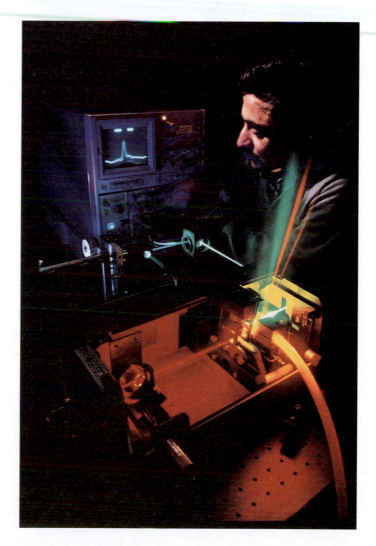

As we have seen in the preceding chapters, the search for the ultimate structure of matter led to the conclusion that matter is not infinitely divisible but is composed of tiny particles, called atoms. Rutherford demonstrated that the atom itself has component parts: a heavy, positively charged nucleus and lighter, negatively charged electrons. Moseley's experiments with x rays showed that the nuclei of atoms possess positive charge in integral multiples of the electronic charge e.

The explanations of blackbody radiation and the photoelectric effect introduced the concept of quantization of electromagnetic radiation. These ideas, in turn, led Bohr to a remarkably successful description of the hydrogen atom, which utilized the idea of energy quantization. These developments, in the early decades of the twentieth century, showed that the classical laws of physics, which worked so successfully when applied to ordinary-sized objects, were not appropriate for the description of matter at the atomic level.

1923 A. H. Compton shows that x-ray photons display particle-like properties and obey the law of conservation of momentum.

1924 Louis de Broglie proposes that all moving particles have a wavelength, a property previously associated only with waves.

1925 Wolfgang Pauli states the exclusion principle: No two electrons in the same atom can have the same four quantum numbers.

Despite the successes of the Bohr theory, it had limitations. For example, although it could predict the wavelength of the spectral lines of the hydrogen atom, the theory did not work for all atoms. Bohr's theory was one of the great milestones in physics, but needed further refinement and development. In this chapter, we trace the development of what is commonly called quantum mechanics or, sometimes, wave mechanics. We start by setting forth some of the important differences between classical physics and quantum physics.

One caution should be added. Keep in mind that our description is a model. We hope that each improved conception is a more accurate description of reality—in the present case, of the nature of the atom. But our description is not reality itself. A model or theory does not make an electron behave in a particular way. However, if a model is complete enough and accurate enough, we may use it to understand certain features of electron behavior and perhaps to predict new effects. ■

28.1 Classical and Quantum Mechanics

Before we discuss the development of quantum mechanics and present some of its key results and applications, let's try to clarify some of the important differences between quantum mechanics and classical mechanics. We will expand on these ideas again later in this chapter, but here we preview some of the ways in which quantum mechanics has changed our view of the physical world.

According to classical physics, there are no limits to how accurately we can measure physical quantities. Certainly, we have practical limitations on the precision and accuracy of our instruments. Also, the task of keeping track of some things—say, the speed and position of every molecule in a gas—is beyond doing. These are practical obstacles, however. In the Newtonian view of things we can, in principle, make all the measurements needed to predict accurately the future state of the entire universe. Even special relativity did not change this perception.

1926 Erwin Schrödinger formulates a wave equation that describes the behavior of particles on the atomic scale.

1927 Werner Heisenberg formulates his uncertainty principle, which states that we cannot measure position and momentum simultaneously with arbitrary precision.

1927 C. J. Davisson (shown) and L. H. Germer observe electron diffraction, as does G. P. Thomson in an independent experiment, confirming that particles act as waves.

Quantum mechanics changes all this. Even in principle, it is not possible to know all physical quantities simultaneously with complete accuracy. This idea has its clearest statement in the uncertainty principle (Section 28.5), which gives specific limitations to how accurately we can know certain properties of a system at the same time. This is not a limitation of our measuring apparatus; this is an inherent property of the physical world.

How, then, can we measure and calculate properties of atoms and other systems? The answer given by quantum mechanics is that we can assign probabilities for values of physical properties. We cannot say that the radius of an electron's orbit in hydrogen is exactly equal to the Bohr radius; however, we can say that the Bohr radius is the distance from the nucleus at which the electron is most likely to be found at a given time. The probability of finding an electron farther than five times the Bohr radius from the nucleus is about 1 chance in 2000. The probabilities involved are calculated using the mathematics of quantum mechanics.

Note that the use of probability in quantum mechanics differs from our use of probability in kinetic theory, where statistics describes the behavior of large numbers of particles, each behaving in accord with Newtonian dynamics. In quantum mechanics, an individual particle does not follow Newtonian dynamics, but has probabilities associated with such properties as position, momentum, and energy. However, the behavior of a large number of particles is well defined.

The interpretation of quantum mechanical results in terms of probabilities is sometimes called the "Copenhagen interpretation," referring to the home of Niels Bohr, who pioneered this view. After having developed the first quantum model of the hydrogen atom in 1913, he went on to make significant contributions to the newer quantum theory, developing new theories and insights. Bohr, more than anyone else, provided the philosophical framework for our present understanding of quantum mechanics.

Bohr also influenced the development of quantum mechanics by his *correspondence principle*. According to this principle, a more general theory, such as relativity or quantum mechanics, must give the same results as a more restricted theory, such as classical mechanics, where the latter is appropriate. For example, for velocities that are small compared with the speed of light, the special theory of relativity gives the same results as does Newtonian mechanics. Moreover, when we apply the quantization of energy and angular momentum to large orbits, say with $n = 10,000$, then an orbit corresponding to $n \pm 1$ is essentially indistinguishable from the orbit for *n*. In this case, we have the classical result that energy and angular momentum vary smoothly. The indistinct boundaries between the domains of applicability of the theories shown in Fig. 4.29 reflect this view.

An important difference between classical and quantum mechanics has to do with the role of the observer—that is, the person making the measurements involved in any experiment. It is a basic idea of quantum mechanics that making a measurement unavoidably affects the object being measured. As a result, we can no longer look at the laws of physics as furnishing a perfectly objective description of the physical universe. Instead, we must consider our role in making these observations. Of course, these observer-induced changes are too small to be noticed in our ordinary experience, so the laws that govern our surroundings seem completely deterministic to us. However, on the atomic and subatomic scales, the observation itself can change, in an unpredictable way, the result of an observation.

In classical mechanics, most of the ideas have clear intuitive meanings that are extrapolated from our own experience. In contrast, many of the ideas of quantum mechanics are not intuitively clear and do not correspond to anything in our prior experience. The introduction of probability into quantum mechanics is one of its most significant departures from classical physics. Even Albert Einstein had difficulty accepting this idea, as reflected by his statement that has been frequently paraphrased as "God does not play dice with the world." Nevertheless, quantum mechanics has had such great success in explaining observed behavior and predicting new behavior that it is universally accepted today. Understanding the laws of quantum mechanics has led to advances in many areas of technology (Fig. 28.1). The development of such things as computer memory chips, lasers, and the tunneling microscope is directly related to the quantum mechanical interpretation of matter. Similarly, knowledge of quantum mechanics has increased our understanding of phenomena ranging from the molecular basis of biology to the life cycles of stars.

Figure 28.1
The diode laser pointer is made possible by an understanding of the quantum behavior of solids. Within the diode, a layer of gallium arsenide (GaAs) only a few atoms thick acts as a quantum potential well to narrow the range of electron energies and thus increase the light emission. This behavior is beyond the explanations of classical physics.

28.2 The Compton Effect

Roentgen's discovery of x rays in 1895 opened the way for numerous advances in physics. For example, the discovery of x-ray diffraction and its explanation (Section 26.5) led to knowledge of the detailed atomic structure of crystalline solids. The study of x-ray spectra helped confirm Bohr's theory and led to a clearer understanding of nuclear charge in the elements.

In 1923, the American physicist Arthur Holly Compton (1892–1962) carried out experiments on the scattering of x rays by matter. The interpretation of these experiments had an impact on the developing quantum theory that was comparable to the impact of Einstein's theory of the photoelectric effect.

According to classical theory, as a wave such as an x ray penetrates matter, it affects every electron in the region through which it travels. The observed scattering of the x ray is due to the combined effect of all these electrons. On the other hand, in the photoelectric effect a single quantum of light—a photon—causes the ejection of a single electron. Thus, Compton reasoned that from the quantum point of view, a single "x-ray photon" should interact with a single electron. In doing so, the photon could scatter by imparting energy and momentum to the electron.

The deflection of an x ray can occur only with a change in its momentum. Consequently, by conservation of momentum, the struck electron must recoil with a momentum equal to the x ray's change in momentum. By conservation of energy, the energy of the scattered photon must equal the energy of the incident photon minus the recoil kinetic energy of the electron. Since the scattered photon necessarily has less energy than the incident photon, according to the Planck relation $E = hc/\lambda$, the scattered photon has a correspondingly increased wavelength. Compton's experiment confirmed this change in wavelength.

We can analyze this collision using conservation laws just as we analyzed elastic collisions in Chapter 8. Imagine an x ray of frequency f scattered by a stationary free electron* of mass m (Fig. 28.2a). Since the x ray moves at the speed of light, the momentum of the incident photon is

$$p_\gamma = \frac{E}{c} = \frac{hf}{c} = \frac{h}{\lambda}.$$

Similarly, the momentum of the scattered photon is

$$p'_\gamma = \frac{E'}{c} = \frac{h}{\lambda'},$$

where λ' is the wavelength of the scattered photon and its direction makes an angle θ with the incident beam.

The momentum of the electron recoiling with speed v is

$$p_e = \frac{mv}{\sqrt{1 - v^2/c^2}},$$

where we use the relativistic formula for momentum developed in Chapter 25. With Compton's assumption that we can treat the x-ray photon as a particle, then the direction of the electron's momentum is determined by conservation of momentum. Figure 28.2(b) shows the momentum diagram for the collision of Fig. 28.2(a). Conservation of momentum requires the three momentum vectors to be coplanar. Furthermore, p'_γ and p_e must equal p_γ. Thus, the three momentum vectors form a triangle as shown in Fig. 28.2(b). The statement of momentum conservation can be put into an equation by using the law of cosines (see Appendix A):

$$p_e^2 = p_\gamma^2 + p'^2_\gamma - 2p_\gamma p'_\gamma \cos\theta.$$

Expressing the momenta in terms of the preceding equations, we get

$$\left(\frac{mv}{\sqrt{1 - v^2/c^2}}\right)^2 = \frac{h^2}{\lambda^2} + \frac{h^2}{\lambda'^2} - \frac{2h^2 \cos\theta}{\lambda\lambda'} \qquad \text{(momentum)}.$$

According to the law of energy conservation, the energy of the incident photon equals the energy of the scattered photon and the recoil electron. Thus

$$\frac{hc}{\lambda} = \frac{hc}{\lambda'} + mc^2\left(\frac{1}{\sqrt{1 - v^2/c^2}} - 1\right) \qquad \text{(energy)}.$$

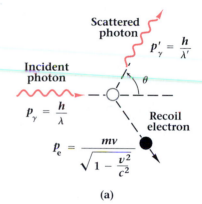

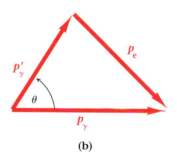

Figure 28.2
Compton scattering of an x-ray photon by an electron. (a) Diagram of the collision. (b) Momentum vector diagram for a photon of energy hf scattered by a free electron. The three momentum vectors form a triangle.

---

*The assumption of a stationary electron is reasonable because the x ray moves with a velocity c. The assumption of a free electron is valid if the recoil energy is very much greater than the energy binding the electron to the atom in the scattering material.

We now have two equations for the two unknown quantities λ' and v, similar to our earlier analysis of elastic collisions. We can obtain the scattered wavelength λ' by solving these equations simultaneously. The result is

$$\lambda' - \lambda = \frac{h}{mc}(1 - \cos\theta). \tag{28.1}$$

The wavelength of the scattered photon is greater than that of the incident photon, as expected. The scattered wavelength is angle-dependent and is greatest for scattering in the backward direction ($\theta = 180°$). The quantity h/mc is known as the **Compton wavelength** of the electron and has the value 2.426×10^{-12} m.

The Compton effect is very important because it points directly to the wave-particle duality of light. Planck's idea of quantized radiation did not fit the classical wave description, and Einstein's explanation of the photoelectric effect supported Planck's quantum hypothesis. But Compton's discovery goes even further. His analysis treats the photon as a particle, having momentum and energy, that collides with another particle, the electron. On one hand, we treat the event as a two-body collision problem; on the other hand, we quantize the photon energy in terms of its wavelength. The same x rays that are scattered like particles by the electrons in the target material are diffracted by the crystal in the spectrometer that measures their wavelength.

Example 28.1

A 17.2-keV x ray from molybdenum is Compton-scattered through an angle of 90°. What is the energy of the x ray after scattering?

Strategy We begin by finding the wavelength of the incident x ray. Then we apply the Compton scattering equation to find the wavelength of the scattered x ray. From that wavelength we can compute the energy of the scattered x ray.

Solution The wavelength of the incident x-ray photon may be calculated using the relationship between wavelength and energy found in Example 27.6:

$$\lambda = \frac{1240 \text{ eV} \cdot \text{nm}}{E} = \frac{1240 \text{ eV} \cdot \text{nm}}{(17.2 \text{ keV})(10^3 \text{ eV/keV})} = 0.0721 \text{ nm}.$$

The wavelength of the scattered x ray is found from Eq. (28.1) to be

$$\lambda' = \lambda + \frac{h}{mc}(1 - \cos\theta).$$

For $\theta = 90°$, $\cos\theta = 0$. The Compton wavelength (h/mc) has the value 0.002426 nm. When these values are all combined, we get

$$\lambda' = 0.0721 \text{ nm} + 0.0024 \text{ nm} = 0.0745 \text{ nm}.$$

Finally, we compute the energy of the scattered x ray from its wavelength:

$$E = \frac{1240 \text{ eV} \cdot \text{nm}}{0.0745 \text{ nm}} = 16{,}600 \text{ eV} = 16.6 \text{ keV}.$$

The missing energy, 0.6 keV, is carried off by the recoil electron.

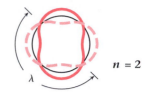

28.3 De Broglie Waves

By 1920 the dual nature of electromagnetic radiation was generally accepted, although it was not understood. As physicists gradually accumulated experimental data, they became accustomed to using a particle theory or a wave theory of light, depending on the nature of the particular experiment to be interpreted. Into this setting came Louis de Broglie, a graduate student at the University of Paris. De Broglie was strongly influenced by conversations about the photoelectric effect, which gave particle properties to waves. He began thinking about the reverse effect: that particles might have some wavelike properties.

In 1924, using arguments from both electrodynamics and relativity, de Broglie proposed that a particle of mass m traveling with speed v should have associated with it a wavelength λ, given by

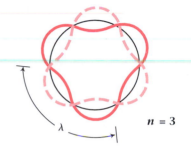

$$\lambda = \frac{h}{mv} = \frac{h}{p}, \tag{28.2}$$

where $p = mv$, the momentum of the particle. (This proposal was part of de Broglie's doctoral dissertation.) The wavelength given by Eq. (28.2) is known as the **de Broglie wavelength.** The de Broglie equation (Eq. 28.2) also gives the correct momentum for a photon of wavelength λ.

The idea of de Broglie waves provided an alternative way for deriving Bohr's condition for quantization of atomic electron orbits. If we describe the electron as a wave, its stable orbits in an atom are those that meet the conditions for a standing wave (Fig. 28.3). For a standing wave to be set up, an integral number of wavelengths must coincide exactly with the circumference of the orbit. Thus, we get $n\lambda = 2\pi r$, where n is an integer. When the de Broglie relation is used for the wavelength, we get $nh/p = 2\pi r$, which, upon rearrangement, gives the quantization condition for angular momentum:

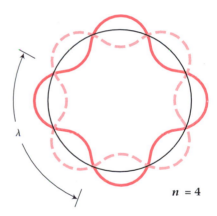

$$L = pr = n\frac{h}{2\pi}. \tag{28.3}$$

Thus, if a wavelength is associated with the electron, the quantization of atomic orbits evolves as a natural consequence of the behavior of the electron. Bohr's principal quantum number n becomes the number of integral wavelengths that fit into each orbit.

Figure 28.3
Standing waves on a circle. The circumference is shown successively equal to two, three, and four wavelengths.

Example 28.2

In a research laboratory, electrons are accelerated to speeds of 6.0×10^6 m/s. Nearby, at the same laboratory, a 1.0×10^{-9} kg speck of dust falls through the air at a speed of 0.020 m/s. Calculate the de Broglie wavelength in both cases.

Solution For the electron,

$$\lambda_e = \frac{h}{p} = \frac{h}{mv} = \frac{(6.63 \times 10^{-34} \text{ J} \cdot \text{s})}{(9.11 \times 10^{-31} \text{ kg})(6.0 \times 10^6 \text{ m/s})}$$

$$\lambda_e = 1.2 \times 10^{-10} \text{ m}.$$

For the dust speck,

$$\lambda_{\text{dust}} = \frac{h}{p} = \frac{h}{mv} = \frac{(6.63 \times 10^{-34} \text{ J} \cdot \text{s})}{(1.0 \times 10^{-9} \text{ kg})(0.020 \text{ m/s})}$$

$$\lambda_{\text{dust}} = 3.3 \times 10^{-23} \text{ m}.$$

Discussion The de Broglie wavelength of the electron is the same order of magnitude as the distance between atomic planes in a crystal. From our knowledge of diffraction and interference, we expect to be able to detect the wavelike behavior of such electrons by their diffraction from crystals. The momentum of the speck of dust is very small for ordinary matter, but is enormously larger than the momentum of the electron just described. Consequently, the de Broglie wavelength of the dust speck is many orders of magnitude smaller than that of the electron, even smaller than nuclear dimensions (10^{-15} m). It is so small that we do not observe its wavelike behavior.

Example 28.3

Find an expression for the de Broglie wavelength, in nanometers, of an electron when the kinetic energy of the electron is given in terms of its accelerating potential difference V.

Solution If the electron is moving slowly enough, we can use the classical expression for kinetic energy,

$$\text{KE} = \frac{1}{2}mv^2 = \frac{p^2}{2m}.$$

Then the wavelength becomes

$$\lambda = \frac{h}{p} = \frac{h}{\sqrt{2m(\text{KE})}}.$$

If an electron is accelerated from rest through an electric potential difference V, its kinetic energy is $\text{KE} = eV$ and its wavelength becomes

$$\lambda = \frac{h}{\sqrt{2meV}}.$$

Inserting numerical values for h, m, and e and measuring V in volts, we get

$$\lambda = \sqrt{\frac{1.50}{V}} \text{ nm}.$$

Discussion This equation we have just derived is a very handy formula for computing electron wavelengths at low energies—that is, at low voltages. At energies above a few kiloelectronvolts, the electron velocity becomes so great that its momentum must be calculated relativistically. As a result, the wavelength is reduced from the value predicted classically. For example, at 50,000 eV the relativistic correction is about 2.5%.

Three years after de Broglie asserted that particles of matter could possess wavelike properties, the diffraction of electrons from the surface of a solid crystal was experimentally observed by Clinton J. Davisson and Lester H. Germer of the Bell Telephone Laboratory. In 1927, they reported their investigation of the angular distribution of electrons scattered from nickel. With careful analysis, they showed that the electron beam was scattered by the surface atoms of the nickel in the exact manner predicted for the diffraction of waves according to Bragg's formula, with a wavelength given by the de Broglie equation. A striking characteristic of diffracted electron beams is their similarity to the Laue pattern resulting from the same crystal when the incident beam is a beam of x rays (Fig. 28.4).

Also in 1927, G. P. Thomson, the son of J. J. Thomson, reported his experiments in which a beam of energetic electrons was diffracted by a thin foil. Thomson found patterns that resembled the x-ray patterns made with powdered (polycrystalline) samples. This kind of diffraction, by many randomly oriented crystalline grains, produces rings (Fig. 28.5). If the wavelength of the electrons is changed by changing their incident energy, the diameters of the diffraction rings change proportionally, as expected from Bragg's equation. These experiments by Davisson and Germer and by Thomson proved that de Broglie waves are not simply mathematical conveniences, but have observable physical effects. Just as Compton showed that waves could act as particles, Davisson and Germer showed that particles could act as waves.

The concept of particles as waves and waves as particles is incompatible with the ideas of classical physics. However, this wave-particle duality, as it is sometimes called, is an integral part of quantum mechanics. To help explain this seeming contradiction, Bohr postulated the *principle of complementarity*. It states that it is neither possible nor necessary to choose between the wave and particle descriptions, but that both are essential for a complete description of nature.

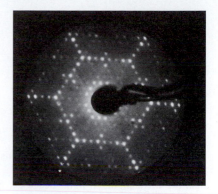

Figure 28.4
A diffraction pattern of low-energy electrons (approximately 48 eV) from the surface of silicon.

Figure 28.5
Transmission electron diffraction patterns produced from a gold foil. Ring diameters decrease with increasing beam voltage: (a) 60 kV, (b) 80 kV, (c) 100 kV.

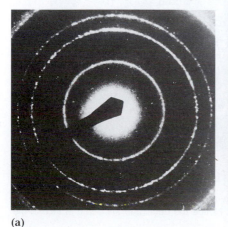

(a)

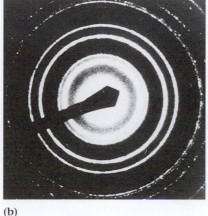

(b)

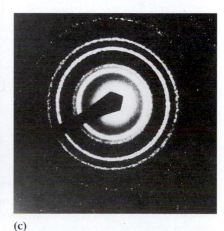

(c)

ELECTRON MICROSCOPES

The first electron microscope was built only a few years after the discovery that high-speed electrons had wavelengths many times smaller than the wavelengths of light. Because of the smaller wavelengths, electron microscopes are capable of much higher resolution than optical microscopes. The beams of electrons can be focused by suitably shaped electric or magnetic fields. For example, the magnetic field of a solenoid acts like a converging lens for electrons (Fig. B28.1). In transmission electron microscopes (TEM), an energetic beam of electrons focused by a magnetic condenser lens illuminates the specimen being examined and passes through to a magnetic objective lens, which forms an intermediate image (Fig. B28.2). A magnetic projector lens then magnifies a portion of the intermediate image to form the final image on a fluorescent screen, a photographic plate, or a semiconductor

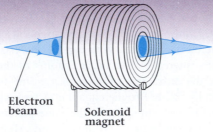

Figure B28.1 The field of a solenoid magnet acts as a converging lens to focus a beam of electrons.

Electron beam

Solenoid magnet

dectector connected to a computer. Just as in the case of an optical microscope, the overall magnification is the product of the magnifications of the objective and projector lenses. With the proper choice of magnetic lens currents, the overall magnification can be as high as 200,000×. The critical limitation is the resolution. Modern instruments are capable of resolving objects smaller than 0.5 nm.

Scanning electron microscopes (SEM), which came into wide usage in the 1970s and 1980s, operate on a different principle from the TEM. In the scanning electron microscope, a finely focused electron beam scans across the surface of the specimen being examined. As the beam scans across the specimen, the incident electrons (primaries) knock out other electrons (secondaries) that come from the area on the specimen where the primary beam is focused. These secondary electrons are collected at a positive electrode. The intensity of the secondary current changes as the primary beam sweeps across the specimen, since more secondaries are generated when the beam strikes a sharply curved edge or sloping surface than when it strikes a flat surface. The intensity information is used to generate a televisionlike image on a cathode ray tube, which gives an impression of three-dimensional surface relief (Fig. B28.3). The size of the beam (typically less than 10 nm) limits the resolution of the instrument. In both types of electron microscopes, the extremely small de Broglie wavelengths of high-speed electrons permit imaging with high resolution.

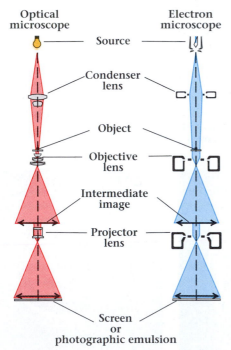

Figure B28.2 Comparison of optical and electron microscopes. The components of the microscopes are arranged side by side to show the similarity of the two systems.

Optical microscope — Electron microscope

Source

Condenser lens

Object

Objective lens

Intermediate image

Projector lens

Screen or photographic emulsion

Figure B28.3 A close-up view of a midge made with a scanning electron microscope. Notice the detail and the depth of field in the image.

28.4 Schrödinger's Equation

The discovery of the wavelike behavior of electrons created the need for a wave theory describing the behavior of matter on the atomic scale. Such a theory was proposed by the Austrian physicist Erwin Schrödinger (1887–1961) only two years after de Broglie formulated the idea of particle waves.

Schrödinger reasoned that if an electron behaves as a wave, then it should be possible to mathematically describe the electron's behavior in space and time as a wave. In 1926, from his knowledge of the general theory of wave phenomena and of de Broglie's work, Schrödinger deduced a fundamental equation that now bears his name. We can write the **Schrödinger equation** in the form

$$\frac{\Delta(\Delta\psi/\Delta x)}{\Delta x} + \frac{8\pi^2 m}{h^2}(E - V)\psi = 0, \qquad (28.4)$$

where E is the total energy of the particle, V is its potential energy, m is its mass, and h is Planck's constant. The symbol ψ (the Greek letter psi) is called the **wave function.** The Schrödinger equation as written here is for only one dimension; the complete equation is written to include three dimensions.

What does ψ, the symbol for the wave function, represent? Its mathematical role in the Schrödinger equation is analogous to the description given by classical mechanics of a wave, such as a transverse wave propagating along a string (Chapter 15). What do we mean by the amplitude of the wave function that arises in Schrödinger's equation?

We cannot measure a property of an electron that directly corresponds to the "amplitude" of the wave in Schrödinger's equation. Nonetheless, we can use Schrödinger's equation to calculate wave functions and predict the results of experiments. We will show several examples of such calculations later in this chapter. We will also discuss the significance of the square of the wave function, which relates not to a physical property of the particle, but to the probability of a given behavior. However, the wave function itself is not a physically observable quantity. From a classical point of view, this is a somewhat unsatisfactory situation. In quantum mechanics, we calculate wave functions in various circumstances (for example, multielectron atoms, liquids, gases, etc.), and then, from the results, we predict the properties and behavior of the corresponding particles. Quantum mechanics allows us to obtain results and predictions that our experimental observations confirm.

With his equation, Schrödinger obtained the wavelength of a free electron, in agreement with de Broglie's formula. For a free electron, the potential $V = 0$ in Eq. (28.4). The Schrödinger equation then predicts that the electron wave can take on any wavelength $\lambda = h/p$ associated with any energy $E = p^2/2m$, in the limit of nonrelativistic velocities. Since classical physics allows a free particle to have any energy whatsoever, all energies are allowed for free particles and the results of the Schrödinger equation are identical with those of classical physics. But, as we shall see on the following page, differences do occur between the two theories when the particle is no longer free.

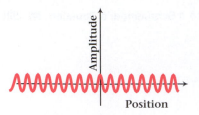

Figure 28.6

A harmonic wave of wavelength λ_0 extending infinitely far in both the $+x$ and $-x$ directions.

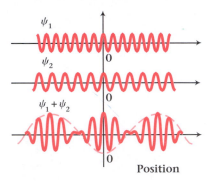

Figure 28.7

Two harmonic waves ψ_1 and ψ_2, of wavelengths λ_1 and λ_2, interfere to produce a new wave $\psi_1 + \psi_2$.

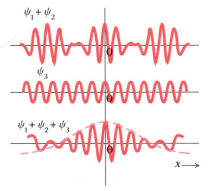

Figure 28.8

The addition of wave ψ_3 to $\psi_1 + \psi_2$ gives a new wave.

28.5 The Uncertainty Principle

We can write the equation for a traveling harmonic wave, such as we described in Chapter 15, as

$$\psi(x,t) = \psi_0 \sin 2\pi\left(\frac{x}{\lambda} - \frac{t}{T}\right). \tag{28.5}$$

Such a wave extends infinitely far in both positive and negative directions (Fig. 28.6). Here ψ_0 is the amplitude of the wave and T is its period. The wave is characterized by a single wavelength λ. However, if we expect a wave to represent an electron or other particle, then it seems unreasonable to have the wave extending to infinity. Instead, the wave should be localized in the vicinity of the electron. Indeed, we see the need for localizing a wave to a small region of space even in the case of light waves. For example, in the photoelectric effect and in Compton scattering, the photon interacts with a single electron, not with all of them.

Fortunately, there is a relatively simple answer to our dilemma. Any complicated wave can be built up from a superposition of harmonic waves of different wavelengths. For example, if we add together two waves ψ_1 and ψ_2 of slightly different wavelengths, a new wave results (Fig. 28.7). Notice that this new wave does not have an appreciable amplitude at all positions, but only over certain repetitive ranges of x. If we add a third wave ψ_3, the waves combine with even greater amplitude in some regions and less in others (Fig. 28.8). By combining a large number of waves of different wavelengths, we can build up a localized wave disturbance like the one shown in Fig. 28.9. We call this kind of resultant wave a **wave packet.** It has an appreciable amplitude only in a small region of space Δx, which corresponds to the location of the particle it represents. Because its component waves are traveling waves, the wave packet moves with time. We identify the velocity of the wave packet with the velocity of the particle.

To construct a wave packet like that of Fig. 28.9, the resultant wave amplitude must be vanishingly small everywhere outside of the small region Δx. An infinite number of pure harmonic waves of different frequencies are needed to completely eliminate the resultant amplitude beyond Δx. However, not all of these pure harmonic waves have large amplitudes. Most of the waves with appreciable amplitude lie in a wavelength range $\Delta \lambda$ about the average wavelength λ_0.

The spread in wavelength can be associated with the spread in momenta through de Broglie's equation. The properties of the wave packet can be combined with the de Broglie equation to find an important relationship between position and momentum. This result is

$$\Delta x \Delta p \geq \frac{h}{2\pi}. \tag{28.6}$$

This inequality is the mathematical statement of the **uncertainty principle,** which was first advanced in 1927 by Werner Heisenberg (1901–1976). The uncertainty principle specifies the limits within which we can apply the particle model to atomic (or even macroscopic) objects. The quantity Δx represents the uncertainty in the position of the particle. In other words, we know the particle is confined within a region Δx. Similarly, Δp represents the uncertainty in the momentum of the particle.

The uncertainty principle limits how well we can know simultaneous values of the position and linear momentum of a particle or wave. It means that *we cannot measure both the position and the related momentum simultaneously with arbitrarily great precision.* For example, suppose that we know the momentum of a certain free electron precisely, but do not know its position. Then, according to the uncertainty principle, any attempt to measure its position will alter its momentum. Consequently, after any experiment to determine its position, our knowledge of its momentum is limited by Eq. (28.6).

In more concrete terms, consider the act of measuring the position of an electron. You might imagine trying to see the electron through a microscope, which would involve a photon of light coming from a source, bouncing off the electron, and then traveling through the microscope to your eye (Fig. 28.10). The photon imparts momentum $p \approx h/\lambda$ to the electron, and so you cannot locate the electron without changing its momentum. Now, how accurately can a photon measure the position of the electron? Only as accurately as the wavelength of the photon, as we saw back in Chapter 24. If we use a photon of shorter wavelength, like an x-ray photon, we will know the position more precisely, but the more energetic photon will impart more momentum to the electron. Consequently, the product of Δx and Δp always remains of the order of Planck's constant.

When we consider motion in three dimensions, three uncertainty relations are required, one for each of the three perpendicular directions. However, because of the independent nature of motion in three directions, there is no uncertainty relation involving other products of these quantities. For example, a measurement of the x position of an object does not limit our knowledge of the object's y position or its momentum in the y direction.

There is another version of the uncertainty principle that relates energy and time. Suppose we try to measure the frequency of a wave (perhaps a photon) by counting the number of cycles that occur in a time interval Δt. The longer we make Δt, the better our measurement of the frequency. Since we have a minimum uncertainty of one count per time interval, we have a spread, or uncertainty, in frequency of $\Delta f \geq 1/\Delta t$. An uncertainty in frequency implies a corresponding uncertainty in energy, since according to the Planck relation the energy is proportional to the frequency. Consequently, we find an uncertainty relation between time and energy:

$$\Delta E \Delta t \geq \frac{h}{2\pi}.$$

(28.7)

There is an uncertainty in the energy of a photon (or any particle or wave) that is related to the time required to define the energy.

The uncertainty principle helped overthrow the ideas of strict determinism in physics. Regardless of our measuring instruments, we cannot predict the motion of an electron beyond an unavoidable uncertainty. According to quantum mechanics, this is the way nature is.

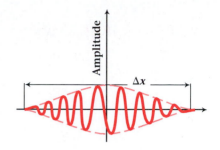

Figure 28.9
A wave packet of average wavelength λ_0 and spatial extent Δx.

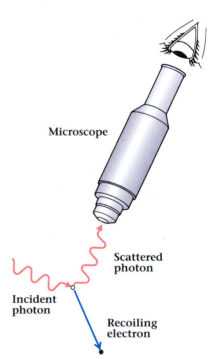

Figure 28.10
A thought experiment for viewing an electron with a microscope. The photon that carries the information of the position of the electron imparts momentum to it, so you cannot locate the electron without disturbing its momentum.

Master the Concept

Heisenberg Uncertainty Principle

Question: Does the uncertainty principle play any noticeable role in large-scale phenomena, such as driving a fast-moving car?

Answer: The uncertainty principle says that we cannot measure exactly both position and momentum of a car at the same time. A limit on the measurement of momentum implies a limit on the measurement of speed. We can express Heisenberg's relationship as the uncertainty in speed times the uncertainty in position. For an automobile of 1,500-kg mass, that product of uncertainties is greater than 4.4×10^{-37} m · m/s. Thus, if we were to determine the car's position to within the size of an atom (about 10^{-10} m), the minimum uncertainty in its measured speed would be 4.4×10^{-27} m/s. This level of uncertainty is far smaller than the uncertainties we would encounter in measuring the speed with the best of instruments. Therefore, in our day-to-day activities we do not directly observe the effects of the Heisenberg uncertainty principle.

Example 28.4

In an experiment, an electron is determined to be within 0.10 mm of a particular point. If we try to measure the electron's velocity, what will be the minimum uncertainty?

Solution Using the Heisenberg relation, we get

$$\Delta v_x = \frac{\Delta p_x}{m} \geq \frac{h/2\pi}{m\Delta x}$$

$$\Delta v_x \geq \frac{6.63 \times 10^{-34} \text{ J} \cdot \text{s}}{(9.11 \times 10^{-31} \text{ kg})(1.0 \times 10^{-4} \text{ m})(2\pi)}$$

$$\Delta v_x \geq 1.2 \text{ m/s}.$$

Discussion The uncertainty in the electron's velocity is 1.2 m/s. Thus, we can predict its velocity only to within 1.2 m/s. Locating the electron at one position affects our ability to know where it will be at later times.

Example 28.5

A grain of sand with a mass of 1.00×10^{-3} g appears to be at rest on a smooth surface. We locate its position to within 0.010 mm. What velocity limit is implied by our measurement of its position?

Solution As before,

$$\Delta v_x = \frac{\Delta p_x}{m} \geq \frac{h/2\pi}{m\Delta x}$$

$$\Delta v_x \geq \frac{6.63 \times 10^{-34} \text{ J} \cdot \text{s}}{(1.0 \times 10^{-6} \text{ kg})(1.0 \times 10^{-5} \text{ m})(2\pi)}$$

$$\Delta v_x \geq 1.1 \times 10^{-23} \text{ m/s}.$$

Discussion The uncertainty in momentum leads to an uncertainty in velocity of 1.1×10^{-23} m/s. This velocity is so small that we do not observe it. The grain of sand may still be considered at rest, as our experience says it should.

Example 28.6

Calculate the approximate energy needed to confine an electron to within a space of the order of nuclear dimensions, about 1.0×10^{-14} m.

Strategy We can estimate the momentum of the electron from the uncertainty relationship, $\Delta p \Delta x \geq h/2\pi$. Then we can use the relativistic expression for total energy to find the minimum energy needed to confine the electron.

Solution If the electron is confined to within $\Delta x = 1.0 \times 10^{-14}$ m, then the uncertainty in momentum is

$$\Delta p \geq \frac{h/2\pi}{\Delta x} = \frac{6.63 \times 10^{-34} \text{ J} \cdot \text{s}}{(1.0 \times 10^{-14} \text{ m})(2\pi)} = 1.06 \times 10^{-20} \text{ J} \cdot \text{s/m}.$$

The average momentum must be at least as great as the uncertainty in momentum Δp. The minimum energy is obtained from the relativistic expression for the total energy that we had in Chapter 25 when we insert the value just obtained for p,

$$E^2 = c^2 p^2 + m_0^2 c^4$$

$$E^2 = (3.00 \times 10^8 \text{ m/s})^2 (1.06 \times 10^{-20} \text{ J} \cdot \text{s/m})^2$$
$$+ (9.11 \times 10^{-31} \text{ kg})^2 (3.00 \times 10^8 \text{ m/s})^4.$$

The energy is

$$E = 3.2 \times 10^{-12} \text{ J} \approx 20 \text{ MeV}.$$

Discussion The energy of an electron confined within a nucleus would be of the order of 20 MeV. However, the observed energies of the electrons ejected from the nucleus in beta decay are of the order of 2 or 3 MeV. Because of this discrepancy, we believe that the nucleus is not composed of and does not contain electrons, even though electrons are emitted during nuclear beta decay. The resolution to this seeming paradox will be discussed in Chapter 29.

28.6 Interpretation of the Wave Function

The Schrödinger wave theory of 1927 was very successful for calculating quantum behavior; we shall examine some of these successes soon. However, at first there was no physical interpretation of the wave function ψ. Initially, Schrödinger tried to associate the wave amplitude with a distribution of electric charge in space, but this idea was soon found unacceptable. Shortly after the publication of Schrödinger's work, the German physicist Max Born (1882–1970) began to apply quantum mechanics to the analysis of collisions between particles. As an outgrowth of this work, Born proposed a surprising interpretation of the physical significance of the wave function.

Born recognized that the quantity ψ could not be measured directly. Instead, he concluded that the square of the absolute value of the wave function $|\psi|^2$ is physically meaningful. Born interpreted this quantity as the **probability density** for the electron (or other particle). That is, the probability, or likelihood, of finding an electron in a particular small volume of space ΔV is given by

$$P = |\psi|^2 \, \Delta V. \tag{28.8}$$

Thus, the particle is most likely to be found where the wave function is largest and is not likely to be found where the wave function is small.

Furthermore, Born pointed out that in collision problems, for example, quantum mechanics does not describe precisely the state of the particles following the collision, as classical mechanics does. Instead, *the quantum theory gives the probabilities of various possible outcomes.* In other words, *quantum mechanics is a statistical theory.* It does not precisely describe a single event, but rather it describes the distribution of outcomes of a large number of identical events. The wave function carries statistical information about the outcome of any particular experimental measurement designed to provide information about the particle.

Suppose, for example, we imagine a beam of electrons traveling along the positive x direction (Fig. 28.11). If the beam passes through a narrow slit whose length is perpendicular to the y direction, we expect the beam to spread out in the y direction by diffraction. (This discussion is valid for photons as well as electrons.) If the diffracted electron beam strikes a fluorescent screen, we expect an intensity pattern of fluorescent light like that shown in Fig. 28.11.

Figure 28.11

Diffraction pattern observed on a fluorescent screen for an electron beam diffracted by a narrow slit.

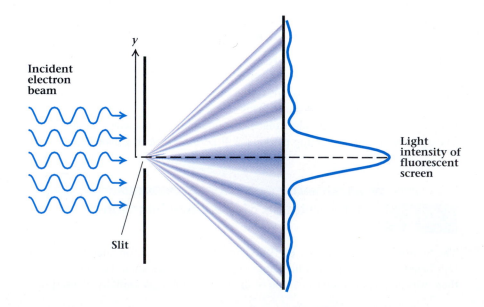

Now suppose that we limited the beam to such a small number of electrons that they passed through the slit one at a time. We could then observe scintillations on the screen due to individual electrons. We would no longer see an interference pattern such as that of Fig. 28.11. Instead, we would see only the tiny flashes of light indicating where each electron hit the screen (Fig. 28.12a). We would be helpless to predict where successive electrons would strike the screen. However, we could calculate the *probability* that an electron would strike any particular position on the screen. If we then waited long enough, until a large number of electrons had passed through the slit and struck the screen (Fig. 28.12b,c,d), the distribution of the observed individual impacts would take on the shape of the theoretical probability curve for electrons striking the screen (Fig. 28.12e). This curve is identical to the diffraction pattern calculated from simple wave theory.

Here again the wave-particle duality appears. The electrons (or photons) are diffracted as waves by the slit. They travel to the screen (or other detector), where they are recorded as individual, particlelike events. Our theory cannot predict exactly where any particular electron will strike, but it can describe the probability of the electron's striking the screen at various points. The sum of a large number of events conforms closely with the probabilistic prediction given by the theory.

We can extend our discussion of single-slit diffraction to include Young's double-slit experiment. As in the case of the single slit, the same general diffraction pattern results from either photons (light) or electrons. You are already familiar with the details of the double-slit experiment with light (Chapter 24). Now let us consider the quantum mechanical interpretation of that experiment. In accordance with Bohr's correspondence principle, quantum mechanics predicts that for high-intensity light (that is, large numbers of photons), the pattern will be the same as predicted by the wave theory (Fig. 28.13a). Just as in the case of the single slit, we cannot calculate where a given photon will strike the screen, but we can calculate the probability that it will strike a particular place on the screen. The resulting probability distribution has the same pattern as that given by the classical theory. These results have been confirmed by experiments in which the light intensity was so low that only one photon at a time passed through the slits.

An interesting point arises when we ask what happens, not at the screen, but at the slits themselves. Considering incident photons one at a time, we may cover

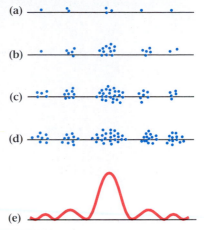

Figure 28.12
A diffraction pattern can be formed from a succession of single-particle events: (a) few events; (b), (c), (d), more events. (e) Theoretical pattern for the sum of a large number of single events, matching the diffraction pattern of Fig. 28.11.

Figure 28.13
(a) Quantum mechanics predicts the intensity distribution for light diffracted by a pair of narrow slits. (b) If one of the slits is covered, the resulting pattern is the usual single-slit pattern.

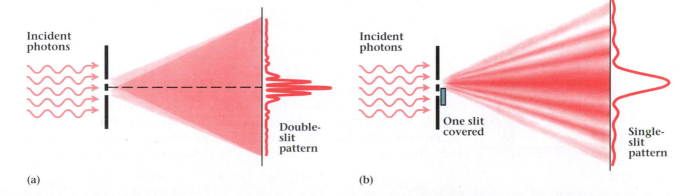

either of the slits and, after a sufficiently long time, observe the buildup of a single-slit pattern (Fig. 28.13b). However, without changing anything else, if we uncover both slits, the pattern built up after a long time will be the usual double-slit one, easily distinguished from that of a single slit. The probability that a photon will arrive at a given place on the screen depends on whether one slit or two slits are open.

If we regard the photons strictly as particles, then individually they must go through one slit or the other, not both. If they go through only one slit, how do they "know" if the other slit is open or closed, so that they generate the appropriate pattern?

The answer to this seeming paradox is not in the description of the experiment, but in our attempt to model the photon as being either a particle or a wave. The solution frequently offered, that the photon (or electron) interferes with itself, is difficult to understand. One way to resolve the difficulty is to say that the wave function ψ must include information about the apparatus as well as about the particle itself. The wave functions for the cases of one slit only and of both slits are different because the apparatus is physically different. With this view of the experiment, there should be no surprise that the diffraction patterns satisfy the correspondence principle.

This experiment points out one of the underlying problems with quantum mechanics. The mathematical rules for predicting results are clear, but interpreting what happens in terms of a physical model is sometimes difficult.

28.7 The Particle in a Box

One of the natural consequences of wave mechanics is the quantization of energy. For example, the stationary electronic states of atoms are intimately related to the wavelike nature of electrons. To illustrate how quantum mechanics reveals this relationship, we examine the case of a particle bound within a one-dimensional box. For simplicity, we assume that the particle is confined between rigid, unyielding walls separated by a distance L (Fig. 28.14). Beyond this allowed region, the potential energy V is enormously large and the wave function must therefore be zero. Within the box, the potential V may be taken as zero and the particle has a wave function ψ and energy E. We often call this kind of situation a potential well, since it resembles a well sunk into the ground. Although this is a somewhat artificial situation, it serves as a good example of a simple quantum mechanical system. With it we obtain quantization of energy.

Within the box, the Schrödinger equation for the particle becomes

$$\frac{\Delta(\Delta\psi/\Delta x)}{\Delta x} + \frac{8\pi^2 m}{h^2} E\psi = 0. \tag{28.9}$$

The condition of the rigid boundaries demands that the wave function should vanish for $x = 0$ and for $x = L$ because the probability density $|\psi|^2$ is zero outside the box and, hence, is also zero right at the boundaries. These conditions are the same as the boundary conditions for a wave along a string fixed at both ends. Wave solutions that are consistent with these boundary conditions are

$$\psi_n(x) = \begin{cases} \psi_0 \sin(n\pi x/L), & n = 1, 2, 3, \ldots & 0 < x < L \\ 0 & \text{everywhere else.} \end{cases} \tag{28.10}$$

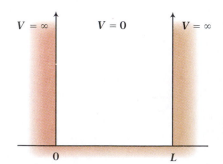

$V = \infty$ $V = 0$ $V = \infty$

0 L

Figure 28.14
A one-dimensional potential well.

The integer n is the quantum number that describes the state of the system. The wave functions $\psi_n(x)$ are zero at the boundaries $x = 0$ and $x = L$ as required. The first four standing wave solutions are shown in Fig. 28.15. We can now use these wave functions in the Schrödinger equation, Eq. (28.9), to find the energy associated with each quantum state (or wave function). We may conclude from our discussion of harmonic motion in Chapter 15 that the rate of change of a cosine function is a negative sine function and that the rate of change of a sine function is a cosine function. Using this information, we can calculate the first term in Eq. (28.9) to be

$$\frac{\Delta(\Delta \psi_n / \Delta x)}{\Delta x} = -\left(\frac{n\pi}{L}\right)^2 \psi_n.$$

Equating the coefficients of ψ_n and solving for E_n, we find a value for the energy of

$$E_n = \frac{h^2}{8mL^2}n^2, \quad n = 1, 2, 3, \ldots. \tag{28.11}$$

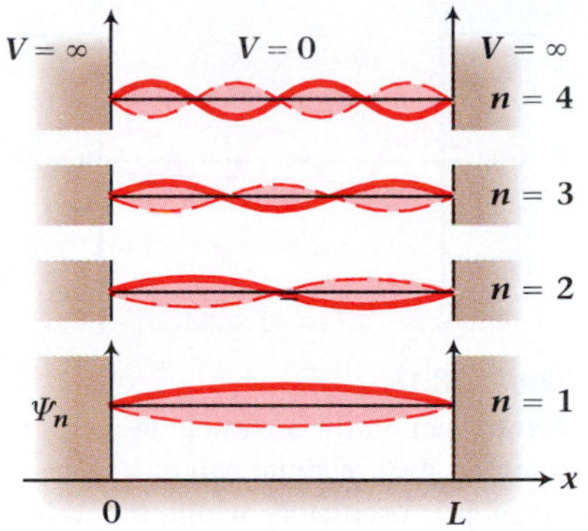

The energy levels are quantized according to the square of the quantum number n. Figure 28.16 shows the first four levels. Particles in a one-dimensional potential well can exist only in states of definite energy E_n, described by the wave functions ψ_n. These states are labeled by the quantum number n. All other energy values are excluded from consideration because the wave functions corresponding to such energies do not satisfy the boundary conditions and consequently are forbidden.

The energy E_n given above is entirely kinetic energy because the potential energy is zero. An important consequence of the wave theory is that zero kinetic energy is not allowed. If we choose $n = 0$, so that $E = 0$, we find that the wave function vanishes everywhere, since $\sin 0 = 0$. Thus, there is no wave function corresponding to zero energy. The lowest energy corresponds to $n = 1$ and is not zero. This result is typical for quantum systems in that, even for a temperature of absolute zero, where classical motions cease, a quantum system still possesses a small amount of kinetic energy. This residual energy is called the **zero-point energy.**

The zero-point energy is observed in solid-state, atomic, and nuclear physics. It is especially apparent in the low-temperature behavior of helium. Classical thermodynamics predicts that the kinetic energy of molecules goes to zero as the absolute temperature goes to zero. This prediction, however, does not take into account the quantum mechanical prediction of zero-point energy. For the case of helium, the zero-point energy is so great that at atmospheric pressure helium does not freeze even when cooled to below 1 K. Instead, helium remains a liquid at temperatures far below the freezing points of all other substances.

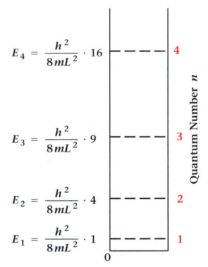

Figure 28.15

The first four standing wave solutions describing the wave function of a particle in a one-dimensional square potential well.

Figure 28.16

Left: the first four energy levels of a one-dimensional potential well of width L; right: the corresponding quantum numbers.

$$E_4 = \frac{h^2}{8mL^2} \cdot 16$$

$$E_3 = \frac{h^2}{8mL^2} \cdot 9$$

$$E_2 = \frac{h^2}{8mL^2} \cdot 4$$

$$E_1 = \frac{h^2}{8mL^2} \cdot 1$$

28.8 Tunneling or Barrier Penetration

A startling prediction of quantum mechanics is the tunneling of particles through a potential-energy barrier. This behavior is unknown in classical physics, where particles cannot penetrate such barriers and can cross them only if the particles have sufficient kinetic energy. The quantum mechanical case is analogous to a football player running headlong into a solid brick wall and passing right through

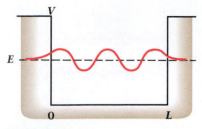

Figure 28.17
Wave function for a particle in a potential well of finite height V. The function is oscillatory within the well, but decreases exponentially outside the well.

to the other side without making a hole in the wall or going over the top. Nevertheless, tunneling is a widespread occurrence throughout the atomic and subatomic world.

In our preceding discussion of the particle in a box, we demanded that the wave function vanish at the boundary. But suppose we relax that requirement, realizing that in actual situations the potential outside is not truly infinite. Then the wave function need not vanish exactly at the boundary. We need only insist that the wave function inside and outside the box merge smoothly at the boundary (Fig. 28.17).

In the region where the potential V exceeds the energy E, the Schrödinger equation becomes

$$\frac{\Delta(\Delta\psi/\Delta x)}{\Delta x} = \frac{8\pi^2 m}{h^2}(V - E)\psi,$$

and the term on the right-hand side is positive. Wave solutions are no longer possible. However, the equation does have exponential solutions of the form

$$\psi(x) = \psi_0 e^{-\alpha x}, \tag{28.12}$$

where α is a constant and x is measured from the boundary into the region of high potential. In the allowed region $0 < x < L$, the particle is still described by an oscillatory function of sines and cosines. In the region beyond the boundary, the wave function for the particle decreases exponentially with distance into the barrier (Fig. 28.17). There is thus a finite though small probability of finding the particle outside the allowed region. This probability diminishes with increasing potential V or increasing distance x into the barrier.

Now suppose the barrier has a finite thickness (Fig. 28.18). Then the amplitude of the wave function does not decrease completely to zero across the width of the barrier. Instead, a wave solution exists on the opposite side of the barrier, but with a reduced amplitude. The probability that the electron (or other particle) will appear on the other side of the barrier is nonzero. This effect is known as **barrier penetration** or **tunneling.**

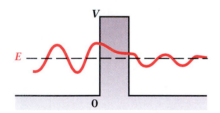

Figure 28.18
Potential energy and wave functions for a finite potential barrier. The presence of ψ on the right-hand side indicates a reduced but finite probability of finding the particle on that side of the barrier.

Tunneling of particles is not predicted in classical mechanics. A particle confined in a gravitational potential cannot escape unless its total energy E exceeds the potential-energy barrier V. Quantum mechanically, a particle such as an electron can tunnel out of a finite potential well even though its total energy E is less than that of the potential barrier. Tunneling effects are important in the radioactive decay of nuclei, in semiconductors, and in superconductors.

*28.9 Wave Theory of the Hydrogen Atom

If we extend our simple example of the particle moving in a rigid box to two dimensions, we find that the wave function must satisfy boundary requirements in both dimensions simultaneously. However, since the two motions are perpendicular, they are independent. Mathematically, this means the solutions to the Schrödinger equation must contain two independent quantum numbers. If we choose the two dimensions of the box as x and y, then the wave must satisfy the condition for standing waves in each direction.

For motion in three dimensions, we need three quantum numbers to specify the complete wave function. Thus, we could describe a particle in a three-dimensional box in terms of the energy states corresponding to various values of the three quantum numbers n_x, n_y, and n_z.

When we apply wave mechanics to the description of a real three-dimensional atomic system, such as the hydrogen atom, it is more natural and more convenient to use a system of spherical coordinates. We still have three quantum numbers, although they are no longer n_x, n_y, and n_z. We shall not attempt here to solve the Schrödinger equation for the hydrogen atom because of the mathematical complexity. However, we will discuss the three quantum numbers and point out some results from the quantum mechanical treatment of the hydrogen atom. As we will see in the rest of this chapter, most of these results, with some modification and extension, underlie the quantum mechanical description of all other atoms.

The electron wave functions that are solutions of the Schrödinger equation for the hydrogen atom are specified by three quantum numbers, one for each of the three coordinates. The first of these is the **principal quantum number** n, which is associated with the radial part of the wave function. This quantum number plays much the same role as the quantum number introduced by Bohr. The energy levels of the atom are roughly arranged into levels, or shells, associated with the principal quantum number n, which takes on integer values 1, 2, 3, 4, ... just as in the Bohr theory. Frequently, we designate each shell by a capital letter:

Principal quantum number n: 1   2   3   4 ...

Shell letter designation: K   L   M   N ...

The second quantum number is the azimuthal or **orbital quantum number** l, which characterizes the orbital angular momentum. This quantum number can take on integer values of 0, 1, 2, ... up to $n - 1$. Each of the principal atomic shells is composed of subshells, characterized by their orbital quantum number. These subshells are labeled with lowercase letters as follows:

Orbital quantum number l: 0   1   2   3   4   5 ...

Subshell letter designation: s   p   d   f   g   h ...

Finally, there is the **magnetic quantum number** m_l, which describes the magnitude of the component of the angular momentum along a particular spatial direction. This quantum number can be any integer value between $-l$ and $+l$, including 0. Each subshell consists of several states, each of which can accommodate two electrons.

To summarize, the energy levels of the hydrogen atom are characterized by three quantum numbers, which can assume integer values according to the following scheme:

$$n = 1, 2, 3, 4, \ldots$$
$$l = 0, 1, 2, 3, \ldots, n - 1 \qquad (28.13)$$
$$m_l = 0, \pm 1, \pm 2, \pm 3, \ldots, \pm l$$

Table 28.1 illustrates the labeling of the first few levels.

When no external fields are present, the energy associated with each wave function depends only on n and is given by

$$E_n = -\frac{me^4}{8\epsilon_0^2 h^2 n^2}.$$

Table 28.1	Designations of the Lowest Atomic Energy Levels		
Principal Quantum Number n	Shell Designation	Orbital Quantum Number l	Subshell Designation
1	K	0	$1s$
2	L	0	$2s$
		1	$2p$
3	M	0	$3s$
		1	$3p$
		2	$3d$

This result is exactly the same as the result obtained from the Bohr theory (Eq. 27.11).

When we compute the radial probability densities from the wave functions, using $|\psi|^2$, they have the forms shown graphically in Fig. 28.19. The probability density for the lowest-energy wave function, labeled by $n = 1$ and $l = 0$, has a maximum at $r = 0.53 \times 10^{-10}$ m, which is the same radial distance from the nucleus as the first Bohr orbit. Similarly, other wave functions for n and $l = n - 1$ have a probability maximum corresponding to the radius of the appropriate Bohr orbit.

When the atom is in a definite quantum state, the wave function is constant in time, just as it was for the earlier example of the particle in a box. Therefore, $|\psi|^2$ is also constant. Thus, the Schrödinger description of the atom replaces the Bohr model of a point electron racing in eternal orbit about the nucleus with a more abstract model of a standing wave, whose amplitude squared measures the probability that the electron is at any particular distance from the nucleus. It is often convenient to imagine the atom as consisting of a point nucleus surrounded by a diffuse cloud of electric charge of density ρ, given by

$$\rho = -e|\psi|^2.$$

In many cases, this model can help us understand the behavior of an atom with its surroundings. Figure 28.20 shows a few examples of the charge density clouds for several wave functions.

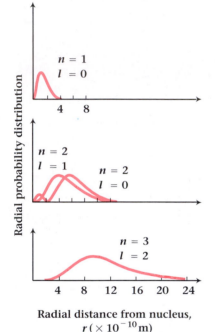

Figure 28.19
Radial probability distributions for several wave functions for hydrogen.

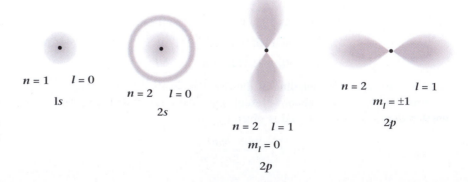

Figure 28.20
Electron charge density clouds for the lowest energy states of hydrogen. The first two clouds are spherically symmetric.

The Zeeman Effect and Space Quantization

We have shown that the principal quantum number n has a clear physical meaning in quantum mechanics, corresponding to distinct energy levels in the atom. What physical characteristics of the atom are associated with the quantum numbers l and m_l? Several experiments conducted before the development of Schrödinger's theory help provide the answer.

After the initial success of the Bohr model of the atom and prior to the development of wave mechanics, Arnold Sommerfeld (1868–1951) extended the Bohr theory to include elliptical orbits. Sommerfeld then argued that a charged electron moving about an orbit is equivalent to a small current loop and therefore behaves like a tiny magnet. When no external magnetic field is present, the spatial orientations of the orbital magnetic moments are all equivalent. But when an external magnetic field is present, the orientation of the orbital magnetic moment contributes to the energy of the atom according to

$$E_m = -\mu B \cos \theta,$$

where μ is the magnitude of the magnetic moment, B is the magnitude of the applied magnetic field, and θ is the angle between them. (In Chapter 16, we discussed a similar equation for the potential energy of an electric dipole in an electric field.)

According to Sommerfeld's theory, these current loops can only orient themselves in directions such that the component of the orbital angular momentum along the direction of the field is an integer multiple of $h/2\pi$ (Fig. 28.21). This effect is known as **space quantization.** To keep track of the possible orientations, Sommerfeld introduced the magnetic quantum number m_l, where

$$m_l = 0, \pm 1, \pm 2, \ldots, \pm l.$$

Notice that the quantum number m_l of the Sommerfeld theory is also found in the Schrödinger wave theory (Eq. 28.13).

The magnetic moment associated with the electronic orbital motion depends on the orbital angular momentum L of the electron:

$$\mu = \frac{-e}{2m}L. \tag{28.14}$$

We can replace the angular momentum L with its quantized form $lh/2\pi$:

$$\mu = \frac{-eh}{4\pi m}l.$$

We define the quantity $eh/4\pi m$ as the **Bohr magneton** μ_B, which has the value 9.274×10^{-24} J/T. In terms of μ_B, the orbital magnetic moment is

$$\mu = -\mu_B l.$$

In the presence of an applied magnetic field B, an electronic energy level with orbital angular momentum labeled by l and having a projection m_l along the field direction is shifted in energy by an amount

$$E_m = m_l \mu_B B. \tag{28.15}$$

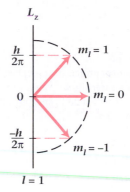

(a)

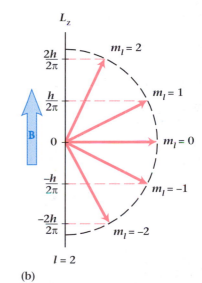

(b)

Figure 28.21
Projections of the angular momentum vector along the direction of an applied field **B**. The projections are quantized in units of $h/2\pi$. (a) $l = 1$; (b) $l = 2$.

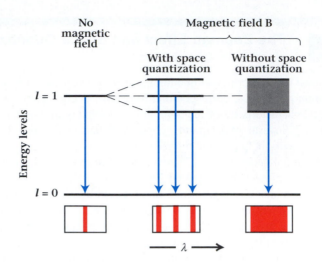

Figure 28.22

In the Zeeman effect, a single spectral line due to a transition from an $l = 1$ state to an $l = 0$ state is split into three lines by an applied magnetic field because the different m_l levels have different energies. Without space quantization, a broadened line would be expected.

This separation in energy of otherwise-identical energy levels provides an explanation for the **Zeeman effect,** which is the observed splitting of spectral lines into separate components by applying a magnetic field to the atom.

For the case of an electronic transition from an $l = 1$ state to an $l = 0$ state, light is emitted. The presence of an external magnetic field splits the expected single spectral line into three closely spaced lines, called a Zeeman triplet (Fig. 28.22). With no applied field, the various m_l states correspond to the same energy. When the magnetic field is applied, the initially identical levels separate according to Eq. (28.15). Transitions from these slightly different states lead to the distinct lines seen in the Zeeman spectra. Classically, only a broadening of the original line would occur, since a classical magnet has no restrictions on the orientations it can assume with respect to the applied field.

Example 28.7

Show that the relationship between an electron's orbital magnetic moment μ and its angular momentum L is given by

$$\mu = \frac{-e}{2m}L.$$

Strategy We can use classical electrodynamics to find the magnetic moment of a current loop. Then we interpret the current in terms of a circulating charge of an orbiting electron, making use of quantities we learned about in rotational mechanics.

Solution The magnetic moment μ of a current loop enclosing an area A and carrying a current I is

$$\mu = IA.$$

Here the current is the charge $-e$ divided by the time it takes to make one complete loop around the nucleus. This time is the period $T = 2\pi r/v$, where v

is the tangential speed of the electron and r is the radius of its motion. Thus

$$I = -\frac{e}{T} = -\frac{ev}{2\pi r}.$$

The area of the loop is πr^2, so

$$\mu = \left(-\frac{ev}{2\pi r}\right)(\pi r^2) = -\frac{evr}{2}.$$

But the definition of angular momentum is $L = mvr$, where m is the mass of the electron. Thus, the magnetic moment becomes

$$\mu = \frac{-e}{2m}L.$$

While the explanation of the Zeeman effect in terms of space quantization of orbital angular momentum is satisfying, it did not unequivocally confirm space quantization. An experiment designed to directly test space quantization was proposed by Otto Stern in 1921 and was subsequently carried out by Stern and W. Gerlach in late 1921 and early 1922.

An electric dipole in a nonuniform electric field experiences a net translational force because the forces acting on the positive and negative charges are unequal. A magnetic dipole in a nonuniform magnetic field also experiences a translational force. Suppose, then, that we direct a narrow beam of atoms, each possessing a magnetic moment, between the poles of a magnet shaped to produce a strong magnetic field gradient perpendicular to the path of the beam. Those atoms with a component of their magnetic moment along the direction of the field experience a deflection. If the component of the magnetic moment lies along the direction of the field, the atoms are deflected one way; if it is opposite to the field, they are deflected oppositely.

In their experiment, Stern and Gerlach generated a beam of silver atoms by evaporation from a heated oven (Fig. 28.23). The atoms escaped through a small aperture into a region of high vacuum. The emerging beam was first shaped by

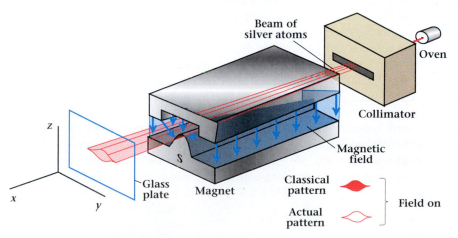

Figure 28.23
Schematic arrangement of the apparatus for the Stern-Gerlach experiments. A beam of atoms was directed along the positive x direction between the specially shaped magnetic pole pieces that produced a strong gradient in B_z. Splitting of the atomic beam was observed at the glass plate.

a series of apertures to a narrow cross section and then directed down the path between the pole pieces of the magnet. After passing through the magnet, the silver atoms struck a glass plate where, after several hours' exposure, they accumulated to form a thin film that was then darkened by chemical treatment. The resulting exposures all had one prominent characteristic: The atomic beam was split into two separate beams by the magnetic field and no atoms passed through undeflected. This experiment was considered another triumph of the early quantum theory and was taken as direct experimental verification of space quantization. If there were no quantization of the magnetic moments, the Stern-Gerlach experiment would give a smeared pattern instead of the two distinct lines observed. The existence of the two distinct beams is proof of two (and only two) unique orientations of the atomic magnetic moment relative to the direction of the applied magnetic field.

*28.11   The Pauli Exclusion Principle

In 1925, before publication of Schrödinger's equation, Wolfgang Pauli (1900–1958) recognized that the shell structure of atoms could be explained naturally if each electron state was labeled by four quantum numbers and if each individual state could be occupied by only a single electron. Pauli combined the four quantum numbers according to the then-current atomic model to give an understanding of the periodic table and of the multiple line structures observed in optical spectra.

As part of this development, Pauli formulated the now famous **exclusion principle: No two electrons in an atom may occupy the same quantum state; that is, no two electrons in an atom may have the same four quantum numbers.** If an electron in an atom exists in a quantum state labeled by a set of four distinct quantum numbers, then that state is occupied and all other electrons in the atom are excluded from that particular state.

In studying Pauli's work on the exclusion principle, Samuel Goudsmit and George Uhlenbeck were struck by the absence of any concrete picture or physical meaning to connect with the fourth quantum number. It was introduced as just part of the formalism of Pauli's atomic model. Goudsmit and Uhlenbeck were familiar with the idea that each quantum number corresponded to a particular coordinate. The prevailing idea in 1925 of a point electron allowed for only three coordinates and, hence, only three quantum numbers.

Goudsmit showed that the four quantum numbers used by Pauli could be replaced by the quantum numbers n, l, m_l, and m_s. The first three quantum numbers are the same three already discussed. The quantum number m_s was observed to take on only the values $+\frac{1}{2}$ and $-\frac{1}{2}$. Goudsmit and Uhlenbeck proposed that this fourth quantum number is associated with an additional property—the angular momentum of the electron or, as it has become known, the electron's **spin.** According to their idea, the electron possesses an intrinsic angular momentum $s = \frac{1}{2}$ whose projection m_s along a given direction could take on only the value of $m_s = +\frac{1}{2}$, or $m_s = -\frac{1}{2}$.* They also postulated that the ratio of the intrinsic

---

*An intrinsic property is a permanent property of the body itself, such as mass or charge. Quantities that depend on the body's circumstances, such as velocity, kinetic energy, and ordinary angular momentum, are not intrinsic properties.

magnetic moment to the intrinsic angular momentum had to be twice as great as for orbital motion. That is,

$$\mu_s = 2\frac{e}{2m}\frac{h}{2\pi}m_s = 2\mu_B m_s. \tag{28.16}$$

Although they did not realize it at the time of their 1922 experiment, Stern and Gerlach were really observing the intrinsic magnetic moment of the electron and not the effects of orbital motion. Had they used other atoms, their results could have been quite different. Nevertheless, their experiment and many different experiments performed since that time established the fact that, in the presence of a magnetic field, atomic magnetic moments arising from both spin and orbital contributions are quantized.

Example 28.8

Calculate the energy difference between two electron states of $m_s = \pm\frac{1}{2}$ in an applied magnetic field of $B = 0.50$ T.

Solution By analogy with Eq. (28.15), we can determine the energy of each state from

$$E = \mu_s B = 2\mu_B m_s B.$$

The quantum number m_s takes on values of $\pm\frac{1}{2}$. Thus

$$\Delta E = E_+ - E_- = 2\mu_B B\left[\frac{1}{2} - \left(-\frac{1}{2}\right)\right] = 2\mu_B B$$
$$\Delta E = 2(9.27 \times 10^{-24} \text{ J/T})(0.50 \text{ T})$$
$$\Delta E = 9.27 \times 10^{-24} \text{ J} = 5.8 \times 10^{-5} \text{ eV}.$$

This energy is nearly 30,000 times smaller than the energy corresponding to the least energetic Balmer transition in hydrogen.

*28.12 Understanding the Periodic Table

With the help of the Pauli exclusion principle, we can bring together the information about quantum states in the atom and their associated quantum numbers to finally understand the electronic structure of the elements and their orderly arrangement in the periodic table. We visualize the lowest energy level, or ground state, of an atom by considering a simple model of energy levels and placing electrons into those levels. In the ground state, we place electrons into the lowest electronic energy levels and fill successive levels until all electrons are accounted for. The remaining energy levels of higher energy are unoccupied. An illustration of such a scheme is shown in Table 28.2.

The first element ($Z = 1$) in the periodic table is hydrogen, which has only one electron. The second element, helium ($Z = 2$), has two electrons, corresponding to the first two states in Table 28.2. Two electrons exhaust all possibilities for energy states with $n = 1$. Together they make up the first shell of electrons about the nucleus, which is called the K shell.

Table 28.2					Quantum Numbers of the Electron States for the First Three Principal Energy Levels	
					Number of Electrons	
Shell	n	l	m_l	m_s	In Subshell	In Shell
K	1	0	0	$+\frac{1}{2}$		2
	1	0	0	$-\frac{1}{2}$		
L	2	0	0	$+\frac{1}{2}$	2	8
	2	0	0	$-\frac{1}{2}$		
	2	1	$+1$	$+\frac{1}{2}$	6	
	2	1	$+1$	$-\frac{1}{2}$		
	2	1	0	$+\frac{1}{2}$		
	2	1	0	$-\frac{1}{2}$		
	2	1	-1	$+\frac{1}{2}$		
	2	1	-1	$-\frac{1}{2}$		
M	3	0	0	$+\frac{1}{2}$	2	18
	3	0	0	$-\frac{1}{2}$		
	3	1	$+1$	$+\frac{1}{2}$	6	
	3	1	$+1$	$-\frac{1}{2}$		
	3	1	0	$+\frac{1}{2}$		
	3	1	0	$-\frac{1}{2}$		
	3	1	-1	$+\frac{1}{2}$		
	3	1	-1	$-\frac{1}{2}$		
	3	2	$+2$	$+\frac{1}{2}$	10	
	3	2	$+2$	$-\frac{1}{2}$		
	3	2	$+1$	$+\frac{1}{2}$		
	3	2	$+1$	$-\frac{1}{2}$		
	3	2	0	$+\frac{1}{2}$		
	3	2	0	$-\frac{1}{2}$		
	3	2	-1	$+\frac{1}{2}$		
	3	2	-1	$-\frac{1}{2}$		
	3	2	-2	$+\frac{1}{2}$		
	3	2	-2	$-\frac{1}{2}$		

The element for $Z = 3$ is lithium. Two of its electrons occupy the K shell, while the third occupies a state for which $n = 2$, beginning the L shell. This configuration of one electron outside a closed shell is similar to hydrogen and accounts for the position of lithium in the periodic table. The fourth element, beryllium, has four electrons, which in the ground state fill the K shell and the subshell of $n = 2$, $l = 0$. The L shell can be considered to be made up of two subshells with $l = 0$ and $l = 1$. The maximum number of electrons in the L shell is 8, corresponding to the element neon ($Z = 10$). See Fig. 28.24. In the periodic table, neon is placed below the element helium, which also has a closed shell of electrons.

The periodicity of chemical behavior of the elements was the basis for Mendeleev's original periodic table (Chapter 26). The additional ordering of these elements according to atomic number Z came as a result of Moseley's x-ray experiments. Finally, with the development of quantum mechanics, the electronic structure of the atoms explained the periodicity of the elements.

Further details about the electronic structure of atoms are beyond the scope of this text. However, the principal point to remember is that the behavior of the elements and their arrangement into the periodic table are well explained within the framework of quantum mechanics.

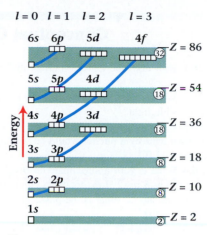

Figure 28.24
Schematic diagram of the electronic states of atoms. Each small square represents two electron states (one each for $m_s = \pm\frac{1}{2}$). Separated groups of squares correspond to subshells. The total number of electrons corresponding to a closed shell is shown in the circle at the right. The number at the far right is the atomic number of the element for which the shells are filled. States having the same value of n are connected with the heavy line.

Summary

Useful Concepts

■ The change in wavelength of Compton-scattered photons is

$$\lambda' - \lambda = \frac{h}{mc}(1 - \cos\theta).$$

■ The de Broglie wavelength of a particle is

$$\lambda = \frac{h}{p}.$$

■ The Heisenberg uncertainty relations can be written in the form

$$\Delta x \Delta p_x \geq \frac{h}{2\pi}$$

and

$$\Delta E \Delta t \geq \frac{h}{2\pi}.$$

■ The Pauli exclusion principle states that no two electrons in any atom may have the same quantum numbers.

Important Terms

You should be able to write the definition or meaning of each of the following:

Compton wavelength	principal quantum number
de Broglie wavelength	orbital quantum number
Schrödinger equation	magnetic quantum number
wave function	space quantization
wave packet	Bohr magneton
uncertainty principle	Zeeman effect
probability density	exclusion principle
zero-point energy	spin
barrier penetration or	
tunneling	

Conceptual Questions

28.1 How does the Compton effect differ from the photoelectric effect?

28.2 Why is the Compton effect observed for x rays but not for visible light?

28.3 Estimate what the de Broglie wavelength of a baseball thrown hard by a pitcher would be if Planck's constant had a value of $1.0 \text{ J} \cdot \text{s}$.

28.4 What is your approximate de Broglie wavelength when you run at 6 m/s? How does that compare with the size of a hydrogen atom? What is the physical significance of such a wavelength?

28.5 Is a baseball diffracted when thrown through a picket fence? Explain.

28.6 How does the uncertainty principle affect our understanding of the principle of conservation of energy?

28.7 How is the interpretation of the Schrödinger wave amplitude similar to the interpretation of the amplitude of an electromagnetic wave?

28.8 Decide, on the basis of some numerical estimates,

whether the quantized energy states of a person confined to a room can be observationally distinguished from the smooth range allowed by classical physics.

28.9 Is the probability that a book lying on a table will tunnel or penetrate through the table exactly zero or is it just extremely small? Could you estimate an upper limit to this probability, based on an estimate of the number of tables in the world and the number of tunneling events reported each year?

28.10 Explain the effects on tunneling of large potentials $(E \ll V)$ and of wide barriers $(x \gg 0)$.

28.11 Why is m_l called the magnetic quantum number?

28.12 How would the atomic shell structure be affected if electrons did not obey the Pauli exclusion principle?

28.13 What are the quantum numbers for the outermost electron of a sodium atom in the ground state? What would be the quantum numbers if the outermost electron were raised to the first excited state?

28.14 How would atomic shell structures be affected if electrons did not obey the Pauli exclusion principle?

Problems

Section 28.2 The Compton Effect

Hints for Solving Problems

Remember that for speeds at which relativistic effects are important you must use the relativistic relationships for kinetic energy and momentum. For a given energy, a lighter particle goes faster than a heavier one.

28.1 A 0.025-nm x ray is scattered through an angle of 60°. What is the frequency of the emergent x ray?

28.2 Compute the value of the Compton wavelength of (a) an electron (mass 9.11×10^{-31} kg), (b) a proton (mass 1.67×10^{-27} kg), and (c) a muon (mass 1.88×10^{-28} kg).

28.3 Show that h/mc has the dimensions of length.

28.4• A 100-keV x ray is Compton-scattered through an angle of 90°. What is the energy of the x ray after scattering?

28.5• Work Example 28.1 for the case in which the x ray is scattered through an angle of 64°.

28.6• A 19.68 keV photon is Compton-scattered so that its final energy is 19.62 keV. Determine the angle through which it is scattered.

28.7• What is the maximum recoil velocity of a free electron that is scattered by an x ray of wavelength $\lambda = 7.1 \times 10^{-11}$ m?

28.8•• An x ray is scattered through an angle of 47°. What is the fractional change in its energy if the incident wavelength is ten times the Compton wavelength of the electron?

28.9•• An x-ray photon of wavelength 0.0023 nm loses 0.024 of its initial energy during a Compton scattering. Through what angle is it scattered?

28.10•• Perform the algebraic steps necessary for eliminating v and f from the equations for conservation of momentum and energy for the Compton effect, and thereby obtain Eq. (28.1).

Section 28.3 De Broglie Waves

28.11 Compute the de Broglie wavelength of a baseball of mass 0.145 kg moving at a speed of 38 m/s.

28.12 The de Broglie wavelength of an electron is 6.0×10^{-9} m. What is the electron's speed?

28.13 The de Broglie wavelength of a charged particle moving with a speed of $0.100c$ is 1.32×10^{-14} m. Is the particle an electron or a proton?

28.14 What is the de Broglie wavelength of a particle moving at a speed of 1.00×10^{6} m/s if it is (a) an electron, (b) a proton? ($m_e = 9.11 \times 10^{-31}$ kg and $m_p = 1.67 \times 10^{-27}$ kg.)

28.15 Compute the de Broglie wavelength of an electron accelerated from rest through a potential of 900 V.

28.16• Compute the de Broglie wavelength of the earth in its orbit.

28.17• What is the kinetic energy of an electron of wavelength 0.30 nm?

28.18• What is the de Broglie wavelength of a 100-MeV proton? (*Hint:* Use relativistic mechanics.)

28.19• What is the wavelength of an electron whose total energy is 0.87 MeV? (*Hint:* Use relativistic mechanics.)

28.20• A 0.25-kg ball is swung on a string in a 0.75-m-radius horizontal circle once every 1.0 s. (a) How many de Broglie wavelengths of the moving ball fit into the circle? (b) By approximately what factor would Planck's constant have to change if the answer to part (a) were 1000 de Broglie wavelengths?

28.21• Calculate the Bragg angle θ for 400-eV electrons diffracted from the planes of a nickel crystal of spacing $d = 0.203$ nm.

28.22• What is the average de Broglie wavelength of oxygen molecules in the air at a temperature of 27°C? Use the results of the kinetic theory of gases. The mass of an oxygen molecule is 5.31×10^{-26} kg.

28.23• In the Davisson and Germer experiment, 54-eV electrons were diffracted from nickel so that the angle between the incident and the diffracted beams was 50°. What is the spacing of the crystal planes?

28.24•• Find the expression requested in Example 28.3 for the case in which relativistic effects must be taken into account.

28.25•• The speed of 440-Hz sound wave traveling through oxygen gas at a temperature of 23°C is 331 m/s. Compare the de Broglie wavelength of oxygen molecules at the rms speed with the wavelength of the sound wave in the gas by finding their ratio.

Section 28.5 The Uncertainty Principle

Hints for Solving Problems

The uncertainty principle can be used to estimate the value of a quantity: For instance, the momentum of a particle in the nucleus can be estimated by assuming that its uncertainty in momentum is not larger than its momentum.

28.26 A slowly moving electron is localized to be within 0.50 mm of a particular point. What is the minimum uncertainty with which its velocity can be simultaneously determined?

28.27 Photons of energy 14.4 keV can be produced with a relative uncertainty in energy of one part in 10^{11}. What is the uncertainty in the lifetime of the state that emits such photons?

28.28 In radiating a photon of green light ($\lambda = 546$ nm), an atom radiates for 1.00×10^{-9} s. (a) What is the energy of the photon? (b) What is the uncertainty in its energy?

28.29• An electron is localized to within a distance of 1.0×10^{-10} m (approximately the diameter of a hydrogen atom). (a) Treat this as a one-dimensional problem and determine the uncertainty in the electron's momentum. (b) What is the kinetic energy associated with this momentum?

28.30• A proton has a kinetic energy of 10.0 MeV. If the x component of its momentum is measured with a relative uncertainty of 1.00%, what is the uncertainty in its position

along the x axis? (*Hint:* A 10.0-MeV proton may be treated nonrelativistically.)

28.31• Show that for two waves of wavelengths λ_1 and λ_2, their spread in momenta can be expressed as

$$\Delta p = \frac{h\Delta\lambda}{\lambda_0^2},$$

where $\Delta\lambda = \lambda_2 - \lambda_1$ and λ_0 is the average wavelength. Use the approximation that when $\Delta\lambda$ is small, $\lambda_0^2 \approx \lambda_1\lambda_2$.

28.32• What is the uncertainty in momentum of an electron with de Broglie wavelength $\lambda = 5.00 \times 10^{-2}$ nm if the uncertainty in wavelength is $\Delta\lambda = 1.00 \times 10^{-3}$ nm? (*Hint:* Use the relationship given in Problem 28.31.)

28.33• An electron with de Broglie wavelength $\lambda = 0.10$ nm has an uncertainty in momentum of $\Delta p = 6.6 \times 10^{-25}$ kg · m/s. What is the uncertainty in the electron's wavelength? (*Hint:* Use the relationship given in Problem 28.31.)

28.34• An experiment measures the energy of a photon to one part in 10^7. What is the minimum uncertainty in locating the photon if it is (a) an x ray of $\lambda = 0.071$ nm, (b) visible light of $\lambda = 550$ nm, and (c) an AM broadcast wave of $\lambda = 300$ m?

28.35•• Show that for angles less than 2π the uncertainty principle can be expressed in the form

$$\Delta L\Delta\theta \geq \frac{h}{2\pi},$$

where ΔL is the uncertainty in the angular momentum and $\Delta\theta$ is the uncertainty in the angular position. (*Hint:* Use a particle moving in a circle.)

Section 28.7 The Particle in a Box

28.36 Imagine an electron confined in a one-dimensional potential box of width $L = 0.12$ nm. Compute the energy of the first three energy levels.

28.37 Imagine an electron confined in a one-dimensional potential box of width $L = 1.0$ nm. Compute the energy of the first three energy levels.

28.38 Imagine a proton of mass 1.67×10^{-27} kg confined in a one-dimensional potential box of width $L = 1.00 \times 10^{-14}$ m, roughly the size of an atomic nucleus. (a) Compute the energy of the first energy level in joules. (b) Give the energy of the first level in units of MeV.

28.39 A particle confined along the x axis to the dimensions of a nucleus ($\approx 1.0 \times 10^{-14}$ m) has a ground-state energy of 3.28×10^{-13} J. What is the particle's mass?

28.40 The ground-state energy of an electron confined along the x axis is 6.03×10^{-18} J. What is the approximate length of the region to which the electron is confined?

28.41• A 75-kg student is confined to a one-dimensional box one meter long. Is the energy difference between the two lowest energy levels great enough to correspond to lifting the student by a distance equal to the diameter of a hydrogen atom?

Section 28.8 Tunneling or Barrier Penetration

28.42 A wave function within a barrier is described by $\psi(x) = \psi_0 e^{-\alpha x}$. What is the probability density for finding the wave at $x = 3/\alpha$? Give your answer as a multiple of ψ_0^2.

28.43 A wave function within a barrier is described by $\psi(x) = \psi_0 e^{-\alpha x}$. What is the ratio of the value of $\psi(x)$ for $x = 1/\alpha$ to that for $x = 10/\alpha$?

28.44•• In an experiment, particles penetrate a rectangular barrier similar to that shown in Fig. 28.18 at the rate of 1.25 $\times 10^4$ particles per second when the barrier is 1.35×10^{-5} m thick. When the barrier thickness is doubled, the number of particles penetrating per second is decreased by a factor of 4. How many particles per second would penetrate the barrier if its thickness were reduced to one-half its initial thickness?

28.45•• The probability that a particle will penetrate a barrier of thickness Δx is 0.0010. What is the probability of penetrating a barrier that is twice as thick? Use the wave function of Eq. (28.12) and assume that everything else is the same.

*Section 28.9 Wave Theory of the Hydrogen Atom

28.46 A given atomic level is characterized by principal quantum number $n = 3$. How many orbital subshells are possible?

28.47 How many orbital subshells are associated with the L electronic shell?

28.48 How many magnetic substates are possible in a p subshell?

28.49 How many magnetic substates are possible in an f subshell?

*Section 28.10 The Zeeman Effect and Space Quantization

28.50 Calculate the value of the Bohr magneton from its definition. Give your answer in SI units.

28.51 Find the shift in energy of a particular orbital angular momentum level having a value $m_l = 1$ due to the presence of an applied magnetic field of 1.00 T. Express this energy in units of electron volts.

28.52 In the presence of a magnetic field B, an electronic p state splits into three energy levels corresponding to $m_l = 0$, $+1$, -1. Calculate the energy separation between adjacent levels.

28.53• In the presence of a magnetic field of $B = 0.10$ T, an electronic d state is split into sublevels of different energies. What is the energy separation between the highest and lowest sublevels?

*Section 28.11 The Pauli Exclusion Principle

28.54 Find the energy difference between states of $m_s = +\frac{1}{2}$ and $m_s = -\frac{1}{2}$ in an applied magnetic field $B = 0.100$ T. Express this energy in electron volts.

28.55• (a) Find the energy difference between electron states of $m_s = +\frac{1}{2}$ and $m_s = -\frac{1}{2}$ in an applied magnetic field of $B = 0.800$ T. Express this energy in electron volts. (b) What is the wavelength of a photon having this same energy?

*Section 28.12 Understanding the Periodic Table

28.56 Predict the electronic states for neon ($Z = 10$) using Tables 28.1 and 28.2.

28.57 Predict the electronic states for sodium ($Z = 11$) using Tables 28.1 and 28.2.

28.58 Predict the electronic states for boron ($Z = 5$) using Tables 28.1 and 28.2.

28.59 Extend Table 28.2 to the N shell for which $n = 4$.

Additional Problems

28.60• A photon whose wavelength is 0.0700 nm is Compton-scattered through an angle of 60°. (a) What is the wavelength and (b) what is the energy of the scattered photon? (c) How much energy is given to the recoil electron?

28.61• A 17.2-keV x ray from molybdenum is Compton-scattered through an angle of 120°. (a) What is the energy of the scattered x ray? (b) What is the kinetic energy of the recoil electron?

28.62• (a) What is the maximum change in wavelength of a photon that undergoes Compton scattering? (b) What is the ratio of this maximum change in wavelength to the wavelength of a 50-kV x ray?

28.63• Calculate the Bragg angle θ for 300-eV electrons diffracted from the atomic planes of silver spaced at $d = 0.236$ nm.

28.64• Diffraction rings of the type shown in Fig. 28.5 are made by electrons diffracted from a polycrystalline sample located a distance L from a fluorescent screen. Prove that the ring diameter is given by

$$D = \frac{2L\lambda}{d}$$

for electrons of kiloelectronvolt energies for which the Bragg angle θ is small and the wavelength λ is much less than the spacing d between atomic planes.

28.65• Show that when the uncertainty in the position of a particle is equal to its de Broglie wavelength, the uncertainty in its momentum is equal to its momentum divided by 2π.

28.66• An atom is excited to an energy level 2.4 eV above its ground state, where it remains an average of 2.0×10^{-8} s before making a transition to the ground state. What is the ratio of the spread in the frequency of the light emitted to the average frequency of this transition?

28.67• A particle of mass m is confined to within a length Δx. Find an expression for its lowest possible energy.

28.68• An electronic state is characterized only by its intrinsic magnetic moment, given by Eq. (28.16). In a magnetic field B, the state is split into two states and the system resonantly absorbs microwave photons of $\lambda = 2.00$ cm. What is the magnitude of B?

28.69•• Prove that the emergent photon in Compton scattering has a frequency that lies between f and $mc^2/2h$ when the incident x ray is very energetic. (*Hint:* By "very energetic" we mean a photon for which $f \gg mc^2/h$.)

28.70•• Electrons are accelerated to 30 GeV for experiments at the DESY laboratory in Hamburg, Germany. What is the de Broglie wavelength of these electrons? (*Hint:* Use relativistic mechanics.)

28.71•• A particle in a two-dimensional square potential box has wave functions that must simultaneously satisfy boundary conditions in both directions. Consequently, two quantum numbers are needed and the energy is given by

$$E = \frac{h^2}{8mL^2}(n_x^2 + n_y^2), \quad \text{for } n_x, n_y = 1, 2, 3, \ldots.$$

Compute the first six energy levels for the two-dimensional square potential box of edge length L. Give the quantum numbers n_x and n_y for each level.

The Nucleus

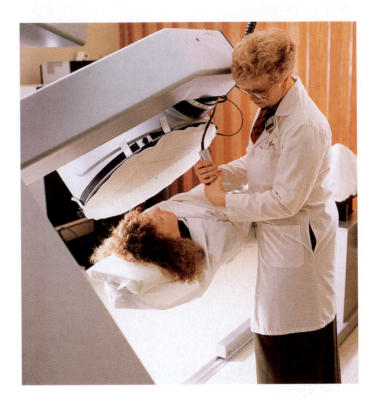

We began our investigation of the atom and its constituent parts in Chapter 26. There we discussed evidence for the existence of the atom and the atomic nucleus. Now we examine the composition and properties of the nucleus itself. We will not introduce these new ideas in the order in which they were discovered. Instead, we have organized this chapter to effectively explain both the properties of the nucleus and the experimental evidence for those properties.

Our understanding of the forces and structure within the nucleus is still incomplete, but present research is advancing along exciting paths. Today's discoveries are of interest to the public as well as to scientists, for they have led to a wealth of practical devices and techniques. Fission and fusion are topics of current commercial, political, and scientific interest. Many people think immediately of nuclear weapons and nuclear power, but knowledge of the nucleus has also revolutionized many areas of medicine and biological science. Radioactive isotopes have been important tools for many years in diagnosing and treating diseases. Radiation plays a major role in cancer therapy. ■

29.1 Radioactivity

We have already discussed the experimental evidence for our nuclear model of the atom. By now, you should be familiar with that material and the associated exercises. Here we will briefly review those particulars that are especially useful in investigating the composition and structure of the nucleus itself.

After Henri Becquerel's 1896 discovery of a penetrating natural radiation, it was found that there are actually three kinds of natural radioactive emissions: alpha (α), beta (β), and gamma (γ) radiation. The alphas are helium nuclei, the betas are electrons, and the gammas are short-wavelength electromagnetic radiation, similar in nature to x rays, radio waves, and light. Table 26.2 lists the masses and charges of these radiations.

Radioactivity is measured by the number of disintegrations per second, called the activity. The SI unit of activity is the becquerel (Bq), which is one disintegration per second. If N_0 radioactive atoms are present at time zero, then the number remaining at some later time t is

$$N = N_0 e^{-0.693t/t_{1/2}},$$

where $t_{1/2}$ is the half-life, which is different for each type of radioactive atom (Fig. 29.1). Several examples of half-life calculations are found in Section 26.8, along with a discussion of how radioactive isotopes can be used in archaeological dating.

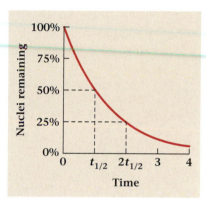

Figure 29.1

The number of radioactive atoms present decreases exponentially with time. In a time interval of one half-life, the number of atoms present decreases by one half.

29.2 Chadwick's Discovery of the Neutron

After Rutherford's discovery of the nucleus in 1911, physicists slowly realized that the nucleus itself must be made of smaller constituent parts. Initially, it seemed reasonable to postulate that the nuclei of atoms heavier than hydrogen consisted of hydrogen nuclei, called **protons,** and electrons. The protons would provide the mass and the positive charge, while the electrons, which were presumed to be in the nucleus, would neutralize some of the charge. This theory could satisfactorily account for a nucleus of atomic mass A and atomic number Z by assuming that the nucleus contained A protons and $A - Z$ electrons. The proton-electron model of the nucleus gave the correct charge and approximately the correct mass for any nucleus. Furthermore, this model accounted for the emission of alpha particles, which were assumed to be composed of four protons and two electrons, and for the emission of beta particles (electrons). According to this theory, gamma radiation was emitted from an excited nuclear state in a manner analogous to the way light was emitted from an excited state of the electrons of an atom.

However, the proton-electron model of the nucleus had several difficulties. For instance, if you take account of the observed spin of particles emitted during decay processes, the proton-electron model of the nucleus does not conserve angular momentum. This fact alone was a strong reason for discarding or drastically revising the theory. Furthermore, as we showed in the previous chapter (Example 28.6), the energy needed to confine an electron to a region comparable in size to the nucleus greatly exceeds the electron energies observed in beta decay. For these reasons, Rutherford suggested as early as 1920 that the nucleus should contain a neutral particle of approximately the same mass as the proton. The search for such a particle was made difficult by the fact that the radiation-detecting devices of the time were sensitive only to charged particles. Not until 1932 did James Chadwick (1891–1974) conclusively demonstrate the existence of such a neutral particle.

In Chadwick's experimental apparatus, alpha particles were allowed to bombard a beryllium target (Fig. 29.2). The beryllium then emitted an unknown radiation, which was unaffected by electric or magnetic fields and thus was electrically neutral. This radiation struck a paraffin sheet. Chadwick used paraffin because it contains a large proportion of hydrogen. The paraffin ejected protons (hydrogen nuclei) when hit by the unknown radiation, and these protons were detected by an ionization chamber.

The maximum energy of the protons emerging from the paraffin was measured to be 5.7 MeV. Then Chadwick showed, by an analysis like that for Compton scattering, that if the radiation that knocked these protons out of the paraffin was electromagnetic radiation, such as x rays or gamma rays, it would have to have an energy of about 55 MeV. This was a surprisingly large amount of energy, much greater than ordinary nuclear decay energies. On the other hand, if the unknown radiation was a neutral particle with no electric charge but with a specific mass, then its interaction with the protons in paraffin could be analyzed by applying the laws of conservation of momentum and energy to a collision between two particles, just as we did back in Chapter 8. (See Problem 29.7.) The unknown particle's velocity and mass could not both be determined in one experiment, so Chadwick repeated the experiment, substituting a nitrogen target for the paraffin.

Figure 29.2
Schematic diagram of Chadwick's experimental apparatus. Chadwick discovered neutrons by analyzing their interaction with hydrogen-rich paraffin in which protons were ejected.

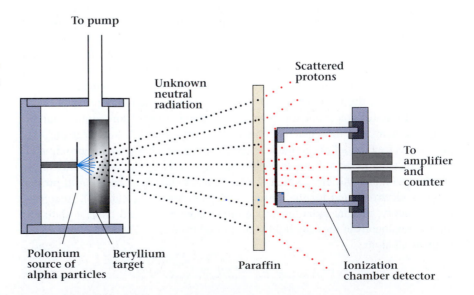

We denote the unknown incident particle with the subscript U, the hydrogen atoms ejected by the paraffin with the subscript H, and the nitrogen with the subscript N. Then, by applying the laws of conservation of momentum and energy to a head-on collision, we obtain the following relationships:

$$v'_H = \frac{2m_U}{m_U + 1m_H}v_U \quad \text{and} \quad v'_N = \frac{2m_U}{m_U + 14m_H}v_U. \tag{29.1}$$

The quantity v_U indicates the speed of the unknown particle before the collision, and the quantities v'_H and v'_N indicate recoil speeds after the collision. Also we have substituted $14m_H$ for m_N. Dividing the first of these equations by the other and inserting the measured values of the final speeds, we get

$$\frac{m_U + 14m_H}{m_U + 1m_H} = \frac{v'_H}{v'_N} = \frac{3.3 \times 10^7 \text{ m/s}}{4.7 \times 10^6 \text{ m/s}} = 7.0.$$

Upon solving for m_U, we find

$$m_U = 1.15m_H.$$

The mass of the unknown particle is approximately the same as that of the proton. Within the precision of Chadwick's experiment, the masses of the two particles were considered the same.

Later, more refined measurements showed the mass to be even closer to the proton mass than indicated here. If the mass of the unknown particle is the same as the mass of the proton, then, assuming an elastic collision, the energy of the incoming particle is transferred to the outgoing target particle, just as for billiard ball collisions. The point is that the model of a neutral particle with the mass of a proton does not give rise to the large energies required if the unknown radiation were electromagnetic. Instead, the radiation could have an energy of 5.7 MeV, well within the range of nuclear decay energies, and still produce the observed effects.

Chadwick's analysis thereby demonstrated the existence of a neutral particle, which was given the name **neutron.** This name had been proposed twelve years earlier by Rutherford for a particle thought to consist of a proton and an electron. Note that Chadwick demonstrated the neutron's existence without directly observing neutrons. Instead, he measured the effects of neutron bombardment on target atoms and applied conservation of energy and momentum to determine the neutron's properties. This is an example of how contemporary physics research uses models and conservation laws to determine new results. Since Chadwick's discovery, the masses of the neutron and proton have been carefully measured in many independent experiments. They are listed in Table 29.1.

Table 29.1	Mass and Charge of Some Subnuclear Particles				
Name	Symbol	Electric Charge	Mass (m_e)	Mass (MeV)	Mass (kg)
Photon (gamma)	γ	0	0	0	0
Electron	e^- (β^-)	$-e$	1	0.5110	9.109×10^{-31}
Positron	e^+ (β^+)	$+e$	1	0.5110	9.109×10^{-31}
Proton	p	$+e$	1836	938.272	1.673×10^{-27}
Neutron	n	0	1839	939.566	1.675×10^{-27}

29.3 Composition and Size of the Nucleus

Because protons and neutrons both appear in the nucleus, they are collectively called **nucleons.** According to today's accepted proton-neutron model of the nucleus, the number of protons in a nucleus is Z, the *atomic number* of the element, and the number of protons plus the number of neutrons is A, the element's *atomic mass number.* This leads to the following way of designating the nucleus of an element X, where X stands for the chemical symbol for the element:

$$^A_Z X. \tag{29.2}$$

For example, the nucleus of hydrogen (one proton) is ^1_1H, the nucleus of helium (two protons and two neutrons) is ^4_2He, and so on. The number of neutrons in a nucleus is simply $A - Z$.

Nuclei that have the same number of protons (Z) but different numbers of neutrons (different A) are called **isotopes.** Isotopes of the same element have the same chemical properties because they have the same number and arrangement of electrons. (Remember, the number of electrons in a neutral atom equals its number of protons.) However, isotopes may not have the same nuclear properties. For instance, some isotopes of an element may be stable while others may be radioactive. This is the case for carbon, whose isotope with $A = 12$ is stable but whose isotope with $A = 14$ is radioactive, as we saw in Chapter 26 (p. 841). The naturally occurring elements collectively have about 280 stable isotopes. However, there are more than 800 known radioactive isotopes, both naturally occurring and laboratory produced. Figure 29.3 shows a chart of the known stable isotopes. Note that for elements of low atomic numbers, the number of protons and neutrons in stable nuclei is about the same. As we go to higher atomic numbers, stable nuclei have a small excess of neutrons, yet the general tendency remains for a stable nucleus to have about half protons and half neutrons.

Very often an element has more than one stable isotope. For example, stable boron occurs in the form $^{10}_5B$, with five protons and five neutrons, and in the form $^{11}_5B$, with five protons and six neutrons. The isotope $^{10}_5B$ has a relative atomic abundance of 19.8%, and $^{11}_5B$ has a relative atomic abundance of 80.2%. An ordinary chemical sample contains a mixture of the two isotopes in the same proportions as their natural abundance, although in some cases a slight variation occurs, depending on the origin of the sample. An element's *chemical atomic mass* (sometimes called *atomic weight*) is a weighted average of its stable isotopes, counting each isotope according to its natural abundance.

Example 29.1

Calculate the expected atomic mass of boron from knowledge of its two naturally occurring isotopes and their relative abundances.

Solution We just mentioned that a sample of natural boron contains isotopes $^{10}_5B$ and $^{11}_5B$ in relative abundances of 19.8% and 80.2%, respectively. Thus,

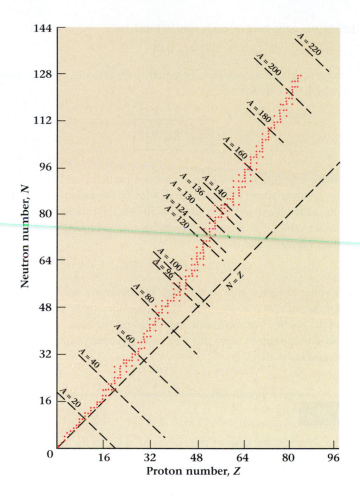

Figure 29.3
Chart of the known stable isotopes. Note how the light elements lie close to the line $N = Z$, indicating equal numbers of neutrons and protons.

19.8% of the sample has atomic mass of approximately 10, and 80.2% of the sample has atomic mass of approximately 11. The averaged mass is then

$$10 \times 0.198 = 1.98$$
$$11 \times 0.802 = \underline{8.82}$$
$$10.80$$

The sum 10.80 represents the approximate average atomic mass of the combined sample. Compare the result of this simple calculation with the value given in the periodic table on the inside back cover.

To calculate an approximate radius for the nucleus, let us make a simple model and assume that the nucleons fill a nucleus in a manner similar to the way marbles fill a bag. Then we can write that the volume V of the nucleus is proportional to the number of nucleons A,

$$V \propto A.$$

If we further assume that the nucleus is spherical (Fig. 29.4), then the volume is proportional to the cube of the radius,

$$r^3 \propto A.$$

Figure 29.4
The nucleus is modeled as a cluster of hard spheres packed together.

Upon taking the cube root of each side, we get

$$r \propto A^{1/3}.$$

We can express this relationship as an equality by introducing a proportionality constant, which we call R_0:

$$r = R_0 A^{1/3}. \tag{29.3}$$

Scattering experiments, both with charged particles and with neutral particles such as neutrons, have been used to determine the radius of the nucleus. The experiments generally agree with the conclusion that nuclear radii increase as $A^{1/3}$. The exact value of R_0 depends somewhat on the type of experiment used to measure it. A typical value is

$$R_0 \approx 1.2 \times 10^{-15} \text{ m.}$$

Using this value of R_0 for gold, which has 197 nucleons, we find that Eq. (29.3) predicts a value of $r = 7 \times 10^{-15}$ m. (Compare this value with the upper limit estimate of 40×10^{-15} m made in Chapter 26 on the basis of Rutherford scattering.) You should keep in mind that this is an extremely simple model and only roughly represents reality.

Example 29.2

What is the approximate ratio of the radius of $^{252}_{98}\text{Cf}$ to the radius of $^{56}_{26}\text{Fe}$?

Solution We can use the relationship between the number of nucleons A and the radius given in Eq. (29.3) to get

$$\frac{r_{\text{Cf}}}{r_{\text{Fe}}} = \frac{R_0 A_{\text{Cf}}^{1/3}}{R_0 A_{\text{Fe}}^{1/3}} = \left(\frac{A_{\text{Cf}}}{A_{\text{Fe}}}\right)^{1/3}$$

$$\frac{r_{\text{Cf}}}{r_{\text{Fe}}} = \left(\frac{252}{56}\right)^{1/3} = (4.5)^{1/3} = 1.65.$$

29.4 Nuclear Forces and Binding Energy

The forces holding the nucleus together must be extremely strong. This is evident from the fact that the positively charged protons remain confined to the small volume of the nucleus. If the nuclear force were weaker, the electrostatic (Coulomb) repulsion of the positively charged protons would predominate and the nucleus would fly apart or, at the very least, the nucleus would occupy a much larger volume.

Many experiments have shown that the attractive force between nucleons—whether between protons and protons, neutrons and neutrons, or neutrons and protons—is the same, once we subtract the effect of the electrostatic Coulomb repulsion between protons. We call this fundamental attractive force the *strong nuclear force*. The strong nuclear force between nucleons in a nucleus is approximately 10^2 times as strong as the Coulomb repulsion between protons, and approximately 10^{39} times as strong as the attractive gravitational force between nucleons. The general form of the strong force between individual nucleons as a function of their separation is sketched in Fig. 29.5. A word of caution is necessary here: This curve does *not* represent the nuclear force in the same precise way that a graph of the gravitational or electrical force represents those forces. We have drawn it only to gain some insight into the nature of the nucleon-nucleon interaction. For example, the force between nucleons is not necessarily directed along the line joining them. However, we can argue for the general shape of the curve on several points. We do not observe nuclear forces at distances much greater than the size of the nucleus. Thus, there should be essentially zero force at distances greater than about R_0. The nucleus is tightly bound, so the force must be large and attractive at distances less than the nuclear radius. However, the nucleus does have a finite size, which depends on the number of particles in the nucleus. This observation indicates that two or more nucleons cannot occupy the same space. Therefore, there must also be a strong repulsive force at very short distances.

Strictly speaking, at distances of nuclear dimensions we must use only quantum mechanical considerations. In quantum mechanics we do not speak of the force between two particles, but instead describe the energy of their interaction. We have greatly oversimplified the picture of the nuclear force in order to gain some understanding of nuclear behavior. The forces among many nucleons are not easily described and are quite different from the forces we have studied so far. The nature of the nuclear force is not fully understood in any simple way, even though a great deal is known about it. Attempts to solve this problem have led to many interesting ideas and results, some of which are discussed in Chapter 32.

Even without knowing the exact nature of the nuclear force, we can, in part, understand the way a nucleus is bound together by considering its energy. Let us use the Einstein mass-energy relationship $E = mc^2$. The masses of the proton and neutron are well known, as are the nuclear masses of several hundred nuclei. Table 29.2 lists the masses of several light atoms. Nuclear and atomic masses are often measured in terms of atomic mass units u.* Through the use of the Einstein energy equation we can show that 1 u is equivalent to 931.5 MeV. In nuclear physics, masses are frequently given in units of MeV.

Let us compare the mass of an atom, such as the $^4_2$He atom, to that of its constituent parts. We begin by computing the total mass of the components of the atom, consisting of two neutrons, two protons, and two electrons. Rather than add together the masses of the six particles individually, we will add the mass of two hydrogen atoms to the mass of two neutrons. We can do this because the energy that binds the electrons to the atoms is small, a few electron volts, in comparison with the mass of the particles involved. (Remember that energy and mass are connected through $E = mc^2$.) The result is a somewhat shorter calculation. Using the values from Table 29.2, we have

$$2m_H = 2(1.007825) = 2.015650 \text{ u}$$
$$+2m_n = 2(1.008665) = \underline{2.017330 \text{ u}}$$
$$4.032980 \text{ u}$$

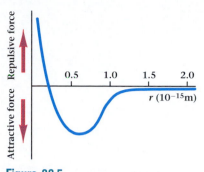

Figure 29.5

General shape of the force between any two nucleons as a function of the distance between them.

---

*The unified atomic mass unit u, also abbreviated as amu, is equal to 1/12 the mass of an atom of $^{12}_6$C; 1 u = 1.66054 × 10⁻²⁷ kg.

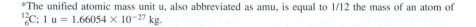

	Table 29.2	*Atomic Mass of Some Light Elements* *The Atomic Mass is Given for the Neutral Atom,* *Including its Electrons.*		
Symbol		**Mass (u)**	**Mass (MeV)**	**Mass (10^{-27} kg)**
p		1.007276	938.272	1.67262
n		1.008665	939.566	1.67493
H		1.007825	938.783	1.67353
$^2_1$H (D)		2.014102	1876.125	3.34450
$^3_1$H (T)		3.016049	2809.433	5.00827
$^3_2$He		3.016029	2809.414	5.00824
$^4_2$He		4.002603	3728.402	6.64648
$^5_2$He		5.012220	4668.854	8.32299
$^5_3$Li		5.012538	4669.151	8.32352
$^6_3$Li		6.015121	5603.051	9.98835
$^7_3$Li		7.016003	6535.367	11.65036
$^9_4$Be		9.012182	8394.796	14.96509

Note: The masses of the proton and neutron have been included as an aid in working problems.

The mass of the $^4_2$He atom is 4.002603 u, a value that is significantly smaller than the sum of the masses of the constituent particles. Using the Einstein relation, we may say that the mass energy of the components is greater than the mass energy of the atom. This difference in energy is what holds the nucleus together. It is called the **binding energy** of the nucleus. The binding energy is the difference in mass energy of the constituent particles when bound together in the nucleus and when considered as separate, isolated particles. Therefore, it is the energy needed to separate the nucleus into its component parts. For $^4_2$He, the mass difference is

$$\Delta m = (2m_H + 2m_n) - m(^4_2He) = 0.030377 \text{ u}.$$

The binding energy is

$$\Delta E = \Delta mc^2 = (0.030377 \text{ u}) (931.5 \text{ MeV/u})$$

$$\Delta E = 28.3 \text{ MeV}.$$

The binding energy of $^4_2$He is rather large for a light nucleus, corresponding to the observation that the alpha particle is very stable. We can calculate the binding energy of any nucleus whose mass and composition are known in the same way that we calculated the binding energy of the alpha particle. Figure 29.6 shows the binding energy per nucleon for a large number of nuclei. Notice that the binding energy per nucleon is greatest for nuclei with A between 40 and 80. The general shape of this curve is important for analyzing some kinds of nuclear reactions. We will discuss its significance further in Sections 29.11 and 29.12.

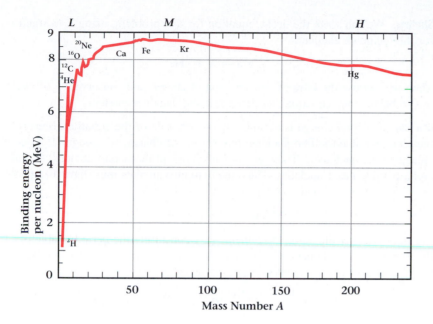

Figure 29.6
Graph of the binding energy per nucleon (in MeV) versus the atomic mass number A.

29.5

Conservation Rules: Radioactive Decay and Nuclear Stability

As we saw in Chapter 26, some nuclei are unstable and decay, most commonly by emitting alpha, beta, or gamma particles. Such decays must be consistent with the law of conservation of energy. However, before considering the energies involved in nuclear decay, we will first describe the notation used to discuss decay and introduce some of the other conservation laws that determine the outcome of nuclear decay.

We can write equations for the radioactive decay of nuclei in a manner analogous to the way in which chemical equations are written. The radiation that Becquerel first observed in 1896 was the emission of alpha particles by the decay of uranium into thorium. The equation for this decay is written

$$^{238}_{92}\text{U} \rightarrow \,^{234}_{90}\text{Th} + \,^{4}_{2}\text{He}.$$

The rules that govern the "balancing" of such equations are *conservation of charge* and *conservation of the total number of nucleons*. Notice that the total number of nucleons (the sum of the protons and neutrons) is the same before and after the decay. As we have seen, this does not mean that mass is conserved in nuclear reactions. We refer to the original nucleus as the *parent nucleus* and the decay product as the *daughter nucleus*.

Example 29.3

The unstable nucleus $^{147}_{62}\text{Sm}$ decays by the emission of alpha particles. What is the daughter nucleus?

Strategy   We can write the decay equation for this problem, using X to stand for the unknown element:

$$^{147}_{62}\text{Sm} \rightarrow X + {}^4_2\text{He}.$$

Then we can use the laws of conservation of charge and conservation of nucleons to balance the equation and determine the daughter nucleus.

Solution   The total charge is 62 units on the left side of the equation. Because the alpha particle (helium nucleus) has 2 units of charge, we assign 60 units of charge to the nucleus X. There are 147 nucleons in the parent nucleus, and the alpha particle has 4 nucleons. Thus, the daughter nucleus must have 143 nucleons to conserve the total number of nucleons. We have

$$^{147}_{62}\text{Sm} \rightarrow {}^{143}_{60}X + {}^4_2\text{He}.$$

We may discover the name of element X by consulting the periodic table of the elements. The element with atomic number 60 is neodymium (Nd). We insert the symbol Nd in the above equation to give

$$^{147}_{62}\text{Sm} \rightarrow {}^{143}_{60}\text{Nd} + {}^4_2\text{He}.$$

Example 29.4

The function of the thyroid gland can be tested using radioactive iodine. A patient drinks a solution containing a small amount of $^{131}_{53}\text{I}$, which is taken up by a properly functioning thyroid. From the reading of a radiation detector placed near the patient's throat region, it is possible to determine if the thyroid is taking in iodine at a normal rate. What element remains after the iodine emits a negative β particle?

Strategy   As in the preceding example, we write down the decay equation, using X for the unknown nucleus:

$$^{131}_{53}\text{I} \rightarrow X + {}^0_{-1}\beta + \bar{\nu}.$$

Then we apply the laws of conservation of charge and conservation of nucleon number to balance the equation. (For completeness, we have included an additional uncharged particle of negligible mass that is also present in beta decay. This particle is called the antineutrino $\bar{\nu}$ and is discussed in Section 29.7.)

Solution   For electric charge to be conserved, the charge of the daughter nucleus must be one greater than that of the parent. Conservation of nucleons requires that the total number of nucleons be the same for the parent nucleus as for the daughter nucleus. The result is equivalent to the conversion of one neutron within the parent into a proton and an electron that is ejected. From the periodic table, we identify the daughter nucleus as xenon (Xe):

$$^{131}_{53}\text{I} \rightarrow {}^{131}_{54}\text{Xe} + {}^0_{-1}\beta + \bar{\nu}.$$

Discussion   Radioactive isotopes used in medicine often must have a short half-life in order to minimize the radiation exposure of the patient. For example, $^{131}_{53}\text{I}$ has a half-life of 8 days, so, from radioactive decay alone, only 9% of the initial activity remains after one month. You should calculate for yourself the amount remaining after one year. In addition, iodine is naturally excreted from the body, so the radioactive material that remains in the body is much less than the value determined here.

The emission of negative electrons from a nucleus—which contains only positive and neutral particles—also presents a conceptual problem. We will discuss the explanation in Section 29.7. Here we only point out that the transformation within the nucleus of a neutron into a proton requires the emission of a negative charge.

■

We can use energy considerations to investigate the important question of stability: Under what conditions can a nucleus spontaneously decay into particular products, and under what conditions is it stable? We answer the question by considering the relative mass energies of the parent nucleus and the decay products. For the decay to take place, the mass energy of the parent nucleus must be greater than the total mass energy of the decay products. If the decay products have more mass than the parent nucleus, additional energy is required from some other source in order to accomplish the reaction. Thus, radioactive decay is possible only if

$$m(\text{parent}) > m(\text{daughter}) + m(\text{decay particle}).$$

The difference between the initial and final mass energies is called the **Q value.** Thus, using $E = mc^2$, we have

$$Q = [m(\text{parent}) - m(\text{daughter}) - m(\text{decay particle})]c^2. \qquad (29.4)$$

A positive Q value indicates that a spontaneous reaction may occur, but a negative Q value indicates that a spontaneous reaction cannot take place. The positive Q-value energy is available to the decay products as kinetic energy.

Example 29.5

Determine whether $^{210}_{84}\text{Po}$ can decay by emission of an alpha particle, and if so, find the kinetic energy released in the process.

Solution   Using conservation of charge and nucleons, we can write the reaction as

$$^{210}_{84}\text{Po} \rightarrow\ ^{206}_{82}\text{Pb} +\ ^{4}_{2}\text{He}.$$

The observed masses are

$$m(^{210}_{84}\text{Po}) = 209.98285 \text{ u},$$

$$m(^{206}_{82}\text{Pb}) = 205.97440 \text{ u},$$

$$m(^{4}_{2}\text{He}) = 4.00260 \text{ u}.$$

The total mass of the decay products is

$$m(^{206}_{82}\text{Pb}) + m(^{4}_{2}\text{He}) = 209.97700 \text{ u}.$$

This is less than the mass of $^{210}_{84}\text{Po}$, so the decay may occur.

The total kinetic energy available is the Q value:

$$KE = Q = [m(^{210}_{84}Po) - m(^{206}_{82}Pb) - m(^4_2He)]c^2$$

$$KE = (209.98285 \text{ u} - 209.97700 \text{ u})c^2 = (0.00585 \text{ u})c^2$$

$$KE = (0.00585 \text{ u})(931.5 \text{ MeV/u}) = 5.45 \text{ MeV}.$$

This kinetic energy is shared between the alpha particle and the daughter nucleus.

29.6 Natural Radioactive Decay Series

In the years following the discovery of radioactivity, scientists found many radioactive elements, most of them relatively heavy. Eventually, they realized that these radioactive elements form decay chains or series. That is, when one nucleus decays into another, the resulting daughter nucleus is also radioactive and subsequently decays, and so on. The end product of each decay chain is a stable isotope of lead ($Z = 82$).

All known naturally radioactive nuclei can be grouped into one of three decay series, which originate with long-lived heavy elements. Each series is named after the longest-lived member of the series: uranium, thorium, and actinium. A fourth series, the neptunium series, consists of isotopes that no longer exist naturally in the earth's crust; it terminates in $^{209}_{83}Bi$. We can categorize each radioactive nucleus into one of these four series according to its mass number A. If n is an integer, then the series are labeled as follows:

| Thorium Series | $A = 4n$ | Uranium Series | $A = 4n + 2$ |
| Neptunium Series | $A = 4n + 1$ | Actinium Series | $A = 4n + 3$ |

A radioactive decay that begins with a member of one of these series produces daughter nuclei that are also members of the same series. Subsequent decay products remain in the same series because alpha decay involves a mass number change of four, while beta and gamma decay involve no change in the mass number A.

We can display these decay series by plotting the number of neutrons N ($N = A - Z$) against the number of protons Z for each nucleus in the series. An alpha decay leads to the lower left of the graph in Fig. 29.7, toward lower N and lower Z. A beta decay leads to the lower right, toward lower N but higher Z. Starting in the upper right-hand corner in each case, you can trace out the sequential decays as the process proceeds. Note that some radioactive nuclei have alternative decay modes, decaying by either alpha emission or beta emission.

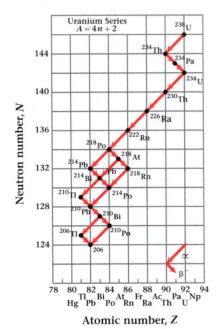

Figure 29.7
Plot of the number of neutrons $N = A - Z$ versus Z for the uranium series. Alpha decays proceed toward the lower left; beta decays proceed toward the lower right.

29.7 Models for Alpha, Beta, and Gamma Decay

We have seen that many naturally occurring nuclei are unstable and decay to other nuclei, with the emission of radioactivity. You may recall from our discussion of radioactive half-life (Sections 26.8 and 29.1) that the time elapsed before any individual nucleus undergoes decay is not subject to exact prediction, but is subject to statistical laws.

The details of the decay process are complex, but we will examine some of the major features, using our knowledge of conservation laws and quantum mechanics. In doing this, we will learn some important facts about nuclear structure and the nature of the subatomic particles themselves. We first classified nuclear decay processes into three types, corresponding to the emission of alpha, beta, or gamma radiation. Now let's look at the major features of each process.

Alpha Decay

A characteristic feature observed in alpha decay is that all alpha particles from a given type of nuclear reaction have essentially the same kinetic energy. This observation allows us to consider alpha decay as a simple breakup of a single parent nucleus into two pieces: the daughter nucleus and the alpha particle.

One approach to understanding alpha decay is to consider the α particle to be formed inside the nucleus before it is emitted. This seems reasonable in view of the energetic stability of the combination of two neutrons and two protons in one bound state (an alpha particle). However, the observed energies of emitted alpha particles are not great enough for them to have escaped the nucleus by overcoming the potential barrier at the nuclear surface (Fig. 29.8). The explanation of how the alpha particle gets out of the nucleus lies in quantum mechanical tunneling. As discussed in Chapter 28, even though the barrier height is greater than the particle's energy, the particle can still get through. The probability of tunneling through the barrier depends on the energy of the particle and the height of the barrier. After working out the details, one gets a prediction that agrees with the experimental observations: The more energetic alpha particles come from isotopes with shorter half-lives.

If the parent nucleus is initially at rest, then conservation of momentum requires that the alpha particle and the daughter nucleus move apart with equal but opposite momenta (Fig. 29.9). Their kinetic energies are provided by the Q value for the decay,

$$Q = \tfrac{1}{2}m_\alpha v_\alpha^2 + \tfrac{1}{2}m_d v_d^2,$$

where subscripts α and d refer to the alpha particle and the daughter nucleus, respectively. Any decay involves a specific energy Q, and in such a two-body breakup it is uniquely shared by the two decay products. For a daughter nucleus of mass number A, the masses m_α and m_d are essentially 4 and A, respectively, in atomic mass units. It can be shown that the kinetic energy of the emitted alpha particle is approximately

$$\boxed{\mathrm{KE}_\alpha \approx \frac{A}{A+4}Q.} \qquad (29.5)$$

The only naturally occurring nuclei that undergo spontaneous alpha decay are those with $Z > 82$. For these heavy nuclei, the fraction $A/(A+4)$ is approximately unity, and the alpha particle takes away almost all of the reaction energy Q.

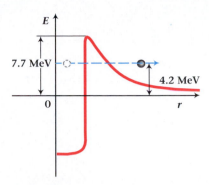

Figure 29.8
The potential-energy diagram of an α particle in a $^{235}_{92}$U nucleus. The observed kinetic energy (4.2 MeV) of the emitted alpha particle is not large enough to overcome the nuclear Coulomb potential barrier, which is greater than 7 MeV. However, the alpha particle can escape by quantum mechanical barrier penetration.

(a)

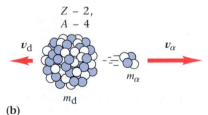

(b)

Figure 29.9
The decay of (a) an unstable nucleus Z, A into (b) a daughter nucleus $Z - 2$, $A - 4$ by the emission of an alpha particle. The momentum of the daughter nucleus is equal in magnitude and opposite in direction to the momentum of the alpha particle.

Example 29.6

Radium-223 decays via alpha emission to radon ($^{219}_{86}$Rn) with an energy release of $Q = 5.864$ MeV. Find the approximate kinetic energy of the alpha particle and its speed.

Strategy The approximate kinetic energy of the alpha can be found with the aid of Eq. (29.5). Then the speed of the alpha can be determined from the non-relativistic definition of kinetic energy, $KE = \frac{1}{2}m_\alpha v_\alpha^2$.

Solution The approximate kinetic energy of the alpha particle is

$$ KE_\alpha \approx \frac{A}{A+4}Q = \frac{219}{223}(5.864 \text{ MeV}) $$

$$ KE_\alpha \approx 5.76 \text{ MeV} = 9.23 \times 10^{-13} \text{ J}. $$

The mass of the alpha is 4.00 u, or 6.65×10^{-27} kg. The speed of the alpha is determined from the expression for kinetic energy,

$$ v_\alpha = \sqrt{\frac{2KE_\alpha}{m}} = \sqrt{\frac{2(9.23 \times 10^{-13} \text{ J})}{6.65 \times 10^{-27} \text{ kg}}} = 1.67 \times 10^7 \text{ m/s}. $$

Discussion This speed just computed is about 5% of the speed of light. Thus, the classical expression for energy is valid, and relativistic theory is not required, unless very high precision is called for.

Beta Decay

In 1932, Carl W. Anderson (1905-1991) discovered the **positron** or **beta plus** particle (β^+) in cosmic rays. Subsequent experiments confirmed that the positron is similar to an ordinary negative electron except for its charge. A positron has the same mass as an electron but carries one unit (e) of positive, rather than negative, electric charge. For clarity, the (negative) electron is often called the **beta minus** (β^-) particle. Positron emission has been observed in some nuclear decays. In a related decay process, known as **electron capture,** the nucleus absorbs an orbital electron from the atom. All of these processes in which the nucleus spontaneously emits or absorbs an electron or positron are called *beta decay.*

If we assume that beta decay is a breakup process in which the end products are a nucleus and an electron, then we should be able to calculate the final energy of the electron by using the laws of conservation of energy and momentum. Furthermore, because all nuclei of a given species have the same mass and all electrons have the same mass, the outgoing electrons should all have the same kinetic energy. Such behavior would be similar to alpha decay. However, this is not what we observe experimentally. Figure 29.10 shows the results of measurements made of the energies of electrons in beta decay. You should note two things about the data:

1. The electrons do not all have the same energy.

2. The energy predicted on the basis of conservation of energy and momentum for a two-body final state (marked by the arrow) is the upper limit for the observed energies.

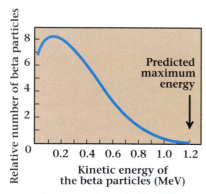

Figure 29.10

The distribution of kinetic energies of beta particles from the decay of bismuth-210. The arrow indicates the energy that the electron would have if the decay were a two-body process, leaving only the daughter nucleus and the electron.

This type of result has been obtained for both β^- and β^+ decay. The large discrepancy between what was predicted and what is observed gave rise to serious problems. In the early 1930s, it became apparent that either conservation of momentum had to be abandoned or some revision of the theory was necessary.

In 1934, Wolfgang Pauli proposed a solution that was so radical as to be almost unbelievable. However, it became widely accepted because it restored conservation of momentum to full standing. Pauli suggested that an additional particle is emitted in beta decay. This particle must be electrically neutral, because charge is already conserved without it. The mass of the particle must be zero or near zero, because some electrons in beta decay do have energies almost as high as the predicted electron energy. Other considerations indicated that the particle should interact only weakly with matter and that it was not electromagnetic in nature (that is, it could not be a photon). The Italian physicist Enrico Fermi (1901–1954) named this particle the **neutrino,** which means little neutral one. The symbol for the neutrino is ν (nu). Neutrinos were not observed by interaction with matter until 1953, when they were first detected by Frederick Reines and Clyde L. Cowan.

Fermi used the neutrino hypothesis to develop a theory of beta decay. His theory accurately predicts the shape of beta decay energy curves like that shown in Fig. 29.10. Close examination of the data in the region of the upper energy limits of these curves reveals that the mass of the neutrino is zero or very nearly so. According to Fermi's theory, any beta decay process is characterized by the change of a neutron into a proton or vice versa, with the simultaneous emission of a neutrino. Thus, the fundamental beta decay processes are

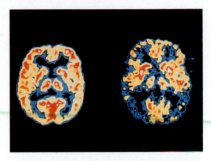

Positron emission tomography (PET) images of the brain of a normal patient (left) and of an Alzheimer's patient (right). To obtain the PET scans a radioactive tracer is injected into the bloodstream to reveal the metabolic activity of the brain.

$$
\begin{aligned}
& \text{n} \rightarrow \text{p} + \beta^- + \bar{\nu} && \text{(beta emission),} && \text{(29.6a)} \\
& \text{p} \rightarrow \text{n} + \beta^+ + \nu && \text{(positron emission),} && \text{(29.6b)} \\
& \text{p} + \text{e}^- \rightarrow \text{n} + \nu && \text{(electron capture).} && \text{(29.6c)}
\end{aligned}
$$

The Greek letter ν with a line over it (pronounced nu-bar) denotes an antineutrino. The distinction between neutrino and antineutrino need not concern us here; it is discussed in Chapter 32.

As a simple example of Eq. (29.6a), the decay of free neutrons into protons occurs with a half-life of about fourteen minutes. Such a decay is energetically favored, for it has a positive Q value. The decay of an isolated proton into a neutron is not allowed, since the neutron mass exceeds the proton mass.

Example 29.7

Calculate the Q value in the beta decay of a free neutron.

Solution If we assume that the antineutrino has no mass, we need only consider the mass of the proton, neutron, and electron:

$$Q = m_\text{n} - m_\text{p} - m_\text{e}.$$

Using the values from Table 29.1, we find that

$$Q = 939.566 \text{ MeV} - 938.272 \text{ MeV} - 0.511 \text{ MeV} = 0.783 \text{ MeV}.$$

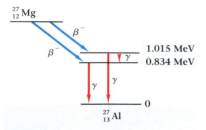

Figure 29.11

Nuclear energy level diagram for gamma decay of excited $^{27}_{13}$Al following the beta decay of $^{27}_{12}$Mg. The $^{27}_{12}$Mg decays to either of the excited $^{27}_{13}$Al states, which then eventually decay to the ground state by emitting gamma-ray photons, in a process analogous to the emission of photons of light by an excited atom.

In the beta decay of a free neutron, the upper limit of the kinetic energy of the beta particle is 0.783 MeV.

■

Gamma Decay

Frequently, the daughter nucleus resulting from alpha or beta decay is left in an excited energy state. These energy states are energy levels of the nucleus, analogous to the electronic energy levels of an atom. Nuclei in excited states can give up their excess energy by emitting electromagnetic radiation. That is, an excited nucleus can return to its lowest, or ground, state through the emission of a photon (Fig. 29.11). The photons associated with nuclear energy level changes are called gamma rays. Typical energies of gamma rays are in the range from tens of keV to a few MeV.

Gamma rays and x rays are both very energetic photons. The distinction between them has to do with their origins: X rays come from electronic excitations, and gamma rays come from nuclear excitations. In general, the gamma rays from nuclear decay are more energetic than the x rays from most x-ray machines, though there is some overlap (Fig. 22.2).

*29.8 Detectors of Radiation

So far, we have discussed the general ideas involved in nuclear stability and nuclear decay. Later we will discuss the units used for radiation measurement and the biological effects of radiation. However, first let's pause to look at the instruments we use to detect the products of nuclear decay.

By observing the flashes, or scintillations, made when radiation strikes a fluorescent screen, physicists gathered much of the early data concerning nuclear physics. However, looking at flashes through a microscope in a darkened room was not easy. (The story is told that Rutherford picked his assistants in part on the basis of how sharp their eyesight was for this task.) Two assistants in Rutherford's laboratory, Hans Geiger and W. Müller, developed a different technique for detecting particles. They invented what was called the Geiger-Müller tube, but is now more frequently referred to as simply a *Geiger tube*. This tube is the detecting element in a Geiger counter.

A typical Geiger tube consists of a cylindrical conducting tube with a wire electrode placed along the axis (Fig. 29.12). The ends of the tube are covered,

Figure 29.12

Schematic diagram of a Geiger counter. A charged particle entering the tube ionizes atoms of the gas, releasing a shower of electrons to the wire. The counter keeps track of each resulting current pulse.

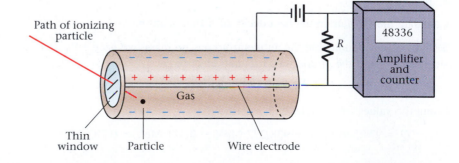

and the tube is filled with a special gas mixture, often argon with a small admixture of polyatomic gases, at a pressure below atmospheric pressure. A dc voltage is applied between the wire and the tube. This voltage must be higher than some threshold value, depending on the gas and the geometry of the tube, but is not so high that discharge takes place between the wire and the tube.

When a charged particle passes through the gas in the tube, it ionizes, or knocks some electrons out of, an atom of gas. The strong electric field between the tube and wire accelerates the freed electrons and gives them enough energy to ionize other atoms, freeing yet more electrons. This multiplication of charge allows a large number of electrons to be collected at the positive wire electrode. A large current then flows through the resistor, R, causing a large potential drop across the resistor and thereby reducing the potential difference between the central wire and the tube. This effect extinguishes the discharge, and the potential of the central wire returns to its original value. The tube is then ready to detect the next charged particle. The positive gas ions are attracted to the negative tube wall, but are prevented from cascading as the electrons do by the presence of the polyatomic gas. Each time a current pulse flows through the resistor, it is counted by an electronic counting circuit, which displays the number of counts on some numerical indicator. Frequently an audio circuit is included, which gives a "click" for each ionizing particle.

The Geiger tube may be provided with thin walls and with especially thin end windows for the detection of low-energy alpha and beta rays. X rays and gamma rays may produce an electron-positron pair in the walls and give rise to particles that will be counted.* However, the efficiency for detection of x rays and gamma rays is quite low.

Although the Geiger tube, or some variation of it, is useful for many purposes, another frequently used radiation detector is a modern descendant of Rutherford's fluorescent scintillating screen. A *scintillation counter* (Fig. 29.13) contains a material that emits a flash of light when radiation passes through it. A typical material for use as a scintillator is a single crystal of sodium iodide with a small amount of thallium; however, many plastics and even some liquids are commonly used. The scintillator is coated with a light-tight layer everywhere except where it is in optical contact with the front of a photomultiplier tube. The tube detects weak flashes of light.

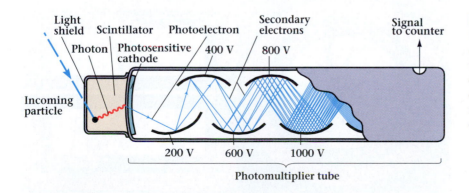

Figure 29.13
Schematic diagram of a scintillation counter. An incoming particle causes the scintillator to emit photons, which cause an electrode to emit photoelectrons. The chain of electrodes magnifies this effect into a sizable current pulse.

*The creation of particle-antiparticle pairs is discussed in Section 32.1.

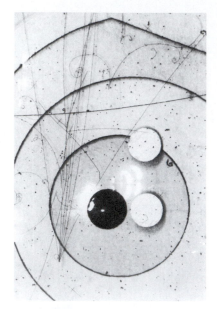

Figure 29.14

A neutrino-induced reaction in a freon-filled bubble chamber. The tracks are tiny bubbles generated by moving charged particles. The particles' paths are curved because of the magnetic field present.

The photomultiplier is a vacuum tube containing several highly sensitive electrodes. The potential of the electrodes increases with position along the length of the tube. When a photon from the scintillator strikes the electrode at the top of the tube, an electron is ejected by the photoelectric effect. This electron is then accelerated toward the next electrode, arriving with enough kinetic energy to release several secondary electrons from it. These electrons, in turn, accelerate toward the next electrode along the tube, triggering an avalanche effect. The end result is a sizable electric pulse, which is registered by an electronic counter. For many scintillation materials, the output of the photomultiplier tube is proportional to the energy of the incident radiation.

One advantage of scintillation counters is that they can be made very sensitive to the detection of x rays and gamma rays, radiation for which the Geiger counter has an efficiency of only a few percent. In addition, they can be ready to record a new event in as little as 10^{-9} s, while a typical Geiger tube requires the order of 10^{-4} s.

Another modern type of detector is the solid-state, or semiconductor, detector. (The physics of semiconductors is discussed in Chapter 31.) It has the advantages of being quite small, having a fast response, and not requiring a high-voltage power supply. Solid-state detectors are frequently useful when good energy resolution is required. However, they are not very efficient detectors; almost all incident particles pass right through the sensitive region of the device.

There are also several kinds of "track-visualization" devices. These instruments record the path in space of an individual particle. They include photographic film, the very first detector used by Becquerel in the discovery of radiation, and bubble chambers developed in the 1950s. In a bubble chamber, the paths of charged particles through a liquid are made visible by the formation of a trail of bubbles (Fig. 29.14).

Newer instruments for visualizing particle tracks are made from huge numbers of small detectors similar to the ones we have just described. By interfacing all of these small individual detectors to a large computer, the paths taken by the particles can be reconstructed in three dimensions (Fig. 29.15). In addition, by applying the laws of basic physics (including relativity), the momentum, energy, and identity of the individual particles can be determined.

Figure 29.15

Particle track reconstructions from the ALEPH detector at CERN. These particles are created in the collision of very high energy particles at the center of the detector.

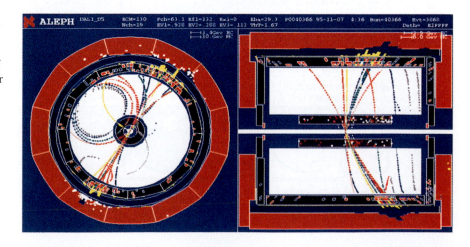

*29.9 Radiation Measurement and Biological Effects

Everyone knows that radioactivity can be dangerous to your health. But before we can examine these biological effects, we need to establish units for measuring how much radioactivity a body is exposed to and how much of this radioactivity a body absorbs. The study of radioactivity measurement, both physical and biological, is called dosimetry.

As we mentioned previously, the number of radioactive disintegrations per unit time is the *activity* of the substance. The SI unit of activity is the *becquerel* (Bq), which is defined as one disintegration per second. The *curie* (Ci), an earlier unit of activity that is sometimes still used, is defined to be 3.7×10^{10} disintegrations per second; this is the activity of one gram of pure radium. The curie turns out to be a large unit of activity, so mCi and μCi are more commonly used.

For discussing the effects of radiation, other quantities are often more useful than activity. The ionizing ability of radiation is important because radiation damage in biological cells is primarily due to excess ionization in the cell. One measure of ionizing ability is the electric charge released when air is ionized by x rays or gamma radiation. The unit is the *roentgen* (R), defined to be 2.58×10^{-4} C/kg of air. However, since the roentgen is defined only for x rays and gamma rays in air, it is not widely used today.

An important quantity for measuring physical effects due to absorption of radiation in matter is called the absorbed dose D, or simply the **dose.** The dose is defined as the energy deposited per unit mass by absorbed radiation. The SI unit of dose is the *gray* (Gy), defined as 1 J/kg. The gray can be used for measuring the energy absorbed from any type of radiation in any material. Another common unit for dose is the *rad* (radiation absorbed dose). One rad equals 0.01 Gy.

Radiation gives rise to two types of biological effects: genetic and somatic. Genetic effects cause mutations in the reproductive cells of an organism and so affect subsequent generations. Since genetic damage occurs only when the reproductive cells are irradiated, the gonads should be shielded if possible any time you are being x-rayed. The younger you are, the more important this is.

Somatic effects (sometimes called "radiation sickness") harm individuals directly, primarily by ionizing molecules in cells. The extent of the damage depends not only on the type of radiation, but also on the part of the body irradiated and on the individual's age. Again, generally speaking, the younger you are, the more hazardous the radiation. In fact, radiation doses before birth are the most dangerous, and pregnant women often require special precautions against radiation exposure. Some somatic effects include reddening of the skin, loss of hair, ulceration, fibrosis of the lungs, formation of holes in tissues, reduction of white blood cells, and induction of cataracts in the eyes. Perhaps one of the most feared effects of radiation is the occurrence of cancer on the skin and in various other organs of the body.

Radiation is also used to destroy cancers. For example, prostate cancer can be treated by inserting tiny pellets of radioactive ^{125}I directly into the prostate. The pellets are carefully placed with the aid of magnetic resonance imaging (MRI) and ultrasonic imaging to deliver doses as big as 150 Gy to the cancer with minimal dose to the surrounding tissues.

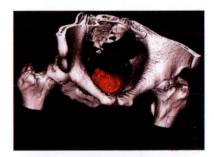

Computed image of the pelvic region. The red portion is the envelope of the dose from radioactive pellets implanted in a cancerous prostate.

Table 29.3	Quality Factor for Several Types of Radiation	
Radiation Type		**Quality Factor**
X rays, gamma rays, beta particles		1
Neutrons, energy < 10 keV		3
Neutrons, energy > 10 keV		10
Protons		10
Alpha particles, fission fragments, recoil nuclei		20

The extent of damage to biological organisms due to radiation depends on the particular type of radiation as well as on the energy deposited. Thus, the dose, which measures only the energy deposited per unit mass, is not sufficient for indicating damage to living tissue. Instead, we use the **dose equivalent** H, which measures the product of the dose (in grays) times a dimensionless number, called the **quality factor** Q, which takes account of the biological effect of each radiation. The quality factor is a judgment based on experiments and experience and is not the direct result of experimentation alone. The result is an equivalent dose, measured in the SI unit called the sievert (Sv):

$$H \text{ (in Sv)} = D \text{ (in Gy)} \times Q. \qquad (29.7)$$

Table 29.3 lists the quality factors for several types of radiation. The equivalent dose is also measured in *rem* (roentgen equivalent man), where 1 rem = 0.01 Sv.

To guard against the health hazards of radiation, the U.S. Nuclear Regulatory Commission (NRC) has established upper limits of acceptable equivalent doses for human exposure to radiation. Table 29.4 gives the present recommended maximum permissible dose for radiation workers and for the general population. The NRC reviews and revises these standards periodically. We should also point out that other countries establish their own standards for maximum permissible equivalent dose, often different from U.S. standards.

People who are likely to receive radiation exposure in their occupation or who enter a high-radiation area must be provided with a personal dosimeter to monitor their exposure. A dosimeter is any device or instrument capable of measuring radiation dose and is used to show compliance with governmental regu-

An OSL dosimeter. External appearance (above) and inside view (below). The large white square is the aluminum oxide detector. Filters of copper and tin are used to discriminate different types and energies of radiation.

Table 29.4	Recommended Maximum Permissible Doses, Excluding Intentional Medical Exposures	
Type of Exposure		**Maximum Permissible Dose (mSv/y)**
Radiation workers:		
Whole body, gonads, or lenses of eyes		50
Skin of whole body		300
Hands and feet		750
General population:		
Whole body		5
Gonads		1.7

Table 29.5	Estimated Annual Effective Dose Equivalent to Individuals in the United States	
Source		**Average Annual Effective Dose Equivalent in the U.S. Population* (mSv)**
Natural radiation		
Cosmic rays		0.280
Terrestrial		0.280
In the body		0.390
Inhaled radon		≈2.000
Radiation from human activity		
Occupational		0.009
Nuclear fuel cycle		0.0005
Consumer products		0.05–0.13
Environmental sources		0.0006
Medical		
Diagnostic x rays		0.39
Nuclear medicine		0.14
Rounded Total		3.6

Source: Taken with permission from "Ionizing Radiation Exposure of the Population of the United States," NCRP Report No. 93 (National Council on Radiation Protection and Measurements, Bethesda, MD, 1987).

*The effective dose equivalent is essentially the dose equivalent when the whole body is irradiated uniformly.

lations. Typical dosimeters are small enough to be worn clipped to a shirt pocket or lapel where they are used to measure the body's exposure to radiation.

Although several types of dosimeters are available, optically stimulated luminescent (OSL) dosimeters are widely used today. They measure radiation exposure due to x-ray, beta, and gamma radiation through a thin layer of aluminum oxide. Some energy from the radiation is retained in the aluminum oxide in the form of electrons excited to long-lived metastable states. When stimulated with laser light, the aluminum oxide becomes luminescent in proportion to the amount of radiation exposure. The emitted light (luminescence) is measured and the radiation exposure is computed from the amount of light released.

In our daily environment, we are subject to radiation exposure from two natural sources: cosmic rays and naturally occurring radioactive isotopes on the earth's surface. This exposure is known as background radiation and varies somewhat from place to place. For example, cosmic ray exposure is much greater at high elevations than at sea level. Consequently, airplane passengers receive a tiny radiation dose while flying that they would not receive on the ground. Additionally, there are several common sources of radiation that are the result of human activity, some of which are listed in Table 29.5. Figure 29.16 shows the relative contributions of the major sources of radiation to the exposure of the population of the United States.

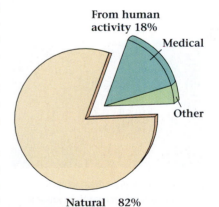

Figure 29.16
The percentage contribution of various radiation sources to the total average effective dose equivalent in the U.S. population. Data are from the source cited in Table 29.5.

Although there is some uncertainty in the exact value, it is clear that radon, a colorless odorless gas, is a major contributor to the equivalent dose in the United States. Radon is part of the natural decay chain of the uranium found in some rocks, and it seeps into houses and buildings through cracks in the foundations. The inhaled radon decays to alpha emitters in the lungs and causes damage and tumor growth. The Environmental Protection Agency estimates that radon causes between 5,000 and 20,000 of the 140,000 annual lung cancer deaths in the United States.

The amount of radon released in the soil varies markedly from place to place because of the underlying geological formations, so not every area receives the same dose. Furthermore, how the foundation of a building is sealed and how the building is ventilated make a difference. You can now buy inexpensive radon monitors for your home, and the Environmental Protection Agency publishes several helpful booklets.

There are two basic methods of shielding against any radiation: distance and mass. The simpler is distance. Most radiation is emitted equally in all directions from its source, so its intensity decreases as $1/r^2$. For example, increasing your distance from a radioactive source by three times reduces your received dose by a factor of nine. The other means of shielding is to place absorbing material around the source. In most cases, for a given thickness of shielding, the more massive it is, the more effective it is. Thus, lead is best for most cases. However, lead is sometimes prohibitively expensive, and so thicker pieces of other material, such as special types of concrete, are used. Because neutrons interact with matter differently from the way other radiation does, shielding against them is best done with a material such as cadmium, which has a high probability of capturing a neutron, or with material containing light nuclei, such as hydrogen, which can absorb the neutron's kinetic energy. Thus, water and paraffin are among the materials commonly used for neutron shielding.

Example 29.8

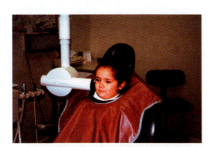

X-ray images allow dentists to see developing teeth before they emerge.

A patient receives a dose equivalent of 1.0 mSv in 0.20 kg of tissue from a dental x-ray machine operating at 90 keV. (a) What is the total energy deposited in the patient? (b) How many x-ray photons contribute to the dose if you assume that each photon gives up all its energy?

Solution (a) The total energy absorbed, E_T, is the dose times the mass of the sample. Thus, to find the total energy we must first find the dose (in grays). X rays have a quality factor of 1, so we get

$$D = \frac{H}{Q} = \frac{1.0 \text{ mSv}}{1} = 1.0 \times 10^{-3} \text{ Gy}$$
$$D = 1.0 \times 10^{-3} \text{ J/kg}.$$

The total energy absorbed, E_T, is

$$E_T = (1.0 \times 10^{-3} \text{ J/kg})(0.20 \text{ kg}) = 2.0 \times 10^{-4} \text{ J}.$$

(b) Each x-ray photon has an energy E_γ of

$$E_\gamma = (90,000 \text{ eV})(1.60 \times 10^{-19} \text{ J/eV}) = 1.44 \times 10^{-14} \text{ J}.$$

The number of photons N is

$$N = \frac{\text{total energy}}{\text{energy per photon}} = \frac{E_T}{E_\gamma} = \frac{2.0 \times 10^{-4} \text{ J}}{1.44 \times 10^{-14} \text{ J/photon}}$$

$$N = 1.4 \times 10^{10} \text{ photons.}$$

29.10 Induced Transmutation and Reactions

In 1902, Rutherford and Soddy proposed that when a radioactive element decayed, the remaining material was a different chemical element. This view was accepted only after several years of experimental verification. With the discovery of radioactivity, the dream of the alchemists had come true: One element was changed into another. Yet it seemed that there was no way to control this transmutation process. No way was known of speeding up or slowing down the rates of radioactive decay or of causing nuclei to decay if they did not already do so naturally.

However, in 1919, Rutherford made the remarkable discovery that nuclear transmutation can be induced, or caused to happen. Rutherford's apparatus is shown in Fig. 29.17. A radioactive source was placed at D within a chamber that could be either evacuated or filled with some gas through the tubes A. In the first part of the experiment, the chamber was filled with hydrogen. When alpha particles from the radioactive source collided head-on with the protons of hydrogen gas, the protons acquired a speed greater than that of the alpha particles. The faster, lighter protons were more penetrating than the alpha particles. When the hydrogen was replaced by air, which contains nitrogen, a surprising effect was observed. Rutherford expected that the head-on collisions of alphas with any of the heavier components of the air would yield slowly moving particles. Instead, he observed rapidly moving particles. He concluded that these rapidly moving particles were protons and that the only explanation was that the nitrogen was disintegrated in the collision with the rapidly moving alpha particle. Rutherford described the reaction as

$$_2^4\text{He} + {}_7^{14}\text{N} \rightarrow {}_8^{17}\text{O} + {}_1^1\text{H}.$$

The resultant product must be ${}_8^{17}\text{O}$ in order to conserve both charge and nucleon number.

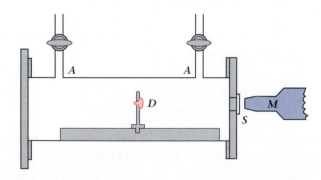

Figure 29.17

Rutherford's schematic drawing of the apparatus used in the discovery of induced transmutation. D = radiation source, A = gas inlet and outlet, S = zinc sulfide screen, and M = microscope.

Figure 29.18
One of Van de Graaff's early electrostatic generators. The first electrostatic generator announced by Van de Graaff in 1931 was capable of producing a potential difference of 1,500,000 V. A later machine was capable of reliably accelerating electrons or ions to energies of 2.75 MeV.

Other elements were soon investigated to see if similar results could be obtained. Two reactions that were found early were

$$\,^4_2\text{He} + \,^9_4\text{Be} \rightarrow \,^{12}_6\text{C} + \,^1_0\text{n}$$

and

$$\,^4_2\text{He} + \,^{11}_5\text{B} \rightarrow \,^{14}_7\text{N} + \,^1_0\text{n}.$$

In both cases, alpha particles were used as the bombarding projectiles. However, the alpha particles from naturally radioactive materials have only enough energy to overcome the repulsive Coulomb potential barrier (Fig. 29.8) of a limited number of nuclei. The need to experiment with particles of higher energy led to the development of what are called particle accelerators. An early accelerator is shown in Fig. 29.18. More recent ones are discussed in Chapter 32.

The first reaction observed with machine-accelerated nuclear particles was produced by the bombardment of lithium foil with protons:

$$\,^1_1\text{H} + \,^7_3\text{Li} \rightarrow \,^4_2\text{He} + \,^4_2\text{He}.$$

Now we can bombard almost any nucleus with almost any other one. We may write a general nuclear reaction in the form

$$A + B \rightarrow C + D. \tag{29.8}$$

The total energy of the particles on the left in Eq. (29.8) must be the same as the total energy of the particles on the right. By "total energy" we mean the sum of the kinetic energies and the rest-mass energies. There is no potential-energy term to be included because the short-range nuclear forces act only during the brief time of the interaction itself.

A given reaction, such as the general one in Eq. (29.8), is characterized by the masses of the nuclei involved. We can calculate the difference in the initial and final masses, and therefore the Q value, for any reaction we care to consider. If the Q value is positive, the kinetic energy of the products is greater than that of the reactants. If the Q value is negative, we must supply energy to the reactant nuclei for the reaction to actually take place.

For a reaction with a negative Q value, the reactant nuclei must have a kinetic energy equal to or greater than the magnitude of the Q value. If the target particle (B) is at rest, the kinetic energy of the incoming particle (A) must be high enough to supply the energy needed for the reaction. Otherwise the reaction will not occur. However, all of the incoming kinetic energy is not available for the reaction, as some must be transferred to the other particles in order to conserve momentum. The lowest energy for the incoming particle at which the reaction will take place is called the **threshold energy.** We can calculate the relationship between the threshold energy KE_{th} and the absolute magnitude of the Q value for the reaction, using conservation of energy and momentum. The result of the somewhat lengthy calculation is

$$\boxed{\text{KE}_{\text{th}} = (1 + m_A/m_B)\,|Q|.} \tag{29.9}$$

We must emphasize that among the principal tools used in investigating nuclear reactions are the conservation rules of momentum and energy. Reactions like those we have just described are two-body interactions, in which two particles come in and two particles go out. Their collision is handled in the same manner as are the collisions between larger bodies (Chapter 8). However, here we also use the Einstein mass-energy relationship. Examples of threshold energy calculations follow.

Example 29.9

Determine the minimum kinetic energy a proton must have to make the following reaction occur:

$$p + {}^{7}_{3}\text{Li} \rightarrow {}^{4}_{2}\text{He} + {}^{4}_{2}\text{He}$$

Solution The appropriate masses are atomic masses, so we use the mass of ${}^{1}_{1}\text{H}$ for the incident proton:

$$m({}^{1}_{1}\text{H}) = 1.007825 \text{ u},$$

$$m({}^{7}_{3}\text{Li}) = 7.016003 \text{ u},$$

$$m({}^{4}_{2}\text{He}) = 4.002603 \text{ u}.$$

We use these masses to find the Q value.

$$Q = \text{initial mass energies} - \text{final mass energies}$$
$$= [m({}^{1}_{1}\text{H}) + m({}^{7}_{3}\text{Li}) - 2m({}^{4}_{2}\text{He})]c^2$$
$$= (0.018622 \text{ u})(931.5 \text{ MeV/u})$$
$$Q = 17.3 \text{ MeV}.$$

The Q value for the reaction is positive, which means that energy is given off in the proposed reaction. Thus, there is no minimum threshold energy. If, on the other hand, the Q value had been negative, the reaction would not proceed by itself and kinetic energy would be required.

Example 29.10

Determine the threshold energy of the alpha particle in the reaction

$${}^{4}_{2}\text{He} + {}^{14}_{7}\text{N} \rightarrow {}^{17}_{8}\text{O} + {}^{1}_{1}\text{H}.$$

Solution The appropriate masses are

$$m({}^{4}_{2}\text{He}) = 4.002603 \text{ u},$$

$$m({}^{14}_{7}\text{N}) = 14.003074 \text{ u},$$

$$m({}^{17}_{8}\text{O}) = 16.999130 \text{ u},$$

$$m({}^{1}_{1}\text{H}) = 1.007825 \text{ u};$$

$$Q = \text{initial mass energy} - \text{final mass energy}$$
$$= (-0.001278 \text{ u})(931.5 \text{ MeV/u}) = -1.190 \text{ MeV}.$$

The Q value is negative, so kinetic energy must be supplied. The threshold, or minimum energy required, is

$$KE_{th} = [1 + m(_2^4H)/m(_7^{14}N)]\,|Q|$$
$$KE_{th} = [(1 + (4.0026)/(14.003)](1.190 \text{ MeV})$$
$$KE_{th} = 1.53 \text{ MeV}.$$

Thus, if the target $_7^{14}N$ nucleus is initially at rest, the alpha particle must have a kinetic energy of at least 1.53 MeV for the reaction to take place.

29.11 Nuclear Fission

In 1939, Otto Hahn and Fritz Strassman published their work on a special kind of nuclear reaction of both theoretical and practical interest. It was already known that many products of artificially induced transmutations were themselves radioactive. Hahn and Strassman, along with Lise Meitner, had bombarded uranium with neutrons. Their experiments were similar to other experiments being done at that time in an effort to create elements of atomic number greater than 92 (uranium).

The products of the reaction were radioactive, but there was great difficulty in identifying the actual nuclei. Finally, a series of extremely careful chemical analyses by Hahn showed that the reaction products contained both $_{56}^{139}Ba$ and $_{57}^{140}La$, two materials much lighter than uranium. Meitner and O. R. Frisch, working in Sweden, soon proposed the explanation that the uranium nucleus had split into two nuclei of smaller mass. The reaction, as we now understand, can lead to several such splittings, one of which is

$$_{92}^{238}U + _0^1n \rightarrow _{56}^{139}Ba + _{36}^{97}Kr + 3_0^1n.$$

Such a process is called **fission.** It is different from other nuclear reactions in that the unstable nucleus splits into two relatively heavy parts plus a few neutrons. Furthermore, the same products do not always result from the same initial bombardment. For instance, the following reaction is also observed:

$$_{92}^{238}U + _0^1n \rightarrow _{57}^{140}La + _{35}^{97}Br + 2_0^1n.$$

This explains why Hahn found both barium and lanthanum among the reaction products. Many other reactions are observed as well. Figure 29.19 shows the relative distribution of the products of the fission of $_{92}^{238}U$ by neutrons. Most fission fragments are highly excited and go through many decays before reaching a stable isotope. Other products accompany the nuclei, as shown in Fig. 29.20. Several neutrons are usually released in the fission process. For $_{92}^{238}U$, the average number of neutrons released per fission is 2.5. Other isotopes of uranium, as well as other heavy nuclei, can undergo fission.

Nuclei that undergo fission are of such a large size that we can describe the breakup in terms of a model, treating the nucleus as a liquid drop. A liquid drop that is disturbed oscillates and, if the disturbance is great enough, eventually breaks apart (Fig. 29.21). The fission process, in which the neutron provides the initial disturbance, is analogous to this behavior.

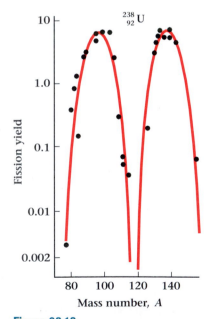

Figure 29.19
Fission yields from $_{92}^{238}U$. The fission yield is the relative number of times, here plotted in arbitrary units, that a nucleus of atomic number A is a product of fission.

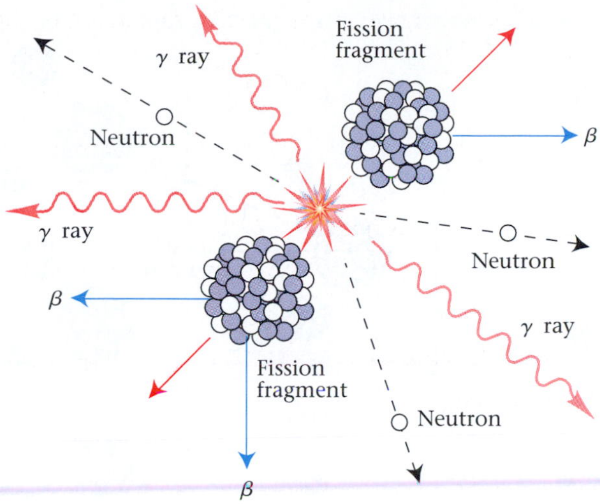

Figure 29.20
Schematic drawing of fission fragments.

Fission processes have the possibility of being of some practical use because they have positive Q values; that is, the kinetic energy of the products is greater than the kinetic energy of the initial reactants. The positive Q value occurs because the heavier elements have a smaller binding energy per nucleon than do the medium-mass elements (Fig. 29.6). The heavy elements can then break up into two components that are more tightly bound and therefore have less total mass energy. On the average, about 200 MeV of energy are released per fission. Compare this value with the average energy in chemical reactions, which is less than 1 eV per molecule.

The large energy produced per fission would be of no practical use if it were not for the fact that each fission releases an average of 2.5 neutrons. If we place the fissioning nuclei in the proper configuration, these neutrons can cause other fissions. The neutrons from those fissions can cause still other fissions, each fission releasing more neutrons and more energy. Such a process is called a **chain reaction** (Fig. 29.22). If the process occurs in an uncontrolled fashion, the result is an explosion. If the process is carried out in a manner that limits the number of neutrons causing successive fission events, then the heat due to the kinetic energy can be extracted and made to do useful work, as in nuclear-fueled power plants. In either case, the products of the fission process are extremely radioactive. The products are mostly long-lived gamma-ray emitters, which offer considerable health hazard.

Figure 29.23 shows the basic components of a fission power plant. The first successful self-sustaining, controlled nuclear chain reaction was accomplished by Enrico Fermi and his colleagues at the University of Chicago in 1942. Although the design of the reactor in Fig. 29.23 differs from that of Fermi's original reactor, all reactors have several things in common.

Most reactors use uranium for their nuclear fuel. Natural uranium is mostly $^{238}_{92}U$ (about 99.3%) with a small amount of $^{235}_{92}U$ (about 0.7%). It is the $^{235}_{92}U$ isotope that plays the important role in the chain reaction.* Because the amount of

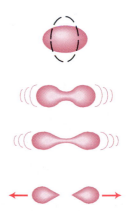

Figure 29.21
Oscillation and breakup of a liquid drop. This process can serve as a model for the fission of a large unstable nucleus.

---

*The $^{238}_{92}U$ isotope does not readily fission when struck by a neutron; instead it captures the neutron to become $^{239}_{92}U$, which, after two beta decays, becomes $^{239}_{94}Pu$. The isotope $^{235}_{92}U$ fissions readily when struck by a slowly moving neutron.

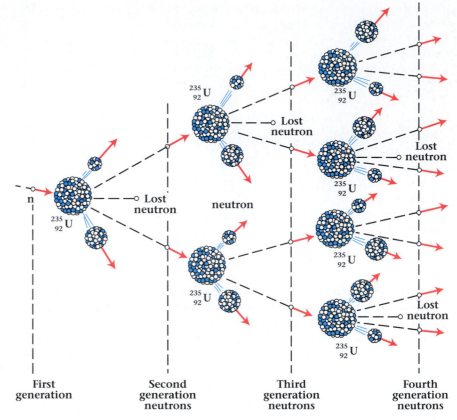

Figure 29.22

Schematic representation of a chain reaction. Here in each step the number of fissioning nuclei increases. In a controlled process, this number is made to stay constant by absorbing the excess neutrons in a moderator.

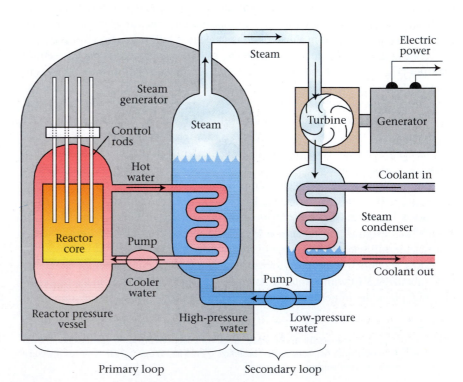

Figure 29.23

The basic components of a fission power plant. The primary loop is kept in a containment structure to seal in radioactivity.

$^{235}_{92}$U in natural uranium is so small, the uranium used for nuclear fuel is enriched to about 3.5% of this isotope.

The probability that an incident neutron will cause $^{235}_{92}$U to fission is significant only for low-energy neutrons. The fast neutrons released by fission must be slowed down so that they can cause other fissions to occur. This is accomplished by a *moderator,* sometimes carbon, ordinary water, or heavy water (D_2O). The neutrons lose kinetic energy by collisions with the moderator. The neutrons eventually slow down until their kinetic energies are of the order of kT, where k is the Boltzmann constant and T is the temperature of the moderator. Slow neutrons of these energies are called *thermal neutrons.*

The placement of the nuclear fuel, which is normally made into the form of long rods, must be carefully designed. The uranium and the moderator must be arranged so that for every fission at least one neutron goes on to cause another fission. There also must be a way to control the rate at which fissions occur. It is necessary to be able to start the chain reaction slowly, regulate it when in progress, and stop it when desired. This control is achieved with control rods of neutron-absorbing material such as cadmium. When the control rods are fully inserted among the fuel rods, so many neutrons are absorbed that the chain reaction is blocked. The rods are slowly withdrawn to allow the reaction to begin and to build up to the desired level for a self-sustaining chain reaction.

The kinetic energy of the fission fragments and neutrons is converted to thermal energy. Thus, the reactor is a large heat source capable of extremely high temperatures. The thermal energy is removed by a liquid coolant that flows through pipes in the reactor core. This primary coolant passes through a heat exchanger, where it gives up heat to a secondary coolant, which may be used to drive a turbine that turns an electric generator. Thus, the energy of fission can be used to generate electricity.

Nuclear reactors are reasonably efficient sources of electrical energy, but they have at least two principal problems. The first problem is with the safety of the reactor itself. The chain reaction must not be allowed to get out of control so that the fuel melts or pressures build up and cause a break in the surrounding containment vessel, because such a break could release radioactive materials into the environment. The solutions to this problem are well within current technology, but require large expenditures of money and effort. A more serious and long-term problem involves the disposal of the radioactive waste from the spent fuel. The solution of this problem is less well agreed upon, and to a great extent involves political and economic decisions, as well as scientific ones.

Special nuclear reactors are designed for research. The High Flux Isotope Reactor (HFIR) at the Oak Ridge National Laboratory is able to produce an extremely high flux of neutrons in its core. These neutrons bombard various target materials to produce elements heavier than uranium. Beams of thermal neutrons can also be used for neutron diffraction studies of magnetic solids.

HFIR, a water-cooled and water-moderated reactor operating at 85 MW, produces intense radiation. A blue glow in the water surrounding the radioactive core is caused by high-speed beta particles emitted in the decay of fission products (Fig. 29.24). This glow, called **Cherenkov radiation,** is analogous to the shock wave caused by an object moving faster than the speed of sound. Here the shock wave is electromagnetic, caused by charged particles exceeding the speed of light in the water. (The speed of the particles is less than c, the speed of light in a vacuum, in accord with the theory of relativity.)

Figure 29.24
Cherenkov radiation is seen as the blue glow in the water surrounding spent fuel as it is removed from the High Flux Isotope Reactor.

LISE MEITNER AND NUCLEAR FISSION

In 1907, Lise Meitner, the second woman to receive a doctorate in physics in Vienna, came to Berlin to attend the lectures of Max Planck. At a colloquium she met the chemist Otto Hahn (1879–1968), who became her closest working colleague for more than 30 years. Hahn, who had studied with Rutherford, had his laboratory at the University of Berlin. Meitner was given a position at the university only after some initial objections due to her being a woman, with the restrictions that she not go to the student laboratories on the upper floors and that she confine her work to the converted carpentry workshop that was Hahn's laboratory.

Rutherford's 1919 discovery that nuclear reactions could be caused by bombarding an element with an alpha particle gave rise to many similar experiments. A typical experiment was to bombard an element with some radioactive decay particle and then determine what element was left and what particle was given off. Often only a few thousand such reactions would take place, giving insufficient products to be analyzed by the methods of ordinary chemistry. This gave rise to the science of radiochemistry, in which elements could be identified from the properties of their radioactive emissions. The team of Hahn and Meitner was one of the most skillful practitioners of radiochemistry, and discovered the element protactinium in 1917.

Chadwick's discovery of the neutron in 1932, gave researchers another atomic projectile with which to work. In Rome, Enrico Fermi and his colleagues made a specialty of the neutron bombardment technique. In these studies the atomic number of the end product was not very different from that of the initial nucleus. However, Fermi noticed that this was not the case with the neutron bombardment of uranium.

In Berlin, Hahn and Meitner were joined in 1935 by Fritz Strassman (1902–1980) and these three did experiments similar to those of Fermi. Initially they agreed with Fermi that transuranic elements (elements with atomic number greater than 92, which are not found in nature) were being created.

Figure B29.1   Lise Meitner (1878–1968).

However, in 1938 they began an extremely careful radiochemical analysis of the products resulting from the neutron bombardment of uranium. Just at this time, Nazi Germany invaded Austria, and Meitner, an Austrian Jew, became unsafe. She fled by train to Holland and then to Sweden, but kept in touch by letter with the work continued by Hahn and Strassman.

Hahn's conclusions, relayed by letter to Meitner, were that the decay products were significantly lighter than uranium, and included barium, which has an atomic number of 56. Hahn wrote that as a "chemist" he was sure that there was barium, but that as a "nuclear chemist" he could not believe it. Meitner invited her nephew Otto Frisch, a physicist at Bohr's institute in Copenhagen, to visit her in the town of Kungälv on Sweden's west coast, where they discussed Hahn's results. Using Bohr's liquid drop model of the nucleus, Meitner and Frisch were able to explain Hahn and Strassman's observation as a breaking apart of the nucleus into two more-or-less equal pieces. Meitner calculated that the mass defect would indeed provide enough energy for such a reaction, and Frisch calculated that the "surface tension" of the uranium nucleus would indeed be very small. When they wrote their paper explaining the experimental observations, they called the break-up "fission" after the biological process of cell division.

Upon returning to Copenhagen, Frisch told Bohr of his and Meitner's work. Shortly after, Bohr inadvertently announced this result during a meeting in the United States, before Frisch and Meitner's paper was published. The interest in fission became so intense that it resulted in more than 100 papers on fission being published in the year after Bohr's talk. The large energy release per nuclear reaction was so striking that the potential of fission to produce large amounts of energy was obvious. Eventually this led to the establishment in the United States of the Manhattan Project, which was the code name for the military development of a fission bomb and which was directed in part by Fermi.

Otto Hahn was awarded the Nobel Prize in chemistry in 1944 for the experimental discovery of fission. Lise Meitner and Otto Frisch are generally credited with explaining what happens during fission.

29.12   Nuclear Fusion

Inspection of Fig. 29.6 suggests another energy-releasing process. When undergoing fission, heavy nuclei release energy in going from H to M on the graph. We should also be able to obtain energy by combining two elements in the region L, forming nuclei of greater binding energy per nucleon in the region M. The process of combining two nuclei to form a heavier one is called **fusion.** An example is

$$\,^2_1\text{H} + \,^2_1\text{H} \rightarrow \,^1_0\text{n} + \,^3_2\text{He},$$

where $\,^2_1\text{H}$ represents the naturally occurring heavy isotope of hydrogen, known as deuterium (D). Deuterium is a naturally occurring component of water and can be extracted from the sea in large amounts. This reaction releases approximately 4 MeV, which is about the same energy per nucleon as is released by fission. An attractive feature of this reaction is that the product $\,^3_2\text{He}$ is stable and not radioactive.

In order for energy to be released in such a fusion reaction, the two deuterium nuclei must get close enough together for the nuclear forces to act. At distances larger than the range of the nuclear force, the repulsive Coulomb force dominates. (Recall Fig. 29.8, which depicts the nuclear Coulomb potential experienced by an alpha particle.) The energy necessary to accomplish the reaction is the energy needed to overcome the Coulomb potential energy barrier at a distance of about 10^{-14} m, the approximate distance at which the reaction will occur. This energy is

$$E = k\frac{e^2}{r} = \frac{(9 \times 10^9\ \text{N} \cdot \text{m}^2/\text{C}^2)(1.6 \times 10^{-19}\ \text{C})^2}{10^{-14}\ \text{m}}$$

$$E = (2.3 \times 10^{-14}\ \text{J})(1\ \text{MeV}/1.6 \times 10^{-13}\ \text{J}) = 0.14\ \text{MeV}.$$

For the fusion reaction to be self-sustaining, its energy must be released in the vicinity of other deuterium nuclei so that they can subsequently interact. In effect, we need a hot gas of deuterium. The temperature of a deuterium gas in which the particles would have an average kinetic energy of 0.14 MeV can be determined from

$$\tfrac{3}{2}kT = \overline{\text{KE}}.$$

The temperature is

$$T = \tfrac{2}{3}\overline{\text{KE}}/k \approx 1 \times 10^9\ \text{K}.$$

The atoms in a gas at such high temperatures are completely ionized. Such a high-temperature gas of positive and negative charged particles is called a *plasma.*

The energy released from fusion is what keeps the sun and other stars hot. The extremely high temperatures and densities, in turn, provide the conditions necessary for a self-sustaining reaction. Even though the reaction will actually take place at temperatures that are lower by a factor of 100, the principal barrier to practical self-sustaining fusion reactions in the laboratory is clear: The high-density fusion fuel must be maintained at an extremely high temperature.

One of the most likely ways of accomplishing a self-sustaining fusion reaction on earth is to compress a high-temperature plasma using magnetic fields.

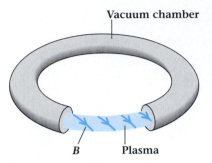

Figure 29.25
Schematic diagram of the toroidal vacuum chamber of a tokamak fusion chamber. The plasma is confined by the helical magnetic field.

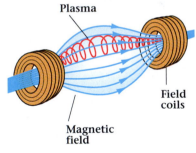

Figure 29.26
The magnetic mirror principle of plasma confinement.

Two methods of doing this are currently under study: the tokamak and the magnetic mirror configuration.

The tokamak, first devised in the former Soviet Union, has a toroidal-shaped chamber with a complex arrangement of coils to generate a magnetic field that spirals around the torus (Fig. 29.25). The circulating plasma is compressed and kept away from the walls by the magnetic field. Research using the tokamak principle is under way in the United States, Great Britain, Japan, and the Commonwealth of Independent States (formerly the USSR).

An alternative method of containing the plasma uses a differently shaped magnetic field that reflects the charged particles back and forth between "magnetic mirrors" (Fig. 29.26). Charged particles move along helical paths in the region where the field is nearly uniform. In the region where the field lines come closer together, the field has a radial component. As the particles move into that region, they encounter a force that retards their motion into the high-field region. Eventually their motion toward the high-field region is reversed, and they are reflected back toward the region of lower field.

In addition to the magnetic confinement methods, other approaches to achieving fusion-produced power are possible. One approach is to bombard small solid pellets containing fusionable material with a burst of high intensity from many lasers at once. This pulsed technique attempts to achieve the required temperatures and densities for a sufficiently long time. Another, entirely different approach is to use muons as part of a lower-temperature fusion reaction. Muons are subatomic particles (discussed in Chapter 32) that are not stable and therefore must be produced by an accelerator.

There is hope that fusion reactors will become practical energy sources, especially since the deuterium fuel is relatively inexpensive and the direct waste products are not radioactive. Furthermore, there is no possibility of the fusion reaction running out of control and exploding. However, the early expectations of the 1950s and 1960s that nuclear fusion would provide practical power before the end of the century were too optimistic. It seems probable now that another 25 to 50 years of research and development will be required before fusion power becomes a practical reality. Despite their virtues, fusion reactors will not be the perfect source of energy. They will be extremely expensive to build, they will be extremely radioactive as a result of their high flux of neutrons, and they will be sources of thermal pollution.

Summary

Useful Concepts

■ A nucleus of the element X with charge Z and total number of nucleons A is written

$$^{A}_{Z}X.$$

■ The radius of a nucleus of mass number A is given approximately by

$$r = R_0 A^{1/3}.$$

■ Nuclear reaction equations at low energies can be "balanced"

using conservation of charge and conservation of nucleon number.

■ The Q value of a decay reaction is written

$$Q = [m(\text{parent}) - m(\text{daughter}) - m(\text{decay products})]c^2.$$

■ In alpha decay processes, the kinetic energy of the emitted alpha is

$$\text{KE}_\alpha \approx \left[\frac{A}{A+4}\right]Q.$$

■ The fundamental beta decay processes are

$$n \rightarrow p + \beta^- + \overline{\nu} \quad \text{(beta emission)},$$

$$p \rightarrow n + \beta^+ + \nu \quad \text{(positron emission)},$$

$$p + e^- \rightarrow n + \nu \quad \text{(electron capture)}.$$

■ The threshold energy for a nuclear reaction is

$$KE_{th} = (1 + m_A/m_B)|Q|.$$

■ The release of energy in both fission and fusion reactions is accomplished through a change in the binding energy per nucleon.

Important Terms

You should be able to write the definition or meaning of each of the following:

protons	neutrino
neutron	dose
nucleons	dose equivalent
isotopes	quality factor
binding energy	threshold energy
Q value	fission
positron	chain reaction
beta plus	Cherenkov radiation
beta minus	fusion
electron capture	

Conceptual Questions

29.1 What are the properties of a nucleus that is likely to undergo (a) alpha decay, (b) beta decay, and (c) gamma decay?

29.2 What would the distribution of electron kinetic energies be like if no neutrinos were emitted in beta decay? What would it look like if a very massive neutral particle were always emitted?

29.3 Why is the decay rate of radioactive nuclei not changed by chemical environment, heating, or pressure?

29.4 Why is radium still found in nature? Its half-life is 1600 years, and the estimated age of the universe is five billion years.

29.5 How can isotopes of an element be separated, since they have the same chemical properties?

29.6 An inventor claims to have invented a device with an input power of a few watts and an output power of many times the input power. He says it is not to be considered a "perpetual motion machine," because it converts mass into energy. The machine gives off no radiation and leaves no radioactive by-products. What tests would you suggest to test his claim, and what questions would you ask based on your knowledge of electricity, relativity, and nuclear physics?

29.7 Can you suggest any reasons why natural radioactivity contains helium nuclei but neither hydrogen nor lithium nuclei? (*Hint:* One key can be found in the binding energy of the decay products.)

29.8 What could be some of the reasons why stable nuclei have about the same number of protons and neutrons, with an excess number of neutrons as A increases?

29.9 A Geiger counter, or any other detection device, registers a finite, though perhaps small, count rate even when no clearly radioactive source is present. Name as many possible sources of this "background" radiation as you can.

29.10 What are some of the advantages and disadvantages of fusion power generation?

29.11 Over which of the radiation sources that affect you do you have any control? By completely eliminating all nonnatural sources, by what fraction could the average U.S. citizen reduce the received dose of radiation?

Problems

Section 29.1 Radioactivity

29.1 A student observes a count rate of 8576 counts/min in a specimen of radioactive Ba that has a half-life of 2.5 min. How long must the student wait before the count rate drops to below 100 counts/min?

29.2 A wooden archeological artifact is reported to be 2000 years old on the basis of carbon-14 dating. What is the fraction of carbon-14 found in the specimen to that found in new wood? The half-life of carbon-14 is 5700 years.

29.3 Radioactive iodine-131 has a half-life of 8 days and an initial activity of 3.6×10^7 Bq. What will be the activity after 45 days have passed?

Section 29.2 Chadwick's Discovery of the Neutron

29.4 Using the fact that Chadwick's unknown particle is now known to be a neutron of mass 1, find the expected ratio of the speeds of the outgoing protons from the bombardment of paraffin and of the recoiling nitrogen ions in Chadwick's experiment.

29.5 If Chadwick had used paraffin and carbon targets instead of paraffin and a nitrogen-containing substance in his neutron-scattering experiments, what value would he have obtained for the ratio of the final velocities for the two cases? Assume the masses of the neutron and the proton to be the same.

29.6 Imagine an experiment similar to Chadwick's in which magnesium, with a relative mass of 24, is used instead of the nitrogen target. What would be the expected ratio of the recoil velocities of the protons relative to the magnesium ions?

29.7•• Derive Eq. (29.1) from the principles of conservation of energy and momentum. Assume head-on collisions.

Section 29.3 Composition and Size of the Nucleus

Hints for Solving Problems

Remember that the atomic number Z is the number of protons in the nucleus. The number of neutrons is $A - Z$, where A is the atomic mass number.

29.8 Identify the isotope of atomic mass number 29 whose nucleus has one more neutron than protons.

29.9 A nucleus of neon contains ten neutrons. What is the atomic mass number, A, of this nucleus?

29.10 How many protons and how many neutrons are in the nucleus of each atom: $^{200}_{80}Hg$, $^{16}_{8}O$, and $^{232}_{90}Th$?

29.11 How many protons and how many neutrons are needed to make the following elements: $^{126}_{53}I$, $^{56}_{26}Fe$, and $^{207}_{82}Pb$?

29.12 Nuclei with the same number of nucleons but with the numbers of protons and neutrons interchanged are called mirror nuclei and are useful for studying nuclear forces. How many protons and how many neutrons would be needed to make the following mirror nuclei: (a) $^{39}_{19}K$, $^{39}_{20}Ca$, and (b) $^{23}_{11}Na$, $^{23}_{12}Mg$?

29.13 Natural chlorine occurs as a mixture of two isotopes. The isotope $^{35}_{17}Cl$ has a relative abundance of 75.5%, and the isotope $^{37}_{17}Cl$ has a relative abundance of 24.5%. Calculate the atomic mass of a natural sample of chlorine.

29.14 Natural copper occurs as a mixture of two isotopes. The isotope $^{63}_{29}Cu$ has a relative abundance of 69%, while the isotope $^{65}_{29}Cu$ has a relative abundance of 31%. What is the atomic mass of natural copper?

29.15 Natural magnesium occurs as a mixture of three isotopes: $^{24}_{12}Mg$ at 78.7%, $^{25}_{12}Mg$ at 10.1%, and $^{26}_{12}Mg$ at 11.2%. Calculate the atomic mass of a natural sample of magnesium.

29.16 What is the approximate radius of the nucleus of $^{235}_{92}U$?

29.17 What is the approximate volume of the nucleus of $^{138}_{56}Ba$?

29.18 What is the approximate mass number of a nucleus whose radius is measured to be 6.0×10^{-15} m?

29.19• A nucleus has 126 neutrons and a volume of approximately 1.51×10^{-42} m^3. Assume that Eq. (29.3) and the value given for R_0 are correct and identify the element.

29.20• Use the mass of the proton and Eq. (29.3) to estimate the density of nuclear matter.

Section 29.4 Nuclear Forces and Binding Energy

29.21 Show that 1 u = 931.5 MeV. (*Hint:* 1 u = 1.66054×10^{-27} kg, 1.6022×10^{-19} J = 1 eV.)

29.22 Using Fig. 29.6, estimate the approximate total binding energy of $^{20}_{10}Ne$ and $^{200}_{80}Hg$.

29.23 Using Fig. 29.6, estimate the approximate total binding energy of $^{58}_{28}Ni$ and $^{232}_{90}Th$.

29.24 Calculate the binding energy of $^{9}_{4}Be$ using the data in Table 29.2.

29.25 Calculate the binding energy of $^{5}_{2}He$ using the data in Table 29.2.

29.26• Using the data in Table 29.2, calculate the ratio of binding energy per nucleon of $^{4}_{2}He$ to that of $^{6}_{3}Li$.

29.27• (a) What is the ratio of the strength of the gravitational force to the strength of the Coulomb force between two protons separated by a distance of the order of nuclear dimensions (10^{-14} m)? (b) Is gravity important in the structure of the nucleus?

Section 29.5 Conservation Rules: Radioactive Decay and Nuclear Stability

Hints for Solving Problems

In decay reactions, charge is conserved and the number of nucleons is conserved.

29.28 When $^{6}_{3}Li$ is bombarded with neutrons, tritium ($^{3}_{1}H$) is emitted. What is the remaining nucleus?

29.29 Identify the daughter nucleus in the decay of $^{234}_{92}U$ by alpha particle emission.

29.30 What is the name of the element that results from the alpha decay of polonium?

29.31 What particle is emitted in the decay of $^{14}_{6}C$ to $^{14}_{7}N$?

29.32 What nucleus decays by negative beta emission to $^{90}_{39}Y$?

29.33 May the reaction $^{9}_{4}Be \rightarrow \alpha + ^{5}_{2}He$ take place naturally?

29.34• Identify the nucleus designated by X in each of the following reactions:

$$\text{(a) } ^{226}_{88}Ra \rightarrow X + \alpha.$$

$$\text{(b) } ^{233}_{91}Pa \rightarrow X + \beta^-.$$

$$\text{(c) } ^{59}_{26}Fe \rightarrow X + \gamma.$$

29.35• (a) What nucleus results from the negative beta decay of $^{66}_{29}Cu$? (b) What nucleus results from the positive beta decay of $^{38}_{19}K$?

29.36• Determine the approximate velocity of the alpha particle in Example 29.5.

29.37• (a) Determine whether $^{223}_{88}Ra$ can decay by the emission of an alpha particle, and if so, find the kinetic energy released in the process. (b) Determine whether $^{64}_{30}Zn$ can decay by the emission of an alpha particle, and if so, find the kinetic energy released in the process. The observed masses are $m(^{223}_{88}\text{Ra}) = 223.01850$ u, $m(^{219}_{86}\text{Rn}) = 219.00948$ u, $m(^{64}_{30}\text{Zn}) = 63.92915$ u, and $m(^{60}_{28}\text{Ni}) = 59.93079$ u.

Section 29.6 Natural Radioactive Decay Series

29.38 The isotope $^{234}_{92}\text{U}$ undergoes five successive alpha decays. Identify the daughter nucleus at each decay step.

29.39 (a) In which decay series do you find the isotope $^{235}_{92}\text{U}$? (b) In which do you find $^{228}_{88}\text{Ra}$?

29.40 (a) Which uranium isotope belongs to the neptunium decay series? (b) To the actinium decay series?

29.41 (a) In which decay series do you find the isotope $^{234}_{91}\text{Pa}$? (b) In which do you find $^{231}_{91}\text{Pa}$?

Section 29.7 Models for Alpha, Beta, and Gamma Decay

> **Hints for Solving Problems**
>
> The Q value is the difference between the initial mass and the final mass. It may be expressed as an energy. A positive Q value means that a decay can take place spontaneously.

29.42 Show that an alpha particle with a kinetic energy of 5 MeV may be treated as a nonrelativistic particle.

29.43 Calculate the Q value for the beta decay of tritium (^3_1H).

29.44 Calculate the approximate Q value of the decay $^{218}_{84}\text{Po} \rightarrow {}^{214}_{82}\text{Pb} + \alpha$, given that the measured kinetic energy of the alpha is 5.998 MeV.

29.45● The atomic mass m_a of a given isotope $^A_Z X$ can be considered to be the sum of the nuclear mass m_N and the mass of Z electrons; that is,

$$m_a(^A_Z X) = m_N(^A_Z X) + Z m_e.$$

Show that nucleus $^A_Z X$ can decay via emission of a negative beta particle if

$$m_a(^A_Z X) > m_a(^A_{Z+1} Y).$$

29.46● (a) Determine the kinetic energy available to products of alpha decay of $^{235}_{92}\text{U}$. (b) Also find the kinetic energy of the alpha particle. The observed masses are $m(^{235}_{92}\text{U}) = 235.04392$ u and $m(^{231}_{90}\text{Th}) = 231.03630$ u.

29.47● (a) Determine whether $^{226}_{88}\text{Ra}$ can decay by the emission of an alpha particle, and if so, find the kinetic energy released in the process. (b) What is the kinetic energy of the alpha particle? The observed masses are $m(^{226}_{88}\text{Ra}) = 226.02540$ u and $m(^{222}_{86}\text{Rn}) = 222.01757$ u.

29.48●● Using conservation of energy and momentum, derive an expression for the kinetic energy of recoil of the daughter nucleus of mass number A resulting from an alpha decay with a given Q value.

*Section 29.8 Detectors of Radiation

29.49 A proton released in a nuclear reaction moves in a circular path perpendicular to a uniform magnetic field of 0.43 T. The kinetic energy of the particle is 0.66 MeV. What is the radius of the circle?

29.50● A small radioactive source is monitored by a detector 10 cm away. The sensitive part of the detector is a disk of 1.50-cm radius and faces directly at the source. If the average count rate of the detector is 4320 counts per minute, what is the approximate activity of the sample in becquerels?

*Section 29.9 Radiation Measurement and Biological Effects

29.51 A patient undergoing therapy is to receive an equivalent dose of 5.0 Sv. What dose (in grays) is this of (a) x rays and (b) fast neutrons?

29.52 A hospital patient receives an x-ray dose of 0.5 m rad. What is the dose equivalent in sieverts?

29.53 A worker at a nuclear plant receives a dose of 0.06 rad of low-energy neutrons. What is the dose equivalent in sieverts?

29.54 What fraction of the annual effective dose equivalent to the U.S. population is due to naturally occurring sources of radiation in the environment?

29.55● A 90-keV x-ray equivalent dose of 1.0 Sv is absorbed by 2.0 kg of tissue in a person's leg. How many photons are absorbed?

Section 29.10 Induced Transmutation and Reactions

> **Hints for Solving Problems**
>
> The Q value of a reaction is the difference between the initial mass and the final mass. It may be expressed as an energy. A positive Q value means that a reaction can take place spontaneously. If the Q value is negative, then energy must be supplied from outside to get the reaction to take place.

29.56 Determine the identity of the nucleus X in the reaction $^{13}_6\text{C} + \text{n} \rightarrow X + \gamma$.

29.57 What element is formed in the reaction $^{10}_5\text{B} + {}^4_2\text{He} \rightarrow X$?

29.58 Identify the particle Y in the reaction $^2_1\text{H} + Y \rightarrow {}^4_2\text{He} + \text{n}$.

29.59● Calculate the threshold energy of the proton for the reaction $^1_1\text{H} + {}^{14}_6\text{C} \rightarrow {}^{14}_7\text{N} + {}^1_0\text{n}$. The masses are $m(^{14}_6\text{C}) = 14.003241$ u and $m(^{14}_7\text{N}) = 14.003074$ u.

Section 29.11 Nuclear Fission

29.60 What is the kinetic energy in MeV of a room-temperature thermal neutron? By "thermal neutron" we mean one with kinetic energy kT. Take room temperature to be 300 K.

29.61 Suppose a nucleus of $^{234}_{92}$U splits exactly in half in a fission reaction in which two neutrons are ejected. What would be the resultant nuclei?

29.62● How much mass is lost in the fission of the nuclear fuel in a 1000-MW power plant in one year? (*Hint:* Assume that the power plant is 33% efficient.)

Section 29.12 Nuclear Fusion

29.63 Calculate the energy released in the fusion reaction $^2_1H + ^2_1H \rightarrow ^1_0n + ^3_2He$.

29.64 Calculate the energy released in the reaction $^3_2He + ^3_2He \rightarrow ^4_2He + ^1_1H + ^1_1H$.

29.65 How much energy is released in the reaction $^{12}_6C + p \rightarrow ^{13}_7N + \gamma$? The masses are $m(^{12}_6C) = 12.00000$ u and $m(^{13}_7N) = 13.005738$ u.

Additional Problems

29.66● What is the ratio of the approximate nuclear radius of $^{232}_{90}$Th to that of $^{57}_{26}$Fe?

29.67● Determine the kinetic energy available to products of alpha decay of $^{238}_{92}$U. Give your answer in units of MeV. The observed masses are $m(^{238}_{92}U) = 238.05078$ u and $m(^{234}_{90}Th) = 234.04359$ u.

29.68● Show that nucleus ^A_ZX can decay by emission of a positron only if

$$m_a(^A_ZX) > m_a(_{Z-1}^A Y) + 2m_e,$$

where m_a is the atomic mass and m_e is the mass of the positron.

29.69● Bismuth-212 decays via alpha emission to $^{208}_{81}$Tl with an energy release of $Q = 6.170$ MeV. Find the approximate kinetic energy of the alpha and its speed.

29.70● (a) Estimate the kinetic energy needed to bring two $^3_2$He nuclei close enough for a reaction to occur ($\approx 10^{-14}$ m). (b) What is the equivalent temperature?

29.71●● Radioactive materials used for pharmaceutical purposes are usually given in amounts of a few millicuries. What would be the mass of radioactive iodine (^{131}I) having activity of 20 mCi? The half-life of ^{131}I is 8 days, and its atomic mass is 131. You may want to refer to Section 26.8.

29.72●● According to a newspaper account, a nuclear power plant had an accidental release of 33,000 curies of tritium. How many grams of tritium were released? The half-life of tritium is 12.3 years. You may want to refer to Section 26.8.

29.73●● Using conservation of energy and momentum, show that the kinetic energy of an alpha particle in a decay with a given Q value is

$$KE_\alpha = \frac{Q}{1 + m_\alpha/m_d},$$

where m_α is the mass of the alpha particle and m_d is the mass of the daughter nucleus.

29.74●● A beta particle emitted from a radioactive nucleus moves in a circular path perpendicular to a uniform magnetic field of 0.37 T. The kinetic energy of the particle is 0.66 MeV. What is the radius of the circle? (*Hint:* First decide if the speed is relativistic.)

29.75●● An alpha particle collides head-on with a stationary proton in an elastic collision. Show that for an initial alpha speed v_α, the struck proton will move off with a velocity $1.6v_\alpha$.

29.76●● The elimination of radioactive material from the human body follows an exponential law so that the amount of material remaining in the body decreases with a characteristic half-life t_{body}. The radioactive material has a natural half-life t_{rad} of its own. Show that the effective biological half-life of the radioactivity in the body is given by

$$\frac{1}{t_{Biol}} = \frac{1}{t_{body}} + \frac{1}{t_{rad}}.$$

Lasers, Holography, and Color

Throughout much of our discussion of optics in Chapters 22–24 we assumed monochromatic light emitted coherently (in phase). In reality, this is a very rare kind of light. Most light, whether from the sun, from fluorescent lamps, or from incandescent bulbs, comes from large numbers of atoms randomly radiating photons. However, with the invention of the laser in 1960, a practical source of coherent light became available, and physicists took a renewed interest in optics. The coherence of laser light made possible many new experiments. For example, laser light can be focused into beams of extremely high intensity, a feature that has led to basic discoveries concerning the nature of the interaction between light and matter. It has also led to practical applications such as drills for making microscopic holes in the hardest substances and surgical instruments for delicate and precise operations.

One of the better-known applications of lasers is the production of three-dimensional images, called holograms. Holography is a rapidly growing field, with scientific and commercial applications. In this chapter, we present the basic ideas of holography, in which the modern tool of laser light is combined with the classical principles of diffraction.

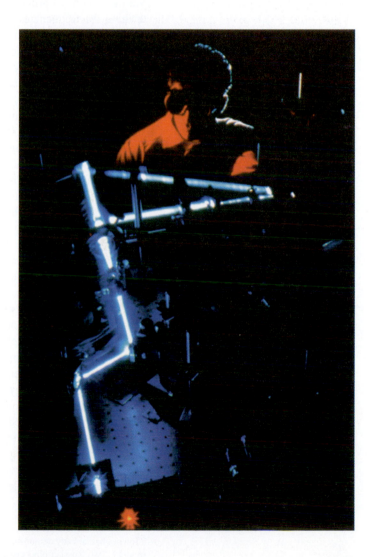

The development of the laser would not have been possible without an understanding of quantum mechanics and atomic physics. Similarly, our understanding of other optical phenomena, such as color and vision, has also advanced through our knowledge of quantum physics. The importance of color and how we see it has grown along with the explosive growth of information in the late twentieth century. The use of color enhances our ability to assimilate information rapidly, as false-color images have demonstrated. Color has become an important tool for presenting information in all areas of life. ■

30.1 Stimulated Emission of Light

We have seen in earlier chapters that the electronic energy levels of an atom determine the frequencies of the light emitted or absorbed by the atom. We have also seen that the likelihood that an atom exists in a particular energy state decreases with the energy of that state. It can be shown that the probability that an atom exists in an energy state E is proportional to $e^{-E/kT}$, called the Boltzmann factor, where E is the energy, k is Boltzmann's constant, and T is the temperature. Thus, for example, a cool (room temperature) gas exists predominantly in the lowest, or ground, state. When the energy difference between the first excited state E_1 and the ground state E_0 is large compared with kT, almost all of the gas atoms will be in the state E_0 (Fig. 30.1). Illuminating such a gas with light of energy $hf_1 = E_1 - E_0$ resonantly excites the atoms from the ground state E_0 to the excited state E_1, with the consequent absorption of the light. For each atom excited from E_0 to E_1, a photon of energy hf_1 is absorbed.

Atoms in excited states are no longer in thermodynamic equilibrium with their surroundings. They are unstable and after a short time return to the ground state, emitting a photon during this process. The average lifetime of an excited atomic state is designated by τ. The emission of photons that occurs this way—through the random de-excitation of excited atoms—is known as **spontaneous emission.**

In 1917, Albert Einstein showed that incident light of the proper frequency could trigger the emission of light by excited atoms. This process, called **stimulated emission,** takes place when the frequency of the incident photon equals the resonant frequency of the transition between energy levels. Thus, when light of frequency f_1 is incident on an atom with an electron in energy level E_1 (Fig. 30.2), a photon of energy hf_1 can stimulate a transition to the ground state, with the simultaneous emission of another photon, also of frequency f_1. In such a case, instead of being absorbed, the incident photon continues its travel, accompanied now by a second photon of the same frequency and phase.

In 1958, some forty years after Einstein's theoretical discovery of stimulated emission, Charles H. Townes and Arthur L. Schawlow published a paper that discussed how this principle could be used in a practical light source.* Two years

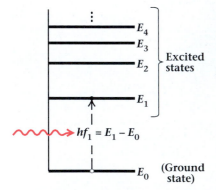

Figure 30.1
Energy levels of an atom. When light of frequency f_1 is resonantly absorbed, the atom goes from ground state E_0 to excited state E_1.

---

*Townes had already demonstrated the stimulated emission of radiofrequency radiation (microwaves) in 1954. Two Russian physicists, Nikolai Basov and Alexander Prokhorov, independently described the laser theoretically.

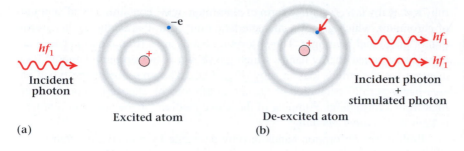

Figure 30.2
Stimulated emission of light. (a) A photon of resonant energy hf_1 incident upon an excited atom. (b) The stimulated atom emits a photon of energy hf_1 in phase with the continuing photon.

later, T. H. Maiman succeeded in operating the first light source of this type.* This device soon became known as a **laser,** an acronym of Light Amplification by Stimulated Emission of Radiation.

30.2 Lasers

To use stimulated emission in a practical laser, we must satisfy several basic conditions. First, we need an **active medium**—that is, a material containing atoms or molecules that can be made to emit light. Second, we need a method for adding energy to this medium in order to promote a sufficient number of its atoms to an excited state. The process of exciting the laser medium is called **pumping.** Finally, we need a way of confining the light so that we can trigger many stimulated emissions before the light escapes from the medium, thus building up an intense output beam. We achieve this result by placing the active medium in an **optical resonator,** which consists of mirrors that reflect the light back and forth. Light emitted along the axis is reflected, causing many additional stimulated emissions, while light directed off the axis is quickly lost from the system.

As an example of a laser system, let us consider a ruby laser similar to the first laser built by Maiman (Fig. 30.3). The active medium of this laser is a small

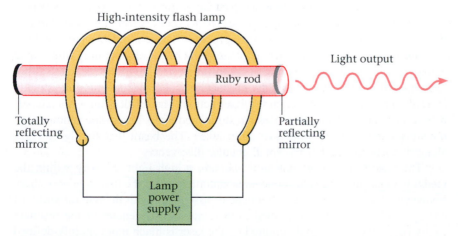

Figure 30.3
Schematic representation of a ruby laser, containing a ruby rod with mirrored ends and a flash lamp.

---

*Gordon Gould also conceived a laser in 1957 and was eventually awarded patents for his ideas, but it is generally believed that the first operating laser was built by Maiman.

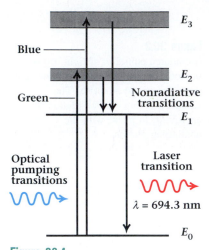

Figure 30.4

Energy-level diagram for chromium ions in ruby. Absorption bands E_2 and E_3 are excited by optical pumping from the ground state E_0. Fast nonradiative transitions to E_1 create the inverted population needed for laser transition from E_1 to E_0.

ruby rod. Ruby is a crystalline form of aluminum oxide containing small amounts of chromium, which give it a characteristic pink color that is due to the absorption of green and blue light. The absorption bands arise from closely spaced energy levels, called energy bands, which are shown in the simplified energy-level diagram for chromium ions in ruby (Fig. 30.4). The presence of the broad energy bands permits excitation of the chromium ions by exposure to bright white light from a flash lamp. Pumping of the active medium by the absorption of light is known as *optical pumping.*

Both of the absorption bands in ruby de-excite by nonradiative transitions, in which the excitation energy is given up to the crystal instead of being carried off by photons. These transitions lead to the two closely spaced states labeled E_1. These energy states are also unstable and give up their energy by emitting light at wavelengths of 694.3 and 692.7 nm. The laser transition corresponds to the 694.3-nm line, which is dominant in the presence of stimulated emission.

Normally, for atoms in thermodynamic equilibrium, upper energy levels are less densely populated than lower ones. In this condition, absorption of photons dominates over stimulated emission. That is, because there are more ground-state ions available to be excited than there are excited ions to be stimulated to emission, there is more absorption than emission. In contrast, when more ions exist in an upper state than in a lower state, we have an **inverted population.** Laser action cannot take place unless we can generate an inverted population of atoms in the active medium.

In a ruby laser, the flash lamp gives out sudden bursts of light, pumping the chromium ions into their excited states. The excited states quickly de-excite, populating the state E_1. Although E_1 is also unstable, its lifetime is long compared with those of the absorption bands. Thus, the pumping leads to an inverted population in state E_1. (Such a state is called a **metastable state.**) When more ions are in the excited state E_1 than in the ground state E_0, the probability that a single photon will cause stimulated emission exceeds the probability that the photon will be absorbed. As a result, the system emits a brief flash of light—laser light. The ruby laser gives out a pulse of laser light each time the flash lamp is activated. Such a laser is called, for obvious reasons, a pulsed laser.

The ends of the ruby crystal are polished and silvered to form a pair of parallel mirrors—the optical resonator. One mirror is totally reflecting, while the other mirror is partially reflecting and partially transmitting. The mirrors reflect the light back and forth through the crystal many times before it escapes through the partially transmitting mirror. This process of extending the path of light through the active medium increases the probability of obtaining stimulated emissions caused by the photons that are traveling parallel to the rod's axis. Photons emitted in other directions, the products of spontaneous emission, quickly leave the rod and play no role in the laser action. The result is an avalanche effect, as each emitted photon causes the stimulated emission of other photons, all traveling parallel to the laser axis (Fig. 30.5). The beam that finally emerges along the axis has a relatively small angular divergence.

The reflecting mirrors not only maintain a high light intensity within the medium, but also serve as the optical resonator that tunes the laser to a very sharp frequency. The narrow band of frequencies associated with the natural width of energy level E_1 is further restricted to the resonant frequencies of the resonant cavity (Fig. 30.6). Thus, light emitted by the laser is much more sharply defined than we would otherwise expect. In this way the laser becomes a source of essentially monochromatic light. However, in many lasers it is not uncommon to

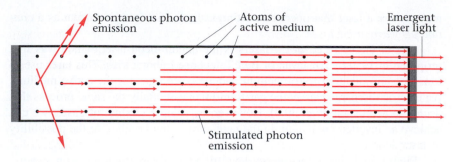

Figure 30.5
As light travels through the active medium, the intensity increases due to stimulated emission.

find several cavity resonances within the envelope of the normal linewidth. In those cases, the laser light is composed of several very closely spaced wavelengths.

We can determine the separation in frequency of the resonator modes in the same way we found resonant frequencies in Chapter 15. For constructive interference within the resonator, the path length between mirrors must be an integral number of half-wavelengths of the light. So, for mirrors separated by a length L, the resonant frequencies are

$$f = \frac{mc}{2L}, \qquad (30.1)$$

where m is an integer and c is the speed of light. For a frequency f' corresponding to $m' = m + 1$, we see that

$$\Delta f = f' - f = \frac{c}{2L}. \qquad (30.2)$$

The spacing of the resonator modes is illustrated in Fig. 30.6.

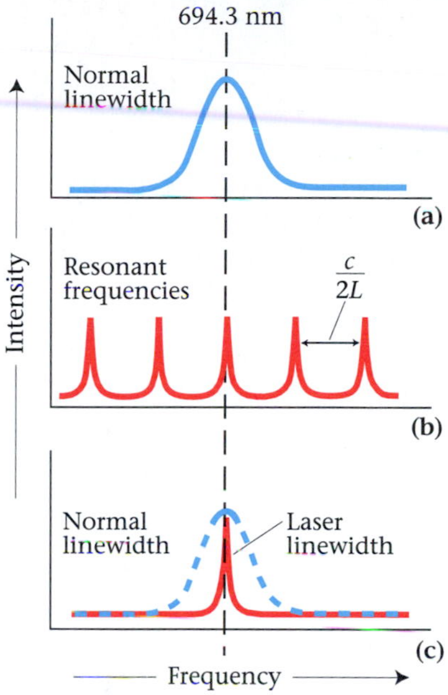

Example 30.1

A laser has mirrors 10 cm apart. If the natural width of the emission line for the laser transition is $\Delta f_0 = 10^8$ Hz, can the laser have more than one frequency?

Solution The separation between the frequency modes is found from Eq. (30.2),

$$\Delta f = \frac{c}{2L} = \frac{3.0 \times 10^8 \text{ m/s}}{2 \times 0.10 \text{ m}} = 1.5 \times 10^9 \text{ Hz}.$$

The linewidth is smaller than the separation of the cavity modes. Thus, there can be only one mode, or frequency, for this laser.

Figure 30.6
Effect of the laser cavity on the frequency. (a) The natural spread of frequencies in the emission line. (b) The resonant frequencies of the optical resonator created by the mirrors. These frequencies are separated by an amount $c/2L$, where c is the speed of light and L is the separation between the mirrors. (c) The product of linewidth with the cavity resonance determines the laser output, here shown with only one resonant frequency.

30.3 The Helium-Neon Laser

Pulsed lasers can deliver a large amount of energy in a very short time interval. Such a laser is particularly appropriate for applications requiring high power in the laser beam, such as drilling or welding. On the other hand, other applications

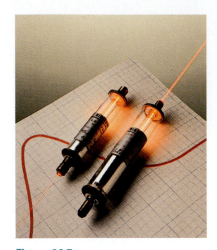

Figure 30.7
A helium-neon laser in operation.

may require a laser that operates continuously. Such a laser is known as a *continuous wave* or *cw laser.*

One of the best-known cw lasers is the helium-neon laser, which was first proposed in 1959 by Ali Javan of Bell Telephone Laboratories. This laser uses a low-pressure mixture of 90% helium and 10% neon gas for the active medium. An electric discharge causes the gas to glow in essentially the same process that occurs in neon signs (Fig. 30.7). However, by using a mixture of gases, we can achieve an inverted population in the neon atoms, thus leading to the possibility of laser action.

Figure 30.8 shows the energy-level diagram for the helium-neon system. The helium atoms are easily excited by electron collisions in the electric discharge. Two of the lowest excited states of helium, labeled 2^1S and 2^3S, are sufficiently long lived that they are considered metastable. These two states are at almost the same energy as two of the excited states of the neon atoms, the 4S and 5S states. Because the energies match so closely, a helium atom can readily transfer its excitation energy to a neon atom during a collision between atoms in the gas. After the collision, the helium atom returns to its ground state while the neon atom is raised to one of its excited states. This process of pumping by collision creates an inverted population distribution in the neon. Atoms in these two excited levels can then radiate energy by stimulated emission to the states labeled 4p and 3p in Fig. 30.8, which quickly de-excite to still lower states, maintaining an inverted population in the excited upper levels. Thus, the laser action is continuous.

In order to select one laser wavelength over others, it is common to use special mirrors that reflect only near the wavelength of the desired transition. By using mirrors that reflect at 632.8 nm but transmit in the infrared, manufacturers construct helium-neon lasers that emit red light at 632.8 nm. Low-power helium-neon lasers of this type are relatively inexpensive. For this reason they are used

Figure 30.8
Energy-level diagram of the helium-neon laser. The lasing transitions are labeled with their wavelengths.

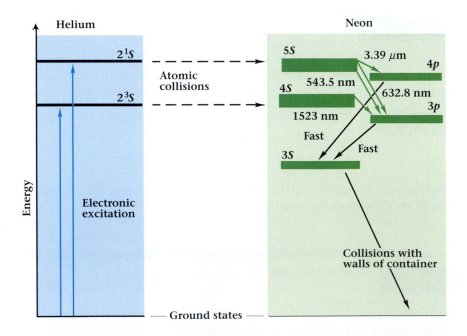

in many high school and college laboratories. Other mirrors permit the construction of helium-neon lasers operating in the green at 543.5 nm and in the infrared at 1.523 nm.

Ruby lasers and helium-neon lasers are only two of a growing number of types of lasers. Other gases such as argon and krypton are also used as laser media, as are metal vapors and molecular gases such as carbon dioxide. Other materials are used in the same manner as ruby in optically pumped lasers. Still other lasers are made from semiconductor diode junctions. Dye lasers, containing fluorescent dyes as the active medium, can be made to emit a continuously variable wavelength. Thus, we can obtain laser light at almost any wavelength desired.

30.4 Properties of Laser Light

When light from a laser passes through a prism or diffraction grating, no noticeable dispersion occurs—an indication that the laser light is monochromatic. In contrast, sunlight is dispersed by a grating into a continuous spectrum of colors, indicating that the sunlight actually consists of a mixture of many different wavelengths. Related effects due to interference of white light often go unnoticed, since the minima and maxima for each separate wavelength occur at different positions and the pattern is washed out. However, monochromatic laser light produces interference patterns that are quite pronounced.

When laser light reflects off a surface, its unusual character immediately becomes visible. The reflected light seems to sparkle with bright regions separated by dark areas. This so-called *speckle pattern* is a characteristic of laser light that is due to coherence. What you see is actually an interference pattern in the reflected light. As you move about, the speckle pattern also moves. Whether it moves in a direction that is the same as or opposite to your motion depends on whether you are nearsighted or farsighted. A nearsighted observer sees the pattern move in the opposite direction to his motion; a farsighted observer sees the pattern move in the same direction. You may wish to try this for yourself.

In our discussion of interference in Chapter 24, we noted that Thomas Young's double-slit experiment was crucial in establishing the wave behavior of light. In Young's experiment, white light from the sun fell on two narrow, closely spaced slits. The light passing through the slits formed an interference pattern of colored stripes on a screen. We can limit the white light to predominantly one color by first passing it through a prism, which spreads the beam into a spectrum, and then directing the spectrum to a narrow slit. We can use the relatively narrow band of wavelengths that pass the slit to generate a double-slit interference pattern. This pattern is more distinct than that made by white light and appears as a set of alternating bright and dark bands.

Now, suppose we illuminate the slits with light from a laser. Since only one wavelength is present, the pattern becomes even more pronounced. If the illumination falling on the two slits is uniform, then we expect that where the waves interfere destructively, the effect should be total; that is, we should observe no light intensity. Then we can take the visibility of the interference lines, or fringes, as a quantitative measure of how well the light interferes with itself. That is, the fringe visibility measures the *coherence* of the light.

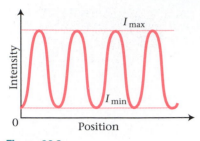

Figure 30.9
Intensity profile of an interference pattern. The visibility is defined in terms of the maximum and minimum intensities, I_{max} and I_{min}.

We define the fringe visibility V to be the ratio

$$V \equiv \frac{I_{max} - I_{min}}{I_{max} + I_{min}}, \qquad (30.3)$$

where I_{max} is the maximum intensity in the pattern and I_{min} is the minimum intensity (Fig. 30.9). If $I_{max} = I_{min}$, no fringes are seen, the visibility is zero, and the light is completely incoherent. If $I_{min} = 0$, the visibility is 1 and the light is completely coherent. Finally, partial coherence results for V between 0 and 1. If the illumination of the slits is equal, the degree of coherence is identical with the fringe visibility. If the illumination is unequal, then the interference pattern has reduced visibility even for perfectly coherent light. In that case, the coherence and the fringe visibility are still proportional. The coherence γ is then defined as

$$\gamma \equiv \frac{I_1 + I_2}{2\sqrt{I_1 I_2}} V, \qquad (30.4)$$

where I_1 and I_2 are the intensities due to each slit independently.

Example 30.2

A double slit, evenly illuminated by a light source, produces interference fringes that are barely visible. Measurement of the intensities reveals that $I_{max} = 2I_{min}$. What is the degree of coherence of the light source?

Solution The fringe visibility is

$$V = \frac{I_{max} - I_{min}}{I_{max} + I_{min}} = \frac{2I_{min} - I_{min}}{2I_{min} + I_{min}} = \frac{1}{3} = 0.33.$$

Since the slits are evenly illuminated, the degree of coherence and the visibility are identical: $\gamma = 0.33$. ■

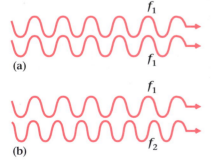

Figure 30.10
(a) Two waves of the same wavelength retain a constant relative phase as they travel through space. (b) Two waves of different wavelength have a changing relative phase as they travel through space.

As light is made more nearly monochromatic, its coherence increases. Thus, two light waves of the same wavelength traveling in the same direction maintain a constant phase between them (Fig. 30.10a). On the other hand, two waves of different wavelengths change their relative phase as they move through space (Fig. 30.10b). As these waves pass a fixed point in space, they go in and out of phase with time. As a result, their time-averaged effect is one of incoherence. A high degree of coherence can occur only if the light is very nearly monochromatic. This coherence is referred to as **temporal coherence.** But even waves of exactly the same wavelength will not be coherent unless their phases are constant in time.

Besides temporal coherence, there is a degree of coherence arising from the spatial extent of the light source, which is called **spatial coherence.** We can associate spatial coherence with the coherence of light from different positions on the wavefront. This complication was taken into account by Young, who passed light through a single pinhole before allowing it to fall on the double slit. In consequence, the light falling on the double slit came from a very small region of the initial wavefront. Light from farther along the wavefront was not sufficiently coherent to produce the interference pattern. Laser light has a high degree of both spatial and temporal coherence, which accounts for its remarkable diffraction and interference effects.

Because of coherence, laser light can be focused into tiny intense beams for surgical use to correct visual defects. Pulses of ultraviolet light from excimer lasers sculpt the cornea into the proper shape to correctly focus light on the retina. Computer-controlled laser beams break molecular bonds between corneal cells with an accuracy up to 0.25 μm. Often, as little as 50 μm of corneal tissue must be removed to achieve the proper amount of correction.

Photorefractive keratectomy (PRK) corrects refractive errors by removing tissue from the corneal surface. In practice, the laser removes the proper amount of tissue to reshape the cornea is less than a minute. PRK is often used to treat myopia and astigmatism.

Laser in-situ keratomileusis (LASIK) corrects refractive errors by removing tissue from beneath the corneal surface. An instrument called a microkeratome enables the surgeon to fold back a thin flap of the cornea. Then the laser removes the proper amount of corneal tissue and the flap is folded back where it bonds in a matter of minutes. LASIK corrects for hyperopia and myopia.

PRK removes tissue from the surface of the cornea.

After PRK, the flatter cornea causes images to focus on the retina.

Master the Concept

Interference and Laser Speckle

Question: Where is the interference pattern that we observe as laser speckle?

Answer: Laser speckle is observed when the spatially coherent light from a laser is reflected from a surface such as a wall or sheet of paper. Because of the microscopic roughness of the surface, the light is diffusely scattered, creating a stationary interference pattern that fills the space surrounding the surface. We can think of this interference pattern as forming a real image in space from diverging rays. Thus, any region of this image can be seen directly by focusing our eyes accordingly. When we look toward the reflecting surface, we see the speckle because our eye focuses the interference pattern on the retina. We see the interference pattern that is in the space between us and the scattering surface.

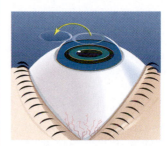

In LASIK, a thin layer of cornea is folded back with a microkeratome.

Light from the excimer laser reshapes the internal cornea and the flap is folded back.

30.5 Holography

Most common optical devices, such as cameras, projectors, and binoculars, use lenses to refract light and form images. As we showed in Chapter 23, we can explain the formation of images by applying the techniques of ray tracing to determine their position and size. However, we have firmly established that light is a wave phenomenon. In fact, only by treating light as a wave can we explain many observed effects. Therefore, we should expect the wave nature of light to be important in understanding image formation. Indeed, it is very important, as our discussion of resolution in Section 24.7 made clear. Further investigation of image formation by waves led to one of the most exciting areas of modern optics: holography.

In 1947, the British physicist Dennis Gabor (1900-1979) began experiments to overcome problems of image formation that were due to aberrations. Gabor recognized that for an object illuminated with coherent light, the pattern generated by interference between the wave scattered off the object and a coherent reference wave contains all of the visual information about the object. This interference pattern can be recorded photographically. Then, if the reference wave alone is used to illuminate the developed photographic film, it is scattered by the film in such a way as to generate a new wave, essentially identical to the original wave reflected from the object. The result is the formation of an image without the use of a lens.

Gabor coined the word **hologram** for this photographic recording of the interference pattern produced by the combination of reference and object beams. This word is derived from the Greek words *holos,* meaning whole, and *gramme,* meaning what is written or drawn. The making and study of holograms is called **holography.** Gabor successfully demonstrated the effect in 1948, more than a decade before the invention of the laser. In the ten years immediately following his discovery, holography received little attention and its development was hampered because of the lack of a suitable source of coherent light.

After the invention of the laser, Emmett N. Leith used coherent laser light to create images in the manner first demonstrated by Gabor. With Leith's introduction of the laser and use of a reference beam at an angle to the object beam, holography became a practical method of photographically recording optical information.

The images created from holograms contain more information than images produced with ordinary photographic techniques. Holographic images are three-dimensional and have both depth and parallax. As you move your head when viewing a holographic image, it seems as if the original objects are present within the image. Figure 30.11 shows three images reconstructed from the same hologram. Parts (a) and (b) show the effects of parallax when the holographic image is viewed from different angles. Part (c) illustrates the depth in the image, revealed when the camera is focused on different portions of the image. Both these effects are missing in ordinary photographs. Even matched stereo pairs of photographs do not contain the visual information of a hologram.

Figure 30.11

(a), (b) Two photographs of the same hologram taken at slightly different vertical positions. Note the effect of parallax in the relative positions of the lens and ruler. The camera was focused on the standing figure. (c) A photograph of the same hologram taken from the same position as (a) but with the camera focused on the rim of the magnifying glass.

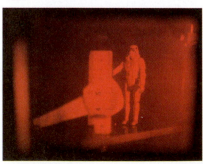

(a)

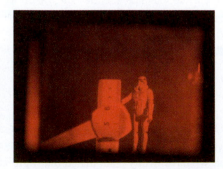

(b)

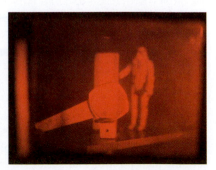

(c)

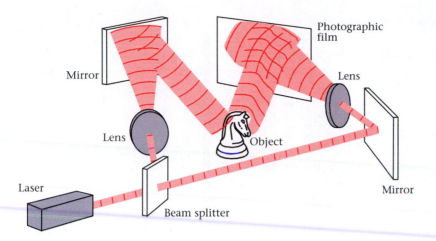

Figure 30.12
Producing a hologram. Light from the laser is split into two beams. The reference beam, here shown to be parallel, is reflected directly at the photographic film. The other beam is scattered off the object onto the film.

The production of a hologram requires two coherent beams of light (Fig. 30.12). These two beams are generated from a single laser beam by a half-silvered mirror or other beam-splitting device. One beam, called the reference beam, reflects directly on the photographic film. Mirrors direct the other beam to the object whose image is to be recorded. Light scattered from the object to the photographic film is called the object beam. No lenses are used to image the object onto the film; instead, the object and reference beam both illuminate the film. Since these beams are coherent, they interfere and the photographic film records that interference pattern.

You can see a holographic image of the object by illuminating the developed film with a laser beam similar to the reference beam. In general, two images are formed (Fig. 30.13). If the reference beam used in making the hologram was a beam of parallel light and the reconstructing beam is also parallel, then one of these two images is a virtual image. You can observe the virtual image by looking through the hologram as indicated in Fig. 30.13. The three-dimensional image appears behind the film. The second image is real and can be formed on a screen. If you are farther away from the film than the image is, your eyes can focus the rays diverging from the real image.

Notice that in producing the hologram, light from each point on the object is scattered over the entire photographic film. This behavior is in sharp contrast to the one-to-one correspondence between points on the object and points on the image in ordinary photography. (See, for example, Figs. 22.18 and 23.5.) Thus, each tiny area of the hologram contains information about the entire object. Even if a large part of the hologram is destroyed, we can still reconstruct the entire image from the part that remains. There is no way to reconstruct part of an ordinary photograph that has been destroyed.

Whether the holographic image is real or virtual depends on both the initial reference beam and the reconstructing beam. These beams need not be parallel, as shown in Fig. 30.13, but can be diverging or converging. Therefore, it is possible to make a hologram and reconstruct it so that either of these two images is real or virtual.

The transmission holograms that we have been describing produce sharp images only when illuminated with laser light or light that is nearly monochromatic. If the light contains a band of wavelengths, as in white light, each wavelength creates its own holographic image. These individual images overlap,

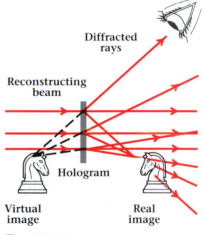

Figure 30.13
Reconstructing a holographic image. The virtual image may be seen with the unaided eye.

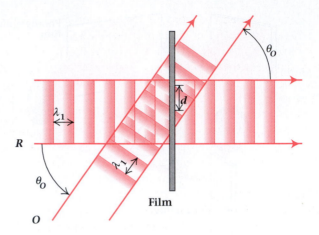

Figure 30.14
Interference pattern formed in the plane of the film by two coherent light beams R and O. The spacing d is given by $d = \lambda_1/\sin \theta_O$.

resulting in a blurred effect. Furthermore, we need the light to come from a point source. If a large or diffuse source is used, each source point produces its own image, again resulting in a smeared image, even if the light is monochromatic. These requirements are equivalent to demanding that the light be both spatially and temporally coherent.

To analyze holography, we must think in terms of diffraction and interference. For simplicity, let's consider the interference of two beams of parallel coherent light striking a photographic film (Fig. 30.14). The object beam O makes an angle θ_O with the reference beam R. The two beams combine to produce an interference pattern, with spacing d between maxima given by

$$d = \lambda_1/\sin \theta_O, \tag{30.5}$$

where λ_1 is the wavelength of the light. A stationary interference pattern occurs only if the two beams are coherent. Furthermore, the relative positions of the light sources, the film, and any other components (such as beam splitters and mirrors) must be stationary and free from vibrations; otherwise the interference pattern in the plane of the film is averaged or washed out. Vibrations with amplitudes as small as $\lambda/4$ are sufficient to prevent recording of the interference pattern.

When the film is developed, it becomes a linear diffraction grating. However, instead of displaying alternating opaque and transparent lines, the film varies in darkness sinusoidally across the width of the film. Such a film is known as a sinusoidal diffraction grating. The spacing d between successive lines on the film is given by Eq. (30.5). When a beam of parallel light of wavelength λ_2 is perpendicularly incident on this linear grating, we expect first-order diffraction at the angle θ given by

$$\sin \theta = \lambda_2/d.$$

Upon inserting Eq. (30.5) into this equation, we find that

$$\sin \theta = \frac{\lambda_2}{\lambda_1} \sin \theta_O.$$

If the reconstructing wavelength λ_2 and the recording wavelength λ_1 are the same, then $\theta = \theta_O$ and one of the diffracted beams is directed in exactly the same direction as the original object beam. The other diffracted beam is scattered on the opposite side of the normal from the direction of the original object beam (Fig. 30.15). In addition, some of the light passes through undeflected, just as in any other grating. However, in contrast with the gratings described in Chapter 23, the sinusoidal grating does not produce any higher-order beams. That is, because of the sinusoidal variation of the intensity in the grating, only zero- and first-order beams are generated.

If the object beam of Fig. 30.14 were a diverging beam from a point source, the pattern recorded on the film would be more complex than that of the linear sinusoidal grating. However, it would still have some of the same properties. As before, only zero- and first-order diffraction would occur. One of the first-order beams would be an expanding wave, emerging in the same direction as the original object wave. This scattered wavefront would have the same shape as the original object wave. For this reason, holography has also been called wavefront reconstruction. The other first-order beam is scattered on the other side of the zero-order beam, as shown in Fig. 30.15. But this wave has a converging wavefront that comes to a point, forming a real image.

We can consider a hologram of a complex object as a superposition of point source holograms from all possible source points on the object. Even in this case, it is still true that only zero- and first-order diffracted beams are generated during holographic reconstruction. One of these first-order beams, called the primary beam, re-creates the waveform as it was scattered from the object. This beam produces the virtual image indicated in Fig. 30.13. The other first-order diffracted beam, called the conjugate beam, converges to form the real image in Fig. 30.13.

If the hologram is illuminated with an undiverged beam of light from a laser, both images can be seen simultaneously on a screen placed in the path of the light (Fig. 30.16). The images remain in focus even when the screen is moved. This behavior is similar to that of a pinhole camera, and for the same reason. The laser beam is scattered by such a small portion of the hologram that the light reaching one point on the screen comes from only one point on the object.

We can apply Eq. (30.5) to compute the resolution requirements for photographic film used in holography. The inverse of the line spacing d gives the number of lines per unit length—the **spatial frequency** of the interference pattern. If the film is to record this pattern, it must be capable of recording lines of at least this spatial frequency. For example, suppose that the angle between the incident beams is 45°. If the coherent light source is a helium-neon laser with $\lambda = 632.8$ nm, then the spatial frequency F of the interference pattern will be

$$F = \frac{1}{d} = \frac{\sin 45°}{632.8 \text{ nm}} = 1117/\text{mm}.$$

Thus, in order to record the interference pattern and serve as a hologram, the film must be capable of recording more than 1117 lines/mm. Most ordinary photographic films have resolutions an order of magnitude less than this and consequently are unsuitable for holography. Fortunately, special holographic films with resolutions in excess of 2000 lines/mm are available.

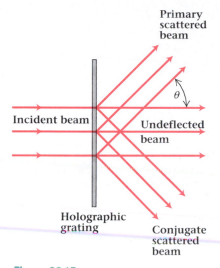

Figure 30.15

A light beam incident on a holographic grating, formed as in Fig. 30.14, is broken into three beams.

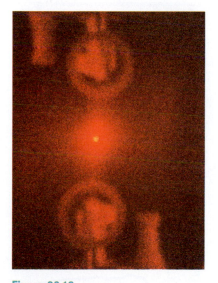

Figure 30.16

When a hologram is illuminated with the undiverged beam from a laser, both images can be seen on a screen at the same time.

Example 30.3

A hologram is made with film whose resolution is 1000 lines/mm. Light from a helium-neon laser, $\lambda = 632.8$ nm, is used to record and view the hologram.

The reference beam makes normal incidence with the hologram. What is the maximum angular field of view of this hologram?

Solution The spatial frequency due to light incident at an angle θ from the normal is

$$F = \frac{\sin \theta}{\lambda}.$$

The highest resolution of the film is 1000 lines/mm, so this must be the maximum spatial frequency. Substituting into the equation, we get

$$\sin \theta = (1000/\text{mm})\lambda$$

$$\sin \theta = \frac{1000}{10^{-3} \text{ m}} 632.8 \times 10^{-9} \text{ m} = 0.6328,$$

$$\theta = 39.3°.$$

The field of view is limited to a cone making an angle of 39.3° from the reference beam.

30.6 Light and Color

Table 30.1 *Colors in the Visible Spectrum*

Color	Wavelength Range (nm)
Red	630–700
Orange	590–630
Yellow	570–590
Green	500–570
Blue	450–500
Violet	400–450

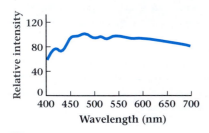

Figure 30.17
Spectral distribution of the intensity of sunlight at the earth's surface at noon.

Isaac Newton's first scientific paper, published in 1672, described his experiments with light and color. Newton passed a beam of sunlight through a prism, spreading the light into a spectrum of colors. A second prism turned the opposite way converged the spectrum back into a narrow beam of white light. Later development of the diffraction grating led to measurements of the wavelengths of light. Although each color merges smoothly into the next across the spectrum, it is possible to assign approximate wavelength ranges to each color (Table 30.1).

Although white light contains all the colors of the spectrum, they need not be present with the same intensity. For example, sunlight, often used as a reference standard for white light, does not have the same intensity at all wavelengths. Instead, it has the intensity distribution shown in Fig. 30.17. The intensity of sunlight reaching the earth's surface peaks near a wavelength of 475 nm and falls off sharply in the ultraviolet. A comparison of the solar data with the Planck theory of blackbody radiation can be used to determine the sun's temperature, giving a result of about 6000 K.

Incandescent lamps are good approximations to blackbody radiators and give off white light; however, this light is deficient in the blue and violet. The spectrum is skewed toward the red because of the relatively low operating temperature of incandescent lamps, about 2900 K. We describe the spectral distribution of the light in terms of the corresponding temperature of a blackbody, termed the *color temperature*. At a color temperature of 2900 K, only 3% of the energy dissipated in the lamp emerges as visible light. Special high-temperature lamps designed for photography and television usually operate at one of two reference temperatures, 3200 K or 3400 K. They produce considerably more blue than do ordinary household lamps. Table 30.2 shows some of the characteristics of a few common light sources.

WHITE-LIGHT HOLOGRAMS

So far we have been describing holograms that produce sharp images only when viewed with light that is nearly monochromatic. Reflection holograms can be viewed in white light. In a *reflection hologram,* the object and reference beams (which *do* need to be monochromatic, coherent light) are brought together from opposite sides of the photographic film (Fig. B30.1a). The resulting interference pattern has structure perpendicular to the plane of the film. If the thickness of the emulsion is greater than about 15 μm, then the interference pattern recorded in the film is truly three-dimensional, with twenty or more layers within the emulsion. The structure of the resulting hologram is analogous to a thin crystal twenty or thirty atomic layers thick. When the developed film is illuminated by a white light beam in the same direction as the original reference beam, the hologram diffracts some of the light backward (Fig. B30.1b). The resulting wave re-creates the original object wave. For each particular angle of incidence, only a narrow range of wavelengths is reflected into the viewer's eye. This diffraction is analogous to the effect of Bragg scattering of x rays. The hologram automatically selects the proper wavelength for each angle of incidence. Thus, we can view reflection holograms with white light, provided it comes from a point source.

Another type of white-light hologram is the *rainbow hologram,* invented by Steven Benton of the Polaroid Corporation in 1969. The rainbow hologram may be viewed with transmitted light from an incandescent lamp. The different wavelengths are dispersed so that each forms only a small portion of the total image, resulting in a rainbowlike change of color across the image. This effect is achieved in

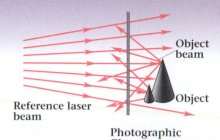

(a)

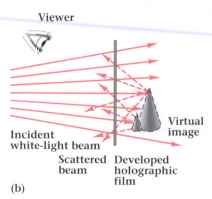

(b)

Figure B30.1 Making a white-light reflection hologram. (a) The reference beam of light passes through the photographic plate to also act as illumination for the object. (b) When illuminated with white light similar to the original reference beam, the developed hologram scatters light backward as if it came from the object. For each angle of incidence, the viewer sees only a narrow range of wavelengths, resulting in a sharp image.

Figure B30.2 Photograph of an embossed rainbow hologram.

a two-step process. First we make a transmission hologram of the object in the usual way. Then we make a second hologram, using the real image from the first hologram as the object. A narrow horizontal slit is placed next to the first hologram during this process. This eliminates the vertical parallax in the second hologram, but the second hologram still records the image of the entire object.

When the second hologram is illuminated with a white light source, each wavelength forms an image of the slit. Each of these images occurs at a different vertical angle, as expected from the laws of diffraction. When looking for the image of the object, an observer actually looks through one of the colored images of the slit. The original object appears with horizontal parallax. As the observer moves up and down, the image remains constant, showing no vertical parallax—but it changes color depending on the vertical position.

In recent years, holograms have been made using a special photographic emulsion called photoresist. Instead of variations in opacity, as in normal photographic emulsion, the developed photoresist has variations in thickness in response to the intensity of the light incident upon it. By using electrolysis, we can deposit a thin layer of nickel on the developed photoresist, making a three-dimensional mold. Duplicates of the interference pattern, made from the mold, can be stamped into plastic and coated with a thin layer of aluminum. The aluminum functions like a mirror, reflecting incident white light back through the interference layers to produce the holographic image. The result is called an *embossed hologram.* Because they can be reproduced inexpensively, embossed holograms are finding their way onto magazine covers, buttons, and novelties (Fig. B30.2). Since 1984, credit card manufacturers have been incorporating embossed holograms into their cards in an effort to thwart counterfeiting.

Table 30.2	Approximate Color Temperature of Some Common Light Sources	
Source		**Color Temperature (K)**
Mercury arc		6000
Daylight		5500
High-intensity carbon arc		5500
Cool white fluorescent		4200
Incandescent tungsten (photoflood 3400)		3400
Incandescent tungsten (photoflood 3200)		3200
Warm white fluorescent		3000
Incandescent tungsten (100W)		2900
Incandescent tungsten (40W)		2650
High pressure sodium arc		2200

Example 30.4

Where is the peak in the spectral distribution of light from a photoflood lamp operating at 3400 K?

Solution   We can compute the wavelength with the greatest intensity from the Wien displacement law (Chapter 27):

$$\lambda_m T = 2.90 \times 10^{-3} \text{ m} \cdot \text{K}.$$

Here T is the temperature of our light source, treated as a blackbody radiator. Thus

$$\lambda_m = \frac{2.90 \times 10^{-3} \text{ m} \cdot \text{K}}{3400 \text{ K}} = 8.53 \times 10^{-7} \text{ m} = 853 \text{ nm}.$$

The peak of the spectral distribution lies beyond the visible range, in the infrared region of the electromagnetic spectrum.

Figure 30.18
Sodium vapor lights are recognized by their characteristic yellow glow.

The spectral distribution of a light source depends on more than just the operating temperature. As we saw in Chapters 27 and 28, characteristic emission spectra of elements are due to electronic transitions between atomic energy levels. Thus, the light from each element has its own characteristic set of emission wavelengths and intensities, which give it a characteristic color. An electric discharge in hydrogen produces light of the Balmer series. We perceive this mixture of wavelengths as pink. Sodium vapor lamps produce a characteristic yellow light that is due to the intensity of the D lines (Fig. 30.18). Similarly, neon lamps produce a red-orange light.

The fluorescent lamp is a gas-discharge tube containing mercury vapor and a coating of fluorescent powder on the inside of the tube. Fluorescent materials absorb light at some wavelengths and then radiate light at longer wavelengths. Generally, both the absorption and emission states are broad bands rather than sharply defined energy levels of the type observed in free atoms. When the mercury vapor in a fluorescent lamp is excited by an electric discharge, it emits its characteristic spectral radiation. Part of this radiation extends beyond the visible range into the ultraviolet. The fluorescent material absorbs the ultraviolet radiation and then radiates light in the visible spectrum. Fluorescent lamps of this type are considerably more efficient than incandescent lamps, converting about 20% of the electrical energy into visible light.

Objects can absorb, reflect, or transmit light, and some objects can do a combination of all three processes. Thus, in addition to color that depends on the detailed nature of the source, light may take on a particular color as it passes through a material that selectively absorbs some wavelengths. For example, a piece of red glass is a filter that selectively passes red light but absorbs shorter wavelengths. Similarly, a blue filter transmits blue but not green, yellow, or red. Thus, color may be produced by selective absorption.

Color is also produced by reflection as well as transmission. The perceived color of most objects is due to the selective reflection of light. Thus, a red object reflects red light but absorbs green and blue. A red apple appears a vivid red when illuminated with red light, but it looks dark under blue light because it does not reflect blue light well.

We should caution you that while we can characterize light according to its spectral distribution, this is *not* the same as describing its color. Color is a *visual* sensation. The biological mechanisms of vision play an important role in our determination of color. Experiments by Edwin Land have demonstrated convincingly that the sensation of color depends on more than the spectral intensities of light reaching our eyes.

30.7 Color by Addition and Subtraction

The visible spectrum contains a continuous range of color, varying smoothly from violet on one end to deep red on the other. However, we can approximate the spectrum using only three separate bands of color, representing equal intervals of wavelength. These bands are called the three **additive primary colors** of light: red, green, and blue. By combining these three colors in various combinations, we can create a wide range of other colors. For instance, the combination of red and green light produces yellow. The addition of blue light to red light gives magenta, while blue added to green gives cyan (Fig. 30.19). If we mix the three primary colors in various proportions, we can obtain a full range of other colors, including purple, brown, and orange. If we add all three primaries with equal intensity, the result is white light. The production of colors in this manner is called *color mixing by addition.*

Color television generates color pictures through the additive mixing of color. If you look closely at the front of a television picture tube, you will see a pattern of closely spaced dots arranged in groups of three.* Each dot contains a fluorescent material that glows when excited by the electron beam. The system uses three different materials to produce red, green, and blue light. By controlling the intensity of the light from the separate dots, a wide range of colors can be produced (Fig. 30.20).

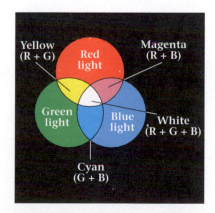

Figure 30.19
The primary colors of light, also called the additive primaries, can combine to generate a full range of colors.

---

*Some manufacturers use closely spaced lines instead of the dots described here.

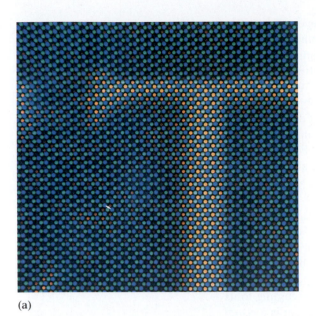

(a)

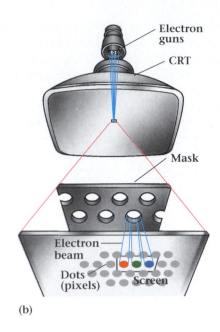

(b)

Figure 30.20

(a) Three different phosphors generate tiny dots of primary colored light on the face of a television screen. The colors blend together to make a full range of colors. (b) A shadow mask limits each of three electron beams to strike only dots of a single color.

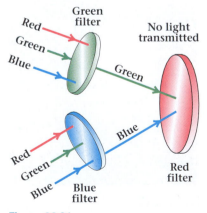

Figure 30.21

Filters in the color of the additive primaries block all other colors of light, so they cannot be used to generate other colors.

We can also produce color by subtraction. In this process we generate colors by selectively filtering light, thus removing certain wavelengths from the beam. For example, a red filter removes green and blue from white light, resulting in a beam of red. Similarly, a green filter transmits green light and a blue filter transmits blue light. However, we cannot combine such filters to produce other colors. Since the red filter passes only red, the combination of red plus green filter or red plus blue filter allows no light to pass (Fig. 30.21). Thus red, green, and blue filters are not appropriate for producing other colors by subtraction.

We use the term complementary colors for any pair of colors of light that can be combined to produce white light. Thus, the color complementary to red is cyan, the color complementary to blue is yellow, and the color complementary to green is magenta. These three colors—cyan, magenta, and yellow—are the **subtractive primary colors.** Filters of these colors *can* be used in succession to produce other colors. For example, passing white light successively through yellow and magenta filters produces red light (Fig. 30.22). The yellow filter removes blue light, while the magenta filter removes the green. The remaining light is red.

Color photography relies on the subtractive method of color production. Photographic films and papers are made with three layers of photosensitive materials, each of which responds to one of the primary colors of light. When the film is developed, dye images in one of the subtractive primaries form in each layer (Fig. 30.23). The varying densities of these filters control the color of the light that passes through. These same subtractive primaries are also used in color printing.

Although both addition and subtraction of colors give good results, as observed in color television and photography, these methods cannot accurately reproduce pure spectral colors. To a television camera, a spectral red of some particular wavelength is detected the same as any other spectral red. The resulting

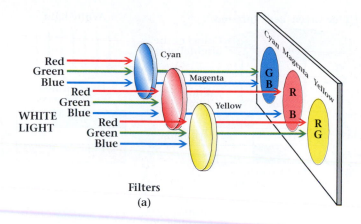

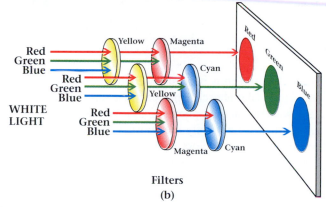

Figure 30.22

Use of subtractive color filters. (a) Cyan, magenta, and yellow filters block the passage of their complementary colors—respectively, red, green, and blue. (b) By using two subtractive filters in succession, we can block the passage of two of the additive primaries and allow the other to pass.

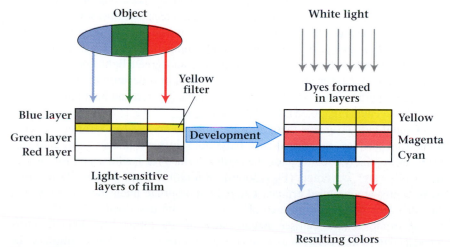

Figure 30.23

Color formation in color slide film.

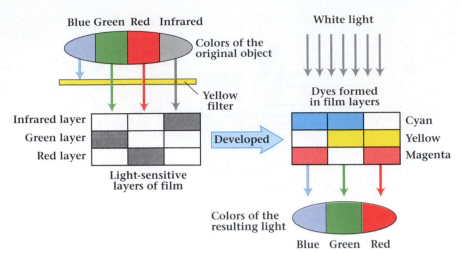

Figure 30.24
Color formation in infrared color film.

(a)

(b)

Figure 30.25
Two photographs of the same flowers with (a) Kodachrome color film and (b) Ektachrome infrared color film. The red flowers reflect both the red and the infrared light, causing the yellow image in the infrared photo. The green leaves reflect both green and infrared, causing the leaves to look magenta in the infrared photo.

red displayed on the screen is necessarily limited to whatever red light is characteristic of the particular phosphor placed in the tube by the manufacturer. Why, then, does color television (and photography) work so well? The answer is that most of the light that we normally see consists of mixtures of wavelengths and not just spectral lines. Most colors are due to broad-banded absorptions or reflections. A yellow lemon photographs as yellow because it reflects both green and red light over a broad range. Since most colors are broad banded, the limitations in color reproductions usually go unnoticed.

We can use our knowledge of color to see beyond the visible range with the help of false-color infrared photography (Fig. 30.24). Special color films are designed with sensitivity that extends into the near-infrared region of the spectrum. They differ from usual color films in that the three image layers are sensitized to green, red, and infrared instead of blue, green, and red (Fig. 30.24) light. During exposure, a yellow filter is used on the camera to block blue light, to which the layers are also sensitive. When processed, the green-sensitive layer has a magenta image, the red-sensitive layer has a yellow image, and the infrared-sensitive layer has a cyan image. These layers, when combined, filter the light to produce an image in which the original infrared, red, and green light appears as red, green, and blue. Leaves appear magenta in such photos because they reflect both infrared and green light, which the film translates into red and blue that combine to give magenta (Fig. 30.25).

Digital video cameras that see in the infrared are used in a variety of locations from satellites to hospital labs. False-color infrared images from high altitudes are routinely used to study the earth's surface for land use and for vegetation and water coverage (Fig. 30.26). Radiation from bands of different wavelengths are assigned primary colors in appropriate intensities. The result is similar to the false-color infrared photographic film described above. However, the electronic cameras can see much farther into the infrared for their detectors are sensitive to the radiation associated with temperatures as low as those of the human body. Color images can be generated for radiation that is entirely beyond the visible (Fig. 30.27).

Figure 30.26
False color digital image of the Nile Delta near Cairo, Egypt. The colors have been enhanced to indicate desert (white) and agricultural (red) features.

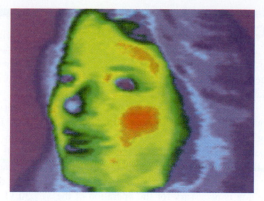

Figure 30.27
Color image of a human face derived from thermal radiation. Warmer regions are red and cooler regions such as the tip of the nose are blue.

Summary

Useful Concepts

■ To create a laser, you need an active medium of atoms or molecules with an inverted population of excited states enclosed in an optical resonator. The laser light is generated by the stimulated emission of photons by the excited medium. Laser light is characterized by its monochromaticity and its coherence.

■ The fringe visibility is

$$V = \frac{I_{max} - I_{min}}{I_{max} + I_{min}}.$$

■ The coherence is

$$\gamma = \frac{I_1 + I_2}{2\sqrt{I_1 I_2}} V.$$

■ Holograms are made by recording the interference pattern of a reference beam and an object beam. The holographic images are reproduced by illuminating the hologram with a beam of light similar to the original reference beam.

■ A full range of colors can be produced by the addition of primary colors of light (red, green, and blue). Colors can also be produced by passing white light through a combination of subtractive filters (the subtractive primaries of yellow, magenta, and cyan).

Important Terms

You should be able to write the definition or meaning of each of the following:

spontaneous emission	temporal coherence
stimulated emission	spatial coherence
laser	hologram
active medium	holography
pumping	spatial frequency
optical resonator	additive primary colors
inverted population	subtractive primary colors
metastable state	

Conceptual Questions

30.1 A photographer has a photoflood lamp whose temperature is too low for the film available. If she uses a color filter to correct the light balance to better match the film, what color should it be?

30.2 What effect would there be if both mirrors of a laser were partly reflecting?

30.3 Give a semiclassical explanation for stimulated emission. Does it explain why the radiation is in the same direction as the incident light?

30.4 Explain why helium is necessary for the operation of the helium-neon laser.

30.5 Sketch a setup to make a hologram. Identify all components shown in your drawing. What special requirements must be met to make a hologram?

30.6 When a hologram is cut in two, can a complete image of the object still be reconstructed from each piece? What differences will there be in the images from the two pieces?

30.7 Explain how viewing the virtual image of a hologram can be thought of as looking through a window. Use this idea to explain the pinhole camera behavior of the images that re-

sult when the hologram is illuminated with the undiverged beam from a laser.

30.8 Why does uneven illumination falling on a pair of slits affect the fringe visibility?

30.9 Describe the image(s) that result when the undiverged beam of a laser is passed through a transmission hologram.

30.10 Why do colored fabrics viewed in sunlight sometimes appear different from the way they look under incandescent light or fluorescent light?

30.11 Name as many commercial systems as you can that create color images using subtractive primary colors.

30.12 Examine a Polaroid instant color slide with a microscope. Is the color reproduction in the image an additive or subtractive process?

30.13 How can a color television reproduce yellow light when the available phosphors are limited to red, green, and blue?

30.14 Explain the photographer's rule that a colored filter lightens the image of objects of its own color and darkens the image of those of the complementary color.

Problems

Section 30.2 Lasers

30.1 Calculate the number of half-wavelengths in the laser cavity of a helium-neon laser whose wavelength is exactly 632.8 nm and whose mirrors are 30.0 cm apart. Assume that the index of refraction is 1.00.

30.2 Calculate the number of half-wavelengths in a ruby laser operating at a wavelength of 694.3 nm in air, given that the ruby rod, silvered on the ends, is exactly 10.0 cm long. The index of refraction of ruby is 1.69.

30.3 Calculate the separation between modes in a helium-neon laser whose cavity length is 35.5 cm. Assume that the index of refraction is 1.00.

30.4 Calculate the separation between modes in a short laser whose cavity length is only 5.00 cm. Assume that the index of refraction is 1.00.

30.5 What is the energy of the transition giving rise to laser light at 694.3 nm? Give your answer in eV.

30.6 What is the energy of the photons in the light from a diode laser at 650 nm? Give your answer in eV.

Section 30.3 The Helium-Neon Laser

30.7 (a) What is the energy separation of the states that give rise to the red 632.8-nm laser line in the helium-neon laser? (b) What is the energy separation of the states that give rise to the infrared line at 1150 nm? Give your answers in electron volts.

30.8 The helium-neon laser has a line with a wavelength of 543 nm. What is the energy, in electron volts, of photons at that wavelength? What color is the light?

30.9 The helium-neon laser is designed to emit a wavelength of 1533 nm. What is the energy, in electron volts, of photons at that wavelength? What color is the light?

Section 30.4 Properties of Laser Light

Hints for Solving Problems

The degree of coherence of light is the same as the visibility of the double-slit pattern when the slits are evenly illuminated.

30.10 A double-slit fringe pattern is observed to have intensity maxima that are three times greater than the intensity minima. What is the visibility of these fringes?

30.11 A certain laser produces a fringe visibility of 0.60. What is the ratio of I_{max}/I_{min} for these fringes?

30.12 An interference pattern produces fringes with a visibility of 0.40. What is the ratio of I_{max}/I_{min} for this pattern?

30.13• A double-slit interference pattern is observed for which the intensity maxima are ten times as great as the minima. Assuming even illumination on the slits, determine the degree of coherence of the incident light.

30.14• The interference pattern of a double slit has intensity maxima that are 7.7 times greater than the minima. It is known that the intensity due to one slit alone is twice as great as the intensity due to the other slit. What is the coherence of the incident light?

30.15• The interference pattern of a double slit has intensity maxima that are 5.5 times greater than the minima. It is known that the intensity due to one slit alone is three times as great as the intensity due to the other slit. What is the coherence of the light source?

30.16• A partially coherent light source, $\gamma = 0.70$, illuminates a double slit unevenly so that one slit receives 1.4 times more light than the other one. Compute the visibility of the fringes and the ratio I_{max}/I_{min} in the interference pattern.

30.17•• Completely coherent laser light is used to illuminate a double slit. The intensity of the light reaching one slit is four times as great as that of the light reaching the other one. (a) What will be the visibility of the resulting fringes? (b) What would be the visibility if the coherence were only 0.90? (c) What would be the ratio of I_{max} to I_{min} if the coherence were only 0.90?

Section 30.5 Holography

Hints for Solving Problems

The maximum spatial frequency that a hologram can have is limited by the resolution of the film.

30.18 In making a particular hologram, the maximum angle between the reference and object beams is 34°. What is the minimum resolution required of the holographic film if the wavelength of the light used is 693.4 nm?

30.19 A hologram is made with film whose resolution is 1500 lines/mm. Light from an argon laser, $\lambda = 488$ nm, was used in recording and viewing the hologram. The reference beam strikes the hologram at normal incidence. What is the angular field of view of this hologram?

30.20 A holographic image subtends an angle of 34° at the center of the holographic film when illuminated with a parallel beam of light of wavelength 488 nm. If the illumination is changed to light from a helium-neon laser at 632.8 nm, what angle will be subtended by the image?

30.21• Two beams of parallel laser light with $\lambda = 632.8$ nm are brought together on a photographic plate to form a sinusoidal holographic grating. One beam strikes the plate at normal incidence. The other beam makes an angle of 28° with the normal. (See Fig. 30.14.) The processed hologram is illuminated with light of wavelength 546.1 nm. If this light strikes the hologram at normal incidence, how many beams emerge and at what angles?

30.22• A grating spectroscope is used to study the emission spectrum from hydrogen. A grating of 1500 lines/mm is illuminated at normal incidence. (a) Determine the angles at which the Balmer lines are found in first order. The wave-

lengths are 656.3, 486.1, 434.0, and 410.2 nm. (b) Can any of the lines be seen in second order?

30.23• A holographic grating of 1500 lines/mm is used as the dispersing element in a spectroscope. In the first-order spectrum, what is the angular separation of the sodium D lines, $\lambda = 589.0$ and 589.6 nm? Assume normal incidence.

Section 30.6 Light and Color

30.24 What color corresponds to the spectral wavelength range (a) 400–450 nm, (b) 570–590 nm, (c) 700–900 nm, and (d) 630–680 nm?

30.25 What is the energy in eV of a photon with a wavelength of (a) 700 nm and (b) 400 nm?

30.26 Radiation from the sun has its maximum intensity at a wavelength of 475 nm. Compute the temperature of the sun using the Wien law. (For comparison, a value of 6000 K is obtained from the Planck law.)

30.27 The blackbody spectrum radiated by a certain lamp has a maximum around the wavelength 967 nm. What is the approximate temperature of this lamp?

30.28 (a) What is the peak wavelength in the spectral distribution of light from a 100-W incandescent lamp? (b) Compare this wavelength with that from the photoflood lamp of Example 30.4 by finding the ratio of the two wavelengths. Consider the lamps to be ideal blackbody radiators.

30.29 What is the peak wavelength of a blackbody radiator that has the same color temperature as (a) a cool white fluorescent lamp and (b) a warm white fluorescent lamp?

30.30 A large room is illuminated with 40 fluorescent lights, each rated at 40 W. How many 100-W incandescent lamps would be required to produce the same level of illumination in the room? Assume that the efficiency of the fluorescent lamps for converting electrical energy to visible radiation is 20%, while that of the incandescents is 3%.

30.31 A new office building is designed to be illuminated with fluorescent lamps that are 20% efficient in converting electrical energy into light (visible radiation). By what fraction would the energy costs for illumination be changed if the building were illuminated at the same level by incandescent lamps of 3.5% efficiency?

Section 30.7 Color by Addition and Subtraction

Hints for Solving Problems

The additive primary colors of light are red, green, and blue. Their complementary colors, cyan, magenta, and yellow, are the subtractive primaries.

30.32 White light passes through a cyan filter, which is, in turn, followed by a second filter. What color emerges if the second filter is (a) yellow, (b) magenta, (c) blue, (d) green?

30.33 Simultaneous excitation in equal intensities of the red and green phosphors on a TV screen produces a yellow color. (a) What color results from full-intensity red and half-intensity green? (b) What color results from full-intensity blue and half-intensity red?

30.34 What color results from the addition of equal intensities of (a) magenta and green light, and (b) blue and yellow light?

30.35 White light passes through two overlapping, full-strength color filters. What is the color of light that emerges if the two filters are (a) red and green, (b) magenta and yellow, (c) red and yellow, or (d) magenta and cyan?

30.36 White light passes through a yellow filter, which is, in turn, followed by a second filter. What color light emerges if the second filter is (a) green, (b) cyan, (c) magenta, or (d) blue?

30.37 The Balmer lines in the hydrogen spectrum are at wavelengths of 410.2, 434.0, 486.1, and 656.3 nm. What are the colors of these lines?

30.38 The principal lines in the emission spectrum of mercury lie at wavelengths of 365.5, 435.8, 546.1, 577.0, and 579.1 nm. Characterize these wavelengths according to their color.

30.39• Violet light ($\lambda = 410$ nm) and orange light ($\lambda = 615$ nm) are shined simultaneously on a pair of slits with a separation of 0.50 mm. The diffraction pattern is formed on a screen 1.0 m away. Describe the location and color of the first five bright fringes.

Additional Problems

30.40 What is the ratio of the peak spectral wavelength of daylight to that of light from a 100-W tungsten lamp?

30.41• A partially coherent light source, $\gamma = 0.80$, illuminates a double slit unevenly so that one slit receives 1.5 times more light than the other one. Compute the visibility of the fringes and the ratio I_{max}/I_{min} in the interference pattern.

30.42•• Completely coherent laser light is used to illuminate a double slit. The intensity of the light reaching one slit is five times as great as that of the light reaching the other one. (a) What will be the visibility of the resulting fringes? (b) What would be the visibility if the coherence were only 0.80? (c) What would be the ratio of I_{max} to I_{min} if the coherence were only 0.80?

30.43•• An advertisement for a commercial helium-neon laser says that the wavelength is 633 nm and the bandwidth of the line is 2×10^9 Hz. This bandwidth has the effect of smearing out the interference pattern that would be due to an absolutely monochromatic source. For a double-slit experiment using this laser, determine the position of the first maximum in the pattern from a double slit with a spacing of 0.20 mm that is 2.50 m from a screen. By how many millimeters is the position of the first maximum smeared out by the bandwidth of the laser?

Condensed Matter

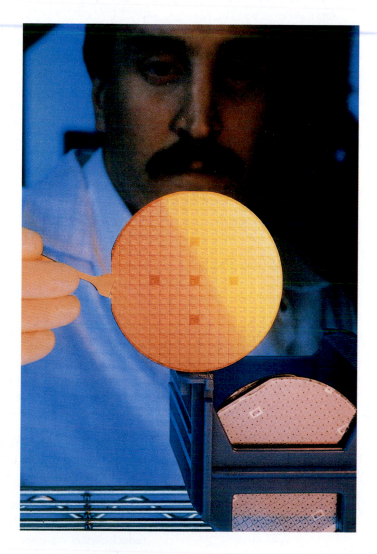

In Chapter 26, we examined some of the early research on crystal structure, which provided evidence for the atomic theory of matter. Although the early crystallographers made progress in categorizing different types of crystal structures, they did not get very far in understanding the behavior of solids. From our present viewpoint, this limited progress is not too surprising. After all, as we pointed out in Chapter 11, atoms in solids are much closer together than atoms in gases. Thus, the kinetic theory model of a gas that works well in explaining the gas laws is inappropriate for a solid or a liquid. The kinetic theory of gases assumes Newtonian mechanics, but a description of solids or liquids must be based on quantum mechanics because neighboring atoms are so close together. The development of quantum mechanics, in the early part of the twentieth century, and the development of x-ray diffraction techniques were key steps along the way to our contemporary understanding of solids.

Our presentation of condensed-matter physics is necessarily brief and introductory. We focus primarily on crystalline solids. However, without getting into the mathematical complexities of quantum mechanics, we can still examine many of the important concepts needed to understand solids. We will develop a descriptive model of the electronic structure of solids, called band theory. This model has been especially successful in explaining the behavior of metals and semi-conductors. ■

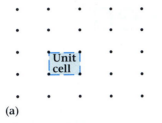

(a)

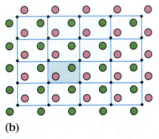

(b)

Figure 31.1

Schematic representation of two-dimensional crystals. (a) The basic repeating unit, the unit cell, is shown by the dashed lines; the points represent the lattice. (b) A crystal is formed when atoms or molecules are associated with each lattice point.

31.1 Types of Condensed Matter

The average separation between molecules in a liquid is comparable to their diameters. Individual molecules are free to move about, but they are constrained to move so that the average separation between near neighbors remains constant. As a result, a given mass of liquid is virtually incompressible and has a definite volume, although its shape can change to match the shape of its container.

In solids, the interatomic separations are also comparable to atomic diameters. But, unlike those in liquids, the atoms in a solid are not free to move about. Instead, atoms of a solid are rigidly fixed and do not move except for small oscillations about their average positions. Thus, the solid has not only a definite volume but a definite shape as well.

As we mentioned in Chapter 26, when the atoms of a solid are arranged in a regular manner, forming a three-dimensional array or lattice, we say that the solid is a **crystalline solid.** We can construct this array by repeating a basic unit pattern in three dimensions, in much the same way that bricks stacked in an orderly fashion form a neat pile. A schematic illustration of a two-dimensional lattice is shown in Fig. 31.1(a). The basic repeating unit, or *unit cell,* of this rectangular lattice is shown by the dashed lines. The entire lattice is built up by adding more unit cells in the adjacent space. In Fig. 31.1(a), we have drawn the unit cell by connecting neighboring points of the lattice with the dashed line. If we associate an atom or molecule with each lattice point, the result is a crystal. If two atoms are associated with each lattice point of Fig. 31.1(a), the resulting crystal is more complex, but the shape of the unit cell is unchanged. Figure 31.1(b) shows a rectangular crystal structure in which two atoms are associated with each lattice point. The periodicity of the atoms in the crystal lattice affects all of the physical properties of crystals.

The scanning tunneling microscope, which we first mentioned in Chapter 26, reveals three-dimensional images of solid surfaces. These images show vertical position differences as small as 0.01 nm and horizontal position differences as small as 0.6 nm. This microscopy technique utilizes the quantum mechanical tunneling of electrons between two conducting solids separated by a narrow layer of insulator or vacuum.

The tunneling microscope takes advantage of the strong exponential dependence of the tunneling current on the width of the insulating gap between the two conducting solids. When a sharp probe tip (Fig. 31.2) is scanned across the surface to be investigated, the vertical position of the tip is changed to keep the tunnel current constant. In this manner, the tip-to-surface distance is kept constant so that the probe follows the surface contour as it scans across. By monitoring the vertical position of the probe, a two-dimensional representation of the surface contours can be made for each scan. A three-dimensional image may be obtained from a sequence of such scans. Figure 31.3 shows one such image of a surface. The regular positions of the surface atoms are easily seen.

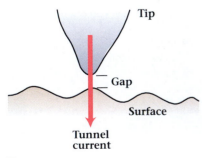

Figure 31.2

The tunneling of electrons across a narrow vacuum gap between two conductors is constant for a constant separation between them.

When the regularity of the pattern extends over the entire crystal, it is called a single crystal or monocrystal (Fig. 31.4). (Other examples of single crystals are shown in Figs. 26.1 and 26.2.) However, frequently the periodicity is disrupted so that the entire solid is made up of numerous individual crystalline subunits, called *grains.* Such material is said to be a **polycrystalline solid.** Figure 31.5(a) shows a possible arrangement of atoms near a grain boundary dividing two otherwise-perfect crystalline regions within a solid. Figure 31.5(b) is a micrograph of a metal, revealing the individual crystalline grains.

Not all solids have the long-range order of crystals. **Amorphous solids** possess only a short-range order, similar to that of liquids. Glass is one example. The arrangement of molecules in a glass is much like that of a liquid in which the atomic positions have become rigid. Schematic representations of crystalline and glassy solids (Fig. 31.6) show an orderly arrangement between neighboring atoms (short-range order) in a glass, but no long-range periodicity of the atomic positions.

A class of material in which the regular periodic arrangement exists in only one or two dimensions is known as a **liquid crystal.** Liquid crystals can flow, form droplets, and maintain long-range orientational and spatial order. Because of these special properties, some physicists refer to liquid crystals as a fourth state of matter. Liquid crystals usually consist of large molecules whose geometric shape forces the periodic structure.

In the remainder of this chapter, we will examine only crystalline solids. In most instances, we will consider the solid to be a perfect single crystal. However, remember that real crystals are never perfect and the presence of imperfections alters the properties of the solid. These imperfections are especially noticeable in the semiconductors, as we shall see later.

The electronic forces binding the atoms in a crystal are not all the same, and differences arise as a result of the geometric arrangement of charges. We often use specialized terms to categorize different crystal bindings, depending on the dominant type of interaction. Thus, we speak of ionic, covalent, and metallic crystals, which correspond to chemical bonding of those three types.

Ionic crystals are made up of positive and negative ions. In ordinary table salt (sodium chloride, NaCl), each sodium atom gives up its outermost electron,

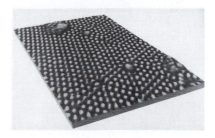

Figure 31.3
Tunneling microscope picture of germanium, showing the location of atoms on the surface. Impurity areas are present in the picture.

Figure 31.4
A laboratory-grown single crystal of aluminum alum.

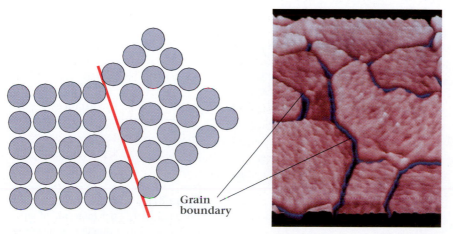

Figure 31.5
(a) Schematic drawing of a grain boundary between two otherwise-perfect crystalline grains.
(b) Tunneling micrograph of a polycrystalline solid shows crystal grains.

Grain boundary

(a)

(b)

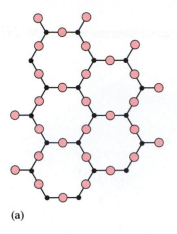

(a)

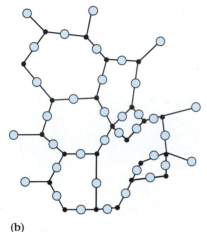

(b)

Figure 31.6

(a) Two-dimensional analog of a crystal. (b) Two-dimensional analog of a glass.

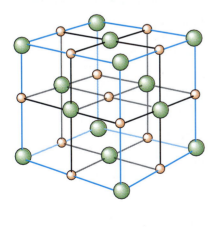

◯ = Chlorine ion
◯ = Sodium ion

Figure 31.7

The crystal structure of sodium chloride. The sodium ions are depicted as smaller than the chlorine ions. The cubic unit cell is shown. This cell is face-centered cubic.

becoming a positively charged ion, and each chlorine atom absorbs one of these electrons to become a negatively charged ion. These oppositely charged ions are then attracted to each other by a Coulomb force that is stronger than the Coulomb repulsion between ions of the same sign. In sodium chloride, the ions arrange themselves in what is called a face-centered cubic lattice (Fig. 31.7). In this crystalline form, the energy of the system is reduced well below the energy of the free ions. If this were not so, the crystal would not be so stable. The large binding energy is a direct result of the fact that each ion is surrounded by six nearest neighbors of opposite charge, thus reducing the electrostatic potential energy. Natural crystals of sodium chloride have a cubic shape suggestive of their inner crystal structure (Fig. 31.8).

Since the electrostatic potential gets lower (more negative) as the separation between ions decreases, you might conclude that the crystal would collapse until it disappeared. However, we know that this does not happen. As the ions are brought closer and closer together, their electronic distributions begin to overlap. At these distances a repulsive force associated with the Pauli exclusion principle arises (Fig. 31.9). We can then think of the ions as hard spheres that pack together tightly. Then the size of the crystal unit cell is governed by the dimensions of the individual ionic radii (Fig. 31.10).

Other ionic crystals have different crystal structures with different numbers of nearest neighbors and different nearest-neighbor distances. The occurrence of a particular crystal structure for a given ionic material depends on the Coulomb potential, the ionic sizes, and the repulsive term in the potential in a complicated way. As an example, cesium chloride, CsCl, which is close chemically to NaCl, has the structure shown in Fig. 31.11.

Covalent crystals result from atoms that exhibit covalent chemical bonding, in which electrons are shared by two closely spaced atoms. Carbon (in the form of diamond), silicon, and germanium all have the same type of crystal structure, known as the diamond structure, in which each atom is bonded to four nearest neighbors in a tetrahedral arrangement (Fig. 31.12). There appears to be a continuous range of crystal types between the extremes of the purely ionic and purely covalent limits. Often the bonding is something intermediate and is characterized for convenience as being mainly ionic or mainly covalent.

Metallic crystals are characterized by high electrical and thermal conductivity and a lustrous appearance. The bonding in a metal may be likened to a

Figure 31.8

A natural crystal of sodium chloride.

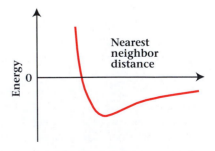

Figure 31.9

Potential energy per molecule of an ionic crystal as a function of nearest-neighbor distance.

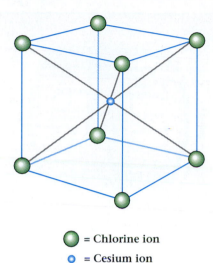

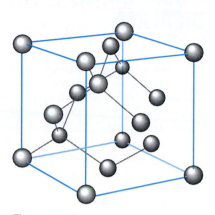

= Chlorine ion

= Cesium ion

Figure 31.10
A hard-sphere model of the sodium chloride crystal.

Figure 31.11
The crystal structure of cesium chloride. The space lattice is simple cubic with one chlorine ion on the corner of the cube and one cesium ion in the center.

Figure 31.12
The diamond structure. The edges of the cubic unit cell are shown. Each atom is tetrahedrally bonded to four nearest neighbors.

form of covalent bonding in which there are more potential bonding sites than there are electrons to be shared. So the shared electrons may move through the crystal, traveling from one bonding site to another. Since these outer electrons are free to move through the solid, it is better to describe them as belonging to the entire solid rather than to particular ions. We consider their behavior further in the next section.

31.2 The Free-Electron Model of Metals

One of the best-known characteristics of metals is their large electrical conductivity. Since most metals are solid at room temperature, their ions are stationary in the crystal lattice except for small-amplitude oscillations about their equilibrium positions. Thus, the large electrical conductivity is not due to motion of these positive ions but results from the motion of some of the negative electrons, which are not tightly bound to the ions.

We can construct a model of a metallic crystal that accounts for these observations. For simplicity, we consider a metal from the first column of the periodic table of the elements, such as sodium. Elements from the first column characteristically have one electron outside of a completely filled electron shell. This outermost electron is only weakly bound to the ion core. In fact, in the solid phase, these outermost electrons are no longer bound to any particular ion, but are free to move throughout the metallic crystal. They belong to the crystal as a whole rather than to an individual ion. In this model, known as the *free-electron model of metals,* we do not worry about details of interactions of the electrons with each other or with the ions. Instead, we treat them as a gas of particles confined to the volume of the crystal.

LIQUID CRYSTAL DISPLAYS

Look around you, wherever you are. Chances are you are not too far from a liquid crystal display (LCD). Adapted in the 1970s for use as display devices in digital watches (Fig. B31.1), they are now used in making digital meters, fever thermometers, and screens for lap-top computers. LCDs are even used to make the display screens for pocket televisions. In all of these applications, the alignment of molecules within the liquid crystal, and hence the optical behavior of the crystal, is changed by applying an electric field. Since very little current is required to activate the LCDs, they are excellent choices for all kinds of battery-operated electronic equipment.

A typical LCD is made from liquid crystal materials that have rodlike molecules with electric dipole moments along the molecular axes. A thin layer of this material is sandwiched between transparent electrodes and polarizing filters (Fig. B31.2).

Figure B31.1 An electric digital watch with a liquid crystal display.

In the structure shown, the polarizers are aligned at right angles to each other. The two surfaces that contact the liquid crystals are treated so that the molecules align with their axes parallel to the surfaces, but with a 90° rotation from the top to bottom surface when no voltage is applied between the electrodes (Fig. B31.3a). Light incident on the display is linearly polarized by the first polarizer. As the light passes through the layer of liquid crystal, the plane of polarization rotates by 90°. The light passes through the second polarizer, strikes a reflector and is reflected back through the system.

When a voltage is applied to the liquid crystal material, the molecules rotate to align with the resulting electric field (Fig. B31.3b). The plane of polarization of the incident linearly polarized light no longer rotates as the light traverses the liquid crystal material. Consequently, the second polarizer absorbs this light. If a segmented display is used, the number segments to which the voltage is applied will appear dark on a light background. If the orientation of the two polarizers is made parallel, then the display is reversed, with light numbers appearing on a dark background.

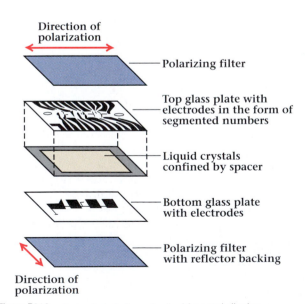

Figure B31.2 An exploded view of a liquid crystal display.

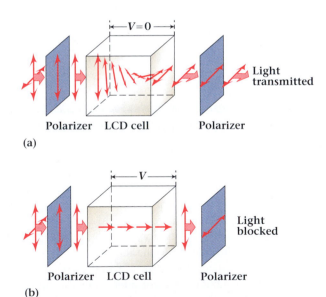

Figure B31.3 The action of an LCD device on polarized light passing through it. (a) The plane of polarization is rotated when no voltage is applied to the cell. (b) When a voltage is applied across the cell, the molecules of the liquid crystal rotate so that light passes through with no change in its polarization.

978

We know, however, that we cannot describe the electrons using Newtonian mechanics, as we did for the classical kinetic model of gases. Instead, we must formulate our model in terms of quantum mechanics. We have already laid the groundwork for this model in our discussion in Chapter 28 of the particle in a box. In the present case, however, we need to treat the problem in three dimensions. Suppose that the electrons are confined to a cube of edge length L. The wave function that satisfies the Schrödinger equation for the three-dimensional box with rigid walls is

$$\Psi_n(x, y, z) = \psi_0 \sin \frac{n_x \pi x}{L} \sin \frac{n_y \pi y}{L} \sin \frac{n_z \pi z}{L},$$

where ψ_0 is the amplitude and n_x, n_y, and n_z are integers. These integers are the quantum numbers that describe the wave function in the x, y, and z directions.

We already established that the essence of the crystal structure is its periodicity. Thus, the environment of each metallic ion must be identical to that of any other similar ion in the crystal. This fact leads us to consider other boundary conditions. Instead of demanding that the electron wave function vanish at the walls, as for the equation above, we need only require that the wave function be the same at points where the potential is the same. As a result, we get a somewhat different wave function, a periodic traveling wave with a period L. Using that wave function and the Schrödinger equation, we have the following expression for the electron energy levels:

$$E_n = \frac{h^2}{2mV^{2/3}}n^2, \tag{31.1}$$

where $n^2 = n_x^2 + n_y^2 + n_z^2$ and V is the volume L^3. In this case, we are not restricted to positive integers for the quantum numbers n_x, n_y, and n_z as before. Instead, they may be positive or negative integers or even zero:

$$n_x, n_y, n_z = 0, \pm 1, \pm 2, \pm 3, \ldots.$$

Each energy state described by a particular set of quantum numbers n_x, n_y, n_z can hold no more than two electrons, one each with spin quantum numbers of $+\frac{1}{2}$ and $-\frac{1}{2}$ as required by the Pauli exclusion principle. Of course, a given energy state could be empty. At very low temperatures, we would expect the lower energy states to be filled, beginning with the lowest state and continuing into successively higher states until all of the free electrons were accounted for. The last filled level would define a cutoff energy. The levels above this cutoff would be vacant (Fig. 31.13). The cutoff energy is known as the **Fermi energy** E_F. The Fermi energy* depends on the electron density—that is, on the number of free electrons N divided by the volume V—and is given by

$$E_F = \frac{h^2}{2m}\left(\frac{3N}{8\pi V}\right)^{2/3}. \tag{31.2}$$

For good conductors, the free-electron density N/V is of the order of 5×10^{28} electrons/m^3. Fermi energies corresponding to electron densities of this

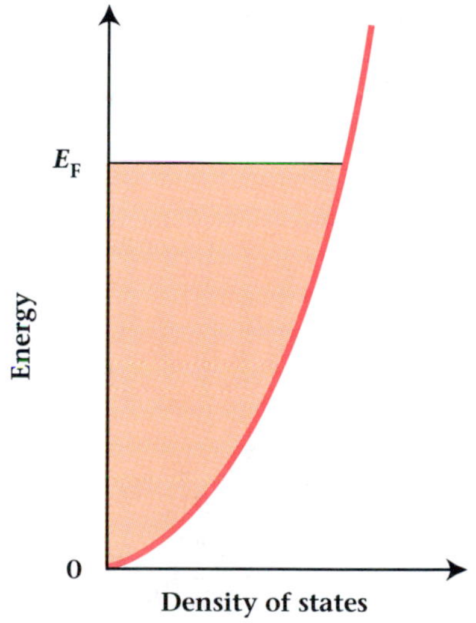

Figure 31.13
Density of electron states (as abscissa) plotted against electron energy (as ordinate). The shaded area represents the occupied states for temperature near 0 K. The cutoff energy is the Fermi energy E_F. Only electrons with energy near E_F interact with applied fields.

---

*The Fermi energy is named after the Italian physicist Enrico Fermi (1901–1954), whose pioneer efforts produced the first nuclear chain reaction. Fermi is also well known for his work on the statistical (thermal) behavior of particles that obey the Pauli principle. These particles, including electrons and protons among others, are collectively known as fermions.

magnitude are of the order of 5 eV. Note that although we have described the lowest, or ground, state of the system, individual electrons possess kinetic energies from 0 to E_F. At temperatures on the order of ordinary room temperatures, where $T \approx 300$ K, the kinetic energy of an electron near the Fermi energy far exceeds the energy (kT) associated with thermal processes. (At room temperature, $kT \approx 1/40$ eV.) Consequently, the Fermi energy of a metal is essentially unaffected by changes in temperature of tens of kelvins, and properties such as electrical conductivity change slowly with change in temperature.

Example 31.1

Calculate the Fermi energy in lithium.

Strategy We need to know the electron density (number of electrons per unit volume) in lithium. So we begin by calculating the number of lithium atoms per unit volume in the solid. This number can be determined from the mass density of lithium (0.53 g/cm^3), its atomic mass (6.94 g/mol), and Avogadro's number. Dividing the density by the atomic mass and multiplying the result by Avogadro's number, we get

$$\frac{\text{density} \times N_A}{\text{atomic mass}} = \frac{(0.53 \text{ g/cm}^3)(6.02 \times 10^{23} \text{ atoms/mol})}{6.94 \text{ g/mol}},$$

$$\text{atomic density} = 4.6 \times 10^{22} \text{ atoms/cm}^3 = 4.6 \times 10^{28} \text{ atoms/m}^3.$$

Since lithium belongs to the first column of the periodic table, we expect each atom to contribute one free electron to the solid. Thus, the electron density is the same as the atomic density, 4.6×10^{28} electrons/m^3.

Solution Now we can use Eq. (31.2) to find the Fermi energy of the electrons in lithium:

$$E_F = \frac{h^2}{2m}\left(\frac{3N}{8\pi V}\right)^{2/3}.$$

Using the free-electron density just determined for N/V, we get

$$E_F = \frac{(6.63 \times 10^{-34} \text{ J} \cdot \text{s})^2}{2(9.11 \times 10^{-31} \text{ kg})}\left(\frac{3(4.6 \times 10^{28}/\text{m}^3)}{8\pi}\right)^{2/3}.$$

$$E_F = 7.51 \times 10^{-19} \text{ J} = 4.7 \text{ eV}.$$

We can summarize our model of a metal as follows. The outermost electrons of the metal atoms are not bound tightly to the ions but are bound only by the potential function of the whole solid. Nonetheless, subject to the rules of quantum mechanics and the Pauli exclusion principle, the electrons are characterized by wave functions that are like free-electron wave functions, except for the restriction on the allowed wavelengths and, hence, energies. In the absence of electric fields applied to the metal, there is no electric current in the crystal even though the electrons are in motion. For each electron characterized by a momentum $\mathbf{p}_i$ there is another electron of equal and opposite momentum $-\mathbf{p}_i$. The net transfer of electric charge within the metal is zero.

The presence of an applied electric field within the metal changes the momentum distribution of the electrons so that the average momentum is nonzero and an electric current flows. However, the momentum distribution can change only if energy levels are available to which the electrons can be promoted by the field. These available levels are present in metals; indeed, they are the chief characteristic of metals, as we shall see below. The effect of the applied field is to give each electron an additional momentum $\Delta \mathbf{p}$. The entire electron gas then has a net momentum $N \Delta \mathbf{p}$. Associated with this net momentum is an average drift velocity $\mathbf{v}$ of the electron gas. (Recall our discussion of drift velocity and current in Section 18.1.)

When an electric field is applied to a metal, the current reaches a finite steady-state value. The electrons are no longer accelerated by the field, and there is some limit to the magnitude of their drift velocity. This limit arises because the electrons are not really free. They do interact with defects, impurities, and thermal distortions of their host crystal. The resulting collisions cause changes in the electron momentum that limit the magnitude of the drift velocity.

We are now in position to reexamine the photoelectric effect to see what it can tell us about metals. In Chapter 27, we found that the relation between the maximum kinetic energy of photoelectrons and the incident light frequency f is given by the Einstein equation:

$$KE_{max} = hf - \phi.$$

The work function ϕ represents the minimum energy required to just remove an electron from the surface of the metal.

We can now interpret the photoelectric effect to mean that, in the metal, the highest occupied electronic energy levels lie at an energy ϕ below the top of the potential well of the metallic crystal. Furthermore, the potential well describing the crystal potential must have finite depth. Our schematic picture of the metallic potential is revised along the lines of Fig. 31.14, where we see the work function ϕ as the energy required to take an electron from the Fermi energy E_F to a state of zero kinetic energy in the vacuum. The difference between the zero kinetic energy state in the vacuum and the bottom of the potential well is $E_0 = E_F + \phi$.

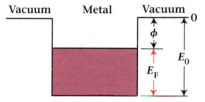

Figure 31.14
The potential well of a metal. The energy levels are filled for $E < E_F$ and empty for $E > E_F$. The photoelectric work function ϕ is the energy needed to take an electron from the Fermi energy E_F in the metal to the top of the potential well.

31.3 Electrical Conductivity and Ohm's Law

Let us now reexamine electrical conductivity in more detail than we did in Chapter 18. We need to introduce some new quantities that will help us to understand the nature of metals and semiconductors.

We learned earlier that when we apply an electric potential difference to the ends of a conducting wire, a current flows through the wire. This current is due to the net drift velocity of the electrons in the wire. A related quantity is the current density $\mathbf{j}$, which we define to be the vector whose magnitude is given by the current divided by the cross-sectional area A of the conductor through which it passes,

$$j = I/A. \tag{31.3}$$

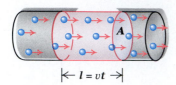

Figure 31.15
Section of a conducting wire. The arrows indicate only the drift motion.

$\leftarrow l = vt \rightarrow$

The direction of the current density vector is always the same as the direction of the applied electric field that drives the current. It is, therefore, the direction in which a positive charge would move under the influence of the field.

To understand the relation between current density and drift velocity, consider a segment of a uniform conducting wire (Fig. 31.15). In a time t, electrons moving with drift velocity $\mathbf{v}$ travel a distance $l = vt$. Thus, in a time t, all the electrons in the volume lA pass through cross section A. If the electron density is n, then the current is given by the electric charge in the volume lA divided by the time,

$$i = \frac{n(-e)lA}{t} = -nevA.$$

The current density becomes

$$\mathbf{j} = -ne\mathbf{v}. \tag{31.4}$$

In the solid, the electrons suffer collisions with impurities and defects in the crystals, including distortions of the potential that result from thermal vibrations. These collisions occur with a characteristic time τ and serve to limit the drift velocity of the electrons as they are accelerated by an applied field. The presence of an electric field causes the electrons to move with an acceleration $\mathbf{a} = -e\mathbf{E}/m$. They acquire an average drift velocity $\mathbf{v}$, limited by the average collision time, $\mathbf{v} = \mathbf{a}\tau$. When these two equations are combined with Eq. (31.4) for current density, we get an expression for current density in terms of the field,

$$\mathbf{j} = \frac{ne^2\tau}{m}\mathbf{E} = \sigma\,\mathbf{E}. \tag{31.5}$$

The quantity σ is the **electrical conductivity.**

Electrical conductivity is an intrinsic characteristic of the material from which the wire is made and is independent of the shape and size of the wire. The SI units of conductivity are $(\text{ohm} \cdot \text{m})^{-1}$. A large conductivity means that a given applied field can produce a large current density in the material. Good conductors like copper and silver have large conductivities.

The reciprocal of the conductivity is the resistivity ρ. Thus, a large conductivity corresponds to a small resistivity. (Table 18.1 lists the resistivities of a number of materials.) The resistivity is

$$\rho = \frac{1}{\sigma} = \frac{E}{j}.$$

For a particular conductor of length l and cross section A, the electric field E within the conductors depends on the applied voltage across it through $E = V/l$. We can then express the resistivity as

$$\rho = \frac{V/l}{I/A},$$

which can be rearranged to give

$$\frac{V}{I} = \frac{\rho l}{A}.$$

However, we earlier defined the ratio of voltage to current to be the electrical resistance R. Thus, the resistance is

$$R = \rho \frac{l}{A}.$$

This is the same expression for resistance in terms of resistivity that we had in Chapter 18. If the resistivity depends only on the microscopic properties of the material and not on the applied field, then the resistance is independent of voltage and the material obeys Ohm's law.

Another quantity related to the drift velocity is the **mobility,** defined as the magnitude of the drift velocity per unit field,

$$\mu = v/E. \tag{31.6}$$

The mobility changes slowly with temperature and, for constant temperature, is constant for a given crystal. By combining Eqs. (31.4) and (31.5), we find that the mobility is primarily determined by the electron collision processes through the collision time τ since

$$\mu = \frac{e\tau}{m}.$$

As a matter of convention, μ is always taken to be a positive quantity.

The conductivity can also be expressed in terms of the mobility through

$$\sigma = ne\mu. \tag{31.7}$$

This equation for conductivity allows us to compute the conductivity from knowledge of the electron density n and the mobility. If there are collections of electrons with different mobilities, we can obtain the total conductivity by adding the contributions due to the different mobilities. We find this approach especially useful when dealing with semiconductors.

Example 31.2

The electrical conductivity of lithium is $1.08 \times 10^7 \ \Omega^{-1}\text{m}^{-1}$. Calculate the mobility of electrons in lithium using the electron density found in Example 31.1.

Solution We can find the mobility from the relation

$$\sigma = ne\mu,$$

which can be rearranged to get

$$\mu = \frac{\sigma}{ne} = \frac{1.08 \times 10^7 \ \Omega^{-1}\text{m}^{-1}}{(4.6 \times 10^{28} \ \text{m}^{-3})(1.6 \times 10^{-19} \ \text{C})}$$

$$\mu = 1.5 \times 10^{-3} \ \text{m}^2 \ \text{V}^{-1}\text{s}^{-1}.$$

The mobility of electrons in lithium is nearly 100 times smaller than the mobility of electrons in silicon.

Band Theory of Solids

In the preceding sections, we developed a model for metals, consisting of nearly free electrons contained within a uniform potential well that extends over the entire crystal. We now examine a somewhat more realistic model in which the potential varies periodically with distance through the crystal. The period of the potential is the interatomic spacing between the ions that make up the crystal. One such potential is illustrated in Fig. 31.16.

We can gain some insight by imagining a solid with interatomic separations so large that the electronic energy levels of the crystal are just those of the isolated atoms. This would create an enormous number of states with the same energy. The number of independent states of a particular energy is the product of the number of atoms in the crystal and the number of distinct states of that energy in an isolated atom. Thus, a crystal of N widely spaced atoms that possess m states of energy E will have Nm independent states belonging to the same energy level E.

If we reduce the interatomic spacing in our imaginary crystal, the mutual interaction of the atoms causes a shift in the energy levels, so that the states no longer have the same energy. The result is that each original energy level is split into a large number of distinct levels (Fig. 31.17). The total number of distinct levels is Nm, which is the total number of states that belong to the same basic atomic level. For a real crystal with a mass of one milligram, $N \approx 10^{19}$. Thus, the number of levels in any macroscopic sample is very great. These levels occur within an energy spread on the order of one electron volt. Thus, the separation between levels is about 10^{-19} eV. For this reason, we often speak of the electronic levels of a crystal as making up a continuous band of states. However, the band is not truly continuous; it contains a very large but finite number of distinct levels.

It is helpful to illustrate the energy bands with a picture like that of Fig. 31.18, which combines the crystal potential of Fig. 31.16 with information on the electronic bands obtained from Fig. 31.17. The low-lying bands correspond to atomiclike wave functions that have appreciable amplitude only near the ion cores. The higher energy states are represented as being more evenly spread throughout the crystal. The nonlocalization of the higher-lying states means that although the electrons are bound to the entire solid, they are not associated with particular atoms. The relatively free motion of some of the electrons through the metal crystal is due to this nonlocalization.

Figure 31.16

Potential energy versus position along a one-dimensional lattice of evenly spaced nuclei.

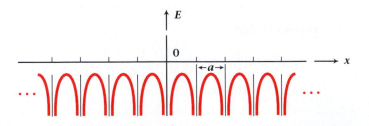

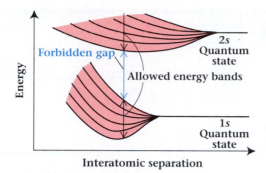

Figure 31.17
Dependence of energy levels upon interatomic separation.

The effect of the periodic crystal potential on the electronic energy levels is to group the higher states into **allowed bands,** separated by regions of energy in which no electron states are allowed, known as **forbidden zones** or gaps. An electron within an allowed band can be described by a wave packet in a manner similar to the treatment of the free-electron model. Each of these wave packets corresponds to a particular electron momentum.

Because of the symmetry of the crystal and, hence, of the energy bands, when all the energy states are filled there is no net momentum in the electron system. Consequently, a completely filled band carries no electric current because the net velocity of the electrons in a filled band is always zero. In addition, the restriction imposed by the forbidden zones means that there are no accessible energy states nearby. Thus, there can be no current in a material with filled energy bands even in the presence of an applied electric field. In other words, we have just described the electronic structure of an insulator.

But how do we describe a metal with the band model? Remember that a set of N regularly spaced potential wells can give rise to a band of N independent states. If each atomic energy level corresponds to two states associated with the electron spin, then the energy band contains $2N$ independent states. Also, we have seen that a completely filled band does not conduct a current any better than an empty band. This fact suggests that the best conductors are those with a half-filled allowed band, such as the alkali metals of the first column in the periodic table. For example, sodium has a single electron above a core of filled atomic levels. Thus, the energy band corresponding to this atomic level in metallic sodium is only half filled (Fig. 31.19a). In fact, sodium does have a high conductivity and is considered a good metal, as would be expected from the band model.

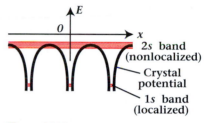

Figure 31.18
Energy bands superposed on a diagram of potential energy versus distance in a crystal.

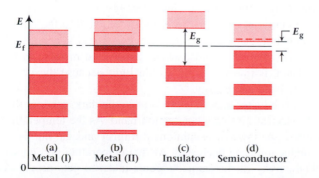

Figure 31.19
Band diagrams for (a) type I metal (partially filled band), (b) type II metal (band overlap), (c) insulator, and (d) semiconductor. The darker color represents occupied electron states, while the lighter color represents allowed but empty electron states. The regions of energy between the allowed bands are forbidden.

Table 31.1	Energy Gap in Some Common Semiconductors at Room Temperature	
Material		**Energy Gap (eV)**
Indium antimonide (InSb)		0.16
Germanium (Ge)		0.67
Silicon (Si)		1.14
Gallium arsenide (GaAs)		1.4
Cadmium sulfide (CdS)		2.42
Zinc sulfide (ZnS)		3.6

At this point, you might expect that elements in the second column of the periodic table, such as beryllium and magnesium, would be insulators, because each atom contributes two electrons to the highest occupied band. Thus the band would be filled, as in the case of the insulator previously described. However, elements from the second column are not insulators, but are metals with conductivities nearly as good as those of the elements in the first column. Their conductivities arise from the overlap of a higher-energy band with an otherwise-filled band (Fig. 31.19b). The noble metals (gold and silver) contribute one electron per atom. They are indeed good metals, as might be predicted from this model.

In some materials the bands do not overlap. In these crystals, if the number of electrons is just sufficient to fill several of these energy bands, the material may be an insulator. In particular, an insulator occurs when the energy gap E_g of the forbidden zone, separating the uppermost filled band from the nearest empty band of allowed states, is so large that thermal excitation or photoexcitation of electrons across the gap is insignificant (Fig. 31.19c). An example of such a material is diamond, which has a large energy gap (5.33 eV) between the filled energy band and the next allowed band.

When the energy gap E_g is sufficiently small (Fig. 31.19d), thermal excitation may become significant and the crystal will no longer be an insulator. Instead, it becomes a **semiconductor,** having a conductivity intermediate between those of an insulator and a metal. In addition, its conductivity may increase in the presence of light as a result of photoexcitation of electrons from the uppermost filled band to the next higher allowed band. Silicon and germanium have crystal structures like that of diamond, but their band gap energies are much lower: 1.14 eV in silicon and 0.67 eV in germanium. They are both semiconductors. Table 31.1 lists the gap energies of some other semiconductors.

Silicon and germanium crystals both look metallic (Fig. 31.20). We explain their appearance by noting that the photon energy in visible light is sufficient to cause photoexcitation of electrons from the filled band to the next higher band, called the conduction band. Thus, the presence of light generates a high density of free electrons within the bands, which creates the metallic luster. Diamond, on the other hand, is transparent to visible light because the photon energy is smaller than the band gap energy.

When semiconductor materials are kept in the dark, random thermal excitations promote electrons into the conduction band. As the temperature of the material is lowered, the thermal excitations diminish and the number of electrons within the conduction band decreases. As the temperature gets closer to absolute zero, the conductivity of the semiconductor material gets progressively poorer until it becomes an insulator.

Figure 31.20
A silicon crystal wafer of the type used in the manufacture of integrated circuits looks like a shiny piece of metal.

Master the Concept

The Energy Gap

Question: Although the band structures of diamond and silicon are quite similar, the band gaps between the valence and conduction bands are different, being 5.33 eV for diamond and 1.14 eV for silicon. How can this difference in band gaps lead to a simple explanation of why silicon has a metallic appearance but diamond is transparent?

Answer: The energies of visible-light photons range from about 1.77 eV to 3.10 eV. Because the band gap energy for silicon is smaller than the energy of the visible-light photons, electrons from the valence band can be photoexcited into the conduction band in sufficient quantities to make the silicon metallic and give it a lustrous appearance. On the other hand, the band gap in diamond is much larger than the photon energies, so the visible photons pass through without interacting with the electrons in the diamond crystal. Thus, diamond is transparent.

31.5 Pure Semiconductors

Semiconductors are just what their name implies; that is, they conduct to some extent, with conductivities intermediate between those of good conductors and good insulators. The conductivities of semiconductors range from about 10^{-6} to $10^4/\Omega \cdot m$ (see Table 18.1). The electrical properties of semiconductors are greatly influenced by impurities, present in amounts as small as one part per million or even less. Semiconductors that are very pure and whose behavior is not dominated by impurity atoms are called **pure (intrinsic) semiconductors.**

At temperatures near absolute zero, an intrinsic semiconductor is characterized by a completely empty energy band, separated by a small energy gap (≈ 1 eV) from a completely filled band (Fig. 31.21). The filled band is made up of energy states of the valence electrons of the atoms and is called the *valence band.*

As the temperature rises, some electrons are excited across the forbidden zone into the empty upper band. Then a continuous process of electron transfer occurs, both into the upper band from below and out of the upper band into the valence band. At any given temperature, the number of electrons in the upper band changes until the rates of transfer into and out of the band become equal. Since the electrons in this nearly empty band have many available states nearby, they can carry a current. For this reason, the upper band is called the *conduction band.*

The transfer of electrons out of the valence band leaves empty states within that band, allowing for a nonzero distribution of momentum within the band. Thus, the valence band can also carry a current. The effective number of charge carriers is equal to the number of empty electron states. Because of the effects of the crystal potential on the momentum of the electrons, we can treat these empty states as positive charge carriers called *holes.* Although the actual mechanism of charge transport is due to electrons, in a nearly filled band we commonly describe the transport in terms of the holes.

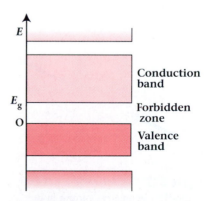

Figure 31.21

Energy band diagram near the valence band in a semiconductor. The top of the valence band is arbitrarily chosen to be the zero level. At low temperature, the valence band is filled and the conduction band is empty.

In an intrinsic semiconductor, all of the electrons excited into the conduction band leave holes in the valence band: Their numbers must be exactly equal. If we let n be the electron density in the conduction band and p be the hole density in the valence band, then $n = p$ in an intrinsic material. In determining the overall conductivity of the semiconductor, we must add the contributions due to both kinds of carriers. Thus, we write

$$\sigma = |e|(n\mu_n + p\mu_p), \tag{31.8}$$

where μ_n and μ_p represent the electron and hole mobilities, respectively. In the following section, we describe an experimental technique that allows us to measure the density, as well as the sign, of the charge carriers.

Example 31.3

Calculate the conductivity of pure silicon at 300 K where the density of carriers is $n = p = 1.45 \times 10^{16}$ m^{-3}. The mobilities are $\mu_n = 1350$ cm^2/V · s and $\mu_p = 480$ cm^2/V · s.

Solution The conductivity can be computed directly from the definition of Eq. (31.8):

$$\sigma = |e|(n\mu_n + p\mu_p).$$

Since $n = p$, we can write the conductivity as

$\sigma = ne(\mu_n + \mu_p)$

$\sigma = (1.45 \times 10^{16}$ m$^{-3})(1.60 \times 10^{-19}$ C$)[(1350 + 480)(\text{cm}^2/\text{V} \cdot \text{s})](10^{-4}$ m$^2/\text{cm}^2)$

$\sigma = 4.25 \times 10^{-4}/\Omega \cdot$ m.

31.6 The Hall Effect

Before we consider the behavior of impure semiconductors, let's examine one experiment that will help us establish the reality of charge transport by holes. In 1879, E. H. Hall observed that a voltage could be generated across a current-carrying conductor by placing the conductor in a magnetic field. When the direction of the current is perpendicular to the direction of the magnetic field, a voltage develops across the conductor at right angles to both the current and the magnetic field. This voltage is known as the Hall voltage, and its occurrence is called the **Hall effect.**

Let's examine the geometry of the Hall effect more closely (Fig. 31.22). An electric field applied to the conducting material in the x direction gives rise to an electric current density $\mathbf{j}_x$. As the charge carriers move through the material, they are deflected by the magnetic field $\mathbf{B}_z$. The result is a net separation of the charge, with an accumulation of negative charge on one edge and positive charge on the other. The separated charges establish an electric field whose effect is to oppose the magnetic deflection on the moving charges. Thus, in the steady state

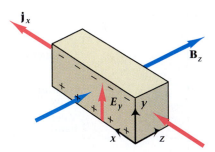

Figure 31.22

Hall-effect geometry: current in the x direction, magnetic field in the z direction, and Hall field in the y direction.

no current exists in the y direction. Subsequent charges pass through without deflection, since the opposing forces are balanced.

A charge e moving in the x direction in Fig. 31.22 experiences a force in the negative y direction due to the magnetic field, given by $F_{\mathrm{m}} = ev_x B_z$. This deflection causes a separation of charge in the material, which generates an electric field **E** in the positive y direction. This field, called the Hall field, exerts a force $F_{\mathrm{e}} = eE_y$ on the charge carriers. Since there is no steady-state current in the y direction, the magnitude of these two forces must be equal. Thus, $F_{\mathrm{e}} = F_{\mathrm{m}}$ or

$$eE_y = ev_x B_z.$$

Using the relation for the current density, $j_x = nev_x$ (where n is the carrier density), we can rewrite this equation as

$$eE_y = \frac{j_x B_z}{n}.$$

Upon rearrangement, the density of charge carriers is given by

$$n = \frac{j_x B_z}{eE_y}. \tag{31.9}$$

Thus, by measuring the current density j_x, the magnetic field B_z, and the Hall field E_y, it is possible to determine the density of mobile charge carriers within the material.

The direction of the Hall field depends on the sign of the charge carrier. Thus, the Hall effect is a relatively simple means for determining the sign of charge carriers in a material. For alkali metals and noble metals, the experimentally determined values of n are close to the values calculated from free-electron theory and the direction of the Hall field is that expected for the motion of negative charges. For many other materials, the agreement with free-electron theory is poorer. In some cases, like metallic bismuth, the sign of the Hall field is reversed, indicating an effective positive charge on the current carriers. Thus, the conductivity of bismuth is described in terms of the motion of positively charged holes.

The transport of electric charge through metals and semiconductors is due to the motion of the electrons. However, when the energy bands are more than half filled, these materials behave as if the charge were being carried by positive carriers, the holes. The Hall effect is one of several experiments that are most simply interpreted in terms of electrons and holes. As you will see in the following section, the concept of charge transport by holes is a great help in understanding the behavior of semiconductors.

Hall probes are Hall-effect devices used for measuring magnetic fields. When a current passes through a small semiconductor crystal in the presence of a magnetic field, a voltage develops just as we have already seen. The magnitude of this voltage depends on the magnitude of the magnetic field and the orientation of the probe relative to the direction of the field. Hall probes are routinely used to measure fields from 10^{-4} T to 10 T.

Example 31.4

A rectangular ribbon of silver, having dimensions $l_z = l_y = 1.0$ mm, is suspended in a magnetic field $B_z = 1.0$ T. When the current through the ribbon is 1.0 A, the Hall field is 1.1×10^{-4} V/m. What is the density of free carriers in the silver?

Solution The density of charge carriers is readily computed from Eq. (31.9). The current density j_x is the current divided by the cross-sectional area of the ribbon, which is $l_z l_y = (1.0 \text{ mm})(1.0 \text{ mm}) = 1.0 \times 10^{-6}$ m^2. Thus

$$n = \frac{j_x B_z}{e E_y} = \frac{(1.0 \text{ A}/1.0 \times 10^{-6} \text{ m}^2)(1.0 \text{ T})}{(1.6 \times 10^{-19} \text{ C})(1.1 \times 10^{-4} \text{ V/m})} = 5.7 \times 10^{28} \text{ m}^{-3}.$$

Example 31.5

A Hall probe used to measure a magnetic field is constructed from a crystal of indium arsenide with dimension $l_y = 2.5$ mm (Fig. 31.22) and a carrier density of 4.2×10^{22}/m^3. If the voltage due to the Hall field across the crystal is 8.9 μV when the current density j_x is 160 A/m^2, what is the magnitude of the magnetic field transverse to the crystal—that is, B_z?

Solution Equation (31.9) can be rewritten to give the magnetic field component B_z:

$$B_z = \frac{n e V_y}{j_x l_y}.$$

When the numerical values are inserted, we get

$$B_z = \frac{(4.2 \times 10^{22}/\text{m}^3)(1.6 \times 10^{-19} \text{ C})(8.9 \times 10^{-6} \text{ V})}{(160 \text{ A/m}^2)(2.5 \times 10^{-3} \text{ m})} = 0.15 \text{ T}.$$

31.7 Impure Semiconductors

Impure (or **extrinsic**) **semiconductors** are made from pure semiconducting material by adding small numbers of impurity atoms of higher or lower valence. In silicon and germanium, the addition of impurities with valence higher than four results in materials whose conductivities are dominated by electron carriers. Addition of atoms with fewer than four valence electrons leads to materials in which the dominant carriers are holes.

The controlled addition of impurities, which is known as *doping,* allows us to control the conductivity, as well as the sign, of the majority of the carriers. The reason we can do so is that the concentration of carriers in a pure semiconductor at room temperature is quite small compared with the density of atoms. For example, the carrier concentration in pure silicon is about 1.5×10^{10} cm^{-3}. The addition of impurity concentrations on the order of parts per million can increase the number of available charge carriers by a factor of a million. Thus, the conductivity of a semiconductor material is extremely sensitive to the impurity concentrations.

Addition of an impurity atom with five valence electrons, such as arsenic, into a crystal of silicon results in a set of new electronic energy levels localized at the impurity atom. The new localized level corresponding to the conduction band state of the host crystal actually lies slightly below the bottom of the conduction band, in the forbidden zone. Such impurity levels usually lie only a small fraction of an electron volt below the band edge (Fig. 31.23a). The energy required to transfer an electron from an impurity level to the conduction band is so small that most impurity sites are empty (or ionized) at room temperature. Because of the ready availability of these electrons for transitions to empty states in the conduction band, this kind of "doped" semiconductor is one in which the negative carriers dominate; that is, $n \gg p$, where n is the free-electron density and p is the density of holes. Impurity atoms that increase the number of negative carriers are called *donor impurities* because they donate electrons to the band. The resulting material is called a negative or **_n_-type semiconductor.**

In a similar manner, impurities with only three valence electrons, such as gallium, lead to localized electron states only slightly above the valence band (Fig. 31.23b). These states would be empty at a temperature of 0 K. However, at room temperature, thermal excitations promote electrons out of the valence band and into these impurity states. This results in an excess of holes, or positive carriers. These impurities are known as *acceptor impurities* because they accept or trap electrons out of the valence band. The resulting materials have a predominance of positive carriers and are known as **_p_-type semiconductors.**

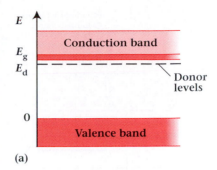

(a)

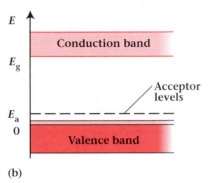

(b)

Figure 31.23
Detail of the band gap in an impurity semiconductor. (a) Donor impurity levels, lying near the bottom of the conduction band, contribute electrons into the band. (b) Acceptor levels, lying near the top of the valence band, trap electrons from the band, leaving holes free to move through the crystal.

31.8 The *pn* Junction

Many contemporary electronic devices owe their existence to junctions where *n*-type and *p*-type semiconductor materials are joined together. Junctions of this type, known as **_pn_ junctions,** have impurity concentrations that vary in such a way that the semiconductor crystal goes from *n*-type to *p*-type across a relatively narrow transition region. The change from *n*-type to *p*-type material may be abrupt, as when regions of virtually constant impurity concentrations contact each other, or the change may be graded so that the impurity concentration varies gradually across the transition region. In either case, the crystal structure is maintained across the transition region.

As an example of a *pn* junction, let's imagine an abrupt junction between *p*-type material containing N_a acceptor impurities per unit volume and an *n*-type material containing N_d donor impurities per unit volume (Fig. 31.24). Since many practical junctions are asymmetric in their impurity concentrations, we let $N_a > N_d$. At equilibrium, with zero applied voltage, electrons and holes diffuse across the junction and recombine. As a result, they create a region around the junction in which the electron and hole concentrations are greatly reduced, a region known as the *depletion zone*. There is a dipolar charge layer in the depletion zone because of the ionized impurities that are no longer neutralized by the mobile charges. This dipolar layer alters the electrostatic potential near the junction, creating an electric field that opposes the flow of electrons out of the *n*-type material and the flow of holes out of the *p*-type material.

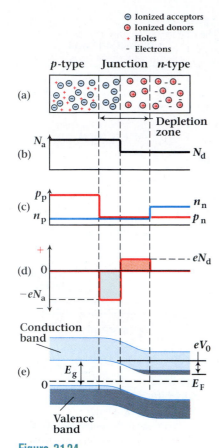

⊖ Ionized acceptors
⊕ Ionized donors
+ Holes
− Electrons

p-type   Junction   *n*-type

(a)

Depletion zone

(b)

N_a

N_d

(c)

p_p
n_p

n_n
p_n

(d)   0

$+$

$−$

eN_d

$−eN_a$

(e)

Conduction band

E_g

0

eV_0

E_F

Valence band

Figure 31.24

(a) Schematic representation of a *pn* junction in equilibrium; (b) impurity concentration, (c) mobile charge concentration, (d) charge density, and (e) energy band diagram showing the effect of the junction potential V_0.

The diffusion of electrons from the *n* side to the *p* side is limited by the electric potential V_0 resulting from the dipolar charge layer. Diffusion of electrons in the other direction, from the *p* side to the *n* side, is limited by their availability because of the small density of free electrons in the *p*-type material. In the same way, the potential inhibits the diffusion of holes from the *p* side to the *n* side. For this reason, we refer to it as a potential barrier.

A *pn* junction **diode** consists of a *pn* junction with two conducting contacts applied, one to the *p* material and one to the *n* material. With no external voltage applied, there is an electric potential across the junction region (Fig. 31.24e). At equilibrium, that potential difference is V_0 and the net current through the junction is zero. When an external voltage is applied to the diode, it passes current in one direction and blocks it in the opposite direction. An ideal diode, or **rectifier,** would have zero resistance for one polarity of applied voltage and infinite resistance for the opposite polarity. The graph of current versus voltage would consist of two straight-line segments (Fig. 31.25a). In practice, we cannot achieve this ideal situation, but real diodes can come quite close (Fig. 31.25b). The direction associated with the low resistance is known as the *forward* direction, and the direction for high resistance is the *reverse,* or *backward,* direction (Fig. 31.26).

We can determine the current-voltage relationship for a *pn* junction diode by considering the currents across the junction. Considering first the electrons, we allow for motion of electrons both ways across the junction. For a density n_n of electrons in the *n*-type material, the diffusion current of electrons across the potential barrier and into the *p*-type material is

$$I_f = Cn_n e^{-eV_0/kT}.$$

The factor $e^{-eV_0/kT}$ gives the probability that an electron in the *n*-type material has a kinetic energy eV_0 and thus is able to get over the potential barrier. C is a constant that depends on the mobility, the junction area, and the electronic charge.

There is also a current resulting from the diffusion of electrons out of the *p*-type material. When these electrons diffuse to the edge of the depletion zone, they are swept across by the junction potential, generating a current

$$I_s = Cn_p,$$

where n_p is the density of electrons in the *p*-type material. At equilibrium, these two opposing currents must be equal for the net current to be zero. Then

$$n_p = n_n e^{-eV_0/kT}. \tag{31.10}$$

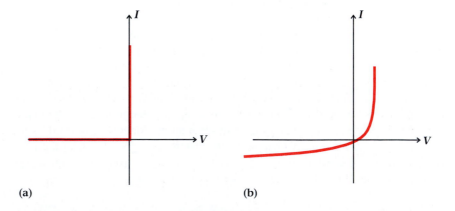

(a)

(b)

Figure 31.25

The current-voltage characteristic curves for (a) an ideal diode and (b) an actual semiconductor diode.

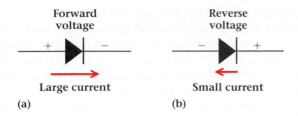

Forward voltage

Large current

(a)

Reverse voltage

Small current

(b)

Figure 31.26
The circuit symbol for a diode. (a) The symbol arrow points in the direction in which a large current can pass (forward voltage). (b) Polarity for reverse voltage of a diode, resulting in a very small current in the reverse direction.

When we apply a potential difference V in the forward direction, the potential barrier is lowered (Fig. 31.27) and the current I_f becomes

$$I_f = Cn_n e^{-e(V - V_0)/kT}.$$

The current I_s is unaffected by the added potential, since all electrons reaching the junction from the p-side still contribute to the current. Thus, the net current becomes

$$I_n = I_f - I_s = I_s(e^{eV/kT} - 1),$$

where we have used Eq. (31.10) to eliminate n_n.

The current I_p due to the motion of holes has the same form as the net current due to the electrons. The total current, which is the sum of the electron current and the hole current, is

$$I = I_n + I_p = I_0(e^{eV/kT} - 1), \tag{31.11}$$

where I_0 represents the maximum current in the reverse direction, known as the *saturation current*. Equation (31.11) is known as the *rectifier equation*. The current-voltage curve of Fig. 31.25(b) is described by the rectifier equation. Note that for strongly doped materials, the densities of holes p_n in the n-type material and electrons n_p in the p-type material are low and the saturation current I_0 is necessarily small.

When an electric potential difference is applied across a pn junction making the p side positive with respect to the n side, a large current results, corresponding to a positive V in the rectifier equation. When the potential difference is reversed, corresponding to a negative V in Eq. (31.11), there is very little current.

The importance of the pn junction in semiconductor devices cannot be overemphasized. Whether used individually as a rectifying junction in a diode or in combination to make transistors or other devices, pn junctions play a paramount role in contemporary electronics. The integrated circuit memory chips that are used in modern digital computers consist of millions of pn junctions assembled together on a single silicon surface no larger than a dime (Fig. 31.28).

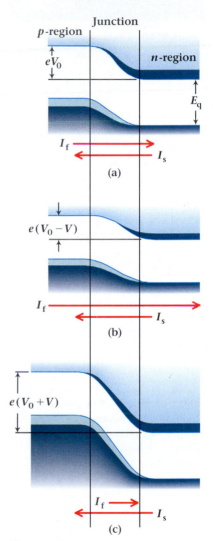

Figure 31.27
Schematic representation of a pn junction under conditions of (a) zero bias, (b) forward bias, and (c) reverse bias. Relative forward and reverse currents are indicated.

*31.9   Rectifier Circuits

A major application of diodes is the rectification of alternating voltages to provide direct voltages suitable for the operation of radios, tape recorders, calculators, etc. The circuit that converts an ac input voltage to a dc output is called a

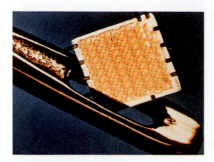

Figure 31.28
An integrated circuit chip.

dc power supply. In many applications, the power supply is designed to operate directly from standard ac power lines. Such circuits range from very simple rectifier circuits to elaborately regulated power supplies that maintain a precise dc voltage level regardless of input fluctuations or changes in the load circuit.

The simplest rectifier circuit is the *half-wave rectifier,* consisting of a single diode in series with a transformer and a load R (Fig. 31.29a). By choosing the proper transformer, we can have the output voltage of the power supply suit any particular need. In the half-wave rectifier shown in Fig. 31.29, the transformer output is $v = V_0 \sin 2\pi ft$. During the positive half cycle, the diode conducts and delivers the positive half of the sine wave voltage to the load. During the negative half cycle, the diode blocks the current and the voltage across the load drops to zero. All of the transformer voltage then appears as a reverse voltage across the diode. As a result, the output waveform across the load is the half-wave rectified signal shown in Fig. 31.29(b).

Half-wave rectifier circuits deliver only half of the available power, since the circuit conducts only half the time. To use more of the available power and also achieve a waveform that is more easily smoothed to a constant value, we need a *full-wave rectifier.* One circuit for achieving this is the full-wave bridge shown in Fig. 31.30(a). When point A of the diagram is positive with respect to point B, the voltage across diodes D_2 and D_3 is in the forward direction. Thus, D_2 and D_3 provide a path for current through the load resistor in the circuit. Diodes D_1 and D_4 block the current because the voltage across them is in their backward direction. When the input voltage reverses polarity, D_1 and D_4 conduct while D_2 and D_3 block (Fig. 31.30b). In both cases, the current through the load resistor is in the same direction. The resulting voltage across the load is the full rectified sine wave signal shown in Fig. 31.30(c).

The maximum current in the bridge circuit is $I_0 = V_0/R$. Similarly, the dc, or average, value of the current is $I_{dc} = \bar{I} = \bar{V}/R$. The average values of the current and voltage can also be expressed in terms of the maximum values

Figure 31.29
(a) A half-wave rectifier circuit. The diode conducts current only during the positive half cycle of the oscillating voltage. (b) Input and output waveforms.

Diode

Input
$v = V_0 \sin 2\pi ft$

R v Output

Iron-core
transformer

(a)

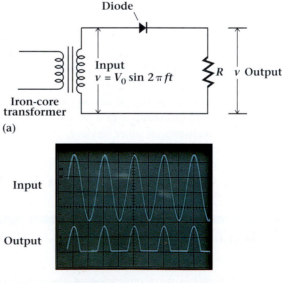

Input

Output

(b)

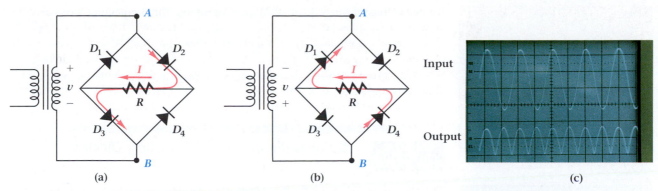

(a)   (b)   (c)

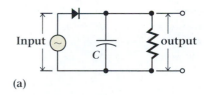

Figure 31.30
A full-wave rectifier circuit. The direction of the voltage across the load (and thus the current) is the same during both the positive (a) and negative (b) half cycles of the input voltage. (c) Input and output waveforms.

(Fig. 31.31). We can use graphical analysis to show that

$$\bar{I} = \frac{2I_0}{\pi} \quad \text{and} \quad \bar{V} = V_{dc} = \frac{2V_0}{\pi} \quad \text{(full rectified sine wave).}$$

The rms value is obtained from the square root of the average of the square of the wave. (See Section 21.3.) Thus, it is the same as for the sine wave itself:

$$I_{rms} = \frac{I_0}{\sqrt{2}} \quad \text{and} \quad V_{rms} = \frac{V_0}{\sqrt{2}} \quad \text{(full rectified sine wave).}$$

The pulsating voltage (and current) of a rectifier circuit can be smoothed out to produce a nearly constant direct voltage. This is usually done with suitable arrangements of capacitors and inductors that are known as **filters** because they filter out the oscillating component of the voltage. The simplest filter circuit, the shunt-capacitor filter, consists of a capacitor connected across the output of the rectifier, in parallel with the load (Fig. 31.32a). When the voltage is first applied, the capacitor charges rapidly with rising input voltage (Fig. 31.32b). As the input voltage begins to drop, the potential difference across the diode reverses and the diode stops conducting. The charge stored on the capacitor then flows off through the load resistor. During this part of the cycle, the voltage across the load resistor decreases, following the exponential decay of the RC discharge. The cycle repeats when the input voltage again gets high enough for the diode to conduct. As the product RC is made larger, the time constant of the discharge becomes larger and the waveform is smoother; that is, the ripple (or alternating part of the waveform) decreases. At the same time, the average (dc) value of the waveform increases toward the peak value of the input wave.

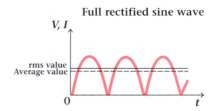

Figure 31.31
A full rectifier sine wave. The lines indicate the average and effective values of the wave.

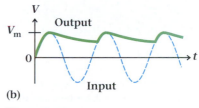

(a)

(b)

Figure 31.32
(a) A capacitively filtered half-wave rectifier. (b) The output voltage wave.

Example 31.6

Let's consider a practical example of a full-wave rectifier. Suppose a transformer in your radio has an rms output of 6.3 V. Assuming that no filter is used, what direct voltage is produced when the transformer is coupled to a full-wave rectifier like that of Fig. 31.30(a)?

Solution We know that for a full-wave rectifier, the dc level of the output voltage is $V_{dc} = 2V_0/\pi$. The value of V_0 is related to the rms value of the transformer voltage through $V_0 = \sqrt{2}V_{\text{rms}}$. Combining these equations, we get

$$V_{\text{dc}} = \frac{2\sqrt{2}V_{\text{rms}}}{\pi}.$$

When we insert 6.3 V for the rms voltage of the transformer, the output dc voltage becomes $V_{\text{dc}} = 5.7$ V.

*31.10 Solar Cells and Light-Emitting Diodes

The absorption of light in semiconductors generates electron-hole pairs in excess of the number already present by virtue of thermal agitation. This effect, called photoabsorption, was mentioned earlier in connection with the photoexcitation of electrons from the valence band to the conduction band. The increased conductivity associated with the excess carriers is called photoconductivity. If the protective material surrounding a diode, transistor, or other semiconductor device has a transparent window, that device may then be controlled with light. Devices of this kind are known as *optoelectronic* devices.

A *photodiode* is a *pn*-junction diode fabricated so that light can fall on the region near the junction. When the diode is operated with a reverse voltage, a large change in its resistance can be caused by photoabsorption. Photons of energy greater than the energy gap E_g generate excess electron-hole pairs, which contribute to an increased reverse current proportional to the incident light intensity. A *solar cell* is a photodiode with zero applied voltage. It produces a current that is due to carriers generated near the junction and carried across by the junction field. Under illumination, the solar cell can be used as a source of dc current to replace batteries.

In addition to the numerous light-activated devices, there are other devices that emit radiation. This radiation, associated with the direct recombination of electrons from the conduction band with holes in the valence band, is seen in *light-emitting diodes* (LEDs). Because the photon energy is essentially that of the band gap E_g, most LEDs emit light in the infrared or red portion of the spectrum (Fig. 31.33). However, blue, green and yellow LEDs are now available.

Infrared LEDs, which emit radiation that is invisible to the eye, are commonly used in television and VCR remote controls (Fig. 31.34). In this application they emit pulses of infrared radiation that are detected by a photodiode or phototransistor in the TV or VCR. These IR detectors are shielded from stray room light by filters that pass infrared but block visible light. Thus, they respond only to the signals from the remote unit. The radiation from the control spreads out in a wide cone so that precise aiming of the radiation is not necessary.

In normal operation, an LED emits light when a forward voltage of about 1.5 to 2.0 V is applied. The light output is roughly proportional to the current in the diode. Because an LED operates in the forward conducting mode, it requires much more power to operate than does an LCD. For this reason, a pocket calculator with an LED display requires larger batteries than the equivalent calculator with a liquid crystal display.

Figure 31.33
A digital voltmeter display made with light-emitting diodes.

Figure 31.34
The invisible radiation from a television remote control is made visible by a sensor that fluoresces in the visible range when illuminated with infrared.

Laser diodes are also available. They are light-emitting diodes that have been especially prepared with polished faces transverse to the depletion zone that act like the mirrors normally used for a laser. Most LEDs and laser diodes are made from alloys of the intermetallic compounds gallium arsenide and gallium phosphide because they have band gaps of the appropriate size for the emission of visible light. In contrast to the narrow beam of light from a helium-neon laser, the light from a diode laser spreads out in a large cone because of the small size of the resonant cavity and effects of diffraction by the tiny apertures (Fig. 31.35). Nevertheless, the light from the diode laser is coherent.

Laser diodes are used in compact disc digital audio systems because they can be brought to ultrasharp focus with a lens (Fig. 31.36). A sharply focused beam, less than 1 μm in diameter, is necessary for detecting the audio information, which is stored as tiny pits only 0.5 μm wide, 0.1 μm deep, and approximately 1 to 3 μm long in a spiral track with a pitch of only 1.6 μm. The pits are pressed into a flat, reflective surface. The beam optics direct the beam to the surface of the disc, where it is reflected and sent to a photodetector. When the beam encounters a pit, it is scattered so that very little light reaches the detector. The sequence of "on" and "off" pulses of light reflected from the disc is converted to digital electrical signals, which are sent to a computer for decoding to reproduce the original sound that was encoded on the disc in the form of the pits.

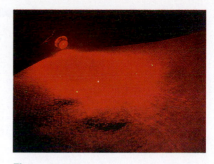

Figure 31.35

A laser diode emits a diverging beam of light.

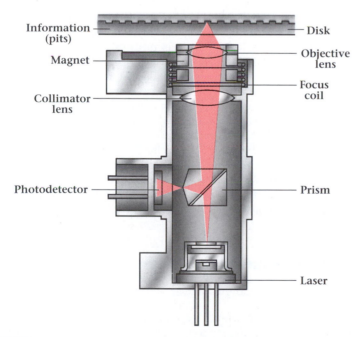

Figure 31.36

Compact disc optical system. The laser beam passes through a semireflecting prism to the lens system, which keeps it in sharp focus on the disc. The light passes through the transparent protective layer and strikes the pitted reflective surface. The reflected light returns through the lens system to the prism, where it is deflected into a photodetector.

Summary

Useful Concepts

■ The current density is

$$j = I/A = nev.$$

■ The mobility of the charge carriers in a conductor is defined by

$$\mu = v/E.$$

■ The conductivity is related to the mobility by

$$\sigma = ne\mu.$$

■ In the Hall effect, the density of charge carriers is

$$n = \frac{j_x B_z}{eE_y}.$$

■ The current in a diode is given by the rectifier equation

$$I = I_0(e^{eV/kT} - 1).$$

Important Terms

You should be able to write the definition or meaning of each of the following:

crystalline solid	pure (intrinsic) semiconductors
polycrystalline solid	Hall effect
amorphous solids	impure (extrinsic)
liquid crystal	semiconductors
Fermi energy	n-type semiconductor
electrical conductivity	p-type semiconductor
mobility	pn junction
allowed bands	diode
forbidden zones	rectifier
semiconductor	filter

Conceptual Questions

31.1 Carefully distinguish between crystals and glasses.

31.2 What keeps the atoms in a crystal apart?

31.3 What are the forces that hold atoms together in solids? In liquids?

31.4 Why is iron a solid at room temperature but mercury is a liquid?

31.5 What gives metals their lustrous appearance?

31.6 What affects the distribution of kinetic energies of the electrons emitted in a photoelectric process when the energy of the photons exceeds the work function by a few electron volts?

31.7 Estimate the drift velocity of the electrons in a wire carrying current to a 100-W lamp. Why does the lamp come on so quickly after the switch is closed?

31.8 Compare semiconductors and insulators. How are they alike and how are they different?

31.9 How can you tell a semiconductor from a metal?

31.10 What evidence is there for the conduction of electricity by holes? Do the holes really exist?

31.11 If a half-wave rectifier is connected to an ordinary electric lamp in a household circuit, the lamp is observed to be dimmer than before; however, it does not blink off and on as you might expect from examining Fig. 31.29(b). Can you give two reasons for the failure to observe such flickering?

31.12 Suppose a large capacitor is placed in parallel with the load resistance in the circuit of Fig. 31.30. What effect would it have on the current and voltage across R? Sketch a diagram of the resulting voltage waveform.

31.13 In what ways does a filter capacitor improve the performance of a dc power supply?

Problems

Section 31.1 Types of Condensed Matter

31.1 The cesium chloride unit cell has an edge length of 0.411 nm. What is the distance between a cesium ion and its nearest cesium neighbor?

31.2 The interatomic distance along the cube edge in cesium chloride is 0.411 nm. What is the interatomic distance between the nearest-neighbor cesium and chlorine ions? (All distances are measured center to center.)

31.3 The nearest-neighbor distance in a crystal with the cesium chloride structure is one-half the corner-to-corner distance along the body diagonal of the cubic unit cell. What is the near-neigh-

bor distance in rubidium chloride, which has the cesium chloride structure with a unit cell edge length of 0.374 nm?

31.4● The nearest-neighbor distance in sodium chloride is 0.281 nm. What is the distance between closest neighbors of the same kind; that is, what is the shortest Na-Na distance?

31.5● The nearest-neighbor distance in diamond is one-fourth the corner-to-corner distance along the body diagonal of the cubic unit cell. Calculate the distance between nearest neighbors in the diamond crystal in terms of the edge length of the cubic unit cell, a_0.

31.6•• Estimate the distance between nearest-neighbor ions in a sodium chloride crystal from knowledge of the molecular weight and the density $\rho = 2.17$ g/cm^3. Compare your result with the distance of 0.281 nm determined experimentally by x-ray diffraction.

Section 31.2 The Free-Electron Model of Metals

Hints for Solving Problems

Be especially careful to use consistent units in calculating Fermi energy, conductivity, and mobility of electrons. Remember that the Fermi energy is the energy of the highest occupied energy state for electrons in a metal, and that the electrons near the Fermi energy are the ones affected by an applied field.

31.7 Calculate the equivalent temperature T_F of the electrons at the Fermi surface of lithium. T_F is defined by $E_F = kT_F$. Use the result of Example 31.1.

31.8 Calculate the free-electron density in potassium. Assume the mass density to be 0.82 g/cm^3 and the atomic mass to be 39.1 g/mol.

31.9 Calculate the free-electron density in sodium, which has a mass density of 0.97 g/cm^3 and an atomic mass of 23.0 g/mol.

31.10• (a) Calculate the Fermi energy for the electrons in rubidium, which has a free-electron density of 1.08×10^{22} cm^{-3}. (b) Calculate the equivalent temperature T_F of the electrons at the Fermi surface of rubidium.

31.11• The photoelectric work function of copper is 4.94 eV. The Fermi energy is 7.00 eV. (a) What is the least energetic photon required to remove an electron from the bottom of the potential well? (b) What is the wavelength of this photon? (c) What is the maximum kinetic energy for photoelectrons ejected by photons of this wavelength?

31.12• Metallic sodium has a photoelectric work function $\phi = 2.75$ eV and a Fermi energy $E_F = 3.10$ eV. (a) What is the least energetic photon required to remove an electron from the bottom of the potential well? (b) What is the wavelength of this photon?

31.13• The threshold wavelength for photoelectric emission from tin is 340 nm. (a) What is the value of the work function? (b) What is the color of the threshold light? (c) Will red light from a helium-neon laser generate photoelectrons from tin?

Section 31.3 Electrical Conductivity and Ohm's Law

Hints for Solving Problems

Conductivity is the reciprocal of resistivity.

31.14 What is the drift velocity of electrons in lithium in a region of $E = 10^{-3}$ V/cm? Use the result of Example 31.2.

31.15 Estimate the average collision time for electrons in lithium from the mobility that was calculated in Example 31.2.

31.16 Estimate the average collision time for electrons in silicon, given that the mobility is 1300 cm^2/V · s.

31.17 Compute the average electron mobility in silver, given that the electrical resistivity is 1.59×10^{-8} Ω · m and the electron density is 5.85×10^{28} electrons/m^3.

31.18 Compute the average electron mobility in copper, given that the electrical resistivity is 1.72×10^{-8} Ω · m and the electron density is 8.50×10^{28} electrons/m^3.

31.19• (a) Calculate the mobility of electrons in gold, given that the electron density is 5.90×10^{28} m^{-3} and the resistivity is 2.4×10^{-8} Ω · m. (b) What is the average collision time of electrons in gold?

31.20• The free electron density in silver is 5.76×10^{28} m^{-3}, and the electrical conductivity is 6.17×10^7 Ω^{-1}m^{-1}. (a) What is the electronic mobility μ? (b) What is the collision time τ?

31.21• Current to a 100-W lamp is carried by a wire with a cross section of 1.00 mm^2. The lamp operates on 24 V-dc. (a) Calculate the drift velocity of the electrons in the wire, assuming that it is made of copper with a carrier density of 8.5×10^{28}/m^3. (b) How long will it take a particular electron to get from the switch to the lamp if the wire from the switch to the lamp is 3.0 m long?

31.22• Current to a 5.0-W lamp is carried by a wire with a cross section of 1.00 mm^2. The lamp operates on 12 V-dc battery. (a) Calculate the drift velocity of the electrons in the wire, assuming that it is made of copper with a carrier density of 8.5×10^{28}/m. (b) How long will it take a particular electron to get from the battery to the lamp if the wire from the battery to the lamp is 1.25 m long?

Section 31.4 Band Theory of Solids

31.23 Calculate the wavelength of the radiation just sufficient to cause electron transitions from the valence band to the conduction band in diamond. To what kind of radiation does this correspond? The energy gap in diamond is $E_g = 5.33$ eV.

31.24 Calculate the wavelength of the radiation just sufficient to cause electron transitions from the valence band to the conduction band in germanium. To what kind of radiation does this correspond? The energy gap is $E_g = 0.67$ eV.

31.25 Silicon can be used as a window for infrared radiation. What is the shortest wavelength that will pass through silicon without causing electron excitation from valence band to conduction band?

Section 31.5 Pure Semiconductors

Hints for Solving Problems

The electrical conductivity of semiconductors depends on the concentrations and mobilities of both electrons and holes.

31.26 Calculate the conductivity of pure germanium at a temperature where the density of electrons is $n = p = 2.4 \times 10^{19}$ m^{-3} if the mobilities are $\mu_n = 3900$ cm^2/V · s and $\mu_p = 1900$ cm^2/V · s.

31.27 Calculate the conductivity of intrinsic gallium arsenide at a temperature where the density of carriers is $n = p = 9.1 \times 10^{12}$ m^{-3}, given that the mobilities are $\mu_n = 8600$ cm^2/V · s and $\mu_p = 250$ cm^2/V · s.

31.28• The mobility of electrons in silicon is 1350 cm^2/V · s. (a) Calculate the drift velocity of an electron in silicon moving in a field of 100 V/cm. (b) Calculate the final velocity of a free electron accelerated from rest through a potential of 100 V.

Section 31.6 The Hall Effect

31.29 A ribbon of copper having dimensions $l_z = 2.0$ mm and $l_y = 5.0$ mm is suspended in a magnetic field $B_z = 1.25$ T. When the current through the ribbon is 10 A, the Hall field is $E_y = 9.3 \times 10^{-5}$ V/m. What is the density of free carriers in this material?

31.30 A ribbon of n-type silicon having dimensions $l_z = 0.20$ cm and $l_y = 0.50$ cm is suspended in a magnetic field $B_z = 1.0$ T. The free-carrier density in this material is 4.6×10^{22} m^{-3}. What is the Hall field E_y when a current of 125 mA is passed through the sample? What is the Hall voltage across the sample?

31.31 (a) Verify that the direction of E_y in Fig. 31.22 is correct for positive charge carriers. (b) In what direction is E_y when the current density j_x is due to negative carriers?

Section 31.7 Impure Semiconductors

31.32 Repeat Problem 31.26 under conditions that $n = 7.2 \times 10^{19}$ m^{-3} and $p = 8 \times 10^{18}$ m^{-3}.

31.33 Calculate the conductivity of n-type gallium arsenide for which $n = 5.1 \times 10^{18}$ m^{-3} and $p = 1.6 \times 10^7$ m^{-3}. The mobilities are $\mu_n = 8600$ cm^2/V · s and $\mu_p = 250$ cm^2/V · s.

Section 31.8 The *pn* Junction

31.34 Calculate the junction potential V_0 of a silicon *pn* junction at room temperature (300 K), given that the materials are doped so that the concentration of electrons in the n-type material is 1.0×10^{14} cm^{-3} while the concentration of electrons in the p-type material is 1.0×10^5 cm^{-3}.

31.35 Calculate the junction potential V_0 of a silicon *pn* junction at room temperature (300 K), given that the materials are doped so that the concentration of electrons in the n-type material is 5.0×10^{14} cm^{-3} while the concentration of electrons in the p-type material is 1.0×10^5 cm^{-3}.

31.36 A diode has a saturation current of 0.130 μA. What current would it carry at a reverse voltage of 1.40 V if it exactly obeyed the rectifier equation?

31.37 A diode has a saturation current of 0.10 μA. What current would it carry at a forward voltage of 0.40 V if it exactly obeyed the rectifier equation?

31.38• Using the rectifier equation, calculate the current through a diode at reverse voltages of 0.10 V and 1.0 V and forward voltage of 0.30 V, given that the saturation current is 1.0×10^{-7} A. Assume that $T = 300$ K.

31.39 A transformer in a TV set has an rms output of 6.3 V. (a) What is the average value of the voltage produced when the transformer is coupled to a half-wave rectifier like that of Fig. 31.29(a)? (b) What is the rms voltage?

31.40 A transformer in a stereo amplifier has an rms output of 15 V. (a) What is the average value of the voltage produced when the transformer is coupled to a full-wave rectifier like that of Fig. 31.30(a)? (b) What is the rms voltage?

31.41• The transformer in a full-wave rectifier circuit provides a sinusoidal voltage with a peak of 9.3 V. (a) What is the average (dc) current if the load resistance is 470 Ω? (b) What is the rms current? (c) What power is dissipated in the resistor?

31.42• Sketch the output signal across R_L from the full-wave rectifier of Fig. 31.28(a) as a function of time if the transformer provides a sinusoidal voltage with a peak of 5.0 V. Consider the situations: for (a) ideal diodes with no resistance, and (b) real diodes with a forward voltage drop of 0.50 V.

*Section 31.10 Solar Cells and Light-Emitting Diodes

31.43 An LED is made from material with a band gap of 1.4 eV. At what wavelength will it radiate?

31.44 Light is emitted by an LED in a process in which an electron makes a direct transition from the conduction band to the valence band, where it recombines with a hole. If the light has a wavelength of 650 nm, what is the energy of the band gap?

31.45• Suppose that you have solar cells that are 10% efficient; that is, they convert 10% of incident sunlight into electrical energy. (a) Assuming an incident solar flux of 1.0 kW/m^2, determine how large a collector is needed to provide a home with electrical power at a rate of 40 kW. (b) How large a collector is needed if the sunlight makes an incident angle of 45° with the surface of the collector?

Additional Problems

31.46• Calculate the Fermi energy for the electrons in cesium, which has a free-electron density of 0.86×10^{22} cm^{-3}.

31.47• (a) Calculate the Fermi energy for the electrons in potassium, which has a free-electron density of 1.34×10^{22} cm^{-3}. (b) Calculate the equivalent temperature T_F of the electrons at the Fermi surface of potassium.

31.48• The photoelectric work function of silver is 4.74 eV. The Fermi energy is 5.51 eV. (a) What is the least energetic photon required to remove an electron from the bottom of the potential well? (b) What is the wavelength of this photon? (c) What is the maximum kinetic energy for photoelectrons ejected by photons of this wavelength?

31.49• A copper wire 10 cm long has a cross section of 0.50 mm^2. (a) Calculate the resistance of the wire, given that copper has a resistivity of 1.69×10^{-8} Ω · m. (b) Calculate the mobility of the electrons, given that the electron density in copper is 8.5×10^{28} m^{-3}. (c) What is the drift velocity of the electrons when the electric potential difference across the length of the wire is one volt?

31.50• (a) If a sinusoidal voltage is applied to the transformer shown in Fig. 31.37, what voltage appears across the load R? That is, make a sketch of the voltage from A to B versus time. Assume the transformer output is $v = 12.0$ V sin $2\pi f t$. (b) What is the average voltage across the load?

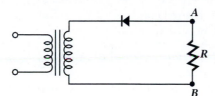

Figure 31.37
Problem 31.50.

31.51• Trace the path of the current through the circuit shown in Fig. 31.38 when (a) the voltage at point A is positive with

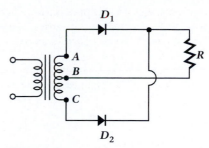

Figure 31.38
Problem 31.51.

respect to point C and (b) the voltage at point C is positive with respect to A. (c) Make a graph of the output voltage across the load resistor. (d) Could this circuit be used as a practical full-wave rectifier?

31.52•• A silicon diode has a saturation current of 0.10 μA. At what voltage would it carry a forward current of 10 mA if the diode exactly obeyed the rectifier equation?

Elementary Particle Physics

In our study of physics, we have discussed many different models of the physical world, from the kinetic theory of gases, through the atomic theory of matter, to subnuclear structure. Most of the topics we have covered are those for which the underlying principles have been known for some time. That is not the case for the material in this chapter, for now we inquire about the nature of the fundamental particles that make up the atoms and nuclei themselves. Although we have achieved considerable understanding, the behavior of the fundamental building blocks of matter remains one of the contemporary frontiers of physics research.

After Chadwick's discovery of the neutron in 1932, it seemed that physicists had found all the subnuclear particles needed as the elementary building blocks for atoms. In appropriate combinations, the proton, neutron, and electron could be used to build up every known element, from hydrogen to the most complex atom. Atoms, in turn, made up molecules and molecules made up ordinary matter. Such a scheme accounted nicely for the masses and sizes of atoms and, in conjunction with quantum mechanics, explained their chemical properties. Thus, the proton, neutron, and electron, along with the photon, were called **elementary,** or **fundamental, particles.**

Soon, however, this seemingly simple and satisfactory picture was complicated by new discoveries. Shortly after the period when Schrödinger was developing his wave formulation of quantum mechanics, other people discovered several new elementary particles. As time passed, literally hundreds of these seemingly elementary particles were discovered. Attempts to understand these particles, their compositions, and their place in nature continues to be one of the most exciting and dynamic areas of modern physics.

The study of elementary particles and their interactions, often known as high-energy physics, has created a number of new terms and classifications that we have not used before. We will try to keep the introduction of new vocabulary to a manageable level. We will focus on those topics now recognized as central to our current understanding and will avoid the many interesting, and often unanswered, questions that have arisen in recent years. As the material in this chapter shows, contemporary physics continues to place great importance on conservation laws, and progress in understanding fundamental particles has frequently come by identifying conserved quantities and trying to explain their origin. ■

32.1   Particles and Antiparticles

One of those involved in the early work on quantum mechanics was Paul A. M. Dirac (1902–1984). At the time, Dirac was Lucasian Professor of Mathematics at Cambridge, a post formerly held by Isaac Newton. One of Dirac's particular concerns was to incorporate relativity into quantum mechanics. To do so required that space and time be treated on an equal footing in quantum theory, as they already were in relativity.

With considerable insight and mathematical elegance, Dirac formulated a relativistically correct quantum wave equation for the electron. The solutions to this equation contained several unexpected results. First, the solutions indicated that electrons have spin, a fact already known from experiment. The second surprising prediction was that electrons could exist in both positive *and* negative energy states. The positive energy states correspond to the ordinary electrons that we already know about. Since energy states correspond to a particle's charge, Dirac thought at first that the negative electron states, which would have positive charge, corresponded to protons. But this could not be the case, in part because the proton is almost two thousand times as massive as the electron. So, in 1930, Dirac postulated the existence of positively charged particles having the same mass as the electron. His prediction was confirmed in 1932, the same year that Chadwick discovered the neutron, when Carl Anderson discovered in cosmic rays a positively charged particle with the mass of an electron. These particles are now called **positrons.**

One major consequence of Dirac's work was the realization that to any particle, not just the electron, there corresponds a similar particle of the same mass but with opposite charge. These oppositely charged particles are called **antiparticles.** The positron was the first antiparticle discovered. Later, in 1955, the ex-

(a)

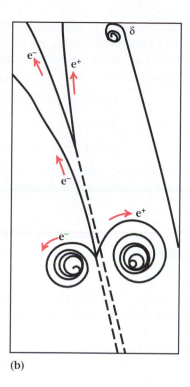

(b)

Figure 32.1
(a) Electron-positron pairs are produced in a bubble chamber. The tracks curve in opposite directions because the oppositely charged particles move in a magnetic field. The nearly straight tracks are left by the heavier charged particles of the incident beam. (b) Schematic view. The dashed lines represent the paths of photons, which, like all neutral particles, leave no tracks in a bubble chamber. In both examples, a photon is converted into an electron-positron pair. In the lower event, a recoil electron is knocked out of a hydrogen atom.

istence of the antiproton was confirmed experimentally by Emilio Segrè and Owen Chamberlain. Further discoveries led to the conclusion that neutral particles also have antiparticles and that some, though not all, neutral particles are their own antiparticles.

The concept of antiparticles is well established in physics today. Symbolically, we denote the antiparticle of a given particle by placing a bar over the letter representing the particle. Thus, a proton is represented by p and the antiproton is denoted by $\bar{p}$ (pee-bar).

When a particle and its antiparticle come together at rest, they annihilate each other. In doing so they produce gamma rays whose total energy equals the total available rest-mass energy. The annihilation of particles and antiparticles has given rise to a large number of interesting science fiction stories. However, producing an antiparticle in the first place requires an amount of energy equal to its rest mass, a relatively large amount of energy on a nuclear scale. Furthermore, antiparticles are not produced alone but only in particle-antiparticle pairs. The dream of making antimatter remains unfulfilled except in the sense of laboratory experiments (Fig. 32.1).

Dual-head gamma scanner adapted for coincidence detection of the two gamma rays resulting from the annihilation of a positron with an electron.

Example 32.1

A particle known as the B meson has a mass slightly more than $5\frac{1}{2}$ times that of a proton. The neutral state, B^0 (bee-naught), with a mass of 9.41×10^{-27} kg, can be produced as one of a particle-antiparticle pair. What is the minimum energy needed to produce a B^0 - $\bar{B}^0$ pair?

Solution Because particle and antiparticle have the same mass, we need a minimum energy equivalent to twice the mass of one of the particles. This energy can be determined using a conversion factor based on Einstein's relationship between mass and energy, $E = mc^2$ (see Problem 32.1):

$$1 \text{ MeV is equivalent to } 1.78 \times 10^{-30} \text{ kg.}$$

The total energy is then

$$2(9.41 \times 10^{-27} \text{ kg})(1 \text{ MeV}/1.78 \times 10^{-30} \text{ kg}) = 10,600 \text{ MeV} = 10.6 \text{ GeV.}$$

This is the minimum energy that an accelerator would have to provide in order to produce a pair of neutral B mesons.

32.2 Pions and the Strong Nuclear Force

Knowing the building blocks of an atom is not the same as knowing what holds them together. The Coulomb electrostatic force binds the negative electrons to the positive nucleus, but what force holds the nucleus together? The Coulomb force between protons in the nucleus is repulsive, and the gravitational force of attraction between nucleons is much weaker than this repulsive force. So, as we saw in Chapter 29, there must be a nuclear force that is quite strong.

Our modern understanding of forces—and fields—is that every fundamental force has a particle associated with it. Although we did not put it in just those terms, we have already seen one example of this type of connection. The electromagnetic force between charged particles is described by Maxwell's equations, of which Coulomb's law of electrostatic force is a part. However, Maxwell's laws also predict the existence of electromagnetic waves, which we also interpret as being photons. Thus, we associate the photon with the electromagnetic force. We say that the photon is the field particle of the electromagnetic force, or that the photon carries, or mediates, the electromagnetic force. We can think of the force between charged bodies as corresponding to an exchange of photons between the bodies. In an analogous fashion, we can describe the strong nuclear force between two nucleons as mediated by a new particle.

In 1932, the Japanese physicist Hideki Yukawa (1907–1981) proposed a mathematical form for the strong nuclear force. His potential energy function (Fig. 32.2) has the mathematical form

$$\text{PE} = -K\frac{e^{-\alpha r}}{r}, \tag{32.1}$$

where the constant α has the dimensions of reciprocal length and K is a positive constant that must be determined from experiments. The force corresponding to the Yukawa potential has a much shorter range than does the Coulomb force. It has appreciable magnitude only over a distance comparable to nuclear dimensions—that is, about 10^{-15} m.

Yukawa predicted the approximate mass of a new elementary particle associated with the nuclear force field, basing his argument on the Heisenberg

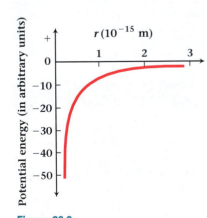

Figure 32.2
The shape of the Yukawa strong nuclear potential, drawn for the case $1/\alpha = 1.4 \times 10^{-15}$ m.

COSMIC RAYS

If you charge an isolated conductor—say, an electroscope on an insulating base—it does not keep its charge indefinitely. Instead, the conductor gradually loses its charge through the air, an effect noticed as far back as 1787 by Charles Coulomb. By the beginning of the twentieth century, scientists had come to suspect that this loss of charge was due to a slight ionization of air caused by radiation from radioactive material in the earth's crust. However, the effect persisted even when an electroscope was placed on a high tower or carried out to sea.

During the early decades of the twentieth century, a series of balloon ascents established the cause of the conductivity of air (Fig. B32.1). Scientists discovered that the amount of ionization decreased from sea level up to heights of about 2000 meters, but surprisingly, began to increase at higher elevations. They concluded that the ionizing radiation came from outer space and gave it the name cosmic radiation, or cosmic rays. By the late 1920s, subsequent experiments established that the observed radiation consisted of charged particles and gamma rays and was indeed of extraterrestrial origin.

Today we believe that all of outer space contains rapidly moving particles known as primary cosmic rays. This radiation consists mostly of protons, with decreasing proportions of heavier nuclei. These particles strike the earth's atmosphere at a rate of approximately one per cm^2 per second, with kinetic energies as high as 10^{20} eV. This maximum energy is approximately ten joules and is a tremendous energy for a single elementary particle to have. It is some 10^8 times the energy that protons have in the Tevatron at Fermi National Laboratory (Section 32.4). An average-size apple will have roughly this energy when dropped from a height of about 6 m.

The origin of the primary cosmic rays is difficult to determine with certainty because the moving charged particles are deflected by the magnetic field present in our galaxy. The extremely energetic ones all appear to come from a single source, Cygnus X3, a neutron star left from the explosion of a supernova. Many of those of intermediate energy are thought to have been accelerated by the explosions of other supernovas.

Figure B32.1   A high-altitude balloon of the type used in cosmic ray research.

In addition to the primaries from deep space, our atmosphere is struck by lower energy protons and electrons from the sun. Known as the solar wind, these particles move toward the polar regions as a result of the earth's magnetic field. Under the right conditions, the energetic electrons ionize the atmosphere and cause it to glow. This glow is called the *aurora borealis* at the North Pole, and the *aurora australis* at the South Pole (Fig. B32.2).

Most energetic primary cosmic radiation interacts with atoms in the upper atmosphere, creating additional high-energy particles called secondary cosmic rays, which consist of almost every known elementary particle. The number of secondary cosmic rays reaching the ground varies with time, elevation, and location, but at sea level an average of about two secondary cosmic rays pass through each square centimeter of the earth's surface every minute. Many of the secondary rays are quite penetrating, making it necessary to put some sensitive experiments in the deepest mines in order to shield against them. This same flux of cosmic rays passes through your body as well. Biologists have estimated that about 2% of genetic mutations are due to changes in the molecular structure of genetic material caused by cosmic rays.

Because of the extreme high energy of some cosmic rays, they have been used as a source of bombarding radiation since the early development of experimental particle physics. However, contemporary particle physics experiments are primarily carried out using accelerators capable of producing particle energies in the GeV to TeV range.

Figure B32.2   The aurora australis as photographed by the crew from the space shuttle *Discovery*.

uncertainty principle. According to Yukawa's theory, the nuclear force field can be represented by a cloud of these new particles that surround a nucleon and that are quickly emitted and reabsorbed by the nucleon. However, if we consider the attraction between nucleons to be mediated by a particle with mass, as Yukawa proposed, then a problem arises with the conservation of energy. The appearance of a new particle with a finite mass seems at first to violate the law of energy conservation; after all, the particle's rest mass must come from somewhere. Nevertheless, if we think of each mediating particle as being emitted from a nucleon and then reabsorbed almost immediately, energy fails to be conserved over only a short period of time. The amount by which the conservation of energy law is uncertain, ΔE, and the time during which the particle exists outside the nucleon, Δt, are related through the uncertainty relationship

$$\Delta E \, \Delta t \approx \frac{h}{2\pi},$$

The length of the particle's path away from a nucleon is of the same order of magnitude as the range of the nuclear force, R_0. If we assume that the particles move with the speed of light, then the uncertainty in time is the ratio of the distance to the speed of light. If the uncertainty in the energy is taken to be the particle's mass energy, the above equation becomes

$$mc^2 \, \frac{R_0}{c} \approx \frac{h}{2\pi},$$

or

$$m \approx \frac{h}{2\pi R_0 c}. \tag{32.2}$$

Inserting the values h, c, and R_0, we find that

$$m \approx 2.93 \times 10^{-28} \text{ kg,}$$

or

$$m \approx 320 m_e.$$

This is not a precise calculation of the particle's mass, only an order-of-magnitude calculation. We would not expect to observe particles of this kind free in nature because they should interact strongly with all nuclear matter. However, it should be possible to produce them in collisions between nucleons when the incoming nucleon has enough kinetic energy to provide the required mass energy.

Within two years of Yukawa's prediction, researchers observed a charged particle of approximately the right mass. However, subsequent investigation showed that the new particle's interaction with matter was so weak that it could not possibly be the Yukawa particle. This new particle was named the **muon,** represented by the symbol μ. Muons are observed with positive or negative charge, both forms having the same mass. The intrinsic angular momentum, or spin, of the muon is in units of $h/2\pi$. This is the same value as the spin of the electron, introduced in Chapter 28. The spin is an important property for classifying elementary particles and for understanding their reactions.

In 1947, C. F. Powell (1903–1969) and his coworkers finally discovered the strongly interacting Yukawa particle while doing experiments involving cosmic rays at high altitudes. This particle was named the pi-meson, or **pion,** and is represented by the symbol π. The pion has a mass of about $270 m_e$, somewhat greater than the mass of the muon, $205 m_e$. Pions occur in positive, negative, and uncharged forms, all with approximately the same mass. The spin of the pion is zero.

Charged pions are unstable and primarily decay with a half-life of about 2.8×10^{-8} s by the process

$$\pi^+ \rightarrow \mu^+ + \nu$$
$$\downarrow$$
$$e^+ + \nu + \bar{\nu}.$$

This notation means that the μ^+ (positively charged muon) subsequently decays into a positron and a neutrino-antineutrino pair. The π^- decays, by a similar process, into a μ^- and a neutrino, and then the μ^- decays into an electron and a neutrino-antineutrino pair. The neutral pion (π^0) decays primarily into two photons, with a half-life of about 0.8×10^{-16} s.

32.3   More and More Particles

In the same year that C. F. Powell and his collaborators discovered the pion, investigators observed other particles that were about a thousand times as massive as an electron. After several years of further research, it became apparent that these were actually two kinds of particles: **hyperons,** whose decay products always included a proton, and **kaons,** or K mesons, whose decay products were mesons only. The **mesons** (from the Greek *mesos,* which means intermediate) form a group of particles whose first members have masses between the masses of the electron and the proton.* Like pions, the hyperons and kaons seemed to be produced as a result of the strong nuclear force between two protons colliding with high energy. However, neither of these newly found particles seemed to be an elementary building block of ordinary matter, nor did it seem necessary for explaining the nuclear force.

The discovery of the muon, pion, hyperon, and kaon created an interest in higher-energy accelerators, capable of enough energy to produce other new particles. Following the construction of such accelerators, other new particles were indeed found, a result that stimulated another increase in the energy of the accelerators. This leapfrog development eventually led to the discovery of a large number of "elementary" particles, some of which are listed in Table 32.1. (We will explain the classification terms *lepton, meson,* and *baryon* in Section 32.5.) We have not listed all the properties of the particles, nor even shown how the unstable ones decay. A few typical accelerator reactions that give rise to some of these elementary particles are as follows (Fig. 32.3):

$$p + p \rightarrow p + p + \pi^+ + \pi^-$$
$$p + p \rightarrow p + n + \pi^+$$
$$p + p \rightarrow p + \Lambda^0 + K^+ \qquad (32.3)$$
$$\pi^- + p \rightarrow \pi^- + p + \pi^0$$
$$\pi^- + p \rightarrow \pi^- + p + \pi^+ + \pi^-$$
$$\pi^- + p \rightarrow \pi^- + n + \pi^+$$

*Other mesons—discovered later—have masses outside this range. See Table 32.1.

		Table 32.1	*Partial List of Elementary Particles*			

Classification	Name	Symbol	Charge	Mass (MeV/c²)	Lifetime (s)
	Photon	γ	0	0	∞
	Neutrino	$\nu_e, \bar{\nu}_e$	0	0	∞
		$\nu_\mu, \bar{\nu}_\mu$	0	0	∞
		$\nu_\tau, \bar{\nu}_\tau$	0	0	∞
Leptons	Electron	$e^\pm$	$\pm e$	0.5110	∞
	Muon	$\mu^\pm$	$\pm e$	105.7	2.20×10^{-6}
	Tau	$\tau^\pm$	$\pm e$	1777	2.91×10^{-13}
Hadrons	Pion	$\pi^\pm$	$\pm e$	139.6	2.60×10^{-8}
		π^0	0	135.0	0.84×10^{-16}
Mesons	Kaon	$K^\pm$	$\pm e$	493.7	1.24×10^{-8}
		K^0	0	497.7	0.89×10^{-10}
	D meson	$D^\pm$	$\pm e$	1869	10.6×10^{-13}
		D^0	0	1865	4.2×10^{-13}
	Proton	p	$+e$	938.3	$>1.6 \times 10^{32}$ y
	Neutron	n	0	939.6	887
	Lambda	Λ^0	0	1116	2.63×10^{-10}
	Sigma	Σ^+	$+e$	1189	0.80×10^{-10}
Baryons		Σ^0	0	1193	7.4×10^{-20}
		Σ^-	$-e$	1197	1.48×10^{-10}
	Xi	Ξ^0	0	1315	2.90×10^{-10}
		Ξ^-	$-e$	1321	1.64×10^{-10}
	Omega	Ω^-	$-e$	1672	0.82×10^{-10}

The usefulness of high-energy accelerators can be seen from at least two standpoints. First, to produce a particle of a given mass m, the accelerated particle must have at least enough kinetic energy so that mc^2 of energy is available for the reaction. Thus, since the first accelerators were of lower energy than later accelerators, they could produce only lower-mass particles, such as pions.

A second way of looking at the advantage of higher energies is to consider the accelerator as a microscope, enabling us to probe deeper into the structure of matter than we can do using only visible light. We have seen that a microscope's resolution, or ability to discern small details, increases as the wavelength of light decreases. The electron microscope was constructed to take advantage of this effect. The de Broglie relationship between a particle's wavelength λ and momentum p is $\lambda = h/p$, where h is Planck's constant. In an electron microscope, electrons are accelerated to such high speeds that their momentum is large enough to make λ much less than the wavelength of visible light. As a result, the electron microscope has much greater resolution than the light microscope. The even

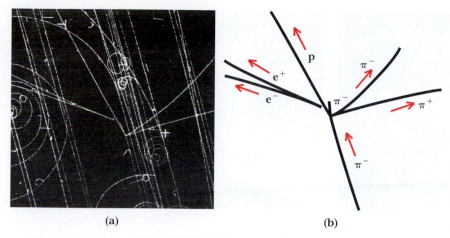

(a)   (b)

Figure 32.3
(a) Bubble chamber photo of the reaction $\pi^- + p \rightarrow \pi^- + p + \pi^- + \pi^+$. (b) Schematic diagram of the reaction. One of the emergent π^- interacts with another proton to give two neutral particles, a π^0 and a neutron. The π^0 decays into two gamma rays. One of the gamma rays creates the electron-positron pair.

greater energies of modern particle accelerators, in the range of tens of GeV and even many TeV, provide still-greater resolution for probing the structure of the nucleus and of the nucleons themselves. It is the decrease in wavelength with increased momentum and energy, and the resultant increase in the resolution and fineness of detail that can be probed, that makes the terms *high-energy physics* and *elementary particle physics* equivalent.

*32.4   Accelerators and Detectors

As we probe deeper into the structure of matter, we require finer and finer probes; that is, we need projectiles of shorter wavelength or, equivalently, higher momentum. Construction of accelerators of increasingly higher energies, and of devices to detect the particles emerging from these high-energy interactions, utilizes our most advanced technology. Reciprocally, the research and development put into the design of particle physics facilities has advanced the development of technology used in other fields.

The Van de Graaff accelerator (Sections 17.2 and 29.10) and the cyclotron (Section 19.5) have features that prevent them from being scaled up to extremely high energies. The Van de Graaff accelerator is limited by the breakdown potential of its insulators. The basic principle of the cyclotron, that of constant period for the outwardly spiraling particles, becomes invalid at higher energies. As the particle's speed increases, relativistic effects become important and the conditions for constant period fail.

Several alternative schemes to overcome these difficulties have been used in constructing accelerators of increasingly higher energies. Linear accelerators use pulses of radio waves to push particles to higher speeds. Circular machines use the same general principles of acceleration but guide the particles through nearly circular orbits, sending them through the same accelerating sections many times (Fig. 32.4a). Magnets placed around the periphery of the donut-shaped beam tube determine the path the particles travel (Fig. 32.4b).

Conservation of momentum requires that for each particle collision, the sum of the momenta of the outgoing particles must equal the momenta of the incoming particles. This requirement limits the energy available for reactions when a fixed target (zero initial momentum) is bombarded with the beam. The

(a)   (b)

Figure 32.4

(a) Aerial view of the site of HERA at DESY in Hamburg, Germany. HERA is the world's only electron-proton colliding-beam facility. (b) View inside the accelerator tunnel, showing a bending magnet (white) and a focusing magnet (orange) around the beam pipe.

outgoing particles necessarily carry away some kinetic energy. This means that the energy available for reactions or production of new particles is not the energy of the incoming particle, but the incoming energy minus the outgoing energy. For highly relativistic velocities, the available energy depends on the square root of the energy of the incoming particle.

We can overcome some of the difficulties of using fixed targets by using colliding beams. Two beams of particles with the same momentum are shot directly at each other, so that the net momentum of the particle collision is zero and all of the energy of both beams is available for the interaction. Colliding beams are usually achieved with storage rings, in which particles circulate in opposite directions in each of two donut-shaped tubes. The tubes intersect, creating places where the countercirculating bunches of particles can collide. In some cases, particles and antiparticles can circulate in the same storage ring.

Even for storage rings, an increase in beam energy means an increase in accelerator size. This requirement is largely due to the relationship between a particle's momentum and the radius of curvature of its path in a magnetic field. The magnetic field of any particular accelerator is limited by the strength of the magnets available. Confinement of faster-moving particles requires a larger circle.

The Tevatron at the Fermi National Accelerator Laboratory near Chicago, Illinois, is an example of an accelerator that serves as a proton-antiproton storage ring. The 2-km-diameter storage ring provides protons an energy of almost 1 TeV for collisions with 1-TeV antiprotons (Fig. 32.5). At the CERN laboratory near Geneva, Switzerland, the storage ring known as LEP has an 8.5-km diameter and allows 161-GeV electrons to collide with 161-GeV positrons (antielectrons). In Hamburg, Germany, at the 2-km diameter storage ring HERA, 30-GeV electrons collide with 820-GeV protons. At these and other sites around the world, both colliding-beam and fixed-target experiments are being conducted in the search for a deeper understanding of the structure of matter.

Experiments in particle physics require not only accelerators to provide particles for the interactions, but also detectors to determine what happens in the interactions. At high energies much of the initial energy may be converted into mass, and typically many particles emerge from the collision between two particles. In a colliding-beam experiment, the net momentum is zero and the particles emerge in all directions from the interaction point. To detect and measure the particles, it is necessary to surround the interaction region completely with

Figure 32.5

Aerial view of the 2-km diameter Tevatron at the Fermi National Accelerator Laboratory.

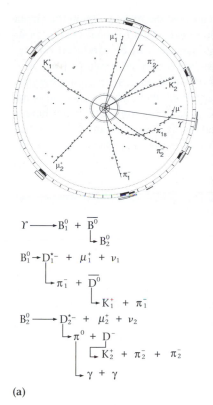

$$\Upsilon \longrightarrow B_1^0 + \overline{B}{}^0$$
$$\qquad \searrow B_2^0$$

$$B_1^0 \to D_1^{*-} + \mu_1^+ + \nu_1$$
$$\qquad \searrow \pi_1^- + \overline{D}{}^0$$
$$\qquad\qquad \searrow K_1^+ + \pi_1^-$$

$$B_2^0 \longrightarrow D_2^{*-} + \mu_2^+ + \nu_2$$
$$\qquad \searrow \pi^0 + D^-$$
$$\qquad\qquad \searrow K_2^+ + \pi_2^- + \pi_2^-$$
$$\qquad \searrow \gamma + \gamma$$

(a)   (b)

Figure 32.6

(a) The emerging particles from a single electron-positron annihilation at 10 GeV, recorded in the central tracking chamber of the ARGUS detector. A B^0-$\overline{B}{}^0$ pair produced in the decay of an upsilon meson Y is identified from the tracks of its decay products. (b) Participants from six different countries assemble the ARGUS detector at the German national laboratory DESY, in Hamburg.

detectors. The objective is to determine the angle at which each particle leaves the interaction region, the charge and energy of the particle, and, when possible, its identity. This is accomplished by using layers of different types of detecting devices, often including a magnetic field as part of the detector. Figure 32.6 shows a detector for colliding-beam experiments and an event recorded by it.

Neither the analysis of data nor the actual running of an experiment can be accomplished without the aid of powerful and fast computers. The many thousands of pieces of information recorded for each event and the large number of components to be operated and monitored make computers as much a part of the detector as are the gas-filled counting tubes and the scintillation counters.

32.5   Classification of Elementary Particles

The large number of elementary particles makes it important to have a classification scheme based on similar properties, as described in Table 32.1. Photons, which interact only by the electromagnetic force, occupy a category of their own. All other particles are classified as either hadrons or leptons. **Hadrons,** after the Greek word *hadros,* which means strong, take part in nuclear interactions through the strong nuclear force. **Leptons,** named from a Greek word meaning light, interact by the weak nuclear force. (Hadrons also experience the weak nuclear force, but the strong force has a greater range and magnitude and so predominates in determining hadron behavior.)

Hadrons are further subdivided into two groups: the mesons (from the Greek for intermediate) and the **baryons** (from the Greek word for heavy). The best-known baryons are the proton and neutron. Baryons can be produced only in baryon-antibaryon pairs. Each baryon is assigned a baryon number of $+1$, and each antibaryon is assigned a baryon number of -1. In nuclear reactions, baryons obey a conservation rule: The sum of the baryon numbers after a reaction is the same as it was before the reaction. Mesons do not obey this rule: They have a baryon number of zero and may be produced singly. Examination of Table 32.1 shows that mesons are no longer confined to the mass range between the electron and the proton.

▼

Example 32.2

Use Table 32.1 to determine which of the following reactions and decays can take place. (a) $p + \bar{p} \rightarrow n + \pi^0 + \pi^- + \pi^+$, (b) $p + p \rightarrow p + n + \pi^+$, (c) $\Lambda \rightarrow p + \pi^-$.

Strategy We can test the proposed reactions to see if both baryon number and charge are conserved. If not, then the reaction cannot take place. If both baryon number and charge are conserved, we should examine other conserved quantities. However, in order to keep our presentation of the main ideas of particle physics as simple and to the point as possible, we have not introduced all of the conserved quantities. In this example, and in the problems, the other conservation rules have already been satisfied.

Solution (a) The proton has a baryon number of $+1$ and the antiproton has a baryon number of -1, making the baryon number of the initial system 0. The right-hand side has a neutron, with a baryon number of $+1$, and three pions, with baryon number 0. Therefore, the reaction cannot occur, even though charge is conserved, because it would mean that baryon number was not conserved.
(b) The total baryon number on the left-hand side is 2, and the charge, in units of the elementary charge e, is equal to 2. Looking at the right-hand side, we see that the only charged particles are p and π^+, for a total of $+2$, indicating that charge is conserved. The proton and neutron both have $+1$ baryon number and the pions have 0 baryon number. Therefore, baryon number is conserved also, and the reaction can, and does, occur.
(c) Table 32.1 lists the Λ as a neutral baryon. Examination of the decay products shows conservation of both charge and baryon number, so the reaction occurs.

■

The best-known lepton is the electron. The electron, the muon, and the tau, together with their respective neutrinos, constitute the known leptons. Since the leptons do not experience the strong nuclear force, their behavior is determined by the weak nuclear force, in addition to the much weaker gravitational force and the powerful electrostatic (Coulomb) force in the case of charged particles. In fact, these same forces are experienced by all particles. Beta decay is associated primarily with the weak nuclear force. The weak force is not only weaker than the strong force by several orders of magnitude, it is also shorter in range,

Table 32.2	The Fundamental Forces of Nature	▼
Name	**Relative Strength**	**Associated Particle**
Strong nuclear	1	Meson
Electromagnetic	10^{-2}	Photon
Weak nuclear	10^{-13}	$W^{\pm}$, Z^0
Gravitational	10^{-40}	Graviton

Note: Because the forces don't vary in the same way with distance, the table is for approximate comparison only.

indicating that any particle involved in mediating that force must be more massive than the pion, which is involved in the nuclear force.

Table 32.2 lists the four fundamental forces of nature in order of decreasing strength, along with the particle (or quantum) that is associated with each force. Experiments have not yet verified the particle associated with the gravitational field, the graviton. The particles associated with the weak nuclear field, the $W^{\pm}$ and the Z^0, were detected in 1983 by the experimental group headed by Carlo Rubbia.

Other properties in addition to those mentioned here are associated with elementary particles. However, in order to outline the current understanding of the basic composition of matter in briefest terms, we have omitted mention of some of these properties. Although these other parameters are necessary for a complete understanding of particle physics, such details are beyond the scope of this book.

Electrons and the other leptons seem to be pointlike particles. That is, they appear to have no observable spatial extent or structure. On the other hand, hadrons have a finite size and seem to have an internal structure. In the next section, we discuss a model for this structure.

32.6 The Quark Model of Matter

The large number of known hadrons (over 200) presents a formidable barrier to understanding the basic composition of matter in any simple way. One approach toward understanding them would be some scheme or technique for classifying the properties of hadrons. The most successful of such schemes, called the *quark model,* was introduced in 1963 by Murray Gell-Mann and George Zweig. The quark model proposes that each hadron consists of a proper combination of a few elementary components called **quarks.** The properties of these quarks are somewhat unusual, such as an electric charge that is a fraction of the elementary unit of charge. At first, many people thought that the quark model did not represent physical reality—that it was only an extremely good mathematical arrangement. But experimental evidence has made clear that quarks are real constituents of matter. Before describing this evidence, we must introduce the elementary properties of quarks.

The earliest form of the quark theory postulated three quarks, all of fractional charge, called the up (u), down (d), and strange (s) quarks (Table 32.3).

Table 32.3	Some Properties of the u, d and s Quarks		
Quark	Charge	Spin	Baryon Number
u	$+\frac{2}{3}$	$\frac{1}{2}$	$\frac{1}{3}$
d	$-\frac{1}{3}$	$\frac{1}{2}$	$\frac{1}{3}$
s	$-\frac{1}{3}$	$\frac{1}{2}$	$\frac{1}{3}$

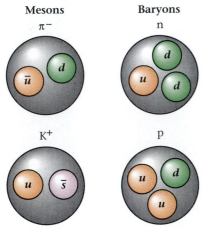

Mesons Baryons

π^- n

K^+ p

Figure 32.7

Quark composition of several particles. Mesons consist of quark-antiquark pairs, and baryons consist of three quarks.

We refer to these names as the different "flavors" of quarks. The quarks also have a spin quantum number of $\frac{1}{2}$ and a baryon number of $\frac{1}{3}$. Each quark has its own antiquark, which differs from it in the sign of the charge and baryon number. For constructing hadrons from quarks, there is a very simple rule: **Mesons are composed of quark-antiquark pairs, and baryons consist of three quarks in proper combination.** For instance, we can represent a π^+ meson by a u quark and an anti-d quark ($\bar{d}$), written as $u\bar{d}$. The proton is made up of a uud combination (Fig. 32.7). Table 32.4 shows some other combinations. No particle consisting of more than three quarks has been observed.

There are other properties, or quantum numbers, that we have not mentioned. They often bear humorous names such as "charm" and "strangeness." In high-energy physics, terms like "charm," "strangeness," "color," and "flavor" do not have their usual everyday meaning, but are only fanciful names for certain quantum numbers. As is the rule with other quantum numbers, these new quantum numbers obey conservation rules.

Example 32.3

Can the following combinations of quarks occur in nature: (a) ud, (b) $\bar{u}d$, and (c) uuu?

Strategy   There are at least two tests that you can apply to see if a given combination of quarks is allowed. The first is based on the fact that all elementary particles have either a whole number or zero charge in units of the elementary charge. Therefore, the quark content must have the same restriction. Likewise the quark composition of a particle must give an integer or zero for the baryon number.

Solution   (a) You should already recognize this combination as impossible because it is not a quark-antiquark pair. Why this is necessary can be seen from Table 32.3. The total charge of the proposed combination is $+\frac{1}{3}$, and the baryon number is $\frac{2}{3}$. So this is not a combination that occurs.
(b) For this combination, remembering that antiquarks have charge and baryon number opposite to those of the quarks, we get a charge of -1 and a baryon number of 0. This is a possible combination and corresponds to the π^-.
(c) The total charge of the combination is 2, and the baryon number is 1. This is a possible combination and is the quark content of the Δ^{++}.

Table 32.4	Quark Composition of Several Particles	
	Particle	Quark Composition
Mesons	π^+	$u\bar{d}$
	π^-	$\bar{u}d$
	K^+	$u\bar{s}$
	K^-	$\bar{u}s$
Baryons	p	uud
	n	udd

To verify the unusual properties of quarks, experimental searches for fractionally charged particles have been made. None of them has demonstrated the existence of a *free* quark in nature. However, other experiments have pointed conclusively to the existence of quarks that are bound together.

It is possible to probe the structure of a hadron, say the proton, using electrons. Such experiments are similar in many ways to Rutherford's scattering experiments to probe atomic structure (Chapter 26). Very rapidly moving electrons have de Broglie wavelengths, $\lambda = h/p$, that are much smaller than the dimensions of a proton. Analyzing the scattering of these electrons from protons tells us much about proton structure. These "deep inelastic" scattering experiments reveal that the proton has an internal structure made up of other particles. The properties of these other particles identify them as quarks. Thus, experiments have demonstrated the existence of quarks in the same way that Rutherford's experiments demonstrated the existence of the nucleus.

To explain the fact that free quarks are not observed, the quark theory proposes an attractive force between quarks that is small or zero when the distance between quarks is of the order of nucleon dimensions and that becomes extremely large when their separation is increased. This behavior, which is known as quark confinement, is quite different from that of other forces, which decrease with increasing distance. These assumptions have been incorporated into theories that successfully explain quark confinement and the properties of hadrons.

The three quarks listed in Table 32.3 were sufficient for the original quark model. However, subsequent discoveries required the introduction of additional quarks. These quarks—the *c*, or charm, quark, the *b*, or bottom, quark, and the *t*, or top, quark—bring the number of experimentally justified quarks to six. The six quarks *u*, *d*, *s*, *c*, *b*, and *t* have now all been experimentally confirmed. You might conclude that the number of quarks is increasing to some large number, analogous to the large number of hadrons. However, experiments now indicate that there are only six quark flavors.

Figure 32.8
The constituents of matter as we now know them. All matter, in the most fundamental sense, consists of various combinations of quarks and leptons.

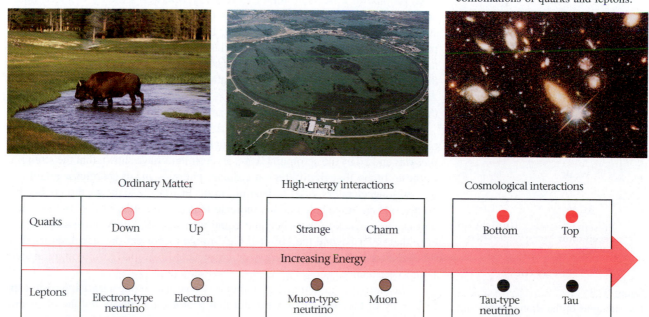

Various combinations of quarks and leptons make up all known constituents of ordinary matter, as well as those particles seen only in high-energy interactions. This classification is shown diagrammatically in Fig. 32.8. The ordinary world—that is, the world of neutrons, protons, and electrons—is composed from the quartet of quarks and leptons at the lowest energy level. To explain the existence of particles observed at higher energies, such as the Λ^0, we need the next higher energy level of quarks and leptons. Finally, the highest level corresponds to the great energies observed in galactic events or in very high energy accelerators. However, as is frequently the case, things are not as simple as they seemed at first glance. Recent experiments—as well as theoretical predictions—give hints that in some sense the quarks themselves may be composite particles.

32.7 Unified Theories

The recognition of only four basic forces in nature has naturally led to the question: Are any or all of these forces related, or might they even be the same? The idea of explaining these forces with one comprehensive theory, called a **unified theory,** had its modern beginning with Einstein. During his later years, he tried, without success, to unify his theory of gravitation with the other long-range force, electromagnetism. To this date, gravitation has not been incorporated with any of the other forces, but efforts to unify the other three have met with considerable success.

In 1967, S. L. Glashow, A. Salam, and S. Weinberg introduced a theory unifying the descriptions of the weak and electromagnetic forces. These two forces have several things in common, one being that they act on quarks and leptons equally. Since quarks and leptons are the basic constituents of matter in the quark model, perhaps it is not surprising that one theory can describe both forces. The Glashow-Salam-Weinberg model has successfully described a large number of observations. In a manner similar to the way Maxwell's theory of electromagnetism predicted electromagnetic waves (photons) and Yukawa's theory of the strong nuclear force predicted the pion, this theory of the "electroweak force" predicts the existence of several particles rather than one. These particles are named the $W^{\pm}$, the Z^0, and the Higgs particle. The most striking verification of this theory was the subsequent confirmation of the existence of the $W^{\pm}$ and Z^0. As of this writing, the Higgs particle has not yet been found, but there is hope that it will be observed within the next several years.

A further step of unification would be to combine the strong, weak, and electromagnetic forces in one theory. Such theories are called *grand unification theories* and go by the acronym GUTs. Experiments have shown that the strengths of these forces, though different at ordinary energies and at energies available today with accelerators, tend to become more nearly the same as the energy becomes higher. Some calculations indicate that the strengths of the strong, weak, and electromagnetic forces become equal at about 10^{15} GeV, an incredible energy that is far beyond the 10^2–10^3 GeV range of present-day accelerators (Fig. 32.9). Nevertheless, GUTs have been constructed that give results in agreement with experiments.

The simplest grand unification theories give rise to one surprising result: the instability of the proton. They predict that protons may decay by one of several modes, such as

$$p \rightarrow \pi^0 + e^+ \qquad \text{or} \qquad p \rightarrow \gamma + e^+.$$

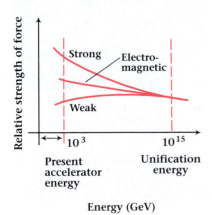

Figure 32.9

The strengths of the strong, weak, and electromagnetic forces as a function of energy.

Such reactions could not happen with a short half-life or the universe as we know it would have vanished long ago, if indeed it could have been created. However, it is possible to calculate the approximate proton lifetime for GUTs. The results of the simplest theories predict a proton lifetime of the order of 10^{30} years, perhaps somewhat longer. Since astronomers believe the approximate lifetime of the universe is of the order of 10^{10} years, the finite lifetime of the proton presents no real danger. Nevertheless, it is still an important prediction that can be tested.

Instead of looking at one proton for 10^{30} years, one collects a large number of protons (many times Avogadro's number of hydrogen atoms) and observes them for a few years. Such experiments often involve placing large volumes of hydrogen-containing liquids deep underground in mines or tunnels to shield them against spurious reactions due to cosmic rays (Fig. 32.10). To date, no one has found conclusive evidence for proton decay, and a lower limit of about 10^{32} years has been placed on the experimentally determined proton lifetime. In spite of the failure of the simplest GUTs, some revised versions of the theories give a proton lifetime consistent with the observations. At the present, it seems that grand unification theories are viable theories.

There are two other consequences of grand unified theories that are of particular interest. One of these is the predicted existence of magnetic monopoles, particles that have an isolated magnetic pole. Such particles would have magnetic fields that would be inconsistent with the predictions of classical electromagnetism (Maxwell's equations). So far, magnetic monopoles have not been observed up to the energies available with present-day accelerators.

Another consequence of grand unification theories is that they allow for the neutrino to have mass. It is clear from many experiments that if neutrinos do indeed have mass, that mass is very small. On first thought, it might not seem significant whether the neutrino mass is zero or just very small. However, the ramifications for the universe as a whole are extremely profound. There are about 100 neutrinos per cm^3 throughout the entire universe, according to reliable estimates. Therefore, even a small mass per neutrino would contribute appreciably to the mass of the universe because of the enormous number of neutrinos involved. The mass (or the density) of the universe is one of the important quantities that is used in the study of cosmology. Thus, we have an important connection between the study of the world of fundamental particles and the study of the universe and its origins.

Figure 32.10
A diver inside the IMB water scintillation detector used in a proton decay experiment.

32.8　Cosmology

Physics and astronomy have been closely connected ever since Newton showed that planetary orbits are determined by the same laws of mechanics and gravitation that operate on the earth. In recent years physicists and astronomers have demonstrated a connection between the very small scale of elementary particles and quarks and the very large scale of the entire universe. This connection arises through the study of the earliest origins of the universe. We can actually observe aspects of the early universe because most stars are so distant from us that their light takes many thousands and even millions of years to reach the earth (Fig. 32.11). Thus studying the properties of far-away stars is the same as studying their state in the distant past.

As the result of many different types of observations and model-building, most astronomers now agree that the universe does not exist in some sort of steady state.

Figure 32.11
The distant galaxy NGC5236. The light from the galaxy takes approximately 10 million years to reach the earth.

Instead, it originated at some particular point in time in the distant past. This model of the early universe, called the **Big Bang,** is supported by considerable evidence, indicating that the age of the universe is about 10–20×10^9 years.

There are at least three important experimental observations supporting the Big Bang theory. One of these observations is the fact that all clusters of galaxies are moving away from one other. We can therefore draw the conclusion that at some time in the past the galaxies were closer together and the universe was much denser than at present—so dense that at one early time, shortly after the Big Bang event, the density of the entire universe was comparable to or greater than the density of nuclear matter. Here is a connection between the study of the origins of the universe and the study of the elementary properties of matter. A second observation supporting the Big Bang model is the observed relative abundance of the lighter elements. These observed abundances can be predicted with a theoretical model that assumes that the early universe was very hot and very dense.

A third experimental observation in support of the Big Bang is the cosmic background radiation that bombards us from all directions. This electromagnetic radiation is believed to be the residual radiation from the early hot universe (Fig. 1.5). By now the universe has expanded and cooled until the background radiation corresponds to a blackbody at 2.735 K. Until 1992, the apparent uniformity of this background radiation represented a problem: If the early universe was uniform, what was the origin of its present structure? In other words, what caused the uniformly distributed matter of the early universe to clump into galaxies, stars, and planets? Small fluctuations in the background radiation temperature were observed by the COBE satellite in 1992 (Fig. 32.12). These fluctuations, corresponding to a time of about 3×10^5 years after the Big Bang, suggest that the mechanism for structure in the universe was the irregularity in the primordial radiation itself.

Understanding the nature of elementary particles sheds light on the behavior of the early universe. Until the quark model and modern unified theories were developed, there was no way in which physics could help explain how matter as we observe it today could have developed from the Big Bang model. Steven Weinberg, one of the authors of the Glashow-Salam-Weinberg unified theory

Figure 32.12

All-sky image from the COBE differential microwave radiometers. The map is in galactic coordinates, with the plane of the Milky Way Galaxy horizontal across the middle and the galactic center at the center. The Doppler effects seen in Fig. 1.5 have been subtracted. The nonuniformities correspond to temperature variation of about 3×10^{-5} K, with the red regions hotter than the blue regions.

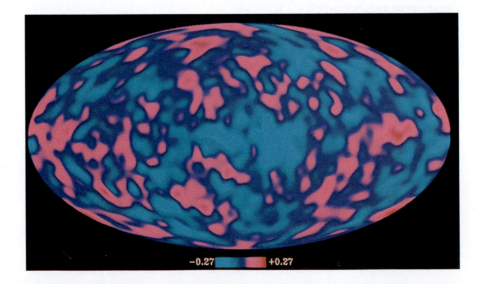

mentioned in the previous section, has written in his book, *The First Three Minutes:* *

> I remember that during the time that I was a student and then began my own research (on other problems) in the 1950's, the study of the early universe was widely regarded as not the sort of thing to which a respectable scientist would devote his time. Nor was this judgment unreasonable. Throughout most of the history of modern physics and astronomy, there simply has not existed an adequate observational and theoretical foundation on which to build a history of the early universe.
>
> Now, in just the past decade, all this has changed. A theory of the early universe has become so widely accepted that astronomers often call it "the standard model." It is more or less the same as what is sometimes called the "big bang" theory, but supplemented with a much more specific recipe for the contents of the universe. . . .

The theory outlines the early history of the universe as follows. The universe began in an explosion—not an ordinary explosion, but a tremendous burst of energy filling all of space. This energy corresponds to an extremely high temperature, of the order of 10^{32} K. At this high energy, no difference exists between the strengths of the fundamental forces. However, as the universe cooled, with consequent decrease of energy, the strengths of the four forces separated until they became as we now observe them in nature. This entire separation process was completed by the end of about 10^{-4} s.

During this initial phase of the universe, the temperature, or energy, was so high that ordinary matter, such as molecules, atoms, and even nucleons, could not exist. Instead, shortly after the big bang, the universe was a soup of photons (radiation), leptons, antileptons, quarks, and antiquarks, though not necessarily in equal numbers. Because the density of this soup was high and the forces between quarks are small when they are close together, the particles all moved very much like free particles. But as the soup cooled, it began to "condense" into elementary particles, including protons and neutrons. As the temperature dropped further, these nucleons condensed into nuclei. Further condensation gave rise to molecules and matter in bulk.

This brief outline skips over several interesting and important points. However, two things are especially worth noting. First, although many details of the big bang model are still uncertain, the same sort of physical theories that successfully describe the subnuclear world are also useful in describing the early history of the universe. Second, these particle physics theories have helped cosmologists deal with one of the more puzzling questions about the origin of the universe: If the initial big bang resulted in equal amounts of matter and antimatter, as our present understanding indicates, why does the universe today not have equal amounts of matter and antimatter? Astronomers have studied this question extensively, but no evidence exists for large amounts of antimatter located far away from us in the universe. One of the most impressive successes of the grand unification theory has been its prediction that, even though the universe starts with a symmetry between matter and antimatter, the end result is a

---

*S. Weinberg, *The First Three Minutes: A Modern View of the Origin of the Universe* (New York: Bantam Books, 1979).

preponderance of one over the other. The theories predict that among the first particles produced in the Big Bang was a particle named the X-boson. The X and its antiparticle the $\overline{X}$ are produced with equal probability, but their decay rates are not the same. As a result, the final decay products are matter rather than antimatter. This effect would have tipped the balance of matter in the universe during the first 10^{-35} s, even before the quark stage had been reached.

The experimental evidence that the universe is expanding, together with the evidence of the Big Bang, leads us to ask: will the universe continue to expand forever, or will it stop and perhaps even contract? The answer to this question depends in part on how much mass there is in the universe. If the density of mass is greater than some critical value, then the effects of gravitation will eventually stop the expansion and cause the universe to collapse on itself. On the other hand, if the density is less than the critical value, the gravitational effects will not be sufficient to stop the expansion and it will continue forever. This critical density is equivalent to about 10 hydrogen atoms per cubic meter. That is, of course, a very small density, but remember that most of the universe is relatively empty space.

In recent years, astronomers and physicists have come to realize that much of the mass of the universe is not luminous and is not detected by ordinary means. This so-called *dark matter* does not reside in stars, intergalactic dust, the known cosmic rays, or any other known component. The experimental evidence for this dark matter comes from several sources. The primary evidence comes from the comparison of measurements of the rotational behavior of the galaxies with the theoretical predictions based on the gravitational effects of the observed mass. These observations lead to the conclusion that between 75% and 90% of the mass of the universe may be made up of this invisible dark matter.

At present, the nature of the dark matter is still unknown. There could be one source or many sources. One proposed explanation for the dark matter is that the unobserved mass may be due to neutrino mass. Because space is filled with neutrinos, even a small mass for each neutrino would give a large contribution to the mass of the universe. Nevertheless, the answer is not known at this time.

Current observations indicate that the mass of the universe is somewhere between one-tenth and two times the critical density. Thus, the answer to our question regarding the continued expansion of the universe is not known. Consequently, the answer to the related question about the mass of the neutrino clearly plays an important role in our understanding the final outcome.

Summary

Useful Concepts

■ The positron was the first antiparticle found, and all particles have antiparticles.

■ Yukawa's expression for the nuclear potential energy is

$$PE = -K\frac{e^{-\alpha r}}{r}.$$

■ The use of colliding beams of particles in an accelerator allows all the energy of the particles to be available for interaction.

■ According to the quark model of matter, mesons can be represented as a quark-antiquark pair; baryons can be represented as a combination of three quarks.

■ Unified theories try to combine all the forces of nature into one type of force.

■ The quark theory has helped explain features of the Big Bang model of the origin of the universe.

Important Terms

You should be able to write the definition or meaning of each of the following:

elementary, or fundamental, particles
positrons
antiparticles
muon
pion
hyperons
kaons

mesons
hadrons
leptons
baryons
quarks
unified theory
Big Bang

Conceptual Questions

32.1 Would you believe someone who told you she had a laboratory flask full of antiwater? (Antiwater is the same as ordinary water except that each ordinary particle is replaced by its antiparticle.)

32.2 If all the particles in the universe were changed into their antiparticles overnight, while you were asleep, could you tell when you awoke that anything had happened?

32.3 What criteria could be used for classifying whether particles are fundamental?

32.4 A π^0 meson never decays into a single photon. Would doing so violate any conservation laws? If so, which ones?

32.5 Discuss the similarities and differences between the photon and the neutrino.

32.6 What is meant by the statement that "a muon is just a fat electron"?

32.7 A free neutron decays into a proton plus an electron plus an antineutrino. Why do we not also observe its decay into a proton-antiproton pair? Why does it not decay into an electron-positron pair?

32.8 Numerous experiments have searched for single particles with a charge that is a fraction of the electron charge. If such particles are never found, would it mean that the quark model is incorrect?

32.9 Defend or attack the following statement, giving reasons or historical analogies: "Theories and investigations into the structure of the proton and the neutron and into the first few minutes of the existence of the universe are great intellectual achievements, but have no practical use."

32.10 What criteria must be satisfied for one elementary particle to decay into another?

32.11 Beams of negative pions have been used to replace x rays in the treatment of cancer. What might be some of the advantages of using pions rather than x rays?

32.12 The weakest of the four forces in nature, gravitation, seems to play the most important role in our daily lives. Is this really true? Explain.

32.13 Assuming that most cosmic rays are charged particles, give some reasons why you would expect the intensity to vary with position on the earth and time of day.

Problems

Section 32.1 Particles and Antiparticles

32.1 Show that the conversion factor between energy and mass can be put in the form

1.000 MeV is equivalent to 1.783×10^{-30} kg.

32.2 How much energy is released when a proton and an antiproton annihilate at rest?

32.3 What is the minimum energy required to produce an electron-positron pair?

32.4 What is the minimum energy required to produce a $D^0 - \overline{D}^0$ pair?

32.5• How much energy would be required to produce a person-antiperson pair with individual masses of 65.0 kg? How

long would the Hoover Dam hydroelectric plant have to operate to produce this much energy? (See Table 6.2.)

32.6• How many proton-antiproton pairs could be produced if all of the kinetic energy gained by a 0.250-kg apple dropped from a height of 0.500 m were to go into producing the particles?

32.7• An electron and a positron come together at rest and then annihilate. Two photons are given off in the process. What can you say about the energy and direction of the photons?

32.8• An electron and a positron collide head-on and annihilate. Each has a kinetic energy of $\frac{1}{2} m_e c^2$ with respect to a stationary observer. What is the total energy released in the collision?

Section 32.2 Pions and the Strong Nuclear Force

32.9 What would be the approximate mass of the Yukawa particle if the range of the nuclear force were twice as great as it is and all of the fundamental constants were unchanged?

32.10 What value of R_0 in Eq. (32.2) corresponds to the actual mass of the π^+ meson?

32.11 The neutral pion decays most frequently into two photons. Show that when the decaying pion is at rest, the photons must have the same energy.

32.12● What is the maximum kinetic energy available to the decay products in the decay of a charged pion into a charged muon?

32.13● What is the maximum kinetic energy available to the decay products in the decay of a μ^+ into an e^+?

32.14● Plot the Yukawa potential and the Coulomb potential, and note any differences that you observe.

Section 32.3 More and More Particles

32.15● (a) On the average, how far will a neutral K meson go before it decays if its speed is $0.90c$? (b) Answer the same question for the charged kaons.

32.16● (a) On the average, how far will a neutron go before it decays if its speed is $0.90c$? (b) Answer the same question for a charged pion.

32.17●● Approximately how many nuclear diameters ($A \approx 100$) will a π^0 meson traverse at a speed of $0.95c$ before it decays?

*Section 32.4 Accelerators and Detectors

32.18 The proton beam of a small accelerator is directed at thin aluminum foil. If the beam current is $1.0 \ \mu A$, how many protons strike the foil per second?

32.19 What momentum must a particle have for its wavelength to be of the order of nuclear dimensions?

32.20 What is the kinetic energy in TeV of an 0.250-kg apple that has fallen through a distance of 1 m?

32.21● Assume that all of the kinetic energy of a 2.0 TeV proton goes in lifting a 0.5-mm-square piece of typing paper straight up. Could you detect this change in elevation? Make a reasonable estimate of the mass of the piece of paper.

32.22● (a) What is the wavelength of a 10-GeV electron? (b) How does this compare with the dimensions of a proton? Take the radius of a proton to be R_0. (For extremely relativistic particles, the mass term in the expression for the total energy may be neglected.)

32.23● (a) What is the wavelength of a 20-TeV proton? (b) How does this compare with R_0? (For extremely relativistic particles, the mass term in the expression for the total energy may be neglected.)

32.24●● What is the wavelength of a proton that has been accelerated to a kinetic energy of (a) 1.00 MeV, (b) 1.00 GeV, and (c) 1.00 TeV?

32.25●● The vacuum in the beam pipe of a storage ring is about 1.5×10^{-6} Pa. (a) Approximately how many air mole-cules are there per cubic meter when the temperature is 15°C? (b) To what height above the surface of the earth does this correspond? For simplicity, assume the atmosphere to be composed entirely of oxygen. (Use the barometric formula given in Chapter 12.)

Section 32.5 Classification of Elementary Particles

Hints for Solving Problems

For elementary particle interactions, the most important thing is to take into account conserved quantities. This means using not only conservation of momentum and energy (keeping in mind relativity), but also such things as conservation of charge and baryon number.

32.26 The upsilon meson has a mass of 9460 MeV and a spread in mass of 52 keV. What is the approximate lifetime of this particle? (Try using the Heisenberg uncertainty relationship.)

32.27● Which of the following reactions can take place? What conservation rules are violated in those that are not allowed?

$$\text{(a)} \ \Sigma^0 \rightarrow \pi^0 + \pi^0 + \gamma,$$

$$\text{(b)} \ K^+ \rightarrow \pi^- + e^+ + e^+,$$

$$\text{(c)} \ \Lambda^0 \rightarrow p + \pi^+ + \pi^-.$$

32.28● Which of the following reactions or decays cannot occur and why?

$$\text{(a)} \ p + p \rightarrow p + n + \pi^+,$$

$$\text{(b)} \ \pi^+ + p \rightarrow n + \pi^+ + \pi^-,$$

$$\text{(c)} \ n \rightarrow e^- + e^+ + \nu,$$

$$\text{(d)} \ p + p \rightarrow \mu^+ + \mu^-.$$

Section 32.6 The Quark Model of Matter

32.29 Write out the quark composition of the antiproton.

32.30 Verify that the quark composition of the particles listed in Table 32.4 gives rise to the correct charge and baryon number.

32.31● Which of the following quark combinations can occur: (a) udd, (b) $\bar{u}ud$, (c) dsu, and (d) $d\bar{u}$? Explain your answer.

32.32● What is the charge of a particle made up of (a) $c\bar{c}$ (the Ψ meson) or (b) uds (the Λ hyperon)?

Additional Problems

32.33● In the reaction

$$\pi^- + p \rightarrow \Lambda^0 + K^0,$$

what is the rest-mass difference between the initial and final products? Does the reaction release energy or require it?

32.34● A hydrogenlike atom can be made by replacing the electron of normal hydrogen with a muon. (a) What wavelength radiation corresponds to the first three Balmer spectral lines? (b) In what part of the electromagnetic spectrum do they lie?

32.35• One estimate of the lower limit of the proton lifetime is 2.5×10^{32} years. Some of the experiments to search for proton decay use a large volume of water. (a) How many water molecules contain 2.5×10^{32} protons? (b) What mass of water is this and what volume does it occupy?

32.36•• Approximately how many nuclear diameters ($A \approx 100$) will a D^0 meson traverse at a speed of $0.92c$ before it decays?

32.37•• A Λ^0 decays at rest into a π^- and a proton. (a) How much energy is available as kinetic energy to the decay products? (b) How much momentum does each particle have, in units of MeV/c?

Appendices

Appendix A: Formulas from Algebra, Geometry, and Trigonometry

BINOMIAL THEOREM

For any positive integer n, the quantity $(a + b)^n$ may be represented by a series of $n + 1$ terms:

$$(a + b)^n = a^n + na^{n-1}b + \frac{n(n - 1)}{2!} a^{n-2}b^2 + \frac{n(n - 1)(n - 2)}{3!} a^{n-3}b^3 + \cdots + nab^{n-1} + b^n.$$

BINOMIAL SERIES

For any n and $|x| < 1$, the quantity $(1 + x)^n$ may be represented by the infinite series:

$$(1 + x)^n = 1 + nx + \frac{n(n - 1)}{2!} x^2 + \frac{n(n - 1)(n - 2)}{3!} x^3 + \cdots,$$

where $m! = m(m - 1)(m - 2)(m - 3) \cdots (1)$.

When $|x| \ll 1$, each successive term is so much smaller than the preceding term that the higher order terms may be neglected and the series may be approximated by the first two terms only. For example, when $|x| \ll 1$:

$$(1 + x)^2 \approx 1 + 2x$$

$$\frac{1}{1 + x} = (1 + x)^{-1} \approx 1 - x$$

$$\sqrt{(1 + x)} = (1 + x)^{\frac{1}{2}} \approx 1 + \frac{1}{2}x$$

$$\frac{1}{\sqrt{(1 + x)}} = (1 + x)^{-\frac{1}{2}} \approx 1 - \frac{1}{2}x$$

AREA AND VOLUME

Area of a rectangle of sides a and b:

$$A = ab$$

Area of a triangle with base b and height h:

$$A = \tfrac{1}{2}bh$$

A-1

Area of a circle of radius r:

$A = \pi r^2$

Area of a sector of a circle when θ is the angle between the radii (in radians):

$A = \frac{1}{2} r^2 \theta$

Area of an ellipse with semiaxes a and b:

$A = \pi ab$

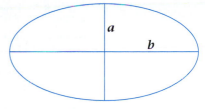

Area of the curved surface of a spherical segment of height h, radius of sphere r:

$A = 2\pi rh$

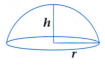

Area of the surface of a sphere of radius r:

$A = 4\pi r^2$

Volume of a sphere of radius r:

$V = \frac{4}{3}\pi r^3$

Volume of a cylinder of radius r and height h:

$V = \pi r^2 h$

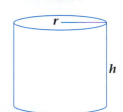

CONIC SECTIONS

Ellipse

Definition: A plane cutting a cone at an angle to the symmetry axis makes an elliptical cross section.

Description: At any point on an ellipse, the sum of distances to two fixed points (foci) is the same. The semimajor axis is labeled a and the semiminor axis is labeled b.

$$\frac{x^2}{a^2} + \frac{y^2}{b^2} = 1$$

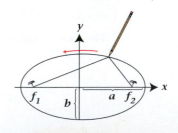

Parabola

Definition: A plane cutting a cone parallel to a side makes a parabolic cross section.

Description: At any point on a parabola, the distance to a fixed point (focus) equals the distance to a fixed line (the directrix).

$$y = -\frac{1}{2d}x^2$$

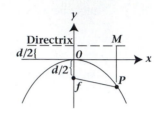

Hyperbola

Definition: A plane cutting a cone parallel to its axis makes a hyperbolic cross section.

Description: At any point on a hyperbola, the difference between the distances to two fixed points (foci) is the same.

$$\frac{x^2}{a^2} - \frac{y^2}{b^2} = 1$$

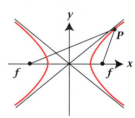

TRIGONOMETRIC RELATIONSHIPS

$\sin^2 x + \cos^2 x = 1$

$\sin(x \pm y) = \sin x \cos y \pm \cos x \sin y$

$\cos(x \pm y) = \cos x \cos y \mp \sin x \sin y$

$\tan(x \pm y) = \dfrac{\tan x \pm \tan y}{1 \mp \tan x \tan y}$

$\sin 2x = 2 \sin x \cos x$

$\cos 2x = \cos^2 x - \sin^2 x$

The law of cosines: In any triangle with angles A, B, and C and opposite sides a, b, and c, respectively,

$$a^2 = b^2 + c^2 - 2bc \cos A$$

Appendix B: The International System of Units

DEFINITIONS OF THE SI BASE UNITS

1. **unit of length (meter)** The 17th General Conference on Weights and Measures (CGPM, 1983) abolished the former definition of the meter and adopted a new definition, which reads: *The meter is the length of the path traveled by light in vacuum during a time interval of 1/299 792 458 of a second.* The old prototype of the meter, which was legalized by the 1st CGPM in 1889, is still kept at the International Bureau of Weights and Measures (BIPM).

2. **unit of mass (kilogram)** The 1st CGPM (1889) legalized the international prototype of the kilogram. The 3d CGPM (1901) declared: *The kilogram is the unit of mass; it is equal to the mass of the international prototype of the kilogram.* This prototype made of platinum-iridium is kept at the BIPM under conditions specified by the 1st CGPM in 1889.

3. **unit of time (second)** The second was defined originally as 1/86 400 of the mean solar day. Because a more precise definition was needed, the 13th CGPM (1967) replaced the astronomical definition of the second by the following: *The second is the duration of 9 192 631 770 periods of the radiation corresponding to the transition between the two hyperfine levels of the ground state of the cesium-133 atom.*

4. **unit of electric current (ampere)** The 9th CGPM (1948) adopted the ampere for the unit of electric current, with the following definition: *The ampere is that constant current which, if maintained in two straight parallel conductors of infinite length, of negligible circular cross section, and placed 1 meter apart in vacuum, would produce between these conductors a force equal to 2×10^{-7} newton per meter of length.*

5. **unit of thermodynamic temperature (kelvin)** The 13th CGPM (1967) adopted the name kelvin (symbol K), and defined the unit as follows: *The kelvin, unit of thermodynamic temperature, is the fraction 1/273.16 of the thermodynamic temperature of the triple point of water.* In addition to the thermodynamic temperature (symbol T), expressed in kelvins, use is also made of Celsius temperature (symbol t) defined by the equation $t = T - T_0$, where $T_0 = 273.15$ K by definition.

6. **unit of amount of substance (mole)** The following is the definition of the mole, adopted by the 14th CGPM (1971): *The mole is the amount of substance of a system that contains as many elementary entities as there are atoms in 0.012 kilogram of carbon 12. When the mole is used, the elementary entities must be specified and may be atoms, molecules, ions, electrons, other particles, or specified groups of such particles.*

7. **unit of luminous intensity (candela)** The unit based on flame or incandescent filament standards before 1948 was replaced initially by a unit based on a Planckian radiator (a blackbody) at the temperature of freezing platinum. Because of the difficulties, and new experimental techniques, the 16th CGPM (1979) adopted the following definition: *The candela is the luminous intensity, in a given direction, of a source that emits monochromatic radiation of frequency 540×10^{12} hertz and that has a radiant intensity in that direction of (1/683) watt per steradian.*

Appendix C: Alphabetical List of Elements

Name	Symbol	Atomic number	Name	Symbol	Atomic number	Name	Symbol	Atomic number
Actinium	Ac	89	Hassium	Hs	108	Radium	Ra	88
Aluminum	Al	13	Helium	He	2	Radon	Rn	86
Americium	Am	95	Holmium	Ho	67	Rhenium	Re	75
Antimony	Sb	51	Hydrogen	H	1	Rhodium	Rh	45
Argon	Ar	18	Indium	In	49	Rubidium	Rb	37
Arsenic	As	33	Iodine	I	53	Ruthenium	Ru	44
Astatine	At	85	Iridium	Ir	77	Rutherfordium	Rf	104
Barium	Ba	56	Iron	Fe	26	Samarium	Sm	62
Berkelium	Bk	97	Krypton	Kr	36	Scandium	Sc	21
Beryllium	Be	4	Lanthanum	La	57	Seaborgium	Sg	106
Bismuth	Bi	83	Lawrencium	Lr	103	Selenium	Se	34
Bohrium	Bh	107	Lead	Pb	82	Silicon	Si	14
Boron	B	5	Lithium	Li	3	Silver	Ag	47
Bromine	Br	35	Lutetium	Lu	71	Sodium	Na	11
Cadmium	Cd	48	Magnesium	Mg	12	Strontium	Sr	38
Calcium	Ca	20	Manganese	Mn	25	Sulfur	S	16
Californium	Cf	98	Meitnerium	Mt	109	Tantalum	Ta	73
Carbon	C	6	Mendelevium	Md	101	Technetium	Tc	43
Cerium	Ce	58	Mercury	Hg	80	Tellurium	Te	52
Cesium	Cs	55	Molybdenum	Mo	42	Terbium	Tb	65
Chlorine	Cl	17	Neodymium	Nd	60	Thallium	Tl	81
Chromium	Cr	24	Neon	Ne	10	Thorium	Th	90
Cobalt	Co	27	Neptuniun	Np	93	Thulium	Tm	69
Copper	Cu	29	Nickel	Ni	28	Tin	Sn	50
Curium	Cm	96	Niobium	Nb	41	Titanium	Ti	22
Dubnium	Db	105	Nitrogen	N	7	Tungsten	W	74
Dysprosium	Dy	66	Nobelium	No	102	(Ununbium)	Uub	112
Einsteinium	Es	99	Osmium	Os	76	(Ununnilium)	Uun	110
Erbium	Er	68	Oxygen	O	8	(Unununium)	Uuu	111
Europium	Eu	63	Palladium	Pd	46	Uranium	U	92
Fermium	Fm	100	Phosphorus	P	15	Vanadium	V	23
Fluorine	F	9	Platinum	Pt	78	Xenon	Xe	54
Francium	Fr	87	Plutonium	Pu	94	Ytterbium	Yb	70
Gadolinium	Gd	64	Polonium	Po	84	Yttrium	Y	39
Gallium	Ga	31	Potassium	K	19	Zinc	Zn	30
Germanium	Ge	32	Praseodymium	Pr	59	Zirconium	Zr	40
Gold	Au	79	Promethium	Pm	61			
Hafnium	Hf	72	Protactinium	Pa	91			

Answers to Odd-Numbered Problems

1.1 (a) Uniform spherical ball on a level floor. (b) Uniform spherical ball on a tilted floor. (c) For (a) ball has a flat spot or is nonuniform and for (b) ball is asymmetric and begins to roll toward the heavier side.
1.3 Die is weighted toward the 2 spot.
1.5 Cube may be hollow if floating in water. Alternately, cube is solid but floats in a liquid that is denser than the cube.
1.7 2.5 cm^3
1.9 30 cm
1.11 1.6×10^8
1.13 m/s^2
1.15 180 cm
1.17 64.7 km
1.19 80.8 miles/h
1.21 \$22.66/m^2
1.23 \$0.282/L
1.25 768 cm^2
1.27 6.6
1.29 $0.549°$
1.31 1.58×10^6 revolutions
1.33 (a) 5.50×10^3 kg/m^3, (b) 2.30×10^{17} kg/m^3, (c) 184 m
1.35 5.94×10^3 cm^3
1.37 (100 ± 2) cm^2
1.39 3.27×10^3
1.41 28 people/km^2
1.43 1.02×10^{-4} m or 0.102 mm
1.45 0.040 mm
1.47 $A = xy \pm xy\left(\dfrac{\Delta x}{x} + \dfrac{\Delta y}{y}\right)$
1.49 0.06 mm
1.51 3.3×10^3 bricks
1.53 6×10^{10} gallons
1.55 1.64×10^4 mm^3
1.57 8.7 L/100 km
1.59 761 mi/h
1.61 (a) 1 min = 60,000 ms, (b) 1 century = 3.16 Gs
1.63 (a) $h = \dfrac{4}{3}R$
1.65 $\pi/6$
1.67 $V = \pi r^2 h \pm 2\pi rh\Delta r \pm \pi r^2 \Delta h$

2.1 4.7 km south of the base
2.3 19 cm from the left edge
2.5 (a) C, (b) A, (c) C, (d) 0 m, (e) A, (f) 2 m

2.7 $x_1 = 9.09$ km, $x_2 = -5.30$ km, $x_3 = -2.52$ km
2.9 64.1 km/h
2.11 1.0 h
2.13 3.77×10^8 m
2.15 (a) 8 km/h, (b) 264 km/h
2.17 (a) 12 km/h north, (b) 140 km/h
2.19 90 m
2.21 -48 m/s
2.23 56 km/h
2.25 at $x = 5$ the slope ≈ 0.70, at $x = 10$ the slope ≈ 1.4
2.29 A_1 is the distance graph and A_2 is the velocity graph. B_1 is the velocity graph and B_2 is the distance graph. C_1 is the velocity graph and C_2 is the distance graph. D_1 is the distance graph and D_2 is the velocity graph.
2.31 750 m
2.33 7 s; The slope is greatest at 7 s.
2.35 30 m
2.37 15 m/s
2.39 5.4 m/s^2
2.41 At 10 s, a $= 1.25$ m/s^2, at 30 s, a $= 0.28$ m/s^2
2.43 (a) A and F, B and E, C and G, (b) A and I, B and K, C and H.
2.45 (a) 3.0 m/s^2, (b) 38 m
2.47 (a) zero, (b) 400 m
2.49 31.7 m/s
2.51 At 20 mi/h, 42 ft; at 40 mi/h, 124 ft; at 50 mi/h, 180 ft
2.53 (a) 19.4 m/s, (b) 27.9 m/s
2.55 $a = \dfrac{v^2}{\sqrt{2}L}$
2.57 3.34 s
2.59 0.929 m/s^2
2.61 20 m/s
2.63 1.6 rev/s
2.65 45.6 m
2.67 68 km/h
2.69 (a) 1900 km/h, (b) 11 km/h, (c) 170
2.71 -6.6 m/s
2.73 (a) $144°$ W of N, (b) $72°$ E of N
2.75 1.2 s
2.77 13 m
2.79 At $t = 1$, $v = -0.4$ cm/s; at $t = 3$, $v = -1.5$ cm/s
2.81 39 m

3.1 (a) $C = 5$ blocks, (b) $\theta = 53°$ N of E
3.3 (a) 40 km, (b) $117°$ from E

3.5 6.2 units at 136°

3.7 (a) A at 0°, (b) 0, (c) 1.73A at 90°, (d) 2A at 180°

3.9 50 m, 100 m, $\theta = 63°$

3.11 $x = 10$, $y = 12$

3.13 $A = 7.8$, $\theta = -52°$

3.15 891 m at 11.7° above the horizontal

3.17 magnitude = 20.4, x component = -9.57

3.19 $D = 10$, $\theta = 97°$

3.21 (a) 13 at $-8.7°$, (b) 17 at 114°

3.23 (a) yes, (b) 13.7 km

3.25 Relative speed of attendant is 2 km/h. Relative speed of the passenger is 6 km/h.

3.27 (a) 30 min, (b) 6.5 km

3.29 (a) 160 km/h, (b) -160 km/h, (c) -10 km/h and -170 km/h

3.31 273 km/h at 4.6° with respect to E

3.33 400 km, 2.9° W of S

3.35 (a) 0.495 s, (b) directly beneath where it was dropped, (c) 34.4 m

3.37 3.9 m

3.39 1.0 km

3.41 2.9 m/s

3.45 3.1 m/s

3.47 (a) 8.86 m, (b) 71.7°

3.49 (a) 18.8 m, (b) 7.68°

3.51 (a) 56 cm, (b) 26.6°, (c) 56°

3.53 (a) 1.4 s, (b) 11 m

3.55 (a) 6.12 m, (b) 5.94 m

3.57 14 at 69°

3.59 635 m

3.61 (a) 15.6 m, (b) 14.7 m

3.63 (a) 29.4 m/s, (b) 44.1 m

3.65 22.0 m

3.67 $\frac{1}{4}R \tan \theta$

3.69 $y_{max} = \frac{1}{4}R_{max}$

4.1 334 N

4.3 0.908F

4.5 22.8 m/s^2

4.7 2.40×10^5 N

4.9 8.2 m

4.13 10.1 kg

4.15 2.27 kg

4.17 480 N

4.19 0.1%

4.21 (a) 1420 N, (b) 235 N, (c) 145 kg

4.23 (a) 1.4 kg, (b) 18 N

4.25 16 N

4.27 (a) 5.8×10^2 N, (b) 6.8×10^2 N, (c) 5.8×10^2 N, (d) 4.1×10^2 N, (e) 6.7×10^2 N

4.29 (a) The reaction force is the force of the table on the book and is equal to the weight of the book. (b) The reaction force is the force of the book on the earth and is equal to the weight of the book. (c) The reaction force is the force of the table on the book and is less than the weight of the book. The

reaction force to the gravity force is the force of the book on the earth and is equal to the weight of the book.

4.31 650 N before and after braking, 780 N during braking

4.33 (a) 167 N, (b) 150 N, (c) 115 N

4.35 8.5 cm

4.37 (a) 2.0 m/s^2, (b) 1.0 N

4.39 accelerating upward

4.41 0.485

4.43 (a) -4.00 m/s^2, (b) 0.408

4.45 1.4 m/s^2

4.47 0.185

4.49 (a) 3.7 m/s^2, (b) 31 N

4.51 (a) 0.64 m/s^2, (b) $T_1 = 24$ N, $T_2 = 37$ N

4.53 30°

4.55 284 N

4.57 35 lb $\approx$ 154 N

4.59 49°

4.61 1.3 kg

4.63 $T_1 = 16.3$ N, $T_2 = 18.4$ N

4.65 (a) 30°, (b) 0.67 m/s^2

4.67 0.58

4.69 0.51

4.71 3.3 m/s^2

4.73 (a) 100 N, (b) 1100 N, (c) 1600 N

4.75 (a) 4.5 s, (b) 61%

4.77 (a) 4.9 m/s^2, (b) 320 N

4.79 0.786 m/s^2 to the left

4.81 (a) 1.30 m/s^2, (b) $T_1 = 41.3$ N, $T_2 = 40.2$ N

4.83 (a) no, (b) 2.0 m, (c) both monkey and mirror fall

4.85 0.81 kg

5.1 1.3 mm

5.3 (a) $\pi/2$ rad, (b) 9.4 in.

5.5 6.0×10^{-3} m/s^2 toward the sun

5.9 (a) 4.0 Hz, (b) 25 rad/s

5.11 (a) 6.28 rad/s, (b) 12.6 m/s, (c) 79.0 m/s^2

5.13 41 km/h

5.15 2.2×10^3 N

5.17 49 m/s

5.19 13.9°

5.21 73.9 N

5.23 (a) 5.12 N, (b) 33.3°

5.25 (a) 2.69 m/s, (b) 3.27 N

5.27 (a) 2.11g, (b) 3.11

5.29 2.1 rad/s

5.31 9.54 AU

5.33 4.23×10^7 m

5.35 0.91%

5.37 0.165

5.39 0.900

5.41 1.7%

5.43 17 kg

5.45 In an inertial frame the accelerations are $a_{moon} = 2.70 \times 10^{-3}$ m/s^2, $a_{earth} = 3.32 \times 10^{-5}$ m/s^2.

5.47 1.90×10^{27} kg

5.49 6.72×10^{-11} N · m^2/kg^2

5.51 (a) $F_{near} = 5.90 \times 10^{-3}$ N, $F_{far} = 5.90 \times 10^{-3}$ N, (b) $F_{near} = 3.44 \times 10^{-5}$ N, $F_{far} = 3.22 \times 10^{-5}$ N,

(c) Force of the sun is about the same, but force of the moon changes by 7% from near to far.

5.53 (a) 4.23×10^7 m. (b) No, only equatorial orbits are allowed.

5.55 267 N

5.57 (a) $v = R\sqrt{\dfrac{4}{3}G\rho\pi}$, (b) $R = 8.1 \times 10^3$ m

5.59 9.80 m/s^2 directed radially toward center of earth.

5.61 1.2×10^{-10} m/s^2 toward the midpoint of the line joining the two masses

5.63 2.2 AU

5.65 3.08 km

5.67 0.60 Hz

5.69 (a) 6.02×10^{24} kg, (b) 5.54×10^3 kg/m^3

5.71 (a) $0.94g$, (b) $2.9g$

5.73 0.58 Hz

5.75 37 rev/min

5.77 (a) $4.4 \times 10^{-3}g$, (b) $1.1 \times 10^{-3}g$, (c) $4.4 \times 10^{-4}g$

6.1 1.4×10^3 J

6.3 (a) 64 N, (b) 1.3×10^3 J

6.5 6.7 m/s

6.7 (a) 35 N, (b) 35 N

6.9 2.63 J

6.11 4.0 J

6.13 3.53 N at 26.6° from the x axis

6.15 3×10^8 gallons

6.17 (a) 1.6×10^{15} W, (b) 2.0×10^{10} m^2, (c) 0.20%

6.19 38 m/s

6.21 4

6.23 (a) 3.7×10^4 J, (b) 7.4×10^4 J

6.27 196 J

6.29 4.7×10^4 N/m

6.31 3.0×10^3 N/m

6.33 (a) 2.3×10^2 J, (b) 7.5×10^2 J

6.35 11.3 m/s

6.37 3.7 m/s

6.39 6.70 m/s

6.41 (a) 11.7 m/s, (b) 1.6 m/s^2

6.43 1.7 m/s

6.45 223 m

6.47 (a) 4.43 m/s, (c) 29.4 N

6.49 $3R$

6.51 (a) 2.9×10^4 J, (b) 44 m/s

6.53 6.02 m/s

6.55 (a) $-mgL\sin 34°$, (b) $mgL\sin 34°$, (c) No change in KE because speed is constant. From conservation of energy we find that $-\Delta PE = W_{friction}$ in agreement with the results computed in parts (a) and (b).

6.57 200

6.59 no

6.61 $0.25

6.63 3.1 GW

6.65 42 kW

6.67 8.0°

6.69 0.688 J

6.71 (a) 91%, (b) 170

6.73 1.25 kg

6.75 0.0033 J

6.77 0.0740 m

6.79 3.5 m/s

6.81 60°

7.1 4.79×10^4 kg · m/s

7.3 The Mazda's momentum is greater.

7.5 (a) 4.9×10^6, (b) 1.2×10^8

7.7 (a) 20.0 km/s, (b) 10.9 m/s

7.9 0.238 kg · m/s

7.11 (a) 0.490 kg · m/s directed horizontally, (b) 1.32 kg · m/s at 68.3° below the horizontal

7.13 (a) 17.4 m/s, (b) 17.4 m/s

7.15 4.17 N

7.17 (a) 60 N, (b) 60 N in the opposite direction

7.21 Work $= \overline{F}\,d = mgh = PE_{initial}$.

7.23 (a) -0.607 kg · m/s, (b) -1.21 kg · m/s, -304 N, -607 N

7.25 0.12 kN

7.27 0.095 s

7.29 -2.47 m/s

7.31 0.43 m/s

7.33 -22.5 cm/s, 13.1 cm/s

7.35 1.13 km/s

7.37 (a) 2120 m/s, (b) 1980 m/s

7.39 6.8×10^2 N/m

7.41 $\dfrac{\sqrt{2gh}}{3}$

7.43 2.78 m/s at 70° from the initial direction of travel of the 1300-kg car

7.45 7 m/s at an angle of 1.56° from the direction of the train

7.47 14 km/h at 28° from the x axis

7.49 $v_1 = 2.22$ m/s and $v_2 = 1.26$ m/s

7.51 390 m/s

7.53 9.2 mi/h

7.55 (a) 2.7 m/s^2, (b) 4.1 m/s^2

7.57 1.8 m/s

7.59 0.37 m/s

7.61 4.2 N · s

7.63 (a) 1.61 m, (b) 3.16 m

7.65 0.94

7.67 $m_2/m_1 = 1.1$

7.69 (a) 7.35 m/s, (b) 3.51 m

7.71 Throwing sequentially is best.

7.73 (a) 1.45 m/s, (b) 0.735 J, (c) $KE_{final} = \frac{1}{2}KE_{initial}$.

7.75 230 N

8.1 no

8.3 2.8 m/s

8.5 yes

8.7 1.01 m

8.9 $v_1 = -v_0$, $v_2 = +v_0$

8.11 $v_1 = -2.0$ m/s, $v_2 = +2.0$ m/s

8.13 (a) 23 m/s, (b) yes

8.15 no

8.17 (a) 5.35×10^4 J, (b) Because of momentum conservation, some of the energy is needed to keep the cars in motion.

8.19 $H = \dfrac{8}{9}\dfrac{v_0^2}{g} - \dfrac{7}{9}h$

8.25 0.75

8.27 Incident deuteron moves at 45° to initial direction, struck deuteron moves at −45° to initial direction.

8.29 0.79 v_0

8.31 -7.63×10^{28} J

8.33 6.26×10^7 J

8.35 0.02 R_E

8.37 7.91×10^3 m/s

8.39 $v_0 = \sqrt{\dfrac{3GM_E}{2R_E}}$

8.41 -1.4×10^6 J

8.43 1.1 R_E

8.45 5.04 km/s

8.47 1.05

8.49 (a) 8.86×10^{-3} m, (b) 2.05×10^{30} kg/m³

8.55 390 N, 13 m

8.57 7.2×10^6 J

8.59 Choose the positive direction as the direction of motion of mass m_1. Then, $v_1' = 1.15$ m/s and $v_2' = -0.495$ m/s.

8.61 (b) $r = 0.90\, R_M$

8.63 $2\sqrt{\dfrac{2h + v^2/g}{g}}$

8.65 3/8

8.67 60°

8.69 −135° from the y axis

8.71 0.100 g

9.1 ω(second hand) = 0.105 rad/s, ω(minute hand) = 1.75×10^{-3} rad/s, ω(hour hand) = 1.45×10^{-4} rad/s

9.3 (a) 20 m/s, (b) 2.1×10^5 rad/s²

9.5 (a) 524 rad/s − 1885 rad/s, (b) 19.7 m/s − 70.7 m/s, (c) 6.7×10^2 rad/s²

9.7 (a) 139 rad/s, (b) 359 m

9.9 2.6 m

9.11 (a) 83.8 rad/s, 356 rad/s, (b) 227 rad/s², (c) 42.0 rev

9.13 32 N

9.15 41.5°

9.17 500 N

9.19 (a) 330 N, (b) 270 N

9.21 2.1 kg

9.23 (a) 1.01 kN, (b) 794 N

9.25 $F_{hands} = 460$ N, $F_{feet} = 276$ N

9.27 930 N

9.29 3.5 m

9.31 4.4×10^4 N

9.33 $W = \dfrac{4F^2 L_0}{\pi Y D^2}$

9.35 9.4 rad/s

9.37 (a) $\dfrac{\tau}{I}t$, (b) $\dfrac{\tau R}{I}t$

9.39 0.58 rad

9.41 (a) 0.047 kg · m², (b) 0.68 m

9.43 6.48 m/s

9.45 7.08×10^{33} kg · m²/s

9.47 29.0 rev/min

9.49 1.9×10^{-2} J

9.51 35 J

9.53 22 rad/s

9.55 Hoop: 5.0 m/s, disk: 5.7 m/s

9.57 $v_{rolling} = \dfrac{1}{\sqrt{2}}\, v_{sliding}$

9.61 (a) 2.34 m/s, (b) 1/3

9.65 $a = \dfrac{g}{1 + \dfrac{I}{mR^2}}$

9.67 9.8 N · m

9.69 2.4 kg

9.71 100 N

9.73 (a) 200 N, (b) 700 N

9.77 2.5 cm

9.79 (a) KE $= \frac{1}{2}|PE|$, (b) $-\frac{1}{2}G\dfrac{Mm}{r}$

9.81 (b) $\sqrt{\dfrac{4}{3}\, gh}$

9.83 3.5 m/s

10.1 (a) 0.752×10^3 kg/m³, (b) oak

10.3 2.17 MPa

10.5 10.0 km

10.7 0.0150 m²

10.9 (a) 410 kPa, 2.7 MPa, (b) 82 kPa

10.11 11.1 kPa

10.13 1.26×10^5 N

10.15 147 kPa

10.17 18.5 cm

10.19 0.70 g/cm³

10.21 1650 kg

10.23 48.5 m³

10.25 76%

10.27 0.0918 N

10.29 (a) 1.6 kg, (b) 1.8 kg

10.31 (b) Net force on m_1 is $F_{1net} = F_{b1} - m_1g - T$, net force on m_2 is $F_{2net} = T + m_2g - F_{b2}$, where T is the tension in the string, (c) $a = \frac{1}{2}\rho g\left(\dfrac{1}{\rho_1} - \dfrac{1}{\rho_2}\right)$

10.33 73×10^{-3} N/m

10.35 0.29 m

10.37 9

10.39 −3.1 kPa

10.45 3.8 kPa

10.47 19%

10.49 1:4:9

10.51 (a) 65 Pa · s, (b) 7.2 Pa · s

10.53 0.66 m/s

10.55 1.14×10^5 N, 6.40×10^4 N, fuel consumption proportional to v^2

10.57 19 Hz

10.63 0.150 mm

10.65 0.11 m³/h

10.67 3.2×10^5 Pa

10.69 6.7×10^{-3} m², 1.25×10^4 Pa

10.71 2.95×10^6 N

10.73 0.91 m/s

10.75 13 cm

10.77 (a) 0.50×10^{-3} m³, (b) 18.7×10^3 kg/m³, (c) 13.6×10^3 kg/m³, (d) mercury

10.79 0.11

10.81 85°C

11.1 (a) 136°F, (b) −129°F

11.3 −269°C = −452°F

11.5 77.35 K = −320.4°F

11.7 +2.67°, +0.44°, −1.22°, −2.33°

11.9 (a) $T(°N) = \frac{5}{14}[T(°F) - 70]$, (b) $T(°N) = \frac{9}{14}T(°C) - 13.6$

11.11 2.9×10^5/°C

11.13 2.1 cm

11.15 1.6 L

11.17 86°C

11.21 1.6 L

11.23 1.2 cal

11.25 0.15 kcal

11.27 6.83×10^3 cal

11.29 0.36°C/s

11.31 0.019°C

11.33 101 g

11.35 89°C

11.37 1.2°C

11.39 45.5°C

11.41 4.28×10^5 J

11.43 2.92×10^5 cal

11.45 100°C

11.47 5.4 cm³/min

11.49 0.5 g of ice, 1.0 g of water

11.51 0°C

11.53 8.8 kW

11.55 459 W/m², 5.67×10^4 W/m², 4.59×10^6 W/m², 5.95×10^6 W/m²

11.57 yes

11.59 5760 K

11.61 18.3 h

11.63 5980 K

11.65 9.0 mm

11.67 6.04 W

11.69 2.99×10^6 J

11.71 (a) 8.6×10^5 J, (b) 5.7×10^4 W, (c) 2.7×10^2 °C

11.73 10 cm

11.75 48.3 g

11.79 0.50 cal/g°C

11.81 All the water remains.

12.1 13.0 m

12.3 107 cm

12.5 5.2×10^{19} N

12.7 1.26×10^5 Pa

12.11 10.3 m

12.13 0.943 L

12.15 (a) 546 K, (b) 746 K, (c) 2546 K

12.17 40°C

12.19 22.5 L

12.21 (a) 1.00, (b) in the same place (exactly in the center)

12.23 4.1×10^7 Pa

12.25 1.7×10^{-3} m³

12.27 4.75, 5.55

12.29 2.18×10^{-6} m

12.31 1.04×10^{-20} J

12.33 1.5 kg/m³

12.35 1.05

12.37 (a) 1.24×10^{-24} kg, (b) No

12.39 3030 J

12.41 1.1 kJ

12.43 $0.50 \, P_0$

12.45 (a) 5.54 km, (b) 11.1 km

12.47 $f(v_m)\Delta v = \frac{4}{e}\sqrt{m/2\pi kT}\,\Delta v$

12.49 1.46

12.51 (a) 0.63 mol, (b) 18 g

12.53 1.6 cm

12.55 (a) 2.44×10^{25}, (b) 1.52×10^5 J, (c) 1.34×10^3 m/s

12.57 (a) 1.41×10^5 K, (b) 1.60×10^5 K. (c) Ratio of nitrogen to oxygen is slightly smaller at upper regions.

12.59 $d = \dfrac{P_0 x}{\rho g(L - x)} + x$

A12.1 −0.061

A12.5 49 particles/min

A12.7 19.38

13.1 22.0 kJ

13.3 29.2 kJ

13.5 impossible

13.7 6.7 kJ

13.9 260 J

13.13 (a) 60 J, (b) heat is given out, (c) −65 J

13.15 (a) 0.502, (b) 0.424

13.17 (a) 26 J, (b) 9 J

13.19 0.020

13.21 $P_0 V_0$

13.23 111 K

13.25 $\frac{1}{2}P_1(V_1 - V_0)$

13.27 actual efficiency = 0.175, theoretical efficiency = 0.308

13.29 0.80

13.31 (a) 0.98, (b) 0.19

13.33 14

13.35 (a) 21, (b) 8.9

13.37 −32°C

13.39 (a) 143 J, (b) 200 J

13.41 68 W

13.43 3.03 kJ/K

13.45 8.8 MW/K

13.47 643 J/K

13.49 (a) 550 J/K, (b) −510 J/K, (c) 40 J/K

13.51 0.22

13.53 15 mJ/K
13.55 42 J
13.57 (a) 97 W, (b) 0.33 W/K
13.59 56% more
13.63 (b) 875 J, (c) 500 J, (d) 375 J
13.65 0.64%

14.1 5.36 cm
14.3 17 cm
14.5 23 cm
14.7 0.575 m
14.9 longer than your arm
14.11 10 cm, 10 cm, 3.1 cm
14.13 2.8 s
14.15 (a) 6.7 cm, (b) 8.1 cm
14.17 (a) 6.0 m/s^2, (b) -4.2 m/s^2, (c) 10.6 cm
14.19 1.2×10^{-2} J
14.21 64 N/m
14.23 (a) 0.067 m, (b) 2.2×10^{-2} J
14.27 (a) 0.318 Hz, (b) 3.14 s
14.29 0.99 m
14.31 0.46 s
14.33 (a) 37.7 cm/s, (b) 118 cm/s^2
14.37 0.408 kg
14.39 4.4 m/s
14.41 (a) 4.0 s, (b) 3.5 s
14.43 2.75 s
14.45 (a) Tarzan saves Jane 2.3 s before the bomb explodes. (b) They are over the middle of the river when the bomb explodes.
14.49 (a) 3.0×10^{-3}/s, (b) 0.069
14.51 $-6.76°$
14.53 8.2 cm
14.55 (a) 2.14 Hz, (b) 39.6 m/s
14.57 (a) 0.140 m, (b) 0.994 m
14.59 6.0 cm
14.61 2.05 Hz
14.63 3.92 m/s^2
14.67 yes
14.69 $5.36\sqrt{L/g}$
14.71 $T = 2\pi\sqrt{L/g}$
14.73 2.0 m

15.1 $y(x,t) = y_0 \sin 2\pi\left(\dfrac{x}{\lambda} + \dfrac{t}{T}\right)$

15.3 4.0 Hz
15.5 4.3×10^{14} Hz $- 7.5 \times 10^{14}$ Hz
15.7 1.17 s
15.9 $y(x) = 8.0$ cm $\cos 2\pi\left(\dfrac{x}{0.28 \text{ m}}\right)$

15.11 The child must triple the frequency.
15.15 (a) 0.170 m, (b) 0.489 m, (c) 0.750 m
15.17 8.5 mm
15.19 20 m
15.23 0.041 m
15.25 48 dB
15.27 50.8 dB
15.29 2.51×10^7 W/m^2
15.31 -12 dB

15.33 89.3 dB
15.35 (a) 916 Hz, (b) 724 Hz
15.37 (a) 44 Hz, (b) You will not hear it.
15.39 2656 Hz
15.41 (a) 1210 Hz, (b) 968 Hz
15.43 Mach 1.49
15.45 28°
15.47 14.2°
15.49 510 m/s
15.51 1.01 g/m
15.53 (a) yes, (b) no
15.55 (a) 2.0 m, (b) 1025 Hz, (c) 0.33 m
15.57 19.6 Hz
15.59 (a) third harmonic, (b) 77 Hz, (c) 307 Hz
15.61 470 Hz, 940 Hz, 1420 Hz, 1890 Hz
15.63 340 Hz, 1020 Hz, 1700 Hz, 2380 Hz
15.65 70, 74, 103, 140, 148, 206, 210, 222, 280, 296, 309, and 412 Hz
15.67 4.6 Hz
15.69 293 Hz
15.71 316
15.73 $v = [1052.2 + 1.1T(°\text{F})]$ ft/s
15.75 (a) 29 m/s, (b) 104 km/h
15.77 $f' = (800$ Hz$)(1 + 0.04 \cos 6\pi t)$
15.79 (a) 10, (b) 360, (c) 120, (d) 96 m
15.81 1% at 100°C
15.83 0.018

16.1 90 N
16.3 1.8 km
16.5 4.2 nC
16.7 0.29
16.9 (a) 1.24×10^{36}, (b) No
16.11 (a) -320 N, attractive, (b) $+40$ N, repulsive
16.13 1.50 μC, 4.50 μC
16.15 2.3 nC/h
16.17 -0.84 N (toward the origin)

16.19 $\left(\sqrt{2} + \dfrac{1}{2}\right)k\dfrac{q^2}{L^2}$ diagonally outward

16.21 Force on each end charge is 7.7 N outward. Force on center charge is zero.
16.23 2.34×10^{-14} N
16.25 8.00×10^5 N/C
16.27 zero
16.31 3.13×10^4 N/C
16.33 7.50×10^{-3} N at 36.9°
16.35 3.29 μC
16.39 1.7×10^5 N/C in the $-x$ direction
16.41 (a) 0, (b) 1.3×10^5 N/C
16.43 (a) -3.95×10^5 N · m^2/C, (b) inward
16.45 0.206 N · m^2/C

16.49 (a) $F = -\dfrac{\rho q r}{3\epsilon_0}$, (b) $r = r_0 \cos \omega t$, with $\omega = \sqrt{\dfrac{\rho q}{3\epsilon_0 m}}$

16.51 8.5×10^{-27} N · m
16.53 3.89 mm
16.55 -1.1×10^{-25} J
16.57 (a) zero, (b) No, because the field is not constant over the surface.

16.59 2.50×10^4 N/C
16.61 (a) negative y direction, (b) $\dfrac{p}{4\pi\epsilon_0 x^3}$
16.63 0.119 m
16.65 2.3 N diagonally inward
16.67 1.49×10^3 C
16.69 zero, $\dfrac{Q(r^3 - a^3)}{4\pi\epsilon_0 r^2(b^3 - a^3)}$, $\dfrac{Q}{4\pi\epsilon_0 r^2}$

16.73 (a) $-y$ direction, (b) $\dfrac{p^2}{2\pi\epsilon_0 r^3}$

17.1 6.14×10^{-18} J
17.3 0.11 pC
17.5 2×10^{-3} C
17.7 (a) 36 V, (b) 3.6×10^{-8} J
17.9 7.58×10^6 m/s
17.11 -1.18 V
17.13 (a) 1.2 MV/m, (b) 64 kV/m, (c) 1.8×10^5 V
17.15 (a) 9.2×10^6 V, (b) 9.4×10^6 V
17.17 (a) 15.3 J, (b) 15.3 J
17.19 (a) 2.40×10^5 m/s, (b) 1.03×10^7 m/s
17.21 (a) 2.50×10^5 eV, (b) 4.00×10^{-14} J, (c) 6.92×10^6 m/s
17.23 $Q/2$
17.25 (a) first sphere, (b) second sphere
17.27 60 V
17.29 375 μC
17.31 120 V
17.33 1.33 μF
17.35 0.83 nF
17.37 0.869 μC
17.39 1.84×10^{-20} J
17.41 0.192 N
17.43 (a) 3.2×10^{-17} J, (b) 200 eV, (c) 8.4×10^6 m/s
17.45 15 kV
17.47 208 V
17.49 2.8×10^{-3} m^2
17.51 5.6
17.53 $CV = \kappa\epsilon_0 AE$, where E is the dielectric strength
17.55 0.50 nF
17.57 (a) 0.050 μC, (b) 1.25×10^{-7} J
17.59 (a) 2.1×10^{-4} J, (b) 1.3×10^{-3} J
17.61 45 V
17.63 5000 eV $= 8.01 \times 10^{-16}$ J
17.65 4.6×10^6 V
17.67 (a) $\sigma/\epsilon_0 E$, (b) 3.8
17.69 $\Delta W = \frac{1}{2}C_0 V_0^2(\kappa - 1)$
17.71 (a) 0, (b) 11 J
17.73 0.70 nF

18.1 (a) 0.50 C, (b) 30 C
18.3 13.9 A
18.5 4.16 nA
18.7 53.2 mA
18.9 5.0 Ω
18.11 0.54 Ω
18.13 no
18.15 0.00 A, 0.10 A, 0.20 A, 0.30 A, 0.40 A, 0.50 A, 0.60 A
18.17 9.8×10^{-8} $\Omega \cdot$ m, iron

18.19 6.4 Ω
18.21 0.969
18.23 (a) 1.25 A, (b) 96.0 Ω
18.25 6.67 A
18.27 \$1.44
18.29 \$1.80
18.31 (a) 420 Ω, (b) higher, (c) resistance increases with temperature
18.33 (a) 0.090 A in left resistor, zero in right resistor, (b) 0.090 A in each resistor
18.35 16 Ω
18.37 400 Ω
18.39 (a) 58 mA, (b) 0.87 V
18.41 (a) 45.2 V, (b) 34.0 V
18.43 3.662 Ω
18.45 8.5 μF
18.47 5.9 μF
18.49 (a) 160 pF, (b) 106 pF
18.51 6.5 μF
18.53 seven combinations: 5, 7.5, 10, 15, 22.5, 30, 45 μF
18.55 (a) 12 V, (b) 180 μC, 300 μC, (c) 1.1×10^{-3} J, 1.8×10^{-3} J
18.57 0.94 Ω
18.59 (a) 5.9 Ω, (b) 8.0 V
18.61 (a) yes, (b) no
18.63 (a) no, (b) The lamp can be operated with either the toaster or the iron.
18.65 11 m
18.67 1.6¢
18.69 23 kg
18.71 5.9 Ω
18.73 (a) 60.0 mA, (b) 0 A in series with switch, 60.0 mA for others
18.75 (a) 9.0 V, (b) 2.4 W
18.77 53 mA
18.79 (a) $R_1 = 22$ Ω, $R_2 = 80$ Ω, $R_3 = 100$ Ω, (b) $I_1 = 0.36$ A, $I_2 = 0.20$ A, $I_3 = 0.16$ A, (c) $P_1 = 2.9$ W, $P_2 = 3.2$ W, $P_3 = 2.56$ W, (d) 66 Ω, (e) 0.36 A, (f) 8.6 W

19.7 80 A $\cdot$ m^2
19.9 2.9×10^{-9} T
19.11 1.11×10^{-4} Wb
19.13 2 μB
19.15 20 A
19.17 0.0228 N
19.19 0.0457 T
19.21 $F = VBA/\rho$
19.23 1.41×10^{13} m/s^2
19.25 The motion is helical with a radius of 0.879 cm.
19.27 8.5×10^{-21} kg $\cdot$ m/s
19.29 (a) 4.79×10^5 m/s, (b) 1.92×10^{-16} J
19.31 11.4 MHz
19.33 34 GHz
19.37 14 cm
19.39 (a) 0.016 N/m, (b) repel
19.41 1.25 A
19.43 8.0 A
19.45 2.6 mT
19.47 0.14 A $\cdot$ m^2

19.49 1.2×10^{-4} N · m
19.51 50 V
19.53 2.5×10^{-2} Ω in parallel with the galvanometer
19.55 $\theta = NIAB/\kappa$
19.57 7.72 mT
19.61 (a) $B = \mu_0 NI/2\pi r$, (b) B varies with position within the torus, (c) zero
19.63 0.27 T
19.65 8.31 turns/m
19.67 The maximum moment occurs for $N = 1$.
19.69 1.58×10^5 m/s

20.1 0.11 V
20.5 0.56 mV
20.7 0.18 T
20.9 (a) from c to d, (b) from d to c, (c) from d to c, (d) from c to d
20.11 (a) 4.8 mV, (b) 19 mV, (c) 9.6 mV, (d) 19 mV
20.13 0.17 mV
20.15 (a) Blv, (b) Blv, (c) Blv/R, (d) $B^2 l^2 v/R$
20.17 0.53 mV
20.19 0.16 mV
20.21 2.3×10^{-5} V
20.23 190 V
20.25 300 V
20.27 10
20.29 3.6
20.31 10.0
20.33 (a) 2.5 A, (b) 0.50 A, (c) 60 W
20.35 37.5 mA
20.37 −0.26 mV, The induced polarity opposes the current.
20.39 0.54 A/s
20.41 208
20.45 5.8×10^{-5} J
20.47 (a) 0.11 J, (b) 0.24 W
20.49 8.6 J/m^3
20.51 0.15 m
20.53 4.740×10^6 m/s
20.55 (a) $E_0 = 0.87$ kV/m, $B_0 = 2.9 \times 10^{-6}$ T, (b) The amplitude of the magnetic field is 5.8% of the earth's field.
20.57 400 V
20.59 (a) 15 mV, (b) 9.4×10^{-5} W
20.61 (a) 520 kW, (b) 440 kV, (c) $360,000
20.63 8.8 A
20.65 40 V
20.69 850 mH
20.71 0.32 H
20.73 $L_1 = 0.056$ H, $L_2 = 0.11$ H

21.3 (a) 30 ms, (b) 0.63
21.5 1.1 ms
21.7 (a) 10 s, (b) 80 mΩ
21.11 12 μs
21.13 39 μF
21.15 (a) $5\tau_C$, (b) 0.24 s
21.17 50.0 Hz
21.19 170 V
21.23 19.4 V

21.25 (a) 160 Ω, (b) 56 mA
21.27 (a) 69 Ω, (b) 72 mA
21.29 6.10 Ω
21.31 33 μF
21.33 (a) 42 mA, (b) 60 mA
21.35 0.15 A
21.37 (a) 0.27 A, (b) 42°, (c) 24 W
21.39 (a) −2.8°, (b) −90°, (c) 0°
21.41 1.13 μF or 1.07 μF
21.43 32 kHz
21.45 (a) 690 Hz, (b) 1500 Ω
21.47 (0.24 A) cos(10,000t)
21.49 (a) 0.14 A, (b) 75 Hz
21.51 (a) 521 Ω, (b) 274 Hz, (c) 470 Ω
21.53 0.693τ
21.55 (a) $V_m/2$, (b) $V_m/\sqrt{2}$
21.57 (a) V/R, (b) V/R

22.1 (a) 32°, (b) 28°
22.3 parallel to the symmetry axis
22.5 2.249×10^8 m/s, 2.29×10^8 m/s, 1.240×10^8 m/s
22.7 72°
22.9 no
22.11 31 cm
21.13 (a) 12.8°, (b) 1.1 mm
22.15 (b) $d = \dfrac{t \sin(\theta_1 - \theta_2)}{\cos\theta_2}$
22.17 42.2°
22.19 97.2°
22.21 (a) 2.18 cm, (b) 0.599
22.23 $86.8° \leq \alpha \leq 99.6°$
22.25 (a) 17 cm, (b) +0.33
22.29 0.86 m
22.31 (a) at the 30 cm mark, (b) 15 cm
22.33 14.4 cm
22.35 44.4 cm, −4.1×
22.37 0.46 m
22.39 0.56 cm
22.41 (a) 13.7 cm, (b) 175 cm
22.43 9 cm to the left of the diverging lens
22.45 +3
22.47 (a) 60 cm, the image is real and inverted, (b) −15 cm, the image is virtual and erect.
22.49 (a) −6.67 cm, (b) −5.45 cm, (c) −3.75 cm
22.51 −0.32 m
22.55 ethyl alcohol
22.57 3
22.63 0.39R from the back surface of the sphere
22.65 0.34 m
22.67 $c = ND\omega/\pi$
22.69 8 cm to the left of the diverging lens
22.71 (a) approximately 0.40 m, (b) 0.43 m

23.1 29 diopters
23.3 0.33 m
23.5 −1.00 diopter
23.7 0.40 m
23.9 −3.4 diopters

23.11 $+12$ diopters
23.13 $20.6°$
23.15 1.4 mm
23.17 $2.5\times$, $1.6\times$, and $4.1\times$
23.19 (a) 4.2 cm, (b) $3\times$
23.21 $f/4.3$
23.23 9.0 cm
23.25 $f/3.4$
23.27 2.2 cm
23.29 (a) no, (b) 1/115 s
23.31 $85°$, $44°$, $10°$
23.33 $f/5.6$ for background alone and $f/11$ for second exposure
23.35 (a) 33.3 cm, (b) 54.4 cm
23.37 1.6 mm
23.39 2.9 mm
23.41 2.03 mm from the objective lens
23.43 $-14\times$
23.45 $-10\times$
23.47 0.27 rad
23.49 (a) $-8.12\times$, (b) 40.6 cm
23.51 15.1 cm
23.53 (a) $+10$ cm, (b) $+2.5$ cm
23.55 200 cm
23.57 7.0 diopters
23.59 7.8 cm

23.61 (a) 15 cm, (b) $-\dfrac{2}{3}\times$

23.63 $f/1.6$
23.67 9.3 m
23.69 (a) 0.54 m, (b) 1.85 diopters, (c) 0.54 m
23.71 (c) 1.33

24.3 (a) 1.82×10^8 m/s, (b) 384 nm
24.5 (a) 180 nm, (b) 319 nm
24.7 Waves converge at $R/2$ in front of the mirror, where $R =$ radius of curvature of the mirror.
24.9 $25°$
24.11 589 nm
24.13 (a) 7.4×10^{-3} rad, (b) 4.65 cm
24.15 1.75×10^{-4}
24.17 No wavelength of visible light is possible.
24.19 1.51 mm
24.21 90 nm
24.23 (a) 143 nm, (b) 71 nm
24.25 714 nm (for red), 556 nm (green), 455 nm (blue)
24.27 12 mm
24.29 630 nm
24.31 12 cm
24.33 3
24.35 seven beams exist, corresponding to four orders
24.37 86.7 μm
24.39 3.36×10^{-4} rad
24.41 31 km
24.43 1.3 km
24.45 0.34 mm
24.47 320 m
24.51 $0.88\, I_{\mathrm{m}}$
24.53 $56.7°$

24.55 (a) 1.66, (b) flint glass
24.57 (a) $0.50 I_0$, (b) $0.50 I_0$, (c) $0.25 I_0$
24.59 (a) 20%, (b) 15%
24.61 680 nm
24.63 590 nm
24.65 480 nm
24.67 (a) no overlap, (b) overlap
24.69 for $m = +1$, $\theta = -10.6°$, for $m = -1$, $\theta = -54.7°$
24.71 (a) along a north-south line, (b) none
24.73 830 lines/mm

25.1 (a) -2 km/h, (b) 14 km/h
25.3 0
25.5 $-0.34c$
25.7 $0.89c$
25.9 (a) c, (b) c
25.13 4
25.15 6:00:01.28
25.17 66.8 ms
25.19 1.21 y
25.21 $0.93\ c$
25.23 1.8×10^{-7} s
25.25 150 m
25.27 $0.60c$
25.29 1.4 m
25.31 70 min
25.33 (a) -7.7×10^{-11} m, (b) 3800 km/h
25.35 1.64×10^{-13} J
25.37 14 years
25.39 4.50×10^{-20} kg $\cdot$ m/s
25.41 5.55×10^7 m/s
25.43 $0.87c$
25.45 $0.87c$
25.47 $0.996c$
25.49 (a) $0.51\ m_0 c^2$, (b) $1.5\ m_0 c^2$
25.51 78 MHz
25.53 (a) 98 MHz, (b) 151 MHz
25.55 (a) moving away, (b) $0.11c$
25.57 $3.8 \times 10^{-13}\ f_0$
25.59 1.2×10^{-3} Hz, positive
25.61 18 ns
25.63 $0.50c$ away
25.65 -245 ns
25.67 (a) -3.76 ms, (b) younger

26.3 (a) No, results are probably untrue. (b) Yes, the results are probably true.
26.5 0.751 g
26.7 singly charged
26.9 2.1 L
26.17 $5.86 \times 10^{22}/\mathrm{cm}^3$
26.19 1.76×10^{-10} m
26.21 1.69×10^{-10} m
26.23 $31.0°$, $64.7°$, and $107°$
26.25 9.54×10^{-5} m
26.27 760 V
26.31 2.9 m
26.33 1.25×10^7 m/s

26.35 0.50

26.37 15 d

26.39 1.22×10^4 Bq

26.41 (b) $t_{1/2} = 3.5$ h, 0.20/h, (c) 2600 disintegrations/min

26.43 (a) 2.6×10^{12} Bq, (b) 71 Ci

26.45 3.2×10^{-14} m

26.47 4.11×10^{-10} m

26.49 1040/min

26.51 For incident particle, recoil the velocity is $-0.82v_0$ and its energy is 0.67KE_0. For the struck particle the velocity is $0.18v_0$ and the energy is 0.33KE_0.

27.1 656.21 nm, 486.08 nm, 434.00 nm, and 410.13 nm

27.3 396.97 nm in the ultraviolet

27.5 7630 K

27.7 569 nm

27.9 2.08×10^{-19} J

27.11 413 nm

27.13 (a) 906 nm, (b) 2.20×10^{-19} J

27.15 (a) $p = \dfrac{hf}{c}$, (b) 1.76×10^{24}/s, (c) 737 kW, (d) not practical

27.17 689 nm

27.19 3.48×10^{-11} m

27.21 2.104 eV

27.23 0.11 eV

27.25 (a) 1.38×10^{-11} m, (b) 2.18×10^{19} Hz

27.27 (a) 2.14 eV, (b) no

27.29 (a) 853 nm, (b) 1.45 eV

27.31 4.76×10^{-10} m

27.33 364.6 nm

27.35 (a) 12.8 eV, (b) 96.9 nm

27.37 1460 nm

27.39 91.16 nm

27.41 1.097×10^7/m

27.43 (a) 10.2 eV, (b) 122 nm

27.45 (a) 1.09×10^6 m/s, (b) $v/c = 3.65 \times 10^{-3}$

27.47 (a) 40.5 keV, (b) 80.8 keV

27.49 (a) 1.67×10^{-10} m, (b) 1.94×10^{-10} m

27.51 25.9 kV

27.53 copper

27.55 zinc

27.57 122 nm, 91.1 nm

27.59 3.10×10^3

27.61 (a) 1.51 eV, (b) 0, (c) 3.65×10^{14} Hz

27.63 1.05×10^5 K

28.1 1.14×10^{19} Hz

28.5 16.9 keV

28.7 2.0×10^7 m/s

28.9 12.4°

28.11 1.2×10^{-34} m

28.13 proton

28.15 4.09×10^{-11} m

28.17 16.8 eV

28.19 1.8×10^{-12} m

28.21 8.68°

28.23 0.20 nm

28.25 3.45×10^{-11}

28.27 4.58 ns

28.29 (a) 1.1×10^{-24} kg · m/s, (b) 3.8 eV

28.33 1.0×10^{-11} m

28.37 $E_1 = 6.0 \times 10^{-20}$ J, $E_2 = 2.4 \times 10^{-19}$ J, $E_3 = 5.4 \times 10^{-19}$ J

28.39 1.7×10^{-27} kg

28.41 no

28.43 8100

28.45 1.0×10^{-6}

28.47 2 subshells

28.49 7

28.51 5.79×10^{-5} eV

28.53 3.7×10^{-24} J

28.55 (a) 9.27×10^{-5} eV, (b) 1.34 cm

28.57 Two $1s$ states, two $2s$ states, six $2p$ states, and one $3s$ state. In this notation, the number is the value of n and the letter labels the value of l.

28.59 For the N shell, $n = 4$. The possible values of l are 0, 1, 2, and 3. Each of these subshells has $2(2l + 1)$ states or $2 + 6 + 10 + 14 = 32$ total states.

28.61 (a) 16.4 keV, (b) 0.8 keV

28.63 8.62°

28.67 $h^2/8\pi^2 m(\Delta x)^2$

28.71 Lowest energies for (n_x, n_y) of (1,1), (1,2) = (2,1), (2,2), (1,3) = (3,1), (3,2) = (2,3), (1,4) = (4,1).

$$E_{1,1} = \frac{2\,h^2}{8mL^2}, E_{1,2} = \frac{5\,h^2}{8mL^2}, E_{2,2} = \frac{8\,h^2}{8mL^2}, E_{1,3} = \frac{10\,h^2}{8mL^2},$$

$$E_{2,3} = \frac{13\,h^2}{8mL^2}, E_{1,4} = \frac{17\,h^2}{8mL^2}.$$

29.1 16 min

29.3 7.3×10^5 Bq

29.5 6.5

29.9 20

29.11 53 protons and 73 neutrons, 26 protons and 30 neutrons, 82 protons and 125 neutrons

29.13 35.5

29.15 24.3

29.17 1.0×10^{-42} m^3

29.19 $^{209}_{83}$Bi

29.23 510 MeV, 1740 MeV

29.25 27.410 MeV

29.27 (a) 8×10^{-37}, (b) no

29.29 $^{230}_{90}$Th

29.31 electron

29.33 reaction cannot take place

29.35 (a) $^{66}_{30}$Zn, (b) $^{38}_{18}$Ar

29.37 (a) yes, 5.98 MeV, (b) no, $Q = -3.95$ MeV

29.39 (a) actinium series, (b) thorium series

29.41 (a) uranium series, (b) actinium series

29.43 0.019 MeV

29.47 (a) yes, 4.87 MeV, (b) 4.78 MeV

29.49 0.27 m

29.51 (a) 5.0 Gy, (b) 0.50 Gy

29.53 1.8 mSv

29.55 1.4×10^{14}

29.57 $^{14}_{7}$N
29.59 0.672 MeV
29.61 $^{116}_{46}$Pd
29.63 3.27 MeV
29.65 1.944 MeV
29.67 4.27 MeV
29.69 6.05 MeV, 1.71×10^7 m/s
29.71 0.16 μg

30.1 9.48×10^5
30.3 423 MHz
30.5 1.786 eV
30.7 (a) 1.960 eV, (b) 1.078 eV
30.9 0.809 eV, infrared
30.11 4.0
30.13 0.82
30.15 0.80
30.17 (a) 0.80, (b) 0.72, (c) 6.1
30.19 47.1°
30.21 Three beams at 0° and $\pm$ 23.9°
30.23 0.11°
30.25 (a) 1.77 eV, (b) 3.10 eV
30.27 3000 K
30.29 (a) 690 nm, (b) 967 nm
30.31 5.7
30.33 (a) orange, (b) purple
30.35 (a) no light emerges, (b) red, (c) red, (d) blue
30.37 violet, violet, blue, and red
30.39 Measuring from the central maximum, maxima of the violet light occur at distances of 0.82 mm, 1.64 mm, 2.46 mm, 3.28 mm, 4.10 mm. Maxima of the orange light occur at 1.23 mm, 2.46 mm, and 3.69 mm. The first five color fringes are: violet; orange; violet; orange + violet (= magenta); and violet.
30.41 V = 0.78, I_{max}/I_{min} = 8.1
30.43 3.3×10^{-5} mm

31.1 0.411 nm
31.3 0.324 nm
31.5 $\sqrt{3}a_0/4$
31.7 5.5×10^4 K
31.9 2.54×10^{28} electrons/m^3
31.11 (a) 11.94 eV, (b) 103.6 nm, (c) 7.00 eV
31.13 (a) 3.65 eV, (b) ultraviolet, (c) no

31.15 8.37×10^{-15} s
31.17 6.72×10^{-3} m^2/V · s
31.19 (a) 4.4×10^{-3} m^2/V · s, (b) 2.5×10^{-14} s
31.21 (a) 3.1×10^{-4} m/s, (b) 2.7 h
31.23 233 nm, ultraviolet
31.25 1090 nm
31.27 1.3×10^{-6}/Ω · m
31.29 8.4×10^{28}/m^3
31.31 (b) $-y$ direction
31.33 0.70/Ω · m
31.35 0.58 V
31.37 + 0.52 A
31.39 (a) 2.8 V, (b) 4.5 V
31.41 (a) 13 mA, (b) 14 mA, (c) 92 mW
31.43 890 nm
31.45 (a) 400 m^2, (b) 570 m^2
31.47 (a) 2.06 eV, (b) 2.39×10^4 K
31.49 (a) 3.4×10^{-3} Ω, (b) 4.35×10^{-3} m^2/V · s, (c) 4.35×10^{-2} m/s
31.51 (a) No current passes through the lower diode. (b) No current passes through the upper diode. (c) yes

32.3 1.022 MeV
32.5 1.17×10^{19} J, 585 years
32.7 Photons emerge in opposite directions, each with an energy of 0.511 MeV.
32.9 160 m_e
32.13 105.2 MeV
32.15 (a) 0.055 m, (b) 7.68 m
32.17 6.9×10^6
32.19 7×10^{-20} kg · m · s^{-1}
32.21 1.7 m
32.23 (a) 6.2×10^{-20} m, (b) 5.2×10^{-5} R_0
32.25 (a) 3.8×10^{14}/m^3, (b) 190 km
32.27 (a) cannot take place: reaction does not conserve baryon number, (b) allowed, (c) cannot take place: reaction does not conserve charge.
32.29 $\overline{u}\,\overline{u}\,\overline{d}$
32.31 (a) can occur, (b) cannot occur, (c) can occur, (d) can occur
32.33 -536 MeV; The reaction requires energy.
32.35 (a) 2.5×10^{31} molecules, 7.5×10^5 kg, 750 m^3
32.37 (a) 38 MeV, (b) 96 MeV/c

Photo Credits

Chapter 14—Opener: © Unicorn Stock Photos/Pamela Whiting; 14.13: © Digital Instruments, Inc.

Chapter 15—Opener: © Stone/John Turner; 15.9: © Jones & Childers; 15.10b: © FPG International/Visual Horizons; 15.10c: © Barry Blanchard; 15.10d: © AT&T Archives; 15.11a,b: © Prof. Bill Melton, UNC Charlotte; 15.13a: © Photo Researchers, Inc.; 15.13b: © AllSport USA/Kirk Schlea; © 15.15a: © The Harold E. Edgerton Foundation, 1997, Courtesy of Palm Press, Inc.; 15.15b: © NASA Langley Research Center/Dr. Leonard Weinstein; 15.19: © Jones & Childers.

Chapter 16—Opener: © Tom Stack & Associates/Wm. L. Wantland; 16.1: © Tom Pantages; 16.2: © FPG International/Michael Stoklos; 16.13c, 16.16b, 16.17b: © Tom Pantages; 16.23: © Museum of Science, Boston.

Chapter 17—Opener: © Museum of Science, Boston; 17.6: © Tom Pantages; 17.7: © National Electrostatics Corporation; 17.8: © Photo Researchers, Inc./George Bernard; 17.12a,b: © Jones & Childers; 17.14b: © Tom Pantages; 17.21: © Jones & Childers; Box 17.2: © The Granger Collection/Currier & Ives.

Chapter 18—Opener: © Stone/Paul Dance; 18.1: © Stock Montage; 18.2: © Jones & Childers; 18.3: © Jones & Childers; 18.10: © Tom Pantages; 18.13a, 18.15a,b: © Jones & Childers; Box 18.2: © Westlight/Chuck O'Rear; Box 18.4: © Bruce Ayres/Tony Stone Images.

Chapter 19—Opener: © Visuals Unlimited/C.P. George; 19.1: © Tom Stack & Associates/Jon Feingersh; 19.3c: © Jones & Childers; 19.4a: © Jones & Childers; 19.4b,c: © Jones & Childers; 19.6a: © Danmarks Tekniske Museum; 19.8b: © Jones & Childers; 19.33: © Argonne National Laboratory; 19.32a,b: © G. E. Medical Systems; Box 19.1: © Photo Researchers, Inc./CNRI/Science Photo Library; Box 19.2: © G. E. Medical Systems.

Chapter 20—Opener: © Visuals Unlimited/Milton H. Tierney, Jr.; 20.1: © The Granger Collection/Thomas Phillips; 20.15: © Jones & Childers; 20.16a,b: DENSO Research Laboratories; 20.18a,b: © Jones & Childers; 20.20: © Cavendish Laboratory, University of Cambridge Department of Physics; 20.21a: © Jones & Childers; Box 20.2: © Varian.

Chapter 21—Opener: © Stone/Jon Riley; Pg. 665: © David J. Sams/Stock, Boston.

Chapter 22—Opener: © Stone/Pete Saloutos; 22.1: © IBM Corporation/Almaden Research; 22.3: © Jones & Childers; 22.6, 22.8a, 22.10b: © Jones & Childers; 22.12b: © Stone/Franz Edson; 22.13a: © Jones & Childers; 22.13b: © Photo Researchers, Inc./Deep Light Productions; 22.14b,c,e, 22.36a,b: © Jones & Childers.

Chapter 23—Opener: © PhotoDisc; 23.6a,b,c: © Jones & Childers; 23.8a: © Nikon; 23.13: © Canon U.S.A.; 23.15: © Jones & Childers; 23.16: © Jones & Childers; Box 23.1: © MMT Observatory; Box 23.2: © European Southern Observatory.

Chapter 24—Opener: © Visuals Unlimited/John D. Cunningham; 24.1: © Jones & Childers; 24.5: © Bill Melton, UNC Charlotte; 24.6: © Bill Melton, UNC Charlotte; 24.7a: © Bill Melton, UNC Charlotte; 24.9a,b, 24.10: © Bill Melton, UNC Charlotte; 24.11b: © Jones & Childers; 24.12: © University of South Carolina/Thomas Cooper Library; 24.15: © Jones & Childers; 24.18a,b: © Bill Melton, UNC Charlotte; 24.19: © Fundamental Photographs; 24.20: © Jones & Childers; 24.24a, 24.25: © Jones & Childers; 24.28, 24.29: © Space Telescope Science Institute; 24.30: © NASA; 24.31: © Electron Microscopy Center, The University of South Carolina;

24.42, 24.45: © Jones & Childers; 24.47: © University of Notre Dame/Paul Chagnon.

Chapter 25—Opener: © Stock Montage; 25.14: © Palomar Observatory/California Institute of Technology; 25.21a: © Space Telescope Science Institute; 25.21b: © Space Telescope Science Institute; Box 25.1: © Photo Researchers, Inc./Science Photo Library; Box 25.2: © Anglo-Australian Observatory/David Malin; Box 25.4C, Box 25.4L, Box 25.4R: © Dr. Ping Kang Hsiung;

Chapter 26—Opener: © Tom Stack & Associates/Brian Parker; Pg. 824: © Bettmann/Corbis; Pg. 824: © American Institute of Physics; Pg. 824: © American Institute of Physics; Pg. 825: © American Institute of Physics; Pg. 825: © American Institute of Physics; Pg. 825: © American Institute of Physics; Pg. 835: © Will & Deni McIntyre/Photo Researchers; 26.1a,b: © Jones & Childers; 26.1c,d: © University of South Carolina/Thomas Cooper Library; Pg. 843: © Tim Dominick/The State Newspaper; 26.6: © Jones & Childers; 26.8b, 26.9: © Reynolds Historical Library; © Jones & Childers; 26.10: © University of South Carolina/X-Ray Laboratory; 26.14a: © National Gallery of Art; 26.14b: © National Gallery of Art; 26.17a,b: © Tom Pantages; Box 26.2a: © Erwin W. Mueller; Box 26.2b: © Erwin W. Mueller; Box 26.3: © Park Scientific Instruments.

Chapter 27—Opener: © Sony; Pg. 854: © Bettmann/Corbis; Pg. 854: © American Institute of Physics; Pg. 854: © Stock Montage; Pg. 855: © Stock Montage; Pg. 855: © Stock Montage; Pg. 855: © American Institute of Physics; 27.1: © Jones & Childers; 27.2a: © Fundamental Photographs/Wabash Instrument Corporation; 27.2b,c,d,e: © Fundamental Photographs/Wabash Instrument Corporation; 27.4a: © Jones & Childers; 27.4b: © Jones & Childers; Box 27.2: © Anglo-Australian Observatory/Royal Observatory Edinburgh.

Chapter 28—Opener: © Stone/John McDermott; © Bettmann/Corbis; 28.1: © Edmund Scientific Company; 28.4: © Omicron Associates; 28.5a,b,c: © University of South Carolina, Electron Microscopy Center; Box 28.3: © University of South Carolina, Electron Microscopy Center.

Chapter 29—Opener: © Photo Researchers, Inc./Stevie Grand/Science Photo Library; Pg. 927: © Dr. Robert Friedland/Science Photo Library/Photo Researchers, Inc.; Pg. 931: © Tom White/Palmetto Baptist Medical Center; Pg. 932 and Pg. 934: © Jones & Childers; U29.4: © Jones & Childers; 29.14; © University of Tennessee/Edward L. Hart; 29.15: © CERN; 29.18: © Smithsonian Institution/MIT/Science Services; 29.24: © Oak Ridge National Laboratory; Box 29.1: © American Institute of Physics.

Chapter 30—Opener: © American Bank Note; 30.7: © Siemens Components, Inc., Isalin, NJ; 30.11a,b,c, 30.16, 30.18, 30.20a, 30.25a,b: © Jones & Childers; 30.26: © ERIM; 30.27: © Electro Optical Industries; Box 30.2: © Jones & Childers.

Chapter 31—Opener: © Stone/Rene Sheret; 31.3: © Oak Ridge National Lab; 31.4: © Jones & Childers; 31.5b: © Digital Instruments Inc., Santa Barbara, CA; 31.8, 31.20: © Jones & Childers; 31.28: © AT&T Archives; 31.33, 31.34, 31.35: © Jones & Childers.

Chapter 32—Opener: © Stone/John Warden; 32.1a, 32.3a: © Lawrence Berkeley Laboratory, University of California; Pg. 1005: © Jones & Childers; 32.4a,b: © DESY, Hamburg, Germany; 32.5: © Aerometric; 32.6b: © DESY, Hamburg, Germany; 32.8a: © PhotoDisc; 32.8b: © Fermilab; 32.8c: © NASA; 32.10: © National Geographic Society Image Collection/Joe Stancampiano; 32.11, 32.12: © NASA; Box 32.1: © NASA Goddard Flight Center; Box 32.2: © NASA.

Index

Note: Page numbers followed by *def.* refer to definitions; page numbers followed by *tab* refer to tables.